Spon's Architects' and Builders' Price Book

2002

Key titles from Spon Press:

Building Regulations Explained
J. Stephenson

Construction Scheduling, Cost Optimisation and Management
H. Adeli and A. Karim

Dictionary of Architectural and Building Technology
H. Cowan and P. Smith

Energy & Environment in Architecture
N. Baker and K. Steemers

Fire From First Principles
P. Stollard and J. Abrahams

Green Building Handbook Volumes 1 & 2
T. Woolley et al

Hazardous Building Materials 2nd Edition
S. Curwell and C. March

Profiled Sheet Roofing and Cladding
National Federation of Roofing Contractors

Refurbishment and Upgrading of Existing Buildings
D. Highfield

Spon's Building Costs Guide for Educational Premises
Barnsley and Partners

Spon's Construction Resource Handbook
B. Spain

Spon's House Improvement Price Book
B. Spain

Spon's Irish Construction Price Book
Franklin and Andrews

Spon's Railway Construction Price Book
Franklin and Andrews

Timber Decay in Buildings
B. Ridout

Understanding the Building Regulations
S. Polley

Understanding JCT Standard Building Contracts
D. Chappell

Understanding Quality Assurance in Construction
H. W. Chung

To order or obtain further information on any of the above or receive a full catalogue please contact:
The Marketing Department, Spon Press, 11 New Fetter Lane, London, EC4P 4EE.
Tel: 020 7583 9855; Fax: 020 7842 2298
For a complete listing of all our titles please visit www.sponpress.com

Spon's Architects' and Builders' Price Book

Edited by
DAVIS LANGDON & EVEREST

2002

One hundred and twenty seventh edition

SPON PRESS
Taylor & Francis Group

London and New York

First edition published 1873
One hundred and twenty seventh edition published 2002
by Spon Press
11 New Fetter Lane, London EC4P 4EE

Simultaneously published in the USA and Canada
by Spon Press
29 West 35th Street, New York, NY 10001

Spon Press is an imprint of the Taylor & Francis Group

© 2002 Spon Press

Printed and bound in Great Britain by
TJ International Ltd, Padstow, Cornwall

All rights reserved. No part of this book may be reprinted or
reproduced or utilised in any form or by any electronic,
mechanical, or other means, now known or hereafter
iinvented, including photocopying and recording, or in any
information storage or retrieval system, without permission in
writing from the publishers.

Publishers note
This book has been produced from camera-ready copy
supplied by the authors

British Library Cataloguing in Publication Data
A catalogue record for this book is available from the British
Library

Library of Congress Cataloging in Publication Data
A catalogue record has been requested

ISBN 0-415-26216-X
ISSN 0-306-3046

Preface to the One hundred and Twenty-seventh edition

The second government term won in June by Labour provides a platform for the promised huge new investment in health, education and transport, education and transport to continue. The 2000 Comprehensive Spending Review, announced by Gordon Brown in July 2000, set out a £43 billion rise in departmental spending by the financial year 2003-04. This included an increase in transport spending of an extra £10-11 billion, much of it on civil engineering construction. In addition, a vast school refurbishment programme was put in place and an increase in spending of £2.50 billion over the three years on social housing was announced. In spite of notorious delays often in translating public sector spending plans into site activity, work in these sectors has begun, over the last year, to fill the gap left by a decline in private commercial work, following the completion of most Millennium projects and the reduction in other large scale lottery funded projects.

Market Conditions

The volume of total construction output (including repair and maintenance) in the year to the first quarter 2001 was unchanged compared to the previous twelve month period but the volume of new work fell slightly. However the volume of new orders obtained by contractors over this period was 7% higher than the previous year. Over these comparable periods, new orders for infrastructure work were 15% higher and public non-housing 5% higher. DETR statistics show that the volume of new orders in the private commercial sector was 16% higher over the same period but this sector now includes privately funded education and health projects, where the volume of new orders rose by 38%.

The Construction Confederation's Quarterly 2001 Construction Trends Survey found that 61% of companies across the country were working at full or near full capacity. Recruitment difficulties were at their highest for 10 years. More than 80% of companies reported difficulty in securing bricklayers, plasterers and carpenters and joiners. The labour problem will continue to worsen if workload increases further. The Construction Industry Training Board has forecast demand for 376,000 extra construction workers over the next 5 years, on top of the 1.50 million employed in the industry. The traditional craft shortages have led to a greater use of prefabrication but this, in turn, has come up against resource restraints in areas such as cladding and curtain walling.

Forecasts

Construction Forecasting and Research Limited in their 'Spring 2001 Construction Forecasts 2001 - 2002' anticipated 2½% growth in 2001 in total construction output compared to 2000, to be followed by a further 3% increase in 2002. Repair and maintenance is forecast to increase considerably more (4% in 2001, 4½% in 2002) than new work (0.50% and 2.60%). Public housing repair and maintenance and infrastructure are the sectors that should see the greatest over these two years.

Workload in the private commercial, industrial and housing sectors will be influenced by events affecting the general economy. At the beginning of 2001 the slowdown in the U.S. economy threatened to spill over into Europe and the UK Nervousness regarding the economic outlook caused some construction clients to slow or even halt progress on construction developments in the pipeline. Some U.S. clients withdrew altogether. The collapse in technology stocks scuppered other schemes on the drawing board. This caused a sudden halt to the climb in construction prices that had been prevalent since the fourth quarter of 1999. At the time of going to press, fear of a sharp slowdown appear to have dissipated. Three interest rate cuts in February, April and May have helped.

Prices

DLE's Tender Price Index shows that prices rose by 7% in the year to the first quarter 2001. Tender prices are forecast to rise by between 3 and 5% in the year to first quarter 2002 and by 3 to 6% in the year to first quarter 2003. The price level of the 2002 edition of Spon's A&B has been indexed at 388, an increase of 6% over the 2001 edition.

From 25 June 2001, the second part of the Construction Industry Joint Council's three year agreement on pay and conditions came into effect, raising wage rates for all grades by 5.5%. The revised labour rates, providing basic hourly rates of £5.04 for general operatives and £6.70 for the craft rate have been taken into account in the preparation of this year's book along with material prices as to June 2001.

Readers of Spon's A&B are reminded that Spon's is the only price book in which key rates are checked against current tender prices. Therefore although "officially quoted" materials and labour prices have increased by 20% over the four years between first quarter 1997 and first quarter 2001, tender prices have risen by 36% over this period in London, though slightly less in most other regions. Before that, the change in tender prices over the six years between the fourth quarter 1989 and the fourth quarter 1996 was measured, in London, at -22%, while the movement in building costs was measured at +28%. Spon's continual monitoring of true tender prices ensures the prices in Spon's A&B accurately reflect real market prices.

Climate Change Levy

Officially inflation remains fairly subdued. Over the year to May 2001, the headline retail price inflation rose by 2.10%. Underlying inflation, excluding mortgage interest payments, rose to 2.40%, just about exactly in line with the Bank of England's target of 2.50%. The input price index for materials and fuel purchased by manufacturing industry rose by 4.70% over the year to May 2001. This statistic excludes the effect of the Climate Change Levy.

The Climate Change Levy came into effect on 1 April 2001. The Levy consists of different levies on various fuels but is generally regarded as equating to a levy of approximately 10-15% on the cost of gas and electricity, to be paid by all UK businesses and public sector organisations.

ONS statistics currently exclude the effect of the Climate Change Levy. However, based on Customs and Excise modelling, DTI Energy Statistics have estimated that the Levy might have added 0.80% to the input (materials and fuel) index in May 2001. The effect on construction materials prices may be yet to feed through.

Aggregates Tax

Readers are reminded that the Aggregates Tax will come into force from April 2002, adding £1.60 a tonne to the price of aggregates. Press Reports suggest that the end-user price may increase significantly more.

Profits and Overheads

For a number of years Spon's A&B has excluded profit and overheads in the build-up of the Measured Rates. The 2002 edition includes a 7½% mark-up on labour and materials for profit and overheads.

Preliminaries

The overall level of preliminaries experienced on projects over the last year has shown no significant change. An 11 to 13% addition for preliminaries is a typical figure to be expected on new work projects. The Prices for Measured Works sections of this book exclude any allowance for preliminaries. In the Approximate Estimating section (Part iV), the Building Prices per Functional Units and the Building Prices per Square Metre do include for preliminaries but the Approximate Estimates (incorporating Comparative Prices) do not.

Dayworks

The Royal Institution of Chartered Surveyors (RICS) have published a new Schedule of Basic Plant Charges for use in connection with daywork under a building contract. The schedule has been revised and extended to reflect the types of plant likely to be available on a building site. Known as the '2001 edition', the schedule became effective from 1 May 2001. The document is reproduced in full in Part 1 - Fees and Daywork - by kind permission of the RICS.

Value Added Tax

Since 1989 all non-domestic building work, refurbishment and alterations, has been subject VAT, currently at 17½%. In the March 2001 Budget VAT payable on residential conversions was cut from 17½% to 5%.

Prices included within this edition do not include for VAT, which must be added if appropriate.

The book is divided into five parts as follows:

Part I: Fees and Daywork

This section contains Fees for Professional Services, Daywork and Prime Cost.

Part II: Rates of Wages

This section includes authorised wage agreements applicable to the Building and associated Industries.

Part III: Prices for Measured Work

This section contains Prices for Measured Work - Major Works, and Prices for Measured Work - Minor Works (on coloured paper).

Part IV: Approximate Estimating

This section contains the Building Cost and Tender Price Index, information on regional price variations, prices per functional unit and square metre for various types of buildings, approximate estimates, cost limits and allowances for "public sector" building work and a procedure for valuing property for insurance purposes.

Part V: Tables and Memoranda

This section contains general formulae, weights and quantities of materials, other design criteria and useful memoranda associated with each trade, and a list of useful Trade Associations.

New Items

A comprehensive list of new sections and items included in this year's edition of Spon's A&B can be found on pages 977 to 985.

While every effort is made to ensure the accuracy of the information given in this publication, neither the Editors nor Publishers in any way accept liability for loss of any kind resulting from the use made by any person of such information.

CD-Rom

Included in the price of each book is a free CD Rom, containing the complete contents of the book itself and an invaluable estimating programme of immense use to all users.

DAVIS LANDON & EVEREST
Princes House
39 Kingsway
London
WC2B 6TP

SPON'S PRICE BOOKS 2002

with free CD-ROM

Free CD-ROM when you order any Spon's 2002 Price Book.
Use the CD-ROM to:
- produce tender documents
- customise data
- keyword search
- export to other major packages
- perform simple calculations.

Spon's Architects' and Builders' Price Book 2002
Davis Langdon & Everest

"Spon's Price Books have always been a 'Bible' in my work - now they have got even better! The CDs are not only quick but easy to use. The CD ROMs will really help me to get the most from my Spon's in my role as a Freelance Surveyor."
Martin Taylor, Isle of Lewis

New Features for 2002 include:
- A new section on Captial Allowances
- Inclusion of new items within a seperate Measured Works section

September 2001: 1024 pages
Hb & CD-ROM: 0-415-26216-X: £110.00

Spon's Landscapes and External Works Price Book 2002
Davis Langdon & Everest, in association with Landscape Projects

New Features for 2002 include:
- Fees for professional services
- Revised and updated sections on Cost Information and how to use this book
- Revisions and expansions of the Approximate Estimating section, together with direct links into the Measured Works Section

September 2001: 484 pages
Hb & CD-ROM: 0-415-26220-8: £80.00

Spon's Mechanical and Electrical Services Price Book 2002
Mott Green & Wall

"An essential reference for everybody concerned with the calculation of costs of mechanical and electrical works." *Cost Engineer*

New Features for 2002 include:
- New sections on modular wiring, emergency lighting, lighting control, sprinkler pre fabricated pipework, UPVC rainwater and gutters, carbon steel pipework and fittings

September 2001: 584 pages
Hb & CD-ROM: 0-415-26222-4: £110.00

Spon's Civil Engineering and Highway Works Price Book 2002
Davis Langdon & Everest

New Features for 2002 include:
- A revised and extended section on Land Remediation
- The Rail Track section now includes data on Permanent Way work with fully reviewed pricing
- Fully reviewed pricing for the Geotextiles section

September 2001: 688 pages
Hb & CD-ROM: 0-415-26218-6: £120.00

updates available to download from the web
www.pricebooks.co.uk

Return your orders to: Spon Press Customer Service Department, ITPS, Cheriton House, North Way, Andover, Hampshire, SP10 5BE · Tel: +44 (0) 1264 343071 · Fax: + 44 (0) 1264 343005 · Email: book.orders@tandf.co.uk
Postage & Packing: 5% of order value (min. charge £1, max. charge £10) for 3–5 days delivery · Option of next day delivery at an additional £6.50.

SPON PRESS · Taylor & Francis Group

The Landfill Tax

The Tax

The Landfill tax came into operation on 1 October 1996. It is levied on operators of licensed landfill sites at the following rates:

  £2 per tonne - Inactive or inert wastes.
           Included are soil, stones, brick, plain and reinforced concrete, plaster and glass.

  £12 per tonne - All other taxable wastes.
           Included are timber, paint and other organic wastes generally found in demolition work, builders skips etc.

From 1 April 2001 the rate of for "all other taxable wastes" was increased from £11 to £12 per tonne, whilst the rate for "inactive or inert wastes" remained at £2 per tonne.

It is intended to raise the standard rate of landfill tax for "all other taxable wastes" in future years by an additional £1 per tonne each year. These increases will take place at least until 2004 when the standard rate will become £15 per tonne.

Mixtures containing wastes not classified as inactive or inert will not qualify for the lower rate of tax unless the amount of non-qualifying material is small and there is no potential for pollution. Water can be ignored and the weight discounted.

Calculating the Weight of Waste

There are two options:

* If licensed sites have a weighbridge, tax will be levied on the actual weight of waste.

* If licensed sites do not have a weighbridge, tax will be levied on the permitted weight of the lorry based on an alternative method of calculation based on volume to weight factors for various categories of waste.

Effect on Prices

The tax is paid by Landfill site operators only. Tipping charges reflect this additional cost.

As an example, Spon's A & B rates for mechanical disposal will be affected as follows:

*	Inactive waste	Spon's A & B 2002 net rate	£10.48 per m^3
		Tax, 2 t per m^3 (unbulked) @ £2.00	£4.00 per m^3
		Spon's rate including tax (page 125)	£14.48 per m^3

Effect on Prices - cont'd

* Active waste — Active waste will normally be disposed of by skip and will probably be mixed inactive waste. The tax levied will depend on the weight of the materials in the skip which can vary significantly.

Exemptions

The following disposals are exempt from Landfill Tax:

* dredgings which arise from the maintenance of inland waterways and harbours.
* naturally occurring materials arising from mining or quarrying operations.
* waste resulting from the cleaning up of historically contaminated land, although to obtain an exemption it is necessary to first obtain a contaminated land certificate from Customs and Excise.
* waste removed from one site to be used on another or to be recycled or incinerated.

An additional exemption was introduced from the 1st October 1999 for inert waste used to restore landfill sites and to fill working and old quarries with a planning condition or obligation in existence to fill the void.

For further information contact the Landfill Tax Helpdesk, Telephone: 0645 128484.

Land Remediation

Statutory framework

In July 1999 new contaminated land provisions, contained in part IIa of the Environmental Protection Act 1990 were introduced. A primary objective of the measures is to encourage the recycling of brownfield land.

Under the Act action to remediate land is required only where there are unacceptable actual or potential risks to health or the environment. Sites that have been polluted from previous land use may not need remediating until the land use is changed. In addition, it may be necessary to take action only where there are appropriate, cost-effective remediation processes that take the use of the site into account.

The Environment Act 1995 amended the Environmental Protection Act 1990 by introducing a new regime designed to deal with the remediation of sites which have been seriously contaiminated by historic activities. The regime became operational on 1 April 2000. Local authorities and/or the Environment Agency will regulate seriously contaminated sites which will be known as "special sites". The risks involved in the purchase of potentially contaminated sites is high, particularly considering that a transaction can result in the transfer of liability for historic contamination from the vendor to the purchaser.

The contaminated land provisions of the Environmental Protection Act 1990 are only one element of a series of statutory measures dealing with pollution and land remediation that are to be introduced this year. Others include:

* groundwater regulations, including pollution prevention measures

* an integrated prevention and control regime for pollution

* selections of the Water Resources Act 1991, which deals with works notices for site controls, restoration and clean up.

The contaminated land measures incorporate statutory guidance on the inspection, definition, remediation, apportionment of liabilities and recovery of costs of remediation. The measures are to be applied in accordance with the following criteria:

* the standard of remediation should relate to the present use

* the costs of remediation should be reasonable in relation to the seriousness of the potential harm

* the proposals should be practical in relation to the availability of remediation technology, impact of site constraints and the effectiveness of the proposed clean-up method.

Liability for the costs of remediation rests with either the party that "caused or knowingly permitted" contamination, or with the current owners or occupiers of the land.

Apportionment of liability, where shared, is determined by the local authority. Although owners or occupiers become liable only if the polluter cannot be identified, the liability for contamination is commonly passed on when land is sold.

The ability to forecast the extent and cost of remedial measures is essential for both parties, so that they can be accurately reflected in the price of the land. If neither the polluter nor owner can be found, the clean-up is funded from public resources.

Land remediation techniques

There are two principal approaches to remediation - dealing with the contamination in situ or off site. The selection of the approach will be influenced by factors such as: initial and long term cost, timeframe for remediation, types of contamination present, depth and distribution of contamination, the existing and planned topography, adjacent land uses, patterns of surface drainage, the location of existing on-site services, depth of excavation necessary for foundations and below-ground services, environmental impact and safety, prospects for future changes in land use and long-term monitoring and maintenance of in situ treatment.

In situ techniques

A range of in situ techniques is available for dealing with contaminants, including:

* Dilution - the reduction of the concentrations of contaminants to below trigger levels by on-site mixing with cleaner material.

* Clean cover - a layer of clean soil is used to segregate contamination from receptor. This technique is best suited to sites with widely dispersed contamination. Costs will vary according to the need for barrier layers to prevent migration of the contaminant.

* On-site encapsulation - the physical containment of contaminants using barriers such as slurry trench cut-off walls. The cost of on-site encapsulation varies in relation to the type and extent of barriers required, the costs of which range from £35/m² to more than £100/m².

There are also in situ techniques for treating more specific contaminants, including:

* Bio-remediation - for removal of oily, organic contaminants through natural digestion by micro-organisms. The process is slow, taking from one to three years, and is particularly effective for the long-term improvement of a site, prior to a change of use.

* Soil washing - involving the separation of a contaminated soil fraction or oily residue through a washing process. The dewatered contaminant still requires disposal to landfill. In order to be cost effective, 70 - 90% of soil mass needs to be recovered.

* Vacuum extraction - involving the extraction of liquid and gas contaminants from soil by vacuum.

* Thermal treatment - the incineration of contaminated soils on site. The uncontaminated soil residue can be recycled. By-products of incineration can create air pollution and exhaust air treatment may be necessary.

* Stabilisation - cement or lime, is used to physically or chemically bound oily or metal contaminants to prevent leaching or migration. Stabilisation can be used in both in situ and off-site locations.

Off-site techniques

Removal is the most common and cost-effective approach to remediation in the UK, providing a broad spectrum solution by dealing with all contaminants. Removal is suited to sites where sources of contamination can be easily identified.

If used in combination with material-handling techniques such as soil washing, the volume of material disposed at landfill sites can be significantly reduced. The disadvantages of the technique include the fact that the contamination is not destroyed, there are risks of pollution during excavation and transfer; road haulage may also cause a local nuisance.

Cost drivers

Cost drivers relate to the selected remediation technique, site conditions and the size and location of a project.

The wide variation of indicative costs of land remediation techniques shown below is largely because of differing site conditions.

Indicative costs of land remediation techniques (excluding landfill tax)		
Remediation technique	Unit	Rate (£/unit)
Removal	disposed material (m^3)	55 - 160
Clean cover	surface area of site (m^2)	21 - 55
On-site encapsulation	encapsulated material (m^3)	21 - 85
Bio-remediation	treated material (tonne)	32 - 90
Soil washing	treated material (tonne)	42 - 90
Soil flushing	treated material (tonne)	65 - 115
Vacuum extraction	treated material (tonne)	55 - 115
Thermal treatment	treated material (tonne)	775 - 1275

Factors that need to be considered include:

* waste classification of the material

* underground obstructions, pockets of contamination and live services

* ground water flows and the requirement for barriers to prevent the migration of contaminants

* health and safety requirements and environmental protection measures

* location, ownership and land use of adjoining sites

* distance from landfill tips, capacity of the tip to accept contaminated materials, and transport restrictions

Other project related variables include size, access to disposal sites and tipping charges; the interaction these factors can have a substantial impact on overall unit rates.

The table below sets out the costs of remediation using *dig-and-dump* methods for different sizes of project. Variation in site establishment and disposal cost accounts for 60 - 70% of the range in cost.

Cost drivers - cont'd

Variation in the costs of land remediation by removal			
Item	Disposal Volume (less than 3000 m³) (£/m³)	Disposal Volume (3000 - 10 000 m³) (£/m³)	Disposal Volume (more than 10 000 m³) (£/m³)
General items and site organisation costs	42 - 75	21 - 37	5 - 16
Site investigation and testing	3 - 13	2 - 5	1 - 3
Excavation and backfill	13 - 27	10 - 21	7 - 14
Disposal costs (including tipping charges but not landfill tax)	16 - 32	21 - 37	27 - 48
Haulage	11 - 27	10 - 14	9 - 18
Total (£/m³)	85 - 174	64 - 114	49 - 99
Allowance for site abnormals	0 - 5 +	0 - 15 +	0 - 10 +

The strict health and safety requirements of remediation can push up the overall costs of site organisation to as much as 50% of the overall project cost. A high proportion of these costs are fixed and, as a result, the unit costs of site organisation increase disproportionally on smaller projects.

Haulage costs are largely determined by the distances to a licensed tip. Current average haulage rates, based on a return journey range from £1.40 to £1.85 a mile. Short journeys to tips, which involve proportionally longer standing times, typically incur higher mileage rates.

A further source of cost variation relates to tipping charges. The table below summarises typical tipping charges for 2000, exclusive of landfill tax:

Typical 2001 tipping charges (excluding landfill tax)	
Waste classification	Charges (£/tonne)
Non-hazardous wastes	6 - 11
Hazardous wastes	11 - 22
Special waste	13 - 27
Contaminated liquid	13 - 38
Contaminated sludge	38 - 50

Tipping charges fluctuate in relation to the grades of material a tip can accept at any point in time. This fluctuation is a further source of cost risk.

Over the past two years, prices at licensed tips have varied by as much as 50%. In addition, a special waste regulation charge of £15 per load, equivalent to 80 p a tonne, is also payable.

Landfill tax, currently charged at £12 a tonne for active waste, is also payable, although exemptions are currently available for the disposal of historically contaminated material (refer also to *"Landfill Tax"* on page ix).

Contents

	page
Preface	v
The Landfill Tax	ix
Land Remediation	xi
Acknowledgements	xvii
How to use this Book	xxxv

PART I - FEES AND DAYWORK

Fees for Professional Services	4
Architects' Fees	4
Quantity Surveyors' Fees	9
Consulting Engineers' Fees	45
The Town and Country Planning Regulations 1997	58
The Building (Local Authority Charges) Regulations 1998	65
Daywork and Prime Cost	71

PART II - RATES OF WAGES

Rates of Wages	91
Building Industry - England, Wales and Scotland	91
Building and Allied Trades Joint Industrial Council	94
Road Haulage Workers employed in the Building Industry	94
Plumbing and Mechanical Engineering Services Industry	95

PART III - PRICES FOR MEASURED WORK

Prices for Measured Work - Major Works	97
Introduction	99
A Preliminaries	105
D Groundwork	120
E In situ concrete/Large precast concrete	135
F Masonry	161
G Structural/Carcassing Metal/Timber	194
H Cladding/Covering	220
J Waterproofing	243
K Linings/Sheathing/Dry partitioning	250
L Windows/Doors/Stairs	267
M Surface finishes	295
N Furniture/Equipment	328
P Building fabric sundries	340
Q Paving/Planting/Fencing/Site furniture	357
R Disposal systems	369
S Piped supply systems	399
T Mechanical heating/Cooling/Refrigeration systems	405
V Electrical systems	406
W Security systems	408

PART III - PRICES FOR MEASURED WORK - cont'd

Prices for Measured Work - Minor Works 409

    Introduction 409
    A Preliminaries/Contract conditions for minor works 410
    C Demolition/Alteration/Renovation 411
    D Groundwork 432
    E In situ concrete/Large precast concrete 445
    F Masonry 463
    G Structural/Carcassing Metal/Timber 493
    H Cladding/Covering 509
    J Waterproofing 526
    K Linings/Sheathings/Dry partitioning 533
    L Windows/Doors/Stairs 548
    M Surface finishes 576
    N Furniture/Equipment 609
    P Building fabric sundries 622
    Q Paving/Planting/Fencing/Site furniture 639
    R Disposal systems 652
    S Piped supply systems 679
    T Mechanical heating/Cooling/Refrigeration systems 685
    V Electrical systems 686
    W Security systems 688

PART IV - APPROXIMATE ESTIMATING

Building Costs and Tender Prices 691
Building Prices per Functional Units 697
Building Prices per Square Metre 703
Approximate Estimates (incorporating Comparative Prices) 711
Cost Limits and Allowances 831
Property Insurance 877
Capital Allowances 881

PART V - TABLES AND MEMORANDA

Conversion Tables 885
Formulae 886
Design Loadings for Buildings 888
Planning Parameters 895
Sound Insulation 913
Thermal Insulation 914
Weights of Various Materials 918
Memoranda for each Trade 919
Useful Addresses for Further Information 956
Natural Sines 971
Natural Cosines 973
Natural Tangents 975
New Items 977
Index 987

Acknowledgements

The Editors wish to record their appreciation of the assistance given by many individuals and organisations in the compilation of this edition.

Material suppliers and sub-contractors who have contributed this year include:

Abbey Building Supplies Company
213 Stourbridge Road
Halesowen
West Midlands
B63 3QY
Building Fixings and Fastenings
Tel: 0121 550 7674
Fax: 0121 585 5031

Ace Minimix
British Rail Goods Yard
York Way
Kings Cross
London
N1 0AU
Ready mixed concrete
Tel: 0207 837 7011
Fax: 0207 278 5333

ACO Technologies Ltd
Hitchin Road
Shefford
Bedfordshire
SG17 5TE
Drainage Channels
Tel: 01462 816 666
Fax: 01462 815 895
E-mail: technologies@aco.co.uk
Website: www.aco.co.uk

Akzo Nobel Decorative Coatings Ltd
PO Box 37
Crown House
Hollins Road
Darwen
Lancashire
BB3 0BG
Paints, Varnishes etc.
Tel: 01254 704 951
Fax: 01245 870 155

Akzo Nobel Woodcare
Meadow Lane
St Ives
Cambridgeshire
PE17 4UY
"Sadolin" Stains and Preservatives
Tel: 01480 496 868
Fax: 01480 496 801

Albion Concrete Products Ltd
Pipe House Wharf
Morfa Road
Swansea
West Glamorgan
SA1 1TD
Precast Concrete Pipes and Manhole Rings, etc.
Tel: 01792 655 968
Fax: 01792 644 461
E-mail: sales@albionc.co.uk

Alexander Adams Ltd
Hylton Bridge Farm
West Boldon
Tyne and Wear
NE36 0BB
Raised Access Floors and Accessories
Tel: 0191 519 0707
Fax: 0191 519 2121

Alfred McAlpine Slate Ltd
Penrhyn Quarry
Bethesda
Bangor
Gwynedd
LL57 4YG
Natural Welsh Slates
Tel: 01248 600 656
Fax: 01248 602 447
E-mail: slate@alfred-mcalpine.com
Website: www.amslate.com

Altro Floors
Works Road
Letchworth
Hertfordshire
SG6 1NW
"Altro" Sheet and Tile Flooring and Stair Nosings
Tel: 0870 548 0480
Fax: 0870 511 3388
E-mail: info@altro.co.uk
Website: www.altro.co.uk

Alumasc Exterior Building Products Ltd
Burton Latimer
Kettering
Northamptonshire
NN15 5JP
Aluminium Rainwater Goods
Tel: 01536 383 838
Fax: 01536 383 830
E-mail: info@alumasc-exteriors.co.uk

Alumasc Interior Building Products Ltd
White House Works
Bold Road
Sutton
Saint Helens
Merseyside
WA9 4JG
Skirting Trunking and Casing Profiles
Tel: 01744 648 400
Fax: 01744 648 401
E-mail: info@almascinteriors.co.uk

Aluminium RW Supplies Ltd
Ryan House
Unit 6
Dumballs Road
Cardiff
South Glamorgan
CF10 5DF
Aluminium Rainwater Goods
Tel: 02920 390 576
Fax: 02920 238 410

AN Wallis & Company Ltd
Greasley Street
Bulwell
Nottingham
Nottinghamshire
NG6 8NJ
Lightning Conductors
Tel: 0115 927 1721
Fax: 0115 977 0069
E-mail: mike@anwallis.emnet.co.uk

Ancon Building Products
President Way
President Park
Sheffield
South Yorkshire
S4 7UR
Steel Reinforcement for Concrete
Tel: 0114 275 5224
Fax: 0114 276 8543
E-mail: info@anconccl.com
Website: www.anconccl.com

Ardex UK Ltd
Homefield Road
Haverhill
Suffolk
CB9 8QP
Building Concrete Admixtures
Tel: 01440 714 939
Fax: 01440 703 424
Website: ardex.co.uk

Armitage Shanks Group Ltd
Armitage
Rugeley
Staffordshire
WS15 4BT
Sanitary Fittings
Tel: 01543 490 253
Fax: 01543 491 677

Armstrong World Industries Ltd
Building Products Operation
38 Market Square
Uxbridge
Middlesex
UB8 1NG
Tiles and Sheet Flooring
Tel: 0800 371 849
Fax: 01895 274 287
Website: www.armstrong.com

Ash Resources
West Burton
Retford
Nottinghamshire
DN22 9BL
Lytag
Tel: 01427 881 234
Fax: 01427 881 000
E-mail: sales@ashresources.co.uk

Ashworth Frazer (Northern) Ltd
Station Road
Hebburn
Tyne and Wear
NE31 1BD
Cast Iron Disposal Goods
Tel:	0191 428 0077
Fax:	0191 483 3628
E-mail:	info@ashworth-frazer.co.uk
Website:	www.ashforth-frazer.co.uk

Balcas Kildare Ltd
Naas
County Kildare
Ireland
Medium Density Fibreboard Profiles
Tel:	00 353 45 877 671
Fax:	00 353 45 877 073

Beacon Hill Brick Company Ltd
Wareham Road
Corfe Mullen
Wimborne
Dorset
BH21 3RX
Facing Bricks
Tel:	01202 697 633
Fax:	01202 605 141

Binns Fencing Ltd
Harvest House
Cranborne Road
Potters Bar
Hertfordshire
EN6 3JF
Fencing
Tel:	01707 855 555
Fax:	01707 857 565

Bison Concrete Products Ltd
Millennium Court
First Avenue
Burton Upon Trent
DE14 2WR
Precast Concrete Flooring Systems
Tel:	01283 495 000
Fax:	01283 544 900
E-mail:	concrete@bison.co.uk

Blue Circle Cement
The Shore
Northfleet
Kent
DA11 9AN
Cement, Lime and Flintag
Tel:	01474 564 344
Fax:	01474 531 281
Website:	www.bluecircle.co.uk

Bolton Gate Company
Waterloo Street
Bolton
Lancashire
BL1 2SP
Roller Shutters
Tel:	01204 871 000
Fax:	01204 871 049

Bonar Floors (Nuway)
High Holburn Road
Ripley
Derbyshire
DE5 3NT
Entrance Matting and Aluminium Matwell Frames
Tel:	01773 744 121
Fax:	01773 744 142
Website:	www.bonarfloors.com

Bourne Steel Ltd
St Clements House
St Clements Road
Poole
Dorset
BH12 4GP
Structural Steelwork
Tel:	01202 746 666
Fax:	01202 732 002
E-mail:	sales@bourne-steel.demon.co.uk

BRC Building Products
Carver Road
Astonfields Industrial Estate
Staffordshire
ST16 3BP
Brick Reinforcement
Tel:	01785 222 288
Fax:	01785 240 029
E-mail:	email@brc-building-products.co.uk
Website:	www.brc-building-products.co.uk

Brickhouse Access Covers & Gratings
Ponymister Industrial Estate
Risca
Newport
Monmouthshire
NP1 6YL
Manhole and Duct Covers
Tel:	01633 612 833
Fax:	01633 601 593
E-mail:	brickhse@celtic.co.uk

British Gypsum Ltd
Technical Services Department
East Leake
Loughborough
Leicestershire
LE12 6HX
Plasterboard and Plaster Products
Tel: 0870 545 6123
Fax   0870 545 6356
E-mail: BGTechnical.Enquiries@bpb.com
Website: www.british-gypsum.bpb.com

Builders Iron & Zincwork Ltd
Millmarsh Lane
Brimsdown
Enfield
Middlesex
EN3 7QA
Roof Trims and Accessories
Tel: 0208 443 3300
Fax: 0208 804 6672

Burlington Slate Ltd
Cavendish House
Kirkby-in-Furness
Cumbria
LA17 7UN
"Westmoreland" Slating
Tel: 01229 889 681
Fax: 01229 889 466

Callenders Construction Products
Harvey Road
Burnt Mills Industrial Estate
Basildon
Essex
SS13 1EJ
Flooring, Roofing and Damp Proofing Products
Tel: 01268 591 155
Fax: 01268 591 165
E-mail: email@callenders.co.uk
Website: www.callenders.co.uk

Camas Building Materials
Hulland Ward
Ashbourne
Derbyshire
DE6 3ET
Reconstructed Stone Walling and Roofing
Tel: 01335 372 222
Fax: 01335 370 074

Cape Calsil Systems Ltd
Iver Lane
Uxbridge
Middlesex
UB8 2JQ
Boarding and TRADA Firecheck Panelling
Tel: 01895 463 400
Fax: 01895 463 401
Website: www.capecalisil.com

Catnic Ltd
Pontygwindy Estate
Caerphilly
Mid Glamorgan
CF8 2WJ
Steel Lintels and "Garador" Products
Tel: 02920 337 900
Fax: 02920 867 796

Cavity Trays Ltd
Administration Centre
Lufton Trading Estate
Yeovil
Somerset
BA22 8HU
Cavity Trays, Closers and Associated Products
Tel: 01935 474 769
Fax: 01935 428 223
E-mail: enquieries@cavitytrays.co.uk

CB Watson Stonemasonry Ltd
The Stoneyard
Atlantic Crescent
No 2 Dock
Barry
Vale of Glamorgan
CF63 3RA
Stonework
Tel: 01446 732 965
Fax: 01446 736 679

CCS Scotseal Ltd
Unit 3
Lyon Road
Linwood Industrial Estate
Paisley
Scotland
PA3 3BQ
External insulation of external walls
Tel: 01505 324 262
Fax: 01505 323 618

CEP Ceilings Ltd
Verulam Road
Common Road Industrial Estate
Stafford
Staffordshire
ST16 3EA
Ceiling Tiles
Tel: 01785 223 435
Fax: 01785 251 309

CF Anderson & Son Ltd
228 London Road
Marks Tey
Colchester
Essex
C06 1HD
Wrought Softwood Boarding and Panel Products
Tel: 01206 211 666
Fax: 01206 212 450
E-mail: gec@ndirect.co.uk

Cordek Ltd
Spring Copse Business Park
Slinfold
West Sussex
RH13 7SZ
Trough Moulds
Tel: 01403 799 600
Fax: 01403 791 718
E-mail: sales@cordek.com

Corus UK Ltd
PO Box 1
Scunthorpe
South Humberside
DN16 1BP
Structural Steel
Tel: 01724 404 040
Fax: 01724 282 599

Cox Building Products
Icknield Way
Tring
Hertfordshire
HP23 4RF
Rooflights
Tel: 01442 820 500
Fax: 01442 820 550
E-mail: coxdome@btinternet.com

Crittal Steel Windows
Springwood Drive
Braintree
Essex
CM7 2YN
Aluminium and Steel Windows
Tel: 01376 324 106
Fax: 01376 349 662
E-mail: hp@crittal-windows.co.uk
Website: www.crittal-windows.co.uk

Daniel Platt Ltd
Brownhills Tileries
Tunstall
Stoke-on-Trent
Staffordshire
ST6 4NY
Quarry & Ceramic Floor Tiles
Tel: 01782 577 187
Fax: 01782 577 877
E-mail: sale@danielplatt.co.uk
Website: www.danielplatt.co.uk

Decra Roof Systems (UK) Ltd
Unit 7
Church Road Industrial Estate
Lowfield Heath
Crawley
West Sussex
RH11 0YA
"Decra Stratos" Roofing System and Accessories
Tel: 01293 545 058
Fax: 01293 562 709
E-mail: technical@decra.co.uk
Website: www.decra.co.uk

Dennis Ruabon Ltd
Hafod Tileries
Ruabon
Wrexham
Clywdd
LL14 6ET
Quarry Floor Tiles
Tel: 01978 843 484
Fax: 01978 843 276
E-mail: sales@dennisruabon.co.uk

Dow Construction Products
2 Heathrow Boulevard
284 Bath Road
West Drayton
Middlesex
UB7 0DQ
Insulation Products
Tel: 0208 917 5050
Fax: 0208 917 5413

DSS C&L Sisalkraft
Commissioners Road
Strood
Rochester
Kent
ME2 4ED
Roof Finish Underlays and Insulation
Tel:		01634 292 700
Fax:		01634 291 029

Dufaylite Developments Ltd
Cromwell Road
St Neots
Huntingdon
Cambridgeshire
PE19 1QW
"Dufaylite Clayboard"
Tel:		01480 215 000
Fax:		01480 405 526
E-mail:	andy.quincey@dufaylite.com
Website:	www.dufaylite.com

Durabella Ltd
2nd Floor
Talisman Square
Kenilworth
Warwickshire
CV8 1J8
"Westbourne" flooring
Tel:		01926 514 000
Fax:		01926 850 359

Envirodoor Ltd
Viking Close
Willerby
East Yorkshire
HU10 6BS
Sliding and Folding Doors
Tel:		01482 659 375
Fax:		01482 655 131
E-mail:	sales@envirodoor.com

Eternit (UK) Ltd
Meldreth
Royston
Hertfordshire
SG8 5RL
Sheeting, Boards and Tiles
Tel:		01763 264 686
Fax:		01763 262 338
E-mail:	marketing@eternit.co.uk
Website:	www.eternit.co.uk

Eurobrick Systems Ltd
Unit 6
Bonville Business Centre
Dixon Road
Brislington
Bristol
BS4 5QW
Insulated Brick Cladding System
Tel:		0117 971 7117
Fax:		0117 971 7217
Website:	www.eurobrick.co.uk

Eurocom Enterprise Ltd
75 Tithe Barn Drive
Bray
Maidenhead
Berkshire
SL6 2DD
Terne Coated Stainless Steel Roofing
Tel:		01628 687 022
Fax:		01628 687 023
E-mail:	terne@dial.pipex.com
Website:	www.enox.co.uk

Evode Ltd
Common Road
Stafford
Staffordshire
ST16 3EH
"Evo-Stik" and "Flashband" Products
Tel:		01785 257 236
Fax:		01785 241 615
E-mail:	evo-stik@evode.co.uk
Website:	www.evode.co.uk

Expamet Building Products
PO Box 52
Longhill Industrial Estate (North)
Hartlepool
Cleveland
TS25 1PR
Expanded Metal Building Products
Tel:		01429 866 611
Fax:		01429 866 633
E-mail:	expamet@compuserve.com

Fendor-Hansen
William Street
Felling
Gateshead
NE10 0TP
Industrial Doors
Tel:		0191 438 3222
Fax:		0191 438 1686

Forbo-Nairn Ltd
PO Box 1
Den Road
Kirkcaldy
Fife
KY1 2SB
Tiles and Sheet Flooring
Tel: 01592 643 777
Fax: 01592 643 999
E-mail: headoffice@forbo-nairn.co.uk
Website: www.forbo-nairn.co.uk

Forticrete Ltd
Bridle Way
Merseyside
L30 4UA
Concrete and Stone Faced Blocks
Tel: 0151 521 3545
Fax: 0151 524 1265
E-mail: Technical@Forticrete.co.uk
Website: www.Forticrete.co.uk

Forticrete Ltd
Thornley Station Industrial Estate
Salters Lane
Shotton Colliery
County Durham
DH6 2QA
"Astra" Glazed Blocks
Tel: 01429 838 001
Fax: 01429 836 206
E-mail: Technical@Forticrete.co.uk
Website: www.Forticrete.co.uk

Forticrete Roofing Products
Heath Road
Leighton Buzzard
Bedfordshire
LU7 8ER
"Hardrow" Slating
Tel: 01525 244 900
Fax: 01525 850 432
E-mail: Technical@Forticrete.co.uk
Website: www.Forticrete.co.uk

Fosroc Ltd
Coleshill Road
Tamworth
Staffordshire
B78 3TL
Admixtures, Expansion Joint Fillers etc
Tel: 01827 262 222
Fax: 01827 262 444
Website: FosrocUK.com

Garador Ltd
Bunford Lane
Yeovil
Somerset
BA20 2YA
Garage Doors
Tel: 01935 443 700
Fax: 01935 443 744
Website: www.garador.co.uk

Geberit Ltd
New Hythe Business Park
New Hythe Lane
Aylesford
Kent
ME20 7PJ
Above and Below Ground Drainage Goods
Tel: 01622 717 811
Fax: 01622 716 920

Gerflor Ltd
Rothwell Road
Warwick
Warwickshire
CV34 5PY
Sheet and Tile Flooring
Tel: 01926 401 500
Fax: 01926 401 650
E-mail: gerflor@btinternet.com

Gil Air Ltd
Artex Avenue
Rustington
West Sussex
BN16 3LN
Aluminium Louvres
Tel: 01903 782 060
Fax: 01903 859 213
E-mail: sales@gill-air.co.uk

Glenigan Cost Information Services
41 - 47 Seabourne Road
Bournemouth
Dorset
BH5 2HU
The Heavyside Building Materials Price Guide
Tel: 01202 432 121
Fax: 01202 421 807
E-mail: info@glenigan.emap.co.uk

GRAB Industrial Flooring Ltd
Park Road
South Wigston
Leicester
Leicestershire
LE18 4QD
Resin Based Flooring
Tel: 0116 277 0024
Fax: 0116 277 4038

Grace Servicised
Expansion Jointing and Waterproofing Division
Ajax Avenue
Slough
Berkshire
SL1 4BH
Expansion Joint Fillers and Waterbars
Tel: 01753 692 929
Fax: 01753 691 623
E-mail: sales@grace-construction-products.co.uk
Website: www.gcp-grace.com

Granwood Flooring Ltd
PO Box 60
Alfreton
Derbyshire
DE55 4ZX
Sports Flooring
Tel: 01773 606 060
Fax: 01773 606 030
Website: www.granwood.co.uk

H&H Celcon Ltd
Celocn House
Ightham
Sevenoaks
Kent
TN15 9JB
Concrete Blocks
Tel: 01732 886 333
Fax: 01732 886 810
E-mail: marketing@celcon.co.uk
Website: www.celcon.co.uk

Halfen Ltd
Humphrys Road
Woodside Estate
Dunstable
Bedfordshire
LU5 4TP
Building Fixings and Fastenings
Tel: 01582 470 300
Fax: 08705 316 304
E-mail: info@halfen.co.uk
Website: www.halfen.co.uk

Hanson Brick Ltd
Stewartby
Bedford
MK43 9LZ
"London Brand" Facing Bricks
Tel: 0870 525 8258
Fax: 01234 762 040

Hanson Bath & Portland Stone
Bumpers Lane
Wakeham
Portland
Dorset
DT5 1HY
Natural Stone
Tel: 01305 820 207
Fax: 01305 860 275
Website: www.hanson-quarryproduct.com

Hanson Conbloc
PO Box 14
Appleford Road
Sutton Courtenay
Abingdon
Oxfordshire
OX14 4UB
Building Block Products
Tel: 01235 848 877
Fax: 01235 848 767
Website: www.hanson-conbloc.com

Hanson Southern
Garston Road
Frome
Somerset
BA11 1RS
Cotswold Stone
Tel: 01373 453 333
Fax: 01373 452 964

Harcros Timber & Building Supplies Ltd
Harcros House
1 Central Road
Worcester Park
Surrey
KT4 8DN
Structural Timber
Tel: 0208 337 6666
Fax: 0208 255 2280
E-mail: feedback@harcros.co.uk

Harvey Fabrication Ltd
Hancock Road
London
E3 3DA
Galvanised Tank & Water Cylinders
Tel: 0208 981 7811
Fax: 0208 981 7815

HCC Protective Coatings Ltd
Suite F5 Bates Business Centre
Church Road
Harold Wood
Romford
Essex
RM3 0JF
Paints and Primers
Tel: 01708 378 666
Fax: 01708 378 868
E-mail: 101712.2170@compuserve.com

Hepworth Building Products Ltd
Hazelhead
Crow Edge
Sheffield
South Yorkshire
S36 4HG
Clay Drainage Goods
Tel: 01226 763 561
Fax: 01226 764 827
Website: www.hepworthdrainage.co.uk

Hillaldam Coburn Ltd
Unit 6
Wyvern Estate
Beverly Way
New Malden
Surrey
KT3 4PH
Sliding and Folding Door Gear
Tel: 0208 336 1515
Fax: 0208 336 1414
E-mail: sales@hillaldam.co.uk
Website: www.coburn.co.uk

Hilti (Great Britain) Ltd
1 Trafford Wharf Road
Trafford Park
Manchester
M17 1BY
Fixings
Tel: 0800 886 100
Fax: 0800 886 200
Website: www.hilti.com

Hinton, Perry & Davenhill Ltd
Dreadnought Works
Pensnett
Brierley Hill
Staffordshire
DY5 4TH
Overlap Roof Tiling
Tel: 01384 774 05
Fax: 01384 745 53

HSS Hire Shops
Group Office
25 Willow Lane
Mitcham
Surrey
CR4 4TS
Paint Guns
Tel: 0208 260 3100
Fax: 0208 687 5005
Website: www.hss.co.uk

Hudevad Britain
Bridge House
Bridge Street
Walton-on-Thames
Surrey
KT12 1AL
Hot Water Radiators and Fittings
Tel: 01932 247 835
Fax: 01932 247 694

Hunter Plastics Ltd
Nathan Way
London
SE28 0AE
Plastic Rainwater Goods
Tel: 0208 855 9851
Fax: 0208 317 7764
E-mail: hunplas@hunterplastics.co.uk

Huntley & Sparks Ltd
Building Products Division
Sterling House
Crewkerne
Somerset
TA18 8LL
"Rigifix" Column Guards
Tel: 01460 722 22
Fax: 01460 764 02
E-mail: email@sterling-hydraulics.co.uk

Ibstock Building Products Ltd
Leicester Road
Ibstock
Leicestershire
LE67 6HS
Facing Bricks
Tel: 01530 261 999
Fax: 01530 261 888
E-mail: marketing@ibstock.co.uk
Website: www.ibstock.co.uk

Acknowledgements

IG Ltd
Avondale Road
Cwmbran
Gwent
NP44 1XY
"Weatherbeater" Doors & Frames
Tel: 01633 486 486
Fax: 01633 486 465
E-mail: info@igltd.co.uk
Website: www.igltd.co.uk

IMI Yorkshire Copper Tube Ltd
East Lancashire Road
Kirkby
Liverpool
Merseyside
L33 7TU
Copper Piping
Tel: 0151 546 2700
Fax: 0151 549 2139

IMI Yorkshire Fittings Ltd
PO Box 166
Haigh Park Road
Leeds
LS1 1RD
"Yorkshire" & "Kuterlite" Fittings
Tel: 0113 270 1104
Fax: 0113 271 5275
E-mail: yorkshire.fittings@imiyf.com
Website: www.imiyf.com

Industrial Textiles & Plastics Ltd
Providence Hill
Husthwaite
York
North Yorkshire
YO61 4PN
Roof Finish Underlays and Insulation
Tel: 01347 868 767
Fax: 01347 868 737
E-mail: info@indtex.co.uk
Website: www.indtex.co.uk

Interface Europe Ltd
Ashlyns Hall
Chestham Road
Berkhamstead
Hertfordshire
HP4 2ST
Non-woven Carpet Tiling
Tel: 01442 285 000
Fax: 01442 876 053
Website: www.interfaceinc.com

James Halstead Ltd
PO Box 3
Radcliffe New Road
Whitefield
Manchester
M45 7NR
"Polyflor" Contract Flooring
Tel: 0161 767 1111
Fax: 0161 767 1128

James Latham (Western)
Badminton Road
Yate
Bristol
Avon
BS37 5JX
Softwood, Hardwood and Panel Products
Tel: 01454 315 421
Fax: 01454 323 488
E-mail: plywood.west@lathams.co.uk
Website: www.lathamtimber-co.uk

Jewson Ltd
Sutherland House
Matlock Road
Foleshill
Coventry
West Midlands
CV1 4JQ
Standard Joinery Products
Tel: 02476 669 100
Fax: 02476 669 101

John Brash & Co Ltd
The Old Shipyard
Gainsborough
Lincolnshire
DN21 1NG
Cedar Shingling
Tel: 01427 613 858
Fax: 01427 810 218
E-mail: email@johnbrash.co.uk

John Newton & Co Ltd
12 Verney Road
London
SE16 3DH
Waterproofing Lathing
Tel: 0207 237 1217
Fax: 0207 252 2769

Jonathan James Ltd
15 - 17 New Road
Rainham
Essex
RM13 8DJ
Plastering and Screeding
Tel: 01708 556 921
Fax: 01708 520 751

Junckers Ltd
Wheaton Court Commercial Centre
Wheaton Road
Whitham
Essex
CM8 3UJ
Hardwood Flooring
Tel: 01376 517 512
Fax: 01376 514 401

Kelsey Roofing Industries Ltd
Paper Mill Drive
Church Hill South
Redditch
Worcestershire
B98 8QJ
Sheet Roof Claddings
Tel: 01527 594 400
Fax: 01527 594 444

Keymer Tiles Ltd
Nye Road
Burgess Hill
West Sussex
RH15 0LZ
Roof Tiling
Tel: 01444 232 931
Fax: 01444 871 852
E-mail: info@keymer.co.uk
Website: www.ketmer.co.uk

Kitsons Insulation Products Ltd
Unit A389
Western Avenue
Team Valley Trading Estate
Gateshead
Tyne and Wear
NE11 0SZ
Pipe Insulation Products
Tel: 0191 482 1261
Fax: 0191 491 0700

Klargester Environmental Engineering Ltd
College Road
Aston Clinton
Aylesbury
Buckinghamshire
HP22 5EW
Petrol Interceptors and Septic Tanks
Tel: 01296 633 000
Fax: 01296 633 001
E-mail: home@klargester.co.uk
Website: www.klargester.co.uk

Kriscut Concrete Drilling & Sawing Ltd
15 Dawley Road
Hayes
Middlesex
UB3 1LT
Diamond Drilling
Tel: 0208 561 8378
Fax: 0208 573 9776

Kvaerner Cementation Foundations Ltd
Maple Cross House
Denham Way
Maple Cross
Rickmansworth
Hertfordshire
WD3 2SW
Piling
Tel: 01923 423 100
Fax: 01923 777 834

Lafarge Redland Precast Ltd
Six Hills
Melton Mowbray
Leicestershire
LE14 3PD
Precast Concrete Paving Units
Tel: 01509 882 121
Fax: 01509 880 799
Website: www.uk.lafarge-aggregates.com

Laidlaw Architectual Hardware
Centurion House
Dakota Avenue
Salford
Manchester
M5 2PU
Ironmongery
Tel: 0161 848 0101
Fax: 0161 848 1750

Lamatherm Products Ltd
Forge Factory Estate
Maesteg
Mid Glamorgan
CF34 0AU
Under Floor Cavity Barriers
Tel: 01656 730 833
Fax: 01656 730 115

Acknowledgements

Langley Waterproofing Systems Ltd
Bishop Crewe House
North Street
Daventry
Northamptonshire
NN11 5PN
Roof Tiling
Tel: 01327 704 778
Fax: 01327 704 845
E-mail: langleyuk@btinternet.com
Website: www.langleyuk.co.uk

Leaderflush Doors + Shapland
PO Box 5404
Nottingham
Nottinghamshire
NG16 4BU
Internal Doors
Tel: 0870 240 0666
Fax: 0870 240 0777
E-mail: marketing@leaderflushshapland.co.uk
Website: www.leaderflush.co.uk

Light Alloy Ltd
Dales Road
Ipswich
Suffolk
IP1 4RJ
"Zig Zag" Loft Ladders
Tel: 01473 740 445
Fax: 01473 240 002

Llewellyn Timber Engineering
Bleak Hall
Milton Keynes
Buckinghamshire
MK6 1LA
Roof Trusses
Tel: 01908 679 222
Fax: 01908 678 631

Luxaflex Blinds (Hunter Douglas Ltd)
Swanscombe Business Centre
17 London Road
Swanscombe
Kent
DA1 0LH
Roller and Vertical Blinds
Tel: 01322 624 580
Fax: 01322 624 558
Website: www.luxaflex.com/uk/projects

Luxcrete Ltd
Premier House
Disraeli Road
Park Royal
Harlesden
London
NW10 7BT
Pavement and Roof Lighting
Tel: 0208 965 7292
Fax: 0208 961 6337

Maccaferri (Ltd)
7400 The Quorum
Oxford Business Park
Garsington Road
Oxford
Oxfordshire
OX4 2JZ
Gabions
Tel: 01865 770 555
Fax: 01865 774 550
E-mail: oxford@maccaferri.co.uk

Manthorpe Building Products Ltd
Manthorpe House
Brittain Drive
Codnor Gate Industrial Estate
Ripley
Derbyshire
DE5 3ND
Cavity Closers
Tel: 01773 743 555
Fax: 01773 748 736
E-mail: sales@manthorpe.co.uk
Website: www.manthorpe.co.uk

Marley Building Materials Ltd
Station Road
Coleshill
Birmingham
B46 1HP
Roof Tiles, Pavings and "Thermalite" Blocks
Tel: 01675 468 400
Fax: 01675 468 485
Website: www.marley.co.uk

Marley Floors Ltd
Dickley Lane
Lenham
Maidstone
Kent
ME17 2DE
Sheet and Tile Flooring
Tel: 01622 854 000
Fax: 01622 854 219
E-mail: info@marfaw.u-net.com
Website: www.marley.co.uk

Marley Plumbing and Drainage
Dickley Lane
Lenham
Maidstone
Kent
ME17 2DE
Drainage Goods
Tel: 01622 858 888
Fax: 01622 858 725
E-mail: marketing@marleyext.com
Website: www.marley.co.uk

Marshalls Mono Ltd
Brier Lodge
Southowram
Halifax
West Yorkshire
HX3 9SY
Concrete Block Paving
Tel: 01422 306 000
Fax: 01422 330 185
Website: www.marshall.co.uk

Metsec Plc
Broadwell Road
Oldbury
Warley
West Midlands
B69 4HE
Lattice Beams
Tel: 0121 601 6000
Fax: 0121 601 6119
E-mail: metsecframing@metsec.com
Website: www.metsec.com

Moelven Laminated Timber Structures Ltd
Unit 10
Vicarage Farm
Winchester Road
Fair Oak
Eastleigh
Hampshire
SO50 7HD
"Glulam" Laminated Timber Beams
Tel: 02380 695 566
Fax: 02380 695 577

Mustang Metal Products
Unit 6
Worcester Trading Estate
Blackpole
Worcester
Worcestershire
WR3 8HR
Rainwater Goods
Tel: 01905 455 151
Fax: 01905 755 055

Myson Radiators
Eastern Avenue
Team Valley Trading Estate
Gateshead
Tyne and Wear
NE11 0PG
Hot Water Radiators and Fittings
Tel: 0191 491 7500
Fax: 0191 491 7568
E-mail: ksmith@myson.co.uk
Website: www.myson.co.uk

Newmond Building Products Ltd
Pioneer House
Lichfield Road Industrial Estate
Tamworth
Staffordshire
B79 7TF
Claddings, Fascias, Bargeboards and Soffits
Tel: 01827 317 200
Fax: 01827 317 201
E-mail: marketing@swishbp.co.uk

Owens Alcopor (UK) Ltd
PO Box 10
Stafford Road
St Helens
Merseyside
WA10 3NS
Building Insulation Products
Tel: 0800 627 465
Fax: 01744 612 007
E-mail: IASUK@ownscorning.com
Website: www.owenscorning.com/worldwide

Parkwood Mellowes Ltd
Ridgacre Road
West Bromwich
West Midlands
B71 1BB
Patent Glazing
Tel: 0121 553 4011
Fax: 0121 553 4019

PD Edenhall Ltd
Danycraig Works
Danycraig Road
Risca
Newport
Monmouthshire
NP1 6DP
Flue Linings and Blocks
Tel: 01633 612 671
Fax: 01633 601 280

Acknowledgements

Phi Group Ltd
Harcourt House
Royal Crescent
Cheltenham
Gloucestershire
GL50 3DA
Crib Walling
Tel: 0870 333 4122
Fax: 0870 333 4121
E-mail: info@phigroup.co.uk
Website: www.phigroup.co.uk

Pilkington United Kingdon Ltd
Prescot Road
St Helens
Merseyside
WA10 3TT
Glazing
Tel: 01744 692 000
Fax: 01744 613 049
E-mail: info@pilkington.com
Website: www.pilkington.co.uk

Pilkington Tiles Ltd
PO Box 4
Rake Lane
Clifton Junction
Swinton
Manchester
M27 8LP
Glazed Wall/Floor Tiles
Tel: 0161 727 1000
Fax: 0161 727 1066

Pressalit Ltd
Riverside Business Park
Dansk Way
Leeds Road
Ilkley
West Yorkshire
LS29 8JZ
Grab Rails
Tel: 01943 607 651
Fax: 01943 607 214

Promat Fire Protection Ltd
Whaddon Road
Meldreth
Near Royston
Hertfordshire
SG8 5RL
Fireproofing Materials
Tel: 01763 262 310
Fax: 01763 262 342

Protim Solignum Ltd
Fieldhouse Lane
Marlow
Buckinghamshire
SL7 1LS
Stains and glazes
Tel: 01628 486 644
Fax: 01628 481 276
Website: www.protimsolignum.com

Quelfire
PO Box 35
Caspian Road
off Atlantic Street
Altrincham
Cheshire
WA14 5QA
Fire Protection Compound
Tel: 0161 928 7308
Fax: 0161 941 2635

Quiligotti Terrazzo Ltd
PO Box 4
Clifton Junction
Manchester
M27 8LP
Terrazzo Products
Tel: 0161 727 1187
Fax: 0161 727 1066

Redland Roofing Systems Ltd
Regent House
Station Approach
Dorking
Surrey
RH4 1TG
Roof Tiling
Tel: 01306 872 000
Fax: 01306 872 111
E-mail: roofing@redland.co.uk
Website: www.redland.co.uk

Rentokil Ltd
Industrial Products
Felcourt
East Grinstead
West Sussex
RH19 2JY
Flame Retardant Products
Tel: 01342 833 022
Fax: 01342 326 229

Rex Bousfield Ltd
Holland Road
Hurst Green
Oxted
Surrey
RH8 9BD
Fire Retardant Chipboarding
Tel:	01883 717 033
Fax:	01883 717 890
E-mail:	sales@bousfield.com
Website:	www.bousfield.com

Richard Lees Steel Decking Ltd
Moor Farm Road West
The Airfield
Ashbourne
Derbyshire
DE6 1HD
Structural Steel Deck Flooring
Tel:	01335 300 999
Fax:	01335 300 888
E-mail:	RLSD-General.E-mail@kvaerner.com

RMC Aggregates (Northern) Ltd
RMC House
8-9 Blezard Business Park
Brenkley Way
Seaton Burn
Newcastle-upon-Tyne
Tyne & Wear
NE13 6DR
Aggregates
Tel:	0191 217 0299
Fax:	0191 217 0298

RMC Concrete Products Ltd
London Road
Wick
Bristol
BS30 5SJ
Paving Blocks
Tel:	0117 937 3740
Fax:	0117 937 3002

RMC Panel Products Ltd
Waldorf Way
Denby Dale Road
Wakefield
West Yorkshire
WF2 8DH
Plasterboard Angle Support System
Tel:	01924 362 081
Fax:	01924 290 126
E-mail:	thermabate@dial.pipex.com
Website:	www.thermabate.co.uk

Rockwool Ltd
Pencoed
Bridgend
Mid Glamorgan
CF35 6NY
Cavity Wall Batts
Tel:	01656 862 621
Fax:	01656 862 302

Rom Ltd
Meadowhall Road
Sheffield
S9 1ED
Reinforcement
Tel:	0114 256 0152
Fax:	0114 256 1011

Ruberoid Building Products Ltd
Business Development Manager
Tewin Road
Welwyn Garden City
Hertfordshire
AL7 1BP
Roofing Felts, DPC's
Tel:	01707 822 222
Fax:	01707 375 060
E-mail:	rbp-wpc@ruberoid.co.uk

Ryton's Building Products Ltd
Design House
Orion Way
Kettering Business Park
Kettering
Northamptonshire
NN15 6NL
Roof Ventilation Products
Tel:	01536 511 874
Fax:	01536 310 455
E-mail:	vents@rytons.com
Website:	www.rytons.com

Saint Gobain
Sinclair Works
PO Box 3
Ketley
Telford
Shropshire
TF1 5AD
Drainage Pipes and Fittings
Tel:	01952 262 500
Fax:	01952 262 555

Sandtoft Roof Tiles Ltd
Sandtoft Tileries
Sandtoft
Doncaster
South Yorkshire
DN8 5SY
Clay Roof Tiling
Tel: 01427 871 200
Fax: 01427 871 222
E-mail: support@sandtoft.co.uk
Website: www.sandtoft.co.uk

Screeduct Ltd
PO Box 7000
Erdington
Birmingham
B23 6TR
Trunking Systems and Conduits
Tel: 0121 384 6200
Fax: 0121 377 8938

Sealmaster
Brewery Road
Pampisford
Cambridge
Cambridgeshire
CB2 4HG
Intumescent Strips and Glazing Compound
Tel: 01223 832 851
Fax: 01223 837 215
E-mail: sales@sealmater.co.uk
Website: www.sealmaster.co.uk

Sealocrete Pla Ltd
Greenfield Lane
Rochdale
Lancashire
OL11 2LD
Building Aids, Repair and Maintenance Materials
Tel: 01706 352 255
Fax: 01706 860 880
E-mail: bestproducts@sealocrete.co.uk

Sheffield Insulation Ltd
Sanderson Street
off Scotswood Road
Newcastle upon Tyne
Tyne and Wear
NE4 7LW
Cavity Wall Insulation
Tel: 0191 2735 843
Fax: 0191 2739 213

Simpson Strong-Tie
Winchester Road
Cardinal Point
Tamworth
Staffordshire
B78 3HG
Expanded Metal Arch Formers, Wall Ties etc
Tel: 01827 255 600
Fax: 01827 255 616
E-mail: tsimons@strongtie.com
Website: www.strongtie.com

Stanton Plc
Lows Lane
Stanton-by-Dale
Ilkeston
Derbyshire
DE7 4QU
Ductile Iron Pipes and Fittings
Tel: 0115 930 5000
Fax: 0115 930 0737

Steel Foundations Ltd
Stevin House
Springwell Road
Gateshead
Tyne and Wear
NE9 7SP
Piling Services
Tel: 0191 417 3545
Fax: 0191 416 2894

Stressline Ltd
Station Road
Stoney Stanton
Leicester
Leicestershire
LE9 4LX
Concrete Lintels
Tel: 01455 272 457
Fax: 01455 274 564

Tarmac Topblock Ltd
Wergs Hall
Wergs Hall Road
Wolverhampton
West Midlands
WV8 2HZ
Concrete Blocks
Tel: 01902 754 131
Fax: 01902 743 171

Spon Press Marketing
Freepost SN926
11 New Fetter Lane
LONDON EC4B 4FH

Affix stamp
here if
posting
from outside
the UK

FREE UPDATES

REGISTER TODAY

Available to download from the web: www.pricebooks.co.uk

All four Spon Price Books - Architects' and Builders', Civil Engineering and Highway Works, Landscape and External Works and Mechanical and Electrical Services - are supported by an updating service. Three updates are issued during the year, in November, February and May. Each gives details of changes in prices of materials, wage rates and other significant items, with regional price level adjustments for Northern Ireland, Scotland and Wales and regions of England.

As a purchaser of a Spon Price Book you are entitled to this updating service for the 2002 edition - free of charge. Simply complete this registration card and return it to us - or register via the website **www.pricebooks.co.uk**. The updates will also be available to download from the website.

The Updating service terminates with the publication of the next annual edition.

REGISTRATION CARD for Spon's Price Book Update   **2002**
Please print your details clearly

Name..
(Please indicate membership of professional bodies e.g. RICS, RIBA, CioB etc.)
Position/Department..
Organisation..
Address...
....
.... Postcode.......................................
Tel:... Fax:...
E-mail address... (Please print clearly)

FIND OUT MORE ABOUT SPON BOOKS
Visit www.sponpress.com for more details.

Spon's Price Books Online
☐ Please tick the box if you are interested in subscribing to Spon's price data online. We will contact you if the service becomes available.
Please tick the other areas of interest
☐ International Price Books   ☐ Occupational Safety and Health
☐ Architecture   ☐ Landscape
☐ Planning   ☐ Journals - inc. Journal of Architecture;
☐ Civil Engineering   Building Research and Information;
☐ Building & Construction Management   Construction Management and Economics
☐ Environmental Engineering

☐ Other - note any other areas of interest...
....

Tarmac Topmix Ltd
Wergs Hall
Wergs Hall Road
Wolverhampton
West Midlands
WV8 2HZ
Ready mixed Concrete
Tel: 01902 353 522
Fax: 01902 742 262
Website: www.Topmix.com

Terminix Ltd
Heritage House
234 High Street
Sutton
Surrey
SM1 1NX
Chemical and Damp-Proofing Products
Tel: 0208 661 6600
Fax: 0208 642 0677
Website: www.terminix.co.uk

Terram Ltd
Mamhilad Park
Pontypool
Gwent
NP4 0YR
Soil Reinforcement Materials
Tel: 01495 757 722
Fax: 01495 762 383
Website: www.terram.co.uk

The Colour Centre
Portland Street
Newport
Gwent
NP9 2DP
Paints, Varnishes and Stains
Tel: 01633 212 481
Fax: 01633 254 801

The Rawlplug Company Ltd
Skibo Drive
Thornliebank Industrial Estate
Glasgow
Scotland
G46 8JR
"Kemfix" Chemical Anchors
Tel: 0141 638 7961
Fax: 0141 273 2333

Tilcon (North) Ltd
PO Box 5
Fell Bank
Birtley
Chester-le-Street
County Durham
DH3 2ST
Mortars
Tel: 0191 492 4000
Fax: 0191 410 8489

Tremco Ltd
393 Edinburgh Avenue
Slough
Berkshire
SL1 4UF
Damp Proofing
Tel: 01753 691 696
Fax: 01753 822 640
Website: www.tremcoeurope.com

United Wire Ltd
Jenna House
101 Blackhorse Lane
London
E17 6DJ
Balloon Guards
Tel: 0208 523 2135
Fax: 0208 531 1633

Uponor Ltd
PO Box 1
Hillcote Plant
Blackwell
Alfreton
Derbyshire
DE55 5JD
MDPE Piping and Fittings
Tel: 01773 811 112
Fax: 01773 812 343
Website: www.uponor.com

USG (UK) Ltd
1 Swan Road
Southwest Industrial Estate
Peterlee
County Durham
SR8 2HS
Ceiling Grid Systems, Panels and Tiles
Tel: 0191 586 1121
Fax: 0191 586 0097

Acknowledgements

Vandex (UK) Ltd
PO Box 88
Leatherhead
Surrey
KT22 7YF
"Vandex Super" & "Premix"
Tel: 01372 363 040
Fax: 01372 363 373

Velux Company Ltd
Woodside Way
Glenrothes East
Fife
KY7 4ND
"Velux" Roof Windows & Flashings
Tel: 01592 772 211
Fax: 01592 771 839
E-mail: enquiries@velux.co.uk
Website: www.velux.co.uk

Vencel Resil Ltd
Infinity House
Anderson Way
Belvedere
Kent
DA17 6BG
"Jablite" Products
Tel: 0208 320 9100
Fax: 0208 320 9110
E-mail: technical@vencel.co.uk

Vigers Hardwood Flooring Systems Ltd
Esgors Farm
Thornwood Common
Epping
Essex
CM16 6LY
Hardwood Flooring Systems
Tel: 01992 572 653
Fax: 01992 572 663

Walter Llewellyn & Sons Ltd
19 North Street
Portslade
Brighton
Sussex
BN41 1ED
Purpose Made Joinery
Tel: 01273 439 494
Fax: 01273 430 741

Wavin Building Products Ltd
Parsonage Way
Chippenham
Wiltshire
SN15 5PN
uPVC Drainage Goods
Tel: 01249 654 121
Fax: 01249 443 286

Welconstruct Company
Woodgate Business Park
Kettles Wood Drive
Birmingham
B32 3GH
Lockers and ShelvingSystems
Tel: 0121 421 9000
Fax: 0121 421 9888
E-mail: mail@welconstruct.co.uk

Wellfill Insulation Ltd
284 High Road
North Weald
Essex
CM16 6EG
Cavity Wall Insulation
Tel: 01992 522 026
Fax: 01992 523 376

Wicanders
Amorim House
Star Road
Partridge Green
Horsham
West Sussex
RH13 8RA
Floor Coverings
Tel: 01403 710 001
Fax: 01403 710 003

William Blyth Ltd
Hoe Hill
Barton-on-Humber
North Lincolnshire
DN18 5ET
Roof Tiles
Tel: 01652 632 175
Fax: 01652 660 966

WP Metals
Westgate
Aldridge
West Midlands
WS9 8DJ
Rainwater Goods
Tel: 01922 743 111
Fax: 01922 743 344
E-mail: sales@wpmetals.co.uk

Yeoman Aggregates Ltd
Stone Terminal
Horn Lane
Acton
London
W3 9EH
Hardcore, Gravels etc
Tel: 0208 896 6800
Fax: 0208 896 6811

How to Use this Book

First-time users of *Spon's Architects' and Builders' Price Book* and others who may not be familiar with the way in which prices are compiled may find it helpful to read this section before starting to calculate the costs of building work. The level of information on a scheme and availability of detailed specifications will determine which section of the book and which level of prices users should refer to.

For preliminary estimates/indicative costs before drawings are prepared, refer to the average overall *Building Prices per Functional Units* (see Part IV) and multiply this by the proposed number of units to be contained within the building (ie. number of bedrooms etc.) or *Building Prices per Square Metre* rates (see Part IV) and multiply this by the gross internal floor area of the building (the sum of all floor areas measured within external walls) to arrive at an overall preliminary cost. These rates include Preliminaries (the Contractors' costs) but make no allowance for the cost of External Works or VAT.

For budget estimates where preliminary drawings are available, one should be able to measure approximate quantities for all the major components of a building and multiply these by individual rates contained in the *Approximate Estimating* section (see Part IV). This should produce a more accurate estimate of cost than using overall prices per square metre. Labour and other incidental associated items, although normally measured separately within Bills of Quantities, are deemed included within approximate estimate rates.

For more detailed estimates or documents such as Bills of Quantities (Quantities of supplied and fixed components in a building, measured from drawings), either use rates from *Prices for Measured Work - Major Works* or *Prices for Measured Work - Minor Works*, depending upon the overall value of the contract. All such prices used may need adjustment for size, site constraints, local conditions location and time, etc., and users are referred to page 694 for an example of how to adjust an estimate for some of these factors. Items within the Measured Works sections are made up of many components: the cost of the material or product; any additional materials needed to carry out the work; the labour involved in unloading and fixing, etc. These components are usually broken down into:

Prime Cost

Commonly known as the "PC", Prime Cost is the actual price of the material such as bricks, blocks, tiles or paint, as sold by suppliers. Prime Cost is given "per square metre", "per 100 bags" or "each" according to the way the supplier sells the product. Unless otherwise stated, prices in *Spon's Architects' and Builders' Price Book* (hereafter referred to as *Spon's A & B*), are deemed to be "delivered to site", (in which case transport costs will be included) and also take account of trade and quantity discounts. Part loads generally cost more than whole loads but, unless otherwise stated, Prime Cost figures are based on average prices for full loads delivered to a hypothetical site in Acton, London W3. Actual prices for "live" tenders will depend on the distance from the supplier, the accessibility of the site, whether the whole quantity ordered is to be supplied in one delivery or at specified dates and market conditions prevailing at the time. Prime Cost figures for commonly-used alternative materials are supplied in tabular form at the beginning of some work sections although these are quoted at "list" prices, readers should deduct their appropriate discount before substituting them as alternatives in rate build ups.

Labour

This figure covers the cost of the operation and is calculated on the gang wage rate (skilled or unskilled) and the time needed for the job. A full explanation and build-up is provided on page 101. Particular shortages have been noted in the Bricklaying, Carpentry, Joinery and Shuttering trades. In response to these pressures and corresponding hikes in daily rates/measured prices, basic labour rates for Bricklayers, Carpenters and Joiners have been enhanced by using the plus 15% and plus 25% bonus payment figures for these trades as outlined on page 102. Extras such as highly skilled craft work, difficult access, intermittent working and the need for additional labourers to back up the craftsman (e.g. for handling heavy concrete blocks etc.) all add to the cost. Large regular or continuous areas of work are cheaper to install than smaller intricate areas, since less labour time is wasted moving from one area to another.

Materials

Material prices include the cost of any ancillary materials, nails, screws, waste, etc., which may be needed in association with the main material product/s. If the material being priced varies from a standard measured rate, then identify the difference between the original PC price and the material price and add this to your alternative material price before adding to the labour cost to produce a new overall Total rate. Alternative material prices, where given, are quoted before discount whereas PC values of materials are after deduction of appropriate trade and quantity discounts, therefore straight substitution of alternative material prices for PC values will result in the actual rate for those alternative materials.

Example:

	PC £	Labour hours	Labour £	Material £	Unit	Total Rate £
100 mm Thermalite Turbo block (see page 177)	6.48	0.46	8.04	8.04	m²	**16.08**
100 mm Toplite standard block (£7.11 see page 177 - less 20% discount)	5.69					
Sundries £8.04 - 6.48 = 1.56 + 5.69 = £7.25						
100 mm Toplite block	5.69	0.46	8.04	7.25	m²	**15.29**

Plant

Plant covers the use of machinery ranging from JCBs to shovels and static plant including running costs such as fuel, water supply (which is metered on a construction site), electricity and rubbish disposal. Some items of plant are included within the "Groundwork" section; other items are included within the Preliminaries section.

Overheads and profit

The general overheads of the Contractor's business, the head office overheads and any profit sought on capital and turnover employed, is usually covered under a general item of overheads and profit which is applied either to all measured rates as a percentage, or alternatively added to the tender summary or included within Preliminaries (site specific overhead costs). At the present time, we are including an allowance of 2% for profit and 5½% for overheads on measured rates to reflect the current market.

Preliminaries

Site specific overheads on a contract, such as insurance, site huts, security, temporary roads and the statutory health and welfare of the labour force, are not directly assignable to individual items so they are generally added as a percentage or calculated allowance after all building component items have been costed and summed. Preliminaries will vary from contract to contract according to the type of construction, difficulties of the site, labour shortage, inclement weather or involvement with other contractors, etc. The overall Preliminary addition for a scheme should be adjusted to allow for these factors. As in the last edition we have retained Preliminary costs at between 11 to 13%.

The first five of these six items combine to form the **Price for Measured Work.** It will be appreciated that a variation in any one item in any group will affect the final Measured Work price. Any cost variation must be weighed against the total cost of the contract, and a small variation in Prime Cost where the items are ordered in thousands may have more effect on the total cost than a large variation on a few items, while a change in design which introduces the need to use, e.g. earth-moving equipment, which must be brought to the site for that one task, will cause a dramatic rise in the contract cost. Similarly, a small saving on multiple items will provide a useful reserve to cover unforeseen extras.

DAVIS LANGDON & EVEREST
Authors of Spon's Price Books

www.davislangdon.com

Davis Langdon & Everest is an independent practice of Chartered Quantity Surveyors, with some 1000 staff in 20 UK offices and, through Davis Langdon & Seah International, some 2,500 staff in 85 offices worldwide.

DLE manages client requirements, controls risk, manages cost and maximises value for money, throughout the course of construction projects, always aiming to be - and to deliver - the best.

TYPICAL PROJECT STAGES, DLE INTEGRATED SERVICES AND THEIR EFFECT:

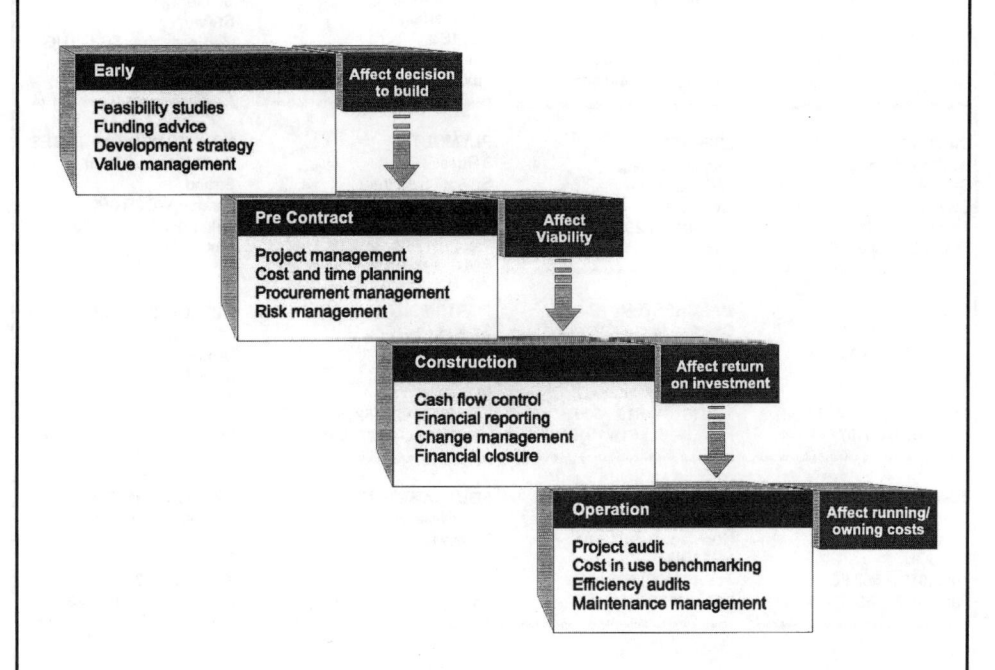

Early
- Feasibility studies
- Funding advice
- Development strategy
- Value management

→ Affect decision to build

Pre Contract
- Project management
- Cost and time planning
- Procurement management
- Risk management

→ Affect Viability

Construction
- Cash flow control
- Financial reporting
- Change management
- Financial closure

→ Affect return on investment

Operation
- Project audit
- Cost in use benchmarking
- Efficiency audits
- Maintenance management

→ Affect running/owning costs

EUROPE ⇨ ASIA - AUSTRALIA - AFRICA - AMERICA
GLOBAL REACH - LOCAL DELIVERY

DAVIS LANGDON & EVEREST

www.davislangdon.com

LONDON
Princes House
39 Kingsway
London
WC2B 6TP
Tel : (020) 7497 9000
Fax : (020) 7497 8858
Email:rob.smith@davislangdon-uk.com

BIRMINGHAM
29 Woodbourne Road
Harborne
Birmingham
B17 8BY
Tel : (0121) 4299511
Fax : (0121) 4292544
Email:richard.d.taylor@davislangdon-uk.com

BRISTOL
St Lawrence House
29/31 Broad Street
Bristol
BS1 2HF
Tel : (0117) 9277832
Fax : (0117) 9251350
Email:alan.trolley@davislangdon-uk.com

CAMBRIDGE
36 Storey's Way
Cambridge
CB3 0DT
Tel : (01223) 351258
Fax : (01223) 321002
Email:stephen.bugg@davislangdon-uk.com

CARDIFF
4 Pierhead Street
Capital Waterside
Cardiff
CF10 4QP
Tel : (029) 20497497
Fax : (029) 20497111
Email:paul.edwards@davislangdon-uk.com

EDINBURGH
74 Great King Street
Edinburgh
EH3 6QU
Tel : (0131) 557 5306
Fax : (0131) 557 5704
Email:ian.mcandie@davislangdon-uk.com

GATESHEAD
11 Regent Terrace
Gateshead
Tyne and Wear
NE8 1LU
Tel : (0191) 477 3844
Fax : (0191) 490 1742
Email:gary.lockey@davislangdon-uk.com

GLASGOW
Cumbrae House
15 Carlton Court
Glasgow
G5 9JP
Tel : (0141) 429 6677
Fax : (0141) 429 2255
Email:hugh.fisher@davislangdon-uk.com

LEEDS
Duncan House
14 Duncan Street
Leeds
LS1 6DL
Tel : (0113) 2432481
Fax : (0113) 2424601
Email:tony.brennan@davislangdon-uk.com

LIVERPOOL
Cunard Building
Water Street
Liverpool L3 1JR
Tel : (0151) 2361992
Fax : (0151) 2275401
Email:john.davenport@davislangdon-uk.com

MANCHESTER
Cloister House
Riverside
New Bailey Street
Manchester M3 5AG
Tel : (0161) 819 7600
Fax : (0161) 819 1818
Email:paul.stanion@davislangdon-uk.com

NORWICH
63 Thorpe Road
Norwich
NR1 1UD
Tel : (01603) 628194
Fax : (01603) 615928
Email:michael.ladbrook@davislangdon-uk.com

MILTON KEYNES
Everest House
Rockingham Drive
Linford Wood
Milton Keynes MK14 6LY
Tel : (01908) 304700
Fax : (01908) 660059
Email:kevin.sims@davislangdon-uk.com

OXFORD
Avalon House
Marcham Road
Abingdon
Oxford OX14 1TZ
Tel : (01235) 555025
Fax : (01235) 554909
Email:paul.coomber@davislangdon-uk.com

PETERBOROUGH
Charterhouse
66 Broadway
Peterborough
PE1 1SU
Tel: (01733) 343625
Fax: (01733) 349177
Email:colin.harrison@davislangdon-uk.com

PLYMOUTH
3 Russell Court
St Andrew Street
Plymouth PL1 2AX
Tel : (01752) 668372
Fax : (01752) 221219
Email:gareth.steventon@davislangdon-uk.com

PORTSMOUTH
St Andrews Court
St Michaels Road
Portsmouth
PO1 2PR
Tel : (023) 92815218
Fax : (023) 92827156
Email:brian.bartholomew@davislangdon-uk.com

SOUTHAMPTON
Brunswick House
Brunswick Place
Southampton SO15 2AP
Tel : (023) 80333438
Fax : (023) 80226099
Email:richard.pitman@davislangdon-uk.com

DAVIS LANGDON CONSULTANCY
Princes House
39 Kingsway
London WC2B 6TP
Tel : (020) 7379 3322
Fax : (020) 7379 3030
Email:jim.meikle@davislangdon-uk.com

MOTT GREEN & WALL
Africa House
64-78 Kingsway
London
WC2B 6NN
Tel : (020) 7836 0836
Fax : (020) 7242 0394
Email:barry.nugent@mottgreenwall.co.uk

SCHUMANN SMITH
14th Flr, Southgate House
St Georges Way
Stevenage
Hertfordshire SG1 1HG
Tel : (01438) 742642
Fax : (01438) 742632
Email:nschumann@schumannsmith.com

NBW CROSHER & JAMES
1-5 Exchange Court
Strand
London WC2R 0PQ
Tel : (020) 7845 0600
Fax : (020) 7845 0601
Email:sanderer@nbwcrosherjames.com

NBW CROSHER & JAMES
102 New Street
Birmingham
B2 4HQ
Tel : (0121) 632 3600
Fax : (0121) 632 3601
Email:whittakerr@bir.nbwcrosherjames.com

NBW CROSHER & JAMES
5 Coates Crescent
Edinburgh
EH3 9AL
Tel : (0131) 220 4225
Fax : (0131) 220 4226
Email:mcfarlanel@nbwcrosherjames.com

with offices throughout Europe, the Middle East, Asia, Australia, Africa and the USA
which together form

DAVIS LANGDON & SEAH INTERNATIONAL

PART I

Fees and Daywork

This part of the book contains the following sections:

Fees for Professional Services, *page* 3
Daywork and Prime Cost, *page* 71

NEW FROM SPON PRESS

The Architectural Expression of Environmental Control Systems

George Baird, Victoria University of Wellington, New Zealand

The Architectural Expression of Environmental Control Systems examines the way project teams can approach the design and expression of both active and passive thermal environmental control systems in a more creative way. Using seminal case studies from around the world and interviews with the architects and environmental engineers involved, the book illustrates innovative responses to client, site and user requirements, focusing upon elegant design solutions to a perennial problem.

This book will inspire architects, building scientists and building services engineers to take a more creative approach to the design and expression of environmental control systems - whether active or passive, whether they influence overall building form or design detail.

March 2001: 276x219: 304pp
135 b+w photos, 40 colour, 90 line illustrations
Hb: 0-419-24430-1: £49.95

To Order: Tel: +44 (0) 8700 768853, or +44 (0) 1264 343071 Fax: +44 (0) 1264 343005, or
Post: Spon Press Customer Services, ITPS Andover, Hants, SP10 5BE, UK Email: book.orders@tandf.co.uk.

Postage & Packing: UK: 5% of order value (min. charge £1, max. charge £10) for 3-5 days delivery. Option of next day delivery at an additional £6.50. Europe: 10% of order value (min. charge £2.95, max. charge £20) for delivery surface post. Option of airmail at an additional £6.50. ROW: 15% of order value (min. charge£6.50, max. charge £30) for airmail delivery.

For a complete listing of all our titles visit: www.sponpress.com

Fees for Professional Services

Extracts from the scales of fees for architects, quantity surveyors and consulting engineers are given together with extracts from the Town and Country Planning Regulations 1993 and Building Regulation Charges. These extracts are reproduced by kind permission of the bodies concerned, in the case of Building Regulation Charges, by kind permission of the London Borough of Ealing. Attention is drawn to the fact that the full scales are not reproduced here and that the extracts are given for guidance only. The full authority scales should be studied before concluding any agreement and the reader should ensure that the fees quoted here are still current at the time of reference.

ARCHITECTS' FEES

Standard Form of Agreement for the Appointment of an Architect (SFA/99), pages 4 and 5
Conditions of Engagement for the Appointment of an Architect (CE/99), page 4 (brief notes only)
Small Works (SW/99), page 4 (brief notes only)
Employer's Requirements (DB1/99), page 4 (brief notes only)
Contractor's Proposals (DB2/99), page 4 (brief notes only)
Form of Appointment as Planning Supervisor (PS/99), page 4 (brief notes only)
Form of Appointment as Sub-Consultant (SC/99), page 4 (brief notes only)
Form of Appointment as Project Manager (PM/99), page 4 (brief notes only)

QUANTITY SURVEYORS' FEES

Scale 36, inclusive scale of professional charges, page 9
Scale 37, itemised scale of professional charges, page 13
Scale 40, professional charges for housing schemes for Local Authorities, page 30
Scale 44, professional charges for improvements to existing housing and environmental improvement works, page 35
Scale 45, professional charges for housing schemes financed by the Housing Corporation, page 37
Scale 46, professional charges for the assessment of damage to buildings from fire etc., page 42
Scale 47, professional charges for the assessment of replacement costs for insurance purposes, page 44

CONSULTING ENGINEERS' FEES

Guidance on Fees, page 45
Basis of Fee Calculations, page 46
Agreement A(1) - Lead Consultant, Civil/Structural, page 49
Agreement A(2) - Lead Consultant, Building Services, page 50
Agreement B(1) - Non-Lead Consultant, Civil/Structural, page 51
Agreement B(2) - Non-Lead Consultant, Building Services, page 52
Agreement C - Design and Construct Project Designer, page 53
Graphical Illustration of fee scales, page 54

PLANNING REGULATION FEES

Part I: General provisions, page 58
Part II: Scale of Fees, page 62

THE BUILDING (LOCAL AUTHORITY CHARGES) REGULATIONS 1998

Charge Schedules, page 65
Charges for erection of one or more small new domestic buildings and connected work, page 67
Charges for erection of certain small domestic buildings, garages, carports and extensions, page 68
Charges for building work other than to which tables 1 and 2 apply, page 69

ARCHITECTS' FEES

RIBA Forms of Appointment 1999
- include forms to cover all aspects of an Architect's practice;
- identify the traditional roles of an Architect;
- reflect or take into account recent legislation such as the Housing Grants, Construction and Regeneration Act 1996, the Arbitration Act 1996 (*applies to England and Wales only*), the Unfair Terms in Consumer Contracts Regulations 1994, Latham issues etc;
- update the Outline Plan of Work Stage titles and descriptions;
- are sufficiently flexible for use with English or Scottish Law [1]; the building projects of any size (except small works); other professional services [2]; different project procedures.

  [1] The Royal Incorporation of Architects in Scotland will publish appropriate replacement pages for use where the law of Scotland applies to the Agreement.
  [2] Excluding acting as adjudicator, arbitrator, expert witness, conciliator, party wall surveyor etc.

All forms require the Architect to agree with the Client the amount of professional indemnity insurance cover for the project.

Standard Form of Agreement for the Appointment of an Architect (SFA/99)
The core document from which all other forms are derived. Suitable for use where an Architect is to provide professional services for a fully designed building project in a wide range of size or complexity and/or to provide other professional services. Used with Articles of Agreement. Includes optional Services Supplement.

Conditions for Engagement for the Appointment of an Architect (CE/99)
Suitable for use where an Architect is to provide services for a fully designed building project and/or to provide other professional services where a Letter of Appointment is preferred in lieu of Articles of Agreement. Includes optional (modified) Services Supplement and draft Model letter.

Small Works (SW/99)
Suitable for the provision of professional services where the cost of construction is not expected to exceed £150,000 and use of the JCT Agreement for Minor Works is appropriate. The appointment is made by a specially drafted Letter of Appointment, the Conditions and an optional Schedule of Services for Small Works.

Employer's Requirements (DB1/99)
A supplement to amend SFA/99 and CE/99 where an Architect is appointed by the Employer Client to prepare Employer's Requirements for a Design and Build contract. Includes replacement Services Supplement and notes on completion for initial appointment, where a change to Design and Build occurs later or where "consultant switch" is contemplated.

Contractor's Proposals (DB2/99)
A supplement to amend SFA/99 where an Architect is appointed by the Contractor Client to prepare Contractor's Proposals under a Design and Build contract. Includes replacement Services Supplement and notes on completion for initial appointment and for "consultant switch".

Form of Appointment as Planning Supervisor (PS/99)
Used with Articles of Agreement. Provides for the Planning Supervisor to prepare the Health and Safety File and the pre-tender Health and Safety Plan.

Form of Appointment as Sub-Consultant (SC/99)
Suitable for use where a Consultant wishes (another Consultant (Sub-Consultant) or Specialist) to perform a part of his responsibility but not for use where the intention is for the Client to appoint Consultants or Specialists directly. Used with Articles of Agreement. Includes draft form of Warranty to the Client.

Form of Appointment as Project Manager (PM/99)
Suitable for a wide range of projects where the Client wishes to appoint a Project Manager to provide a management service and/or other professional service. Used with Articles of Agreement. Does not duplicate or conflict with the Architect's services under other RIBA forms.

ARCHITECTS' FEES

Guides

Matters connected with using and completing the forms are covered in "The Architect's Contract: A guide to RIBA Forms of Appointment 1999 and other Architect's appointments". The guidance covers the Standard Form (SFA/99) including a worked example; the options for calculating fees ie. percentages, calculated or fixed lump sums, time charges and other methods; the other forms in the suite; the Design and Build amendments including advice on 'consultant switch' and 'novation' agreements; and topics for other appointments connected with dispute resolution, etc. Notes on appointments for Historic Buildings and Community Architecture projects are also included.

A series of guides "Engaging an Architect", addressed directly to clients, is also published on topics associated with the appointment of an Architect. These guides include graphs showing indicative percentage fees for the architect's normal services in stages C-L under SFA/99, CE/99 or SW/99 for works to new and existing buildings. Indicative hourly rates are also shown. Tables 2, 3 and 4 on page 8 show by interpolation of the graphs the indicative fees for different classes of building adjusted to reflect current tender price indices. Table 1 gives the classification of buildings. Indicative fees do not include expenses, disbursements or VAT fees.

The appointing documents and guides are published by:
RIBA Publications, Construction House, 56-64 Leonard Street, London. EC2A 4LT.
Telephone: 0207 251 0791   Fax: 0207 608 2375

The Standard Form of Agreement for the Appointment of an Architect (SFA/99)

SFA/99 is the core document, which is used as the basis for all the documents in the RIBA 1999 suite. It should be suitable for any Project to be procured in the 'traditional' manner. Supplements are available for use with Design and Build procurement. It is used with Articles of Agreement and formal attestation underhand or as a deed.

There are some changes from SFA/92 in presentation, in some of the responsibilities and liabilities of the parties, to some definitions and clauses and their arrangement for greater clarity or flexibility, for compliance with the Construction Act; and to ensure that each role of the Architect is separately identified.

It comprises:

- the formal declaration of intent - Recitals and Articles;

- Appendix to the Conditions;

- Schedule 1 - Project Description;

- Schedule 2 - Services indicating the roles the Architect is to perform, which Work Stages apply and any other services required. A table of 'Other activities' identifies items not included in 'Normal Services';

- Schedule 3 - Fess and Expenses;

- Schedule 4 - Other appointments;

- the Conditions governing performance of the contract and obligations of the parties;

- a Services Supplement describing the Architect's roles and activities' which can be edited to suit the requirements of the Project or removed if it is not suitable for the Project, and if appropriate, another inserted in its place; and

- notes on completion.

ARCHITECTS' FEES

Table 1   Classification by building type

Type	Class 1	Class 2
Industrial	Storage sheds	Speculative factories and warehouses Assembly and machine workshops Transport garages
Agricultural	Barns and sheds	Stables
Commercial	Speculative shops Surface parks	Multi-storey and underground car parks
Community		Community halls
Residential		Dormitory hostels
Education		
Recreation		
Medical/Social services		

ARCHITECTS' FEES

Class 3	Class 4	Class 5
Purpose-built factories and warehouses		
Animal breeding units		
Supermarkets Banks Purpose-built shops Office developments Retail warehouses Garages/showrooms	Department stores Shopping centres Food processing units Breweries Telecommunications and computer buildings Restaurants Public houses	High risk research and production buildings Research and development laboratories Radio, TV and recording studios
Community centres Branch libraries Ambulances and fire stations Bus stations Railway stations Airports Police stations Prisons Postal buildings Broadcasting	Civic centres Churches and crematoria Specialist libraries Museums and art galleries Magistrates/County Courts	Theatres Opera houses Concert halls Cinemas Crown Courts
Estates housing and flats Barracks Sheltered housing Housing for single people Student housing	Parsonages/manses Apartment blocks Hotels Housing for the handicapped Housing for the frail elderly	Houses and flats for individual clients
Primary/nursery/first schools	Other schools, including middle and secondary University complexes	University laboratories
Sports halls Squash courts	Swimming pools Leisure complexes	Leisure pools Specialised complexes
Clinics	Health centres General hospitals Nursing homes Surgeries	Teaching hospitals Hospital laboratories Dental surgeries

Table 2   Indicative percentage fees for new works

Construction Cost £	Class 1 %	Class 2 %	Class 3 %	Class 4 %	Class 5 %
50,000	7.90	8.70	-	-	-
75,000	7.25	7.80	8.40	-	-
100,000	7.10	7.60	8.20	8.90	9.60
250,000	6.20	6.70	7.20	7.80	8.40
500,000	5.75	6.25	6.75	7.25	7.90
1,000,000	5.40	5.90	6.20	6.80	7.50
2,500,000	5.15	5.60	6.10	6.60	7.10
5,000,000	-	-	5.97	6.50	7.00
over 10,000,000	-	-	5.95	6.45	6.97

Table 3   Indicative percentage fees for works to existing buildings

Construction Cost £	Class 1 %	Class 2 %	Class 3 %	Class 4 %	Class 5 %
50,000	11.60	12.60	-	-	-
75,000	10.70	11.60	12.40	-	-
100,000	10.40	11.30	12.20	13.15	14.10
250,000	9.30	10.10	10.85	11.75	12.55
500,000	8.70	9.45	10.20	11.05	11.80
1,000,000	8.25	9.00	9.70	10.55	11.30
2,500,000	-	-	9.25	10.00	10.75
5,000,000	-	-	9.10	9.85	10.55
over 10,000,000	-	-	9.00	9.75	10.45

Table 4   Indicative hourly rates

Type of work	General	Complex	Specialist
Partner/Director or equivalent	£95	£140	£180
Senior Architect	£75	£105	£140
Architect	£55	£75	£95

QUANTITY SURVEYORS' FEES

Author's Note:

The Royal Institution of Chartered Surveyors formally abolished the standard Quantity Surveyors' fee scales with effect from 31 December 1998. To date the Institution has not published a guidance booklet to assist practitioners in compiling fee proposals and we believe that they are unlikley to do so. The last standard Quantity Surveyors' fee scales have therefore been reproduced here in order to assist the reader.

SCALE 36 INCLUSIVE OF PROFESSIONAL CHARGES FOR QUANTITY SURVEYING SERVICES FOR BUILDING WORKS ISSUED BY THE ROYAL INSTITUTION OF CHARTERED SURVEYORS. The scale is recommended and not mandatory.

EFFECTIVE FROM JULY 1988

1.0. GENERALLY

1.1 This scale is for the use when an inclusive scale of professional charges is considered to appropriate by mutual agreement between the employer and the quantity surveyor.

1.2. This scale does not apply to civil engineering works, housing schemes financed by local authorities and the Housing Corporation and housing improvement work for which separate scales of fees have been published.

1.3. The fees cover quantity surveying services as may be required in connection with a building project irrespective of the type of contract from initial appointment to final certification of the contractor's account such as:

(a) Budget estimating; cost planning and advice on tendering procedures and contract arrangements.
(b) Preparing tendering documents for main contract and specialist sub-contracts; examining tenders received and reporting thereon or negotiating tenders and pricing with a selected contractor and/or sub-contractors.
(c) Preparing recommendations for interim payments on account to the contractor; preparing periodic assessments of anticipated final cost and reporting thereon; measuring work and adjusting variations in accordance with the terms of the contract and preparing final account, pricing same and agreeing totals with the contractor.
(d) Providing a reasonable number of copies of bills of quantities and other documents; normal travelling and other expenses. Additional copies of documents, abnormal travelling and other expenses (e.g. in remote areas or overseas) and the provision of checkers on site shall be charged in addition by prior arrangement with the employer.

1.4. If any of the materials used in the works are supplied by the employer or charged at a preferential rate, then the actual or estimated market value thereof shall be included in the amounts upon which fees are to be calculated.

1.5. If the quantity surveyor incurs additional costs due to exceptional delays in building operations or any other cause beyond the control of the quantity surveyor then the fees may be adjusted by agreement between the employer and the quantity surveyor to cover the reimbursement of these additional costs.

1.6. The fees and charges are in all cases exclusive of value added tax which will be applied in accordance with legislation.

1.7. Copyright in bills of quantities and other documents prepared by the quantity surveyor is reserved to the quantity surveyor.

QUANTITY SURVEYORS' FEES

2.0. INCLUSIVE SCALE

2.1. The fees for the services outlined in para.1.3, subject to the provision of para. 2.2, shall be as follows:

(a) Category A: Relatively complex works and/or works with little or no repetition.

Examples:
Ambulance and fire stations; banks; cinemas; clubs; computer buildings; council offices; crematoria; fitting out of existing buildings; homes for the elderly; hospitals and nursing homes; laboratories; law courts; libraries; "one off" houses; petrol stations; places of religious worship; police stations; public houses, licensed premises; restaurants; sheltered housing; sports pavilions; theatres; town halls; universities, polytechnics and colleges of further education (other than halls of residence and hostels); and the like.

Value of work	£	Category A fee £	£
Up to	150 000	380 + 6.0% (Minimum fee £3 380)	
150 000 -	300 000	9 380 + 5.0% on balance over	150 000
300 000 -	600 000	16 880 + 4.3% on balance over	300 000
600 000 -	1 500 000	29 780 + 3.4% on balance over	600 000
1 500 000 -	3 000 000	60 380 + 3.0% on balance over	1 500 000
3 000 000 -	6 000 000	105 380 + 2.8% on balance over	3 000 000
Over	6 000 000	189 380 + 2.4% on balance over	6 000 000

(b) Category B: Less complex works and/or works with some element of repetition.

Examples:
Adult education facilities; canteens; church halls; community centres; departmental stores; enclosed sports stadia and swimming baths; halls of residence; hostels; motels; offices other than those included in Categories A and C; railway stations; recreation and leisure centres; residential hotels; schools; self-contained flats and maisonettes; shops and shopping centres; supermarkets and hypermarkets; telephone exchanges; and the like.

Value of work	£	Category B fee £	£
Up to	150 000	360 + 5.8% (Minimum fee £3 260)	
150 000 -	300 000	9 060 + 4.7% on balance over	150 000
300 000 -	600 000	16 110 + 3.9% on balance over	300 000
600 000 -	1 500 000	27 810 + 2.8% on balance over	600 000
1 500 000 -	3 000 000	53 010 + 2.6% on balance over	1 500 000
3 000 000 -	6 000 000	92 010 + 2.4% on balance over	3 000 000
Over	6 000 000	164 010 + 2.0% on balance over	6 000 000

(c) Category C: Simple works and/or works with a substantial element of repetition.

Examples:
Factories; garages; multi-storey car parks; open-air sports stadia; structural shell offices not fitted out; warehouses; workshops; and the like.

QUANTITY SURVEYORS' FEES

Value of work £		Category C fee	£
Up to	150 000	300 + 4.9% (Minimum fee £2 750)	
150 000 -	300 000	7 650 + 4.1% on balance over	150 000
300 000 -	600 000	13 800 + 3.3% on balance over	300 000
600 000 -	1 500 000	23 700 + 2.5% on balance over	600 000
1 500 000 -	3 000 000	46 200 + 2.2% on balance over	1 500 000
3 000 000 -	6 000 000	79 200 + 2.0% on balance over	3 000 000
Over	6 000 000	139 200 + 1.6% on balance over	6 000 000

(d) Fees shall be calculated upon the total of the final account for the whole of the work including all nominated sub-contractors' and nominated supplier's accounts. When work normally included in a building contract is the subject of a separate contract for which the quantity surveyor has not been paid fees under any other clause hereof, the value of such work shall be included in the amount upon which fees are charged.

(e) When a contract comprises buildings which fall into more than one category, the fee shall be calculated as follows:

    (i) The amount upon which fees are chargeable shall be allocated to the categories of work applicable and the amounts so allocated expressed as percentages of the total amount upon which fees are chargeable.

    (ii) Fees shall then be calculated for each category on the total amount upon which fees are chargeable.

    (iii) The fee chargeable shall then be calculated by applying the percentages of work in each category to the appropriate total fee and adding the resultant amounts.

    (iv) A consolidated percentage fee applicable to the total value of the work may be charged by prior agreement between the employer and the quantity surveyor. Such a percentage shall be based on this scale and on the estimated cost of the various categories of work and calculated in accordance with the principles stated above.

(f) When a project is subject to a number of contracts then, for the purpose of calculating fees, the values of such contracts shall not be aggregated but each contract shall be taken separately and the scale of charges (paras. 2.1 (a) to (e)) applied as appropriate.

2.2. Air conditioning, heating, ventilating and electrical services

(a) When the services outlined in para. 1.3 are provided by the quantity surveyor for the air conditioning, heating, ventilating and electrical services there shall be a fee for these services in addition to the fee calculated in accordance with para. 2.1 as follows:

Value of work £		Additional fee £	£
Up to	120 000	5.0%	
120 000 -	240 000	6 000 + 4.7% on balance over	120 000
240 000 -	480 000	11 640 + 4.0% on balance over	240 000
480 000 -	750 000	21 240 + 3.6% on balance over	480 000
750 000 -	1 000 000	30 960 + 3.0% on balance over	750 000
1 000 000 -	4 000 000	38 460 + 2.7% on balance over	1 000 000
Over	4 000 000	119 200 + 2.4% on balance over	4 000 000

(b) The value of such services, whether the subject of separate tenders or not, shall be aggregated and the total value of work so obtained used for the purpose of calculating the additional fee chargeable in accordance with para. (a). (Except that when more than one firm of consulting engineers is engaged on the design of these services, the separate values for which each such firm is responsible shall be aggregated and the additional fees charged shall be calculated independently on each such total value so obtained.)

QUANTITY SURVEYORS' FEES

(c) Fees shall be calculated upon the basis of the account for the whole of the air conditioning, heating, ventilating and electrical services for which bills of quantities and final accounts have been prepared by the quantity surveyor.

2.3 Works of alteration
On works of alteration or repair, or on those sections of the work which are mainly works of alteration or repair, there shall be a fee of 1.0% in addition to the fee calculated in accordance with paras. 2.1 and 2.2.

2.4. Works of redecoration and associated minor repairs
On works of redecoration and associated minor repairs, there shall be a fee of 1.5% in addition to the fee calculated in accordance with paras. 2.1 and 2.2.

2.5. Generally
If the works are substantially varied at any stage or if the quantity surveyor is involved in an excessive amount of abortive work, then the fees shall be adjusted by agreement between the employer and the quantity surveyor.

3.0. ADDITIONAL SERVICES

3.1. For additional services not normally necessary, such as those arising as a result of the termination of a contract before completion, liquidation, fire damage to the buildings, services in connection with arbitration, litigation and investigation of the validity of contractors' claims, services in connection with taxation matters, and all similar services where the employer specifically instructs the quantity surveyor, the charges shall be in accordance with para. 4.0 below.

4.0. TIME CHARGES

4.1. (a) For consultancy and other services performed by a principal, a fee by arrangement according to the circumstances including the professional status and qualifications of the quantity surveyor.
(b) When a principal does work which would normally be done by a member of staff, the charge shall be calculated as para. 4.2 below.

4.2. (a) For services by a member of staff, the charges for which are to be based on the time involved, such charges shall be calculated on the hourly cost of the individual involved plus 145%.
(b) A member of staff shall include a principal doing work normally done by an employee (as para. 4.1 (b) above), technical and supporting staff, but shall exclude secretarial staff or staff engaged upon general administration.
(c) For the purpose of para. 4.2 (b) above, a principal's time shall be taken at the rate applicable to a senior assistant in the firm.
(d) The supervisory duties of a principal shall be deemed to be included in the addition of 145% as para. 4.2 (a) above and shall not be charged separately.
(e) The hourly cost to the employer shall be calculated by taking the sum of the annual cost of the member of staff of:

(i) Salary and bonus but excluding expenses;
(ii) Employer's contributions payable under any Pension and Life Assurance Schemes;
(iii) Employer's contributions made under the National Insurance Acts, the Redundancy Payments Act and any other payments made in respect of the employee by virtue of any statutory requirements; and
(iv) Any other payments or benefits made or granted by the employer in pursuance of the terms of employment of the member of staff;

and dividing by 1,650.

QUANTITY SURVEYORS' FEES

5.0. INSTALMENT PAYMENTS

5.1 In the absence of agreement to the contrary, fees shall be paid by instalments as follows:

(a) Upon acceptance by the employer of a tender for the works, one half of the fee calculated on the amount of the accepted tender.

(b) The balance by instalments at intervals to be agreed between the date of the first certificate and one month after final certification of the contractor's account.

5.2. (a) In the event of no tender being accepted, one half of the fee shall be paid within three months of completion of the tender documents. The fee shall be calculated upon the basis of the lowest original bona fide tender received. In the event of no tender being received, the fee shall be calculated upon a reasonable valuation of the works based upon the tender documents.

(b) In the event of the project being abandoned at any stage other than those covered by the foregoing, the proportion of fee payable shall be by agreement between the employer and the quantity surveyor.

NOTE: In the foregoing context "bona fide tender" shall be deemed to mean a tender submitted in good faith without major errors of computation and not subsequently withdrawn by the tenderer.

SCALE 37 ITEMISED SCALE OF PROFESSIONAL CHARGES FOR QUANTITY SURVEYING SERVICES FOR BUILDING WORK ISSUED BY THE ROYAL INSTITUTION OF CHARTERED SURVEYORS.
The scale is recommended and not mandatory.

EFFECTIVE FROM JULY 1988

1.0. GENERALLY

1.1. The fees are in all cases exclusive of travelling and other expenses (for which the actual disbursement is recoverable unless there is some prior arrangement for such charges) and of the cost of reproduction of bills of quantities and other documents, which are chargeable in addition at net cost.

1.2. The fees are in all cases exclusive of services in connection with the allocation of the cost of the works for purposes of calculating value added tax for which there shall be an additional fee based on the time involved (see paras. 19.1 and 19.2).

1.3. If any of the materials used in the works are supplied by the employer or charged at a preferential rate, then the actual or estimated market value thereof shall be included in the amounts upon which fees are to be calculated.

1.4. The fees are in all cases exclusive of preparing a specification of the materials to be used and the works to be done, but the fees for preparing bills of quantities and similar documents do include for incorporating preamble clauses describing the materials and workmanship (from instructions given by the architect and/or consulting engineer)

1.5. If the quantity surveyor incurs additional costs due to exceptional delays in building operations or any other cause beyond the control of the quantity surveyor then the fees may be adjusted by agreement between the employer and the quantity surveyor to cover the reimbursement of these additional costs.

1.6. The fees and charges are in all cases exclusive of value added tax which will be applied in accordance with legislation.

QUANTITY SURVEYORS' FEES

1.7. Copyright in bills of quantities and other documents prepared by the quantity surveyor is reserved to the quantity surveyor.

CONTRACTS BASED ON BILLS OF QUANTITIES: PRE-CONTRACT SERVICES

2.0. **BILLS OF QUANTITIES**

   2.1. Basic scale

   For preparing bills of quantities and examining tenders received and reporting thereon.

   (a) Category A: Relatively complex works and/or works with little or no repetition.

   Examples:
   Ambulance and fire stations; banks; cinemas; clubs; computer buildings; council offices; crematoria; fitting out of existing buildings; homes for the elderly; hospitals and nursing homes; laboratories; law courts; libraries; "one off" houses; petrol stations; places of religious worship; police stations; public houses; licensed premises; restaurants; sheltered housing; sports pavilions; theatres; town halls; universities, polytechnics and colleges of further education (other than halls of residence and hostels); and the like.

Value of work £	Category A fee £	£
Up to 150 000	230 + 3.0% (Minimum fee £1730)	
150 000 - 300 000	4 730 + 2.3% on balance over	150 000
300 000 - 600 000	8 180 + 1.8% on balance over	300 000
600 000 - 1 500 000	13 580 + 1.5% on balance over	600 000
1 500 000 - 3 000 000	27 080 + 1.2% on balance over	1 500 000
3 000 000 - 6 000 000	45 080 + 1.1% on balance over	3 000 000
Over 6 000 000	78 080 + 1.0% on balance over	6 000 000

   (b) Category B: Less complex works and/or works with some element of repetition.

   Examples:
   Adult education facilities; canteens; church halls; community centres; departmental stores; enclosed sports stadia and swimming baths; halls of residence; hostels; motels; offices other than those included in Categories A and C; railway stations; recreation and leisure centres; residential hotels; schools; self-contained flats and maisonettes; shops and shopping centres; supermarkets and hypermarkets; telephone exchanges; and the like.

Value of work £	Category B fee £	£
Up to 150 000	210 + 2.8% (Minimum fee £1 680)	
150 000 - 300 000	4 410 + 2.0% on balance over	150 000
300 000 - 600 000	7 410 + 1.5% on balance over	300 000
600 000 - 1 500 000	11 910 + 1.1% on balance over	600 000
1 500 000 - 3 000 000	21 810 + 1.0% on balance over	1 500 000
3 000 000 - 6 000 000	36 810 + 0.9% on balance over	3 000 000
Over 6 000 000	63 810 + 0.8% on balance over	6 000 000

   (c) Category C: Simple works and/or works with a substantial element of repetition

   Examples:
   Factories; garages; multi-storey car parks; open-air sports stadia; structural shell offices not fitted out; warehouses; workshops and the like.

QUANTITY SURVEYORS' FEES

Value of work	£	Category C fee	£
Up to	150 000	180 + 2.5% (Minimum fee £1 430)	
150 000 -	300 000	3 930 + 1.8% on balance over	150 000
300 000 -	600 000	6 630 + 1.2% on balance over	300 000
600 000 -	1 500 000	10 230 + 0.9% on balance over	600 000
1 500 000 -	3 000 000	18 330 + 0.8% on balance over	1 500 000
3 000 000 -	6 000 000	30 330 + 0.7% on balance over	3 000 000
Over	6 000 000	51 330 + 0.6% on balance over	6 000 000

(d) The scales of fees for preparing bills of quantities (paras. 2.1 (a) to (c)) are overall scales based upon the inclusion of all provisional and prime cost items, subject to the provision of para. 2.1 (g). When work normally included in a building contract is the subject of a separate contract for which the quantity surveyor has not been paid fees under any other clause hereof, the value of such work shall be included in the amount upon which fees are charged.

(e) Fees shall be calculated upon the accepted tender for the whole of he work subject to the provisions of para. 2.6. In the event of no tender being accepted, fees shall be calculated upon the basis of the lowest original bona fide tender received. In the event of no such tender being received, the fees shall be calculated upon a reasonable valuation of the works based upon the original bills of quantities.

NOTE: In the foregoing context "bona fide tender" shall be deemed to mean a tender submitted in good faith without major errors of computation and not subsequently withdrawn by the tenderer.

(f) In calculating the amount upon which fees are charged the total of any credits and the totals of any alternative bills shall be aggregated and added to the amount described above. The value of any omission or addition forming part of an alternative bill shall not be added unless measurement or abstraction from the original dimension sheets was necessary.

(g) Where the value of the air conditioning, heating, ventilating and electrical services included in the tender documents together exceeds 25% of the amount calculated as described in paras. 2.1 (d) and (e), then, subject to the provisions of para. 2.2, no fee is chargeable on the amount by which the value of these services exceeds the said 25%. In this context the term "value" excludes general contractor's profit, attendance, builder's work in connection with the services, preliminaries and any similar additions.

(h) When a contract comprises buildings which fall into more than one category, the fee shall be calculated as follows:

    (i) The amount upon which fees are chargeable shall be allocated to the categories of work applicable and the amounts so allocated expressed as percentages of the total amount upon which fees are chargeable.

    (ii) Fees shall then be calculated for each category on the total amount upon which fees are chargeable.

    (iii) The fee chargeable shall then be calculated by applying the percentages of work in each category to the appropriate total fee and adding the resultant amounts.

(j) When a project is the subject of a number of contracts then, for the purpose of calculating fees, the values of such contracts shall not be aggregated but each contract shall be taken separately and the scale of charges (paras. 2.1 (a) to (h)) applied as appropriate.

(k) Where the quantity surveyor is specifically instructed to provide cost planning services the fee calculated in accordance with paras. 2.1 (a) to (j) shall be increased by a sum calculated in accordance with the following table and based upon the same value of work as that upon which the aforementioned fee has been calculated:

QUANTITY SURVEYORS' FEES

Categories A & B: (as defined in paras. 2.1 (a) and (b)).

Value of work £	Fee £	£
Up to 600 000	0.70%	
600 000 - 3 000 000	4 200 + 0.40% on balance over	600 000
3 000 000 - 6 000 000	13 800 + 0.35% on balance over	3 000 000
Over 6 000 000	24 300 + 0.30% on balance over	6 000 000

Category C: (as defined in paras. 2.1 (c))

Value of work £	Fee £	£
Up to 600 000	0.50%	
600 000 - 3 000 000	3 000 + 0.30% on balance over	600 000
3 000 000 - 6 000 000	10 200 + 0.25% on balance over	3 000 000
Over 6 000 000	17 700 + 0.20% on balance over	6 000 000

2.2. Air conditioning, heating, ventilating and electrical services

(a) Where bills of quantities are prepared by the quantity surveyor for the air conditioning, heating, ventilating and electrical services there shall be a fee for these services (which shall include examining tenders received and reporting thereon), in addition to the fee calculated in accordance with para. 2.1, as follows:

Value of work £	Additional fee £	£
Up to 120 000	2.50%	
120 000 - 240 000	3 000 + 2.25% on balance over	120 000
240 000 - 480 000	5 700 + 2.00% on balance over	240 000
480 000 - 750 000	10 500 + 1.75% on balance over	480 000
750 000 - 1 000 000	15 225 + 1.25% on balance over	750 000
Over 1 000 000	18 350 + 1.15% on balance over	1 000 000

(b) The values of such services, whether the subject of separate tenders or not, shall be aggregated and the total value of work so obtained used for the purpose of calculating the additional fee chargeable in accordance with para. (a). (Except that when more than one firm of consulting engineers is engaged on the design of these services, the separate values for which each such firm is responsible shall be aggregated and the additional fees charged shall be calculated independently on each such total value so obtained.)

(c) Fees shall be calculated upon the accepted tender for the whole of the air conditioning, heating, ventilating and electrical services for which bills of quantities have been prepared by the quantity surveyor. In the event of no tender being accepted, fees shall be calculated upon the basis of the lowest original bona fide tender received. In the event of no such tender being received, the fees shall be calculated upon a reasonable valuation of the services based upon the original bills of quantities.

NOTE   In the foregoing context "bona fide tender"' shall be deemed to mean a tender submitted in good faith without major errors of computation and not subsequently withdrawn by the tenderer.

(d) When cost planning services are provided by the quantity surveyor for air conditioning, heating, ventilating and electrical services (or for any part of such services) there shall be an additional fee based on the time involved (see paras. 19.1 and 19.2). Alternatively the fee may be on a lump sum or percentage basis agreed between the employer and the quantity surveyor.

QUANTITY SURVEYORS' FEES

NOTE The incorporation of figures for air conditioning, heating, ventilating and electrical services provided by the consulting engineer is deemed to be included in the quantity surveyor's services under para. 2.1.

2.3. Works of alteration
On works of alteration or repair, or on those sections of the works which are mainly works of alteration or repair, there shall be a fee of 1.0% in addition to the fee calculated in accordance with paras. 2.1 and 2.2.

2.4. Works of redecoration and associated minor repairs,
On works of redecoration and associated minor repairs, there shall be a fee of 1.5% in addition to the fee calculated in accordance with paras. 2.1 and 2.2.

2.5. Bills of quantities prepared in special forms
Fees calculated in accordance with paras. 2.1, 2.2, 2.3 and 2.4 include for the preparation of bills of quantities on a normal trade basis. If the employer requires additional information to be provided in the bills of quantities or the bills to be prepared in an elemental, operational or similar form, then the fee may be adjusted by agreement between the employer and the quantity surveyor.

2.6. Reduction of tenders

(a) When cost planning services have been provided by the quantity surveyor and a tender, when received, is reduced before acceptance, and if the reductions are not necessitated by amended instructions of the employer or by the inclusion in the bills of quantities of items which the quantity surveyor has indicated could not be contained within the approved estimate, then in such a case no charge shall be made by the quantity surveyor for the preparation of bills of reductions and the fee for the preparation of the bills of quantities shall be based on the amount of the reduced tender.

(b) When cost planning services have not been provided by the quantity surveyor and if a tender, when received, is reduced before acceptance, fees are to be calculated upon the amount of the unreduced tender. When the preparation of bills of reductions is required, a fee is chargeable for preparing such bills of reductions as follows:

(i) 2.0% upon the gross amount of all omissions requiring measurement or abstraction from original dimensional sheets.
(ii) 3.0% upon the gross amount of all additions requiring measurement.
(iii) 0.5% upon the gross amount of all remaining additions.

NOTE: The above scale for the preparation of bills of reductions applies to work in all categories.

2.7 Generally
If the works are substantially varied at any stage or if the quantity surveyor is involved in an excessive amount of abortive work, then the fees shall be adjusted by agreement between the employer and the quantity surveyor.

3.0. NEGOTIATING TENDERS

3.1. (a) For negotiating and agreeing prices with a contractor:

Value of work	Fee
Up to 150 000	0.5%
150 000 - 600 000	750 + 0.3% on balance over 150 000
600 000 - 1 200 000	2 100 + 0.2% on balance over 600 000
Over 1 200 000	3 300 + 0.1% on balance over 1 200 000

QUANTITY SURVEYORS' FEES

- (b) The fee shall be calculated on the total value of the works as defined in paras. 2.1 (d), (e), (f), (g) and (j).
- (c) For negotiating and agreeing prices with a contractor for air conditioning, heating, ventilating and electrical services there shall be an additional fee as para. 3.1 (a) calculated on the total value of such services as defined in para. 2.2 (b).

4.0. CONSULTATIVE SERVICES AND PRICING BILLS OF QUANTITIES

4.1. Consultative services

Where the quantity surveyor is appointed to prepare approximate estimates, feasibility studies or submissions for the approval of financial grants or similar services, then the fee shall be based on the time involved (see paras. 19.1 and 19.2) or alternatively, on a lump sum or percentage basis agreed between the employer and the quantity surveyor.

4.2. Pricing bills of quantities

- (a) For pricing bills of quantities, if instructed, to provide an estimate comparable with tenders, the fee shall be one-third (33.33%) of the fee for negotiating and agreeing prices with a contractor, calculated in accordance with paras. 3.1 (a) and (b).
- (b) For pricing bills of quantities, if instructed, to provide an estimate comparable with tenders for air conditioning, heating, ventilating and electrical services the fee shall be one-third (33.33%) of the fee calculated in accordance with para. 3.1. (c).

CONTRACTS BASED ON BILLS OF QUANTITIES: POST-CONTRACT SERVICES

Alternative scales (I and II) for post-contract services are set out below to be used at the quantity surveyor's discretion by prior agreement with the employer.

5.0. ALTERNATIVE I: OVERALL SCALE OF CHARGES FOR POST-CONTRACT SERVICES

5.1.
If the quantity surveyor appointed to carry out the post-contract services did not prepare the bills of quantities then the fees in paras. 5.2 and 5.3 shall be increased to cover the additional services undertaken by the quantity surveyor.

5.2. Basic scale

For taking particulars and reporting valuations for interim certificates for payments on account to the contractor, preparing periodic assessments of anticipated final cost and reporting thereon, measuring and making up bills of variations including pricing and agreeing totals with the contractor, and adjusting fluctuations in the cost of labour and materials if required by the contract.

- (a) Category A: Relatively complex works and/or works with little or no repetition.

    Examples:
    Ambulance and fire stations; banks; cinemas; clubs; computer buildings; council offices; crematoria; fitting out existing buildings; homes for the elderly; hospitals and nursing homes; laboratories; law courts; libraries; "one-off" houses; petrol stations; places of religious worship; police stations; public houses; licensed premises; restaurants; sheltered housing; sports pavilions; theatres; town halls; universities, polytechnics and colleges of further education (other than halls of residence and hostels); and the like.

QUANTITY SURVEYORS' FEES

Value of work £		Category A fee	
		£	£
Up to	150 000	150 + 2.0% (Minimum fee £1 150)	
150 000 -	300 000	3 150 + 1.7% on balance over	150 000
300 000 -	600 000	5 700 + 1.6% on balance over	300 000
600 000 -	1 500 000	10 500 + 1.3% on balance over	600 000
1 500 000 -	3 000 000	22 200 + 1.2% on balance over	1 500 000
3 000 000 -	6 000 000	40 200 + 1.1% on balance over	3 000 000
Over	6 000 000	73 200 + 1.0% on balance over	6 000 000

(b) Category B: Less complex works and/or works with some element of repetition.

Examples:
Adult education facilities; canteens; church halls; community centres; departmental stores; enclosed sports stadia and swimming baths; halls of residence; hostels; motels; offices other than those included in Categories A and C; railway stations; recreation and leisure centres; residential hotels; schools; self-contained flats and maisonettes; shops and shopping centres; supermarkets and hypermarkets; telephone exchanges; and the like.

Value of work £		Category B fee	
		£	£
Up to	150 000	150 + 2.0% (Minimum fee £1 150)	
150 000 -	300 000	3 150 + 1.7% on balance over	150 000
300 000 -	600 000	5 700 + 1.5% on balance over	300 000
600 000 -	1 500 000	10 200 + 1.1% on balance over	600 000
1 500 000 -	3 000 000	20 100 + 1.0% on balance over	1 500 000
3 000 000 -	6 000 000	35 100 + 0.9% on balance over	3 000 000
Over	6 000 000	62 100 + 0.8% on balance over	6 000 000

(c) Category C: Simple works and/or works with a substantial element of repetition.

Examples:
Factories; garages; multi-storey car parks; open-air sports stadia; structural shell offices not fitted out; warehouses; workshops; and the like.

Value of work £		Category C fee	
		£	£
Up to	150 000	120 + 1.6% (Minimum fee £920)	
150 000 -	300 000	2 520 + 1.5% on balance over	150 000
300 000 -	600 000	4 770 + 1.4% on balance over	300 000
600 000 -	1 500 000	8 970 + 1.1% on balance over	600 000
1 500 000 -	3 000 000	18 870 + 0.9% on balance over	1 500 000
3 000 000 -	6 000 000	32 370 + 0.8% on balance over	3 000 000
Over	6 000 000	56 370 + 0.7% on balance over	6 000 000

(d) The scales of fees for post-contract services (paras. 5.2 (a) to (c)) are overall scales based upon the inclusion of all nominated sub-contractors' and nominated suppliers' accounts, subject to the provision of para. 5.2 (g). When work normally included in a building contract is the subject of a separate contract for which the quantity surveyor has not been paid fees under any other clause hereof, the value of such work shall be included in the amount on which fees are charged.

(e) Fees shall be calculated upon the basis of the account for the whole of the work, subject to the provisions of para. 5.3.

(f) In calculating the amount on which fees are charged the total of any credits is to be added to the amount described above.

QUANTITY SURVEYORS' FEES

(g) Where the value of air conditioning, heating, ventilating and electrical services included in the tender documents together exceeds 25% of the amount calculated as described in paras. 5.2. (d) and (e) above, then, subject to provisions of para. 5.3, no fee is chargeable on the amount by which the value of these services exceeds the said 25%. In this context the term "value" excludes general contractors' profit, attendance, builders work in connection with the services, preliminaries and other similar additions.

(h) When a contract comprises buildings which fall into more than one category, the fee shall be calculated as follows:

    (i) The amount upon which fees are chargeable shall be allocated to the categories of work applicable and the amounts so allocated expressed as percentages of the total amount upon which fees are chargeable.

    (ii) Fees shall then be calculated for each category on the total amount upon which fees are chargeable.

    (iii) The fee chargeable shall then be calculated by applying the percentages of work in each category to the appropriate total fee and adding the resultant amounts.

(j) When a project is the subject of a number of contracts then, for the purposes of calculating fees, the values of such contracts shall not be aggregated but each contract shall be taken separately and the scale of charges (paras. 5.2 (a) to (h)), applied as appropriate.

(k) When the quantity surveyor is required to prepare valuations of materials or goods off site, an additional fee shall be charged based on the time involved (see paras. 19.1 and 19.2).

(l) The basic scale for post-contract services includes for a simple routine of periodically estimating final costs. When the employer specifically requests a cost monitoring service which involves the quantity surveyor in additional or abortive measurement an additional fee shall be charged based on the time involved (see paras. 19.1 and 19.2), or alternatively on a lump sum or percentage basis agreed between the employer and the quantity surveyor.

(m) The above overall scales of charges for post-contract services assume normal conditions when the bills of quantities are based on drawings accurately depicting the building work the employer requires. If the works are materially varied to the extent that substantial remeasurement is necessary then the fee for post contract services shall be adjusted by agreement between the employer and the quantity surveyor.

5.3. Air conditioning, heating, ventilating and electrical services

(a) Where final accounts are prepared by the quantity surveyor for the air conditioning, heating, ventilating and electrical services there shall be a fee for these services, in addition to the fee calculated in accordance with para. 5.2, as follows:

Value of work £	Additional fee £	£
Up to 120 000	2.00%	
120 000 - 240 000	2 400 + 1.60% on balance over	120 000
240 000 - 1 000 000	4 320 + 1.25% on balance over	240 000
1 000 000 - 4 000 000	13 820 + 1.00% on balance over	1 000 000
Over 4 000 000	43 820 + 0.90% on balance over	4 000 000

(b) The values of such services, whether the subject of separate tenders or not, shall be aggregated and the total value of work so obtained used for the purpose of calculating the additional fee chargeable in accordance with para. (a). (Except that when more than one firm of consulting engineers is engaged on the design of these services the separate values for which each such firm is responsible shall be aggregated and the additional fee charged shall be calculated independently on each such total value so obtained.)

QUANTITY SURVEYORS' FEES

(c) The scope of the services to be provided by the quantity surveyor under para. (a) above shall be deemed to be equivalent to those described for the basic scale for post-contract services.

(d) When the quantity surveyor is required to prepare periodic valuations of materials or goods off site, an additional fee shall be charged based on the time involved (see paras. 19.1 and 19.2).

(e) The basic scale for post-contract services includes for a simple routine of periodically estimating final costs. When the employer specifically requests a cost monitoring service which involves the quantity surveyor in additional or abortive measurement an additional fee shall be based on the time involved (see paras. 19.1 and 19.2), or alternatively on a lump sum or percentage basis agreed between the employer and the quantity surveyor.

(f) Fees shall be calculated upon the basis of the account for the whole of the air conditioning, heating, ventilating and electrical services for which final accounts have been prepared by the quantity surveyor.

6.0. ALTERNATIVE II: SCALE OF CHARGES FOR SEPARATE STAGES OF POST-CONTRACT SERVICES

6.1. If the quantity surveyor appointed to carry out the post-contract services did not prepare the bills of quantities then the fees in paras. 6.2 and 6.3 shall be increased to cover the additional services undertaken by the quantity surveyor.

NOTE: The scales of fees in paras. 6.2 and 6.3 apply to work in all categories (including air conditioning, heating, ventilating and electrical services).

6.2. Valuations for interim certificates

(a) For taking particulars and reporting valuations for interim certificates for payments on account to the contractor.

Total of valuations £	Fee £	£
Up to 300 000	0.5%	
300 000 - 1 000 000	1 500 + 0.4% on balance over	300 000
1 000 000 - 6 000 000	4 300 + 0.3% on balance over	1 000 000
Over 6 000 000	19 300 + 0.2% on balance over	6 000 000

NOTES:

1. Subject to note 2 below, the fees are to be calculated on the total of all interim valuations (i.e. the amount of the final account less only the net amount of the final valuation).

2. When consulting engineers are engaged in supervising the installation of air conditioning, heating, ventilating and electrical services and their duties include reporting valuations for inclusion in interim certificates for payments on account in respect of such services, then valuations so reported shall be excluded from any total amount of valuations used for calculating fees.

(b) When the quantity surveyor is required to prepare valuations of materials or goods off site, an additional fee shall be charged based on the time involved (see paras. 19.1 and 19.2).

6.3. Preparing accounts of variation upon contracts
For measuring and making up bills of variations including pricing and agreeing totals with the contractor:

(a) An initial lump sum of £600 shall be payable on each contract.

QUANTITY SURVEYORS' FEES

(b) 2.0% upon the gross amount of omissions requiring measurement or abstraction from the original dimension sheets.

(c) 3.0% upon the gross amount of additions requiring measurement and upon dayworks.

(d) 0.5% upon the gross amount of remaining additions which shall be deemed to include all nominated sub-contractors' and nominated suppliers' accounts which do not involve measurement or checking of quantities but only checking against lump sum estimates.

(e) 3.0% upon the aggregate of the amounts of the increases and/or decreases in the cost of labour and materials in accordance with any fluctuations clause in the conditions of contract, except where a price adjustment formula applies.

(f) On contracts where fluctuations are calculated by the use of a price adjustment formula method the following scale shall be applied to the account for the whole of the work:

Value of work	£	Fee £	£
Up to	300 000	300 + 0.5%	
300 000 -	1 000 000	1 800 + 0.3% on balance over	300 000
Over	1 000 000	3 900 + 0.1% on balance over	1 000 000

(g) When consulting engineers are engaged in supervising the installation of air conditioning, heating, ventilating and electrical services and their duties include for the adjustment of accounts and pricing and agreeing totals with the sub-contractors for inclusion in the measured account, then any totals so agreed shall be excluded from any amounts used for calculating fees.

6.4. Cost monitoring services

The fee for providing all approximate estimates of final cost and/or a cost monitoring service shall be based on the time involved (see paras. 19.1 and 19.2), or alternatively on a lump sum or percentage basis agreed between the employer and the quantity surveyor.

7.0. BILLS OF APPROXIMATE QUANTITIES, INTERIM CERTIFICATES AND FINAL ACCOUNTS

7.1. Basic scale

For preparing bills of approximate quantities suitable for obtaining competitive tenders which will provide a schedule of prices and a reasonably close forecast of the cost of the works, but subject to complete remeasurement, examining tenders and reporting thereon, taking particulars and reporting valuations for interim certificates for payments on account to the contractor, preparing periodic assessments of anticipated final cost and reporting thereon, measuring and preparing final account, including pricing and agreeing totals with the contractor and adjusting fluctuations in the cost of labour and materials if required by the contract:

(a) Category A: Relatively complex works and/or works with little or no repetition.

Examples:
Ambulance and fire stations; banks; cinemas; clubs; computer buildings; council offices; crematoria; fitting out existing buildings; homes for the elderly; hospitals and nursing homes; laboratories; law courts; libraries; "one-off" houses; petrol stations; places of religious worship; police stations; public houses; licensed premises; restaurants; sheltered housing; sports pavilions; theatres; town halls; universities, polytechnics and colleges of further education (other than halls of residence and hostels); and the like.

QUANTITY SURVEYORS' FEES

Value of work £		Category A fee	
		£	£
Up to	150 000	380 + 5.0% (Minimum fee £2 880)	
150 000 -	300 000	7 880 + 4.0% on balance over	150 000
300 000 -	600 000	13 880 + 3.4% on balance over	300 000
600 000 -	1 500 000	24 080 + 2.8% on balance over	600 000
1 500 000 -	3 000 000	49 280 + 2.4% on balance over	1 500 000
3 000 000 -	6 000 000	85 280 + 2.2% on balance over	3 000 000
Over	6 000 000	151 280 + 2.0% on balance over	6 000 000

(b) Category B: Less complex works and/or works with some element of repetition

Examples:
Adult education facilities; canteens; church halls; community centres; departmental stores; enclosed sports stadia and swimming baths; halls of residence; hostels; motels; offices other than those included in Categories A and C; railway stations; recreation and leisure centres; residential hotels; schools; self-contained flats and maisonettes; shops and shopping centres; supermarkets and hypermarkets; telephone exchanges; and the like.

Value of work £		Category B fee	
		£	£
Up to	150 000	360 + 4.8% (Minimum fee £2 760)	
150 000 -	300 000	7 560 + 3.7% on balance over	150 000
300 000 -	600 000	13 110 + 3.0% on balance over	300 000
600 000 -	1 500 000	22 110 + 2.2% on balance over	600 000
1 500 000 -	3 000 000	41 910 + 2.0% on balance over	1 500 000
3 000 000 -	6 000 000	71 910 + 1.8% on balance over	3 000 000
Over	6 000 000	125 910 + 1.6% on balance over	6 000 000

(c) Category C: Simple works and/or works with a substantial element of repetition.
Examples:
Factories; garages; multi-storey car parks; open air sports stadia; structural shell offices not fitted out; warehouses; workshops; and the like.

Value of work £		Category C fee	
		£	£
Up to	150 000	300 + 4.1% (Minimum fee £2 350)	
150 000 -	300 000	6 450 + 3.3% on balance over	150 000
300 000 -	600 000	11 400 + 2.6% on balance over	300 000
600 000 -	1 500 000	19 200 + 2.0% on balance over	600 000
1 500 000 -	3 000 000	37 200 + 1.7% on balance over	1 500 000
3 000 000 -	6 000 000	62 700 + 1.5% on balance over	3 000 000
Over	6 000 000	107 700 + 1.3% on balance over	6 000 000

(d) The scales of fees for pre-contract and post-contract services (paras. 7.1 (a) to (c)) are overall scales based upon the inclusion of all nominated sub-contractors' and nominated suppliers' accounts, subject to the provision of para. 7.1. (g). When work normally included in a building contract is the subject of a separate contract for which the quantity surveyor has not been paid fees under any other clause hereof, the value of such work shall be included in the amount on which fees are charged.

(e) Fees shall be calculated upon the basis of the account for the whole of the work, subject to the provisions of para. 7.2.

(f) In calculating the amount on which fees are charged the total of any credits is to be added to the amount described above.

QUANTITY SURVEYORS' FEES

(g) Where the value of air conditioning, heating, ventilating and electrical services included in tender documents together exceeds 25% of the amount calculated as described in paras. 7.1. (d) and (e), then, subject to the provisions of para. 7.2 no fee is chargeable on the amount by which the value of these services exceeds the said 25%. In this context the term "value" excludes general contractors' profit, attendance, builders' work in connection with the services, preliminaries and any other similar additions.

(h) When a contract comprises buildings which fall into more than one category, the fee shall be calculated as follows.

    (i) The amount upon which fees are chargeable shall be allocated to the categories of work applicable and the amount so allocated expressed as percentages of the total amount upon which fees are chargeable.

    (ii) Fees shall then be calculated for each category on the total amount upon which fees are chargeable.

    (iii) The fee chargeable shall then be calculated by applying the percentages of work in each category to the appropriate total fee adding the resultant amounts.

(j) When a project is the subject of a number of contracts then, for the purpose of calculating fees, the values of such contracts shall not be aggregated but each contract shall be taken separately and the scale of charges (paras. 7.1(a) to (h)) applied as appropriate.

(k) Where the quantity surveyor is specifically instructed to provide cost planning services, the fee calculated in accordance with paras. 7.1 (a) to (j) shall be increased by a sum calculated in accordance with the following table and based upon the same value of work as that upon which the aforementioned fee has been calculated:

Categories A & B: (as defined in paras. 7.1 (a) and (b))

Value of work £	Fee £	£
Up to 600 000	0.70%	
600 000 - 3 000 000	4 200 + 0.40% on balance over	600 000
3 000 000 - 6 000 000	13 800 + 0.35% on balance over	3 000 000
Over 6 000 000	24 300 + 0.30% on balance over	6 000 000

Category C: (as defined in para. 7.1 (c))

Value of work £	Fee £	£
Up to 600 000	0.50%	
600 000 - 3 000 000	3 000 + 0.30% on balance over	600 000
3 000 000 - 6 000 000	10 200 + 0.25% on balance over	3 000 000
Over 6 000 000	17 700 + 0.20% on balance over	6 000 000

(l) When the quantity surveyor is required to prepare valuations of materials or goods off site, an additional fee shall be charged based on the time involved (see paras. 19.1 and 19.2).

(m) The basic scale for post-contract services includes for a simple routine of periodically estimating final costs. When the employer specifically requests a cost monitoring service which involves the quantity surveyor in additional or abortive measurement an additional fee shall be charged based on the time involved (see paras. 19.1 and 19.2), or alternatively on a lump sum or percentage basis agreed between the employer and the quantity surveyor.

QUANTITY SURVEYORS' FEES

7.2. Air conditioning, heating, ventilating and electrical services

(a) Where bills of approximate quantities and final accounts are prepared by the quantity surveyor for the air conditioning, heating, ventilating and electrical services there shall be a fee for these services in addition to the fee calculated in accordance with para. 7.1 as follows:

Value of work £		Category A fee £	£
Up to	120 000	4.50%	
120 000 -	240 000	5 400 + 1.85% on balance over	120 000
240 000 -	480 000	10 020 + 3.25% on balance over	240 000
480 000 -	750 000	17 820 + 3.00% on balance over	480 000
750 000 -	1 000 000	25 920 + 2.50% on balance over	750 000
1 000 000 -	4 000 000	32 170 + 2.15% on balance over	1 000 000
Over	4 000 000	96 670 + 2.05% on balance over	4 000 000

(b) The value of such services, whether the subject of separate tenders or not, shall be aggregated and the value of work so obtained used for the purpose of calculating the additional fee chargeable in accordance with para. (a). (Except that when more than one firm of consulting engineers is engaged on the design of these services, the separate values for which each such firm is responsible shall be aggregated and the additional fees charged shall be calculated independently on each such total value so obtained.)

(c) The scope of the services to be provided by the quantity surveyor under para. (a) above shall be deemed to be equivalent to those described for the basic scale for pre-contract and post-contract services.

(d) When the quantity surveyor is required to prepare valuations of materials or goods off site, an additional fee shall be charged based on the time involved (see paras. 19.1 and 19.2).

(e) The basic scale for post-contract services includes for a simple routine of periodically estimating final costs. When the employer specifically requests a cost monitoring service, which involves the quantity surveyor in additional or abortive measurement, an additional fee shall be charged based on the time involved (see paras. 19.1 and 19.2), or alternatively on a lump sum or percentage basis agreed between the employer and the quantity surveyor.

(f) Fees shall be calculated upon the basis of the account for the whole of the air conditioning, heating, ventilating and electrical services for which final accounts have been prepared by the quantity surveyor.

(g) When cost planning services are provided by the quantity surveyor for air conditioning, heating, ventilating and electrical services (or for any part of such services) there shall be an additional fee based on the time involved (see paras. 19.1 and 19.2) or alternatively on a lump sum or percentage basis agreed between the employer and quantity surveyor.

NOTE: The incorporation of figures for air conditioning, heating, ventilating and electrical services provided by the consulting engineer is deemed to be included in the quantity surveyor's services under para 7.1.

7.3. Works of alteration
On works of alteration or repair, or on those sections of the work which are mainly works of alteration or repair, there shall be a fee of 1.0% in addition to the fee calculated in accordance with paras. 7.1 and 7.2

7.4. Works of redecoration and associated minor repairs
On works of redecoration and associated minor repairs, there shall be a fee of 1.5% in addition to the fee calculated in accordance with paras. 7.1 and 7.2.

QUANTITY SURVEYORS' FEES

7.5. **Bills of quantities and/or final accounts prepared in special forms**
Fees calculated in accordance with paras. 7.1, 7.2, 7.3 and 7.4 include for the preparation of bills of quantities and/or final accounts on a normal trade basis. If the employer requires additional information to be provided in the bills of quantities and/or final accounts or the bills and/or final accounts to be prepared in an elemental, operational or similar form, then the fee may be adjusted by agreement between the employer and the quantity surveyor.

7.6. **Reduction of tenders**

(a) When cost planning services have been provided by the quantity surveyor and a tender, when received, is reduced before acceptance and if the reductions are not necessitated by amended instructions of the employer or by the inclusion in the bills of approximate quantities of items which the quantity surveyor has indicated could not be contained within the approved estimate, then in such a case no charge shall be made by the quantity surveyor for the preparation of bills of reductions and the fee for the preparation of bills of approximate quantities shall be based on the amount of the reduced tender.

(b) When cost planning services have not been provided by the quantity surveyor and if a tender, when received, is reduced before acceptance, fees are to be calculated upon the amount of the unreduced tender. When the preparation of bills of reductions is required, a fee is chargeable for preparing such bills of reductions as follows:

(i) 2.0% upon the gross amount of all omissions requiring measurement or abstraction from original dimension sheets.
(ii) 3.0% upon the gross amount of all additions requiring measurement.
(iii) 0.5% upon the gross amount of all remaining additions.

NOTE: The above scale for the preparation of bills of reductions applies to work in all categories.

7.7. **Generally**
If the works are substantially varied at any stage or if the quantity surveyor is involved in an excessive amount of abortive work, then the fees shall be adjusted by agreement between the employer and the quantity surveyor.

8.0. NEGOTIATING TENDERS

8.1 (a) For negotiating and agreeing prices with a contractor:

Value of work £		Fee £	£
Up to	150 000	0.5%	
150 000 -	600 000	750 + 0.3% on balance over	150 000
600 000 -	1 200 000	2 100 + 0.2% on balance over	600 000
Over	1 200 000	3 300 + 0.1% on balance over	1 200 000

(b) The fee shall be calculated on the total value of the works as defined in paras. 7.1 (d), (e), (f), (g) and (j).

(c) For negotiating and agreeing prices with a contractor for air conditioning, heating, ventilating and electrical services there shall be an additional fee as para. 8.1 (a) calculated on the total value of such services as defined in para. 7.2 (b).

QUANTITY SURVEYORS' FEES

9.0. CONSULTATIVE SERVICES AND PRICING BILLS OF APPROXIMATE QUANTITIES

9.1. Consultative services
Where the quantity surveyor is appointed to prepare approximate estimates, feasibility studies or submissions for the approval of financial grants or similar services, then the fee shall be based on the time involved (see paras. 19.1 and 19.2) or alternatively, on a lump sum or percentage basis agreed between the employer and the quantity surveyor.

9.2. Pricing bills of approximate quantities
For pricing bills of approximate quantities, if instructed, to provide an estimate comparable with tenders, the fees shall be the same as for the corresponding services in paras. 4.2 (a) and (b).

10.0. INSTALMENT PAYMENTS

10.1. For the purpose of instalment payments the fee for preparation of bills of approximate quantities only shall be the equivalent of forty per cent (40%) of the fees calculated in accordance with the appropriate sections of paras. 7.1 to 7.5, and the fee for providing cost planning services shall be in accordance with the appropriate sections of para. 7.1 (k); both fees shall be based on the total value of the bills of approximate quantities ascertained in accordance with the provisions of para. 2.1 (e).

10.2. In the absence of agreement to the contrary, fees shall be paid by instalments as follows:

(a) Upon acceptance by the employer of a tender for the works the above defined fees for the preparation of bills of approximate quantities and for providing cost planning services.
(b) In the event of no tender being accepted, the aforementioned fees shall be paid within three months of completion of the bills of approximate quantities.
(c) The balance by instalments at intervals to be agreed between the date of the first certificate and one month after certification of the contractor's account.

10.3. In the event of the project being abandoned at any stage other than those covered by the foregoing, the proportion of fee payable shall be by agreement between the employer and the quantity surveyor.

11.0. SCHEDULES OF PRICES

11.1. The fee for preparing, pricing and agreeing schedules of prices shall be based on the time involved (see paras. 19.1 and 19.2). Alternatively, the fee may be on a lump sum or percentage basis agreed between the employer and the quantity surveyor.

12.0. COST PLANNING AND APPROXIMATE ESTIMATES

12.1. The fee for providing cost planning services or for preparing approximate estimates shall be based on the time involved (see paras. 19.1 and 19.2). Alternatively, the fee may be on a lump sum or percentage basis agreed between the employer and the quantity surveyor.

CONTRACTS BASED ON SCHEDULES OF PRICES: POST-CONTRACT SERVICES

QUANTITY SURVEYORS' FEES

13.0. FINAL ACCOUNTS

13.1. Basic Scale

(a) For taking particulars and reporting valuations for interim certificates for payments on account to the contractor, preparing periodic assessments of anticipated final cost and reporting thereon, measuring and preparing final account including pricing and agreeing totals with the contractor, and adjusting fluctuations in the cost of labour and materials if required by the contract, the fee shall be equivalent to sixty per cent (60%) of the fee calculated in accordance with paras. 7.1 (a) to (j).

(b) When the quantity surveyor is required to prepare valuations of materials or goods off site, an additional fee shall be charged on the basis of the time involved (see paras. 19.1 and 19.2).

(c) The basic scale for post-contract services includes for a simple routine of periodically estimating final costs. When the employer specifically requests a cost monitoring service which involves the quantity surveyor in additional or abortive measurement an additional fee shall be charged based on the time involved (see paras. 19.1 and 19.2), or alternatively on a lump sum or percentage basis agreed between the employer and the quantity surveyor.

13.2. Air conditioning, heating, ventilating and electrical services
Where final accounts are prepared by the quantity surveyor for the air conditioning, heating, ventilating and electrical services there shall be a fee for these services, in addition to the fee calculated in accordance with para. 13.1, equivalent to sixty per cent (60%) of the fee calculated in accordance with paras. 7.2 (a) to (f).

13.3. Works of alterations
On works of alteration or repair, or on those sections of the work which are mainly works of alteration or repair, there shall be a fee of 1.0% in addition to the fee calculated in accordance with paras. 13.1 and 13.2.

13.4. Works of redecoration and associated minor repairs
On works of redecoration and associated minor repairs, there shall be a fee of 1.5% in addition to the fee calculated in accordance with paras. 13.1 and 13.2.

13.5. Final accounts prepared in special forms
Fees calculated in accordance with paras. 13.1, 13.2, 13.3 and 13.4 include for the preparation of final accounts on a normal trade basis. If the employer requires additional information to be provided in the final accounts or the accounts to be prepared in an elemental, operational or similar form, then the fee may be adjusted by agreement between the employer and the quantity surveyor.

PRIME COST CONTRACTS: PRE-CONTRACT AND POST-CONTRACT SERVICES

14.0. COST PLANNING

14.1. The fee for providing a cost planning service shall be based on the time involved (see paras. 19.1 and 19.2). Alternatively, the fee may be on a lump sum or percentage basis agreed between the employer and the quantity surveyor.

QUANTITY SURVEYORS' FEES

15.0. ESTIMATES OF COST

15.1. (a) For preparing an approximate estimate, calculated by measurement, of the cost of work, and, if required under the terms of the contract, negotiating, adjusting and agreeing the estimate:

Value of work £		Fee £	£
Up to	30 000	1.25%	
30 000 -	150 000	375 + 1.00% on balance over	30 000
150 000 -	600 000	1 575 + 0.75% on balance over	150 000
Over	600 000	4 950 + 0.50% on balance over	600 000

(b) The fee shall be calculated upon the total of the approved estimates.

16.0. FINAL ACCOUNTS

16.1. (a) For checking prime costs, reporting for interim certificates for payments on account to the contractor and preparing final accounts:

Value of work £		Fee £	£
Up to	30 000	2.50%	
30 000 -	150 000	750 + 2.00% on balance over	30 000
150 000 -	600 000	3 150 + 1.50% on balance over	150 000
Over	600 000	9 900 + 1.25% on balance over	600 000

(b) The fee shall be calculated upon the total of the final account with the addition of the value of credits received for old materials removed and less the value of any work charged for in accordance with para. 16.1 (c).

(c) On the value of any work to be paid for on a measured basis, the fee shall be 3%.

(d) When the quantity surveyor is required to prepare valuations of materials or goods off site, an additional fee shall be charged based on the time involved (see paras. 19.1 and 19.2).

(e) The above charges do not include the provision of checkers on the site. If the quantity surveyor is required to provide such checkers an additional charge shall be made by arrangement.

17.0. COST REPORTING AND MONITORING SERVICES

17.1. The fee for providing cost reporting and/or monitoring services (e.g. preparing periodic assessments of anticipated final costs and reporting thereon) shall be based on the time involved (see paras. 19.1 and 19.2) or alternatively, on a lump sum or percentage basis agreed between the employer and the quantity surveyor.

18.0. ADDITIONAL SERVICES

18.1. For additional services not normally necessary, such as those arising as a result of the termination of a contract before completion, liquidation, fire damage to the buildings, services in connection with arbitration, litigation and investigation of the validity of contractors' claims, services in connection with taxation matters and all similar services where the employer specifically instructs the quantity surveyor, the charges shall be in accordance with paras. 19.1 and 19.2.

QUANTITY SURVEYORS' FEES

19.0. TIME CHARGES

19.1. (a) For consultancy and other services performed by a principal, a fee by arrangement according to the circumstances including the professional status and qualifications of the quantity surveyor.

(b) When a principal does work which would normally be done by a member of staff, the charge shall be calculated as para. 19.2 below.

19.2. (a) For services by a member of staff, the charges for which are to be based on the time involved, such charges shall be calculated on the hourly cost of the individual involved plus 145%.

(b) A member of staff shall include a principal doing work normally done by an employee (as para. 19.1 (b) above), technical and supporting staff, but shall exclude secretarial staff or staff engaged upon general administration.

(c) For the purpose of para. 19.2 (b) above, a principal's time shall be taken at the rate applicable to a senior assistant in the firm.

(d) The supervisory duties of a principal shall be deemed to be included in the addition of 145% as para. 19.2 (a) above and shall not be charged separately.

(e) The hourly cost to the employer shall be calculated by taking the sum of the annual cost of the member of staff of:

  (i) Salary and bonus but excluding expenses;
  (ii) Employer's contributions payable under any Pension and Life Assurance Schemes;
  (iii) Employer's contributions made under the National Insurance Acts, the Redundancy Payments Act and any other payments made in respect of the employee by virtue of any statutory requirements; and
  (iv) Any other payments or benefits made or granted by the employer in pursuance of the terms of employment of the member of staff;

and dividing by 1,650.

19.3. The foregoing Time Charges under paras. 19.1 and 19.2 are intended for use where other paras. of the Scale (not related to Time Charges) form a significant proportion of the overall fee. In all other cases an increased time charge may be agreed.

20.0. INSTALMENT PAYMENTS

20.1. In the absence of agreement to the contrary, payments to the quantity surveyor shall be made by instalments by arrangement between the employer and the quantity surveyor.

SCALE 40 PROFESSIONAL CHARGES FOR QUANTITY SURVEYING SERVICES IN CONNECTION WITH HOUSING SCHEMES FOR LOCAL AUTHORITIES
The scale is recommended and not mandatory.

EFFECTIVE FROM FEBRUARY 1983

1.0 GENERALLY

1.1 The scale is applicable to housing schemes of self-contained dwellings regardless of type (e.g. houses, maisonettes, bungalows or flats) and irrespective of the amount of repetition of identical types or blocks within an individual housing scheme and shall also apply to all external works forming part of the contract for the housing scheme. This scale does not apply to improvement to existing dwellings.

QUANTITY SURVEYORS' FEES

1.2 The fees set out below cover the following quantity surveying services as may be required:

(a) Preparing bills of quantities or other tender documents; checking tenders received or negotiating tenders and pricing with a selected contractor; reporting thereon.

(b) Preparing recommendations for interim payments on account to the contractor; measuring work and adjusting variations in accordance with the terms of the contract and preparing the final account; pricing same and agreeing totals with the contractor; adjusting fluctuations in the cost of labour and materials if required by the contract.

(c) Preparing periodic financial statements showing the anticipated final cost by means of a simple routine of estimating final costs and reporting thereon, but excluding cost monitoring (see para. 1.4).

1.3 Where the quantity surveyor is appointed to prepare approximate estimates to establish and substantiate the economic viability of the scheme and to obtain the necessary approvals and consents, or to enable the scheme to be designed and constructed within approved cost criteria an additional fee shall be charged based on the time involved (see para. 7.0) or, alternatively, on a lump sum or percentage basis agreed between the employer and the quantity surveyor. (Cost planning services, see para. 3.0).

1.4 When the employer specifically requests a post-contract cost monitoring service which involves the quantity surveyor in additional or abortive work an additional fee shall be charged based on the time involved (see para. 7.0) or, alternatively, on a lump sum or percentage basis agreed between the employer and the quantity surveyor.

1.5 The fees are in all cases exclusive of travelling and other expenses (for which the actual disbursement is recoverable unless there is some prior arrangement for such charges) and of the cost of reproduction of bills of quantities and other documents, which are chargeable in addition at net cost.

1.6 The fees are in all cases exclusive of services in connection with the allocation of the cost of the works for purposes of calculating value added tax for which there shall be an additional fee based on the time involved (see para. 7.0).

1.7 When work normally included in a building contract is the subject of a separate contract for which the quantity surveyor has not been paid fees under any other clause thereof, the value of such work shall be included in the amount upon which fees are charged.

1.8 If any of the materials used in the works are supplied by the employer or charged at a preferential rate, then the estimated or actual value thereof shall be included in the amount upon which fees are to be calculated.

1.9 The fees are in all cases exclusive of preparing a specification of the materials to be used and the works to be done, but the fees for preparing bills of quantities and similar documents do include for incorporating preamble clauses describing the materials and workmanship (from information given by the architect and/or consulting engineer).

1.10 If the quantity surveyor incurs additional costs due to exceptional delays in building operations or any other cause beyond the control of the quantity surveyor, then the fees shall be adjusted by agreement between the employer and the quantity surveyor to cover the reimbursement of these additional costs.

1.11 When a project is the subject of a number of contracts then for the purposes of calculating fees, the values of such contracts shall not be aggregated but each contract shall be taken separately and the scale of charges applied as appropriate.

1.12 The fees and charges are in all cases exclusive of value added tax which will be applied in accordance with legislation.

QUANTITY SURVEYORS' FEES

1.13 Copyright in bills of quantities and other documents prepared by the quantity surveyor is reserved to the quantity surveyor.

2.0 BASIC SCALE

2.1 The basic fee for the services outlined in para. 1.2 shall be as follows:-

Value of work		Fee	
		£	£
Up to	75 000	250 + 4.6%	
75 000 -	150 000	3 700 + 3.6% on balance over	75 000
150 000 -	750 000	6 400 + 2.3% on balance over	150 000
750 000 -	1 500 000	20 200 + 1.7% on balance over	750 000
Over	1 500 000	32 950 + 1.5% on balance over	1 500 000

2.2 Fees shall be calculated upon the total of the final account for the whole of the work including all nominated sub-contractors' and nominated suppliers' accounts.

2.3 For services in connection with accommodation designed for the elderly or the disabled or other special category occupants for whom special facilities are required an addition of 10% shall be made to the fee calculated in accordance with para. 2.1.

2.4 When additional fees under para. 2.3 are chargeable on a part or parts of a scheme, the value of basic fee to which the additional percentages shall be applied shall be determined by the proportion that the values of the various types of accommodation bear to the total of those values.

2.5 When the quantity surveyor is required to prepare an interim valuation of materials or goods off site, an additional fee shall be charged based on the time involved (see para. 7.0).

2.6 If the works are substantially varied at any stage and if the quantity surveyor is involved in an excessive amount of abortive work, then the fee shall be adjusted by agreement between the employer and the quantity surveyor.

2.7 The fees payable under paras. 2.1 and 2.3 include for the preparation of bills of quantities or other tender documents on a normal trade basis. If the employer requires additional information to be provided in bills of quantities, or bills of quantities to be prepared in an elemental, operational or similar form, then the fee may be adjusted by agreement between the employer and the quantity surveyor.

3.0 COST PLANNING

3.1 When the quantity surveyor is specifically instructed to provide cost planning services, the fee calculated in accordance with paras. 2.1 and 2.3 shall be increased by a sum calculated in accordance with the following table and based upon the amount of the accepted tender.

Value of work	£	Fee	
		£	£
Up to	150 000	0.45%	
150 000 -	750 000	675 + 0.35% on balance over	150 000
Over	750 000	2 775 + 0.25% on balance over	750 000

3.2 Cost planning is defined as the process of ascertaining a cost limit, where necessary, within the guidelines set by any appropriate Authority, and thereafter checking the cost of the project within that limit throughout the design process. It includes the preparation of a cost plan (based upon elemental analysis or other suitable criterion) checking and revising it where required and effecting the necessary liaison with other consultants employed.

QUANTITY SURVEYORS' FEES

    3.3 (a) When cost planning services have been provided by the quantity surveyor and bills of reductions are required, then no charge shall be made by the quantity surveyor for the bills of reductions unless the reductions are necessitated by amended instructions of the employer or by the inclusion in the bills of quantities of items which the quantity surveyor has indicated could not be contained within the approved estimate.

        (b) When cost planning services have not been provided by the quantity surveyor and bills of reductions are required, a fee is chargeable for preparing such bills of reductions as follows:

            (i) 2.0% upon the gross amount of all omissions requiring measurement or abstraction from original dimension sheets.
            (ii) 3.0% upon the gross amount of all additions requiring measurement.
            (iii) 0.5% upon the gross amount of all remaining additions.

4.0 HEATING, VENTILATING AND ELECTRICAL SERVICES

    (a) When bills of quantities and the final account are prepared by the quantity surveyor for the heating, ventilating and electrical services, there shall be a fee for these services in addition to the fee calculated in accordance with paras. 2.1 and 2.3 as follows:

Value of work £		Fee £		£
Up to	60 000	4.50%		
60 000	- 120 000	2 700 + 3.85%	on balance over	60 000
120 000	- 240 000	5 010 + 3.25%	on balance over	120 000
240 000	- 375 000	8 910 + 3.00%	on balance over	240 000
375 000	- 500 000	12 960 + 2.50%	on balance over	375 000
Over	500 000	16 085 + 2.15%	on balance over	500 000

    (b) The value of such services, whether the subject of separate tenders or not shall be aggregated and the total value of work so obtained used for the purpose of calculating the additional fee chargeable in accordance with para. (a). (Except that when more than one firm of consulting engineers is engaged on the design of these services, the separate values for which each such firm is responsible shall be aggregated and the additional fees charged shall be calculated independently on each such total value so obtained).

    (c) The scope of the services to be provided by the quantity surveyor under para. (a) above shall be deemed to be equivalent to those outlined in para. 1.2.

    (d) Fee shall be calculated upon the basis of the account for the whole of the heating, ventilating and electrical services for which final accounts have been prepared by the quantity surveyor.

5.0 INSTALMENT PAYMENTS

    5.1 In the absence of agreement to the contrary, fees shall be paid by instalments as follows:

        (a) Upon receipt by the employer of a tender for the works sixty per cent (60%) of the fees calculated in accordance with paras. 2.0 and 4.0 in the amount of the accepted tender plus the appropriate recoverable expenses and the full amount of the fee for cost planning services if such services have been instructed by the employer.

        (b) The balance of fees and expenses by instalments at intervals to be agreed between the date of the first certificate and one month after final certification of the contractor's account.

    5.2 In the event of no tender being accepted, sixty per cent (60%) of the fees, plus the appropriate recoverable expenses, and the full amount of the fee for cost planning services if such services have been instructed by the employer, shall be paid within three months of the completion of the tender documents. The fee shall be calculated on the amount of the lowest original bona fide tender received. In the event of no tender being received, the fee shall be calculated on a reasonable valuation of the work based upon the tender documents.

QUANTITY SURVEYORS' FEES

> NOTE: In the foregoing context "bona fide tender" shall be deemed to mean a tender submitted in good faith without major errors of computation and not subsequently withdrawn by the tenderer.

5.3 In the event of the project being abandoned at any stage other than those covered by the foregoing, the proportion of fee payable shall be by agreement between the employer and the quantity surveyor.

5.4 When the quantity surveyor is appointed to carry out post-contract services only and has not prepared the bills of quantities then the fees shall be agreed between the employer and the quantity surveyor as a proportion of the scale set out in paras. 2.0 and 4.0 with an allowance for the necessary familiarisation and any additional services undertaken by the quantity surveyor. The percentages stated in paras. 5.1 and 5.2 are not intended to be used as a means of calculating the fees payable for post-contract services only.

6.0 ADDITIONAL SERVICES

6.1 For additional services not normally necessary such as those arising as a result of the termination of a contract before completion, liquidation, fire damage to the buildings, services in connection with arbitration, litigation and investigation of the validity of contractors' claims, services in connection with taxation matters, and all similar services where the employer specifically instructs the quantity surveyor, the charge shall be in accordance with para. 7.0.

7.0 TIME CHARGES

7.1 (a) For consultancy and other services performed by a principal, a fee by arrangement according to the circumstances, including the professional status and qualifications of the quantity surveyor.
 (b) When a principal does work which would normally be done by a member of staff, the charge shall be calculated as para. 7.2.

7.2 (a) For services by a member of staff, the charges for which are to be based on the time involved, such hourly charges shall be calculated on the basis of annual salary (including bonus and any other payments or benefits previously agreed with the employer) multiplied by a factor of 2.5, plus reimbursement of payroll costs, all divided by 1600. Payroll costs shall include inter alia employer's contributions payable under any Pension and Life Assurance Schemes, employer's contributions made under the National Insurance Acts, the Redundancy Payments Act and any other payments made in respect of the employee by virtue of any statutory requirements. In this connection it would not be unreasonable in individual cases to take account of the cost of providing a car as part of the "salary" of staff engaged on time charge work when considering whether the salaries paid to staff engaged on such work are reasonable.
 (b) A member of staff shall include a principal doing work normally done by an employee (as para. 7.1 (b) above), technical and supporting staff, but shall exclude secretarial staff or staff engaged upon general administration.
 (c) For the purpose of para. 7.2 (b) above a principal's time shall be taken at the rate applicable to a senior assistant in the firm.
 (d) The supervisory duties of a principal shall be deemed to be included in the multiplication factor as para. 7.2 (a) above and shall not be charged separately.

7.3 The foregoing Time Charges under paras. 7.1 and 7.2 are intended for use where other paras. of the scale (not related to Time Charges) form a significant proportion of the overall fee. In all other cases an increased Time Charge may be agreed.

QUANTITY SURVEYORS' FEES

SCALE 44   PROFESSIONAL CHARGES FOR QUANTITY SURVEYING SERVICES IN CONNECTION WITH IMPROVEMENTS TO EXISTING HOUSING AND ENVIRONMENTAL IMPROVEMENT WORKS
The scale is recommended and not mandatory.

EFFECTIVE FROM FEBRUARY 1973

1. This scale of charges is applicable to all works of improvement to existing housing for local authorities, development corporations, housing associations and the like and to environmental improvement works associated therewith or of a similar nature.

2. The fees set out below cover such quantity surveying services as may be required in connection with an improvement project irrespective of the type of contract or contract documentation from initial appointment to final certification of the contractor's account such as:

    (a) Preliminary cost exercises and advice on tendering procedures and contract arrangements.
    (b) Providing cost advice to assist the design and construction of the project within approved cost limits.
    (c) Preliminary inspection of a typical dwelling of each type.
    (d) Preparation of tender documents; checking tenders received and reporting thereon or negotiating tenders and agreeing prices with a selected contractor.
    (e) Making recommendations for and, where necessary, preparing bills of reductions except in cases where the reductions are necessitated by amended instructions of the employer or by the inclusion in the bills of quantities of items which the quantity surveyor has indicated could not be contained within the approved estimate.
    (f) Analysing tenders and preparing details for submission to a Ministry or Government Department and attending upon the employer in any negotiations with such Ministry or Government Department.
    (g) Recording the extent of work required to every dwelling before work commences.
    (h) Preparing recommendations for interim payments on account to the contractor; preparing periodic assessments of the anticipated final cost of the works and reporting thereon
    (j) Measurement of work and adjustment of variations and fluctuations in the cost of labour and materials in accordance with the terms of the contract and preparing final account, pricing same and agreeing totals with the contractor.

3. The services listed in para. 2 do not include the carrying out of structural surveys.

4. The fees set out below have been calculated on the basis of experience that all of the services described above will not normally be required and in consequence these scales shall not be abated if, by agreement, any of the services are not required to be provided by the quantity surveyor.

QUANTITY SURVEYORS' FEES

IMPROVEMENT WORKS TO HOUSING

5. The fee for quantity surveying services in connection with improvement works to existing housing and external works in connection therewith shall be calculated from a sliding scale based upon the total number of houses or flats in a project divided by the total number of types substantially the same in design and plan as follows:

Total number of houses or flats divided by total number of types substantially the same in design and plan	Fee
not exceeding 1	see note below
exceeding 1 but not exceeding 2	7.0%
exceeding 2 but not exceeding 3	5.0%
exceeding 3 but not exceeding 4	4.5%
exceeding 4 but not exceeding 20	4.0%
exceeding 20 but not exceeding 50	3.6%
exceeding 50 but not exceeding 100	3.2%
exceeding 100	3.0%

and to the result of the computation shall be added 12.5%

NOTE: For schemes of only one house or flat per type an appropriate fee is to be agreed between the employer and the quantity surveyor on a percentage, lump sum or time basis.

ENVIRONMENTAL IMPROVEMENT WORKS

6. The fee for quantity surveying services in connection with environmental improvement works associated with improvements to existing housing or environmental improvement works of a similar nature shall be as follows:

Value of work £		Fee £		£
Up to	50 000	4.5%		
50 000 -	200 000	2 250 + 3.0%	on balance over	50 000
200 000 -	500 000	6 750 + 2.1%	on balance over	200 000
Over	500 000	13 050 + 2.0%	on balance over	500 000

and to the result of that computation shall be added 12.5%

GENERALLY

7. When tender documents prepared by a quantity surveyor for an earlier scheme are re-used without amendment by the quantity surveyor for a subsequent scheme or part thereof for the same employer, the percentage fee in respect of such subsequent scheme or the part covered by such reused documents shall be reduced by 20%.

8. The foregoing fees shall be calculated upon the separate totals of the final account for improvement works to housing and environmental Government works respectively including all nominated sub-contractors' and nominated suppliers' accounts and (subject to para. 5 above) regardless of the amount of repetition within the scheme. When environmental improvement works are the subject of a number of contracts then for the purpose of calculating fees, the values of such contracts shall not be aggregated but each contract shall be taken separately and the scale of charges in para. 6 above applied as appropriate.

9. In cases where any of the materials used in the works are supplied by the employer, the estimated or actual value thereof is to be included in the total on which the fee is calculated.

QUANTITY SURVEYORS' FEES

10. In the absence of agreement to the contrary, fees shall be paid by instalments as follows:

    (a) Upon acceptance by the employer of a tender for the works, one half of the fee calculated on the amount of the accepted tender.

    (b) The balance by instalments at intervals to be agreed between the date of the first certificate and one month after final certification of the contractor's account.

11. (a) In the event of no tender being accepted, one half of the fee shall be paid within three months of completion of the tender documents. The fee shall be calculated on the amount of the lowest original bona fide tender received. If no such tender has been received, the fee shall be calculated upon a reasonable valuation of the work based upon the tender documents.

    (b) In the event of the project being abandoned at any stage other than those covered by the foregoing, the proportion of fee payable shall be by agreement between the employer and the quantity surveyor.

12. If the works are substantially varied at any stage or if the quantity surveyor is involved in an excessive amount of abortive work, then the fee shall be adjusted by agreement between the employer and the quantity surveyor.

13. When the quantity surveyor is required to perform additional services in connection with the allocation of the costs of the works for purposes of calculating value added tax there shall be an additional fee based on the time involved.

14. For additional services not normally necessary such as those arising as a result of the termination of the contract before completion, liquidation, fire damage to the buildings, services in connection with arbitration, litigation and claims on which the employer specifically instructs the surveyor to investigate and report, there shall be an additional fee to be agreed between the employer and the quantity surveyor.

15. Copyright in the bills of quantities and other documents prepared by the quantity surveyor is reserved to the quantity surveyor.

16. The foregoing fees are in all cases exclusive of travelling expenses and lithography or other charges for copies of documents, the net amount of such expenses and charges to be paid for in addition. Subsistence expenses, if any, to be charged by arrangement with the employer.

17. The foregoing fees and charges are in all cases exclusive of value added tax which shall be applied in accordance with legislation current at the time the account is rendered.

SCALE 45 PROFESSIONAL CHARGES FOR QUANTITY SURVEYING SERVICES IN CONNECTION WITH HOUSING SCHEMES FINANCED BY THE HOUSING CORPORATION

EFFECTIVE FROM JANUARY 1982 - reprinted 1989

1. (a) This scale of charges has been agreed between The Royal Institution of Chartered Surveyors and the Housing Corporation and shall apply to housing schemes of self-contained dwellings financed by the Housing Corporation regardless of type (e.g. houses, maisonettes, bungalows or flats) and irrespective of the amount of repetition of identical types or blocks within an individual housing scheme.

    (b) This scale does not apply to services in connection with improvements to existing dwellings.

2. The fees set out below cover the following quantity surveying services as may be required in connection with the particular project:

    (a) Preparing such estimates of cost as are required by the employer to establish and substantiate the economic viability of the scheme and to obtain the necessary approvals and consents from the Housing Corporation but excluding cost planning services (see para. 10)

    (b) Providing pre-contract cost advice (e.g. approximate estimates on a floor area or similar basis) to enable the scheme to be designed and constructed within the approved cost criteria but excluding cost planning services (see para. 10).

QUANTITY SURVEYORS' FEES

- (c) Preparing bills of quantities or other tender documents; checking tenders received or negotiating tenders and pricing with a selected contractor; reporting thereon.
- (d) Preparing an elemental analysis of the accepted tender (RICS/BCIS Detailed Form of Cost Analysis excluding the specification notes or equivalent).
- (e) Preparing recommendations for interim payments on account to the contractor; measuring the work and adjusting variations in accordance with the terms of the contract and preparing the final account, pricing same and agreeing totals with the contractor; adjusting fluctuations in the cost of labour and materials if required by the contract.
- (f) Preparing periodic post-contract assessments of the anticipated final cost by means of a simple routine of periodically estimating final costs and reporting thereon, but excluding a cost monitoring service specifically required by the employer.

3. The fees set out below are exclusive of travelling and of other expenses (for which the actual disbursement is recoverable unless there is some special prior arrangement for such charges) and the cost of reproduction of bills of quantities and other documents, which are chargeable in addition at net cost.

4. Copyright in the bills of quantities and other documents prepared by the quantity surveyor is reserved to the quantity surveyor.

5. (a) The basic fee for the services outlined in para. 2 (regardless of the extent of services described in para. 2) shall be as follows:

Value of work	Fee	
	£	£
Up to 75 000	210 + 3.8%	
75 000 - 150 000	3 060 + 3.0% on balance over	75 000
150 000 - 750 000	5 310 + 2.0% on balance over	150 000
750 000 - 1 500 000	17 310 + 1.5% on balance over	750 000
Over 1 500 000	28 560 + 1.3% on balance over	1 500 000

(b) (i) For services in connection with Categories 1 and 2 Accommodation designed for Old People in accordance with the standards described in Ministry of Housing and Local Government Circulars 82/69 and 27/70 (Welsh Office Circulars 84/69 & 30/70), there shall be a fee in addition to that in accordance with para. 5 (a), calculated as follows:

Category 1 An addition of five per cent (5%) to the basic fee calculated in accordance with para. 5 (a)
Category 2 An addition of twelve and a half per cent (12.5%) to the basic fee calculated in accordance with para. 5 (a).

(ii) For services in connection with Accommodation designed for the Elderly in Scotland in accordance with the standards described in Scottish Housing Handbook Part 5, Housing for the Elderly, the fee shall be calculated as follows:

Mainstream and Amenity Housing	Basic fee in accordance with para. 5 (a)
Basic Sheltered Housing (i.e. Amenity Housing plus Warden's accommodation and alarm system)	An addition of five per cent (5%) to the basic fee calculated in accordance with para. 5 (a)
Sheltered Housing, including optional facilities	An addition of twelve and a half per cent (12.5%) of the basic fee calculated in accordance with para. 5 (a)

QUANTITY SURVEYORS' FEES

- (c) (i) For services in connection with Accommodation designed for Disabled People in accordance with the standards described in Department of Environment Circular 92/75 (Welsh Office Circular 163/75), there shall be an addition of fifteen per cent (15%) to the fee calculated in accordance with paragraph 5 (a).
  - (ii) For services in connection with Accommodation designed for the Disabled in Scotland in accordance with the standards described in Scottish Housing Handbook Part 6, Housing for the Disabled, there shall be an addition of fifteen per cent (15%) to the fee calculated in accordance with para. 5 (a).
- (d) For services in connection with Accommodation designed for Disabled Old People, the fee shall be calculated in accordance with para. 5 (c).
- (e) For services in connection with Subsidised Fair Rent New Build Housing, there shall be a fee, in addition to that in accordance with paras. 5 (a) to (d), calculated as follows:

Value of work £	Category A fee £	£
Up to 75 000	20 + 0.40%	
75 000 - 150 000	320 + 0.20% on balance over	75 000
150 000 - 500 000	470 + 0.07% on balance over	150 000
Over 500 000	715	

6. (a) Where additional fees under paras. 5 (b) to (d) are chargeable on a part or parts of a scheme, the value of basic fee to which the additional percentages shall be applied shall be determined by the proportion that the values of the various types of accommodation bear to the total of those values.
   (b) Fees shall be calculated upon the total of the final account for the whole of the work including all nominated sub-contractors' and nominated suppliers' accounts.
   (c) If any of the materials used in the works are supplied free of charge to the contractor, the estimated or actual value thereof shall be included in the amount upon which fees are to be calculated.
   (d) When a project is the subject of a number of contracts then, for the purpose of calculating fees, the values of such contracts shall not be aggregated but each contract shall be taken separately and the scale of charges applied as appropriate.

7. If bills of quantities and final accounts are prepared by the quantity surveyor for the heating, ventilating or electrical services, there shall be an additional fee by agreement between the employer and the quantity surveyor subject to the approval of the Housing Corporation.

8. In the absence of agreement to the contrary, fees shall be paid by instalments as follows:
   (a) Upon receipt by the employer of a tender for the works, or when the employer certifies to the Housing Corporation that the tender documents have been completed, a sum on account representing ninety per cent (90%) of the anticipated sum under para. 8 (b) below.
   (b) Upon acceptance by the employer of a tender for the works, sixty per cent (60%) of the fee calculated on the amount of the accepted tender, plus the appropriate recoverable expense
   (c) The balance of fees and expenses by instalments at intervals to be agreed between the date of the first certificate and one month after final certification of the contractor's account.

9. (a) In the event of no tender being accepted, sixty per cent (60%) of the fee and the appropriate recoverable expenses shall be paid within six months of completion of the tender documents. The fee shall be calculated on the amount of the lowest original bona fide tender received. In the event of no tender being received, the fee shall be calculated upon a reasonable valuation of the work based upon the tender documents.

   NOTE: In the foregoing context "bona fide tender" shall be deemed to mean a tender submitted in good faith without major errors of computation and not subsequently withdrawn by the tenderer.

QUANTITY SURVEYORS' FEES

- (b) In the event of part of the project being postponed or abandoned after the preparation of the bills of quantities or other tender documents, sixty per cent (60%) of the fee on this part shall be paid within three months of the date of postponement or abandonment.
- (c) In the event of the project being postponed or abandoned at any stage other than those covered by the foregoing, the proportion of fee payable shall be by agreement between the employer and the quantity surveyor.

10. (a) Where with the approval of the Housing Corporation the employer instructs the quantity surveyor to carry out cost planning services there shall be a fee additional to that charged under para. 5 as follows:

Value of work £	Category A fee £	£
Up to 150 000	0.45%	
150 000 - 750 000	675 + 0.35% on balance over	150 000
Over 750 000	2 775 + 0.25% on balance over	750 000

- (b) Cost planning is defined as the process of ascertaining a cost limit where necessary, within guidelines set by any appropriate Authority, and thereafter checking the cost of the project within that limit throughout the design process. It includes the preparation of a cost plan (based upon elemental analysis or other suitable criterion) checking and revising it where required and effecting the necessary liaison with the other consultants employed.

11. If the quantity surveyor incurs additional costs due to exceptional delays in building operations or any other cause beyond the control of the quantity surveyor, then the fees shall be adjusted by agreement between the employer and the quantity surveyor to cover the reimbursement of these additional costs.

12. When the quantity surveyor is required to prepare an interim valuation of materials or goods off site, an additional fee shall be charged based on the time involved (see paras. 15 and 16) in respect of each such valuation.

13. If the Works are materially varied to the extent that substantial remeasurement is necessary, then the fee may be adjusted by agreement between the employer and the quantity surveyor.

14. For additional services not normally necessary, such as those arising as a result of the termination of a contract before completion, fire damage to the buildings, cost monitoring (see para. 2 (f)), services in connection with arbitration, litigation and investigation of the validity of contractors' claims, services in connection with taxation matters and similar all services where the employer specifically instructs the quantity surveyor, the charges shall be in accordance with paras. 15 & 16.

15. (a) For consultancy and other services performed by a principal, a fee by arrangement according to the circumstances, including the professional status and qualifications of the quantity surveyor.
    (b) When a principal does work which would normally be done by a member of staff, the charge shall be calculated as para. 16.

16. (a) For services by a member of staff, the charges for which are to be based on the time involved, such hourly charges shall be calculated on the basis of annual salary (including bonus and any other payments or benefits previously agreed with the employer) multiplied by a factor of 2.5, plus reimbursement of payroll costs, all divided by 1600. Payroll costs shall include inter alia employer's contributions payable under any Pension and Life Assurance Schemes, employer's contributions made under the National Insurance Acts, the Redundancy Payments Act and any other payments made in respect of the employee by virtue of any statutory requirements in this connection it would not be unreasonable in individual cases to take account of the cost of providing a car as part of the "salary" of staff engaged on time charge work when considering whether the salaries paid to staff engaged on such work are reasonable.

QUANTITY SURVEYORS' FEES

- (b) A member of staff shall include a principal doing work normally done by an employee (as para. 15 (b) above), technical and supporting staff, but shall exclude secretarial staff or staff engaged upon general administration.
- (c) For the purpose of para. 16 (b) above a principal's time shall be taken at the rate applicable to a senior assistant in the firm.
- (d) The supervisory duties of a principal shall be deemed to be included in the multiplication factor as para. 16 (a) above and shall not be charged separately.

17. The foregoing Time Charges under paras. 15 and 16 are intended for use where other paras. of the scale (not related to Time Charges) form a significant proportion of the overall fee. In all other cases an increased time charge may be agreed.

18. (a) In the event of the employment of the contractor being determined due to bankruptcy or liquidation, the fee for the services outlined in para. 2, and for the additional services required, shall be recalculated to the aggregate of the following:

- (i) Fifty per cent (50%) of the fee in accordance with paragraphs 5 and 6 calculated upon the total of the Notional Final Account in accordance with the terms of the original contracts.
- (ii) Fifty per cent (50%) of the fee in accordance with paragraphs 5 and 6 calculated upon the aggregate of the total value (which may differ from the total of interim valuations) of work up to the date of determination in accordance with the terms of the original contract plus the total of the final account for the completion contract;
- (iii) A charge based upon time involved (in accordance with paragraphs 15 and 16) in respect of dealing with those matters specifically generated by the liquidation (other than normal post-contract services related to the completion contract), which may include (inter alia):

Site inspection and (where required) security (initial and until the replacement contractor takes possession);
Taking instructions from and/or advising the employer;
Representing the employer at meeting(s) of creditors;
Making arrangements for the continued employment of sub-contractors and similar related matters;
Preparing bills of quantities or other appropriate documents for the completion contract, obtaining tenders, checking and reporting thereon;
The additional cost (over and above the preparation of the final account for the completion contract) of pre-paring the Notional Final Account; pricing the same
Negotiations with the liquidator (trustee or receiver).

- (b) In calculating fees under para. 18 (a) (iii) above, regard shall be taken of any services carried out by the quantity surveyor for which a fee will ultimately be chargeable under para. 18 (a) (i) and (ii) above in respect of which a suitable abatement shall be made from the fee charged (e.g. measurement of variations for purposes of the completion contract where such would contribute towards the preparation of the contract final account).
- (c) Any interim instalments of fees paid under para. 8 in respect of services outlined in para. 2 shall be deducted from the overall fee computed as outlined herein.
- (d) In the absence of agreement to the contrary fees and expenses in respect of those services outlined in para. 18 (a) (iii) above up to acceptance of a completion tender shall be paid upon such acceptance; the balance of fees and expenses shall be paid in accordance with para. 8 (c).
- (e) For the purpose of this Scale the term "Notional Final Account" shall be deemed to mean an account indicating that which would have been payable to the original contractor had he completed the whole of the works and before deduction of interim payments to him.

19. The fees and charges are in all cases exclusive of Value Added Tax which will be applied in accordance with legislation.

QUANTITY SURVEYORS' FEES

EXPLANATORY NOTE:

(Source: Chartered Quantity Surveyor, August 1986)

For rehabilitation projects the basic fee set out in paragraph 5 (a) of the scale will apply with the addition of a further 1% fee calculated upon the total of the final account for rehabilitation works including all nominated sub-contractors' and nominated suppliers' accounts.
In the case of special housing categories (e.g., elderly people) the additional percentage should be applied before the application of the additional percentage set out in paragraph 5 (b). The provisions of paragraph 6 (a) of the scale will also apply.
There is no longer any distinction between "hostel" and "cluster dwellings" which now have a single category of shared housing.
For shared housing new build projects other than those specified below the fee should be calculated in accordance with paragraph 5 (a) plus an enhancement of 10%.
For shared housing rehabilitation projects other than those specified below the fee should be calculated in accordance with paragraph 5 (a) of the scale plus 1% plus an enhancement of 10%.
For shared housing projects comprising wheelchair accommodation (as described in the Housing Corporation's Design and Contract Criteria) or frail elderly accommodation (as described in Housing Corporation circular HCO1/85) the fee should be calculated in accordance with paragraph 5 (a), (plus 1% for rehabilitation schemes where applicable) plus an enhancement of 15%.
The additional percentage set out in paragraph 5 (b) does not apply to shared housing projects, but the provisions of paragraph 6 (a) are applicable.

SCALE 46   PROFESSIONAL CHARGES FOR QUANTITY SURVEYING SERVICES IN CONNECTION WITH LOSS ASSESSMENT OF DAMAGE TO BUILDINGS FROM FIRE, ETC ISSUED BY THE ROYAL INSTITUTION OF CHARTERED SURVEYORS.
The scale is recommended and not mandatory.

EFFECTIVE FROM JULY 1988

1. This scale of professional charges is for use in assessing loss resulting from damage to buildings by fire etc., under the "building" section of an insurance policy and is applicable to all categories of buildings.

2. The fees set out below cover the following quantity surveying services as may be required in connection with the particular loss assessment:

    (a) Examining the insurance policy.
    (b) Visiting the building and taking all necessary site notes.
    (c) Measuring at site and/or from drawings and preparing itemised statement of claim and pricing same.
    (d) Negotiating and agreeing claim with the loss adjuster.

3. The fees set out below are exclusive of the following:

    (a) Travelling and other expenses (for which the actual disbursement is recoverable unless there is some special prior arrangement for such charge.)
    (b) Cost of reproduction of all documents, which are chargeable in addition at net cost.

4. Copyright in all documents prepared by the quantity surveyor is reserved.

QUANTITY SURVEYORS' FEES

5. (a) The fees for the services outlined in paragraph 2 shall be as follows:

Agreed Amount of Damage £	Fee £	£
Up to 60 000	see note 5(c) below	
60 000 - 180 000	2.5%	
180 000 - 360 000	4 500 + 2.3% on balance over	180 000
360 000 - 720 000	8 640 + 2.0% on balance over	360 000
Over 720 000	15 840 + 1.5% on balance over	720 000

and to the result of that computation shall be added 12.5%

(b) The sum on which the fees above shall be calculated shall be arrived at after having given effect to the following:

   (i) The sum shall be based on the amount of damage, including such amounts in respect of architects', surveyors and other consultants' fees for reinstatement, as admitted by the loss adjuster.

   (ii) When a policy is subject to an average clause, the sum shall be the agreed amount before the adjustment for "average".

   (iii) When, in order to apply the average clause, the reinstatement value of the whole subject is calculated and negotiated an additional fee shall be charged commensurate with the work involved.

(c) Subject to 5 (b) above, when the amount of the sum on which fees shall be calculated is under £60,000 the fee shall be based on time involved as defined in Scale 37 (July 1988) para. 19 or on a lump sum or percentage basis agreed between the building owner and the quantity surveyor.

6. The foregoing scale of charges is exclusive of any services in connection with litigation and arbitration.

7. The fees and charges are in all cases exclusive of value added tax which shall be applied in accordance with legislation.

QUANTITY SURVEYORS' FEES

SCALE 47 PROFESSIONAL CHARGES FOR THE ASSESSMENT OF REPLACEMENT COSTS BUILDINGS FOR INSURANCE, CURRENT COST ACCOUNTING AND OTHER PURPOSES ISSUED BY THE ROYAL INSTITUTION OF CHARTERED SURVEYORS
The scale is recommended and not mandatory.

EFFECTIVE FROM JULY 1988

1.0 GENERALLY

1.1. The fees are in all cases exclusive of travelling and other expenses (for which the actual disbursement is recoverable unless there is some prior arrangement for such charges).
1.2. The fees and charges are in all cases exclusive of value added tax which will be applied in accordance with legislation.

2.0 ASSESSMENT OF REPLACEMENT COSTS OF BUILDINGS FOR INSURANCE PURPOSES

2.1. Assessing the current replacement cost of buildings where adequate drawings for the purpose are available.

Assessed current costs £	Fee £	£
Up to 140 000	0.2%	
140 000 - 700 000	280 + 0.075% on balance over	140 000
700 000 - 4 200 000	700 + 0.025% on balance over	700 000
Over 4 200 000	1 575 + 0.01% on balance over	4 200 000

2.2. Fees to be calculated on the assessed cost, i.e. base value, for replacement purposes including allowances for demolition and the clearance but excluding inflation allowances and professional fees.

2.3. Where drawings adequate for the assessment of costs are not available or where other circumstances require that measurements of the whole or part of the buildings are taken, an additional fee shall be charged based on the time involved or alternatively on a lump sum basis agreed between the employer and the surveyor.

2.4 when the assessment is for buildings of different character or on more than one site, the costs shall not be aggregated for the purpose of calculating fees.

2.5 For current cost accounting purposes this scale refers only to the assessment of replacement cost of buildings.

2.6 The scale is appropriate for initial assessments but for annual review or a regular reassessment the fee should be by arrangement having regard to the scale and to the amount of work involved and the time taken.

2.7. The fees are exclusive of services in connection with negotiations with brokers, accountants or insurance companies for which there shall be an additional fee based upon the time involve

CONSULTING ENGINEERS' FEES

INTRODUCTION

A scale of professional charges for consulting engineering services is published by the Association of Consulting Engineers (ACE)

Copies of the document can be obtained direct from:

> The Association of Consulting Engineers
> 12 Caxton Street
> Westminster
> London
> SW1H 0QL
>
> Tel: 0207 222 6557
> Fax: 0207 222 0750
> E-mail: consult@acenet.co.uk

GUIDANCE ON FEES

The 1995 ACE Conditions of Engagement (2nd Edition 1998) and their Guidance on Completion provide for payment of fees to be calculated by:

- time charges;
- lump sums; or
- ad valorem percentages on the cost of the Project or of the Works.

The Association considers that, in normal circumstances, the level of renumeration represented by the scales of percentage fees and hourly charging rates set out in this guidance is such as to enable the provision by a Consulting Engineer to his client of a full, competent and reliable standard of service.

The levels of fees recommended are based closely on those published in the ACE Conditions of Engagement 1981, modified only as necessary to be compatible with the 1995 Agreements. No account has been taken of factors which have, since 1981, increased the demands on consulting engineers. These include requirements of new and more complex design codes and legislation such as the CDM Regulations. Due regard should be paid to these factors when arriving at the appropriate fee for a specific commission.

Variation of Fees

These scales and rates are not mandatory but are presented as guidelines for work of average complexity which may, by negotiation, be adjusted upwards or downwards to take account of the abnormal complexity or simplicity of design, increase or diminution of the extent of services to be provided, long-standing client relationships or other circumstances. As there is a wide range of construction and of degrees of complexity, the Association does not make specific recommendations but suggests that the range of adjustment may be represented by such factors as 0.75 for highly repetitive new works, 1.00 for normal new works, 1.50 for non-competitive or complex new works and from 1.25 to 1.75 for alterations/additions to existing works. Alternatively, the fee for the design of alterations/additions can be negotiated as a lump sum or time charges.

An approach to the complexity of work which may be helpful is that used by the RIBA, whose Guidance for Clients on Fees classifies building types in various sectors by the amount of work their design requires. For example, industrial buildings may need least work for storage sheds, more for factors and garages and about average for purpose built factories and warehouses. In the commercial sector, surface car parks would be at the low end, multi-storey and underground car parks higher, supermarkets, banks and offices about average, department stores and restaurants higher and research laboratories and radio and television studios highest. For mechanical and electrical services, domestic premises would be at the low end, offices about average and hospitals at the high end. It is difficult to quantify these factors precisely but they provide some guidelines.

CONSULTING ENGINEERS' FEES

Quality

The Association strongly advises clients to satisfy themselves that the level and quality of services they want will be covered if they make appointments based on charges which are appreciably lower than those set out in this guidance note.

Extent of Services

The figures given for percentage (ad valorem) fees are based upon the Normal Services defined in the Conditions of Engagement. They do not include allowance for any Additional Services, such as those so listed in the Agreements or material changes to the brief leading to extra design work, for which a further fee is normally chargeable. Nor do they allow for acting as Planning Supervisor in accordance with the requirements of the Construction (Design and Management) Regulations, for which a separate appointment and fee is applicable. The normal fees take no account of actual expenses, such as printing, reproduction and purchase of maps, records and photographs, courier charges, travelling, hotel and subsistence payments, charges for use of special equipment and any other expenses for which repayment is specifically authorised, which are recoverable in addition to the fees.

Brief

The client's brief for the project is of the utmost importance in the preparation of a fee bid. The more comprehensive and accurate the brief the more reliable will be the fee bids. The ACE issues guidance to clients on the preparation of briefs.

Partial Services

When the client wishes to appoint the Consulting Engineer for partial services only, it is important that both parties recognise the limitation which such an appointment places upon the responsibility of the Consulting Engineer who cannot be held liable for matters that are outside his control.

The terms of reference for the appointment should be carefully drawn up and the relevant ACE Conditions of Engagement should be adapted to suit the scope of the services required.

Professional charges for partial services are usually best calculated on a time basis, but may, in suitable cases, be a commensurate part of the percentage fee for normal services shown in this guidance.

Instalments

Provision is made in the ACE Agreements for payment of fees in instalments. This is a statutory requirement for construction contracts under the Housing Grants, Construction and Regeneration Act if the commission lasts longer than 45 days.

BASIS OF FEE CALCULATIONS

Time Charges

When it is not possible to estimate in advance the construction cost, a percentage of the estimated construction cost cannot be used as a basis of renumeration. The normal method of payment in these cases is a time charge for staff actually employed on the project. This method is also the most usual for feasibility studies, advisory work and small projects.

It may be appropriate to agree a budget fee, to exceed which the Consulting Engineer must seek the client's authorisation. This will introduce a degree of cost control.

Hourly rates are most conveniently calculated by applying a multiplier, which covers overheads and profits, to the staff renumeration cost and then adding the net amount of other payroll costs. The recommended level of multiplier is 2.60 for office based and 1.30 for site staff.

CONSULTING ENGINEERS' FEES

Time Charges - cont'd

The major part of the multiplier is attributable to the Consulting Engineer's overheads which may include, inter alia, the following costs and expenses:

a. rent, rates and other expenses of up-keep of his office, its furnishings, equipment and supplies;

b. insurance premiums other than those recovered in the payroll costs;

c. administrative, accounting, secretarial and financing costs;

d. the cost of ensuring that staff keep abreast of advances in engineering and undertake continuing professional development;

e. the expense of preliminary arrangements for new or prospective projects;

f. loss of productive time of technical staff between assignments.

In this context the following definitions apply:

a. Renumeration Cost. The annualised cost to the Consulting Engineer of the gross renumeration paid to a person employed by him including the cost of all benefits in kind, divided by 1600 (deemed to be the average annual total of effective working hours of an employee).

b. Other Payroll Cost. The annualised cost to the Consulting Engineer of all contributions and payments made directly or indirectly in respect of a person employed by him for pension, life assurance, prolonged disability and other like schemes and also the annual amount for National Insurance contributions and any other tax, charge, levy, impost or payment of any kind whatsoever which the Consulting Engineer is obliged at any time during the performance of this Agreement by law to make in respect of such person, divided by 1600.

When calculating amounts chargeable on a time basis, a Consulting Engineer is entitled to include time spent by staff in travelling in connection with the performance of the services. The time spent by secretarial staff engaged on general accountancy or administration duties in the Consulting Engineers' office is not chargeable unless otherwise agreed.

If time charges are agreed as a stated amount per hour for specified staff, consideration should be given to their periodic review and indexation.

Lump Sum Fees

Lump sums, which may be broken down into components applicable to particular duties or stages of work, have the advantage by negotiation or tender. It is not possible to provide to provide guidance for clients on likely lump sums but comparison with fees calculated by the percentage (ad valorem) method or by estimating time charges will give an indication of appropriate levels. Lump sums will inevitably incorporate an allowance for the additional risk involved in making such arrangements. Great care should be exercised to ensure that the work covered by the lump sum is specified detail to avoid subsequent disputes as to what was or was not included.

If the commission is a lengthy one allowance should be made for cost increases during its term and consideration should be given to provision for adjustment of lump sums to account for significant changes in the value of the work.

CONSULTING ENGINEERS' FEES

Percentage (Ad Valorem) Fees

The recommendations in the graphs (pages 54 to 57) are expressed in terms of 1999 prices for construction work. If they are used at a time when construction prices are significantly different an allowance for this should be made. The equivalent 1999 price with which to enter the graph can be calculated by multiplying the current price or estimate by the 1999 Output Price Index (OPI), which was 371, and dividing by the current OPI.

Current Output Price Index

The Output Price Index at the time of going to press is **394** at first quarter 2001 (1975 = 100).

Keep your figures up to date, free of charge
Download updates from the web: www.pricebooks.co.uk

This section, and most of the other information in this Price Book, is brought up to date every three months with the Price Book Updates, until the next annual edition. The updates are available free to all Price Book purchasers.

To ensure you receive your free copies, either complete the reply card from the centre of the book and return it to us or register via the website www.pricebooks.co.uk

Agreement A(1) - Lead Consultant, Civil/Structural

For works where the Consulting Engineer is lead consultant it is normal for the fees to be based upon a percentage of the Project Cost. If additional duties, such as detailed drawings and bar bending schedules, are undertaken in connection with structures involving reinforced or prestressed concrete, masonry, timber, plastics, steel and other metals then an additional fee of between 1.50% and 6.00% of the cost of such work is usually charged.

For works of average complexity the percentages of the Project Costs for fee calculation may be determined from the graph on page 54 using the estimated Project Cost which should be agreed with the Client. The figure so obtained can then be adjusted to allow for the degrees of complexity, of repetition, and other factors particular to the commission (see Variation on Fees on page 45). Further adjustments may be made if any of the Normal Services are not required or if any of those shown as Additional Services are required to be considered as Normal Services for a commission and paid for accordingly.

Example 1 - Highway Project.

Total Project Cost £5m; cost of structural work in bridges etc. £1.50 m. Allow 10% reduction for overall for repetition.

Initial project cost percentage for £5 m is 5.50%.

Structural addition percentage for £1.50 m is, say, 2.60%.

Overall percentage = $0.90 \left(5.50 + \dfrac{2.60 \times 1.50}{5}\right) = 5.65\%$

Example 2 - Building Project.

Total Project Cost £1.90 m; cost of structural work at say 25%. Allow 25% addition for complex and non-repetitive design.

Initial project cost percentage for £1.90 m is 6.50%.

Structural addition percentage for £1.90 m x 0.25 (£475k) is, say, 3.10%.

Overall percentage = 1.25 (6.50 + 3.10 x 0.25) = 9.10%.

The proportionate amount of the fee for Normal Services to be paid for each work stage is:

CIVIL AND STRUCTURAL WORK STAGE	
Outline Proposals	15%
Detailed Proposals	35%
Final Proposals	60%
Tender Action	85%
Construction and Completion	100%

CONSULTING ENGINEERS' FEES

Agreement A(2) - Lead Consultant, Building Services

For works where the Consulting Engineer is lead consultant it is normal for the fees to be based upon a percentage of the Project Cost.

For works of average complexity the percentages of the Project Cost for fee calculation may be determined from the graph on page 56 using the estimated Project Cost which should be agreed with the Client. The figure so obtained can then be adjusted to allow for the degrees of complexity, of repetition, and other factors particular to the commission (see Variation of Fees on page 45). Further adjustments may be made of any of the Normal Services are not required or if any of those shown as Additional Services are required to be considered as Normal Services for a commission and paid for accordingly.

The Normal Services in the 1995 Agreements for Building Services work approximate to the Abridged Duties in the 1981 Agreements (i.e. partial design by the Consulting Engineer, detailed design by the Contractor) or Performance Duties (i.e. performance specified by the Consulting Engineer, design by the Contractor) is agreed by importing some of the Additional Services or deleting some of the Normal Services the appropriate curve in the graph on page 56 should be used.

Example - Building Services Refurbishment Project.

Total Project Cost £2.00 m; almost all building services work. Allow 25% for complex non-repetitive design.

Percentage Fee for £2.00 m is 8.00% from graph on page 56.

With 25% addition, Fee is 1.25 x 8.00% = 10.00% of Project Cost.

The proportionate amount of the fee for Normal Services to be paid for each work stage is:

CIVIL AND STRUCTURAL WORK	
STAGE	
Outline Proposals	7%
Detailed Proposals	17%
Final Proposals	40%
Tender Action	80%
Construction and Completion	100%

CONSULTING ENGINEERS' FEES

Agreement B(1) - Non-Lead Consultant, Civil/Structural

The fee to be charged can be calculated as a percentage of the Works Cost, i.e. that part of the Project for which the structural engineer takes responsibility, using the upper (B1) graph on page 55. If a fee expressed as a percentage of the Project Cost is preferred, this can be found by using the Project Cost and the lower (B2) curve which corresponds to the ratio of Works Cost to Project Cost (W/P). The figure so obtained can then be adjusted to allow for the degrees of complexity, of repetition, and other factors particular to the commission (see Variation of Fees on page 45). If additional duties, such as detailed drawings and bar bending schedules, are undertaken in connection with structures involving reinforced or pre-stressed concrete, masonry, timber, plastics, steel and other metals then an additional fee (usually 3% of the cost of such work) is charged. Further adjustments may be made if any of the Normal Services are not required or if any of those shown as Additional Services are required to be considered as Normal Services for a commission and paid for accordingly.

The proportionate amount of the fee for Normal Services to be paid for each work stage is:

CIVIL AND STRUCTURAL WORK	
STAGE	
Outline Proposals	15%
Detailed Proposals	35%
Final Proposals	60%
Tender Action	85%
Construction and Completion	100%

CONSULTING ENGINEERS' FEES

Agreement B(2) - Non-Lead Consultant, Building Services

The fee to be charged should be calculated as a percentage of the Works Costs, i.e. that part of the Project for which the Consulting Engineer takes responsibility. This may be expressed as a percentage of the Project Cost by multiplying the percentage of the Works Cost by the ratio of Works Cost to Project Cost (W/P).

For works of average complexity the percentages of the Works Cost for fee calculation may be determined from the graph on page 57 using the estimated Works Cost which should be agreed with the Client. The figure so obtained can then be adjusted to allow for the degrees of complexity, of repetition, and other factors particular to the commission (see Variation of Fees on page 45). Further adjustments may be made if any of the Normal Services are not required or if any of those shown as Additional Services are required to be considered as Normal Services for a commission and paid for accordingly.

The Normal Services in the 1995 Agreements for Building Services work approximate to the Abridged Duties in the 1981 Agreements (i.e. partial design by the Consulting Engineer, detailed design by the Sub-Contractor). If the equivalent of Full Duties (i.e. complete design by the Consulting Engineer) or Performance Duties (i.e. performance specified by the Consulting Engineer, design by the Sub-Contractor) is agreed by importing some of the Additional Services or deleting some of the Normal Services the appropriate curve in the graph in page 57 should be used.

The proportionate amount of the fee for Normal Services to be paid for each work stage is:

CIVIL AND STRUCTURAL WORK STAGE	
Outline Proposals	7%
Detailed Proposals	17%
Final Proposals	40%
Tender Action	80%
Construction and Completion	100%

CONSULTING ENGINEERS' FEES

Agreement C - Design and Construct Project Designer

The fees for design services in a Design and Construct project can be related to the fees for the non-lead consultant carrying out the appropriate type of work, making due allowance for the services actually provided.

Fees for Professional Services

GRAPH SHOWING RELATIONSHIP OF PERCENTAGE FEE WITH PROJECT COST
AGREEMENT A(1), LEAD CONSULTANT, CIVIL/STRUCTURAL

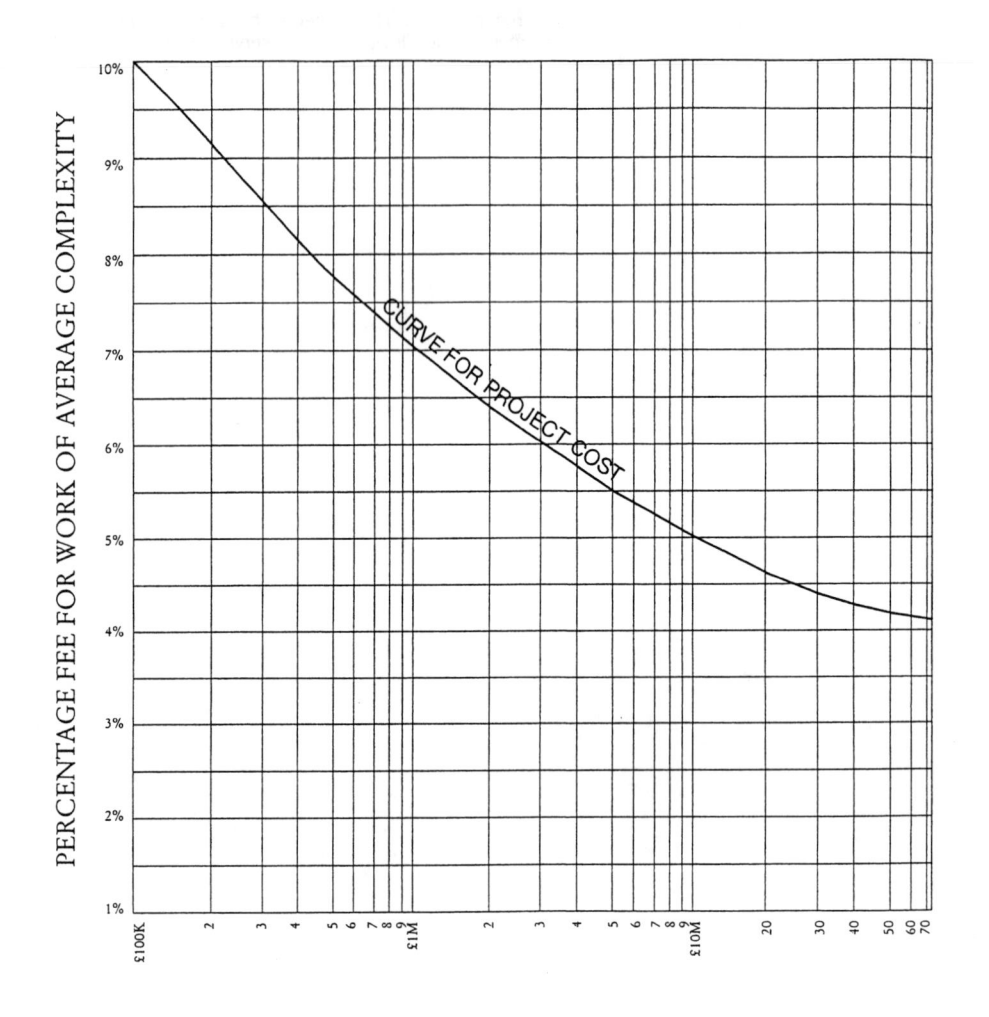

PROJECT COST IN £
(1999 PRICES, OPI 371, 1975 = 100)

Note: When Project Cost is below £100K time charges should be used.

Fees for Professional Services 55

GRAPH SHOWING RELATIONSHIP OF PERCENTAGE FEE WITH WORKS COST/PROJECT COST
AGREEMENT B(1), DIRECTLY ENGAGED CONSULTANT, NOT IN THE LEAD, CIVIL/STRUCTURAL

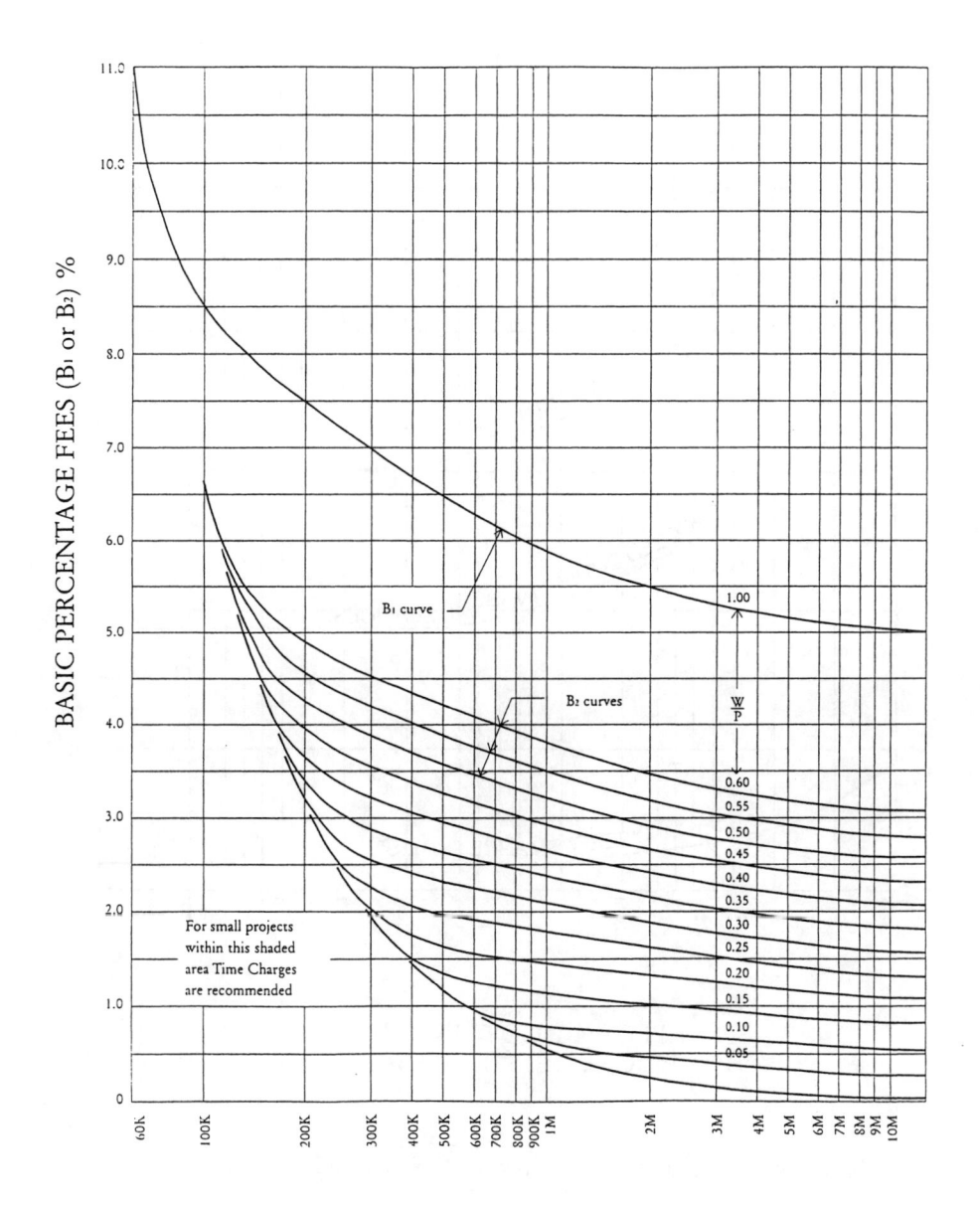

BASIC PERCENTAGE FEES (B_1 or B_2) %

B_1 curve

B_2 curves

$\frac{W}{P}$

1.00
0.60
0.55
0.50
0.45
0.40
0.35
0.30
0.25
0.20
0.15
0.10
0.05

For small projects within this shaded area Time Charges are recommended

PROJECT COST OR WORKS COST IN £
(1999 PRICES, OPI 371, 1975 = 100)

GRAPH SHOWING RELATIONSHIP OF PERCENTAGE FEE WITH PROJECT COST

AGREEMENT A(2), LEAD CONSULTANT, ELECTRICAL AND MECHANICAL ENGINEERING SERVICES

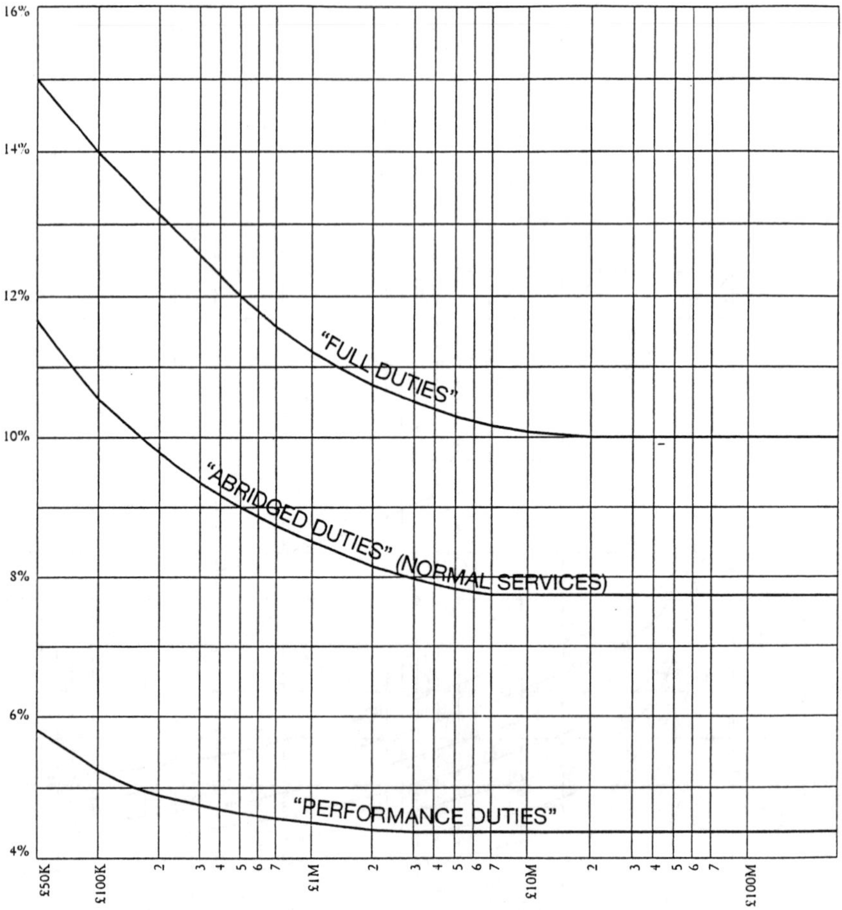
16%
14%
12%
10%
8%
6%
4%
"FULL DUTIES"
"ABRIDGED DUTIES" (NORMAL SERVICES)
"PERFORMANCE DUTIES"
£50K
£100K
2
3
4
5
6
7
£1M
2
3
4
5
6
7
£10M
2
3
4
5
6
7
£100M

PROJECT COST IN £
(1999 PRICES, OPI 371, 1975 = 100)

Note: When Project Cost is below £50K time charges should be used.

GRAPH SHOWING RELATIONSHIP OF PERCENTAGE FEE WITH WORKS COST
AGREEMENT B(2), DIRECTLY ENGAGED CONSULTANT, NOT IN THE LEAD, ELECTRICAL AND MECHANICAL ENGINEERING SERVICES

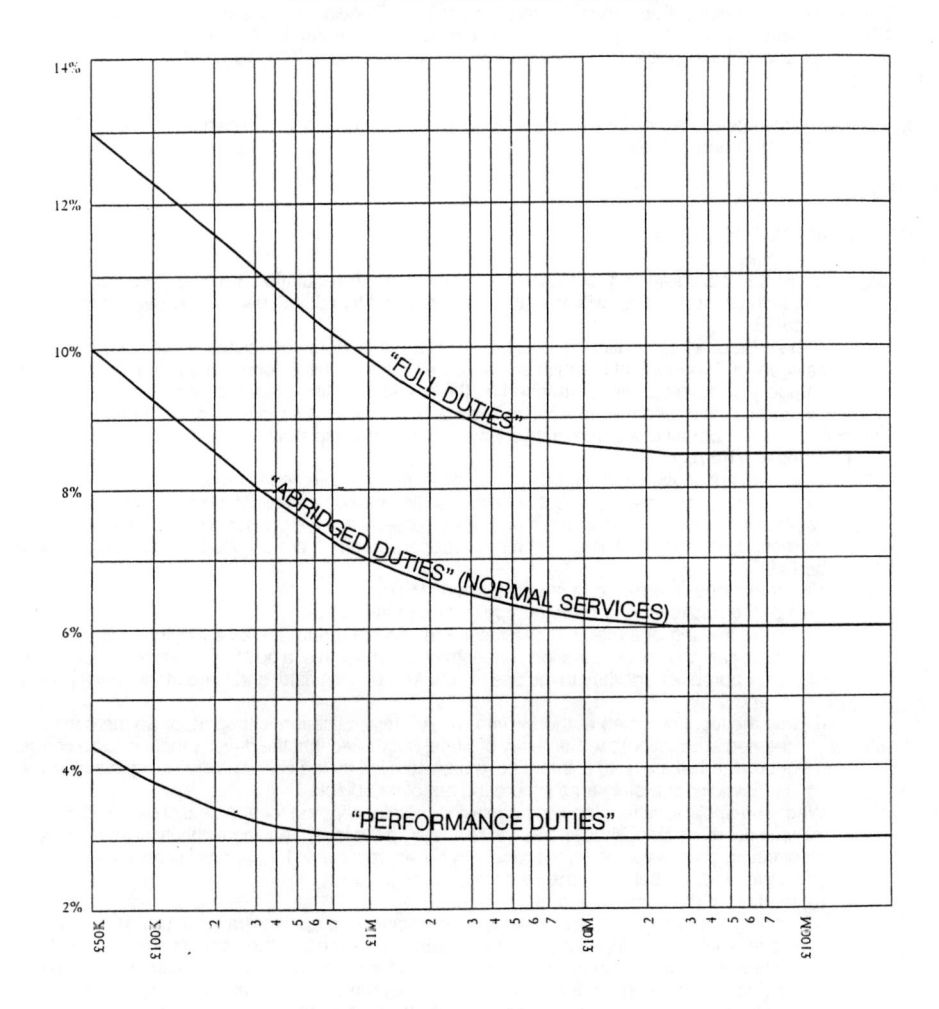

WORKS COST IN £
(1999 PRICES, OPI 371, 1975 = 100)

Note: When Works Cost is below £50K time charges should be used.

Fees for Professional Services

THE TOWN AND COUNTRY PLANNING (FEES FOR APPLICATIONS AND DEEMED APPLICATIONS) (AMENDMENT) REGULATIONS 1997
Operative from 1st October 1997

The following extracts from the Town and Country Planning Fees Regulations, available from HMSO referring to SI 1997:37, relate only to those applications which meet the "deemed to qualify clauses" laid down in regulations 1 to 12 of S.I. No. 1989/193, updated by 1 to 5 of S.I. 1991/2735, 1992/1817, 1992/3052 and 1993/3170.

Further advice on the interpretation of these regulations can be found in the 1990 Town and Country Planning Acts and S.I. 1989/193 and 1990/2473.

SCHEDULE 1

PART I: GENERAL PROVISIONS

1. (1) Subject to paragraphs 3 to 11, the fee payable under regulation 3 or regulation 10 shall be calculated in accordance with the table set out in Part II of this Schedule and paragraphs 2 and 12 to 16.
   (2) In the case of an application for approval of reserved matters, references in this Schedule to the category of development to which an application relates shall be construed as references to the category of development authorised by the relevant outline planning permission.
2. Where an application or deemed application is made or deemed to be made by or on behalf of a parish council or by or on behalf of a community council, the fee payable shall be one-half of the amount as would otherwise be payable.
3. (1) Where an application or deemed application is made or deemed to be made by or on behalf of a club, society or other organisation(including any persons administering a trust) which is not established or conducted for profit and whose objects are the provision of facilities for sport or recreation, and the conditions specified in subparagraph (2) are satisfied, the fee payable shall be £190).
   (2) The conditions referred to in subparagraph (1) are:
      (a) that the application or deemed application relates to:
         (i) the making of a material change in the use of land to use as a playing field; or
         (ii) the carrying out of operations (other than the erection of a building containing floor space) for purposes ancillary to the use of land as a playing field, and to no other development: and
      (b) that the local planning authority with whom the application is lodged, or (in the case of a deemed application) the Secretary of State, is satisfied that the development is to be carried out on land which is, or is intended to be, occupied by the club, society or organisation and used wholly or mainly for the carrying out of its objects.
4. (1) Where an application for planning permission or an application for approval of reserved matters is made not more than 28 days after the lodging with the local planning authority of an application for planning permission or, as the case may be, an application for approval of reserved matters:
      (a) made by or on behalf of the same applicant;
      (b) relating to the same site; and
      (c) relating to the same development or, in the case of an application for approval of reserved matters, relating to the same reserved matters in respect of the same building or buildings authorised by the same outline planning permission, and a fee of the full amount payable in respect of the category or categories of development to which the applications relate has been paid in respect of the earlier application, the fee payable in respect of the later application shall, subject to sub-paragraph (2) be one-quarter of the amount paid in respect of the earlier application.
   (2) Sub-paragraph (1) apply only in respect of one application made by or on behalf of the same applicant in relation to the same development or in relation to the same reserved matters (as the case may be).

THE TOWN AND COUNTRY PLANNING (FEES FOR APPLICATIONS AND DEEMED APPLICATIONS) (AMENDMENT) REGULATIONS 1997

5. (1) This paragraph applies where:
    (a) an application is made for approval of one or more reserved matters ("the current application"); and
    (b) the applicant has previously applied for such approval under the same outline planning permission and paid fees in relation to one or more such applications; and
    (c) no application has been made under that permission other than by or on behalf of the applicant.
  (2) Where the amount paid as mentioned in sub-paragraph (1) (b) is not less than the amount which would be payable if the applicant were by his current application seeking approval of all the matters reserved by the outline permission (and in relation to the whole of the development authorised by the permission), the fee payable in respect of the current application shall be £190.
  (3) Where:
    (a) a fee has been paid as mentioned in sub-paragraph (1) (b) at a rate lower than that prevailing at the date of the current application; and
    (b) sub-paragraph (2) would apply if that fee had been paid at the rate applying at that date, the fee in respect of the current application shall be the amount specified in sub-paragraph (2).

6. Where an application is made pursuant to section 31A of the 1971 Act the fee payable in respect of the application shall be £95.

7A Where an application relates to development to which section 73A(b) of the Town and Country Planning Act 1990 applies, the fee payable in respect of the application shall be:
  (a) where the application relates to development carried out within planning permission, the fee that would be payable if the application were for planning permission to carry out that development;
  (b) £95 in any case

7B. Where an application is made for the renewal of planning permission and:
  (a) a planning permission has previously been granted for development which has not yet begun, and
  (b) a limit as to the time by which the development must be begun was imposed under section 91 (limit of duration of planning permission) or section 92 (outline planning permission) of the Town and Country Planning Act 1990 which has not yet expired, the fee payable in respect of the application shall be £95.

8. (1) This paragraph applies where applications are made for planning permission or for the approval of reserved matters in respect of the development of land lying in the areas of:
    (a) two or more local planning authorities in a metropolitan county or in Greater London; or
    (b) two or more district planning authorities in a non-metropolitan county; or
    (c) one or more such local planning authorities and one or more such district planning authorities.
  (2) A fee shall be payable only to the local planning authority or district planning authority in whose area the largest part of the relevant land is situated: and the amount payable shall be:
    (a) where the applications relate wholly or partly to a county matter within the meaning of paragraph 32 of Schedule 16 to the Local Government Act 1972, and all the land is situated in a single non-metropolitan county, the amount which would have been payable if application had fallen to be made to one authority in relation to the whole development;
    (b) in any other case, one and a half times the amount which would have been payable if application had fallen to be made to a single authority or the sum of the amounts which would have been payable but for this paragraph, whichever is the lesser.

THE TOWN AND COUNTRY PLANNING (FEES FOR APPLICATIONS AND DEEMED APPLICATIONS) (AMENDMENT) REGULATIONS 1997

9. (1) This paragraph applies where application for planning permission is deemed to have been made by virtue of section 88B (3) of the 1971 Act in respect of such land as is mentioned in paragraph 8 (1).
   (2) The fee payable to the Secretary of State shall be the amount which would be payable by virtue of paragraph 8 (2) if application for the like permission had been made to the relevant local or district planning authority on the date on which notice of appeal was given in accordance with section 88 (3) of the 1971 Act.
10. (1) Where:
    (a) application for planning permission is made in respect of two or more alternative proposals for the development of the same land; or
    (b) application for approval of reserved matters is made, in respect of two or more alternative proposals for the carrying out of the development authorised by an outline planning permission, and application is made in respect of all of the alternative proposals on the same date and by or on behalf of the same applicant, a single fee shall be payable in respect of all such alternative proposals, calculated as provided in sub-paragraph (2).
    (2) Calculations shall be made, in accordance with this Schedule, of the fee appropriate to each of the alternative proposals and the single fee payable in respect of all the alternative proposals shall be the sum of:
    (a) an amount equal to the highest of the amounts calculated in respect of each of the alternative proposals; and
    (b) an amount calculated by adding together the amounts appropriate to all of the alternative proposals, other than the amount referred to in subparagraph (a), and dividing that total by the figure of 2.
11. In the case of an application for planning permission which is deemed to have been made by virtue of section 95 (6) of the 1971 Act, the fee payable shall be the sum of £190.
12. Where, in respect of any category of development specified in the table set out in Part II of this Schedule, the fee is to be calculated by reference to the site area:
    (a) that area shall be taken as consisting of the area of land to which the application relates or, in the case of an application for planning permission which is deemed to have been made by virtue of section 88B (3) of the 1971 Act, the area of land to which the relevant enforcement notice relates; and
    (b) where the area referred to in sub-paragraph (a) above is not an exact multiple of the unit of measurement specified in respect of the relevant category of development, the fraction of a unit remaining after division of the total area by the unit of measurement shall be treated as a complete unit.
13. (1) In relation to development within any of the categories 2 to 4 specified in the table in Part II of this Schedule, the area of gross floor space to be created by the development shall be ascertained by external measurement of the floor space, whether or not it is to be bounded (wholly or partly) by external walls of a building.
    (2) In relation to development within category 2 specified in the said table, where the area of gross floor space to be created by the development exceeds 75 sq metres and is not an exact multiple of 75 sq metres, the area remaining after division of the total number of square metres of gross floor space by the figure of 75 shall be treated as being 75 sq metres.
    (3) In relation to development within category 3 specified in the said table, where the area of gross floor space exceeds 540 sq metres and the amount of the excess is not an exact multiple of 75 sq metres, the area remaining after division of the number of square metres of that excess area of gross floor space by the figure of 75 shall be treated as being 75 sq metres.

THE TOWN AND COUNTRY PLANNING (FEES FOR APPLICATIONS AND DEEMED APPLICATIONS) (AMENDMENT) REGULATIONS 1997

14. (1) Where an application (other than an outline application) or a deemed application relates to development which is in part within category 1 in the table set out in Part II of this Schedule and in part within category 2, 3, or 4, the following sub-paragraphs shall apply for the purpose of calculating the fee payable in respect of the application or deemed application.

    (2) An assessment shall be made of the total amount of gross floor space which is to be created by that part of the development which is within category 2, 3 or 4 ("the non-residential floor space"), and the sum payable in respect of the non-residential floor space to be created by the development shall be added to the sum payable in respect of that part of the development which is within category 1, and subject to sub-paragraph (4), the sum so calculated shall be the fee payable in respect of the application or deemed application.

    (3) For the purpose of calculating the fee under sub-paragraph (2)-
        (a) Where any of the buildings is to contain floor space which it is proposed to use for the purposes of providing common access or common services or facilities for persons occupying or using that building for residential purposes and for persons occupying or using it for non-residential purposes ("common floor space"), the amount of non-residential floor space shall be assessed, in relation to that building, as including such proportion of the common floor space as the amount of non-residential floor space in the building bears to the total amount of gross floor space in the building to be created by the development;
        (b) where the development falls within more than one of categories 2, 3 and 4 an amount shall be calculated in accordance with each such category and the highest amount so calculated shall be taken as the sum payable in respect of all of the non-residential floor space.

    (4) Where an application or deemed application to which this paragraph applies relates to development which is also within one or more than one of the categories 5 to 13 in the table set out in Part II of this Schedule, an amount shall be calculated in accordance with each such category and if any of the amounts so calculated exceeds the amount calculated in accordance with sub-paragraph (2) that higher amount shall be the fee payable in respect of all of the development to which the application or deemed application relates.

15. (1) Subject to paragraph 14, and sub-paragraph (2), where an application or deemed application relates to development which is within more than one of the categories specified in the table set out in Part II of this Schedule:
        (a) an amount shall be calculated in accordance with each such category; and
        (b) the highest amount so calculated shall be the fee payable in respect of the application or deemed application.

    (2) Where an application is for outline planning permission and relates to development which is within more than one of the categories specified in the said table, the fee payable in respect of the application shall be £190 for each 0.1 hectares of the site area, subject to a maximum of £4,750.

16. In the case of an application for planning permission which is deemed to have been made by virtue of section 88B (3) of the 1971 Act, references in this Schedule to the development to which an application relates shall be construed as references to the use of land or the operations (as the case may be) to which the relevant enforcement notice relates; references to the amount of floor space or the number of dwelling houses to be created by the development shall be construed as references to the amount of floor space or the number of dwelling houses to which that enforcement notice relates; and references to the purposes for which it is proposed that floor space be used shall be construed as references to the purposes for which floor space was stated to be used in the enforcement notice.

THE TOWN AND COUNTRY PLANNING (FEES FOR APPLICATIONS AND DEEMED APPLICATIONS) (AMENDMENT) REGULATIONS 1997

PART II: SCALE OF FEES

Category of development	Fee Payable
I. Operations	
1. The erection of dwelling houses (other than development within category 6 below).	(a) Where the application is for outline planning permission £190 for each 0.1 hectare of the site area, subject to a maximum of £4,750; (b) in other cases, £190 for each dwelling house to be created by the development, subject to a maximum of £9,500.
2. The erection of buildings (other than buildings coming within categories 1, 3, 4, 5 or 7.)	(a) Where the application is for outline planning permission £190 for each 0.1 hectare of the the site area, subject to a maximum of £4,750; (b) in other cases: (i) where no floor space is to be created by the development, £95; (ii) where the area of gross floor space to be created by the development does not exceed 40 sq metres, £95; (iii) where the area of gross floor space to be created by the development exceeds 40 sq metres but does not exceed 75 sq metres, £190; and (iv) where the area of gross floor space to be created by the development exceeds 75 sq metres, £190 for each 75 sq metres, subject to a maximum of £9,500.
3. The erection, on land used for the purposes of agriculture, of buildings to be used for agricultural purposes (other than buildings coming within category 4).	(a) Where the application is for outline planning permission £190 for each hectare of the site area subject to a maximum of £4,750; (b) in other cases: (i) where the area of gross floor space to be created by the development does not exceed 465 sq metres, £35; (ii) where the area of gross floor space to be created by the development exceeds 465 sq metres but does not exceed 540 sq metres, £190; and (iii) where the area of gross floor space to be created by development exceeds 540 sq metres, £190 for the first 540 sq metres and £190 for each 75 sq metres in excess of that figure, subject to a maximum of £9,500.
4. The erection of glasshouses on land used for the purposes of agriculture.	(a) Where the area of gross floor space to be created by the development does not exceed 465 sq metres, £35; (b) where the area of gross floor space to be created by the development exceeds 465 sq metres, £1,085.
5. The erection, alteration or replacement of plant or machinery.	£190 for each 0.1 hectare of the site area, subject to a maximum of £9,500.

THE TOWN AND COUNTRY PLANNING (FEES FOR APPLICATIONS AND DEEMED APPLICATIONS) (AMENDMENT) REGULATIONS 1997

Category of development	Fee Payable
6. The enlargement, improvement or other alteration of existing dwelling houses	(a) Where the application relates to one dwelling house, £95; (b) where the application relates to 2 or more dwelling houses, £190.
7. (a) The carrying out of operations (including the erection of a building within the curtilage of an existing dwelling house, for purposes ancillary to the enjoyment of the dwelling house as such, or the erection or construction of gates, fences, walls or other means of enclosure along a boundary of the curtilage of an existing dwelling house; or (b) the construction of car parks, service roads and other means of access on land used for the purposes of a single undertaking, where the development is required for a purpose incidental to the existing use of the land.	£95
8. The carrying out of any operations connected with exploratory drilling for oil or natural gas.	£190 for each 0.1 hectare of the site area, subject to a maximum of £14,250.
9. The carrying out of any operations not coming within any of the above categories.	£95 for each 0.1 hectare of the site area, subject to a maximum of: (a) in the case of operations for the winning and working of minerals, £14,250; (b) in other cases, £950.

II. Uses of Land

10. The change of use of a building to use as one or more separate dwelling houses.	(a) Where the change is from a previous use as a single dwelling house to use as a two or more single dwelling houses, £190 for each additional dwelling house to be created by the development, subject to a maximum of £9,500. (b) in other cases, £190 for each dwelling house to be created by the development, subject to a maximum of £9,500.
11. (a) The use of land for the disposal of refuse or waste materials or for the deposit of material remaining after minerals have been extracted from land; or (b) the use of land for the storage of minerals in the open.	£95 for each 0.1 hectare of the site area, subject to a maximum of £14,250.
12. The making of a material change in the use of a building or land (other than a material change of use coming within any of the above categories).	£190.

THE TOWN AND COUNTRY PLANNING (FEES FOR APPLICATIONS AND DEEMED APPLICATIONS) (AMENDMENT) REGULATIONS 1997

SCHEDULE 2

SCALE OF FEES IN RESPECT OF APPLICATIONS FOR CONSENT TO DISPLAY ADVERTISEMENTS

Category of advertisement	Fee payable
1. Advertisements displayed on business premises, on the forecourt of business premises or on other land within the curtilage of business premises, wholly with reference to all or any of the following matters: (a) the nature of the business or other activity carried out on the premises; (b) the goods sold or the services provided on the premises; or (c) the name and qualifications of the person carrying on such business or activity or supplying such goods or services.	£50
2. Advertisements for the purpose of directing members of the public to, or otherwise drawing attention to the existence of, business premises which are in the same locality as the site on which the advertisement is to be displayed but which are not visible from that site.	£50
3. All other advertisements.	£190.

THE BUILDING (LOCAL AUTHORITY CHARGES) REGULATIONS 1998

Author's Note:

On the 31st July 1998 the Minister for Construction, announced his intention of improving the flexibility with which local authorities responsible for building control in England and Wales could respond to competition from the private sector by devolving to individual authorities the setting of charges for building control functions carried out in respect of the Building Regulations 1991.

The Building (Local Authority Charges) Regulations 1998 (the Charges Regulations) require each local authority to prepare a Scheme within which they are to fix their charges. They came into effect on the 1st April 1999. In a number of major cities, uniform levels of fees have been adopted. In some local authorities charges have fallen a third in comparison to those prescribed within the 1991 Regulations. A number of authorities have adopted the Local Government Association (LGA) Model Fee Scheme 2000, which is for local authority distribution only.

Consultation should be made to each local authority for their Charges, however as guidance we have kindly been given permission by the London Borough of Ealing to publish the Charges for their district, which includes Acton W3.

CHARGE SCHEDULES

With effect from 1st April 2001, there are three main charge Schedule Tables:

Table 1, For erection of one or more small new domestic buildings and connected work, ie. houses and flats up to 3 storeys in height with an internal floor area not exceeding 300 m^2;

Table 2, For erection of certain small domestic building, and extensions, ie detached garages and carports not exceeding 40 m^2 and not exempt, and extensions including all new loft conversions up to a total of 60 m^2 at the same time;

Table 3, For building work other than where Tables 1 and 2 apply. Charges relate to estimated cost of the works.

TYPES OF CHARGES

There are four application types and five types of charge:

1) **Building Notice** applications where the full **Charge** must be paid at time of notification (column 10 in Tables 1, 2 and 3). This is mainly used for small domestic alterations. Where structural work is involved calculations need to be provided. Upon satisfactory completion of works on site a Completion Certificate may be issued;

2) **Full Plan** applications where the **Plan Charge** (column 4 in Tables 1, 2 and 3) must be paid in deposit of plans. A subsequent re-submission, further to a Rejection of Plans, will NOT attract an additional fee for essentially the same work. If the inspection charge is not paid upon the deposit of plans, an invoice will be raised after the first inspection on site as this is when **Inspection Charge** (column 7 in Tables 1, 2 and 3) becomes payable. Where work is to be done to Shops, Factories, Offices, Railway Premises, Hotels and Boarding Houses, and Non-domestic Workplaces a Full Application should be made. It is also appropriate for Domestic Loft Conversions and other Extensions and Erections of Domestic Buildings. It is important to start work before 3 years have expired or the Plan Approval may be withdrawn. Upon satisfactory completion of the works on site a Completion Certificate may be issued; and

3) **Regularisation** applications where the full **Charge of 1.2 times the Building Notice Charge** must be paid at time of notification (ie 1.2 times column 8 in Tables 1, 2 and 3). VAT is not chargeable. That is where work started after November 1985 but was not notified to Building Control. It is usually necessary to open up works to show what has been done. A letter may be sent outlining what needs to be done to comply with the Building Regulations. Upon satisfactory completion of works on site a Regularisation Certificate may be issued; and

THE BUILDING (LOCAL AUTHORITY CHARGES) REGULATIONS 1998

TYPES OF CHARGES - cont'd

4) **Reversion** applications where the full **Charge equal to the Building Notice Charge** must be paid at time of notification (column 8 in Tables 1, 2 and 3). VAT is chargeable. This is where an Approved Inspector has supervised Building Regulations but the work has reverted to Building Control. It is usually necessary to open up works to show what has been done. A letter may be sent outlining what needs to be done to comply with the Building Regulations. Upon satisfactory completion of works on site a Reversion Certificate may be issued.

Unless a required fee is included with an application form, it cannot be accepted as a valid application. Work may not legally start on site until the fee is received. If Cheque(s) are Dishonoured the application also cannot be accepted as valid.

EXEMPTIONS

No fees are charged, where we are satisfied **work** is **solely** for the purpose of providing means of access **for disabled persons** or within a building, or for providing facilities designed to secure greater health, safety, welfare or convenience and is carried out in relation to a building to which members of the public are admitted or is a dwelling occupied by a disabled person.

THE BUILDING (LOCAL AUTHORITY CHARGES) REGULATIONS 1998

TABLE 1

CHARGES FOR ERECTION OF ONE OF MORE SMALL NEW DOMESTIC BUILDINGS AND CONNECTED WORK

NOTES

Dwellings in excess of 300 m² in floor area (excluding garage or carport) are to be calculated on estimated cost on accordance with Table 3.
Buildings in excess of 3 storeys (including any basements) to be calculated on estimated cost in accordance with Table 3.
The Charges in this table includes for works of drainage in connection with erection of a building(s), even where those drainage works are commenced in advance of the plans for the building being deposited.
The charges include for an integral garage and where a garage or carport shares at least one wall of the domestic building. Detached garages are not included in this Table (see Notes Table 2).
Where a Plan or Inspection Charge exceeds £4,000.00 (columns 2 and 5) the Council may agree payment in instalments.
Where all dwellings on a site or an estate are substantially the same, it may be possible to offer a discount of 30% reduction on PLAN Charge OR equivalent reduction on the BUILDING NOTICE Charge.

Number of Dwelling	PLAN CHARGE			INSPECTION CHARGE			BUILDING NOTICE		
	CHARGE £	VAT £	TOTAL £	CHARGE £	VAT £	TOTAL £	CHARGE £	VAT £	TOTAL £
(1)	(2)	(3)	(4)	(5)	(6)	(7)	(8)	(9)	(10)
1	140.00	24.50	164.50	265.00	46.38	311.38	405.00	70.88	475.88
2	205.00	35.88	240.88	401.00	70.18	471.18	606.00	106.06	712.06
3	270.00	47.25	317.25	541.00	94.68	635.68	811.00	141.93	952.93
4	335.00	58.63	393.63	666.00	116.55	782.55	1,001.00	175.18	1,176.18
5	405.00	70.86	475.86	766.00	134.05	900.05	1,171.00	204.93	1,375.93
6	475.00	83.13	558.13	911.00	159.43	1070.43	1,386.00	242.56	1,628.56
7	495.00	86.63	581.63	974.00	170.45	1144.45	1,469.00	257.08	1,726.08
8	515.00	90.13	605.13	1,137.00	198.98	1,335.98	1,652.00	289.11	1,941.11
9	535.00	93.63	628.63	1,301.00	227.68	1,528.68	1,836.00	321.31	2,157.31
10	540.00	94.50	634.50	1,480.00	259.00	1,739.00	2,020.00	353.50	2,373.50
11	545.00	95.38	640.38	1,623.00	284.03	1,907.03	2,168.00	379.41	2,547.41
12	550.00	96.25	646.25	1,765.00	308.89	2,073.89	2,315.00	405.14	2,720.14
13	555.00	97.13	652.13	1,908.00	333.90	2,241.90	2,463.00	431.03	2,894.03
14	560.00	98.00	658.00	2,051.00	358.93	2,409.93	2,611.00	456.93	3,067.93
15	565.00	98.88	663.88	2,194.00	383.95	2,577.95	2,759.00	482.83	3,241.83
16	570.00	99.75	669.75	2,337.00	408.98	2,745.98	2,907.00	508.73	3,415.73
17	575.00	100.63	675.63	2,480.00	434.00	2,914.00	3,055.00	534.63	3,589.63
18	580.00	101.50	681.50	2,623.00	459.03	3,082.03	3,203.00	560.53	3,763.53
19	585.00	102.38	687.38	2,766.00	484.05	3,250.05	3,351.00	586.43	3,937.43
20	590.00	103.25	693.25	2,909.00	509.08	3,418.08	3,499.00	612.33	4,111.33
21	600.00	105.00	705.00	3,011.00	520.93	3,537.93	3,611.00	631.93	4,242.93
22	610.00	106.75	716.75	3,113.00	544.78	3,657.78	3,723.00	651.53	4,374.53
23	620.00	108.50	728.50	3,215.00	562.63	3,777.63	3,835.00	671.13	4,506.13
24	630.00	110.25	740.25	3,317.00	580.48	3,897.48	3,947.00	690.73	4,637.73
25	640.00	112.00	752.00	3,420.00	598.50	4,018.50	4,060.00	710.50	4,770.50
26	650.00	113.75	763.75	3,522.00	616.35	4,138.35	4,172.00	730.10	4,902.10
27	660.00	115.50	775.50	3,624.00	634.20	4,258.20	4,284.00	749.70	5,033.70
28	670.00	117.25	787.25	3,726.00	652.05	4,378.05	4,396.00	769.30	5,165.30
29	680.00	119.00	799.00	3,828.00	669.90	4,497.90	4,508.00	788.90	5,296.90
30	690.00	120.75	810.75	3,885.00	679.88	4,564.88	4,575.00	800.63	5,375.63
31	700.00	122.50	822.50	3,940.00	689.50	4,629.50	4,640.00	812.00	5,452.00
31 AND OVER	For each dwelling in excess of 31 add £5 + VAT (= £5.88 inclussive of VAT)			For each dwelling in excess of 31 add £75 + VAT (= £88.13 inclussive of VAT)			For each dwelling in excess of 31 add £80 + VAT (= £94.00 inclussive of VAT)		

THE BUILDING (LOCAL AUTHORITY CHARGES) REGULATIONS 1998

TABLE 2

CHARGES FOR ERECTION OF CERTAIN SMALL DOMESTIC BUILDINGS, GARAGES, CARPORTS AND EXTENSIONS

NOTES

Detached garages and carports having an internal floor area not exceeding 30 m² are 'exempt buildings', providing that in the case of a garage it is sited at least 1.0 m away from the boundary or is constructed substantially of non-combustible materials.
A carport extension having an internal floor area not exceeding 30 m² would be exempt if it is fully open on at least 2 sides.
Detached garages in excess of 40 m² and extensions in excess of 60 m² in floor area use Table 3.
A new Dormer Windows which does not increase the usable floor area would be an alteration so use Table 3.
If the **total floor area** of all extensions being done at the same time **exceeds 60 m² use Table 3.**
Loft conversions with new internal useable floor area in roof space are to be treated as an extension in this Table 2.
Extensions more than three storeys high (including any basement) **should use Table 3.**
Chargeable installations of Cavity Fill Insulation, and Unvented Hot Water Systems should use Table 3 (see Table 3 notes).
Extensions to a building that is **NOT wholly domestic** should use **Table 3.**
Where on an estate erections of garages or extensions are substantially the same, it may be possible to offer a discount of 30% reduction on PLAN Charge OR a 7½% reduction on the BUILDING NOTICE Charge.

Type of Work	PLAN CHARGE			INSPECTION CHARGE			BUILDING NOTICE		
	CHARGE £	VAT £	TOTAL £	CHARGE £	VAT £	TOTAL £	CHARGE £	VAT £	TOTAL £
(1)	(2)	(3)	(4)	(5)	(6)	(7)	(8)	(9)	(10)
1 Erection of a **detached** building which consists of a **garage or carport** or both having a floor area not exceeding 40m² in total and intended to be used in common with an existing building, and which is not an exempt building.	35.00	6.13	41.13	70.00	12.25	82.25	105.00	18.38	123.38
2 Any extension of a dwelling the total floor area of which **does not exceed 10m²**, including means of access and work in connection with that extension.	210.00	36.75	246.75	No inspection charge (included in plan charge)			210.00	36.75	246.75
3 Any extension of a dwelling the total floor area of which **exceeds 10m², but does not exceed 40m²**, including means of access and work in connection with that extension.	80.00	14.00	94.00	235.00	41.13	276.13	315.00	55.13	370.13
4 Any extension of a dwelling the total floor area of which **exceeds 40m², but does not exceed 60m²**, including means of access and work in connection with that extension.	105.00	18.38	123.38	315.00	55.13	370.13	420.00	73.50	493.50

THE BUILDING (LOCAL AUTHORITY CHARGES) REGULATIONS 1998

TABLE 3

CHARGES FOR BUILDING WORK OTHER THAN TO WHICH TABLES 1 AND 2 APPLY.
CHARGES RELATE TO ESTIMATED COST.

NOTES

If some building work is covered by Table 2 and some by Table 3, both fees are payable.

Estimated cost of work should not include any professional fees (e.g. Architect, Quantity Surveyor, etc) nor any VAT.

Installation of cavity fill insulation in accordance with Part D of Schedule 1 to the Principle Regulations where installation is not certified to an approved standard or is not installed by an approved installer, or is part of a larger project this Table Building Notice Charge is payable (columns 8, 9 and 10).

Installation of an unvented hot water system in accordance with Part G3 of Schedule 1 to the Principle Regulations where the installation is not part of a larger project and where the authority carry out an inspection, this Table Building Notice Charge is payable (columns 8, 9 and 10).

Where a Plan or Inspection Charge exceeds £4,000.00 (columns 2 or 5) the Council may agree payment in instalments.

If application is for erection of work substantially the same type under current regulations, it may be possible to offer a discount of PLAN CHARGE by 30% OR BUILDING NOTICE Charge by 7½%.

Use Table 1 for Dwellings up to 300 m² in floor area and up to and including 3 storeys high.

ESTIMATED COST OF WORKS £ (1)		PLAN CHARGE			INSPECTION CHARGE			BUILDING NOTICE		
		CHARGE £ (2)	VAT £ (3)	TOTAL £ (4)	CHARGE £ (5)	VAT £ (6)	TOTAL £ (7)	CHARGE £ (8)	VAT £ (9)	TOTAL £ (10)
0 - 2,000		100.00	17.50	117.50	No inspection charge (included in plan charge)			100.00	17.50	117.50
2,001 - 5,000		165.00	28.88	193.88				165.00	28.88	193.88
5,001 - 6,000		43.50	7.61	51.11	130.50	22.84	153.34	174.00	30.45	204.45
6,001 - 7,000		45.75	8.01	53.76	137.25	24.03	161.28	183.00	32.03	215.03
7,001 - 8,000		48.00	8.40	56.40	144.00	25.20	169.20	192.00	33.60	225.60
8,001 - 9,000		50.25	8.79	59.04	150.75	26.38	177.13	201.00	35.18	236.18
9,001 - 10,000		52.50	9.19	61.69	157.50	27.56	185.06	210.00	36.75	246.75
10,001 - 11,000		54.75	9.58	64.33	164.25	28.74	192.99	219.00	38.33	257.33
11,001 - 12,000		57.00	9.98	66.98	171.00	29.93	200.93	228.00	39.90	267.90
12,001 - 13,000		59.25	10.37	69.62	177.75	31.10	208.86	237.00	41.48	278.48
13,001 - 14,000		61.50	10.76	72.26	184.50	32.29	216.79	246.00	43.05	289.05
14,001 - 15,000		63.75	11.16	74.91	191.25	33.47	224.72	255.00	44.63	299.63
15,001 - 16,000		66.00	11.55	77.55	198.00	34.65	232.65	264.00	46.20	310.20
16,001 - 17,000		68.25	11.94	80.19	204.75	35.83	240.58	273.00	47.78	320.78
17,001 - 18,000		70.50	12.34	82.84	211.50	37.01	248.51	282.00	49.35	331.35
18,001 - 19,000		72.75	12.73	85.48	218.25	38.19	256.44	291.00	50.93	341.93
19,001 - 20,000		75.00	13.12	88.12	225.00	39.38	264.38	300.00	52.50	352.50
20,001 to 100,000	Basic	75.00	+ VAT	-	225.00	+ VAT	-	300.00	52.50	352.50
	+ for each £1,000 or part thereof by which cost exceeds £20,000	2.00	+ 17½% VAT	-	6.00	+ 17½% VAT	-	8.00	1.40	9.40
100,001 to 1,000,000	Basic	235.00	+ VAT	-	705.00	+ VAT	-	940.00	164.50	1,104.50
	+ for each £1,000 or part thereof by which cost exceeds £100,000	0.80	+ 17½% VAT	-	2.40	+ 17½% VAT	-	3.20	0.56	3.76
1,000,001 to 10,000,000	Basic	955.00	+ VAT	-	2,865.00	+ VAT	-	3,820.00	668.50	4,488.50
	+ for each £1,000 or part thereof by which cost exceeds £1 m	0.59	+ 17½% VAT	-	1.76	+ 17½% VAT	-	2.35	0.41	2.76
10,000,001 and Over	Basic	6,242.50	+ VAT	-	18,727.50	+ VAT	-	24,970.00	4369.75	29,339.75
	+ for each £1,000 or part thereof by which cost exceeds £10 m	0.38	+ 17½% VAT	-	1.13	+ 17½% VAT	-	1.50	0.35	2.35

NEW FROM SPON PRESS

Dictionary of Property and Construction Law

Edited by J. Rostron, School of the Built Environment, Liverpool John Moores University, UK

This is a new dictionary containing over 6,000 entries. It provides clear and concise explanations of the terms used in land, property and construction law and management. The four key areas of coverage are: planning/construction law, land law, equity/trusts and finance and administration. It will be a useful reference for property and building professionals and for students of property and construction law. Jack Rostron is an experienced author and editor whose 1997 Spon title *Sick Building Syndrome* has been well received and widely reviewed.

August 2001: 234x156: 160pp:
Pb: 0-419-26110-9: £19.99
Hb: 0-419-26100-1: £50.00

To Order: Tel: +44 (0) 8700 768853, or +44 (0) 1264 343071 Fax: +44 (0) 1264 343005, or
Post: Spon Press Customer Services, ITPS Andover, Hants, SP10 5BE, UK Email: book.orders@tandf.co.uk.

Postage & Packing: UK: 5% of order value (min. charge £1, max. charge £10) for 3-5 days delivery. Option of next day delivery at an additional £6.50. Europe: 10% of order value (min. charge £2.95, max. charge £20) for delivery surface post. Option of airmail at an additional £6.50. ROW: 15% of order value (min. charge £6.50, max. charge £30) for airmail delivery.

For a complete listing of all our titles visit: www.sponpress.com

SPON PRESS
Taylor & Francis Group

Daywork And Prime Cost

When work is carried out which cannot be valued in any other way it is customary to assess the value on a cost basis with an allowance to cover overheads and profit. The basis of costing is a matter for agreement between the parties concerned, but definitions of prime cost for the building industry have been prepared and published jointly by the Royal Institution of Chartered Surveyors and the National Federation of Building Trades Employers (now the Building Employers Confederation) for the convenience of those who wish to use them. These documents are reproduced on the following pages by kind permission of the publishers.

The daywork schedule published by the Civil Engineering Contractors Association is included in the A&B's companion title, *"Spons Civil Engineering and Highway Works Price Book"*.

For larger Prime Cost contracts the reader is referred to the form of contract issued by the Royal Institute of British Architects.

BUILDING INDUSTRY

DEFINITION OF PRIME COST OF DAYWORK CARRIED OUT UNDER A BUILDING CONTRACT (DECEMBER 1975 EDITION)

This definition of Prime Cost is published by the Royal Institution of Chartered Surveyors and the National Federation of Building Trades Employers, for convenience and for use by people who choose to use it. Members of the National Federation of Building Trades Employers are not in any way debarred from defining Prime Cost and rendering their accounts for work carried out on that basis in any way they choose. Building owners are advised to reach agreement with contractors on the Definition of Prime Cost to be used prior to issuing instructions.

SECTION 1 - APPLICATION

1.1. This definition provides a basis for the valuation of daywork executed under such building contracts as provide for its use (e.g. contracts embodying the Standard Forms issued by the Joint Contracts Tribunal).

1.2. It is not applicable in any other circumstances, such as jobbing or other work carried out as a separate or main contract, nor in the case of daywork executed during the Defects Liability Period of contracts embodying the above mentioned Standard Forms.

SECTION 2 - COMPOSITION OF TOTAL CHARGES

2.1. The prime cost of daywork comprises the sum of the following costs:
    (a)   Labour as defined in Section 3.
    (b)   Material and goods as defined in Section 4.
    (c)   Plant as defined in Section 5.

2.2. Incidental costs, overheads and profit as defined in Section 6, as provided in the building contract and expressed therein as percentage adjustments are applicable to each of 2.1 (a)-(c).

BUILDING INDUSTRY

SECTION 3 - LABOUR

3.1. The standard wage rates, emoluments and expenses referred to below and the standard working hours referred to in 3.2 are those laid down for the time being in the rules or decisions of the National Joint Council for the Building Industry and the terms of the Building and Civil Engineering Annual and Public Holiday Agreements applicable to the works, or the rules or decisions or agreements of such body, other than the National Joint Council for the Building Industry, as may be applicable relating to the class of labour concerned at the time when and in the area where the daywork is executed.

3.2. Hourly base rates for labour are computed by dividing the annual prime cost of labour, based upon standard working hours and as defined in 3.4 (a)-(I), by the number of standard working hours per annum.

3.3. The hourly rates computed in accordance with 3.2 shall be applied in respect of the time spent by operatives directly engaged on daywork, including those operating mechanical plant and transport and erecting and dismantling other plant (unless otherwise expressly provided in the building contract).

3.4. The annual prime cost of labour comprises the following:

    (a)   Guaranteed minimum weekly earnings (e.g. Standard Basic Rate of Wages, Joint Board Supplement and Guaranteed Minimum Bonus Payment in the case of NJCBI rules).
    (b)   All other guaranteed minimum payments (unless included in Section 6).
    (c)   Differentials or extra payments in respect of skill, responsibility, discomfort, inconvenience or risk (excluding those in respect of supervisory responsibility - see 3.5).
    (d)   Payments in respect of public holidays.
    (e)   Any amounts which may become payable by the Contractor to or in respect of operatives arising from the operation of the rules referred to in 3.1 which are not provided for in 3.4 (a)-(d) or in Section 6.
    (f)   Employer's National Insurance contributions applicable to 3.4 (a)-(e).
    (g)   Employer's contributions to annual holiday credits.
    (h)   Employer's contributions to death benefit scheme.
    (i)   Any contribution, levy or tax imposed by statute, payable by the contractor in his capacity as an employer.

3.5. **Note:** Differentials or extra payments in respect of supervisory responsibility are excluded from the annual prime cost (see Section 6). The time of principals, foremen, gangers, leading hands and similar categories, when working manually, is admissible under this Section at the appropriate rates for the trades concerned.

SECTION 4 - MATERIALS AND GOODS

4.1. The prime cost of materials and goods obtained from stockists or manufacturers is the invoice cost after deduction of all trade discounts but including cash discounts not exceeding 5 per cent and includes the cost of delivery to site.

4.2. The prime cost of materials and goods supplied from the Contractor's stock is based upon the current market prices plus any appropriate handling charges.

4.3. Any Value Added Tax which is treated, or is capable of being treated, as input tax (as defined in the Finance Act, 1972) by the Contractor is excluded.

BUILDING INDUSTRY

SECTION 5 - PLANT

5.1. The rates for plant shall be as provided in the building contract.

5.2. The costs included in this Section comprise the following:
- (a) Use of mechanical plant and transport for the time employed on daywork.
- (b) Use of non-mechanical plant (excluding non-mechanical hand tools) for the time employed on daywork.

5.3. **Note:** The use of non-mechanical hand tools and of erected scaffolding, staging, trestles or the like is excluded (see Section 6).

SECTION 6 - INCIDENTAL COSTS, OVERHEADS AND PROFIT

6.1. The percentage adjustments provided in the building contract, which are applicable to each of the totals of Sections 3, 4 and 5, comprise the following:
- (a) Head Office charges.
- (b) Site staff, including site supervision.
- (c) The additional cost of overtime (other than that referred to in 6.2).
- (d) Time lost due to inclement weather.
- (e) The additional cost of bonuses and all other incentive payments in excess of any guaranteed minimum included in 3.4. (a).
- (f) Apprentices study time.
- (g) Subsistence and periodic allowances.
- (h) Fares and travelling allowances.
- (i) Sick pay or insurance in respect thereof.
- (j) Third-party and employers' liability insurance.
- (k) Liability in respect of redundancy payments to employees.
- (l) Employers' National Insurance contributions not included in Section 3.4.
- (m) Tool allowances.
- (n) Use, repair and sharpening of non-mechanical hand tools.
- (o) Use of erected scaffolding, staging, trestles or the like.
- (p) Use of tarpaulins, protective clothing, artificial lighting, safety and welfare facilities, storage and the like that may be available on the site.
- (q) Any variation to basic rates required by the Contractor in cases where the building contract provides for the use of a specified schedule of basic plant charges (to the extent that no other provision is made for such variation).
- (r) All other liabilities and obligations whatsoever not specifically referred to in this Section nor chargeable under any other Section.
- (s) Profit.

6.2. **Note:** The additional cost of overtime, where specifically ordered by the Architect/Supervising Officer shall only be chargeable in the terms of prior written agreement between the parties to the building contract.

BUILDING INDUSTRY

Example of calculation of typical standard hourly base rate (as defined in Section 3) for NJCBI building craftsman and labourer in Grade A areas at 1st July, 1975.

		Rate £	Craftsman £	Rate £	Labourer £
Guaranteed minimum weekly earnings					
Standard Basic Rate	49 wks	37.00	1813.00	31.40	1538.60
Joint Board Supplement	49 wks	5.00	245.00	4.20	205.80
Guaranteed Minimum Bonus	49 wks	4.00	196.00	3.60	176.40
			2254.00		1920.00
Employer's National Insurance Contribution at 8.5%			191.59		163.27
			2445.59		2084.07
Employer's Contribution to :					
CITB annual levy			15.00		3.00
Annual holiday credits	49 wks	2.80	137.20	2.80	137.20
Public holidays (included in guaranteed minimum weekly earnings above)					
Death benefit scheme	49 wks	0.10	4.90	0.10	4.90
Annual labour cost as defined in section 3			£ 2602.69		£ 2229.17

Hourly rate of labour as defined in section 3, Clause 3.2 $\frac{2602.69}{1904}$ = £ 1.37 $\frac{2229.17}{1904}$ = £ 1.17

Note:

1. Standard working hours per annum calculated as follows:

52 weeks @ 39 hours		2080
Less		
3 weeks holiday @ 40 hours	120	
7 days public holidays @ 8 hours	56	-176
		1904

2. It should be noted that all labour costs incurred by the Contractor in his capacity as an employer other than those contained in the hourly rate, are to be taken into account under section 6.

3. The above example is for the convenience of users only and does not form part of the Definition; all basic costs are subject to re-examination according to the time when and in the area where the daywork is executed.

Daywork and Prime Cost

BUILDING INDUSTRY

For the convenience of readers the example which appears on the previous page has been updated by the Editors for London and Liverpool rates as at 25 June, 2001.

		Rate £	Craftsman £	Rate £	Labourer £
Guaranteed minimum weekly earnings					
Standard Basic Rate	46.200 wks	261.30	12072.06	196.56	9081.07
Guaranteed Minimum Bonus	46.200 wks	0.00	0.00	0.00	0.00
			12072.06		9081.07
Employer's National Insurance Contribution (11.97% after the first £87.00 per week)			958.27		602.34
			13030.33		9683.41
Employer's Contributions to:					
CITB levy (0.50% of payroll)			67.94		51.11
Holiday Pay	4.200 wks	261.3	1097.46	196.56	825.55
Public Holidays	1.600 wks	261.3	418.08	196.56	314.50
Retirement cover scheme (Death and Accident cover is provided free)	52.000 wks	1.90	98.80	1.90	98.80
Annual labour cost as defined in section 3			**£ 14712.60**		**£ 10973.36**
Hourly Base Rates as defined in Section 3, Clause 3.2 (annual cost of labour/hours per annum)			**£ 8.17**		**£ 6.09**

Note:

1. Calculated following Definition of Prime Cost of Daywork carried out under a Building Contract, published by the Royal Institution of Chartered Surveyors and the Constuction Confederation.

2. Standard basic rates effective from 25 June 2001.

3. Standard working hours per annum calculated as follows:
   52 weeks @ 39 hours = 2028.00
   Less
   4.20 weeks holiday @ 39 hours = 163.80
   8 days public holidays @ 7.80 hours = 62.40
   226.20
   1801.80

4. All labour costs incurred by the contractor in his capacity as an employer, other than those contained in the hourly base rate, are to be taken into account under Section 6.

5. The above example is for guidance only and does not form part of the Definition; all the basic costs are subject to re-examination according to the time when and in the area where the daywork is executed.

6. N.I. payments are at not-contracted out rates applicable from 5 April 2001.

7. Basic rate and GMB number of weeks =
   52 weeks - 4.20 weeks annual hoilday - 1.600 weeks public holiday = 46.20 weeks.

BUILDING INDUSTRY

DEFINITION OF PRIME COST OF BUILDING WORKS OF A JOBBING OR MAINTENANCE CHARACTER (1980 EDITION)

This Definition of Prime Cost is published by the Royal Institution of Chartered Surveyors and the National Federation of Building Trades Employers for convenience and for use by people who choose to use it. Members of the National Federation of Building Trades Employers are not in any way debarred from defining Prime Cost and rendering their accounts for work carried out on that basis in any way they choose. Building owners are advised to reach agreement with contractors on the Definition of Prime Cost to be used prior to issuing instructions.

SECTION 1 - APPLICATION

1.1. This definition provides a basis for the valuation of work of a jobbing or maintenance character executed under such building contracts as provide for its use.

1.2. It is not applicable in any other circumstances, such as daywork executed under or incidental to a building contract.

SECTION 2 - COMPOSITION OF TOTAL CHARGES

2.1. The prime cost of jobbing work comprises the sum of the following costs:

    (a)   Labour as defined in Section 3.
    (b)   Materials and goods as defined in Section 4.
    (c)   Plant, consumable stores and services as defined in Section 5.
    (d)   Sub-contracts as defined in Section 6.

2.2. Incidental costs, overhead and profit as defined in Section 7 and expressed as percentage adjustments are applicable to each of 2.1 (a)-(d).

SECTION 3 - LABOUR

3.1. Labour costs comprise all payments made to or in respect of all persons directly engaged upon the work, whether on or off the site, except those included in Section 7.

3.2. Such payments are based upon the standard wage rates, emoluments and expenses as laid down for the time being in the rules or decisions of the National Joint Council for the Building Industry and the terms of the Building and Civil Engineering Annual and Public Holiday Agreements applying to the works, or the rules of decisions or agreements of such other body as may relate to the class of labour concerned, at the time when and in the area where the work is executed, together with the Contractor's statutory obligations, including:

    (a)   Guaranteed minimum weekly earnings (e.g. Standard Basic Rate of Wages and Guaranteed Minimum Bonus Payment in the case of NJCBI rules).
    (b)   All other guaranteed minimum payments (unless included in Section 7).
    (c)   Payments in respect of incentive schemes or productivity agreements applicable to the works.
    (d)   Payments in respect of overtime normally worked; or necessitated by the particular circumstances of the work; or as otherwise agreed between the parties.
    (e)   Differential or extra payments in respect of skill, responsibility, discomfort, inconvenience or risk.
    (f)   Tool allowance.
    (g)   Subsistence and periodic allowances.
    (h)   Fares, travelling and lodging allowances.
    (j)   Employer's contributions to annual holiday credits.
    (k)   Employer's contributions to death benefit schemes.
    (l)   Any amounts which may become payable by the Contractor to or in respect of operatives arising from the operation of the rules referred to in 3.2 which are not provided for in 3.2 (a)-(k) or in Section 7.

BUILDING INDUSTRY

(m) Employer's National Insurance contributions and any contribution, levy or tax imposed by statute, payable by the Contractor in his capacity as employer.

Note:

Any payments normally made by the Contractor which are of a similar character to those described in 3.2 (a)-(c) but which are not within the terms of the rules and decisions referred to above are applicable subject to the prior agreement of the parties, as an alternative to 3.2 (a)-(c).

3.3. The wages or salaries of supervisory staff, timekeepers, storekeepers, and the like, employed on or regularly visiting site, where the standard wage rates, etc., are not applicable, are those normally paid by the Contractor together with any incidental payments of a similar character to 3.2 © - (k).

3.4. Where principals are working manually their time is chargeable, in respect of the trades practised, in accordance with 3.2.

SECTION 4 - MATERIALS AND GOODS

4.1. The prime cost of materials and goods obtained by the Contractor from stockists or manufacturers is the invoice cost after deduction of all trade discounts but including cash discounts not exceeding 5 per cent, and includes the cost of delivery to site.

4.2. The prime cost of materials and goods supplied from the Contractor's stock is based upon the current market prices plus any appropriate handling charges.

4.3. The prime cost under 4.1 and 4.2 also includes any costs of:

(a) non-returnable crates or other packaging.
(b) returning crates and other packaging less any credit obtainable.

4.4. Any Value Added Tax which is treated, or is capable of being treated, as input tax (as defined in the Finance Act, 1972 or any re-enactment thereof) by the Contractor is excluded.

SECTION 5 - PLANT, CONSUMABLE STORES AND SERVICES

5.1. The prime cost of plant and consumable stores as listed below is the cost at hire rates agreed between the parties or in the absence of prior agreement at rates not exceeding those normally applied in the locality at the time when the works are carried out, or on a use and waste basis where applicable:

(a) Machinery in workshops.
(b) Mechanical plant and power-operated tools.
(c) Scaffolding and scaffold boards.
(d) Non-mechanical plant excluding hand tools.
(e) Transport including collection and disposal of rubbish.
(f) Tarpaulins and dust sheets.
(g) Temporary roadways, shoring, planking and strutting, hoarding, centering, formwork, temporary fans, partitions or the like.
(h) Fuel and consumable stores for plant and power-operated tools unless included in 5.1 (a), (b), (d) or (e) above.
(j) Fuel and equipment for drying out the works and fuel for testing mechanical services.

BUILDING INDUSTRY

5.2. The prime cost also includes the net cost incurred by the Contractor of the following services, excluding any such cost included under Sections 3, 4 or 7:

   (a) Charges for temporary water supply including the use of temporary plumbing and storage.
   (b) Charges for temporary electricity or other power and lighting including the use of temporary installations.
   (c) Charges arising from work carried out by local authorities or public undertakings.
   (d) Fees, royalties and similar charges.
   (e) Testing of materials.
   (f) The use of temporary buildings including rates and telephone and including heating and lighting not charged under (b) above.
   (g) The use of canteens, sanitary accommodation, protective clothing and other provision for the welfare of persons engaged in the work in accordance with the current Working Rule Agreement and any Act of Parliament, statutory instrument, rule, order, regulation or bye-law.
   (h) The provision of safety measures necessary to comply with any Act of Parliament.
   (j) Premiums or charges for any performance bonds or insurances which are required by the Building Owner and which are not referred to elsewhere in this Definition.

SECTION 6 - SUB-CONTRACTS

6.1. The prime cost of work executed by sub-contractors, whether nominated by the Building Owner or appointed by the Contractor, is the amount which is due from the Contractor to the sub-contractors in accordance with the terms of the sub-contracts after deduction of all discounts except any cash discount offered by any sub-contractor to the Contractor not exceeding 2.5%.

SECTION 7 - INCIDENTAL COSTS, OVERHEADS AND PROFIT

7.1. The percentage adjustments provided in the building contract, which are applicable to each of the totals of Sections 3-6, provide for the following:

   (a) Head Office charges.
   (b) Off-site staff including supervisory and other administrative staff in the Contractor's workshops and yard.
   (c) Payments in respect of public holidays.
   (d) Payments in respect of apprentices' study time.
   (e) Sick pay or insurance in respect thereof.
   (f) Third party employer's liability insurance.
   (g) Liability in respect of redundancy payments made to employees.
   (h) Use, repair and sharpening of non-mechanical hand tools.
   (j) Any variations to basic rates required by the Contractor in cases where the building contract provides for the use of a specified schedule of basic plant charges (to the extent that no other provision is made for such variation).
   (k) All other liabilities and obligations whatsoever not specifically referred to in this Section nor chargeable under any other section.
   (l) Profit.

BUILDING INDUSTRY

SPECIMEN ACCOUNT FORMAT

If this Definition of Prime Cost is followed the Contractor's account could be in the following format:

                   £

Labour (as defined in Section 3)
 Add ____ % (see Section 7)

Materials and goods (as defined in Section 4)
 Add ____ % (see Section 7)

Plant, consumable stores and services
(as defined in Section 5)
 Add ____ % (see Section 7)

Sub-contracts (as defined in Section 6)
 Add ____ % (see Section 7)   _____

              £ _____

VAT to be added if applicable.

SCHEDULE OF BASIC PLANT CHARGES (1st MAY 2001 ISSUE)

This Schedule is published by the Royal Institution of Chartered Surveyors and is for use in connection with Dayworks under a Building Contract.

EXPLANATORY NOTES

1. The rates in the Schedule are intended to apply solely to daywork carried out under and incidental to a Building Contract. They are NOT intended to apply to:
   (i) Jobbing or any other work carried out as a main or separate contract; or
   (ii) Work carried out after the date of commencement of the Defects Liability Period.

2. The rates apply only to plant and machinery already on site, whether hired or owned by the Contractor.

3. The rates, unless otherwise stated, include the cost of fuel and power of every description, lubricating oils, grease, maintenance, sharpening of tools, replacement of spare parts, all consumable stores and for licences and insurances applicable to items of plant.

4. The rates, unless otherwise stated, do not include the costs of drivers and attendants.

5. The rates are base costs and may be subject to the overall adjustment for price movement, overheads and profit, quoted by the Contractor prior to the placing of the Contract.

6. The rates should be applied to the time during which the plant is actually engaged in daywork.

7. Whether or not plant is chargeable on daywork depends on the daywork agreement in use and the inclusion of an item of plant in this schedule does not necessarily indicate that the item is chargeable.

8. Rates for plant not included in the Schedule or which is not already on site and is specifically provided or hired for daywork shall be settled at prices which are reasonably related to the rates in the Schedule having regard to any overall adjustment quoted by the Contractor in the Conditions of Contract.

BUILDING INDUSTRY

MECHANICAL PLANT AND TOOLS

Item of plant	Size/Rating	Unit	Rate per hour £
PUMPS			
Mobile Pumps			
Including pump hoses, valves and strainers etc.			
Diaphragm	50 mm diameter	Each	0.87
Diaphragm	76 mm diameter	Each	1.29
Submersible	50 mm diameter	Each	1.18
Induced Flow	50 mm diameter	Each	1.54
Induced Flow	76 mm diameter	Each	2.05
Centifugal, self priming	50 mm diameter	Each	1.96
Centifugal, self priming	102 mm diameter	Each	2.52
Centifugal, self priming	152 mm diameter	Each	3.87
SCAFFOLDING, SHORING, FENCING			
Complete Scaffolding			
Mobile working towers, single width	1.80 m x 0.80 m base x 7.00 m high	Each	2.00
Mobile working towers, single width	1.80 m x 0.80 m base x 9.00 m high	Each	2.80
Mobile working towers, double width	1.80 m x 1.40 m base x 7.00 m high	Each	2.15
Mobile working towers, double width	1.80 m x 1.40 m base x 15.00 m high	Each	5.10
Chimney scaffold, single unit		Each	1.79
Chimney scaffold, twin unit		Each	2.05
Chimney scaffold, four unit		Each	3.59
Trestles			
Trestle, adjustable	Any height	Pair	0.10
Trestle, painters	1.80 m high	Pair	0.21
Trestle, painters	2.40 m high	Pair	0.26
Shoring, Planking and Strutting			
'Acrow' adjustable prop	Sizes up to 4.90 m (open)	Each	0.10
'Strong boy' support attachment		Each	0.15
Adjustable trench struts	Sizes up to 1.67 m (open)	Each	0.10
Trench sheet		Metre	0.01
Backhoe trench box		Each	1.00
Temporary Fencing			
Including block and coupler			
Site fencing steel grid panel	3.50 m x 2.00 m	Each	0.08
Anti-climb site steel grid fence panel	3.50 m x 2.00 m	Each	0.08
LIFTING APPLIANCES AND CONVEYORS			
Cranes			
<u>Mobile cranes</u>			
Rates are inclusive of drivers			
Lorry mounted, telescopic jib			
Two wheel drive	6 tonnes	Each	24.40
Two wheel drive	7 tonnes	Each	25.00
Two wheel drive	8 tonnes	Each	25.62
Two wheel drive	10 tonnes	Each	26.90
Two wheel drive	12 tonnes	Each	28.25
Two wheel drive	15 tonnes	Each	29.66
Two wheel drive	18 tonnes	Each	31.14
Two wheel drive	20 tonnes	Each	32.70
Two wheel drive	25 tonnes	Each	34.33

BUILDING INDUSTRY

Item of plant	Size/Rating		Unit	Rate per hour £

LIFTING APPLIANCES AND CONVEYORS - cont'd
Cranes - cont'd
Mobile cranes - cont'd
Rates are inclusive of drivers
Lorry mounted, telescopic jib - cont'd

Item of plant	Size/Rating		Unit	Rate per hour £
Four wheel drive	10 tonnes		Each	27.44
Four wheel drive	12 tonnes		Each	28.81
Four wheel drive	15 tonnes		Each	30.25
Four wheel drive	20 tonnes		Each	33.35
Four wheel drive	25 tonnes		Each	35.19
Four wheel drive	30 tonnes		Each	37.12
Four wheel drive	45 tonnes		Each	39.16
Four wheel drive	50 tonnes		Each	41.32

Track-mounted tower crane
Rates are inclusive of drivers
Note: Capacity equals maximum lift in tonnes times maximum radius at which it can be lifted

	Capacity (Metre/tonnes) up to	Height under hook above ground (m) up to	Unit	Rate per hour £
Tower crane	10	17	Each	7.99
Tower crane	15	17	Each	8.59
Tower crane	20	18	Each	9.18
Tower crane	25	20	Each	11.56
Tower crane	30	22	Each	13.78
Tower crane	40	22	Each	18.09
Tower crane	50	22	Each	22.20
Tower crane	60	22	Each	24.32
Tower crane	70	22	Each	23.00
Tower crane	80	22	Each	25.91
Tower crane	110	22	Each	26.45
Tower crane	125	30	Each	29.38
Tower crane	150	30	Each	32.35

Static tower cranes
Rates inclusive of driver
To be charged at 90% of the above rates for tower mounted tower cranes

Crane Equipment

Item	Size	Unit	Rate
Mucking tipping skip	up to 0.25 m³	Each	0.56
Muck tipping skip	0.50 m³	Each	0.67
Muck tipping skip	0.75 m³	Each	0.82
Muck tipping skip	1.00 m³	Each	1.03
Muck tipping skip	1.50 m³	Each	1.18
Muck tipping skip	2.00 m³	Each	1.38
Mortar skips	up to 0.38 m³	Each	0.41
Boat skips	1.00 m³	Each	1.08
Boat skips	1.50 m³	Each	1.33
Boat skips	2.00 m³	Each	1.59
Concrete skips, hand levered	0.50 m³	Each	1.00
Concrete skips, hand levered	0.75 m³	Each	1.10
Concrete skips, hand levered	1.00 m³	Each	1.25
Concrete skips, hand levered	1.50 m³	Each	1.50
Concrete skips, hand levered	2.00 m³	Each	1.65

Item of plant	Size/Rating		Unit	Rate per hour £

LIFTING APPLIANCES AND CONVEYORS - cont'd

Static tower cranes - cont'd
Rates inclusive of driver - cont'd
To be charged at 90% of the above rates for tower mounted tower cranes - cont'd

Crane Equipment - cont'd

Item of plant	Size/Rating		Unit	Rate per hour £
Concrete skips, geared	0.50 m³		Each	1.30
Concrete skips, geared	0.75 m³		Each	1.40
Concrete skips, geared	1.00 m³		Each	1.55
Concrete skips, geared	1.50 m³		Each	1.80
Concrete skips, geared	2.00 m³		Each	2.05
Hoists				
Scaffold hoists	200 kg		Each	1.92
Rack and pinion (goods only)	500 kg		Each	3.31
Rack and pinion (goods only)	1100 kg		Each	4.28
Rack and pinion goods and passenger	15 person, 1200 kg		Each	5.62
Wheelbarrow chain sling			Each	0.31
Conveyors				
Belt conveyors				
Conveyor	7.50 m long x 400 mm wide		Each	6.41
Miniveyor, control box and loading hopper	3.00 m unit		Each	3.59
Other Conveying Equipment				
Wheelbarrow			Each	0.21
Hydraulic superlift			Each	2.95
Pavac slab lifter			Each	1.03
Hand pad and hose attachment			Each	0.26
Lifting Trucks				
Fork lift, two wheel drive	Payload	Maximum Lift		
Fork lift, two wheel drive	1100 kg	up to 3.00 m	Each	4.87
Fork lift, two wheel drive	2540 kg	up to 3.70 m	Each	5.12
Fork lift, four wheel drive	1524 kg	up to 6.00 m	Each	6.04
Fork lift, four wheel drive	2600 kg	up to 5.40 m	Each	7.69
Lifting Platforms				
Hydraulic platform (Cherry picker)	7.50 m		Each	4.23
Hydraulic platform (Cherry picker)	13.00 m		Each	9.23
Scissors lift	7.80 m		Each	7.56
Telescopic handlers	7.00 m, 2 tonne		Each	7.18
Telescopic handlers	13.00 m, 3 tonne		Each	8.72
Lifting and Jacking Gear				
Pipe winch including gantry	1.00 tonne		Sets	1.92
Pipe winch including gantry	3.00 tonnes		Sets	3.21
Chain block	1.00 tonne		Each	0.45
Chain block	2.00 tonnes		Each	0.71
Chain block	5.00 tonnes		Each	1.22
Pull lift (Tirfor winch)	1.00 tonne		Each	0.64
Pull lift (Tirfor winch)	1.60 tonnes		Each	0.90
Pull lift (Tirfor winch)	3.20 tonnes		Each	1.15
Brother or chain slings, two legs	not exceeding 4.20 tonnes		Set	0.35
Brother or chain slings, two legs	not exceeding 7.50 tonnes		Set	0.45
Brother or chain slings, four legs	not exceeding 3.10 tonnes		Set	0.41
Brother or chain slings, four legs	not exceeding 11.20 tonnes		Set	1.28

BUILDING INDUSTRY

Item of plant	Size/Rating	Unit	Rate per hour £
CONSTRUCTION VEHICLES			
Lorries			
Plated lorries			
Rates are inclusive of driver			
Platform lorries	7.50 tonnes	Each	19.00
Platform lorries	17.00 tonnes	Each	21.00
Platform lorries	24.00 tonnes	Each	26.00
Platform lorries with winch and skids	7.50 tonnes	Each	21.40
Platform lorries with crane	17.00 tonnes	Each	27.50
Platform lorries with crane	24.00 tonnes	Each	32.10
Tipper Lorries			
Rates are inclusive of driver			
Tipper lorries	15.00/17.00 tonnes	Each	19.50
Tipper lorries	24.00 tonnes	Each	21.40
Tipper lorries	30.00 tonnes	Each	27.10
Dumpers			
Site use only (excluding tax, insurance and extra cost of DEFV etc. when operating on highway)	Makers capacity		
Two wheel drive	0.80 tonnes	Each	1.20
Two wheel drive	1.00 tonne	Each	1.30
Two wheel drive	1.20 tonnes	Each	1.60
Four wheel drive	2.00 tonnes	Each	2.50
Four wheel drive	3.00 tonnes	Each	3.00
Four wheel drive	4.00 tonnes	Each	3.50
Four wheel drive	5.00 tonnes	Each	4.00
Four wheel drive	6.00 tonnes	Each	4.50
Dumper Trucks			
Rates are inclusive of drivers			
Dumper trucks	10.00/13.00 tonnes	Each	20.00
Dumper trucks	18.00/20.00 tonnes	Each	20.40
Dumper trucks	22.00/25.00 tonnes	Each	26.30
Dumper trucks	35.00/40.00 tonnes	Each	36.60
Tractors			
<u>Agricultural Type</u>			
Wheeled, rubber-clad tyred			
Light	48 h.p.	Each	4.65
Heavy	65 h.p.	Each	5.15
<u>Crawler Tractors</u>			
With bull or angle dozer	80/90 h.p.	Each	21.40
With bull or angle dozer	115/130 h.p.	Each	25.10
With bull or angle dozer	130/150 h.p.	Each	26.00
With bull or angle dozer	155/175 h.p.	Each	27.74
With bull or angle dozer	210/230 h.p.	Each	28.00
With bull or angle dozer	300/340 h.p.	Each	31.10
With bull or angle dozer	400/440 h.p.	Each	46.90
With loading shovel	0.80 m^3	Each	25.00
With loading shovel	1.00 m^3	Each	28.00
With loading shovel	1.20 m^3	Each	32.00
With loading shovel	1.40 m^3	Each	36.00
With loading shovel	1.80 m^3	Each	45.00
Light vans			
Ford Escort or the like		Each	4.74
Ford Transit or the like	1.00 tonnes	Each	6.79
Luton Box Van or the like	1.80 tonnes	Each	8.33

BUILDING INDUSTRY

Item of plant	Size/Rating	Unit	Rate per hour £
CONSTRUCTION VEHICLES- cont'd			
Water/Fuel Storage			
Mobile water container	110 litres	Each	0.28
Water bowser	1100 litres	Each	0.55
Water bowser	3000 litres	Each	0.74
Mobile fuel container	110 litres	Each	0.28
Fuel bowser	1100 litres	Each	0.65
Fuel bowser	3000 litres	Each	1.02
EXCAVATORS AND LOADERS			
Excavators			
Wheeled, hydraulic	7.00/10.00 tonnes	Each	12.00
Wheeled, hydraulic	11.00/13.00 tonnes	Each	12.70
Wheeled, hydraulic	15.00/16.00 tonnes	Each	14.80
Wheeled, hydraulic	17.00/18.00 tonnes	Each	16.70
Wheeled, hydraulic	20.00/23.00 tonnes	Each	16.70
Crawler, hydraulic	12.00/14.00 tonnes	Each	12.00
Crawler, hydrualic	15.00/17.50 tonnes	Each	14.00
Crawler, hydraulic	20.00/23.00 tonnes	Each	16.00
Crawler, hydraulic	25.00/30.00 tonnes	Each	21.00
Crawler, hydraulic	30.00/35.00 tonnes	Each	30.00
Mini excavators	1000/1500 kg	Each	4.50
Mini excavators	2150/2400 kg	Each	5.50
Mini excavators	2700/3500 kg	Each	6.50
Mini excavators	3500/4500 kg	Each	8.50
Mini excavators	4500/6000 kg	Each	9.50
Loaders			
Wheeled skip loader		Each	4.50
Shovel loaders, four wheel drive	1.60 m³	Each	12.00
Shovel loaders, four wheel drive	2.40 m³	Each	19.00
Shovel loaders, four wheel drive	3.60 m³	Each	22.00
Shovel loaders, four wheel drive	4.40 m³	Each	23.00
Shovel loaders, crawlers	0.80 m³	Each	11.00
Shovel loaders, crawlers	1.20 m³	Each	14.00
Shovel loaders, crawlers	1.60 m³	Each	16.00
Shovel loaders, crawlers	2.00 m³	Each	17.00
Skid steer loaders wheeled	300/400 kg payload	Each	6.00
Exacavator Loaders			
Wheeled tractor type with back-hoe excavator			
Four wheel drive	2.50/3.50 tonnes	Each	7.00
Four wheel drive, 2 wheel steer	7.00/8.00 tonnes	Each	9.00
Four wheel drive, 4 wheel steer	7.00/8.00 tonnes	Each	10.00
Crawler, hydraulic	12 tonnes	Each	20.00
Crawler, hydraulic	20 tonnes	Each	16.00
Crawler, hydrualic	30 tonnes	Each	35.00
Crawler, hydraulic	40 tonnes	Each	38.00
Attachments			
Breakers for excavators		Each	7.50
Breakers for mini excavators		Each	3.60
Breakers for back-hoe exacavtor/loaders		Each	6.00

BUILDING INDUSTRY

Item of plant	Size/Rating	Unit	Rate per hour £
COMPACTION EQUIPMENT			
Rollers			
Vibrating roller	368 - 420 kg	Each	1.68
Single roller	533 kg	Each	1.92
Single roller	750 kg	Each	2.41
Twin roller	698 kg	Each	1.93
Twin roller	851 kg	Each	2.41
Twin roller with seat end steering wheel	1067 kg	Each	3.03
Twin roller with seat end steering wheel	1397 kg	Each	3.17
Pavement rollers	3.00 - 4.00 tonnes dead weight	Each	3.18
Pavement rollers	4.00 - 6.00 tonnes	Each	4.13
Pavement rollers	6.00 - 10.00 tonnes	Each	4.84
Rammers			
Tamper rammer 2 stroke-petrol	225 mm - 275 mm	Each	1.59
Soil Compactors			
Plate compactor	375 mm - 400 mm	Each	1.20
Plate compactor rubber pad	375 mm - 1400 mm	Each	0.33
Plate compactor reversible plate- petrol	400 mm	Each	2.20
CONCRETE EQUIPMENT			
Concrete/Mortar Mixers			
Open drum without hopper	0.09/0.06 m³	Each	0.62
Open drum without hopper	0.12/0.09 m³	Each	0.68
Open drum without hopper	0.15/0.10 m³	Each	0.72
Open drum with hopper	0.20/0.15 m³	Each	0.80
Concrete/Mortar Transport Equipment			
Concrete pump including hose, valve and couplers			
Lorry mounted concrete pump	23 m maximum distance	Each	36.00
Lorry mounted concrete pump	50 m maximum distance	Each	46.00
Concrete Equipment			
Vibrator, poker, petrol type	up to 75 mm diameter	Each	1.62
Air vibrator (excluding compressor and hose)	up to 75 mm diameter	Each	0.79
Extra poker heads	25/36/60 mm diameter	Each	0.77
Vibrating screed unit with beam	5 .00 m	Each	1.77
Vibrating screed unit with adjustable beam	3.00 m - 5.00 m	Each	2.18
Power float	725 mm - 900 mm	Each	1.72
Power grouter		Each	0.92
TESTING EQUIPMENT			
Pipe Testing Equipment			
Pressure testing pump, electric		Sets	1.87
Pipe pressure testing equipment, hydraulic		Sets	2.46
Pressure test pump		Sets	0.64

BUILDING INDUSTRY

Item of plant	Size/Rating	Unit	Rate per hour £
SITE ACCOMMODATION AND TEMPORARY SERVICES			
Heating Equipment			
Space heaters - propane	80,000 Btu/hr	Each	0.77
Space heaters - propane/electric	125,000 Btu/hr	Each	1.56
Space heaters - propane/electric	250,000 Btu/hr	Each	1.79
Space heaters - propane	125,000 Btu/hr	Each	1.33
Space heaters - propane	260,000 Btu/hr	Each	1.64
Cabinet headers		Each	0.41
Cabinet heater catalytic		Each	0.46
Electric halogen heaters		Each	1.28
Ceramic heaters	3kW	Each	0.79
Fan heaters	3kW	Each	0.41
Cooling fan		Each	1.15
Mobile cooling unit - small		Each	1.38
Mobile cooling unit - large		Each	1.54
Air conditioning unit		Each	2.62
Site Lighting and Equipment			
Tripod floodlight	500W	Each	0.36
Tripod floodlight	1000W	Each	0.34
Towable floodlight	4 x 1000W	Each	2.00
Hand held floodlight	500W	Each	0.22
Recahargeable light		Each	0.62
Inspection light		Each	0.15
Plasterers light		Each	0.56
Lighting mast		Each	0.92
Festoon light string	33.00 m	Each	0.31
Site Electrical Equipment			
Extension leads	240V/14.00 m	Each	0.20
Extension leads	110V/14.00 m	Each	0.20
Cable reel	25.00 m 110V/240V	Each	0.28
Cable reel	50.00 m 110V/240V	Each	0.33
4 way junction box	110V	Each	0.17
Power Generating Units			
Generator - petrol	2kVA	Each	1.08
Generator - silenced petrol	2kVA	Each	1.54
Generator - petrol	3kVA	Each	1.38
Generator - diesel	5kVA	Each	1.92
Generator - silenced diesel	8kVA	Each	3.59
Generator - silenced diesel	15kVA	Each	7.69
Tail adaptor	240V	Each	0.20
Transformers			
Transformer	3kVA	Each	0.36
Transformer	5kVA	Each	0.51
Transformer	7.50 kVA	Each	0.82
Transformer	10 kVA	Each	0.87
Rubbish Collection and Disposal Equipment			
Rubbish Chutes			
Standard plastic module	1.00 m section	Each	0.18
Steel liner insert		Each	0.26
Steel top hopper		Each	0.20
Plastic side entry hopper		Each	0.20
Plastic side entry hopper liner		Each	0.20
Dust Extraction Plant			
Dust extraction unit, light duty		Each	1.03
Dust extraction unit, heavy duty		Each	1.64

BUILDING INDUSTRY

Item of plant	Size/Rating	Unit	Rate per hour £
SITE EQUIPMENT			
Welding Equipment			
Arc-(Electric) Complete With Leads			
Welder generator - petrol	200 amp	Each	2.26
Welder generator - diesel	300/350 amp	Each	3.33
Welder generator - diesel	400 amp	Each	4.74
Extra welding lead sets		Each	0.29
Gas-Oxy Welder			
Welding and cutting set (including oxygen and acetylene, excluding underwater equipment and thermic boring)			
Small		Each	1.41
Large		Each	2.00
Mig welder		Each	1.00
Fume extractor		Each	0.92
Road Works Equipment			
Traffic lights, mains/generator	2-way	Set	4.01
Traffic lights, mains/generator	3-way	Set	7.92
Traffic lights, mains/generator	4-way	Set	9.81
Traffic lights, mains/generator - trailer mounted	2-way	Set	3.98
Flashing light		Each	0.20
Road safety cone	450 mm	10	0.26
Safety cone	750 mm	10	0.38
Safety barrier plank	1.25 m	Each	0.03
Saftey barrier plank	2.00 m	Each	0.04
Road sign		Each	0.26
DPC Equipment			
Damp proofing injection machine		Each	1.49
Cleaning Equipment			
Vacuum cleaner (industrial wet) single motor		Each	0.62
Vacuum cleaner (industrial wet) twin motor		Each	1.23
Vacuum cleaner (industrial wet) triple motor		Each	1.44
Vacuum cleaner (industrial wet) back pack		Each	0.97
Pressure washer, light duty, electric	1450 PSI	Each	0.97
Pressure washer, heavy duty, diesel	2500 PSI	Each	2.69
Cold pressure washer, electric		Each	1.79
Hot pressure washer, petrol		Each	2.92
Cold pressure washer, petrol		Each	2.00
Sandblast attachment to last washer		Each	0.54
Drain cleaning attachment to last washer		Each	0.31
Surface Preparation Equipment			
Rotavators	5 h.p.	Each	1.67
Scabbler, up to three heads		Each	1.15
Scabbler, pole		Each	1.50
Scabbler, multi-headed floor		Each	4.00
Floor preparation machine		Each	2.82

BUILDING INDUSTRY

Item of plant	Size/Rating	Unit	Rate per hour £
SITE EQUIPMENT - cont'd			
Compressors and Equipment			
<u>Portable Compressors</u>			
Compressors - electric	0.23 m³/min	Each	1.59
Compressors - petrol	0.28 m³/min	Each	1.74
Compressors - petrol	0.71 m³/min	Each	2.00
Compressors - diesel	up to 2.83 m³/min	Each	1.24
Compressors - diesel	up to 3.68 m³/min	Each	1.49
Compressors - diesel	up to 4.25 m³/min	Each	1.60
Compressors - diesel	up to 4.81 m³/min	Each	1.92
Compressors - diesel	up to 7.64 m³/min	Each	3.08
Compressors - diesel	up to 11.32 m³/min	Each	4.23
Compressors - diesel	up to 18.40 m³/min	Each	5.73
<u>Mobile Compressors</u>			
Lorry mounted compressors *(machine plus lorry only)*	2.86 - 4.24 m³/min	Each	12.50
Tractor mounted compressors *(machine plus rubber tyred tractor)*	2.86 - 3.40 m³/min	Each	13.50
<u>Accessories (Pneumatic Tools)</u>			
(with and including up to 15.00 m of air hose)			
Demolition pick		Each	1.03
Breakers (with six steels) light	up to 150 kg	Each	0.79
Breakers (with six steels) medium	295 kg	Each	1.08
Breakers (with six steels) heavy	386 kg	Each	1.44
Rock drill (for use with compressor) hand held		Each	0.90
Additional hoses	15.00 m	Each	0.16
Muffler, tool silencer		Each	0.14
Breakers			
Demolition hammer drill, heavy duty, electric		Each	1.00
Road breaker, electric		Each	1.65
Road breaker, 2 stroke, petrol		Each	2.05
Hydraulic breaker unit, light duty, petrol		Each	2.05
Hydraulic breaker unit, heavy duty, petrol		Each	2.60
Hydraulic breaker unit, heavy duty, diesel		Each	2.95
Quarrying and Tooling Equipment			
Block and stone splitter, hydraulic	600 mm x 600 mm	Each	1.35
Block and stone splitter, manual		Each	1.10
Steel Reinforcement Equipment			
Bar bending machine - manual	up to 13 mm diameter rods	Each	0.90
Bar bending machine - manual	up to 20 mm diameter rods	Each	1.28
Bar shearing machine - electric	up to 38 mm diameter rods	Each	2.82
Bar shearing machine - electric	up to 40 mm diameter rods	Each	3.85
Bar cropper machine - electric	up to 13 mm diameter rods	Each	1.54
Bar cropper machine - electric	up to 20 mm diameter rods	Each	2.05
Bar cropper machine - electric	up to 40 mm diameter rods	Each	2.82
Bar cropper machine - 3 phase	up to 40 mm diameter rods	Each	3.85
Dehumidifiers			
110/240v Water	68 litres extraction per 24 hours	Each	1.28
110/240v Water	90 litres extraction per 24 hours	Each	1.85

BUILDING INDUSTRY

Item of plant	Size/Rating	Unit	Rate per hour £
SMALL TOOLS			
Saws			
Masonry saw bench	350 mm - 500 mm diameter	Each	2.80
Floor saw	350 mm diameter, 125 mm max. cut	Each	1.90
Floor saw	450 mm diameter, 150 mm max. cut	Each	2.60
Floor saw, reversible	Max. Cut 300 mm	Each	13.00
Chop/cut off saw, electric	350 mm diameter	Each	1.33
Circular saw, electric	230 mm diameter	Each	0.60
Tyrannosaw		Each	1.20
Reciprocating saw		Each	0.60
Door trimmer		Each	0.90
Chainsaw, petrol	500 mm	Each	2.13
Full chainsaw safety kit		Each	0.50
Worktop jig		Each	0.60
PipeWork Equipment			
Pipe bender	15 mm - 22 mm	Each	0.33
Pipe bender, hydraulic	50 mm	Each	0.60
Pipe bender, electric	50 mm - 150 mm diameter	Each	1.35
Pipe cutter, hydraulic		Each	1.84
Tripod pipe vice		Set	0.40
Ratchet threader	12 mm - 32 mm	Each	0.55
Pipe threading machine, electric	12 mm - 75 mm	Each	2.40
Pipe threading machine, electric	12 mm - 100 mm	Each	3.00
Impact wrench, electric		Each	0.54
Impact wrench, two stroke, petrol		Each	4.49
Impact wrench, heavy duty, electric		Each	1.13
Plumber's furnace, calor gas or similar		Each	2.16
Hand-held Drills and Equipment			
Impact or hammer drill	up to 25 mm diameter	Each	0.50
Impact or hammer drill	35 mm diameter	Each	0.90
Angle head drills		Each	0.70
Stirrer, mixed drills		Each	0.70
Paint, Insulation Application Equipment			
Airless spray unit		Each	4.20
Portaspray unit		Each	1.65
HVPL turbine spray unit		Each	1.65
Compressor and spray gun		Each	2.20
Other Handtools			
Screwing machine	13 mm - 50 mm diameter	Each	0.77
Screwing machine	25 mm - 100 mm diameter	Each	1.57
Staple gun		Each	0.33
Air nail gun	110V	Each	3.33
Cartridge hammer		Each	1.00
Tongue and groove nailer complete with mallet		Each	0.93
Chasing machine	152 mm	Each	1.72
Chasing machine	76 mm - 203 mm	Each	5.99
Floor grinder		Each	3.00
Floor plane		Each	3.67
Diamond concrete planer		Each	2.05
Autofeed screwdriver, electric		Each	1.13
Laminate trimmer		Each	0.64
Biscuit jointer		Each	0.87
Random orbital sander		Each	0.73
Floor sander		Each	1.33

BUILDING INDUSTRY

Item of plant	Size/Rating	Unit	Rate per hour £
SMALL TOOLS - cont'd			
Other Handtools - cont'd			
Palm, delta, flap or belt sander		Each	0.38
Saw cutter, two strokes, petrol	300 mm	Each	1.26
Grinder, angle or cutter	up to 225 mm	Each	0.60
Grinder, angle or cutter	300 mm	Each	1.10
Mortar raking tool attachment		Each	0.15
Floor/polish scrubber	325 mm	Each	1.03
Floor tile stripper		Each	1.74
Wallpaper stripper, electric		Each	0.56
Electric scraper		Each	0.51
Hot air paint stripper		Each	0.38
Electric diamond tile cutter	All sizes	Each	1.38
Hasnd tile cutter		Each	0.36
Electric needle gun		Each	1.08
Needle chipping gun		Each	0.72
Redestrian floor sweeper	1.2 m wide	Each	0.87

Keep your figures up to date, free of charge
Download updates from the web: www.pricebooks.co.uk

This section, and most of the other information in this Price Book, is brought up to date every three months with the Price Book Updates, until the next annual edition. The updates are available free to all Price Book purchasers.

To ensure you receive your free copies, either complete the reply card from the centre of the book and return it to us or register via the website www.pricebooks.co.uk

PART II

Rates of Wages

BUILDING INDUSTRY - ENGLAND, WALES AND SCOTLAND

The Building and Civil Engineering Joint Negotiating Committee have agreed a new three year agreement on pay and conditions for building and civil engineering operatives.

The Working Rule Agreement includes a pay structure with a general operative and additional skilled rates of pay as well as craft rate. Plus rates and additional payments will be consolidated into basic pay to provide the following rates (for a normal 39 hour week) which will come into effect from the following dates:

Effective from 25 June 2001

The following basic rates of pay will apply:

	£
General operative	196.56
Skill Rate 4	211.77
Skill Rate 3	224.64
Skill Rate 2	239.85
Skill Rate 1	248.82
Craft Rate	261.30

Effective from 24 June 2002

The following basic rates of pay will apply:

	£
General operative	214.11
Skill Rate 4	230.49
Skill Rate 3	244.53
Skill Rate 2	261.30
Skill Rate 1	271.05
Craft Rate	284.70

BUILDING INDUSTRY - ENGLAND, WALES AND SCOTLAND - cont'd

Drivers of excavators with a rated bucket capacity over 7.65 cu metres, and mobile cranes over 120 tonnes capacity will receive an additional payment.

Holidays with Pay and Benefits Schemes

The Building and Civil Engineering benefits scheme has unveiled a new holiday pay plan following the introduction of the Working Time Directive. From 2 August 1999 there are no fixed holiday credits, instead employers will calculate appropriate sums to fund operatives' holiday pay entitlement and make regular monthly payments into the B&CE scheme.

For full details contact B&CE on 0345 414142.

Employers contribution towards retirement benefit is paid at £1.90 per week, effective from 10 January 2000.

Death and accident cover is provided free.

Young Labourers

Effective from 25 June 2001

The rates of wages for young labourers shall be the following proportions of the labourers' rates:

> At 16 years of age 50%
> At 17 years of age 70%
> At 18 years of age 100%

Apprentices/Trainees

The Construction Apprenticeship Scheme (CAS) operates throughout Great Britain from 1 August 1998 it is open to all young people from the age of 16 years. For further information telephone CAS helpline - 01485 578 333.

Apprentice rates - effective from 29 November 1999

Please note that the rates below are for guidance only:

	Per Week (39 hours) £	Per Hour £
Year 1 (first six months)	74.88	1.92
Year 1 (second six months)	95.16	2.44
Year 2	129.09	3.31
Year 3	161.85	4.15

BUILDING AND ALLIED TRADES JOINT INDUSTRIAL COUNCIL

Authorised rates of wages in the building industry in England and Wales agreed by the Building and Allied Trades Joint Industrial Council.

Effective from 11 June 2001

Subject to the conditions prescribed in the Working Rule Agreement the standard weekly rates of wages shall be as follows:

	£
Craft operative	261.30
Adult general operative	211.38

ROAD HAULAGE WORKERS EMPLOYED IN THE BUILDING INDUSTRY

Authorised rates of pay for road haulage workers in the building industry recommended by the Builders Employers Confederation.

Effective from 25 June 2001

Employers	Operatives
The Building Employers Confederation 82 New Cavendish Street London W1M BAD	The Transport and General Workers Union Transport House Smith Squares Westminster London SM1P 3JB
Tel: 0207 580 5588	Tel: 0207 828 7788

Lorry Drivers

The rate for Lorry drivers required to hold a LGV licence of either or both classes C & E and Lorry drivers employed whole time to hold Class C & E Licence.

NOTE: This clause does not apply in Scotland.

	£
Lorry Drivers (Skill Rate 1)	248.82

PLUMBING AND MECHANICAL ENGINEERING SERVICES INDUSTRY

Authorised rates of wages agreed by the Joint Industry Board for the Plumbing and Mechanical Engineering Services Industry in England and Wales

Effective from 23 August 1999 and 21 August 2000

The Joint Industry Board for Plumbing and Mechanical Engineering Services in England and Wales
Brook House
Brook Street
St Neots
Huntingdon
Cambridgeshire
PE19 2HW

Tel: 01480 476 925
Fax: 01480 403 081
E-mail: info@jib-pmes.org.uk

	Rate per hour 23/08/1999 £	Rate per hour 21/08/2000 £
Operatives		
Technical plumber and Service technician (Gas)	8.00	8.40
Advanced plumber and Service engineer (Gas)	7.20	7.56
Trained plumber and Service fitter (Gas)	6.17	6.48
Apprentice Plumbers and Apprentice Service fitters		
1st year of training	3.00	3.15
2nd year of training	3.43	3.60
3rd year of training	3.87	4.06
3rd year of training with NVQ Level 2	4.70	4.94
4th year of training	4.76	5.00
4th year of training with NVQ Level 2	5.42	5.69
4th year of training with NVQ Level 3	5.98	6.28
Adult Trainees		
1st - 6 months of employment	4.83	5.07
2nd - 6 months of employment	5.16	5.42
3rd - 6 months of employment	5.39	5.66

NEW FROM SPON PRESS

Inclusive Design
Designing and Developing Accessible Environments

Rob Imrie and **Peter Hall**, both from the Royal Holloway College, University of London, UK

'This is a well written and informative book on an important topic. The authors have taken an interesting perspective on access and the built environment by focussing on the role of the development industry. Their findings are shocking. Anyone interested in urban issues will find a wealth of insightful material in this book.' Nick Oatley, University of the West of England, UK

The reality of the built environment for disabled people is one of social, physical and attitudinal barriers which prevent their ease of mobility, movement and access. In the United Kingdom, most homes cannot be accessed by wheelchair, while accessible transport is the exception rather than the rule. Pavements are littered with street furniture, while most public and commercial buildings provide few design features to permit disabled people ease of access.

Inclusive Design is a documentation of the attitudes, values, and practices of property professionals, including developers, surveyors and architects, in responding to the building needs of disabled people. It looks at the way in which pressure for accessible building design is influencing the policies and practices of property companies and professionals, with a primary focus on commercial developments in the UK. The book also provides comments on, and references to, other countries, particularly Sweden, New Zealand, and the USA.

June 2001: 246x189: 240pp
5 line figures, 34 b+w photos
Pb: 0-419-25620-2: £29.99

To Order: Tel: +44 (0) 8700 768853, or +44 (0) 1264 343071 Fax: +44 (0) 1264 343005, or
Post: Spon Press Customer Services, ITPS Andover, Hants, SP10 5BE, UK Email: book.orders@tandf.co.uk.

Postage & Packing: UK: 5% of order value (min. charge £1, max. charge £10) for 3-5 days delivery. Option of next day delivery at an additional £6.50. Europe: 10% of order value (min. charge £2.95, max. charge £20) for delivery surface post. Option of airmail at an additional £6.50. ROW: 15% of order value (min.charge£6.50, max. charge £30) for airmail delivery.

For a complete listing of all our titles visit: www.sponpress.com

PART III

Prices for Measured Work

This part of the book contains the following sections:

Prices for Measured Work - Major Works, *page* 99

Prices for Measured Work - Minor Works, *page* 409

NEW FROM SPON PRESS

Housing Design Quality
THROUGH POLICY, GUIDANCE AND REVIEW

Matthew Carmona,
University College London, UK

Housing Design Quality directly addresses the major planning debate of our time - the delivery and quality of new housing development.

As pressure for new housing development in England increases, a widespread desire to improve the design of the resulting residential environments becomes ever more apparent with increasing condemnation of the standard products of the volume house builders. In recent years central government has come to accept the need to deliver higher quality living environments, and the important role of the planning system in helping to raise design standards. *Housing Design Quality* focuses on this role and in particular on how the various policy instruments available to public authorities can be used in a positive manner to deliver higher quality residential developments.

March 2001: 246x189: 368pp
90 b+w illustrations
Pb: 0-419-25650-4: £35.00

To Order: Tel: +44 (0) 8700 768853, or +44 (0) 1264 343071 Fax: +44 (0) 1264 343005, or
Post: Spon Press Customer Services, ITPS Andover, Hants, SP10 5BE, UK Email: book.orders@tandf.co.uk.

Postage & Packing: UK: 5% of order value (min. charge £1, max. charge £10) for 3-5 days delivery. Option of next day delivery at an additional £6.50. Europe: 10% of order value (min. charge £2.95, max. charge £20) for delivery surface post. Option of airmail at an additional £6.50. ROW: 15% of order value (min.charge£6.50, max. charge £30) for airmail delivery.

For a complete listing of all our titles visit: www.sponpress.com

Prices for Measured Works - Major Works

INTRODUCTION

The rates contained in "Prices for Measured Work - Major Works" are intended to apply to a project in the outer London area costing about £1,300,000 (excluding Preliminaries) and assume that reasonable quantities of all types of work are required. Similarly it has been necessary to assume that the size of the project warrants the sub-letting of all types of work normally sub-let. Adjustments should be made to standard rates for time, location, local conditions, site constraints and any other factors likely to affect costs of a specific scheme.

The distinction between builders' work and work normally sub-let is stressed because prices for work which can be sub-let may well be quite inadequate for the contractor who is called upon to carry out relatively small quantities of such work himself.

As explained in more detail later the prices are generally based on wage rates which came into force on 25 June 2001, material costs as stated, and include no allowance for overheads or profit. They do not allow for preliminary items which are dealt with under a separate heading (see page 105) or for any Value Added Tax.

The format of this section is so arranged that, in the case of work normally undertaken by the Main Contractor, the constituent parts of the total rate are shown enabling the reader to make such adjustments as may be required in particular circumstances. Similar details have also been given for work normally sub-let although it has not been possible to provide this in all instances.

As explained in the Preface, there is a facility available to readers which enables a comparison to be made between the level of prices in this section and current tenders by means of a tender index. The tender index for this Major Works section of Spon's is 388 (as shown on the front cover) which is 1.31% above our forecast tender price index of 383 for the outer London region for the fourth quarter of 2001.

To adjust prices for other regions/times, the reader is recommended to refer to the explanations and examples on how to apply these tender indices, given on pages 691 - 695.

There follow explanations and definitions of the basis of costs in the "Prices for Measured Work" section under the following headings:

- Overhead charges and profit
- Labour hours and Labour £ column
- Material £ column
- Material/Plant £ columns
- Total rate £ column

OVERHEAD CHARGES AND PROFIT

All rates checked against winning tenders include overhead charges and profit at current levels.

LABOUR HOURS AND LABOUR £ COLUMNS

"Labour rates" are based upon typical gang costs divided by the number of primary working operatives for the trade concerned, and for general building work include an allowance for trade supervision (see below). "Labour hours" multiplied by "Labour rate" with the appropriate addition for overhead charges and profit (currently zero) gives "Labour £". In some instances, due to variations in gangs used, "Labour rate" figures have not been indicated, but can be calculated by dividing "Labour £" by "Labour hours".

Building craft operatives and labourers

From 25 June 2001 guaranteed minimum weekly earnings in the London area for craft operatives and general operatives are £261.30 and £196.56 respectively; to these rates have been added allowances for the items below in accordance with the recommended procedure of the Institute of Building in its "Code of Estimating Practice". The resultant hourly rates on which the "Prices for Measured Work" have generally been based are £9.15 and £6.83 for craft operatives and labourers, respectively.

- Lost time
- Construction Industry Training Board Levy
- Holidays with pay
- Accidental injury, retirement and death benefits scheme
- Sick pay
- National Insurance
- Severance pay and sundry costs
- Employer's liability and third party insurance

NOTE: For travelling allowances and site supervision see "Preliminaries" section.

Prices for Measured Works - Major Works

The table which follows illustrates how the "all-in" hourly rates referred to on page 102 have been calculated. Productive time has been based on a total of 1954 hours worked per year.

		Craft Operatives £	£	General Operatives £	£
Wages at standard basic rate:					
Productive Time	44.300 wks	261.30	11574.52	196.56	8706.80
Lost Time Allowance	0.904 wks	261.30	236.22	196.56	177.69
Overtime	5.800 wks	391.95	2273.31	294.84	1710.07
			14084.04		10594.56
Extra payments under					
National Working Rules	45.200 wks		-		-
Sick Pay	1.000 wks		-		-
CITB Allowance (0.50% of Payroll)	1.000 yr		79.46		59.77
Holiday Pay	4.200 wks	311.59	1308.69	234.39	984.45
Public Holiday	1.600 wks	311.59	498.55	234.39	375.03
Employer's contributions under					
Retirement cover scheme (Death and Accident cover is provided free)	52.000 wks	1.90	98.80	1.90	98.80
National Insurance (Average Weekly Payment)	46.200 wks	25.92	1197.69	16.94	782.44
			17267.23		12895.05
Severance Pay and Sundry Costs	Plus	1.50%	259.01	1.50%	193.43
			17526.24		13088.48
Employer's Liability and Third-Party Insurance	Plus	2.00%	350.52	2.00%	261.77
Annual labour cost as defined in section 3			**£ 17876.76**		**£ 13350.25**
Total cost per hour			**£9.15**		**£6.83**

NOTES:

1. Absence due to sickness has been assumed to be for periods not exceeding 3 days for which no payment is due (Working Rule 20.7.3).

2. Retirement benefit effective from 10 January 2000. Death and accident benefit cover is provided free of charge.

3. All N.I. payments are at not-contracted out rates applicable from 5 April 2001.

Prices for Measured Works - Major Works

The "labour rates" used in the Measured Work sections have been based on the following gang calculations which generally include an allowance for supervision by a foreman or ganger. Alternative labour rates are given showing the effect of various degrees of bonus.

Gang	Total Gang rate £/hour		Productive unit rate £/hour		Alternative labour rates £/hour +25%	+15%	+5%
Groundwork Gang							
1 Ganger	1 x 7.38 =	7.38					
6 General Operatives	6 x 6.83 =	40.98					
		48.36	÷ 6.5 =	7.44	9.36	8.59	7.83
Concreting gang							
1 Ganger	1 x 7.83 =	7.83					
4 General Operatives	4 x 7.29 =	29.16					
		36.99	÷ 4.5 =	8.22	10.33	9.49	8.64
Steelfixing Gang							
1 Foreman	1 x 9.69 =	9.69					
4 Steelfixers	4 x 9.15 =	36.60					
		46.29	÷ 4.5 =	10.29	12.92	11.86	10.82
Formwork Gang							
1 Foreman	1 x 9.69 =	9.69					
10 Carpenters	10 x 9.15 =	91.50					
1 General Operative	1 x 6.83 =	6.83					
		108.02	÷ 10.5 =	10.29	12.92	11.86	10.82
Bricklaying/Lightweight Blockwork Gang							
1 Foreman	1 x 9.69 =	9.69					
6 Bricklayers	6 x 9.15 =	54.90					
4 General Operatives	4 x 6.83 =	27.32					
		91.91	÷ 6.5 =	14.14	17.76	16.31	14.87
Dense Blockwork Gang							
1 Foreman	1 x 9.69 =	9.69					
6 Bricklayers	6 x 9.15 =	54.90					
6 General Operatives	6 x 6.83 =	40.98					
		105.57	÷ 6.5 =	16.24	20.41	18.74	17.08
Carpentry/Joinery Gang							
1 Foreman	1 x 9.69 =	9.69					
5 Carpenters	5 x 9.15 =	45.75					
1 General Operative	1 x 6.83 =	6.83					
		62.27	÷ 5.5 =	11.32	14.22	13.06	11.91
Craft Operative (Painter, Wall and Floor Tiler, Slater)	1 x 9.15 =	9.15	÷ 1 =	9.15	11.49	10.55	9.62
1 and 1 Gang							
1 Craft Operative	1 x 9.15 =	9.15					
1 General Operative	1 x 7.29 =	7.29					
		16.44	÷ 1 =	16.44	20.65	18.97	17.28
2 and 1 Gang							
2 Craft Operatives	2 x 9.15 =	18.30					
1 General Operative	1 x 7.29 =	7.29					
		25.59	÷ 2 =	12.80	16.07	14.76	13.45
Small Labouring Gang (making good)							
1 Ganger	1 x 7.95 =	7.95					
4 General Operatives	4 x 7.29 =	29.16					
		37.11	÷ 4.5 =	8.22	10.33	9.49	8.64
Drain Laying Gang/Clayware							
2 General Operatives	2 x 7.29 =	14.58	÷ 2 =	7.29	9.16	8.42	7.66

Sub-Contractor's operatives
Similar labour rates are shown in respect of sub-let trades where applicable.

Plumbing operatives
From 21 August 2000 the hourly earnings for technical and trained plumbers are £8.40 and £6.48, respectively; to these rates have been added allowances similar to those added for building operatives (see below). The resultant average hourly rate on which the "Prices for Measured Work" have been based is £10.63.

The items referred to above for which allowance has been made are:

Tool allowance
Plumbers' welding supplement
Holidays with pay
Pension and welfare stamp
National Insurance "contracted out" contributions
Severance pay and sundry costs
Employer's liability and third party insurance

No allowance has been made for supervision as we have assumed the use of a team of technical or trained plumbers who are able to undertake such relatively straightforward plumbing works, e.g. on housing schemes, without further supervision.

The table which follows shows how the average hourly rate referred to above has been calculated. Productive time has been based on a total of 1695.00 hours worked per year.

			Technical Plumber £	£	Trained Plumber £	£
Wages at standard basic rate						
productive time	1695.00 hrs	8.40		14,238.00	6.48	10,983.60
Overtime (paid at standard basic rate)	67.80 hrs	8.40		569.52	6.48	439.34
Overtime	158.20 hrs	12.60		1,993.32	9.72	1,537.70
Plumber's welding supplement (gas and arc)	1921.00 hrs	0.46		883.66		0.00
Tool Allowance - now discontinued				0.00		0.00
				17,684.50		12,960.65
Employer's contribution to						
holiday credit/welfare stamps	52.00 wks	42.98		2,234.96	33.16	1,724.32
pension (6.5% of earnings)	46.20 wks	25.00		1,155.00	18.33	846.85
Holiday Top-up funding (provided by Employer)	52.00 wks	1.64		85.28	1.26	65.52
National Insurance	47.80 wks	33.89		1,619.94	22.08	1,055.42
				22,779.68		16,652.76
Severance pay and sundry costs	Plus		1.5%	341.70	1.5%	249.79
				23,121.38		16,902.55
Employer's Liability and Third Party Insurance	Plus		2.0%	462.43	2.0%	338.05
Total cost per annum				**£23,583.81**		**£17,240.60**
Total cost per hour				**£12.28**		**£8.97**
Average all-in rate per hour					**£10.63**	

MATERIAL £ COLUMN

Many items have reference to a "PC" value. This indicates the prime cost of the principal material delivered to site in the outer London area assuming maximum discounts for large quantities. When obtaining material prices from other sources, it is important to identify any discounts that may apply. Some manufacturers only offer 5 to 10% discount for the largest of orders; or "firm" orders (as distinct from quotations). For some materials, discounts of 30% to 40% may be obtainable, depending on value of order, preferential position of the purchaser or state of the market.

The "Material £" column indicates the total materials cost including delivery, waste, sundry materials and an allowance (currently zero) for overhead charges and profit for the unit of work concerned. Alternative material prices are given at the beginning of many sections, by means of which alternative "Total rate £" prices may be calculated. All material prices quoted are exclusive of Value Added Tax.

MATERIAL PLANT £ COLUMN

Plant costs have been based on current weekly hire charges and estimated weekly cost of oil, grease, ropes (where necessary), site servicing and cartage charges. The total amount is divided by 30 (assuming 25% idle time) to arrive at a cost per working hour of plant. To this hourly rate is added one hour fuel consumption and one hour for an operator where indicated; the rate to be calculated in accordance with the principles set out earlier in this section, i.e. with an allowance for plus rates, etc.

For convenience the all-in rates per hour used in the calculations of "Prices for Measured Work" are shown below and where included in Material/Plant £ column.

Plant	Labour	"All-in" rate per hour £
Excavator (4 wheeled - 0.76 m^3 shovel, 0.24 m^3 bucket)	Driver	20.45
Excavator (JCB 3C - 0.24 m^3 bucket)	Driver	15.80
Excavator (JCB 3C off centre - 0.24 m^3 bucket)	Driver	18.55
Excavator (Hitachi EX120 - 0.53 m^3 bucket)	Driver	21.35
Dumper (2.30 m^3)	Driver	14.85
Two tool portable compressor (125 cfm)		
per breaking tool		2.00
per punner foot and stem rammer		1.60
Roller		
Bomag BW75S - pedestrian double drum		2.90
Bomag BW120AD - tandem		6.65
5/3.50 cement mixer		1.10
Kango heavy duty breaker		0.85
Power float		1.85
Light percussion drill		0.42

* Operation of compressor by tool operator

TOTAL RATE £ COLUMN

"Total rate £" column is the sum of "Labour £" and "Material £" columns. This column excludes any allowance for "Preliminaries" which must be taken into account if one is concerned with the total cost of work.
The example of "Preliminaries" in the following section indicates that in the absence of detailed calculations currently 11% should be added to all prices for measured work to arrive at total cost.

A PRELIMINARIES

The number of items priced in the "Preliminaries" section of Bills of Quantities and the manner in which they are priced vary considerably between Contractors. Some Contractors, by modifying their percentage factor for overheads and profit, attempt to cover the costs of "Preliminary" items in their "Prices for Measured Work". However, the cost of "Preliminaries" will vary widely according to job size and complexity, site location, accessibility, degree of mechanisation practicable, position of the Contractor's head office and relationships with local labour/domestic Sub-Contractors. It is therefore usually far safer to price "Preliminary" items separately on their merits according to the job.

In amending the Preliminaries/General Conditions section for SMM7, the Joint Committee stressed that the preliminaries section of a bill should contain two types of cost significant item:

1. Items which are not specific to work sections but which have an identifiable cost which is useful to consider separately in tendering e.g. contractual requirements for insurances, site facilities for the employer's representative and payments to the local authority.
2. Items for fixed and time-related costs which derive from the contractor's expected method of carrying out the work, e.g. bringing plant to and from site, providing temporary works and supervision.

A fixed charge is for work the cost of which is to be considered as independent of duration. A time related charge is for work the cost of which is to be considered as dependent on duration. The fixed and time-related subdivision given for a number of preliminaries items will enable tenderers to price the elements separately should they so desire. Tenderers also have the facility at their discretion to extend the list of fixed and time-related cost items to suit their particular methods of construction.

The opportunity for Tenderers to price fixed and time-related items in A30-A37, A40-A44 and A51-A52 have been noted against the following appropriate items although we have not always provided guidance as costs can only be assessed in the light of circumstances of a particular job.

Works of a temporary nature are deemed to include rates, fees and charges related thereto in Sections A36, A41, A42, and A44, all of which will probably be dealt with as fixed charges.

In addition to the cost significant items required by the method, other preliminaries items which are important from other points of view, e.g. quality control requirements, administrative procedures, may need to be included to complete the Preliminaries/General conditions as a comprehensive statement of the employer's requirements.

Typical clause descriptions from a "Preliminaries/General Conditions" section are given below together with details of those items that are usually priced in detail in tenders.

An example in pricing "Preliminaries" follows, and this assumes the form of contract used is the Standard Form of Building Contract 1998 Edition and the value, excluding "Preliminaries", is £1 300 000. The contract is estimated to take 80 weeks to complete and the value is built up as follows:

	£
Labour value	465 000
Material value	395 000
Provisional sums and all Sub-Contractors	440 000
	£ 1 300 000

At the end of the section the examples are summarised to give a total value of "Preliminaries" for the project.

A PRELIMINARIES

A PRELIMINARIES/GENERAL CONDITIONS

Preliminary particulars

A10 Project particulars - Not priced

A11 Drawings - Not priced

A12 The Site/Existing buildings - Generally not priced

The reference to existing buildings relates only to those buildings which could have an influence on cost. This could arise from their close proximity making access difficult, their heights relative to the possible use of tower cranes or the fragility of, for example, an historic building necessitating special care.

A13 Description of the work - Generally not priced

A20 The Contract/Sub-contract
(The Standard Form of Building Contract 1998 Edition is assumed)

Clause no
1. **Interpretation, definitions, etc.** - Not priced
2. **Contractor's obligations** - Not priced
3. **Contract Sum - adjustment - Interim certificates** - Not priced
4. **Architect's Instructions** - Not priced
5. **Contract documents - other documents - issue of certificates**
   The contract conditions may require a master programme to be prepared. This will normally form part of head office overheads and therefore is not priced separately here.
6. **Statutory obligations, notices, fees and charges** - Not priced. Unless the Contractor is specifically instructed to allow for these items.
7. **Level and setting out of the works** - Not priced
8. **Materials, goods and workmanship to conform to description, testing and inspection** - Not priced
9. **Royalties and patent rights** - Not priced

NOTE: The term "Not priced" or "Generally not priced" where used throughout this section means either that the cost implication is negligible or that it is usually included elsewhere in the tender.

10. **Person-in-charge**
    Under this heading are usually priced any staff that will be required on site. The staff required will vary considerably according to the size, layout and complexity of the scheme, from one foreman-in-charge to a site agent, general foreman, assistants, checkers and storemen, etc. The costs included for such people should include not only their wages, but their total cost to the site including statutory payments, pension, expenses, holiday relief, overtime, etc. Part of the foreman's time, together with that of an assistant, will be spent on setting out the site. Allow, say, £2.15 per day for levels, staff, pegs and strings, plus the assistant's time if not part of the general management team. Most sites usually include for one operative to clean up generally and do odd jobs around the site.
    Cost of other staff, such as buyers, and quantity surveyors, are usually part of head office overhead costs, but alternatively, may now be covered under A40 Management and staff.

A PRELIMINARIES

A20 The Contract/Sub-contract - cont'd

A typical build-up of a foreman's costs might be:

	£
Annual salary	17,600.00
Expenses	1,500.00
Bonus - say	1,800.00
Employer's National Insurance contribution (10.4% on £19 400.00) say	2,018.00
Training levy @ 0.25% say	49.00
Pension scheme (say 6% of salary and bonus) say	1,164.00
Sundries, including Employer's Liability and Third Party (say 3.5% of salary and bonus) say	679.00
	£ 24,810.00
Divide by 47 to allow for holidays: per week	527.87
Say	£ 530.00

Corresponding costs for other site staff should be calculated in a similar manner, e.g.

Site administration	£
General foreman 80 weeks @ £530	42,400.00
Holiday relief 4 weeks @ £530	2,120.00
Assistant foreman 60 weeks @ £265	15,900.00
Storeman/checker 60 weeks @ £195	11,700.00
	£ 72,120.00

11. **Access for Architect to the Works** - Not priced
12. **Clerk of Works** - Not priced
13. **Variations and provisional sums** - Not priced
14. **Contract Sum** - Not priced
15. **Value Added Tax - supplemental provisions**
    Major changes to the VAT status of supplies of goods and services by Contractors came into effect on 1 April 1989. It is clear that on and from that date the majority of work supplied under contracts on JCT Forms will be chargeable on the Contractor at the standard rate of tax. In April 1989, the JCT issued Amendment 8 and a guidance note dealing with the amendment to VAT provisions. This involves a revision to clause 15 and the Supplemental Provisions (the VAT agreement). Although the standard rating of most supplies should reduce the amount of VAT analysing previously undertaken by contractors, he should still allow for any incidental costs and expenses which may be incurred.
16. **Materials and goods unfixed or off-site** - Not priced
17. **Practical completion and Defects Liability**
    Inevitably some defects will arise after practical completion and an allowance will often be made to cover this. An allowance of say 0.25 to 0.50% should be sufficient, e.g. Example

    Defects after completion

    Based on 0.25% of the contract sum

    £1,300,000 x 0.25%   £ 3,250.00

18. **Partial possession by Employer** - Not priced
19. **Assignment and Sub-Contracts** - Not priced

A PRELIMINARIES

A20 The Contract/Sub-contract - cont'd.

    19A **Fair wages** - Not priced
20. **Injury to persons and property and Employers indemnity**
(See Clause no 21)
21. **Insurance against injury to persons and property**
The Contractor's Employer's Liability and Public Liability policies (which would both be involved under this heading) are often in the region of 0.50 to 0.60% on the value of his own contract work (excluding provisional sums and work by Sub-Contractors whose prices should allow for these insurances). However, this allowance can be included in the all-in hourly rate used in the calculation of "Prices for Measured Work" (see page 99).
Under Clause 21.2 no requirement is made upon the Contractor to insure as stated by the clause unless a provisional sum is allowed in the Contract Bills.
22. **Insurance of the works against Clause 22 Perils**
If, at the Contractor's risk, the insurance cover must be sufficient to include the full cost of reinstatement, all increases in cost, professional fees and any consequential costs such as demolition. The average provision for fire risk is 0.10% of the value of the work after adding for increased costs and professional fees.

Contractor's Liability - Insurance of works against fire, etc.	£
Contract value (including "Preliminaries"), say	1,475,000.00
Estimated increased costs during contract period, say 6%	88,500.00
	1,563,500.00
Estimated increased costs incurred during period of reinstatement, say 3%	46,900.00
	1,610,400.00
Professional fees @ 16%	257,700.00
	1,868,100.00
Allow 0.1%	1,868.00

NOTE: Insurance premiums are liable to considerable variation, depending on the Contractor, the nature of the work and the market in which the insurance is placed.

23. **Date of possession, completion and postponement** - Not priced
24. **Damages for non-completion** - Not priced
25. **Extension of time** - Not priced
26. **Loss and expense caused by matters materially affecting regular progress of the works** - Not priced
27. **Determination by Employer** - Not priced
28. **Determination by Contractor** -Not priced
29. **Works by Employer or persons employed or engaged by Employer** - Not priced
30. **Certificates and payments** - Not priced
31. **Finance (No.2) Act 1975 - Statutory tax deduction scheme** - Not priced
32. **Outbreak of hostilities** -Not priced
33. **War damage** - Not priced
34. **Antiquities** - Not priced
35. **Nominated Sub-Contractors** - Not priced here. An amount should be added to the relevant PC sums, if required, for profit and a further sum for special attendance.
36. **Nominated Suppliers** - Not priced here. An amount should be added to the relevant PC sums, if required, for profit.

A PRELIMINARIES

A20 The Contract/Sub-contract - cont'd.

37. **Choice of fluctuation provisions - entry in Appendix**
    The amount which the Contractor may recover under the fluctuations clauses (Clause nos 38, 39 and 40) will vary depending on whether the Contract is "firm", i.e. Clause no 38 is included, or "fluctuating", whether the traditional method of assessment is used, i.e. Clause no 39 or the formula method, i.e. Clause no 40. An allowance should be made for any shortfall in reimbursement under fluctuating contracts.
38. **Contribution Levy and Tax Fluctuations** (see Clause no 37)
39. **Labour and Materials Cost and Tax Fluctuations** (see Clause no 37)
40. **Use of Price Adjustment Formulae** (see Clause no 37)
    Details should include special conditions or amendments to standard conditions, the Appendix insertions and the Employer's insurance responsibilities. Additional obligations may include the provision of a performance bond. If the Contractor is required to provide sureties for the fulfilment of the work the usual method of providing this is by a bond provided by one or more insurance companies. The cost of a performance bond depends largely on the financial standing of the applying Contractor. Figures tend to range from 0.25 to 0.50% of the contract sum.

A30-A37 EMPLOYERS' REQUIREMENTS

These include the following items but costs can only be assessed in the light of circumstances on a particular job. Details should be given for each item and the opportunity for the Tenderer to separately price items related to fixed charges and time related charges.

A30 Tendering/Sub-letting/Supply

A31 Provision, content and use of documents

A32 Management of the works

A33 Quality standards/control

A34 Security/Safety/Protection

This includes noise and pollution control, maintaining adjoining buildings, public and private roads, live services, security and the protection of work in all sections.

(i) Control of noise, pollution and other obligations
The Local Authority, Landlord or Management Company may impose restrictions on the timing of certain operations, particularly noisy of dust-producing operations, which may necessitate the carrying out of these works outside normal working hours or using special tools and equipment. The situation is most likely to occur in built-up areas such as city centres, shopping malls etc., where the site is likely to be in close proximity to offices, commercial or residential property.

(ii) Maintenance of public and private roads
Some additional value or allowance may be required against this item to insure/protect against damage to entrance gates, kerbs or bridges caused by extraordinary traffic in the execution of the works.

A35 Specific limitations on method/sequence/timing

This includes design constraints, method and sequence of work, access, possession and use of the site, use or disposal of materials found, start of work, working hours, employment of labour and sectional possession or partial possession etc.

A PRELIMINARIES

A36 Facilities/Temporary work/Services

This includes offices, sanitary accommodation, temporary fences, hoardings, screens and roofs, name boards, technical and surveying equipment, temperature and humidity, telephone/facsimile installation and rental/maintenance, special lighting and other general requirements, etc. The attainment and maintenance of suitable levels necessary for satisfactory completion of the work including the installation of joinery, suspended ceilings, lift machinery, etc. is the responsibility of the contractor. The following is an example how to price a mobile office for a Clerk of Works

			£
(a)	Fixed charge		
	Haulage to and from site - say		150.00
(b)	Time related charge		
	Hire charge - 15m^2 x 76 weeks @ £2.00/m^2		2,280.00
	Lighting, heating and attendance on office,		
	say 76 weeks @ £25.00		1,900.00
	Rates on temporary building based on £14.00/m^2 per annum		310.00
			4,490.00
(c)	Combined charge		4,640.00

The installation of telephones or facsimiles for the use of the Employer, and all related charges therewith, shall be given as a provisional sum.

A37 Operation/Maintenance of the finished building

A40-A44 CONTRACTORS GENERAL COST ITEMS

For items A41-A44 it shall be clearly indicated whether such items are to be "Provided by the Contractor" or "Made available (in any part) by the Employer".

A40 Management and staff (Provided by the Contractor)

NOTE: The cost of site administrative staff has previously been included against Clause 10 of the Conditions of Contract, where Readers will find an example of management and staff costs.

When required allow for the provision of a watchman or inspection by a security organization. Other general administrative staff costs, e.g. Engineering, Programming and production and Quantity Surveying could be priced as either fixed or time related charges, under this section. For the purpose of this example allow, say, 1% of measured work value for other administrative staff costs.

(a) Time related charge

Based on 1% of £1,300,000 - say £ 13,000.00

A PRELIMINARIES

A41 Site accommodation (Provided by the Contractor or made available by the Employer)

This includes all temporary offices laboratories, cabins, stores, compounds, canteens, sanitary facilities and the like for the Contractor's and his domestic sub-contractors' use (temporary office for a Clerk of Works is covered under obligations and restrictions imposed by the Employer).
Typical costs for jack-type offices are as follows, based upon a twelve months minimum hire period they exclude furniture which could add a further £14.00 - £16.00 per week.

Size	Rate per week £
12 ft x 9 ft (10.03 m^2)	21.00 (£2.09/m^2)
16 ft x 9 ft (13.37 m^2)	24.00 (£1.80/m^2)
24 ft x 9 ft (20.06 m^2)	32.00 (£1.60/m^2)
32 ft x 12 ft (35.67 m^2)	50.00 (£1.40/m^2)

Typical rates for security units are as follows:

Size	Rate per week £
12 ft x 8 ft (8.92 m^2)	21.00 (£2.35/m^2)
16 ft x 8 ft (11.89 m^2)	25.00 (£2.10/m^2)
24 ft x 8 ft (17.84 m^2)	34.00 (£1.91/m^2)
32 ft x 8 ft (23.78 m^2)	43.00 (£1.81/m^2)

The following example is for one Foreman's office and two security units.

			£
(a)	Fixed charge		
	Haulage to and from site - Foreman's Office, say		150.00
	Haulage to and from site - Storage sheds, say		300.00
			450.00
(b)	Time related charge		
	Hire charge - Foreman's office 15 m^2 x 76 weeks @ £1.80/m^2		2,052.00
	Hire charge - Storage sheds 40 m^2 x 70 weeks @ £1.80/m^2		5,040.00
	Lighting, heating and attendance on office, say 76 weeks @ £25.00		1,900.00
	Rates on temporary building based on £14.00/m^2 per annum		1,440.00
			10,432.00
(c)	Combined charge		10,884.00

A PRELIMINARIES

A42 Services and facilities (Provided by the Contractor or made available by the Employer)

This generally includes the provision of all of the Contractor's own services, power, lighting, fuels, water, telephone and administration, safety, health and welfare, storage of materials, rubbish disposal, cleaning, drying out, protection of work in all sections, security, maintaining public and private roads, small plant and tools and general attendance on nominated sub-contractors.
However, this section does not cover fuel for testing and commissioning permanent installations which would be measured under Sections Y51 and Y81. Examples of build-ups/allowances for some of the major items are provided below:

(i) Lighting and power for the works
The Contractor is usually responsible for providing all temporary lighting and power for the works and all charges involved. On large sites this could be expensive and involve sub-stations and the like, but on smaller sites it is often limited to general lighting (depending upon time of year) and power for power operated tools for which a small diesel generator and some transformers usually proves adequate.

Typical costs are:

Low voltage diesel generator	£90.00 - £100.00 per week
2.20 to 10 kVA transformer	£10.00 - £50.00 per week

A typical allowance, including charges, installation and fitting costs could be 1% of contract value.

Example
Lighting and power for the works
Combined charge
The fixed charge would normally represent a proportion of the following allowance applicable to connection and supply charges. The residue would be allocated to time related charges.

Dependant on the nature of the work, time of year and incidence of power operated tools
based on 1% of £1,300,000 - say   £ 13,000.00

(ii) Water for the works

Charges should properly be ascertained from the local Water Authority. If these are not readily available, an allowance of 0.33% of the value of the contract is probably adequate, providing water can be obtained directly from the mains. Failing this, each case must be dealt with on its merits. In all cases an allowance should also be made for temporary plumbing including site storage of water if required.

Useful rates for temporary plumbing include:

Piping	£5.15 per metre
Connection	£150.00
Standpipe	£55.00

Plus an allowance for barrels and hoses.

A PRELIMINARIES

A42 Services and facilities - cont'd

Example

Water for the works
Combined charge
The fixed charge would normally represent a proportion of the following allowance applicable to connection and supply charges. The residue would be allocated to time related charges.

	£
0.33% of £1,300,000	4,290.00
Temporary plumbing, say	400.00
	£ 4,690.00

(iii) Temporary telephones for the use of the Contractor

Against this item should be included the cost of installation, rental and an assessment of the cost of calls made during the contract.

Installation costs	£99.00
Rental	£28.67 Initial Monthly Payment
	£14.33 Monthly Payment Thereafter
Cost of calls	For sites with one telephone allow about £22.00 per week.

Example
Temporary telephones   £

(a) Fixed charge
    Connection charges
    line with socket outlet   99.00
    external ringing device   10.00
                                                                          109.00

(b) Time related charge
    line rental - 6 quarters -   1 Month @ £28.67   28.67
                                17 Months @ £14.33   243.61
    external ringing device rental - 6 quarters @ £2.50   15.00
    call charges - 76 weeks @ £22.00   1,672.00
                                                                        1,959.28

(c) Combined charges, say   £ 2,070.00

(iv) Safety, health and welfare of workpeople
The Contractor is required to comply with the Code of Welfare Conditions for the Building Industry which sets out welfare requirements for the following:

1. Shelter from inclement weather
2. Accommodation for clothing
3. Accommodation and provision for meals
4. Provision of drinking water
5. Sanitary conveniences
6. Washing facilities
7. First aid
8. Site conditions

A PRELIMINARIES

A42 Services and facilities - cont'd

A variety of self-contained mobile or jack-type units are available for hire and a selection of rates is given below:

	£	
Kitchen with cooker, fridge, sink unit, water heater and basin		
32 ft x 10 ft jack-type	75.00	per week
Mess room with water heater, wash basin and seating		
16 ft x 7 ft 6 in mobile	40.00	per week
16 ft x 9 ft jack-type	38.00	per week
Welfare unit with drying rack, lockers, tables, seating, cooker, heater, sink and basin		
22 ft x 7 ft 6 in mobile	55.00	per week
Toilets (mains type)		
One pan unit	15.00	per week
Two pan unit mobile	40.00	per week
Four pan unit jack-type	50.00	per week

Allowance must be made in addition for transport costs to and from site, setting up costs, connection to mains, fuel supplies and attendance

Site first aid kit   £ 3.65 per week

A general provision to comply with the above code is often 0.50 to 0.75% of the measured work value. The costs of safety supervisors (required for firms employing more than 20 people) are usually part of head office overhead costs.

Example
Safety, health and welfare
Combined charge
The fixed charge would normally represent a proportion of the following allowance, with the majority allocated to time related charges.

Based on 0.66% of £1,300,000 - say   £ 8,580.00

(v)   Removing rubbish, protective casings and coverings and cleaning the works on completion. This includes removing surplus materials and final cleaning of the site prior to handover. Allow for sufficient "bins" for the site throughout the contract duration and for some operatives time at the end of the contract for final clearing and cleaning ready for handover.

      Cost of "bins" - approx. £25.00 each
      A general allowance of 0.20% of measured work value is probably sufficient.

      Example
      Removing rubbish, etc., and cleaning
      Combined charge
      The fixed charge would normally represent an allowance for final clearing of the works on completion with the residue for cleaning throughout the contract period

      Based on 0.2% of £1,300,000 - say   £ 2,600.00

A PRELIMINARIES

A42 Services and facilities - cont'd

(vi) Drying the works
Use or otherwise of an installed heating system will probably determine the value to be placed against this item. Dependant upon the time of year, say allow 0.1% to 0.2% of the contract value to cover this item.

Example
Drying of the works
Combined charge
Generally this cost is likely to be related to a fixed charge only

Based on 0.1% of £1,300,000 - say   £ 1,300.00

(vii) Protecting the works from inclement weather
In areas likely to suffer particularly inclement weather, some nominal allowance should be included for tarpaulins, polythene sheeting, battening, etc., and the effect of any delays in concreting or brickwork by such weather.

(viii) Small plant and tools
Small plant and hand tools are usually assessed as between 0.5% and 1.5% of total labour value.
Combined charge

Based on 1% of labour cost of £465,000 - say   £ 4,650.00

(ix) General attendance on nominated sub-contractors
In the past this item was located after each PC sum. Under SMM7 it is intended that two composite items (one for fixed charges and the other for time related charges) should be provided in the preliminaries bill for general attendance on all nominated sub-contractors.

A43 Mechanical plant

This includes for cranes, hoists, personnel transport, transport, earthmoving plant, concrete plant, piling plant, paving and surfacing plant, etc. SMM6 required that items for protection or for plant be given in each section, whereas SMM7 provides for these items to be covered under A34, A42 and A43, as appropriate.

(i) Plant
Quite often, the Contractors own plant and plant employed by sub-contractors are included in measured rates, (e.g. for earthmoving, concrete or piling plant) and the Editors have adopted this method of pricing where they believe it to be appropriate. As for other items of plant e.g. cranes, hoists, these tend to be used by a variety of trades. An example of such an item might be:

A PRELIMINARIES

A43 Mechanical Plant - cont'd

    Example
    Tower crane - static 30 m radius - 4/5 tonne max. load

			£
(a)	Fixed charge		
	Haulage of crane to and from site, erection, testing, commissioning & dismantling - say		5,000.00
(b)	Time related charge		
	Hire of crane, say 30 weeks @ 600.00		18,000.00
	electricity fuel and oil - say		5,000.00
	operator 30 weeks x 40 hours @ £10.00		12,000.00
			35,000.00
(c)	Combined charge, say		£ 40,000.00

For the purpose of this example in pricing of Preliminaries, the Editors have assumed that the costs of the above tower crane are included in with "Measured Rates" items and have not therefore carried this total to the Sample Summary.

    (ii)   Personnel transport
    The labour rates per hour on which "Prices for Measured Work" have been based do not cover travel and lodging allowances which must be assessed according to the appropriate working rule agreement.

    Example
    Personnel transport

    Assuming all labour can be found within the London region, the labour value of £465 000 represents approximately 2050 man weeks. Assume each man receives an allowance of £1.54 per day or £7.70 per week of five days.

    Combined/time related charge

    2050 man weeks @ £7.70   £ 15,785.00

A44 Temporary works (Provided by the Contractor or made available by the Employer)

This includes for temporary roads, temporary walkways, access scaffolding, support scaffolding and propping, hoardings, fans, fencing etc., hardstanding and traffic regulations etc. The Contractor should include maintaining any temporary works in connection with the items, adapting, clearing away and making good, and all notices and fees to Local Authorities and public undertakings. On fluctuating contracts, i.e. where Clause 39 or 40 is incorporated there Is no allowance for fluctuations in respect of plant and temporary works and in such instances allowances must be made for any increases likely to occur over the contract period.

Examples of build-ups/allowances for some items are provided below:

    (i)   Temporary roads, hardstandings, crossings and similar items
    Quite often consolidated bases of eventual site roads are used throughout a contract to facilitate movement of materials around the site. However, during the initial setting up of a site, with drainage works outstanding, this is not always possible and occasionally temporary roadways have to be formed and ground levels later reinstated.

A PRELIMINARIES

A44 Temporary works - cont'd

Typical costs are:

Removal of topsoil and provision of a 225 mm thick stone/hardcore base blinded with ashes as a temporary roadway 3.50 m wide and subsequent reinstatement:

on level ground	£22.00 per metre
on sloping ground (including 1m of cut or fill)	£27.50 per metre

Removal of topsoil and provision of 225 mm thick stone/hardcore base blinded with ashes as a temporary hardstanding £6.45 per m^2

Provision of "Aluminium Trakway"; Eve Trakway or equal and approved; portable road system laid directly onto existing surface or onto compacted removable granular base to protect existing surface (Note: most systems are based on a weekly hire charge and laid and removed by the supplier):

Heavy Duty Trakpanel (3.00 m x 2.50 m)	£3.93 per m²/per week
Delivery charge for Trakpanel	£1.83 per m²
Outrigger mats for use in conjunction with Heavy Duty Trakpanel - per set of 4 mats	£80.00 per set/per week
Single Trak Roadway (3.90 m x 1.22 m)	£2.53 per m²/per week
LD20 Eveolution - Light Duty Trakway (roll out system - minimum delivery 75 m)	£4.60 per m²/per first week £1.50 per m²/per week thereafter
Terraplas Walkways (turf protection system) - per section (1.00 m x 1.00 m)	£5.00 per m²/per week

NOTE: Any allowance for special hardcore hardstandings for piling Sub-Contractors is usually priced against the "special attendance" clause after the relevant Prime Cost sum.

(ii) Scaffolding
The General Contractor's standing scaffolding is usually undertaken by specialist Sub-Contractors who will submit quotations based on the specific requirements of the works. It is not possible to give rates here for the various types of scaffolding that may be required but for the purposes of this section it is assumed that the cost of supplying, erecting, maintaining and subsequently dismantling the scaffolding required would amount to £10,000.00 inclusive of overheads and profit.

(iii) Temporary fencing, hoarding, screens, fans, planked footways, guardrails, gantries, and similar items.
This item must be considered in some detail as it is dependant on site perimeter, phasing of the work, work within existing buildings, etc.

Useful rates include:

A PRELIMINARIES

A44 Temporary works - cont'd

   (iii) Temporary fencing, hoarding, screens, fans, planked footways, guardrails, gantries, and similar items - cont'd

Hoarding 2.30 m high of 18 mm thick plywood with 50 mm x 100 mm sawn softwood studding, rails and posts, including later dismantling

undecorated	£55.00/m (£23.90/m^2)
decorated one side	£61.50/m (£26.75/m^2)
Pair of gates for hoarding	extra £225.00 per pair
Cleft Chestnut fencing 1.20 m high including dismantling	£6.75/m
Morarflex "T-Plus" scaffold sheeting	£3.25/m^2

Example
Temporary hoarding
Combined fixed charge

	£
Decorated plywood hoarding 100m @ £61.50	6,150.00
extra for one pair of gates	225.00
	£ 6,375.00

   (iv) Traffic regulations

Waiting and unloading restrictions can occasionally add considerably to costs, resulting in forced overtime or additional weekend working. Any such restrictions must be carefully assessed for the job in hand.

A50 Work/Materials by the Employer

A description shall be given of works by others directly engaged by the Employer and any attendance that is required shall be priced in the same way as works by nominated sub-contractors.

A51 Nominated sub-contractors

This section governs how nominated sub-contractors should be covered in the bills of quantities for main contracts. Bills of quantities used for inviting tenders from potential nominated sub-contractors should be drawn up in accordance with SMM7 as a whole as if the work was main contractor's work. This means, for example, that bills issued to potential nominated sub-contractors should include preliminaries and be accompanied by the drawings.

As much information as possible should be given in respect of nominated sub-contractors' work in order that tenderers can make due allowance when assessing the overall programme and establishing the contract period if not already laid down. A simple list of the component elements of the work might not be sufficient, but a list describing in addition the extent and possible value of each element would be more helpful. The location of the main plant, e.g. whether in the basement or on the roof, would clearly have a bearing on tenderers' programmes. It would be good practice to seek programme information when obtaining estimates from sub-contractors so that this can be incorporated in the bills of quantities for the benefit of tenderers.

A percentage should be added for the main contractor's profit together with items for fixed and time related charges for special attendances required by nominated sub-contractors. Special attendances to include scaffolding (additional to the Contractor's standing scaffolding), access roads, hardstandings, positioning (including unloading, distributing, hoisting or placing in position items of significant weight or size), storage, power, temperature and humidity etc.

A52 Nominated suppliers

Goods and materials which are required to be obtained from a nominated supplier shall be given as a prime cost sum to which should be added, if required, a percentage for profit.

A PRELIMINARIES

A53 Work by statutory authorities

Works which are to be carried out by a Local Authority or statutory undertakings shall be given as provisional sums.

A54 Provisional work

One of the more significant revisions in SMM7 is the requirement to identify provisional sums as being for either defined or undefined work.
The rules require that each sum for defined work should be accompanied in the bills of quantities by a description of the work sufficiently detailed for the tenderer to make allowance for its effect in the pricing of relevant preliminaries. The information should also enable the length of time required for execution of the work to be estimated and its position in the sequence of construction to be determined and incorporated into the programme. Where Provisional Sums are given for undefined work the Contractor will be deemed not to have made any allowance in programming, planning and pricing preliminaries.
Any provision for Contingencies shall be given as a provisional sum for undefined work.

A55 Dayworks

To include provisional sums for:

Labour
Materials and goods
Plant

SAMPLE SUMMARY

Item		£
A20.10	Site administration	72,120.00
A20.17	Defects after completion	3,250.00
A20.22	Insurance of the works against fire, etc.	1,868.00
A36	Clerk of Work's Office	4,640.00
A40	Additional management and staff	13,000.00
A41	Contractor's accommodation	10,884.00
A42(i)	Lighting and power for the works	13,000.00
A42(ii)	Water for the works	4,690.00
A42(iii)	Temporary telephones	1,850.00
A42(iv)	Safety, health and welfare	8,580.00
A42(v)	Removing rubbish, etc., and cleaning	2,600.00
A42(vi)	Drying the works	1,300.00
A42(viii)	Small plant and tools	4,650.00
A43(ii)	Personnel transport	15,785.00
A44(ii)	Scaffolding	10,000.00
A44(iii)	Temporary hoarding	6,375.00
	TOTAL £	174,592.00

It is emphasized that the above is an example only of the way in which Preliminaries may be priced and it is essential that for any particular contract or project the items set out in Preliminaries should be assessed on their respective values.

Preliminaries as a percentage of a total contract will vary considerably according to each scheme and each Contractors' estimating practice. The value of the Preliminaries items in the above example represents approximately a 13% addition to the value of measured work. Current trends are for Preliminaries to vary from 11 to 13% on tenders. For the purposes of an estimate based on the "Prices for Measured Works - Major Works" section in the current edition of Spons the authors recommend that a 11% addition for preliminaries be applied.

D GROUNDWORK

D20 EXCAVATING AND FILLING

NOTE: Prices are applicable to excavation in firm soil. Multiplying factors for other soils are as follows:

	Mechanical	Hand
Clay	x 2.00	x 1.20
Compact gravel	x 3.00	x 1.50
Soft chalk	x 4.00	x 2.00
Hard rock	x 5.00	x 6.00
Running sand or silt	x 6.00	x 2.00

	PC £	Labour hours	Labour £	Material/ Plant £	Unit	Total rate £
Site preparation						
Removing trees						
girth 600 mm - 1.50 m	-	18.50	147.96	-	nr	147.96
girth 1.50 - 3.00 m	-	32.50	259.94	-	nr	259.94
girth exceeding 3.00 m	-	46.50	371.91	-	nr	371.91
Removing tree stumps						
girth 600 mm - 1.50 m	-	0.93	7.44	36.00	nr	43.44
girth 1.50 - 3.00 m	-	0.93	7.44	52.46	nr	59.90
girth exceeding 3.00 m	-	0.93	7.44	72.00	nr	79.44
Clearing site vegetation						
bushes, scrub, undergrowth, hedges and trees and tree stumps not exceeding 600 mm girth	-	0.03	0.24	-	m²	0.24
Lifting turf for preservation						
stacking	-	0.32	2.56	-	m²	2.56
Excavating; by machine						
Topsoil for preservation						
average depth 150 mm	-	0.02	0.16	0.99	m²	1.15
add or deduct for each 25 mm variation in average depth	-	0.01	0.08	0.26	m²	0.34
To reduce levels						
maximum depth not exceeding 0.25 m	-	0.05	0.40	1.30	m³	1.70
maximum depth not exceeding 1.00 m	-	0.03	0.24	0.91	m³	1.15
maximum depth not exceeding 2.00 m	-	0.05	0.40	1.30	m³	1.70
maximum depth not exceeding 4.00 m	-	0.07	0.56	1.64	m³	2.20
Basements and the like; commencing level exceeding 0.25 m below existing ground level						
maximum depth not exceeding 1.00 m	-	0.07	0.56	1.23	m³	1.79
maximum depth not exceeding 2.00 m	-	0.05	0.40	0.92	m³	1.32
maximum depth not exceeding 4.00 m	-	0.07	0.56	1.23	m³	1.79
maximum depth not exceeding 6.00 m	-	0.09	0.72	1.54	m³	2.26
maximum depth not exceeding 8.00 m	-	0.12	0.96	1.84	m³	2.80
Pits						
maximum depth not exceeding 0.25 m	-	0.31	2.48	4.92	m³	7.40
maximum depth not exceeding 1.00 m	-	0.33	2.64	4.31	m³	6.95
maximum depth not exceeding 2.00 m	-	0.39	3.12	4.92	m³	8.04
maximum depth not exceeding 4.00 m	-	0.47	3.76	5.53	m³	9.29
maximum depth not exceeding 6.00 m	-	0.49	3.92	5.84	m³	9.76
Extra over pit excavating for commencing level exceeding 0.25 m below existing ground level						
1.00 m below	-	0.03	0.24	0.62	m³	0.86
2.00 m below	-	0.05	0.40	0.92	m³	1.32
3.00 m below	-	0.06	0.48	1.23	m³	1.71
4.00 m below	-	0.09	0.72	1.54	m³	2.26

D GROUNDWORK

	PC £	Labour hours	Labour £	Material/ Plant £	Unit	Total rate £
Trenches; width not exceeding 0.30 m						
maximum depth not exceeding 0.25 m	-	0.26	2.08	4.00	m³	6.08
maximum depth not exceeding 1.00 m	-	0.28	2.24	3.38	m³	5.62
maximum depth not exceeding 2.00 m	-	0.33	2.64	4.00	m³	6.64
maximum depth not exceeding 4.00 m	-	0.40	3.20	4.92	m³	8.12
maximum depth not exceeding 6.00 m	-	0.46	3.68	5.84	m³	9.52
Trenches; width exceeding 0.30 m						
maximum depth not exceeding 0.25 m	-	0.23	1.84	3.69	m³	5.53
maximum depth not exceeding 1.00 m	-	0.25	2.00	3.07	m³	5.07
maximum depth not exceeding 2.00 m	-	0.30	2.40	3.69	m³	6.09
maximum depth not exceeding 4.00 m	-	0.35	2.80	4.31	m³	7.11
maximum depth not exceeding 6.00 m	-	0.43	3.44	5.53	m³	8.97
Extra over trench excavating for commencing level exceeding 0.25 m below existing ground level						
1.00 m below	-	0.03	0.24	0.62	m³	0.86
2.00 m below	-	0.05	0.40	0.92	m³	1.32
3.00 m below	-	0.06	0.48	1.23	m³	1.71
4.00 m below	-	0.09	0.72	1.54	m³	2.26
For pile caps and ground beams between piles						
maximum depth not exceeding 0.25 m	-	0.39	3.12	6.46	m³	9.58
maximum depth not exceeding 1.00 m	-	0.35	2.80	5.84	m³	8.64
maximum depth not exceeding 2.00 m	-	0.39	3.12	6.46	m³	9.58
To bench sloping ground to receive filling						
maximum depth not exceeding 0.25 m	-	0.09	0.72	1.54	m³	2.26
maximum depth not exceeding 1.00 m	-	0.07	0.56	1.84	m³	2.40
maximum depth not exceeding 2.00 m	-	0.09	0.72	1.54	m³	2.26
Extra over any types of excavating irrespective of depth						
excavating below ground water level	-	0.13	1.04	2.15	m³	3.19
next existing services	-	2.55	20.39	1.23	m³	21.62
around existing services crossing excavation	-	5.80	46.39	3.38	m³	49.77
Extra over any types of excavating irrespective of depth for breaking out existing materials						
rock	-	2.95	23.59	12.83	m³	36.42
concrete	-	2.55	20.39	9.24	m³	29.63
reinforced concrete	-	3.60	28.79	13.76	m³	42.55
brickwork, blockwork or stonework	-	1.85	14.80	7.01	m³	21.81
Extra over any types of excavating irrespective of depth for breaking out existing hard pavings, 75 mm thick						
coated macadam or asphalt	-	0.19	1.52	0.50	m²	2.02
Extra over any types of excavating irrespective of depth for breaking out existing hard pavings, 150 mm thick						
concrete	-	0.39	3.12	1.50	m²	4.62
reinforced concrete	-	0.58	4.64	1.95	m²	6.59
coated macadam or asphalt and hardcore	-	0.26	2.08	0.54	m²	2.62
Working space allowance to excavations						
reduce levels, basements and the like	-	0.07	0.56	1.23	m²	1.79
pits	-	0.19	1.52	3.38	m²	4.90
trenches	-	0.18	1.44	3.07	m²	4.51
pile caps and ground beams between piles	-	0.20	1.60	3.38	m²	4.98
Extra over excavating for working space for backfilling in with special materials						
hardcore	-	0.13	1.04	11.39	m²	12.43
sand	-	0.13	1.04	17.18	m²	18.22
40 mm - 20 mm gravel	-	0.13	1.04	17.51	m²	18.55
plain in situ ready mixed designated concrete						
C7.5 - 40 mm aggregate	-	0.93	8.22	39.40	m²	47.62

D GROUNDWORK

	PC £	Labour hours	Labour £	Material/ Plant £	Unit	Total rate £
D20 EXCAVATING AND FILLING - cont'd						
Excavating; by hand						
Topsoil for preservation						
average depth 150 mm	-	0.23	1.84	-	m²	1.84
add or deduct for each 25 mm variation in						
average depth	-	0.03	0.24	-	m²	0.24
To reduce levels						
maximum depth not exceeding 0.25 m	-	1.44	11.52	-	m³	11.52
maximum depth not exceeding 1.00 m	-	1.63	13.04	-	m³	13.04
maximum depth not exceeding 2.00 m	-	1.80	14.40	-	m³	14.40
maximum depth not exceeding 4.00 m	-	1.99	15.92	-	m³	15.92
Basements and the like; commencing level exceeding 0.25 m below existing ground level						
maximum depth not exceeding 1.00 m	-	1.90	15.20	-	m³	15.20
maximum depth not exceeding 2.00 m	-	2.04	16.32	-	m³	16.32
maximum depth not exceeding 4.00 m	-	2.73	21.83	-	m³	21.83
maximum depth not exceeding 6.00 m	-	3.33	26.63	-	m³	26.63
maximum depth not exceeding 8.00 m	-	4.02	32.15	-	m³	32.15
Pits						
maximum depth not exceeding 0.25 m	-	2.13	17.04	-	m³	17.04
maximum depth not exceeding 1.00 m	-	2.75	21.99	-	m³	21.99
maximum depth not exceeding 2.00 m	-	3.30	26.39	-	m³	26.39
maximum depth not exceeding 4.00 m	-	4.18	33.43	-	m³	33.43
maximum depth not exceeding 6.00 m	-	5.17	41.35	-	m³	41.35
Extra over pit excavating for commencing level exceeding 0.25 m below existing ground level						
1.00 m below	-	0.42	3.36	-	m³	3.36
2.00 m below	-	0.88	7.04	-	m³	7.04
3.00 m below	-	1.30	10.40	-	m³	10.40
4.00 m below	-	1.71	13.68	-	m³	13.68
Trenches; width not exceeding 0.30 m						
maximum depth not exceeding 0.25 m	-	1.85	14.80	-	m³	14.80
maximum depth not exceeding 1.00 m	-	2.76	22.07	-	m³	22.07
maximum depth not exceeding 2.00 m	-	3.24	25.91	-	m³	25.91
maximum depth not exceeding 4.00 m	-	3.96	31.67	-	m³	31.67
maximum depth not exceeding 6.00 m	-	5.10	40.79	-	m³	40.79
Trenches; width exceeding 0.30 m						
maximum depth not exceeding 0.25 m	-	1.80	14.40	-	m³	14.40
maximum depth not exceeding 1.00 m	-	2.46	19.68	-	m³	19.68
maximum depth not exceeding 2.00 m	-	2.88	23.03	-	m³	23.03
maximum depth not exceeding 4.00 m	-	3.66	29.27	-	m³	29.27
maximum depth not exceeding 6.00 m	-	4.68	37.43	-	m³	37.43
Extra over trench excavating for commencing level exceeding 0.25 m below existing ground level						
1.00 m below	-	0.42	3.36	-	m³	3.36
2.00 m below	-	0.88	7.04	-	m³	7.04
3.00 m below	-	1.30	10.40	-	m³	10.40
4.00 m below	-	1.71	13.68	-	m³	13.68
For pile caps and ground beams between piles						
maximum depth not exceeding 0.25 m	-	2.78	22.23	-	m³	22.23
maximum depth not exceeding 1.00 m	-	2.96	23.67	-	m³	23.67
maximum depth not exceeding 2.00 m	-	3.52	28.15	-	m³	28.15
To bench sloping ground to receive filling						
maximum depth not exceeding 0.25 m	-	1.30	10.40	-	m³	10.40
maximum depth not exceeding 1.00 m	-	1.48	11.84	-	m³	11.84
maximum depth not exceeding 2.00 m	-	1.67	13.36	-	m³	13.36
Extra over any types of excavating irrespective of depth						
excavating below ground water level	-	0.32	2.56	-	m³	2.56
next existing services	-	0.93	7.44	-	m³	7.44
around existing services crossing excavation	-	1.85	14.80	-	m³	14.80

D GROUNDWORK

	PC £	Labour hours	Labour £	Material/ Plant £	Unit	Total rate £
Extra over any types of excavating irrespective of depth for breaking out existing materials						
rock	-	4.63	37.03	5.60	m³	42.63
concrete	-	4.16	33.27	4.67	m³	37.94
reinforced concrete	-	5.55	44.39	6.53	m³	50.92
brickwork, blockwork or stonework	-	2.78	22.23	2.80	m³	25.03
Extra over any types of excavating irrespective of depth for breaking out existing hard pavings, 60 mm thick						
precast concrete paving slabs	-	0.28	2.24	-	m²	2.24
Extra over any types of excavating irrespective of depth for breaking out existing hard pavings, 75 mm thick						
coated macadam or asphalt	-	0.37	2.96	0.37	m²	3.33
Extra over any types of excavating irrespective of depth for breaking out existing hard pavings, 150 mm thick						
concrete	-	0.65	5.20	0.65	m²	5.85
reinforced concrete	-	0.83	6.64	0.93	m²	7.57
coated macadam or asphalt and hardcore	-	0.46	3.68	0.47	m²	4.15
Working space allowance to excavations						
reduce levels, basements and the like	-	2.13	17.04	-	m²	17.04
pits	-	2.22	17.76	-	m²	17.76
trenches	-	1.94	15.52	-	m²	15.52
pile caps and ground beams between piles	-	2.31	18.48	-	m²	18.48
Extra over excavation for working space for backfilling with special materials						
hardcore	-	0.74	5.92	9.57	m²	15.49
sand	-	0.74	5.92	15.36	m²	21.28
40 mm - 20 mm gravel	-	0.74	5.92	15.69	m²	21.61
plain in situ concrete ready mixed designated concrete; C7.5 - 40 mm aggregate	-	1.02	9.01	37.58	m²	46.59
Earthwork support (average "risk" prices)						
Maximum depth not exceeding 1.00 m						
distance between opposing faces not exceeding 2.00 m	-	0.10	0.80	0.33	m²	1.13
distance between opposing faces 2.00 - 4.00 m	-	0.11	0.88	0.39	m²	1.27
distance between opposing faces exceeding 4.00 m	-	0.12	0.96	0.49	m²	1.45
Maximum depth not exceeding 2.00 m						
distance between opposing faces not exceeding 2.00 m	-	0.12	0.96	0.39	m²	1.35
distance between opposing faces 2.00 - 4.00 m	-	0.13	1.04	0.49	m²	1.53
distance between opposing faces exceeding 4.00 m	-	0.14	1.12	0.62	m²	1.74
Maximum depth not exceeding 4.00 m						
distance between opposing faces not exceeding 2.00 m	-	0.16	1.28	0.49	m³	1.77
distance between opposing faces 2.00 - 4.00 m	-	0.16	1.28	0.62	m²	1.90
distance between opposing faces exceeding 4.00 m	-	0.18	1.44	0.78	m²	2.22
Maximum depth not exceeding 6.00 m						
distance between opposing faces not exceeding 2.00 m	-	0.18	1.44	0.59	m²	2.03
distance between opposing faces 2.00 - 4.00 m	-	0.19	1.52	0.78	m²	2.30
distance between opposing faces exceeding 4.00 m	-	0.22	1.76	0.98	m²	2.74
Maximum depth not exceeding 8.00 m						
distance between opposing faces not exceeding 2.00 m	-	0.23	1.84	0.78	m²	2.62
distance between opposing faces 2.00 - 4.00 m	-	0.28	2.24	0.98	m²	3.22
distance between opposing faces exceeding 4.00 m	-	0.33	2.64	1.17	m²	3.81

Prices for Measured Works - Major Works
D GROUNDWORK

	PC £	Labour hours	Labour £	Material/ Plant £	Unit	Total rate £
D20 EXCAVATING AND FILLING - cont'd						
Earthwork support (open boarded)						
Maximum depth not exceeding 1.00 m						
distance between opposing faces not exceeding 2.00 m	-	0.28	2.24	0.69	m²	2.93
distance between opposing faces 2.00 - 4.00 m	-	0.31	2.48	0.78	m²	3.26
distance between opposing faces exceeding 4.00 m	-	0.35	2.80	0.98	m²	3.78
Maximum depth not exceeding 2.00 m						
distance between opposing faces not exceeding 2.00 m	-	0.35	2.80	0.78	m²	3.58
distance between opposing faces 2.00 - 4.00 m	-	0.39	3.12	0.94	m²	4.06
distance between opposing faces exceeding 4.00 m	-	0.44	3.52	1.17	m²	4.69
Maximum depth not exceeding 4.00 m						
distance between opposing faces not exceeding 2.00 m	-	0.44	3.52	0.89	m²	4.41
distance between opposing faces 2.00 - 4.00 m	-	0.50	4.00	1.09	m²	5.09
distance between opposing faces exceeding 4.00 m	-	0.56	4.48	1.37	m²	5.85
Maximum depth not exceeding 6.00 m						
distance between opposing faces not exceeding 2.00 m	-	0.56	4.48	0.98	m²	5.46
distance between opposing faces 2.00 - 4.00 m	-	0.61	4.88	1.23	m²	6.11
distance between opposing faces exceeding 4.00 m	-	0.70	5.60	1.56	m²	7.16
Maximum depth not exceeding 8.00 m						
distance between opposing faces not exceeding 2.00 m	-	0.74	5.92	1.28	m²	7.20
distance between opposing faces 2.00 - 4.00 m	-	0.83	6.64	1.47	m²	8.11
distance between opposing faces exceeding 4.00 m	-	0.97	7.76	1.96	m²	9.72
Earthwork support (close boarded)						
Maximum depth not exceeding 1.00 m						
distance between opposing faces not exceeding 2.00 m	-	0.74	5.92	1.37	m²	7.29
distance between opposing faces 2.00 - 4.00 m	-	0.81	6.48	1.56	m²	8.04
distance between opposing faces exceeding 4.00 m	-	0.90	7.20	1.96	m²	9.16
Maximum depth not exceeding 2.00 m						
distance between opposing faces not exceeding 2.00 m	-	0.93	7.44	1.56	m²	9.00
distance between opposing faces 2.00 - 4.00 m	-	1.02	8.16	1.87	m²	10.03
distance between opposing faces exceeding 4.00 m	-	1.11	8.88	2.35	m²	11.23
Maximum depth not exceeding 4.00 m						
distance between opposing faces not exceeding 2.00 m	-	1.16	9.28	1.76	m²	11.04
distance between opposing faces 2.00 - 4.00 m	-	1.30	10.40	2.19	m²	12.59
distance between opposing faces exceeding 4.00 m	-	1.43	11.44	2.74	m²	14.18
Maximum depth not exceeding 6.00 m						
distance between opposing faces not exceeding 2.00 m	-	1.44	11.52	1.96	m²	13.48
distance between opposing faces 2.00 - 4.00 m	-	1.57	12.56	2.46	m²	15.02
distance between opposing faces exceeding 4.00 m	-	1.76	14.08	3.13	m²	17.21

D GROUNDWORK

	PC £	Labour hours	Labour £	Material/ Plant £	Unit	Total rate £
Maximum depth not exceeding 8.00 m						
distance between opposing faces not exceeding 2.00 m	-	1.76	14.08	2.54	m^2	**16.62**
distance between opposing faces 2.00 - 4.00 m	-	1.94	15.52	2.93	m^2	**18.45**
distance between opposing faces exceeding 4.00 m	-	2.22	17.76	3.52	m^2	**21.28**
Extra over earthwork support for						
Curved	-	0.02	0.16	0.33	m^2	**0.49**
Below ground water level	-	0.28	2.24	0.30	m^2	**2.54**
Unstable ground	-	0.46	3.68	0.59	m^2	**4.27**
Next to roadways	-	0.37	2.96	0.49	m^2	**3.45**
Left in	-	0.60	4.80	13.69	m^2	**18.49**
Earthwork support (average "risk" prices - inside existing buildings)						
Maximum depth not exceeding 1.00 m						
distance between opposing faces not exceeding 2.00 m	-	0.18	1.44	0.49	m^2	**1.93**
distance between opposing faces 2.00 - 4.00 m	-	0.19	1.52	0.56	m^2	**2.08**
distance between opposing faces exceeding 4.00 m	-	0.22	1.76	0.69	m^2	**2.45**
Maximum depth not exceeding 2.00 m						
distance between opposing faces not exceeding 2.00 m	-	0.22	1.76	0.56	m^2	**2.32**
distance between opposing faces 2.00 - 4.00 m	-	0.24	1.92	0.75	m^2	**2.67**
distance between opposing faces exceeding 4.00 m	-	0.32	2.56	0.83	m^2	**3.39**
Maximum depth not exceeding 4.00 m						
distance between opposing faces not exceeding 2.00 m	-	0.28	2.24	0.75	m^2	**2.99**
distance between opposing faces 2.00 - 4.00 m	-	0.31	2.48	0.89	m^2	**3.37**
distance between opposing faces exceeding 4.00 m	-	0.34	2.72	1.04	m^2	**3.76**
Maximum depth not exceeding 6.00 m						
distance between opposing faces not exceeding 2.00 m	-	0.34	2.72	0.84	m^2	**3.56**
distance between opposing faces 2.00 - 4.00 m	-	0.38	3.04	1.04	m^2	**4.08**
distance between opposing faces exceeding 4.00 m	-	0.43	3.44	1.23	m^2	**4.67**
Disposal; by machine						
Excavated material						
off site; to tip not exceeding 13 km (using lorries); including Landfill Tax based on inactive waste	-	-	-	14.48	m^3	**14.48**
on site; depositing in spoil heaps; average 25 m distance	-	-	-	0.78	m^3	**0.78**
on site; spreading; average 25 m distance	-	0.20	1.60	0.52	m^3	**2.12**
on site; depositing in spoil heaps; average 50 m distance	-	-	-	1.30	m^3	**1.30**
on site; spreading; average 50 m distance	-	0.20	1.60	1.04	m^3	**2.64**
on site; depositing in spoil heaps; average 100 m distance	-	-	-	2.34	m^3	**2.34**
on site; spreading; average 100 m distance	-	0.20	1.60	1.56	m^3	**3.16**
on site; depositing in spoil heaps; average 200 m distance	-	-	-	2.93	m^3	**2.93**
on site; spreading; average 200 m distance	-	0.20	1.60	2.08	m^3	**3.68**

D20 EXCAVATING AND FILLING - cont'd

	PC £	Labour hours	Labour £	Material/ Plant £	Unit	Total rate £
Disposal; by hand						
Excavated material						
off site; to tip not exceeding 13 km (using lorries); including Landfill Tax based on inactive waste	-	0.74	5.92	23.47	m³	29.39
on site; depositing in spoil heaps; average 25 m distance	-	1.02	8.16	-	m³	8.16
on site; spreading; average 25 m distance	-	1.34	10.72	-	m³	10.72
on site; depositing in spoil heaps; average 50 m distance	-	1.34	10.72	-	m³	10.72
on site; spreading; average 50 m distance	-	1.62	12.96	-	m³	12.96
on site; depositing in spoil heaps; average 100 m distance	-	1.94	15.52	-	m³	15.52
on site; spreading; average 100 m distance	-	2.22	17.76	-	m³	17.76
on site; depositing in spoil heaps; average 200 m distance	-	2.87	22.95	-	m³	22.95
on site; spreading; average 200 m distance	-	3.15	25.19	-	m³	25.19
Filling to excavations; by machine						
Average thickness not exceeding 0.25 m						
arising from the excavations	-	0.17	1.36	2.08	m³	3.44
obtained off site; hardcore	7.36	0.19	1.52	21.13	m³	22.65
obtained off site; granular fill type one	10.46	0.19	1.52	30.49	m³	32.01
obtained off site; granular fill type two	10.06	0.19	1.52	29.47	m³	30.99
Average thickness exceeding 0.25 m						
arising from the excavations	-	0.14	1.12	1.56	m³	2.68
obtained off site; hardcore	7.36	0.16	1.28	17.89	m³	19.17
obtained off site; granular fill type one	10.46	0.16	1.28	29.71	m³	30.99
obtained off site; granular fill type two	10.06	0.16	1.28	28.69	m³	29.97
Filling to make up levels; by machine						
Average thickness not exceeding 0.25 m						
arising from the excavations	-	0.24	1.92	2.25	m³	4.17
obtained off site; imported topsoil	8.83	0.24	1.92	12.26	m³	14.18
obtained off site; hardcore	7.36	0.28	2.24	21.15	m³	23.39
obtained off site; granular fill type one	10.46	0.28	2.24	30.51	m³	32.75
obtained off site; granular fill type two	10.06	0.28	2.24	29.49	m³	31.73
obtained off site; sand	10.46	0.28	2.24	30.51	m³	32.75
Average thickness exceeding 0.25 m						
arising from the excavations	-	0.20	1.60	1.62	m³	3.22
obtained off site; imported topsoil	8.83	0.20	1.60	11.63	m³	13.23
obtained off site; hardcore	7.36	0.24	1.92	17.80	m³	19.72
obtained off site; granular fill type one	10.46	0.24	1.92	29.62	m³	31.54
obtained off site; granular fill type two	10.06	0.24	1.92	28.60	m³	30.52
obtained off site; sand	10.46	0.24	1.92	29.62	m³	31.54
Filling to excavations; by hand						
Average thickness not exceeding 0.25 m						
arising from the excavations	-	1.16	9.28	-	m³	9.28
obtained off site; hardcore	7.36	1.25	10.00	19.05	m³	29.05
obtained off site; granular fill type one	10.46	1.48	11.84	26.59	m³	38.43
obtained off site; granular fill type two	10.06	1.48	11.84	25.58	m³	37.42
obtained off site; sand	10.46	1.48	11.84	26.59	m³	38.43

D GROUNDWORK

	PC £	Labour hours	Labour £	Material/ Plant £	Unit	Total rate £
Average thickness exceeding 0.25 m						
arising from the excavations	-	0.93	7.44	-	m³	7.44
obtained off site; hardcore	7.36	1.02	8.16	16.33	m³	24.49
obtained off site; granular fill type one	10.46	1.20	9.60	26.59	m³	36.19
obtained off site; granular fill type two	10.06	1.20	9.60	25.58	m³	35.18
obtained off site; sand	10.46	1.20	9.60	26.59	m³	36.19
Filling to make up levels; by hand						
Average thickness not exceeding 0.25 m						
arising from the excavations	-	1.25	10.00	3.58	m³	13.58
obtained off site; imported soil	8.83	1.25	10.00	13.07	m³	23.07
obtained off site; hardcore	7.36	1.39	11.12	23.03	m³	34.15
obtained off site; granular fill type one	10.46	1.54	12.32	30.99	m³	43.31
obtained off site; granular fill type two	10.06	1.54	12.32	29.97	m³	42.29
obtained off site; sand	10.46	1.54	12.32	30.99	m³	43.31
Average thickness exceeding 0.25 m						
arising from the excavations	-	1.02	8.16	2.91	m³	11.07
arising from on site spoil heaps; average 25 m distance; multiple handling	-	2.22	17.76	6.36	m³	24.12
obtained off site; imported soil	8.83	1.02	8.16	12.41	m³	20.57
obtained off site; hardcore	7.36	1.34	10.72	20.17	m³	30.89
obtained off site; granular fill type one	10.46	1.43	11.44	30.70	m³	42.14
obtained off site; granular fill type two	10.06	1.43	11.44	29.68	m³	41.12
obtained off site; sand	10.46	1.43	11.44	30.70	m³	42.14
Surface packing to filling						
To vertical or battered faces	-	0.17	1.36	0.18	m²	1.54
Surface treatments						
Compacting						
filling; blinding with sand	-	0.04	0.32	1.46	m²	1.78
bottoms of excavations	-	0.04	0.32	0.03	m²	0.35
Trimming						
sloping surfaces	-	0.17	1.36	-	m²	1.36
sloping surfaces; in rock	-	0.93	7.29	1.31	m²	8.60
Filter membrane; one layer; laid on earth to receive granular material						
"Terram 500" filter membrane or other equal and approved; one layer; laid on earth	-	0.04	0.32	0.34	m²	0.66
"Terram 700" filter membrane or other equal and approved; one layer; laid on earth	-	0.04	0.32	0.36	m²	0.68
"Terram 1000"; filter membrane or other equal and approved; one layer; laid on earth	-	0.04	0.32	0.38	m²	0.70
"Terram 2000"; filter membrane or other equal and approved; one layer; laid on earth	-	0.04	0.32	0.81	m²	1.13
D30 CAST IN PLACE PILING						
NOTE: The following approximate prices, for the quantities of piling quoted, are for work on clear open sites with reasonable access. They are based on 500 mm nominal diameter piles, normal concrete mix 20.00 N/mm² reinforced for loading up to 40,000 kg depending upon ground conditions and include up to 0.16 m of projecting reinforcement at top of pile. The prices do not allow for removal of spoil.						
(* indicates work normally carried out by the Main Contractor)						

Prices for Measured Works - Major Works
D GROUNDWORK

	PC £	Labour hours	Labour £	Material/ Plant £	Unit	Total rate £
D30 CAST IN PLACE PILING - cont'd						
Tripod bored cast-in-place concrete piles						
Provision of all plant; including bringing to and removing from site; maintenance, erection and dismantling at each pile position for 100 nr piles	-	-	-	-	item	7250.38
Bored piles						
500 mm diameter piles; reinforced; 10 m long	-	-	-	-	nr	709.06
add for additional piles length up to 15 m	-	-	-	-	m	74.90
deduct for reduction in pile length	-	-	-	-	m	24.97
Cutting off tops of piles*	-	1.20	16.51	-	m	16.51
Blind bored piles						
500 mm diameter	-	-	-	-	m	49.93
Delays						
rig standing time	-	-	-	-	hour	89.88
Extra over piling						
breaking through obstructions	-	-	-	-	hour	109.85
Pile tests						
working to 600 kN/t; using tension piles as reaction; first pile	-	-	-	-	nr	4344.24
working to 600 kN/t; using tension piles as reaction; subsequent piles	-	-	-	-	nr	3994.70
Rotary bored cast-in-place concrete piles						
Provision of all plant; including bringing to and removing from site; maintenance, erection and dismantling at each pile position for 100 nr piles	-	-	-	-	item	8688.47
Bored piles						
500 mm diameter piles; reinforced; 10 m long	-	-	-	-	nr	349.54
add for additional piles length up to 15 m	-	-	-	-	m	37.45
deduct for reduction in pile length	-	-	-	-	m	16.97
Cutting off tops of piles*	-	1.20	16.51	-	m	16.51
Blind bored piles						
500 mm diameter	-	-	-	-	m	17.97
Delays						
rig standing time	-	-	-	-	hour	149.80
Extra over piling						
breaking through obstructions	-	-	-	-	hour	174.76
Pile tests						
working to 600 kN/t; using tension piles as reaction; first pile	-	-	-	-	nr	4094.57
working to 600 kN/t; using tension piles as reaction; subsequent piles	-	-	-	-	nr	3495.36
D32 STEEL PILING						
"Frodingham" steel sheeting piling or other equal and approved; grade 43A; pitched and driven						
Provision of all plant; including bringing to and removing from site; maintenance, erection and dismantling; assuming one rig for 1500 m² of piling	-	-	-	-	item	3645.16
Driven shell piles						
type 1N; 99.10 kg/m²	-	-	-	-	m²	82.39
type 2N; 112.30 kg/m²	-	-	-	-	m²	94.87
type 3N; 137.10 kg/m²	-	-	-	-	m²	104.87
type 4N; 170.80 kg/m²	-	-	-	-	m²	129.83
type 5N; 236.90 kg/m²	-	-	-	-	m²	169.77
Burning off tops of piles	-	-	-	-	m	10.99

Prices for Measured Works - Major Works
D GROUNDWORK

	PC £	Labour hours	Labour £	Material/ Plant £	Unit	Total rate £
"Frodingham" steel sheet piling or other equal or approved; extract only						
Provision of all plant; including bringing to and removing from site; maintenance, erection and dismantling; assuming one rig as before	-	-	-	-	nr	3245.69
Driven sheet piling; extract only						
type 1N; 99.10 kg/m²	-	-	-	-	m²	15.97
type 2N; 112.30 kg/m²	-	-	-	-	m²	15.97
type 3N; 137.10 kg/m²	-	-	-	-	m²	15.97
type 4N; 170.80 kg/m²	-	-	-	-	m²	15.97
type 5N; 236.90 kg/m²	-	-	-	-	m²	16.97
D40 EMBEDDED RETAINING WALLING						
Diaphragm walls; contiguous panel construction; panel lengths not exceeding 5 m						
Provision of all plant; including bringing to and removing from site; maintenance, erection and dismantling; assuming one rig for 1000 m² of walling	-	-	-	-	item	69907.25
Excavation for diaphragm wall; excavated material removed from site; Bentonite slurry supplied and disposed of						
600 mm thick walls	-	-	-	-	m³	214.72
1000 mm thick walls	-	-	-	-	m³	149.80
Ready mixed reinforced in situ concrete; normal portland cement; C30 - 10 mm aggregate in walls	-	-	-	-	m³	94.87
Reinforcement bar; BS 4449 cold rolled deformed square high yield steel bars; straight or bent						
25 mm - 40 mm diameter	-	-	-	-	t	649.14
20 mm diameter	-	-	-	-	t	649.14
16 mm diameter	-	-	-	-	t	649.14
Formwork 75 mm thick to form chases	-	-	-	-	m²	54.92
Construct twin guide walls in reinforced concrete; together with reinforcement and formwork along the axis of the diaphragm wall	-	-	-	-	m	224.71
Delays						
rig standing	-	-	-	-	hour	379.50
D41 CRIB WALLS/GABIONS/REINFORCED EARTH						
Crib walls						
Precast concrete crib units; "Anda-crib Mini" doweless system or other equal and approved; Phi-O-Group; dry joints; machine filled with crushed rock (excavating and foundations not included)						
125 mm thick	-	-	-	-	m²	112.35
Precast concrete crib units; "Anda-crib Maxi" doweless system or other equal and approved; Phi-O-Group; dry joints; machine filled with crushed rock (excavating and foundations not included)						
125 mm thick	-	-	-	-	m²	142.81
Precast concrete crib units; "Anda-crib Super Maxi" doweless system or other equal and approved; Phi-O-Group; dry joints; machine filled with crushed rock (excavating and foundations not included)						
125 mm thick	-	-	-	-	m²	164.78

D GROUNDWORK

	PC £	Labour hours	Labour £	Material/ Plant £	Unit	Total rate £
D41 CRIB WALLS/GABIONS/REINFORCED EARTH - cont'd						
Gabion baskets						
Wire mesh gabion baskets; Maccaferri Ltd or other equal and approved; galvanised mesh 80 mm x 100 mm; filling with broken stones 125 mm - 200 mm size						
2.00 x 1.00 x 0.50 m	-	1.46	20.09	63.08	nr	83.17
2.00 x 1.00 x 1.00 m	-	1.39	19.13	112.69	nr	131.82
3.00 x 1.00 x 0.50 m	-	1.94	26.69	82.52	nr	109.21
3.00 x 1.00 x 1.00 m	-	3.89	53.53	152.55	nr	206.08
"Reno" mattress gabion baskets or other equal and approved; Maccaferri Ltd; filling with broken stones 125 mm - 200 mm size						
6.00 x 2.00 x 0.17 m	-	0.46	6.33	147.82	nr	154.15
6.00 x 2.00 x 0.23 m	-	0.62	8.53	177.54	nr	186.07
6.00 x 2.00 x 0.30 m	-	0.69	9.49	208.75	nr	218.24
D50 UNDERPINNING						
Excavating; by machine						
Preliminary trenches						
maximum depth not exceeding 1.00 m	-	0.23	1.84	5.87	m^3	7.71
maximum depth not exceeding 2.00 m	-	0.28	2.24	7.05	m^3	9.29
maximum depth not exceeding 4.00 m	-	0.32	2.56	8.22	m^3	10.78
Extra over preliminary trench excavating for breaking out existing hard pavings, 150 mm thick						
concrete	-	0.65	5.20	0.65	m^2	5.85
Excavating; by hand						
Preliminary trenches						
maximum depth not exceeding 1.00 m	-	2.68	21.43	-	m^3	21.43
maximum depth not exceeding 2.00 m	-	3.05	24.39	-	m^3	24.39
maximum depth not exceeding 4.00 m	-	3.93	31.43	-	m^3	31.43
Extra over preliminary trench excavating for breaking out existing hard pavings, 150 mm thick						
concrete	-	0.28	2.24	1.64	m^2	3.88
Underpinning pits; commencing from 1.00 m below existing ground level						
maximum depth not exceeding 0.25 m	-	4.07	32.55	-	m^3	32.55
maximum depth not exceeding 1.00 m	-	4.44	35.51	-	m^3	35.51
maximum depth not exceeding 2.00 m	-	5.32	42.55	-	m^3	42.55
Underpinning pits; commencing from 2.00 m below existing ground level						
maximum depth not exceeding 0.25 m	-	5.00	39.99	-	m^3	39.99
maximum depth not exceeding 1.00 m	-	5.37	42.95	-	m^3	42.95
maximum depth not exceeding 2.00 m	-	6.24	49.91	-	m^3	49.91
Underpinning pits; commencing from 4.00 m below existing ground level						
maximum depth not exceeding 0.25 m	-	5.92	47.35	-	m^3	47.35
maximum depth not exceeding 1.00 m	-	6.29	50.31	-	m^3	50.31
maximum depth not exceeding 2.00 m	-	7.17	57.35	-	m^3	57.35
Extra over any types of excavating irrespective of depth						
excavating below ground water level	-	0.32	2.56	-	m^3	2.56

D GROUNDWORK

	PC £	Labour hours	Labour £	Material/ Plant £	Unit	Total rate £
Earthwork support to preliminary trenches (open boarded - in 3.00 m lengths)						
Maximum depth not exceeding 1.00 m						
distance between opposing faces not exceeding 2.00 m	-	0.37	2.96	1.28	m²	4.24
Maximum depth not exceeding 2.00 m						
distance between opposing faces not exceeding 2.00 m	-	0.46	3.68	1.56	m²	5.24
Maximum depth not exceeding 4.00 m						
distance between opposing faces not exceeding 2.00 m	-	0.59	4.72	1.96	m²	6.68
Earthwork support to underpinning pits (open boarded - in 3.00 m lengths)						
Maximum depth not exceeding 1.00 m						
distance between opposing faces not exceeding 2.00 m	-	0.41	3.28	1.37	m²	4.65
Maximum depth not exceeding 2.00 m						
distance between opposing faces not exceeding 2.00 m	-	0.51	4.08	1.76	m²	5.84
Maximum depth not exceeding 4.00 m						
distance between opposing faces not exceeding 2.00 m	-	0.65	5.20	2.15	m²	7.35
Earthwork support to preliminary trenches (closed boarded - in 3.00 m lengths)						
Maximum depth not exceeding 1.00 m						
1.00 m deep	-	0.93	7.44	2.15	m²	9.59
Maximum depth not exceeding 2.00 m						
distance between opposing faces not exceeding 2.00 m	-	1.16	9.28	2.74	m²	12.02
Maximum depth not exceeding 4.00 m						
distance between opposing faces not exceeding 2.00 m	-	1.43	11.44	3.32	m²	14.76
Earthwork support to underpinning pits (closed boarded - in 3.00 m lengths)						
Maximum depth not exceeding 1.00 m						
distance between opposing faces not exceeding 2.00 m	-	1.02	8.16	2.35	m²	10.51
Maximum depth not exceeding 2.00 m						
distance between opposing faces not exceeding 2.00 m	-	1.28	10.24	2.93	m²	13.17
Maximum depth not exceeding 4.00 m						
distance between opposing faces not exceeding 2.00 m	-	1.57	12.56	3.72	m²	16.28
Extra over earthwork support for						
Left in	-	0.69	5.52	13.69	m²	19.21
Cutting away existing projecting foundations						
Concrete						
maximum width 150 mm; maximum depth 150 mm	-	0.15	1.20	0.13	m	1.33
maximum width 150 mm; maximum depth 225 mm	-	0.22	1.76	0.19	m	1.95
maximum width 150 mm; maximum depth 300 mm	-	0.30	2.40	0.26	m	2.66
maximum width 300 mm; maximum depth 300 mm	-	0.58	4.64	0.50	m	5.14
Masonry						
maximum width one brick thick; maximum depth one course high	-	0.04	0.32	0.04	m	0.36
maximum width one brick thick; maximum depth two courses high	-	0.13	1.04	0.11	m	1.15

Prices for Measured Works - Major Works
D GROUNDWORK

	PC £	Labour hours	Labour £	Material/ Plant £	Unit	Total rate £
D50 UNDERPINNING - cont'd						
Cutting away existing projecting foundations - cont'd						
Masonry - cont'd						
maximum width one brick thick; maximum depth three courses high	-	0.25	2.00	0.22	m	2.22
maximum width one brick thick; maximum depth four courses high	-	0.42	3.36	0.36	m	3.72
Preparing the underside of existing work to receive the pinning up of the new work						
Width of existing work						
380 mm wide	-	0.56	4.48	-	m	4.48
600 mm wide	-	0.74	5.92	-	m	5.92
900 mm wide	-	0.93	7.44	-	m	7.44
1200 mm wide	-	1.11	8.88	-	m	8.88
Disposal; by hand						
Excavated material						
off site; to tip not exceeding 13 km (using lorries); including Landfill Tax based on inactive waste	-	0.74	5.92	29.33	m³	35.25
Filling to excavations; by hand						
Average thickness exceeding 0.25 m						
arising from the excavations	-	0.93	7.44	-	m³	7.44
Surface treatments						
Compacting						
bottoms of excavations	-	0.04	0.32	0.03	m²	0.35
Plain in situ ready mixed designated concrete; C10 - 40 mm aggregate; poured against faces of excavation						
Underpinning						
thickness not exceeding 150 mm	-	3.42	30.22	69.36	m³	99.58
thickness 150 - 450 mm	-	2.87	25.36	69.36	m³	94.72
thickness exceeding 450 mm	-	2.50	22.09	69.36	m³	91.45
Plain in situ ready mixed designated concrete; C20 - 20 mm aggregate; poured against faces of excavation						
Underpinning						
thickness not exceeding 150 mm	-	3.42	30.22	71.75	m³	101.97
thickness 150 - 450 mm	-	2.87	25.36	71.75	m³	97.11
thickness exceeding 450 mm	-	2.50	22.09	71.75	m³	93.84
Extra for working around reinforcement	-	0.28	2.47	-	m³	2.47
Sawn formwork; sides of foundations in underpinning						
Plain vertical						
height exceeding 1.00 m	-	1.48	20.46	4.89	m²	25.35
height not exceeding 250 mm	-	0.51	7.05	1.40	m²	8.45
height 250 - 500 mm	-	0.79	10.92	2.61	m²	13.53
height 500 mm - 1.00 m	-	1.20	16.59	4.89	m²	21.48
Reinforcement bar; BS4449; hot rolled plain round mild steel bars						
20 mm diameter nominal size						
bent	258.26	27.00	291.99	319.89	t	611.88
16 mm diameter nominal size						
bent	258.26	29.00	314.12	323.44	t	637.56

D GROUNDWORK

	PC £	Labour hours	Labour £	Material/ Plant £	Unit	Total rate £
12 mm diameter nominal size						
bent	261.05	31.00	336.24	330.13	t	**666.37**
10 mm diameter nominal size						
bent	265.69	32.00	347.30	341.42	t	**688.72**
8 mm diameter nominal size						
bent	275.91	34.00	367.20	356.50	t	**723.70**
Reinforcement bar; BS 4461; cold worked deformed square high yield steel bars						
20 mm diameter nominal size						
bent	258.26	27.00	291.99	319.89	t	**611.88**
16 mm diameter nominal size						
bent	258.26	29.00	314.12	323.44	t	**637.56**
12 mm diameter nominal size						
bent	261.05	31.00	336.24	330.13	t	**666.37**
10 mm diameter nominal size						
bent	265.69	32.00	347.30	341.42	t	**688.72**
8 mm diameter nominal size						
bent	275.91	34.00	367.20	356.50	t	**723.70**
Common bricks; PC £175.80/1000; in cement mortar (1:3)						
Walls in underpinning						
one brick thick	-	2.22	38.80	26.09	m²	**64.89**
one and a half brick thick	-	3.05	53.31	38.81	m²	**92.12**
two brick thick	-	3.79	66.25	54.22	m²	**120.47**
Class A engineering bricks; PC £258.30/1000; in cement mortar (1:3)						
Walls in underpinning						
one brick thick	-	2.22	38.80	36.73	m²	**75.53**
one and a half brick thick	-	3.05	53.31	54.77	m²	**108.08**
two brick thick	-	3.79	66.25	75.50	m²	**141.75**
Class B engineering bricks; PC £184.00/1000; in cement mortar (1:3)						
Walls in underpinning						
one brick thick	-	2.22	38.80	27.15	m²	**65.95**
one and a half brick thick	-	3.05	53.31	40.39	m²	**93.70**
two brick thick	-	3.79	66.25	56.33	m²	**122.58**
Add or deduct for variation of £10.00/1000 in PC of bricks						
one brick thick	-	-	-	1.29	m²	**1.29**
one and a half bricks thick	-	-	-	1.94	m²	**1.94**
two bricks thick	-	-	-	2.58	m²	**2.58**
"Pluvex" (hessian based) damp proof course or similar; 200 mm laps; in cement mortar (1:3)						
Horizontal						
width exceeding 225 mm	5.62	0.23	4.02	6.98	m²	**11.00**
width not exceeding 225 mm	5.62	0.46	8.04	7.14	m²	**15.18**
"Hyload" (pitch polymer) damp proof course or similar; 150 mm laps; in cement mortar (1:3)						
Horizontal						
width exceeding 225 mm	4.96	0.23	4.02	6.16	m²	**10.18**
width not exceeding 225 mm	4.96	0.46	8.04	6.31	m²	**14.35**

D GROUNDWORK

	PC £	Labour hours	Labour £	Material/ Plant £	Unit	Total rate £
D50 UNDERPINNING - cont'd						
"Ledkore" grade A (bitumen based lead cored) damp proof course or other equal and approved; 200 mm laps; in cement mortar (1:3)						
Horizontal						
width exceeding 225 mm	14.75	0.31	5.42	18.32	m²	**23.74**
width not exceeding 225 mm	14.75	0.61	10.66	18.75	m²	**29.41**
Two courses of slates in cement mortar (1:3)						
Horizontal						
width exceeding 225 mm	-	1.39	24.30	21.62	m²	**45.92**
width not exceeding 225 mm	-	2.31	40.38	22.11	m²	**62.49**
Wedging and pinning						
To underside of existing construction with slates in cement mortar (1:3)						
width of wall - half brick thick	-	1.02	17.83	5.18	m	**23.01**
width of wall - one brick thick	-	1.20	20.98	10.36	m	**31.34**
width of wall - one and a half brick thick	-	1.39	24.30	15.54	m	**39.84**

E05 IN SITU CONCRETE CONSTRUCTION GENERALLY

MIXED CONCRETE PRICES (£/m³)

Designed mixes

Definition: "Mix for which the purchaser is responsible for specifying the required performances and the producer is responsible for selecting the mix proportions to produce the required performance".

NOTE: The following prices are for designed mix concrete ready for placing excluding any allowance for waste, discount or overheads and profit. Prices are based upon delivery to site within a 5 mile (8 km) radius of concrete mixing plant, using full loads.

Designed mix	Unit	Aggregate		
		10 mm £	20 mm £	40 mm £
Grade C7.5; cement to BS12	m³	66.33	64.31	63.66
Grade C7.5; sulphate resisting cement	m³	69.31	67.19	66.51
Grade C10; cement to BS12	m³	67.34	65.29	64.15
Grade C10; sulphate resisting cement	m³	72.44	68.22	67.03
Grade C15; cement to BS12	m³	67.87	65.80	64.63
Grade C15; sulphate resisting cement	m³	70.93	68.75	67.53
Grade C20; cement to BS12	m³	68.38	66.29	65.13
Grade C20; sulphate resisting cement	m³	71.47	69.27	68.06
Grade C25; cement to BS12	m³	68.90	66.80	65.62
Grade C25; sulphate resisting cement	m³	72.01	69.81	68.57
Grade C30; cement to BS12	m³	69.42	67.30	66.13
Grade C30; sulphate resisting cement	m³	72.56	70.33	69.10
Grade C40; cement to BS12	m³	69.96	68.31	-
Grade C40; sulphate resisting cement	m³	73.12	70.88	-
Grade C50; cement to BS12	m³	70.49	68.33	-
Grade C50; sulphate resisting cement	m³	73.68	71.41	-

E05 IN SITU CONCRETE CONSTRUCTION GENERALLY - cont'd

MIXED CONCRETE PRICES (£/m³)

Prescribed mixes

Definition: "Mix for which the purchaser specifies the proportions of the constituents and is responsible for ensuring that these proportions will produce a concrete with the performance required".

NOTE: The following prices are for prescribed mix concrete ready for placing excluding any allowance for waste, discount or overheads and profit. Prices are based upon delivery to site within a 5 mile (8 km) radius of concrete mixing plant, using full loads.

Prescribed mix	Unit	Aggregate		
		10 mm £	20 mm £	40 mm £
Grade C7.5; cement to BS12	m³	68.74	66.74	66.10
Grade C7.5; sulphate resisting cement	m³	71.69	69.59	68.92
Grade C10; cement to BS12	m³	69.74	67.71	66.58
Grade C10; sulphate resisting cement	m³	74.79	70.61	69.43
Grade C15; cement to BS12	m³	70.26	68.21	67.06
Grade C15; sulphate resisting cement	m³	73.29	71.14	69.93
Grade C20; cement to BS12	m³	70.77	68.70	67.55
Grade C20; sulphate resisting cement	m³	73.83	71.65	70.45
Grade C25; cement to BS12	m³	71.28	69.20	68.04
Grade C25; sulphate resisting cement	m³	74.36	72.18	70.96
Grade C30; cement to BS12	m³	71.80	69.70	68.54
Grade C30; sulphate resisting cement	m³	74.91	72.70	71.45
Grade C40; cement to BS12	m³	72.33	70.21	-
Grade C40; sulphate resisting cement	m³	75.46	73.24	-
Grade C50; cement to BS12	m³	72.86	70.72	-
Grade C50; sulphate resisting cement	m³	76.02	73.77	-

E IN SITU CONCRETE/LARGE PRECAST CONCRETE

MIXED CONCRETE PRICES (£/m³)

Designated mixes

Definition: "Mix produced in accordance with the specification given in section 5 of BS 5328: 2 : 1991 and requiring the producer to hold current product conformity certification based on product testing and surveillance coupled with approval of his quality system to BS 5750 : 1 (En 29001)".

NOTE: The following prices are for designated mix concrete ready for placing excluding any allowance for waste, discount or overheads and profit. Prices are based upon delivery to site within a 5 mile (8 km) radius of concrete mixing plant, using full loads.

Designated mix	Unit	Aggregate		
		10 mm £	20 mm £	40 mm £
Grade C7.5; cement to BS12	m³	65.74	63.74	63.10
Grade C7.5; sulphate resisting cement	m³	68.69	66.59	65.92
Grade C10; cement to BS12	m³	66.74	64.71	63.58
Grade C10; sulphate resisting cement	m³	71.79	67.61	66.43
Grade C15; cement to BS12	m³	67.26	65.21	64.06
Grade C15; sulphate resisting cement	m³	70.29	68.14	66.93
Grade C20; cement to BS12	m³	67.76	65.70	64.55
Grade C20; sulphate resisting cement	m³	70.83	68.65	67.45
Grade C25; cement to BS12	m³	68.28	66.20	65.04
Grade C25; sulphate resisting cement	m³	71.36	69.18	67.96
Grade C30; cement to BS12	m³	68.80	66.70	65.54
Grade C30; sulphate resisting cement	m³	71.91	69.70	68.48
Grade C40; cement to BS12	m³	69.33	67.21	-
Grade C40; sulphate resisting cement	m³	72.46	70.24	-
Grade C50; cement to BS12	m³	69.86	67.72	-
Grade C50; sulphate resisting cement	m³	73.02	70.77	-

E05 IN SITU CONSTRUCTION GENERALLY - cont'd

MIXED CONCRETE PRICES (£/m³)

Standard mixes

Definition: "Mix selected from the restricted list given in section 4 of BS 5328 : 2 : 1991 and made with a a restricted range of materials".

NOTE: The following prices are for standard mix concrete ready for placing excluding any allowance for waste, discount or overheads and profit. Prices are based upon delivery to site within a 5 mile (8 km) radius of concrete mixing plant, using full loads.

Standard mix	Unit	£
GEN 0	m³	55.48
GEN 1	m³	56.75
GEN 2	m³	58.02
GEN 3	m³	59.29
GEN 4	m³	60.56
ST 1	m³	58.02
ST 2	m³	59.55
ST 3	m³	61.07
ST 4	m³	62.59
ST 5	m³	64.62

Lightweight concrete

Mix	Unit	Aggregate		
		10 mm £	20 mm £	40 mm £
Grade 15; Lytag Medium and Natural Sand	m³	-	-	80.10
Grade 20; Lytag Medium and Natural Sand	m³	-	81.05	-
Grade 25; Lytag Medium and Natural Sand	m³	-	81.45	-
Grade 30; Lytag Medium and Natural Sand	m³	-	83.65	-

SITE MIXED CONCRETE

Site mixed	Unit	Aggregate		
		10 mm £	20 mm £	40 mm £
Mix 7.50 N/mm²; cement to BS12 (1:8)	m³	-	-	72.40
Mix 7.50 N/mm²; sulphate resisting cement (1:8)	m³	-	-	74.50
Mix 10.00 N/mm²; cement to BS12 (1:8)	m³	-	-	72.40
Mix 10.00 N/mm²; sulphate resisting cement (1:8)	m³	-	-	74.50
Mix 20.00 N/mm²; cement to BS12 (1:2:4)	m³	-	75.55	-
Mix 20.00 N/mm²; sulphate resisting cement (1:2:4)	m³	-	78.05	-
Mix 25.00 N/mm²; cement to BS12 (1:1:5:3)	m³	-	82.70	-
Mix 25.00 N/mm²; sulphate resisting cement (1:1:5:3)	m³	-	86.10	-

E05 IN SITU CONCRETE CONSTRUCTION GENERALLY - cont'd

MIXED CONCRETE PRICES (£/m³)

Add to the preceding prices for:	Unit	£
Rapid-hardening cement to BS 12	m³	3.79
Polypropylene fibre additive	m³	1.07
Air entrained concrete	m³	3.79
Water repellent additive	m³	4.04
Distance per mile in excess of 5 miles (8 km)	m³	0.85
Part loads per m³ below full load	m³	19.48

CEMENTS

Cement type:	Bulk	£
Ordinary Portland to BS12	t	91.07
Lightning high alumina	t	272.25
Sulfacrete sulphate resisting	t	108.03
Ferrocrete rapid hardening	t	102.39
Snowcrete white cement	t	160.14

CEMENT ADMIXTURES

Cement Admixture:	Unit	£
Febtone colorant - red, marigold, yellow, brown, black	kg	9.63
Febproof waterproof	5 litres	17.63
Febond PVA bonding agent	5 litres	17.58
Febspeed frostproofer and hardener	5 litres	6.25

NOTE: A discount of 7% has been applied to the above Mixed Concrete Prices before shrinkage factors (at plus 2½% (or 5% where poured on or against earth or unblinded hardcore)) and waste (at plus 7½%) have been added to arrive at the following material prices for measured works.

E IN SITU CONCRETE/LARGE PRECAST CONCRETE

	PC £	Labour hours	Labour £	Material £	Unit	Total rate £
E IN SITU CONCRETE/LARGE PRECAST						
E05 IN SITU CONCRETE CONSTRUCTING GENERALLY						
Plain in situ ready mixed designated concrete; C10 - 40 mm aggregate						
Foundations	-	1.20	10.60	64.63	m³	75.23
Isolated foundations	-	1.39	12.28	64.63	m³	76.91
Beds						
thickness not exceeding 150 mm	-	1.62	14.32	64.63	m³	78.95
thickness 150 - 450 mm	-	1.16	10.25	64.63	m³	74.88
thickness exceeding 450 mm	-	0.93	8.22	64.63	m³	72.85
Filling hollow walls						
thickness not exceeding 150 mm	-	3.15	27.84	64.63	m³	92.47
Plain in situ ready mixed designated concrete; C10 - 40 mm aggregate; poured on or against earth or unblinded hardcore						
Foundations	-	1.25	11.05	66.21	m³	77.26
Isolated foundations	-	1.48	13.08	66.21	m³	79.29
Beds						
thickness not exceeding 150 mm	-	1.71	15.11	66.21	m³	81.32
thickness 150 - 450 mm	-	1.25	11.05	66.21	m³	77.26
thickness exceeding 450 mm	-	0.97	8.57	66.21	m³	74.78
Plain in situ ready mixed designated concrete; C20 - 20 mm aggregate						
Foundations	-	1.20	10.60	66.86	m³	77.46
Isolated foundations	-	1.39	12.28	66.86	m³	79.14
Beds						
thickness not exceeding 150 mm	-	1.76	15.55	66.86	m³	82.41
thickness 150 - 450 mm	-	1.20	10.60	66.86	m³	77.46
thickness exceeding 450 mm	-	0.93	8.22	66.86	m³	75.08
Filling hollow walls						
thickness not exceeding 150 mm	-	3.15	27.84	66.86	m³	94.70
Plain in situ ready mixed designated concrete; C20 - 20 mm aggregate; poured on or against earth or unblinded hardcore						
Foundations	-	1.25	11.05	68.49	m³	79.54
Isolated foundations	-	1.48	13.08	68.49	m³	81.57
Beds						
thickness not exceeding 150 mm	-	1.85	16.35	68.49	m³	84.84
thickness 150 - 450 mm	-	1.30	11.49	68.49	m³	79.98
thickness exceeding 450 mm	-	0.97	8.57	68.49	m³	77.06
Reinforced in situ ready mixed designated concrete; C20 - 20 mm aggregate						
Foundations	-	1.30	11.49	66.86	m³	78.35
Ground beams	-	2.59	22.89	66.86	m³	89.75
Isolated foundations	-	1.57	13.87	66.86	m³	80.73
Beds						
thickness not exceeding 150 mm	-	2.04	18.03	66.86	m³	84.89
thickness 150 - 450 mm	-	1.48	13.08	66.86	m³	79.94
thickness exceeding 450 mm	-	1.20	10.60	66.86	m³	77.46
Slabs						
thickness not exceeding 150 mm	-	3.24	28.63	66.86	m³	95.49
thickness 150 - 450 mm	-	2.59	22.89	66.86	m³	89.75
thickness exceeding 450 mm	-	2.31	20.41	66.86	m³	87.27

Prices for Measured Works - Major Works
E IN SITU CONCRETE/LARGE PRECAST CONCRETE

	PC £	Labour hours	Labour £	Material £	Unit	Total rate £
E05 IN SITU CONCRETE CONSTRUCTING GENERALLY - cont'd						
Reinforced in situ ready mixed designated concrete; C20 - 20 mm aggregate - cont'd						
Coffered and troughed slabs						
thickness 150 - 450 mm	-	2.96	26.16	66.86	m³	93.02
thickness exceeding 450 mm	-	2.59	22.89	66.86	m³	89.75
Extra over for sloping						
not exceeding 15 degrees	-	0.23	2.03	-	m³	2.03
over 15 degrees	-	0.46	4.06	-	m³	4.06
Walls						
thickness not exceeding 150 mm	-	3.42	30.22	66.86	m³	97.08
thickness 150 - 450 mm	-	2.73	24.12	66.86	m³	90.98
thickness exceeding 450 mm	-	2.41	21.30	66.86	m³	88.16
Beams						
isolated	-	3.70	32.70	66.86	m³	99.56
isolated deep	-	4.07	35.96	66.86	m³	102.82
attached deep	-	3.70	32.70	66.86	m³	99.56
Beam casings						
isolated	-	4.07	35.96	66.86	m³	102.82
isolated deep	-	4.44	39.23	66.86	m³	106.09
attached deep	-	4.07	35.96	66.86	m³	102.82
Columns	-	4.44	39.23	66.86	m³	106.09
Column casings	-	4.90	43.30	66.86	m³	110.16
Staircases	-	5.55	49.04	66.86	m³	115.90
Upstands	-	3.56	31.46	66.86	m³	98.32
Reinforced in situ ready mixed designated concrete; C25 - 20 mm aggregate						
Foundations	-	1.30	11.49	67.36	m³	78.85
Ground beams	-	2.59	22.89	67.36	m³	90.25
Isolated foundations	-	1.57	13.87	67.36	m³	81.23
Beds						
thickness not exceeding 150 mm	-	1.85	16.35	67.36	m³	83.71
thickness 150 - 450 mm	-	1.39	12.28	67.36	m³	79.64
thickness exceeding 450 mm	-	1.20	10.60	67.36	m³	77.96
Slabs						
thickness not exceeding 150 mm	-	3.05	26.95	67.36	m³	94.31
thickness 150 - 450 mm	-	2.54	22.44	67.36	m³	89.80
thickness exceeding 450 mm	-	2.31	20.41	67.36	m³	87.77
Coffered and troughed slabs						
thickness 150 - 450 mm	-	2.87	25.36	67.36	m³	92.72
thickness exceeding 450 mm	-	2.59	22.89	67.36	m³	90.25
Extra over for sloping						
not exceeding 15 degrees	-	0.23	2.03	-	m³	2.03
over 15 degrees	-	0.46	4.06	-	m³	4.06
Walls						
thickness not exceeding 150 mm	-	3.33	29.43	67.36	m³	96.79
thickness 150 - 450 mm	-	2.68	23.68	67.36	m³	91.04
thickness exceeding 450 mm	-	2.41	21.30	67.36	m³	88.66
Beams						
isolated	-	3.70	32.70	67.36	m³	100.06
isolated deep	-	4.07	35.96	67.36	m³	103.32
attached deep	-	3.70	32.70	67.36	m³	100.06
Beam casings						
isolated	-	4.07	35.96	67.36	m³	103.32
isolated deep	-	4.44	39.23	67.36	m³	106.59
attached deep	-	4.07	35.96	67.36	m³	103.32
Columns	-	4.44	39.23	67.36	m³	106.59
Column casings	-	4.90	43.30	67.36	m³	110.66

E IN SITU CONCRETE/LARGE PRECAST CONCRETE

	PC £	Labour hours	Labour £	Material £	Unit	Total rate £
Staircases	-	5.55	49.04	67.36	m³	116.40
Upstands	-	3.56	31.46	67.36	m³	98.82
Reinforced in situ ready mixed designated concrete; C30 -20 mm aggregate						
Foundations	-	1.30	11.49	67.87	m³	79.36
Ground beams	-	2.59	22.89	67.87	m³	90.76
Isolated foundations	-	1.57	13.87	67.87	m³	81.74
Beds						
thickness not exceeding 150 mm	-	2.04	18.03	67.87	m³	85.90
thickness 150 - 450 mm	-	1.48	13.08	67.87	m³	80.95
thickness exceeding 450 mm	-	1.20	10.60	67.87	m³	78.47
Slabs						
thickness not exceeding 150 mm	-	3.24	28.63	67.87	m³	96.50
thickness 150 - 450 mm	-	2.59	22.89	67.87	m³	90.76
thickness exceeding 450 mm	-	2.31	20.41	67.87	m³	88.28
Coffered and troughed slabs						
thickness 150 - 450 mm	-	2.96	26.16	67.87	m³	94.03
thickness exceeding 450 mm	-	2.59	22.89	67.87	m³	90.76
Extra over for sloping						
not exceeding 15 degrees	-	0.23	2.03	-	m³	2.03
over 15 degrees	-	0.46	4.06	-	m³	4.06
Walls						
thickness not exceeding 150 mm	-	3.42	30.22	67.87	m³	98.09
thickness 150 - 450 mm	-	2.73	24.12	67.87	m³	91.99
thickness exceeding 450 mm	-	2.41	21.30	67.87	m³	89.17
Beams						
isolated	-	3.70	32.70	67.87	m³	100.57
isolated deep	-	4.07	35.96	67.87	m³	103.83
attached deep	-	3.70	32.70	67.87	m³	100.57
Beam casings						
isolated	-	4.07	35.96	67.87	m³	103.83
isolated deep	-	4.44	39.23	67.87	m³	107.10
attached deep	-	4.07	35.96	67.87	m³	103.83
Columns	-	4.44	39.23	67.87	m³	107.10
Column casings	-	4.90	43.30	67.87	m³	111.17
Staircases	-	5.55	49.04	67.87	m³	116.91
Upstands	-	3.56	31.46	67.87	m³	99.33
Extra over vibrated concrete for						
Reinforcement content over 5%	-	0.51	4.51	-	m³	4.51
Grouting with cement mortar (1:1)						
Stanchion bases						
10 mm thick	-	0.93	8.22	0.11	nr	8.33
25 mm thick	-	1.16	10.25	0.29	nr	10.54
Grouting with epoxy resin						
Stanchion bases						
10 mm thick	-	1.16	10.25	7.87	nr	18.12
25 mm thick	-	1.39	12.28	19.68	nr	31.96
Grouting with "Conbextra GP" cementitious grout						
Stanchion bases						
10 mm thick	-	1.16	10.25	1.09	nr	11.34
25 mm thick	-	1.39	12.28	2.91	nr	15.19

E IN SITU CONCRETE/LARGE PRECAST CONCRETE

	PC £	Labour hours	Labour £	Material £	Unit	Total rate £
E05 IN SITU CONCRETE CONSTRUCTING GENERALLY - cont'd						
Filling; plain in situ designated concrete; C20 - 20 mm aggregate						
Mortices	-	0.09	0.80	0.37	nr	1.17
Holes	-	0.23	2.03	78.22	m^3	80.25
Chases exceeding 0.01 m^2	-	0.19	1.68	78.22	m^3	79.90
Chases not exceeding 0.01 m^2	-	0.14	1.24	0.78	m	2.02
Sheeting to prevent moisture loss						
Building paper; lapped joints						
subsoil grade; horizontal on foundations	-	0.02	0.18	0.43	m^2	0.61
standard grade; horizontal on slabs	-	0.04	0.35	0.61	m^2	0.96
Polythene sheeting; lapped joints; horizontal on slabs						
250 microns; 0.25 mm thick	-	0.04	0.35	0.29	m^2	0.64
"Visqueen" sheeting or other equal and approved; lapped joints; horizontal on slabs						
250 microns; 0.25 mm thick	-	0.04	0.35	0.24	m^2	0.59
300 microns; 0.30 mm thick	-	0.05	0.44	0.33	m^2	0.77
E20 FORMWORK FOR IN SITU CONCRETE						
NOTE: Generally all formwork based on four uses unless otherwise stated.						
Sides of foundations; basic finish						
Plain vertical						
height exceeding 1.00 m	-	1.48	20.46	7.98	m^2	28.44
height exceeding 1.00 m; left in	-	1.30	17.97	19.25	m^2	37.22
height not exceeding 250 mm	-	0.42	5.81	3.17	m	8.98
height not exceeding 250 mm; left in	-	0.42	5.81	5.60	m	11.41
height 250 - 500 mm	-	0.79	10.92	6.21	m	17.13
height 250 - 500 mm; left in	-	0.69	9.54	12.51	m	22.05
height 500 mm - 1.00 m	-	1.11	15.35	7.98	m	23.33
height 500 mm - 1.00 m; left in	-	1.06	14.65	19.25	m	33.90
Sides of foundations; polystyrene sheet formwork; Cordek "Claymaster" or other equal and approved; 50 mm thick						
Plain vertical						
height exceeding 1.00 m; left in	-	0.30	4.15	4.67	m^2	8.82
height not exceeding 250 mm; left in	-	0.09	1.24	1.17	m	2.41
height 250 - 500 mm; left in	-	0.16	2.21	2.34	m	4.55
height 500 mm - 1.00 m; left in	-	0.24	3.32	4.67	m	7.99
Sides of foundations; polystyrene sheet formwork; Cordek "Claymaster" or other equal and approved; 100 mm thick						
Plain vertical						
height exceeding 1.00 m; left in	-	0.32	4.42	9.33	m^2	13.75
height not exceeding 250 mm; left in	-	0.10	1.38	2.33	m	3.71
height 250 - 500 mm; left in	-	0.18	2.49	4.67	m	7.16
height 500 mm - 1.00 m; left in	-	0.27	3.73	9.33	m	13.06
Sides of ground beams and edges of beds; basic finish						
Plain vertical						
height exceeding 1.00 m	-	1.53	21.15	7.93	m^2	29.08
height not exceeding 250 mm	-	0.46	6.36	3.12	m	9.48
height 250 - 500 mm	-	0.83	11.47	6.16	m	17.63
height 500 mm - 1.00 m	-	1.16	16.04	7.93	m	23.97

E IN SITU CONCRETE/LARGE PRECAST CONCRETE

	PC £	Labour hours	Labour £	Material £	Unit	Total rate £
Edges of suspended slabs; basic finish						
Plain vertical						
height not exceeding 250 mm	-	0.69	9.54	3.22	m	12.76
height 250 - 500 mm	-	1.02	14.10	5.19	m	19.29
height 500 mm - 1.00 m	-	1.62	22.40	8.03	m	30.43
Sides of upstands; basic finish						
Plain vertical						
height exceeding 1.00 m	-	1.85	25.58	9.58	m^2	35.16
height not exceeding 250 mm	-	0.58	8.02	3.33	m	11.35
height 250 - 500 mm	-	0.93	12.86	6.37	m	19.23
height 500 mm - 1.00 m	-	1.62	22.40	9.58	m	31.98
Steps in top surfaces; basic finish						
Plain vertical						
height not exceeding 250 mm	-	0.46	6.36	3.38	m	9.74
height 250 - 500 mm	-	0.74	10.23	6.42	m	16.65
Steps in soffits; basic finish						
Plain vertical						
height not exceeding 250 mm	-	0.51	7.05	2.65	m	9.70
height 250 - 500 mm	-	0.81	11.20	4.77	m	15.97
Machine bases and plinths; basic finish						
Plain vertical						
height exceeding 1.00 m	-	1.48	20.46	7.93	m^2	28.39
height not exceeding 250 mm	-	0.46	6.36	3.12	m	9.48
height 250 - 500 mm	-	0.79	10.92	6.16	m	17.08
height 500 mm - 1.00 m	-	1.16	16.04	7.93	m	23.97
Soffits of slabs; basic finish						
Slab thickness not exceeding 200 mm						
horizontal; height to soffit not exceeding 1.50 m	-	1.67	23.09	7.36	m^2	30.45
horizontal; height to soffit 1.50 - 3.00 m	-	1.62	22.40	7.46	m^2	29.86
horizontal; height to soffit 1.50 - 3.00 m (based on 5 uses)	-	1.53	21.15	6.18	m^2	27.33
horizontal; height to soffit 1.50 - 3.00 m (based on 6 uses)	-	1.48	20.46	5.32	m^2	25.78
horizontal; height to soffit 3.00 - 4.50 m	-	1.57	21.70	7.72	m^2	29.42
horizontal; height to soffit 4.50 - 6.00 m	-	1.67	23.09	7.98	m^2	31.07
Slab thickness 200 - 300 mm						
horizontal; height to soffit 1.50 - 3.00 m	-	1.67	23.09	9.42	m^2	32.51
Slab thickness 300 - 400 mm						
horizontal; height to soffit 1.50 - 3.00 m	-	1.71	23.64	10.40	m^2	34.04
Slab thickness 400 - 500 mm						
horizontal; height to soffit 1.50 - 3.00 m	-	1.80	24.88	11.38	m^2	36.26
Slab thickness 500 - 600 mm						
horizontal; height to soffit 1.50 - 3.00 m	-	1.94	26.82	11.38	m^2	38.20
Extra over soffits of slabs for						
sloping not exceeding 15 degrees	-	0.19	2.63	-	m^2	2.63
sloping exceeding 15 degrees	-	0.37	5.12	-	m^2	5.12
Soffits of slabs; Expamet "Hy-Rib" permanent shuttering and reinforcement or other equal and approved; ref. 2411						
Slab thickness not exceeding 200 mm						
horizontal; height to soffit 1.50 - 3.00 m	-	1.39	19.22	15.23	m^2	34.45

E IN SITU CONCRETE/LARGE PRECAST CONCRETE

	PC £	Labour hours	Labour £	Material £	Unit	Total rate £
E20 FORMWORK FOR IN SITU CONCRETE - cont'd						
Soffits of slabs; Richard Lees "Ribdeck AL" permanent shuttering or other equal and approved; 0.90 mm gauge; shot-fired to frame (not included)						
Slab thickness not exceeding 200 mm						
horizontal; height to soffit 1.50 - 3.00 m	-	0.65	8.99	11.85	m²	20.84
horizontal; height to soffit 3.00 - 4.50 m	-	0.74	10.23	11.90	m²	22.13
horizontal; height to soffit 4.50 - 6.00 m	-	0.83	11.47	11.96	m²	23.43
Soffits of slabs; Richard Lees "Ribdeck AL" permanent shuttering or other equal and approved; 1.20 mm gauge; shot-fired to frame (not included)						
Slab thickness not exceeding 200 mm						
horizontal; height to soffit 1.50 - 3.00 m	-	0.74	10.23	13.72	m²	23.95
horizontal; height to soffit 3.00 - 4.50 m	-	0.83	11.47	13.51	m²	24.98
horizontal; height to soffit 4.50 - 6.00 m	-	0.97	13.41	13.60	m²	27.01
Soffits of slabs; Richard Lees "Super Holorib" permanent shuttering or other equal and approved; 0.90 mm gauge						
Slab thickness not exceeding 200 mm						
horizontal; height to soffit 1.50 - 3.00 m	-	0.65	8.99	15.71	m²	24.70
horizontal; height to soffit 3.00 - 4.50 m	-	0.74	10.23	16.23	m²	26.46
horizontal; height to soffit 4.50 - 6.00 m	-	0.83	11.47	16.75	m²	28.22
Soffits of slabs; Richard Lees "Super Holorib" permanent shuttering or other equal and approved; 1.20 mm gauge						
Slab thickness not exceeding 200 mm						
horizontal; height to soffit 1.50 - 3.00 m	-	0.74	10.23	17.99	m²	28.22
horizontal; height to soffit 3.00 - 4.50 m	-	0.74	10.23	18.77	m²	29.00
horizontal; height to soffit 4.50 - 6.00 m	-	0.97	13.41	19.29	m²	32.70
Soffits of landings; basic finish						
Slab thickness not exceeding 200 mm						
horizontal; height to soffit 1.50 - 3.00 m	-	1.67	23.09	7.85	m²	30.94
Slab thickness 200 - 300 mm						
horizontal; height to soffit 1.50 - 3.00 m	-	1.76	24.33	10.01	m²	34.34
Slab thickness 300 - 400 mm						
horizontal; height to soffit 1.50 - 3.00 m	-	1.80	24.88	11.09	m²	35.97
Slab thickness 400 - 500 mm						
horizontal; height to soffit 1.50 - 3.00 m	-	1.90	26.27	12.17	m²	38.44
Slab thickness 500 - 600 mm						
horizontal; height to soffit 1.50 - 3.00 m	-	2.04	28.20	12.17	m²	40.37
Extra over soffits of landings for						
sloping not exceeding 15 degrees	-	0.19	2.63	-	m²	2.63
sloping exceeding 15 degrees	-	0.37	5.12	-	m²	5.12
Soffits of coffered or troughed slabs; basic finish Cordek "Correx" trough mould or other equal and approved; 300 mm deep; ribs of mould at 600 mm centres and cross ribs at centres of bay; slab thickness 300 - 400 mm						
horizontal; height to soffit 1.50 - 3.00 m	-	2.31	31.93	12.03	m²	43.96
horizontal; height to soffit 3.00 - 4.50 m	-	2.41	33.32	12.29	m²	45.61
horizontal; height to soffit 4.50 - 6.00 m	-	2.50	34.56	12.45	m²	47.01
Top formwork; basic finish						
Sloping exceeding 15 degrees	-	1.39	19.22	5.76	m²	24.98

E IN SITU CONCRETE/LARGE PRECAST CONCRETE

	PC £	Labour hours	Labour £	Material £	Unit	Total rate £
Walls; basic finish						
Vertical	-	1.67	23.09	9.42	m²	32.51
Vertical; height exceeding 3.00 m above floor level	-	2.04	28.20	9.68	m²	37.88
Vertical; interrupted	-	1.94	26.82	9.68	m²	36.50
Vertical; to one side only	-	3.24	44.79	11.91	m²	56.70
Battered	-	2.59	35.81	10.15	m²	45.96
Beams; basic finish						
Attached to slabs						
regular shaped; square or rectangular; height to soffit 1.50 - 3.00 m	-	2.04	28.20	9.11	m²	37.31
regular shaped; square or rectangular; height to soffit 3.00 - 4.50 m	-	2.13	29.45	9.42	m²	38.87
regular shaped; square or rectangular; height to soffit 4.50 - 6.00 m	-	2.22	30.69	9.68	m²	40.37
Attached to walls						
regular shaped; square or rectangular; height to soffit 1.50 - 3.00 m	-	2.13	29.45	9.11	m²	38.56
Isolated						
regular shaped; square or rectangular; height to soffit 1.50 - 3.00 m	-	2.22	30.69	9.11	m²	39.80
regular shaped; square or rectangular; height to soffit 3.00 - 4.50 m	-	2.31	31.93	9.42	m²	41.35
regular shaped; square or rectangular; height to soffit 4.50 - 6.00 m	-	2.41	33.32	9.68	m²	43.00
Extra over beams for						
regular shaped; sloping not exceeding 15 degrees	-	0.28	3.87	0.93	m²	4.80
regular shaped; sloping exceeding 15 degrees	-	0.56	7.74	1.86	m²	9.60
Beam casings; basic finish						
Attached to slabs						
regular shaped; square or rectangular; height to soffit 1.50 - 3.00 m	-	2.13	29.45	9.11	m²	38.56
regular shaped; square or rectangular; height to soffit 3.00 - 4.50 m	-	2.22	30.69	9.42	m²	40.11
Attached to walls						
regular shaped; square or rectangular; height to soffit 1.50 - 3.00 m	-	2.22	30.69	9.11	m²	39.80
Isolated						
regular shaped; square or rectangular; height to soffit 1.50 - 3.00 m	-	2.31	31.93	9.11	m²	41.04
regular shaped; square or rectangular; height to soffit 3.00 - 4.50 m	-	2.41	33.32	9.42	m²	42.74
Extra over beam casings for						
regular shaped; sloping not exceeding 15 degrees	-	0.28	3.87	0.93	m²	4.80
regular shaped; sloping exceeding 15 degrees	-	0.56	7.74	1.86	m²	9.60
Columns; basic finish						
Attached to walls						
regular shaped; square or rectangular; height to soffit 1.50 - 3.00 m	-	2.04	28.20	7.98	m²	36.18
Isolated						
regular shaped; square or rectangular; height to soffit 1.50 - 3.00 m	-	2.13	29.45	7.98	m²	37.43
regular shaped; circular; not exceeding 300 mm diameter; height to soffit 1.50 - 3.00 m	-	3.70	51.15	12.83	m²	63.98
regular shaped; circular; 300 - 600 mm diameter; height to soffit 1.50 - 3.00 m	-	3.47	47.97	11.38	m²	59.35
regular shaped; circular; 600 - 900 mm diameter; height to soffit 1.50 - 3.00 m	-	3.24	44.79	11.12	m²	55.91

E20 FORMWORK FOR IN SITU CONCRETE - cont'd

	PC £	Labour hours	Labour £	Material £	Unit	Total rate £
Column casings; basic finish						
Attached to walls						
regular shaped; square or rectangular; height to soffit 1.50 - 3.00 m	-	2.13	29.45	7.98	m²	37.43
Isolated						
regular shaped; square or rectangular; height to soffit 1.50 - 3.00 m	-	2.22	30.69	7.98	m²	38.67
Recesses or rebates						
12 x 12 mm	-	0.06	0.83	0.08	m	0.91
25 x 25 mm	-	0.06	0.83	0.13	m	0.96
25 x 50 mm	-	0.06	0.83	0.16	m	0.99
50 x 50 mm	-	0.06	0.83	0.27	m	1.10
Nibs						
50 x 50 mm	-	0.51	7.05	0.99	m	8.04
100 x 100 mm	-	0.72	9.95	1.31	m	11.26
100 x 200 mm	-	0.96	13.27	2.37	m	15.64
Extra over a basic finish for fine formed finishes						
Slabs	-	0.32	4.42	-	m²	4.42
Walls	-	0.32	4.42	-	m²	4.42
Beams	-	0.32	4.42	-	m²	4.42
Columns	-	0.32	4.42	-	m²	4.42
Add to prices for basic formwork for						
Curved radius 6.00 m - 50%						
Curved radius 2.00 m - 100%						
Coating with retardant agent	-	0.01	0.14	0.46	m²	0.60
Wall kickers; basic finish						
Height 150 mm	-	0.46	6.36	2.30	m	8.66
Height 225 mm	-	0.60	8.29	2.83	m	11.12
Suspended wall kickers; basic finish						
Height 150 mm	-	0.58	8.02	2.32	m	10.34
Wall ends, soffits and steps in walls; basic finish						
Plain						
width exceeding 1.00 m	-	1.76	24.33	9.42	m²	33.75
width not exceeding 250 mm	-	0.56	7.74	2.39	m	10.13
width 250 - 500 mm	-	0.88	12.17	5.23	m	17.40
width 500 mm - 1.00 m	-	1.39	19.22	9.42	m	28.64
Openings in walls						
Plain						
width exceeding 1.00 m	-	1.94	26.82	9.42	m²	36.24
width not exceeding 250 mm	-	0.60	8.29	2.39	m	10.68
width 250 - 500 mm	-	1.02	14.10	5.23	m	19.33
width 500 mm - 1.00 m	-	1.57	21.70	9.42	m	31.12
Stairflights						
Width 1.00 m; 150 mm waist; 150 mm undercut risers						
string, width 300 mm	-	4.63	64.01	20.77	m	84.78
Width 2.00 m; 200 mm waist; 150 mm undercut risers						
string, width 350 mm	-	8.33	115.16	36.34	m	151.50

Prices for Measured Works - Major Works
E IN SITU CONCRETE/LARGE PRECAST CONCRETE

	PC £	Labour hours	Labour £	Material £	Unit	Total rate £
Mortices						
Girth not exceeding 500 mm						
depth not exceeding 250 mm; circular	-	0.14	1.94	0.79	nr	**2.73**
Holes						
Girth not exceeding 500 mm						
depth not exceeding 250 mm; circular	-	0.19	2.63	1.04	nr	**3.67**
depth 250 - 500 mm; circular	-	0.28	3.87	2.50	nr	**6.37**
Girth 500 mm - 1.00 m						
depth not exceeding 250 mm; circular	-	0.23	3.18	1.75	nr	**4.93**
depth 250 - 500 mm; circular	-	0.35	4.84	4.64	nr	**9.48**
Girth 1.00 - 2.00 m						
depth not exceeding 250 mm; circular	-	0.42	5.81	4.64	nr	**10.45**
depth 250 - 500 mm; circular	-	0.62	8.57	9.70	nr	**18.27**
Girth 2.00 - 3.00 m						
depth not exceeding 250 mm; circular	-	0.56	7.74	9.34	nr	**17.08**
depth 250 - 500 mm; circular	-	0.83	11.47	16.71	nr	**28.18**
E30 REINFORCEMENT FOR IN SITU CONCRETE						
Bar; BS 4449; hot rolled plain round mild steel bars						
40 mm diameter nominal size						
straight or bent	290.78	20.00	214.56	353.05	t	**567.61**
curved	319.85	20.00	214.56	385.86	t	**600.42**
32 mm diameter nominal size						
straight or bent	268.48	21.00	225.62	328.93	t	**554.55**
curved	295.33	21.00	225.62	359.23	t	**584.85**
25 mm diameter nominal size						
straight or bent	263.84	23.00	247.75	323.66	t	**571.41**
curved	290.22	23.00	247.75	353.44	t	**601.19**
20 mm diameter nominal size						
straight or bent	261.05	25.00	269.87	323.04	t	**592.91**
curved	287.15	25.00	269.87	352.50	t	**622.37**
16 mm diameter nominal size						
straight or bent	261.05	27.00	291.99	326.58	t	**618.57**
curved	287.15	27.00	291.99	356.05	t	**648.04**
12 mm diameter nominal size						
straight or bent	265.69	29.00	314.12	335.37	t	**649.49**
curved	292.26	29.00	314.12	365.36	t	**679.48**
10 mm diameter nominal size						
straight or bent	271.27	31.00	336.24	347.71	t	**683.95**
curved	298.39	31.00	336.24	378.33	t	**714.57**
8 mm diameter nominal size						
straight or bent	280.56	33.00	356.14	361.75	t	**717.89**
links	280.56	36.00	389.32	363.84	t	**753.10**
curved	308.61	33.00	356.14	393.41	t	**749.55**
6 mm diameter nominal size						
straight or bent	305.64	37.00	400.39	390.06	t	**790.45**
links	305.64	40.00	433.57	390.06	t	**823.63**
curved	336.21	37.00	400.39	424.56	t	**824.95**
Bar; BS 4449; hot rolled deformed high steel bars grade 460						
40 mm diameter nominal size						
straight or bent	265.69	20.00	214.56	324.73	t	**539.29**
curved	292.26	20.00	214.56	354.72	t	**569.28**
32 mm diameter nominal size						
straight or bent	261.05	21.00	225.62	320.54	t	**546.16**
curved	287.15	21.00	225.62	350.00	t	**575.62**

E IN SITU CONCRETE/LARGE PRECAST CONCRETE

	PC £	Labour hours	Labour £	Material £	Unit	Total rate £
E30 REINFORCEMENT FOR IN SITU CONCRETE - cont'd						
Bar; BS 4449; hot rolled deformed high steel bars grade 460 - cont'd						
25 mm diameter nominal size						
straight or bent	256.40	23.00	247.75	315.27	t	563.02
curved	282.04	23.00	247.75	344.21	t	591.96
20 mm diameter nominal size						
straight or bent	258.26	25.00	269.87	319.89	t	589.76
curved	284.09	25.00	269.87	349.04	t	618.91
16 mm diameter nominal size						
straight or bent	258.26	27.00	291.99	323.44	t	615.43
curved	284.09	27.00	291.99	352.59	t	644.58
12 mm diameter nominal size						
straight or bent	261.05	29.00	314.12	330.13	t	644.25
curved	287.15	29.00	314.12	359.59	t	673.71
10 mm diameter nominal size						
straight or bent	265.69	31.00	336.24	341.42	t	677.66
curved	292.26	31.00	336.24	371.41	t	707.65
8 mm diameter nominal size						
straight or bent	275.91	33.00	356.14	356.50	t	712.64
links	275.91	36.00	389.32	358.60	t	747.92
curved	303.50	33.00	356.14	387.64	t	743.78
Bar; stainless steel						
32 mm diameter nominal size						
straight or bent	2100.85	21.00	225.62	2284.29	t	2509.91
curved	2181.25	21.00	225.62	2370.72	t	2596.34
25 mm diameter nominal size						
straight or bent	2100.85	23.00	247.75	2284.27	t	2532.02
curved	2181.25	23.00	247.75	2370.70	t	2618.45
20 mm diameter nominal size						
straight or bent	2100.85	25.00	269.87	2286.79	t	2556.66
curved	2181.25	25.00	269.87	2373.22	t	2643.09
16 mm diameter nominal size						
straight or bent	1714.93	27.00	291.99	1875.47	t	2167.46
curved	1795.33	27.00	291.99	1961.90	t	2253.89
12 mm diameter nominal size						
straight or bent	1714.93	29.00	314.12	1879.02	t	2193.14
curved	1821.86	29.00	314.12	1993.97	t	2308.09
10 mm diameter nominal size						
straight or bent	1714.93	31.00	336.24	1885.07	t	2221.31
curved	1835.53	31.00	336.24	2014.71	t	2350.95
8 mm diameter nominal size						
straight or bent	1714.93	33.00	356.14	1888.62	t	2244.76
curved	1862.06	33.00	356.14	2046.78	t	2402.92
Fabric; BS 4483						
Ref A98 (1.54 kg/m^2)						
400 mm minimum laps	0.68	0.12	1.33	0.80	m^2	2.13
strips in one width; 600 mm width	0.68	0.15	1.66	0.80	m^2	2.46
strips in one width; 900 mm width	0.68	0.14	1.55	0.80	m^2	2.35
strips in one width; 1200 mm width	0.68	0.13	1.44	0.80	m^2	2.24
Ref A142 (2.22 kg/m^2)						
400 mm minimum laps	0.71	0.12	1.33	0.84	m^2	2.17
strips in one width; 600 mm width	0.71	0.15	1.66	0.84	m^2	2.50
strips in one width; 900 mm width	0.71	0.14	1.55	0.84	m^2	2.39
strips in one width; 1200 mm width	0.71	0.13	1.44	0.84	m^2	2.28

E IN SITU CONCRETE/LARGE PRECAST CONCRETE

	PC £	Labour hours	Labour £	Material £	Unit	Total rate £
Ref A193 (3.02 kg/m²)						
400 mm minimum laps	0.97	0.12	1.33	1.15	m²	**2.48**
strips in one width; 600 mm width	0.97	0.15	1.66	1.15	m²	**2.81**
strips in one width; 900 mm width	0.97	0.14	1.55	1.15	m²	**2.70**
strips in one width; 1200 mm width	0.97	0.13	1.44	1.15	m²	**2.59**
Ref A252 (3.95 kg/m²)						
400 mm minimum laps	1.26	0.13	1.44	1.49	m²	**2.93**
strips in one width; 600 mm width	1.26	0.16	1.77	1.49	m²	**3.26**
strips in one width; 900 mm width	1.26	0.15	1.66	1.49	m²	**3.15**
strips in one width; 1200 mm width	1.26	0.14	1.55	1.49	m²	**3.04**
Ref A393 (6.16 kg/m²)						
400 mm minimum laps	1.93	0.15	1.66	2.28	m²	**3.94**
strips in one width; 600 mm width	1.93	0.18	1.99	2.28	m²	**4.27**
strips in one width; 900 mm width	1.93	0.17	1.88	2.28	m²	**4.16**
strips in one width; 1200 mm width	1.93	0.16	1.77	2.28	m²	**4.05**
Ref B196 (3.05 kg/m²)						
400 mm minimum laps	1.00	0.12	1.33	1.18	m²	**2.51**
strips in one width; 600 mm width	1.00	0.15	1.66	1.18	m²	**2.84**
strips in one width; 900 mm width	1.00	0.14	1.55	1.18	m²	**2.73**
strips in one width; 1200 mm width	1.00	0.13	1.44	1.18	m²	**2.62**
Ref B283 (3.73 kg/m²)						
400 mm minimum laps	1.21	0.12	1.33	1.43	m²	**2.76**
strips in one width; 600 mm width	1.21	0.15	1.66	1.43	m²	**3.09**
strips in one width; 900 mm width	1.21	0.14	1.55	1.43	m²	**2.98**
strips in one width; 1200 mm width	1.21	0.13	1.44	1.43	m²	**2.87**
Ref B385 (4.53 kg/m²)						
400 mm minimum laps	1.48	0.13	1.44	1.75	m²	**3.19**
strips in one width; 600 mm width	1.48	0.16	1.77	1.75	m²	**3.52**
strips in one width; 900 mm width	1.48	0.15	1.66	1.75	m²	**3.41**
strips in one width; 1200 mm width	1.48	0.14	1.55	1.75	m²	**3.30**
Ref B503 (5.93 kg/m²)						
400 mm minimum laps	1.93	0.15	1.66	2.28	m²	**3.94**
strips in one width; 600 mm width	1.93	0.18	1.99	2.28	m²	**4.27**
strips in one width; 900 mm width	1.93	0.17	1.88	2.28	m²	**4.16**
strips in one width; 1200 mm width	1.93	0.16	1.77	2.28	m²	**4.05**
Ref B785 (8.14 kg/m²)						
400 mm minimum laps	2.65	0.17	1.88	3.13	m²	**5.01**
strips in one width; 600 mm width	2.65	0.20	2.21	3.13	m²	**5.34**
strips in one width; 900 mm width	2.65	0.19	2.10	3.13	m²	**5.23**
strips in one width; 1200 mm width	2.65	0.18	1.99	3.13	m²	**5.12**
Ref B1131 (10.90 kg/m²)						
400 mm minimum laps	3.45	0.18	1.99	4.08	m²	**6.07**
strips in one width; 600 mm width	3.45	0.24	2.65	4.08	m²	**6.73**
strips in one width; 900 mm width	3.45	0.22	2.43	4.08	m²	**6.51**
strips in one width; 1200 mm width	3.45	0.20	2.21	4.08	m²	**6.29**
Ref C385 (3.41 kg/m²)						
400 mm minimum laps	1.14	0.12	1.33	1.35	m²	**2.68**
strips in one width; 600 mm width	1.14	0.15	1.66	1.35	m²	**3.01**
strips in one width; 900 mm width	1.14	0.14	1.55	1.35	m²	**2.90**
strips in one width; 1200 mm width	1.14	0.13	1.44	1.35	m²	**2.79**
Ref C503 (4.34 kg/m²)						
400 mm minimum laps	1.46	0.13	1.44	1.73	m²	**3.17**
strips in one width; 600 mm width	1.46	0.16	1.77	1.73	m²	**3.50**
strips in one width; 900 mm width	1.46	0.15	1.66	1.73	m²	**3.39**
strips in one width; 1200 mm width	1.46	0.14	1.55	1.73	m²	**3.28**
Ref C636 (5.55 kg/m²)						
400 mm minimum laps	1.86	0.14	1.55	2.20	m²	**3.75**
strips in one width; 600 mm width	1.86	0.17	1.88	2.20	m²	**4.08**
strips in one width; 900 mm width	1.86	0.16	1.77	2.20	m²	**3.97**
strips in one width; 1200 mm width	1.86	0.15	1.66	2.20	m²	**3.86**

E IN SITU CONCRETE/LARGE PRECAST CONCRETE

	PC £	Labour hours	Labour £	Material £	Unit	Total rate £
E30 REINFORCEMENT FOR IN SITU CONCRETE - cont'd						
Fabric; BS 4483 - cont'd						
Ref C785 (6.72 kg/m^2)						
400 mm minimum laps	2.26	0.14	1.55	2.67	m^2	4.22
strips in one width; 600 mm width	2.26	0.17	1.88	2.67	m^2	4.55
strips in one width; 900 mm width	2.26	0.16	1.77	2.67	m^2	4.44
strips in one width; 1200 mm width	2.26	0.15	1.66	2.67	m^2	4.33
Ref D49 (0.77 kg/m^2)						
100 mm minimum laps; bent	0.79	0.24	2.65	0.93	m^2	3.58
Ref D98 (1.54 kg/m^2)						
200 mm minimum laps; bent	0.59	0.24	2.65	0.70	m^2	3.35
E40 DESIGNED JOINTS IN IN SITU CONCRETE						
Formed; Fosroc Expandite "Flexcell" impregnated fibreboard joint filler or other equal and approved						
Width not exceeding 150 mm						
12.50 mm thick	-	0.14	1.94	1.26	m	3.20
20 mm thick	-	0.19	2.63	1.83	m	4.46
25 mm thick	-	0.19	2.63	2.08	m	4.71
Width 150 - 300 mm						
12.50 mm thick	-	0.19	2.63	1.95	m	4.58
20 mm thick	-	0.23	3.18	2.82	m	6.00
25 mm thick	-	0.23	3.18	3.25	m	6.43
Width 300 - 450 mm						
12.50 mm thick	-	0.23	3.18	2.93	m	6.11
20 mm thick	-	0.28	3.87	4.22	m	8.09
25 mm thick	-	0.28	3.87	4.88	m	8.75
Formed; Grace Servicised "Kork-pak" waterproof bonded cork joint filler board or other equal and approved						
Width not exceeding 150 mm						
10 mm thick	-	0.14	1.94	2.47	m	4.41
13 mm thick	-	0.14	1.94	2.50	m	4.44
19 mm thick	-	0.14	1.94	3.19	m	5.13
25 mm thick	-	0.14	1.94	3.60	m	5.54
Width 150 - 300 mm						
10 mm thick	-	0.19	2.63	4.57	m	7.20
13 mm thick	-	0.19	2.63	4.63	m	7.26
19 mm thick	-	0.19	2.63	6.01	m	8.64
25 mm thick	-	0.19	2.63	6.84	m	9.47
Width 300 - 450 mm						
10 mm thick	-	0.23	3.18	6.96	m	10.14
13 mm thick	-	0.23	3.18	7.05	m	10.23
19 mm thick	-	0.23	3.18	9.12	m	12.30
25 mm thick	-	0.23	3.18	10.37	m	13.55
Sealants; Fosroc Expandite "Pliastic 77" hot poured rubberized bituminous compound or other equal and approved						
Width 10 mm						
25 mm depth	-	0.17	2.35	0.79	m	3.14
Width 12.50 mm						
25 mm depth	-	0.18	2.49	0.97	m	3.46
Width 20 mm						
25 mm depth	-	0.19	2.63	1.57	m	4.20
Width 25 mm						
25 mm depth	-	0.20	2.76	1.93	m	4.69

E IN SITU CONCRETE/LARGE PRECAST CONCRETE

	PC £	Labour hours	Labour £	Material £	Unit	Total rate £
Sealants; Fosroc Expandite "Thioflex 600" gun grade two part polysulphide or other equal and approved						
Width 10 mm						
25 mm depth	-	0.05	0.69	2.99	m	**3.68**
Width 12.50 mm						
25 mm depth	-	0.06	0.83	3.74	m	**4.57**
Width 20 mm						
25 mm depth	-	0.07	0.97	5.98	m	**6.95**
Width 25 mm						
25 mm depth	-	0.08	1.11	7.47	m	**8.58**
Sealants; Grace Servicised "Paraseal" polysulphide compound or other equal and approved; priming with Grace Servicised "Servicised P" or other equal and approved						
Width 10 mm						
25 mm depth	-	0.19	1.68	1.64	m	**3.32**
Width 13 mm						
25 mm depth	-	0.19	1.68	2.11	m	**3.79**
Width 19 mm						
25 mm depth	-	0.23	2.03	3.03	m	**5.06**
Width 25 mm						
25 mm depth	-	0.23	2.03	3.95	m	**5.98**
Waterstops						
PVC water stop; flat dumbbell type; heat welded joints; cast into concrete						
170 mm wide	2.02	0.23	2.54	2.39	m	**4.93**
flat angle	6.45	0.28	3.10	5.85	nr	**8.95**
vertical angle	9.27	0.28	3.10	8.95	nr	**12.05**
flat three way intersection	8.22	0.37	4.09	8.29	nr	**12.38**
vertical three way intersection	11.62	0.37	4.09	12.02	nr	**16.11**
four way intersection	8.98	0.46	5.09	9.59	nr	**14.68**
210 mm wide	2.64	0.23	2.54	3.08	m	**5.62**
flat angle	8.00	0.28	3.10	7.45	nr	**10.55**
vertical angle	8.82	0.28	3.10	8.35	nr	**11.45**
flat three way intersection	10.44	0.37	4.09	10.91	nr	**15.00**
vertical three way intersection	8.43	0.37	4.09	8.71	nr	**12.80**
four way intersection	11.43	0.46	5.09	12.77	nr	**17.86**
250 mm wide	3.49	0.28	3.10	4.05	m	**7.15**
flat angle	9.62	0.32	3.54	9.16	nr	**12.70**
vertical angle	9.92	0.32	3.54	9.49	nr	**13.03**
flat three way intersection	12.43	0.42	4.65	13.49	nr	**18.14**
vertical three way intersection	9.47	0.42	4.65	10.21	nr	**14.86**
four way intersection	13.80	0.61	5.64	16.19	nr	**21.83**
PVC water stop; centre bulb type; heat welded joints; cast into concrete						
160 mm wide	2.19	0.23	2.54	2.57	m	**5.11**
flat angle	6.64	0.28	3.10	5.94	nr	**9.04**
vertical angle	9.60	0.28	3.10	9.20	nr	**12.30**
flat three way intersection	8.42	0.37	4.09	8.43	nr	**12.52**
vertical three way intersection	9.20	0.37	4.09	9.29	nr	**13.38**
four way intersection	9.20	0.46	5.09	9.80	nr	**14.89**
210 mm wide	3.13	0.23	2.54	3.64	m	**6.18**
flat angle	8.44	0.28	3.10	7.67	nr	**10.77**
vertical angle	10.68	0.28	3.10	10.14	nr	**13.24**
flat three way intersection	10.82	0.37	4.09	11.20	nr	**15.29**
vertical three way intersection	10.22	0.37	4.09	10.55	nr	**14.64**
four way intersection	11.82	0.46	5.09	13.21	nr	**18.30**

E IN SITU CONCRETE/LARGE PRECAST CONCRETE

	PC £	Labour hours	Labour £	Material £	Unit	Total rate £
E40 DESIGNED JOINTS IN IN SITU CONCRETE - cont'd						
Waterstops - cont'd						
260 mm wide	3.66	0.28	3.10	4.23	m	7.33
flat angle	9.83	0.32	3.54	9.33	nr	12.87
vertical angle	10.05	0.32	3.54	9.57	nr	13.11
flat three way intersection	12.68	0.42	4.65	13.74	nr	18.39
vertical three way intersection	9.66	0.42	4.65	10.41	nr	15.06
four way intersection	14.04	0.51	5.64	16.50	nr	22.14
325 mm wide	8.12	0.32	3.54	9.27	m	12.81
flat angle	20.09	0.37	4.09	20.58	nr	24.67
vertical angle	14.91	0.37	4.09	14.87	nr	18.96
flat three way intersection	22.70	0.46	5.09	27.14	nr	32.23
vertical three way intersection	14.46	0.46	5.09	18.07	nr	23.16
four way intersection	25.06	0.56	6.19	33.43	nr	39.62
E41 WORKED FINISHES/CUTTING TO IN SITU CONCRETE						
Worked finishes						
Tamping by mechanical means	-	0.02	0.18	0.09	m²	0.27
Power floating	-	0.16	1.41	0.29	m²	1.70
Trowelling	-	0.31	2.74	-	m²	2.74
Hacking						
by mechanical means	-	0.31	2.74	0.26	m²	3.00
by hand	-	0.65	5.74	-	m²	5.74
Lightly shot blasting surface of concrete	-	0.37	3.27	-	m²	3.27
Blasting surface of concrete						
to produce textured finish	-	0.65	5.74	0.56	m²	6.30
Wood float finish	-	0.12	1.06	-	m²	1.06
Tamped finish						
level or to falls	-	0.06	0.53	-	m²	0.53
to falls	-	0.09	0.80	-	m²	0.80
Spade finish	-	0.14	1.24	-	m²	1.24
Cutting chases						
Depth not exceeding 50 mm						
width 10 mm	-	0.31	2.74	0.73	m	3.47
width 50 mm	-	0.46	4.06	0.86	m	4.92
width 75 mm	-	0.61	5.39	0.99	m	6.38
Depth 50 mm - 100 mm						
width 75 mm	-	0.83	7.33	1.57	m	8.90
width 100 mm	-	0.93	8.22	1.65	m	9.87
width 100 mm; in reinforced concrete	-	1.39	12.28	2.55	m	14.83
Depth 100 mm - 150 mm						
width 100 mm	-	1.20	10.60	1.98	m	12.58
width 100 mm; in reinforced concrete	-	1.85	16.35	3.23	m	19.58
width 150 mm	-	1.48	13.08	2.22	m	15.30
width 150 mm; in reinforced concrete	-	2.22	19.62	3.61	m	23.23
Cutting rebates						
Depth not exceeding 50 mm						
width 50 mm	-	0.46	4.06	0.86	m	4.92
Depth 50 mm - 100 mm						
width 100 mm	-	0.93	8.22	1.65	m	9.87

E IN SITU CONCRETE/LARGE PRECAST CONCRETE

	PC £	Labour hours	Labour £	Material £	Unit	Total rate £
NOTE: The following rates for cutting mortices and holes in reinforced concrete allow for diamond drilling.						
Cutting mortices						
Depth not exceeding 100 mm						
cross sectional size 20 mm diameter; making good	-	0.14	1.24	0.08	nr	**1.32**
cross sectional size 50 mm diameter; making good	-	0.16	1.41	0.11	nr	**1.52**
cross sectional size 150 mm x 150 mm; making good	-	0.32	2.83	0.29	nr	**3.12**
cross sectional size 300 mm x 300 mm; making good	-	0.65	5.74	0.61	nr	**6.35**
cross sectional size 50 mm diameter; in reinforced concrete; making good	-	-	-	-	nr	**7.48**
cross sectional size 50 mm - 75 mm diameter; in reinforced concrete; making good	-	-	-	-	nr	**8.75**
cross sectional size 75 mm - 100 mm diameter; in reinforced concrete; making good	-	-	-	-	nr	**9.36**
cross sectional size 100 mm - 125 mm diameter; in reinforced concrete; making good	-	-	-	-	nr	**9.99**
cross sectional size 125 mm - 150 mm diameter; in reinforced concrete; making good	-	-	-	-	nr	**11.23**
Depth 100 mm - 200 mm						
cross sectional size 50 mm diameter; in reinforced concrete; making good	-	-	-	-	nr	**10.47**
cross sectional size 50 mm - 75 mm diameter; in reinforced concrete; making good	-	-	-	-	nr	**11.98**
cross sectional size 75 mm - 100 mm diameter; in reinforced concrete; making good	-	-	-	-	nr	**13.13**
cross sectional size 100 mm - 125 mm diameter; in reinforced concrete; making good	-	-	-	-	nr	**14.94**
cross sectional size 125 mm - 150 mm diameter; in reinforced concrete; making good	-	-	-	-	nr	**16.82**
Depth 200 mm - 300 mm						
cross sectional size 50 mm diameter; in reinforced concrete; making good	-	-	-	-	nr	**20.95**
cross sectional size 50 mm - 75 mm diameter; in reinforced concrete; making good	-	-	-	-	nr	**23.94**
cross sectional size 75 mm - 100 mm diameter; in reinforced concrete; making good	-	-	-	-	nr	**26.20**
cross sectional size 100 mm - 125 mm diameter; in reinforced concrete; making good	-	-	-	-	nr	**29.93**
cross sectional size 125 mm - 150 mm diameter; in rcinforocd oonorotc; making good	-	-	-	-	nr	**33.09**
Depth exceeding 300 mm; 400 mm depth						
cross sectional size 50 mm diameter; in reinforced concrete; making good	-	-	-	-	nr	**27.92**
cross sectional size 50 mm - 75 mm diameter; in reinforced concrete; making good	-	-	-	-	nr	**31.92**
cross sectional size 75 mm - 100 mm diameter; in reinforced concrete; making good	-	-	-	-	nr	**34.91**
cross sectional size 100 mm - 125 mm diameter; in reinforced concrete; making good	-	-	-	-	nr	**39.89**
cross sectional size 125 mm - 150 mm diameter; in reinforced concrete; making good	-	-	-	-	nr	**44.89**

Prices for Measured Works - Major Works
E IN SITU CONCRETE/LARGE PRECAST CONCRETE

	PC £	Labour hours	Labour £	Material £	Unit	Total rate £
E41 WORKED FINISHES/CUTTING TO IN SITU CONCRETE - cont'd						
Cutting mortices- cont'd						
Depth exceeding 300 mm; 500 mm depth						
cross sectional size 50 mm diameter; in reinforced concrete; making good	-	-	-	-	nr	34.91
cross sectional size 50 mm - 75 mm diameter; in reinforced concrete; making good	-	-	-	-	nr	39.89
cross sectional size 75 mm - 100 mm diameter; in reinforced concrete; making good	-	-	-	-	nr	43.65
cross sectional size 100 mm - 125 mm diameter; in reinforced concrete; making good	-	-	-	-	nr	49.87
cross sectional size 125 mm - 150 mm diameter; in reinforced concrete; making good	-	-	-	-	nr	56.13
cross sectional size 200 mm x 150 mm; in reinforced concrete; making good	-	-	-	-	nr	149.60
Depth exceeding 300 mm; 600 mm depth						
cross sectional size 50 mm diameter; in reinforced concrete; making good	-	-	-	-	nr	41.90
cross sectional size 50 mm - 75 mm diameter; in reinforced concrete; making good	-	-	-	-	nr	47.88
cross sectional size 75 mm - 100 mm diameter; in reinforced concrete; making good	-	-	-	-	nr	52.37
cross sectional size 100 mm - 125 mm diameter; in reinforced concrete; making good	-	-	-	-	nr	59.85
cross sectional size 125 mm - 150 mm diameter; in reinforced concrete; making good	-	-	-	-	nr	67.34
Cutting holes						
Depth not exceeding 100 mm						
cross sectional size 50 mm diameter; making good	-	0.32	2.83	0.29	nr	3.12
cross sectional size 100 mm diameter; making good	-	0.37	3.27	0.34	nr	3.61
cross sectional size 150 mm x 150 mm; making good	-	0.42	3.71	0.38	nr	4.09
cross sectional size 300 mm x 300 mm; making good	-	0.51	4.51	0.46	nr	4.97
Depth 100 mm - 200 mm						
cross sectional size 50 mm diameter; making good	-	0.46	4.06	0.42	nr	4.48
cross sectional size 50 mm diameter; in reinforced concrete; making good	-	0.69	6.10	0.63	nr	6.73
cross sectional size 100 mm diameter; making good	-	0.56	4.95	0.51	nr	5.46
cross sectional size 100 mm diameter; in reinforced concrete; making good	-	0.83	7.33	0.76	nr	8.09
cross sectional size 150 mm x 150 mm; making good	-	0.69	6.10	0.63	nr	6.73
cross sectional size 150 mm x 150 mm; in reinforced concrete; making good	-	1.06	9.37	0.97	nr	10.34
cross sectional size 300 mm x 300 mm; making good	-	0.88	7.78	0.80	nr	8.58
cross sectional size 300 mm x 300 mm; in reinforced concrete; making good	-	1.34	11.84	1.22	nr	13.06
cross sectional size not exceeding 0.10 m²; in reinforced concrete; making good	-	-	-	-	nr	87.28
cross sectional size 0.10 m² - 0.20 m²; in reinforced concrete; making good	-	-	-	-	nr	174.56
cross sectional size 0.20 m² - 0.30 m²; in reinforced concrete; making good	-	-	-	-	nr	197.00

E IN SITU CONCRETE/LARGE PRECAST CONCRETE

	PC £	Labour hours	Labour £	Material £	Unit	Total rate £
Depth 100 mm - 200 mm - cont'd						
cross sectional size 0.30 m² - 0.40 m²; in reinforced concrete; making good	-	-	-	-	nr	229.48
cross sectional size 0.40 m² - 0.50 m²; in reinforced concrete; making good	-	-	-	-	nr	263.08
cross sectional size 0.50 m² - 0.60 m²; in reinforced concrete; making good	-	-	-	-	nr	306.74
cross sectional size 0.60 m² - 0.70 m²; in reinforced concrete; making good	-	-	-	-	nr	350.40
cross sectional size 0.70 m² - 0.80 m²; in reinforced concrete; making good	-	-	-	-	nr	399.45
Depth 200 mm - 300 mm						
cross sectional size 50 mm diameter; making good	-	0.69	6.10	0.63	nr	6.73
cross sectional size 50 mm diameter; in reinforced concrete; making good	-	1.06	9.37	0.97	nr	10.34
cross sectional size 100 mm diameter; making good	-	0.83	7.33	0.76	nr	8.09
cross sectional size 100 mm diameter; in reinforced concrete; making good	-	1.25	11.05	1.14	nr	12.19
cross sectional size 150 mm x 150 mm; making good	-	1.02	9.01	0.93	nr	9.94
cross sectional size 150 mm x 150 mm; in reinforced concrete; making good	-	1.53	13.52	1.39	nr	14.91
cross sectional size 300 mm x 300 mm; making good	-	1.30	11.49	1.18	nr	12.67
cross sectional size 300 mm x 300 mm; in reinforced concrete; making good	-	1.94	17.14	1.77	nr	18.91
E42 ACCESSORIES CAST INTO IN SITU CONCRETE						
Foundation bolt boxes						
Temporary plywood; for group of 4 nr bolts						
75 mm x 75 mm x 150 mm	-	0.42	5.81	1.01	nr	6.82
75 mm x 75 mm x 250 mm	-	0.42	5.81	1.25	nr	7.06
Expanded metal; Expamet Building Products Ltd or other equal and approved						
75 mm diameter x 150 mm long	-	0.28	3.87	0.93	nr	4.80
75 mm diameter x 300 mm long	-	0.28	3.87	1.28	nr	5.15
100 mm diameter x 450 mm long	-	0.28	3.87	2.28	nr	6.15
Foundation bolts and nuts						
Black hexagon						
10 mm diameter x 100 mm long	-	0.23	3.18	0.52	nr	3.70
12 mm diameter x 120 mm long	-	0.23	3.18	0.79	nr	3.97
16 mm diameter x 160 mm long	-	0.28	3.87	2.18	nr	6.05
20 mm diameter x 180 mm long	-	0.28	3.87	2.54	nr	6.41
Masonry slots						
Galvanised steel; dovetail slots; 1.20 mm thick; 18G						
75 mm long	-	0.07	0.97	0.18	nr	1.15
100 mm long	-	0.07	0.97	0.20	nr	1.17
150 mm long	-	0.08	1.11	0.24	nr	1.35
225 mm long	-	0.09	1.24	0.34	nr	1.58
Galvanised steel; metal insert slots; Halfen Ltd or other equal and approved; 2.50 mm thick; end caps and foam filling						
41 mm x 41 mm; ref P3270	-	0.37	5.12	8.08	m	13.20
41 mm x 41 mm x 75 mm; ref P3249	-	0.09	1.24	1.61	nr	2.85
41 mm x 41 mm x 100 mm; ref P3250	-	0.09	1.24	2.00	nr	3.24
41 mm x 41 mm x 150 mm; ref P3251	-	0.09	1.24	2.41	nr	3.65

Prices for Measured Works - Major Works
E IN SITU CONCRETE/LARGE PRECAST CONCRETE

	PC £	Labour hours	Labour £	Material £	Unit	Total rate £
E42 ACCESSORIES CAST INTO IN SITU CONCRETE - cont'd						
Cramps						
Mild steel; once bent; one end shot fired into concrete; other end fanged and built into brickwork joint						
200 mm girth	-	0.14	1.96	0.45	nr	2.41
Column guards						
White nylon coated steel; "Rigifix" or other equal and approved; Huntley and Sparks Ltd; plugging; screwing to concrete; 1.50 mm thick						
75 mm x 75 mm x 1000 mm	-	0.74	10.23	15.41	nr	25.64
Galvanised steel; "Rigifix" or other equal and approved; Huntley and Sparks Ltd; 3 mm thick						
75 mm x 75 mm x 1000 mm	-	0.56	7.74	10.82	nr	18.56
Galvanised steel; "Rigifix" or other equal and approved; Huntley and Sparks Ltd; 4.50 mm thick						
75 mm x 75 mm x 1000 mm	-	0.56	7.74	14.55	nr	22.29
Stainless steel; "HKW" or other equal and approved; Halfen Ltd; 5 mm thick						
50 mm x 50 mm x 1500 mm	-	0.93	12.86	42.20	nr	55.06
50 mm x 50 mm x 2000 mm	-	1.11	15.35	60.84	nr	76.19
Channels						
Stainless steel; Halfen Ltd or other equal and approved						
ref 38/17/HTA	-	0.32	4.42	12.18	m	16.60
ref 41/22/HZA; 80 mm long; including "T" headed bolts and plate washers	-	0.09	1.24	8.49	nr	9.73
Channel ties						
Stainless steel; Halfen Ltd or other equal and approved						
ref HTS - B12; 150 mm projection; including insulation retainer	-	0.03	0.52	0.37	nr	0.89
ref HTS - B12; 200 mm projection; including insulation retainer	-	0.03	0.52	0.52	nr	1.04
E60 PRECAST/COMPOSITE CONCRETE DECKING						
Prestressed precast flooring planks; Bison "Drycast" or other equal and approved; cement and sand (1:3) grout between planks and on prepared bearings						
100 mm thick suspended slabs; horizontal						
600 mm wide planks	-	-	-	-	m^2	34.34
1200 mm wide planks	-	-	-	-	m^2	32.56
150 mm thick suspended slabs; horizontal						
1200 mm wide planks	-	-	-	-	m^2	33.96

Prices for Measured Works - Major Works
E IN SITU CONCRETE/LARGE PRECAST CONCRETE

	PC £	Labour hours	Labour £	Material £	Unit	Total rate £
Prestressed precast concrete beam and block floor; Bison "Housefloor" or other equal and approved; in situ concrete 30.00 N/mm² - 10 mm aggregate in filling at wall abutments; cement and sand (1:6) grout brushed in between beams and blocks						
155 mm thick suspended slab at ground level; 440 mm x 215 mm x 100 mm blocks; horizontal						
beams at 510 mm centres; up to 3.30 m span with a superimposed load of 5.00 kN/m²	-	-	-	-	m²	18.60
beams at 285 mm centres; up to 4.35 m span with a superimposed load of 5.00 kN/m²	-	-	-	-	m²	21.73
Prestressed precast concrete structural suspended floors; Bison "Hollowcore" or other equal and approved; supplied and fixed on hard level bearings, to areas of 500 m² per site visit; top surface screeding and ceiling finishes by others						
Floors to dwellings, offices, car parks, shop retail floors, hospitals, school teaching rooms, staff rooms and the like; superimposed load of 5.00 kN/m²						
floor spans up to 3.00 m; 1200 mm x 150 mm	-	-	-	-	m²	33.51
floor spans 3.00 m - 6.00 m; 1200 mm x 150 mm	-	-	-	-	m²	33.96
floor spans 6.00 m - 7.50 m; 1200 mm x 200 mm	-	-	-	-	m²	34.88
floor spans 7.50 m - 9.50 m; 1200 mm x 250 mm	-	-	-	-	m²	39.56
floor spans 9.50 m - 12.00 m; 1200 mm x 300 mm	-	-	-	-	m²	41.91
floor spans 12.00 m - 13.00 m; 1200 mm x 350 mm	-	-	-	-	m²	44.12
floor spans 13.00 m - 14.00 m; 1200 mm x 400 mm	-	-	-	-	m²	47.87
floor spans 14.00 m - 15.00 m; 1200 mm x 450 mm	-	-	-	-	m²	48.63
Floors to shop stockrooms, light warehousing, schools, churches or similar places of assembly, light factory accommodation, laboratories and the like; superimposed load of 8.50 kN/m²						
floor spans up to 3.00 m; 1200 mm x 150 mm	-	-	-	-	m²	33.55
floor spans 3.00 m - 6.00 m; 1200 mm x 200 mm	-	-	-	-	m²	34.95
floor spans 6.00 m - 7.50 m; 1200 mm x 250 mm	-	-	-	-	m²	39.60
Floors to heavy warehousing, factories, stores and the like; superimposed load of 12.50 kN/m²						
floor spans up to 3.00 m; 1200 mm x 150 mm	-	-	-	-	m²	33.62
floor spans 3.00 m - 6.00 m; 1200 mm x 250 mm	-	-	-	-	m²	39.66
Prestressed precast concrete staircase, supplied and fixed in conjunction with Bison "Hollowcore" flooring system or similar; comprising 2 nr 1100 mm wide flights with 7 nr 275 mm treads, 8 nr 185 mm risers and 150 mm waist; 1 nr 2200 mm x 1400 mm x 150 mm half landing and 1 nr top landing						
3.00 m storey height	-	-	-	-	nr	1524.20

E IN SITU CONCRETE/LARGE PRECAST CONCRETE

	PC £	Labour hours	Labour £	Material £	Unit	Total rate £
E60 PRECAST/COMPOSITE CONCRETE DECKING - cont'd						
Composite floor comprising reinforced in situ ready-mixed concrete 30.00 N/mm^2; on and including 1.20 mm thick "Super Holorib" steel deck permanent shutting; complete with reinforcement to support imposed loading and A142 anti-crack mesh						
150 mm thick suspended slab; 5.00 kN/m^2 loading						
1.50 m - 3.00 m high to soffit	-	1.43	16.23	28.61	m^2	44.84
3.00 m - 4.50 m high to soffit	-	1.43	16.23	29.76	m^2	45.99
4.50 m - 6.00 m high to soffit	-	1.67	19.55	30.20	m^2	49.75
200 mm thick suspended slab; 7.50 kN/m^2 loading						
1.50 m - 3.00 m high to soffit	-	1.47	16.58	31.92	m^2	48.50
3.00 m - 4.50 m high to soffit	-	1.47	16.58	33.07	m^2	49.65
4.50 m - 6.00 m high to soffit	-	1.70	19.76	33.50	m^2	53.26

F MASONRY

F10 BRICK/BLOCK WALLING

BASIC MORTAR PRICES

Coloured mortar materials (£/tonne); (excluding cement)
light	39.85	medium	41.44
dark	50.14	extra dark	50.14

Mortar materials (£/tonne)
cement	77.65	lime	123.43
sand	13.84	white cement	144.01

Mortar materials (£/5 litres)
"Cemplas Super" mortar plasticiser   4.25

Common bricks; PC £175.80/1000; in cement mortar (1:3)

	PC £	Labour hours	Labour £	Material £	Unit	Total rate £
Walls						
half brick thick	-	0.93	16.26	13.31	m²	29.57
half brick thick; building against other work; concrete	-	1.02	17.83	14.62	m²	32.45
half brick thick; building overhand	-	1.16	20.28	13.31	m²	33.59
half brick thick; curved; 6.00 m radii	-	1.20	20.98	13.31	m²	34.29
half brick thick; curved; 1.50 m radii	-	1.57	27.44	15.20	m²	42.64
one brick thick	-	1.57	27.44	26.61	m²	54.05
one brick thick; curved; 6.00 m radii	-	2.04	35.66	28.50	m²	64.16
one brick thick; curved; 1.50 m radii	-	2.54	44.40	29.16	m²	73.56
one and a half brick thick	-	2.13	37.23	39.92	m²	77.15
one and a half brick thick; battering	-	2.45	42.82	39.92	m²	82.74
two brick thick	-	2.59	45.27	53.23	m²	98.50
two brick thick; battering	-	3.05	53.31	53.23	m²	106.54
337 mm average thick; tapering, one side	-	2.68	46.85	39.92	m²	86.77
450 mm average thick; tapering, one side	-	3.47	60.65	53.23	m²	113.88
337 mm average thick; tapering, both sides	-	3.10	54.19	39.92	m²	94.11
450 mm average thick; tapering, both sides	-	3.89	68.00	53.88	m²	121.88
facework one side, half brick thick	-	1.02	17.83	13.31	m²	31.14
facework one side, one brick thick	-	1.67	29.19	26.61	m²	55.80
facework one side, one and a half brick thick	-	2.22	38.80	39.92	m²	78.72
facework one side, two brick thick	-	2.68	46.85	53.23	m²	100.08
facework both sides, half brick thick	-	1.11	19.40	13.31	m²	32.71
facework both sides, one brick thick	-	1.76	30.76	26.61	m²	57.37
facework both sides, one and a half brick thick	-	2.31	40.38	39.92	m²	80.30
facework both sides, two brick thick	-	2.78	48.59	53.23	m²	101.82
Isolated piers						
one brick thick	-	2.36	41.25	26.61	m²	67.86
two brick thick	-	3.70	64.67	53.88	m²	118.55
three brick thick	-	4.67	81.63	81.15	m²	162.78
Isolated casings						
half brick thick	-	1.20	20.98	13.31	m²	34.29
one brick thick	-	2.04	35.66	26.61	m²	62.27
Chimney stacks						
one brick thick	-	2.36	41.25	26.61	m²	67.86
two brick thick	-	3.70	64.67	53.88	m²	118.55
three brick thick	-	4.67	81.63	81.15	m²	162.78
Projections						
225 mm width; 112 mm depth; vertical	-	0.28	4.89	2.75	m	7.64
225 mm width; 225 mm depth; vertical	-	0.56	9.79	5.50	m	15.29
327 mm width; 225 mm depth; vertical	-	0.83	14.51	8.25	m	22.76
440 mm width; 225 mm depth; vertical	-	0.93	16.26	11.00	m	27.26

Prices for Measured Works - Major Works
F MASONRY

	PC £	Labour hours	Labour £	Material £	Unit	Total rate £
F10 BRICK/BLOCK WALLING - cont'd						
Common bricks; PC £175.80/1000; in cement mortar (1:3) - cont'd						
Closing cavities						
width of cavity 50 mm, closing with common brickwork half brick thick; vertical	-	0.28	4.89	0.68	m	**5.57**
width of cavity 50 mm, closing with common brickwork half brick thick; horizontal	-	0.28	4.89	2.03	m	**6.92**
width of cavity 50 mm, closing with common brickwork half brick thick; including damp proof course; vertical	-	0.37	6.47	1.39	m	**7.86**
width of cavity 50 mm, closing with common brickwork half brick thick; including damp proof course; horizontal	-	0.32	5.59	2.55	m	**8.14**
width of cavity 75 mm, closing with common brickwork half brick thick; vertical	-	0.28	4.89	0.99	m	**5.88**
width of cavity 75 mm, closing with common brickwork half brick thick; horizontal	-	0.28	4.89	2.97	m	**7.86**
width of cavity 75 mm, closing with common brickwork half brick thick; including damp proof course; vertical	-	0.37	6.47	1.70	m	**8.17**
width of cavity 75 mm, closing with common brickwork half brick thick; including damp proof course; horizontal	-	0.32	5.59	3.49	m	**9.08**
Bonding to existing						
half brick thick	-	0.28	4.89	0.73	m	**5.62**
one brick thick	-	0.42	7.34	1.46	m	**8.80**
one and a half brick thick	-	0.65	11.36	2.19	m	**13.55**
two brick thick	-	0.88	15.38	2.91	m	**18.29**
ADD or DEDUCT to walls for variation of £10.00/1000 in PC of common bricks						
half brick thick	-	-	-	0.65	m²	**0.65**
one brick thick	-	-	-	1.29	m²	**1.29**
one and a half brick thick	-	-	-	1.94	m²	**1.94**
two brick thick	-	-	-	2.58	m²	**2.58**
Common bricks; PC £175.80/1000; in gauged mortar (1:1:6)						
Walls						
half brick thick	-	0.93	16.26	13.10	m²	**29.36**
half brick thick; building against other work; concrete	-	1.02	17.83	14.28	m²	**32.11**
half brick thick; building overhand	-	1.16	20.28	13.10	m²	**33.38**
half brick thick; curved; 6.00 m radii	-	1.20	20.98	13.10	m²	**34.08**
half brick thick; curved; 1.50 m radii	-	1.57	27.44	14.99	m²	**42.43**
one brick thick	-	1.57	27.44	26.20	m²	**53.64**
one brick thick; curved; 6.00 m radii	-	2.04	35.66	28.09	m²	**63.75**
one brick thick; curved; 1.50 m radii	-	2.54	44.40	28.68	m²	**73.08**
one and a half brick thick	-	2.13	37.23	39.31	m²	**76.54**
one and a half brick thick; battering	-	2.45	42.82	39.31	m²	**82.13**
two brick thick	-	2.59	45.27	52.41	m²	**97.68**
two brick thick; battering	-	3.05	53.31	52.41	m²	**105.72**
337 average thick; tapering, one side	-	2.68	46.85	39.31	m²	**86.16**
450 average thick; tapering, one side	-	3.47	60.65	52.41	m²	**113.06**
337 average thick; tapering, both sides	-	3.10	54.19	39.31	m²	**93.50**
450 average thick; tapering, both sides	-	3.89	68.00	53.00	m²	**121.00**
facework one side, half brick thick	-	1.02	17.83	13.10	m²	**30.93**
facework one side, one brick thick	-	1.67	29.19	26.20	m²	**55.39**
facework one side, one and a half brick thick	-	2.22	38.80	39.31	m²	**78.11**
facework one side, two brick thick	-	2.68	46.85	52.41	m²	**99.26**

F MASONRY

	PC £	Labour hours	Labour £	Material £	Unit	Total rate £
Walls - cont'd						
facework both sides, half brick thick	-	1.11	19.40	13.10	m²	32.50
facework both sides, one brick thick	-	1.76	30.76	26.20	m²	56.96
facework both sides, one and a half brick thick	-	2.31	40.38	39.31	m²	79.69
facework both sides, two brick thick	-	2.78	48.59	52.41	m²	101.00
Isolated piers						
one brick thick	-	2.36	41.25	26.20	m²	67.45
two brick thick	-	3.70	64.67	53.00	m²	117.67
three brick thick	-	4.67	81.63	79.79	m²	161.42
Isolated casings						
half brick thick	-	1.20	20.98	13.10	m²	34.08
one brick thick	-	2.04	35.66	26.20	m²	61.86
Chimney stacks						
one brick thick	-	2.36	41.25	26.20	m²	67.45
two brick thick	-	3.70	64.67	53.00	m²	117.67
three brick thick	-	4.67	81.63	79.79	m²	161.42
Projections						
225 mm width; 112 mm depth; vertical	-	0.28	4.89	2.73	m	7.62
225 mm width; 225 mm depth; vertical	-	0.56	9.79	5.45	m	15.24
337 mm width; 225 mm depth; vertical	-	0.83	14.51	8.18	m	22.69
440 mm width; 225 mm depth; vertical	-	0.93	16.26	10.90	m	27.16
Closing cavities						
width of cavity 50 mm, closing with common brickwork half brick thick; vertical	-	0.28	4.89	0.67	m	5.56
width of cavity 50 mm, closing with common brickwork half brick thick; horizontal	-	0.28	4.89	2.01	m	6.90
width of cavity 50 mm, closing with common brickwork half brick thick; including damp proof course; vertical	-	0.37	6.47	1.38	m	7.85
width of cavity 50 mm, closing with common brickwork half brick thick; including damp proof course; horizontal	-	0.32	5.59	2.53	m	8.12
width of cavity 75 mm, closing with common brickwork half brick thick; vertical	-	0.28	4.89	0.97	m	5.86
width of cavity 75 mm, closing with common brickwork half brick thick; horizontal	-	0.28	4.89	2.96	m	7.85
width of cavity 75 mm, closing with common brickwork half brick thick; including damp proof course; vertical	-	0.37	6.47	1.69	m	8.16
width of cavity 75 mm, closing with common brickwork half brick thick; including damp proof course; horizontal	-	0.32	5.59	3.48	m	9.07
Bonding to existing						
half brick thick	-	0.28	4.89	0.72	m	5.61
one brick thick	-	0.42	7.34	1.44	m	8.78
one and a half brick thick	-	0.65	11.36	2.15	m	13.51
two brick thick	-	0.88	15.38	2.87	m	18.25
Arches						
height on face 102 mm, width of exposed soffit 102 mm, shape of arch - segmental, one ring	-	1.57	20.56	7.79	m	28.35
height on face 102 mm, width of exposed soffit 215 mm, shape of arch - segmental, one ring	-	2.04	28.78	9.39	m	38.17
height on face 102 mm, width of exposed soffit 102 mm, shape of arch - semi-circular, one ring	-	1.99	27.91	7.79	m	35.70
height on face 102 mm, width of exposed soffit 215 mm, shape of arch - semi-circular, one ring	-	2.50	36.82	9.39	m	46.21
height on face 215 mm, width of exposed soffit 102 mm, shape of arch - segmental, two ring	-	1.99	27.91	9.17	m	37.08
height on face 215 mm, width of exposed soffit 215 mm, shape of arch - segmental, two ring	-	2.45	35.95	12.14	m	48.09

F10 BRICK/BLOCK WALLING - cont'd

	PC £	Labour hours	Labour £	Material £	Unit	Total rate £
Common bricks; PC £175.80/1000; in gauged mortar (1:1:6) - cont'd						
Arches - cont'd						
height on face 215 mm, width of exposed soffit						
102 mm, shape of arch - semi-circular, two ring	-	2.68	39.97	9.17	m	49.14
height on face 215 mm, width of exposed soffit						
215 mm, shape of arch - semi-circular, two ring	-	3.05	46.43	12.14	m	58.57
ADD or DEDUCT to walls for variation of						
£10.00/1000 in PC of common bricks						
half brick thick	-	-	-	0.65	m²	0.65
one brick thick	-	-	-	1.29	m²	1.29
one and a half brick thick	-	-	-	1.94	m²	1.94
two brick thick	-	-	-	2.58	m²	2.58
Class A engineering bricks; PC £258.30/1000; in cement mortar (1:3)						
Walls						
half brick thick	-	1.02	17.83	18.63	m²	36.46
one brick thick	-	1.67	29.19	37.26	m²	66.45
one brick thick; building against other work	-	1.99	34.78	39.22	m²	74.00
one brick thick; curved; 6.00 m radii	-	2.22	38.80	37.26	m²	76.06
one and a half brick thick	-	2.22	38.80	55.88	m²	94.68
one and a half brick thick; building against other work	-	2.68	46.85	55.88	m²	102.73
two brick thick	-	2.78	48.59	74.51	m²	123.10
337 mm average thick; tapering, one side	-	2.87	50.17	55.88	m²	106.05
450 mm average thick; tapering, one side	-	3.70	64.67	74.51	m²	139.18
337 mm average thick; tapering, both sides	-	3.33	58.21	55.88	m²	114.09
450 mm average thick; tapering, both sides	-	4.21	73.59	75.17	m²	148.76
facework one side, half brick thick	-	1.11	19.40	18.63	m²	38.03
facework one side, one brick thick	-	1.76	30.76	37.26	m²	68.02
facework one side, one and a half brick thick	-	2.31	40.38	55.88	m²	96.26
facework one side, two brick thick	-	2.87	50.17	74.51	m²	124.68
facework both sides, half brick thick	-	1.20	20.98	18.63	m²	39.61
facework both sides, one brick thick	-	1.85	32.34	37.26	m²	69.60
facework both sides, one and a half brick thick	-	2.41	42.13	55.88	m²	98.01
facework both sides, two brick thick	-	2.96	51.74	74.51	m²	126.25
Isolated piers						
one brick thick	-	2.59	45.27	37.26	m²	82.53
two brick thick	-	4.07	71.14	75.17	m²	146.31
three brick thick	-	5.00	87.40	113.08	m²	200.48
Isolated casings						
half brick thick	-	1.30	22.72	18.63	m²	41.35
one brick thick	-	2.22	38.80	37.26	m²	76.06
Projections						
225 mm width; 112 mm depth; vertical	-	0.32	5.59	3.93	m	9.52
225 mm width; 225 mm depth; vertical	-	0.60	10.49	7.86	m	18.35
337 mm width; 225 mm depth; vertical	-	0.88	15.38	11.80	m	27.18
440 mm width; 225 mm depth; vertical	-	1.02	17.83	15.73	m	33.56
Bonding to existing						
half brick thick	-	0.32	5.59	1.02	m	6.61
one brick thick	-	0.46	8.04	2.05	m	10.09
one and a half brick thick	-	0.65	11.36	3.07	m	14.43
two brick thick	-	0.97	16.96	4.10	m	21.06
ADD or DEDUCT to walls for variation of						
£10.00/1000 in PC of bricks						
half brick thick	-	-	-	0.65	m²	0.65
one brick thick	-	-	-	1.29	m²	1.29
one and a half brick thick	-	-	-	1.94	m²	1.94
two brick thick	-	-	-	2.58	m²	2.58

F MASONRY

	PC £	Labour hours	Labour £	Material £	Unit	Total rate £
Class B engineering bricks; PC £184.00/1000; in cement mortar (1:3)						
Walls						
half brick thick	-	1.02	17.83	13.84	m²	**31.67**
one brick thick	-	1.67	29.19	27.67	m²	**56.86**
one brick thick; building against other work	-	1.99	34.78	29.64	m²	**64.42**
one brick thick; curved; 6.00 m radii	-	2.22	38.80	27.67	m²	**66.47**
one and a half brick thick	-	2.22	38.80	41.51	m²	**80.31**
one and a half brick thick; building against other work	-	2.68	46.85	41.51	m²	**88.36**
two brick thick	-	2.78	48.59	55.34	m²	**103.93**
337 mm thick; tapering, one side	-	2.87	50.17	41.51	m²	**91.68**
450 mm thick; tapering, one side	-	3.70	64.67	55.34	m²	**120.01**
337 mm thick; tapering, both sides	-	3.33	58.21	41.51	m²	**99.72**
450 mm thick; tapering, both sides	-	4.21	73.59	56.00	m²	**129.59**
facework one side, half brick thick	-	1.11	19.40	13.84	m²	**33.24**
facework one side, one brick thick	-	1.76	30.76	27.67	m²	**58.43**
facework one side, one and a half brick thick	-	2.31	40.38	41.51	m²	**81.89**
facework one side, two brick thick	-	2.87	50.17	55.34	m²	**105.51**
facework both sides, half brick thick	-	1.20	20.98	13.84	m²	**34.82**
facework both sides, one brick thick	-	1.85	32.34	27.67	m²	**60.01**
facework both sides, one and a half brick thick	-	2.41	42.13	41.51	m²	**83.64**
facework both sides, two brick thick	-	2.96	51.74	55.34	m²	**107.08**
Isolated piers						
one brick thick	-	2.59	45.27	27.67	m²	**72.94**
two brick thick	-	4.07	71.14	56.00	m²	**127.14**
three brick thick	-	5.00	87.40	84.33	m²	**171.73**
Isolated casings						
half brick thick	-	1.30	22.72	13.84	m²	**36.56**
one brick thick	-	2.22	38.80	27.67	m²	**66.47**
Projections						
225 mm width; 112 mm depth; vertical	-	0.32	5.59	2.87	m	**8.46**
225 mm width; 225 mm depth; vertical	-	0.60	10.49	5.73	m	**16.22**
337 mm width; 225 mm depth; vertical	-	0.88	15.38	8.60	m	**23.98**
440 mm width; 225 mm depth; vertical	-	1.02	17.83	11.47	m	**29.30**
Bonding to existing						
half brick thick	-	0.32	5.59	0.76	m	**6.35**
one brick thick	-	0.46	8.04	1.52	m	**9.56**
one and a half brick thick	-	0.65	11.36	2.27	m	**13.63**
two brick thick	-	0.97	16.96	3.03	m	**19.99**
ADD or DEDUCT to walls for variation of £10.00/1000 in PC of bricks						
half brick thick	-	-	-	0.65	m²	**0.65**
one brick thick	-	-	-	1.29	m²	**1.29**
one and a half brick thick	-	-	-	1.94	m²	**1.94**
two brick thick	-	-	-	2.58	m²	**2.58**

F10 BRICK/BLOCK WALLING - cont'd

ALTERNATIVE FACING BRICK PRICES (£/1000)

Ibstock facing bricks; 215 x 102.50 x 65 mm

	£		£
Aldridge Brown Blend	254.70	Leicester Red Stock	228.70
Aldridge Leicester Anglican Red Rustic	206.70	Roughdales Red Multi Rustic	215.80
Ashdown Cottage Mixture	370.60	Roughdales Trafford Multi Rustic	227.70
Ashdown Crowbridge Multi	451.40	Stourbridge Himley Mixed Russet	201.80
Ashdown Pevensey Multi	454.40	Stourbridge Kenilworth Multi	329.60
Cattybrook Gloucestershire Golden	237.70	Stourbridge Pennine Pastone	323.60
Chailey Stock	492.40	Stratfford Red Rustic	244.70
Dorking Multi	383.50	Swanage Handmade Restoration	540.30
Funton Second Hand Stock	508.40	Tonbridge Handmade Multi	585.20
Holbrook Smooth Red	426.50		

Hanson Brick Ltd, London brand; 215 x 102.50 x 65 mm

	£		£
Harvest Series		Country Wood Series	
- Autumn leaf	184.00	- Claydon Red Multi	192.20
- Burgundy Red	191.10	- Delph Autumn	185.10
- Dawn Red	192.20	- Longville Stone	186.20
- Hawthron Blend	182.90	- Morteyne Russet	195.50
- Honey Buff	185.10	- Nene Valley Stone	186.20
- Sunset Red	194.50	- Orton Multi Buff	186.20
Sovereign Series		Ridgeway Series	
- Georgian	192.20	- Brecken Grey	189.00
- Regency	206.00	- Ironstone	192.20
- Saxon Gold	208.80	- Milton Buff	201.60
- Tudor	220.30	- Windsor	194.50
Shire Series			
- Chiltern	220.30		
- Heather	220.30		
- Hereward Light	194.50		
- Sandfaced	215.30		

Breacon Hill Brick Company; 215 x 102.50 x 65 mm

	£		£
Brown Flint Rustic	324.50	Buff Multi CS Textured	339.55
Charcoal Flint Textured	250.40	Dorset Buff Smooth	204.72
Dorset Golden Smooth	240.40	Golden Buff Flint Textured	259.66
Grey Black Flint Textured	250.40	Minster Autumn Gold	214.72
Minster Bracken Mixture	208.80	Minster Mixed Red	234.68
Oatmeal CS Plain	160.50	Oatmeal Flint Textured	209.72
Red Black Multi Flint Plain	339.55	Smoke Grey Multi CS Textured	339.55
Straw Flint Textured	259.66	Warm Brown Multi CS Textured	339.55
Wessex Brindle Buff	234.68	Wessex Brindle Red	234.68
Wessex Red Blend	359.52		

F MASONRY

	PC £	Labour hours	Labour £	Material £	Unit	Total rate £
Facing bricks; sand faced; PC £130.00/1000 (unless otherwise stated); in gauged mortar (1:1:6)						
Walls						
facework one side, half brick thick; stretcher bond	-	1.20	20.98	10.15	m²	31.13
facework one side, half brick thick; flemish bond with snapped headers	-	1.39	24.30	10.15	m²	34.45
facework one side, half brick thick; stretcher bond; building against other work; concrete	-	1.30	22.72	11.32	m²	34.04
facework one side, half brick thick; flemish bond with snapped headers; building against other work; concrete	-	1.48	25.87	11.32	m²	37.19
facework one side, half brick thick; stretcher bond; building overhand	-	1.48	25.87	10.15	m²	36.02
facework one side, half brick thick; flemish bond with snapped headers; building overhand	-	1.67	29.19	10.15	m²	39.34
facework one side, half brick thick; stretcher bond; curved; 6.00 m radii	-	1.76	30.76	10.15	m²	40.91
facework one side, half brick thick; flemish bond with snapped headers; curved; 6.00 m radii	-	1.99	34.78	10.15	m²	44.93
facework one side, half brick thick; stretcher bond; curved; 1.50 m radii	-	2.22	38.80	10.15	m²	48.95
facework one side, half brick thick; flemish bond with snapped headers; curved; 1.50 m radii	-	2.59	45.27	10.15	m²	55.42
facework both sides, one brick thick; two stretcher skins tied together	-	2.08	36.36	21.32	m²	57.68
facework both sides, one brick thick; flemish bond	-	2.13	37.23	20.30	m²	57.53
facework both sides, one brick thick; two stretcher skins tied together; curved; 6.00 m radii	-	2.87	50.17	22.71	m²	72.88
facework both sides, one brick thick; flemish bond; curved; 6.00 m radii	-	2.96	51.74	21.69	m²	73.43
facework both sides, one brick thick; two stretcher skins tied together; curved; 1.50 m radii	-	3.56	62.23	24.70	m²	86.93
facework both sides, one brick thick; flemish bond; curved; 1.50 m radii	-	3.70	64.67	23.68	m²	88.35
Isolated piers						
facework both sides, one brick thick; two stretcher skins tied together	-	2.45	42.82	21.90	m²	64.72
facework both sides, one brick thick; flemish bond	-	2.50	43.70	21.90	m²	65.60
Isolated casings						
facework one side, half brick thick; stretcher bond	-	1.85	32.34	10.15	m²	42.49
facework one side, half brick thick; flemish bond with snapped headers	-	2.04	35.66	10.15	m²	45.81
Projections						
225 mm width; 112 mm depth; stretcher bond; vertical	-	0.28	4.89	2.07	m²	6.96
225 mm width; 112 mm depth; flemish bond with snapped headers; vertical	-	0.37	6.47	2.07	m²	8.54
225 mm width; 225 mm depth; flemish bond; vertical	-	0.60	10.49	4.14	m²	14.63
328 mm width; 112 mm depth; stretcher bond; vertical	-	0.56	9.79	3.11	m²	12.90
328 mm width; 112 mm depth; flemish bond with snapped headers; vertical	-	0.65	11.36	3.11	m²	14.47
328 mm width; 225 mm depth; flemish bond; vertical	-	1.11	19.40	6.18	m²	25.58

Prices for Measured Works - Major Works
F MASONRY

	PC £	Labour hours	Labour £	Material £	Unit	Total rate £
F10 BRICK/BLOCK WALLING - cont'd						
Facing bricks; sand faced; PC £130.00/1000 (unless otherwise stated); in gauged mortar (1:1:6) - cont'd						
Projections - cont'd						
440 mm width; 112 mm depth; stretcher bond; vertical	-	0.83	14.51	4.14	m²	18.65
440 mm width; 112 mm depth; flemish bond with snapped headers; vertical	-	0.88	15.38	4.14	m²	19.52
440 mm width; 225 mm depth; flemish bond; vertical	-	1.62	28.32	8.28	m²	36.60
Arches						
height on face 215 mm, width of exposed soffit 102 mm, shape of arch - flat	-	0.93	12.82	3.67	m	16.49
height on face 215 mm, width of exposed soffit 215 mm, shape of arch - flat	-	1.39	20.86	5.85	m	26.71
height on face 215 mm, width of exposed soffit 102 mm, shape of arch - segmental, one ring	-	1.76	23.12	8.51	m	31.63
height on face 215 mm, width of exposed soffit 215 mm, shape of arch - segmental, one ring	-	2.13	29.59	10.60	m	40.19
height on face 215 mm, width of exposed soffit 102 mm, shape of arch - semi-circular, one ring	-	2.68	39.20	8.51	m	47.71
height on face 215 mm, width of exposed soffit 215 mm, shape of arch - semi-circular, one ring	-	3.61	55.46	10.60	m	66.06
height on face 215 mm, width of exposed soffit 102 mm, shape of arch - segmental, two ring	-	2.07	28.54	8.51	m	37.05
height on face 215 mm, width of exposed soffit 215 mm, shape of arch - segmental, two ring	-	2.82	41.65	10.60	m	52.25
height on face 215 mm, width of exposed soffit 102 mm, shape of arch - semi-circular, two ring	-	3.61	55.46	8.51	m	63.97
height on face 215 mm, width of exposed soffit 215 mm, shape of arch - semi-circular, two ring	-	5.00	79.75	10.60	m	90.35
Arches; cut voussoirs; PC £162.00/100						
height on face 215 mm, width of exposed soffit 102 mm, shape of arch - segmental, one ring	-	1.80	23.82	32.58	m	56.40
height on face 215 mm, width of exposed soffit 215 mm, shape of arch - segmental, one ring	-	2.27	32.04	58.74	m	90.78
height on face 215 mm, width of exposed soffit 102 mm, shape of arch - semi-circular, one ring	-	2.04	28.01	32.58	m	60.59
height on face 215 mm, width of exposed soffit 215 mm, shape of arch - semi-circular, one ring	-	2.59	37.63	58.74	m	96.37
height on face 320 mm, width of exposed soffit 102 mm, shape of arch - segmental, one and a half ring	-	2.41	34.48	58.63	m	93.11
height on face 320 mm, width of exposed soffit 215 mm, shape of arch - segmental, one and a half ring	-	3.15	47.42	117.42	m	164.84
Arches; bullnosed specials; PC £103.50/100						
height on face 215 mm, width of exposed soffit 102 mm, shape of arch - flat	-	0.97	13.90	18.99	m	32.89
height on face 215 mm, width of exposed soffit 215 mm, shape of arch - flat	-	1.43	21.94	36.86	m	58.80
Bullseye windows; 600 mm diameter						
height on face 215 mm, width of exposed soffit 102 mm, two rings	-	4.63	73.29	6.84	nr	80.13
height on face 215 mm, width of exposed soffit 215 mm, two rings	-	6.48	105.62	11.48	nr	117.10

Prices for Measured Works - Major Works
F MASONRY

	PC £	Labour hours	Labour £	Material £	Unit	Total rate £
Bullseye windows; 600 mm diameter; cut voussoirs; PC £162.00/100						
height on face 215 mm, width of exposed soffit 102 mm, one ring	-	3.89	60.35	71.00	nr	**131.35**
height on face 215 mm, width of exposed soffit 215 mm, one ring	-	5.37	86.22	139.80	nr	**226.02**
Bullseye windows; 1200 mm diameter						
height on face 215 mm, width of exposed soffit 102 mm, two rings	-	7.22	118.56	17.54	nr	**136.10**
height on face 215 mm, width of exposed soffit 215 mm, two rings	-	10.36	173.44	27.35	nr	**200.79**
Bullseye windows; 1200 mm diameter; cut voussoirs PC £162.00/100						
height on face 215 mm, width of exposed soffit 102 mm, one ring	-	6.11	99.16	127.00	nr	**226.16**
height on face 215 mm, width of exposed soffit 215 mm, one ring	-	8.70	144.43	245.09	nr	**389.52**
ADD or DEDUCT for variation of £10.00/1000 in PC of facing bricks in 102 mm high arches with 215 mm soffit	-	-	-	0.29	m	**0.29**
Facework sills						
150 mm x 102 mm; headers on edge; pointing top and one side; set weathering; horizontal	-	0.51	8.91	2.07	m	**10.98**
150 mm x 102 mm; cant headers on edge; PC £103.50/100; pointing top and one side; set weathering; horizontal	-	0.56	9.79	17.38	m	**27.17**
150 mm x 102 mm; bullnosed specials; PC £103.50/100; headers on flat; pointing top and one side; horizontal	-	0.46	8.04	17.38	m	**25.42**
Facework copings						
215 mm x 102 mm; headers on edge; pointing top and both sides; horizontal	-	0.42	7.34	2.16	m	**9.50**
260 mm x 102 mm; headers on edge; pointing top and both sides; horizontal	-	0.65	11.36	3.17	m	**14.53**
215 mm x 102 mm; double bullnose specials headers on edge PC £103.50/1000; pointing top and both sides; horizontal	-	0.46	8.04	17.47	m	**25.51**
260 mm x 102 mm; single bullnose specials headers on edge PC £103.50/1000; pointing top and both sides; horizontal	-	0.65	11.36	34.72	m	**46.08**
ADD or DEDUCT for variation of £10.00/1000 in PC of facing bricks in copings 215 mm wide, 102 mm high	-	-	-	0.14	m	**0.14**
Extra over facing bricks for; facework ornamental bands and the like, plain bands; PC £150.00/1000						
flush; horizontal, 225 mm width; entirely of stretchers	-	0.19	3.32	0.29	m	**3.61**
Extra over facing bricks for; facework quoins; PC £150.00/1000						
flush; mean girth 320 mm	-	0.28	4.89	0.29	m	**5.18**
Bonding to existing						
facework one side, half brick thick; stretcher bond	-	0.46	8.04	0.55	m	**8.59**
facework one side, half brick thick; flemish bond with snapped headers	-	0.46	8.04	0.55	m	**8.59**
facework both sides, one brick thick; two stretcher skins tied together	-	0.65	11.36	1.11	m	**12.47**
facework both sides, one brick thick; flemish bond	-	0.65	11.36	1.11	m	**12.47**
ADD or DEDUCT for variation of ú10.00/1000 in PC of facing bricks; in walls built entirely of facings; in stretcher or flemish bond						
half brick thick	-	-	-	0.65	m^2	**0.65**
one brick thick	-	-	-	1.29	m^2	**1.29**

Prices for Measured Works - Major Works
F MASONRY

	PC £	Labour hours	Labour £	Material £	Unit	Total rate £
F10 BRICK/BLOCK WALLING - cont'd						
Facing bricks; white sandlime; PC £165.00/1000 in gauged mortar (1:1:6)						
Walls						
facework one side, half brick thick; stretcher bond	-	1.20	20.98	12.41	m²	33.39
facework both sides, one brick thick; flemish bond	-	2.13	37.23	24.81	m²	62.04
ADD or DEDUCT for variation of £10.00/1000 in PC of facing bricks; in walls built entirely of facings; in stretcher or flemish bond						
half brick thick	-	-	-	0.65	m²	0.65
one brick thick	-	-	-	1.29	m²	1.29
Facing bricks; machine made facings; PC £275.00/1000 (unless otherwise stated); in gauged mortar (1:1:6)						
Walls						
facework one side, half brick thick; stretcher bond	-	1.20	20.98	19.50	m²	40.48
facework one side, half brick thick, flemish bond with snapped headers	-	1.39	24.30	19.50	m²	43.80
facework one side, half brick thick, stretcher bond; building against other work; concrete	-	1.30	22.72	20.68	m²	43.40
facework one side, half brick thick; flemish bond with snapped headers; building against other work; concrete	-	1.48	25.87	20.68	m²	46.55
facework one side, half brick thick; stretcher bond; building overhand	-	1.48	25.87	19.50	m²	45.37
facework one side, half brick thick; flemish bond with snapped headers; building overhand	-	1.67	29.19	19.50	m²	48.69
facework one side, half brick thick; stretcher bond; curved; 6.00 m radii	-	1.76	30.76	19.50	m²	50.26
facework one side, half brick thick; flemish bond with snapped headers; curved; 6.00 m radii	-	1.99	34.78	19.50	m²	54.28
facework one side, half brick thick; stretcher bond; curved; 1.50 m radii	-	2.22	38.80	22.46	m²	61.26
facework one side, half brick thick; flemish bond with snapped headers; curved; 1.50 m radii	-	2.59	45.27	22.46	m²	67.73
facework both sides, one brick thick; two stretcher skins tied together	-	2.08	36.36	40.02	m²	76.38
facework both sides, one brick thick; flemish bond	-	2.13	37.23	39.00	m²	76.23
facework both sides, one brick thick; two stretcher skins tied together; curved; 6.00 m radii	-	2.87	50.17	42.98	m²	93.15
facework both sides, one brick thick; flemish bond; curved; 6.00 m radii	-	2.96	51.74	41.96	m²	93.70
facework both sides, one brick thick; two stretcher skins tied together; curved; 1.50 m radii	-	3.56	62.23	46.52	m²	108.75
facework both sides, one brick thick; flemish bond; curved; 1.50 m radii	-	3.70	64.67	45.50	m²	110.17
Isolated piers						
facework both sides, one brick thick; two stretcher skins tied together	-	2.45	42.82	40.60	m²	83.42
facework both sides, one brick thick; flemish bond	-	2.50	43.70	40.60	m²	84.30

F MASONRY

	PC £	Labour hours	Labour £	Material £	Unit	Total rate £
Isolated casings						
facework one side, half brick thick; stretcher bond	-	1.85	32.34	19.50	m²	51.84
facework one side, half brick thick; flemish bond with snapped headers	-	2.04	35.66	19.50	m²	55.16
Projections						
225 mm width; 112 mm depth; stretcher bond; vertical	-	0.28	4.89	4.15	m	9.04
225 mm width; 112 mm depth; flemish bond with snapped headers; vertical	-	0.37	6.47	4.15	m	10.62
225 mm width; 225 mm depth; flemish bond; vertical	-	0.60	10.49	8.29	m	18.78
328 mm width; 112 mm depth; stretcher bond; vertical	-	0.56	9.79	6.22	m	16.01
328 mm width; 112 mm depth; flemish bond with snapped headers; vertical	-	0.65	11.36	6.22	m	17.58
328 mm width; 225 mm depth; flemish bond; vertical	-	1.11	19.40	12.41	m	31.81
440 mm width; 112 mm depth; stretcher bond; vertical	-	0.83	14.51	8.29	m	22.80
440 mm width; 112 mm depth; flemish bond with snapped headers; vertical	-	0.88	15.38	8.29	m	23.67
440 mm width; 225 mm depth; flemish bond; vertical	-	1.62	28.32	16.59	m	44.91
Arches						
height on face 215 mm, width of exposed soffit 102 mm, shape of arch - flat	-	0.93	13.20	5.75	m	18.95
height on face 215 mm, width of exposed soffit 215 mm, shape of arch - flat	-	1.39	21.24	10.06	m	31.30
height on face 215 mm, width of exposed soffit 102 mm, shape of arch - segmental, one ring		1.76	23.12	10.59	m	33.71
height on face 215 mm, width of exposed soffit 215 mm, shape of arch segmental, one ring	-	2.13	29.59	14.75	m	44.34
height on face 215 mm, width of exposed soffit 102 mm, shape of arch - semi-circular, one ring	-	2.68	39.20	10.59	m	49.79
height on face 215 mm, width of exposed soffit 215 mm, shape of arch - semi-circular, one ring		3.61	55.46	14.75	m	70.21
height on face 215 mm, width of exposed soffit 102 mm, shape of arch - segmental, two ring	-	2.17	30.29	10.59	m	40.88
height on face 215 mm, width of exposed soffit 215 mm, shape of arch - segmental; two ring		2.82	41.65	14.75	m	56.40
height on face 215 mm, width of exposed soffit 102 mm, shape of arch - semi-circular, two ring	-	3.61	55.46	10.59	m	66.05
height on face 215 mm, width of exposed soffit 215 mm, shape of arch - semi-circular, two ring		5.00	79.75	14.75	m	94.50
Arches; cut voussoirs; PC £2570.00/1000						
height on face 215 mm, width of exposed soffit 102 mm, shape of arch - segmental, one ring	-	1.80	23.82	43.48	m	67.30
height on face 215 mm, width of exposed soffit 215 mm, shape of arch - segmental, one ring	-	2.27	32.04	80.54	m	112.58
height on face 215 mm, width of exposed soffit 102 mm, shape of arch - semi-circular, one ring	-	2.04	28.01	43.48	m	71.49
height on face 215 mm, width of exposed soffit 215 mm, shape of arch - semi-circular, one ring	-	2.59	37.63	80.54	m	118.17
height on face 320 mm, width of exposed soffit 102 mm, shape of arch - segmental, one and a half ring	-	2.41	34.48	80.44	m	114.92
height on face 320 mm, width of exposed soffit 215 mm, shape of arch - segmental, one and a half ring	-	3.15	47.42	161.03	m	208.45

Prices for Measured Works - Major Works
F MASONRY

	PC £	Labour hours	Labour £	Material £	Unit	Total rate £
F10 BRICK/BLOCK WALLING - cont'd						
Facing bricks; machine made facings; PC £275.00/1000 (unless otherwise stated); in gauged mortar (1:1:6) - cont'd						
Arches; bullnosed specials; PC £1186.00/1000						
height on face 215 mm, width of exposed soffit 102 mm, shape of arch - flat	-	0.97	13.90	18.81	m	32.71
height on face 215 mm, width of exposed soffit 215 mm, shape of arch - flat	-	1.43	21.94	36.50	m	58.44
Bullseye windows; 600 mm diameter						
height on face 215 mm, width of exposed soffit 102 mm, two rings	-	4.63	73.29	11.21	nr	84.50
height on face 215 mm, width of exposed soffit 215 mm, two rings	-	6.48	105.62	20.21	nr	125.83
Bullseye windows; 600 mm; cut voussoirs; PC £2570.00/1000						
height on face 215 mm, width of exposed soffit 102 mm, one ring	-	3.89	60.35	99.63	nr	159.98
height on face 215 mm, width of exposed soffit 215 mm, one ring	-	5.37	86.22	197.05	nr	283.27
Bullseye windows; 1200 mm diameter						
height on face 215 mm, width of exposed soffit 102 mm, two rings	-	7.22	118.56	26.27	nr	144.83
height on face 215 mm, width of exposed soffit 215 mm, two rings	-	10.36	173.44	44.81	nr	218.25
Bullseye windows; 1200 mm diameter; cut voussoirs; PC £2570.00/1000						
height on face 215 mm, width of exposed soffit 102 mm, one ring	-	6.11	99.16	176.07	nr	275.23
height on face 215 mm, width of exposed soffit 215 mm, one ring	-	8.70	144.43	343.23	nr	487.66
ADD or DEDUCT for variation of £10.00/1000 in PC of facing bricks in 102 mm high arches with 215 mm soffit	-	-	-	0.29	m	0.29
Facework sills						
150 mm x 102 mm; headers on edge; pointing top and one side; set weathering; horizontal	-	0.51	8.91	4.15	m	13.06
150 mm x 102 mm; cant headers on edge; PC £1186.00/1000; pointing top and one side; set weathering; horizontal	-	0.56	9.79	17.21	m	27.00
150 mm x 102 mm; bullnosed specials; PC £1186.00/1000; headers on flat; pointing top and one side; horizontal	-	0.46	8.04	17.21	m	25.25
Facework copings						
215 mm x 102 mm; headers on edge; pointing top and both sides; horizontal	-	0.42	7.34	4.24	m	11.58
260 mm x 102 mm; headers on edge; pointing top and both sides; horizontal	-	0.65	11.36	6.28	m	17.64
215 mm x 102 mm; double bullnose specials; PC £1186.00/1000; headers on edge; pointing top and both sides; horizontal	-	0.46	8.04	17.29	m	25.33
260 mm x 102 mm; single bullnose specials; PC £1186.00/1000; headers on edge; pointing top and both sides; horizontal	-	0.65	11.36	34.37	m	45.73
ADD or DEDUCT for variation of £10.00/1000 in PC of facing bricks in copings 215 mm wide, 102 mm high	-	-	-	0.14	m	0.14

F MASONRY

	PC £	Labour hours	Labour £	Material £	Unit	Total rate £
Extra over facing bricks for; facework ornamental bands and the like, plain bands; PC £305.00/1000						
flush; horizontal; 225 mm width; entirely of stretchers	-	0.19	3.32	0.44	m	3.76
Extra over facing brick for; facework quoins; PC £305.00/1000						
flush; mean girth 320 mm	-	0.28	4.89	0.43	m	5.32
Bonding to existing						
facework one side, half brick thick; stretcher bond	-	0.46	8.04	1.07	m	9.11
facework one side, half brick thick; flemish bond with snapped headers	-	0.46	8.04	1.07	m	9.11
facework both sides, one brick thick; two stretcher skins tied together	-	0.65	11.36	2.15	m	13.51
facework both sides, one brick thick; flemish bond	-	0.65	11.36	2.15	m	13.51
ADD or DEDUCT for variation of £10.00/1000 in PC of facing bricks; in walls built entirely of facings; in stretcher or flemish bond						
half brick thick	-	-	-	0.65	m²	0.65
one brick thick	-	-	-	1.29	m²	1.29
Facing bricks; hand made; PC £450.00/1000 (unless otherwise stated); in gauged mortar (1:1:6)						
Walls						
facework one side, half brick thick; stretcher bond	-	1.20	20.98	30.79	m²	51.77
facework one side, half brick thick; flemish bond with snapped headers		1.39	24.30	30.79	m²	55.09
facework one side; half brick thick; stretcher bond; building against other work; concrete	-	1.30	22.72	31.96	m²	54.68
facework one side, half brick thick; flemish bond with snapped headers; building against other work; concrete	-	1.48	25.87	31.96	m²	57.83
facework one side, half brick thick; stretcher bond; building overhand	-	1.48	25.87	30.79	m²	56.66
facework one side, half brick thick; flemish bond with snapped headers; building overhand		1.67	29.19	30.79	m²	59.98
facework one side, half brick thick; stretcher bond; curved; 6.00 m radii	-	1.76	30.76	30.79	m²	61.55
facework one side, half brick thick; flemish bond with snapped headers; curved; 6.00 m radii	-	1.99	34.78	34.42	m²	69.20
facework one side, half brick thick; stretcher bond; curved 1.50 m radii	-	2.22	38.80	30.79	m²	69.59
facework one side, half brick thick; flemish bond with snapped headers; curved; 1.50 m radii	-	2.59	45.27	36.84	m²	82.11
facework both sides, one brick thick; two stretcher skins tied together	-	2.08	36.36	62.60	m²	98.96
facework both sides, one brick thick; flemish bond	-	2.13	37.23	61.58	m²	98.81
facework both sides; one brick thick; two stretcher skins tied together; curved; 6.00 m radii	-	2.87	50.17	67.43	m²	117.60
facework both sides, one brick thick; flemish bond; curved; 6.00 m radii	-	2.96	51.74	66.41	m²	118.15
facework both sides, one brick thick; two stretcher skins tied together; curved; 1.50 m radii	-	3.56	62.23	72.86	m²	135.09
facework both sides, one brick thick; flemish bond; curved; 1.50 m radii	-	3.70	64.67	71.84	m²	136.51

Prices for Measured Works - Major Works
F MASONRY

	PC £	Labour hours	Labour £	Material £	Unit	Total rate £
F10 BRICK/BLOCK WALLING - cont'd						
Facing bricks; hand made; PC £450.00/1000 (unless otherwise stated); in gauged mortar (1:1:6) - cont'd						
Isolated piers						
facework both sides, one brick thick; two stretcher skins tied together	-	2.45	42.82	63.18	m²	106.00
facework both sides, one brick thick; flemish bond	-	2.50	43.70	63.18	m²	106.88
Isolated casings						
facework one side, half brick thick; stretcher bond	-	1.85	32.34	30.79	m²	63.13
facework one side, half brick thick; flemish bond with snapped headers	-	2.04	35.66	30.79	m²	66.45
Projections						
225 mm width; 112 mm depth; stretcher bond; vertical	-	0.28	4.89	6.66	m	11.55
225 mm width; 112 mm depth; flemish bond with snapped headers; vertical	-	0.37	6.47	6.66	m	13.13
225 mm width; 225 mm depth; flemish bond; vertical	-	0.60	10.49	13.31	m	23.80
328 mm width; 112 mm depth; stretcher bond; vertical	-	0.56	9.79	9.99	m	19.78
328 mm width; 112 mm depth; flemish bond with snapped headers; vertical	-	0.65	11.36	9.99	m	21.35
328 mm width; 225 mm depth; flemish bond; vertical	-	1.11	19.40	19.94	m	39.34
440 mm width; 112 mm depth; stretcher bond; vertical	-	0.83	14.51	13.31	m	27.82
440 mm width; 112 mm depth; flemish bond with snapped headers; vertical	-	0.88	15.38	13.31	m	28.69
440 mm width; 225 mm depth; flemish bond; vertical	-	1.62	28.32	26.62	m	54.94
Arches						
height on face 215 mm, width of exposed soffit 102 mm, shape of arch - flat	-	0.93	13.20	8.26	m	21.46
height on face 215 mm, width of exposed soffit 215 mm, shape of arch - flat	-	1.39	21.24	15.13	m	36.37
height on face 215 mm, width of exposed soffit 102 mm, shape of arch - segmental, one ring	-	1.76	23.12	13.10	m	36.22
height on face 215 mm, width of exposed soffit 215 mm, shape of arch - segmental, one ring	-	2.13	29.59	19.77	m	49.36
height on face 215 mm, width of exposed soffit 102 mm, shape of arch - semi-circular, one ring	-	2.68	39.20	13.10	m	52.30
height on face 215 mm, width of exposed soffit 215 mm, shape of arch - semi-circular, one ring	-	3.61	55.46	19.77	m	75.23
height on face 215 mm, width of exposed soffit 102 mm, shape of arch - segmental, two ring	-	2.17	30.29	13.10	m	43.39
height on face 215 mm, width of exposed soffit 215 mm, shape of arch - segmental, two ring	-	2.82	41.65	19.77	m	61.42
height on face 215 mm, width of exposed soffit 102 mm, shape of arch - semi-circular, two ring	-	3.61	55.46	13.10	m	68.56
height on face 215 mm, width of exposed soffit 215 mm, shape of arch - semi-circular, two ring	-	5.00	79.75	19.77	m	99.52
Arches; cut voussoirs; PC £2520.00/1000						
height on face 215 mm, width of exposed soffit 102 mm, shape of arch - segmental, one ring	-	1.80	23.82	42.77	m	66.59
height on face 215 mm, width of exposed soffit 215 mm, shape of arch - segmental, one ring	-	2.27	32.04	79.11	m	111.15
height on face 215 mm, width of exposed soffit 102 mm, shape of arch - semi-circular, one ring	-	2.04	28.01	42.77	m	70.78
height one face 215 mm, width of exposed soffit						

Prices for Measured Works - Major Works

F MASONRY

	PC £	Labour hours	Labour £	Material £	Unit	Total rate £
Arches; cut voussoirs; PC £2520.00/1000 - cont'd						
215 mm, shape of - arch semi-circular, one ring height on face 320 mm, width of exposed soffit	-	2.59	37.63	79.11	m	116.74
102 mm, shape of arch - segmental, one and a half ring	-	2.41	34.48	79.00	m	113.48
height on face 320 mm, width of exposed soffit 215 mm, shape of arch - segmental, one and a half ring	-	3.15	47.42	158.17	m	205.59
Arches; bullnosed specials; PC £1160.00/1000						
height on face 215 mm, width of exposed soffit 102 mm, shape of arch - flat	-	0.97	13.90	18.44	m	32.34
height on face 215 mm, width of exposed soffit 215 mm, shape of arch - flat	-	1.43	21.94	35.74	m	57.68
Bullseye windows; 600 mm diameter						
height on face 215 mm, width of exposed soffit 102 mm, two ring	-	4.63	73.29	16.47	nr	89.76
height on face 215 mm, width of exposed soffit 215 mm, two ring	-	6.48	105.62	41.33	nr	146.95
Bullseye windows; 600 mm diameter; cut voussoirs; PC £2520.00/1000						
height on face 215 mm, width of exposed soffit 102 mm, one ring	-	3.89	60.35	97.74	nr	158.09
height on face 215 mm, width of exposed soffit 215 mm, one ring	-	5.37	86.22	193.29	nr	279.51
Bullseye windows; 1200 mm diameter						
height on face 215 mm, width of exposed soffit 102 mm, two ring	-	7.22	118.56	36.80	nr	155.36
height on face 215 mm, width of exposed soffit 215 mm, two ring	-	10.36	173.44	65.88	nr	239.32
Bullseye windows; 1200 mm diameter; cut voussoirs; PC £2520.00/1000						
height on face 215 mm, width of exposed soffit 102 mm, one ring	-	6.11	99.16	172.84	nr	272.00
height on face 215 mm, width of exposed soffit 215 mm, one ring	-	8.70	144.43	336.78	nr	481.21
ADD or DEDUCT for variation of £10.00/1000 in PC of facing bricks in 102 mm high arches with 215 mm soffit	-	-	-	0.29	m	0.29
Facework sills						
150 mm x 102 mm; headers on edge; pointing top and one side; set weathering; horizontal	-	0.51	8.91	6.66	m	15.57
150 mm x 102 mm; cant headers on edge; PC £1160.00/1000; pointing top and one side; set weathering; horizontal	-	0.56	9.79	16.83	m	26.62
150 mm x 102 mm; bullnosed specials; PC £1160.00/1000; headers on edge; pointing top and one side; horizontal	-	0.46	8.04	16.83	m	24.87
Facework copings						
215 mm x 102 mm; headers on edge; pointing top and both sides; horizontal	-	0.42	7.34	6.74	m	14.08
260 mm x 102 mm; headers on edge; pointing top and both sides; horizontal	-	0.65	11.36	10.05	m	21.41
215 mm x 102 mm; double bullnose specials; PC £1160.00/1000; headers on edge; pointing top and both sides; horizontal	-	0.46	8.04	16.92	m	24.96
260 mm x 102 mm; single bullnose specials; PC £1160.00/1000; headers on edge; pointing top and both sides; horizontal	-	0.65	11.36	33.62	m	44.98
ADD or DEDUCT for variation of £10.00/1000 in PC of facing bricks in copings 215 mm wide, 102 mm high	-	-	-	0.14	m	0.14

F MASONRY

	PC £	Labour hours	Labour £	Material £	Unit	Total rate £
F10 BRICK/BLOCK WALLING - cont'd						
Facing bricks; hand made; PC £450.00/1000 (unless otherwise stated); in gauged mortar (1:1:6) - cont'd						
Extra over facing bricks for; facework ornamental bands and the like, plain bands; PC £490.00/1000						
flush; horizontal; 225 mm width; entirely of stretchers	-	0.19	3.32	0.57	m	3.89
Extra over facing bricks for; facework quoins; PC £490.00/1000						
flush mean girth 320 mm	-	0.28	4.89	0.57	m	5.46
Bonding ends to existing						
facework one side, half brick thick; stretcher bond	-	0.46	8.04	1.70	m	9.74
facework one side, half brick thick; flemish bond with snapped headers	-	0.46	8.04	1.70	m	9.74
facework both sides, one brick thick; two stretcher skins tied together	-	0.65	11.36	3.40	m	14.76
facework both sides, one brick thick; flemish bond	-	0.65	11.36	3.40	m	14.76
ADD or DEDUCT for variation of £10.00/1000 in PC of facing bricks; in walls built entirely of facings; in stretcher or flemish bond						
half brick thick	-	-	-	0.65	m²	0.65
one brick thick	-	-	-	1.29	m²	1.29
Facing bricks; slips 50 mm thick; PC £1050.00/1000; in gauged mortar (1:1:6) built up against concrete including flushing up at back (ties not included)						
Walls	-	1.85	32.34	69.44	m²	101.78
Edges of suspended slabs; 200 mm wide	-	0.56	9.79	13.89	m	23.68
Columns; 400 mm wide	-	1.11	19.40	27.78	m	47.18
Engineering bricks; PC £258.30/1000; and specials at PC £1186.00/1000; in cement mortar (1:3)						
Facework steps						
215 mm x 102 mm; all headers-on-edge; edges set with bullnosed specials; pointing top and one side; set weathering; horizontal	-	0.51	8.91	17.23	m	26.14
returned ends pointed	-	0.14	2.45	3.18	nr	5.63
430 mm x 102 mm; all headers-on-edge; edges set with bullnosed specials; pointing top and one side; set weathering; horizontal	-	0.74	12.93	21.16	m	34.09
returned ends pointed	-	0.19	3.32	4.06	nr	7.38

F MASONRY

ALTERNATIVE BLOCK PRICES (£/m²)						
	£		£			£
Aerated concrete Durox "Supablocs"; 630 x 225 mm						
100 mm	7.54	140 mm	12.29	215 mm		17.97
130 mm	10.63	150 mm	12.59			
Hanson Conbloc blocks; 450 x 225 mm						
Cream fair faced	£		£			£
100 mm hollow	6.74	140 mm solid	11.89	215 mm hollow		14.20
100 mm solid	7.30	190 mm hollow	14.23			
140 mm hollow	9.99	190 mm solid	15.68			
"Fenlite"						
100 mm solid; 3.5 N/mm²	5.61	100 mm solid; 7.00 N/mm²	5.91	140 mm solid; 3.50 N/mm²		8.38
Standard Dense						
100 mm solid	5.24	140 mm hollow	7.78	190mm solid		10.96
140 mm solid	7.83	190 mm hollow	11.19	215mm hollow		11.74
Celcon "Standard" blocks; 450 x 225 mm						
100 mm	9.89	140 mm	13.99			
125 mm	12.18	215 mm	23.47			
Forticrete painting quality blocks; 450 x 225 mm						
100 mm solid	10.49	140 mm hollow	13.01	215 mm solid		21.79
100 mm hollow	9.43	190 mm solid	10.83	215 mm hollow		18.12
140 mm solid	15.19	190 mm hollow	17.28			
Tarmac "Topblocks"; 450 x 225 mm						
3.50 N/mm² "Hemelite" blocks						
100 mm solid	4.62	190 mm solid	9.48	215 mm solid		10.20
7.00 N/mm² "Hemelite" blocks						
100 mm solid	4.97	190 mm solid	10.17			
140mm solid	7.11	215mm solid	10.93			
"Toplite" standard blocks						
100 mm	7.11	150 mm	10.64			
140mm	9.93	215mm	15.28			
"Toplite" GTI (thermal) blocks						
115 mm	8.42	130 mm	9.51	150 mm		10.99
125 mm	9.16	140 mm	10.26	215 mm		15.75

Discount of 0 - 20% available depending on quantity / status.

	PC £	Labour hours	Labour £	Material £	Unit	Total rate £
Lightweight aerated concrete blocks; Thermalite "Turbo" blocks or other equal and approved; in gauged mortar (1:2:9)						
Walls						
100 mm thick	6.48	0.46	8.04	8.04	m²	**16.08**
115 mm thick	7.45	0.46	8.04	9.24	m²	**17.28**
125 mm thick	8.11	0.46	8.04	10.06	m²	**18.10**
130 mm thick	8.43	0.46	8.04	10.46	m²	**18.50**
140 mm thick	9.08	0.51	8.91	11.27	m²	**20.18**
150 mm thick	9.72	0.51	8.91	12.06	m²	**20.97**
190 mm thick	12.32	0.56	9.79	15.29	m²	**25.08**
200 mm thick	12.97	0.56	9.79	16.09	m²	**25.88**
215 mm thick	13.94	0.56	9.79	17.30	m²	**27.09**

Prices for Measured Works - Major Works
F MASONRY

	PC £	Labour hours	Labour £	Material £	Unit	Total rate £
F10 BRICK/BLOCK WALLING - cont'd						
Lightweight aerated concrete blocks;						
Thermalite "Turbo" blocks or other equal and						
approved; in gauged mortar (1:2:9) - cont'd						
Isolated piers or chimney stacks						
190 mm thick	-	0.83	14.51	15.29	m²	29.80
215 mm thick	-	0.83	14.51	17.30	m²	31.81
Isolated casings						
100 mm thick	-	0.51	8.91	8.04	m²	16.95
115 mm thick	-	0.51	8.91	9.24	m²	18.15
125 mm thick	-	0.51	8.91	10.06	m²	18.97
140 mm thick	-	0.56	9.79	11.27	m²	21.06
Extra over for fair face; flush pointing						
walls; one side	-	0.04	0.70	-	m²	0.70
walls; both sides	-	0.09	1.57	-	m²	1.57
Closing cavities						
width of cavity 50 mm, closing with lightweight blockwork 100 mm thick; vertical	-	0.23	4.02	0.45	m	4.47
width of cavity 50 mm, closing with lightweight blockork 100 mm thick; including damp proof course; vertical	-	0.28	4.89	1.17	m	6.06
width of cavity 75 mm, closing with lightweight blockwork 100 mm thick; vertical	-	0.23	4.02	0.65	m	4.67
width of cavity 75 mm, closing with lightweight blockwork 100 mm thick; including damp proof course; vertical	-	0.28	4.89	1.36	m	6.25
Bonding ends to common brickwork						
100 mm thick	-	0.14	2.45	0.93	m	3.38
115 mm thick	-	0.14	2.45	1.06	m	3.51
125 mm thick	-	0.23	4.02	1.16	m	5.18
130 mm thick	-	0.23	4.02	1.21	m	5.23
140 mm thick	-	0.23	4.02	1.31	m	5.33
150 mm thick	-	0.23	4.02	1.39	m	5.41
190 mm thick	-	0.28	4.89	1.76	m	6.65
200 mm thick	-	0.28	4.89	1.86	m	6.75
215 mm thick	-	0.32	5.59	2.00	m	7.59
Lightweight aerated concrete blocks;						
Thermalite "Shield 2000" blocks or other						
equal and approved; in gauged mortar (1:2:9)						
Walls						
75 mm thick	4.69	0.42	7.34	5.33	m²	12.67
90 mm thick	5.63	0.42	7.34	6.39	m²	13.73
100 mm thick	6.25	0.46	8.04	7.10	m²	15.14
140 mm thick	8.75	0.51	8.91	9.94	m²	18.85
150 mm thick	9.38	0.51	8.91	10.65	m²	19.56
190 mm thick	11.88	0.56	9.79	13.49	m²	23.28
200 mm thick	12.50	0.56	9.79	14.19	m²	23.98
Isolated piers or chimney stacks						
190 mm thick	-	0.83	14.51	13.49	m²	28.00
Isolated casings						
75 mm thick	-	0.51	8.91	5.33	m²	14.24
90 mm thick	-	0.51	8.91	6.39	m²	15.30
100 mm thick	-	0.51	8.91	7.10	m²	16.01
140 mm thick	-	0.56	9.79	9.94	m²	19.73
Extra over for fair face; flush pointing						
walls; one side	-	0.04	0.70	-	m²	0.70
walls; both sides	-	0.09	1.57	-	m²	1.57

F MASONRY

	PC £	Labour hours	Labour £	Material £	Unit	Total rate £
Closing cavities						
width of cavity 50 mm, closing with lightweight blockwork 100 mm thick; vertical	-	0.23	4.02	0.40	m	4.42
width of cavity 50 mm, closing with lightweight blockork 100 mm thick; including damp proof course; vertical	-	0.28	4.89	1.12	m	6.01
width of cavity 75 mm, closing with lightweight blockwork 100 mm thick; vertical	-	0.23	4.02	0.58	m	4.60
width of cavity 75 mm, closing with lightweight blockwork 100 mm thick; including damp proof course; vertical	-	0.28	4.89	1.29	m	6.18
Bonding ends to common brickwork						
75 mm thick	-	0.09	1.57	0.62	m	2.19
90 mm thick	-	0.09	1.57	0.74	m	2.31
100 mm thick	-	0.14	2.45	0.82	m	3.27
140 mm thick	-	0.23	4.02	1.16	m	5.18
150 mm thick	-	0.23	4.02	1.23	m	5.25
190 mm thick	-	0.28	4.89	1.56	m	6.45
200 mm thick	-	0.28	4.89	1.65	m	6.54
Lightweight smooth face aerated concrete blocks; Thermalite "Smooth Face" blocks or other equal and approved; in gauged mortar (1:2:9); flush pointing one side						
Walls						
100 mm thick	10.74	0.56	9.79	13.08	m^2	22.87
140 mm thick	15.04	0.65	11.36	18.31	m^2	29.67
150 mm thick	16.12	0.65	11.36	19.63	m^2	30.99
190 mm thick	20.41	0.74	12.93	24.85	m^2	37.78
200 mm thick	21.49	0.74	12.93	26.17	m^2	39.10
215 mm thick	23.10	0.74	12.93	28.13	m^2	41.06
Isolated piers or chimney stacks						
190 mm thick	-	0.93	16.26	24.85	m^2	41.11
200 mm thick	-	0.93	16.26	26.17	m^2	42.43
215 mm thick	-	0.93	16.26	28.13	m^2	44.39
Isolated casings						
100 mm thick	-	0.69	12.06	13.08	m^2	25.14
140 mm thick	-	0.74	12.93	18.31	m^2	31.24
Extra over for fair face flush pointing						
walls; both sides	-	0.04	0.70	-	m^2	0.70
Bonding ends to common brickwork						
100 mm thick	-	0.23	4.02	1.48	m	5.50
140 mm thick	-	0.23	4.02	2.09	m	6.11
150 mm thick	-	0.28	4.89	2.23	m	7.12
190 mm thick	-	0.32	5.59	2.82	m	8.41
200 mm thick	-	0.32	5.59	2.98	m	8.57
215 mm thick	-	0.32	5.59	3.21	m	8.80
Lightweight smooth face aerated concrete blocks; Thermalite "Party Wall" blocks (650 kg/m³) or other equal and approved; in gauged mortar (1:2:9); flush pointing one side						
Walls						
100 mm thick	6.35	0.56	9.79	7.21	m^2	17.00
215 mm thick	13.65	0.74	12.93	15.49	m^2	28.42
Isolated piers or chimney stacks						
215 mm thick	-	0.93	16.26	15.49	m^2	31.75
Isolated casings						
100 mm thick	-	0.69	12.06	7.21	m^2	19.27
Extra over for fair face flush pointing						
walls; both sides	-	0.04	0.70	-	m^2	0.70

Prices for Measured Works - Major Works
F MASONRY

	PC £	Labour hours	Labour £	Material £	Unit	Total rate £
F10 BRICK/BLOCK WALLING - cont'd						
Lightweight smooth face aerated concrete blocks; Thermalite "Party Wall" blocks (650 kg/m³) or other equal and approved; in gauged mortar (1:2:9); flush pointing one side - cont'd						
Bonding ends to common brickwork						
100 mm thick	-	0.23	4.02	0.83	m	4.85
215 mm thick	-	0.32	5.59	1.80	m	7.39
Lightweight smooth face aerated concrete blocks; Thermalite "Party Wall" blocks (880 kg/m³) or other equal and approved; in gauged mortar (1:2:9); flush pointing one side						
Walls						
215 mm thick	18.69	0.88	15.38	20.91	m²	36.29
Isolated piers or chimney stacks						
215 mm thick	-	1.16	20.28	20.91	m²	41.19
Extra over for fair face flush pointing						
walls; both sides	-	0.04	0.70	-	m²	0.70
Bonding ends to common brickwork						
215 mm thick	-	0.37	6.47	2.40	m	8.87
Lightweight aerated high strength concrete blocks (7.00 N/mm²); Thermalite "High Strength" blocks or other equal and approved; in cement mortar (1:3)						
Walls						
100 mm thick	7.86	0.46	8.04	9.73	m²	17.77
140 mm thick	11.00	0.51	8.91	13.62	m²	22.53
150 mm thick	11.79	0.51	8.91	14.60	m²	23.51
190 mm thick	14.93	0.56	9.79	18.48	m²	28.27
200 mm thick	15.72	0.56	9.79	19.46	m²	29.25
215 mm thick	16.90	0.56	9.79	20.92	m²	30.71
Isolated piers or chimney stacks						
190 mm thick	-	0.83	14.51	18.48	m²	32.99
200 mm thick	-	0.83	14.51	19.46	m²	33.97
215 mm thick	-	0.83	14.51	20.92	m²	35.43
Isolated casings						
100 mm thick	-	0.51	8.91	9.73	m²	18.64
140 mm thick	-	0.56	9.79	13.62	m²	23.41
150 mm thick	-	0.56	9.79	14.60	m²	24.39
190 mm thick	-	0.69	12.06	18.48	m²	30.54
200 mm thick	-	0.69	12.06	19.46	m²	31.52
215 mm thick	-	0.69	12.06	20.92	m²	32.98
Extra over for flush pointing						
walls; one side	-	0.04	0.70	-	m²	0.70
walls; both sides	-	0.09	1.57	-	m²	1.57
Bonding ends to common brickwork						
100 mm thick	-	0.23	4.02	1.12	m	5.14
140 mm thick	-	0.23	4.02	1.58	m	5.60
150 mm thick	-	0.28	4.89	1.68	m	6.57
190 mm thick	-	0.32	5.59	2.13	m	7.72
200 mm thick	-	0.32	5.59	2.24	m	7.83
215 mm thick	-	0.32	5.59	2.42	m	8.01

Prices for Measured Works - Major Works

F MASONRY

	PC £	Labour hours	Labour £	Material £	Unit	Total rate £
Lightweight aerated high strength concrete blocks; Thermalite "Trenchblock" blocks or other equal and approved; in cement mortar (1:4)						
Walls						
255 mm thick	16.19	0.74	12.93	18.41	m²	31.34
275 mm thick	17.47	0.79	13.81	19.86	m²	33.67
305 mm thick	19.37	0.83	14.51	22.02	m²	36.53
355 mm thick	22.54	0.93	16.26	25.58	m²	41.84
Dense aggregate concrete blocks; Hanson "Conbloc" or other equal and approved; in cement mortar (1:2:9)						
Walls or partitions or skins of hollow walls						
75 mm thick; solid	4.16	0.56	9.78	5.20	m²	14.98
100 mm thick; solid	4.59	0.69	12.05	5.81	m²	17.86
140 mm thick; solid	9.04	0.83	14.49	11.22	m²	25.71
140 mm thick; hollow	8.68	0.74	12.92	10.79	m²	23.71
190 mm thick; hollow	10.24	0.93	16.24	12.83	m²	29.07
215 mm thick; hollow	10.68	1.02	17.81	13.44	m²	31.25
Isolated piers or chimney stacks						
140 mm thick; hollow	-	1.02	17.81	10.79	m²	28.60
190 mm thick; hollow	-	1.34	23.39	12.83	m²	36.22
215 mm thick; hollow	-	1.53	26.71	13.44	m²	40.15
Isolated casings						
75 mm thick; solid	-	0.69	12.05	5.20	m²	17.25
100 mm thick; solid	-	0.74	12.92	5.81	m²	18.73
140 mm thick; solid	-	0.93	16.24	11.22	m²	27.46
Extra over for fair face; flush pointing						
walls; one side	-	0.09	1.57	-	m²	1.57
walls; both sides	-	0.14	2.44	-	m²	2.44
Bonding ends to common brickwork						
75 mm thick solid	-	0.14	2.44	0.60	m	3.04
100 mm thick solid	-	0.23	4.02	0.68	m	4.70
140 mm thick solid	-	0.28	4.89	1.30	m	6.19
140 mm thick hollow	-	0.28	4.89	1.25	m	6.14
190 mm thick hollow	-	0.32	5.59	1.49	m	7.08
215 mm thick hollow	-	0.37	6.46	1.57	m	8.03
Dense aggregate concrete blocks; (7.00 N/mm²) Forticrete "Shepton Mallet Common" blocks or other equal and approved; in cement mortar (1:3)						
Walls						
75 mm thick; solid	5.94	0.56	9.78	7.35	m²	17.13
100 mm thick; hollow	6.93	0.69	12.05	8.03	m²	20.08
100 mm thick; solid	4.47	0.69	12.05	5.73	m²	17.78
140 mm thick; hollow	9.57	0.74	12.92	11.93	m²	24.85
140 mm thick; solid	6.86	0.83	14.49	8.72	m²	23.21
190 mm thick; hollow	12.71	0.93	16.24	15.86	m²	32.10
190 mm thick; solid	9.37	1.02	17.81	11.91	m²	29.72
215 mm thick; hollow	10.11	1.02	17.81	12.89	m²	30.70
215 mm thick; solid	10.00	1.16	20.25	12.76	m²	33.01
Dwarf support wall						
140 mm thick; solid	-	1.16	20.25	8.72	m²	28.97
190 mm thick; solid	-	1.34	23.39	11.91	m²	35.30
215 mm thick; solid	-	1.53	26.71	12.76	m²	39.47
Isolated piers or chimney stacks						
140 mm thick; hollow	-	1.02	17.81	11.93	m²	29.74
190 mm thick; hollow	-	1.34	23.39	15.86	m²	39.25
215 mm thick; hollow	-	1.53	26.71	12.89	m²	39.60

Prices for Measured Works - Major Works
F MASONRY

	PC £	Labour hours	Labour £	Material £	Unit	Total rate £
F10 BRICK/BLOCK WALLING - cont'd						
Dense aggregate concrete blocks; (7.00 N/mm²) Forticrete "Shepton Mallet Common" blocks or other equal and approved; in cement mortar (1:3) - cont'd						
Isolated casings						
75 mm thick; solid	-	0.69	12.05	7.35	m²	19.40
100 mm thick; solid	-	0.74	12.92	5.73	m²	18.65
140 mm thick; solid	-	0.93	16.24	8.72	m²	24.96
Extra over for fair face; flush pointing						
walls; one side	-	0.09	1.57	-	m²	1.57
walls; both sides	-	0.14	2.44	-	m²	2.44
Bonding ends to common brickwork						
75 mm thick solid	-	0.14	2.44	0.84	m	3.28
100 mm thick solid	-	0.23	4.02	0.66	m	4.68
140 mm thick solid	-	0.28	4.89	1.01	m	5.90
190 mm thick solid	-	0.32	5.59	1.37	m	6.96
215 mm thick solid	-	0.37	6.46	1.48	m	7.94
Dense aggregate coloured concrete blocks; Forticrete "Shepton Mallet Bathstone" or other equal and approved; in coloured gauged mortar (1:1:6); flush pointing one side						
Walls						
100 mm thick hollow	15.48	0.74	12.92	17.96	m²	30.88
100 mm thick solid	18.75	0.74	12.92	21.65	m²	34.57
140 mm thick hollow	20.37	0.83	14.49	23.67	m²	38.16
140 mm thick solid	27.32	0.93	16.24	31.52	m²	47.76
190 mm thick hollow	25.89	1.02	17.81	30.14	m²	47.95
190 mm thick solid	37.51	1.16	20.25	43.26	m²	63.51
215 mm thick hollow	27.33	1.16	20.25	31.89	m²	52.14
215 mm thick solid	39.75	1.20	20.95	45.91	m²	66.86
Isolated piers or chimney stacks						
140 mm thick solid	-	1.25	21.82	31.52	m²	53.34
190 mm thick solid	-	1.39	24.27	43.26	m²	67.53
215 mm thick solid	-	1.57	27.41	45.91	m²	73.32
Extra over blocks for						
100 mm thick half lintel blocks; ref D14	-	0.23	4.02	16.54	m	20.56
140 mm thick half lintel blocks; ref H14	-	0.28	4.89	20.95	m	25.84
140 mm thick quoin blocks; ref H16	-	0.32	5.59	24.05	m	29.64
140 mm thick cavity closer blocks; ref H17	-	0.32	5.59	28.90	m	34.49
140 mm thick cill blocks; ref H21	-	0.28	4.89	17.15	m	22.04
190 mm thick half lintel blocks; ref A14	-	0.32	5.59	24.04	m	29.63

F MASONRY

	PC £	Labour hours	Labour £	Material £	Unit	Total rate £
"Astra-Glaze" satin-gloss glazed finish blocks or other equal and approved; Forticrete Ltd; standard colours; in gauged mortar (1:1:6); joints raked out; gun applied latex grout to joints						
Walls or partitions or skins of hollow walls						
100 mm thick; glazed one side	65.03	0.93	16.24	73.80	m²	90.04
extra; glazed square end return	8.83	0.37	6.46	24.83	m	31.29
150 mm thick; glazed one side	74.32	1.02	17.81	84.48	m²	102.29
extra; glazed square end return	14.12	0.46	8.03	50.12	m	58.15
200 mm thick; glazed one side	83.61	1.16	20.25	95.16	m²	115.41
extra; glazed square end return	-	0.51	8.90	70.76	m	79.66
100 mm thick; glazed both sides	92.90	1.11	19.38	105.25	m²	124.63
150 mm thick; glazed both sides	102.19	1.20	20.95	115.93	m²	136.88
100 mm thick lintel 200 mm high; glazed one side	-	0.83	10.68	20.61	m	31.29
150 mm thick lintel 200 mm high; glazed one side	-	0.93	12.43	25.19	m	37.62
200 mm thick lintel 200 mm high; glazed one side	-	1.25	16.11	23.96	m	40.07
F11 GLASS BLOCK WALLING						
NOTE: The following specialist prices for glass block walling assume standard blocks; panels of 50 m²; no fire rating; work in straight walls at ground level; and all necessary ancillary fixing; strengthening; easy access; pointing and expansion materials etc.						
Hollow glass block walling; Luxcrete sealed "Luxblocks" or other equal and approved; in cement mortar "Luxfix" joints or other equal and approved; reinforced with 6 mm diameter stainless steel rods; with "Luxfibre" or other equal and approved at head and jambs; pointed both sides with "Luxseal" mastic or other equal and approved						
Walls; facework both sides						
115 mm x 115 mm x 80 mm flemish blocks	-	-	-	-	m²	505.53
190 mm x 190 mm x 80 mm flemish; cross reeded or clear blocks	-	-	-	-	m²	231.50
240 mm x 240 mm x 80 mm flemish; cross reeded or clear blocks	-	-	-	-	m²	222.05
240 mm x 115 mm x 80 mm flemish or clear blocks	-	-	-	-	m²	359.07
F20 NATURAL STONE RUBBLE WALLING						
Cotswold Guiting limestone or other equal and approved ; laid dry						
Uncoursed random rubble walling						
275 mm thick	-	2.07	32.70	28.56	m²	61.26
350 mm thick	-	2.46	38.53	36.34	m²	74.87
425 mm thick	-	2.81	43.65	44.13	m²	87.78
500 mm thick	-	3.15	48.60	51.92	m²	100.52

	PC £	Labour hours	Labour £	Material £	Unit	Total rate £
F20 NATURAL STONE RUBBLE WALLING - cont'd						
Cotswold Guiting limestone or other equal and approved; bedded; jointed and pointed in cement:lime mortar (1:2:9)						
Uncoursed random rubble walling; faced and pointed; both sides						
275 mm thick	-	1.98	31.10	32.00	m²	63.10
350 mm thick	-	2.18	33.58	40.72	m²	74.30
425 mm thick	-	2.39	36.23	49.45	m²	85.68
500 mm thick	-	2.59	38.70	58.17	m²	96.87
Coursed random rubble walling; rough dressed; faced and pointed one side						
114 mm thick	-	1.48	24.83	39.79	m²	64.62
150 mm thick	-	1.76	29.69	43.98	m²	73.67
Fair returns on walling						
114 mm wide	-	0.02	0.35	-	m	0.35
150 mm wide	-	0.03	0.52	-	m	0.52
275 mm wide	-	0.06	1.05	-	m	1.05
350 mm wide	-	0.08	1.40	-	m	1.40
425 mm wide	-	0.10	1.75	-	m	1.75
500 mm wide	-	0.12	2.10	-	m	2.10
Fair raking cutting or circular cutting						
114 mm wide	-	0.20	3.50	5.75	m	9.25
150 mm wide	-	0.25	4.38	6.32	m	10.70
Level uncoursed rubble walling for damp proof courses and the like						
275 mm wide	-	0.19	3.36	2.13	m	5.49
350 mm wide	-	0.20	3.53	2.66	m	6.19
425 mm wide	-	0.21	3.71	3.24	m	6.95
500 mm wide	-	0.22	3.89	3.83	m	7.72
Copings formed of rough stones; faced and pointed all round						
275 mm x 200 mm (average) high	-	0.56	9.10	7.68	m	16.78
350 mm x 250 mm (average) high	-	0.75	12.02	10.67	m	22.69
425 mm x 300 mm (average) high	-	0.97	15.38	14.71	m	30.09
500 mm x 300 mm (average) high	-	1.23	19.26	19.83	m	39.09
F22 CAST STONE ASHLAR WALLING DRESSINGS						
Reconstructed limestone walling; Bradstone 100 bed weathered "Cotswold" or "North Cerney" masonry blocks or other equal and approved; laid to pattern or course recommended; bedded; jointed and pointed in approved coloured cement:lime mortar (1:2:9)						
Walls; facing and pointing one side						
masonry blocks; random uncoursed	-	1.04	18.18	22.43	m²	40.61
extra; returned ends	-	0.37	6.47	23.96	m	30.43
extra; plain L shaped quoins	-	0.12	2.10	8.18	m	10.28
traditional walling; coursed squared	-	1.30	22.72	22.39	m²	45.11
squared random rubble	-	1.30	22.72	24.83	m²	47.55
squared coursed rubble (large module)	-	1.20	20.98	24.68	m²	45.66
squared coursed rubble (small module)	-	1.25	21.85	24.98	m²	46.83
squared and pitched rock faced walling; coursed	-	1.34	23.42	23.27	m²	46.69
rough hewn rockfaced walling; random	-	1.39	24.30	23.12	m²	47.42
extra; returned ends	-	0.15	2.62	-	m	2.62

F MASONRY

	PC £	Labour hours	Labour £	Material £	Unit	Total rate £
Isolated piers or chimney stacks; facing and pointing one side						
masonry blocks; random uncoursed	-	1.43	25.00	22.43	m²	**47.43**
traditional walling; coursed squared	-	1.80	31.46	22.39	m²	**53.85**
squared random rubble	-	1.80	31.46	24.83	m²	**56.29**
squared coursed rubble (large module)	-	1.67	29.19	24.68	m²	**53.87**
squared coursed rubble (small module)	-	1.76	30.76	24.98	m²	**55.74**
squared and pitched rock faced walling; coursed	-	1.90	33.21	23.27	m²	**56.48**
rough hewn rockfaced walling; random	-	1.94	33.91	23.12	m²	**57.03**
Isolated casings; facing and pointing one side						
masonry blocks; random uncoursed	-	1.25	21.85	22.43	m²	**44.28**
traditional walling; coursed squared	-	1.57	27.44	22.39	m²	**49.83**
squared random rubble	-	1.57	27.44	24.83	m²	**52.27**
squared coursed rubble (large module)	-	1.43	25.00	24.68	m²	**49.68**
squared coursed rubble (small module)	-	1.53	26.74	24.98	m²	**51.72**
squared and pitched rock faced walling; coursed	-	1.62	28.32	23.27	m²	**51.59**
rough hewn rockfaced walling; random	-	1.67	29.19	23.12	m²	**52.31**
Fair returns 100 mm wide						
masonry blocks; random uncoursed	-	0.11	1.92	-	m²	**1.92**
traditional walling; coursed squared	-	0.14	2.45	-	m²	**2.45**
squared random rubble	-	0.14	2.45	-	m²	**2.45**
squared coursed rubble (large module)	-	0.13	2.27	-	m²	**2.27**
squared coursed rubble (small module)	-	0.13	2.27	-	m²	**2.27**
squared and pitched rock faced walling; coursed	-	0.14	2.45	-	m²	**2.45**
rough hewn rockfaced walling; random	-	0.15	2.62	-	m²	**2.62**
Fair raking cutting or circular cutting						
100 mm wide	-	0.17	2.97	-	m	**2.97**
Reconstructed limestone dressings; "Bradstone Architectural" dressings in weathered "Cotswold" or "North Cerney" shades or other equal and approved; bedded, jointed and pointed in approved coloured cement:lime mortar (1:2:9)						
Copings; twice weathered and throated						
152 mm x 76 mm; type A	-	0.31	5.42	9.62	m	**15.04**
178 mm x 64 mm; type B	-	0.31	5.42	9.34	m	**14.76**
305 mm x 76 mm; type A	-	0.37	6.47	18.24	m	**24.71**
Extra for						
fair end	-	-	-	3.97	nr	**3.97**
returned mitred fair end	-	-	-	3.97	nr	**3.97**
Copings; once weathered and throated						
191 mm x 76 mm	-	0.31	5.42	10.37	m	**15.79**
305 mm x 76 mm	-	0.37	6.47	17.96	m	**24.43**
365 mm x 76 mm	-	0.37	6.47	19.23	m	**25.70**
Extra for						
fair end	-	-	-	3.97	nr	**3.97**
returned mitred fair end	-	-	-	3.97	nr	**3.97**
Chimney caps; four times weathered and throated; once holed						
553 mm x 553 mm x 76 mm	-	0.37	6.47	22.39	nr	**28.86**
686 mm x 686 mm x 76 mm	-	0.37	6.47	36.67	nr	**43.14**
Pier caps; four times weathered and throated						
305 mm x 305 mm	-	0.23	4.02	8.06	nr	**12.08**
381 mm x 381 mm	-	0.23	4.02	11.18	nr	**15.20**
457 mm x 457 mm	-	0.28	4.89	15.16	nr	**20.05**
533 mm x 533 mm	-	0.28	4.89	21.39	nr	**26.28**

Prices for Measured Works - Major Works
F MASONRY

	PC £	Labour hours	Labour £	Material £	Unit	Total rate £
F22 CAST STONE ASHLAR WALLING DRESSINGS - cont'd						
Reconstructed limestone dressings; "Bradstone Architectural" dressings in weathered "Cotswold" or "North Cerney" shades or other equal and approved; bedded, jointed and pointed in approved coloured cement:lime mortar (1:2:9) - cont'd						
Splayed corbels						
479 mm x 100 mm x 215 mm	-	0.14	2.45	10.42	nr	12.87
665 mm x 100 mm x 215 mm	-	0.19	3.32	16.72	nr	20.04
Air bricks						
300 mm x 140 mm x 76 mm	-	0.07	1.22	5.93	nr	7.15
100 mm x 140 mm lintels; rectangular; reinforced with mild steel bars						
not exceeding 1.22 m long	-	0.22	3.85	16.35	m	20.20
1.37 m - 1.67 m long	-	0.24	4.20	16.35	m	20.55
1.83 m - 1.98 m long	-	0.26	4.54	16.35	m	20.89
100 mm x 215 mm lintels; rectangular; reinforced with mild steel bars						
not exceeding 1.67 m long	-	0.24	4.20	22.02	m	26.22
1.83 m - 1.98 m long	-	0.26	4.54	22.02	m	26.56
2.13 m - 2.44 m long	-	0.28	4.89	22.02	m	26.91
2.59 m - 2.90 m long	-	0.30	5.24	22.02	m	27.26
197 mm x 67 mm sills to suit standard softwood windows; stooled at ends						
not exceeding 2.54 m long	-	0.28	4.89	25.64	m	30.53
Window surround; traditional with label moulding; for single light; sill 146 mm x 133 mm; jambs 146 mm x 146 mm; head 146 mm x 105 mm; including all dowels and anchors						
overall size 508 mm x 1479 mm	120.20	0.83	14.51	130.69	nr	145.20
Window surround; traditional with label moulding; three light; for windows 508 mm x 1219 mm; sill 146 mm x 133 mm; jambs 146 mm x 146 mm; head 146 mm x 103 mm; mullions 146 mm x 108 mm; including all dowels and anchors						
overall size 1975 mm x 1479 mm	256.99	2.17	37.93	279.95	nr	317.88
Door surround; moulded continuous jambs and head with label moulding; including all dowels and anchors						
door 839 mm x 1981 mm in 102 mm x 64 mm frame	219.69	1.53	26.74	236.90	nr	263.64
F30 ACCESSORIES/SUNDRY ITEMS FOR BRICK/BLOCK/STONE WALLING						
Forming cavities						
In hollow walls						
width of cavity 50 mm; polypropylene ties; three wall ties per m^2	-	0.05	0.87	0.18	m^2	1.05
width of cavity 50 mm; galvanised steel butterfly wall ties; three wall ties per m^2	-	0.05	0.87	0.16	m^2	1.03
width of cavity 50 mm; galvanised steel twisted wall ties; three wall ties per m^2	-	0.05	0.87	0.32	m^2	1.19
width of cavity 50 mm; stainless steel butterfly wall ties; three wall ties per m^2	-	0.05	0.87	0.33	m^2	1.20
width of cavity 50 mm; stainless steel twisted wall ties; three wall ties per m^2	-	0.05	0.87	0.65	m^2	1.52

F MASONRY

	PC £	Labour hours	Labour £	Material £	Unit	Total rate £
In hollow walls - cont'd						
width of cavity 75 mm; polypropylene ties; three wall ties per m²	-	0.05	0.87	0.18	m²	1.05
width of cavity 75 mm; galvanised steel butterfly wall ties; three wall ties per m²	-	0.05	0.87	0.20	m²	1.07
width of cavity 75 mm; galvanised steel twisted wall ties; three wall ties per m²	-	0.05	0.87	0.35	m²	1.22
width of cavity 75 mm; stainless steel butterfly wall ties; three wall ties per m²	-	0.05	0.87	0.47	m²	1.34
width of cavity 75 mm; stainless steel twisted wall ties; three wall ties per m²	-	0.05	0.87	0.71	m²	1.58
Damp proof courses						
"Pluvex No 1" (hessian based) damp proof course or other equal and approved; 200 mm laps; in gauged morter (1:1:6)						
width exceeding 225 mm; horizontal	5.62	0.23	4.02	6.49	m²	10.51
width exceeding 225 mm; forming cavity gutters in hollow walls; horizontal	-	0.37	6.47	6.49	m²	12.96
width not exceeding 225 mm; horizontal	-	0.46	8.04	6.49	m²	14.53
width not exceeding 225 mm; vertical	-	0.69	12.06	6.49	m²	18.55
"Pluvex No 2" (fibre based) damp proof course or other equal and approved; 200 mm laps; in gauged mortar (1:1:6)						
width exceeding 225 mm; horizontal	3.78	0.23	4.02	4.37	m²	8.39
width exceeding 225 mm wide; forming cavity gutters in hollow walls; horizontal	-	0.37	6.47	4.37	m²	10.84
width not exceeding 225 mm; horizontal	-	0.46	8.04	4.37	m²	12.41
width not exceeding 225 mm; vertical	-	0.69	12.06	4.37	m²	16.43
"Ruberthene" polythylene damp proof course or other equal and approved; 200 mm laps; in gauged mortar (1:1:6)						
width exceeding 225 mm; horizontal	0.88	0.23	4.02	1.02	m²	5.04
width exceeding 225 mm; forming cavity gutters in hollow walls; horizontal	-	0.37	6.47	1.02	m²	7.49
width not exceeding 225 mm; horizontal	-	0.46	8.04	1.02	m²	9.06
width not exceeding 225 mm; vertical	-	0.69	12.06	1.02	m²	13.08
"Permabit" bitumen polymer damp proof course or other equal and approved; 150 mm laps; in gauged mortar (1:1:6)						
width exceeding 225 mm; horizontal	4.96	0.23	4.02	5.73	m²	9.75
width exceeding 225 mm; forming cavity gutters in hollow walls; horizontal	-	0.37	6.47	5.73	m²	12.20
width not exceeding 225 mm; horizontal	-	0.46	8.04	5.73	m²	13.77
width not exceeding 225 mm; vertical	-	0.69	12.06	5.73	m²	17.79
"Hyload" (pitch polymer) damp proof course or other equal and approved; 150 mm laps; in gauged mortar (1:1:6)						
width exceeding 225 mm; horizontal	4.96	0.23	4.02	5.73	m²	9.75
width exceeding 225 mm; forming cavity gutters in hollow walls; horizontal	-	0.37	6.47	5.73	m²	12.20
width not exceeding 225 mm; horizontal	-	0.46	8.04	5.73	m²	13.77
width not exceeding 225 mm	-	0.69	12.06	5.73	m²	17.79
"Ledkore" grade A (bitumen based lead cored); damp proof course or other equal and approved; 200 mm laps; in gauged mortar (1:1;6)						
width exceeding 225 mm; horizontal	14.75	0.31	5.42	17.05	m²	22.47
width exceeding 225 mm; forming cavity gutters in hollow walls; horizontal	-	0.49	8.56	17.05	m²	25.61
width not exceeding 225 mm; horizontal	-	0.60	10.49	17.05	m²	27.54
width not exceeding 225 mm; horizontal	-	0.83	14.51	17.05	m²	31.56

Prices for Measured Works - Major Works
F MASONRY

	PC £	Labour hours	Labour £	Material £	Unit	Total rate £
F30 ACCESSORIES/SUNDRY ITEMS FOR BRICK/BLOCK/STONE WALLING - cont'd						
Damp proof courses - cont'd						
Milled lead damp proof course; PC £20.85/m^2; BS 1178; 1.80 mm thick (code 4), 175 mm laps; in cement:lime mortar (1:2:9)						
width exceeding 225 mm; horizontal	-	1.85	32.34	15.52	m^2	47.86
width not exceeding 225 mm; horizontal	-	2.78	48.59	15.52	m^2	64.11
Two courses slates in cement:mortar (1:3)						
width exceeding 225 mm; horizontal	-	1.39	24.30	9.71	m^2	34.01
width exceeding 225 mm; vertical	-	2.08	36.36	9.71	m^2	46.07
"Synthaprufe" damp proof membrane or other equal and approved; PC £53.10/25 litres; three coats brushed on						
width not exceeding 150 mm; vertical	-	0.31	2.43	5.71	m^2	8.14
width 150 mm - 225 mm; vertical	-	0.30	2.35	5.71	m^2	8.06
width 225 mm - 300 mm; vertical	-	0.28	2.19	5.71	m^2	7.90
width exceeding 300 mm wide; vertical	-	0.26	2.04	5.71	m^2	7.75
Joint reinforcement						
"Brickforce" galvanised steel joint reinforcement or other equal and approved						
width 60 mm; ref GBF35W60B	-	0.02	0.35	0.32	m	0.67
width 100 mm; ref GBF35W100B	-	0.03	0.52	0.39	m	0.91
width 150 mm; ref GBF35W150B	-	0.04	0.70	0.49	m	1.19
width 175 mm; ref GBF35W175B	-	0.05	0.87	0.58	m	1.45
"Brickforce" stainless steel joint reinforcement or other equal and approved						
width 60 mm; ref SBF35W60BSC	-	0.02	0.35	0.89	m	1.24
width 100 mm; ref SBF35W100BSC	-	0.03	0.52	0.92	m	1.44
width 150 mm; ref SBF35W150BSC	-	0.04	0.70	0.95	m	1.65
width 175 mm; ref SBF35W175BSC	-	0.05	0.87	1.02	m	1.89
"Wallforce" stainless steel joint reinforcement or other equal and approved						
width 210 mm; ref SWF35W210	-	0.05	0.87	2.51	m	3.38
width 235 mm; ref SWF35W235	-	0.06	1.05	2.58	m	3.63
width 250 mm; ref SWF35W250	-	0.07	1.22	2.66	m	3.88
width 275 mm; ref SWF35W275	-	0.08	1.40	2.73	m	4.13
Weather fillets						
Weather fillets in cement:mortar (1:3)						
50 mm face width	-	0.11	1.92	0.04	m	1.96
100 mm face width	-	0.19	3.32	0.16	m	3.48
Angle fillets						
Angle fillets in cement:mortar (1:3)						
50 mm face width	-	0.11	1.92	0.04	m	1.96
100 mm face width	-	0.19	3.32	0.16	m	3.48
Pointing in						
Pointing with mastic						
wood frames or sills	-	0.09	0.89	0.43	m	1.32
Pointing with polysulphide sealant						
wood frames or sills	-	0.09	0.89	1.31	m	2.20
Wedging and pinning						
To underside of existing construction with slates in cement mortar (1:3)						
width of wall - one brick thick	-	0.74	12.93	2.07	m	15.00
width of wall - one and a half brick thick	-	0.93	16.26	4.13	m	20.39
width of wall - two brick thick	-	1.11	19.40	6.20	m	25.60

F MASONRY

	PC £	Labour hours	Labour £	Material £	Unit	Total rate £
Joints						
Hacking joints and faces of brickwork or blockwork						
to form key for plaster	-	0.24	1.88	-	m²	1.88
Raking out joint in brickwork or blockwork for turned-in edge of flashing						
horizontal	-	0.14	2.45	-	m	2.45
stepped	-	0.19	3.32	-	m	3.32
Raking out and enlarging joint in brickwork or blockwork for nib of asphalt						
horizontal	-	0.19	3.32	-	m	3.32
Cutting grooves in brickwork or blockwork						
for water bars and the like	-	0.23	1.80	0.47	m	2.27
for nib of asphalt; horizontal	-	0.23	1.80	0.47	m	2.27
Preparing to receive new walls						
top existing 215 mm wall	-	0.19	3.32	-	m	3.32
Cleaning and priming both faces; filling with pre-formed closed cell joint filler and pointing one side with polysulphide sealant; 12 mm deep						
expansion joints; 12 mm wide	-	0.23	3.26	2.62	m	5.88
expansion joints; 20 mm wide	-	0.28	3.75	3.82	m	7.57
expansion joints; 25 mm wide	-	0.32	4.06	4.61	m	8.67
Fire resisting horizontal expansion joints; filling with joint filler; fixed with high temperature slip adhesive; between top of wall and soffit						
wall not exceeding 215 mm wide; 10 mm wide joint with 30 mm deep filler (one hour fire seal)	-	0.23	4.02	4.49	m	8.51
wall not exceeding 215 mm wide; 10 mm wide joint with 30 mm deep filler (two hour fire seal)	-	0.23	4.02	4.49	m	8.51
wall not exceeding 215 mm wide; 20 mm wide joint with 45 mm deep filler (two hour fire seal)	-	0.28	4.89	6.87	m	11.76
wall not exceeding 215 mm wide; 30 mm wide joint with 75 mm deep filler (three hour fire seal)		0.32	5.59	16.95	m	22.54
Fire resisting vertical expansiojn joints; filling with joint filler; fixed with high temperature slip adhesive; with polysulphide sealant one side; between end of wall and concrete						
wall not exceeding 215 mm wide; 20 mm wide joint with 45 mm deep filler (two hour fire seal)	-	0.37	5.70	9.48	m	15.18
Slate and tile sills						
Sills; two courses of machine made plain roofing tiles						
set weathering; bedded and pointed	-	0.56	9.79	3.41	m	13.20
Flue linings						
Flue linings; Marflex "ML 200" square refractory concrete flue linings or other equal and approved; rebated joints in flue joint mortar mix						
linings; ref MLS1U	-	0.28	4.89	9.23	m	14.12
bottom swivel unit; ref MLS2U	-	0.09	1.57	7.43	nr	9.00
45 degree bend; ref MLS4U	-	0.09	1.57	5.43	nr	7.00
90 mm offset liner; ref MLS5U	-	0.09	1.57	3.60	nr	5.17
70 mm offset liner; ref MLS6U	-	0.09	1.57	3.24	nr	4.81
57 mm offset liner; ref MLS7U	-	0.09	1.57	7.88	nr	9.45
pot; ref MLS9UR	-	0.09	1.57	13.98	nr	15.55
single cap unit; ref MLS10U	-	0.23	4.02	17.81	nr	21.83
double cap unit; ref MLS11U	-	0.28	4.89	16.39	nr	21.28
combined gather lintel; ref MLSUL4	-	0.28	4.89	16.64	nr	21.53

Prices for Measured Works - Major Works
F MASONRY

	PC £	Labour hours	Labour £	Material £	Unit	Total rate £
F30 ACCESSORIES/SUNDRY ITEMS FOR BRICK/BLOCK/STONE WALLING - cont'd						
Air bricks						
Air bricks; red terracotta; building into prepared openings						
215 mm x 65 mm	-	0.07	1.22	0.98	nr	2.20
215 mm x 140 mm	-	0.07	1.22	1.37	nr	2.59
215 mm x 215 mm	-	0.07	1.22	3.69	nr	4.91
Gas flue blocks						
Gas flue system; Marflex "HP"or other equal and approved; concrete blocks built in; in flue joint mortar mix; cutting brickwork or blockwork around						
recess unit; ref HP1	-	0.09	1.57	1.53	nr	3.10
cover block; ref HP2	-	0.09	1.57	3.41	nr	4.98
222 mm standard block with nib; ref HP3	-	0.09	1.57	2.41	nr	3.98
112 mm standard block with nib; ref HP3112	-	0.09	1.57	1.89	nr	3.46
72 mm standard block with nib; ref HP372	-	0.09	1.57	1.89	nr	3.46
222 mm standard block without nib; ref HP4	-	0.09	1.57	2.35	nr	3.92
112 mm standard block without nib; ref HP4112	-	0.09	1.57	1.89	nr	3.46
72 mm standard block without nib; ref HP472	-	0.09	1.57	1.89	nr	3.46
120 mm side offset block; ref HP5	-	0.09	1.57	2.54	nr	4.11
70 mm back offset block; ref HP6	-	0.09	1.57	7.43	nr	9.00
vertical exit block; ref HP7	-	0.09	1.57	4.86	nr	6.43
angled entry/exit block; ref HP8	-	0.09	1.57	4.86	nr	6.43
reverse rebate block; ref HP9	-	0.09	1.57	3.49	nr	5.06
corbel block; ref HP10	-	0.09	1.57	4.68	nr	6.25
lintel unit; ref HP11	-	0.09	1.57	4.41	nr	5.98
Proprietary items						
"Thermabate" cavity closers or other equal and approved; RMC Panel Products Ltd						
closing cavities; width of cavity 50 mm	-	0.14	2.45	4.73	m	7.18
closing cavities; width of cavity 75 mm	-	0.14	2.45	5.06	m	7.51
"Westbrick" cavity closers or other equal and approved; Manthorpe Building Products Ltd						
closing cavities; width of cavity 50 mm	-	0.14	2.45	4.40	m	6.85
"Type H cavicloser" or other equal and approved; uPVC universal cavity closer, insulator and damp proof course by Cavity Trays Ltd; built into cavity wall as work proceeds, complete with face closer and ties						
closing cavities; width of cavity 50 mm - 100 mm	-	0.07	1.22	4.75	m	5.97
"Type L" durropolyethelene lintel stop ends or other equal and approved; Cavity Trays Ltd; fixing with butyl anchoring strip; building in as the work proceeds						
adjusted to lintel as required	-	0.04	0.70	0.43	nr	1.13
"Type W" polypropylene weeps/vents or other equal and approved; Cavity Trays Ltd; built into cavity wall as work proceeds						
100/115 mm x 65 mm x 10 mm including lock fit wedges	-	0.04	0.70	0.36	nr	1.06
extra; extension duct 200/225 mm x 65 mm x 10 mm	-	0.07	1.22	0.62	nr	1.84

Prices for Measured Works - Major Works
F MASONRY

	PC £	Labour hours	Labour £	Material £	Unit	Total rate £
"Type X" polypropylene abutment cavity tray or other equal and approved; Cavity Trays Ltd; built into facing brickwork as the work proceeds; complete with Code 4 flashing; intermediate/catchment tray with short leads (requiring soakers); to suit roof of						
17 - 20 degree pitch	-	0.05	0.87	4.03	nr	4.90
21 - 25 degree pitch	-	0.05	0.87	3.75	nr	4.62
26 - 45 degree pitch	-	0.05	0.87	3.58	nr	4.45
"Type X" polypropylene abutment cavity tray or other equal and approved; Cavity Trays Ltd; built into facing brickwork as the work proceeds; complete with Code 4 flashing; intermediate/catchment tray with long leads (suitable only for corrugated roof tiles); to suit roof of						
17 - 20 degree pitch	-	0.05	0.87	5.44	nr	6.31
21 - 25 degree pitch	-	0.05	0.87	5.00	nr	5.87
26 - 45 degree pitch	-	0.05	0.87	4.62	nr	5.49
"Type X" polypropylene abutment cavity tray or other equal and approved; Cavity Trays Ltd; built into facing brickwork as the work proceeds; complete with Code 4 flashing; ridge tray with short/long leads; to suit roof of						
17 - 20 degree pitch	-	0.05	0.87	9.17	nr	10.04
21 - 25 degree pitch	-	0.05	0.87	8.51	nr	9.38
26 - 45 degree pitch	-	0.05	0.87	7.56	nr	8.43
Servicised "Bitu-thene" self-adhesive cavity flashing or other equal and approved; type "CA"; well lapped at joints; in gauged mortar (1:1:6)						
width exceeding 225 mm wide	-	0.79	13.81	16.17	m^2	29.98
"Expamet" stainless steel wall starters or other equal and approved; plugged and screwed						
to suit walls 60 mm - 75 mm thick	-	0.23	1.80	5.49	m	7.29
to suit walls 100 mm - 115 mm thick	-	0.23	1.80	6.04	m	7.84
to suit walls 125 mm - 180 mm thick	-	0.37	2.90	8.08	m	10.98
to suit walls 190 mm - 260 mm thick	-	0.46	3.60	10.32	m	13.92
Brickwork support angle welded to bracket reference HC6C or other equal and approved; Halfen Ltd; to suit 75 mm cavity, support to brickwork 6000 mm high						
6 mm thick; bolting with M12 x 50 mm T head bolts to cast in channel (not included)	-	0.32	4.48	40.67	m	45.15
Slotted frame cramp; Halfen Ltd or other equal and approved; fixing by bolting (bolts measured elsewhere)						
ref. HTS - FH12; 150 mm projection	-	0.07	0.62	0.37	nr	0.99
Single expansion bolt; Halfen Ltd or other equal and approved; including washer						
8 mm diameter; ref. SEB 8	-	0.11	1.54	1.31	nr	2.85
Head restraint fixings; sliding brick anchors with 500 mm long stem; 2 nr 100 mm projection HST brick anchor ties or other equal and approved; Halfen Ltd; fixing with bolts to concrete soffit (bolts not included)						
ref. SBA/L	-	0.19	2.66	4.29	nr	6.95
Ties in walls; 200 mm long butterfly type; building into joints of brickwork or blockwork						
galvanised steel or polypropylene	-	0.02	0.35	0.07	nr	0.42
stainless steel	-	0.02	0.35	0.13	nr	0.48

F MASONRY

	PC £	Labour hours	Labour £	Material £	Unit	Total rate £
F30 ACCESSORIES/SUNDRY ITEMS FOR BRICK/BLOCK/STONE WALLING - cont'd						
Proprietary items - cont'd						
Ties in walls; 20 mm x 3 mm x 200 mm long twisted wall type; building into joints of brickwork or blockwork						
galvanised steel	-	0.02	0.35	0.13	nr	0.48
stainless steel	-	0.02	0.35	0.26	nr	0.61
Anchors in walls; 25 mm x 3 mm x 100 mm long; one end dovetailed; other end building into joints of brickwork or blockwork						
galvanised steel	-	0.05	0.87	0.18	nr	1.05
stainless steel	-	0.05	0.87	0.28	nr	1.15
Fixing cramps; 25 mm x 3 mm x 250 mm long; once bent; fixed to back of frame; other end building into joints of brickwork or blockwork						
galvanised steel	-	0.05	0.87	0.14	nr	1.01
Chimney pots; red terracotta; plain or cannon-head; setting and flaunching in cement mortar (1:3)						
185 mm diameter x 300 mm long	18.89	1.67	29.19	21.62	nr	50.81
185 mm diameter x 600 mm long	33.64	1.85	32.34	37.47	nr	69.81
185 mm diameter x 900 mm long	61.31	1.85	32.34	67.22	nr	99.56
Galvanised steel lintels; "Catnic" or other equal and approved; built into brickwork or blockwork; "CN7" combined lintel; 143 mm high; for standard cavity walls						
750 mm long	16.84	0.23	4.02	18.14	nr	22.16
900 mm long	20.39	0.28	4.89	21.95	nr	26.84
1200 mm long	26.43	0.32	5.59	28.45	nr	34.04
1500 mm long	30.21	0.37	6.47	32.51	nr	38.98
1800 mm long	42.46	0.42	7.34	45.68	nr	53.02
2100 mm long	48.74	0.46	8.04	52.43	nr	60.47
Galvanised steel lintels; "Catnic" or other equal and approved; built into brickwork or blockwork; "CN8" combined lintel; 219 mm high; for standard cavity walls						
2400 mm long	59.69	0.56	9.79	64.20	nr	73.99
2700 mm long	70.37	0.65	11.36	75.68	nr	87.04
3000 mm long	93.82	0.74	12.93	100.89	nr	113.82
3300 mm long	104.72	0.83	14.51	112.61	nr	127.12
3600 mm long	114.91	0.93	16.26	123.56	nr	139.82
3900 mm long	149.50	1.02	17.83	160.75	nr	178.58
4200 mm long	163.95	0.46	8.04	176.28	nr	184.32
Galvanised steel lintels; "Catnic" or other equal and approved; built into brickwork or blockwork; "CN92" single lintel; for 75 mm internal walls						
900 mm long	2.52	0.28	4.89	2.73	nr	7.62
1050 mm long	3.03	0.28	4.89	3.27	nr	8.16
1200 mm long	3.43	0.32	5.59	3.70	nr	9.29
Galvanised steel lintels; "Catnic" or other equal and approved; built into brickwork or blockwork; "CN102" single lintel; for 100 mm internal walls						
900 mm long	3.11	0.28	4.89	3.36	nr	8.25
1050 mm long	3.84	0.28	4.89	4.15	nr	9.04
1200 mm long	4.24	0.32	5.59	4.58	nr	10.17

F MASONRY

	PC £	Labour hours	Labour £	Material £	Unit	Total rate £
F31 PRECAST CONCRETE SILLS/LINTELS/ COPINGS/FEATURES						
Mix 20.00 N/mm^2 - 20 mm aggregate (1:2:4)						
Lintels; plate; prestressed bedded						
100 mm x 70 mm x 750 mm long	2.74	0.37	6.47	2.97	nr	9.44
100 mm x 70 mm x 900 mm long	3.25	0.37	6.47	3.51	nr	9.98
100 mm x 70 mm x 1050 mm long	3.86	0.37	6.47	4.17	nr	10.64
100 mm x 70 mm x 1200 mm long	4.37	0.37	6.47	4.72	nr	11.19
150 mm x 70 mm x 900 mm long	4.46	0.46	8.04	4.83	nr	12.87
150 mm x 70 mm x 1050 mm long	5.20	0.46	8.04	5.62	nr	13.66
150 mm x 70 mm x 1200 mm long	5.95	0.46	8.04	6.43	nr	14.47
220 mm x 70 mm x 900 mm long	6.87	0.56	9.79	7.43	nr	17.22
220 mm x 70 mm x 1200 mm long	9.15	0.56	9.79	9.88	nr	19.67
220 mm x 70 mm x 1500 mm long	11.47	0.65	11.36	12.38	nr	23.74
265 mm x 70 mm x 900 mm long	7.34	0.56	9.79	7.94	nr	17.73
265 mm x 70 mm x 1200 mm long	9.80	0.56	9.79	10.59	nr	20.38
265 mm x 70 mm x 1500 mm long	12.22	0.65	11.36	13.19	nr	24.55
265 mm x 70 mm x 1800 mm long	14.63	0.74	12.93	15.78	nr	28.71
Lintels; rectangular; reinforced with mild steel bars; bedded						
100 mm x 145 mm x 900 mm long	6.64	0.56	9.79	7.16	nr	16.95
100 mm x 145 mm x 1050 mm long	7.80	0.56	9.79	8.40	nr	18.19
100 mm x 145 mm x 1200 mm long	8.83	0.56	9.79	9.51	nr	19.30
225 mm x 145 mm x 1200 mm long	9.99	0.74	12.93	10.79	nr	23.72
225 mm x 225 mm x 1800 mm long	19.36	1.39	24.30	20.86	nr	45.16
Lintels; boot; reinforced with mild steel bars; bedded						
250 mm x 225 mm x 1200 mm long	16.04	1.11	19.40	17.29	nr	36.69
275 mm x 225 mm x 1800 mm long	21.51	1.67	29.19	23.17	nr	52.36
Padstones						
300 mm x 100 mm x 75 mm	3.27	0.28	4.89	3.53	nr	8.42
225 mm x 225 mm x 150 mm	4.97	0.37	6.47	5.38	nr	11.85
450 mm x 450 mm x 150 mm	13.14	0.56	9.79	14.26	nr	24.05
Mix 30.00 N/mm^2 - 20 mm aggregate (1:1:2)						
Copings; once weathered; once throated; bedded and pointed						
152 mm x 76 mm	3.98	0.65	11.36	4.56	m	15.92
178 mm x 64 mm	4.39	0.65	11.36	5.02	m	16.38
305 mm x 76 mm	7.42	0.74	12.93	8.53	m	21.46
extra for fair ends	-	-	-	4.10	nr	4.10
extra for angles	-	-	-	4.64	nr	4.64
Copings; twice weathered; twice throated; bedded and pointed						
152 mm x 76 mm	3.98	0.65	11.36	4.56	m	15.92
178 mm x 64 mm	4.35	0.65	11.36	4.98	m	16.34
305 mm x 76 mm	7.42	0.74	12.93	8.53	m	21.46
extra for fair ends	-	-	-	4.10	nr	4.10
extra for angles	-	-	-	4.64	nr	4.64

G10 STRUCTURAL STEEL FRAMING

NOTE: The following basic prices are for basic quantities of BS EN10025 1993 grade 275 JR steel (over 10 tonnes of one quality, one serial size and one thickness in lengths between 6 m and 18.5 m, for delivery to one destination). Transport charges are shown in a separate schedule.

See page 195 for other extra charges.

Based on delivery Middlesbrough/Scunthorpe/Stoke-on-Trent Railway Stations - refer Corus section availability at each location.

See page 196 for delivery charges.

Universal beams (kg/m)	£/tonne	Universal beams (kg/m)	£/tonne
1016 x 305 mm (222,249,272,314, 349,393,438,487)	485.00	356 x 171 mm (45,51,57,67)	435.00
914 x 419 mm (343,388)	475.00	356 x 127 mm (33,39)	435.00
914 x 305 mm (201,224,253,289)	470.00	305 x 165 mm (40,46,54)	430.00
838 x 292 mm (176,194,226)	465.00	305 x 127 mm (37,42,48)	430.00
762 x 267 mm (134,147,173,197)	465.00	305 x 102 mm (25,28,33)	430.00
686 x 254 mm (125,140,152,170)	465.00	254 x 146 mm (31,37,43)	415.00
610 x 305 mm (149,179,238)	455.00	254 x 102 mm (22,25,28)	415.00
610 x 229 mm (101,113,125,140)	455.00	203 x 133 mm (25,30)	375.00
533 x 210 mm (82,92,101,109,122)	440.00	203 x 102 mm (23)	395.00
457 x 191 mm (67,74,82,89,98)	430.00	178 x 102 mm (19)	365.00
457 x 152 mm (52,60,67,74,82)	430.00	152 x 89 mm (16)	405.00
406 x 178 mm (54,60,67,74)	435.00	127 x 76 mm (13)	425.00
406 x 140 mm (39,46)	435.00		
Universal columns (kg/m)	**£/tonne**	**Universal columns (kg/m)**	**£/tonne**
356 x 406 mm (235,287,340,393,467, 551, 634)	475.00	254 x 254 mm (73,89,107,132,167)	435.00
356 x 368 mm (129,153,177,202)	475.00	203 x 203 mm (46,52,60,71,86)	425.00
305 x 305 mm (97,118,137,158,198, 240,283)	455.00	152 x 152 mm (23,30,37)	395.00
Joists (kg/m)	**£/tonne**	**Joists (kg/m)**	**£/tonne**
254 x 203 mm (82.0)	420.00	114 x 114 mm (26.9)	400.00
203 x 152 mm (60.2)	435.00	102 x 102 mm (23.0)	400.00
203 x 152 mm (52.3)	410.00	102 x 44 mm (7.5)	360.00
152 x 127 mm (37.3)	410.00	89 x 89 mm (19.5)	400.00
127 x 114 mm (26.9,29.3)	400.00	76 x 76 mm (12.8)	450.00
Channels (kg/m)	**£/tonne**	**Channels (kg/m)**	**£/tonne**
430 x 100 mm (64.4)	480.00	200 x 90 mm (29.7)	450.00
380 x 100 mm (54.0)	480.00	200 x 75 mm (23.4)	370.00
300 x 100 mm (45.5)	450.00	180 x 90 mm (26.1)	440.00
300 x 90 mm (41.4)	450.00	180 x 75 mm (20.3)	370.00
260 x 90 mm (34.8)	450.00	150 x 90 mm (23.9)	440.00
260 x 75 mm (27.6)	450.00	150 x 75 mm (17.9)	370.00
230 x 90 mm (32.2)	450.00	125 x 65 mm (14.8)	370.00
230 x 75 mm (25.7)	450.00	100 x 50 mm (10.2)	365.00

G STRUCTURAL/CARCASSING METAL/TIMBER

Equal angles	£/tonne	Equal angles	£/tonne	Equal angles	£/tonne
200 x 200 x 16 mm	410.00	150 x 150 x 18 mm	405.00	100 x 100 x 12 mm	345.00
200 x 200 x 18 mm	410.00	120 x 120 x 8 mm	380.00	100 x 100 x 15 mm	345.00
200 x 200 x 20 mm	410.00	120 x 120 x 10 mm	380.00	90 x 90 x 6 mm	345.00
200 x 200 x 24 mm	410.00	120 x 120 x 12 mm	380.00	90 x 90 x 7 mm	345.00
150 x 150 x 10 mm	405.00	120 x 120 x 15 mm	380.00	90 x 90 x 8 mm	345.00
150 x 150 x 12 mm	405.00	100 x 100 x 8 mm	345.00	90 x 90 x 10 mm	345.00
150 x 150 x 15 mm	405.00	100 x 100 x 10 mm	345.00	90 x 90 x 12 mm	345.00

Unequal angles	£/tonne	Unequal angles	£/tonne	Unequal angles	£/tonne
200 x 150 x 12 mm	425.00	150 x 90 x 12 mm	405.00	125 x 75 x 12 mm	360.00
200 x 150 x 15 mm	425.00	150 x 90 x 15 mm	405.00	100 x 75 x 8 mm	355.00
200 x 150 x 18 mm	425.00	150 x 75 x 10 mm	395.00	100 x 75 x 10 mm	355.00
200 x 100 x 10 mm	420.00	150 x 75 x 12 mm	395.00	100 x 75 x 12 mm	355.00
200 x 100 x 12 mm	420.00	150 x 75 x 15 mm	395.00	100 x 65 x 7 mm	355.00
200 x 100 x 15 mm	420.00	125 x 75 x 8 mm	360.00	100 x 65 x 8 mm	355.00
150 x 90 x 10 mm	405.00	125 x 75 x 10 mm	360.00	100 x 65 x 10 mm	355.00

Add to the aforementioned prices for:

	Unit	£
Universal beams, columns, joists, channels and angles non-standard size	t	25.00
Quantity under 10 tonnes to 5 tonnes	t	10.00
under 5 tonnes to 2 tonnes	t	25.00
under 2 tonnes	t	50.00
Size lengths 3,000 mm to under 6,000 mm in 100 mm increments	t	15.00
lengths over 18,000 mm to 24,000 mm in 100 mm increments	t	5.00
lengths over 24,000 to 27,000 mm in 100 mm increments	t	10.00
lengths over 27,000 to 29,500 mm in 100 mm increments	t	60.00
Tees cut from universal beams and columns and joists		
weight per metre of rolled section before splitting		
up to 25 kg per metre	t	150.00
25 - 40 kg per metre	t	120.00
40 - 73 kg per metre	t	100.00
73 - 125 kg per metre	t	90.00
125 kg + per metre	t	85.00
impact testing within the specification	t	10.00
Shotblasting and priming		
Epoxy Zinc Phosphate primer to universal beams and columns	m²	2.60
Epoxy Zinc Phosphate primer to channels and angles	m²	2.70
Zinc rich epoxy primer to universal beams and columns	m²	3.60
Zinc rich epoxy primer to channels and angles	m²	3.50
Surface quality		
Specification of class D in respect of EN 10163-3: 1991	t	40.00

Prices for Measured Works - Major Works
G STRUCTURAL/CARCASSING METAL/TIMBER

G10 STRUCTRURAL STEEL FRAMING - cont'd

RADIAL DISTANCES MILES FROM BASING POINT	0. under 5 tonnes £/tonne	5. under 10 tonnes £/tonne	10. under 20 tonnes £/tonne	20. under and over £/tonne
Schedule IX - Iron and Steel				
up to 10	7.05	4.90	3.70	3.35
over 10 up to 15	7.75	5.30	4.15	3.70
over 15 up to 20	8.25	5.80	4.60	4.10
over 20 up to 25	8.75	6.20	4.95	4.55
over 25 up to 30	9.15	6.70	5.40	4.90
over 30 up to 35	9.60	7.10	5.75	5.30
over 35 up to 40	10.10	7.55	6.15	5.65
over 40 up to 45	10.50	8.00	6.55	6.05
over 45 up to 50	10.85	8.45	6.95	6.40
over 50 up to 60	11.40	9.05	7.50	6.95
over 60 up to 70	11.90	9.85	8.25	7.70
over 70 up to 80	12.70	10.60	8.95	8.45
over 80 up to 90	13.60	11.35	9.65	9.10
over 90 up to 100	14.40	12.10	10.35	9.85
over 100 up to 110	15.05	12.80	11.10	10.50
over 110 up to 120	15.60	13.50	11.75	11.20
over 120 up to 130	16.10	14.20	12.45	11.90
over 130 up to 140	16.70	14.90	13.10	12.55
over 140 up to 150	16.85	15.55	13.80	13.20
over 150 up to 160	17.25	16.25	14.45	13.90
over 160 up to 170	17.85	16.90	15.15	14.55
over 170 up to 180	18.30	17.55	15.75	15.20
over 180 up to 190	18.95	18.20	16.45	15.80
over 190 up to 200	19.55	18.85	17.10	16.20
over 200 up to 210	20.10	19.45	17.70	16.85
over 210 up to 220	20.70	20.10	18.35	17.40
over 220 up to 230	21.20	20.70	18.85	18.10
over 230 up to 240	22.00	21.30	19.35	18.60
over 240 up to 250	22.45	21.95	19.90	19.00
over 250 up to 275	23.35	22.95	20.75	20.00
over 275 up to 300	24.70	24.45	22.00	21.20
over 300 up to 325	25.95	25.90	23.25	22.55
over 325 up to 350	27.30	27.35	24.55	23.80
over 350 up to 375	28.65	28.75	25.60	25.20
over 375 up to 400	30.00	30.15	26.70	26.30

NOTE: The minimum charge under this schedule will be as for 2 tonnes.
Lengths over 14 metres and up to 18 metres will be subject to a surcharge of 20%.
Lengths over 18 metres and up to 24 metres will be subject to a surcharge of 40%.
Lengths over 24 metres will be subect to a surcharge of 70%.

Collection by customer from works will incur an extra cost of £3.00/tonne in addition to any additional transport charges between the basing point and the producing works.

G STRUCTURAL/CARCASSING METAL/TIMBER

NOTE: The following basic prices are for basic quantities of 10 tonnes and over in one size, thickness, length, steelgrade and surface finish and include delivery to mainland of Great Britain to one destination. Additional costs for variations to these factors vary between sections and should be ascertained from the supplier.

The following lists are not fully comprehensive and for alternative sections Corus price lists should be consulted.

Hot formed structural hollow section	Approx metres per tonne (m)	S276J2H Grade 43D £/100m	S355J2H Grade 50D £/100m
Circular (kg/m)			
21.30 x 3.20 mm (1.43)	700.00	82.13	-
26.90 x 3.20 mm (1.87)	535.00	107.40	-
33.70 x 3.20 mm (2.41)	415.00	138.41	-
33.70 x 4.00 mm (2.93)	342.00	169.90	-
42.40 x 3.20 mm (3.09)	324.00	176.61	-
42.40 x 4.00 mm (3.79)	264.00	218.72	-
48.30 x 3.20 mm (3.56)	281.00	203.48	223.82
48.30 x 4.00 mm (4.37)	229.00	252.19	277.41
48.30 x 5.00 mm (5.34)	188.00	308.17	338.98
60.30 x 3.20 mm (4.51)	222.00	257.77	283.55
60.30 x 4.00 mm (5.55)	181.00	332.78	366.06
60.30 x 5.00 mm (6.82)	147.00	408.93	449.82
76.10 x 3.20 mm (5.75)	174.00	328.65	361.51
76.10 x 4.00 mm (7.11)	141.00	426.32	468.95
76.10 x 5.00 mm (8.77)	115.00	525.85	578.43
88.90 x 3.20 mm (6.76)	148.00	386.37	425.01
88.90 x 4.00 mm (8.38)	120.00	502.46	552.71
88.90 x 5.00 mm (10.30)	97.10	617.59	679.35
114.30 x 3.60 mm (9.83)	102.00	589.41	648.35
114.30 x 5.00 mm (13.50)	74.10	809.46	890.41
114.30 x 6.30 mm (16.80)	59.60	1007.33	1108.06
139.70 x 5.00 mm (16.60)	60.30	1029.81	1132.80
139.70 x 6.30 mm (20.70)	48.40	1284.17	1412.59
139.70 x 8.00 mm (26.00)	38.50	1612.96	1774.27
139.70 x 10.00 mm (32.00)	31.30	2053.54	2258.88
168.30 x 5.00 mm (20.10)	49.80	1246.94	1371.64
168.30 x 6.30 mm (25.20)	39.70	1563.33	1719.67
168.30 x 8.00 mm (31.60)	31.70	1960.37	2156.42
168.30 x 10.00 mm (39.00)	25.70	2502.75	2753.01
168.30 x 12.50 mm (48.00)	20.80	3736.75	4110.43
193.70 x 5.00 mm (23.30)	42.90	1445.46	1590.02
193.70 x 6.30 mm (29.10)	34.40	1805.28	1985.81
193.70 x 8.00 mm (36.60)	27.30	2270.55	2497.62
193.70 x 10.00 mm (45.30)	22.10	2907.04	3197.73
193.70 x 12.50 mm (55.90)	17.90	4351.76	4786.94
219.10 x 5.00 mm (26.40)	37.90	1938.42	2132.28
219.10 x 6.30 mm (33.10)	30.20	2430.37	2673.42
219.10 x 8.00 mm (41.60)	24.10	3054.48	3359.95
219.10 x 10.00 mm (51.60)	19.40	3788.73	4167.63
219.10 x 12.50 mm (63.70)	15.70	5096.57	5606.24

G10 STRUCTURAL STEEL FRAMING - cont'd

Hot formed structural hollow section	Approx metres per tonne (m)	S276J2H Grade 43D £/100m	S355J2H Grade 50D £/100m
Circular (kg/m) - cont'd			
219.10 x 16.00 mm (80.10)	12.50	6408.72	7049.60
244.50 x 6.30 mm (37.00)	27.10	2716.72	2988.42
244.50 x 8.00 mm (46.70)	21.50	3428.95	3771.87
244.50 x 10.00 mm (57.80)	17.40	4243.96	4668.39
244.50 x 12.50 mm (71.50)	14.00	5541.89	6096.09
244.50 x 16.00 mm (90.20)	11.10	7216.81	7938.50
273.00 x 6.30 mm (41.40)	24.20	3039.80	3343.80
273.00 x 8.00 mm (52.30)	19.10	3840.13	4224.17
273.00 x 10.00 mm (64.90)	15.40	4765.28	5241.84
273.00 x 12.50 mm (80.30)	12.50	6223.97	6846.38
273.00 x 16.00 mm (101.00)	9.91	8080.91	8889.01
323.90 x 6.30 mm (49.30)	20.30	3619.85	3981.86
323.90 x 8.00 mm (62.30)	16.10	4574.38	5031.85
323.90 x 10.00 mm (77.40)	12.90	5683.10	6251.44
323.90 x 12.50 mm (96.00)	10.40	7440.86	8184.96
323.90 x 16.00 mm (121.00)	8.27	9681.09	10649.21
355.60 x 8.00 mm (68.60)	14.60	5036.96	5540.68
355.60 x 10.00 mm (85.30)	11.80	6255.81	6881.43
355.60 x 12.50 mm (106.00)	9.44	8215.95	9037.56
355.60 x 16.00 mm (134.00)	7.47	10721.21	11793.34
406.40 x 10.00 mm (97.80)	10.20	7180.96	7899.11
406.40 x 12.50 mm (121.00)	8.27	9378.59	10316.46
406.40 x 16.00 mm (154.00)	6.50	12321.39	13553.54
457.00 x 10.00 mm (110.00)	9.09	8076.75	8884.48
457.00 x 12.50 mm (137.00)	7.30	10618.73	11680.62
457.00 x 16.00 mm (174.00)	5.75	13921.57	15313.74
508.00 x 10.00 mm (123.00)	8.13	9031.28	9934.46
508.00 x 12.50 mm (153.00)	6.54	11858.88	13044.78
508.00 x 16.00 mm (194.00)	5.16	15521.75	17073.94
Square (kg/m)	346.00	169.30	186.23
40 x 40 x 2.50 mm (2.89)	293.30	200.03	220.03
40 x 40 x 3.00 mm (3.41)	277.00	212.21	233.43
40 x 40 x 3.20 mm (3.61)	227.80	258.59	284.45
40 x 40 x 4.00 mm (4.39)	189.40	313.09	344.40
40 x 40 x 5.00 mm (5.28)	271.70	215.11	236.62
50 x 50 x 2.50 mm (3.68)	229.90	254.53	279.99
50 x 50 x 3.00 mm (4.35)	216.50	270.19	297.21
50 x 50 x 3.20 mm (4.62)	177.30	331.65	364.81
50 x 50 x 4.00 mm (5.64)	146.00	404.12	444.53
50 x 50 x 5.00 mm (6.85)	120.30	492.25	541.48
50 x 50 x 6.30 mm (8.31)	189.00	309.61	340.57
60 x 60 x 3.00 mm (5.29)			

Hot formed structural hollow section	Approx metres per tonne (m)	S276J2H Grade 43D £/100m	S355J2H Grade 50D £/100m
Square (kg/m) - cont'd			
60 x 60 x 3.20 mm (5.62)	177.90	328.75	361.62
60 x 60 x 4.00 mm (6.90)	145.00	423.94	466.33
60 x 60 x 5.00 mm (8.42)	118.80	519.44	571.38
60 x 60 x 6.30 mm (10.30)	97.10	638.65	702.51
60 x 60 x 8.00 mm (12.50)	80.00	778.55	856.40
70 x 70 x 3.00 mm (6.24)	160.30	353.46	388.81
70 x 70 x 3.60 mm (7.40)	135.10	440.45	484.49
70 x 70 x 5.00 mm (9.99)	100.10	614.32	675.75
70 x 70 x 6.30 mm (12.30)	81.30	760.30	836.32
70 x 70 x 8.00 mm (15.00)	66.70	930.61	1023.66
80 x 80 x 3.60 mm (8.53)	117.20	507.16	557.88
80 x 80 x 5.00 mm (11.60)	86.20	711.64	782.80
80 x 80 x 6.30 mm (14.20)	70.40	875.87	963.45
80 x 80 x 8.00 mm (17.50)	57.10	1082.67	1190.93
90 x 90 x 3.60 mm (9.66)	103.50	573.88	631.27
90 x 90 x 5.00 mm (13.10)	76.30	808.96	889.85
90 x 90 x 6.30 mm (16.20)	61.70	997.51	1097.26
90 x 90 x 8.00 mm (20.10)	49.80	1240.81	1364.88
100 x 100 x 4.00 mm (11.90)	84.00	688.84	757.72
100 x 100 x 5.00 mm (14.70)	68.00	900.20	990.21
100 x 100 x 6.30 mm (18.20)	54.90	1119.16	1231.07
100 x 100 x 8.00 mm (22.60)	44.20	1461.87	1608.06
100 x 100 x 10.00 mm (27.40)	36.50	1816.51	1998.17
120 x 120 x 4.00 mm (14.40)	69.40	912.60	1003.86
120 x 120 x 5.00 mm (17.80)	56.20	1132.88	1246.18
120 x 120 x 6.30 mm (22.20)	45.00	1403.52	1543.87
120 x 120 x 8.00 mm (27.60)	36.20	1755.97	1931.57
120 x 120 x 10.00 mm (33.70)	29.70	2226.69	2449.37
120 x 120 x 12.50 mm (40.90)	24.40	2708.49	2979.35
140 x 140 x 5.00 mm (21.00)	47.60	1327.99	1460.80
140 x 140 x 6.30 mm (26.10)	38.30	1655.27	1820.80
140 x 140 x 8.00 mm (32.60)	30.70	2070.66	2277.73
140 x 140 x 10.00 mm (40.00)	25.00	2630.36	2893.41
140 x 140 x 12.50 mm (48.70)	20.50	4411.34	4852.48
150 x 150 x 5.00 mm (22.60)	44.20	1442.24	1586.48
150 x 150 x 6.30 mm (28.10)	35.60	1798.04	1977.86
150 x 150 x 8.00 mm (35.10)	28.50	2249.14	2474.07
150 x 150 x 10.00 mm (43.10)	23.20	2865.65	3152.24
150 x 150 x 12.50 mm (52.70)	19.00	4758.90	5234.80
150 x 150 x 16.00 mm (65.20)	15.30	7079.20	8067.19
160 x 160 x 5.00 mm (24.10)	41.50	1820.20	2002.21
160 x 160 x 6.30 mm (30.10)	33.20	2279.01	2506.90
160 x 160 x 8.00 mm (37.60)	26.60	2850.65	3135.69
160 x 160 x 10.00 mm (46.30)	21.60	3512.54	3863.77
160 x 160 x 12.50 mm (56.30)	17.70	4549.39	5004.35

G10 STRUCTURAL STEEL FRAMING - cont'd

Hot formed structural hollow section	Approx metres per tonne (m)	S276J2H Grade 43D £/100m	S355J2H Grade 50D £/100m
Square (kg/m) - cont'd			
160 x 160 x 16.50 mm (70.20)	14.20	6362.81	6999.06
180 x 180 x 5.00 mm (27.30)	36.60	2060.89	2266.97
180 x 180 x 6.30 mm (34.00)	29.40	2572.35	2829.57
180 x 180 x 8.00 mm (42.70)	23.40	3234.24	3557.65
180 x 180 x 10.00 mm (52.50)	19.00	3986.40	4385.01
180 x 180 x 12.50 mm (64.40)	15.50	5176.62	5694.31
180 x 180 x 16.00 mm (80.20)	12.50	7253.96	7979.32
200 x 200 x 5.00 mm (30.40)	32.90	2294.06	2523.45
200 x 200 x 6.30 mm (38.00)	26.30	2873.21	3160.52
200 x 200 x 8.00 mm (47.70)	21.00	3610.32	3971.33
200 x 200 x 10.00 mm (58.80)	17.00	4460.25	4906.24
200 x 200 x 12.50 mm (72.30)	13.80	5795.91	6375.53
200 x 200 x 16.00 mm (90.30)	11.10	8154.02	8969.38
250 x 250 x 6.30 mm (47.90)	20.90	3617.84	3979.60
250 x 250 x 8.00 mm (60.30)	16.60	4550.51	5005.53
250 x 250 x 10.00 mm (74.50)	13.40	5641.12	6205.20
250 x 250 x 12.50 mm (91.90)	10.90	7352.07	8087.31
250 x 250 x 16.00 mm (115.00)	8.70	10426.46	11469.04
300 x 300 x 6.30 mm (57.80)	17.30	4354.95	4790.41
300 x 300 x 8.00 mm (72.80)	13.70	5498.22	6048.00
300 x 300 x 10.00 mm (90.20)	11.10	6822.00	7504.16
300 x 300 x 12.50 mm (112.00)	8.93	8892.35	9781.63
300 x 300 x 16.00 mm (141.00)	7.09	12654.33	13919.69
350 x 350 x 8.00 mm (85.40)	11.70	6445.93	7090.48
350 x 350 x 10.00 mm (106.00)	9.43	7972.79	8770.02
350 x 350 x 12.50 mm (131.00)	7.63	10480.27	11528.35
350 x 350 x 16.00 mm (166.00)	6.02	14882.20	16370.34
400 x 400 x 10.00 mm (122.00)	8.20	9176.23	10093.79
400 x 400 x 12.50 mm (151.00)	6.62	12068.19	13275.07
400 x 400 x 16.00 mm (191.00)	5.24	17110.08	18820.99
Rectangular (kg/m)			
50 x 30 x 2.50 mm (2.89)	346.00	169.30	186.23
50 x 30 x 3.00 mm (3.41)	293.30	200.03	220.03
50 x 30 x 3.20 mm (3.61)	277.00	212.21	233.43
50 x 30 x 4.00 mm (4.39)	227.80	258.59	284.45
50 x 30 x 5.00 mm (5.28)	189.40	313.09	344.40
60 x 40 x 2.50 mm (3.68)	271.70	215.11	236.62
60 x 40 x 3.00 mm (4.35)	229.90	254.53	279.99
60 x 40 x 3.20 mm (4.62)	216.50	270.19	297.21
60 x 40 x 4.00 mm (5.64)	177.30	331.65	364.81
60 x 40 x 5.00 mm (6.85)	146.00	404.12	444.53
60 x 40 x 6.30 mm (8.31)	120.30	492.25	541.48
76.20 x 50.80 x 3.20 mm (5.97)	167.50	334.04	378.44

G STRUCTURAL/CARCASSING METAL/TIMBER

Hot formed structural hollow section	Approx metres per tonne (m)	S276J2H Grade 43D £/100m	S355J2H Grade 50D £/100m
Rectangular (kg/m) - cont'd			
76.20 x 50.80 x 4.00 mm (7.34)	136.20	444.21	488.63
76.20 x 50.80 x 6.30 mm (11.00)	90.90	698.25	768.07
80 x 40 x 3.00 mm (5.29)	189.00	309.61	340.57
80 x 40 x 3.20 mm (5.62)	177.90	328.75	361.62
80 x 40 x 4.00 mm (6.90)	144.90	423.94	466.33
80 x 40 x 5.00 mm (8.42)	118.80	519.44	571.38
80 x 40 x 6.30 mm (10.30)	97.10	638.65	702.51
80 x 40 x 8.00 mm (12.50)	80.00	778.55	856.40
90 x 50 x 3.00 mm (6.24)	160.30	353.46	388.81
90 x 50 x 3.60 mm (7.40)	135.10	440.45	484.49
90 x 50 x 5.00 mm (9.99)	100.10	614.32	675.75
90 x 50 x 6.30 mm (12.30)	81.30	760.30	836.32
90 x 50 x 8.00 mm (15.00)	66.70	930.61	1023.66
100 x 50 x 3.00 mm (6.71)	149.00	379.92	417.91
100 x 50 x 3.20 mm (7.13)	140.30	404.12	444.53
100 x 50 x 4.00 mm (8.78)	113.90	523.10	575.41
100 x 50 x 5.00 mm (10.80)	92.60	662.98	729.28
100 x 50 x 6.30 mm (13.30)	75.20	815.04	896.54
100 x 50 x 8.00 mm (16.30)	61.30	1009.68	1110.64
100 x 60 x 3.00 mm (7.18)	139.30	406.37	447.00
100 x 60 x 3.60 mm (8.53)	117.20	507.16	557.88
100 x 60 x 5.00 mm (11.60)	86.20	711.64	782.80
100 x 60 x 6.30 mm (14.20)	70.40	875.87	963.45
100 x 60 x 8.00 mm (17.50)	57.10	1082.67	1190.93
120 x 60 x 3.60 mm (9.66)	103.50	573.88	631.27
120 x 60 x 5.00 mm (13.10)	76.30	808.96	889.85
120 x 60 x 6.30 mm (16.20)	61.70	997.51	1097.26
120 x 60 x 8.00 mm (20.10)	49.80	1240.81	1364.88
120 x 80 x 5.00 mm (14.70)	68.00	900.20	990.21
120 x 80 x 6.30 mm (18.20)	54.90	1119.16	1231.07
120 x 80 x 8.00 mm (22.60)	44.20	1461.87	1608.06
120 x 80 x 10.00 mm (27.40)	36.50	1816.51	1998.17
150 x 100 x 4.00 mm (15.10)	66.20	950.36	1045.40
150 x 100 x 5.00 mm (18.60)	53.80	1176.94	1294.64
150 x 100 x 6.30 mm (23.10)	43.30	1466.46	1613.11
150 x 100 x 8.00 mm (28.90)	34.60	1831.50	2014.65
150 x 100 x 10.00 mm (35.30)	28.30	2324.36	2556.80
150 x 100 x 12.50 mm (42.80)	23.40	2838.71	3122.59
160 x 80 x 4.00 mm (14.40)	69.40	912.60	1003.86
160 x 80 x 5.00 mm (17.80)	50.20	1132.08	1246.18
160 x 80 x 6.30 mm (22.20)	45.00	1403.52	1543.87
160 x 80 x 8.00 mm (27.60)	36.20	1755.97	1931.57
160 x 80 x 10.00 mm (33.70)	29.70	2226.69	2449.37
160 x 80 x 12.50 mm (40.90)	24.40	2708.49	2979.35
200 x 100 x 5.00 mm (22.60)	44.20	1442.24	1586.48

G10 STRUCTURAL STEEL FRAMING - cont'd

Hot formed structural hollow section	Approx metres per tonne (m)	S276J2H Grade 43D £/100m	S355J2H Grade 50D £/100m
Rectangular (kg/m) - cont'd			
200 x 100 x 6.30 mm (28.10)	35.60	1798.04	1977.86
200 x 100 x 8.00 mm (35.10)	28.50	2249.14	2474.07
200 x 100 x 10.00 mm (43.10)	23.20	2865.65	3152.24
200 x 100 x 12.50 mm (52.70)	19.00	4758.90	5234.80
200 x 100 x 16.00 mm (65.20)	15.30	7879.20	8667.19
250 x 150 x 5.00 mm (30.40)	32.90	2294.06	2523.45
250 x 150 x 6.30 mm (38.00)	26.30	2873.21	3160.52
250 x 150 x 8.00 mm (47.70)	21.00	3610.32	3971.33
250 x 150 x 10.00 mm (58.80)	17.00	4460.25	4906.24
250 x 150 x 12.50 mm (72.30)	13.80	5795.91	6375.53
250 x 150 x 16.00 mm (90.30)	11.10	8154.02	8969.38
300 x 200 x 6.30 mm (47.90)	20.90	3617.84	3979.60
300 x 200 x 8.00 mm (60.30)	16.60	4550.51	5005.53
300 x 200 x 10.00 mm (74.50)	13.40	5641.12	6205.20
300 x 200 x 12.50 mm (91.90)	10.90	7352.07	8087.31
300 x 200 x 16.00 mm (115.00)	8.70	10426.46	11469.04
400 x 200 x 6.30 mm (57.80)	17.30	4354.95	4790.41
400 x 200 x 8.00 mm (72.80)	13.70	5498.22	6048.00
400 x 200 x 10.00 mm (90.20)	11.10	6822.00	7504.16
400 x 200 x 12.50 mm (112.00)	8.93	8892.35	9781.63
400 x 200 x 16.00 mm (141.00)	7.09	12654.33	13919.69
450 x 250 x 8.00 mm (85.40)	11.70	6445.93	7090.48
450 x 250 x 10.00 mm (106.00)	9.43	7972.79	8770.02
450 x 250 x 12.50 mm (131.00)	7.63	10480.27	11528.35
450 x 250 x 16.00 mm (166.00)	6.02	14882.20	16370.34
500 x 300 x 10.00 mm (122.00)	8.20	9176.23	1009 3.79
500 x 300 x 12.50 mm (151.00)	6.62	12068.19	13275.07
500 x 300 x 16.00 mm (191.00)	5.24	17110.08	18820.99

Add to the aforementioned prices for:

	Percentage extra %
Finish Self colour is suplied unless otherwise specified. Transit primer painted (All sections except for circular hollow sections over 200 mm diameter).	5.00

Test Certificates

Test certificates will be charged at a rate of £25 per certificate.

G STRUCTURAL CARCASSING/METAL/TIMBER
Excluding overheads and profit

Add to the aforementioned prices for:

	Percentage extra %
Quantity	
- **Work despatches**	
Orders for the following hollow sections of one size, thickness, length, steel grade and surface finish.	
a) circular hollow sections over 200 mm diameter	
b) square hollow sections over 150 x 150 mm	
c) rectangular hollow sections over 600 mm girth	
Order Quantity:	
4 tonnes to under 10 tonnes	15.00
2 tonnes to under 4 tonnes	20.00
1 tonne to under 2 tonnes	25.00
orders under 1 tonne are not supplied	
- **Warehouse despatches**	
Orders are for the following hollow sections in one steel grade, for delivery to one destination in one assignment.	
a) circular hollow sections less than 200 mm diameter	
b) square hollow sections up to and including 150 x 150 mm	
c) rectangular hollow sections up to and including 600 mm girth	
Order Quantity:	
10 tonnes and over	7.50
4 tonnes to under 10 tonnes	10.00
2 tonnes to under 4 tonnes	12.50
1 tonne to under 2 tonnes	17.50
500 kg to under 1 tonne	22.50
250 kg to under 500 kg	35.00
100 kg to under 250 kg	50.00
under 100 kg	100.00
a) circular hollow sections over 200 mm diameter	
b) square hollow sections over 150 x 150 mm	
c) rectangular hollow sections over 600 mm girth	
Order Quantity:	
10 tonnes and over	12.50
4 tonnes to under 10 tonnes	15.00
2 tonnes to under 4 tonnes	20.00
1 tonne to under 2 tonnes	25.00
500 kg to under 1 tonne	45.00
250 kg to under 500 kg	75.00

G10 STRUCTURAL STEEL FRAMING

Framing, fabrication; weldable steel; BS EN 10025: 1993 Grade S275; hot rolled structural steel sections; welded fabrication

	PC £	Labour hours	Labour £	Material £	Unit	Total rate £
Columns						
weight not exceeding 40 kg/m	-	-	-	-	t	858.86
weight not exceeding 40 kg/m; castellated	-	-	-	-	t	963.73
weight not exceeding 40 kg/m; curved	-	-	-	-	t	1198.41
weight not exceeding 40 kg/m; square hollow section	-	-	-	-	t	1283.30
weight not exceeding 40 kg/m; circular hollow section	-	-	-	-	t	1423.12
weight 40 - 100 kg/m	-	-	-	-	t	838.89
weight 40 - 100 kg/m; castellated	-	-	-	-	t	938.75
weight 40 - 100 kg/m; curved	-	-	-	-	t	1158.46
weight 40 - 100 kg/m; square hollow section	-	-	-	-	t	1258.33
weight 40 - 100 kg/m; circular hollow section	-	-	-	-	t	1398.14
weight exceeding 100 kg/m	-	-	-	-	t	823.90
weight exceeding 100 kg/m; castellated	-	-	-	-	t	923.78
weight exceeding 100 kg/m; curved	-	-	-	-	t	1128.50
weight exceeding 100 kg/m; square hollow section	-	-	-	-	t	1238.36
weight exceeding 100 kg/m; circular hollow section	-	-	-	-	t	1378.17
Beams						
weight not exceeding 40 kg/m	-	-	-	-	t	898.81
weight not exceeding 40 kg/m; castellated	-	-	-	-	t	1023.65
weight not exceeding 40 kg/m; curved	-	-	-	-	t	1313.25
weight not exceeding 40 kg/m; square hollow section	-	-	-	-	t	1398.14
weight not exceeding 40 kg/m; circular hollow section	-	-	-	-	t	1597.88
weight 40 - 100 kg/m	-	-	-	-	t	878.83
weight 40 - 100 kg/m; castellated	-	-	-	-	t	988.69
weight 40 - 100 kg/m; curved	-	-	-	-	t	1288.29
weight 40 - 100 kg/m; square hollow section	-	-	-	-	t	1368.18
weight 40 - 100 kg/m; circular hollow section	-	-	-	-	t	1572.91
weight exceeding 100 kg/m	-	-	-	-	t	868.85
weight exceeding 100 kg/m; castellated	-	-	-	-	t	968.71
weight exceeding 100 kg/m; curved	-	-	-	-	t	1268.32
weight exceeding 100 kg/m; square hollow section	-	-	-	-	t	1348.21
weight exceeding 100 kg/m; circular hollow section	-	-	-	-	t	1552.94
Bracings						
weight not exceeding 40 kg/m	-	-	-	-	t	1153.46
weight not exceeding 40 kg/m; square hollow section	-	-	-	-	t	1597.88
weight not exceeding 40 kg/m; circular hollow section	-	-	-	-	t	1797.62
weight 40 - 100 kg/m	-	-	-	-	t	1123.51
weight 40 - 100 kg/m; square hollow section	-	-	-	-	t	1577.91
weight 40 - 100 kg/m; circular hollow section	-	-	-	-	t	1767.65
weight exceeding 100 kg/m	-	-	-	-	t	1098.54
weight exceeding 100 kg/m; square hollow section	-	-	-	-	t	1557.93
weight exceeding 100 kg/m; circular hollow section	-	-	-	-	t	1747.68

G STRUCTURAL/CARCASSING METAL/TIMBER

	PC £	Labour hours	Labour £	Material £	Unit	Total rate £
Purlins and cladding rails						
weight not exceeding 40 kg/m	-	-	-	-	t	1018.65
weight not exceeding 40 kg/m; square hollow section	-	-	-	-	t	1448.08
weight not exceeding 40 kg/m; circular hollow section	-	-	-	-	t	1597.88
weight 40 - 100 kg/m	-	-	-	-	t	988.69
weight 40 - 100 kg/m; square hollow section	-	-	-	-	t	1418.12
weight 40 - 100 kg/m; circular hollow section	-	-	-	-	t	1577.91
weight exceeding 100 kg/m	-	-	-	-	t	963.73
weight exceeding 100 kg/m; square hollow section	-	-	-	-	t	1398.14
weight exceeding 100 kg/m; circular hollow section	-	-	-	-	t	1557.93
Grillages						
weight not exceeding 40 kg/m	-	-	-	-	t	699.07
weight 40 - 100 kg/m	-	-	-	-	t	739.02
weight exceeding 100 kg/m	-	-	-	-	t	699.07
Trestles, towers and built up columns						
straight	-	-	-	-	t	1338.22
Trusses and built up girders						
straight	-	-	-	-	t	1338.22
curved	-	-	-	-	t	1712.72
Fittings	-	-	-	-	t	973.70
Framing, erection						
Trial erection	-	-	-	-	t	239.68
Permanent erection on site	-	-	-	-	t	194.75
Surface preparation						
At works						
blast cleaning	-	-	-	-	m²	1.80
Surface treatment						
At works						
galvanising	-	-	-	-	m²	10.99
grit blast and one coat zinc chromate primer	-	-	-	-	m²	2.64
touch up primer and one coat of two pack epoxy zinc phosphate primer	-	-	-	-	m²	3.25
intumescent paint fire protection, 30 minutes; spray applied	-	-	-	-	m²	8.00
intumescent paint fire protection, 60 minutes; spray applied	-	-	-	-	m²	12.00
Extra over intumescent paint for; decorative sealer top	-	-	-	-	m²	2.00
On site						
intumescent paint fire protection, 30 minutes; spray applied	-	-	-	-	m²	6.00
intumescent paint fire protection, 60 minutes; spray applied	-	-	-	-	m²	8.00
Extra over intumescent paint for; decorative sealer top	-	-	-	-	m²	2.00
G12 ISOLATED STRUCTURAL METAL MEMBERS						
Isolated structural member; weldable steel; BS EN 10025: 1993 Grade S275; hot rolled structural steel sections						
Plain member; beams						
weight not exceeding 40 kg/m	-	-	-	-	t	624.18
weight 40 - 100 kg/m	-	-	-	-	t	599.21
weight exceeding 100 kg/m	-	-	-	-	t	584.23

G STRUCTURAL/CARCASSING METAL/TIMBER	PC £	Labour hours	Labour £	Material £	Unit	Total rate £
G12 ISOLATED STRUCTURAL METAL MEMBERS - cont'd						
Metsec open web steel lattice beams or other equal and approved; in single members; raised 3.50 m above ground; ends built in						
Beams; one coat zinc phosphate primer at works						
200 mm deep; to span 5.00 m (7.69 kg/m); ref B22	-	0.19	3.36	27.89	m	31.25
200 mm deep; to span 5.00 m (7.64 kg/m); ref B27	-	0.19	3.36	25.14	m	28.50
300 mm deep; to span 7.00 m (10.26 kg/m); ref B30	-	0.23	4.06	30.64	m	34.70
350 mm deep; to span 8.50 m (10.60 kg/m); ref B35	-	0.23	4.06	33.41	m	37.47
350 mm deep; to span 10.00 m (12.76 kg/m); ref D35	-	0.28	4.95	47.19	m	52.14
450 mm deep; to span 11.50 m (17.08 kg/m); ref D50	-	0.32	5.66	52.70	m	58.36
450 mm deep; to span 13.00 m (25.44 kg/m); ref G50	-	0.46	8.13	76.93	m	85.06
Beams; galvanised						
200 mm deep; to span 5.00 m (7.69 kg/m); ref B22	-	0.19	3.36	28.79	m	32.15
200 mm deep; to span 5.00 m (7.64 kg/m); ref B27	-	0.19	3.36	25.95	m	29.31
300 mm deep; to span 7.00 m (10.26 kg/m); ref B30	-	0.23	4.06	31.63	m	35.69
350 mm deep; to span 8.50 m (10.60 kg/m); ref B35	-	0.23	4.06	34.48	m	38.54
350 mm deep; to span 10.00 m (12.76 kg/m); ref D35	-	0.28	4.95	48.71	m	53.66
450 mm deep; to span 11.50 m (17.08 kg/m); ref D50	-	0.32	5.66	54.41	m	60.07
450 mm deep; to span 13.00 m (25.44 kg/m); ref G50	-	0.46	8.13	79.44	m	87.57
G20 CARPENTRY/TIMBER FRAMING/FIRST FIXING						
BASIC TIMBER PRICES						

	£		£		£
Hardwood; Joinery quality (£/m³)					
Agba	953.00	American White Ash	925.00	Beech	564.00
Brazilian Mahogany	1465.00	European Oak	2000.00	Iroko	914.00
Maple	1790.00	Sapele	728.00	Teak	2000.00
Utile	1005.00	W.A. Mahogany	713.00		
Softwood; Carcassing quality (£/m³)					
2.00 - 4.80 m lengths	236.78	6 .00 - 9.00 m lengths	248.05	S. S. Grade	27.44
4.80 - 6.00 m lengths	225.50	G. S. Grade	14.28		
Joinery quality - (£/m³)	290.00				

"Treatment" - (£/m³)

Pre-treatment of timber by vacuum/pressure impregnation, excluding transport costs and any subsequent seasoning:		
interior work; minimum. salt retention 4.00 kg/m³	43.60	
exterior work; minimum salt retention 5.30 kg/m³	49.80	
Pre-treatment of timber including flame proofing all purposes; minimum salt retention 36.00 kg/m³	115.00	

"Aquaseal" timber treatments - (£/25 litres)

"Timbershield"	26.10
"Longlife Wood Protector"	70.51

Prices for Measured Works - Major Works
G STRUCTURAL/CARCASSING METAL/TIMBER

	PC £	Labour hours	Labour £	Material £	Unit	Total rate £
Sawn softwood; untreated						
Floor members						
38 mm x 100 mm	-	0.11	1.54	0.99	m	**2.53**
38 mm x 150 mm	-	0.13	1.82	1.45	m	**3.27**
50 mm x 75 mm	-	0.11	1.54	0.99	m	**2.53**
50 mm x 100 mm	-	0.13	1.82	1.26	m	**3.08**
50 mm x 125 mm	-	0.13	1.82	1.53	m	**3.35**
50 mm x 150 mm	-	0.14	1.96	1.86	m	**3.82**
50 mm x 175 mm	-	0.14	1.96	2.15	m	**4.11**
50 mm x 200 mm	-	0.15	2.10	2.52	m	**4.62**
50 mm x 225 mm	-	0.15	2.10	2.92	m	**5.02**
50 mm x 250 mm	-	0.16	2.24	3.31	m	**5.55**
75 mm x 125 mm	-	0.15	2.10	2.31	m	**4.41**
75 mm x 150 mm	-	0.15	2.10	2.77	m	**4.87**
75 mm x 175 mm	-	0.15	2.10	3.25	m	**5.35**
75 mm x 200 mm	-	0.16	2.24	3.78	m	**6.02**
75 mm x 225 mm	-	0.16	2.24	4.36	m	**6.60**
75 mm x 250 mm	-	0.17	2.38	4.97	m	**7.35**
100 mm x 150 mm	-	0.20	2.80	3.78	m	**6.58**
100 mm x 200 mm	-	0.21	2.94	5.18	m	**8.12**
100 mm x 250 mm	-	0.23	3.22	7.17	m	**10.39**
100 mm x 300 mm	-	0.25	3.50	8.86	m	**12.36**
Wall or partition members						
25 mm x 25 mm	-	0.06	0.84	0.30	m	**1.14**
25 mm x 38 mm	-	0.06	0.84	0.38	m	**1.22**
25 mm x 75 mm	-	0.08	1.12	0.52	m	**1.64**
38 mm x 38 mm	-	0.08	1.12	0.44	m	**1.56**
38 mm x 50 mm	-	0.08	1.12	0.50	m	**1.62**
38 mm x 75 mm	-	0.11	1.54	0.75	m	**2.29**
38 mm x 100 mm	-	0.14	1.96	0.99	m	**2.95**
50 mm x 50 mm	-	0.11	1.54	0.70	m	**2.24**
50 mm x 75 mm	-	0.14	1.96	1.01	m	**2.97**
50 mm x 100 mm	-	0.17	2.38	1.28	m	**3.66**
50 mm x 125 mm	-	0.18	2.52	1.55	m	**4.07**
75 mm x 75 mm	-	0.17	2.38	1.69	m	**4.07**
75 mm x 100 mm	-	0.19	2.66	1.90	m	**4.56**
100 mm x 100 mm	-	0.19	2.66	2.51	m	**5.17**
Roof members; flat						
38 mm x 75 mm	-	0.13	1.82	0.75	m	**2.57**
38 mm x 100 mm	-	0.13	1.82	0.99	m	**2.81**
38 mm x 125 mm	-	0.13	1.82	1.24	m	**3.06**
38 mm x 150 mm	-	0.13	1.82	1.45	m	**3.27**
50 mm x 100 mm	-	0.13	1.82	1.26	m	**3.08**
50 mm x 125 mm	-	0.13	1.82	1.53	m	**3.35**
50 mm x 150 mm	-	0.14	1.96	1.86	m	**3.82**
50 mm x 175 mm	-	0.14	1.96	2.15	m	**4.11**
50 mm x 200 mm	-	0.15	2.10	2.52	m	**4.62**
50 mm x 225 mm	-	0.15	2.10	2.92	m	**5.02**
50 mm x 250 mm	-	0.16	2.24	3.31	m	**5.55**
75 mm x 150 mm	-	0.15	2.10	2.77	m	**4.87**
75 mm x 175 mm	-	0.15	2.10	3.25	m	**5.35**
75 mm x 200 mm	-	0.16	2.24	3.78	m	**6.02**
75 mm x 225 mm	-	0.16	2.24	4.36	m	**6.60**
75 mm x 250 mm	-	0.17	2.38	4.97	m	**7.35**
Roof members; pitched						
25 mm x 100 mm	-	0.11	1.54	0.68	m	**2.22**
25 mm x 125 mm	-	0.11	1.54	0.83	m	**2.37**
25 mm x 150 mm	-	0.14	1.96	1.06	m	**3.02**
25 mm x 175 mm	-	0.16	2.24	1.38	m	**3.62**
25 mm x 200 mm	-	0.17	2.38	1.61	m	**3.99**
38 mm x 100 mm	-	0.14	1.96	0.99	m	**2.95**

G20 CARPENTRY/TIMBER FRAMING/FIRST FIXING - cont'd

	PC £	Labour hours	Labour £	Material £	Unit	Total rate £
Sawn softwood; untreated						
Roof members; pitched						
38 mm x 125 mm	-	0.14	1.96	1.24	m	3.20
38 mm x 150 mm	-	0.14	1.96	1.45	m	3.41
50 mm x 50 mm	-	0.11	1.54	0.68	m	2.22
50 mm x 75 mm	-	0.14	1.96	0.99	m	2.95
50 mm x 100 mm	-	0.17	2.38	1.26	m	3.64
50 mm x 125 mm	-	0.17	2.38	1.53	m	3.91
50 mm x 150 mm	-	0.19	2.66	1.86	m	4.52
50 mm x 175 mm	-	0.19	2.66	2.15	m	4.81
50 mm x 200 mm	-	0.19	2.66	2.52	m	5.18
50 mm x 225 mm	-	0.19	2.66	2.92	m	5.58
75 mm x 100 mm	-	0.23	3.22	1.86	m	5.08
75 mm x 125 mm	-	0.23	3.22	2.31	m	5.53
75 mm x 150 mm	-	0.23	3.22	2.77	m	5.99
100 mm x 150 mm	-	0.28	3.92	3.81	m	7.73
100 mm x 175 mm	-	0.28	3.92	4.42	m	8.34
100 mm x 200 mm	-	0.28	3.92	5.18	m	9.10
100 mm x 225 mm	-	0.31	4.34	6.31	m	10.65
100 mm x 250 mm	-	0.31	4.34	7.17	m	11.51
Plates						
38 mm x 75 mm	-	0.11	1.54	0.75	m	2.29
38 mm x 100 mm	-	0.14	1.96	0.99	m	2.95
50 mm x 75 mm	-	0.14	1.96	0.99	m	2.95
50 mm x 100 mm	-	0.17	2.38	1.26	m	3.64
75 mm x 100 mm	-	0.19	2.66	1.86	m	4.52
75 mm x 125 mm	-	0.22	3.08	2.28	m	5.36
75 mm x 150 mm	-	0.25	3.50	2.75	m	6.25
Plates; fixing by bolting						
38 mm x 75 mm	-	0.20	2.80	0.75	m	3.55
38 mm x 100 mm	-	0.23	3.22	0.99	m	4.21
50 mm x 75 mm	-	0.23	3.22	0.99	m	4.21
50 mm x 100 mm	-	0.26	3.64	1.26	m	4.90
75 mm x 100 mm	-	0.29	4.06	1.86	m	5.92
75 mm x 125 mm	-	0.31	4.34	2.28	m	6.62
75 mm x 150 mm	-	0.34	4.76	2.75	m	7.51
Joist strutting; herringbone						
50 mm x 50 mm; depth of joist 150 mm	-	0.46	6.44	1.62	m	8.06
50 mm x 50 mm; depth of joist 175 mm	-	0.46	6.44	1.65	m	8.09
50 mm x 50 mm; depth of joist 200 mm	-	0.46	6.44	1.68	m	8.12
50 mm x 50 mm; depth of joist 225 mm	-	0.46	6.44	1.72	m	8.16
50 mm x 50 mm; depth of joist 250 mm	-	0.46	6.44	1.75	m	8.19
Joist strutting; block						
50 mm x 150 mm; depth of joist 150 mm	-	0.28	3.92	2.13	m	6.05
50 mm x 175 mm; depth of joist 175 mm	-	0.28	3.92	2.42	m	6.34
50 mm x 200 mm; depth of joist 200 mm	-	0.28	3.92	2.79	m	6.71
50 mm x 225 mm; depth of joist 225 mm	-	0.28	3.92	3.20	m	7.12
50 mm x 250 mm; depth of joist 250 mm	-	0.28	3.92	3.59	m	7.51
Cleats						
225 mm x 100 mm x 75 mm	-	0.19	2.66	0.49	nr	3.15
Extra for stress grading to above timbers						
general structural (GS) grade	-	-	-	15.90	m^3	15.90
special structural (SS) grade	-	-	-	30.31	m^3	30.31
Extra for protecting and flameproofing timber with "Celgard CF" protection or other equal and approved						
small sections	-	-	-	94.62	m^3	94.62
large sections	-	-	-	90.51	m^3	90.51

Prices for Measured Works - Major Works
G STRUCTURAL/CARCASSING METAL/TIMBER

	PC £	Labour hours	Labour £	Material £	Unit	Total rate £
Wrot surfaces						
plain; 50 mm wide	-	0.02	0.28	-	m	0.28
plain; 100 mm wide	-	0.03	0.42	-	m	0.42
plain; 150 mm wide	-	0.04	0.56	-	m	0.56
Sawn softwood; "Tanalised"						
Floor members						
38 mm x 75 mm	-	0.11	1.54	0.86	m	2.40
38 mm x 100 mm	-	0.11	1.54	1.14	m	2.68
38 mm x 150 mm	-	0.13	1.82	1.68	m	3.50
50 mm x 75 mm	-	0.11	1.54	1.14	m	2.68
50 mm x 100 mm	-	0.13	1.82	1.46	m	3.28
50 mm x 125 mm	-	0.13	1.82	1.78	m	3.60
50 mm x 150 mm	-	0.14	1.96	2.16	m	4.12
50 mm x 175 mm	-	0.14	1.96	2.50	m	4.46
50 mm x 200 mm	-	0.15	2.10	2.91	m	5.01
50 mm x 225 mm	-	0.15	2.10	3.37	m	5.47
50 mm x 250 mm	-	0.16	2.24	3.81	m	6.05
75 mm x 125 mm	-	0.15	2.10	2.68	m	4.78
75 mm x 150 mm	-	0.15	2.10	3.22	m	5.32
75 mm x 175 mm	-	0.15	2.10	3.77	m	5.87
75 mm x 200 mm	-	0.16	2.24	4.38	m	6.62
75 mm x 225 mm	-	0.16	2.24	5.04	m	7.28
75 mm x 250 mm	-	0.17	2.38	5.72	m	8.10
100 mm x 150 mm	-	0.20	2.80	4.38	m	7.18
100 mm x 200 mm	-	0.21	2.94	5.98	m	8.92
100 mm x 250 mm	-	0.23	3.22	8.16	m	11.38
100 mm x 300 mm	-	0.25	3.50	10.05	m	13.55
Wall or partition members						
25 mm x 25 mm	-	0.06	0.84	0.32	m	1.16
25 mm x 38 mm	-	0.06	0.84	0.42	m	1.26
25 mm x 75 mm	-	0.08	1.12	0.59	m	1.71
38 mm x 38 mm	-	0.08	1.12	0.49	m	1.61
38 mm x 50 mm	-	0.08	1.12	0.58	m	1.70
38 mm x 75 mm	-	0.11	1.54	0.86	m	2.40
38 mm x 100 mm	-	0.14	1.96	1.14	m	3.10
50 mm x 50 mm	-	0.11	1.54	0.80	m	2.34
50 mm x 75 mm	-	0.14	1.96	1.16	m	3.12
50 mm x 100 mm	-	0.17	2.38	1.48	m	3.86
50 mm x 125 mm	-	0.18	2.52	1.81	m	4.33
75 mm x 75 mm	-	0.17	2.38	1.91	m	4.29
75 mm x 100 mm	-	0.19	2.66	2.20	m	4.86
100 mm x 100 mm	-	0.19	2.66	2.91	m	5.57
Roof members; flat						
38 mm x 75 mm	-	0.13	1.82	0.86	m	2.68
38 mm x 100 mm	-	0.13	1.82	1.14	m	2.06
38 mm x 125 mm	-	0.13	1.82	1.44	m	3.26
38 mm x 150 mm	-	0.13	1.82	1.68	m	3.50
50 mm x 100 mm	-	0.13	1.82	1.46	m	3.28
50 mm x 125 mm	-	0.13	1.82	1.78	m	3.60
50 mm x 150 mm	-	0.14	1.96	2.16	m	4.12
50 mm x 175 mm	-	0.14	1.96	2.50	m	4.46
50 mm x 200 mm	-	0.15	2.10	2.91	m	5.01
50 mm x 225 mm	-	0.15	2.10	3.37	m	5.47
50 mm x 250 mm	-	0.16	2.24	3.81	m	6.05
75 mm x 150 mm	-	0.15	2.10	3.22	m	5.32
75 mm x 175 mm	-	0.15	2.10	3.77	m	5.87
75 mm x 200 mm	-	0.16	2.24	4.38	m	6.62
75 mm x 225 mm	-	0.16	2.24	5.04	m	7.28
75 mm x 250 mm	-	0.17	2.38	5.72	m	8.10

Prices for Measured Works - Major Works
G STRUCTURAL/CARCASSING METAL/TIMBER

	PC £	Labour hours	Labour £	Material £	Unit	Total rate £
G20 CARPENTRY/TIMBER FRAMING/FIRST FIXING - cont'd						
Sawn softwood; "Tanalised" - cont'd						
Roof members; pitched						
25 mm x 100 mm	-	0.11	1.54	0.78	m	2.32
25 mm x 125 mm	-	0.11	1.54	0.95	m	2.49
25 mm x 150 mm	-	0.14	1.96	1.21	m	3.17
25 mm x 175 mm	-	0.16	2.24	1.56	m	3.80
25 mm x 200 mm	-	0.17	2.38	1.81	m	4.19
38 mm x 100 mm	-	0.14	1.96	1.14	m	3.10
38 mm x 125 mm	-	0.14	1.96	1.44	m	3.40
38 mm x 150 mm	-	0.14	1.96	1.68	m	3.64
50 mm x 50 mm	-	0.11	1.54	0.78	m	2.32
50 mm x 75 mm	-	0.14	1.96	1.14	m	3.10
50 mm x 100 mm	-	0.17	2.38	1.46	m	3.84
50 mm x 125 mm	-	0.17	2.38	1.78	m	4.16
50 mm x 150 mm	-	0.19	2.66	2.16	m	4.82
50 mm x 175 mm	-	0.19	2.66	2.50	m	5.16
50 mm x 200 mm	-	0.19	2.66	2.91	m	5.57
50 mm x 225 mm	-	0.19	2.66	3.37	m	6.03
75 mm x 100 mm	-	0.23	3.22	2.16	m	5.38
75 mm x 125 mm	-	0.23	3.22	2.68	m	5.90
75 mm x 150 mm	-	0.23	3.22	3.22	m	6.44
100 mm x 150 mm	-	0.28	3.92	4.40	m	8.32
100 mm x 175 mm	-	0.28	3.92	5.11	m	9.03
100 mm x 200 mm	-	0.28	3.92	5.98	m	9.90
100 mm x 225 mm	-	0.31	4.34	7.21	m	11.55
100 mm x 250 mm	-	0.31	4.34	8.16	m	12.50
Plates						
38 mm x 75 mm	-	0.11	1.54	0.90	m	2.44
38 mm x 100 mm	-	0.14	1.96	1.14	m	3.10
50 mm x 75 mm	-	0.14	1.96	1.14	m	3.10
50 mm x 100 mm	-	0.17	2.38	1.46	m	3.84
75 mm x 100 mm	-	0.19	2.66	2.16	m	4.82
75 mm x 125 mm	-	0.22	3.08	2.66	m	5.74
75 mm x 150 mm	-	0.25	3.50	3.19	m	6.69
Plates; fixing by bolting						
38 mm x 75 mm	-	0.20	2.80	0.86	m	3.66
38 mm x 100 mm	-	0.23	3.22	1.14	m	4.36
50 mm x 75 mm	-	0.23	3.22	1.14	m	4.36
50 mm x 100 mm	-	0.26	3.64	1.46	m	5.10
75 mm x 100 mm	-	0.29	4.06	2.16	m	6.22
75 mm x 125 mm	-	0.31	4.34	2.66	m	7.00
75 mm x 150 mm	-	0.34	4.76	3.19	m	7.95
Joist strutting; herringbone						
50 mm x 50 mm; depth of joist 150 mm	-	0.46	6.44	1.82	m	8.26
50 mm x 50 mm; depth of joist 175 mm	-	0.46	6.44	1.86	m	8.30
50 mm x 50 mm; depth of joist 200 mm	-	0.46	6.44	1.90	m	8.34
50 mm x 50 mm; depth of joist 225 mm	-	0.46	6.44	1.93	m	8.37
50 mm x 50 mm; depth of joist 250 mm	-	0.46	6.44	1.97	m	8.41
Joist strutting; block						
50 mm x 150 mm; depth of joist 150 mm	-	0.28	3.92	2.43	m	6.35
50 mm x 175 mm; depth of joist 175 mm	-	0.28	3.92	2.77	m	6.69
50 mm x 200 mm; depth of joist 200 mm	-	0.28	3.92	3.19	m	7.11
50 mm x 225 mm; depth of joist 225 mm	-	0.28	3.92	3.64	m	7.56
50 mm x 250 mm; depth of joist 250 mm	-	0.28	3.92	4.08	m	8.00
Cleats						
225 mm x 100 mm x 75 mm	-	0.19	2.66	0.55	nr	3.21

Prices for Measured Works - Major Works
G STRUCTURAL/CARCASSING METAL/TIMBER

	PC £	Labour hours	Labour £	Material £	Unit	Total rate £
Extra for stress grading to above timbers						
general structural (GS) grade	-	-	-	15.90	m³	15.90
special structural (SS) grade	-	-	-	30.31	m³	30.31
Extra for protecting and flameproofing timber with "Celgard CF" protection or other equal and approved						
small sections	-	-	-	94.62	m³	94.62
large sections	-	-	-	90.51	m³	90.51
Wrot surfaces						
plain; 50 mm wide	-	0.02	0.28	-	m	0.28
plain; 100 mm wide	-	0.03	0.42	-	m	0.42
plain; 150 mm wide	-	0.04	0.56	-	m	0.56
Trussed rafters, stress graded sawn softwood pressure impregnated; raised through two storeys and fixed in position						
"W" type truss (Fink); 22.5 degree pitch; 450 mm eaves overhang						
5.00 m span	-	1.48	20.71	24.68	nr	45.39
7.60 m span	-	1.62	22.67	31.11	nr	53.78
10.00 m span	-	1.85	25.89	40.50	nr	66.39
"W" type truss (Fink); 30 degree pitch; 450 mm eaves overhang						
5.00 m span	-	1.48	20.71	24.50	nr	45.21
7.60 m span	-	1.62	22.67	31.96	nr	54.63
10.00 m span	-	1.85	25.89	45.81	nr	71.70
"W" type truss (Fink); 45 degree pitch; 450 mm eaves overhang						
4.60 m span	-	1.48	20.71	26.32	nr	47.03
7.00 m span	-	1.62	22.67	35.42	nr	58.09
"Mono" type truss; 17.5 degree pitch; 450 mm eaves overhang						
3.30 m span	-	1.30	18.20	19.93	nr	38.13
5.60 m span	-	1.48	20.71	28.20	nr	48.91
7.00 m span	-	1.71	23.93	35.42	nr	59.35
"Mono" type truss; 30 degree pitch; 450 mm eaves overhang						
3.30 m span	-	1.30	18.20	21.40	nr	39.60
5.60 m span	-	1.48	20.71	30.52	nr	51.23
7.00 m span	-	1.71	23.93	40.22	nr	64.15
"Attic" type truss; 45 degree pitch; 450 mm eaves overhang						
5.00 m span	-	2.91	40.73	53.30	nr	94.03
7.60 m span	-	3.05	42.69	86.45	nr	129.14
9.00 m span	-	3.24	45.35	110.69	nr	156.04
"Moelven Toreboda" glulam timber beams or other equal and approved; Moelven Laminated Timber Structures; LB grade whitewood; pressure impregnated; phenbol resorcinal adhesive; clean planed finish; fixed						
Laminated roof beams						
56 mm x 225 mm	-	0.51	7.14	4.89	m	12.03
66 mm x 315 mm	-	0.65	9.10	9.56	m	18.66
90 mm x 315 mm	-	0.83	11.62	13.05	m	24.67
90 mm x 405 mm	-	1.06	14.84	16.77	m	31.61
115 mm x 405 mm	-	1.34	18.76	21.44	m	40.20
115 mm x 495 mm	-	1.67	23.37	26.20	m	49.57
115 mm x 630 mm	-	2.04	28.55	33.34	m	61.89

G STRUCTURAL/CARCASSING METAL/TIMBER

	PC £	Labour hours	Labour £	Material £	Unit	Total rate £
G20 CARPENTRY/TIMBER FRAMING/FIRST FIXING - cont'd						
"Masterboard" or other equal and approved; 6 mm thick						
Eaves, verge soffit boards, fascia boards and the like						
over 300 mm wide	4.92	0.65	9.10	6.11	m²	15.21
75 mm wide	-	0.19	2.66	0.47	m	3.13
150 mm wide	-	0.22	3.08	0.92	m	4.00
225 mm wide	-	0.26	3.64	1.36	m	5.00
300 mm wide	-	0.28	3.92	1.81	m	5.73
Plywood; external quality; 15 mm thick						
Eaves, verge soffit boards, fascia boards and the like						
over 300 mm wide	9.31	0.76	10.64	11.18	m²	21.82
75 mm wide	-	0.23	3.22	0.85	m	4.07
150 mm wide	-	0.27	3.78	1.68	m	5.46
225 mm wide	-	0.31	4.34	2.51	m	6.85
300 mm wide	-	0.34	4.76	3.33	m	8.09
Plywood; external quality; 18 mm thick						
Eaves, verge soffit boards, fascia boards and the like						
over 300 mm wide	11.12	0.76	10.64	13.27	m²	23.91
75 mm wide	-	0.23	3.22	1.01	m	4.23
150 mm wide	-	0.27	3.78	1.99	m	5.77
225 mm wide	-	0.31	4.34	2.98	m	7.32
300 mm wide	-	0.34	4.76	3.96	m	8.72
Plywood; marine quality; 18 mm thick						
Gutter boards; butt joints						
over 300 mm wide	8.28	0.86	12.04	10.00	m²	22.04
150 mm wide	-	0.31	4.34	1.50	m	5.84
225 mm wide	-	0.34	4.76	2.26	m	7.02
300 mm wide	-	0.38	5.32	3.00	m	8.32
Eaves, verge soffit boards, fascias boards and the like						
over 300 mm wide	-	0.76	10.64	10.00	m²	20.64
75 mm wide	-	0.23	3.22	0.76	m	3.98
150 mm wide	-	0.27	3.78	1.50	m	5.28
225 mm wide	-	0.31	4.34	2.24	m	6.58
300 mm wide	-	0.34	4.76	2.98	m	7.74
Plywood; marine quality; 25 mm thick						
Gutter boards; butt joints						
over 300 mm wide	11.51	0.93	13.02	13.72	m²	26.74
150 mm wide	-	0.32	4.48	2.06	m	6.54
225 mm wide	-	0.37	5.18	3.10	m	8.28
300 mm wide	-	0.42	5.88	4.12	m	10.00
Eaves, verge soffit boards, fascia baords and the like						
over 300 mm wide	-	0.81	11.34	13.72	m²	25.06
75 mm wide	-	0.24	3.36	1.04	m	4.40
150 mm wide	-	0.29	4.06	2.06	m	6.12
225 mm wide	-	0.29	4.06	3.08	m	7.14
300 mm wide	-	0.37	5.18	4.10	m	9.28

G STRUCTURAL/CARCASSING METAL/TIMBER

	PC £	Labour hours	Labour £	Material £	Unit	Total rate £
Sawn softwood; untreated						
Gutter boards; butt joints						
19 mm thick; sloping	-	1.16	16.24	5.83	m²	22.07
19 mm thick; 75 mm wide	-	0.32	4.48	0.44	m	4.92
19 mm thick; 150 mm wide	-	0.37	5.18	0.85	m	6.03
19 mm thick; 225 mm wide	-	0.42	5.88	1.47	m	7.35
25 mm thick; sloping	-	1.16	16.24	7.52	m²	23.76
25 mm thick; 75 mm wide	-	0.32	4.48	0.54	m	5.02
25 mm thick; 150 mm wide	-	0.37	5.18	1.10	m	6.28
25 mm thick; 225 mm wide	-	0.42	5.88	1.95	m	7.83
Cesspools with 25 mm thick sides and bottom						
225 mm x 225 mm x 150 mm	-	1.11	15.54	1.50	nr	17.04
300 mm x 300 mm x 150 mm	-	1.30	18.20	1.89	nr	20.09
Individual supports; firrings						
50 mm wide x 36 mm average depth	-	0.14	1.96	0.96	m	2.92
50 mm wide x 50 mm average depth	-	0.14	1.96	1.23	m	3.19
50 mm wide x 75 mm average depth	-	0.14	1.96	1.58	m	3.54
Individual supports; bearers						
25 mm x 50 mm	-	0.09	1.26	0.44	m	1.70
38 mm x 50 mm	-	0.09	1.26	0.55	m	1.81
50 mm x 50 mm	-	0.09	1.26	0.72	m	1.98
50 mm x 75 mm	-	0.09	1.26	1.03	m	2.29
Individual supports; angle fillets						
38 mm x 38 mm	-	0.09	1.26	0.48	m	1.74
50 mm x 50 mm	-	0.09	1.26	0.72	m	1.98
75 mm x 75 mm	-	0.11	1.54	1.68	m	3.22
Individual supports; tilting fillets						
19 mm x 38 mm	-	0.09	1.26	0.22	m	1.48
25 mm x 50 mm	-	0.09	1.26	0.32	m	1.58
38 mm x 75 mm	-	0.09	1.26	0.54	m	1.80
50 mm x 75 mm	-	0.09	1.26	0.67	m	1.93
75 mm x 100 mm	-	0.14	1.96	1.25	m	3.21
Individual supports; grounds or battens						
13 mm x 19 mm	-	0.04	0.56	0.27	m	0.83
13 mm x 32 mm	-	0.04	0.56	0.30	m	0.86
25 mm x 50 mm	-	0.04	0.56	0.40	m	0.96
Individual supports; grounds or battens; plugged and screwed						
13 mm x 19 mm	-	0.14	1.96	0.29	m	2.25
13 mm x 32 mm	-	0.14	1.96	0.32	m	2.28
25 mm x 50 mm	-	0.14	1.96	0.42	m	2.38
Framed supports; open-spaced grounds or battens; at 300 mm centres one way						
25 mm x 50 mm	-	0.14	1.96	1.32	m²	3.28
25 mm x 50 mm; plugged and screwed	-	0.42	5.88	1.40	m²	7.28
Framed supports; at 300 mm centres one way and 600 mm centres the other way						
25 mm x 50 mm	-	0.69	9.66	1.98	m²	11.64
38 mm x 50 mm	-	0.69	9.66	2.50	m²	12.16
50 mm x 50 mm	-	0.69	9.66	3.37	m²	13.03
50 mm x 75 mm	-	0.69	9.66	4.93	m²	14.59
75 mm x 75 mm	-	0.69	9.66	8.34	m²	18.00
Framed supports; at 300 mm centres one way and 600 mm centres the other way; plugged and screwed						
25 mm x 50 mm	-	1.16	16.24	2.24	m²	18.48
38 mm x 50 mm	-	1.16	16.24	2.76	m²	19.00
50 mm x 50 mm	-	1.16	16.24	3.63	m²	19.87
50 mm x 75 mm	-	1.16	16.24	5.19	m²	21.43
75 mm x 75 mm	-	1.16	16.24	8.59	m²	24.83

Prices for Measured Works - Major Works
G STRUCTURAL/CARCASSING METAL/TIMBER

	PC £	Labour hours	Labour £	Material £	Unit	Total rate £
G20 CARPENTRY/TIMBER FRAMING/FIRST FIXING - cont'd						
Sawn softwood; untreated - cont'd						
Framed supports; at 500 mm centres both ways						
25 mm x 50 mm; to bath panels	-	0.83	11.62	2.58	m^2	14.20
Framed supports; as bracketing and cradling around steelwork						
25 mm x 50 mm	-	1.30	18.20	2.80	m^2	21.00
50 mm x 50 mm	-	1.39	19.46	4.75	m^2	24.21
50 mm x 75 mm	-	1.48	20.71	6.93	m^2	27.64
Sawn softwood; "Tanalised"						
Gutter boards; butt joints						
19 mm thick; sloping	-	1.16	16.24	6.58	m^2	22.82
19 mm thick; 75 mm wide	-	0.32	4.48	0.50	m	4.98
19 mm thick; 150 mm wide	-	0.37	5.18	0.96	m	6.14
19 mm thick; 225 mm wide	-	0.42	5.88	1.64	m	7.52
25 mm thick; sloping	-	1.16	16.24	8.52	m^2	24.76
25 mm thick; 75 mm wide	-	0.32	4.48	0.61	m	5.09
25 mm thick; 150 mm wide	-	0.37	5.18	1.25	m	6.43
25 mm thick; 225 mm wide	-	0.42	5.88	2.18	m	8.06
Cesspools with 25 mm thick sides and bottom						
225 mm x 225 mm x 150 mm	-	1.11	15.54	1.70	nr	17.24
300 mm x 300 mm x 150 mm	-	1.30	18.20	2.16	nr	20.36
Individual supports; firrings						
50 mm wide x 36 mm average depth	-	0.14	1.96	1.04	m	3.00
50 mm wide x 50 mm average depth	-	0.14	1.96	1.33	m	3.29
50 mm wide x 75 mm average depth	-	0.14	1.96	1.73	m	3.69
Individual supports; bearers						
25 mm x 50 mm	-	0.09	1.26	0.49	m	1.75
38 mm x 50 mm	-	0.09	1.26	0.62	m	1.88
50 mm x 50 mm	-	0.09	1.26	0.82	m	2.08
50 mm x 75 mm	-	0.09	1.26	1.18	m	2.44
Individual supports; angle fillets						
38 mm x 38 mm	-	0.09	1.26	0.51	m	1.77
50 mm x 50 mm	-	0.09	1.26	0.77	m	2.03
75 mm x 75 mm	-	0.11	1.54	1.79	m	3.33
Individual supports; tilting fillets						
19 mm x 38 mm	-	0.09	1.26	0.24	m	1.50
25 mm x 50 mm	-	0.09	1.26	0.34	m	1.60
38 mm x 75 mm	-	0.09	1.26	0.59	m	1.85
50 mm x 75 mm	-	0.09	1.26	0.75	m	2.01
75 mm x 100 mm	-	0.14	1.96	1.40	m	3.36
Individual supports; grounds or battens						
13 mm x 19 mm	-	0.04	0.56	0.28	m	0.84
13 mm x 32 mm	-	0.04	0.56	0.31	m	0.87
25 mm x 50 mm	-	0.04	0.56	0.45	m	1.01
Individual supports; grounds or battens; plugged and screwed						
13 mm x 19 mm	-	0.14	1.96	0.30	m	2.26
13 mm x 32 mm	-	0.14	1.96	0.33	m	2.29
25 mm x 50 mm	-	0.14	1.96	0.47	m	2.43
Framed supports; open-spaced grounds or battens; at 300 mm centres one way						
25 mm x 50 mm	-	0.14	1.96	1.49	m^2	3.45
25 mm x 50 mm; plugged and screwed	-	0.42	5.88	1.57	m^2	7.45

Prices for Measured Works - Major Works

G STRUCTURAL/CARCASSING METAL/TIMBER

	PC £	Labour hours	Labour £	Material £	Unit	Total rate £
Framed supports; at 300 mm centres one way and 600 mm centres the other way						
25 mm x 50 mm	-	0.69	9.66	2.23	m²	**11.89**
38 mm x 50 mm	-	0.69	9.66	2.88	m²	**12.54**
50 mm x 50 mm	-	0.69	9.66	3.86	m²	**13.52**
50 mm x 75 mm	-	0.69	9.66	5.68	m²	**15.34**
75 mm x 75 mm	-	0.69	9.66	9.45	m²	**19.11**
Framed supports; at 300 mm centres one way and 600 mm centres the other way; plugged and screwed						
25 mm x 50 mm	-	1.16	16.24	2.49	m²	**18.73**
38 mm x 50 mm	-	1.16	16.24	3.14	m²	**19.38**
50 mm x 50 mm	-	1.16	16.24	4.12	m²	**20.36**
50 mm x 75 mm	-	1.16	16.24	5.93	m²	**22.17**
75 mm x 75 mm	-	1.16	16.24	9.71	m²	**25.95**
Framed supports; at 500 mm centres both ways						
25 mm x 50 mm; to bath panels	-	0.83	11.62	2.90	m²	**14.52**
Framed supports; as bracketing and cradling around steelwork						
25 mm x 50 mm	-	1.30	18.20	3.16	m²	**21.36**
50 mm x 50 mm	-	1.39	19.46	5.44	m²	**24.90**
50 mm x 75 mm	-	1.48	20.71	7.98	m²	**28.69**
Wrought softwood						
Gutter boards; tongued and grooved joints						
19 mm thick; sloping	-	1.39	19.46	7.52	m²	**26.98**
19 mm thick; 75 mm wide	-	0.37	5.18	0.88	m	**6.06**
19 mm thick; 150 mm wide	-	0.42	5.88	1.10	m	**6.98**
19 mm thick; 225 mm wide	-	0.46	6.44	1.94	m	**8.38**
25 mm thick; sloping	-	1.39	19.46	9.20	m²	**28.66**
25 mm thick; 75 mm wide	-	0.37	5.18	1.23	m	**6.41**
25 mm thick; 150 mm wide	-	0.42	5.88	1.30	m	**7.18**
25 mm thick; 225 mm wide	-	0.46	6.44	2.47	m	**8.91**
Eaves, verge soffit boards, fascia boards and the like						
19 mm thick; over 300 mm wide	-	1.15	16.10	7.52	m²	**23.62**
19 mm thick; 150 mm wide; once grooved	-	0.19	2.66	1.34	m	**4.00**
25 mm thick; 150 mm wide; once grooved	-	0.19	2.66	1.62	m	**4.28**
25 mm thick; 175 mm wide; once grooved	-	0.19	2.66	1.86	m	**4.52**
32 mm thick; 225 mm wide; once grooved	-	0.23	3.22	3.27	m	**6.49**
Wrought softwood; "Tanalised"						
Gutter boards; tongued and grooved joints						
19 mm thick; sloping	-	1.39	19.46	8.28	m²	**27.74**
19 mm thick; 75 mm wide	-	0.37	5.18	0.94	m	**6.12**
19 mm thick; 150 mm wide	-	0.42	5.88	1.22	m	**7.10**
19 mm thick; 225 mm wide	-	0.46	6.44	2.11	m	**8.55**
25 mm thick; sloping	-	1.39	19.46	10.19	m²	**29.65**
25 mm thick; 75 mm wide	-	0.37	5.18	1.30	m	**6.48**
25 mm thick; 150 mm wide	-	0.42	5.88	1.45	m	**7.33**
25 mm thick; 225 mm wide	-	0.46	6.44	2.69	m	**9.13**
Eaves, verge soffit boards, fascia boards and the like						
19 mm thick; over 300 mm wide	-	1.15	16.10	8.27	m²	**24.37**
19 mm thick; 150 mm wide; once grooved	-	0.19	2.66	1.46	m	**4.12**
25 mm thick; 150 mm wide; once grooved	-	0.19	2.66	1.77	m	**4.43**
25 mm thick; 175 mm wide; once grooved	-	0.20	2.80	2.04	m	**4.84**
32 mm thick; 225 mm wide; once grooved	-	0.23	3.22	3.56	m	**6.78**

Prices for Measured Works - Major Works
G STRUCTURAL/CARCASSING METAL/TIMBER

	PC £	Labour hours	Labour £	Material £	Unit	Total rate £
G20 CARPENTRY/TIMBER FRAMING/FIRST FIXING - cont'd						
Straps; mild steel; galvanised						
Standard twisted vertical restraint; fixing to softwood and brick or blockwork						
30 mm x 2.5 mm x 400 mm girth	-	0.23	3.22	2.02	nr	5.24
30 mm x 2.5 mm x 600 mm girth	-	0.24	3.36	2.99	nr	6.35
30 mm x 2.5 mm x 800 mm girth	-	0.25	3.50	4.12	nr	7.62
30 mm x 2.5 mm x 1000 mm girth	-	0.28	3.92	5.35	nr	9.27
30 mm x 2.5 mm x 1200 mm girth	-	0.29	4.06	6.32	nr	10.38
Hangers; mild steel; galvanised						
Joist hangers 0.90 mm thick; The Expanded Metal Company Ltd "Speedy" or other equal and approved; for fixing to softwood; joist sizes						
50 mm wide; all sizes to 225 mm deep	0.60	0.11	1.54	0.75	nr	2.29
75 mm wide; all sizes to 225 mm deep	0.63	0.14	1.96	0.82	nr	2.78
100 mm wide; all sizes to 225 mm deep	0.68	0.17	2.38	0.91	nr	3.29
Joist hangers 2.50 mm thick; for building in; joist sizes						
50 mm x 100 mm	1.17	0.07	1.05	1.37	nr	2.42
50 mm x 125 mm	1.17	0.07	1.05	1.37	nr	2.42
50 mm x 150 mm	1.10	0.09	1.33	1.31	nr	2.64
50 mm x 175 mm	1.16	0.09	1.33	1.38	nr	2.71
50 mm x 200 mm	1.28	0.11	1.61	1.54	nr	3.15
50 mm x 225 mm	1.36	0.11	1.61	1.63	nr	3.24
75 mm x 150 mm	1.70	0.09	1.33	1.99	nr	3.32
75 mm x 175 mm	1.60	0.09	1.33	1.88	nr	3.21
75 mm x 200 mm	1.70	0.11	1.61	2.01	nr	3.62
75 mm x 225 mm	1.82	0.11	1.61	2.15	nr	3.76
75 mm x 250 mm	1.93	0.13	1.89	2.30	nr	4.19
100 mm x 200 mm	2.12	0.11	1.61	2.49	nr	4.10
Metal connectors; mild steel; galvanised						
Round toothed plate; for 10 mm or 12 mm diameter bolts						
38 mm diameter; single sided	-	0.01	0.14	0.18	nr	0.32
38 mm diameter; double sided	-	0.01	0.14	0.20	nr	0.34
50 mm diameter; single sided	-	0.01	0.14	0.20	nr	0.34
50 mm diameter; double sided	-	0.01	0.14	0.22	nr	0.36
63 mm diameter; single sided	-	0.01	0.14	0.29	nr	0.43
63 mm diameter; double sided	-	0.01	0.14	0.32	nr	0.46
75 mm diameter; single sided	-	0.01	0.14	0.42	nr	0.56
75 mm diameter; double sided	-	0.01	0.14	0.44	nr	0.58
framing anchor	-	0.14	1.96	0.33	nr	2.29
Bolts; mild steel; galvanised						
Fixing only bolts; 50 mm - 200 mm long						
6 mm diameter	-	0.03	0.42	-	nr	0.42
8 mm diameter	-	0.03	0.42	-	nr	0.42
10 mm diameter	-	0.04	0.56	-	nr	0.56
12 mm diameter	-	0.04	0.56	-	nr	0.56
16 mm diameter	-	0.05	0.70	-	nr	0.70
20 mm diameter	-	0.05	0.70	-	nr	0.70
Bolts						
Expanding bolts; "Rawlbolt" projecting type or other equal and approved; The Rawlplug Company; plated; one nut; one washer						
6 mm diameter; ref M6 10P	-	0.09	1.26	0.49	nr	1.75
6 mm diameter; ref M6 25P	-	0.09	1.26	0.55	nr	1.81

G STRUCTURAL/CARCASSING METAL/TIMBER

	PC £	Labour hours	Labour £	Material £	Unit	Total rate £
Expanding bolts; "Rawlbolt" projecting type or other equal and approved; The Rawlplug Company; plated; one nut; one washer - cont'd						
6 mm diameter; ref M6 60P	-	0.09	1.26	0.58	nr	1.84
8 mm diameter; ref M8 25P	-	0.09	1.26	0.65	nr	1.91
8 mm diameter; ref M8 60P	-	0.09	1.26	0.70	nr	1.96
10 mm diameter; ref M10 15P	-	0.09	1.26	0.85	nr	2.11
10 mm diameter; ref M10 30P	-	0.09	1.26	0.89	nr	2.15
10 mm diameter; ref M10 60P	-	0.09	1.26	0.93	nr	2.19
12 mm diameter; ref M12 15P	-	0.09	1.26	1.35	nr	2.61
12 mm diameter; ref M12 30P	-	0.10	1.40	0.13	nr	1.53
12 mm diameter; ref M12 75P	-	0.09	1.26	1.81	nr	3.07
16 mm diameter; ref M16 35P	-	0.09	1.26	3.45	nr	4.71
16 mm diameter; ref M16 75P	-	0.09	1.26	3.62	nr	4.88
Expanding bolts; "Rawlbolt" loose bolt type or other equal and approved; The Rawlplug Company; plated; one bolt; one washer						
6 mm diameter; ref M6 10L	-	0.09	1.26	0.49	nr	1.75
6 mm diameter; ref M6 25L	-	0.09	1.26	0.52	nr	1.78
6 mm diameter; ref M6 40L	-	0.09	1.26	0.52	nr	1.78
8 mm diameter; ref M8 25L	-	0.09	1.26	0.64	nr	1.90
8 mm diameter; ref M8 40L	-	0.09	1.26	0.68	nr	1.94
10 mm diameter; ref M10 10L	-	0.09	1.26	0.82	nr	2.08
10 mm diameter; ref M10 25L	-	0.09	1.26	0.84	nr	2.10
10 mm diameter; ref M10 50L	-	0.09	1.26	0.89	nr	2.15
10 mm diameter; ref M10 75L	-	0.09	1.26	0.92	nr	2.18
12 mm diameter; ref M12 10L	-	0.09	1.26	1.22	nr	2.48
12 mm diameter; ref M12 25L	-	0.09	1.26	1.35	nr	2.61
12 mm diameter; ref M12 40L	-	0.09	1.26	1.41	nr	2.67
12 mm diameter; ref M12 60L	-	0.09	1.26	1.48	nr	2.74
16 mm diameter; ref M16 30L	-	0.09	1.26	3.28	nr	4.54
16 mm diameter; ref M16 60L	-	0.09	1.26	3.55	nr	4.81
Truss clips						
Truss clips; fixing to softwood; joist size						
38 mm wide	0.21	0.14	1.96	0.38	nr	2.34
50 mm wide	0.22	0.14	1.96	0.40	nr	2.36
Sole plate angles; mild steel galvanised						
Sole plate angle; fixing to softwood and concrete						
112 mm x 40 mm x 76 mm	0.27	0.19	2.66	1.01	nr	3.67
Chemical anchors						
"Kemfix" capsules and standard studs or other equal and approved; The Ramplug Company; with nuts and washers; drilling masonry						
capsule ref 60-408; stud ref 60-448	-	0.25	3.50	1.19	nr	4.69
capsule ref 60-410; stud ref 60-454	-	0.28	3.92	1.34	nr	5.26
capsule ref 60-412; stud ref 60-460	-	0.31	4.34	1.63	nr	5.97
capsule ref 60-416; stud ref 60-472	-	0.34	4.76	2.56	nr	7.32
capsule ref 60-420; stud ref 60-478	-	0.36	5.04	3.91	nr	8.95
capsule ref 60-424; stud ref 60-484	-	0.40	5.60	5.81	nr	11.41
"Kemfix" capsules and stainless steel studs or other equal and approved; The Rawlplug Company; with nuts and washers; drilling masonry						
capsule ref 60-408; stud ref 60-905	-	0.25	3.50	2.10	nr	5.60
capsule ref 60-410; stud ref 60-910	-	0.28	3.92	2.81	nr	6.73
capsule ref 60-412; stud ref 60-915	-	0.31	4.34	3.86	nr	8.20
capsule ref 60-416; stud ref 60-920	-	0.34	4.76	6.57	nr	11.33
capsule ref 60-420; stud ref 60-925	-	0.36	5.04	10.14	nr	15.18
capsule ref 60-424; stud ref 60-930	-	0.40	5.60	16.40	nr	22.00

Prices for Measured Works - Major Works
G STRUCTURAL/CARCASSING METAL/TIMBER

	PC £	Labour hours	Labour £	Material £	Unit	Total rate £
G20 CARPENTRY/TIMBER FRAMING/FIRST FIXING - cont'd						
Chemical anchors						
"Kemfix" capsules and standard internal threaded sockets or other equal and approved; The Rawlpug Company; drilling masonry						
capsule ref 60-408; socket ref 60-650	-	0.25	3.50	1.23	nr	4.73
capsule ref 60-410; socket ref 60-656	-	0.28	3.92	1.49	nr	5.41
capsule ref 60-412; socket ref 60-662	-	0.31	4.34	1.86	nr	6.20
capsule ref 60-416; socket ref 60-668	-	0.34	4.76	2.55	nr	7.31
capsule ref 60-420; socket ref 60-674	-	0.36	5.04	3.07	nr	8.11
capsule ref 60-424; socket ref 60-676	-	0.40	5.60	4.98	nr	10.58
"Kemfix" capsules and stainless steel internal threaded sockets or other equal and approved; The Rawlplug Company; drilling masonry						
capsule ref 60-408; socket ref 60-943	-	0.25	3.50	2.33	nr	5.83
capsule ref 60-410; socket ref 60-945	-	0.28	3.92	2.64	nr	6.56
capsule ref 60-412; socket ref 60-947	-	0.31	4.34	3.03	nr	7.37
capsule ref 60-416; socket ref 60-949	-	0.34	4.76	4.32	nr	9.08
capsule ref 60-420; socket ref 60-951	-	0.36	5.04	5.21	nr	10.25
capsule ref 60-424; socket ref 60-955	-	0.40	5.60	9.53	nr	15.13
"Kemfix" capsules, perforated sleeves and standard studs or other equal and approved; The Rawlplug Company; in low density material; with nuts and washers; drilling masonry						
capsule ref 60-408; sleeve ref 60-538; stud ref 60-448	-	0.25	3.50	1.75	nr	5.25
capsule ref 60-410; sleeve ref 60-544; stud ref 60-454	-	0.28	3.92	1.93	nr	5.85
capsule ref 60-412; sleeve ref 60-550; stud ref 60-460	-	0.31	4.34	2.31	nr	6.65
capsule ref 60-416; sleeve ref 60-562; stud ref 60-472	-	0.34	4.76	3.40	nr	8.16
"Kemfix" capsules, perforated sleeves and stainless steel studs or other equal and approved; The Rawlplug Company; in low density material; with nuts and washers; drilling masonry						
capsule ref 60-408; sleeve ref 60-538; stud ref 60-905	-	0.25	3.50	2.66	nr	6.16
capsule ref 60-410; sleeve ref 60-544; stud ref 60-910	-	0.28	3.92	3.40	nr	7.32
capsule ref 60-412; sleeve ref 60-550; stud ref 60-915	-	0.31	4.34	4.54	nr	8.88
capsule ref 60-416; sleeve ref 60-562; stud ref 60-920	-	0.34	4.76	7.40	nr	12.16
"Kemfix" capsules, perforated sleeves and standard internal threaded sockets or other equal and approved; The Rawlplug Company; in low density material; with nuts and washers; drilling masonry						
capsule ref 60-408; sleeve ref 60-538; socket ref 60-650	-	0.25	3.50	1.79	nr	5.29
capsule ref 60-410; sleeve ref 60-544; socket ref 60-656	-	0.28	3.92	2.08	nr	6.00
capsule ref 60-412; sleeve ref 60-550; socket ref 60-662	-	0.31	4.34	2.54	nr	6.88
capsule ref 60-416; sleeve ref 60-562; socket ref 60-668	-	0.34	4.76	3.38	nr	8.14

	PC £	Labour hours	Labour £	Material £	Unit	Total rate £
"Kemfix" capsules, perforated sleeves and stainless steel internal threaded sockets or other equal and approved; The Rawlplug Company; in low density material; drilling masonry						
capsule ref 60-408; sleeve ref 60-538; socket ref 60-943	-	0.25	3.50	2.90	nr	6.40
capsule ref 60-410; sleeve ref 60-544; socket ref 60-945	-	0.28	3.92	3.23	nr	7.15
capsule ref 60-412; sleeve ref 60-550; socket ref 60-947	-	0.31	4.34	3.71	nr	8.05
capsule ref 60-416; sleeve ref 60-562; socket ref 60-949	-	0.34	4.76	5.15	nr	9.91
G32 EDGE SUPPORTED/REINFORCED WOODWOOL SLAB DECKING						
Woodwool interlocking reinforced slabs; Torvale "Woodcelip" or other equal and approved; natural finish; fixing to timber or steel with galvanized nails or clips; flat or sloping						
50 mm thick slabs; type 503; maximum span 2100 mm						
1800 mm - 2100 mm lengths	13.59	0.46	6.44	15.53	m²	21.97
2400 mm lengths	14.24	0.46	6.44	16.27	m²	22.71
2700 mm - 3000 mm lengths	14.47	0.46	6.44	16.53	m²	22.97
75 mm thick slabs; type 751; maximum span 2100 mm						
1800 mm - 2400 mm lengths	20.37	0.51	7.14	23.22	m²	30.36
2700 mm - 3000 mm lengths	20.47	0.51	7.14	23.34	m²	30.48
75 mm thick slabs; type 752; maximum span 2100 mm						
1800 mm - 2400 mm lengths	20.31	0.51	7.14	23.16	m²	30.30
2700 mm - 3000 mm lengths	20.35	0.51	7.14	23.20	m²	30.34
75 mm thick slabs; type 753; maximum span 3600 mm						
2400 mm lengths	19.71	0.51	7.14	22.48	m²	29.62
2700 mm - 3000 mm lengths	20.56	0.51	7.14	23.44	m²	30.58
3300 mm - 3900 mm lengths	24.78	0.51	7.14	28.20	m²	35.34
extra for holes for pipes and the like	-	0.12	1.68	-	nr	1.68
100 mm thick slabs; type 1001; maximum span 3600 mm						
3000 mm lengths	27.12	0.56	7.84	30.88	m²	38.72
3300 mm - 3600 mm lengths	29.35	0.56	7.84	33.40	m²	41.24
100 mm thick slabs; type 1002; maximum span 3600 mm						
3000 mm lengths	26.48	0.56	7.84	30.16	m²	38.00
3300 mm - 3600 mm lengths	28.12	0.56	7.84	32.01	m²	39.85
100 mm thick slabs; type 1003; maximum span 4000 mm						
3000 mm - 3600 mm lengths	25.15	0.56	7.84	28.65	m²	36.49
3900 mm - 4000 mm lengths	25.15	0.66	7.84	28.65	m²	36.49
125 mm thick slabs; type 1252; maximum span 3000 mm						
2400 - 3000 lengths	28.62	0.56	7.84	32.57	m²	40.41
Extra over slabs for						
pre-screeded deck	-	-	-	1.02	m²	1.02
pre-screeded soffit	-	-	-	2.62	m²	2.62
pre-screeded deck and soffit	-	-	-	3.42	m²	3.42
pre-screeded and proofed deck	-	-	-	2.11	m²	2.11
pre-screeded and proofed deck plus pre-screeded soffit	-	-	-	4.90	m²	4.90
pre-felted deck (glass fibre)	-	-	-	2.88	m²	2.88
pre-felted deck plus pre-screeded soffit	-	-	-	5.50	m²	5.50

H CLADDING/COVERING

	PC £	Labour hours	Labour £	Material £	Unit	Total rate £
H CLADDING/COVERING						
H10 PATENT GLAZING						
Patent glazing; aluminium alloy bars 2.55 m long at 622 mm centres; fixed to supports						
Roof cladding						
glazing with 7 mm thick Georgian wired cast glass	-	-	-	-	m²	109.85
Associated code 4 lead flashings						
top flashing; 210 mm girth	-	-	-	-	m	34.96
bottom flashing; 240 mm girth	-	-	-	-	m	28.47
end flashing; 300 mm girth	-	-	-	-	m	64.91
Wall cladding						
glazing with 7 mm thick Georgian wired cast glass	-	-	-	-	m²	114.84
glazing with 6 mm thick plate glass	-	-	-	-	m²	104.87
Extra for aluminium alloy members						
38 mm x 38 mm x 3 mm angle jamb	-	-	-	-	m	14.99
pressed cill member	-	-	-	-	m	32.45
pressed channel head and PVC case	-	-	-	-	m	27.47
"Kawneer" window frame system or other equal and approved; polyester powder coated glazing bars; glazed with double hermetically sealed units in toughened safety glass; one 6 mm thick air space; overall 18 mm thick						
Vertical surfaces						
single tier; aluminium glazing bars at 850 mm centres x 890 mm long; timber supports at 890 mm centres	-	-	-	-	m²	189.75
"Kawneer" window frame system or other equal and approved; polyester powder coated glazing bars; glazed with clear toughened safety glass; 10.70 mm thick						
Vertical surfaces						
single tier; aluminium glazing bars at 850 mm centres x 890 mm long; timber supports at 890 mm centres	-	-	-	-	m²	179.76
H14 CONCRETE ROOFLIGHTS/PAVEMENT LIGHTS						
Reinforced concrete rooflights/pavement lights; "Luxcrete" or other equal and approved; with glass lenses; supplied and fixed complete						
Rooflights						
2.50 kN/m² loading; ref R254/125	-	-	-	-	m²	503.51
2.50 kN/m² loading; ref R200/90	-	-	-	-	m²	534.71
2.50 kN/m² loading; home office; double glazed	-	-	-	-	m²	548.07
Pavement lights						
20.00 kN/m²; pedestrian traffic; ref. P150/100	-	-	-	-	m²	441.14
60.00 kN/m²; vehicular traffic; ref. P165/165	-	-	-	-	m²	454.50
brass terrabond	-	-	-	-	m²	9.35
150 x 75 identification plates	-	-	-	-	m²	22.28
escape hatch	-	-	-	-	m²	4558.37

H CLADDING/COVERING

	PC £	Labour hours	Labour £	Material £	Unit	Total rate £
H20 RIGID SHEET CLADDING						
"Resoplan" sheet or other equal and approved; Eternit UK Ltd; flexible neoprene gasket joints; fixing with stainless steel screws and coloured caps						
6 mm thick cladding to walls						
over 300 mm wide	-	1.94	27.15	46.04	m²	73.19
not exceeding 300 mm wide	-	0.65	9.10	16.11	m	25.21
Eternit 2000 "Glasal" sheet or other equal and approved; Eternit UK Ltd; flexible neoprene gasket joints; fixing with stainless steel screws and coloured caps						
7.50 mm thick cladding to walls						
over 300 mm wide	-	1.94	27.15	42.07	m²	69.22
not exceeding 300 mm wide	-	0.65	9.10	14.92	m	24.02
external angle trim	-	0.09	1.26	7.70	m	8.96
7.50 mm thick cladding to eaves, verge soffit boards, fascia boards or the like						
100 mm wide	-	0.46	6.44	7.22	m	13.66
200 mm wide	-	0.56	7.84	11.07	m	18.91
300 mm wide	-	0.65	9.10	14.92	m	24.02
H30 FIBRE CEMENT PROFILED SHEET CLADDING/COVERING/SIDING						
Asbestos-free corrugated sheets; Eternit "2000" or other equal and approved						
Roof cladding; sloping not exceeding 50 degrees; fixing to timber purlins with drive screws						
"Profile 3"; natural grey	8.78	0.19	3.36	11.83	m²	15.19
"Profile 3"; coloured	10.53	0.19	3.36	13.80	m²	17.16
"Profile 6"; natural grey	7.74	0.23	4.06	10.59	m²	14.65
"Profile 6"; coloured	9.29	0.23	4.06	12.34	m²	16.40
"Profile 6"; natural grey; insulated 60 glass fibre infill; lining panel	-	0.42	7.42	25.95	m²	33.37
Roof cladding; sloping not exceeding 50 degrees; fixing to steel purlins with hook bolts						
"Profile 3"; natural grey	-	0.23	4.06	12.79	m²	16.85
"Profile 3"; coloured	-	0.23	4.06	14.77	m²	18.83
"Profile 6"; natural grey	-	0.28	4.95	11.55	m²	16.50
"Profile 6"; coloured	-	0.28	4.95	13.30	m²	18.25
"Profile 6"; natural grey; insulated 60 glass fibre infill; lining panel	-	0.46	8.13	23.65	m²	31.78
Wall cladding; vertical; fixing to steel rails with hook bolts						
"Profile 3"; natural grey	-	0.28	4.95	12.79	m²	17.74
"Profile 3"; coloured	-	0.28	4.95	14.77	m²	19.72
"Profile 6"; natural grey	-	0.32	5.66	11.55	m²	17.21
"Profile 6"; coloured	-	0.32	5.66	13.30	m²	18.96
"Profile 6"; natural grey; insulated 60 glass fibre infill; lining panel	-	0.51	9.01	23.65	m²	32.66
raking cutting	-	0.14	2.47	-	m	2.47
holes for pipe and the like	-	0.14	2.47	-	nr	2.47
Accessories; to "Profile 3" cladding; natural grey						
eaves filler	-	0.09	1.59	9.07	m	10.66
vertical corrugation closer	-	0.11	1.94	9.07	m	11.01
apron flashing	-	0.11	1.94	9.13	m	11.07
plain wing or close fitting two piece adjustable capping to ridge	-	0.16	2.83	8.53	m	11.36
ventilating two piece adjustable capping to ridge	-	0.16	2.83	12.87	m	15.70

Prices for Measured Works - Major Works
H CLADDING/COVERING

	PC £	Labour hours	Labour £	Material £	Unit	Total rate £
H30 FIBRE CEMENT PROFILED SHEET CLADDING/COVERING/SIDING - cont'd						
Asbestos-free corrugated sheets; Eternit "2000" or other equal and approved - cont'd						
Accessories; to "Profile 3" cladding; coloured						
eaves filler	-	0.09	1.59	11.35	m	12.94
vertical corrugation closer	-	0.11	1.94	11.35	m	13.29
apron flashing	-	0.11	1.94	11.39	m	13.33
plain wing or close fitting two piece adjustable capping to ridge	-	0.16	2.83	10.28	m	13.11
ventilating two piece adjustable capping to ridge	-	0.16	2.83	16.07	m	18.90
Accessories; to "Profile 6" cladding; natural grey						
eaves filler	-	0.09	1.59	6.73	m	8.32
vertical corrugation closer	-	0.11	1.94	6.73	m	8.67
apron flashing	-	0.11	1.94	7.25	m	9.19
underglazing flashing	-	0.11	1.94	6.92	m	8.86
plain cranked crown to ridge	-	0.16	2.83	13.58	m	16.41
plain wing or close fitting two piece adjustable capping to ridge	-	0.16	2.83	13.27	m	16.10
ventilating two piece adjustable capping to ridge	-	0.16	2.83	16.94	m	19.77
Accessories; to "Profile 6" cladding; coloured						
eaves filler	-	0.09	1.59	8.41	m	10.00
vertical corrugation closer	-	0.11	1.94	8.41	m	10.35
apron flashing	-	0.11	1.94	9.06	m	11.00
underglazing flashing	-	0.11	1.94	8.66	m	10.60
plain cranked crown to ridge	-	0.16	2.83	17.00	m	19.83
plain wing or close fitting two piece adjustable capping to ridge	-	0.16	2.83	16.59	m	19.42
ventilating two piece adjustable capping to ridge	-	0.16	2.83	21.21	m	24.04
H31 METAL PROFILED/FLAT SHEET CLADDING/COVERING/SIDING						
Galvanised steel strip troughed sheets; PMF Strip Mill Products or other equal and approved; colorcoat "Plastisol" finish						
Roof cladding; sloping not exceeding 50 degrees; fixing to steel purlins with plastic headed self-tapping screws						
0.70 mm thick type 12.5; 75 mm corrugated	6.36	0.32	5.66	10.03	m²	15.69
0.70 mm thick Long Rib 1000; 35 mm deep	6.41	0.37	6.54	10.09	m²	16.63
Wall cladding; vertical; fixing to steel rails with plastic headed self-tapping screws						
0.70 mm thick type 12.5; 75 mm corrugated	-	0.37	6.54	10.03	m²	16.57
0.70 mm thick Scan Rib 1000; 19 mm deep	-	0.42	7.42	9.98	m²	17.40
raking cutting	-	0.20	3.53	-	m	3.53
holes for pipes and the like	-	0.37	6.54	-	nr	6.54
Accessories; colorcoat silicone polyester finish 0.90 thick standard flashings; bent to profile						
250 mm girth	-	0.19	3.36	3.72	m²	7.08
375 mm girth	-	0.20	3.53	4.98	m²	8.51
500 mm girth	-	0.22	3.89	6.33	m²	10.22
625 mm girth	-	0.28	4.95	7.67	m²	12.62
Galvanised steel profile sheet cladding; Kelsey Roofing Industries Ltd or other equal and approved						
Roof cladding; sloping not exceeding 50 degrees; fixing to steel purlins with plastic headed self-tapping screws						
0.70 mm thick	-	-	-	-	m²	13.93
0.70 mm thick; PVF2 coated soffit finish	-	-	-	-	m²	20.22

H CLADDING/COVERING

	PC £	Labour hours	Labour £	Material £	Unit	Total rate £
Galvanised steel profile sheet cladding; Plannja Ltd or other equal and approved; metallack coated finish						
Roof cladding; sloping not exceeding 50 degrees; fixing to steel purlins with plastic headed self-tapping screws						
0.72 mm thick "profile 20B"	-	-	-	-	m^2	15.60
0.72 mm thick "profile TOP 40"	-	-	-	-	m^2	16.49
0.72 mm thick "profile 45"	-	-	-	-	m^2	17.82
extra over for 80 mm insulation and 0.40 mm thick coated inner lining sheet	-	-	-	-	m^2	12.03
Wall cladding; vertical fixing to steel rails with plastic headed self-tapping screws						
0.60 mm thick "profile 20B"; corrugations vertical	-	-	-	-	m^2	15.15
0.60 mm thick "profile 30"; corrugations vertical	-	-	-	-	m^2	16.04
0.60 mm thick "profile TOP 40"; corrugations vertical	-	-	-	-	m^2	17.07
0.60 mm thick "profile 60B"; corrugations vertical	-	-	-	-	m^2	18.89
0.60 mm thick "profile 30"; corrugations horizontal	-	-	-	-	m^2	16.49
0.60 mm thick "profile 60B"; corrugations horizontal	-	-	-	-	m^2	20.32
extra for 80 mm insulation and 0.40 mm thick coated inner lining sheet	-	-	-	-	m^2	12.48
Accessories for roof/vertical cladding; PVF2 coated finish; 0.60 mm thick flashings; once bent						
250 mm girth	-	-	-	-	m	11.14
375 mm girth	-	-	-	-	m	13.81
500 mm girth	-	-	-	-	m	15.60
625 mm girth	-	-	-	-	m	17.38
extra bends	-	-	-	-	nr	0.19
profile fillers	-	-	-	-	m	0.71
Lightweight galvanised steel roof tiles; Decra Roof Systems (UK) Ltd or other equal and approved; coated finish						
Roof coverings	-	0.23	4.06	12.53	m^2	16.59
Accessories for roof cladding						
pitched "D" ridge	-	0.09	1.59	6.73	m	8.32
barge cover (handed)	-	0.09	1.59	7.23	m	8.82
in line air vent	-	0.09	1.59	33.16	nr	34.75
in line soil vent	-	0.09	1.59	51.97	nr	53.56
gas flue terminal	-	0.19	3.36	68.59	nr	71.95
Standing seam aluminium roof cladding; "Kal-zip" Hoogovens Aluminium Building Systems Ltd or other equal and approved; ref AA 3004 A1 Mn1 Mg1; standard natural aluminium stucco embossed finish						
Roof coverings (lining sheets not included); sloping not exceeding 50 degrees; 305 mm wide units						
0.90 mm thick	-	-	-	-	m^2	25.81
1.00 mm thick	-	-	-	-	m^2	28.26
1.20 mm thick	-	-	-	-	m^2	33.41
extra over for						
polyester coating	-	-	-	-	m^2	5.84
PVF2 coating	-	-	-	-	m^2	8.23
smooth curved	-	-	-	-	m^2	30.96
factory formed tapered sheets	-	-	-	-	m^2	37.10
non standard lengths	-	-	-	-	m^2	1.55
raking cutting	-	-	-	-	m	6.44

Prices for Measured Works - Major Works
H CLADDING/COVERING

	PC £	Labour hours	Labour £	Material £	Unit	Total rate £
H31 METAL PROFILED/FLAT SHEET CLADDING/COVERING/SIDING - cont'd						
Standing seam aluminium roof cladding; "Kal-zip" Hoogovens Aluminium Building Systems Ltd or other equal and approved; ref AA 3004 A1 Mn1 Mg1; standard natural aluminium stucco embossed finish - cont'd						
Accessories for roof coverings						
single skin PVC rooflights; 305 mm cover	-	-	-	-	m²	93.37
double skin PVC rooflights; 305 mm cover	-	-	-	-	m²	115.35
thermal insulation quilt; 60 mm thick	-	-	-	-	m²	1.49
thermal insulation quilt; 80 mm thick	-	-	-	-	m²	1.70
thermal insulation quilt; 100 mm thick	-	-	-	-	m²	1.85
thermal insulation quilt; 150 mm thick	-	-	-	-	m²	2.85
semi-rigid insulation slab; 30 mm thick; tissue faced	-	-	-	-	m²	5.79
semi-rigid insulation slab; 50 mm thick	-	-	-	-	m²	10.99
"Kal-Foil" vapour check	-	-	-	-	m²	3.10
Lining sheets; "Kal-bau" aluminium ref TR 30/150; natural aluminium stucco embossed finish; fixed to steel purlins with stainless steel screws						
0.70 mm gauge	-	-	-	-	m²	14.53
0.80 mm gauge	-	-	-	-	m²	16.33
0.90 mm gauge	-	-	-	-	m²	18.12
1.00 mm gauge	-	-	-	-	m²	19.77
1.20 mm gauge	-	-	-	-	m²	23.42
extra over for						
crimped curve	-	-	-	-	m²	6.19
perforated sheet	-	-	-	-	m²	3.34
Lining sheets; "Kal-bau" aluminium ref TR 35/200; natural aluminium stucco embossed finish; fixed to steel purlins with stainless steel screws						
0.70 mm gauge	-	-	-	-	m²	14.33
0.80 mm gauge	-	-	-	-	m²	16.08
0.90 mm gauge	-	-	-	-	m²	17.63
1.00 mm gauge	-	-	-	-	m²	19.58
1.20 mm gauge	-	-	-	-	m²	23.07
extra over for						
crimped curve	-	-	-	-	m²	6.10
perforated sheet	-	-	-	-	m²	3.34
Lining sheets; profiled steel; 915 mm cover; bright white polyester paint finish; fixed to steel purlins with stainless steel screws						
0.40 mm gauge	-	-	-	-	m²	4.64
0.55 mm gauge	-	-	-	-	m²	8.04
0.70 mm gauge	-	-	-	-	m²	9.34
extra over for						
crimped curve	-	-	-	-	m²	3.49
perforated sheet	-	-	-	-	m²	3.34
tapered 0.70 mm plain galvanised steel liner	-	-	-	-	m²	20.18
plastisol coating on external liners	-	-	-	-	m²	2.00
Lining sheets; profiled steel; 1000 mm cover; bright white polyester paint finish; fixed to steel purlins with stainless steel screws						
0.40 mm gauge	-	-	-	-	m²	6.69
0.55 mm gauge	-	-	-	-	m²	8.04
0.70 mm gauge	-	-	-	-	m²	9.34
extra over for						
crimped curve	-	-	-	-	m²	3.49
perforated sheet	-	-	-	-	m²	3.34
tapered 0.70 mm plain galvanised steel liner	-	-	-	-	m²	20.18

	PC £	Labour hours	Labour £	Material £	Unit	Total rate £
Eaves detail for 305 mm wide "Kal-zip" roof cladding units; including high density polythylene foam fillers; 2 mm extruded alloy drip angle; fixed to Kalzip sheet using stainless steel blind rivets						
40 mm x 20 mm angle: single skin	-	-	-	-	m	6.54
70 mm x 30 mm angle: single skin	-	-	-	-	m	8.44
40 mm x 20 mm angle: double skin	-	-	-	-	m	8.04
70 mm x 30 mm angle: double skin	-	-	-	-	m	9.89
Verge detail for 305 mm wide "Kal-zip" roof cladding units 2 mm extruded aluminium alloy gable closure section; fixed with stainless steel blind sealed rivets; gable end hook/verge clip fixed to ST clip with stainless steel screws and 2 mm extruded aluminium tolerance clip	-	-	-	-	m	8.84
Ridge detail for 305 mm wide "Kal-zip" roof cladding units abutment ridge including natural aluminium stucco embossed "U" type ridge closures fixed with stainless steel blind steel rivets; "U" type polythylene ridge fillers and 2 mm extruded aluminium aluminium alloy support Zed fixed with stainless steel blind sealed rivets; fixing with rivets through small seam of "Kal-zip" into ST clip using stainless steel blind sealed rivets	-	-	-	-	m	14.18
duo ridge including natural aluminium stucco embossed "U" type ridge closures fixed with stainless steel blind sealed rivets; "U" type polythylene ridge fillers and 2 mm extruded aluminium alloy support Zed fixed with stainless steel blind sealed rivets; fixing with rivets through small seam of "Kal-zip" into ST clip using stainless steel blind steel rivets	-	-	-	-	m	28.26
H32 PLASTICS PROFILED SHEET CLADDING/COVERING/SIDING						
Extended, hard skinned, foamed PVC-UE profiled sections; Swish Celuka or other equal and approved; Class 1 fire rated to BS 476; Part 7; 1987; in white finish						
Wall cladding; vertical; fixing to timber						
100 mm shiplap profiles; Code 001	-	0.35	4.90	32.71	m²	37.61
150 mm shiplap profiles; Code 002	-	0.32	4.48	29.08	m²	33.56
125 mm feather-edged profiles; Code C208	-	0.34	4.76	31.67	m²	36.43
Vertical angles	-	0.19	2.66	3.50	m	6.16
Raking cutting	-	0.14	1.96	-	m	1.96
Holes for pipes and the like	-	0.03	0.42	-	nr	0.42
H41 GLASS REINFORCED PLASTICS PANEL CLADDING/FEATURES						
Glass fibre translucent sheeting grade AB class 3						
Roof cladding; sloping not exceeding 50 degrees; fixing to timber purlins with drive screws; to suit						
"Profile 3" or other equal and approved	9.44	0.18	3.18	12.03	m²	15.21
"Profile 6" or other equal and approved	8.81	0.23	4.06	11.32	m²	15.38
Roof cladding; sloping not exceeding 50 degrees; fixing to timber purlins with hook bolts; to suit						
"Profile 3" or other equal and approved	9.44	0.23	4.06	13.00	m²	17.06
"Profile 6" or other equal and approved	8.81	0.28	4.95	12.28	m²	17.23
"Longrib 1000" or other equal and approved	8.65	0.28	4.95	12.10	m²	17.05

H51 NATURAL STONE SLAB CLADDING FEATURES

BASIC NATURAL STONE BLOCK PRICES

Block prices

	(£/m3)		(£/m3)
Monks Park bed ht n.e. 300 mm	235.00	Beer Stone - all sizes	545.00
Monks Park bed ht 300 - 599 mm	335.00	Portland Stone n.e. 0.60 m³	220.00
Monks Park bed ht 600 mm +	335.00	Portland Stone 0.60 - 0.99 m³	415.00
Westwood Ground n.e. 0.60 m³	230.00	Portland Stone 1.00 m³ +	495.00
Westwood Ground 0.60 - 0.99 m³	335.00		
Westwood Ground 1.00 m³ +	335.00		

	PC £	Labour hours	Labour £	Material £	Unit	Total rate £
Portland Whitbed limestone bedded and jointed in cement - lime - mortar (1:2:9); slurrying with weak lime and stone dust mortar; flush pointing and cleaning on completion (cramps etc. measured separately)						
Facework; one face plain and rubbed; bedded against backing						
50 mm thick stones	-	-	-	-	m²	162.28
63 mm thick stones	-	-	-	-	m²	209.72
75 mm thick stones	-	-	-	-	m²	209.72
100 mm thick stones	-	-	-	-	m²	234.68
Fair returns on facework						
50 mm wide	-	-	-	-	m	1.05
63 mm wide	-	-	-	-	m	1.49
75 mm wide	-	-	-	-	m	1.49
100 mm wide	-	-	-	-	m	1.95
Fair raking cutting on facework						
50 mm thick	-	-	-	-	m	15.48
63 mm thick	-	-	-	-	m	17.58
75 mm thick	-	-	-	-	m	19.67
100 mm thick	-	-	-	-	m	23.62
Copings; once weathered; and throated; rubbed; set horizontal or raking						
250 mm x 50 mm	-	-	-	-	m	109.85
extra for external angle	-	-	-	-	nr	16.78
extra for internal angle	-	-	-	-	nr	25.71
300 mm x 50 mm	-	-	-	-	m	114.84
extra for external angle	-	-	-	-	nr	20.97
extra for internal angle	-	-	-	-	nr	31.45
350 mm x 75 mm	-	-	-	-	m	159.79
extra for external angle	-	-	-	-	nr	23.62
extra for internal angle	-	-	-	-	nr	35.66
400 mm x 100 mm	-	-	-	-	m	184.76
extra for external angle	-	-	-	-	nr	27.26
extra for internal angle	-	-	-	-	nr	40.89
450 mm x 100 mm	-	-	-	-	m	199.74
extra for external angle	-	-	-	-	nr	31.45
extra for internal angle	-	-	-	-	nr	47.19
500 mm x 125 mm	-	-	-	-	m	249.67
extra for external angle	-	-	-	-	nr	34.60
extra for internal angle	-	-	-	-	nr	52.43

Prices for Measured Works - Major Works
H CLADDING/COVERING

	PC £	Labour hours	Labour £	Material £	Unit	Total rate £
Band courses; plain; rubbed; horizontal						
225 mm x 112 mm	-	-	-	-	m	78.65
300 mm x 112 mm	-	-	-	-	m	94.37
extra for stopped ends	-	-	-	-	nr	0.70
extra for external angles	-	-	-	-	nr	0.70
Band courses; moulded 100 mm girth on face; rubbed; horizontal						
125 mm x 75 mm	-	-	-	-	m	99.87
extra for stopped ends	-	-	-	-	nr	0.70
extra for external angles	-	-	-	-	nr	9.49
extra for internal angles	-	-	-	-	nr	18.97
150 mm x 75 mm	-	-	-	-	m	109.85
extra for stopped ends	-	-	-	-	nr	0.70
extra for external angles	-	-	-	-	nr	12.59
extra for internal angles	-	-	-	-	nr	25.47
200 mm x 100 mm	-	-	-	-	m	139.81
extra for stopped ends	-	-	-	-	nr	0.70
extra for external angles	-	-	-	-	nr	21.47
extra for internal angles	-	-	-	-	nr	40.95
250 mm x 150 mm	-	-	-	-	m	179.76
extra for stopped ends	-	-	-	-	nr	0.70
extra for external angles	-	-	-	-	nr	28.32
extra for internal angles	-	-	-	-	nr	50.43
300 mm x 250 mm	-	-	-	-	m	264.65
extra for stopped ends	-	-	-	-	nr	0.70
extra for external angles	-	-	-	-	nr	38.45
extra for internal angles	-	-	-	-	nr	68.16
Coping apex block; two sunk faces; rubbed						
650 mm x 450 mm x 225 mm	-	-	-	-	nr	284.62
Coping kneeler block; three sunk faces; rubbed						
350 mm x 350 mm x 375 mm	-	-	-	-	nr	209.72
450 mm x 450 mm x 375 mm	-	-	-	-	nr	299.60
Corbel; turned and moulded; rubbed						
225 mm x 225 mm x 375 mm	-	-	-	-	nr	149.8
Slab surrounds to openings; one face splayed; rubbed						
75 mm x 100 mm	-	-	-	-	m	57.67
75 mm x 200 mm	-	-	-	-	m	78.65
100 mm x 100 mm	-	-	-	-	m	68.16
125 mm x 100 mm	-	-	-	-	m	73.40
125 mm x 150 mm	-	-	-	-	m	89.13
175 mm x 175 mm	-	-	-	-	m	114.84
225 mm x 175 mm	-	-	-	-	m	129.83
300 mm x 175 mm	-	-	-	-	m	159.79
300 mm x 225 mm	-	-	-	-	m	199.74
Slab surrounds to openings; one face sunk splayed; rubbed						
75 mm x 100 mm	-	-	-	-	m	69.66
75 mm x 200 mm	-	-	-	-	m	90.63
100 mm x 100 mm	-	-	-	-	m	80.14
125 mm x 100 mm	-	-	-	-	m	85.39
125 mm x 150 mm	-	-	-	-	m	101.11
175 mm x 175 mm	-	-	-	-	m	126.83
225 mm x 175 mm	-	-	-	-	m	141.81
300 mm x 175 mm	-	-	-	-	m	171.77
300 mm x 225 mm	-	-	-	-	m	211.72
extra for throating	-	-	-	-	m	6.04
extra for rebates and grooves	-	-	-	-	m	18.48
extra for stooling	-	-	-	-	m	16.97

Prices for Measured Works - Major Works
H CLADDING/COVERING

	PC £	Labour hours	Labour £	Material £	Unit	Total rate £
H51 NATURAL STONE SLAB CLADDING FEATURES - cont'd						
Sundries - stone walling						
Coating backs of stones with brush applied cold bitumen solution; two coats						
limestone facework	-	0.19	1.68	1.90	m²	3.58
Cutting grooves in limestone masonry for						
water bars or the like	-	-	-	-	m	5.79
Mortices in limestone masonry for						
metal dowel	-	-	-	-	nr	0.55
metal cramp	-	-	-	-	nr	1.05
"Eurobrick" insulated brick cladding systems or other equal and approved ; extruded polystyrene foam insulation; brick slips bonded to insulation panels with "Eurobrick" gun applied adhesive or other equal and approved; pointing with formulated mortar grout						
25 mm insulation to walls						
over 300 mm wide; fixing with proprietary screws and plates to timber	-	1.39	22.55	39.80	m²	62.35
50 mm insulation to walls						
over 300 mm wide; fixing with proprietary screws and plates; to timber	-	1.39	22.55	43.99	m²	66.54
Cramps and dowels; Harris and Edgar's "Hemax" range or other equal and approved; one end built into brickwork or set in slot in concrete; stainless steel						
Dowel						
8 mm diameter x 75 mm long	0.18	0.04	0.70	0.20	nr	0.90
10 mm diameter x 150 mm long	0.52	0.04	0.70	0.59	nr	1.29
Pattern "J" tie						
25 mm x 3 mm x 100 mm	0.39	0.06	1.05	0.44	nr	1.49
Pattern "S" cramp; with two 20 mm turndowns (190 mm girth)						
25 mm x 3 mm x 150 mm	0.54	0.06	1.05	0.61	nr	1.66
Pattern "B" anchor; with 8 mm x 75 mm loose dowel						
25 mm x 3 mm x 150 mm	0.67	0.09	1.57	0.76	nr	2.33
Pattern "Q" tie						
25 mm x 3 mm x 200 mm	0.67	0.06	1.05	0.76	nr	1.81
38 mm x 3 mm x 250 mm	1.37	0.06	1.05	1.54	nr	2.59
Pattern "P" half twist tie						
25 mm x 3 mm x 200 mm	0.73	0.06	1.05	0.82	nr	1.87
38 mm x 3 mm x 250 mm	1.15	0.06	1.05	1.30	nr	2.35

H60 CLAY/CONCRETE ROOF TILING

ALTERNATIVE TILE PRICES (£/1000)

CLAY TILES; INTERLOCKING AND PANTILES

Langley's "Sterreberg" pantiles
	£		£		£
Anthracite	1098.50	Glazed Brown	1166.50	Natural Red/Rustic	808.90

Sandtoft pantiles
	£		£		£
The Barrow Bold Roman	1368.20	Gaelic	1033.60	County Interlocking	704.10

William Blyth pantiles
	£		£
Barco Bold Roll	565.89	Celtic (French)	641.65

CONCRETE TILES; PLAIN AND INTERLOCKING

Marley roof tiles
	£		£
Anglia Plus	632.00	Ludlow Plus	514.00
Ludlow Major	724.00	Plain	318.00

Redland roof tiles
	£		£
Redland 49	437.00	50 Double Roman	623.20
Mini Stoneworld	675.10	Grovebury	655.10

Discounts of 2½ - 15% available depending on quantity/status

	PC £	Labour hours	Labour £	Material £	Unit	Total rate £
NOTE: The following items of tile roofing unless otherwise described, include for conventional fixing assuming normal exposure with appropriate nails and/or rivets or clips to pressure impregnated softwood battens fixed with galvanised nails; prices also include for all bedding and pointing at verges, beneath ridge tiles, etc.						
Clay interlocking pantiles; Sandtoft Goxhill "Tudor" red sand faced or other equal and approved; PC £933.70/1000; 470 mm x 285 mm; to 100 mm lap; on 25 mm x 38 mm battens and type 1F reinforced underlay						
Roof coverings	-	0.37	6.54	12.32	m^2	18.86
Extra over coverings for						
fixing every tile	-	0.07	1.24	1.07	m^2	2.31
eaves course with plastic filler	-	0.28	4.95	5.12	m	10.07
verges; extra single undercloak course of plain tiles	-	0.28	4.95	7.23	m	12.18
open valleys; cutting both sides	-	0.17	3.00	3.92	m	6.92
ridge tiles	-	0.56	9.90	10.63	m	20.53
hips; cutting both sides	-	0.69	12.19	14.55	m	26.74
holes for pipes and the like	-	0.19	3.36	-	nr	3.36

H CLADDING/COVERING

H60 PLAIN ROOF TILING - cont'd

	PC £	Labour hours	Labour £	Material £	Unit	Total rate £
Clay pantiles; Sandtoft Goxhill "Old English"; red sand faced or other equal and approved; PC £628.10/1000; 342 mm x 241 mm; to 75 mm lap; on 25 mm x 38 mm battens and type 1F reinforced underlay						
Roof coverings	-	0.42	7.42	14.04	m²	21.46
Extra over coverings for						
fixing every tile	-	0.02	0.35	2.68	m²	3.03
other colours	-	-	-	0.49	m²	0.49
double course at eaves	-	0.31	5.48	3.53	m	9.01
verges; extra single undercloak course of plain tiles	-	0.28	4.95	7.62	m	12.57
open valleys; cutting both sides	-	0.17	3.00	2.64	m	5.64
ridge tiles; tile slips	-	0.56	9.90	15.90	m	25.80
hips; cutting both sides	-	0.69	12.19	18.54	m	30.73
holes for pipes and the like	-	0.19	3.36	-	nr	3.36
Clay pantiles; William Blyth's "Lincoln" natural or other equal and approved; 343 mm x 280 mm; to 75 mm lap; PC £802.10/1000; on 19 mm x 38 mm battens and type 1F reinforced underlay						
Roof coverings	-	0.42	7.42	16.54	m²	23.96
Extra over coverings for						
fixing every tile	-	0.02	0.35	2.68	m²	3.03
other colours	-	-	-	1.45	m²	1.45
double course at eaves	-	0.31	5.48	4.26	m	9.74
verges; extra single undercloak course of plain tiles	-	0.28	4.95	10.42	m	15.37
open valleys; cutting both sides	-	0.17	3.00	3.37	m	6.37
ridge tiles; tile slips	-	0.56	9.90	17.77	m	27.67
hips; cutting both sides	-	0.69	12.19	21.14	m	33.33
holes for pipes and the like	-	0.19	3.36	-	nr	3.36
Clay plain tiles; Hinton, Perry and Davenhill "Dreadnought" smooth red machine-made or other equal and approved; PC £198.80/1000; 265 mm x 165 mm; on 19 mm x 38 mm battens and type 1F reinforced underlay						
Roof coverings; to 64 mm lap	-	0.97	17.14	18.68	m²	35.82
Wall coverings; to 38 mm lap	-	1.16	20.50	15.96	m²	36.46
Extra over coverings for						
other colours	-	-	-	6.86	m²	6.86
ornamental tiles	-	-	-	10.29	m²	10.29
double course at eaves	-	0.23	4.06	2.30	m	6.36
verges	-	0.28	4.95	1.25	m	6.20
swept valleys; cutting both sides	-	0.60	10.60	25.97	m	36.57
bonnet hips; cutting both sides	-	0.74	13.08	26.04	m	39.12
external vertical angle tiles; supplementary nail fixings	-	0.37	6.54	29.09	m	35.63
half round ridge tiles	-	0.56	9.90	8.76	m	18.66
holes for pipes and the like	-	0.19	3.36	-	nr	3.36
Clay plain tiles; Keymer best hand-made sand-faced tiles or other equal and approved; PC £498.30/1000; 265 mm x 165 mm; on 25 mm x 38 mm battens and type 1F reinforced underlay						
Roof coverings; to 64 mm lap	-	0.97	17.14	37.55	m²	54.69
Wall coverings; to 38 mm lap	-	1.16	20.50	32.63	m²	53.13
Extra over coverings for						
ornamental tiles	-	-	-	6.86	m²	6.86
double course at eaves	-	0.23	4.06	4.19	m	8.25
verges	-	0.28	4.95	1.82	m	6.77
swept valleys; cutting both sides	-	0.60	10.60	29.70	m	40.30

H CLADDING/COVERING

	PC £	Labour hours	Labour £	Material £	Unit	Total rate £
Extra over coverings for - cont'd						
bonnet hips; cutting both sides	-	0.74	13.08	29.76	m	**42.84**
external vertical angle tiles; supplementary nail fixings	-	0.37	6.54	40.75	m	**47.29**
half round ridge tiles	-	0.56	9.90	9.18	m	**19.08**
holes for pipes and the like	-	0.19	3.36	-	nr	**3.36**
Concrete interlocking tiles; Marley "Bold Roll" granule finish tiles or other equal and approved; PC £639.80/ 1000; 420 mm x 330 mm; to 75 mm lap; on 25 mm x 38 mm battens and type 1F reinforced underlay						
Roof coverings	-	0.32	5.66	9.04	m^2	**14.70**
Extra over coverings for						
fixing every tile	-	0.02	0.35	0.45	m^2	**0.80**
eaves; eaves filler	-	0.04	0.71	0.86	m	**1.57**
verges; 150 mm wide asbestos free strip undercloak	-	0.21	3.71	1.60	m	**5.31**
valley trough tiles; cutting both sides	-	0.51	9.01	16.47	m	**25.48**
segmental ridge tiles; tile slips	-	0.51	9.01	9.07	m	**18.08**
segmental hip tiles; tile slips; cutting both sides	-	0.65	11.49	11.08	m	**22.57**
dry ridge tiles; segmental including batten sections; unions and filler pieces	-	0.28	4.95	13.20	m	**18.15**
segmental mono-ridge tiles	-	0.51	9.01	12.73	m	**21.74**
gas ridge terminal	-	0.46	8.13	46.82	nr	**54.95**
holes for pipes and the like	-	0.19	3.36	-	nr	**3.36**
Concrete interlocking tiles; Marley "Ludlow Major" granule finish tiles or other equal and approved; PC £606.30/1000; 420 mm x 330 mm; to 75 mm lap; on 25 mm x 38 mm battens and type 1F reinforced underlay						
Roof coverings	-	0.32	5.66	8.77	m^2	**14.43**
Extra over coverings for						
fixing every tile	-	0.02	0.35	0.45	m^2	**0.80**
eaves; eaves filler	-	0.04	0.71	0.34	m	**1.05**
verges; 150 mm wide asbestos free strip undercloak	-	0.21	3.71	1.60	m	**5.31**
dry verge system; extruded white pvc	-	0.14	2.47	6.75	m	**9.22**
segmental ridge cap to dry verge	-	0.02	0.35	2.46	m	**2.81**
valley trough tiles; cutting both sides	-	0.51	9.01	16.40	m	**25.41**
segmental ridge tiles	-	0.46	8.13	6.32	m	**14.45**
segmental hip tiles; cutting both sides	-	0.60	10.60	8.23	m	**18.83**
dry ridge tiles; segmental including batten sections; unions and filler pieces	-	0.28	4.95	13.20	m	**18.15**
segmental mono-ridge tiles	-	0.46	8.13	11.13	m	**19.26**
gas ridge terminal	-	0.46	8.13	46.82	nr	**54.95**
holes for pipes and the like	-	0.19	3.36	-	nr	**3.36**
Concrete interlocking tiles; Marley "Mendip" granule finish double pantiles or other equal and approved; PC £651.60/1000; 420 mm x 330 mm; to 75 mm lap; on 22 mm x 38 mm battens and type 1F reinforced underlay						
Roof coverings	-	0.32	5.66	9.16	m^2	**14.82**
Extra over coverings for						
fixing every tile	-	0.02	0.35	0.45	m^2	**0.80**
verges; 150 mm wide asbestos free strip undercloak	-	0.21	3.71	1.60	m	**5.31**
dry verge system; extruded white pvc	-	0.14	2.47	6.75	m	**9.22**
valley trough tiles; cutting both sides	-	0.51	9.01	16.49	m	**25.50**
segmental ridge tiles	-	0.51	9.01	9.07	m	**18.08**
segmental hip tiles; cutting both sides	-	0.65	11.49	11.12	m	**22.61**
dry ridge tiles; segmental including batten sections; unions and filler pieces	-	0.28	4.95	13.20	m	**18.15**
segmental mono-ridge tiles	-	0.46	8.13	12.50	m	**20.63**
gas ridge terminal	-	0.46	8.13	46.82	nr	**54.95**
holes for pipes and the like	-	0.19	3.36	-	nr	**3.36**

H CLADDING/COVERING

	PC £	Labour hours	Labour £	Material £	Unit	Total rate £
H60 PLAIN ROOF TILING - cont'd						
Concrete interlocking tiles; Marley "Modern" smooth finish tiles or other equal and approved; PC £606.30/1000; 420 mm x 220 mm; to 75 mm lap; on 25 mm x 38 mm battens and type 1F reinforced underlay						
Roof coverings	-	0.32	5.66	8.93	m²	14.59
Extra over coverings for						
fixing every tile	-	0.02	0.35	0.45	m²	0.80
verges; 150 wide asbestos free strip undercloak	-	0.21	3.71	1.60	m	5.31
dry verge system; extruded white pvc	-	0.19	3.36	6.75	m	10.11
"Modern" ridge cap to dry verge	-	0.02	0.35	2.46	m	2.81
valley trough tiles; cutting both sides	-	0.51	9.01	16.40	m	25.41
"Modern" ridge tiles	-	0.46	8.13	6.64	m	14.77
"Modern" hip tiles; cutting both sides	-	0.60	10.60	8.55	m	19.15
dry ridge tiles; "Modern"; including batten sections; unions and filler pieces	-	0.28	4.95	13.51	m	18.46
"Modern" mono-ridge tiles	-	0.46	8.13	11.13	m	19.26
gas ridge terminal	-	0.46	8.13	46.82	nr	54.95
holes for pipes and the like	-	0.19	3.36	-	nr	3.36
Concrete interlocking tiles; Marley "Wessex" smooth finish tiles or other equal and approved; PC £1003.20/1000; 413 mm x 330 mm; to 75 mm lap; on 25 mm x 38 mm battens and type 1F reinforced underlay						
Roof coverings	-	0.32	5.66	13.06	m²	18.72
Extra over coverings for						
fixing every tile	-	0.02	0.35	0.45	m²	0.80
verges; 150 mm wide asbestos free strip undercloak	-	0.21	3.71	1.60	m	5.31
dry verge system; extruded white pvc	-	0.19	3.36	6.75	m	10.11
"Modern" ridge cap to dry verge	-	0.02	0.35	2.46	m	2.81
valley trough tiles; cutting both sides	-	0.51	9.01	17.23	m	26.24
"Modern" ridge tiles	-	0.46	8.13	6.64	m	14.77
"Modern" hip tiles; cutting both sides	-	0.60	10.60	9.80	m	20.40
dry ridge tiles; "Modern"; including batten sections; unions and filler pieces	-	0.28	4.95	13.51	m	18.46
"Modern" mono-ridge tiles	-	0.46	8.13	11.13	m	19.26
gas ridge terminal	-	0.46	8.13	46.82	nr	54.95
holes for pipes and the like	-	0.19	3.36	-	nr	3.36
Concrete interlocking slates; Redland "Richmond" smooth finish tiles or other equal and approved; PC £753.70/1000; 430 x 380; to 75 mm lap; on 25 mm x 38 mm battens and type 1F reinforced underlay						
Roof coverings	-	0.32	5.66	9.02	m²	14.68
Extra over coverings for						
fixing every tile	-	0.02	0.35	0.45	m²	0.80
eaves; eaves filler	-	0.02	0.35	0.50	m	0.85
verges; extra single undercloak course of plain tiles	-	0.23	4.06	2.69	m	6.75
ambi-dry verge system	-	0.19	3.36	7.36	m	10.72
ambi-dry verge eave/ridge end piece	-	0.02	0.35	2.64	m	2.99
valley trough tiles; cutting both sides	-	0.56	9.90	20.99	m	30.89
"Delta" ridge tiles	-	0.46	8.13	7.83	m	15.96
"Delta" hip tiles; cutting both sides	-	0.60	10.60	10.21	m	20.81
"Delta" mono-ridge tiles	-	0.46	8.13	15.41	m	23.54
gas ridge terminal	-	0.46	8.13	61.77	nr	69.90
ridge vent with 110 mm diameter flexible adaptor	-	0.46	8.13	54.31	nr	62.44
holes for pipes and the like	-	0.19	3.36	-	nr	3.36

Prices for Measured Works - Major Works
H CLADDING/COVERING

	PC £	Labour hours	Labour £	Material £	Unit	Total rate £
Concrete interlocking tiles; Redland "Norfolk" smooth finish pantiles or other equal and approved; PC £383.60/1000; 381 mm x 229 mm; to 75 mm lap; on 25 mm x 38 mm battens and type 1F reinforced underlay						
Roof coverings	-	0.42	7.42	10.07	m^2	17.49
Extra over coverings for						
fixing every tile	-	0.04	0.71	0.25	m^2	0.96
eaves; eaves filler	-	0.04	0.71	0.81	m	1.52
verges; extra single undercloak course of plain tiles	-	0.28	4.95	4.69	m	9.64
valley trough tiles; cutting both sides	-	0.56	9.90	19.84	m	29.74
universal ridge tiles	-	0.46	8.13	8.35	m	16.48
universal hip tiles; cutting both sides	-	0.60	10.60	10.36	m	20.96
universal gas flue ridge tile	-	0.46	8.13	62.42	nr	70.55
universal ridge vent tile with 110 mm diameter adaptor	-	0.50	8.84	56.45	nr	65.29
holes for pipes and the like	-	0.19	3.36	-	nr	3.36
Concrete interlocking tiles; Redland "Regent" granule finish bold roll tiles or other equal and approved; PC £549.30/1000; 418 mm x 332 mm; to 75 mm lap; on 25 mm x 38 mm battens and type 1F reinforced underlay						
Roof coverings	-	0.32	5.66	8.29	m^2	13.95
Extra over coverings for						
fixing every tile	-	0.03	0.53	0.55	m^2	1.08
eaves; eaves filler	-	0.04	0.71	0.61	m	1.32
verges; extra single undercloak course of plain tiles	-	0.23	4.06	2.23	m	6.29
cloaked verge system	-	0.14	2.47	4.75	m	7.22
valley trough tiles; cutting both sides	-	0.51	9.01	19.55	m	28.56
universal ridge tiles	-	0.46	8.13	8.35	m	16.48
universal hip tiles; cutting both sides	-	0.60	10.60	10.08	m	20.68
dry ridge system; universal ridge tiles	-	0.23	4.06	26.80	m	30.86
universal half round mono-pitch ridge tiles	-	0.51	9.01	17.58	m	26.59
universal gas flue ridge tile	-	0.46	8.13	62.42	nr	70.55
universal ridge vent tile with 110 mm diameter adaptor	-	0.46	8.13	56.45	nr	64.58
holes for pipes and the like	-	0.19	3.36	-	nr	3.36
Concrete interlocking tiles; Redland "Renown" granule finish tiles or other equal and approved; PC £522.60/1000; 418 mm x 330 mm; to 75 mm lap; on 25 mm x 38 mm battens and type 1F reinforced underlay						
Roof coverings	-	0.32	5.66	8.02	m^2	13.68
Extra over coverings for						
fixing every tile	-	0.02	0.35	0.30	m^2	0.65
verges; extra single undercloak course of plain tiles	-	0.23	4.06	2.69	m	6.75
cloaked verge system	-	0.14	2.47	4.83	m	7.30
valley trough tiles; cutting both sides	-	0.51	9.01	19.47	m	28.48
universal ridge tiles	-	0.46	8.13	8.35	m	16.48
universal hip tiles; cutting both sides	-	0.60	10.60	9.99	m	20.59
dry ridge system; universal ridge tiles	-	0.23	4.06	26.80	m	30.86
universal half round mono-pitch ridge tiles	-	0.51	9.01	17.58	m	26.59
universal gas flue ridge tile	-	0.46	8.13	62.42	nr	70.55
universal ridge vent tile with 110 mm diameter adaptor	-	0.46	8.13	56.45	nr	64.58
holes for pipes and the like	-	0.19	3.36	-	nr	3.36

H CLADDING/COVERING

	PC £	Labour hours	Labour £	Material £	Unit	Total rate £
H60 PLAIN ROOF TILING - cont'd						
Concrete interlocking tiles; Redland "Stonewold II" smooth finish tiles or other equal and approved; PC £1064.40/1000; 430 mm x 380 mm; to 75 mm lap; on 25 mm x 38 mm battens and type 1F reinforced underlay						
Roof coverings	-	0.32	5.66	13.64	m^2	19.30
Extra over coverings for						
fixing every tile	-	0.02	0.35	0.84	m^2	1.19
verges; extra single undercloak course of plain tiles	-	0.28	4.95	2.69	m	7.64
ambi-dry verge system	-	0.19	3.36	7.36	m	10.72
ambi-dry verge eave/ridge end piece	-	0.02	0.35	2.64	m	2.99
valley trough tiles; cutting both sides	-	0.51	9.01	21.18	m	30.19
universal angle ridge tiles	-	0.46	8.13	6.21	m	14.34
universal hip tiles; cutting both sides	-	0.60	10.60	9.56	m	20.16
dry ridge system; universal angle ridge tiles	-	0.23	4.06	16.26	m	20.32
universal mono-pitch angle ridge tiles	-	0.51	9.01	12.01	m	21.02
universal gas flue angle ridge tile	-	0.46	8.13	62.42	nr	70.55
universal angle ridge vent tile with 110 mm diameter adaptor	-	0.46	8.13	56.45	nr	64.58
holes for pipes and the like	-	0.19	3.36	-	nr	3.36
Concrete plain tiles; EN 490 and 491 group A; PC £228.70/1000; 267 mm x 165 mm; on 25 mm x 38 mm battens and type 1F reinforced underlay						
Roof coverings; to 64 mm lap	-	0.97	17.14	20.56	m^2	37.70
Wall coverings; to 38 mm lap	-	1.16	20.50	17.62	m^2	38.12
Extra over coverings for						
ornamental tiles	-	-	-	12.18	m^2	12.18
double course at eaves	-	0.23	4.06	2.49	m	6.55
verges	-	0.31	5.48	0.98	m	6.46
swept valleys; cutting both sides	-	0.60	10.60	20.52	m	31.12
bonnet hips; cutting both sides	-	0.74	13.08	20.59	m	33.67
external vertical angle tiles; supplementary nail fixings	-	0.37	6.54	14.48	m	21.02
half round ridge tiles	-	0.46	8.13	5.31	m	13.44
third round hip tiles; cutting both sides	-	0.46	8.13	6.75	m	14.88
holes for pipes and the like	-	0.19	3.36	-	nr	3.36
Sundries						
Hip irons						
galvanised mild steel; fixing with screws	-	0.09	1.59	2.43	nr	4.02
"Rytons Clip strip" or other equal and approved; continuous soffit ventilator						
51 mm wide; plastic; code CS351	-	0.28	4.95	0.98	m	5.93
"Rytons over fascia ventilator" or other equal and approved; continuous eaves ventilator						
40 mm wide; plastic; code OFV890	-	0.09	1.59	1.66	m	3.25
"Rytons roof ventilator" or other equal and approved; to suit rafters at 600 mm centres						
250 mm deep x 43 mm high; plastic; code TV600	-	0.09	1.59	1.57	m	3.16
"Rytons push and lock ventilators" or other equal and approved; circular						
83 mm diameter; plastic; code PL235	-	0.04	0.56	0.26	nr	0.82
Fixing only						
lead soakers (supply cost not included)	-	0.07	0.86	-	nr	0.86
Pressure impregnated softwood counter battens; 25 mm x 50 mm						
450 mm centres	-	0.06	1.06	1.07	m^2	2.13
600 mm centres	-	0.04	0.71	0.81	m^2	1.52

Prices for Measured Works - Major Works

H CLADDING/COVERING

	PC £	Labour hours	Labour £	Material £	Unit	Total rate £
Underlay; BS 747 type 1B; bitumen felt weighing 14 kg/10 m^2; 75 mm laps						
To sloping or vertical surfaces	0.47	0.02	0.35	1.03	m^2	1.38
Underlay; BS 747 type 1F; reinforced bitumen felt weighing 22.50 kg/10 m^2; 75 mm laps						
To sloping or vertical surfaces	0.88	0.02	0.35	1.49	m^2	1.84
Underlay; Visqueen "Tilene 200P" or other equal and approved; micro-perforated sheet; 75 mm laps						
To sloping or vertical surfaces	0.46	0.02	0.35	1.05	m^2	1.40
Underlay; "Powerlon 250 BM" or other equal and approved; reinforced breather membrane; 75 mm laps						
To sloping or vertical surfaces	0.56	0.02	0.35	1.16	m^2	1.51
Underlay; "Anticon" or other equal and approved sarking membrane; Euroroof Ltd; polyethylene; 75 mm laps						
To sloping or vertical surfaces	-	0.02	0.35	1.39	m^2	1.74
H61 FIBRE CEMENT SLATING						
Asbestos-free artificial slates; Eternit "2000" or other equal and approved; to 75 mm lap; on 19 mm x 50 mm battens and type 1F reinforced underlay						
Coverings; 400 mm x 200 mm slates						
roof coverings	-	0.74	13.08	23.37	m^2	36.45
wall coverings	-	0.97	17.14	23.37	m^2	40.51
Coverings; 500 mm x 250 mm slates						
roof coverings	-	0.60	10.60	21.15	m^2	31.75
wall coverings	-	0.74	13.08	21.15	m^2	34.23
Coverings; 600 mm x 300 mm slates						
roof coverings	-	0.46	8.13	18.58	m^2	26.71
wall coverings	-	0.60	10.60	18.58	m^2	29.18
Extra over slate coverings for						
double course at eaves	-	0.23	4.06	4.61	m	8.67
verges; extra single undercloak course	-	0.31	5.48	1.10	m	6.58
open valleys; cutting both sides	-	0.19	3.36	4.39	m	7.75
valley gutters; cutting both sides	-	0.51	9.01	18.79	m	27.80
half round ridge tiles	-	0.46	8.13	16.69	m	24.82
stop end	-	0.09	1.59	5.71	nr	7.30
roll top ridge tiles	-	0.56	9.90	18.16	m	28.06
stop end	-	0.09	1.59	8.50	nr	10.09
mono-pitch ridge tiles	-	0.46	8.13	21.33	m	29.46
stop end	-	0.09	1.59	25.04	nr	27.23
duo-pitch ridge tiles	-	0.46	8.13	17.24	m	25.37
stop end	-	0.09	1.59	18.83	nr	20.42
mitred hips; cutting both sides	-	0.19	3.36	4.39	m	7.75
half round hip tiles; cutting both sides	-	0.19	3.36	21.08	m	24.44
holes for pipes and the like	-	0.19	3.36	-	nr	3.36
H62 NATURAL SLATING						
NOTE: The following items of slate roofing unless otherwise described, include for conventional fixing assuming "normal exposure" with appropriate nails and/or rivets or clips to pressure impregnated softwood battens fixed with galvanised nails; prices also include for all bedding and pointing at verges; beneath verge tiles etc.						

H CLADDING/COVERING

	PC £	Labour hours	Labour £	Material £	Unit	Total rate £
H62 NATURAL SLATING - cont'd						
Natural slates; BS 680 Part 2; Welsh blue; uniform size; to 75 mm lap; on 25 mm x 50 mm battens and type 1F reinforced underlay						
Coverings; 405 mm x 255 mm slates						
roof coverings	-	0.83	14.67	32.34	m²	47.01
wall coverings	-	1.06	18.73	32.34	m²	51.07
Coverings; 510 mm x 255 mm slates						
roof coverings	-	0.69	12.19	39.00	m²	51.19
wall coverings	-	0.83	14.67	39.00	m²	53.67
Coverings; 610 mm x 305 mm slates						
roof coverings	-	0.56	9.90	50.01	m²	59.91
wall coverings	-	0.69	12.19	50.01	m²	62.20
Extra over coverings for						
double course at eaves	-	0.28	4.95	13.14	m	18.09
verges; extra single undercloak course	-	0.39	6.89	7.65	m	14.54
open valleys; cutting both sides	-	0.20	3.53	30.25	m	33.78
blue/black glazed ware 152 mm half round ridge tiles	-	0.46	8.13	6.70	m	14.83
blue/black glazed ware 125 mm x 125 mm plain angle ridge tiles	-	0.46	8.13	6.70	m	14.83
mitred hips; cutting both sides	-	0.20	3.53	30.25	m	33.78
blue/black glazed ware 152 mm half round hip tiles; cutting both sides	-	0.65	11.49	36.94	m	48.43
blue/black glazed ware 125 mm x 125 mm plain angle hip tiles; cutting both sides	-	0.65	11.49	36.94	m	48.43
holes for pipes and the like	-	0.19	3.36	-	nr	3.36
Natural slates; Westmoreland green; PC £1700.75/t; random lengths; 457 mm - 229 mm proportionate widths to 75 mm lap; in diminishing courses; on 25 mm x 50 mm battens and type 1F underlay						
Roof coverings	-	1.06	18.73	101.07	m²	119.80
Wall coverings	-	1.34	23.68	101.07	m²	124.75
Extra over coverings for						
double course at eaves	-	0.61	10.78	18.66	m	29.44
verges; extra single undercloak course slates 152 mm wide	-	0.69	12.19	16.24	m	28.43
holes for pipes and the like	-	0.28	4.95	-	nr	4.95
H63 RECONSTRUCTED STONE SLATING/TILING						
Reconstructed stone slates; "Hardrow Slates" or other equal and approved; standard colours; or similar; PC £773.80/1000 75 mm lap; on 25 mm x 50 mm battens and type 1F reinforced underlay						
Coverings; 457 mm x 305 mm slates						
roof coverings	-	0.74	13.08	18.75	m²	31.83
wall coverings	-	0.93	16.44	18.75	m²	35.19
Coverings; 457 mm x 457 mm slates						
roof coverings	-	0.60	10.60	18.50	m²	29.10
wall coverings	-	0.79	13.96	18.50	m²	32.46
Extra over 457 mm x 305 mm coverings for						
double course at eaves	-	0.28	4.95	3.39	m	8.34
verges; pointed	-	0.39	6.89	0.06	m	6.95
open valleys; cutting both sides	-	0.20	3.53	8.12	m	11.65
ridge tiles	-	0.46	8.13	21.31	m	29.44
hip tiles; cutting both sides	-	0.65	11.49	17.62	m	29.11
holes for pipes and the like	-	0.19	3.36	-	nr	3.36

Prices for Measured Works - Major Works

H CLADDING/COVERING

	PC £	Labour hours	Labour £	Material £	Unit	Total rate £
Reconstructed stone slates; "Hardrow Slates" or other equal and approved; green/oldstone colours; PC £773.80/1000; to 75 mm lap; on 25 mm x 50 mm battens and type 1F reinforced underlay						
Coverings; 457 mm x 305 mm slates						
roof coverings	-	0.74	13.08	18.75	m²	**31.83**
wall coverings	-	0.93	16.44	18.75	m²	**35.19**
Coverings; 457 mm x 457 mm slates						
roof coverings	-	0.60	10.60	18.50	m²	**29.10**
wall coverings	-	0.79	13.96	18.50	m²	**32.46**
Extra over 457 mm x 305 mm coverings for						
double course at eaves	-	0.28	4.95	3.39	m	**8.34**
verges; pointed	-	0.39	6.89	0.06	m	**6.95**
open valleys; cutting both sides	-	0.20	3.53	8.12	m	**11.65**
ridge tiles	-	0.46	8.13	21.31	m	**29.44**
hip tiles; cutting both sides	-	0.65	11.49	17.62	m	**29.11**
holes for pipes and the like	-	0.19	3.36	-	nr	**3.36**
Reconstructed stone slates; Bradstone "Cotswold" stye or other equal and approved; PC £19.87/m²; random lengths 550 mm - 300 mm; proportional widths; to 80 mm lap; in diminishing courses; on 25 mm x 50 mm battens and type 1F reinforced underlay						
Roof coverings (all-in rate inclusive of eaves and verges)	-	0.97	17.14	25.65	m²	**42.79**
Extra over coverings for						
open valleys/mitred hips; cutting both sides	-	0.42	7.42	9.39	m²	**16.81**
ridge tiles	-	0.61	10.78	12.56	m	**23.34**
hip tiles; cutting both sides	-	0.97	17.14	21.25	m	**38.39**
holes for pipes and the like	-	0.28	4.95	-	nr	**4.95**
Reconstructed stone slates; Bradstone "Moordale" style or other equal and approved; PC £18.72/m²; random lengths 550 mm - 450 mm; proportional widths; to 80 mm lap; in diminishing course; on 25 mm x 50 mm battens and type 1F reinforced underlay						
Roof coverings (all-in rate inclusive of eaves and verges)	-	0.97	17.14	24.43	m²	**41.57**
Extra over coverings for						
open valleys/mitred hips; cutting both sides	-	0.42	7.42	8.84	m²	**16.26**
ridge tiles	-	0.61	10.78	12.56	m	**23.34**
holes for pipes and the like	-	0.28	4.95	-	nr	**4.95**
H64 TIMBER SHINGLING						
Red cedar sawn shingles preservative treated; PC £25.55/bundle; uniform length 400 mm; to 125 mm gauge; on 25 mm x 38 mm battens and type 1F reinforced underlay						
Roof coverings; 125 mm gauge, 2.28 m²/bundle	-	0.97	17.14	16.72	m²	**33.86**
Wall coverings; 190 mm gauge, 3.47 m²/bundle	-	0.74	13.08	11.39	m²	**24.47**
Extra over coverings for						
double course at eaves	-	0.19	3.36	1.52	m	**4.88**
open valleys; cutting both sides	-	0.19	3.36	2.88	m	**6.24**
pre-formed ridge capping	-	0.28	4.95	5.12	m	**10.07**
pre-formed hip capping; cutting both sides	-	0.46	8.13	8.00	m	**16.13**
double starter course to cappings	-	0.09	1.59	0.53	m	**2.12**
holes for pipes and the like	-	0.14	2.47	-	nr	**2.47**

H CLADDING/COVERING

	PC £	Labour hours	Labour £	Material £	Unit	Total rate £
H71 LEAD SHEET COVERINGS/FLASHINGS						
Milled lead; BS 1178; PC £729.10/t						
1.32 mm thick (code 3) roof coverings						
flat	-	2.50	30.56	11.38	m^2	41.94
sloping 10 - 50 degrees	-	2.78	33.98	11.38	m^2	45.36
vertical or sloping over 50 degrees	-	3.05	37.28	11.38	m^2	48.66
1.80 mm thick (code 4) roof coverings						
flat	-	2.68	32.76	15.52	m^2	48.28
sloping 10 - 50 degrees	-	2.96	36.18	15.52	m^2	51.70
vertical or sloping over 50 degrees	-	3.24	39.60	15.52	m^2	55.12
1.80 mm thick (code 4) dormer coverings						
flat	-	3.15	38.50	15.52	m^2	54.02
sloping 10 - 50 degrees	-	3.61	44.12	15.52	m^2	59.64
vertical or sloping over 50 degrees	-	3.89	47.55	15.52	m^2	63.07
2.24 mm thick (code 5) roof coverings						
flat	-	2.87	35.08	20.69	m^2	55.77
sloping 10 - 50 degrees	-	3.15	38.50	20.69	m^2	59.19
vertical or sloping over 50 degrees	-	3.42	41.80	20.69	m^2	62.49
2.24 mm thick (code 5) dormer coverings						
flat	-	3.42	41.80	20.69	m^2	62.49
sloping 10 - 50 degrees	-	3.79	46.32	20.69	m^2	67.01
vertical or sloping over 50 degrees	-	4.16	50.85	20.69	m^2	71.54
2.65 mm thick (code 6) roof coverings						
flat	-	3.05	37.28	22.84	m^2	60.12
sloping 10 - 50 degrees	-	3.33	40.70	22.84	m^2	63.54
vertical or sloping over 50 degrees	-	3.61	44.12	22.84	m^2	66.96
2.65 mm thick (code 6) dormer coverings						
flat	-	3.70	45.22	22.84	m^2	68.06
sloping 10 - 50 degrees	-	3.98	48.65	22.84	m^2	71.49
vertical or sloping over 50 degrees	-	4.35	53.17	22.84	m^2	76.01
3.15 mm thick (code 7) roof coverings						
flat	-	3.24	39.60	27.15	m^2	66.75
sloping 10 - 50 degrees	-	3.52	43.02	27.15	m^2	70.17
vertical or sloping over 50 degrees	-	3.79	46.32	27.15	m^2	73.47
3.15 mm thick (code 7) dormer coverings						
flat	-	3.89	47.55	27.15	m^2	74.70
sloping 10 - 50 degrees	-	4.16	50.85	27.15	m^2	78.00
vertical or sloping over 50 degrees	-	4.53	55.37	27.15	m^2	82.52
3.55 mm thick (code 8) roof coverings						
flat	-	3.42	41.80	30.60	m^2	72.40
sloping 10 - 50 degrees	-	3.70	45.22	30.60	m^2	75.82
vertical or sloping over 50 degrees	-	3.98	48.65	30.60	m^2	79.25
3.55 mm thick (code 8) dormer coverings						
flat	-	4.16	50.85	30.60	m^2	81.45
sloping 10 - 50 degrees	-	4.44	54.27	30.60	m^2	84.87
vertical or sloping over 50 degrees	-	4.81	58.79	30.60	m^2	89.39
Dressing over glazing bars and glass	-	0.31	3.79	-	m	3.79
Soldered dot	-	0.14	1.71	-	nr	1.71
Copper nailing; 75 mm centres	-	0.19	2.32	0.27	m	2.59
1.80 mm thick (code 4) lead flashings, etc.						
Flashings; wedging into grooves						
150 mm girth	-	0.74	9.04	2.42	m	11.46
240 mm girth	-	0.83	10.14	3.86	m	14.00
Stepped flashings; wedging into grooves						
180 mm girth	-	0.83	10.14	2.88	m	13.02
270 mm girth	-	0.93	11.37	4.31	m	15.68
Linings to sloping gutters						
390 mm girth	-	1.11	13.57	6.28	m	19.85
450 mm girth	-	1.20	14.67	7.19	m	21.86

H CLADDING/COVERING

	PC £	Labour hours	Labour £	Material £	Unit	Total rate £
Cappings to hips or ridges						
450 mm girth	-	1.39	16.99	7.19	m	24.18
600 mm girth	-	1.48	18.09	9.60	m	27.69
Soakers						
200 mm x 200 mm	-	0.14	1.71	0.62	nr	2.33
300 mm x 300 mm	-	0.19	2.32	1.40	nr	3.72
Saddle flashings; at intersections of hips and ridges; dressing and bossing						
450 mm x 600 mm	-	1.67	20.41	4.31	nr	24.72
Slates; with 150 mm high collar						
450 mm x 450 mm; to suit 50 mm diameter pipe	-	1.57	19.19	3.72	nr	22.91
450 mm x 450 mm; to suit 100 mm diameter pipe	-	1.85	22.61	4.08	nr	26.69
2.24 mm thick (code 5) lead flashings, etc.						
Flashings; wedging into grooves						
150 mm girth	-	0.74	9.04	3.00	m	12.04
240 mm girth	-	0.83	10.14	4.76	m	14.90
Stepped flashings; wedging into grooves						
180 mm girth	-	0.83	10.14	3.59	m	13.73
270 mm girth	-	0.93	11.37	5.36	m	16.73
Linings to sloping gutters						
390 mm girth	-	1.11	13.57	7.78	m	21.35
450 mm girth	-	1.20	14.67	8.94	m	23.61
Cappings to hips or ridges						
450 mm girth	-	1.39	16.99	8.94	m	25.93
600 mm girth	-	1.48	18.09	11.95	m	30.04
Soakers						
200 mm x 200 mm	-	0.14	1.71	0.83	nr	2.54
300 mm x 300 mm	-	0.19	2.32	1.86	nr	4.18
Saddle flashings; at intersections of hips and ridges; dressing and bossing						
450 mm x 600 mm	-	1.67	20.41	5.37	nr	25.78
Slates; with 150 mm high collar						
450 mm x 450 mm; to suit 50 mm diameter pipe	-	1.57	19.19	4.63	nr	23.82
450 mm x 450 mm; to suit 100 mm diameter pipe	-	1.85	22.61	5.08	nr	27.69
H72 ALUMINIUM SHEET COVERINGS/FLASHINGS						
Aluminium roofing; commercial grade						
0.90 mm thick roof coverings						
flat	5.38	2.78	33.98	6.07	m^2	40.05
sloping 10 - 50 degrees	-	3.05	37.28	6.07	m^2	43.35
vertical or sloping over 50 degrees	-	3.33	40.70	6.07	m^2	46.77
0.90 mm thick dormer coverings						
flat	-	3.33	40.70	6.07	m^2	46.77
sloping 10 - 50 degrees	-	3.70	45.22	6.07	m^2	51.29
vertical or sloping over 50 degrees	-	4.07	49.75	6.07	m^2	55.82
Aluminium nailing; 75 spacing	-	0.19	2.32	0.25	m	2.57
0.90 mm commercial grade aluminium flashings, etc.						
Flashings; wedging into grooves						
150 mm girth	-	0.74	9.04	0.95	m	9.99
240 mm girth	-	0.82	10.02	1.54	m	11.56
300 mm girth	-	0.97	11.86	1.86	m	13.72
Stepped flashings; wedging into grooves						
180 mm girth	-	0.83	10.14	1.12	m	11.26
270 mm girth	-	0.93	11.37	1.74	m	13.11

H CLADDING/COVERING

	PC £	Labour hours	Labour £	Material £	Unit	Total rate £
H72 ALUMINIUM SHEET COVERINGS/FLASHINGS - cont'd						
1.20 mm commercial grade aluminium flashings; polyester powder coated						
Flashings; fixed with self tapping screws						
170 mm girth	-	0.83	10.14	1.30	m	11.44
200 mm girth	-	0.88	10.76	1.72	m	12.48
280 mm girth	-	1.02	12.47	2.26	m	14.73
H73 COPPER STRIP SHEET COVERINGS/FLASHINGS						
Copper roofing; BS 2870						
0.56 mm thick (24 swg) roof coverings						
flat	16.95	2.96	36.18	19.13	m^2	55.31
sloping 10 - 50 degrees	-	3.24	39.60	19.13	m^2	58.73
vertical or sloping over 50 degrees	-	3.52	43.02	19.13	m^2	62.15
0.56 mm thick (24 swg) dormer coverings						
flat	16.95	3.52	43.02	19.13	m^2	62.15
sloping 10 - 50 degrees	-	3.89	47.55	19.13	m^2	66.68
vertical or sloping over 50 degrees	-	4.26	52.07	19.13	m^2	71.20
0.61 mm thick (23 swg) roof coverings						
flat	18.53	2.96	36.18	20.92	m^2	57.10
sloping 10 - 50 degrees	-	3.24	39.60	20.92	m^2	60.52
vertical or sloping over 50 degrees	-	3.52	43.02	20.92	m^2	63.94
0.61 mm thick (23 swg) dormer coverings						
flat	18.53	3.52	43.02	20.92	m^2	63.94
sloping 10 - 50 degrees	-	3.89	47.55	20.92	m^2	68.47
vertical or sloping over 50 degrees	-	4.26	52.07	20.92	m^2	72.99
Copper nailing; 75 mm spacing	-	0.19	2.32	0.27	m	2.59
0.56 mm thick copper flashings, etc.						
Flashings; wedging into grooves						
150 mm girth	-	0.74	9.04	3.01	m	12.05
240 mm girth	-	0.83	10.14	4.82	m	14.96
300 mm girth	-	0.97	11.86	6.05	m	17.91
Stepped flashings; wedging into grooves						
180 mm girth	-	0.83	10.14	3.63	m	13.77
270 mm girth	-	0.93	11.37	5.43	m	16.80
0.61 mm thick copper flashings, etc.						
Flashings; wedging into grooves						
150 mm girth	-	0.74	9.04	3.27	m	12.31
240 mm girth	-	0.83	10.14	5.26	m	15.40
300 mm girth	-	0.97	11.86	6.57	m	18.43
Stepped flashings; wedging into grooves						
180 mm girth	-	0.83	10.14	3.94	m	14.08
270 mm girth	-	0.93	11.37	5.91	m	17.28
H74 ZINC STRIP SHEET COVERINGS/FLASHINGS						
Zinc; BS 849						
0.80 mm thick roof coverings						
flat	12.22	2.96	36.18	13.79	m^2	49.97
sloping 10 - 50 degrees	-	3.24	39.60	13.79	m^2	53.39
vertical or sloping over 50 degrees	-	3.52	43.02	13.79	m^2	56.81
0.80 mm thick (23 swg) dormer coverings						
flat	-	3.52	43.02	13.79	m^2	56.81
sloping 10 - 50 degrees	-	3.89	47.55	13.79	m^2	61.34
vertical or sloping over 50 degrees	-	4.26	52.07	13.79	m^2	65.86

Prices for Measured Works - Major Works
H CLADDING/COVERING

	PC £	Labour hours	Labour £	Material £	Unit	Total rate £
0.80 mm thick zinc flashings, etc.						
Flashings; wedging into grooves						
150 mm girth	-	0.74	9.04	2.14	m	11.18
240 mm girth	-	0.83	10.14	3.47	m	13.61
300 mm girth	-	0.97	11.86	4.36	m	16.22
Stepped flashings; wedging into grooves						
180 mm girth	-	0.83	10.14	2.61	m	12.75
270 mm girth	-	0.93	11.37	3.91	m	15.28
H75 STAINLESS STEEL SHEET COVERINGS/ FLASHINGS						
Terne coated stainless steel roofing; Eurocom Enterprise Ltd or other equal and approved						
Roof coverings; "Uginox AME"; 0.40 mm thick						
flat; fixing with stainless steel screws to timber	13.01	0.93	16.44	19.23	m²	35.67
flat; including batten rolls; fixing with stainless steel screws to timber	-	0.93	16.44	20.66	m²	37.10
Roof coverings; "Uginox AE"; 0.40 mm thick						
flat; fixing with stainless steel screws to timber	10.92	0.93	16.44	16.49	m²	32.93
flat; including batten rolls; fixing with stainless steel screws to timber	-	0.93	16.44	17.87	m²	34.31
Wall coverings; "Uginox AME"; 0.40 mm thick						
vertical; coulisseau joints; fixing with stainless steel screws; to timber	-	0.83	14.67	18.12	m²	32.79
Wall coverings; "Uginox AE"; 0.40 mm thick						
vertical; coulisseau joints; fixing with stainless steel screws; to timber	-	0.83	14.67	15.44	m²	30.11
Flashings						
head abutments; swiss fold and cover flashings 150 mm girth	-	0.37	6.54	3.82	m	10.36
head abutments; manchester fold and cover flashings 150 mm girth	-	0.46	8.13	3.82	m	11.95
head abutments; soldered saddle flashings 150 mm girth	-	0.46	8.13	4.33	m	12.46
head abutments; into brick wall with timber infill; cover flashings 150 mm girth	-	0.74	13.08	5.99	m	19.07
head abutments; into brick wall with soldered apron; cover flashings 150 mm girth	-	0.88	15.55	6.71	m	22.26
side abutments; "Uginox" stepped flashings 170 mm girth	-	1.48	26.16	4.80	m	30.96
eaves flashings 140 mm girth; approned with single lock welts	-	0.32	5.66	6.89	m	12.55
eaves flashings 190 mm girth; approned with flashings and soldered strips	-	0.46	8.13	8.44	m	16.57
verge flashings 150 mm girth	-	0.23	4.06	7.09	m	11.15
Aprons						
fan aprons 250 mm girth	-	0.28	4.95	5.92	m	10.87
Ridges						
ridge with coulisseau closures	-	0.56	9.90	2.78	m	12.68
monoridge flanged apex (AME)	-	0.28	4.95	4.10	m	9.05
ridge with tapered timber infill and flat saddles	-	0.93	16.44	5.60	m	22.04
monoridge with timber infill and flat saddles	-	0.51	9.01	8.77	m	17.78
ventilated ridge with timber kerb and flat saddles	-	1.11	19.62	14.94	m	34.56
ventilated ridge with stainless steel brackets	-	0.83	14.67	25.87	m	40.54
Edges						
downstand	-	0.28	4.95	5.51	m	10.46
flanged	-	0.23	4.06	5.92	m	9.98

H CLADDING/COVERING

	PC £	Labour hours	Labour £	Material £	Unit	Total rate £
H75 STAINLESS STEEL SHEET COVERINGS/ FLASHINGS - cont'd						
Terne coated stainless steel roofing; Eurocom Enterprise Ltd or other equal and approved - cont'd						
Gutters						
double lock welted valley gutters	-	0.97	17.14	9.11	m	26.25
eaves gutters; against masonry work	-	1.11	19.62	12.37	m	31.99
recessed valley gutters	-	0.83	14.67	17.51	m	32.18
recessed valley gutters with laplocks	-	1.02	18.03	20.62	m	38.65
secret	-	0.93	16.44	21.51	m	37.95
saddle expansion joints in gutters	-	0.93	16.44	7.40	m	23.84
H76 FIBRE BITUMEN THERMOPLASTIC SHEET COVERINGS/FLASHINGS						
Glass fibre reinforced bitumen strip slates; "Ruberglas 105" or other equal and approved; 1000 mm x 336 mm mineral finish; to external quality plywood boarding (boarding not included)						
Roof coverings	9.27	0.23	4.06	10.92	m²	14.98
Wall coverings	-	0.37	6.54	10.92	m²	17.46
Extra over coverings for						
double course at eaves; felt soaker	-	0.19	3.36	7.34	m	10.70
verges; felt soaker	-	0.14	2.47	6.09	m	8.56
valley slate; cut to shape; felt soaker and cutting both sides	-	0.42	7.42	9.64	m	17.06
ridge slate; cut to shape	-	0.28	4.95	6.09	m	11.04
hip slate; cut to shape; felt soaker and cutting both sides	-	0.42	7.42	9.54	m	16.96
holes for pipes and the like	-	0.48	8.48	-	nr	8.48
"Evode Flashband Original" sealing strips and flashings or other equal and approved; special grey finish						
Flashings; wedging at top if required; pressure bonded; flashband primer before application; to walls						
75 mm girth	-	0.17	2.08	0.89	m	2.97
100 mm girth	-	0.23	2.81	1.22	m	4.03
150 mm girth	-	0.31	3.79	1.82	m	5.61
225 mm girth	-	0.37	4.52	2.86	m	7.38
300 mm girth	-	0.42	5.13	3.59	m	8.72

J WATERPROOFING

	PC £	Labour hours	Labour £	Material £	Unit	Total rate £
J WATERPROOFING						
J10 SPECIALIST WATERPROOF RENDERING						
"Sika" waterproof rendering or other equal and approved; steel trowelled						
20 mm work to walls; three coat; to concrete base						
over 300 mm wide	-	-	-	-	m²	32.96
not exceeding 300 mm wide	-	-	-	-	m²	49.93
25 mm work to walls; three coat; to concrete base						
over 300 mm wide	-	-	-	-	m²	38.95
not exceeding 300 mm wide	-	-	-	-	m²	59.92
40 mm work to walls; four coat; to concrete base						
over 300 mm wide	-	-	-	-	m²	57.43
not exceeding 300 mm wide	-	-	-	-	m²	89.88
J20 MASTIC ASPHALT TANKING/DAMP PROOF MEMBRANES						
Mastic asphalt to BS 6925 Type T 1097						
13 mm thick one coat coverings to concrete base; flat; subsequently covered						
over 300 mm wide	-	-	-	-	m²	8.41
225 mm - 300 mm wide	-	-	-	-	m²	19.94
150 mm - 225 mm wide	-	-	-	-	m²	21.92
not exceeding 150 mm wide	-	-	-	-	m²	27.36
20 mm thick two coat coverings to concrete base; flat; subsequently covered						
over 300 mm wide	-	-	-	-	m²	12.14
225 mm - 300 mm wide	-	-	-	-	m²	22.19
150 mm - 225 mm wide	-	-	-	-	m²	24.22
not exceeding 150 mm wide	-	-	-	-	m²	29.61
30 mm thick three coat coverings to concrete base; flat; subsequently covered						
over 300 mm wide	-	-	-	-	m²	17.36
225 mm - 300 mm wide	-	-	-	-	m²	25.21
150 mm - 225 mm wide	-	-	-	-	m²	27.36
not exceeding 150 mm wide	-	-	-	-	m²	32.63
13 mm thick two coat coverings to brickwork base; vertical; subsequently covered						
over 300 mm wide	-	-	-	-	m²	33.72
225 mm - 300 mm wide	-	-	-	-	m²	41.14
150 mm - 225 mm wide	-	-	-	-	m²	48.61
not exceeding 150 mm wide	-	-	-	-	m²	63.01
20 mm thick three coat coverings to brickwork base; vertical; subsequently covered						
over 300 mm wide	-	-	-	-	m²	52.78
225 mm - 300 mm wide	-	-	-	-	m²	56.79
150 mm - 225 mm wide	-	-	-	-	m²	60.31
not exceeding 150 mm wide	-	-	-	-	m²	77.99
Turning into groove 20 mm deep	-	-	-	-	m	0.66
Internal angle fillets; subsequently covered	-	-	-	-	m	3.90

	PC £	Labour hours	Labour £	Material £	Unit	Total rate £
J21 MASTIC ASPHALT ROOFING/INSULATION/ FINISHES						
Mastic asphalt to BS 6925 Type R 988						
20 mm thick two coat coverings; felt isolating membrane; to concrete (or timber) base; flat or to falls or slopes not exceeding 10 degrees from horizontal						
over 300 mm wide	-	-	-	-	m^2	11.98
225 mm - 300 mm wide	-	-	-	-	m^2	21.97
150 mm - 225 mm wide	-	-	-	-	m^2	24.34
not exceeding 150 mm wide	-	-	-	-	m^2	29.39
Add to the above for covering with:						
10 mm thick limestone chippings in hot bitumen	-	-	-	-	m^2	2.37
coverings with solar reflective paint	-	-	-	-	m^2	2.14
300 mm x 300 mm x 8 mm g.r.p. tiles in hot bitumen	-	-	-	-	m^2	31.37
Cutting to line; jointing to old asphalt	-	-	-	-	m	5.28
13 mm thick two coat skirtings to brickwork base						
not exceeding 150 mm girth	-	-	-	-	m	9.40
150 mm - 225 mm girth	-	-	-	-	m	10.83
225 mm - 300 mm girth	-	-	-	-	m	12.14
13 mm thick three coat skirtings; expanded metal lathing reinforcement nailed to timber base						
not exceeding 150 mm girth	-	-	-	-	m	14.89
150 mm - 225 mm girth	-	-	-	-	m	17.58
225 mm - 300 mm girth	-	-	-	-	m	20.49
13 mm thick two coat fascias to concrete base						
not exceeding 150 mm girth	-	-	-	-	m	9.40
150 mm - 225 mm girth	-	-	-	-	m	10.83
20 mm thick two coat linings to channels to concrete base						
not exceeding 150 mm girth	-	-	-	-	m	22.96
150 mm - 225 mm girth	-	-	-	-	m	24.34
225 mm - 300 mm girth	-	-	-	-	m	27.47
20 mm thick two coat lining to cesspools						
250 mm x 150 mm x 150 mm deep	-	-	-	-	nr	22.30
Collars around pipes, standards and like members	-	-	-	-	nr	10.83
Accessories						
Eaves trim; extruded aluminium alloy; working asphalt into trim						
"Alutrim"; type A roof edging or other equal and approved	-	-	-	-	m	10.44
extra; angle	-	-	-	-	nr	5.83
Roof screed ventilator - aluminium alloy "Extr-aqua-vent" or other equal and approved; set on screed over and including dished sinking; working collar around ventilator	-	-	-	-	nr	20.21
J30 LIQUID APPLIED TANKING/DAMP PROOF MEMBRANES						
Tanking and damp proofing						
"Synthaprufe" or other equal and approved; blinding with sand; horizontal on slabs						
two coats	-	0.19	1.68	2.64	m^2	4.32
three coats	-	0.26	2.30	3.91	m^2	6.21

J WATERPROOFING

	PC £	Labour hours	Labour £	Material £	Unit	Total rate £
"Tretolastex 202T" or other equal and approved; on vertical surfaces of concrete						
two coats	-	0.19	1.68	0.44	m²	2.12
three coats	-	0.26	2.30	0.66	m²	2.96
One coat Vandex "Super" 0.75 kg/m² slurry or other equal and approved; one consolidating coat of Vandex "Premix" 1 kg/m² slurry or other equal and approved; horizontal on beds						
over 225 mm wide	-	0.32	2.83	2.31	m²	5.14
J40 FLEXIBLE SHEET TANKING/DAMP PROOF MEMBRANES						
Tanking and damp proofing						
"Bituthene" sheeting or other equal and approved; lapped joints; horizontal on slabs						
500 grade	-	0.09	0.80	2.19	m²	2.99
1000 grade	-	0.10	0.88	3.00	m²	3.88
heavy duty grade	-	0.12	1.06	4.35	m²	5.41
"Bituthene" sheeting or other equal and approved; lapped joints; dressed up vertical face of concrete						
1000 grade	-	0.17	1.50	3.00	m²	4.50
"Servi-pak" protection board or other equal and approved; butt jointed; taped joints; to horizontal surfaces;						
3 mm thick	-	0.14	1.24	4.14	m²	5.38
6 mm thick	-	0.14	1.24	5.72	m²	6.96
12 mm thick	-	0.19	1.68	8.95	m²	10.63
"Servi-pak" protection board or other equal and approved; butt jointed; taped joints; to vertical surfaces						
3 mm thick	-	0.19	1.68	4.14	m²	5.82
6 mm thick	-	0.19	1.68	5.72	m²	7.40
12 mm thick	-	0.23	2.03	8.95	m²	10.98
"Bituthene" fillet or other equal and approved						
40 mm x 40 mm	-	0.09	0.80	3.94	m	4.74
"Bituthene" reinforcing strip or other equal and approved; 300 mm wide						
1000 grade	-	0.09	0.80	0.90	m	1.70
Expandite "Famflex" hot bitumen bonded waterproof tanking or other equal and approved; 150 mm laps						
horizontal; over 300 mm wide	-	0.37	3.27	11.66	m²	14.93
vertical; over 300 mm wide	-	0.60	5.30	11.66	m²	16.96
J41 BUILT UP FELT ROOF COVERINGS						
NOTE: The following items of felt roofing, unless otherwise described, include for conventional lapping, laying and bonding between layers and to base; and laying flat or to falls, crossfalls or to slopes not exceeding 10 degrees - but exclude any insulation etc.						
Felt roofing; BS 747; suitable for flat roofs						
Three layer coverings first layer type 3G; subsequent layers type 3B bitumen glass fibre based felt	-	-	-	-	m²	7.29
Extra over felt for covering with and bedding in hot bitumen						
13 mm thick stone chippings	-	-	-	-	m²	2.34
300 mm x 300 mm x 8 mm g.r.p. tiles	-	-	-	-	m²	40.15
working into outlet pipes and the like	-	-	-	-	m²	9.89
Skirtings; three layer; top layer mineral surfaced; dressed over tilting fillet; turned into groove						
not exceeding 200 mm girth	-	-	-	-	m	7.34
200 mm - 400 mm girth	-	-	-	-	m	8.79

	PC £	Labour hours	Labour £	Material £	Unit	Total rate £
J41 BUILT UP FELT ROOF COVERINGS - cont'd						
Felt roofing; BS 747; suitable for flat roofs - cont'd						
Coverings to kerbs; three layer						
400 mm - 600 mm girth	-	-	-	-	m	18.52
Linings to gutters; three layer						
400 mm - 600 mm girth	-	-	-	-	m	12.59
Collars around pipes and the like; three layer mineral surface; 150 mm high						
not exceeding 55 mm nominal size	-	-	-	-	nr	8.23
55 mm - 110 mm nominal size	-	-	-	-	nr	8.23
Three layer coverings; two base layers type 5U bitumen polyester based felt; top layer type 5B polyester based mineral surfaced felt; 10 mm stone chipping covering; bitumen bonded	-	-	-	-	m²	15.33
Coverings to kerbs						
not exceeding 200 mm girth	-	-	-	-	m	9.69
200 mm - 400 mm girth	-	-	-	-	m	12.08
Outlets and dishing to gullies						
300 mm diameter	-	-	-	-	nr	9.89
"Andersons" high performance polyester-based roofing system or other equal and approved						
Two layer coverings; first layer HT 125 underlay; second layer HT 350; fully bonded to wood; fibre or cork base	-	-	-	-	m²	10.99
Extra over for						
top layer mineral surfaced	-	-	-	-	m²	2.15
13 mm thick stone chippings	-	-	-	-	m²	2.34
third layer of type 3B as underlay for concrete or screeded base	-	-	-	-	m²	2.15
working into outlet pipes and the like	-	-	-	-	nr	9.89
Skirtings; two layer; top layer mineral surfaced; dressed over tilting fillet; turned into groove						
not exceeding 200 mm girth	-	-	-	-	m	7.79
200 mm - 400 mm girth	-	-	-	-	m	9.38
Coverings to kerbs; two layer						
400 mm - 600 mm girth	-	-	-	-	m	21.88
Linings to gutters; three layer						
400 mm - 600 mm girth	-	-	-	-	m	15.97
Collars around pipes and the like; two layer; 150 mm high						
not exceeding 55 mm nominal size	-	-	-	-	nr	8.23
55 mm - 110 mm nominal size	-	-	-	-	nr	8.23
"Ruberglas 120 GP" high performance roofing or other equal and approved						
Two layer coverings; first and second layers "Ruberglas 120 GP"; fully bonded to wood, fibre or cork base	-	-	-	-	m²	6.79
Extra over for						
top layer mineral surfaced	-	-	-	-	m²	2.15
13 mm thick stone chippings	-	-	-	-	m²	2.34
third layer of "Rubervent 3G" as underlay for concrete or screeded base	-	-	-	-	m²	2.15
working into outlet pipes and the like	-	-	-	-	nr	9.89
Skirtings; two layer; top layer mineral surfaced; dressed over tilting fillet; turned into groove						
not exceeding 200 mm girth	-	-	-	-	m	6.79
200 mm - 400 mm girth	-	-	-	-	m	7.89
Coverings to kerbs; two layer						
400 mm - 600 mm girth	-	-	-	-	m	13.58
Linings to gutters; three layer						
400 mm - 600 mm girth	-	-	-	-	m	11.29

J WATERPROOFING

	PC £	Labour hours	Labour £	Material £	Unit	Total rate £
Collars around pipes and the like; two layer, 150 mm high						
not exceeding 55 mm nominal size	-	-	-	-	nr	8.23
55 mm - 110 mm nominal size	-	-	-	-	nr	8.23
"Ruberfort HP 350" high performance roofing or other equal and approved						
Two layer coverings; first layer "Ruberfort HP 180"; second layer "Ruberfort HP 350"; fully bonded; to wood; fibre or cork base	-	-	-	-	m²	10.54
Extra over for						
top layer mineral surfaced	-	-	-	-	m²	2.15
13 mm thick stone chippings	-	-	-	-	m²	2.34
third layer of "Rubervent 3G"; as underlay for concrete or screeded base	-	-	-	-	m²	2.19
working into outlet pipes and the like	-	-	-	-	nr	9.89
Skirtings; two layer; top layer mineral surface; dressed over tilting fillet; turned into groove						
not exceeding 200 mm girth	-	-	-	-	m	7.49
200 mm - 400 mm girth	-	-	-	-	m	8.89
Coverings to kerbs; two layer						
400 mm - 600 mm girth	-	-	-	-	m	21.03
Linings to gutters; three layer						
400 mm - 600 mm girth	-	-	-	-	m	15.18
Collars around pipes and the like; two layer; 150 mm high						
not exceeding 55 mm nominal size	-	-	-	-	nr	8.23
55 mm - 110 mm nominal size	-	-	-	-	nr	8.23
"Polybit 350" elastomeric roofing or other equal and approved						
Two layer coverings; first layer "Polybit 180"; second layer "Polybit 350"; fully bonded to wood; fibre or cork base	-	-	-	-	m²	10.78
Extra over for						
top layer mineral surfaced	-	-	-	-	m²	2.15
13 mm thick stone chippings	-	-	-	-	m²	2.34
third layer of "Rubervent 3G" as underlay for concrete or screeded base	-	-	-	-	m²	2.15
working into outlet pipes and the like	-	-	-	-	nr	9.89
Skirtings; two layer; top layer mineral surfaced; dressed over tilting fillet; turned into groove						
not exceeding 200 mm girth	-	-	-	-	m	7.69
200 mm - 400 mm girth	-	-	-	-	m	10.29
Coverings to kerbs; two layer						
400 mm - 600 mm girth	-	-	-	-	m	21.52
Linings to gutters; three layer						
400 mm - 600 mm girth	-	-	-	-	m	15.58
Collars around pipes and the like; two layer; 150 mm high						
not exceeding 55 mm nominal size	-	-	-	-	nr	8.23
55 mm - 110 mm nominal size	-	-	-	-	nr	8.23
"Hyload 150 E" elastomeric roofing or other equal and approved						
Two layer coverings; first layer "Ruberglas 120 GHP"; second layer "Hyload 150 E" fully bonded to wood; fibre or cork base	-	-	-	-	m²	21.97
Extra over for						
13 mm thick stone chippings	-	-	-	-	m²	2.34
third layer of "Rubervent 3G" as underlay for concrete or screeded base	-	-	-	-	m²	2.15
working into outlet pipes and the like	-	-	-	-	nr	9.84

Prices for Measured Works - Major Works
J WATERPROOFING

	PC £	Labour hours	Labour £	Material £	Unit	Total rate £
J41 BUILT UP FELT ROOF COVERINGS - cont'd						
"Hyload 150 E" elastomeric roofing or other equal and approved - cont'd						
Skirtings; two layer; dressed over tilting fillet; turned into groove						
200 mm - 400 mm girth	-	-	-	-	m	20.12
Coverings to kerbs; two layer						
400 mm - 600 mm girth	-	-	-	-	m	41.00
Linings to gutters; three layer						
400 mm - 600 mm girth	-	-	-	-	m	27.11
Collars around pipes and the like; two layer; 150 mm high						
not exceeding 55 mm nominal size	-	-	-	-	nr	8.23
55 mm - 110 mm nominal size	-	-	-	-	nr	8.23
Accessories						
Eaves trim; extruded aluminium alloy; working felt into trim						
"Alutrim"; type F roof edging or other equal and approved	-	-	-	-	m	7.34
extra over for; external angle	-	-	-	-	nr	9.84
Roof screed ventilator - aluminium alloy						
"Extr-aqua-vent" or other equal and approved - set on screed over and including dished sinking and collar	-	-	-	-	nr	21.88
Insulation board underlays						
Vapour barrier						
reinforced; metal lined	-	-	-	-	m²	10.54
Cork boards; density 112 - 125 kg/m³						
60 mm thick	-	-	-	-	m²	13.73
Rockwool; slab RW4 or other equal and approved						
60 mm thick	-	-	-	-	m²	16.73
Perlite boards or other equal and approved; density 170 - 180 kg/m³						
60 mm thick	-	-	-	-	m²	21.88
Polyurethane boards; density 32 kg/m³						
30 mm thick	-	-	-	-	m²	8.64
35 mm thick	-	-	-	-	m²	9.23
50 mm thick	-	-	-	-	m²	10.04
Wood fibre boards; impregnated; density 220 - 350 kg/m³						
12.70 mm thick	-	-	-	-	m²	3.85
Insulation board overlays						
Dow "Roofmate SL" extruded polystyrene foam boards or other equal and approved						
50 mm thick	-	0.28	4.95	12.56	m²	17.51
75 mm thick	-	0.28	4.95	17.18	m²	22.13
Dow "Roofmate LG" extruded polystyrene foam boards or other equal and approved						
60 mm thick	-	0.28	4.95	22.54	m²	27.49
90 mm thick	-	0.28	4.95	25.58	m²	30.53
110 mm thick	-	0.28	4.95	28.26	m²	33.21
Dow "Roofmate PR" extruded polystyrene foam boards or other equal and approved						
90 mm thick	-	0.28	4.95	19.36	m²	24.31
120 mm thick	-	0.28	4.95	25.78	m²	30.73

	PC £	Labour hours	Labour £	Material £	Unit	Total rate £
J42 SINGLE LAYER PLASTICS ROOF COVERINGS						
"Trocal S" PVC roofing or other equal and approved						
Coverings	-	-	-	-	m²	14.99
Skirtings; dressed over metal upstands						
not exceeding 200 mm girth	-	-	-	-	m	11.63
200 mm - 400 mm girth	-	-	-	-	m	14.28
Coverings to kerbs						
400 mm - 600 mm girth	-	-	-	-	m	26.11
Collars around pipes and the like; 150 mm high						
not exceeding 55 mm nominal size	-	-	-	-	nr	7.99
55 mm - 110 mm nominal size	-	-	-	-	nr	7.99
Accessories						
"Trocal" metal upstands or other equal and approved						
not exceeding 200 mm girth	-	-	-	-	m	8.49
200 mm - 400 mm girth	-	-	-	-	m	10.99
J43 PROPRIETARY ROOF DECKING WITH FELT FINISH						
"Bitumetal" flat roof construction or other equal and approved; fixing to timber, steel or concrete; flat or sloping; vapour check; 32 mm thick polyurethane insulation; 3G perforated felt underlay; two layers of glass fibre base felt roofing; stone chipping finish						
0.70 mm thick galvanized steel						
35 mm deep profiled decking; 2.38 m span	-	-	-	-	m²	31.65
46 mm deep profiled decking; 2.96 m span	-	-	-	-	m²	32.00
60 mm deep profiled decking; 3.74 m span	-	-	-	-	m²	33.00
100 mm deep profiled decking; 5.13 m span	-	-	-	-	m²	34.05
0.90 mm thick aluminium; mill finish						
35 mm deep profiled decking; 1.79 m span	-	-	-	-	m²	35.39
60 mm deep profiled decking; 2.34 m span	-	-	-	-	m²	36.84
"Bitumetal" flat roof construction or other equal and approved; fixing to timber, steel or concrete; flat or sloping; vapour check; 32 mm polyurethane insulation; 3G perforated felt underlay; two layers of polyester based roofing; stone chipping finish						
0.70 mm thick galvanised steel						
35 mm deep profiled decking; 2.38 m span	-	-	-	-	m²	35.35
46 mm deep profiled decking; 2.96 m span	-	-	-	-	m²	35.70
60 mm deep profiled decking; 3.74 m span	-	-	-	-	m²	36.70
100 mm deep profiled decking; 5.13 m span	-	-	-	-	m²	37.74
0.90 mm thick aluminium; mill finish						
35 mm deep profiled decking; 1.79 m span	-	-	-	-	m²	39.09
60 mm deep profiled decking; 2.34 m span	-	-	-	-	m²	40.54
"Plannja" flat roof construction or other equal and approved; fixing to timber; steel or concrete; flat or sloping; 3B vapour check; 32 mm thick polyurethene insulation; 3G perforated felt underlay; two layers of glass fibre bitumen felt roofing type 3B; stone chipping finish						
0.72 mm thick galvanised steel						
45 mm deep profiled decking; 3.12 m span	-	-	-	-	m²	36.45
70 mm deep profiled decking; 4.40 m span	-	-	-	-	m²	38.25

K10 PLASTERBOARD DRY LINING

ALTERNATIVE SHEET LINING MATERIAL PRICES

	£		£		£
Blockboard: Gaboon faced (£/10 m²)					
16 mm	124.84	22 mm	149.8	25 mm	164.78
18 mm Decorative faced (£/10 m²)					
Ash	162.28	Beech	144.81	Oak	154.80
Teak	169.77				
Edgings; self adhesive (£/50 m roll)					
22 mm Ash	11.63	22 mm Oak	11.63		
22 mm Beech	11.63				
Chipboard Standard Grade (£/10 m²)					
12 mm	25.96	16 mm	31.71	22 mm	44.94
25 mm	49.93				
Melamine faced (£/10 m²)					
12 mm	44.94	18 mm	52.93		
Laminboard; Birch faced (£/10 m²)					
18 mm	199.74	25 mm	259.66		
Medium density fibreboard (£/10 m²)					
6 mm	30.45	9 mm	39.95	19 mm	69.91
25 mm	89.88				
Plasterboard (£/100 m²)					
Wallboard plank					
9.50 mm	186.75	12.50 mm	211.72	19 mm	241.68
Industrial board					
12.50 mm	390.48				
Fireline board					
12.50 mm	241.68	15 mm	294.61		
Moisture Resistant Board					
9.50 mm	261.65				

Discounts of 0 - 20% available depending on quantity/status

	PC £	Labour hours	Labour £	Material £	Unit	Total rate £
"Gyproc" laminated proprietary partitions or other equal and approved; two skins of gypsum plasterboard bonded to a centre core of plasterboard square edge plank 19 mm thick; fixing with nails to softwood perimeter battens (not included); joints filled with filler and joint tape; to receive direct decoration						
50 mm partition; two outer skins of 12.50 mm thick tapered edge wallboard						
height 2.10 m - 2.40 m	-	1.85	25.70	15.81	m	41.51
height 2.40 m - 2.70 m	-	2.41	33.45	18.07	m	51.52
height 2.70 m - 3.00 m	-	2.68	37.18	19.64	m	56.82
height 3.00 m - 3.30 m	-	2.96	41.05	21.23	m	62.28
height 3.30 m - 3.60 m	-	3.24	44.92	22.88	m	67.80
65 mm partition; two outer skins of 19 mm thick tapered edge wallboard						
height 2.10 m - 2.40 m	-	2.04	28.36	19.68	m	48.04
height 2.40 m - 2.70 m	-	2.59	35.97	22.18	m	58.15
height 2.70 m - 3.00 m	-	2.87	39.84	24.00	m	63.84
height 3.00 m - 3.30 m	-	3.19	44.27	25.83	m	70.10
height 3.30 m - 3.60 m	-	3.47	48.14	27.72	m	75.86

K LININGS/SHEATHING/DRY PARTITIONING

	PC £	Labour hours	Labour £	Material £	Unit	Total rate £
"Gyproc" metal stud proprietary partitions or other equal and approved; comprising 48 mm wide metal stud frame; 50 mm wide floor channel plugged and screwed to concrete through 38 mm x 48 mm tanalised softwood sole plate						
Tapered edge panels; joints filled with joint filler and joint tape to receive direct decoration; 80 mm thick partition; one hour; one layer of 15 mm thick "Fireline" board or other equal and approved each side						
height 2.10 m - 2.40 m	-	3.89	54.26	18.61	m	72.87
height 2.40 m - 2.70 m	-	4.49	62.56	21.45	m	84.01
height 2.70 m - 3.00 m	-	5.00	69.65	23.60	m	93.25
height 3.00 m - 3.30 m	-	5.78	80.52	25.96	m	106.48
height 3.30 m - 3.60 m	-	6.34	88.31	27.98	m	116.29
height 3.60 m - 3.90 m	-	7.59	105.76	30.18	m	135.94
height 3.90 m - 4.20 m	-	8.14	113.41	32.39	m	145.80
angles	-	0.19	2.64	1.14	m	3.78
T-junctions	-	0.09	1.26	-	m	1.26
fair ends	-	0.19	2.64	0.40	m	3.04
Tapered edge panels; joints filled with joint filler and joint tape to receive direct decoration; 100 mm thick partition; two hour; two layers of 12.50 mm thick "Fireline" board or other equal and approved both sides						
height 2.10 m - 2.40 m	-	4.81	67.13	25.85	m	92.98
height 2.40 m - 2.70 m	-	5.53	77.12	29.62	m	106.74
height 2.70 m - 3.00 m	-	6.15	85.75	32.66	m	118.41
height 3.00 m - 3.30 m	-	6.13	85.42	35.93	m	121.35
height 3.30 m - 3.60 m	-	7.72	107.63	38.84	m	146.47
height 3.60 m - 3.90 m	-	7.59	105.76	42.00	m	147.76
height 3.90 m - 4.20 m	-	9.76	136.09	45.05	m	181.14
angles	-	0.28	3.90	1.24	m	5.14
T-junctions	-	0.09	1.26	-	m	1.26
fair ends	-	0.28	3.90	0.50	m	4.40
Gypsum plasterboard; BS 1230; plain grade tapered edge wallboard; fixing with nails; joints left open to receive "Artex" finish or other equal and approved; to softwood base						
9.50 mm board to ceilings						
over 300 mm wide	-	0.23	4.20	1.54	m^2	5.74
9.50 mm board to beams						
girth not exceeding 600 mm	-	0.28	5.11	0.94	m^2	6.05
girth 600 mm - 1200 mm	-	0.37	6.75	1.87	m^2	8.62
12.50 mm board to ceilings						
over 300 mm wide	-	0.31	5.66	1.81	m^2	7.47
12.50 mm board to beams						
girth not exceeding 600 mm	-	0.28	5.11	1.11	m^2	6.22
girth 600 mm - 1200 mm	-	0.37	6.75	2.17	m^2	8.92
Gypsum plasterboard to BS 1230; fixing with nails; joints filled with joint filler and joint tape to receive direct decoration; to softwood base						
Plain grade tapered edge wallboard						
9.50 mm board to walls						
wall height 2.40 m - 2.70 m	-	0.93	12.87	5.14	m	18.01
wall height 2.70 m - 3.00 m	-	1.06	14.67	5.72	m	20.39
wall height 3.00 m - 3.30 m	-	1.20	16.61	6.29	m	22.90
wall height 3.30 m - 3.60 m	-	1.39	19.24	6.90	m	26.14

K10 PLASTERBOARD DRY LINING/ PARTITIONS/CEILINGS - cont'd

	PC £	Labour hours	Labour £	Material £	Unit	Total rate £
Plain grade tapered edge wallboard - cont'd						
9.50 mm board to reveals and soffits of openings and recesses						
not exceeding 300 mm wide	-	0.19	2.64	1.03	m	3.67
300 mm - 600 mm wide	-	0.37	5.13	1.48	m	6.61
9.50 mm board to faces of columns - 4 nr						
not exceeding 600 mm total girth	-	0.46	6.38	2.09	m	8.47
600 mm - 1200 mm total girth	-	0.93	12.89	3.00	m	15.89
1200 mm - 1800 mm total girth	-	1.20	16.61	3.91	m	20.52
9.50 mm board to ceilings						
over 300 mm wide	-	0.39	5.41	1.91	m^2	7.32
9.50 mm board to faces of beams - 3 nr						
not exceeding 600 mm total girth	-	0.56	7.77	2.06	m	9.83
600 mm - 1200 mm total girth	-	1.02	14.13	2.97	m	17.10
1200 mm - 1800 mm total girth	-	1.30	17.99	3.87	m	21.86
add for "Duplex" insulating grade or other equal and approved	-	-	-	0.46	m^2	0.46
12.50 mm board to walls						
wall height 2.40 m - 2.70 m	-	0.97	13.43	5.79	m	19.22
wall height 2.70 m - 3.00 m	-	1.11	15.37	6.44	m	21.81
wall height 3.00 m - 3.30 m	-	1.25	17.31	7.08	m	24.39
wall height 3.30 m - 3.60 m	-	1.43	19.80	7.76	m	27.56
12.50 mm board to reveals and soffits of openings and recesses						
not exceeding 300 mm wide	-	0.19	2.64	1.12	m	3.76
300 mm - 600 mm wide	-	0.37	5.13	1.64	m	6.77
12.50 mm board to faces of columns - 4 nr						
not exceeding 600 mm total girth	-	0.46	6.38	2.28	m	8.66
600 mm - 1200 mm total girth	-	0.93	12.89	3.32	m	16.21
1200 mm - 1800 mm total girth	-	1.20	16.61	4.36	m	20.97
12.50 mm board to ceilings						
over 300 mm wide	-	0.41	5.69	2.15	m^2	7.84
12.50 mm board to faces of beams - 3 nr						
not exceeding 600 mm total girth	-	0.56	7.77	2.23	m	10.00
600 mm - 1200 mm total girth	-	1.02	14.13	3.27	m	17.40
1200 mm - 1800 mm total girth	-	1.30	17.99	4.31	m	22.30
external angle; with joint tape bedded and covered with "Jointex" or other equal and approved	-	0.11	1.51	0.27	m	1.78
add for "Duplex" insulating grade or other equal and approved	-	-	-	0.46	m^2	0.46
Tapered edge plank						
19 mm plank to walls						
wall height 2.40 m - 2.70 m	-	1.02	14.13	7.97	m	22.10
wall height 2.70 m - 3.00 m	-	1.20	16.63	8.86	m	25.49
wall height 3.00 m - 3.30 m	-	1.30	18.01	9.75	m	27.76
wall height 3.30 m - 3.60 m	-	1.53	21.20	10.67	m	31.87
19 mm plank to reveals and soffits of openings and recesses						
not exceeding 300 mm wide	-	0.20	2.78	1.37	m	4.15
300 mm - 600 mm wide	-	0.42	5.83	2.12	m	7.95
19 mm plank to faces of columns - 4 nr						
not exceeding 600 mm total girth	-	0.51	7.08	2.77	m	9.85
600 mm - 1200 mm total girth	-	0.97	13.45	4.29	m	17.74
1200 mm - 1800 mm total girth	-	1.25	17.31	5.82	m	23.13
19 mm plank to ceilings						
over 300 mm wide	-	0.43	5.97	2.95	m^2	8.92

K LININGS/SHEATHING/DRY PARTITIONING

	PC £	Labour hours	Labour £	Material £	Unit	Total rate £
19 mm plank to faces of beams - 3 nr						
not exceeding 600 mm total girth	-	0.60	8.33	2.71	m	11.04
600 mm - 1200 mm total girth	-	1.06	14.69	4.24	m	18.93
1200 mm - 1800 mm total girth	-	1.34	18.55	5.77	m	24.32
Thermal Board						
27 mm board to walls						
wall height 2.40 m - 2.70 m	-	1.06	14.69	14.61	m	29.30
wall height 2.70 m - 3.00 m	-	1.23	17.05	16.23	m	33.28
wall height 3.00 m - 3.30 m	-	1.34	18.58	17.86	m	36.44
wall height 3.30 m - 3.60 m	-	1.62	22.46	19.52	m	41.98
27 mm board to reveals and soffits of openings and recesses						
not exceeding 300 mm wide	-	0.21	2.92	2.10	m	5.02
300 mm - 600 mm wide	-	0.43	5.97	3.60	m	9.57
27 mm board to faces of columns - 4 nr						
not exceeding 600 mm total girth	-	0.52	7.22	4.24	m	11.46
600 mm - 1200 mm total girth	-	1.02	14.15	7.24	m	21.39
1200 mm - 1800 mm total girth	-	1.30	18.01	10.24	m	28.25
27 mm board to ceilings						
over 300 mm wide	-	0.46	6.39	5.41	m^2	11.80
27 mm board to faces of beams - 3 nr						
not exceeding 600 mm total girth	-	0.56	7.78	4.19	m	11.97
600 mm - 1200 mm total girth	-	1.06	14.71	7.19	m	21.90
1200 mm - 1800 mm total girth	-	1.43	19.81	10.19	m	30.00
50 mm board to walls						
wall height 2.40 m - 2.70 m	-	1.06	14.69	14.54	m	29.23
wall height 2.70 m - 3.00 m	-	1.30	18.03	16.18	m	34.21
wall height 3.00 m - 3.30 m	-	1.43	19.84	17.80	m	37.64
wall height 3.30 m - 3.60 m	-	1.71	23.72	19.45	m	43.17
50 mm board to reveals and soffits of openings and recesses						
not exceeding 300 mm wide	-	0.23	3.20	2.10	m	5.30
300 mm - 600 mm wide	-	0.46	6.39	3.59	m	9.98
50 mm board to faces of columns - 4 nr						
not exceeding 600 mm total girth	-	0.56	7.78	4.27	m	12.05
600 mm - 1200 mm total girth	-	1.11	15.41	7.26	m	22.67
1200 mm - 1800 mm total girth	-	1.43	19.83	10.24	m	30.07
50 mm board to ceilings						
over 300 mm wide	-	0.49	6.81	5.39	m^2	12.20
50 mm board to faces of beams - 3 nr						
not exceeding 600 mm total girth	-	0.58	8.06	4.31	m	12.37
600 mm - 1200 mm total girth	-	1.17	16.25	7.33	m	23.58
1200 mm - 1800 mm total girth	-	1.57	21.77	10.34	m	32.11
White plastic faced gypsum plasterboard to BS 1230; industrial grade square edge wallboard; fixing with screws; butt joints; to softwood base						
9.50 mm board to walls						
wall height 2.40 m - 2.70 m	-	0.65	9.10	7.95	m	17.05
wall height 2.70 m - 3.00 m	-	0.79	11.06	8.82	m	19.88
wall height 3.00 m - 3.30 m	-	0.93	13.02	9.69	m	22.71
wall height 3.30 m - 3.60 m	-	1.06	14.84	10.56	m	25.40
9.50 mm board to reveals and soffits of openings and recesses						
not exceeding 300 mm wide	-	0.14	1.96	0.91	m	2.87
300 mm - 600 mm wide	-	0.28	3.92	1.77	m	5.69
9.50 mm board to faces of columns - 4 nr						
not exceeding 600 mm total girth	-	0.37	5.18	1.91	m	7.09
600 mm - 1200 mm total girth	-	0.74	10.36	3.74	m	14.10
1200 mm - 1800 mm total girth	-	0.97	13.58	5.52	m	19.10

K LININGS/SHEATHING/DRY PARTITIONING

	PC £	Labour hours	Labour £	Material £	Unit	Total rate £
K10 PLASTERBOARD DRY LINING/ PARTITIONS/CEILINGS - cont'd						
White plastic faced gypsum plasterboard to BS 1230; industrial grade square edge wallboard; fixing with screws; butt joints; to softwood base - cont'd						
12.50 mm board to walls						
wall height 2.40 m - 2.70 m	-	0.69	9.66	8.44	m	18.10
wall height 2.70 m - 3.00 m	-	0.83	11.62	9.37	m	20.99
wall height 3.00 m - 3.30 m	-	0.97	13.58	10.30	m	23.88
wall height 3.30 m - 3.60 m	-	1.11	15.54	11.23	m	26.77
12.50 mm board to reveals and soffits of openings and recesses						
not exceeding 300 mm wide	-	0.15	2.10	0.97	m	3.07
300 mm - 600 mm wide	-	0.30	4.20	1.88	m	6.08
12.50 mm board to faces of columns - 4 nr						
not exceeding 600 mm total girth	-	0.39	5.46	2.04	m	7.50
600 mm - 1200 mm total girth	-	0.78	10.92	3.98	m	14.90
1200 mm - 1800 mm total girth	-	1.02	14.28	5.87	m	20.15
Plasterboard jointing system; filling joint with jointing compounds						
To ceilings						
to suit 9.50 mm or 12.50 mm thick boards	-	0.09	1.26	1.33	m	2.59
Angle trim; plasterboard edge support system						
To ceilings						
to suit 9.50 mm or 12.50 mm thick boards	-	0.09	1.26	1.25	m	2.51
Two layers of gypsum plasterboard to BS 1230; plain grade square and tapered edge wallboard; fixing with nails; joints filled with joint filler and joint tape; top layer to receive direct decoration; to softwood base						
19 mm two layer board to walls						
wall height 2.40 m - 2.70 m	-	1.30	18.05	9.42	m	27.47
wall height 2.70 m - 3.00 m	-	1.48	20.55	10.48	m	31.03
wall height 3.00 m - 3.30m	-	1.67	23.18	11.53	m	34.71
wall height 3.30 m - 3.60m	-	1.94	26.94	12.61	m	39.55
19 mm two layer board to reveals and soffits of openings and recesses						
not exceeding 300 mm wide	-	0.28	3.90	1.55	m	5.45
300 mm - 600 mm wide	-	0.56	7.79	2.46	m	10.25
19 mm two layer board to faces of columns - 4 nr						
not exceeding 600 mm total girth	-	0.69	9.60	3.18	m	12.78
600 mm - 1200 mm total girth	-	1.34	18.63	5.01	m	23.64
1200 mm - 1800 mm total girth	-	1.67	23.18	6.84	m	30.02
25 mm two layer board to walls						
wall height 2.40 m - 2.70 m	-	1.39	19.31	10.59	m	29.90
wall height 2.70 m - 3.00 m	-	1.57	21.81	11.77	m	33.58
wall height 3.00 m - 3.30 m	-	1.76	24.44	12.96	m	37.40
wall height 3.30 m - 3.60 m	-	2.04	28.34	14.17	m	42.51
25 mm two layer board to reveals and soffits of openings and recesses						
not exceeding 300 mm wide	-	0.28	3.90	1.71	m	5.61
300 mm - 600 mm wide	-	0.56	7.79	2.73	m	10.52
25 mm two layer board to faces of columns - 4 nr						
not exceeding 600 mm total girth	-	0.69	9.60	3.48	m	13.08
600 mm - 1200 mm total girth	-	1.34	18.63	5.56	m	24.19
1200 mm - 1800 mm total girth	-	1.67	23.18	7.65	m	30.83

Prices for Measured Works - Major Works
K LININGS/SHEATHING/DRY PARTITIONING

	PC £	Labour hours	Labour £	Material £	Unit	Total rate £
Gyproc Dri-Wall dry lining system or other equal or approved; plain grade tapered edge wallboard; fixed to walls with adhesive; joints filled with joint filler and joint tape; to receive direct decoration						
9.50 mm board to walls						
wall height 2.40 m - 2.70 m	-	1.11	15.39	6.37	m	21.76
wall height 2.70 m - 3.00 m	-	1.28	17.75	7.07	m	24.82
wall height 3.00 m - 3.30 m	-	1.43	19.83	7.77	m	27.60
wall height 3.30 m - 3.60 m	-	1.67	23.16	8.50	m	31.66
9.50 mm board to reveals and soffits of openings and recesses						
not exceeding 300 mm wide	-	0.23	3.20	1.13	m	4.33
300 mm - 600 mm wide	-	0.46	6.39	1.71	m	8.10
9.50 mm board to faces of columns - 4 nr						
not exceeding 600 mm total girth	-	0.58	8.06	2.26	m	10.32
600 mm - 1200 mm total girth	-	1.14	15.83	3.53	m	19.36
1200 mm - 1800 mm total girth	-	1.43	19.83	4.60	m	24.43
Angle; with joint tape bedded and covered with "Jointex" or other equal and approved						
internal	-	0.05	0.69	0.27	m	0.96
external	-	0.11	1.51	0.27	m	1.78
Gyproc Dri-Wall M/F dry lining system or other equal or approved; mild steel furrings fixed to walls with adhesive; tapered edge wallboard screwed to furrings; joints filled with joint filler and joint tape						
12.50 mm board to walls						
wall height 2.40 m - 2.70 m	-	1.48	20.57	11.48	m	32.05
wall height 2.70 m - 3.00 m	-	1.69	23.49	12.75	m	36.24
wall height 3.00 m - 3.30 m	-	1.90	26.40	14.03	m	40.43
wall height 3.30 m - 3.60 m	-	2.22	30.86	15.34	m	46.20
12.50 mm board to reveals and soffits of openings and recesses						
not exceeding 300 mm wide	-	0.23	3.20	1.02	m	4.22
300 mm - 600 mm wide	-	0.46	6.39	1.49	m	7.88
add for one coat of "Drywall" top coat or other equal and approved	-	0.04	0.55	0.66	m^2	1.21
Vermiculite gypsum cladding; "Vicuclad 900R" board or other equal and approved; fixed with adhesive; joints pointed in adhesive						
25 mm thick column linings, faces - 4; 2 hour fire protection rating						
not exceeding 600 mm girth	-	0.69	9.66	12.32	m	21.98
600 mm - 1200 mm girth	-	0.83	11.62	24.42	m	36.04
1200 mm - 1800 mm girth	-	1.16	16.24	36.53	m	52.77
30 mm thick beam linings, faces - 3; 2 hour fire protection rating						
not exceeding 600 mm girth	-	0.56	7.84	14.52	m	22.36
600 mm - 1200 mm girth	-	0.69	9.66	28.82	m	38.48
1200 mm - 1800 mm girth	-	0.83	11.62	43.13	m	54.75
Vermiculite gypsum cladding; "Vicuclad 1050R" board or other equal and approved; fixed with adhesive; joints pointed in adhesive						
55 mm thick column linings, faces - 4 ; 4 hour fire protection rating						
not exceeding 600 mm girth	-	0.83	11.62	32.43	m	44.05
600 mm - 1200 mm girth	-	1.02	14.28	64.65	m	78.93
1200 mm - 1800 mm girth	-	1.39	19.46	96.87	m	116.33

K LININGS/SHEATHING/DRY PARTITIONING

	PC £	Labour hours	Labour £	Material £	Unit	Total rate £
K10 PLASTERBOARD DRY LINING/ PARTITIONS/CEILINGS - cont'd						
Vermiculite gypsum cladding; "Vicuclad 1050R" board or other equal and approved; fixed with adhesive; joints pointed in adhesive - cont'd						
60 mm thick beam linings, faces - 3; 4 hour fire protection rating						
not exceeding 600 mm girth	-	0.69	9.66	35.54	m	45.20
600 mm - 1200 mm girth	-	0.83	11.62	70.87	m	82.49
1200 mm - 1800 mm girth	-	1.02	14.28	106.20	m	120.48
Add to the above for						
plus 3% for work 3.50 m - 5.00 m high						
plus 6% for work 5.00 m - 6.50 m high						
plus 12% for work 6.50 m - 8.00 m high						
plus 18% for work over 8.00 m high						
Cutting and fitting around steel joints, angles, trunking, ducting, ventilators, pipes, tubes, etc						
over 2 m girth	-	0.42	5.88	-	m	5.88
not exceeding 0.30 m girth	-	0.28	3.92	-	nr	3.92
0.30 m - 1 m girth	-	0.37	5.18	-	nr	5.18
1 m - 2 m girth	-	0.51	7.14	-	nr	7.14
K11 RIGID SHEET FLOORING/SHEATHING/ LININGS/CASING						
Blockboard (Birch faced)						
Linings to walls 12 mm thick						
over 300 mm wide	-	0.43	6.02	9.39	m²	15.41
not exceeding 300 mm wide	-	0.29	4.06	2.83	m	6.89
holes for pipes and the like	-	0.03	0.42	-	nr	0.42
Two-sided 12 mm thick pipe casing; to softwood framing (not included)						
300 mm girth	-	0.69	9.66	2.89	m	12.55
600 mm girth	-	0.83	11.62	5.66	m	17.28
Three-sided 12 mm thick pipe casing; to softwood framing (not included)						
450 mm girth	-	0.93	13.02	4.34	m	17.36
900 mm girth	-	1.11	15.54	8.51	m	24.05
extra for 400 mm x 400 mm removable access panel; brass cups and screws; additional framing	-	1.00	14.00	0.77	nr	14.77
Lining to walls 18 mm thick						
over 300 wide	11.15	0.46	6.44	13.06	m²	19.50
not exceeding 300 wide	-	0.30	4.20	3.93	m	8.13
holes for pipes and the like	-	0.04	0.56	-	nr	0.56
Lining to walls 25 mm thick						
over 300 wide	14.52	0.50	7.00	16.99	m²	23.99
not exceeding 300 wide	-	0.32	4.48	5.10	m	9.58
holes for pipes and the like	-	0.05	0.70	-	nr	0.70
Chipboard (plain)						
Lining to walls 12 mm thick						
over 300 mm wide	1.78	0.35	4.90	2.23	m²	7.13
not exceeding 300 mm wide	-	0.20	2.80	0.68	m	3.48
holes for pipes and the like	-	0.02	0.28	-	nr	0.28
Lining to walls 15 mm thick						
over 300 mm wide	2.04	0.37	5.18	2.53	m²	7.71
not exceeding 300 mm wide	-	0.22	3.08	0.77	m	3.85
holes for pipes and the like	-	0.03	0.42	-	nr	0.42

Prices for Measured Works - Major Works
K LININGS/SHEATHING/DRY PARTITIONING

	PC £	Labour hours	Labour £	Material £	Unit	Total rate £
Two-sided 15 mm thick pipe casing; to softwood framing (not included)						
300 mm girth	-	0.56	7.84	0.83	m	8.67
600 mm girth	-	0.65	9.10	1.54	m	10.64
Three-sided 15 mm thick pipe casing; to softwood framing (not included)						
450 mm girth	-	1.16	16.24	1.25	m	17.49
900 mm girth	-	1.39	19.46	2.33	m	21.79
extra for 400 mm x 400 mm removable access panel; brass cups and screws; additional framing	-	0.93	13.02	0.77	nr	13.79
Lining to walls 18 mm thick						
over 300 mm wide	2.45	0.39	5.46	3.04	m²	8.50
not exceeding 300 mm wide	-	0.25	3.50	0.91	m	4.41
holes for pipes and the like	-	0.04	0.56	-	nr	0.56
Fire-retardent chipboard; Antivlam or other equal and approved; Class 1 spread of flame						
Lining to walls 12 mm thick						
over 300 mm wide	-	0.35	4.90	7.42	m²	12.32
not exceeding 300 mm wide	-	0.20	2.80	2.24	m	5.04
holes for pipes and the like	-	0.02	0.28	-	nr	0.28
Lining to walls 18 mm thick						
over 300 mm wide	-	0.39	5.46	9.79	m²	15.25
not exceeding 300 mm wide	-	0.25	3.50	2.95	m	6.45
holes for pipes and the like	-	0.04	0.56	-	nr	0.56
Lining to walls 22 mm thick						
over 300 mm wide	-	0.41	5.74	12.43	m²	18.17
not exceeding 300 mm wide	-	0.28	3.92	3.74	m	7.66
holes for pipes and the like	-	0.05	0.70	-	nr	0.70
Chipboard Melamine faced; white matt finish; laminated masking strips						
Lining to walls 15 mm thick						
over 300 mm wide	3.29	0.97	13.58	4.13	m²	17.71
not exceeding 300 mm wide	-	0.63	8.82	1.31	m	10.13
Chipboard boarding and flooring						
Boarding to floors; butt joints						
18 mm thick	2.39	0.28	3.92	2.97	m²	6.89
22 mm thick	3.10	0.31	4.34	3.80	m²	8.14
Boarding to floors; tongued and grooved joints						
18 mm thick	2.52	0.30	4.20	3.13	m²	7.33
22 mm thick	3.10	0.32	4.48	3.80	m²	8.28
Boarding to roofs; butt joints						
18 mm thick	3.14	0.31	4.34	3.84	m²	8.18
Durabella "Westbourne" flooring system or other equal and approved; comprising 19 mm thick tongued and grooved chipboard panels secret nailed to softwood MK 10X-profiled foam backed battens at 600 mm centres; on concrete floor						
Flooring tongued and grooved joints						
63 mm thick overall; 44 mm x 54 mm nominal size battens	-	-	-	-	m²	17.97
75 mm thick overall; 56 mm x 54 mm nominal size battens	-	-	-	-	m²	19.73
Plywood flooring						
Boarding to floors; tongued and grooved joints						
15 mm thick	14.20	0.37	5.18	16.63	m²	21.81
18 mm thick	16.49	0.41	5.74	19.27	m²	25.01

Prices for Measured Works - Major Works
K LININGS/SHEATHING/DRY PARTITIONING

	PC £	Labour hours	Labour £	Material £	Unit	Total rate £
K11 RIGID SHEET FLOORING/SHEATHING/ LININGS/CASING - cont'd						
Plywood; external quality; 18 mm thick						
Boarding to roofs; butt joints						
flat to falls	11.12	0.37	5.18	13.06	m²	18.24
sloping	11.12	0.40	5.60	13.06	m²	18.66
vertical	11.12	0.53	7.42	13.06	m²	20.48
Plywood; external quality; 12 mm thick						
Boarding to roofs; butt joints						
flat to falls	7.42	0.37	5.18	8.78	m²	13.96
sloping	7.42	0.40	5.60	8.78	m²	14.38
vertical	7.42	0.53	7.42	8.78	m²	16.20
Glazed hardboard to BS 1142; on and including 38 mm x 38 mm sawn softwood framing 3.20 mm thick panel						
to side of bath	-	1.67	23.37	4.06	nr	27.43
to end of bath	-	0.65	9.10	1.13	nr	10.23
Insulation board to BS 1142						
Lining to walls 12 mm thick						
over 300 mm wide	1.35	0.22	3.08	1.72	m²	4.80
not exceeding 300 mm wide	-	0.13	1.82	0.53	m	2.35
holes for pipes and the like	-	0.01	0.14	-	nr	0.14
Lining to walls 18 mm thick						
over 300 mm wide	2.12	0.24	3.36	2.62	m²	5.98
not exceeding 300 mm wide	-	0.15	2.10	0.80	m	2.90
holes for pipes and the like	-	0.01	0.14	-	nr	0.14
Laminboard (Birch Faced); 18 mm thick						
Lining to walls						
over 300 mm wide	13.92	0.49	6.86	16.26	m²	23.12
not exceeding 300 mm wide	-	0.31	4.34	4.89	m	9.23
holes for pipes and the like	-	0.05	0.70	-	nr	0.70
Non-asbestos board; "Masterboard" or other equal and approved; sanded finish						
Lining to walls 6 mm thick						
over 300 mm wide	4.92	0.31	4.34	5.81	m²	10.15
not exceeding 300 mm wide	-	0.19	2.66	1.75	m	4.41
Lining to ceilings 6 mm thick						
over 300 mm wide	4.92	0.41	5.74	5.81	m²	11.55
not exceeding 300 mm wide	-	0.25	3.50	1.75	m	5.25
holes for pipes and the like	-	0.02	0.28	-	nr	0.28
Lining to walls 9 mm thick						
over 300 mm wide	10.38	0.33	4.62	12.12	m²	16.74
not exceeding 300 mm wide	-	0.19	2.66	3.64	m	6.30
Lining to ceilings 9 mm thick						
over 300 mm wide	10.38	0.42	5.88	12.12	m²	18.00
not exceeding 300 mm wide	-	0.27	3.78	3.64	m	7.42
holes for pipes and the like	-	0.03	0.42	-	nr	0.42
Non-asbestos board; "Supalux" or other equal and approved; sanded finish						
Lining to walls 6 mm thick						
over 300 mm wide	7.56	0.31	4.34	8.86	m²	13.20
not exceeding 300 mm wide	-	0.19	2.66	2.66	m	5.32

K LININGS/SHEATHING/DRY PARTITIONING

	PC £	Labour hours	Labour £	Material £	Unit	Total rate £
Lining to ceilings 6 mm thick						
over 300 mm wide	7.56	0.41	5.74	8.86	m²	14.60
not exceeding 300 mm wide	-	0.25	3.50	2.66	m	6.16
holes for pipes and the like	-	0.03	0.42	-	nr	0.42
Lining to walls 9 mm thick						
over 300 mm wide	11.25	0.33	4.62	13.13	m²	17.75
not exceeding 300 mm wide	-	0.19	2.66	3.94	m	6.60
Lining to ceilings 9 mm thick						
over 300 mm wide	11.25	0.42	5.88	13.13	m²	19.01
not exceeding 300 mm wide	-	0.27	3.78	3.94	m	7.72
holes for pipes and the like	-	0.03	0.42	-	nr	0.42
Lining to walls 12 mm thick						
over 300 mm wide	14.89	0.37	5.18	17.33	m²	22.51
not exceeding 300 mm wide	-	0.22	3.08	5.20	m	8.28
Lining to ceilings 12 mm thick						
over 300 mm wide	14.89	0.49	6.86	17.33	m²	24.19
not exceeding 300 mm wide	-	0.30	4.20	5.20	m	9.40
holes for pipes and the like	-	0.04	0.56	-	nr	0.56
Non-asbestos board; "Monolux 40" or other equal and approved; 6 mm x 50 mm "Supalux" cover fillets or other equal and approved one side						
Lining to walls 19 mm thick						
over 300 mm wide	28.23	0.65	9.10	33.98	m²	43.08
not exceeding 300 mm wide	-	0.46	6.44	11.37	m	17.81
Lining to walls 25 mm thick						
over 300 mm wide	33.83	0.69	9.66	40.46	m²	50.12
not exceeding 300 mm wide	33.83	0.49	6.86	13.31	m	20.17
Plywood (Far Eastern); internal quality						
Lining to walls 4 mm thick						
over 300 mm wide	1.59	0.34	4.76	2.00	m²	6.76
not exceeding 300 mm wide	-	0.22	3.08	0.61	m	3.69
Lining to ceilings 4 mm thick						
over 300 mm wide	1.59	0.46	6.44	2.00	m²	8.44
not exceeding 300 mm wide	-	0.30	4.20	0.61	m	4.81
holes for pipes and the like	-	0.02	0.28	-	nr	0.28
Lining to walls 6 mm thick						
over 300 mm wide	1.88	0.37	5.18	2.34	m²	7.52
not exceeding 300 mm wide	-	0.24	3.36	0.71	m	4.07
Lining to ceilings 6 mm thick						
over 300 mm wide	1.88	0.49	6.86	2.34	m²	9.20
not exceeding 300 mm wide	-	0.32	4.48	0.71	m	5.19
holes for pipes and the like	-	0.02	0.28	-	nr	0.28
Two-sided 6 mm thick pipe casings; to softwood framing (not included)						
300 mm girth	-	0.74	10.36	0.78	m	11.14
600 mm girth	-	0.93	13.02	1.43	m	14.45
Three-sided 6 mm thick pipe casing; to softwood framing (not included)						
450 mm girth	-	1.06	14.84	1.17	m	16.01
900 mm girth	-	1.25	17.50	2.16	m	19.66
Lining to walls 9 mm thick						
over 300 mm wide	2.75	0.40	5.60	3.34	m²	8.94
not exceeding 300 mm wide	-	0.26	3.64	1.02	m	4.66
Lining to ceilings 9 mm thick						
over 300 mm wide	2.75	0.53	7.42	3.34	m²	10.76
not exceeding 300 mm wide	-	0.34	4.76	1.02	m	5.78
holes for pipes and the like	-	0.03	0.42	-	nr	0.42

K LININGS/SHEATHING/DRY PARTITIONING

	PC £	Labour hours	Labour £	Material £	Unit	Total rate £
K11 RIGID SHEET FLOORING/SHEATHING/ LININGS/CASING - cont'd						
Plywood (Far Eastern); internal quality - cont'd						
Lining to walls 12 mm thick						
over 300 mm wide	3.54	0.43	6.02	4.26	m²	10.28
not exceeding 300 mm wide	-	0.28	3.92	1.29	m	5.21
Lining to ceilings 12 mm thick						
over 300 mm wide	3.54	0.56	7.84	4.26	m²	12.10
not exceeding 300 mm wide	-	0.37	5.18	1.29	m	6.47
holes for pipes and the like	-	0.03	0.42	-	nr	0.42
Plywood (Far Eastern); external quality						
Lining to walls 4 mm thick						
over 300 mm wide	2.86	0.34	4.76	3.47	m²	8.23
not exceeding 300 mm wide	-	0.22	3.08	1.05	m	4.13
Lining to ceilings 4 mm thick						
over 300 mm wide	2.86	0.46	6.44	3.47	m²	9.91
not exceeding 300 mm wide	-	0.30	4.20	1.05	m	5.25
holes for pipes and the like	-	0.02	0.28	-	nr	0.28
Lining to walls 6 mm thick						
over 300 mm wide	3.76	0.37	5.18	4.52	m²	9.70
not exceeding 300 mm wide	-	0.24	3.36	1.37	m	4.73
Lining to ceilings 6 mm thick						
over 300 mm wide	3.76	0.49	6.86	4.52	m²	11.38
not exceeding 300 mm wide	-	0.32	4.48	1.37	m	5.85
holes for pipes and the like	-	0.02	0.28	-	nr	0.28
Two-sided 6 mm thick pipe casings; to softwood framing (not included)						
300 mm girth	-	0.74	10.36	1.43	m	11.79
600 mm girth	-	0.93	13.02	2.74	m	15.76
Three-sided 6 mm thick pipe casing; to softwood framing (not included)						
450 mm girth	-	1.06	14.84	2.15	m	16.99
900 mm girth	-	1.25	17.50	4.13	m	21.63
Lining to walls 9 mm thick						
over 300 mm wide	5.01	0.40	5.60	5.96	m²	11.56
not exceeding 300 mm wide	-	0.26	3.64	1.80	m	5.44
Lining to ceilings 9 mm thick						
over 300 mm wide	5.01	0.53	7.42	5.96	m²	13.38
not exceeding 300 mm wide	-	0.34	4.76	1.80	m	6.56
holes for pipes and the like	-	0.03	0.42	-	nr	0.42
Lining to walls 12 mm thick						
over 300 mm wide	6.65	0.43	6.02	7.86	m²	13.88
not exceeding 300 mm wide	-	0.28	3.92	2.37	m	6.29
Lining to ceilings 12 mm thick						
over 300 mm wide	6.65	0.56	7.84	7.86	m²	15.70
not exceeding 300 mm wide	-	0.37	5.18	2.37	m	7.55
holes for pipes and the like	-	0.03	0.42	-	nr	0.42
Extra over wall linings fixed with nails for screwing	-	-	-	-	m²	1.56
Preformed white melamine faced plywood casings; Pendock Profiles Ltd or other equal and approved; to softwood battens (not included)						
Skirting trunking profile; plain butt joints in the running length						
45 mm x 150 mm; ref TK150	-	0.11	1.54	6.36	m	7.90
extra for stop end	-	0.04	0.56	5.28	nr	5.84
extra for external corner	-	0.09	1.26	6.52	nr	7.78
extra for internal corner	-	0.09	1.26	4.87	nr	6.13

K LININGS/SHEATHING/DRY PARTITIONING

	PC £	Labour hours	Labour £	Material £	Unit	Total rate £
Casing profiles						
150 mm x 150 mm; ref MX150/150; 5 mm thick	-	0.11	1.54	10.12	m	11.66
extra for stop end	-	0.04	0.56	4.14	nr	4.70
extra for external corner	-	0.09	1.26	15.38	nr	16.64
extra for internal corner	-	0.09	1.26	4.87	nr	6.13
Woodwool unreinforced slabs; Torvale "Woodcemair" or other equal and approved; BS 1105 type SB; natural finish; fixing to timber or steel with galvanized nails or clips; flat or sloping						
50 mm thick slabs; type 500; 600 mm maximum span						
1800 mm - 2400 mm lengths	5.55	0.37	5.18	6.50	m²	11.68
2700 mm - 3000 mm lengths	5.69	0.37	5.18	6.66	m²	11.84
75 mm thick slabs; type 750; 900 mm maximum span						
2100 mm lengths	6.62	0.42	5.88	7.75	m²	13.63
2400 mm - 2700 mm lengths	6.71	0.42	5.88	7.85	m²	13.73
3000 mm lengths	6.72	0.42	5.88	7.86	m²	13.74
100 mm thick slabs; type 1000; 1200 mm maximum span						
3000 mm - 3600 mm lengths	9.16	0.46	6.44	10.57	m²	17.01
Internal quality American Cherry veneered plywood; 6 mm thick						
Lining to walls						
over 300 mm wide	4.76	0.41	5.74	5.63	m²	11.37
not exceeding 300 mm wide	-	0.27	3.78	1.72	m	5.50
"Tacboard" or other equal and approved; Eternit UK Ltd; fire resisting boards; butt joints; to softwood base						
Lining to walls; 6 mm thick						
over 300 mm wide	-	0.31	4.34	8.33	m²	12.67
not exceeding 300 mm wide	-	0.19	2.66	2.53	m	5.19
Lining to walls; 9 mm thick						
over 300 mm wide	-	0.33	4.62	15.26	m²	19.88
not exceeding 300 mm wide	-	0.20	2.80	4.60	m	7.40
Lining to walls; 12 mm thick						
over 300 wide	-	0.37	5.18	19.84	m²	25.02
not exceeding 300 mm wide	-	0.22	3.08	5.98	m	9.06
"Tacfire" or other equal and approved; Eternit UK Ltd; fire resisting boards						
Lining to walls; 6 mm thick						
over 300 mm wide	-	0.31	4.34	11.17	m²	15.51
not exceeding 300 mm wide	-	0.19	2.66	3.38	m	6.04
Lining to walls; 9 mm thick						
over 300 mm wide	-	0.33	4.62	17.04	m²	21.66
not exceeding 300 mm wide	-	0.20	2.80	5.14	m	7.94
Lining to walls; 12 mm thick						
over 300 mm wide	-	0.37	5.18	22.42	m²	27.60
not exceeding 300 mm wide	-	0.22	3.08	6.75	m	9.83

Prices for Measured Works - Major Works
K LININGS/SHEATHING/DRY PARTITIONING

	PC £	Labour hours	Labour £	Material £	Unit	Total rate £
K14 GLASS REINFORCED GYPSUM LININGS/ PANELLING/CASINGS/MOULDINGS						
Glass reinforced gypsum Glasroc Multi-board or other equal and approved; fixing with nails; joints filled with joint filler and joint tape; finishing with "Jointex" or other equal and approved to receive decoration; to softwood base						
10 mm board to walls						
wall height 2.40 m - 2.70 m	-	0.93	12.87	35.32	m	48.19
wall height 2.70 m - 3.00 m	-	1.06	14.67	39.25	m	53.92
wall height 3.00 m - 3.30 m	-	1.20	16.61	43.17	m	59.78
wall height 3.30 m - 3.60 m	-	1.39	19.24	47.14	m	66.38
12.50 mm board to walls						
wall height 2.40 m - 2.70 m	-	0.97	13.43	46.17	m	59.60
wall height 2.70 m - 3.00 m	-	1.11	15.37	51.30	m	66.67
wall height 3.00 m - 3.30 m	-	1.25	17.31	56.44	m	73.75
wall height 3.30 m - 3.60 m	-	1.43	19.80	61.60	m	81.40
K20 TIMBER BOARD FLOORING/SHEATHING/ LININGS/CASINGS						
Sawn softwood; untreated						
Boarding to roofs; 150 mm wide boards; butt joints						
19 mm thick; flat; over 300 mm wide	-	0.42	5.88	5.51	m²	11.39
19 mm thick; flat; not exceeding 300 mm wide	-	0.28	3.92	1.67	m	5.59
19 mm thick; sloping; over 300 mm wide	-	0.46	6.44	5.51	m²	11.95
19 mm thick; sloping; not exceeding 300 mm wide	-	0.31	4.34	1.67	m	6.01
19 mm thick; sloping; laid diagonally; over 300 mm wide	-	0.58	8.12	5.51	m²	13.63
19 mm thick; sloping; laid diagonally; not exceeding 300 mm wide	-	0.37	5.18	1.67	m	6.85
25 mm thick; flat; over 300 mm wide	-	0.42	5.88	7.20	m²	13.08
25 mm thick; flat; not exceeding 300 mm wide	-	0.28	3.92	2.18	m	6.10
25 mm thick; sloping; over 300 mm wide	-	0.46	6.44	7.20	m²	13.64
25 mm thick; sloping; not exceeding 300 mm wide	-	0.31	4.34	2.18	m	6.52
25 mm thick; sloping; laid diagonally; over 300 mm wide	-	0.58	8.12	7.20	m²	15.32
25 mm thick; sloping; laid diagonally; not exceeding 300 mm wide	-	0.37	5.18	2.18	m	7.36
Boarding to tops or cheeks of dormers; 150 mm wide boards; butt joints						
19 mm thick; laid diagonally; over 300 mm wide	-	0.74	10.36	5.51	m²	15.87
19 mm thick; laid diagonally; not exceeding 300 mm wide	-	0.46	6.44	1.67	m	8.11
19 mm thick; laid diagonally; area not exceeding 1.00 m² irrespective of width	-	0.93	13.02	5.23	nr	18.25
Sawn softwood; "Tanalised"						
Boarding to roofs; 150 wide boards; butt joints						
19 mm thick; flat; over 300 mm wide	-	0.42	5.88	6.26	m²	12.14
19 mm thick; flat; not exceeding 300 mm wide	-	0.28	3.92	1.90	m	5.82
19 mm thick; sloping; over 300 mm wide	-	0.46	6.44	6.26	m²	12.70
19 mm thick; sloping; not exceeding 300 mm wide	-	0.31	4.34	1.90	m	6.24
19 mm thick; sloping; laid diagonally; over 300 mm wide	-	0.58	8.12	6.26	m²	14.38
19 mm thick; sloping; laid diagonally; not exceeding 300 mm wide	-	0.37	5.18	1.90	m	7.08
25 mm thick; flat; over 300 mm wide	-	0.42	5.88	8.20	m²	14.08
25 mm thick; flat; not exceeding 300 mm wide	-	0.28	3.92	2.48	m	6.40
25 mm thick; sloping; over 300 mm wide	-	0.46	6.44	8.20	m²	14.64

K LININGS/SHEATHING/DRY PARTITIONING

	PC £	Labour hours	Labour £	Material £	Unit	Total rate £
Boarding to roofs; 150 wide boards; butt joints - cont'd						
25 mm thick; sloping; not exceeding 300 mm wide	-	0.31	4.34	2.48	m	6.82
25 mm thick; sloping; laid diagonally; over 300 mm wide	-	0.58	8.12	8.20	m²	16.32
25 mm thick; sloping; laid diagonally; not exceeding 300 mm wide	-	0.37	5.18	2.48	m	7.66
Boarding to tops or cheeks of dormers; 150 mm wide boards; butt joints						
19 mm thick; laid diagonally; over 300 mm wide	-	0.74	10.36	6.26	m²	16.62
19 mm thick; laid diagonally; not exceeding 300 mm wide	-	0.46	6.44	1.90	m	8.34
19 mm thick; laid diagonally; area not exceeding 1.00 m² irrespective of width	-	0.93	13.02	5.99	nr	19.01
Wrought softwood						
Boarding to floors; butt joints						
19 mm x 75 mm boards	-	0.56	7.84	6.34	m²	14.18
19 mm x 125 mm boards	-	0.51	7.14	6.98	m²	14.12
22 mm x 150 mm boards	-	0.46	6.44	7.48	m²	13.92
25 mm x 100 mm boards	-	0.51	7.14	8.07	m²	15.21
25 mm x 150 mm boards	-	0.46	6.44	8.66	m²	15.10
Boarding to floors; tongued and grooved joints						
19 mm x 75 mm boards	-	0.65	9.10	6.93	m²	16.03
19 mm x 125 mm boards	-	0.60	8.40	7.56	m²	15.96
22 mm x 150 mm boards	-	0.56	7.84	8.07	m²	15.91
25 mm x 100 mm boards	-	0.60	8.40	8.66	m²	17.06
25 mm x 150 mm boards	-	0.56	7.84	9.24	m²	17.08
Boarding to internal walls; tongued and grooved and V-jointed						
12 mm x 100 mm boards	-	0.74	10.36	6.08	m²	16.44
16 mm x 100 mm boards	-	0.74	10.36	8.14	m²	18.50
19 mm x 100 mm boards	-	0.74	10.36	9.13	m²	19.49
19 mm x 125 mm boards	-	0.69	9.66	9.59	m²	19.25
19 mm x 125 mm boards; chevron pattern	-	1.11	15.54	9.59	m²	25.13
25 mm x 125 mm boards	-	0.69	9.66	12.35	m²	22.01
12 mm x 100 mm boards; knotty pine	-	0.74	10.36	5.02	m²	15.38
Boarding to internal ceilings						
12 mm x 100 mm boards	-	0.93	13.02	6.08	m²	19.10
16 mm x 100 mm boards	-	0.93	13.02	8.14	m²	21.16
19 mm x 100 mm boards	-	0.93	13.02	9.13	m²	22.15
19 mm x 125 mm boards	-	0.88	12.32	9.59	m²	21.91
19 mm x 125 mm boards; chevron pattern	-	1.30	18.20	9.59	m²	27.79
25 mm x 125 mm boards	-	0.88	12.32	12.35	m²	24.67
12 mm x 100 mm boards; knotty pine	-	0.93	13.02	5.02	m²	18.04
Boarding to roofs; tongued and grooved joints						
19 mm thick; flat to falls	-	0.51	7.14	7.20	m²	14.34
19 mm thick; sloping	-	0.56	7.84	7.20	m²	15.04
19 mm thick; sloping; laid diagonally	-	0.72	10.08	7.20	m²	17.28
25 mm thick; flat to falls	-	0.51	7.14	8.67	m²	15.81
25 mm thick; sloping	-	0.56	7.84	8.67	m²	16.51
Boarding to tops or cheeks of dormers; tongued and grooved joints						
19 mm thick; laid diagonally	-	0.93	13.02	7.20	m²	20.22
Wrought softwood; "Tanalised"						
Boarding to roofs; tongued and grooved joints						
19 mm thick; flat to falls	-	0.51	7.14	7.96	m²	15.10
19 mm thick; sloping	-	0.56	7.84	7.96	m²	15.80
19 mm thick; sloping; laid diagonally	-	0.72	10.08	7.96	m²	18.04
25 mm thick; flat to falls	-	0.51	7.14	9.66	m²	16.80
25 mm thick; sloping	-	0.56	7.84	9.66	m²	17.50

Prices for Measured Works - Major Works
K LININGS/SHEATHING/DRY PARTITIONING

	PC £	Labour hours	Labour £	Material £	Unit	Total rate £
K20 TIMBER BOARD FLOORING/SHEATHING/ LININGS/CASINGS - cont'd						
Wrought softwood; "Tanalised" - cont'd						
Boarding to tops or cheeks of dormers; tongued and grooved joints						
19 mm thick; laid diagonally	-	0.93	13.02	7.96	m²	20.98
Wood strip; 22 mm thick; "Junckers" pre-treated or other equal and approved; tongued and grooved joints; pre-finished boards; level fixing to resilient battens; to cement and sand base						
Strip flooring; over 300 mm wide						
beech; prime	-	-	-	-	m²	54.92
beech; standard	-	-	-	-	m²	47.44
beech; sylvia squash	-	-	-	-	m²	49.93
oak; quality A	-	-	-	-	m²	59.92
Wrought hardwood						
Strip flooring to floors; 25 mm thick x 75 mm wide; tongue and grooved joints; secret fixing; surface sanded after laying						
american oak	-	-	-	-	m²	39.95
canadian maple	-	-	-	-	m²	42.44
gurjun	-	-	-	-	m²	37.45
iroko	-	-	-	-	m²	34.96
K32 FRAMED PANEL CUBICLE PARTITIONS						
Cubicle partitions; Armitage Shanks "Rapide" range or other equal and approved; standard colours; standard ironmongery; assembling; screwing to floor and wall						
Cubicle sets complete with doors						
range of 1 nr open ended cubicles; 1 nr panel; 1 nr door	-	2.08	29.11	259.66	nr	288.77
range of 3 nr open ended cubicles; 3 nr panel; 3 nr door	-	6.24	87.34	699.07	nr	786.41
range of 6 nr open ended cubicles; 6 nr panel; 6 nr door	-	12.50	174.96	1348.21	nr	1523.17
K33 CONCRETE/TERRAZZO PARTITIONS						
Terrazzo faced partitions; polished on two faces						
Pre-cast reinforced terrazzo faced WC partitions						
38 mm thick; over 300 mm wide	-	-	-	-	m²	179.76
50 mm thick; over 300 mm wide	-	-	-	-	m²	194.75
Wall post; once rebated						
64 mm x 102 mm	-	-	-	-	m	104.87
64 mm x 152 mm	-	-	-	-	m	114.84
Centre post; twice rebated						
64 mm x 102 mm	-	-	-	-	m	114.84
64 mm x 152 mm	-	-	-	-	m	119.84
Lintel; once rebated						
64 mm x 102 mm	-	-	-	-	m	104.87
Pair of brass topped plates or sockets cast into posts for fixings (not included)	-	-	-	-	nr	26.47
Brass indicator bolt lugs cast into posts for fixings (not included)	-	-	-	-	nr	12.73

K LININGS/SHEATHING/DRY PARTITIONING

	PC £	Labour hours	Labour £	Material £	Unit	Total rate £
K40 DEMOUNTABLE SUSPENDED CEILINGS						
Suspended ceilings; Donn Products exposed suspended ceiling system or other equal and approved; hangers plugged and screwed to concrete soffit; 600 mm x 600 mm x 15 mm Cape TAP Ceilings Ltd; "Solitude" tegular fissured tile						
Lining to ceilings; hangers average 400 mm long						
over 300 mm wide	-	0.32	4.03	7.84	m²	11.87
Suspended ceilings, Gyproc M/F suspended ceiling system or other equal and approved; hangers plugged and screwed to concrete soffit, 900 mm x 1800 mm x 12.50 mm tapered edge wallboard infill; joints filled with joint filler and taped to receive direct direction						
Lining to ceilings; hangers average 400 mm long						
over 300 mm wide	-	-	-	-	m²	23.82
not exceeding 300 mm wide in isolated strips	-	-	-	-	m	9.63
300 mm - 600 mm wide in isolated strips	-	-	-	-	m	15.18
Vertical bulkhead; including additional hangers						
over 300 mm wide	-	-	-	-	m²	26.07
not exceeding 300 mm wide in isolated strips	-	-	-	-	m	14.68
300 mm - 600 mm wide in isolated strips	-	-	-	-	m	20.47
Suspended ceilings; "Slimline" exposed suspended ceiling system or other equal and approved; hangers plugged and screwed to concrete soffit, 600 mm x 600 mm x 19 mm Treetex "Glacier" mineral tile infill; PC £11.00/m²						
Lining to ceilings; hangers average 400 mm long						
over 300 mm wide	-	-	-	-	m²	24.37
not exceeding 300 mm wide in isolated strips	-	-	-	-	m	10.23
300 mm - 600 mm wide in isolated strips	-	-	-	-	m	19.27
Edge detail 24 mm x 19 mm white finished angle edge trim	-	-	-	-	m	1.90
Cutting and fitting around pipes; not exceeding 0.30 m girth	-	-	-	-	nr	1.34
Suspended ceilings; "Z" demountable suspended ceiling system or other equal and approved; hangers plugged and screwed to concrete soffit, 600 mm x 600 mm x 19 mm "Echostop" glass reinforced fibrous plaster lightweight plain bevelled edge tiles; PC £16.00/m²						
Lining to ceilings; hangers average 400 mm long						
over 300 mm wide	-	-	-	-	m²	51.03
not exceeding 300 mm wide in isolated strips	-	-	-	-	m	17.58
Suspended ceilings; concealed galvanised steel suspension system; hangers plugged and screwed to concrete soffit, "Burgess" white stove enamelled perforated mild steel tiles 600 mm x 600 mm; PC £15.00/m²						
Lining to ceilings; hangers average 400 mm long						
over 300 mm wide	-	-	-	-	m²	41.44
not exceeding 300 mm wide; in isolated strips	-	-	-	-	m	14.68

Prices for Measured Works - Major Works
K LININGS/SHEATHING/DRY PARTITIONING

	PC £	Labour hours	Labour £	Material £	Unit	Total rate £
K40 DEMOUNTABLE SUSPENDED CEILINGS - cont'd						
Suspended ceilings; concealed galvanised steel "Trulok" suspension system or other equal and approved; hangers plugged and screwed to concrete; Armstrong "Travertone" 300 mm x 300 mm x 18 mm mineral ceiling tiles PC £17.50/m²						
Linings to ceilings; hangers average 700 mm long						
over 300 mm wide	-	-	-	-	m²	49.29
over 300 mm wide; 3.50 m - 5.00 m high	-	-	-	-	m²	49.84
over 300 mm wide; in staircase areas or plant rooms	-	-	-	-	m²	51.03
not exceeding 300 mm wide; in isolated strips	-	-	-	-	m	16.97
300 mm - 600 mm wide; in isolated strips	-	-	-	-	m	30.06
Extra for cutting and fitting around modular downlighter including yoke	-	-	-	-	nr	4.49
Vertical bulkhead; including additional hangers						
over 300 mm wide	-	-	-	-	m²	51.03
not exceeding 300 mm wide; in isolated strips	-	-	-	-	m	22.67
300 mm - 600 mm wide; in isolated strips	-	-	-	-	m	36.21
24 mm x 19 mm white finished angle edge trim	-	-	-	-	m	1.85
Cutting and fitting around pipes; not exceeding 0.30 m girth	-	-	-	-	nr	1.34
Suspended ceilings; galvanised steel suspension system; hangers plugged and screwed to concrete soffit, "Luxalon" stove enamelled aluminium linear panel ceiling, type 80B or other equal and approved, complete with mineral insulation; PC £20.00/m²						
Linings to ceilings; hangers average 700 mm long						
over 300 mm wide	-	-	-	-	m²	51.03
not exceeding 300 mm wide; in isolated strips	-	-	-	-	m	17.58
K41 RAISED ACCESS FLOORS						
USG (U.K.) Ltd raised flooring systems or other equal and approved; laid on or fixed to concrete floor						
Full access system; 150 mm high overall; pedestal supports						
PSA light grade; steel finish	-	-	-	-	m²	38.55
PSA medium grade; steel finish	-	-	-	-	m²	44.04
PSA heavy grade; steel finish	-	-	-	-	m²	49.54
Extra for						
factory applied needlepunch carpet	-	-	-	-	m²	12.69
factory applied anti-static vinyl	-	-	-	-	m²	27.51
factory applied black PVC edge strips	-	-	-	-	m	3.30
ramps; 3.00 m x 1.40 m (no finish)	-	-	-	-	nr	604.19
steps (no finish)	-	-	-	-	m	33.11
forming cut-out for electrical boxes	-	-	-	-	nr	7.44

L WINDOWS/DOORS/STAIRS

L10 WINDOWS/ROOFLIGHTS/SCREENS/LOUVRES

Standard windows; "treated" wrought softwood; Rugby Joinery or other equal and approved
Side hung casement windows without glazing bars; with 140 mm wide softwood sills; opening casements and ventilators hung on rustproof hinges; fitted with aluminized lacquered finish casement stays and fasteners

	PC £	Labour hours	Labour £	Material £	Unit	Total rate £
500 mm x 750 mm; ref N07V	41.30	0.65	9.10	44.61	nr	**53.71**
500 mm x 900 mm; ref N09V	42.68	0.74	10.36	46.09	nr	**56.45**
600 mm x 750 mm; ref 107V	50.22	0.74	10.36	54.19	nr	**64.55**
600 mm x 750 mm; ref 107C	51.24	0.74	10.36	55.29	nr	**65.65**
600 mm x 900 mm; ref 109V	50.38	0.83	11.62	54.37	nr	**65.99**
600 mm x 900 mm; ref 109C	54.36	0.74	10.36	58.65	nr	**69.01**
600 mm x 1050 mm; ref 109V	51.57	0.74	10.36	55.72	nr	**66.08**
600 mm x 1050 mm; ref 110V	54.59	0.93	13.02	58.96	nr	**71.98**
915 mm x 900 mm; ref 2NO9W	61.48	1.02	14.28	66.30	nr	**80.58**
915 mm x 1050 mm; ref 2N1OW	62.33	1.06	14.84	67.28	nr	**82.12**
915 mm x 1200 mm; ref 2N12W	67.25	1.11	15.54	72.57	nr	**88.11**
915 mm x 1350 mm; ref 2N13W	68.60	1.25	17.50	74.02	nr	**91.52**
915 mm x 1500 mm; ref 2N15W	69.52	1.30	18.20	75.08	nr	**93.28**
1200 mm x 750 mm; ref 2O7C	66.89	1.06	14.84	72.18	nr	**87.02**
1200 mm x 750 mm; ref 2O7CV	83.63	1.06	14.84	90.18	nr	**105.02**
1200 mm x 900 mm; ref 2O9C	68.44	1.11	15.54	73.85	nr	**89.39**
1200 mm x 900 mm; ref 2O9W	73.13	1.11	15.54	78.89	nr	**94.43**
1200 mm x 900 mm; ref 2O9CV	84.45	1.11	15.54	91.06	nr	**106.60**
1200 mm x 1050 mm; ref 210C	68.51	1.25	17.50	74.00	nr	**91.50**
1200 mm x 1050 mm; ref 210W	74.41	1.25	17.50	80.34	nr	**97.84**
1200 mm x 1050 mm; ref 210T	86.71	1.25	17.50	93.56	nr	**111.06**
1200 mm x 1050 mm; ref 210CV	84.56	1.25	17.50	91.25	nr	**108.75**
1200 mm x 1200 mm; ref 212C	71.16	1.34	18.76	76.91	nr	**95.67**
1200 mm x 1200 mm; ref 212W	75.82	1.34	18.76	81.92	nr	**100.68**
1200 mm x 1200 mm; ref 212TX	93.68	1.34	18.76	101.12	nr	**119.88**
1200 mm x 1200 mm; ref 212CV	87.24	1.34	18.76	94.20	nr	**112.96**
1200 mm x 1350 mm; ref 213W	77.36	1.43	20.02	83.58	nr	**103.60**
1200 mm x 1350 mm; ref 213CV	96.60	1.43	20.02	104.26	nr	**124.28**
1200 mm x 1500 mm; ref 215W	79.18	1.57	21.97	85.60	nr	**107.57**
1770 mm x 750 mm; ref 307CC	98.47	1.30	18.20	106.27	nr	**124.47**
1770 mm x 900 mm; ref 309CC	100.83	1.57	21.97	108.81	nr	**130.78**
1770 mm x 1050 mm; ref 310C	84.88	1.62	22.67	91.66	nr	**114.33**
1770 mm x 1050 mm; ref 310T	101.02	1.57	21.97	109.01	nr	**130.98**
1770 mm x 1050 mm; ref 310CC	101.22	1.30	18.20	100.23	nr	**127.43**
1770 mm x 1050 mm; ref 310WW	116.75	1.30	18.20	125.92	nr	**144.12**
1770 mm x 1200 mm; ref 312C	87.37	1.67	23.37	94.41	nr	**117.78**
1770 mm x 1200 mm; ref 312T	103.85	1.67	23.37	112.12	nr	**135.49**
1770 mm x 1200 mm; ref 312CC	104.90	1.67	23.37	113.25	nr	**136.62**
1770 mm x 1200 mm; ref 312WW	119.73	1.67	23.37	129.20	nr	**152.57**
1770 mm x 1200 mm; ref 312CVC	127.90	1.67	23.37	137.98	nr	**161.35**
1770 mm x 1350 mm; ref 313CC	118.32	1.76	24.63	127.68	nr	**152.31**
1770 mm x 1350 mm; ref 313WW	122.49	1.76	24.63	132.16	nr	**156.79**
1770 mm x 1350 mm; ref 313CVC	135.38	1.76	24.63	146.02	nr	**170.65**
1770 mm x 1500 mm; ref 315T	114.12	1.85	25.89	123.16	nr	**149.05**
2340 mm x 1050 mm; ref 410CWC	144.73	1.80	25.19	156.07	nr	**181.26**
2340 mm x 1200 mm; ref 412CWC	149.37	1.90	26.59	161.13	nr	**187.72**
2340 mm x 1350 mm; ref 413CWC	157.56	2.04	28.55	170.00	nr	**198.55**

Prices for Measured Works - Major Works
L WINDOWS/DOORS/STAIRS

	PC £	Labour hours	Labour £	Material £	Unit	Total rate £
L10 WINDOWS/ROOFLIGHTS/SCREENS/LOUVRES - cont'd						
Standard windows; "treated" wrought softwood; Rugby Joinery or other equal and approved - cont'd						
Top hung casement windows; with 140 mm wide softwood sills; opening casements and ventilators hung on rustproof hinges; fitted with aluminized lacquered finish casement stays						
600 mm x 750 mm; ref 107A	57.05	0.74	10.36	61.54	nr	71.90
600 mm x 900 mm; ref 109A	58.79	0.83	11.62	63.41	nr	75.03
600 mm x 1050 mm; ref 110A	61.01	0.93	13.02	65.86	nr	78.88
915 mm x 750 mm; ref 2N07A	68.04	0.97	13.58	73.35	nr	86.93
915 mm x 900 mm; ref 2N09A	72.87	1.02	14.28	78.54	nr	92.82
915 mm x 1050 mm; ref 2N10A	75.46	1.06	14.84	81.40	nr	96.24
915 mm x 1350 mm; ref 2N13AS	85.34	1.25	17.50	92.09	nr	109.59
1200 mm x 750 mm; ref 207A	76.38	1.06	14.84	82.39	nr	97.23
1200 mm x 900 mm; ref 209A	80.78	1.11	15.54	87.12	nr	102.66
1200 mm x 1050 mm; ref 210A	83.34	1.25	17.50	89.94	nr	107.44
1200 mm x 1200 mm; ref 212A	85.80	1.34	18.76	92.58	nr	111.34
1200 mm x 1350 mm; ref 213AS	93.68	1.43	20.02	101.12	nr	121.14
1200 mm x 1500 mm; ref 215AS	96.20	1.57	21.97	103.83	nr	125.80
1770 mm x 1050 mm; ref 310AE	97.87	1.57	21.97	105.63	nr	127.60
1770 mm x 1200 mm; ref 312A	100.76	1.67	23.37	108.73	nr	132.10
High performance single light with canopy sash windows; ventilators; weather stripping; opening sashes and fanlights hung on rustproof hinges; fitted with aluminized lacquered espagnolette bolts						
600 mm x 900 mm; ref AL0609	116.88	0.83	11.62	125.85	nr	137.47
900 mm x 900 mm; ref AL0909	121.25	1.02	14.28	130.55	nr	144.83
900 mm x 1050 mm; ref AL0910	121.90	1.06	14.84	131.25	nr	146.09
900 mm x 1200 mm; ref AL0912	124.31	1.16	16.24	133.91	nr	150.15
900 mm x 1500 mm; ref AL0915	130.99	1.30	18.20	141.09	nr	159.29
1200 mm x 1050 mm; ref AL1210	141.28	1.25	17.50	152.15	nr	169.65
1200 mm x 1200 mm; ref AL1212	142.18	1.34	18.76	153.19	nr	171.95
1200 mm x 1500 mm; ref AL1215	160.53	1.57	21.97	172.99	nr	194.96
1500 mm x 1050 mm; ref AL1510	141.52	1.53	21.41	152.48	nr	173.89
1500 mm x 1200 mm; ref AL1512	148.61	1.57	21.97	160.10	nr	182.07
1500 mm x 1500 mm; ref AL1515	160.04	1.67	23.37	172.53	nr	195.90
1500 mm x 1650 mm; ref AL1516	165.19	1.76	24.63	178.00	nr	202.63
High performance double hung sash windows with glazing bars; solid frames; 63 mm x 175 mm softwood sills; standard flush external linings; spiral spring balances and sash catch						
635 mm x 1050 mm; ref VS0610B	185.47	1.85	25.89	199.66	nr	225.55
635 mm x 1350 mm; ref VS0613B	200.88	2.04	28.55	216.29	nr	244.84
635 mm x 1650 mm; ref VS0616B	214.60	2.27	31.77	231.11	nr	262.88
860 mm x 1050 mm; ref VS0810B	202.33	2.13	29.81	217.78	nr	247.59
860 mm x 1350 mm; ref VS0813B	219.00	2.41	33.73	235.77	nr	269.50
860 mm x 1650 mm; ref VS0816B	235.64	2.78	38.91	253.73	nr	292.64
1085 mm x 1050 mm; ref VS1010B	220.06	2.41	33.73	236.84	nr	270.57
1085 mm x 1350 mm; ref VS1013B	241.38	2.78	38.91	259.83	nr	298.74
1085 mm x 1650 mm; ref VS1016B	255.37	3.42	47.87	274.94	nr	322.81
1725 mm x 1050 mm; ref VS1710B	402.93	3.42	47.87	433.50	nr	481.37
1725 mm x 1350 mm; ref VS1713B	439.13	4.26	59.63	472.48	nr	532.11
1725 mm x 1650 mm; ref VS1716B	471.17	4.35	60.88	506.99	nr	567.87

Prices for Measured Works - Major Works
L WINDOWS/DOORS/STAIRS

	PC £	Labour hours	Labour £	Material £	Unit	Total rate £
Standard windows; Premdor Crosby; Meranti; factory applied preservative stain base coat						
Side hung casement windows; 45 mm x 140 mm hardwood sills; weather stripping; opening sashes on canopy hinges; fitted with fasteners; brown finish ironmongery						
630 mm x 750 mm; ref 107C	91.12	0.88	12.32	98.16	nr	110.48
630 mm x 900 mm; ref 109C	95.06	1.11	15.54	102.40	nr	117.94
630 mm x 900 mm; ref 109V	90.59	0.88	12.32	97.59	nr	109.91
630 mm x 1050 mm; ref 110V	92.89	1.20	16.80	100.07	nr	116.87
915 mm x 900 mm; ref 2N09W	113.21	1.39	19.46	121.98	nr	141.44
915 mm x 1050 mm; ref 2N10W	115.03	1.48	20.71	124.00	nr	144.71
915 mm x 1200 mm; ref 2N12W	117.91	1.57	21.97	127.10	nr	149.07
915 mm x 1350 mm; ref 2N13W	120.59	1.67	23.37	129.98	nr	153.35
915 mm x 1500 mm; ref 2N15W	123.23	1.76	24.63	132.89	nr	157.52
1200 mm x 900 mm; ref 209C	123.05	1.57	21.97	132.49	nr	154.46
1200 mm x 900 mm; ref 209W	129.26	1.57	21.97	139.16	nr	161.13
1200 mm x 1050 mm; ref 210C	127.52	1.67	23.37	137.29	nr	160.66
1200 mm x 1050 mm; ref 210W	131.75	1.67	23.37	141.84	nr	165.21
1200 mm x 1200 mm; ref 212C	132.88	1.80	25.19	143.12	nr	168.31
1200 mm x 1200 mm; ref 212W	134.39	1.80	25.19	144.75	nr	169.94
1200 mm x 1350 mm; ref 213W	137.33	1.94	27.15	147.91	nr	175.06
1200 mm x 1500 mm; ref 215W	139.95	2.04	28.55	150.72	nr	179.27
1770 mm x 1050 mm; ref 310C	150.00	2.08	29.11	161.60	nr	190.71
1770 mm x 1050 mm; ref 310CC	193.98	2.08	29.11	208.88	nr	237.99
1770 mm x 1200 mm; ref 312C	154.31	2.22	31.07	166.23	nr	197.30
1770 mm x 1200 mm ; ref 312CC	201.60	2.22	31.07	217.07	nr	248.14
2339 mm x 1200 mm; ref 412CMC	251.67	2.41	33.73	270.96	nr	304.69
Top hung casement windows; 45 mm x 140 mm hardwood sills; weather stripping; opening sashes on canopy hinges; fitted with fasteners; brown finish ironmongery						
630 mm x 900 mm; ref 109A	86.28	0.88	12.32	92.96	nr	105.28
630 mm x 1050 mm; ref 110A	91.68	1.20	16.80	98.83	nr	115.63
915 mm x 900 mm; ref 2N09A	105.22	1.39	19.46	113.32	nr	132.78
915 mm x 1050 mm; ref 2N10A	110.61	1.48	20.71	119.18	nr	139.89
915 mm x 1200 mm; ref 2N12A	116.54	1.57	21.97	125.56	nr	147.53
915 mm x 1350 mm; ref 2N13D	140.40	1.67	23.37	151.28	nr	174.65
1200 mm x 900 mm; ref 209A	118.81	1.39	19.46	127.93	nr	147.39
1200 mm x 1050 mm; ref 210A	124.24	1.57	21.97	133.84	nr	155.81
1200 mm x 1200 mm; ref 212A	130.16	1.67	23.37	140.20	nr	163.57
1770 mm x 1050 mm; ref 310AE	150.53	1.80	25.19	162.17	nr	187.36
1770 mm x 1200 mm; ref 312AE	158.25	1.85	25.89	170.47	nr	196.36
Purpose made window casements; "treated" wrought softwood						
Casements; rebated; moulded - supply only						
38 mm thick	-	-	-	22.71	m²	22.71
50 mm thick	-	-	-	27.01	m²	27.01
Casements; rebated; moulded; in medium panes - supply only						
38 mm thick	-	-	-	33.21	m²	33.21
50 mm thick	-	-	-	41.05	m²	41.05
Casements; rebated; moulded; with semi-circular head - supply only						
38 mm thick	-	-	-	41.46	m²	41.46
50 mm thick	-	-	-	51.35	m²	51.35

Prices for Measured Works - Major Works
L WINDOWS/DOORS/STAIRS

	PC £	Labour hours	Labour £	Material £	Unit	Total rate £
L10 WINDOWS/ROOFLIGHTS/SCREENS/LOUVRES - cont'd						
Purpose made window casements; "treated" wrought softwood - cont'd						
Casements; rebated; moulded; to bullseye window - supply only						
38 mm thick; 600 mm diameter	-	-	-	74.57	nr	74.57
38 mm thick; 900 mm diameter	-	-	-	110.62	nr	110.62
50 mm thick; 600 mm diameter	-	-	-	85.78	nr	85.78
50 mm thick; 900 mm diameter	-	-	-	127.17	nr	127.17
Fitting and hanging casements						
square or rectangular	-	0.46	6.44	-	nr	6.44
semi-circular	-	1.16	16.24	-	nr	16.24
bullseye	-	1.85	25.89	-	nr	25.89
Purpose made window casements; selected West African Mahogany						
Casements; rebated; moulded - supply only						
38 mm thick	-	-	-	38.25	m^2	38.25
50 mm thick	-	-	-	49.86	m^2	49.86
Casements; rebated; moulded; in medium panes - supply only						
38 mm thick	-	-	-	46.42	m^2	46.42
50 mm thick	-	-	-	62.10	m^2	62.10
Casements; rebated; moulded with semi-circular head - supply only						
38 mm thick	-	-	-	58.03	m^2	58.03
50 mm thick	-	-	-	77.69	m^2	77.69
Casements; rebated; moulded; to bullseye window - supply only						
38 mm thick; 600 mm diameter	-	-	-	125.60	nr	125.60
38 mm thick; 900 mm diameter	-	-	-	187.96	nr	187.96
50 mm thick; 600 mm diameter	-	-	-	144.43	nr	144.43
50 mm thick; 900 mm diameter	-	-	-	215.34	nr	215.34
Fitting and hanging casements						
square or rectangular	-	0.65	9.10	-	nr	9.10
semi-circular	-	1.57	21.97	-	nr	21.97
bullseye	-	2.50	34.99	-	nr	34.99
Purpose made window casements; American White Ash						
Casements; rebated; moulded - supply only						
38 mm thick	-	-	-	45.21	m^2	45.21
50 mm thick	-	-	-	55.51	m^2	55.51
Casements; rebated; moulded; in medium panes - supply only						
38 mm thick	-	-	-	58.60	m^2	58.60
50 mm thick	-	-	-	70.42	m^2	70.42
Casements; rebated; moulded with semi-circular head - supply only						
38 mm thick	-	-	-	73.26	m^2	73.26
50 mm thick	-	-	-	88.05	m^2	88.05
Casements; rebated; moulded; to bullseye window - supply only						
38 mm thick; 600 mm diameter	-	-	-	148.52	nr	148.52
38 mm thick; 900 mm diameter	-	-	-	221.90	nr	221.90
50 mm thick; 600 mm diameter	-	-	-	171.31	nr	171.31
50 mm thick; 900 mm diameter	-	-	-	255.41	nr	255.41

L WINDOWS/DOORS/STAIRS

	PC £	Labour hours	Labour £	Material £	Unit	Total rate £
Fitting and hanging casements						
square or rectangular	-	0.65	9.10	-	nr	9.10
semi-circular	-	1.57	21.97	-	nr	21.97
bullseye	-	2.50	34.99	-	nr	34.99
Purpose made window frames; "treated" wrought softwood; prices indicated are for supply only as part of a complete window frame; the reader is referred to the previous pages for fixing costs for frames based on the overall window size						
Frames; rounded; rebated check grooved						
25 mm x 120 mm	-	-	-	6.92	m	6.92
50 mm x 75 mm	-	-	-	8.19	m	8.19
50 mm x 100 mm	-	-	-	10.05	m	10.05
50 mm x 125 mm	-	-	-	12.19	m	12.19
63 mm x 100 mm	-	-	-	12.51	m	12.51
75 mm x 150 mm	-	-	-	21.02	m	21.02
90 mm x 140 mm	-	-	-	27.10	m	27.10
Mullions and transoms; twice rounded, rebated and check grooved						
50 mm x 75 mm	-	-	-	10.07	m	10.07
50 mm x 100 mm	-	-	-	11.92	m	11.92
63 mm x 100 mm	-	-	-	14.38	m	14.38
75 mm x 150 mm	-	-	-	22.88	m	22.88
Sill; sunk weathered, rebated and grooved						
75 mm x 100 mm	-	-	-	21.06	m	21.06
75 mm x 150 mm	-	-	-	27.59	m	27.59
Add 5% to the above material prices for "selected" softwood for staining						
Purpose made window frames; selected West African Mahogany; prices given are for supply only as part of a complete window frame; the reader is referred to the previous pages for fixing costs for frames based on the overall window size						
Frames; rounded; rebated check grooved						
25 mm x 120 mm	-	-	-	11.94	m	11.94
50 mm x 75 mm	-	-	-	14.27	m	14.27
50 mm x 100 mm	-	-	-	17.66	m	17.66
50 mm x 125 mm	-	-	-	21.54	m	21.54
63 mm x 100 mm	-	-	-	22.03	m	22.03
75 mm x 150 mm	-	-	-	37.32	m	37.32
90 mm x 140 mm	-	-	-	48.39	m	48.39
Mullions and transoms; twice rounded, rebated and check grooved						
50 mm x 75 mm	-	-	-	17.06	m	17.06
50 mm x 100 mm	-	-	-	20.45	m	20.45
63 mm x 100 mm	-	-	-	24.83	m	24.83
75 mm x 150 mm	-	-	-	40.10	m	40.10
Sill; sunk weathered, rebated and grooved						
75 mm x 100 mm	-	-	-	35.47	m	35.47
75 mm x 150 mm	-	-	-	47.25	m	47.25

L10 WINDOWS/ROOFLIGHTS/SCREENS/LOUVRES - cont'd

	PC £	Labour hours	Labour £	Material £	Unit	Total rate £
Purpose made window frames; American White Ash; prices indicated are for supply only as part of a complete window frame; the reader is referred to the previous pages for fixing costs for frames based on the overall window size						
Frames; rounded; rebated check grooved						
25 mm x 120 mm	-	-	-	13.30	m	13.30
50 mm x 75 mm	-	-	-	15.96	m	15.96
50 mm x 100 mm	-	-	-	19.93	m	19.93
50 mm x 125 mm	-	-	-	24.38	m	24.38
63 mm x 100 mm	-	-	-	24.91	m	24.91
75 mm x 150 mm	-	-	-	42.43	m	42.43
90 mm x 140 mm	-	-	-	55.25	m	55.25
Mullions and transoms; twice rounded, rebated and check grooved						
50 mm x 75 mm	-	-	-	18.75	m	18.75
50 mm x 100 mm	-	-	-	22.72	m	22.72
63 mm x 100 mm	-	-	-	27.71	m	27.71
75 mm x 150 mm	-	-	-	45.22	m	45.22
Sill; sunk weathered, rebated and grooved						
75 mm x 100 mm	-	-	-	38.89	m	38.89
75 mm x 150 mm	-	-	-	52.39	m	52.39
Purpose made double hung sash windows; "treated" wrought softwood						
Cased frames of 100 mm x 25 mm grooved inner linings; 114 mm x 25 mm grooved outer linings; 125 mm x 38 mm twice rebated head linings; 125 mm x 32 mm twice rebated grooved pulley stiles; 150 mm x 13 mm linings; 50 mm x 19 mm parting slips; 25 mm x 19 mm inside beads; 150 mm x 75 mm Oak twice sunk weathered throated sill; 50 mm thick rebated and moulded sashes; moulded horns						
over 1.25 m² each; both sashes in medium panes; including spiral spring balances	154.74	2.08	29.11	230.86	m²	259.97
As above but with cased mullions	163.27	2.31	32.33	240.03	m²	272.36
Purpose made double hung sash windows; selected West African Mahogany						
Cased frames of 100 mm x 25 mm grooved inner linings; 114 mm x 25 mm grooved outer linings; 125 mm x 38 mm twice rebated head linings; 125 mm x 32 mm twice rebated grooved pulley stiles; 150 mm x 13 mm linings; 50 mm x 19 mm parting slips; 25 mm x 19 mm inside beads; 150 mm x 75 mm Oak twice sunk weathered throated sill; 50 mm thick rebated and moulded sashes; moulded horns						
over 1.25 m² each; both sashes in medium panes; including spiral sash balances	247.00	2.78	38.91	330.03	m²	368.94
As above but with cased mullions	260.57	3.08	43.11	344.62	m²	387.73

L WINDOWS/DOORS/STAIRS

	PC £	Labour hours	Labour £	Material £	Unit	Total rate £
Purpose made double hung sash windows; American White Ash						
Cased frames of 100 mm x 25 mm grooved inner linings; 114 mm x 25 mm grooved outer linings; 125 mm x 38 mm twice rebated head linings; 125 mm x 32 mm twice rebated grooved pulley stiles; 150 mm x 13 mm linings; 50 mm x 19 mm parting slips; 25 mm x 19 mm inside beads; 150 mm x 75 mm Oak twice sunk weathered throated sill; 50 mm thick rebated and moulded sashes; moulded horns over 1.25 m² each; both sashes in medium panes; including spiral spring balances	267.87	2.78	38.91	352.47	m²	391.38
As above but with cased mullions	-	3.08	43.11	368.28	m²	411.39
Galvanised steel fixed, casement and fanlight windows; Crittal "Homelight" range or other equal and approved; site glazing not included; fixed in position including lugs plugged and screwed to brickwork or blockwork						
Basic fixed lights; including easy-glaze beads						
628 mm x 292 mm; ref ZNG5	16.04	1.11	15.54	17.38	nr	32.92
628 mm x 923 mm; ref ZNC5	22.02	1.11	15.54	23.88	nr	39.42
628 mm x 1513 mm; ref ZNDV5	31.45	1.11	15.54	34.16	nr	49.70
1237 mm x 292 mm; ref ZNG13	25.16	1.11	15.54	27.19	nr	42.73
1237 mm x 923 mm; ref ZNC13	31.45	1.62	22.67	34.02	nr	56.69
1237 mm x 1218 mm; ref ZND13	37.74	1.62	22.67	40.85	nr	63.52
1237 mm x 1513 mm; ref ZNDV13	40.88	1.62	22.67	44.29	nr	66.96
1846 mm x 292 mm; ref ZNG14	31.45	1.11	15.54	34.02	nr	49.56
1846 mm x 923 mm; ref ZNC14	37.74	1.62	22.67	40.92	nr	63.59
1846 mm x 1513 mm; ref ZNDV14	47.17	2.04	28.55	51.19	nr	79.74
Basic opening lights; including easy-glaze beads and weatherstripping						
628 mm x 292 mm; ref ZNG1	40.88	1.11	15.54	44.08	nr	59.62
1237 mm x 292 mm; ref ZNG13G	59.76	1.11	15.54	64.37	nr	79.91
1846 mm x 292 mm; ref ZNG4	100.64	1.11	15.54	108.47	nr	124.01
One piece composites; including easy-glaze beads and weatherstripping						
628 mm x 923 mm; ref ZNC5F	56.61	1.11	15.54	61.06	nr	76.60
628 mm x 1513 mm; ref ZNDV5F	66.04	1.11	15.54	71.35	nr	86.89
1237 mm x 923 mm; ref ZNC2F	106.93	1.62	22.67	115.16	nr	137.83
1237 mm x 1218 mm; ref ZND2F	125.80	1.62	22.67	135.51	nr	158.18
1237 mm x 1513 mm; ref ZNDV2V	147.81	1.62	22.67	159.24	nr	181.91
1846 mm x 923 mm; ref ZNC4F	163.54	1.62	22.67	176.15	nr	198.82
1846 mm x 1218 mm; ref ZND10F	157.25	2.04	28.55	169.46	nr	198.01
Reversible windows; including easy-glaze beads						
997 mm x 923 mm; ref NC13R	141.52	1.43	20.02	152.35	nr	172.37
997 mm x 1067 mm; ref NCO13R	147.81	1.43	20.02	159.10	nr	179.12
1237 mm x 923 mm; ref ZNC13R	154.10	2.13	29.81	165.87	nr	195.68
1237 mm x 1218 mm; ref ZND13R	166.69	2.13	29.81	179.47	nr	209.28
1237 mm x 1513 mm; ref ZNDV13RS	188.70	2.13	29.81	203.20	nr	233.01
Pressed steel sills; to suit above window widths						
628 mm long	9.75	0.32	4.48	10.48	nr	14.96
997 mm long	12.58	0.42	5.88	13.52	nr	19.40
1237 mm long	16.04	0.51	7.14	17.24	nr	24.38
1846 mm long	22.02	0.69	9.66	23.67	nr	33.33

Prices for Measured Works - Major Works
L WINDOWS/DOORS/STAIRS

	PC £	Labour hours	Labour £	Material £	Unit	Total rate £
L10 WINDOWS/ROOFLIGHTS/SCREENS/LOUVRES - cont'd						
Factory finished steel fixed light; casement and fanlight windows; Crittall polyester powder coated "Homelight" range or other equal and approved; site glazing not included; fixed in position; including lugs plugged and screwed to brickwork or blockwork						
Basic fixed lights; including easy-glaze beads						
628 mm x 292 mm; ref ZNG5	20.44	1.11	15.54	22.11	nr	37.65
628 mm x 923 mm; ref ZNC5	28.30	1.11	15.54	30.63	nr	46.17
628 mm x 1513 mm; ref ZNDV5	40.88	1.11	15.54	44.29	nr	59.83
1237 mm x 292 mm; ref ZNG13	31.45	1.11	15.54	33.95	nr	49.49
1237 mm x 923 mm; ref ZNC13	40.88	1.62	22.67	44.15	nr	66.82
1237 mm x 1218 mm; ref ZND13	47.17	1.62	22.67	50.99	nr	73.66
1237 mm x 1513 mm; ref ZNDV13	50.32	1.62	22.67	54.44	nr	77.11
1846 mm x 292 mm; ref ZNG14	40.88	1.11	15.54	44.08	nr	59.62
1846 mm x 923 mm; ref ZNC14	50.32	1.62	22.67	54.30	nr	76.97
1846 mm x 1513 mm; ref ZNDV14	59.76	2.04	28.55	64.58	nr	93.13
Basic opening lights; including easy-glaze beads and weatherstripping						
628 mm x 292 mm; ref ZNG1	48.75	1.11	15.54	52.55	nr	68.09
1237 mm x 292 mm; ref ZNG13G	69.19	1.11	15.54	74.52	nr	90.06
1846 mm x 292 mm; ref ZNG4	119.51	1.11	15.54	128.61	nr	144.15
One piece composites; including easy-glaze beads and weatherstripping						
628 mm x 923 mm; ref ZNC5F	66.04	1.11	15.54	71.21	nr	86.75
628 mm x 1513 mm; ref ZNDV5F	81.77	1.11	15.54	88.25	nr	103.79
1237 mm x 923 mm; ref ZNC2F	132.09	1.62	22.67	142.20	nr	164.87
1237 mm x 1218 mm; ref ZND2F	150.96	1.62	22.67	162.49	nr	185.16
1237 mm x 1513 mm; ref ZNDV2V	176.12	1.62	22.67	189.68	nr	212.35
1846 mm x 923 mm; ref ZNC4F	198.13	1.62	22.67	213.20	nr	235.87
1846 mm x 1218 mm; ref ZND10F	188.70	2.04	28.55	203.20	nr	231.75
Reversible windows; including easy-glaze beads						
997 mm x 923 mm; ref NC13R	182.41	1.43	20.02	196.30	nr	216.32
997 mm x 1218 mm; ref NCO13R	191.84	1.43	20.02	206.44	nr	226.46
1237 mm x 923 mm; ref ZNC13R	198.13	2.13	29.81	213.20	nr	243.01
1237 mm x 1218 mm; ref ZND13R	217.00	2.13	29.81	233.55	nr	263.36
1237 mm x 1513 mm; ref ZNDV13RS	242.16	2.13	29.81	260.67	nr	290.48
Pressed steel sills; to suit above window widths						
628 mm long	12.58	0.32	4.48	13.52	nr	18.00
997 mm long	16.98	0.42	5.88	18.25	nr	24.13
1237 mm long	20.13	0.51	7.14	21.64	nr	28.78
1846 mm long	28.30	0.69	9.66	30.42	nr	40.08
uPVC windows to BS 2782; reinforced where appropriate with aluminium alloy; including standard ironmongery; cills and glazing; fixed in position; including lugs plugged and screwed to brickwork or blockwork						
Fixed light; including e.p.d.m. glazing gaskets and weather seals						
600 mm x 900 mm; single glazed	35.30	1.39	19.46	38.23	nr	57.69
600 mm x 900 mm; double glazed	45.52	1.39	19.46	49.21	nr	68.67

L WINDOWS/DOORS/STAIRS

	PC £	Labour hours	Labour £	Material £	Unit	Total rate £
Casement/fixed light; including e.p.d.m. glazing gaskets and weather seals						
600 mm x 1200 mm; single glazed	76.18	1.62	22.67	82.24	nr	104.91
600 mm x 1200 mm; double glazed	91.04	1.62	22.67	98.21	nr	120.88
1200 mm x 1200 mm; single glazed	101.26	1.85	25.89	109.27	nr	135.16
1200 mm x 1200 mm; double glazed	142.14	1.85	25.89	153.22	nr	179.11
1800 mm x 1200 mm; single glazed	111.48	1.85	25.89	120.26	nr	146.15
1800 mm x 1200 mm; double glazed	152.36	1.85	25.89	164.20	nr	190.09
"Tilt & Turn" light; including e.p.d.m. glazing gaskets and weather seals						
1200 mm x 1200 mm; single glazed	105.91	1.85	25.89	114.27	nr	140.16
1200 mm x 1200 mm; double glazed	131.92	1.85	25.89	142.23	nr	168.12
Rooflights, skylights, roof windows and frames; pre-glazed; "treated" Nordic Red Pine and aluminium trimmed "Velux" windows or other equal and approved; type U flashings and soakers (for tiles and pantiles), and sealed double glazing unit (trimming opening not included)						
Roof windows						
550 mm x 780 mm; ref GGL-3000-102	100.51	1.85	25.89	108.29	nr	134.18
550 mm x 980 mm; ref GGL-3000-104	111.06	2.08	29.11	119.62	nr	148.73
660 mm x 1180 mm; ref GGL-3000-206	130.91	2.31	32.33	141.03	nr	173.36
780 mm x 980 mm; ref GGL-3000-304	125.95	2.31	32.33	135.69	nr	168.02
780 mm x 1400 mm; ref GGL-3000-308	151.38	2.31	32.33	163.14	nr	195.47
940 mm x 1600 mm; ref GGL-3000-410	181.79	2.78	38.91	195.87	nr	234.78
1140 mm x 1180 mm; ref GGL-3000-606	167.51	2.78	38.91	180.51	nr	219.42
1340 mm x 980 mm; ref GGL-3000-804	173.10	2.78	38.91	186.52	nr	225.43
1340 mm x 1400 mm; ref GGL-3000-808	205.36	2.78	38.91	221.20	nr	260.11
Rooflights, skylights, roof windows and frames; galvanised steel; plugged and screwed to concrete; or screwed to timber						
Rooflight; "Coxdome TPX" or other equal and approved; dome; galvanised steel; UV protected polycarbonate glazing; GRP splayed upstand; hit and miss vent						
600 mm x 600 mm	127.45	1.62	22.67	137.22	nr	159.89
900 mm x 600 mm	179.32	1.76	24.63	193.02	nr	217.65
900 mm x 900 mm	203.93	1.90	26.59	219.51	nr	246.10
1200 mm x 900 mm	238.21	2.04	28.55	256.40	nr	284.95
1200 mm x 1200 mm	281.28	2.17	30.37	302.73	nr	333.10
1800 mm x 1200 mm	370.94	2.31	32.33	399.15	nr	431.48
Rooflights, skylights, roof windows and frames; uPVC; plugged and screwed to concrete; or screwed to timber						
Rooflight; "Coxdome Universal Dome" or other equal and approved; acrylic double skin; dome						
600 mm x 600 mm	82.63	1.39	19.46	89.04	nr	108.50
900 mm x 600 mm	153.82	1.43	20.02	165.61	nr	185.63
900 mm x 900 mm	153.82	1.67	23.37	165.64	nr	189.01
1200 mm x 900 mm	186.35	1.80	25.19	200.65	nr	225.84
1200 mm x 1200 mm	235.57	1.94	27.15	253.59	nr	280.74
1500 mm x 1050 mm	349.84	2.04	28.55	376.47	nr	405.02
Louvres and frames; polyester powder coated aluminium; fixing in position including brackets						
Louvre; Gil Airvac "Plusaire 75SP" weatherlip or other equal and approved						
1250 mm x 675 mm	-	-	-	-	nr	249.67
2000 mm x 530 mm	-	-	-	-	nr	279.63

L WINDOWS/DOORS/STAIRS

	PC £	Labour hours	Labour £	Material £	Unit	Total rate £
L20 DOORS/SHUTTERS/HATCHES						
Doors; standard matchboarded; wrought softwood						
Matchboarded, ledged and braced doors; 25 mm thick ledges and braces; 19 mm thick tongued, grooved and V-jointed boarding; one side vertical boarding						
762 mm x 1981mm	33.80	1.39	19.46	36.34	nr	55.80
838 mm x 1981mm	33.80	1.39	19.46	36.34	nr	55.80
Matchboarded, framed, ledged and braced doors; 44 mm thick overall; 19 mm thick tongued, grooved and V-jointed boarding; one side vertical boarding						
762 mm x 1981mm	41.42	1.67	23.37	44.53	nr	67.90
838 mm x 1981mm	41.42	1.67	23.37	44.53	nr	67.90
Doors; standard flush; softwood composition						
Flush door; internal quality; skeleton or cellular core; hardboard faced both sides; Premdor Crosby "Primaseal" or other equal and approved						
457 mm x 1981 mm x 35 mm	13.52	1.16	16.24	14.53	nr	30.77
533 mm x 1981 mm x 35 mm	13.52	1.16	16.24	14.53	nr	30.77
610 mm x 1981 mm x 35 mm	13.52	1.16	16.24	14.53	nr	30.77
686 mm x 1981 mm x 35 mm	13.52	1.16	16.24	14.53	nr	30.77
762 mm x 1981 mm x 35 mm	13.52	1.16	16.24	14.53	nr	30.77
838 mm x 1981 mm x 35 mm	14.20	1.16	16.24	15.27	nr	31.51
526 mm x 2040 mm x 40 mm	14.63	1.16	16.24	15.73	nr	31.97
626 mm x 2040 mm x 40 mm	14.63	1.16	16.24	15.73	nr	31.97
726 mm x 2040 mm x 40 mm	14.63	1.16	16.24	15.73	nr	31.97
826 mm x 2040 mm x 40 mm	14.63	1.16	16.24	15.73	nr	31.97
Flush door; internal quality; skeleton or cellular core; chipboard veneered; faced both sides; lipped on two long edges; Premdor Crosby "Popular" or other equal and approved						
457 mm x 1981 mm x 35 mm	19.44	1.16	16.24	20.90	nr	37.14
533 mm x 1981 mm x 35 mm	19.44	1.16	16.24	20.90	nr	37.14
610 mm x 1981 mm x 35 mm	19.44	1.16	16.24	20.90	nr	37.14
686 mm x 1981 mm x 35 mm	19.44	1.16	16.24	20.90	nr	37.14
762 mm x 1981 mm x 35 mm	19.44	1.16	16.24	20.90	nr	37.14
838 mm x 1981 mm x 35 mm	20.34	1.16	16.24	21.87	nr	38.11
526 mm x 2040 mm x 40 mm	21.37	1.16	16.24	22.97	nr	39.21
626 mm x 2040 mm x 40 mm	21.37	1.16	16.24	22.97	nr	39.21
726 mm x 2040 mm x 40 mm	21.37	1.16	16.24	22.97	nr	39.21
826 mm x 2040 mm x 40 mm	21.37	1.16	16.24	22.97	nr	39.21
Flush door; internal quality; skeleton or cellular core; Sapele faced both sides; lipped on all four edges; Premdor Crosby "Landscape Sapele" or other equal and approved						
457 mm x 1981 mm x 35 mm	22.06	1.25	17.50	23.71	nr	41.21
533 mm x 1981 mm x 35 mm	22.06	1.25	17.50	23.71	nr	41.21
610 mm x 1981 mm x 35 mm	22.06	1.25	17.50	23.71	nr	41.21
686 mm x 1981 mm x 35 mm	22.06	1.25	17.50	23.71	nr	41.21
762 mm x 1981 mm x 35 mm	22.06	1.25	17.50	23.71	nr	41.21
838 mm x 1981 mm x 35 mm	22.96	1.25	17.50	24.68	nr	42.18
526 mm x 2040 mm x 40 mm	23.78	1.25	17.50	25.56	nr	43.06
626 mm x 2040 mm x 40 mm	23.78	1.25	17.50	25.56	nr	43.06
726 mm x 2040 mm x 40 mm	23.78	1.25	17.50	25.56	nr	43.06
826 mm x 2040 mm x 40 mm	23.78	1.25	17.50	25.56	nr	43.06

L WINDOWS/DOORS/STAIRS

	PC £	Labour hours	Labour £	Material £	Unit	Total rate £
Flush door; half-hour fire check (FD20); hardboard faced both sides; Premdor Crosby "Primaseal Fireshield" or other equal and approved						
762 mm x 1981 mm x 44 mm	32.53	1.62	22.67	34.97	nr	57.64
838 mm x 1981 mm x 44 mm	33.72	1.62	22.67	36.25	nr	58.92
726 mm x 2040 mm x 44 mm	34.54	1.62	22.67	37.13	nr	59.80
826 mm x 2040 mm x 44 mm	34.54	1.62	22.67	37.13	nr	59.80
Flush door; half-hour fire check (30/20); chipboard veneered; faced both sides; lipped on all four edges; Premdor Crosby "Popular Fireshield" or other equal and approved						
457 mm x 1981 mm x 44 mm	32.00	1.62	22.67	34.40	nr	57.07
533 mm x 1981 mm x 44 mm	32.00	1.62	22.67	34.40	nr	57.07
610 mm x 1981 mm x 44 mm	32.00	1.62	22.67	34.40	nr	57.07
686 mm x 1981 mm x 44 mm	32.00	1.62	22.67	34.40	nr	57.07
762 mm x 1981 mm x 44 mm	32.00	1.62	22.67	34.40	nr	57.07
838 mm x 1981 mm x 44 mm	33.33	1.62	22.67	35.83	nr	58.50
526 mm x 2040 mm x 44 mm	34.99	1.62	22.67	37.61	nr	60.28
626 mm x 2040 mm x 44 mm	34.99	1.62	22.67	37.61	nr	60.28
726 mm x 2040 mm x 44 mm	34.99	1.62	22.67	37.61	nr	60.28
826 mm x 2040 mm x 44 mm	34.99	1.62	22.67	37.61	nr	60.28
Flush door; half hour fire resisting "Leaderflush" (type B30) or other equal and approved; chipboard for painting; hardwood lipping two long edges						
526 mm x 2040 mm x 44 mm	79.82	1.71	23.93	85.81	nr	109.74
626 mm x 2040 mm x 44 mm	83.87	1.71	23.93	90.16	nr	114.09
726 mm x 2040 mm x 44 mm	82.76	1.71	23.93	88.97	nr	112.90
826 mm x 2040 mm x 44 mm	83.65	1.71	23.93	89.92	nr	113.85
Flush door; half-hour fire check (FD20); Sapele faced both sides; lipped on all four edges; Premdor Crosby "Landscape Sapele Fireshield" or other equal and approved						
686 mm x 1981 mm x 44 mm	46.42	1.71	23.93	49.90	nr	73.83
762 mm x 1981 mm x 44 mm	46.42	1.71	23.93	49.90	nr	73.83
838 mm x 1981 mm x 44 mm	47.61	1.71	23.93	51.18	nr	75.11
726 mm x 2040 mm x 44 mm	48.43	1.71	23.93	52.06	nr	75.99
826 mm x 2040 mm x 44 mm	48.43	1.71	23.93	52.06	nr	75.99
Flush door; half-hour fire resisting (30/30) Sapele faced both sides; lipped on all four edges						
610 mm x 1981 mm x 44 mm	51.95	1.71	23.93	55.85	nr	79.78
686 mm x 1981 mm x 44 mm	51.95	1.71	23.93	55.85	nr	79.78
762 mm x 1981 mm x 44 mm	51.95	1.71	23.93	55.85	nr	79.78
838 mm x 1981 mm x 44 mm	51.95	1.71	23.93	55.85	nr	79.78
726 mm x 2040 mm x 44 mm	51.95	1.71	23.93	55.85	nr	79.78
826 mm x 2040 mm x 44 mm	51.95	1.71	23.93	55.85	nr	79.78
Flush door; half-hour fire resisting "Leaderflush" (type B30) or other equal and approved; American light oak veneer; hardwood lipping all edges						
526 mm x 2040 mm x 44 mm	113.49	1.71	23.93	122.00	nr	145.93
626 mm x 2040 mm x 44 mm	119.01	1.71	23.93	127.94	nr	151.87
726 mm x 2040 mm x 44 mm	119.35	1.71	23.93	128.30	nr	152.23
826 mm x 2040 mm x 44 mm	121.66	1.71	23.93	130.78	nr	154.71
Flush door; one-hour fire check (60/45); plywood faced both sides; lipped on all four edges; Premdor Crosby "Popular Firemaster" or other equal and approved						
762 mm x 1981 mm x 54 mm	121.17	1.85	25.89	130.26	nr	156.15
838 mm x 1981 mm x 54 mm	127.04	1.85	25.89	136.57	nr	162.46
726 mm x 2040 mm x 54 mm	125.11	1.85	25.89	134.49	nr	160.38
826 mm x 2040 mm x 54 mm	125.11	1.85	25.89	134.49	nr	160.38

Prices for Measured Works - Major Works
L WINDOWS/DOORS/STAIRS

	PC £	Labour hours	Labour £	Material £	Unit	Total rate £
L20 DOORS/SHUTTERS/HATCHES - cont'd						
Doors; standard flush; softwood composition - cont'd						
Flush door; one-hour fire check (60/45); Sapele faced both sides; lipped on all four edges						
762 mm x 1981 mm x 54 mm	200.31	1.94	27.15	215.33	nr	242.48
838 mm x 1981 mm x 54 mm	209.62	1.94	27.15	225.34	nr	252.49
Flush door; one-hour fire resisting "Leaderflush" (type B60) or other equal and approved; Afrormosia veneer; hardwood lipping all edges including groove and "Leaderseal" intumescent strip or other equal and approved						
457 mm x 1981 mm x 54 mm	149.76	1.94	27.15	160.99	nr	188.14
533 mm x 1981 mm x 54 mm	147.19	1.94	27.15	158.23	nr	185.38
610 mm x 1981 mm x 54 mm	166.18	1.94	27.15	178.64	nr	205.79
686 mm x 1981 mm x 54 mm	168.76	1.94	27.15	181.42	nr	208.57
762 mm x 1981 mm x 54 mm	171.32	1.94	27.15	184.17	nr	211.32
838 mm x 1981 mm x 54 mm	180.74	1.94	27.15	194.30	nr	221.45
526 mm x 2040 mm x 54 mm	146.97	1.94	27.15	157.99	nr	185.14
626 mm x 2040 mm x 54 mm	166.62	1.94	27.15	179.12	nr	206.27
726 mmx 2040 mm x 54 mm	169.87	1.94	27.15	182.61	nr	209.76
826 mm x 2040 mm x 54 mm	173.12	1.94	27.15	186.10	nr	213.25
Flush door; external quality; skeleton or cellular core; plywood faced both sides; lipped on all four edges						
762 mm x 1981 mm x 54 mm	32.60	1.62	22.67	35.05	nr	57.72
838 mm x 1981 mm x 54 mm	33.74	1.62	22.67	36.27	nr	58.94
Flush door; external quality with standard glass opening; skeleton or cellular core; plywood faced both sides; lipped on all four edges; including glazing beads						
762 mm x 1981 mm x 54 mm	40.97	1.62	22.67	44.04	nr	66.71
838 mm x 1981 mm x 54 mm	42.04	1.62	22.67	45.19	nr	67.86
Doors; purpose made panelled; wrought softwood						
Panelled doors; one open panel for glass; including glazing beads						
686 mm x 1981 mm x 44 mm	52.00	1.62	22.67	55.90	nr	78.57
762 mm x 1981 mm x 44 mm	53.60	1.62	22.67	57.62	nr	80.29
838 mm x 1981 mm x 44 mm	55.17	1.62	22.67	59.31	nr	81.98
Panelled doors; two open panel for glass; including glazing beads						
686 mm x 1981 mm x 44 mm	71.92	1.62	22.67	77.31	nr	99.98
762 mm x 1981 mm x 44 mm	75.78	1.62	22.67	81.46	nr	104.13
838 mm x 1981 mm x 44 mm	79.54	1.62	22.67	85.51	nr	108.18
Panelled doors; four 19 mm thick plywood panels; mouldings worked on solid both sides						
686 mm x 1981 mm x 44 mm	99.27	1.62	22.67	106.72	nr	129.39
762 mm x 1981 mm x 44 mm	103.03	1.62	22.67	110.76	nr	133.43
838 mm x 1981 mm x 44 mm	108.62	1.62	22.67	116.77	nr	139.44
Panelled doors; six 25 mm thick panels raised and fielded; mouldings worked on solid both sides						
686 mm x 1981 mm x 44 mm	226.80	1.94	27.15	243.81	nr	270.96
762 mm x 1981 mm x 44 mm	236.23	1.94	27.15	253.95	nr	281.10
838 mm x 1981 mm x 44 mm	248.05	1.94	27.15	266.65	nr	293.80
rebated edges beaded	-	-	-	1.29	m	1.29
rounded edges or heels	-	-	-	0.65	m	0.65
weatherboard fixed to bottom rail	-	0.23	3.22	2.47	m	5.69
stopped groove for weatherboard	-	-	-	3.86	m	3.86

L WINDOWS/DOORS/STAIRS

	PC £	Labour hours	Labour £	Material £	Unit	Total rate £
Doors; purpose made panelled; selected West African Mahogany						
Panelled doors; one open panel for glass; including glazing beads						
686 mm x 1981 mm x 50 mm	105.15	2.31	32.33	113.04	nr	**145.37**
762 mm x 1981 mm x 50 mm	108.62	2.31	32.33	116.77	nr	**149.10**
838 mm x 1981 mm x 50 mm	112.09	2.31	32.33	120.50	nr	**152.83**
686 mm x 1981 mm x 63 mm	127.96	2.54	35.55	137.56	nr	**173.11**
762 mm x 1981 mm x 63 mm	132.30	2.54	35.55	142.22	nr	**177.77**
838 mm x 1981 mm x 63 mm	136.61	2.54	35.55	146.86	nr	**182.41**
Panelled doors; 250 mm wide cross tongued intermediate rail; two open panels for glass; mouldings worked on the solid one side; 19 mm x 13 mm beads one side; fixing with brass cups and screws						
686 mm x 1981 mm x 50 mm	143.60	2.31	32.33	154.37	nr	**186.70**
762 mm x 1981 mm x 50 mm	150.70	2.31	32.33	162.00	nr	**194.33**
838 mm x 1981 mm x 50 mm	157.70	2.31	32.33	169.53	nr	**201.86**
686 mm x 1981 mm x 63 mm	172.62	2.54	35.55	185.57	nr	**221.12**
762 mm x 1981 mm x 63 mm	181.18	2.54	35.55	194.77	nr	**230.32**
838 mm x 1981 mm x 63 mm	189.73	2.54	35.55	203.96	nr	**239.51**
Panelled doors; four panels; (19 mm thick for 50 mm doors, 25 mm thick for 63 mm doors); mouldings worked on solid both sides						
686 mm x 1981 mm x 50 mm	182.97	2.31	32.33	196.69	nr	**229.02**
762 mm x 1981 mm x 50 mm	190.58	2.31	32.33	204.87	nr	**237.20**
838 mm x 1981 mm x 50 mm	200.11	2.31	32.33	215.12	nr	**247.45**
686 mm x 1981 mm x 63 mm	216.06	2.54	35.55	232.26	nr	**267.81**
762 mm x 1981 mm x 63 mm	225.09	2.54	35.55	241.97	nr	**277.52**
838 mm x 1981 mm x 63 mm	236.34	2.54	35.55	254.07	nr	**289.62**
Panelled doors; 150 mm wide stiles in one width; 430 mm wide cross tongued bottom rail; six panels raised and fielded one side; (19 mm thick for 50 mm doors, 25 mm thick for 63 mm doors); mouldings worked on solid both sides						
686 mm x 1981 mm x 50 mm	386.02	2.31	32.33	414.97	nr	**447.30**
762 mm x 1981 mm x 50 mm	402.12	2.31	32.33	432.28	nr	**464.61**
838 mm x 1981 mm x 50 mm	422.19	2.31	32.33	453.85	nr	**486.18**
686 mm x 1981 mm x 63 mm	475.40	2.54	35.55	511.06	nr	**546.61**
762 mm x 1981 mm x 63 mm	495.24	2.54	35.55	532.38	nr	**567.93**
838 mm x 1981 mm x 63 mm	519.98	2.54	35.55	558.98	nr	**594.53**
rebated edges beaded	-	-	-	1.92	m	**1.92**
rounded edges or heels	-	-	-	0.97	m	**0.97**
weatherboard fixed to bottom rail	-	0.31	4.34	5.53	m	**9.87**
stopped groove for weatherboard	-	-	-	4.73	m	**4.73**
Doors; purpose made panelled; American White Ash						
Panelled doors; one open panel for glass; including glazing beads						
686 mm x 1981 mm x 50 mm	117.75	2.31	32.33	126.58	nr	**158.91**
762 mm x 1981 mm x 50 mm	121.70	2.31	32.33	130.83	nr	**163.16**
838 mm x 1981 mm x 50 mm	125.64	2.31	32.33	135.06	nr	**167.39**
686 mm x 1981 mm x 63 mm	143.68	2.54	35.55	154.46	nr	**190.01**
762 mm x 1981 mm x 63 mm	148.64	2.54	35.55	159.79	nr	**195.34**
838 mm x 1981 mm x 63 mm	153.56	2.54	35.55	165.08	nr	**200.63**
Panelled doors; 250 mm wide cross tongued intermediate rail; two open panels for glass; mouldings worked on the solid one side; 19 mm x 13 mm beads one side; fixing with brass cups and screws						
686 mm x 1981 mm x 50 mm	160.23	2.31	32.33	172.25	nr	**204.58**

Prices for Measured Works - Major Works
L WINDOWS/DOORS/STAIRS

	PC £	Labour hours	Labour £	Material £	Unit	Total rate £
L20 DOORS/SHUTTERS/HATCHES - cont'd						
Doors; purpose made panelled; American White Ash - cont'd						
Panelled doors; 250 mm wide cross tongued intermediate rail; two open panels for glass; mouldings worked on the solid one side; 19 mm x 13 mm beads one side; fixing with brass cups and screws - cont'd						
762 mm x 1981 mm x 50 mm	168.25	2.31	32.33	180.87	nr	213.20
838 mm x 1981 mm x 50 mm	176.22	2.31	32.33	189.44	nr	221.77
686 mm x 1981 mm x 63 mm	193.39	2.54	35.55	207.89	nr	243.44
762 mm x 1981 mm x 63 mm	203.11	2.54	35.55	218.34	nr	253.89
838 mm x 1981 mm x 63 mm	212.80	2.54	35.55	228.76	nr	264.31
Panelled doors; four panels; (19 mm thick for 50 mm doors, 25 mm thick for 63 mm doors); mouldings worked on solid both sides						
686 mm x 1981 mm x 50 mm	203.50	2.31	32.33	218.76	nr	251.09
762 mm x 1981 mm x 50 mm	211.94	2.31	32.33	227.84	nr	260.17
838 mm x 1981 mm x 50 mm	222.55	2.31	32.33	239.24	nr	271.57
686 mm x 1981 mm x 63 mm	241.82	2.54	35.55	259.96	nr	295.51
762 mm x 1981 mm x 63 mm	251.91	2.54	35.55	270.80	nr	306.35
838 mm x 1981 mm x 63 mm	264.51	2.54	35.55	284.35	nr	319.90
Panelled doors; 150 mm wide stiles in one width; 430 mm wide cross tongued bottom rail; six panels raised and fielded one side; (19 mm thick for 50 mm doors, 25 mm thick for 63 mm doors); mouldings worked on solid						
686 mm x 1981 mm x 50 mm	419.39	2.31	32.33	450.84	nr	483.17
762 mm x 1981 mm x 50 mm	436.87	2.31	32.33	469.64	nr	501.97
838 mm x 1981 mm x 50 mm	458.71	2.31	32.33	493.11	nr	525.44
686 mm x 1981 mm x 63 mm	517.65	2.54	35.55	556.47	nr	592.02
762 mm x 1981 mm x 63 mm	539.21	2.54	35.55	579.65	nr	615.20
838 mm x 1981 mm x 63 mm	566.19	2.54	35.55	608.65	nr	644.20
rebated edges beaded	-	-	-	1.92	m	1.92
rounded edges or heels	-	-	-	0.97	m	0.97
weatherboard fixed to bottom rail	-	0.31	4.34	6.05	m	10.39
stopped groove for weatherboard	-	-	-	4.73	m	4.73
Doors; galvanised steel "up and over" type garage doors; Catnic "Horizon 90" or other equal and approved; spring counter balanced; fixed to timber frame (not included)						
Garage door						
2135 mm x 1980 mm	156.85	3.70	51.79	168.91	nr	220.70
2135 mm x 2135 mm	176.54	3.70	51.79	190.07	nr	241.86
2400 mm x 2135 mm	213.88	3.70	51.79	230.26	nr	282.05
3965 mm x 2135 mm	562.89	5.55	77.68	605.79	nr	683.47
Doorsets; Anti-Vandal Security door and frame units; Bastion Security Ltd or other equal and approved; to BS 5051; factory primed; fixing with frame anchors to masonry; cutting mortices; external						
46 mm thick insulated door with birch grade plywood; sheet steel bonded into door core; 2 mm thick polyester coated laminate finish; hardwood lippings all edges; 95 mm x 65 mm hardwood frame; polyester coated standard ironmongery; weather stripping all round; low projecting aluminium threshold; plugging; screwing						
for 980 mm x 2100 mm structural opening; single door sets; panic bolt	-	-	-	-	nr	1298.28
for 1830 mm x 2100 mm structural opening; double door sets; panic bolt	-	-	-	-	nr	2197.09

L WINDOWS/DOORS/STAIRS

	PC £	Labour hours	Labour £	Material £	Unit	Total rate £
Doorsets; steel door and frame units; Jandor Architectural Ltd or other equal and approved; polyester powder coated; ironmongery						
Single action door set; "Metset MD01" doors and "Metset MF" frames						
900 mm x 2100 mm	-	-	-	-	nr	1597.88
pair 1800 mm x 2100 mm	-	-	-	-	nr	2197.09
Doorsets; factory assembled; Dorma Entrance Systems or equal and approved; space saver automatic double doors; powder coated aluminium framing and base plates; alumnium entrance fanlights; laminated safety glazing; weatherstops; fixing by bolting						
Doorsets						
1485 mm x 2985 mm (co-ordinated size)	-	-	-	-	nr	7000.00
Rolling shutters and collapsible gates; steel counter shutters; Bolton Brady Ltd or other equal and approved; push-up, self-coiling; polyester power coated; fixing by bolting						
Shutters						
3000 mm x 1000 mm	-	-	-	-	nr	918.78
4000 mm x 1000 mm; in two panels	-	-	-	-	nr	1600.00
Rolling shutters and collapsible gates; galvanised steel; Bolton Brady Type 474 or other equal and approved; one hour fire resisting; self-coiling; activated by fusible link; fixing by bolting						
Rolling shutters and collapsible gates						
1000 mm x 2750 mm	-	-	-	-	nr	1100.00
1500 mm x 2750 mm	-	-	-	-	nr	1155.00
2400 mm x 2750 mm	-	-	-	-	nr	1365.00
Rolling shutters and collapsible gates; GRP vertically opening insulated panel shutter doors; electrically operated; Envirodoor Markus Ltd; type HT40 or other equal and approved; manual over-ride, lock interlock, stop and return safety cage, "deadmans" down button, anti-flip device, photo-electric cell and beam deflectors; fixing by bolting						
Shutter doors						
3000 mm x 3000 mm; windows - 2 nr	-	-	-	-	nr	7090.59
4000 mm x 4000 mm; windows - 2 nr	-	-	-	-	nr	8788.34
5000 mm x 5000 mm; windows - 2 nr	-	-	-	-	nr	10486.09
Sliding/folding partitions; aluminium double glazed sliding patio doors; Crittal "Luminaire" or equal and approved; white acrylic finish; with and including 18 mm thick annealed double glazing; fixed in position; including lugs plugging and screwing to brickwork or blockwork						
Patio doors						
1800 mm x 2100 mm; ref PF1821	943.50	2.31	32.33	1015.10	nr	1047.43
2400 m x 2100 mm; ref PF2421	1132.20	2.78	38.91	1217.95	nr	1256.86
2700 mm x 2100 mm; ref PF2721	1258.00	3.24	45.35	1353.18	nr	1398.53

L WINDOWS/DOORS/STAIRS

	PC £	Labour hours	Labour £	Material £	Unit	Total rate £
L20 DOORS/SHUTTERS/HATCHES - cont'd						
Grilles; "Galaxy" nylon rolling counter grille or other equal and approved; Bolton Brady Ltd; colour, off-white; self-coiling; fixing by bolting						
Grilles						
3000 mm x 1000 mm	-	-	-	-	nr	828.90
4000 mm x 1000 mm	-	-	-	-	nr	1248.34
Door frames and door linings, sets; standard joinery; wrought softwood						
Internal door frame or lining composite sets for 686 mm x 1981 mm door; all with loose stops unless rebated; "finished sizes"						
27 mm x 94 mm lining	19.98	0.74	10.36	21.76	nr	32.12
27 mm x 107 mm lining	21.02	0.74	10.36	22.87	nr	33.23
35 mm x 107 mm rebated lining	21.81	0.74	10.36	23.72	nr	34.08
27 mm x 121 mm lining	22.73	0.74	10.36	24.71	nr	35.07
27 mm x 121 mm lining with fanlight over	32.01	0.88	12.32	34.76	nr	47.08
27 mm x 133 mm lining	24.50	0.74	10.36	26.62	nr	36.98
35 mm x 133 mm rebated lining	25.18	0.74	10.36	27.35	nr	37.71
27 mm x 133 mm lining with fanlight over	34.40	0.88	12.32	37.33	nr	49.65
33 mm x 57 mm frame	16.10	0.74	10.36	17.59	nr	27.95
33 mm x 57 mm storey height frame	19.86	0.88	12.32	21.70	nr	34.02
33 mm x 57 mm frame with fanlight over	24.07	0.88	12.32	26.22	nr	38.54
33 mm x 64 mm frame	17.35	0.74	10.36	18.93	nr	29.29
33 mm x 64 mm storey height frame	21.26	0.88	12.32	23.20	nr	35.52
33 mm x 64 mm frame with fanlight over	25.41	0.88	12.32	27.66	nr	39.98
44 mm x 94 mm frame	26.88	0.85	11.90	29.17	nr	41.07
44 mm x 94 mm storey height frame	31.98	1.02	14.28	34.73	nr	49.01
44 mm x 94 mm frame with fanlight over	37.29	1.02	14.28	40.43	nr	54.71
44 mm x 107 mm frame	30.74	0.93	13.02	33.32	nr	46.34
44 mm x 107 mm storey height frame	36.01	1.02	14.28	39.06	nr	53.34
44 mm x 107 mm frame with fanlight over	41.09	1.02	14.28	44.52	nr	58.80
Internal door frame or lining composite sets for 762 mm x 1981 mm door; all with loose stops unless rebated; "finished sizes"						
27 mm x 94 mm lining	19.98	0.74	10.36	21.76	nr	32.12
27 mm x 107 mm lining	21.02	0.74	10.36	22.87	nr	33.23
35 mm x 107 mm rebated lining	21.81	0.74	10.36	23.72	nr	34.08
27 mm x 121 mm lining	22.73	0.74	10.36	24.71	nr	35.07
27 mm x 121 mm lining with fanlight over	32.01	0.88	12.32	34.76	nr	47.08
27 mm x 133 mm lining	24.50	0.74	10.36	26.62	nr	36.98
35 mm x 133 mm rebated lining	25.18	0.74	10.36	27.35	nr	37.71
27 mm x 133 mm lining with fanlight over	34.40	0.88	12.32	37.33	nr	49.65
33 mm x 57 mm frame	16.10	0.74	10.36	17.59	nr	27.95
33 mm x 57 mm storey height frame	19.86	0.88	12.32	21.70	nr	34.02
33 mm x 57 mm frame with fanlight over	24.07	0.88	12.32	26.22	nr	38.54
33 mm x 64 mm frame	17.35	0.74	10.36	18.93	nr	29.29
33 mm x 64 mm storey height frame	21.26	0.88	12.32	23.20	nr	35.52
33 mm x 64 mm frame with fanlight over	25.41	0.88	12.32	27.66	nr	39.98
44 mm x 94 mm frame	26.88	0.85	11.90	29.17	nr	41.07
44 mm x 94 mm storey height frame	31.98	1.02	14.28	34.73	nr	49.01
44 mm x 94 mm frame with fanlight over	37.29	1.02	14.28	40.43	nr	54.71
44 mm x 107 mm frame	30.74	0.93	13.02	33.32	nr	46.34
44 mm x 107 mm storey height frame	36.01	1.02	14.28	39.06	nr	53.34
44 mm x 107 mm frame with fanlight over	41.09	1.02	14.28	44.52	nr	58.80

L WINDOWS/DOORS/STAIRS

	PC £	Labour hours	Labour £	Material £	Unit	Total rate £
Internal door frame or lining composite sets for 726 mm x 2040 mm door; with loose stops						
30 mm x 94 mm lining	23.30	0.74	10.36	25.33	nr	35.69
30 mm x 94 mm lining with fanlight over	33.21	0.88	12.32	36.05	nr	48.37
30 mm x 107 mm lining	27.40	0.74	10.36	29.73	nr	40.09
30 mm x 107 mm lining with fanlight over	37.42	0.88	12.32	40.57	nr	52.89
30 mm x 133 mm lining	30.70	0.74	10.36	33.28	nr	43.64
30 mm x 133 mm lining with fanlight over	40.35	0.88	12.32	43.72	nr	56.04
Internal door frame or lining composite sets for 826 mm x 2040 mm door; with loose stops						
30 mm x 94 mm lining	23.30	0.74	10.36	25.33	nr	35.69
30 mm x 94 mm lining with fanlight over	33.21	0.88	12.32	36.05	nr	48.37
30 mm x 107 mm lining	27.40	0.74	10.36	29.73	nr	40.09
30 mm x 107 mm lining with fanlight over	37.42	0.88	12.32	40.57	nr	52.89
30 mm x 133 mm lining	30.70	0.74	10.36	33.28	nr	43.64
30 mm x 133 mm lining with fanlight over	40.35	0.88	12.32	43.72	nr	56.04
Door frames and door linings, sets; purpose made; wrought softwood						
Jambs and heads; as linings						
32 mm x 63 mm	-	0.16	2.24	3.93	m	6.17
32 mm x 100 mm	-	0.16	2.24	5.03	m	7.27
32 mm x 140 mm	-	0.16	2.24	6.44	m	8.68
Jambs and heads; as frames; rebated, rounded and grooved						
38 mm x 75 mm	-	0.16	2.24	4.99	m	7.23
38 mm x 100 mm	-	0.16	2.24	5.91	m	8.15
38 mm x 115 mm	-	0.16	2.24	6.88	m	9.12
38 mm x 140 mm	-	0.19	2.66	8.00	m	10.66
50 mm x 100 mm	-	0.19	2.66	7.44	m	10.10
50 mm x 125 mm	-	0.19	2.66	8.85	m	11.51
63 mm x 88 mm	-	0.19	2.66	8.53	m	11.19
63 mm x 100 mm	-	0.19	2.66	9.10	m	11.76
63 mm x 125 mm	-	0.19	2.66	10.86	m	13.52
75 mm x 100 mm	-	0.19	2.66	10.43	m	13.09
75 mm x 125 mm	-	0.20	2.80	12.91	m	15.71
75 mm x 150 mm	-	0.20	2.80	15.05	m	17.85
100 mm x 100 mm	-	0.23	3.22	13.57	m	16.79
100 mm x 150 mm	-	0.23	3.22	19.44	m	22.66
Mullions and transoms; in linings						
32 mm x 63 mm	-	0.11	1.54	6.41	m	7.95
32 mm x 100 mm	-	0.11	1.54	7.54	m	9.08
32 mm x 140 mm	-	0.11	1.54	8.87	m	10.41
Mullions and transoms; in frames; twice rebated, rounded and grooved						
38 mm x 75 mm	-	0.11	1.54	7.64	m	9.18
38 mm x 100 mm	-	0.11	1.54	8.56	m	10.10
38 mm x 115 mm	-	0.11	1.54	9.30	m	10.84
38 mm x 140 mm	-	0.13	1.82	10.43	m	12.25
50 mm x 100 mm	-	0.13	1.82	9.87	m	11.69
50 mm x 125 mm	-	0.13	1.82	11.42	m	13.24
63 mm x 88 mm	-	0.13	1.82	11.02	m	12.84
63 mm x 100 mm	-	0.13	1.82	11.59	m	13.41
75 mm x 100 mm	-	0.13	1.82	13.01	m	14.83
Add 5% to the above material prices for selected softwood for staining						

Prices for Measured Works - Major Works
L WINDOWS/DOORS/STAIRS

	PC £	Labour hours	Labour £	Material £	Unit	Total rate £
L20 DOORS/SHUTTERS/HATCHES - cont'd						
Door frames and door linings, sets; purpose made; selected West African Mahogany						
Jambs and heads; as linings						
32 mm x 63 mm	6.72	0.21	2.94	7.29	m	10.23
32 mm x 100 mm	8.74	0.21	2.94	9.46	m	12.40
32 mm x 140 mm	11.09	0.21	2.94	12.06	m	15.00
Jambs and heads; as frames; rebated, rounded and grooved						
38 mm x 75 mm	8.47	0.21	2.94	9.17	m	12.11
38 mm x 100 mm	10.27	0.21	2.94	11.11	m	14.05
38 mm x 115 mm	11.80	0.21	2.94	12.82	m	15.76
38 mm x 140 mm	13.64	0.25	3.50	14.80	m	18.30
50 mm x 100 mm	12.97	0.25	3.50	14.08	m	17.58
50 mm x 125 mm	15.62	0.25	3.50	16.93	m	20.43
63 mm x 88 mm	14.51	0.25	3.50	15.67	m	19.17
63 mm x 100 mm	16.09	0.25	3.50	17.37	m	20.87
63 mm x 125 mm	19.34	0.25	3.50	20.93	m	24.43
75 mm x 100 mm	18.54	0.25	3.50	20.07	m	23.57
75 mm x 125 mm	23.06	0.28	3.92	24.93	m	28.85
75 mm x 150 mm	27.06	0.28	3.92	29.23	m	33.15
100 mm x 100 mm	24.28	0.28	3.92	26.24	m	30.16
100 mm x 150 mm	35.22	0.28	3.92	38.00	m	41.92
Mullions and transoms; in linings						
32 mm x 63 mm	10.30	0.15	2.10	11.07	m	13.17
32 mm x 100 mm	12.31	0.15	2.10	13.23	m	15.33
32 mm x 140 mm	14.67	0.15	2.10	15.77	m	17.87
Mullions and transoms; in frames; twice rebated, rounded and grooved						
38 mm x 75 mm	12.24	0.15	2.10	13.16	m	15.26
38 mm x 100 mm	14.03	0.15	2.10	15.08	m	17.18
38 mm x 115 mm	15.39	0.15	2.10	16.54	m	18.64
38 mm x 140 mm	17.22	0.17	2.38	18.51	m	20.89
50 mm x 100 mm	16.53	0.17	2.38	17.77	m	20.15
50 mm x 125 mm	19.39	0.17	2.38	20.84	m	23.22
63 mm x 88 mm	18.10	0.17	2.38	19.46	m	21.84
63 mm x 100 mm	19.66	0.17	2.38	21.13	m	23.51
75 mm x 100 mm	22.31	0.17	2.38	23.98	m	26.36
Sills; once sunk weathered; once rebated, three times grooved						
63 mm x 175 mm	36.17	0.31	4.34	38.88	m	43.22
75 mm x 125 mm	32.76	0.31	4.34	35.22	m	39.56
75 mm x 150 mm	36.83	0.31	4.34	39.59	m	43.93
Door frames and door linings, sets; purpose made; American White Ash						
Jambs and heads; as linings						
32 mm x 63 mm	7.63	0.21	2.94	8.27	m	11.21
32 mm x 100 mm	10.00	0.21	2.94	10.82	m	13.76
32 mm x 140 mm	12.91	0.21	2.94	14.02	m	16.96

L WINDOWS/DOORS/STAIRS

	PC £	Labour hours	Labour £	Material £	Unit	Total rate £
Jambs and heads; as frames; rebated, rounded and grooved						
38 mm x 75 mm	9.63	0.21	2.94	10.42	m	13.36
38 mm x 100 mm	11.81	0.21	2.94	12.77	m	15.71
38 mm x 115 mm	13.64	0.21	2.94	14.80	m	17.74
38 mm x 140 mm	15.76	0.25	3.50	17.08	m	20.58
50 mm x 100 mm	14.97	0.25	3.50	16.23	m	19.73
50 mm x 125 mm	18.14	0.25	3.50	19.64	m	23.14
63 mm x 88 mm	16.79	0.25	3.50	18.12	m	21.62
63 mm x 100 mm	18.64	0.25	3.50	20.11	m	23.61
63 mm x 125 mm	22.53	0.25	3.50	24.36	m	27.86
75 mm x 100 mm	21.55	0.25	3.50	23.31	m	26.81
75 mm x 125 mm	26.84	0.28	3.92	28.99	m	32.91
75 mm x 150 mm	31.60	0.28	3.92	34.11	m	38.03
100 mm x 100 mm	28.30	0.28	3.92	30.56	m	34.48
100 mm x 150 mm	41.29	0.28	3.92	44.53	m	48.45
Mullions and transoms; in linings						
32 mm x 63 mm	11.20	0.15	2.10	12.04	m	14.14
32 mm x 100 mm	13.59	0.15	2.10	14.61	m	16.71
32 mm x 140 mm	16.49	0.15	2.10	17.73	m	19.83
Mullions and transoms; in frames; twice rebated, rounded and grooved						
38 mm x 75 mm	13.39	0.15	2.10	14.39	m	16.49
38 mm x 100 mm	15.58	0.15	2.10	16.75	m	18.85
38 mm x 115 mm	17.22	0.15	2.10	18.51	m	20.61
38 mm x 140 mm	19.34	0.17	2.38	20.79	m	23.17
50 mm x 100 mm	18.55	0.17	2.38	19.94	m	22.32
50 mm x 125 mm	21.91	0.17	2.38	23.55	m	25.93
63 mm x 88 mm	20.40	0.17	2.38	21.93	m	24.31
63 mm x 100 mm	22.21	0.17	2.38	23.88	m	26.26
75 mm x 100 mm	25.32	0.17	2.38	27.22	m	29.60
Sills; once sunk weathered; once rebated, three times grooved						
63 mm x 175 mm	40.61	0.31	4.34	43.66	m	48.00
75 mm x 125 mm	36.55	0.31	4.34	39.29	m	43.63
75 mm x 150 mm	41.39	0.31	4.34	44.49	m	48.83
Door frames and door linings, sets; European Oak						
Sills; once sunk weathered; once rebated, three times grooved						
63 mm x 175 mm	71.59	0.31	4.34	76.96	m	81.30
75 mm x 125 mm	63.57	0.31	4.34	68.34	m	72.68
75 mm x 150 mm	72.87	0.31	4.34	78.34	m	82.68
Bedding and pointing frames						
Pointing wood frames or sills with mastic						
one side	-	0.09	0.89	0.43	m	1.32
both sides	-	0.19	1.87	0.87	m	2.74
Pointing wood frames or sills with polysulphide sealant						
one side	-	0.09	0.89	1.31	m	2.20
both sides	-	0.19	1.87	2.62	m	4.49
Bedding wood frames in cement mortar (1:3) and point						
one side	-	0.07	1.22	0.07	m	1.29
both sides	-	0.09	1.57	0.09	m	1.66
one side in mortar; other side in mastic	-	0.19	2.63	0.50	m	3.13

L30 STAIRS/WALKWAYS/BALUSTRADES

	PC £	Labour hours	Labour £	Material £	Unit	Total rate £
Standard staircases; wrought softwood (parana pine) Stairs; 25 mm thick treads with rounded nosings; 9 mm thick plywood risers; 32 mm thick strings; bullnose bottom tread; 50 mm x 75 mm hardwood handrail; 32 mm square plain balusters; 100 mm square plain newel posts						
straight flight; 838 mm wide; 2676 mm going; 2600 mm rise; with two newel posts	-	6.48	90.70	335.25	nr	425.95
straight flight with turn; 838 mm wide; 2676 mm going; 2600 mm rise; with two newel posts; three top treads winding	-	6.48	90.70	424.36	nr	515.06
dogleg staircase; 838 mm wide; 2676 mm going; 2600 mm rise; with two newel posts; quarter space landing third riser from top	-	6.48	90.70	357.53	nr	448.23
dogleg staircase; 838 mm wide; 2676 mm going; 2600 mm rise; with two newel posts; half space landing third riser from top	-	7.40	103.57	379.80	nr	483.37
Standard balustrades; wrought softwood Landing balustrade; 50 mm x 75 mm hardwood handrail; three 32 mm x 140 mm balustrade kneerails; two 32 mm x 50 mm stiffeners; one end of handrail jointed to newel post; other end built into wall; (newel post and mortices both not included)						
3.00 m long	-	2.54	35.55	115.85	nr	151.40
Landing balustrade; 50 mm x 75 mm hardwood handrail; 32 mm square plain balusters; one end of handrail jointed to newel post; other end built into wall; balusters housed in at bottom (newel post and mortices both not included)						
3.00 m long	-	3.70	51.79	71.29	nr	123.08
Hardwood staircases; purpose made; assembled at works Fixing only complete staircase including landings, balustrades, etc.						
plugging and screwing to brickwork or blockwork	-	13.88	194.27	1.75	nr	196.02
The following are supply only prices for purpose made staircase components in selected West African Mahogany supplied as part of an assembled staircase and may be used to arrive at a guide price for a complete hardwood staircase Board landings; cross-tongued joints; 100 mm x 50 mm sawn softwood bearers						
25 mm thick	-	-	-	82.81	m²	82.81
32 mm thick	-	-	-	101.53	m²	101.53
Treads; cross-tongued joints and risers; rounded nosings; tongued, grooved, glued and blocked together; one 175 mm x 50 mm sawn softwood carriage						
25 mm treads; 19 mm risers	-	-	-	102.52	m²	102.52
ends; quadrant	-	-	-	47.47	nr	47.47
ends; housed to hardwood	-	-	-	0.88	nr	0.88
32 mm treads; 25 mm risers	-	-	-	123.04	m²	123.04
ends; quadrant	-	-	-	61.01	nr	61.01
ends; housed to hardwood	-	-	-	0.88	nr	0.88

L WINDOWS/DOORS/STAIRS

	PC £	Labour hours	Labour £	Material £	Unit	Total rate £
Winders; cross-tongued joints and risers in one width; rounded nosings; tongued, grooved glued and blocked together; one 175 mm x 50 mm sawn softwood carriage						
25 mm treads; 19 mm risers	-	-	-	112.77	m²	112.77
32 mm treads; 25 mm risers	-	-	-	135.36	m²	135.36
wide ends; housed to hardwood	-	-	-	1.75	nr	1.75
narrow ends; housed to hardwood	-	-	-	1.31	nr	1.31
Closed strings; in one width; 230 mm wide; rounded twice						
32 mm thick	-	-	-	21.74	m	21.74
38 mm thick	-	-	-	26.10	m	26.10
50 mm thick	-	-	-	34.48	m	34.48
Closed strings; cross-tongued joints; 280 mm wide; once rounded						
32 mm thick	-	-	-	38.04	m	38.04
extra for short ramp	-	-	-	19.02	nr	19.02
38 mm thick	-	-	-	43.16	m	43.16
extra for short ramp	-	-	-	21.62	nr	21.62
50 mm thick	-	-	-	53.46	m	53.46
extra for short ramp	-	-	-	26.78	nr	26.78
The following labours are irrespective of timber width						
ends; fitted	-	-	-	1.13	nr	1.13
ends; framed	-	-	-	6.62	nr	6.62
extra for tongued heading joint	-	-	-	3.27	nr	3.27
Closed strings; ramped; crossed tongued joints 280 mm wide; once rounded; fixing with screws; plugging 450 mm centres						
32 mm thick	-	-	-	41.88	m	41.88
38 mm thick	-	-	-	47.52	m	47.52
50 mm thick	-	-	-	58.79	m	58.79
Apron linings; in one width 230 mm wide						
19 mm thick	-	-	-	8.62	m	8.62
25 mm thick	-	-	-	10.43	m	10.43
Handrails; rounded						
40 mm x 50 mm	-	-	-	7.18	m	7.18
50 mm x 75 mm	-	-	-	8.98	m	8.98
63 mm x 87 mm	-	-	-	11.61	m	11.61
75 mm x 100 mm	-	-	-	14.11	m	14.11
Handrails; moulded						
40 mm x 50 mm	-	-	-	7.70	m	7.70
50 mm x 75 mm	-	-	-	9.50	m	9.50
63 mm x 87 mm	-	-	-	12.15	m	12.15
75 mm x 100 mm	-	-	-	14.64	m	14.64
Handrails; rounded; ramped						
40 mm x 50 mm	-	-	-	14.34	m	14.34
50 mm x 75 mm	-	-	-	17.97	m	17.97
63 mm x 87 mm	-	-	-	23.25	m	23.25
75 mm x 100 mm	-	-	-	23.00	m	23.00
Handrails; moulded; ramped						
40 mm x 50 mm	-	-	-	15.39	m	15.39
50 mm x 75 mm	-	-	-	19.03	m	19.03
63 mm x 87 mm	-	-	-	24.31	m	24.31
75 mm x 100 mm	-	-	-	29.27	m	29.27
Heading joints to handrail; mitred or raked						
overall size not exceeding 50 mm x 75 mm	-	-	-	24.56	nr	24.56
overall size not exceeding 75 mm x 100 mm	-	-	-	31.27	nr	31.27
Balusters; stiffeners						
25 mm x 25 mm	-	-	-	3.48	m	3.48
32 mm x 32 mm	-	-	-	4.04	m	4.04
50 mm x 50 mm	-	-	-	4.88	m	4.88
ends; housed	-	-	-	1.11	nr	1.11

L WINDOWS/DOORS/STAIRS

	PC £	Labour hours	Labour £	Material £	Unit	Total rate £
L30 STAIRS/WALKWAYS/BALUSTRADES - cont'd						
The following are supply only prices for purpose made staircase components in selected West African Mahogany supplied as part of an assembled staircase and may be used to arrive at a guide price for a complete hardwood staircase - cont'd						
Sub rails						
32 mm x 63 mm	-	-	-	6.21	m	6.21
ends; framed joint to newel	-	-	-	4.46	nr	4.46
Knee rails						
32 mm x 140 mm	-	-	-	9.71	m	9.71
ends; framed joint to newel	-	-	-	4.46	nr	4.46
Newel posts						
50 mm x 100 mm; half newel	-	-	-	10.38	m	10.38
75 mm x 75 mm	-	-	-	11.21	m	11.21
100 mm x 100 mm	-	-	-	17.32	m	17.32
Newel caps; splayed on four sides						
62.50 mm x 125 mm x 50 mm	-	-	-	6.53	nr	6.53
100 mm x 100 mm x 50 mm	-	-	-	6.53	nr	6.53
125 mm x 125 mm x 50 mm	-	-	-	7.06	nr	7.06
The following are supply only prices for purpose made staircase components in selected Oak; supplied as part of an assembled staircase						
Board landings; cross-tongued joints; 100 mm x 50 mm sawn softwood bearers						
25 mm thick	-	-	-	174.16	m²	174.16
32 mm thick	-	-	-	214.67	m²	214.67
Treads; cross-tongued joints and risers; rounded nosings; tongued, grooved, glued and blocked together; one 175 mm x 50 mm sawn softwood carriage						
25 mm treads; 19 mm risers	-	-	-	189.65	m²	189.65
ends; quadrant	-	-	-	94.85	nr	94.85
ends; housed to hardwood	-	-	-	1.16	nr	1.16
32 mm treads; 25 mm risers	-	-	-	233.46	m²	233.46
ends; quadrant	-	-	-	116.75	nr	116.75
ends; housed to hardwood	-	-	-	1.16	nr	1.16
Winders; cross-tongued joints and risers in one width; rounded nosings; tongued, grooved glued and blocked together; one 175 mm x 50 mm sawn softwood carriage						
25 mm treads; 19 mm risers	-	-	-	208.63	m²	208.63
32 mm treads; 25 mm risers	-	-	-	256.82	m²	256.82
wide ends; housed to hardwood	-	-	-	2.33	nr	2.33
narrow ends; housed to hardwood	-	-	-	1.75	nr	1.75
Closed strings; in one width; 230 mm wide; rounded twice						
32 mm thick	-	-	-	45.82	m	45.82
38 mm thick	-	-	-	54.61	m	54.61
50 mm thick	-	-	-	72.19	m	72.19
Closed strings; cross-tongued joints; 280 mm wide; once rounded						
32 mm thick	-	-	-	73.22	m	73.22
extra for short ramp	-	-	-	36.59	nr	36.59
38 mm thick	-	-	-	83.98	m	83.98
extra for short ramp	-	-	-	41.99	nr	41.99
50 mm thick	-	-	-	105.67	m	105.67
extra for short ramp	-	-	-	52.85	nr	52.85

Prices for Measured Works - Major Works
L WINDOWS/DOORS/STAIRS

	PC £	Labour hours	Labour £	Material £	Unit	Total rate £
Closed strings; ramped; crossed tongued joints 280 mm wide; once rounded; fixing with screws; plugging 450 mm centres						
32 mm thick	-	-	-	80.55	m	80.55
38 mm thick	-	-	-	92.40	m	92.40
50 mm thick	-	-	-	116.24	m	116.24
Apron linings; in one width 230 mm wide						
19 mm thick	-	-	-	20.93	m	20.93
25 mm thick	-	-	-	26.31	m	26.31
Handrails; rounded						
40 mm x 50 mm	-	-	-	14.90	m	14.90
50 mm x 75 mm	-	-	-	20.29	m	20.29
63 mm x 87 mm	-	-	-	28.15	m	28.15
75 mm x 100 mm	-	-	-	35.56	m	35.56
Handrails; moulded						
40 mm x 50 mm	-	-	-	15.60	m	15.60
50 mm x 75 mm	-	-	-	20.98	m	20.98
63 mm x 87 mm	-	-	-	23.65	m	23.65
75 mm x 100 mm	-	-	-	36.26	m	36.26
Handrails; rounded; ramped						
40 mm x 50 mm	-	-	-	29.80	m	29.80
50 mm x 75 mm	-	-	-	40.59	m	40.59
63 mm x 87 mm	-	-	-	56.31	m	56.31
75 mm x 100 mm	-	-	-	71.14	m	71.14
Handrails; moulded; ramped						
40 mm x 50 mm	-	-	-	31.20	m	31.20
50 mm x 75 mm	-	-	-	41.99	m	41.99
63 mm x 87 mm	-	-	-	57.70	m	57.70
75 mm x 100 mm	-	-	-	72.53	m	72.53
Heading joints to handrail; mitred or raked						
overall size not exceeding 50 mm x 75 mm	-	-	-	37.54	nr	37.54
overall size not exceeding 75 mm x 100 mm	-	-	-	45.72	nr	45.72
Balusters; stiffeners						
25 mm x 25 mm	-	-	-	5.67	m	5.67
32 mm x 32 mm	-	-	-	7.30	m	7.30
50 mm x 50 mm	-	-	-	9.80	m	9.80
ends; housed	-	-	-	1.45	nr	1.45
Sub rails						
32 mm x 63 mm	-	-	-	12.51	m	12.51
ends; framed joint to newel	-	-	-	5.94	nr	5.94
Knee rails						
32 mm x 140 mm	-	-	-	22.82	m	22.82
ends; framed joint to newel	-	-	-	5.94	nr	5.94
Newel posts						
50 mm x 100 mm; half newel	-	-	-	24.90	m	24.90
75 mm x 75 mm	-	-	-	27.35	m	27.35
100 mm x 100 mm	-	-	-	45.53	m	45.53
Newel caps; splayed on four sides						
62.50 mm x 125 mm x 50 mm	-	-	-	9.47	nr	9.47
100 mm x 100 mm x 50 mm	-	-	-	9.47	nr	9.47
125 mm x 125 mm x 50 mm	-	-	-	11.09	nr	11.09
Spiral staircases, balustrades and handrails; mild steel; galvanised and polyester powder coated						
Staircase						
2080 mm diameter x 3695 mm high; 18 nr treads; 16 mm diameter intermediate balusters; 1040 mm x 1350 mm landing unit with matching balustrade both sides; fixing with 16 mm diameter resin anchors to masonry at landing and with 12 mm diameter expanding bolts to concrete at base	-	-	-	-	nr	3495.36

Prices for Measured Works - Major Works
L WINDOWS/DOORS/STAIRS

	PC £	Labour hours	Labour £	Material £	Unit	Total rate £
L30 STAIRS/WALKWAYS/BALUSTRADES - cont'd						
Aluminium alloy folding loft ladders; "Zig Zag" stairways, model B or other equal and approved; Light Alloy Ltd; on and including plywood backboard; fixing with screws to timber lining (not included)						
Loft ladders						
ceiling height not exceeding 2500 mm; model 888801	-	0.93	13.02	434.15	nr	447.17
ceiling height not exceeding 2800 mm; model 888802	-	0.93	13.02	476.04	nr	489.06
ceiling height not exceeding 3100 mm; model 888803	-	0.93	13.02	531.28	nr	544.30
Flooring, balustrades and handrails; mild steel						
Chequer plate flooring; over 300 mm wide; bolted to steel supports						
6 mm thick	-	-	-	44.94	m^2	44.94
8 mm thick	-	-	-	59.92	m^2	59.92
Open grid steel flooring; Eurogrid Ltd; "Safeway" type D38 diamond pattern flooring or other equal and approved to steel supports at 1200 mm centres; galvanised; to BS 729 part 1						
30 mm x 5 mm sections; in one mat; fixing to 10 mm thick channels with type F5 clamps	-	-	-	114.84	m^2	114.84
Balustrades; welded construction; 1070 mm high; galvanized; 50 mm x 50 mm x 3.20 mm rhs top rail; 38 mm x 13 mm bottom rail, 50 mm x 50 mm x 3.20 mm rhs standards at 1830 mm centres with base plate drilled and bolted to concrete; 13 mm x 13 mm balusters at 102 mm centres	-	-	-	109.85	m	109.85
Balusters; isolated; one end ragged and cemented in; one 76 mm x 25 mm x 6 mm flange plate welded on; ground to a smooth finish; countersunk drilled and tap screwed to underside of handrail						
19 mm square; 914 mm long bar	-	-	-	14.99	nr	14.99
Core-rails; joints prepared, welded and ground to a smooth finish; fixing on brackets (not included)						
38 mm x 10 mm flat bar	-	-	-	19.97	m	19.97
50 mm x 8 mm flat bar	-	-	-	19.22	m	19.22
Handrails; joints prepared, welded and ground to a smooth finish; fixing on brackets (not included)						
38 mm x 12 mm half oval bar	-	-	-	26.47	m	26.47
44 mm x 13 mm half oval bar	-	-	-	28.96	m	28.96
Handrail bracket; comprising 40 mm x 5 mm plate with mitred and welded angle; one end welded to 100 mm diameter x 5 mm backplate; three times holed and plugged and screwed to brickwork; other end scribed and welded to underside of handrail; 140 mm girth	-	-	-	15.97	nr	15.97
Holes						
Holes; countersunk; for screws or bolts						
6 mm diameter; 3 mm thick	-	-	-	0.80	nr	0.80
6 mm diameter; 6 mm thick	-	-	-	1.19	nr	1.19
8 mm diameter; 6 mm thick	-	-	-	1.19	nr	1.19
10 mm diameter; 6 mm thick	-	-	-	1.25	nr	1.25
12 mm diameter; 8 mm thick	-	-	-	1.60	nr	1.60
Balustrades; stainless steel BS 970 grade 316; TIG welded joints; mirror polished finish						
Raking balustrade; comprising 40 mm handrail on 40 mm standards at 1 m centres with two 16 mm diameter intermediate standards; 1.00 m high	-	-	-	599.21	m	599.21

L40 GENERAL GLAZING

BASIC GLASS PRICES (£/m²)

Trade cut prices - all to limiting sizes

	£		£		£		£
Ordinary transluscent/patterned glass							
2 mm	15.97	5 mm	34.96	12 mm	109.85	25 mm	234.68
3 mm	19.97	6 mm	39.95	15 mm	119.84		
4 mm	21.97	10 mm	74.90	19 mm	169.77		
Obscured ground sheet glass - extra on sheet glass prices							
PC £10.50/m²							
Patterned							
4 mm tinted	28.96	4 mm white	16.48	6 mm white	27.96		
Rough cast							
6 mm	21.97	10 mm	84.89				
Ordinary Georgian wired							
7 mm cast	22.97	6 mm polish	59.92				
Polycarbonate standard sheets							
2 mm	42.44	3 mm	59.92	4 mm	79.89	5 mm	97.37
6 mm	119.84	8 mm	159.79	10 mm	179.76	12 mm	234.68

SPECIAL GLASSES

	£		£		£		£
"Antisun"; float; bronze grey							
4 mm	44.94	6 mm	59.92				
10 mm	139.81	12 mm	174.76				
"Cetuff" toughened							
- float							
4 mm	32.96	5 mm	38.95	6 mm	43.44	10 mm	84.89
12 mm	124.84						
- patterned							
6 mm rough	82.39	10 mm rough	119.84	4 mm white	37.95	6 mm white	48.93
Clear laminated							
- security							
7.50 mm	99.87	9.50 mm	104.87	11.50 mm	109.89		
- safety							
4.40 mm	49.93	5.40 mm	51.93	6.40 mm	54.92	6.80 mm	64.91
8.80 mm	82.39	10.80 mm	99.87				
"Permawal"							
- reflective							
6 mm	82.39	10 mm	119.84				
"Permasol"							
6 mm	114.84						
"Silvered"							
2 mm	29.96	3 mm	33.45	4 mm	36.45	6 mm	49.93
"Silvered tinted"							
4 mm bronze	57.43	6 mm bronze	79.89	4 mm grey	57.92	6 mm grey	79.89
"Venetian striped"							
4 mm	99.87	6 mm	109.85				

Trade discounts of 30 - 50% are usually available off the above prices. The following "measured rates" are provided by a glazing sub-contractor and assume in excess of 500 m², within 20 miles of the suppliers branch. Therefore deduction of "trade prices for cut sizes" from "measured rates" is not a reliable basis for identifying fixing costs.

Prices for Measured Works - Major Works
L WINDOWS/DOORS/STAIRS

	PC £	Labour hours	Labour £	Material £	Unit	Total rate £
L40 GENERAL GLAZING - cont'd						
Standard plain glass; BS 952; clear float; panes area 0.15 m² - 4.00 m²						
3 mm thick; glazed with						
putty or bradded beads	-	-	-	-	m²	21.11
bradded beads and butyl compound	-	-	-	-	m²	21.11
screwed beads	-	-	-	-	m²	23.86
screwed beads and butyl compound	-	-	-	-	m²	23.86
4 mm thick; glazed with						
putty or bradded beads	-	-	-	-	m²	22.49
bradded beads and butyl compound	-	-	-	-	m²	22.49
screwed beads	-	-	-	-	m²	25.24
screwed beads and butyl compound	-	-	-	-	m²	25.24
5 mm thick; glazed with						
putty or bradded beads	-	-	-	-	m²	26.63
bradded beads and butyl compound	-	-	-	-	m²	26.63
screwed beads	-	-	-	-	m²	29.38
screwed beads and butyl compound	-	-	-	-	m²	29.38
6 mm thick; glazed with						
putty or bradded beads	-	-	-	-	m²	28.00
bradded beads and butyl compound	-	-	-	-	m²	28.00
screwed beads	-	-	-	-	m²	30.76
screwed beads and butyl compound	-	-	-	-	m²	30.76
Standard plain glass; BS 952; white patterned; panes area 0.15 m² - 4.00 m²						
4 mm thick; glazed with						
putty or bradded beads	-	-	-	-	m²	25.24
bradded beads and butyl compound	-	-	-	-	m²	25.24
screwed beads	-	-	-	-	m²	28.00
screwed beads and butyl compound	-	-	-	-	m²	28.00
6 mm thick; glazed with						
putty or bradded beads	-	-	-	-	m²	30.76
bradded beads and butyl compound	-	-	-	-	m²	30.76
screwed beads	-	-	-	-	m²	33.51
screwed beads and butyl compound	-	-	-	-	m²	33.51
Standard plain glass; BS 952; rough cast; panes area 0.15 m² - 4.00 m²						
6 mm thick; glazed with						
putty or bradded beads	-	-	-	-	m²	33.51
bradded beads and butyl compound	-	-	-	-	m²	33.51
screwed beads	-	-	-	-	m²	36.26
screwed beads and butyl compound	-	-	-	-	m²	36.26
Standard plain glass; BS 952; Georgian wired cast; panes area 0.15 m² - 4.00 m²						
7 mm thick; glazed with						
putty or bradded beads	-	-	-	-	m²	36.26
bradded beads and butyl compound	-	-	-	-	m²	36.26
screwed beads	-	-	-	-	m²	39.02
screwed beads and butyl compound	-	-	-	-	m²	39.02
Extra for lining up wired glass	-	-	-	-	m²	2.79

L WINDOWS/DOORS/STAIRS

	PC £	Labour hours	Labour £	Material £	Unit	Total rate £
Standard plain glass; BS 952; Georgian wired polished; panes area 0.15 m² - 4.00 m²						
6 mm thick; glazed with						
putty or bradded beads	-	-	-	-	m²	66.55
bradded beads and butyl compound	-	-	-	-	m²	66.55
screwed beads	-	-	-	-	m²	68.85
screwed beads and butyl compound	-	-	-	-	m²	68.85
Extra for lining up wired glass	-	-	-	-	m²	2.79
Special glass; BS 952; toughened clear float; panes area 0.15 m² - 4.00 m²						
4 mm thick; glazed with						
putty or bradded beads	-	-	-	-	m²	25.24
bradded beads and butyl compound	-	-	-	-	m²	25.24
screwed beads	-	-	-	-	m²	28.00
screwed beads and butyl compound	-	-	-	-	m²	28.00
5 mm thick; glazed with						
putty or bradded beads	-	-	-	-	m²	33.51
bradded beads and butyl compound	-	-	-	-	m²	33.51
screwed beads	-	-	-	-	m²	36.26
screwed beads and butyl compound	-	-	-	-	m²	36.26
6 mm thick; glazed with						
putty or bradded beads	-	-	-	-	m²	33.51
bradded beads and butyl compound	-	-	-	-	m²	33.51
screwed beads	-	-	-	-	m²	36.26
screwed beads and butyl compound	?	-	-	-	m²	36.26
10 mm thick; glazed with						
putty or bradded beads	-	-	-	-	m²	66.55
bradded beads and butyl compound	-	-	-	-	m²	66.55
screwed beads	-	-	-	-	m²	68.85
screwed beads and butyl compound	-	-	-	-	m²	68.85
Special glass; BS 952; clear laminated safety glass; panes area 0.15 m² - 4.00m²						
4.40 mm thick; glazed with						
putty or bradded beads	-	-	-	-	m²	55.08
bradded beads and butyl compound	-	-	-	-	m²	55.08
screwed beads	-	-	-	-	m²	56.46
screwed beads and butyl compound	-	-	-	-	m²	56.46
5.40 mm thick; glazed with						
putty or bradded beads	-	-	-	-	m²	66.55
bradded beads and butyl compound	-	-	-	-	m²	66.55
screwed beads	-	-	-	-	m²	69.77
screwed beads and butyl compound	-	-	-	-	m²	69.77
6.40 mm thick; glazed with						
putty or bradded beads	-	-	-	-	m²	73.44
bradded beads and butyl compound	-	-	-	-	m²	73.44
screwed beads	-	-	-	-	m²	80.79
screwed beads and butyl compound	-	-	-	-	m²	80.79
Special glass; BS 952; "Pyran" half-hour fire resisting glass or other equal or approved						
6.50 mm thick rectangular panes; glazed with screwed hardwood beads and Sealmaster "Fireglaze" intumescent compound or other equal and approved to rebated frame						
300 mm x 400 mm pane	-	0.37	6.54	26.48	nr	33.02
400 mm x 800 mm pane	-	0.46	8.13	66.01	nr	74.14
500 mm x 1400 mm pane	-	0.74	13.08	139.73	nr	152.81
600 mm x 1800 mm pane	-	0.93	16.44	237.12	nr	253.56

Prices for Measured Works - Major Works
L WINDOWS/DOORS/STAIRS

	PC £	Labour hours	Labour £	Material £	Unit	Total rate £
L40 GENERAL GLAZING - cont'd						
Special glass; BS 952; "Pyrostop" one-hour fire resisting glass or other equal and approved						
15 mm thick regular panes; glazed with screwed hardwood beads and Sealmaster "Fireglaze" intumescent liner and compound or other equal and approved both sides						
300 mm x 400 mm pane	-	1.11	19.62	59.62	nr	79.24
400 mm x 800 mm pane	-	1.39	24.57	116.38	nr	140.95
500 mm x 1400 mm pane	-	1.85	32.70	229.83	nr	262.53
600 mm x 1800 mm pane	-	2.31	40.82	335.15	nr	375.97
Special glass; BS 952; clear laminated security glass						
7.50 mm thick regular panes; glazed with screwed hardwood beads and Intergens intumescent strip						
300 mm x 400 mm pane	-	0.37	6.54	16.24	nr	22.78
400 mm x 800 mm pane	-	0.46	8.13	39.09	nr	47.22
500 mm x 1400 mm pane	-	0.74	13.08	80.97	nr	94.05
600 mm x 1800 mm pane	-	0.93	16.44	145.67	nr	162.11
Mirror panels; BS 952; silvered; insulation backing						
4 mm thick float; fixing with adhesive						
1000 mm x 1000 mm	-	-	-	-	nr	146.89
1000 mm x 2000 mm	-	-	-	-	nr	275.42
1000 mm x 4000 mm	-	-	-	-	nr	532.47
Glass louvres; BS 952; with long edges ground or smooth						
6 mm thick float						
150 mm wide	-	-	-	-	m	11.29
7 mm thick Georgian wired cast						
150 mm wide	-	-	-	-	m	15.61
6 mm thick Georgian wire polished						
150 mm wide	-	-	-	-	m	22.26
Factory made double hermetically sealed units; to wood or metal with screwed or clipped beads						
Two panes; BS 952; clear float glass; 4 mm thick; 6 mm air space						
2.00 m² - 4.00 m²	-	-	-	-	m²	50.49
1.00 m² - 2.00 m²	-	-	-	-	m²	50.49
0.75 m² - 1.00 m²	-	-	-	-	m²	50.49
0.50 m² - 0.75 m²	-	-	-	-	m²	50.49
0.35 m² - 0.50 m²	-	-	-	-	m²	50.49
0.25 m² - 0.35 m²	-	-	-	-	m²	50.49
not exceeding 0.25 m²	-	-	-	-	nr	12.17
Two panes; BS 952; clear float glass; 6 mm thick; 6 mm air space						
2.00 m² - 4.00 m²	-	-	-	-	m²	59.67
1.00 m² - 2.00 m²	-	-	-	-	m²	59.67
0.75 m² - 1.00 m²	-	-	-	-	m²	59.67
0.50 m² - 0.75 m²	-	-	-	-	m²	59.67
0.35 m² - 0.50 m²	-	-	-	-	m²	59.67
0.25 m² - 0.35 m²	-	-	-	-	m²	59.67
not exceeding 0.25 m²	-	-	-	-	nr	15.15

M SURFACE FINISHES

	PC £	Labour hours	Labour £	Material £	Unit	Total rate £
M SURFACE FINISHES						
M10 CEMENT:SAND/CONCRETE SCREEDS/TOPPINGS						
Cement and sand (1:3); steel trowelled						
Work to floors; one coat; level and to falls not exceeding 15 degrees from horizontal; to concrete base; over 300 mm wide						
32 mm thick	-	0.27	3.72	2.10	m^2	5.82
40 mm thick	-	0.29	3.99	2.62	m^2	6.61
48 mm thick	-	0.31	4.27	3.15	m^2	7.42
50 mm thick	-	0.32	4.40	3.28	m^2	7.68
60 mm thick	-	0.34	4.68	3.94	m^2	8.62
65 mm thick	-	0.37	5.09	4.26	m^2	9.35
70 mm thick	-	0.38	5.23	4.59	m^2	9.82
75 mm thick	-	0.39	5.37	4.92	m^2	10.29
Add to the above for work to falls and crossfalls and to slopes						
not exceeding 15 degrees from horizontal	-	0.02	0.28	-	m^2	0.28
over 15 degrees from horizontal	-	0.09	1.24	-	m^2	1.24
water repellent additive incorporated in the mix	-	0.02	0.28	0.55	m^2	0.83
oil repellent additive incorporated in the mix	-	0.07	0.96	3.60	m^2	4.56
Cement and sand (1:3) beds and backings						
Work to floors; one coat level; to concrete base; screeded; over 300 mm wide						
25 mm thick	-	-	-	-	m^2	8.00
50 mm thick	-	-	-	-	m^2	9.44
75 mm thick	-	-	-	-	m^2	12.31
100 mm thick	-	-	-	-	m^2	15.21
Work to floors; one coat; level; to concrete base; steel trowelled; over 300 mm wide						
25 mm thick	-	-	-	-	m^2	8.00
50 mm thick	-	-	-	-	m^2	9.44
75 mm thick	-	-	-	-	m^2	12.31
100 mm thick	-	-	-	-	m^2	15.21
Fine concrete (1:4) beds and backings						
Work to floors; one coat; level; to concrete base; steel trowelled; over 300 mm wide						
50 mm thick	-	-	-	-	m^2	9.21
75 mm thick	-	-	-	-	m^2	11.80
Granolithic paving; cement and granite chippings 5 to dust (1:1:2); steel trowelled						
Work to floors; one coat; level; laid on concrete while green; over 300 mm wide						
20 mm thick	-	-	-	-	m^2	14.45
25 mm thick	-	-	-	-	m^2	15.60
Work to floors; two coat; laid on hacked concrete with slurry; over 300 mm wide						
38 mm thick	-	-	-	-	m^2	18.61
50 mm thick	-	-	-	-	m^2	21.38
75 mm thick	-	-	-	-	m^2	28.91
Work to landings; one coat; level; laid on concrete while green; over 300 mm wide						
20 mm thick	-	-	-	-	m^2	16.19
25 mm thick	-	-	-	-	m^2	17.34

M SURFACE FINISHES

	PC £	Labour hours	Labour £	Material £	Unit	Total rate £
M10 CEMENT:SAND/CONCRETE SCREEDS/TOPPINGS - cont'd						
Granolithic paving; cement and granite chippings 5 to dust (1:1:2); steel trowelled - cont'd						
Work to landings; two coat; laid on hacked concrete with slurry; over 300 mm wide						
38 mm thick	-	-	-	-	m^2	20.34
50 mm thick	-	-	-	-	m^2	23.13
75 mm thick	-	-	-	-	m^2	30.64
Add to the above over 300 mm wide for						
liquid hardening additive incorporated in the mix	-	0.04	0.55	0.64	m^2	1.19
oil-repellent additive incorporated in the mix	-	0.07	0.96	3.60	m^2	4.56
25 mm work to treads; one coat; to concrete base						
225 mm wide	-	0.83	14.67	7.49	m	22.16
275 mm wide	-	0.83	14.67	8.39	m	23.06
returned end	-	0.17	3.00	-	nr	3.00
13 mm skirtings; rounded top edge and coved bottom junction; to brickwork or blockwork base						
75 mm wide on face	-	0.51	9.01	0.36	m	9.37
150 mm wide on face	-	0.69	12.19	6.59	m	18.78
ends; fair	-	0.04	0.71	-	nr	0.71
angles	-	0.06	1.06	-	nr	1.06
13 mm outer margin to stairs; to follow profile of and with rounded nosing to treads and risers; fair edge and arris at bottom, to concrete base						
75 mm wide	-	0.83	14.67	3.60	m	18.27
angles	-	0.06	1.06	-	nr	1.06
13 mm wall string to stairs; fair edge and arris on top; coved bottom junction with treads and risers; to brickwork or blockwork base						
275 mm (extreme) wide	-	0.74	13.08	6.29	m	19.37
ends	-	0.04	0.71	-	nr	0.71
angles	-	0.06	1.06	-	nr	1.06
ramps	-	0.07	1.24	-	nr	1.24
ramped and wreathed corners	-	0.09	1.59	-	nr	1.59
13 mm outer string to stairs; rounded nosing on top at junction with treads and risers; fair edge and arris at bottom; to concrete base						
300 mm (extreme) wide	-	0.74	13.08	7.79	m	20.87
ends	-	0.04	0.71	-	nr	0.71
angles	-	0.06	1.06	-	nr	1.06
ramps	-	0.07	1.24	-	nr	1.24
ramps and wreathed corners	-	0.09	1.59	-	nr	1.59
19 mm thick skirtings; rounded top edge and coved bottom junction; to brickwork or blockwork base						
75 mm wide on face	-	0.51	9.01	6.59	m	15.60
150 mm wide on face	-	0.69	12.19	10.19	m	22.38
ends; fair	-	0.04	0.55	-	nr	0.55
angles	-	0.06	1.06	-	nr	1.06
19 mm riser; one rounded nosing; to concrete base						
150 mm high; plain	-	0.83	14.67	5.69	m	20.36
150 mm high; undercut	-	0.83	14.67	5.69	m	20.36
180 mm high; plain	-	0.83	14.67	7.79	m	22.46
180 mm high; undercut	-	0.83	14.67	7.79	m	22.46

M SURFACE FINISHES

	PC £	Labour hours	Labour £	Material £	Unit	Total rate £
M11 MASTIC ASPHALT FLOORING/FLOOR UNDERLAYS						
Mastic asphalt flooring to BS 6925 Type F 1076; black						
20 mm thick; one coat coverings; felt isolating membrane; to concrete base; flat						
over 300 mm wide	-	-	-	-	m²	10.78
225 mm - 300 mm wide	-	-	-	-	m²	20.07
150 mm - 225 mm wide	-	-	-	-	m²	22.02
not exceeding 150 mm wide	-	-	-	-	m²	26.96
25 mm thick; one coat coverings; felt isolating membrane; to concrete base; flat						
over 300 mm wide	-	-	-	-	m²	12.53
225 mm - 300 mm wide	-	-	-	-	m²	21.37
150 mm - 225 mm wide	-	-	-	-	m²	23.32
not exceeding 150 mm wide	-	-	-	-	m²	28.26
20 mm three coat skirtings to brickwork base						
not exceeding 150 mm girth	-	-	-	-	m	10.78
150 mm - 225 mm girth	-	-	-	-	m	12.53
225 mm - 300 mm girth	-	-	-	-	m	14.18
Mastic asphalt flooring; acid-resisting; black						
20 mm thick; one coat coverings; felt isolating membrane; to concrete base flat						
over 300 mm wide	-	-	-	-	m²	12.63
225 mm - 300 mm wide	-	-	-	-	m²	23.12
150 mm - 225 mm wide	-	-	-	-	m²	23.86
not exceeding 150 mm wide	-	-	-	-	m²	28.81
25 mm thick; one coat coverings; felt isolating membrane; to concrete base; flat						
over 300 mm wide	-	-	-	-	m²	14.93
225 mm - 300 mm wide	-	-	-	-	m²	23.77
150 mm - 225 mm wide	-	-	-	-	m²	25.71
not exceeding 150 mm wide	-	-	-	-	m²	30.66
20 mm thick; three coat skirtings to brickwork base						
not exceeding 150 mm girth	-	-	-	-	m	11.14
150 mm - 225 mm girth	-	-	-	-	m	12.99
225 mm - 300 mm girth	-	-	-	-	m	14.73
Mastic asphalt flooring to BS 6925 Type F 1451; red						
20 mm thick; one coat coverings; felt isolating membrane; to concrete base; flat						
over 300 mm wide	-	-	-	-	m²	17.67
225 mm - 300 mm wide	-	-	-	-	m²	29.21
150 mm - 225 mm wide	-	-	-	-	m²	31.56
not exceeding 150 mm wide	-	-	-	-	m²	37.75
20 mm thick; three coat skirtings to brickwork base						
not exceeding 150 mm girth	-	-	-	-	m	13.88
150 mm - 225 mm girth	-	-	-	-	m	17.67
M12 TROWELLED BITUMEN/RESIN/RUBBER LATEXFLOORING						
Latex cement floor screeds; steel trowelled						
Work to floors; level; to concrete base; over 300 mm wide						
3 mm thick; one coat	-	-	-	-	m²	3.70
5 mm thick; two coats	-	-	-	-	m²	5.25

M SURFACE FINISHES

	PC £	Labour hours	Labour £	Material £	Unit	Total rate £
M12 TROWELLED BITUMEN/RESIN/RUBBER LATEXFLOORING - cont'd						
Epoxy resin flooring; Altro "Altroflow 3000" or other equal and approved; steel trowelled						
Work to floors; level; to concrete base; over 300 mm wide						
3 mm thick; one coat	-	-	-	-	m^2	22.47
Isocrete K screeds or other equal and approved; steel trowelled						
Work to floors; level; to concrete base; over 300 mm wide						
35 mm thick; plus polymer bonder coat	-	-	-	-	m^2	12.26
40 mm thick	-	-	-	-	m^2	10.81
45 mm thick	-	-	-	-	m^2	11.08
50 mm thick	-	-	-	-	m^2	11.39
Work to floors; to falls or cross-falls; to concrete base; over 300 mm wide						
55 mm (average) thick	-	-	-	-	m^2	12.56
60 mm (average) thick	-	-	-	-	m^2	13.14
65 mm (average) thick	-	-	-	-	m^2	13.71
75 mm (average) thick	-	-	-	-	m^2	14.86
90 mm (average) thick	-	-	-	-	m^2	16.60
Bituminous lightweight insulating roof screeds						
"Bit-Ag" or similar roof screed or other equal and approved; to falls or cross-falls; bitumen felt vapour barrier; over 300 mm wide						
75 mm (average) thick	-	-	-	-	m^2	28.36
100 mm (average) thick	-	-	-	-	m^2	35.95
M20 PLASTERED/RENDERED/ROUGHCAST COATINGS						
Cement and sand (1:3) beds and backings						
10 mm thick work to walls; one coat; to brickwork or blockwork base						
over 300 mm wide	-	-	-	-	m^2	9.02
not exceeding 300 mm wide	-	-	-	-	m	4.48
13 mm thick; work to walls; two coats; to brickwork or blockwork base						
over 300 mm wide	-	-	-	-	m^2	10.78
not exceeding 300 mm wide	-	-	-	-	m	5.39
15 mm thick work to walls; two coats; to brickwork or blockwork base						
over 300 mm wide	-	-	-	-	m^2	11.64
not exceeding 300 mm wide	-	-	-	-	m	5.82
Cement and sand (1:3); steel trowelled						
13 mm thick work to walls; two coats; to brickwork or blockwork base						
over 300 mm wide	-	-	-	-	m^2	10.48
not exceeding 300 mm wide	-	-	-	-	m	5.24
16 mm thick work to walls; two coats; to brickwork or blockwork base						
over 300 mm wide	-	-	-	-	m^2	11.74
not exceeding 300 mm wide	-	-	-	-	m	5.88
19 mm thick work to walls; two coats; to brickwork or blockwork base						
over 300 mm wide	-	-	-	-	m^2	13.00
not exceeding 300 mm wide	-	-	-	-	m	6.50

M SURFACE FINISHES

	PC £	Labour hours	Labour £	Material £	Unit	Total rate £
ADD to above						
over 300 mm wide in water repellant cement	-	-	-	-	m²	2.66
finishing coat in colour cement	-	-	-	-	m²	6.13
Cement-lime-sand (1:2:9); steel trowelled						
19 mm thick work to walls; two coats; to brickwork or blockwork base						
over 300 mm wide	-	-	-	-	m²	13.15
not exceeding 300 mm wide	-	-	-	-	m	6.57
Cement-lime-sand (1:1:6); steel trowelled						
13 mm thick work to walls; two coats; to brickwork or blockwork base						
over 300 mm wide	-	-	-	-	m²	10.64
not exceeding 300 mm wide	-	-	-	-	m	5.32
Add to the above over 300 mm wide for waterproof additive	-	-	-	-	m²	1.82
19 mm thick work to ceilings; three coats; to metal lathing base						
over 300 mm wide	-	-	-	-	m²	16.14
not exceeding 300 mm wide	-	-	-	-	m	8.32
Sto External Render System; StoRend Cote or other equal and approved; CCS Scotseal Ltd						
12 mm thick StoLevell Cote; one coat Sto primer; to brickwork or blockwork base						
over 300 mm wide	-	0.62	14.33	14.45	m²	28.78
Sto External Render System; StoRend Fibre or other equal and approved; CCS Scotseal Ltd						
12 mm thick Sto Fibrecoat; one coat Sto primer; to brickwork or blockwork base						
over 300 mm wide	-	0.62	14.33	20.38	m²	34.71
Sto External Render System; StoRend Fibre Plus or other equal and approved; CCS Scotseal Ltd						
10 mm thick Sto Fibrecoat; glassfibre reinforcing mesh; one coat Sto primer; to brickwork or blockwork base						
over 300 mm wide	-	0.69	15.95	20.25	m²	36.20
Sto External Render System; StoRend Flex or other equal and approved; CCS Scotseal Ltd						
2 mm Sto RFP reinforcing coat; glassfibre reinforcing mesh; one coat Sto primer; to existing rendered brickwork or blockwork base						
over 300 mm wide	-	0.54	12.48	14.01	m²	27.39
Sto External Render System; StoRend Flex Cote or other equal and approved; CCS Scotseal Ltd						
12 mm StoLevell Cote; 2 mm Sto RFPreinforcing coat; glassfibre reinforcing mesh; to brickwork or blockwork base						
over 300 mm wide	-	0.85	19.65	23.36	m²	43.01
Plaster; first and finishing coats of "Carlite" pre-mixed lightweight plaster or other equal and approved; steel trowelled						
13 mm thick work to walls; two coats; to brickwork or blockwork base (or 10 mm thick work to concrete base)						
over 300 mm wide	-	-	-	-	m²	7.65
over 300 mm wide; in staircase areas or plant rooms	-	-	-	-	m²	8.83
not exceeding 300 mm wide	-	-	-	-	m	3.81

Prices for Measured Works - Major Works
M SURFACE FINISHES

	PC £	Labour hours	Labour £	Material £	Unit	Total rate £
M20 PLASTERED/RENDERED/ROUGHCAST COATINGS - cont'd						
Plaster; first and finishing coats of "Carlite" pre-mixed lightweight plaster or other equal and approved; steel trowelled - cont'd						
13 mm thick work to isolated piers or columns; two coats						
over 300 mm wide	-	-	-	-	m²	9.41
not exceeding 300 mm wide	-	-	-	-	m	4.70
10 mm thick work to ceilings; two coats; to concrete base						
over 300 mm wide	-	-	-	-	m²	8.19
over 300 wide; 3.50 m - 5.00 m high	-	-	-	-	m²	8.76
over 300 mm wide; in staircase areas or plant rooms	-	-	-	-	m²	9.35
not exceeding 300 mm wide	-	-	-	-	m	4.10
10 mm thick work to isolated beams; two coats; to concrete base						
over 300 mm wide	-	-	-	-	m²	9.95
over 300 mm wide; 3.50 m - 5.00 m high	-	-	-	-	m²	10.54
not exceeding 300 mm wide	-	-	-	-	m	4.97
Plaster; first coat of "Thistle Hardwall" plaster or other equal and approved; finishing coat of "Thistle Multi Finish" plaster; steel trowelled						
13 mm thick work to walls; two coats; to brickwork or blockwork base						
over 300 mm wide	-	-	-	-	m²	8.33
over 300 mm wide; in staircase areas or plant rooms	-	-	-	-	m²	9.57
not exceeding 300 mm wide	-	-	-	-	m	4.18
13 mm thick work to isolated columns; two coats						
over 300 mm wide	-	-	-	-	m²	10.18
not exceeding 300 mm wide	-	-	-	-	m	5.10
Plaster; one coat "Snowplast" plaster or other equal and approved; steel trowelled						
13 mm thick work to walls; one coat; to brickwork or blockwork base						
over 300 mm wide	-	-	-	-	m²	8.89
over 300 mm wide; in staircase areas or plant rooms	-	-	-	-	m²	10.12
not exceeding 300 mm wide	-	-	-	-	m	4.44
13 thick work to isolated columns; one coat						
over 300 mm wide	-	-	-	-	m²	10.73
not exceeding 300 mm wide	-	-	-	-	m	5.38
Plaster; first coat of cement and sand (1:3); finishing coat of "Thistle" class B plaster or other equal and approved; steel trowelled						
13 mm thick; work to walls; two coats; to brickwork or blockwork base						
over 300 mm wide	-	-	-	-	m²	8.51
over 300 mm wide; in staircase areas or plant rooms	-	-	-	-	m²	9.74
not exceeding 300 mm wide	-	-	-	-	m	4.27
13 mm thick work to isolated columns; two coats						
over 300 mm wide	-	-	-	-	m²	10.98
not exceeding 300 mm wide	-	-	-	-	m	5.48

Prices for Measured Works - Major Works
M SURFACE FINISHES

	PC £	Labour hours	Labour £	Material £	Unit	Total rate £
Plaster; first coat of cement-lime-sand (1:1:6); finishing coat of "Multi Finish" plaster or other equal and approved; steel trowelled						
13 mm thick work to walls; two coats; to brickwork or blockwork base						
over 300 mm wide	-	-	-	-	m²	8.74
over 300 mm wide; in staircase areas or plant rooms	-	-	-	-	m²	9.98
not exceeding 300 mm wide	-	-	-	-	m	4.18
13 mm thick work to isolated columns; two coats						
over 300 mm wide	-	-	-	-	m²	10.18
not exceeding 300 mm wide	-	-	-	-	m	5.10
Plaster; first coat of "Limelite" renovating plaster or other equal and approved; finishing coat of "Limelite" renovating plaster or other equal and approved; finishing coat of "Limelite" finishing plaster or other equal and approved; steel trowelled						
13 mm thick work to walls; two coats; to brickwork or blockwork base						
over 300 mm wide	-	-	-	-	m²	12.28
over 300 mm wide; in staircase areas or plant rooms	-	-	-	-	m²	13.51
not exceeding 300 mm wide	-	-	-	-	m	6.15
Dubbing out existing walls with undercoat plaster; average 6 mm thick						
over 300 mm wide	-	-	-	-	m²	3.69
not exceeding 300 mm wide	-	-	-	-	m	1.85
Dubbing out existing walls with undercoat plaster; average 12 mm thick						
over 300 mm wide	-	-	-	-	m²	7.37
not exceeding 300 mm wide	-	-	-	-	m	3.69
Plaster; first coat of "Thistle X-ray" plaster or other equal and approved; finishing coat of "Thistle X-ray" finishing plaster or other equal and approved; steel trowelled						
17 mm thick work to walls; two coats; to brickwork or blockwork base						
over 300 mm wide	-	-	-	-	m²	39.94
over 300 mm wide; in staircase areas or plant rooms	-	-	-	-	m²	41.16
not exceeding 300 mm wide	-	-	-	-	m	15.97
17 mm thick work to isolated columns; two coats						
over 300 mm wide	-	-	-	-	m²	43.61
not exceeding 300 mm wide	-	-	-	-	m	17.44
Plaster; one coat "Thistle" projection plaster or other equal and approved; steel trowelled						
13 mm thick work to walls; one coat; to brickwork or blockwork base						
over 300 mm wide	-	-	-	-	m²	8.57
over 300 mm wide; in staircase areas or plant rooms	-	-	-	-	m²	9.79
not exceeding 300 mm wide	-	-	-	-	m	4.29
10 mm thick work to isolated columns; one coat						
over 300 mm wide	-	-	-	-	m²	10.42
not exceeding 300 mm wide	-	-	-	-	m	5.20
Plaster; first, second and finishing coats of "Carlite" pre-mixed lightweight plaster or other equal and approved; steel trowelled						
13 mm thick work to ceilings; three coats to metal lathing base						
over 300 mm wide	-	-	-	-	m²	11.07

	PC £	Labour hours	Labour £	Material £	Unit	Total rate £
M20 PLASTERED/RENDERED/ROUGHCAST COATINGS - cont'd						
Plaster; first, second and finishing coats of "Carlite" pre-mixed lightweight plaster or other equal and approved; steel trowelled - cont'd						
13 mm thick work to ceilings; three coats to metal lathing base - cont'd						
over 300 mm wide; in staircase areas or plant rooms	-	-	-	-	m²	12.31
not exceeding 300 mm wide	-	-	-	-	m	5.53
13 mm thick work to swept soffit of metal lathing arch former						
not exceeding 300 mm wide	-	-	-	-	m	7.37
300 mm - 400 mm wide	-	-	-	-	m	9.85
13 mm thick work to vertical face of metal lathing arch former						
not exceeding 0.50 m² per side	-	-	-	-	nr	10.47
0.50 m² - 1 m² per side	-	-	-	-	nr	15.71
Squash court plaster, Prodorite Ltd; first coat "Formula Base" screed or other equal and approved; finishing coat "Formula 90" finishing plaster or other equal and approved; steel trowelled and finished with sponge float						
12 mm thick work to walls; two coats; to brickwork or blockwork base						
over 300 mm wide	-	-	-	-	m²	28.32
not exceeding 300 mm wide	-	-	-	-	m	14.14
demarcation lines on battens	-	-	-	-	m	4.10
"Cemrend" self-coloured render or other equal and approved; one coat; to brickwork or blockwork base						
20 mm thick work to walls; to brickwork or blockwork base						
over 300 mm wide	-	-	-	-	m²	27.96
not exceeding 300 mm wide	-	-	-	-	m	15.97
Tyrolean decorative rendering or similar; 13 mm thick first coat of cement-lime-sand (1:1:6); finishing three coats of "Cullamix'" or other equal and approved; applied with approved hand operated machine external						
To walls; four coats; to brickwork or blockwork base						
over 300 mm wide	-	-	-	-	m²	19.65
not exceeding 300 mm wide	-	-	-	-	m	9.83
Drydash (pebbledash) finish of Derbyshire Spar chippings or other equal and approved on and including cement-lime-sand (1:2:9) backing						
18 mm thick work to walls; two coats; to brickwork or blockwork base						
over 300 mm wide	-	-	-	-	m²	18.08
not exceeding 300 mm wide	-	-	-	-	m	9.05
Plaster; one coat "Thistle" board finish or other equal and approved; steel trowelled (prices included within plasterboard rates)						
3 mm thick work to walls or ceilings; one coat; to plasterboard base						
over 300 mm wide	-	-	-	-	m²	4.35
over 300 mm wide; in staircase areas or plant rooms	-	-	-	-	m²	4.98
not exceeding 300 mm wide	-	-	-	-	m	2.60

M SURFACE FINISHES

	PC £	Labour hours	Labour £	Material £	Unit	Total rate £
Plaster; one coat "Thistle" board finish or other and approved; steel trowelled 3 mm work to walls or ceilings; one coat on and including gypsum plasterboard; BS 1230; fixing with nails; 3 mm joints filled with plaster and jute scrim cloth; to softwood base; plain grade baseboard or lath with rounded edges						
9.50 mm thick boards to walls						
over 300 mm wide	-	0.97	9.04	2.38	m²	11.42
not exceeding 300 mm wide	-	0.37	3.64	0.67	m	4.31
9.50 mm thick boards to walls; in staircase areas or plant rooms						
over 300 mm wide	-	1.06	9.93	2.38	m²	12.31
not exceeding 300 mm wide	-	0.46	4.52	0.67	m	5.19
9.50 mm thick boards to isolated columns						
over 300 mm wide	-	1.06	9.93	2.38	m²	12.31
not exceeding 300 mm wide	-	0.56	5.51	0.67	m	6.18
9.50 mm thick boards to ceilings						
over 300 mm wide	-	0.89	8.25	2.38	m²	10.63
over 300 mm wide; 3.50 m - 5.00 m high	-	1.03	9.63	2.38	m²	12.01
not exceeding 300 mm wide	-	0.43	4.23	0.67	m	4.90
9.50 mm thick boards to ceilings; in staircase areas or plant rooms						
over 300 mm wide	-	0.98	9.14	2.38	m²	11.52
not exceeding 300 mm wide	-	0.47	4.62	0.67	m	5.29
9.50 mm thick boards to isolated beams						
over 300 mm wide	-	1.05	9.83	2.38	m²	12.21
not exceeding 300 mm wide	-	0.50	4.92	0.67	m	5.59
add for "Duplex" insulating grade 12.50 mm thick boards to walls	-	-	-	0.46	m²	0.46
12.50 mm thick boards to walls; in staircase areas or plant rooms						
over 300 mm wide	-	1.12	10.52	2.58	m²	13.10
not exceeding 300 mm wide	-	0.50	4.92	0.73	m	5.65
12.50 mm thick boards to isolated columns						
over 300 mm wide	-	1.12	10.52	2.58	m²	13.10
not exceeding 300 mm wide	-	0.59	5.80	0.73	m	6.53
12.50 mm thick boards to ceilings						
over 300 mm wide	-	0.95	8.84	2.58	m²	11.42
over 300 mm wide; 3.50 m - 5.00 m high	-	1.06	9.93	2.58	m²	12.51
not exceeding 300 mm wide	-	0.45	4.43	0.73	m	5.16
12.50 mm thick boards to ceilings; in staircase areas or plant rooms						
over 300 mm wide	-	1.06	9.93	2.58	m²	12.51
not exceeding 300 mm wide	-	0.51	5.02	0.73	m	5.75
12.50 mm thick boards to isolated beams						
over 300 mm wide	-	1.15	10.81	2.58	m²	13.39
not exceeding 300 mm wide	-	0.56	5.51	0.73	m	6.24
add for "Duplex" insulating grade 12.50 mm thick boards to walls	-	-	-	0.46	m²	0.46
Accessories						
"Expamet" render beads or other equal and approved; white PVC nosings; to brickwork or blockwork base						
external stop bead; ref 573	-	0.07	0.69	1.74	m	2.43
"Expamet" render beads or other equal and approved; stainless steel; to brickwork or blockwork base						
stop bead; ref 546	-	0.07	0.69	1.43	m	2.12
stop bead; ref 547	-	0.07	0.69	1.43	m	2.12

Prices for Measured Works - Major Works
M SURFACE FINISHES

	PC £	Labour hours	Labour £	Material £	Unit	Total rate £
M20 PLASTERED/RENDERED/ROUGHCAST COATINGS - cont'd						
Accessories - cont'd						
"Expamet" plaster beads or other equal and approved; galvanised steel; to brickwork or blockwork base						
angle bead; ref 550	-	0.08	0.79	0.57	m	1.36
architrave bead; ref 579	-	0.10	0.98	0.88	m	1.86
stop bead; ref 562	-	0.07	0.69	0.67	m	1.36
stop beads; ref 563	-	0.07	0.69	0.67	m	1.36
movement bead; ref 588	-	0.09	0.89	3.13	m	4.02
"Expamet" plaster beads or other equal and approved; stainless steel; to brickwork or blockwork base						
angle bead; ref 545	-	0.08	0.79	1.60	m	2.39
stop bead; ref 534	-	0.07	0.69	1.43	m	2.12
stop bead; ref 533	-	0.07	0.69	1.43	m	2.12
"Expamet" thin coat plaster beads or other equal and approved; galvanised steel; to timber base						
angle bead; ref 553	-	0.07	0.69	0.56	m	1.25
angle bead; ref 554	-	0.07	0.69	0.78	m	1.47
stop bead; ref 560	-	0.06	0.59	0.72	m	1.31
stop bead; ref 561	-	0.06	0.59	0.78	m	1.37
M21 INSULATION WITH RENDERED FINISH						
"StoTherm" mineral external wall insulation system or other equal and approved; CCS Scotseal Ltd						
70 mm thick expanded polystyrene; Sto RFP reinforcing coat; glassfibre reinforcing mesh; one coat Sto Primer; mechanical fixing using PVC intermediate tracks and T-splines; to brickwork or blockwork base						
over 300 mm wide	-	1.23	28.43	29.89	m²	58.32
70 mm thick expanded polystyrene; Sto RFP reinforcing coat; glassfibre reinforcing mesh; fixing using adhesive; to brickwork or blockwork base						
over 300 mm wide	-	1.00	23.11	26.37	m²	49.48
70 mm thick mineral wool; Sto RFP reinforcing coat; glassfibre reinforcing mesh; one coat Sto Primer; mechanical fixing using PVC intermediate tracks and T-splines; to brickwork or blockwork base						
over 300 mm wide	-	1.54	35.59	38.77	m²	74.36
70 mm thick mineral wool; StoLevell Uni reinforcing coat; glassfibre reinforcing mesh; one coat Sto Primer; fixing using adhesive; to brickwork or blockwork base						
over 300 mm wide	-	1.39	32.13	33.83	m²	65.96
M22 SPRAYED MINERAL FIBRE COATINGS						
Prepare and apply by spray "Mandolite CP2" fire protection or other equal and approved on structural steel/metalwork						
16 mm thick (one hour) fire protection						
to walls and columns	-	-	-	-	m²	8.29
to ceilings and beams	-	-	-	-	m²	9.14
to isolated metalwork	-	-	-	-	m²	18.22

Prices for Measured Works - Major Works

M SURFACE FINISHES

	PC £	Labour hours	Labour £	Material £	Unit	Total rate £
22 mm thick (one and a half hour) fire protection						
to walls and columns	-	-	-	-	m^2	9.63
to ceilings and beams	-	-	-	-	m^2	10.69
to isolated metalwork	-	-	-	-	m^2	21.37
28 mm thick (two hour) fire protection						
to walls and columns	-	-	-	-	m^2	11.29
to ceilings and beams	-	-	-	-	m^2	12.33
to isolated metalwork	-	-	-	-	m^2	24.67
52 mm thick (four hour) fire protection						
to walls and columns	-	-	-	-	m^2	17.08
to ceilings and beams	-	-	-	-	m^2	19.03
to isolated metalwork	-	-	-	-	m^2	37.85
Prepare and apply by spray; cementitious "Pyrok WF26" render or other equal and approved; on expanded metal lathing (not included)						
15 mm thick						
to ceilings and beams	-	-	-	-	m^2	26.22
M30 METAL MESH LATHING/ANCHORED REINFORCEMENT FOR PLASTERED COATING						
Accessories						
Pre-formed galvanised expanded steel arch-frames; "Simpson's Strong-Tie" or other equal and approved; semi-circular; to suit walls up to 230 mm thick						
375 mm radius; for 800 mm opening; ref SC 750	15.41	0.46	4.06	16.98	nr	21.04
425 mm radius; for 850 mm opening; ref SC 850	14.72	0.46	4.06	16.22	nr	20.28
450 mm radius; for 900 mm opening; ref SC 900	15.22	0.46	4.06	16.77	nr	20.83
600 mm radius; for 1200 mm opening; ref SC 1200	17.61	0.46	4.06	19.40	nr	23.46
Lathing; Expamet "BB" expanded metal lathing or other equal and approved; BS 1369; 50 mm laps						
6 mm thick mesh linings to ceilings; fixing with staples; to softwood base; over 300 mm wide						
ref BB263; 0.500 mm thick	2.70	0.56	4.95	3.05	m^2	8.00
ref BB264; 0.675 mm thick	3.16	0.56	4.95	3.57	m^2	8.52
6 mm thick mesh linings to ceilings; fixing with wire; to steelwork; over 300 mm wide						
ref BB263; 0.500 mm thick	-	0.59	5.22	3.05	m^2	8.27
ref BB264; 0.675 mm thick	-	0.59	5.22	3.57	m^2	8.79
6 mm thick mesh linings to ceilings; fixing with wire; to steelwork; not exceeding 300 mm wide						
ref BB263; 0.500 mm thick	-	0.37	3.26	3.05	m^2	6.31
ref BB264; 0.675 mm thick	-	0.37	3.26	3.57	m^2	6.83
raking cutting	-	0.19	1.87	-	m	1.87
cutting and fitting around pipes; not exceeding 0.30 m girth	-	0.28	2.75	-	nr	2.75
Lathing; Expamet "Riblath" or "Spraylath" or other equal and approved stiffened expanded metal lathing or similar; 50 mm laps						
10 mm thick mesh lining to walls; fixing with nails; to softwood base; over 300 mm wide						
"Riblath" ref 269; 0.30 mm thick	3.39	0.46	4.06	3.92	m^2	7.98
"Riblath" ref 271; 0.50 mm thick	3.90	0.46	4.06	4.50	m^2	8.56
"Spraylath" ref 273; 0.50 mm thick	-	0.46	4.06	5.45	m^2	9.51

Prices for Measured Works - Major Works
M SURFACE FINISHES

	PC £	Labour hours	Labour £	Material £	Unit	Total rate £
M30 METAL MESH LATHING/ANCHORED REINFORCEMENT FOR PLASTERED COATING - cont'd						
Lathing; Expamet "Riblath" or "Spraylath" or other equal and approved stiffened expanded metal lathing or similar; 50 mm laps - cont'd						
10 mm thick mesh lining to walls; fixing with nails; to softwood base; not exceeding 300 mm wide						
"Riblath" ref 269; 0.30 mm thick	-	0.28	2.47	1.21	m²	3.68
"Riblath" ref 271; 0.50 mm thick	-	0.28	2.47	1.38	m²	3.85
"Spraylath" ref 273; 0.50 mm thick	-	0.28	2.47	1.70	m²	4.17
10 mm thick mesh lining to walls; fixing to brick or blockwork; over 300 mm wide						
"Red-rib" ref 274; 0.50 mm thick	4.31	0.37	3.26	5.41	m²	8.67
stainless steel "Riblath" ref 267; 0.30 mm thick	8.55	0.37	3.26	10.19	m²	13.45
10 mm thick mesh lining to ceilings; fixing with wire; to steelwork; over 300 mm wide						
"Riblath" ref 269; 0.30 mm thick	-	0.59	5.22	4.34	m²	9.56
"Riblath" ref 271; 0.50 mm thick	-	0.59	5.22	4.92	m²	10.14
"Spraylath" ref 273; 0.50 mm thick	-	0.59	5.22	5.88	m²	11.10
M31 FIBROUS PLASTER						
Fibrous plaster; fixing with screws; plugging; countersinking; stopping; filling and pointing joints with plaster						
16 mm thick plain slab coverings to ceilings						
over 300 mm wide	-	-	-	-	m²	104.87
not exceeding 300 mm wide	-	-	-	-	m	34.96
Coves; not exceeding 150 mm girth						
per 25 mm girth	-	-	-	-	m	4.99
Coves; 150 mm - 300 mm girth						
per 25 mm girth	-	-	-	-	m	6.10
Cornices						
per 25 mm girth	-	-	-	-	m	6.19
Cornice enrichments						
per 25 mm girth; depending on degree of enrichments	-	-	-	-	m	7.34
Fibrous plaster; fixing with plaster wadding filling and pointing joints with plaster; to steel base						
16 mm thick plain slab coverings to ceilings						
over 300 mm wide	-	-	-	-	m²	104.87
not exceeding 300 mm wide	-	-	-	-	m	34.96
16 mm thick plain casings to stanchions						
per 25 mm girth	-	-	-	-	m	3.10
16 mm thick plain casings to beams						
per 25 mm girth	-	-	-	-	m	3.10
Gyproc cove or other equal and approved; fixing with adhesive; filling and pointing joints with plaster						
Cove						
125 mm girth	-	0.19	1.87	0.75	m	2.62
Angles	-	0.03	0.30	0.47	nr	0.77

M40 STONE/CONCRETE/QUARRY/CERAMIC TILING/MOSAIC

ALTERNATIVE TILE MATERIALS

Dennis Ruabon clay floor quarries (£/1000) excl. VAT

	£		£		£
Heather Brown					
150 x 150 x 12.50 mm	397.70	194 x 94 x 12.50 mm	451.40	194 x 194 x 12.50 mm	698.70
Red					
150 x 150 x 12.50 mm; square	301.00	150 x 150 x 12.50 mm; hexagonal	494.50		

Daniel Platt heavu duty floor tiles (£/m²)

	£		£		£
"Ferrolite" flat; 100 x 100 x 9 mm					
Black	16.18	Cream mingled (M29)	16.18	Red	16.18
"Ferrolite" flat; 152 x 152 x 12 mm					
Black	28.47	Cream mingled (M29)	24.04	Chocolate	28.47
Red	22.65				
"Ferrundum" anti-slip; 152 x 152 x 12 mm					
Black	30.51	Cream mingled (M29)	25.89		

Marley floor tiles (£/m²); 300 x 300 mm

"Europa"	- 2.00 mm	5.71	"Heavy Duty Hi-Tech Dissipative"		14.64
	- 2.50 mm	7.03	"Travertine" - 2.50 mm		9.34
"Marleyflex"			"Vylon Plus"		5.51
series 2	- 2.50 mm	5.08	series 4	- 2.50 mm	5.75

	PC £	Labour hours	Labour £	Material £	Unit	Total rate £
Clay floor quarries; BS 6431; class 1; Daniel Platt "Crown" tiles or other equal and approved; level bedding 10 mm thick and jointing in cement and sand (1:3); butt joints; straight both ways; flush pointing with grout; to cement and sand base						
Work to floors; over 300 mm wide						
150 mm x 150 mm x 12.50 mm thick; red	-	0.74	10.18	16.75	m²	26.93
150 mm x 150 mm x 12.50 mm thick; brown	-	0.74	10.18	19.66	m²	29.84
200 mm x 200 mm x 19 mm thick; brown	-	0.60	8.26	27.72	m²	35.98
Works to floors; in staircase areas or plant rooms						
150 mm x 150 mm x 12.50 mm thick; red	-	0.83	11.42	16.75	m²	28.17
150 mm x 150 mm x 12.50 mm thick; brown	-	0.83	11.42	19.66	m²	31.08
200 mm x 200 mm x 19 mm thick; brown	-	0.69	9.49	27.72	m²	37.21
Work to floors; not exceeding 300 mm wide						
150 mm x 150 mm x 12.50 mm thick; red	-	0.37	5.09	3.71	m	8.80
150 mm x 150 mm x 12.50 mm thick; brown	-	0.37	5.09	4.59	m	9.68
200 mm x 200 mm x 19 mm thick; brown	-	0.31	4.27	7.01	m	11.28
fair square cutting against flush edges of existing finishes	-	0.11	1.02	0.82	m	1.84
raking cutting	-	0.19	1.80	0.92	m	2.72
cutting around pipes; not exceeding 0.30 m girth	-	0.14	1.38	-	nr	1.38
extra for cutting and fitting into recessed manhole cover 600 mm x 600 mm	-	0.93	9.15	-	nr	9.15
Work to sills; 150 mm wide; rounded edge tiles						
150 mm x 150 mm x 12.50 mm thick; red	0.21	0.31	4.27	3.25	m	7.52
150 mm x 150 mm x 12.50 mm thick; brown	-	0.31	4.27	4.13	m	8.40
fitted end	-	0.14	1.38	-	nr	1.38
Coved skirtings; 138 mm high; rounded top edge						
150 mm x 138 mm x 12.50 mm thick; red	0.37	0.23	3.16	3.96	m	7.12
150 mm x 138 mm x 12.50 mm thick; brown	0.39	0.23	3.16	4.10	m	7.26
ends	-	0.04	0.39	-	nr	0.39
angles	-	0.14	1.38	1.43	nr	2.81

M SURFACE FINISHES

	PC £	Labour hours	Labour £	Material £	Unit	Total rate £
M40 STONE/CONCRETE/QUARRY/CERAMIC TILING/MOSAIC - cont'd						
Glazed ceramic wall tiles; BS 6431; fixing with adhesive; butt joints; straight both ways; flush pointing with white grout; to plaster base						
Work to walls; over 300 mm wide						
152 mm x 152 mm x 5.50 mm thick; white	9.94	0.56	9.90	15.47	m²	25.37
152 mm x 152 mm x 5.50 mm thick; light colours	12.13	0.56	9.90	17.94	m²	27.84
152 mm x 152 mm x 5.50 mm thick; dark colours	13.26	0.56	9.90	19.22	m²	29.12
extra for RE or REX tile	-	-	-	5.76	m²	5.76
200 mm x 100 mm x 6.50 mm thick; white and light colours	9.94	0.56	9.90	15.47	m²	25.37
250 mm x 200 mm x 7 mm thick; white and light colours	10.77	0.56	9.90	16.41	m²	26.31
Work to walls; in staircase areas or plant rooms						
152 mm x 152 mm x 5.50 mm thick; white	-	0.62	10.96	15.47	m²	26.43
Work to walls; not exceeding 300 mm wide						
152 mm x 152 mm x 5.50 mm thick; white	-	0.28	4.95	4.59	m	9.54
152 mm x 152 mm x 5.50 mm thick; light colours	-	0.28	4.95	7.36	m	12.31
152 mm x 152 mm x 5.50 mm thick; dark colours	-	0.28	4.95	7.74	m	12.69
200 mm x 100 mm x 6.50 mm thick; white and light colours	-	0.28	4.95	4.59	m	9.54
250 mm x 200 mm x 7 mm thick; white and light colours	-	0.23	4.06	4.87	m	8.93
cutting around pipes; not exceeding 0.30 m girth	-	0.09	0.89	-	nr	0.89
Work to sills; 150 mm wide; rounded edge tiles						
152 mm x 152 mm x 5.50 mm thick; white	-	0.23	4.06	2.30	m	6.36
fitted end	-	0.09	0.89	-	nr	0.89
198 mm x 64.50 mm x 6 mm thick wall tiles; fixing with adhesive; butt joints; straight both ways; flush pointing with white grout; to plaster base						
Work to walls						
over 300 mm wide	20.25	1.67	29.51	27.11	m²	56.62
not exceeding 300 mm wide	-	0.65	11.49	8.08	m	19.57
20 mm x 20 mm x 5.50 mm thick glazed mosaic wall tiles; fixing with adhesive; butt joints; straight both ways; flush pointing with white grout; to plaster base						
Work to walls						
over 300 mm wide	24.85	1.76	31.10	33.05	m²	64.15
not exceeding 300 mm wide	-	0.69	12.19	11.79	m	23.98
50 mm x 50 mm x 5.50 mm thick slip resistant mosaic floor tiles, Series 2 or other equal and approved; Langley London Ltd; fixing with adhesive; butt joints; straight both ways; flush pointing with white grout; to cement and sand base						
Work to floors						
over 300 mm wide	23.55	1.76	24.22	32.46	m²	56.68
not exceeding 300 mm wide	-	0.69	9.49	11.35	m	20.84
Dakota mahogany granite cladding; polished finish; jointed and pointed in coloured mortar (1:2:8)						
20 mm work to floors; level; to cement and sand base						
over 300 mm wide	-	-	-	-	m²	231.50
20 mm x 300 mm treads; plain nosings	-	-	-	-	m	113.39
raking, cutting	-	-	-	-	m	23.63
polished edges	-	-	-	-	m	37.80
birdsmouth	-	-	-	-	m	23.63

	PC £	Labour hours	Labour £	Material £	Unit	Total rate £
Dakota mahogany granite cladding; polished finish; jointed and pointed in coloured mortar (1:2:8) - cont'd						
20 mm thick work to walls; to cement and sand base						
over 300 mm wide	-	-	-	-	m²	236.23
not exceeding 300 mm wide	-	-	-	-	m	99.22
40 mm thick work to walls; to cement and sand base						
over 300 mm wide	-	-	-	-	m²	392.15
not exceeding 300 mm wide	-	-	-	-	m	151.19
Riven Welsh slate floor tiles; level; bedding 10 mm thick and jointing in cement and sand (1:3); butt joints; straight both ways; flush pointing with coloured mortar; to cement and sand base						
Work to floors; over 300 mm wide						
250 mm x 250 mm x 12 mm - 15 mm thick	-	0.56	9.90	32.14	m²	42.04
Work to floors; not exceeding 300 mm wide						
250 mm x 250 mm x 12 mm - 15 mm thick	-	0.28	4.95	9.71	m	14.66
Roman Travertine marble cladding; polished finish; jointed and pointed in coloured mortar (1:2:8)						
20 mm thick work to floors; level; to cement and sand base						
over 300 mm wide	-	-	-	-	m²	151.19
20 mm x 300 mm treads; plain nosings	-	-	-	-	m	80.32
raking cutting	-	-	-	-	m	16.06
polished edges	-	-	-	-	m	16.06
birdsmouth	-	-	-	-	m	26.93
20 mm thick work to walls; to cement and sand base						
over 300 mm wide	-	-	-	-	m²	151.19
not exceeding 300 mm wide	-	-	-	-	m	66.14
40 mm thick work to walls; to cement and sand base						
over 300 mm wide	-	-	-	-	m²	236.23
not exceeding 300 mm wide	-	-	-	-	m	103.94
M41 TERRAZZO TILING/IN SITU TERRAZZO						
Terrazzo tiles; BS 4131; aggregate size random ground grouted and polished to 80's grit finish; standard colour range; 3 mm joints symmetrical layout; bedding in 42 mm cement semi-dry mix (1:4); grouting with neat matching cement						
300 mm x 300 mm x 28 mm (nominal), "Quil-Terra Terrazzo" tile units or other equal and approved; Quiligotti Contracts Ltd; hydraulically pressed, mechanically vibrated, steam cured; to floors on concrete base (not included); sealed with "Quil-Shield" penetrating case hardener or other equal and approved; 2 coats applied immediately after final polishing						
plain; laid level	-	-	-	-	m²	33.06
plain; to slopes exceeding 15 degrees from horizontal	-	-	-	-	m²	37.75
to small areas/toilets	-	-	-	-	m²	73.40
Accessories						
plastic division strips; 6 mm x 38 mm; set into floor tiling above crack inducing joints, to the nearest full tile module	-	-	-	-	m	2.25

M SURFACE FINISHES

	PC £	Labour hours	Labour £	Material £	Unit	Total rate £
M41 TERRAZZO TILING/IN SITU TERRAZZO - cont'd						
Specially made terrazzo precast units; BS 4387; aggregate size random; standard colour range; 3 mm joints; grouting with neat matching cement						
Standard size Quiligotti Tread and Riser square combined terrazzo units (with riser cast down) or other equal and approved; 280 mm wide; 150 mm high; 40 mm thick; machine made; vibrated and fully machine polished; incorporating 1 nr. "Ferodo" anti-slip insert ref.OT/D or other equal and approved cast-in during manufacture; one end polished only						
fixed with cement:sand (1:4) mortar on prepared backgrounds (not included); grouted in neat tinted cement; wiped clean on completion of fixing	-	-	-	-	m	143.66
Standard size Quiligotti Tread square terrazzo units or other equal and approved; 40 mm thick; 280 mm wide; factory polished; incorporating 1 nr. "Ferodo" anti-slip insert ref. OT/D or other equal and approved						
fixed with cement:sand (1:4) mortar on prepared backgrounds (not included); grouted in neat tinted cement; wiped clean on completion of fixing	-	-	-	-	m	83.89
Standard size Quiligotti Riser square terrazzo units or other equal and approved; 40 mm thick; 150 mm high; factory polished						
fixed with cemnt:sand (1:4) mortar on prepared backgrounds (not included); grouted in neat tinted cement; wiped clean on completion of fixing	-	-	-	-	m	62.92
Standard size Quiligotti coved terrazzo skirting units or other equal and approved; 904 mm long; 150 mm high; nominal finish; 23 mm thick; with square top edge						
fixed with cement:sand (1:4) mortar on prepared backgrounds (by others); grouted in neat tinted cement; wiped clean on completion of fixing	-	-	-	-	m	41.95
extra over for special internal/external angle pieces to match	-	-	-	-	m	16.78
extra over for special polished ends	-	-	-	-	nr	5.55
M42 WOOD BLOCK/COMPOSITION BLOCK/PARQUET FLOORING						
Wood block; Vigers, Stevens & Adams "Vigerflex" or other equal and approved; 7.50 mm thick; level; fixing with adhesive; to cement:sand base						
Work to floors; over 300 mm wide						
"Maple 7"	-	0.46	6.33	16.21	m²	22.54
Wood blocks 25 mm thick; tongued and grooved joints; herringbone pattern; level; fixing with adhesive; to cement:sand base						
Work to floors; over 300 mm wide						
iroko	-	-	-	-	m²	36.95
"Maple 7"	-	-	-	-	m²	39.95
french Oak	-	-	-	-	m²	54.92
american oak	-	-	-	-	m²	49.93
fair square cutting against flush edges of existing finishings	-	-	-	-	m	4.74

M SURFACE FINISHES

	PC £	Labour hours	Labour £	Material £	Unit	Total rate £
Wood blocks 25 mm thick; tongued and grooved joints; herringbone pattern; level; fixing with adhesive; to cement:sand base - cont'd						
Work to floors; over 300 mm wide - cont'd						
extra for cutting and fitting into recessed duct covers 450 mm wide; lining up with adjoining work	-	-	-	-	nr	12.18
cutting around pipes; not exceeding 0.30 m girth	-	-	-	-	nr	2.00
extra for cutting and fitting into recessed manhole covers 600 mm x 600 mm; lining up with adjoining work	-	-	-	-	nr	8.99
Add to wood block flooring over 300 wide for						
sanding; one coat sealer; one coat wax polish	-	-	-	-	m²	4.00
sanding; two coats sealer; buffing with steel wool	-	-	-	-	m²	3.89
sanding; three coats polyurethane lacquer; buffing down between coats	-	-	-	-	m²	5.99
M50 RUBBER/PLASTICS/CORK/LINO/CARPET TILING/SHEETING						
Linoleum sheet; BS 6826; Forbo-Nairn Floors or other equal and approved; level; fixing with adhesive; butt joints; to cement and sand base						
Work to floors; over 300 mm wide						
2.50 mm thick; plain	-	0.37	5.09	11.30	m²	16.39
3.20 mm thick; marbled	-	0.37	5.09	11.97	m²	17.06
Linoleum sheet; "Marmoleum Real" or other equal and approved; level; with welded seams; fixing with adhesive; to cement and sand base						
Work to floors; over 300 mm wide						
2.50 mm thick	-	0.46	6.33	9.67	m²	16.00
Vinyl sheet; Altro "Safety" range or other equal and approved; with welded seams; level; fixing with adhesive; to cement and sand base						
Work to floors; over 300 mm wide						
2.00 mm thick; "Marine T20"	-	0.56	7.71	13.27	m²	20.98
2.50 mm thick; "Classic D25"	-	0.65	8.94	14.92	m²	23.86
3.50 mm thick; "Stronghold"	-	0.74	10.18	19.76	m²	29.94
Slip resistant vinyl sheet; Forbo-Nairn "Surestep" or other equal and approved; level with welded seams; fixing with adhesive; to cement and sand base						
Work to floors; over 300 mm wide						
2.00 mm thick	-	0.46	6.33	11.92	m²	18.25
Vinyl sheet; heavy duty; Marley "HD" or other equal and approved; level; with welded seams; fixing with adhesive; level; to cement and sand base						
Work to floors; over 300 mm wide						
2.00 mm thick	-	0.42	5.78	7.28	m²	13.06
2.50 mm thick	-	0.46	6.33	8.16	m²	14.49
2.00 mm thick skirtings						
100 mm high	-	0.11	1.51	1.30	m	2.81
Vinyl sheet; "Gerflex" standard sheet; "Classic" range or other equal and approved; level; with welded seams; fixing with adhesive; to cement and sand base						
Work to floors; over 300 mm wide						
2.00 mm thick	-	0.46	6.33	6.02	m²	12.35

M SURFACE FINISHES

	PC £	Labour hours	Labour £	Material £	Unit	Total rate £
M50 RUBBER/PLASTICS/CORK/LINO/CARPET TILING/SHEETING - cont'd						
Vinyl sheet; "Armstrong Rhino Contract" or other equal and approved; level; with welded seams; fixing with adhesive; to cement and sand base						
Work to floors; over 300 mm wide						
2.50 mm thick	-	0.46	6.33	6.96	m²	13.29
Vinyl tiles; "Accoflex" or other equal and approved; level; fixing with adhesive; butt joints; straight both ways; to cement and sand base						
Work to floors; over 300 mm wide						
300 mm x 300 mm x 2.00 mm thick	-	0.23	3.16	4.13	m²	7.29
Vinyl semi-flexible tiles; "Arlon" or other equal and approved; level; fixing with adhesive; butt joints; straight both ways; to cement and sand base						
Work to floors; over 300 mm wide						
250 mm x 250 mm x 2.00 mm thick	-	0.23	3.16	4.20	m²	7.36
Vinyl semi-flexible tiles; Marley "Marleyflex" or other equal and approved; level; fixing with adhesive; butt joints; straight both ways; to cement and sand base						
Work to floors; over 300 mm wide						
300 mm x 300 mm x 2.00 mm thick	-	0.23	3.16	4.17	m²	7.33
300 mm x 300 mm x 2.50 mm thick	-	0.23	3.16	5.03	m²	8.19
Vinyl semi-flexible tiles; "Vylon" or other equal and approved; level; fixing with adhesive; butt joints; straight both ways; to cement and sand base						
Work to floors; over 300 mm wide						
250 mm x 250 mm x 2.00 mm thick	-	0.26	3.58	4.17	m²	7.75
Vinyl tiles; anti-static; level; fixing with adhesive; butt joints; straight both ways; to cement and sand base						
Work to floors; over 300 mm wide						
457 mm x 457 mm x 2.00 mm thick	-	0.42	5.78	7.92	m²	13.70
Vinyl tiles; "Polyflex" or other equal and approved; level; fixing with adhesive; butt joints; straight both ways; to cement and sand base						
Work to floors; over 300 mm wide						
300 mm x 300 mm x 1.50 mm thick	-	0.23	3.16	3.42	m²	6.58
300 mm x 300 mm x 2.00 mm thick	-	0.23	3.16	3.78	m²	6.94
Vinyl tiles; "Polyflor XL" or other equal and approved; level; fixing with adhesive; butt joints; straight both ways; to cement and sand base						
Work to floors; over 300 mm wide						
300 mm x 300 mm x 2.00 mm thick	-	0.32	4.40	4.71	m²	9.11
Vinyl tiles; Marley "HD" or other equal and approved; level; fixing with adhesive; butt joints; straight both ways; to cement and sand base						
Work to floors; over 300 mm wide						
300 mm x 300 mm x 2.00 mm thick	-	0.32	4.40	7.28	m²	11.68

M SURFACE FINISHES

	PC £	Labour hours	Labour £	Material £	Unit	Total rate £
Thermoplastic tiles; Marley "Marleyflex" or other equal and approved; level; fixing with adhesive; butt joints; straight both ways; to cement and sand base						
Work to floors; over 300 mm wide						
300 mm x 300 mm x 2.00 mm thick; series 2	-	0.21	2.89	3.43	m²	6.32
300 mm x 300 mm x 2.00 mm thick; series 4	-	0.21	2.89	3.95	m²	6.84
Linoleum tiles; BS 6826; Forbo-Nairn Floors or other equal and approved; level; fixing with adhesive; butt joints; straight both ways; to cement and sand base						
Work to floors; over 300 mm wide						
2.50 mm thick (marble pattern)	-	0.28	3.85	11.88	m²	15.73
Cork tiles Wicanders "Cork-Master" or other equal and approved; level; fixing with adhesive; butt joints; straight both ways; to cement and sand base						
Work to floors; over 300 mm wide						
300 mm x 300 mm x 4.00 mm thick	-	0.37	5.09	18.95	m²	24.04
Rubber studded tiles; Altro "Mondopave" or other equal and approved; level; fixing with adhesive; butt joints; straight to cement and sand base						
Work to floors; over 300 mm wide						
500 mm x 500 mm x 2.50 mm thick; type MRB; black	-	0.56	7.71	23.43	m²	31.14
500 mm x 500 mm x 4.00 mm thick; type MRB; black	-	0.56	7.71	22.42	m²	30.13
Work to landings; over 300 mm wide						
500 mm x 500 mm x 4.00 mm thick; type MRB; black	-	0.74	10.18	22.42	m²	32.60
4.00 mm thick to tread						
275 mm wide	-	0.46	6.33	6.66	m	12.99
4.00 mm thick to riser						
180 mm wide	-	0.56	7.71	4.74	m	12.45
Sundry floor sheeting underlays						
For floor finishings; over 300 mm wide						
building paper to BS 1521; class A; 75 mm lap (laying only)	-	0.05	0.39	-	m²	0.39
3.20 mm thick hardboard	-	0.19	3.36	1.11	m²	4.47
6.00 mm thick plywood	-	0.28	4.95	5.72	m²	10.67
Stair nosings						
Light duty hard aluminium alloy stair tread nosings; plugged and screwed in concrete						
57 mm x 32 mm	6.31	0.23	2.26	7.06	m	9.32
84 mm x 32 mm	8.73	0.28	2.75	9.72	m	12.47
Heavy duty aluminium alloy stair tread nosings; plugged and screwed to concrete						
60 mm x 32 mm	7.50	0.28	2.75	8.37	m	11.12
92 mm x 32 mm	10.28	0.32	3.15	11.43	m	14.58

Prices for Measured Works - Major Works
M SURFACE FINISHES

	PC £	Labour hours	Labour £	Material £	Unit	Total rate £
M50 RUBBER/PLASTICS/CORK/LINO/CARPET TILING/SHEETING - cont'd						
Heavy duty carpet tiles; "Heuga 580 Olympic" or other equal and approved; to cement and sand base						
Work to floors						
over 300 mm wide	15.79	0.28	3.85	17.82	m²	21.67
Nylon needlepunch carpet; "Marleytex" or other equal and approved; fixing; with adhesive; level; to cement						
Work to floors						
over 300 mm wide	-	0.23	3.16	5.01	m²	8.17
M51 EDGE FIXED CARPETING						
Fitted carpeting; Wilton wool/nylon or other equal and approved; 80/20 velvet pile; heavy domestic plain						
Work to floors						
over 300 mm wide	33.84	0.37	3.27	40.02	m²	43.29
Work to treads and risers						
over 300 mm wide	-	0.74	6.54	40.02	m²	46.56
Underlay to carpeting						
Work to floors						
over 300 mm wide	3.20	0.07	0.62	3.61	m²	4.23
raking cutting	-	0.07	0.55	-	m	0.55
Sundries						
Carpet gripper fixed to floor; standard edging						
22 mm wide	-	0.04	0.31	0.20	m	0.51
M52 DECORATIVE PAPERS/FABRICS						
Lining paper; PC £1.95/roll; and hanging						
Plaster walls or columns						
over 300 mm girth	-	0.19	1.87	0.33	m²	2.20
Plaster ceilings or beams						
over 300 mm girth	-	0.23	2.26	0.33	m²	2.59
Decorative paper-backed vinyl wallpaper; PC £10.48/roll; and hanging						
Plaster walls or columns						
over 300 mm girth	-	0.23	2.26	1.86	m²	4.12

Prices for Measured Works - Major Works
M SURFACE FINISHES

M60 PAINTING/CLEAR FINISHING

BASIC PAINT PRICES (£/5 LITRE TIN)

Paints
"Dulux"
	£		£		£
- matt emulsion	16.94	oil based undercoat	24.68	"Sandtex" masonry paint	
- gloss	18.14	"Weathershield" gloss	37.39	- brilliant white	14.60
- eggshell gloss	34.81	"Weathershield" undercoat	37.39	- coloured	16.62

Primer/undercoats
acrylic	14.88	water based	19.35	masonry sealer	16.73
red oxide	18.08	zinc phosphate	30.32	knotting solution	37.45

Special paints
solar reflective aluminium	35.45	"Hammerite"		fire retardant	
anti-graffiti	118.70	"Smoothrite"	39.34	- undercoat	48.44
bituminous emulsion	9.09			- top coat	64.51

Stains and Preservatives
	£			£
Cuprinol		Sikkens		
- "Clear"	18.96	- "Cetol HLS"		39.50
- "Boiled linseed oil"	16.82	- "Cetol TS"		58.44
Sadolin		- "Cetol Filter 7"		60.10
- "Extra"	63.41	Protim Soliignum		
- "New base"	35.45	- "Architectural"		54.36
		- "Brown"		16.68
		- "Cedar"		25.86

Varnishes
polyurethene		27.06

	PC £	Labour hours	Labour £	Material £	Unit	Total rate £

NOTE: The following prices include for preparing surfaces. Painting woodwork also includes for knotting prior to applying the priming coat and for all stopping of nail holes.

PAINTING/CLEAR FINISHING - INTERNALLY

One coat primer; on wood surfaces before fixing
General surfaces

	PC £	Labour hours	Labour £	Material £	Unit	Total rate £
over 300 mm girth	-	0.08	0.79	0.44	m^2	1.23
isolated surfaces not exceeding 300 mm girth	-	0.02	0.20	0.16	m	0.36
isolated areas not exceeding 0.50 m^2 irrespective of girth	-	0.06	0.59	0.18	nr	0.77

One coat sealer; on wood surfaces before fixing
General surfaces

	PC £	Labour hours	Labour £	Material £	Unit	Total rate £
over 300 mm girth	-	0.10	0.98	0.65	m^2	1.63
isolated surfaces not exceeding 300 mm girth	-	0.03	0.30	0.23	m	0.53
isolated surfaces not exceeding 0.50 m^2 irrespective of girth	-	0.08	0.79	0.31	nr	1.10

One coat of Sikkens "Cetol HLS" stain or other equal and approved; on wood surfaces before fixing
General surfaces

	PC £	Labour hours	Labour £	Material £	Unit	Total rate £
over 300 mm girth	-	0.11	1.08	0.60	m^2	1.68
isolated surfaces not exceeding 300 mm girth	-	0.03	0.30	0.23	m	0.53
isolated surfaces not exceeding 0.50 m^2 irrespective of girth	-	0.08	0.79	0.29	nr	1.08

Prices for Measured Works - Major Works
M SURFACE FINISHES

	PC £	Labour hours	Labour £	Material £	Unit	Total rate £
M60 PAINTING/CLEAR FINISHING - cont'd						
PAINTING/CLEAR FINISHING - INTERNALLY - cont'd						
One coat of Sikkens "Cetol TS" interior stain or other equal and approved; on wood surfaces before fixing						
General surfaces						
over 300 mm girth	-	0.11	1.08	0.94	m^2	2.02
isolated surfaces not exceeding 300 mm girth	-	0.03	0.30	0.36	m	0.66
isolated areas not exceeding 0.50 m^2; irrespective of girth	-	0.08	0.79	0.45	nr	1.24
One coat Cuprinol clear wood preservative or other equal and approved; on wood surfaces before fixing						
General surfaces						
over 300 mm girth	-	0.08	0.79	0.40	m^2	1.19
isolated surfaces not exceeding 300 mm girth	-	0.02	0.20	0.15	m	0.35
isolated areas not exceeding 0.50 m^2; irrespective of girth	-	0.05	0.49	0.19	nr	0.68
One coat HCC Protective Coatings Ltd "Permacor" urethane alkyd gloss finishing coat or other equal and approved; on previously primed steelwork						
Members of roof trusses						
over 300 mm girth	-	0.06	0.59	1.30	m^2	1.89
Two coats emulsion paint						
Brick or block walls						
over 300 mm girth	-	0.21	2.07	0.74	m^2	2.81
Cement render or concrete						
over 300 mm girth	-	0.20	1.97	0.65	m^2	2.62
isolated surfaces not exceeding 300 mm girth	-	0.10	0.98	0.20	m	1.18
Plaster walls or plaster/plasterboard ceilings						
over 300 mm girth	-	0.18	1.77	0.63	m^2	2.40
over 300 mm girth; in multi colours	-	0.24	2.36	0.76	m^2	3.12
over 300 mm girth; in staircase areas	-	0.21	2.07	0.73	m^2	2.80
cutting in edges on flush surfaces	-	0.08	0.79	-	m	0.79
Plaster/plasterboard ceilings						
over 300 mm girth; 3.50 m - 5.00 m high	-	0.21	2.07	0.64	m^2	2.71
One mist and two coats emulsion paint						
Brick or block walls						
over 300 mm girth	-	0.19	1.87	0.96	m^2	2.83
Cement render or concrete						
over 300 mm girth	-	0.19	1.87	0.89	m^2	2.76
Plaster walls or plaster/plasterboard ceilings						
over 300 mm girth	-	0.18	1.77	0.89	m^2	2.66
over 300 mm girth; in multi colours	-	0.25	2.46	0.90	m^2	3.36
over 300 mm girth; in staircase areas	-	0.21	2.07	0.89	m^2	2.96
cutting in edges on flush surfaces	-	0.09	0.89	-	m	0.89
Plaster/plasterboard ceilings						
over 300 mm girth; 3.50 m - 5.00 m high	-	0.21	2.07	0.89	m^2	2.96
One coat "Tretol No 10 Sealer" or other equal and approved; two coats "Tretol sprayed Supercover Spraytone" emulsion paint or other equal and approved						
Plaster walls or plaster/plasterboard ceilings						
over 300 mm girth	-	-	-	-	m^2	4.29

Prices for Measured Works - Major Works
M SURFACE FINISHES

	PC £	Labour hours	Labour £	Material £	Unit	Total rate £
Textured plastic; "Artex" or other equal and approved finish						
Plasterboard ceilings						
over 300 mm girth	-	0.19	1.87	1.58	m²	3.45
Concrete walls or ceilings						
over 300 mm girth	-	0.23	2.26	1.52	m²	3.78
One coat "Portabond" or other equal and approved; one coat "Portaflek" or other equal and approved; on plaster surfaces; spray applied, masking adjacent surfaces						
General surfaces						
over 300 mm girth	-	-	-	-	m²	8.99
extra for one coat standard HD glaze	-	-	-	-	m²	1.70
not exceeding 300 mm girth	-	-	-	-	m	3.89
Touch up primer; one undercoat and one finishing coat of gloss oil paint; on wood surfaces						
General surfaces						
over 300 mm girth	-	0.23	2.26	1.17	m²	3.43
isolated surfaces not exceeding 300 mm girth	-	0.09	0.89	0.39	m	1.28
isolated areas not exceeding 0.50 m²; irrespective of girth	-	0.18	1.77	0.60	nr	2.37
Glazed windows and screens						
panes; area not exceeding 0.10 m²	-	0.38	3.74	0.85	m²	4.59
panes; area 0.10 m² - 0.50 m²	-	0.31	3.05	0.67	m²	3.72
panes; area 0.50 m² - 1.00 m²	-	0.26	2.56	0.53	m²	3.09
panes; area over 1.00 m²	-	0.23	2.26	0.45	m²	2.71
Touch up primer; two undercoats and one finishing coat of gloss oil paint; on wood surfaces						
General surfaces						
over 300 mm girth	-	0.32	3.15	1.08	m²	4.23
isolated surfaces not exceeding 300 mm girth	-	0.13	1.28	0.44	m	1.72
isolated areas not exceeding 0.50 m²; irrespective of girth	-	0.24	2.36	0.62	nr	2.98
Glazed windows and screens						
panes; area not exceeding 0.10 m²	-	0.54	5.31	1.17	m²	6.48
panes; area 0.10 m² - 0.50 m²	-	0.43	4.23	0.92	m²	5.15
panes; area 0.50 m² - 1.00 m²	-	0.37	3.64	0.76	m²	4.40
panes; area over 1.00 m²	-	0.32	3.15	0.64	m²	3.79
Knot; one coat primer; stop; one undercoat and one finishing coat of gloss oil paint; on wood surfaces						
General surfaces						
over 300 mm girth	-	0.33	3.25	1.06	m²	4.31
isolated surfaces not exceeding 300 mm girth	-	0.13	1.28	0.35	m	1.63
isolated areas not exceeding 0.50 m²; irrespective of girth	-	0.25	2.46	0.69	nr	3.15
Glazed windows and screens						
panes; area not exceeding 0.10 m²	-	0.56	5.51	1.06	m²	6.57
panes; area 0.10 m² - 0.50 m²	-	0.45	4.43	0.89	m²	5.32
panes; area 0.50 m² - 1.00 m²	-	0.40	3.93	0.89	m²	4.82
panes; area over 1.00 m²	-	0.33	3.25	0.65	m²	3.90
Knot; one coat primer; stop; two undercoats and one finishing coat of gloss oil paint; on wood surfaces						
General surfaces						
over 300 mm girth	-	0.43	4.23	1.48	m²	5.71
isolated surfaces not exceeding 300 mm girth	-	0.18	1.77	0.51	m	2.28

Prices for Measured Works - Major Works
M SURFACE FINISHES

	PC £	Labour hours	Labour £	Material £	Unit	Total rate £
M60 PAINTING/CLEAR FINISHING - cont'd						
PAINTING/CLEAR FINISHING - INTERNALLY - cont'd						
Knot; one coat primer; stop; two undercoats and one finishing coat of gloss oil paint; on wood surfaces - cont'd						
General surfaces - cont'd						
isolated areas not exceeding 0.50 m^2; irrespective of girth	-	0.32	3.15	0.83	nr	3.98
Glazed windows and screens						
panes; area not exceeding 0.10 m^2	-	0.71	6.98	1.40	m^2	8.38
panes; area 0.10 m^2 - 0.50 m^2	-	0.56	5.51	1.28	m^2	6.79
panes; area 0.50 m^2 - 1.00 m^2	-	0.50	4.92	1.16	m^2	6.08
panes; area over 1.00 m^2	-	0.43	4.23	0.86	m^2	5.09
One coat primer; one undercoat and one finishing coat of gloss oil paint						
Plaster surfaces						
over 300 mm girth	-	0.30	2.95	1.35	m^2	4.30
One coat primer; two undercoats and one finishing coat of gloss oil paint						
Plaster surfaces						
over 300 mm girth	-	0.40	3.93	1.75	m^2	5.68
One coat primer; two undercoats and one finishing coat of eggshell paint						
Plaster surfaces						
over 300 mm girth	-	0.40	3.93	1.71	m^2	5.64
Touch up primer; one undercoat and one finishing coat of gloss paint; on iron or steel surfaces						
General surfaces						
over 300 mm girth	-	0.23	2.26	0.82	m^2	3.08
isolated surfaces not exceeding 300 mm girth	-	0.09	0.89	0.28	m	1.17
isolated areas not exceeding 0.50 m^2; irrespective of girth	-	0.18	1.77	0.45	nr	2.22
Glazed windows and screens						
panes; area not exceeding 0.10 m^2	-	0.38	3.74	0.85	m^2	4.59
panes; area 0.10 m^2 - 0.50 m^2	-	0.31	3.05	0.67	m^2	3.72
panes; area 0.50 m^2 - 1.00 m^2	-	0.26	2.56	0.51	m^2	3.07
panes; area over 1.00 m^2	-	0.23	2.26	0.43	m^2	2.69
Structural steelwork						
over 300 mm girth	-	0.25	2.46	0.85	m^2	3.31
Members of roof trusses						
over 300 mm girth	-	0.34	3.34	0.97	m^2	4.31
Ornamental railings and the like; each side measured overall						
over 300 mm girth	-	0.40	3.93	1.08	m^2	5.01
Iron or steel radiators						
over 300 mm girth	-	0.23	2.26	0.89	m^2	3.15
Pipes or conduits						
over 300 mm girth	-	0.34	3.34	0.93	m^2	4.27
not exceeding 300 mm girth	-	0.13	1.28	0.30	m	1.58
Touch up primer; two undercoats and one finishing coat of gloss oil paint; on iron or steel surfaces						
General surfaces						
over 300 mm girth	-	0.32	3.15	1.12	m^2	4.27
isolated surfaces not exceeding 300 mm girth	-	0.13	1.28	0.41	m	1.69

M SURFACE FINISHES

	PC £	Labour hours	Labour £	Material £	Unit	Total rate £
General surfaces - cont'd						
isolated areas not exceeding 0.50 m²;						
irrespective of girth	-	0.24	2.36	0.65	nr	**3.01**
Glazed windows and screens						
panes; area not exceeding 0.10 m²	-	0.54	5.31	1.19	m²	**6.50**
panes; area 0.10 m² - 0.50 m²	-	0.43	4.23	0.98	m²	**5.21**
panes; area 0.50 m² - 1.00 m²	-	0.37	3.64	0.84	m²	**4.48**
panes; area over 1.00 m²	-	0.32	3.15	0.66	m²	**3.81**
Structural steelwork						
over 300 mm girth	-	0.36	3.54	1.15	m²	**4.69**
Members of roof trusses						
over 300 mm girth	-	0.48	4.72	1.38	m²	**6.10**
Ornamental railings and the like; each side measured overall						
over 300 mm girth	-	0.55	5.41	1.48	m²	**6.89**
Iron or steel radiators						
over 300 mm girth	-	0.32	3.15	1.25	m²	**4.40**
Pipes or conduits						
over 300 mm girth	-	0.48	4.72	1.40	m²	**6.12**
not exceeding 300 mm girth	-	0.19	1.87	0.45	m	**2.32**
One coat primer; one undercoat and one finishing coat of gloss oil paint; on iron or steel surfaces						
General surfaces						
over 300 mm girth	-	0.23	2.26	0.82	m²	**3.08**
isolated surfaces not exceeding 300 mm girth	-	0.12	1.18	0.45	m	**1.63**
isolated areas not exceeding 0.50 m²;						
irrespective of girth	-	0.23	2.26	0.78	nr	**3.04**
Glazed windows and screens						
panes; area not exceeding 0.10 m²	-	0.50	4.92	1.26	m²	**6.18**
panes; area 0.10 m² - 0.50 m²	-	0.40	3.93	1.00	m²	**4.93**
panes; area 0.50 m² - 1.00 m²	-	0.34	3.34	0.86	m²	**4.20**
panes; area over 1.00 m²	-	0.30	2.95	0.78	m²	**3.73**
Structural steelwork						
over 300 mm girth	-	0.33	3.25	1.23	m²	**4.48**
Members of roof trusses						
over 300 mm girth	-	0.45	4.43	1.31	m²	**5.74**
Ornamental railings and the like; each side measured overall						
over 300 mm girth	-	0.51	5.02	1.55	m²	**6.57**
Iron or steel radiators						
over 300 mm girth	-	0.30	2.95	1.31	m²	**4.26**
Pipes or conduits						
over 300 mm girth	-	0.45	4.43	1.31	m²	**5.74**
not exceeding 300 mm girth	-	0.18	1.77	0.43	m	**2.20**
One coat primer; two undercoats and one finishing coat of gloss oil paint; on iron or steel surfaces						
General surfaces						
over 300 mm girth	-	0.40	3.93	1.52	m²	**5.45**
isolated surfaces not exceeding 300 mm girth	-	0.16	1.57	0.61	m	**2.18**
isolated areas not exceeding 0.50 m²;						
irrespective of girth	-	0.29	2.85	0.86	nr	**3.71**
Glazed windows and screens						
panes; area not exceeding 0.10 m²	-	0.64	6.30	1.56	m²	**7.86**
panes; area 0.10 m² - 0.50 m²	-	0.54	5.31	0.87	m²	**6.18**
panes; area 0.50 m² - 1.00 m²	-	0.45	4.43	1.07	m²	**5.50**
panes; area over 1.00 m²	-	0.40	3.93	0.91	m²	**4.84**
Structural steelwork						
over 300 mm girth	-	0.44	4.33	1.56	m²	**5.89**

Prices for Measured Works - Major Works
M SURFACE FINISHES

	PC £	Labour hours	Labour £	Material £	Unit	Total rate £
M60 PAINTING/CLEAR FINISHING - cont'd						
PAINTING/CLEAR FINISHING - INTERNALLY - cont'd						
One coat primer; two undercoats and one finishing coat of gloss oil paint; on iron or steel surfaces - cont'd						
Members of roof trusses						
over 300 mm girth	-	0.58	5.71	1.87	m²	7.58
Ornamental railings and the like; each side measured overall						
over 300 mm girth	-	0.66	6.49	2.05	m²	8.54
Iron or steel radiators						
over 300 mm girth	-	0.40	3.93	1.68	m²	5.61
Pipes or conduits						
over 300 mm girth	-	0.59	5.80	1.89	m²	7.69
not exceeding 300 mm girth	-	0.24	2.36	0.60	m	2.96
Two coats of bituminous paint; on iron or steel surfaces						
General surfaces						
over 300 mm girth	-	0.23	2.26	0.40	m²	2.66
Inside of galvanized steel cistern						
over 300 mm girth	-	0.34	3.34	0.48	m²	3.82
Two coats bituminous paint; first coat blinded with clean sand prior to second coat; on concrete surfaces						
General surfaces						
over 300 mm girth	-	0.79	7.77	1.19	m²	8.96
Mordant solution; one coat HCC Protective Coatings Ltd "Permacor Alkyd MIO" or other equal and approved; one coat "Permatex Epoxy Gloss" finishing coat on galvanised steelwork						
Structural steelwork						
over 300 mm girth	-	0.44	4.33	2.31	m²	6.64
One coat HCC Protective Coatings Ltd "Epoxy Zinc Primer" or other equal and approved; two coats "Permacor Alkyd MIO" or other equal and approved; one coat "Permacor Epoxy Gloss" finishing coat or other equal and approved on steelwork						
Structural steelwork						
over 300 mm girth	-	0.63	6.20	4.18	m²	10.38
Steel protection; HCC Protective Coatings Ltd "Unitherm" or other equal and approved; two coats to steelwork						
Structural steelwork						
over 300 mm girth	-	0.99	9.74	1.43	m²	11.17
Two coats epoxy resin sealer; HCC Protective Coatings Ltd "Betonol" or other equal and approved; on concrete surfaces						
General surfaces						
over 300 mm girth	-	0.19	1.87	3.22	m²	5.09
"Nitoflor Lithurin" floor hardener and dust proofer or other equal and approved; Fosroc Expandite Ltd; two coats; on concrete surfaces						
General surfaces						
over 300 mm girth	-	0.24	1.88	0.43	m²	2.31

M SURFACE FINISHES

	PC £	Labour hours	Labour £	Material £	Unit	Total rate £
Two coats of boiled linseed oil; on hardwood surfaces						
General surfaces						
over 300 mm girth	-	0.18	1.77	1.14	m^2	**2.91**
isolated surfaces not exceeding 300 mm girth	-	0.07	0.69	0.37	m	**1.06**
isolated areas not exceeding 0.50 m^2; irrespective of girth	-	0.13	1.28	0.66	nr	**1.94**
Two coats polyurethane varnish; on wood surfaces						
General surfaces						
over 300 mm girth	-	0.18	1.77	1.14	m^2	**2.91**
isolated surfaces not exceeding 300 mm girth	-	0.07	0.69	0.42	m	**1.11**
isolated areas not exceeding 0.50 m^2; irrespective of girth	-	0.13	1.28	0.13	nr	**1.41**
Three coats polyurethane varnish; on wood surfaces						
General surfaces						
over 300 mm girth	-	0.26	2.56	1.74	m^2	**4.30**
isolated surfaces not exceeding 300 mm girth	-	0.10	0.98	0.53	m	**1.51**
isolated areas not exceeding 0.50 m^2; irrespective of girth	-	0.19	1.87	0.97	nr	**2.84**
One undercoat; and one finishing coat; of "Albi" clear flame retardant surface coating or other equal and approved; on wood surfaces						
General surfaces						
over 300 mm girth	-	0.34	3.34	2.27	m^2	**5.61**
isolated surfaces not exceeding 300 mm girth	-	0.14	1.38	0.79	m	**2.17**
isolated areas not exceeding 0.50 m^2; irrespective of girth	-	0.19	1.87	1.73	nr	**3.60**
Two undercoats; and one finishing coat; of "Albi" clear flame retardant surface coating or other equal and approved; on wood surfaces						
General surfaces						
over 300 mm girth	-	0.40	3.93	3.20	m^2	**7.13**
isolated surfaces not exceeding 300 mm girth	-	0.20	1.97	1.15	m	**3.12**
isolated areas not exceeding 0.50 m^2; irrespective of girth	-	0.33	3.25	1.76	nr	**5.01**
Seal and wax polish; dull gloss finish on wood surfaces						
General surfaces						
over 300 mm girth	-	-	-	-	m^2	**7.94**
isolated surfaces not exceeding 300 mm girth	-	-	-	-	m	**3.59**
isolated areas not exceeding 0.50m^2; irrespective of girth	-	-	-	-	nr	**5.55**
One coat of "Sadolin Extra" or other equal and approved; clear or pigmented; one further coat of "Holdex" clear interior silk matt lacquer or similar						
General surfaces						
over 300 mm girth	-	0.25	2.46	3.38	m^2	**5.84**
isolated surfaces not exceeding 300 mm girth	-	0.10	0.98	1.58	m	**2.56**
isolated areas not exceeding 0.50 m^2; irrespective of girth	-	0.20	1.97	1.65	nr	**3.62**
Glazed windows and screens						
panes; area not exceeding 0.10 m^2	-	0.42	4.13	1.93	m^2	**6.06**
panes; area 0.10 m^2 - 0.50 m^2	-	0.33	3.25	1.80	m^2	**5.05**
panes; area 0.50 m^2 - 1.00 m^2	-	0.29	2.85	1.67	m^2	**4.52**
panes; area over 1.00 m^2	-	0.25	2.46	1.58	m^2	**4.04**

Prices for Measured Works - Major Works
M SURFACE FINISHES

	PC £	Labour hours	Labour £	Material £	Unit	Total rate £
M60 PAINTING/CLEAR FINISHING - cont'd						
PAINTING/CLEAR FINISHING - INTERNALLY - cont'd						
Two coats of "Sadolin Extra" or other equal and approved; clear or pigmented; two further coats of "Holdex" clear interior silk matt lacquer or other equal and approved						
General surfaces						
over 300 mm girth	-	0.40	3.93	6.24	m²	10.17
isolated surfaces not exceeding 300 mm girth	-	0.16	1.57	3.12	m	4.69
isolated areas not exceeding 0.50 m²;						
irrespective of girth	-	0.30	2.95	3.56	nr	6.51
Glazed windows and screens						
panes; area not exceeding 0.10 m²	-	0.66	6.49	3.82	m²	10.31
panes; area 0.10 m² - 0.50 m²	-	0.52	5.11	3.56	m²	8.67
panes; area 0.50 m² - 1.00 m²	-	0.45	4.43	3.30	m²	7.73
panes; area over 1.00 m²	-	0.40	3.93	3.12	m²	7.05
Two coats of Sikkens "Cetol TS" interior stain or other equal and approved; on wood surfaces						
General surfaces						
over 300 mm girth	-	0.19	1.87	1.69	m²	3.56
isolated surfaces not exceeding 300 mm girth	-	0.08	0.79	0.59	m	1.38
isolated areas not exceeding 0.50 m²;						
irrespective of girth	-	0.13	1.28	0.92	nr	2.20
Body in and wax polish; dull gloss finish; on hardwood surfaces						
General surfaces						
over 300 mm girth	-	-	-	-	m²	9.14
isolated surfaces not exceeding 300 mm girth	-	-	-	-	m	4.15
isolated areas not exceeding 0.50 m²;						
irrespective of girth	-	-	-	-	nr	6.40
Stain; body in and wax polish; dull gloss finish; on hardwood surfaces						
General surfaces						
over 300 mm girth	-	-	-	-	m²	12.63
isolated surfaces not exceeding 300 mm girth	-	-	-	-	m	5.70
isolated areas not exceeding 0.50 m²;						
irrespective of girth	-	-	-	-	nr	8.89
Seal; two coats of synthetic resin lacquer; decorative flatted finish; wire down, wax and burnish; on wood surfaces						
General surfaces						
over 300 mm girth	-	-	-	-	m²	15.18
isolated surfaces not exceeding 300 mm girth	-	-	-	-	m	6.84
isolated areas not exceeding 0.50 m²;						
irrespective of girth	-	-	-	-	nr	10.63
Stain; body in and fully French polish; full gloss finish; on hardwood surfaces						
General surfaces						
over 300 mm girth	-	-	-	-	m²	17.48
isolated surfaces not exceeding 300 mm girth	-	-	-	-	m	7.89
isolated areas not exceeding 0.50 m²;						
irrespective of girth	-	-	-	-	nr	12.23

M SURFACE FINISHES

	PC £	Labour hours	Labour £	Material £	Unit	Total rate £
Stain; fill grain and fully French polish; full gloss finish; on hardwood surfaces						
General surfaces						
over 300 mm girth	-	-	-	-	m²	23.97
isolated surfaces not exceeding 300 mm girth	-	-	-	-	m	10.78
isolated areas not exceeding 0.50 m²; irrespective of girth	-	-	-	-	nr	16.78
Stain black; body in and fully French polish; ebonized finish; on hardwood surfaces						
General surfaces						
over 300 mm girth	-	-	-	-	m²	29.96
isolated surfaces not exceeding 300 mm girth	-	-	-	-	m	12.48
isolated areas not exceeding 0.50 m²; irrespective of girth	-	-	-	-	nr	20.97
PAINTING/CLEAR FINISHING - EXTERNALLY						
Two coats of cement paint, "Sandtex Matt" or other equal and approved						
Brick or block walls						
over 300 mm girth	-	0.26	2.56	1.57	m²	4.13
Cement render or concrete walls						
over 300 mm girth	-	0.23	2.26	1.04	m²	3.30
Roughcast walls						
over 300 mm girth	-	0.40	3.93	1.04	m²	4.97
One coat sealer and two coats of external grade emulsion paint, Dulux "Weathershield" or other equal and approved						
Brick or block walls						
over 300 mm girth	-	0.43	4.23	5.79	m²	10.02
Cement render or concrete walls						
over 300 mm girth	-	0.35	3.44	3.86	m²	7.30
Concrete soffits						
over 300 mm girth	-	0.40	3.93	3.86	m²	7.79
One coat sealer (applied by brush) and two coats of external grade emulsion paint, Dulux "Weathershield" or other equal and approved (spray applied)						
Roughcast						
over 300 mm girth	-	0.29	2.85	7.80	m²	10.65
Two coat sealer and one coat of Anti-Graffiti paint (spray applied)						
Roughcast						
over 300 mm girth	-	0.29	2.85	33.28	m²	36.13
Two coats solar reflective aluminium paint; on bituminous roofing						
General surfaces						
over 300 mm girth	-	0.44	4.33	10.05	m²	14.38
Touch up primer; one undercoat and one finishing coat of gloss oil paint; on wood surfaces						
General surfaces						
over 300 mm girth	-	0.25	2.46	0.81	m²	3.27
isolated surfaces not exceeding 300 mm girth	-	0.11	1.08	0.25	m	1.33
isolated areas not exceeding 0.50 m²; irrespective of girth	-	0.19	1.87	0.49	nr	2.36

M SURFACE FINISHES

	PC £	Labour hours	Labour £	Material £	Unit	Total rate £
M60 PAINTING/CLEAR FINISHING - cont'd						
PAINTING/CLEAR FINISHING - EXTERNALLY - cont'd						
Touch up primer; one undercoat and one finishing coat of gloss oil paint; on wood surfaces - cont'd						
Glazed windows and screens						
panes; area not exceeding 0.10 m^2	-	0.43	4.23	0.89	m^2	5.12
panes; area 0.10 m^2 - 0.50 m^2	-	0.34	3.34	0.75	m^2	4.09
panes; area 0.50 m^2 - 1.00 m^2	-	0.30	2.95	0.59	m^2	3.54
panes; area over 1.00 m^2	-	0.25	2.46	0.49	m^2	2.95
Glazed windows and screens; multi-coloured work						
panes; area not exceeding 0.10 m^2	-	0.46	4.52	0.89	m^2	5.41
panes; area 0.10 m^2 - 0.50 m^2	-	0.38	3.74	0.77	m^2	4.51
panes; area 0.50 m^2 - 1.00 m^2	-	0.32	3.15	0.59	m^2	3.74
panes; area over 1.00 m^2	-	0.29	2.85	0.49	m^2	3.34
Touch up primer; two undercoats and one finishing coat of gloss oil paint; on wood surfaces						
General surfaces						
over 300 mm girth	-	0.35	3.44	1.06	m^2	4.50
isolated surfaces not exceeding 300 mm girth	-	0.15	1.48	0.29	m	1.77
isolated areas not exceeding 0.50 m^2; irrespective of girth	-	0.27	2.66	0.58	nr	3.24
Glazed windows and screens						
panes; area not exceeding 0.10 m^2	-	0.59	5.80	0.94	m^2	6.74
panes; area 0.10 m^2 - 0.50 m^2	-	0.59	5.80	0.79	m^2	6.59
panes; area 0.50 m^2 - 1.00 m^2	-	0.47	4.62	0.70	m^2	5.32
panes; area over 1.00 m^2	-	0.35	3.44	0.58	m^2	4.02
Glazed windows and screens; multi-coloured work						
panes; area not exceeding 0.10 m^2	-	0.68	6.69	0.94	m^2	7.63
panes; area 0.10 m^2 - 0.50 m^2	-	0.55	5.41	0.82	m^2	6.23
panes; area 0.50 m^2 - 1.00 m^2	-	0.47	4.62	0.70	m^2	5.32
panes; area over 1.00 m^2	-	0.41	4.03	0.58	m^2	4.61
Knot; one coat primer; one undercoat and one finishing coat of gloss oil paint; on wood surfaces						
General surfaces						
over 300 mm girth	-	0.38	3.74	1.12	m^2	4.86
isolated surfaces not exceeding 300 mm girth	-	0.16	1.57	0.36	m	1.93
isolated areas not exceeding 0.50 m^2; irrespective of girth	-	0.29	2.85	0.62	nr	3.47
Glazed windows and screens						
panes; area not exceeding 0.10 m^2	-	0.62	6.10	1.16	m^2	7.26
panes; area 0.10 m^2 - 0.50 m^2	-	0.50	4.92	1.04	m^2	5.96
panes; area 0.50 m^2 - 1.00 m^2	-	0.44	4.33	0.81	m^2	5.14
panes; area over 1.00 m^2	-	0.38	3.74	0.57	m^2	4.31
Glazed windows and screens; multi-coloured work						
panes; area not exceeding 0.10 m^2	-	0.68	6.69	1.16	m^2	7.85
panes; area 0.10 m^2 - 0.50 m^2	-	0.55	5.41	1.04	m^2	6.45
panes; area 0.50 m^2 - 1.00 m^2	-	0.48	4.72	0.80	m^2	5.52
panes; area over 1.00 m^2	-	0.41	4.03	0.56	m^2	4.59

M SURFACE FINISHES

	PC £	Labour hours	Labour £	Material £	Unit	Total rate £
Knot; one coat primer; two undercoats and one finishing coat of gloss oil paint; on wood surfaces						
General surfaces						
over 300 mm girth	-	0.46	4.52	1.34	m²	5.86
isolated surfaces not exceeding 300 mm girth	-	0.19	1.87	0.48	m	2.35
isolated areas not exceeding 0.50 m²; irrespective of girth	-	0.35	3.44	0.81	nr	4.25
Glazed windows and screens						
panes; area not exceeding 0.10 m²	-	0.78	7.67	1.50	m²	9.17
panes; area 0.10 m² - 0.50 m²	-	0.62	6.10	1.33	m²	7.43
panes; area 0.50 m² - 1.00 m²	-	0.55	5.41	1.02	m²	6.43
panes; area over 1.00 m²	-	0.46	4.52	0.71	m²	5.23
Glazed windows and screens; multi-coloured work						
panes; area not exceeding 0.10 m²	-	0.89	8.75	1.50	m²	10.25
panes; area 0.10 m² - 0.50 m²	-	0.72	7.08	1.34	m²	8.42
panes; area 0.50 m² - 1.00 m²	-	0.64	6.30	1.02	m²	7.32
panes; area over 1.00 m²	-	0.54	5.31	0.71	m²	6.02
Touch up primer; one undercoat and one finishing coat of gloss oil paint; on iron or steel surfaces						
General surfaces						
over 300 mm girth	-	0.25	2.46	0.81	m²	3.27
isolated surfaces not exceeding 300 mm girth	-	0.11	1.08	0.46	m	1.54
isolated areas not exceeding 0.50 m²; irrespective of girth	-	0.19	1.87	0.43	nr	2.30
Glazed windows and screens						
panes; area not exceeding 0.10 m²	-	0.43	4.23	0.85	m²	5.08
panes; area 0.10 m² - 0.50 m²	-	0.34	3.34	0.74	m²	4.08
panes; area 0.50 m² - 1.00 m²	-	0.30	2.95	0.62	m²	3.57
panes; area over 1.00 m²	-	0.25	2.46	0.46	m²	2.92
Structural steelwork						
over 300 mm girth	-	0.29	2.85	0.85	m²	3.70
Members of roof trusses						
over 300 mm girth	-	0.38	3.74	0.97	m²	4.71
Ornamental railings and the like; each side measured overall						
over 300 mm girth	-	0.43	4.23	1.00	m²	5.23
Eaves gutters						
over 300 mm girth	-	0.45	4.43	1.16	m²	5.59
not exceeding 300 mm girth	-	0.19	1.87	0.38	m	2.25
Pipes or conduits						
over 300 mm girth	-	0.38	3.74	1.16	m²	4.90
not exceeding 300 mm girth	-	0.15	1.48	0.38	m	1.86
Touch up primer; two undercoats and one finishing coat of gloss oil paint; on iron or steel surfaces						
General surfaces						
over 300 mm girth	-	0.35	3.44	1.10	m²	4.54
isolated surfaces not exceeding 300 mm girth	-	0.14	1.38	0.30	m	1.68
isolated areas not exceeding 0.50 m²; irrespective of girth	-	0.26	2.56	0.61	nr	3.17
Glazed windows and screens						
panes; area not exceeding 0.10 m²	-	0.59	5.80	1.11	m²	6.91
panes; area 0.10 m² - 0.50 m²	-	0.47	4.62	0.96	m²	5.58
panes; area 0.50 m² - 1.00 m²	-	0.41	4.03	0.81	m²	4.84
panes; area over 1.00 m²	-	0.35	3.44	0.66	m²	4.10

M SURFACE FINISHES

	PC £	Labour hours	Labour £	Material £	Unit	Total rate £
M60 PAINTING/CLEAR FINISHING - cont'd						
PAINTING/CLEAR FINISHING - EXTERNALLY - cont'd						
Touch up primer; two undercoats and one finishing coat of gloss oil paint; on iron or steel surfaces - cont'd						
Structural steelwork						
over 300 mm girth	-	0.40	3.93	1.15	m²	5.08
Members of roof trusses						
over 300 mm girth	-	0.54	5.31	1.30	m²	6.61
Ornamental railings and the like; each side measured overall						
over 300 mm girth	-	0.60	5.90	1.34	m²	7.24
Eaves gutters						
over 300 mm girth	-	0.64	6.30	1.49	m²	7.79
not exceeding 300 mm girth	-	0.25	2.46	0.63	m	3.09
Pipes or conduits						
over 300 mm girth	-	0.54	5.31	1.49	m²	6.80
not exceeding 300 mm girth	-	0.21	2.07	0.51	m	2.58
One coat primer; one undercoat and one finishing coat of gloss oil paint; on iron or steel surfaces						
General surfaces						
over 300 mm girth	-	0.32	3.15	0.96	m²	4.11
isolated surfaces not exceeding 300 mm girth	-	0.13	1.28	0.25	m	1.53
isolated areas not exceeding 0.50 m²; irrespective of girth	-	0.25	2.46	0.50	nr	2.96
Glazed windows and screens						
panes; area not exceeding 0.10 m²	-	0.55	5.41	0.89	m²	6.30
panes; area 0.10 m² - 0.50 m²	-	0.44	4.33	0.78	m²	5.11
panes; area 0.50 m² - 1.00 m²	-	0.38	3.74	0.66	m²	4.40
panes; area over 1.00 m²	-	0.32	3.15	0.50	m²	3.65
Structural steelwork						
over 300 mm girth	-	0.37	3.64	1.01	m²	4.65
Members of roof trusses						
over 300 mm girth	-	0.48	4.72	1.12	m²	5.84
Ornamental railings and the like; each side measured overall						
over 300 mm girth	-	0.56	5.51	1.12	m²	6.63
Eaves gutters						
over 300 mm girth	-	0.58	5.71	1.36	m²	7.07
not exceeding 300 mm girth	-	0.24	2.36	0.47	m	2.83
Pipes or conduits						
over 300 mm girth	-	0.48	4.72	1.36	m²	6.08
not exceeding 300 mm girth	-	0.19	1.87	0.45	m	2.32
One coat primer; two undercoats and one finishing coat of gloss oil paint; on iron or steel surfaces						
General surfaces						
over 300 mm girth	-	0.43	4.23	1.26	m²	5.49
isolated surfaces not exceeding 300 mm girth	-	0.18	1.77	0.33	m	2.10
isolated areas not exceeding 0.50 m²; irrespective of girth	-	0.32	3.15	0.65	nr	3.80
Glazed windows and screens						
panes; area not exceeding 0.10 m²	-	0.71	6.98	1.15	m²	8.13
panes; area 0.10 m² - 0.50 m²	-	0.56	5.51	1.00	m²	6.51
panes; area 0.50 m² - 1.00 m²	-	0.50	4.92	0.85	m²	5.77
panes; area over 1.00 m²	-	0.43	4.23	0.65	m²	4.88

M SURFACE FINISHES

	PC £	Labour hours	Labour £	Material £	Unit	Total rate £
Structural steelwork						
over 300 mm girth	-	0.48	4.72	1.30	m²	6.02
Members of roof trusses						
over 300 mm girth	-	0.64	6.30	1.45	m²	7.75
Ornamental railings and the like; each side measured overall						
over 300 mm girth	-	0.72	7.08	1.45	m²	8.53
Eaves gutters						
over 300 mm girth	-	0.76	7.48	1.69	m²	9.17
not exceeding 300 mm girth	-	0.31	3.05	0.58	m	3.63
Pipes or conduits						
over 300 mm girth	-	0.64	6.30	1.69	m²	7.99
not exceeding 300 mm girth	-	0.25	2.46	0.56	m	3.02
One coat of Andrews "Hammerite" paint or other equal and approved; on iron or steel surfaces						
General surfaces						
over 300 mm girth	-	0.15	1.48	1.02	m²	2.50
isolated surfaces not exceeding 300 mm girth	-	0.08	0.79	0.31	m	1.10
isolated areas not exceeding 0.50 m²; irrespective of girth	-	0.11	1.08	0.59	nr	1.67
Glazed windows and screens						
panes; area not exceeding 0.10 m²	-	0.25	2.46	0.77	m²	3.23
panes; area 0.10 m² - 0.50 m²	-	0.19	1.87	0.85	m²	2.72
panes; area 0.50 m² - 1.00 m²	-	0.18	1.77	0.76	m²	2.53
panes; area over 1.00 m²	-	0.15	1.48	0.76	m²	2.24
Structural steelwork						
over 300 mm girth	-	0.17	1.67	0.93	m²	2.60
Members of roof trusses						
over 300 mm girth	-	0.23	2.26	1.02	m²	3.28
Ornamental railings and the like; each side measured overall						
over 300 mm girth	-	0.26	2.56	1.02	m²	3.58
Eaves gutters						
over 300 mm girth	-	0.27	2.66	1.10	m²	3.76
not exceeding 300 mm girth	-	0.08	0.79	0.51	m	1.30
Pipes or conduits						
over 300 mm girth	-	0.26	2.56	0.93	m²	3.49
not exceeding 300 mm girth	-	0.08	0.79	0.43	m	1.22
Two coats of creosote; on wood surfaces						
General surfaces						
over 300 mm girth	-	0.16	1.57	0.17	m²	1.74
isolated surfaces not exceeding 300 mm girth	-	0.05	0.49	0.11	m	0.60
Two coats of "Solignum" wood preservative or other equal and approved; on wood surfaces						
General surfaces						
over 300 mm girth	-	0.14	1.38	0.86	m²	2.24
isolated surfaces not exceeding 300 mm girth	-	0.05	0.49	0.26	m	0.75
Three coats of polyurethane; on wood surfaces						
General surfaces						
over 300 mm girth	-	0.29	2.85	1.92	m²	4.77
isolated surfaces not exceeding 300 mm girth	-	0.11	1.08	0.95	m	2.03
isolated areas not exceeding 0.50 m²; irrespective of girth	-	0.21	2.07	1.10	nr	3.17

N FURNITURE/EQUIPMENT

N10/11 GENERAL FIXTURES/KITCHEN FITTINGS

Fixing general fixtures

NOTE: The fixing of general fixtures will vary considerably dependent upon the size of the fixture and the method of fixing employed. Prices for fixing like sized kitchen fittings may be suitable for certain fixtures, although adjustment to those rates will almost invariably be necessary and the reader is directed to section "G20" for information on bolts, plugging brickwork and blockwork, etc. which should prove useful in building up a suitable rate.

The following supply only prices are for purpose made fittings components in various materials supplied as part of an assembled fitting and therefore may be used to arrive at a guide price for a complete fitting.

	PC £	Labour hours	Labour £	Material £	Unit	Total rate £
Fitting components; blockboard						
Backs, fronts, sides or divisions; over 300 mm wide						
12 mm thick	-	-	-	35.41	m²	35.41
19 mm thick	-	-	-	45.70	m²	45.70
25 mm thick	-	-	-	60.03	m²	60.03
Shelves or worktops; over 300 mm wide						
19 mm thick	-	-	-	45.70	m²	45.70
25 mm thick	-	-	-	60.03	m²	60.03
Flush doors; lipped on four edges						
450 mm x 750 mm x 19 mm	-	-	-	24.81	nr	24.81
450 mm x 750 mm x 25 mm	-	-	-	30.40	nr	30.40
600 mm x 900 mm x 19 mm	-	-	-	36.89	nr	36.89
600 mm x 900 mm x 25 mm	-	-	-	45.04	nr	45.04
Fitting components; chipboard						
Backs, fronts, sides or divisions; over 300 mm wide						
6 mm thick	-	-	-	10.49	m²	10.49
9 mm thick	-	-	-	14.91	m²	14.91
12 mm thick	-	-	-	18.39	m²	18.39
19 mm thick	-	-	-	26.52	m²	26.52
25 mm thick	-	-	-	35.80	m²	35.80
Shelves or worktops; over 300 mm wide						
19 mm thick	-	-	-	26.52	m²	26.52
25 mm thick	-	-	-	35.80	m²	35.80
Flush doors; lipped on four edges						
450 mm x 750 mm x 19 mm	-	-	-	18.94	nr	18.94
450 mm x 750 mm x 25 mm	-	-	-	23.03	nr	23.03
600 mm x 900 mm x 19 mm	-	-	-	27.77	nr	27.77
600 mm x 900 mm x 25 mm	-	-	-	33.54	nr	33.54
Fitting components; Melamine faced chipboard						
Backs, fronts, sides or divisions; over 300 mm wide						
12 mm thick	-	-	-	25.26	m²	25.26
19 mm thick	-	-	-	34.25	m²	34.25
Shelves or worktops; over 300 mm wide						
19 mm thick	-	-	-	34.25	m²	34.25

	PC £	Labour hours	Labour £	Material £	Unit	Total rate £
Flush doors; lipped on four edges						
450 mm x 750 mm x 19 mm	-	-	-	23.52	nr	23.52
600 mm x 900 mm x 25 mm	-	-	-	34.58	nr	34.58
Fitting components; "Warerite Xcel" standard colour laminated chipboard type LD2 or other equal and approved						
Backs, fronts, sides or divisions; over 300 mm wide						
13.20 mm thick	-	-	-	67.78	m²	67.78
Shelves or worktops; over 300 mm wide						
13.20 mm thick	-	-	-	67.78	m²	67.78
Flush doors; lipped on four edges						
450 mm x 750 mm x 13.20 mm	-	-	-	33.28	m²	33.28
600 mm x 900 mm x 13.20 mm	-	-	-	50.03	m²	50.03
Fitting components; plywood						
Backs, fronts, sides or divisions; over 300 mm wide						
6 mm thick	-	-	-	17.37	m²	17.37
9 mm thick	-	-	-	23.61	m²	23.61
12 mm thick	-	-	-	29.75	m²	29.75
19 mm thick	-	-	-	43.38	m²	43.38
25 mm thick	-	-	-	58.53	m²	58.53
Shelves or worktops; over 300 mm wide						
19 mm thick	-	-	-	43.38	m²	43.38
25 mm thick	-	-	-	58.53	m²	58.53
Flush doors; lipped on four edges						
450 mm x 750 mm x 19 mm	-	-	-	23.73	nr	23.73
450 mm x 750 mm x 25 mm	-	-	-	29.51	nr	29.51
600 mm x 900 mm x 19 mm	-	-	-	35.35	nr	35.35
600 mm x 900 mm x 25 mm	-	-	-	43.77	nr	43.77
Fitting components; wrought softwood						
Backs, fronts, sides or divisions; cross-tongued joints; over 300 mm wide						
25 mm thick	-	-	-	41.73	m²	41.73
Shelves or worktops; cross-tongued joints; over 300 mm wide						
25 mm thick	-	-	-	41.73	m²	41.73
Bearers						
19 mm x 38 mm	-	-	-	2.37	m	2.37
25 mm x 50 mm	-	-	-	2.97	m	2.97
50 mm x 50 mm	-	-	-	4.33	m	4.33
50 mm x 75 mm	-	-	-	5.79	m	5.79
Bearers; framed						
19 mm x 38 mm	-	-	-	4.63	m	4.63
25 mm x 50 mm	-	-	-	5.24	m	5.24
50 mm x 50 mm	-	-	-	6.60	m	6.60
50 mm x 75 mm	-	-	-	8.06	m	8.06
Framing to backs, fronts or sides						
19 mm x 38 mm	-	-	-	4.63	m	4.63
25 mm x 50 mm	-	-	-	5.24	m	5.24
50 mm x 50 mm	-	-	-	6.60	m	6.60
50 mm x 75 mm	-	-	-	8.06	m	8.06
Flush doors, softwood skeleton or cellular core; plywood facing both sides; lipped on four edges						
450 mm x 750 mm x 35 mm	-	-	-	23.43	nr	23.43
600 mm x 900 mm x 35 mm	-	-	-	37.32	nr	37.32
Add 5% to the above material prices for selected softwood staining						

Prices for Measured Works - Major Works
N FURNITURE/EQUIPMENT

	PC £	Labour hours	Labour £	Material £	Unit	Total rate £
N10/11 GENERAL FIXTURES/KITCHEN FITTINGS - cont'd						
Fitting components; selected West African Mahogany						
Bearers						
19 mm x 38 mm	-	-	-	3.96	m	3.96
25 mm x 50 mm	-	-	-	5.21	m	5.21
50 mm x 50 mm	-	-	-	8.02	m	8.02
50 mm x 75 mm	-	-	-	10.98	m	10.98
Bearers; framed						
19 mm x 38 mm	-	-	-	7.37	m	7.37
25 mm x 50 mm	-	-	-	8.63	m	8.63
50 mm x 50 mm	-	-	-	11.44	m	11.44
50 mm x 75 mm	-	-	-	14.36	m	14.36
Framing to backs, fronts or sides						
19 mm x 38 mm	-	-	-	7.37	m	7.37
25 mm x 50 mm	-	-	-	8.63	m	8.63
50 mm x 50 mm	-	-	-	11.44	m	11.44
50 mm x 75 mm	-	-	-	14.36	m	14.36
Fitting components; Iroko						
Backs, fronts, sides or divisions; cross-tongued joints; over 300 mm wide						
25 mm thick	-	-	-	93.75	m²	93.75
Shelves or worktops; cross-tongued joints; over 300 mm wide						
25 mm thick	-	-	-	93.75	m²	93.75
Draining boards; cross-tongued joints; over 300 mm wide						
25 mm thick	-	-	-	99.07	m²	99.07
stopped flutes	-	-	-	2.99	m	2.99
grooves; cross-grain	-	-	-	0.70	m	0.70
Bearers						
19 mm x 38 mm	-	-	-	4.82	m	4.82
25 mm x 50 mm	-	-	-	6.24	m	6.24
50 mm x 50 mm	-	-	-	9.50	m	9.50
50 mm x 75 mm	-	-	-	12.94	m	12.94
Bearers; framed						
19 mm x 38 mm	-	-	-	9.37	m	9.37
25 mm x 50 mm	-	-	-	10.79	m	10.79
50 mm x 50 mm	-	-	-	14.05	m	14.05
50 mm x 75 mm	-	-	-	17.49	m	17.49
Framing to backs, fronts or sides						
19 mm x 38 mm	-	-	-	9.37	m	9.37
25 mm x 50 mm	-	-	-	10.79	m	10.79
50 mm x 50 mm	-	-	-	14.05	m	14.05
50 mm x 75 mm	-	-	-	17.49	m	17.49
Fitting components; Teak						
Backs, fronts, sides or divisions; cross-tongued joints; over 300 mm wide						
25 mm thick	-	-	-	190.25	m²	190.25
Shelves or worktops; cross-tongued joints; over 300 mm wide						
25 mm thick	-	-	-	190.25	m²	190.25
Draining boards; cross-tongued joints; over 300 mm wide						
25 mm thick	-	-	-	194.25	m²	194.25
stopped flutes	-	-	-	2.99	m	2.99
grooves; cross-grain	-	-	-	0.70	m	0.70

N FURNITURE/EQUIPMENT

	PC £	Labour hours	Labour £	Material £	Unit	Total rate £
Fixing kitchen fittings						
NOTE: Kitchen fittings vary considerably. PC supply prices for reasonable quantities for a moderately priced range of kitchen fittings (Rugby Joinery "Lambeth" range) have been shown but not extended.						
Fixing to backgrounds requiring plugging; including any pre-assembly						
Wall units						
300 mm x 580 mm x 300 mm	33.05	1.11	10.92	0.19	nr	11.11
300 mm x 720 mm x 300 mm	30.95	1.16	11.41	0.19	nr	11.60
600 mm x 580 mm x 300 mm	34.91	1.30	12.79	0.19	nr	12.98
600 mm x 720 mm x 300 mm	38.79	1.48	14.56	0.19	nr	14.75
Floor units with drawers						
600 mm x 900 mm x 500 mm	56.47	1.16	11.41	0.19	nr	11.60
600 mm x 900 mm x 600 mm	58.88	1.30	12.79	0.19	nr	12.98
1200 mm x 900 mm x 600 mm	105.29	1.57	15.44	0.19	nr	15.63
Sink units (excluding sink top)						
1000 mm x 900 mm x 600 mm	87.74	1.48	14.56	0.19	nr	14.75
1200 mm x 900 mm x 600 mm	96.84	1.67	16.43	0.19	nr	16.62
Laminated plastics worktops; single rolled edge; prices include for fixing						
28 mm thick; 600 mm wide	-	0.37	3.64	12.52	m	16.16
38 mm thick; 600 mm wide	-	0.37	3.64	19.86	m	23.50
extra for forming hole for inset sink	-	0.69	6.79	-	nr	6.79
extra for jointing strip at corner intersection of worktops	-	0.14	1.38	8.33	nr	9.71
extra for butt and scribe joint at corner intersection of worktops	-	4.16	40.92	-	nr	40.92
Lockers; The Welconstruct Company or other equal and approved						
Standard clothes lockers; steel body and door within reinforced 19G frame, powder coated finish, cam locks						
1 compartment; placing in position						
305 mm x 305 mm x 1830 mm	-	0.23	1.80	51.33	nr	53.13
305 mm x 460 mm x 1830 mm	-	0.23	1.80	62.32	nr	64.12
460 mm x 460 mm x 1830 mm	-	0.28	2.19	78.40	nr	80.59
610 mm x 460 mm x 1830 mm	-	0.28	2.19	94.47	nr	96.66
Compartment lockers; steel body and door within reinforced 19G frame, powder coated finish, cam locks						
2 compartments; placing in position						
305 mm x 305 mm x 1830 mm	-	0.23	1.80	69.21	nr	71.01
305 mm x 460 mm x 1830 mm	-	0.23	1.80	72.51	nr	74.31
460 mm x 460 mm x 1830 mm	-	0.28	2.19	86.18	nr	88.37
4 compartments; placing in position						
305 mm x 305 mm x 1830 mm	-	0.23	1.80	87.69	nr	89.49
305 mm x 460 mm x 1830 mm	-	0.23	1.80	93.68	nr	95.48
460 mm x 460 mm x 1830 mm	-	0.28	2.19	98.27	nr	100.46
Wet area lockers; galvanised steel 18/22G etched primed coating body, galvanised steel 18/20G reinforced door on non-ferrous hinges, powder coated finish, cam locks						
1 compartment; placing in position						
305 mm x 305 mm x 1830 mm	-	0.23	1.80	72.96	nr	74.76

N FURNITURE/EQUIPMENT

Prices for Measured Works - Major Works

	PC £	Labour hours	Labour £	Material £	Unit	Total rate £
N10/11 GENERAL FIXTURES/KITCHEN FITTINGS - cont'd						
Lockers; The Welconstruct Company or other equal and approved - cont'd						
Wet area lockers; galvanised steel 18/22G etched primed coating body, galvanised steel 18/20G reinforced door on non-ferrous hinges, powder coated finish, cam locks - cont'd						
2 compartments; placing in position						
305 mm x 305 mm x 1830 mm	-	0.23	1.80	94.55	nr	96.35
305 mm x 460 mm x 1830 mm	-	0.23	1.80	99.17	nr	100.97
4 compartments; placing in position						
305 mm x 305 mm x 1830 mm	-	0.23	1.80	119.70	nr	121.50
305 mm x 460 mm x 1830 mm	-	0.23	1.80	128.08	nr	129.88
Extra for						
coin operated lock; coin returned	-	-	-	59.02	nr	59.02
coin operated lock; coin retained	-	-	-	82.79	nr	82.79
Timber clothes lockers; veneered MDF finish, routed door, cam locks						
1 compartment; placing in position						
380 mm x 380 mm x 1830 mm	-	0.28	2.19	224.20	nr	226.39
4 compartments; placing in position						
380 mm x 380 mm x 1830 mm	-	0.28	2.19	336.55	nr	338.74
Shelving support systems; The Welconstruct Company or other equal and approved						
Rolled front shelving support systems; steel body; stove enamelled finish; assembling						
open initial bay; 5 shelves; placing in position						
915 mm x 305 mm x 1905 mm	-	0.69	6.10	73.34	nr	79.44
915 mm x 460 mm x 1905 mm	-	0.69	6.10	95.60	nr	101.70
open extension bay; 5 shelves; placing in position						
915 mm x 305 mm x 1905 mm	-	0.83	7.33	59.46	nr	66.79
915 mm x 460 mm x 1905 mm	-	0.83	7.33	76.71	nr	84.04
closed initial bay; 5 shelves; placing in position						
915 mm x 305 mm x 1905 mm	-	0.69	6.10	91.94	nr	98.04
915 mm x 460 mm x 1905 mm	-	0.69	6.10	114.11	nr	120.21
closed extension bay; 5 shelves; placing in position						
915 mm x 305 mm x 1905 mm	-	0.83	7.33	82.30	nr	89.63
915 mm x 460 mm x 1905 mm	-	0.83	7.33	99.36	nr	106.69
6 mm thick rectangular glass mirrors; silver backed; fixed with chromium plated domed headed screws; to background requiring plugging						
Mirror with polished edges						
365 mm x 254 mm	4.56	0.74	7.28	5.82	nr	13.10
400 mm x 300 mm	5.80	0.74	7.28	7.36	nr	14.64
560 mm x 380 mm	9.95	0.83	8.16	12.49	nr	20.65
640 mm x 460 mm	12.64	0.93	9.15	15.81	nr	24.96
Mirror with bevelled edges						
365 mm x 254 mm	7.88	0.74	7.28	9.93	nr	17.21
400 mm x 300 mm	9.12	0.74	7.28	11.46	nr	18.74
560 mm x 380 mm	14.92	0.83	8.16	18.63	nr	26.79
640 mm x 460 mm	18.65	0.93	9.15	23.24	nr	32.39

N FURNITURE/EQUIPMENT

	PC £	Labour hours	Labour £	Material £	Unit	Total rate £
Door mats						
Entrance mats; "Tuftiguard type C" or other equal and approved; laying in position; 12 mm thick						
900 mm x 550 mm	169.04	0.46	3.60	90.86	nr	94.46
1200 mm x 750 mm	169.04	0.46	3.60	163.55	nr	167.15
2400 mm x 1200 mm	169.04	0.93	7.29	523.35	nr	530.64
Matwells						
Polished aluminium matwell; comprising 34 mm x 26 mm x 6 mm angle rim; with brazed angles and lugs brazed on; to suit mat size						
914 mm x 560 mm	31.22	0.93	7.29	33.56	nr	40.85
1067 mm x 610 mm	34.30	0.93	7.29	36.87	nr	44.16
1219 mm x 762 mm	40.54	0.93	7.29	43.58	nr	50.87
Polished brass matwell; comprising 38 mm x 38 mm x 6 mm angle rim; with brazed angles and lugs welded on; to suit mat size						
914 mm x 560 mm	92.45	0.93	7.29	99.38	nr	106.67
1067 mm x 610 mm	101.55	0.93	7.29	109.17	nr	116.46
1219 mm x 762 mm	120.04	0.93	7.29	129.04	nr	136.33
Internal blinds; Luxaflex Ltd or other equal and approved						
Roller blinds; Luxaflex "Safeweave RB"; fire resisting material; 1200 mm drop; fixing with screws						
1000 mm wide	35.30	0.93	7.29	37.95	nr	45.24
2000 mm wide	68.75	1.45	11.36	73.91	nr	85.27
3000 mm wide	104.05	1.97	15.44	111.85	nr	127.29
Roller blinds; Luxaflex "Plain RB"; plain type material; 1200 mm drop; fixing with screws						
1000 mm wide	22.30	0.93	7.29	23.97	nr	31.26
2000 mm wide	42.73	1.45	11.36	45.93	nr	57.29
3000 mm wide	63.17	1.97	15.44	67.91	nr	83.35
Roller blinds; Luxaflex "Dimout plain RB"; blackout material; 1200 mm drop; fixing with screws						
1000 mm wide	35.30	0.93	7.29	37.95	nr	45.24
2000 mm wide	68.75	1.45	11.36	73.91	nr	85.27
3000 mm wide	104.05	1.97	15.44	111.85	nr	127.29
Roller blinds; Luxaflex "Lite-master Crank Op"; 100% blackout; 1200 mm drop; fixing with screws						
1000 mm wide	160.72	1.96	15.36	172.77	nr	188.13
2000 mm wide	228.53	2.75	21.55	245.67	nr	267.22
3000 mm wide	287.06	3.53	27.66	308.59	nr	336.25
Vertical louvre blinds; 89 mm wide louvres; Luxaflex "Finessa 3430" Group 0; 1200 mm drop; fixing with screws						
1000 mm wide	50.17	0.82	6.43	53.93	nr	60.36
2000 mm wide	86.40	1.30	10.19	92.88	nr	103.07
3000 mm wide	122.63	1.77	13.87	131.83	nr	145.70
Vertical louvre blinds; 127 mm wide louvres; "Finessa 3430" Group 0; 1200 mm drop; fixing with screws						
1000 mm wide	44.59	0.88	6.90	47.93	nr	54.83
2000 mm wide	74.32	1.35	10.58	79.89	nr	90.47
3000 mm wide	106.83	1.81	14.18	114.85	nr	129.03

Prices for Measured Works - Major Works
N FURNITURE/EQUIPMENT

	PC £	Labour hours	Labour £	Material £	Unit	Total rate £
N13 SANITARY APPLIANCES/FITTINGS						
Sinks; Armitage Shanks or equal and approved						
Sinks; white glazed fireclay; BS 1206; pointing all round with Dow Corning Hansil silicone sealant ref 785						
Belfast sink; 455 mm x 380 mm x 205 mm ref 350016S; Nimbus ½" inclined bib taps ref 6610400; ½" wall mounts ref 81460PR; 1½" slotted waste, chain and plug, screw stay ref 70668M8; aluminium alloy build-in fixing brackets ref 7931WD0	-	2.78	33.98	117.33	nr	**151.31**
Belfast sink; 610 mm x 455 mm x 255 mm ref 350086S; Nimbus ½" inclined bib taps ref 6610400; ½" wall mounts ref 81460PR; 1½" slotted waste, chain and plug, screw stay ref 70668M8; aluminium alloy build-in fixing brackets ref 7931VE0	-	2.78	33.98	147.84	nr	**181.82**
Belfast sink; 760 mm x 455 mm x 255 mm ref 3500A6S; Nimbus ½" inclined bib taps ref 6610400; ½" wall mounts ref 81460PR; 1½" slotted waste, chain and plug, screw stay ref 70668M8; aluminium alloy build-in fixing brackets ref 7931VE0	-	2.78	33.98	215.69	nr	**249.67**
Lavatory basins; Armitage Shanks or equal and approved						
Basins; white vitreous china; BS 5506 Part 3; pointing all round with Dow Corning Hansil silcone sealant ref 785						
Portman basin 400 mm x 365 mm ref 117913J; Nuastyle 2 ½" pillar taps with anti-vandal indices ref 6973400; 1½" bead chain waste and plug, 80 mm slotted tail, bolt stay ref 90547N1; 1½" plastics bottle trap with 75 mm seal ref 70237Q4; concealed fixing bracket ref 790002Z; Isovalve servicing valve ref 9060400; screwing	-	2.13	26.03	91.61	nr	**117.64**
Portman basin 500 mm x 420 mm ref 117923S; Nuastyle 2 ½" pillar taps with anti-vandal indices ref 6973400; 1½" bead chain waste and plug, 80 mm slotted tail, bolt stay ref 90547N1; 1½" plastics bottle trap with 75 mm seal ref 70237Q4; concealed fixing bracket ref 790002Z; Isovalve servicing valve ref 9060400; screwing	-	2.13	26.03	111.29	nr	**137.32**
Tiffany basin 560 mm x 455 mm ref 121614A; Tiffany pedestal ref 132408S; Millenia ½" monobloc mixer tap with non-return valves and 1½" pop-up waste ref 6104XXX; Universal Porcelain handwheels ref 63614XXX; Isovalve servicing valve ref 9060400; screwing	-	2.13	26.03	124.24	nr	**150.27**
Montana basin 510 mm x 410 mm ref 120814E; Montana pedestal ref 132408S; Millenia ½" monobloc mixer tap with non-return valves and 1½" pop-up waste ref 6104XXX; Millenia metal handwheels ref 63664XX; Isovalve servicing valve ref 9060400; screwing	-	2.31	28.23	129.27	nr	**157.50**

Prices for Measured Works - Major Works 335
N FURNITURE/EQUIPMENT

	PC £	Labour hours	Labour £	Material £	Unit	Total rate £
Basins; white vitreous china; BS 5506 Part 3; pointing all round with Dow Corning Hansil silcone sealant ref 785 - cont'd						
Montana basin 580 mm x 475 mm ref 120824E; Montana pedestal ref 132408S; Millenia ½" monobloc mixer tap with non-return valves and 1½" pop-up waste ref 6104XXX; Millenia metal handwheels ref 63664XX; Isovalve servicing valve ref 9060400; screwing	-	2.31	28.23	132.43	nr	160.66
Portman basin 600 mm x 480 mm ref 117933S; Nuastyle 2 ½" pillar taps with anti-vandal indices ref 6973400; 1½" bead chain waste and plug, 80 mm slotted tail, bolt stay ref 90547N1; 1½" plastics bottle trap with 75 mm seal ref 70237Q4; concealed fixing bracket ref 790002Z; Isovalve servicing valve ref 9060400; screwing	-	2.13	26.03	145.53	nr	171.56
Cottage basin 560 mm x 430 mm ref 121013S; Cottage pedestal ref 132008S; Cottage ½" pillar taps ref 9290400; 1½" bead chain waste and plug, 80 mm slotted tail, bolt stay ref 90547N1' Isovalve servicing valve ref 9060400; screwing	-	2.31	28.23	206.11	nr	234.34
Cottage basin 625 mm x 500 mm ref 121023S; Cottage pedestal ref 132008S; Cottage ½" pillar taps ref 9290400; 1½" bead chain waste and plug, 80 mm slotted tail, bolt stay ref 90547N1' Isovalve servicing valve ref 9060400; screwing	-	2.31	28.23	243.74	nr	271.97
Cliveden basin 620 mm x 525 mm ref 121223S; Cliveden pedestal ref 132308S; Cliveden ½" mixer tap with 1½" pop-up waste ref 9456400; Isovalve servicing vale ref 9060400; screwing	-	2.31	28.23	532.18	nr	560.41
Drinking fountains; Armitage Shanks or equal and approved						
White vitreous china fountains; pointing all round with Dow Corning Hansil silicone selant ref 785						
Aqualon drinking fountain ref 200100; ½" self closing non-conclussive valve with flow control, plastics strainer waste, concealed hanger ref 0275A00; 1½" plastics bottle trap with 75 mm seal; screwing	-	2.31	28.23	173.04	nr	201.27
Stainless steel fountains; pointing all round with Dow Corning Hansil silicone selant ref 785						
Purita drinking fountain, self closing non-conclussive valve with push button operation, flow control and 1½" strainer waste ref 53400Z5; screwing	-	2.31	28.23	154.99	nr	183.22
Purita drinking fountain, self closing non-conclussive valve with push button operation, flow control and 1½" strainer waste ref 53400Z5; pedestal shroud ref 5341000; screwing	-	2.78	33.98	343.59	nr	377.57

N13 SANITARY APPLIANCES/FITTINGS - cont'd

	PC £	Labour hours	Labour £	Material £	Unit	Total rate £
Baths; Armitage Shanks or equal and approved						
Bath; reinforced acrylic rectangular pattern; chromium plated overflow chain and plug; 40 mm diameter chromium plated waste; cast brass "P" trap with plain outlet and overflow connection; pair 20 mm diameter chromium plated easy clean pillar taps to BS 1010						
1700 mm long; white	171.87	3.50	42.78	184.76	nr	227.54
1700 mm long; coloured	190.44	3.50	42.78	204.72	nr	247.50
Bath; enamelled steel; medium gauge rectangular pattern; 40 mm diameter chromium plated overflow chain and plug; 40 mm diameter chromium plated waste; cast brass "P" trap with plain outlet and overflow connection; pair 20 mm diameter chromium plated easy clean pillar taps to BS 1010						
1700 mm long; white	-	3.50	42.78	284.62	nr	327.40
1700 mm long; coloured	-	3.50	42.78	299.60	nr	342.38
Water closets; Armitage Shanks or equal and approved						
White vitreous china pans and cisterns; pointing all round base with Dow Corning Hansil silicone sealant ref 785						
Seville close coupled washdown pan ref 147001A; Seville plastics seat and cover ref 68780B1; Panekta WC pan P trap connector ref 9013000; Seville 6 litres capacity cistern and cover, bottom supply ball valve, bottom overflow and close coupling fitment ref 17156FR; Seville modern lever ref 7959STR	-	3.05	37.28	156.71	nr	193.99
Extra over for; Panekta WC S trap connector ref 9014000	-	-	-	1.46	nr	1.46
Wentworth close coupled washdown pan ref 150601A; Orion III plastics seat and cover; Panekta WC pan P trap connector ref 9013000; Group 7½ litres capacity cistern and cover, bottom supply ballvalve, bottom overflow, close coupling fitment and lever ref 17650FB	-	3.05	37.28	184.66	nr	221.94
Tiffany back to wall washdown pan ref 154601A; Saturn plastics seat and cover ref 68980B1; Panekta WC pan P trap connector ref 9013000; Conceala 7½ litres capacity cistern and cover, side supply ball valve, side overflow, flushbend and extended lever ref 42350JE	-	3.05	37.28	200.28	nr	237.56
Extra over for; Panekta WC pan S trap connector ref 9014000	-	-	-	1.46	nr	1.46
Tiffany close coupled washdown pan ref 154301A; Saturn plastics seat and cover ref 68980B1; Panekta WC pan P trap connector ref 9013000; Tiffany 7l litres capacity cistern and cover, bottom supply ball valve, bottom overflow, close coupling fitment and side lever ref 17751FN	-	3.05	37.28	213.50	nr	250.78
Extra over for Panekta WC pan S trap connector ref 9014000	-	-	-	1.46	nr	1.46

Prices for Measured Works - Major Works
N FURNITURE/EQUIPMENT

	PC £	Labour hours	Labour £	Material £	Unit	Total rate £
White vitreous china pans and cisterns; pointing all round base with Dow Corning Hansil silicone sealant ref 785 - cont'd						
Cameo close coupled washdown pan ref 154301A; Cameo plastics seat and cover ref 6879NB2; Panekta WC pan P trap connector ref 9013000; Cameo 7½ litres capactity cistern and cover, bottom supply ball valve, bottom overflow and close coupling fitment ref 17831KR; luxury metal lever ref 7968000	-	3.05	37.28	275.34	nr	312.62
Extra over for; Panekta WC pan S trap connector ref 9014000	-	-	-	1.46	nr	1.46
Cottage close coupled washdown pan ref 152301A; mahogany seat and cover ref 68970B2; Panekta WC pan P trap connector ref 9013000; 7½ lites capacity cistern and cover, bottom supply ball valve, bottom overflow and close coupling fitment ref 17700FR; mahogany lever assembly ref S03SN23	-	3.05	37.28	431.13	nr	468.41
Extra over for; Panekta WC pan S trap connector ref 9014000	-	-	-	1.46	nr	1.46
Cliveden close coupled washdown pan ref 153201A; mahogany seat and cover ref 68970B2; Panekta WC pan P trap connector ref 9013000; 7½ litres capacity cistern and cover, bottom supply ball valve, bottom overflow and close coupling fitment ref 17720FR; brass level assembly ref S03SN01	-	3.05	37.28	479.09	nr	516.37
Extra over for; Panekta WC pan S trap connector ref 9014000	-	-	-	1.46	nr	1.46
Concept back to wall washdown pan ref 153301A; Concept plastics seat and cover ref 6896AB2; Panekta WC pan P trap connector ref 9013000; Conceala 7½ litres capacity cistern and cover, side supply ball valve, side overflow, flushbend and extended lever ref 42350JE	-	3.05	37.28	566.60	nr	603.88
Wall urinals; Armitage Shanks or equal and approved						
White vitreous china bowls and cisterns; pointing all round with Dow Corning Hansil silicone sealant ref 785						
single Sanura 400 mm bowls ref 261119E; top inlet spreader ref 74344A1; concealed steel hangers ref 7220000; 1½" plastics domed strainer waste ref 90568N0; 1½" plastics bottle trap with 75 mm seal ref 70238Q4; Conceala 4½ litres capacity cistern and cover ref 4225100; polished stainless steel exposed flushpipes ref 74450A1PO; screwing	-	3.70	45.22	208.21	nr	253.43
single Sanura 500 mm bowls ref 261129E; top inlet spreader ref 74344A1; concealed steel hangers ref 7220000; 1½" plastics domed strainer waste ref 90568N0; 1½" plastics bottle trap with 75 mm seal ref 70238Q4; Conceala 4½ litres capacity cistern and cover ref 4225100; polished stainless steel exposed flushpipes ref 74450A1PO; screwing	-	3.70	45.22	269.77	nr	314.99

Prices for Measured Works - Major Works
N FURNITURE/EQUIPMENT

	PC £	Labour hours	Labour £	Material £	Unit	Total rate £

N13 SANITARY APPLIANCES/FITTINGS - cont'd

Wall urinals; Armitage Shanks or equal and approved - cont'd
White vitreous china bowls and cisterns; pointing all round with Dow Corning Hansil silicone sealant ref 785 - cont'd

	PC £	Labour hours	Labour £	Material £	Unit	Total rate £
range of 2 nr Sanura 400 mm bowls ref 261119E; top inlet spreaders ref 74344A1; concealed steel hangers ref 7220000; 1½" plastics domed strainer wastes ref 90568N0; 1½" plastics bottle traps with 75 mm seal ref 70238Q4; Conceala 9 litres capacity cistern and cover ref 4225200; polished stainless steel exposed flushpipes ref 74450B1PO; screwing	-	6.94	84.83	338.03	nr	422.86
range of 2 nr Sanura 500 mm bowls ref 261129E; top inlet spreaders ref 74344A1; concealed steel hangers ref 7220000; 1½" plastics domed strainer wastes ref 90568N0; 1½" plastics bottle traps with 75 mm seal ref 70238Q4; Conceala 9 litres capacity cistern and cover ref 4225200; polished stainless steel exposed flushpipes ref 74450B1PO; screwing	-	6.94	84.83	461.15	nr	545.98
range of 3 nr Sanura 400 mm bowls ref 261119E; top inlet spreaders ref 74344A1; concealed steel hangersref 7220000; 1½" plastics domed strainer wastes ref 90568N0; 1½" plastics bottle traps with 75 mm seal ref 70238Q4; Conceala 9 litres capacity cistern and cover ref 4225200; polished stainless steel flushpipes ref 74450C1PO; screwing	-	10.18	124.43	456.35	nr	580.78
range of 3 nr Sanura 500 mm bowls ref 261129E; top inlet spreaders ref 74344A1; concealed steel hangers ref 7220000; 1½" plastics domed strainer wastes ref 90568N0; 1½" plastics bottle traps with 75 mm seal ref 70238Q4; Conceala 9 litres capacity cistern and cover ref 4225200; polished stainless steel flushpipes ref 74450C1PO; screwing	-	10.18	124.43	640.82	nr	765.25
range of 4 nr Sanura 400 mm bowls ref 261119E; top inlet spreaders ref 74344A1; concealed steel hangers ref 7220000; 1½" plastics domed strainer wastes ref 90568N0; 1½" plastics bottle traps with 75 mm seal ref 70238Q4; Conceala 13.60 litres capacity cistern and cover ref 4225300; polished stainless steel flushpipes ref 74450D1PO; screwing	-	13.41	163.91	592.52	nr	756.43
range of 4 nr Sanura 500 mm bowls ref 261129E; top inlet spreaders ref 74344A1; concealed steel hangers ref 7220000; 1½" plastics domed strainer wastes ref 90568N0; 1½" plastics bottle traps with 75 mm seal ref 70238Q4; Conceala 13.60 litres capacity cistern and cover ref 4225300; polished stainless steel flushpipes ref 74450D1PO; screwing	-	13.41	163.91	838.75	nr	1002.66
White vitreous china division panels; pointing all round with Dow Corning Hansill silicone sealant ref 785						
625 mm long ref 2605000; screwing	-	0.69	8.43	53.47	nr	61.90

N FURNITURE/EQUIPMENT

	PC £	Labour hours	Labour £	Material £	Unit	Total rate £
Bidets; Armitage Shanks or equal and approved						
Bidet; vitreous china; chromium plated waste; mixer tap with hand wheels						
600 mm x 400 mm x 400 mm; white	-	3.50	42.78	184.76	nr	227.54
Shower trays and fittings; Armitage Shanks or equal and approved						
Shower tray; glazed fireclay with outlet and grated waste; chain and plug; bedding and pointing in waterproof cement mortar						
760 mm x 760 mm x 180 mm; white	-	3.00	36.67	114.85	nr	151.52
760 mm x 760 mm x 180 mm; coloured	-	3.00	36.67	174.76	nr	211.43
Shower fitting; riser pipe with mixing valve and shower rose; chromium plated; plugging and screwing mixing valve and pipe bracket						
15 mm diameter riser pipe; 127 mm diameter shower rose	-	5.00	61.11	224.71	nr	285.82
Miscellaneous fittings; Pressalit or equal and approved						
Raised seats						
Ergosit; 50 mm high; ref R19000	-	0.50	4.92	76.50	nr	81.42
Dania; 50 mm high; ref R23000	-	0.50	4.92	84.50	nr	89.42
Dania; 100 mm high; ref R24000	-	0.50	4.92	129.50	nr	134.42
Dania; sloping 100 mm - 50 mm high; ref R25000	-	0.50	4.92	123.00	nr	127.92
Raised seats and covers						
Ergosit; 50 mm high; ref R20000	-	0.50	4.92	104.50	nr	109.42
Dania; 50 mm high; ref R33000	-	0.50	4.92	91.50	nr	96.42
Dania; 100 mm high; ref R34000	-	0.50	4.92	151.00	nr	155.92
Dania; sloping 100 mm - 50 mm high; ref R35000	-	0.50	4.92	145.50	nr	150.42
Miscellaneous fittings; Pressalit Ltd or equal and approved						
Grab rails						
300 mm long ref RT100000; screwing	-	0.50	4.92	37.43	nr	42.35
450 mm long ref RT101000; screwing	-	0.50	4.92	43.77	nr	48.69
600 mm long ref RT102000; screwing	-	0.50	4.92	50.26	nr	55.18
800 mm long ref RT103000; screwing	-	0.50	4.92	56.25	nr	61.17
1000 mm long ref RT104000; screwing	-	0.50	4.92	64.74	nr	69.66
Angled grab rails						
900 mm long, angled 135 ref RT110000; screwing	-	0.50	4.92	81.22	nr	86.14
1300 mm long, angled 90 ref RT119000; screwing	-	0.75	7.38	127.32	nr	134.70
Hinged grab rails						
600 mm long ref R3016000 ; screwing	-	0.35	3.44	131.71	nr	135.15
600 mm long with spring counter balance ref RF016000 ; screwing	-	0.35	3.44	182.64	nr	186.08
800 mm long ref R3010000 ; screwing	-	0.35	3.44	157.68	nr	161.12
800 mm long with spring counter balance ref RF010000 ; screwing	-	0.35	3.44	194.12	nr	197.56
N15 SIGNS/NOTICES						
Plain script; in gloss oil paint; on painted or varnished surfaces						
Capital letters; lower case letters or numerals						
per coat; per 25 mm high	-	0.09	0.89	-	nr	0.89
Stops						
per coat	-	0.02	0.20	-	nr	0.20

P BUILDING FABRIC SUNDRIES

P10 SUNDRY INSULATION/PROOFING WORK/FIRESTOPS

ALTERNATIVE INSULATION PRICES

Insulation (£/m²)	£		£		£		£
EXPANDED POLYSTYRENE							
"Crown Wallmate"							
25 mm	4.25	30 mm	4.74	40 mm	5.12	50 mm	7.78
"Crown Wool"							
60 mm	2.02	80 mm	2.64	100 mm	3.16		
"Crown Dritherm" - part fill							
30 mm	2.17	50 mm	3.43	75 mm	4.56		
"Crown Sound Deadening Quilt"							
13 mm	2.34	25 mm	4.49				
"Crown Factoryclad"							
80 mm	3.02	90 mm	3.29	100 mm	3.56		

P BUILDING FABRIC SUNDRIES	PC £	Labour hours	Labour £	Material £	Unit	Total rate £
"Sisalkraft" building papers/vapour barriers or other equal and approved						
Building paper; 150 mm laps; fixed to softwood						
"Moistop" grade 728 (class A1F)	-	0.09	1.26	0.86	m²	2.12
Vapour barrier/reflective insulation 150 laps; fixed to softwood						
"Insulex" grade 714; single sided	-	0.09	1.26	1.06	m²	2.32
"Insulex" grade 714; double sided	-	0.09	1.26	1.53	m²	2.79
Mat or quilt insulation						
Glass fibre quilt; "Isowool 1000" or other equal and approved; laid loose between members at 600 mm centres						
60 mm thick	1.90	0.19	2.66	2.04	m²	4.70
80 mm thick	2.49	0.21	2.94	2.68	m²	5.62
100 mm thick	2.98	0.23	3.22	3.20	m²	6.42
150 mm thick	4.54	0.28	3.92	4.88	m²	8.80
Mineral fibre quilt; "Isowool 1200" or other equal and approved; pinned vertically to softwood						
25 mm thick	1.61	0.13	1.82	1.73	m²	3.55
50 mm thick	2.58	0.14	1.96	2.77	m²	4.73
Glass fibre building roll; pinned vertically to softwood						
60 mm thick	1.47	0.14	1.96	1.66	m²	3.62
80 mm thick	1.93	0.15	2.10	2.18	m²	4.28
100 mm thick	2.30	0.16	2.24	2.60	m²	4.84
Glass fibre flanged building roll; paper faces; pinned vertically or to slope between timber framing						
60 mm thick	3.24	0.17	2.38	3.66	m²	6.04
80 mm thick	4.05	0.18	2.52	4.57	m²	7.09
100 mm thick	4.72	0.19	2.66	5.33	m²	7.99
Glass fibre aluminium foil faced roll; pinned to softwood						
80 mm thick	-	0.17	2.38	3.86	m²	6.24

P BUILDING FABRIC SUNDRIES

	PC £	Labour hours	Labour £	Material £	Unit	Total rate £
Board or slab insulation						
Jablite expanded polystyrene board standard grade RD/N or other equal and approved; fixed with adhesive						
25 mm thick	-	0.39	5.46	2.33	m²	7.79
40 mm thick	-	0.42	5.88	3.25	m²	9.13
50 mm thick	-	0.46	6.44	3.88	m²	10.32
Jablite expanded polystyrene board; grade EHD(N) or other equal and approved						
100 mm thick	-	0.46	6.44	1.23	m²	7.67
"Styrofoam Floormate 350" extruded polystyrene foam or other equal and approved						
50 mm thick	-	0.46	6.44	11.60	m²	18.04
Fire stops						
Cape "Firecheck" channel; intumescent coatings on cut mitres; fixing with brass cups and screws						
19 mm x 44 mm or 19 mm x 50 mm	7.57	0.56	7.84	8.88	m	16.72
"Sealmaster" intumescent fire and smoke seals or other equal and approved; pinned into groove in timber						
type N30; for single leaf half hour door	5.01	0.28	3.92	5.65	m	9.57
type N60; for single leaf one hour door	7.63	0.31	4.34	8.61	m	12.95
type IMN or IMP; for meeting or pivot stiles of pair of one hour doors; per stile	7.63	0.31	4.34	8.61	m	12.95
intumescent plugs in timber; including boring	-	0.09	1.26	0.26	nr	1.52
Rockwool fire stops or other equal and approved; between top of brick/block wall and concrete soffit						
25 mm deep x 112 mm wide	-	0.07	0.98	0.77	m	1.75
25 mm deep x 150 mm wide	-	0.09	1.26	1.03	m	2.29
50 mm deep x 225 mm wide	-	0.14	1.96	3.63	m	5.59
Fire protection compound						
Quelfire QF4, fire protection compound or other equal and approved; filling around pipes, ducts and the like; including all necessary formwork						
300 mm x 300 mm x 250 mm; pipes - 2	-	0.93	11.61	10.15	nr	21.76
500 mm x 500 mm x 250 mm; pipes - 2	-	1.16	13.54	30.45	nr	43.99
Fire barriers						
Rockwool fire barrier or other equal and approved; between top of suspended ceiling and concrete soffit						
one 50 mm layer x 900 mm wide; half hour	-	0.56	7.84	5.25	m²	13.09
two 50 mm layers x 900 mm wide; one hour	-	0.83	11.62	10.29	m²	21.91
Lamatherm fire barrier or other equal and approved; to void below raised access floors						
75 mm thick x 300 mm high; half hour	-	0.17	2.30	13.40	m	15.78
75 mm thick x 600 mm high; half hour	-	0.17	2.38	30.12	m	32.50
90 mm thick x 300 mm high; half hour	-	0.17	2.38	16.86	m	19.24
90 mm thick x 600 mm high; half hour	-	0.17	2.38	40.18	m	42.56
Dow Chemicals "Styrofoam 1B" or other equal and approved; cold bridging insulation fixed with adhesive to brick, block or concrete base						
Insulation to walls						
25 mm thick	-	0.31	4.34	5.51	m²	9.85
50 mm thick	-	0.33	4.62	10.13	m²	14.75
75 mm thick	-	0.35	4.90	15.12	m²	20.02
Insulation to isolated columns						
25 mm thick	-	0.38	5.32	5.51	m²	10.83
50 mm thick	-	0.41	5.74	10.13	m²	15.87
75 mm thick	-	0.43	6.02	15.12	m²	21.14

Prices for Measured Works - Major Works
P BUILDING FABRIC SUNDRIES

	PC £	Labour hours	Labour £	Material £	Unit	Total rate £
P10 SUNDRY INSULATION/PROOFING WORK/FIRESTOPS - cont'd						
Dow Chemicals "Styrofoam 1B" or other equal and approved; cold bridging insulation fixed with adhesive to brick, block or concrete base - cont'd						
Insulation to ceilings						
25 mm thick	-	0.33	4.62	5.51	m^2	10.13
50 mm thick	-	0.36	5.04	10.13	m^2	15.17
75 mm thick	-	0.39	5.46	15.12	m^2	20.58
Insulation to isolated beams						
25 mm thick	-	0.41	5.74	5.51	m^2	11.25
50 mm thick	-	0.43	6.02	10.13	m^2	16.15
75 mm thick	-	0.46	6.44	15.12	m^2	21.56
P11 FOAMED/FIBRE/BEAD CAVITY WALL INSULATION						
Injected insulation						
Cavity wall insulation; injecting 65 mm cavity with						
blown EPS granules	-	-	-	-	m^2	2.30
blown mineral wool	-	-	-	-	m^2	2.40
P20 UNFRAMED ISOLATED TRIMS/ SKIRTINGS/SUNDRY ITEMS						
Blockboard (Birch faced); 18 mm thick						
Window boards and the like; rebated; hardwood lipped on one edge						
18 mm x 200 mm	-	0.23	3.22	3.69	m	6.91
18 mm x 250 mm	-	0.26	3.64	4.34	m	7.98
18 mm x 300 mm	-	0.29	4.06	4.98	m	9.04
18 mm x 350 mm	-	0.31	4.34	5.63	m	9.97
returned and fitted ends	-	0.20	2.80	0.37	nr	3.17
Blockboard (Sapele veneered one side); 18 mm thick						
Window boards and the like; rebated; hardwood lipped on one edge						
18 mm x 200 mm	-	0.25	3.50	3.29	m	6.79
18 mm x 250 mm	-	0.28	3.92	3.84	m	7.76
18 mm x 300 mm	-	0.31	4.34	4.38	m	8.72
18 mm x 350 mm	-	0.33	4.62	4.93	m	9.55
returned and fitted ends	-	0.20	2.80	0.37	nr	3.17
Blockboard (American White Ash veneered one side); 18 mm thick						
Window boards and the like; rebated; hardwood lipped on one edge						
18 mm x 200 mm	-	0.25	3.50	3.77	m	7.27
18 mm x 250 mm	-	0.28	3.92	4.43	m	8.35
18 mm x 300 mm	-	0.31	4.34	5.09	m	9.43
18 mm x 350 mm	-	0.33	4.62	5.75	m	10.37
returned and fitted ends	-	0.20	2.80	0.37	nr	3.17
Medium density fibreboard primed profiles; Balcas Kildare Ltd or other equal and approved						
Window boards; rounded and rebated						
25 mm x 220 mm	-	0.28	3.92	3.74	m	7.66
25 mm x 245 mm	-	0.28	3.92	4.05	m	7.97

P BUILDING FABRIC SUNDRIES

Prices for Measured Works - Major Works

	PC £	Labour hours	Labour £	Material £	Unit	Total rate £
Skirtings						
14.50 mm x 45 mm; rounded	-	0.09	1.26	0.71	m	1.97
14.50 mm x 70 mm; rounded	-	0.09	1.26	0.96	m	2.22
14.50 mm x 95 mm; rounded	-	0.09	1.26	1.15	m	2.41
14.50 mm x 95 mm; moulded	-	0.09	1.26	1.15	m	2.41
14.50 mm x 120 mm; moulded	-	0.09	1.26	1.32	m	2.58
18 mm x 70 mm; moulded	-	0.09	1.26	1.06	m	2.32
18 mm x 145 mm; moulded	-	0.09	1.26	1.73	m	2.99
Dado rail						
18 mm x 58 mm; moulded	-	0.09	1.26	1.07	m	2.33
Wrought softwood						
Skirtings, picture rails, dado rails and the like; splayed or moulded						
19 mm x 50 mm; splayed	-	0.09	1.26	1.41	m	2.67
19 mm x 50 mm; moulded	-	0.09	1.26	1.53	m	2.79
19 mm x 75 mm; splayed	-	0.09	1.26	1.65	m	2.91
19 mm x 75 mm; moulded	-	0.09	1.26	1.78	m	3.04
19 mm x 100 mm; splayed	-	0.09	1.26	1.92	m	3.18
19 mm x 100 mm; moulded	-	0.09	1.26	2.04	m	3.30
19 mm x 150 mm; moulded	-	0.11	1.54	2.60	m	4.14
19 mm x 175 mm; moulded	-	0.11	1.54	2.85	m	4.39
22 mm x 100 mm; splayed	-	0.09	1.26	2.11	m	3.37
25 mm x 50 mm; moulded	-	0.09	1.26	1.70	m	2.96
25 mm x 75 mm; splayed	-	0.09	1.26	1.92	m	3.18
25 mm x 100 mm; splayed	-	0.09	1.26	2.26	m	3.52
25 mm x 150 mm; splayed	-	0.11	1.54	2.96	m	4.50
25 mm x 150 mm; moulded	-	0.11	1.54	3.08	m	4.62
25 mm x 175 mm; moulded	-	0.11	1.54	3.44	m	4.98
25 mm x 225 mm; moulded	-	0.13	1.82	4.15	m	5.97
returned ends	-	0.14	1.96	-	nr	1.96
mitres	-	0.09	1.26	-	nr	1.26
Architraves, cover fillets and the like; half round; splayed or moulded						
13 mm x 25 mm; half round	-	0.11	1.54	1.08	m	2.62
13 mm x 50 mm; moulded	-	0.11	1.54	1.38	m	2.92
16 mm x 32 mm; half round	-	0.11	1.54	1.17	m	2.71
16 mm x 38 mm; moulded	-	0.11	1.54	1.34	m	2.88
16 mm x 50 mm; moulded	-	0.11	1.54	1.45	m	2.99
19 mm x 50 mm; splayed	-	0.11	1.54	1.41	m	2.95
19 mm x 63 mm; splayed	-	0.11	1.54	1.54	m	3.08
19 mm x 75 mm; splayed	-	0.11	1.54	1.65	m	3.19
25 mm x 44 mm; splayed	-	0.11	1.54	1.59	m	3.13
25 mm x 50 mm; moulded	-	0.11	1.54	1.70	m	3.24
25 mm x 63 mm; splayed	-	0.11	1.54	1.74	m	3.28
25 mm x 75 mm; splayed	-	0.11	1.54	1.92	m	3.46
32 mm x 88 mm; moulded	-	0.11	1.54	2.66	m	4.20
38 mm x 38 mm; moulded	-	0.11	1.54	1.69	m	3.23
50 mm x 50 mm; moulded	-	0.11	1.54	2.26	m	3.80
returned ends	-	0.14	1.96	-	nr	1.96
mitres	-	0.09	1.26	-	nr	1.26
Stops; screwed on						
16 mm x 38 mm	-	0.09	1.26	1.06	m	2.32
16 mm x 50 mm	-	0.09	1.26	1.15	m	2.41
19 mm x 38 mm	-	0.09	1.26	1.11	m	2.37
25 mm x 38 mm	-	0.09	1.26	1.22	m	2.48
25 mm x 50 mm	-	0.09	1.26	1.42	m	2.68

Prices for Measured Works - Major Works
P BUILDING FABRIC SUNDRIES

	PC £	Labour hours	Labour £	Material £	Unit	Total rate £
P20 UNFRAMED ISOLATED TRIMS/ SKIRTINGS/SUNDRY ITEMS - cont'd						
Wrought softwood - cont'd						
Glazing beads and the like						
13 mm x 16 mm	-	0.04	0.56	0.78	m	1.34
13 mm x 19 mm	-	0.04	0.56	0.80	m	1.36
13 mm x 25 mm	-	0.04	0.56	0.86	m	1.42
13 mm x 25 mm; screwed	-	0.04	0.56	0.91	m	1.47
13 mm x 25 mm; fixing with brass cups and screws	-	0.09	1.26	1.15	m	2.41
16 mm x 25 mm; screwed	-	0.04	0.56	0.96	m	1.52
16 mm quadrant	-	0.04	0.56	0.94	m	1.50
19 mm quadrant or scotia	-	0.04	0.56	1.02	m	1.58
19 mm x 36 mm; screwed	-	0.04	0.56	1.13	m	1.69
25 mm x 38 mm; screwed	-	0.04	0.56	1.24	m	1.80
25 mm quadrant or scotia	-	0.04	0.56	1.13	m	1.69
38 mm scotia	-	0.04	0.56	1.60	m	2.16
50 mm scotia	-	0.04	0.56	2.18	m	2.74
Isolated shelves, worktops, seats and the like						
19 mm x 150 mm	-	0.15	2.10	2.45	m	4.55
19 mm x 200 mm	-	0.20	2.80	2.93	m	5.73
25 mm x 150 mm	-	0.15	2.10	2.93	m	5.03
25 mm x 200 mm	-	0.20	2.80	3.60	m	6.40
32 mm x 150 mm	-	0.15	2.10	3.49	m	5.59
32 mm x 200 mm	-	0.20	2.80	4.38	m	7.18
Isolated shelves, worktops, seats and the like; cross-tongued joints						
19 mm x 300 mm	-	0.26	3.64	11.43	m	15.07
19 mm x 450 mm	-	0.31	4.34	13.08	m	17.42
19 mm x 600 mm	-	0.37	5.18	19.02	m	24.20
25 mm x 300 mm	-	0.26	3.64	12.68	m	16.32
25 mm x 450 mm	-	0.31	4.34	14.73	m	19.07
25 mm x 600 mm	-	0.37	5.18	21.28	m	26.46
32 mm x 300 mm	-	0.26	3.64	13.89	m	17.53
32 mm x 450 mm	-	0.31	4.34	16.56	m	20.90
32 mm x 600 mm	-	0.37	5.18	23.76	m	28.94
Isolated shelves, worktops, seats and the like; slatted with 50 wide slats at 75 mm centres						
19 mm thick	-	1.23	17.22	8.86	m	26.08
25 mm thick	-	1.23	17.22	10.15	m	27.37
32 mm thick	-	1.23	17.22	11.28	m	28.50
Window boards, nosings, bed moulds and the like; rebated and rounded						
19 mm x 75 mm	-	0.17	2.38	1.93	m	4.31
19 mm x 150 mm	-	0.19	2.66	2.78	m	5.44
19 mm x 225 mm; in one width	-	0.24	3.36	3.56	m	6.92
19 mm x 300 mm; cross-tongued joints	-	0.28	3.92	11.80	m	15.72
25 mm x 75 mm	-	0.17	2.38	2.19	m	4.57
25 mm x 150 mm	-	0.19	2.66	3.29	m	5.95
25 mm x 225 mm; in one width	-	0.24	3.36	4.35	m	7.71
25 mm x 300 mm; cross-tongued joints	-	0.28	3.92	13.01	m	16.93
32 mm x 75 mm	-	0.17	2.38	2.50	m	4.88
32 mm x 150 mm	-	0.19	2.66	3.87	m	6.53
32 mm x 225 mm; in one width	-	0.24	3.36	5.21	m	8.57
32 mm x 300 mm; cross-tongued joints	-	0.28	3.92	14.27	m	18.19
38 mm x 75 mm	-	0.17	2.38	2.77	m	5.15
38 mm x 150 mm	-	0.19	2.66	4.37	m	7.03
38 mm x 225 mm; in one width	-	0.24	3.36	5.99	m	9.35
38 mm x 300 mm; cross-tongued joints	-	0.28	3.92	15.40	m	19.32
returned and fitted ends	-	0.14	1.96	-	nr	1.96

P BUILDING FABRIC SUNDRIES

	PC £	Labour hours	Labour £	Material £	Unit	Total rate £
Handrails; mopstick						
50 mm diameter	-	0.23	3.22	3.32	m	6.54
Handrails; rounded						
44 mm x 50 mm	-	0.23	3.22	3.32	m	6.54
50 mm x 75 mm	-	0.25	3.50	4.07	m	7.57
63 mm x 87 mm	-	0.28	3.92	5.50	m	9.42
75 mm x 100 mm	-	0.32	4.48	6.17	m	10.65
Handrails; moulded						
44 mm x 50 mm	-	0.23	3.22	3.70	m	6.92
50 mm x 75 mm	-	0.25	3.50	4.46	m	7.96
63 mm x 87 mm	-	0.28	3.92	5.88	m	9.80
75 mm x 100 mm	-	0.32	4.48	6.54	m	11.02
Add 5% to the above material prices for selected softwood for staining						
Selected West African Mahogany						
Skirtings, picture rails, dado rails and the like; splayed or moulded						
19 mm x 50 mm; splayed	2.99	0.13	1.82	3.46	m	5.28
19 mm x 50 mm; moulded	3.16	0.13	1.82	3.65	m	5.47
19 mm x 75 mm; splayed	3.60	0.13	1.82	4.15	m	5.97
19 mm x 75 mm; moulded	3.76	0.13	1.82	4.33	m	6.15
19 mm x 100 mm; splayed	4.21	0.13	1.82	4.84	m	6.66
19 mm x 100 mm; moulded	4.39	0.13	1.82	5.04	m	6.86
19 mm x 150 mm; moulded	5.62	0.15	2.10	6.43	m	8.53
19 mm x 175 mm; moulded	6.45	0.15	2.10	7.37	m	9.47
22 mm x 100 mm; splayed	4.58	0.13	1.82	5.25	m	7.07
25 mm x 50 mm; moulded	3.55	0.13	1.82	4.09	m	5.91
25 mm x 75 mm; splayed	4.19	0.13	1.82	4.81	m	6.63
25 mm x 100 mm; splayed	4.98	0.13	1.82	5.71	m	7.53
25 mm x 150 mm; splayed	6.86	0.15	2.10	7.83	m	9.93
25 mm x 150 mm; moulded	7.02	0.15	2.10	8.01	m	10.11
25 mm x 175 mm; moulded	7.82	0.15	2.10	8.91	m	11.01
25 mm x 225 mm; moulded	9.42	0.17	2.38	10.72	m	13.10
returned ends	-	0.20	2.80	-	nr	2.80
mitres	-	0.14	1.96	-	nr	1.96
Architraves, cover fillets and the like; half round; splayed or moulded						
13 mm x 25 mm; half round	2.20	0.15	2.10	2.57	m	4.67
13 mm x 50 mm; moulded	2.76	0.15	2.10	3.20	m	5.30
16 mm x 32 mm; half round	2.43	0.15	2.10	2.83	m	4.93
16 mm x 38 mm; moulded	2.70	0.15	2.10	3.13	m	5.23
16 mm x 50 mm; moulded	2.95	0.15	2.10	3.41	m	5.51
19 mm x 50 mm; splayed	2.99	0.15	2.10	3.46	m	5.56
19 mm x 63 mm; splayed	3.31	0.15	2.10	3.82	m	5.92
19 mm x 75 mm; splayed	3.60	0.15	2.10	4.15	m	6.25
25 mm x 44 mm; splayed	3.38	0.15	2.10	3.90	m	6.00
25 mm x 50 mm; moulded	3.55	0.15	2.10	4.09	m	6.19
25 mm x 63 mm; splayed	3.80	0.15	2.10	4.37	m	6.47
25 mm x 75 mm; splayed	4.19	0.15	2.10	4.81	m	6.91
32 mm x 88 mm; moulded	5.90	0.15	2.10	6.74	m	8.84
38 mm x 38 mm; moulded	3.62	0.15	2.10	4.17	m	6.27
50 mm x 50 mm; moulded	4.98	0.15	2.10	5.71	m	7.81
returned ends	-	0.20	2.80	-	nr	2.80
mitres	-	0.14	1.96	-	nr	1.96
Stops; screwed on						
16 mm x 38 mm	2.37	0.14	1.96	2.68	m	4.64
16 mm x 50 mm	2.64	0.14	1.96	2.98	m	4.94
19 mm x 38 mm	2.55	0.14	1.96	2.88	m	4.84
25 mm x 38 mm	2.84	0.14	1.96	3.21	m	5.17
25 mm x 50 mm	3.23	0.14	1.96	3.65	m	5.61

P BUILDING FABRIC SUNDRIES

	PC £	Labour hours	Labour £	Material £	Unit	Total rate £
P20 UNFRAMED ISOLATED TRIMS/ SKIRTINGS/SUNDRY ITEMS - cont'd						
Selected West African Mahogany - cont'd						
Glazing beads and the like						
13 mm x 16 mm	1.86	0.04	0.56	2.10	m	2.66
13 mm x 19 mm	1.91	0.04	0.56	2.16	m	2.72
13 mm x 25 mm	2.03	0.04	0.56	2.29	m	2.85
13 mm x 25 mm; screwed	1.86	0.07	0.98	2.25	m	3.23
13 mm x 25 mm; fixing with brass cups and screws	2.03	0.14	1.96	2.52	m	4.48
16 mm x 25 mm; screwed	2.12	0.07	0.98	2.54	m	3.52
16 mm quadrant	2.11	0.06	0.84	2.38	m	3.22
19 mm quadrant or scotia	2.23	0.06	0.84	2.52	m	3.36
19 mm x 36 mm; screwed	2.55	0.06	0.84	3.03	m	3.87
25 mm x 38 mm; screwed	2.84	0.06	0.84	3.36	m	4.20
25 mm quadrant or scotia	2.57	0.06	0.84	2.90	m	3.74
38 mm scotia	3.62	0.06	0.84	4.09	m	4.93
50 mm scotia	4.98	0.06	0.84	5.62	m	6.46
Isolated shelves; worktops, seats and the like						
19 mm x 150 mm	5.80	0.20	2.80	6.55	m	9.35
19 mm x 200 mm	6.98	0.28	3.92	7.88	m	11.80
25 mm x 150 mm	6.98	0.20	2.80	7.88	m	10.68
25 mm x 200 mm	8.51	0.28	3.92	9.61	m	13.53
32 mm x 150 mm	8.27	0.20	2.80	9.33	m	12.13
32 mm x 200 mm	10.32	0.28	3.92	11.65	m	15.57
Isolated shelves, worktops, seats and the like; cross-tongued joints						
19 mm x 300 mm	19.46	0.35	4.90	21.97	m	26.87
19 mm x 450 mm	23.20	0.42	5.88	26.19	m	32.07
19 mm x 600 mm	33.64	0.51	7.14	37.97	m	45.11
25 mm x 300 mm	22.18	0.35	4.90	25.04	m	29.94
25 mm x 450 mm	27.09	0.42	5.88	30.58	m	36.46
25 mm x 600 mm	38.82	0.51	7.14	43.82	m	50.96
32 mm x 300 mm	24.88	0.35	4.90	28.08	m	32.98
32 mm x 450 mm	31.06	0.42	5.88	35.06	m	40.94
32 mm x 600 mm	44.35	0.51	7.14	50.06	m	57.20
Isolated shelves, worktops, seats and the like; slatted with 50 wide slats at 75 mm centres						
19 mm thick	21.76	1.62	22.67	25.46	m²	48.13
25 mm thick	24.55	1.62	22.67	28.69	m²	51.36
32 mm thick	27.13	1.62	22.67	31.67	m²	54.34
Window boards, nosings, bed moulds and the like; rebated and rounded						
19 mm x 75 mm	4.00	0.22	3.08	4.71	m	7.79
19 mm x 150 mm	5.86	0.25	3.50	6.81	m	10.31
19 mm x 225 mm; in one width	7.69	0.33	4.62	8.87	m	13.49
19 mm x 300 mm; cross-tongued joints	19.36	0.37	5.18	22.04	m	27.22
25 mm x 75 mm	4.60	0.22	3.08	5.38	m	8.46
25 mm x 150 mm	7.08	0.25	3.50	8.18	m	11.68
25 mm x 225 mm; in one width	9.42	0.33	4.62	10.82	m	15.44
25 mm x 300 mm; cross-tongued joints	21.96	0.37	5.18	24.98	m	30.16
32 mm x 75 mm	5.28	0.22	3.08	6.15	m	9.23
32 mm x 150 mm	8.37	0.25	3.50	9.64	m	13.14
32 mm x 225 mm; in one width	11.47	0.33	4.62	13.14	m	17.76
32 mm x 300 mm; cross-tongued joints	24.67	0.37	5.18	28.04	m	33.22
38 mm x 75 mm	5.86	0.22	3.08	6.81	m	9.89
38 mm x 150 mm	9.54	0.25	3.50	10.96	m	14.46
38 mm x 225 mm ; in one width	13.22	0.33	4.62	15.11	m	19.73
38 mm x 300 mm; cross-tongued joints	27.12	0.37	5.18	30.80	m	35.98
returned and fitted ends	-	0.21	2.94	-	nr	2.94

P BUILDING FABRIC SUNDRIES

	PC £	Labour hours	Labour £	Material £	Unit	Total rate £
Handrails; rounded						
44 mm x 50 mm	6.68	0.31	4.34	7.54	m	11.88
50 mm x 75 mm	8.35	0.33	4.62	9.43	m	14.05
63 mm x 87 mm	10.80	0.37	5.18	12.19	m	17.37
75 mm x 100 mm	13.13	0.42	5.88	14.82	m	20.70
Handrails; moulded						
44 mm x 50 mm	7.16	0.31	4.34	8.08	m	12.42
50 mm x 75 mm	8.84	0.33	4.62	9.98	m	14.60
63 mm x 87 mm	11.30	0.37	5.18	12.75	m	17.93
75 mm x 100 mm	13.62	0.42	5.88	15.37	m	21.25
American White Ash						
Skirtings, picture rails, dado rails and the like; splayed or moulded						
19 mm x 50 mm; splayed	3.36	0.13	1.82	3.88	m	5.70
19 mm x 50 mm; moulded	3.52	0.13	1.82	4.06	m	5.88
19 mm x 75 mm; splayed	4.19	0.13	1.82	4.81	m	6.63
19 mm x 75 mm; moulded	4.34	0.13	1.82	4.98	m	6.80
19 mm x 100 mm; splayed	4.98	0.13	1.82	5.71	m	7.53
19 mm x 100 mm; moulded	5.15	0.13	1.82	5.90	m	7.72
19 mm x 150 mm; moulded	6.75	0.15	2.10	7.70	m	9.80
19 mm x 175 mm; moulded	7.80	0.15	2.10	8.89	m	10.99
22 mm x 100 mm; splayed	5.49	0.13	1.82	6.28	m	8.10
25 mm x 50 mm; moulded	4.06	0.13	1.82	4.67	m	6.49
25 mm x 75 mm; splayed	4.94	0.13	1.82	5.66	m	7.48
25 mm x 100 mm; splayed	6.00	0.13	1.82	6.86	m	8.68
25 mm x 150 mm; splayed	8.37	0.15	2.10	9.53	m	11.63
25 mm x 150 mm; moulded	8.53	0.15	2.10	9.71	m	11.81
25 mm x 175 mm; moulded	9.57	0.15	2.10	10.89	m	12.99
25 mm x 225 mm; moulded	11.70	0.17	2.38	13.29	m	15.67
returned end	-	0.20	2.80	-	nr	2.80
mitres	-	0.14	1.96	-	nr	1.96
Architraves, cover fillets and the like; half round; splayed or moulded						
13 mm x 25 mm; half round	2.33	0.15	2.10	2.71	m	4.81
13 mm x 50 mm; moulded	3.01	0.15	2.10	3.48	m	5.58
16 mm x 32 mm; half round	2.61	0.15	2.10	3.03	m	5.13
16 mm x 38 mm; moulded	2.97	0.15	2.10	3.44	m	5.54
16 mm x 50 mm; moulded	3.30	0.15	2.10	3.81	m	5.91
19 mm x 50 mm; splayed	3.36	0.15	2.10	3.88	m	5.98
19 mm x 63 mm; splayed	3.81	0.15	2.10	4.39	m	6.49
19 mm x 75 mm; splayed	4.19	0.15	2.10	4.81	m	6.91
25 mm x 44 mm; splayed	3.89	0.15	2.10	4.48	m	6.58
25 mm x 50 mm; moulded	4.06	0.15	2.10	4.67	m	6.77
25 mm x 63 mm; splayed	4.42	0.15	2.10	5.07	m	7.17
25 mm x 75 mm; splayed	4.94	0.15	2.10	5.66	m	7.76
32 mm x 88 mm; moulded	7.17	0.15	2.10	8.18	m	10.28
38 mm x 38 mm; moulded	4.21	0.15	2.10	4.84	m	6.94
50 mm x 50 mm; moulded	6.00	0.15	2.10	6.86	m	8.96
returned end	-	0.20	2.80	-	nr	2.80
mitres	-	0.14	1.96	-	nr	1.96
Stops; screwed on						
16 mm x 38 mm	2.65	0.14	1.96	2.99	m	4.95
16 mm x 50 mm	2.97	0.14	1.96	3.35	m	5.31
19 mm x 38 mm	2.82	0.14	1.96	3.18	m	5.14
25 mm x 38 mm	3.20	0.14	1.96	3.61	m	5.57
25 mm x 50 mm	3.74	0.14	1.96	4.22	m	6.18

Prices for Measured Works - Major Works
P BUILDING FABRIC SUNDRIES

	PC £	Labour hours	Labour £	Material £	Unit	Total rate £
P20 UNFRAMED ISOLATED TRIMS/ SKIRTINGS/SUNDRY ITEMS - cont'd						
American White Ash - cont'd						
Glazing beads and the like						
13 mm x 16 mm	1.96	0.04	0.56	2.21	m	2.77
13 mm x 19 mm	2.02	0.04	0.56	2.28	m	2.84
13 mm x 25 mm	2.17	0.04	0.56	2.45	m	3.01
13 mm x 25 mm; screwed	2.17	0.07	0.98	2.60	m	3.58
13 mm x 25 mm; fixing with brass cups and screws	2.17	0.14	1.96	2.68	m	4.64
16 mm x 25 mm; screwed	2.28	0.07	0.98	2.72	m	3.70
16 mm quadrant	2.21	0.06	0.84	2.49	m	3.33
19 mm quadrant or scotia	2.38	0.06	0.84	2.69	m	3.53
19 mm x 36 mm; screwed	2.82	0.06	0.84	3.33	m	4.17
25 mm x 38 mm; screwed	3.20	0.06	0.84	3.76	m	4.60
25 mm quadrant or scotia	2.84	0.06	0.84	3.21	m	4.05
38 mm scotia	4.21	0.06	0.84	4.75	m	5.59
50 mm scotia	6.00	0.06	0.84	6.77	m	7.61
Isolated shelves, worktops, seats and the like						
19 mm x 150 mm	6.99	0.20	2.80	7.89	m	10.69
19 mm x 200 mm	8.51	0.28	3.92	9.61	m	13.53
25 mm x 150 mm	8.51	0.20	2.80	9.61	m	12.41
25 mm x 200 mm	10.52	0.28	3.92	11.87	m	15.79
32 mm x 150 mm	10.21	0.20	2.80	11.52	m	14.32
32 mm x 200 mm	12.89	0.28	3.92	14.55	m	18.47
Isolated shelves, worktops, seats and the like; cross-tongued joints						
19 mm x 300 mm	22.00	0.35	4.90	24.83	m	29.73
19 mm x 450 mm	26.91	0.42	5.88	30.37	m	36.25
19 mm x 600 mm	38.71	0.51	7.14	43.69	m	50.83
25 mm x 300 mm	25.55	0.35	4.90	28.84	m	33.74
25 mm x 450 mm	31.98	0.42	5.88	36.10	m	41.98
25 mm x 600 mm	45.50	0.51	7.14	51.36	m	58.50
32 mm x 300 mm	29.10	0.35	4.90	32.85	m	37.75
32 mm x 450 mm	37.23	0.42	5.88	42.02	m	47.90
32 mm x 600 mm	52.77	0.51	7.14	60.98	m	68.12
Isolated shelves, worktops, seats and the like; slatted with 50 wide slats at 75 mm centres						
19 mm thick	24.56	1.62	22.67	28.70	m²	51.37
25 mm thick	28.08	1.62	22.67	32.77	m²	55.44
32 mm thick	31.82	1.62	22.67	37.09	m²	59.76
Window boards, nosings, bed moulds and the like; rebated and rounded						
19 mm x 75 mm	4.57	0.22	3.08	5.35	m	8.43
19 mm x 150 mm	7.00	0.25	3.50	8.09	m	11.59
19 mm x 225 mm; in one width	9.40	0.33	4.62	10.80	m	15.42
19 mm x 300 mm; cross-tongued joints	21.93	0.37	5.18	24.94	m	30.12
25 mm x 75 mm	5.37	0.22	3.08	6.25	m	9.33
25 mm x 150 mm	8.62	0.25	3.50	9.92	m	13.42
25 mm x 225 mm; in one width	11.67	0.33	4.62	13.36	m	17.98
25 mm x 300 mm; cross-tongued joints	25.33	0.37	5.18	28.78	m	33.96
32 mm x 75 mm	6.24	0.22	3.08	7.23	m	10.31
32 mm x 150 mm	10.31	0.25	3.50	11.83	m	15.33
32 mm x 225 mm; in one width	-	0.33	4.62	16.43	m	21.05
32 mm x 300 mm; cross-tongued joints	28.88	0.37	5.18	32.79	m	37.97
38 mm x 75 mm	7.09	0.22	3.08	8.38	m	11.46
38 mm x 150 mm	11.86	0.25	3.50	13.58	m	17.08
38 mm x 225 mm; in one width	16.67	0.33	4.62	19.01	m	23.63
38 mm x 300 mm; cross-tongued joints	32.11	0.37	5.18	36.43	m	41.61
returned and fitted ends	-	0.21	2.94	-	nr	2.94

P BUILDING FABRIC SUNDRIES

	PC £	Labour hours	Labour £	Material £	Unit	Total rate £
Handrails; rounded						
44 mm x 50 mm	7.67	0.31	4.34	8.66	m	13.00
50 mm x 75 mm	9.80	0.33	4.62	11.06	m	15.68
63 mm x 87 mm	14.13	0.37	5.18	15.95	m	21.13
75 mm x 100 mm	16.14	0.42	5.88	18.22	m	24.10
Handrails; moulded						
44 mm x 50 mm	8.17	0.31	4.34	9.22	m	13.56
50 mm x 75 mm	10.30	0.33	4.62	11.63	m	16.25
63 mm x 87 mm	14.62	0.37	5.18	16.50	m	21.68
75 mm x 100 mm	16.62	0.42	5.88	18.76	m	24.64
Pin-boards; medium board						
Sundeala "A" pin-board or other equal and approved; fixed with adhesive to backing (not included); over 300 mm wide						
6.40 mm thick	-	0.56	7.84	11.57	m²	19.41
Sundries on softwood/hardwood						
Extra over fixing with nails for						
gluing and pinning	-	0.01	0.14	0.06	m	0.20
masonry nails	-	-	-	-	m	0.27
steel screws	-	-	-	-	m	0.25
self-tapping screws	-	-	-	-	m	0.26
steel screws; gluing	-	-	-	-	m	0.44
steel screws; sinking; filling heads	-	-	-	-	m	0.56
steel screws; sinking; pellating over	-	-	-	-	m	1.21
brass cups and screws	-	-	-	-	m	1.49
Extra over for						
countersinking	-	-	-	-	m	0.23
pellating	-	-	-	-	m	1.05
Head or nut in softwood						
let in flush	-	-	-	-	nr	0.56
Head or nut; in hardwood						
let in flush	-	-	-	-	nr	0.83
let in over; pellated	-	-	-	-	nr	1.94
Metalwork; mild steel						
Angle section bearers; for building in						
150 mm x 100 mm x 6 mm	-	0.31	4.27	11.69	m	15.96
150 mm x 150 mm x 8 mm	-	0.32	4.40	18.71	m	23.11
200 mm x 200 mm x 12 mm	-	0.37	5.09	35.09	m	40.18
Metalwork; mild steel; galvanized						
Waterbars; groove in timber						
6 mm x 30 mm	-	0.46	6.44	5.01	m	11.45
6 mm x 40 mm	-	0.46	6.44	5.85	m	12.29
6 mm x 50 mm	-	0.46	6.44	7.81	m	14.25
Angle section bearers; for building in						
150 mm x 100 mm x 6 mm	-	0.31	4.27	15.44	m	19.71
150 mm x 150 mm x 8 mm	-	0.32	4.40	24.56	m	28.96
200 mm x 200 mm x 12 mm	-	0.37	5.09	49.12	m	54.21
Dowels; mortice in timber						
8 mm diameter x 100 mm long	-	0.04	0.56	0.09	nr	0.65
10 mm diameter x 50 mm long	-	0.04	0.56	0.24	nr	0.80
Cramps						
25 mm x 3 mm x 230 mm girth; one end bent, holed and screwed; other end fishtailed for building in	-	0.06	0.84	0.56	nr	1.40
Metalwork; stainless steel						
Angle section bearers; for building in						
150 mm x 100 mm x 6 mm	-	0.31	4.27	43.51	m	47.78
150 mm x 150 mm x 8 mm	-	0.32	4.40	70.17	m	74.57

Prices for Measured Works - Major Works
P BUILDING FABRIC SUNDRIES

	PC £	Labour hours	Labour £	Material £	Unit	Total rate £
P21 IRONMONGERY						
NOTE: Ironmongery is largely a matter of selection and prices vary considerably; indicative prices for reasonable quantities of "standard quality" ironmongery are given below.						
Ironmongery; NT Laidlaw Ltd or other equal and approved; standard ranges; to softwood						
Bolts						
barrel; 100 mm x 32 mm; PAA	-	0.31	4.34	5.18	nr	9.52
barrel; 150 mm x 32 mm; PAA	-	0.39	5.46	5.92	nr	11.38
flush; 152 mm x 25 mm; SCP	-	0.56	7.84	12.03	nr	19.87
flush; 203 mm x 25 mm; SCP	-	0.56	7.84	13.25	nr	21.09
flush; 305 mm x 25 mm; SCP	-	0.56	7.84	17.95	nr	25.79
indicating; 76 mm x 41 mm; SAA	-	0.62	8.68	8.79	nr	17.47
indicating; coin operated; SAA	-	0.62	8.68	12.03	nr	20.71
panic; single; SVE	-	2.31	32.33	55.50	nr	87.83
panic; double; SVE	-	3.24	45.35	75.59	nr	120.94
necked tower; 152 mm; BJ	-	0.31	4.34	2.44	nr	6.78
necked tower; 203 mm; BJ	-	0.31	4.34	3.29	nr	7.63
mortice security; SCP	-	0.56	7.84	15.96	nr	23.80
garage door bolt; 305 mm	-	0.56	7.84	5.94	nr	13.78
monkey tail with knob; 305 mm x 19 mm; BJ	-	0.62	8.68	7.27	nr	15.95
monkey tail with bow; 305 mm x 19 mm; BJ	-	0.62	8.68	9.16	nr	17.84
Butts						
63 mm; light steel	-	0.23	3.22	0.60	pr	3.82
100 mm; light steel	-	0.23	3.22	1.04	pr	4.26
102 mm; SC	-	0.23	3.22	3.36	pr	6.58
102 mm double flap extra strong; SC	-	0.23	3.22	2.10	pr	5.32
76 mm x 51 mm; SC	-	0.23	3.22	2.42	pr	5.64
51 mm x 22 mm narrow suite S/D	-	0.23	3.22	2.02	pr	5.24
76 mm x 35 mm narrow suite S/D	-	0.23	3.22	2.83	pr	6.05
51 mm x 29 mm broad suite S/D	-	0.23	3.22	2.02	pr	5.24
76 mm x 41 mm broad suite S/D	-	0.23	3.22	2.73	pr	5.95
102 mm x 60 mm broad suite S/D	-	0.23	3.22	5.72	pr	8.94
102 mm x 69 mm lift-off (R/L hand)	-	0.23	3.22	3.04	pr	6.26
high security heavy butts F/pin; 100 mm x 94 mm; ball bearing; SSS	-	0.37	5.18	17.96	pr	23.14
"Hi-load" butts; 125 mm x 102 mm	-	0.28	3.92	41.17	pr	45.09
S/D rising butts (R/L hand); 102 mm x 67 mm; BRS	-	0.23	3.22	34.40	pr	37.62
ball bearing butts; 102 mm x 67 mm; SSS	-	0.23	3.22	8.05	pr	11.27
Catches						
ball catch; 13 mm diameter; BRS	-	0.31	4.34	0.49	nr	4.83
double ball catch; 50 mm; BRS	-	0.37	5.18	0.65	nr	5.83
57 mm x 38 mm cupboard catch; SCP	-	0.28	3.92	5.60	nr	9.52
14 mm x 35 mm magnetic catch; WHT	-	0.16	2.24	0.56	nr	2.80
adjustable nylon roller catch; WNY	-	0.23	3.22	1.48	nr	4.70
"Bales" catch; Nr 4; 41 mm x 16 mm; self colour brass; mortice	-	0.31	4.34	0.87	nr	5.21
"Bales" catch; Nr 8; 66 mm x 25 mm; self colour brass; mortice	-	0.32	4.48	1.94	nr	6.42
Door closers and furniture						
standard concealed overhead door closer (L/R hand); SIL	-	1.16	16.24	143.63	nr	159.87
light duty surface fixed door closer (L/R hand); SIL	-	0.93	13.02	61.06	nr	74.08
"Perkomatic" concealed door closer; BRS	-	0.62	8.68	70.64	nr	79.32
"Softline" adjustable size 2 - 4 overhead door closer; SIL	-	1.39	19.46	67.70	nr	87.16
"Centurion II" size 3 overhead door closer; SIL	-	1.16	16.24	31.49	nr	47.73

Prices for Measured Works - Major Works
P BUILDING FABRIC SUNDRIES

	PC £	Labour hours	Labour £	Material £	Unit	Total rate £
Door closers and furniture - cont'd						
"Centurion II" size 4 overhead door closer; SIL	-	1.39	19.46	38.85	nr	58.31
backcheck door closer size 3; SAA	-	1.16	16.24	80.65	nr	96.89
backcheck door closer size 4; SAA	-	1.39	19.46	84.31	nr	103.77
door selector; face fixing; SAA	-	0.56	7.84	60.45	nr	68.29
finger plate; 300 mm x 75 mm x 3 mm	-	0.16	2.24	3.48	nr	5.72
kicking plate; 1000 mm x 150 mm x 30 mm; PAA	-	0.23	3.22	10.13	nr	13.35
floor spring; single and double action; ZP	-	2.31	32.33	148.46	nr	180.79
lever furniture; 280 mm x 40 mm	-	0.23	3.22	19.94	pr	23.16
pull handle; 225 mm; back fixing; PAA	-	0.16	2.24	5.33	nr	7.57
pull handle; f/fix with cover rose; PAA	-	0.31	4.34	28.62	nr	32.96
letter plate; 330 mm x 76 mm; aluminium finish	-	1.23	17.22	7.25	nr	24.47
Latches						
102 mm mortice latch; SCP	-	1.16	16.24	6.08	nr	22.32
cylinder rim night latch; SC7	-	0.69	9.66	17.25	nr	26.91
Locks						
drawer or cupboard lock; 63 mm x 32 mm; SC	-	0.39	5.46	7.59	nr	13.05
mortice dead lock; 63 mm x 108 mm; SSS	-	0.69	9.66	9.78	nr	19.44
12 mm rebate conversion set to mortice dead lock	-	0.46	6.44	9.70	nr	16.14
rim lock; 140 mm x 73 mm; GYE	-	0.39	5.46	7.44	nr	12.90
rim dead lock; 92 mm x 74.5 mm; SCP	-	0.39	5.46	19.67	nr	25.13
upright mortice lock; 103 mm x 82 mm; 3 lever	-	0.77	10.78	10.08	nr	20.86
upright mortice lock; 103 mm x 82 mm; 5 lever	-	0.77	10.78	31.50	nr	42.28
Window furniture						
casement stay; 305 mm long; 2 pin; SAA	-	0.16	2.24	6.59	nr	8.83
casement fastener; standard; 113 mm; SAA	-	0.16	2.24	4.27	nr	6.51
cockspur fastener; ASV	-	0.31	4.34	7.21	nr	11.55
sash fastener; 65 mm; SAA	-	0.23	3.22	2.42	nr	5.64
Sundries						
numerals; SAA	-	0.07	0.98	0.68	nr	1.66
rubber door stop; SAA	-	0.07	0.98	1.02	nr	2.00
medium hot pressed cabin hook; 102 mm; CP	-	0.16	2.24	8.32	nr	10.56
medium hot pressed cabin hook; 203 mm; CP	-	0.16	2.24	11.01	nr	13.25
coat hook; SAA	-	0.07	0.98	1.12	nr	2.10
toilet roll holder; CP	-	0.16	2.24	14.59	nr	16.83
Ironmongery; NT Laidlaw Ltd or similar; standard ranges; to hardwood						
Bolts						
barrel; 100 mm x 32 mm; PAA	-	0.41	5.74	5.18	nr	10.92
barrel; 150 mm x 32 mm; PAA	-	0.52	7.28	5.92	nr	13.20
flush; 152 mm x 25 mm; SCP	-	0.74	10.36	12.03	nr	22.39
flush; 203 mm x 25 mm; SCP	-	0.74	10.36	13.25	nr	23.61
flush; 305 mm x 25 mm; SCP	-	0.74	10.36	17.95	nr	28.31
indicating; 76 mm x 41 mm; SAA	-	0.82	11.48	8.79	nr	20.27
indicating; coin operated; SAA	-	0.82	11.48	12.03	nr	23.51
panic; single	-	3.08	43.11	55.50	nr	98.61
panic; double	-	4.32	60.46	75.59	nr	136.05
necked tower; 152 mm; BJ	-	0.41	5.74	2.44	nr	8.18
necked tower; 203 mm; BJ	-	0.41	5.74	3.29	nr	9.03
mortice security; SCP	-	0.74	10.36	15.96	nr	26.32
garage door bolt; 305 mm	-	0.74	10.36	5.94	nr	16.30
monkey tail with knob; 305 mm x 19 mm; BJ	-	0.79	11.06	7.27	nr	18.33
monkey tail with bow; 305 mm x 19 mm; BJ	-	0.79	11.06	9.16	nr	20.22
Butts						
63 mm; light steel	-	0.32	4.48	0.60	pr	5.08
100 mm; light steel	-	0.32	4.48	1.04	pr	5.52
102 mm; SC	-	0.32	4.48	3.36	pr	7.84
102 mm double flap extra strong; SC	-	0.32	4.48	2.10	pr	6.58
76 mm x 51 mm; SC	-	0.32	4.48	2.42	pr	6.90
51 mm x 22 mm narrow suite S/D	-	0.32	4.48	2.02	pr	6.50

P BUILDING FABRIC SUNDRIES

	PC £	Labour hours	Labour £	Material £	Unit	Total rate £
P21 IRONMONGERY - cont'd						
Ironmongery; NT Laidlaw Ltd or similar; standard ranges; to hardwood - cont'd						
Butts - cont'd						
76 mm x 35 mm narrow suite S/D	-	0.32	4.48	2.83	pr	7.31
51 mm x 29 mm broad suite S/D	-	0.32	4.48	2.02	pr	6.50
76 mm x 41 mm broad suite S/D	-	0.32	4.48	2.73	pr	7.21
102 mm x 60 mm broad suite S/D	-	0.32	4.48	5.72	pr	10.20
102 mm x 69 mm lift-off (R/L hand)	-	0.32	4.48	3.04	pr	7.52
high security heavy butts F/pin; 100 mm x 94 mm; ball bearing; SSS	-	0.42	5.88	17.96	pr	23.84
"Hi-load" butts; 125 mm x 102 mm	-	0.37	5.18	41.17	pr	46.35
S/D rising butts (R/L hand); 102 mm x 67 mm; BRS	-	0.32	4.48	34.40	pr	38.88
ball bearing butts; 102 mm x 67 mm; SSS	-	0.32	4.48	8.05	pr	12.53
Catches						
ball catch; 13 mm diameter; BRS	-	0.42	5.88	0.49	nr	6.37
double ball catch; 50 mm; BRS	-	0.43	6.02	0.65	nr	6.67
57 mm x 38 mm cupboard catch; SCP	-	0.37	5.18	5.60	nr	10.78
14 mm x 35 mm magnetic catch; WHT	-	0.21	2.94	0.56	nr	3.50
adjustable nylon roller catch; WNY	-	0.31	4.34	1.48	nr	5.82
"Bales" catch; Nr 4; 41 mm x 16 mm; self colour brass; mortice	-	0.41	5.74	0.87	nr	6.61
"Bales" catch; Nr 8; 66 mm x 25 mm; self colour brass; mortice	-	0.43	6.02	1.94	nr	7.96
Door closers and furniture						
standard concealed overhead door closer (L/R hand); SIL	-	1.54	21.55	143.63	nr	165.18
light duty surface fixed door closer (L/R hand); SIL	-	1.20	16.80	61.06	nr	77.86
"Perkomatic" concealed door closer; BRS	-	0.82	11.48	70.64	nr	82.12
"Softline" adjustable size 2 - 4 overhead door closer; SIL	-	1.85	25.89	67.70	nr	93.59
"Centurion II" size 3 overhead door closer; SIL	-	1.62	22.67	31.49	nr	54.16
"Centurion II" size 4 overhead door closer; SIL	-	1.85	25.89	38.85	nr	64.74
backcheck door closer size 3; SAA	-	1.62	22.67	80.65	nr	103.32
backcheck door closer size 4; SAA	-	1.85	25.89	84.31	nr	110.20
door selector; soffit fixing; SSS	-	0.62	8.68	27.82	nr	36.50
door selector; face fixing; SAA	-	0.74	10.36	60.45	nr	70.81
finger plate; 300 mm x 75 mm x 3 mm	-	0.21	2.94	3.48	nr	6.42
kicking plate; 1000 mm x 150 mm x 30 mm; PAA	-	0.41	5.74	10.13	nr	15.87
floor spring; single and double action; ZP	-	3.08	43.11	148.46	nr	191.57
lever furniture; 280 mm x 40 mm	-	0.31	4.34	19.94	pr	24.28
pull handle; 225 mm; back fixing; PAA	-	0.21	2.94	5.33	nr	8.27
pull handle; f/fix with cover rose; PAA	-	0.41	5.74	28.62	nr	34.36
letter plate; 330 mm x 76 mm; Aluminium finish	-	1.64	22.95	7.25	nr	30.20
Latches						
102 mm mortice latch; SCP	-	1.62	22.67	6.08	nr	28.75
cylinder rim night latch; SC7	-	0.93	13.02	17.25	nr	30.27
Locks						
drawer or cupboard lock; 63 mm x 32 mm; SC	-	0.52	7.28	7.59	nr	14.87
mortice dead lock; 63 mm x 108 mm; SSS	-	0.93	13.02	9.78	nr	22.80
12 mm rebate conversion set to mortice dead lock	-	0.69	9.66	9.70	nr	19.36
rim lock; 140 mm x 73 mm; GYE	-	0.52	7.28	7.44	nr	14.72
rim dead lock; 92 mm x 74.50 mm; SCP	-	0.52	7.28	19.67	nr	26.95
upright mortice lock; 103 mm x 82 mm; 3 lever	-	1.03	14.42	10.08	nr	24.50
upright mortice lock; 103 mm x 82 mm; 5 lever	-	1.03	14.42	31.50	nr	45.92
Window furniture						
casement stay; 305 mm long; 2 pin; SAA	-	0.21	2.94	6.59	nr	9.53
casement fastener; standard; 113 mm; SAA	-	0.21	2.94	4.27	nr	7.21
cockspur fastener; ASV	-	0.41	5.74	7.21	nr	12.95
sash fastener; 65 mm; SAA	-	0.31	4.34	2.42	nr	6.76

P BUILDING FABRIC SUNDRIES

	PC £	Labour hours	Labour £	Material £	Unit	Total rate £
Sundries						
numerals; SAA	-	0.10	1.40	0.68	nr	2.08
rubber door stop; SAA	-	0.10	1.40	1.02	nr	2.42
medium hot pressed cabin hook; 102 mm; CP	-	0.21	2.94	8.32	nr	11.26
medium hot pressed cabin hook; 203 mm; CP	-	0.21	2.94	11.01	nr	13.95
coat hook; SAA	-	0.10	1.40	1.12	nr	2.52
toilet roll holder; CP	-	0.23	3.22	14.59	nr	17.81
Sliding door gear; Hillaldam Coburn Ltd or other equal and approved; Commercial/Light industrial; for top hung timber/metal doors, weight not exceeding 365 kg						
Sliding door gear						
bottom guide; fixed to concrete in groove	12.90	0.46	6.44	13.87	m	20.31
top track	18.08	0.23	3.22	19.44	m	22.66
detachable locking bar and padlock	30.98	0.31	4.34	33.30	nr	37.64
hangers; timber doors	33.44	0.46	6.44	35.95	nr	42.39
hangers; metal doors	23.84	0.46	6.44	25.63	nr	32.07
head brackets; open, side fixing; bolting to masonry	3.57	0.46	6.44	6.01	nr	12.45
head brackets; open, soffit fixing; screwing to timber	2.70	0.32	4.48	2.95	nr	7.43
door guide to timber door	6.71	0.23	3.22	7.21	nr	10.43
door stop; rubber buffers; to masonry	18.37	0.69	9.66	19.75	nr	29.41
drop bolt; screwing to timber	34.41	0.46	6.44	36.99	nr	43.43
bow handle; to timber	7.44	0.23	3.22	8.00	nr	11.22
Sundries						
rubber door stop; plugged and screwed to concrete	4.59	0.09	1.26	4.93	nr	6.19
P30 TRENCHES/PIPEWAYS/PITS FOR BURIED ENGINEERING SERVICES						
Excavating trenches; by machine; grading bottoms; earthwork support; filling with excavated material and compacting; disposal of surplus soil on site; spreading on site average 50 m						
Services not exceeding 200 mm nominal size						
average depth of run not exceeding 0.50 m	-	0.28	2.24	1.20	m	3.44
average depth of run not exceeding 0.75 m	-	0.37	2.96	1.99	m	4.95
average depth of run not exceeding 1.00 m	-	0.79	6.32	3.83	m	10.15
average depth of run not exceeding 1.25 m	-	1.16	9.28	5.22	m	14.50
average depth of run not exceeding 1.50 m	-	1.48	11.84	6.82	m	18.66
average depth of run not exceeding 1.75 m	-	1.85	14.80	8.72	m	23.52
average depth of run not exceeding 2.00 m	-	2.13	17.04	9.97	m	27.01
Excavating trenches; by hand; grading bottoms; earthwork support; filling with excavated material and compacting; disposal; of surplus soil on site; spreading on site average 50 m						
Services not exceeding 200 mm nominal size						
average depth of run not exceeding 0.50 m	-	0.93	7.44	-	m	7.44
average depth of run not exceeding 0.75 m	-	1.39	11.12	-	m	11.12
average depth of run not exceeding 1.00 m	-	2.04	16.32	1.33	m	17.65
average depth of run not exceeding 1.25 m	-	2.87	22.95	1.83	m	24.78
average depth of run not exceeding 1.50 m	-	3.93	31.43	2.23	m	33.66
average depth of run not exceeding 1.75 m	-	5.18	41.43	2.70	m	44.13
average depth of run not exceeding 2.00 m	-	5.92	47.35	2.97	m	50.32

Prices for Measured Works - Major Works
P BUILDING FABRIC SUNDRIES

	PC £	Labour hours	Labour £	Material £	Unit	Total rate £
P30 TRENCHES/PIPEWAYS/PITS FOR BURIED ENGINEERING SERVICES - cont'd						
Stop cock pits, valve chambers and the like; excavating; half brick thick walls in common bricks in cement mortar (1:3); on in situ concrete designated mix C20 - 20 mm aggregate bed; 100 mm thick						
Pits						
100 mm x 100 mm x 750 mm deep; internal holes for one small pipe; polypropylene hinged box cover; bedding in cement mortar (1:3)	-	3.89	68.00	24.58	nr	92.58
P31 HOLES/CHASES/COVERS/SUPPORTS FOR SERVICES						
Builders' work for electrical installations; cutting away for and making good after electrician; including cutting or leaving all holes, notches, mortices, sinkings and chases, in both the structure and its coverings, for the following electrical points						
Exposed installation						
lighting points	-	0.28	2.79	-	nr	2.79
socket outlet points	-	0.46	4.79	-	nr	4.79
fitting outlet points	-	0.46	4.79	-	nr	4.79
equipment points or control gear points	-	0.65	6.87	-	nr	6.87
Concealed installation						
lighting points	-	0.37	3.79	-	nr	3.79
socket outlet points	-	0.65	6.87	-	nr	6.87
fitting outlet points	-	0.65	6.87	-	nr	6.87
equipment points or control gear points	-	0.93	9.66	-	nr	9.66
Builders' work for other services installations						
Cutting chases in brickwork						
for one pipe; not exceeding 55 mm nominal size; vertical	-	0.37	2.90	-	m	2.90
for one pipe; 55 mm - 110 mm nominal size; vertical	-	0.65	5.09	-	m	5.09
Cutting and pinning to brickwork or blockwork; ends of supports						
for pipes not exceeding 55 mm nominal size	-	0.19	3.32	-	nr	3.32
for cast iron pipes 55 mm - 110 mm nominal size	-	0.31	5.42	-	nr	5.42
Cutting holes for pipes or the like; not exceeding 55 mm nominal size						
half brick thick	-	0.31	2.74	-	nr	2.74
one brick thick	-	0.51	4.51	-	nr	4.51
one and a half brick thick	-	0.83	7.33	-	nr	7.33
100 mm blockwork	-	0.28	2.47	-	nr	2.47
140 mm blockwork	-	0.37	3.27	-	nr	3.27
215 mm blockwork	-	0.46	4.06	-	nr	4.06
Cutting holes for pipes or the like; 55 mm - 110 mm nominal size						
half brick thick	-	0.37	3.27	-	nr	3.27
one brick thick	-	0.65	5.74	-	nr	5.74
one and a half brick thick	-	1.02	9.01	-	nr	9.01
100 mm blockwork	-	0.32	2.83	-	nr	2.83
140 mm blockwork	-	0.46	4.06	-	nr	4.06
215 mm blockwork	-	0.56	4.95	-	nr	4.95

Prices for Measured Works - Major Works
P BUILDING FABRIC SUNDRIES

	PC £	Labour hours	Labour £	Material £	Unit	Total rate £
Cutting holes for pipes or the like; over 110 mm nominal size						
half brick thick	-	0.46	4.06	-	nr	4.06
one brick thick	-	0.79	6.98	-	nr	6.98
one and a half brick thick	-	1.25	11.05	-	nr	11.05
100 mm blockwork	-	0.42	3.71	-	nr	3.71
140 mm blockwork	-	0.56	4.95	-	nr	4.95
215 mm blockwork	-	0.69	6.10	-	nr	6.10
Add for making good fair face or facings one side						
pipe; not exceeding 55 mm nominal size	-	0.07	1.22	-	nr	1.22
pipe; 55 mm - 110 mm nominal size	-	0.09	1.57	-	nr	1.57
pipe; over 110 mm nominal size	-	0.11	1.92	-	nr	1.92
Add for fixing sleeve (supply not included)						
for pipe; small	-	0.14	2.45	-	nr	2.45
for pipe; large	-	0.19	3.32	-	nr	3.32
for pipe; extra large	-	0.28	4.89	-	nr	4.89
Cutting or forming holes for ducts; girth not exceeding 1.00 m						
half brick thick	-	0.56	4.95	-	nr	4.95
one brick thick	-	0.93	8.22	-	nr	8.22
one and a half brick thick	-	1.48	13.08	-	nr	13.08
100 mm blockwork	-	0.46	4.06	-	nr	4.06
140 mm blockwork	-	0.65	5.74	-	nr	5.74
215 mm blockwork	-	0.83	7.33	-	nr	7.33
Cutting or forming holes for ducts; girth 1.00 m - 2.00 m						
half brick thick	-	0.65	5.74	-	nr	5.74
one brick thick	-	1.11	9.81	-	nr	9.81
one and a half brick thick	-	1.76	15.55	-	nr	15.55
100 mm blockwork	-	0.56	4.95	-	nr	4.95
140 mm blockwork	-	0.74	6.54	-	nr	6.54
215 mm blockwork	-	0.93	8.22	-	nr	8.22
Cutting or forming holes for ducts; girth 2.00 m - 3.00 m						
half brick thick	-	1.02	9.01	-	nr	9.01
one brick thick	-	1.76	15.55	-	nr	15.55
one and a half brick thick	-	2.78	24.57	-	nr	24.57
100 mm blockwork	-	0.88	7.78	-	nr	7.78
140 mm blockwork	-	1.20	10.60	-	nr	10.60
215 mm blockwork	-	1.53	13.52	-	nr	13.52
Cutting or forming holes for ducts; girth 3.00 m - 4.00 m						
half brick thick	-	1.39	12.28	-	nr	12.28
one brick thick	-	2.31	20.41	-	nr	20.41
one and a half brick thick	-	3.70	32.70	-	nr	32.70
100 mm blockwork	-	1.02	9.01	-	nr	9.01
140 mm blockwork	-	1.39	12.28	-	nr	12.28
215 mm blockwork	-	1.76	15.55	-	nr	15.55
Mortices in brickwork						
for expansion bolt	-	0.19	1.68	-	nr	1.68
for 20 mm diameter bolt; 75 mm deep	-	0.14	1.24	-	nr	1.24
for 20 mm diameter bolt; 150 mm deep	-	0.23	2.03	-	nr	2.03
Mortices in brickwork; grouting with cement mortar (1:1)						
75 mm x 75 mm x 200 mm deep	-	0.28	2.47	0.12	nr	2.59
75 mm x 75 mm x 300 mm deep	-	0.37	3.27	0.19	nr	3.46
Holes in softwood for pipes, bars, cables and the like						
12 mm thick	-	0.03	0.42	-	nr	0.42
25 mm thick	-	0.05	0.70	-	nr	0.70
50 mm thick	-	0.09	1.26	-	nr	1.26
100 mm thick	-	0.14	1.96	-	nr	1.96

P BUILDING FABRIC SUNDRIES

	PC £	Labour hours	Labour £	Material £	Unit	Total rate £
P31 HOLES/CHASES/COVERS/SUPPORTS FOR SERVICES - cont'd						
Builders' work for other services installations - cont'd						
Holes in hardwood for pipes, bars, cables and the like						
12 mm thick	-	0.05	0.70	-	nr	0.70
25 mm thick	-	0.08	1.12	-	nr	1.12
50 mm thick	-	0.14	1.96	-	nr	1.96
100 mm thick	-	0.20	2.80	-	nr	2.80
"SFD Screeduct" or other equal and approved; MDT Ducting Ltd; laid within floor screed; galvanised mild steel						
Floor ducting						
100 mm wide; 40 mm deep; ref SFD40/100/00	5.05	0.19	2.66	5.70	m	8.36
extra for						
45 degree bend	5.49	0.09	1.26	6.20	nr	7.46
90 degree bend	4.39	0.09	1.26	4.96	nr	6.22
tee section	4.39	0.09	1.26	4.96	nr	6.22
cross over	5.71	0.09	1.26	6.45	nr	7.71
reducer; 100 mm - 50 mm	10.33	0.09	1.26	11.66	nr	12.92
connector / stop end	0.44	0.09	1.26	0.50	nr	1.76
divider	1.32	0.09	1.26	1.49	nr	2.75
ply cover 15 mm/16 mm thick WBP exterior grade	0.51	0.09	1.26	0.58	m	1.84
200 mm wide; 65 mm deep; ref SFD65/200/00	5.93	0.19	2.66	6.69	m	9.35
extra for						
45 degree bend	7.69	0.09	1.26	8.68	nr	9.94
90 degree bend	5.93	0.09	1.26	6.69	nr	7.95
tee section	5.93	0.09	1.26	6.69	nr	7.95
cross over	8.13	0.09	1.26	9.18	nr	10.44
reducer; 200 mm - 100 mm	10.33	0.09	1.26	11.66	nr	12.92
connector / stop end	0.57	0.09	1.26	0.64	nr	1.90
divider	1.32	0.09	1.26	1.49	nr	2.75
ply cover 15 mm/16 mm thick WBP exterior grade	0.86	0.09	1.26	0.97	m	2.23

Q PAVING/PLANTING/FENCING/SITE FURNITURE

	PC £	Labour hours	Labour £	Material £	Unit	Total rate £
Q PAVING/PLANTING/FENCING/SITE FURNITURE						
Q10 KERBS/EDGINGS/CHANNELS/PAVING ACCESSORIES						
Excavating; by machine						
Excavating trenches; to receive kerb foundations; average size						
300 mm x 100 mm	-	0.02	0.16	0.30	m	**0.46**
450 mm x 150 mm	-	0.02	0.16	0.60	m	**0.76**
600 mm x 200 mm	-	0.03	0.26	0.84	m	**1.10**
Excavating curved trenches; to receive kerb foundations; average size						
300 mm x 100 mm	-	0.01	0.08	0.50	m	**0.58**
450 mm x 150 mm	-	0.03	0.24	0.70	m	**0.94**
600 mm x 200 mm	-	0.04	0.32	0.90	m	**1.22**
Excavating; by hand						
Excavating trenches; to receive kerb foundations; average size						
150 mm x 50 mm	-	0.02	0.16	-	m	**0.16**
200 mm x 75 mm	-	0.06	0.48	-	m	**0.48**
250 mm x 100 mm	-	0.10	0.80	-	m	**0.80**
300 mm x 100 mm	-	0.13	1.04	-	m	**1.04**
Excavating curved trenches; to receive kerb foundations; average size						
150 mm x 50 mm	-	0.03	0.24	-	m	**0.24**
200 mm x 75 mm	-	0.07	0.56	-	m	**0.56**
250 mm x 100 mm	-	0.11	0.88	-	m	**0.88**
300 mm x 100 mm	-	0.14	1.12	-	m	**1.12**
Plain in situ ready mixed designated concrete; C7.5 - 40 mm aggregate; poured on or against earth or unblinded hardcore						
Foundations	54.20	1.16	10.25	65.77	m^3	**76.02**
Blinding beds						
thickness not exceeding 150 mm	54.20	1.71	15.11	65.77	m^3	**80.88**
Plain in situ ready mixed designated concrete; C10 - 40 mm aggregate; poured on or against earth or unblinded hardcore						
Foundations	54.56	1.16	10.25	66.21	m^3	**76.46**
Blinding beds						
thickness not exceeding 150 mm	54.56	1.71	15.11	66.21	m^3	**81.32**
Plain in situ ready mixed designated concrete; C20 - 20 mm aggregate; poured on or against earth or unblinded hardcore						
Foundations	56.44	1.16	10.25	68.49	m^3	**78.74**
Blinding beds						
thickness not exceeding 150 mm	56.44	1.71	15.11	68.49	m^3	**83.60**
Precast concrete kerbs, channels, edgings, etc.; BS 340; bedded, jointed and pointed in cement mortar (1:3); including haunching up one side with in situ ready mix designated concrete C10 - 40 mm aggregate; to concrete base						
Edgings; straight; square edge, fig 12						
50 mm x 150 mm	-	0.23	3.16	2.58	m	**5.74**
50 mm x 200 mm	-	0.23	3.16	3.17	m	**6.33**
50 mm x 255 mm	-	0.23	3.16	3.37	m	**6.53**

Q PAVING/PLANTING/FENCING/SITE FURNITURE

	PC £	Labour hours	Labour £	Material £	Unit	Total rate £
Q10 KERBS/EDGINGS/CHANNELS/PAVING ACCESSORIES - cont'd						
Precast concrete kerbs, channels, edgings, etc.; BS 340; bedded, jointed and pointed in cement mortar (1:3); including haunching up one side with in situ ready mix designated concrete C10 - 40 mm aggregate; to concrete base - cont'd						
Kerbs; straight						
125 mm x 255 mm; fig 7	-	0.31	4.27	4.89	m	9.16
150 mm x 305 mm; fig 6	-	0.31	4.27	7.15	m	11.42
Kerbs; curved						
125 mm x 255 mm; fig 7	-	0.46	6.33	6.36	m	12.69
150 mm x 305 mm; fig 6	-	0.46	6.33	11.40	m	17.73
Channels; 255 x 125 mm; fig 8						
straight	-	0.31	4.27	4.38	m	8.65
curved	-	0.46	6.33	5.79	m	12.12
Quadrants; fig 14						
305 mm x 305 mm x 150 mm	-	0.32	4.40	6.99	nr	11.39
305 mm x 305 mm x 255 mm	-	0.32	4.40	7.52	nr	11.92
457 mm x 457 mm x 150 mm	-	0.37	5.09	8.25	nr	13.34
457 mm x 457 mm x 255 mm	-	0.37	5.09	8.77	nr	13.86
Q20 HARDCORE/GRANULAR/CEMENT BOUND BASES/SUB-BASES TO ROADS/PAVINGS						
Filling to make up levels; by machine						
Average thickness not exceeding 0.25 m						
obtained off site; hardcore	-	0.28	2.24	20.40	m^3	22.64
obtained off site; granular fill type one	-	0.28	2.24	27.95	m^3	30.19
obtained off site; granular fill type two	-	0.28	2.24	26.93	m^3	29.17
Average thickness exceeding 0.25 m						
obtained off site; hardcore	-	0.24	1.92	17.55	m^3	19.47
obtained off site; granular fill type one	-	0.24	1.92	27.82	m^3	29.74
obtained off site; granular fill type two	-	0.24	1.92	26.80	m^3	28.72
Filling to make up levels; by hand						
Average thickness not exceeding 0.25 m						
obtained off site; hardcore	-	0.61	4.88	20.80	m^3	25.68
obtained off site; sand	-	0.71	5.68	28.63	m^3	34.31
Average thickness exceeding 0.25 m						
obtained off site; hardcore	-	0.51	4.08	17.79	m^3	21.87
obtained off site; sand	-	0.60	4.80	28.31	m^3	33.11
Surface treatments						
Compacting						
filling; blinding with sand	-	0.04	0.32	1.46	m^2	1.78
Q21 IN SITU CONCRETE ROADS/PAVINGS						
Reinforced in situ ready mixed designated concrete; C10 - 40 mm aggregate						
Roads; to hardcore base						
thickness not exceeding 150 mm	54.56	1.85	16.35	63.05	m^3	79.40
thickness 150 mm - 450 mm	54.56	1.30	11.49	63.05	m^3	74.54

Prices for Measured Works - Major Works
Q PAVING/PLANTING/FENCING/SITE FURNITURE

	PC £	Labour hours	Labour £	Material £	Unit	Total rate £
Reinforced in situ ready mixed designated concrete; C20 - 20 mm aggregate						
Roads; to hardcore base						
thickness not exceeding 150 mm	56.44	1.85	16.35	65.22	m³	81.57
thickness 150 mm - 450 mm	56.44	1.30	11.49	65.22	m³	76.71
Reinforced in situ ready mixed designated concrete; C25 - 20 mm aggregate						
Roads; to hardcore base						
thickness ot exceeding 150 mm	56.87	1.85	16.35	65.72	m³	82.07
thickness 150 mm - 450 mm	56.87	1.30	11.49	65.72	m³	77.21
Formwork; sides of foundations; basic finish						
Plain vertical						
height not exceeding 250 mm	-	0.39	5.39	1.42	m	6.81
height 250 mm - 500 mm	-	0.57	7.88	2.47	m	10.35
height 500 mm - 1.00 m	-	0.83	11.47	4.89	m	16.36
add to above for curved radius 6m	-	0.03	0.41	0.19	m	0.60
Reinforcement; fabric; BS 4483; lapped; in roads, footpaths or pavings						
Ref A142 (2.22 kg/m²)						
400 mm minimum laps	0.88	0.14	1.55	1.04	m²	2.59
Ref A193 (3.02 kg/m²)						
400 mm minimum laps	-	0.14	1.55	1.43	m²	2.98
Formed joints; Fosroc Expandite "Flexcell" impregnated joint filler or other equal and approved						
Width not exceeding 150 mm						
12.50 mm thick	-	0.14	1.94	1.67	m	3.61
25 mm thick	-	0.19	2.63	2.48	m	5.11
Width 150 - 300 mm						
12.50 mm thick	-	0.19	2.63	2.76	m	5.39
25 mm thick	-	0.19	2.63	4.06	m	6.69
Width 300 - 450 mm						
12.50 mm thick	-	0.23	3.18	4.15	m	7.33
25 mm thick	-	0.23	3.18	6.09	m	9.27
Sealants; Fosroc Expandite "Pliastic N2" hot poured rubberized bituminous compound or other equal and approved						
Width 25 mm						
25 mm depth	-	0.20	2.76	1.45	m	4.21
Concrete sundries						
Treating surfaces of unset concrete; grading to cambers; tamping with a 75 mm thick steel shod tamper	-	0.23	2.03	-	m²	2.03

Prices for Measured Works - Major Works
Q PAVING/PLANTING/FENCING/SITE FURNITURE

	PC £	Labour hours	Labour £	Material £	Unit	Total rate £
Q22 COATED MACADAM/ASPHALT ROADS/PAVINGS						
In situ finishings; Associated Adphalt or other equal and approved						
NOTE: The prices for all in situ finishings to roads and footpaths include for work to falls, crossfalls or slopes not exceeding 15 degrees from horizontal; for laying on prepared bases (prices not included) and for rolling with an appropriate roller. The following rates are based on black bitumen macadam, except where stated. Red bitumen macadam rates are approximately 50% dearer.						
Fine graded wearing course; BS 4987; clause 2.7.7, tables 34 - 36; 14 mm nominal size pre-coated igneous rock chippings; tack coat of bitumen emulsion						
19 mm work to roads; one coat						
igneous aggregate	-	-	-	-	m^2	11.53
Close graded bitumen macadam; BS 4987; 10 mm nominal size graded aggregate to clause 2.7.4 tables 34 - 36; tack coat of bitumen emulsion						
30 mm work to roads; one coat						
limestone aggregate	-	-	-	-	m^2	11.04
igneous aggregate	-	-	-	-	m^2	11.14
Bitumen macadam; BS 4987; 45 thick base course of 20 mm open graded aggregate to clause 2.6.1 tables 5 - 7; 20 mm thick wearing course of 6 mm nominal size medium graded aggregate to clause 2.7.6 tables 32 - 33						
65 mm work to pavements/footpaths; two coats						
limestone aggregate	-	-	-	-	m^2	12.88
igneous aggregate	-	-	-	-	m^2	13.03
add to last for 10 nominal size chippings; sprinkled into wearing course	-	-	-	-	m^2	0.30
Bitumen macadam; BS 4987; 50 mm nominal size graded aggregate to clause 2.6.2 tables 8 - 10						
75 mm work to roads; one coat						
limestone aggregate	-	-	-	-	m^2	13.63
igneous aggregate	-	-	-	-	m^2	13.88
Dense bitumen macadam; BS 4987; 50 mm thick base course of 20 mm graded aggregate to clause 2.6.5 tables 15 - 16; 200 pen. binder; 30 mm wearing course of 10 mm nominal size graded aggregate to clause 2.7.4 tables 26 - 28						
80 mm work to roads; two coats						
limestone aggregate	-	-	-	-	m^2	13.73
igneous aggregate	-	-	-	-	m^2	13.88

Prices for Measured Works - Major Works
Q PAVING/PLANTING/FENCING/SITE FURNITURE

	PC £	Labour hours	Labour £	Material £	Unit	Total rate £
Bitumen macadam; BS 4987; 50 mm thick base course of 20 mm nominal size graded aggregate to clause 2.6.1 tables 5 - 7; 25 mm thick wearing course of 10 mm nominal size graded aggregate to clause 2.7.2 tables 20 - 22						
75 mm work to roads; two coats						
limestone aggregate	-	-	-	-	m^2	14.28
igneous aggregate	-	-	-	-	m^2	14.43
Dense bitumen macadam; BS 4987; 70 mm thick base course of 20 mm nominal size graded aggregate to clause 2.6.5 tables 15 - 16; 200 pen. binder; 30 mm wearing course of 10 mm nominal size graded aggregate to clause 2.7.4 tables 26 - 28						
100 mm work to roads; two coats						
limestone aggregate	-	-	-	-	m^2	15.48
igneous aggregate	-	-	-	-	m^2	15.73
Dense bitumen macadam; BS 4987; 70 mm thick road base of 28 mm nominal size graded aggregate to clause 2.5.2 tables 3 - 4; 50 mm thick base course of 20 mm nominal size graded aggregate to clause 2.6.5 tables 15 - 16; 30 mm wearing course of 10 mm nominal size graded aggregate to clause 2.7.4 tables 26 - 28						
150 mm work to roads; three coats						
limestone aggregate	-	-	-	-	m^2	25.07
igneous aggregate	-	-	-	-	m^2	25.47
Red bitumen macadam; BS 4987; 70 mm thick base course of 20 mm nominal size graded aggregate to clause 2.6.5 tables 15 - 16; 30 mm wearing course of 10 mm nominal size graded aggregate to clause 2.7.4 tables 26 - 28						
100 mm work to roads; two coats						
limestone base; igneous wearing course	-	-	-	-	m^2	18.48
igneous aggregate	-	-	-	-	m^2	18.73
Q23 GRAVEL/HOGGIN/WOODCHIP ROADS/PAVINGS						
Two coat gravel paving; level and to falls; first layer course clinker aggregate and wearing layer fine gravel aggregate						
Pavings; over 300 mm wide						
50 mm thick	-	0.07	0.96	1.41	m^2	2.37
63 mm thick	-	0.09	1.24	1.86	m^2	3.10
Q25 SLAB/BRICK/BLOCK/SETT/COBBLE PAVINGS						
Artificial stone paving; Redland Aggregates "Texitone" or other equal and approved; to falls or crossfalls; bedding 25 mm thick in cement mortar (1:3); staggered joints; jointing in coloured cement mortar (1:3), brushed in; to sand base						
Pavings; over 300 mm wide						
450 mm x 600 mm x 50 mm thick; grey or coloured	2.71	0.42	5.78	14.02	m^2	19.80
600 mm x 600 mm x 50 mm thick; grey or coloured	3.14	0.39	5.37	12.55	m^2	17.92
750 mm x 600 mm x 50 mm thick; grey or coloured	3.66	0.36	4.95	11.88	m^2	16.83
900 mm x 600 mm x 50 mm thick; grey or coloured	4.19	0.33	4.54	11.46	m^2	16.00

Prices for Measured Works - Major Works
Q PAVING/PLANTING/FENCING/SITE FURNITURE

	PC £	Labour hours	Labour £	Material £	Unit	Total rate £
Q25 SLAB/BRICK/BLOCK/SETT/COBBLE PAVINGS - cont'd						
Brick paviors; 215 mm x 103 mm x 65 mm rough stock bricks; PC £479.30/1000; to falls or crossfalls; bedding 10 mm thick in cement mortar (1:3); jointing in cement mortar (1:3); as work proceeds; to concrete base						
Pavings; over 300 mm wide; straight joints both ways						
bricks laid flat	-	0.74	12.93	22.58	m²	35.51
bricks laid on edge	-	1.04	18.18	33.54	m²	51.72
Pavings; over 300 mm wide; laid to herringbone pattern						
bricks laid flat	-	0.93	16.26	22.58	m²	38.84
bricks laid on edge	-	1.30	22.72	33.54	m²	56.26
Add or deduct for variation of £10.00/1000 in PC of brick paviours						
bricks laid flat	-	-	-	0.45	m²	0.45
bricks laid on edge	-	-	-	0.68	m²	0.68
River washed cobble paving; 50 mm -75 mm; PC £79.89/t; to falls or crossfalls; bedding 13 mm thick in cement mortar (1:3); jointing to a height of two thirds of cobbles in dry mortar (1:3); tightly butted, washed and brushed; to concrete						
Pavings; over 300 mm wide						
regular	-	3.70	50.91	18.72	m²	69.63
laid to pattern	-	4.63	63.71	18.72	m²	82.43
Concrete paving flags; BS 7263; to falls or crossfalls; bedding 25 mm thick in cement and sand mortar (1:4); butt joints straight both ways; jointing in cement and sand (1:3); brushed in; to sand base						
Pavings; over 300 mm wide						
450 mm x 600 mm x 50 mm thick; grey	1.43	0.42	5.78	7.28	m²	13.06
450 mm x 600 mm x 60 mm thick; coloured	1.64	0.42	5.78	8.16	m²	13.94
600 mm x 600 mm x 50 mm thick; grey	1.62	0.39	5.37	6.39	m²	11.76
600 mm x 600 mm x 50 mm thick; coloured	1.87	0.39	5.37	7.17	m²	12.54
750 mm x 600 mm x 50 mm thick; grey	1.91	0.36	4.95	6.10	m²	11.05
750 mm x 600 mm x 50 mm thick; coloured	2.20	0.36	4.95	6.83	m²	11.78
900 mm x 600 mm x 50 mm thick; grey	2.11	0.33	4.54	5.72	m²	10.26
900 mm x 600 mm x 50 mm thick; coloured	2.46	0.33	4.54	6.45	m²	10.99
Concrete rectangular paving blocks; to falls or crossfalls; bedding 50 mm thick in dry sharp sand; filling joints with sharp sand brushed in; on earth base						
Pavings; "Keyblock" or other equal and approved; over 300 mm wide; straight joints both ways						
200 mm x 100 mm x 60 mm thick; grey	6.45	0.69	9.49	9.25	m²	18.74
200 mm x 100 mm x 60 mm thick; coloured	7.07	0.69	9.49	9.95	m²	19.44
200 mm x 100 mm x 80 mm thick; grey	7.04	0.74	10.18	10.08	m²	20.26
200 mm x 100 mm x 80 mm thick; coloured	8.18	0.74	10.18	11.36	m²	21.54
Pavings; "Keyblock" or other equal and approved; over 300 mm wide; laid to herringbone pattern						
200 mm x 100 mm x 60 mm thick; grey	-	0.88	12.11	9.25	m²	21.36
200 mm x 100 mm x 60 mm thick; coloured	-	0.88	12.11	9.95	m²	22.06
200 mm x 100 mm x 80 mm thick; grey	-	0.93	12.80	10.08	m²	22.88
200 mm x 100 mm x 80 mm thick; coloured	-	0.93	12.80	11.36	m²	24.16

Prices for Measured Works - Major Works
Q PAVING/PLANTING/FENCING/SITE FURNITURE

	PC £	Labour hours	Labour £	Material £	Unit	Total rate £
Extra for two row boundary edging to herringbone pavings; 200 mm wide; including a 150 mm high in situ concrete mix C10 - 40 mm aggregate haunching to one side; blocks laid breaking joint						
200 mm x 100 mm x 60 mm; coloured	-	0.28	3.85	1.82	m	5.67
200 mm x 100 mm x 80 mm; coloured	-	0.28	3.85	1.91	m	5.76
Pavings; "Mount Sorrel" or other equal and approved; over 300 mm wide; straight joints both ways						
200 mm x 100 mm x 60 mm thick; grey	5.64	0.69	9.49	8.33	m²	17.82
200 mm x 100 mm x 60 mm thick; coloured	6.49	0.69	9.49	9.29	m²	18.78
200 mm x 100 mm x 80 mm thick; grey	6.71	0.74	10.18	9.70	m²	19.88
200 mm x 100 mm x 80 mm thick; coloured	7.71	0.74	10.18	10.83	m²	21.01
Pavings; "Pedesta" or other equal and approved; over 300 mm wide; straight joints both ways						
200 mm x 100 mm x 60 mm thick; grey	6.47	0.69	9.49	9.27	m²	18.76
200 mm x 100 mm x 60 mm thick; coloured	6.55	0.69	9.49	9.36	m²	18.85
200 mm x 100 mm x 80 mm thick; grey	7.50	0.74	10.18	10.59	m²	20.77
200 mm x 100 mm x 80 mm thick; coloured	7.96	0.74	10.18	11.11	m²	21.29
Pavings; "Intersett" or other equal and approved; over 300 mm wide; straight joints both ways						
200 mm x 100 mm x 60 mm thick; grey	8.08	0.69	9.49	11.09	m²	20.58
200 mm x 100 mm x 60 mm thick; coloured	8.96	0.69	9.49	12.08	m²	21.57
200 mm x 100 mm x 80 mm thick; grey	9.66	0.74	10.18	13.03	m²	23.21
200 mm x 100 mm x 80 mm thick; coloured	10.73	0.74	10.18	14.24	m²	24.42
Concrete rectangular paving blocks; to falls or crossfalls; 6 mm wide joints; symmetrical layout; bedding in 15 mm semi-dry cement mortar (1:4); jointing and pointing in cement and sand (1:4); on concrete base						
Pavings; "Trafica" or other equal and approved; over 300 mm wide						
400 mm x 400 mm x 65 mm; Standard; natural	1.22	0.44	6.05	9.67	m²	15.72
400 mm x 400 mm x 65 mm; Standard; buff	1.76	0.44	6.05	13.42	m²	19.47
400 mm x 400 mm x 65 mm; Saxon textured; natural	2.40	0.44	6.05	17.87	m²	23.92
400 mm x 400 mm x 65 mm; Saxon textured; buff	2.83	0.44	6.05	20.86	m²	26.91
400 mm x 400 mm x 65 mm; Perfecta; natural	2.86	0.44	6.05	21.06	m²	27.11
400 mm x 400 mm x 65 mm; Perfecta; buff	3.40	0.44	6.05	24.82	m²	30.87
450 mm x 450 mm x 70 mm; Standard; natural	1.77	0.43	5.92	10.92	m²	16.84
450 mm x 450 mm x 70 mm; Standard; buff	2.35	0.43	5.92	14.11	m²	20.03
450 mm x 450 mm x 70 mm; Saxon textured; natural	2.97	0.43	5.92	17.52	m²	23.44
450 mm x 450 mm x 70 mm; Saxon textured; buff	3.52	0.43	5.92	20.55	m²	26.47
450 mm x 450 mm x 70 mm; Perfecta; natural	3.56	0.43	5.92	20.77	m²	26.69
450 mm x 450 mm x 70 mm; Perfecta; buff	4.18	0.43	5.92	24.18	m²	30.10
450 mm x 450 mm x 100 mm; Standard; natural	4.72	0.44	6.05	27.15	m²	33.20
450 mm x 450 mm x 100 mm; Standard; buff	5.84	0.44	6.05	33.31	m²	39.36
450 mm x 450 mm x 100 mm; Saxon textured; natural	5.32	0.44	6.05	30.45	m²	36.50
450 mm x 450 mm x 100 mm; Saxon textured; buff	6.29	0.44	6.05	35.79	m²	41.84
450 mm x 450 mm x 100 mm; Perfecta; natural	5.71	0.44	6.05	32.60	m²	38.65
450 mm x 450 mm x 100 mm; Perfecta; buff	6.68	0.44	6.05	37.93	m²	43.98
York stone slab pavings; to falls or crossfalls; bedding 25 mm thick in cement:sand mortar (1:4); 5 mm wide joints; jointing in coloured cement mortar (1:3); brushed in; to sand base						
Pavings; over 300 mm wide						
50 mm thick; random rectangular pattern	55.74	0.69	12.06	63.01	m²	75.07
600 mm x 600 mm x 50 mm thick	69.67	0.39	6.82	78.36	m²	85.18
600 mm x 900 mm x 50 mm thick	72.00	0.33	5.77	80.93	m²	86.70

	PC £	Labour hours	Labour £	Material £	Unit	Total rate £
Q25 SLAB/BRICK/BLOCK/SETT/COBBLE PAVINGS - cont'd						
Granite setts; BS 435; 200 mm x 100 mm x 100 mm; PC £125.42/t; standard "C" dressing; tightly butted to falls or crossfalls; bedding 25 mm thick in cement mortar (1:3); filling joints with dry mortar (1:6); washed and brushed; on concrete base						
Pavings; over 300 mm wide						
straight joints	-	1.48	20.36	40.35	m²	60.71
laid to pattern	-	1.85	25.46	40.35	m²	65.81
Two rows of granite setts as boundary edging; 200 mm wide; including a 150 mm high ready mixed designated concrete C10 - 40 mm aggregate; haunching to one side; blocks laid breaking joint	-	0.65	8.94	9.26	m	18.20
Q26 SPECIAL SURFACINGS/PAVINGS FOR SPORT/GENERAL AMENITY						
Sundries						
Line marking						
width not exceeding 300 mm	-	0.04	0.39	0.31	m	0.70
Q30 SEEDING/TURFING						
Vegetable soil						
Selected from spoil heaps; grading; prepared for turfing or seeding; to general surfaces						
average 75 mm thick	-	0.21	1.68	-	m²	1.68
average 100 mm thick	-	0.23	1.84	-	m²	1.84
average 125 mm thick	-	0.25	2.00	-	m²	2.00
average 150 mm thick	-	0.26	2.08	-	m²	2.08
average 175 mm thick	-	0.27	2.16	-	m²	2.16
average 200 mm thick	-	0.29	2.32	-	m²	2.32
Selected from spoil heaps; grading; prepared for turfing or seeding; to cuttings or embankments						
average 75 mm thick	-	0.24	1.92	-	m²	1.92
average 100 mm thick	-	0.26	2.08	-	m²	2.08
average 125 mm thick	-	0.28	2.24	-	m²	2.24
average 150 mm thick	-	0.30	2.40	-	m²	2.40
average 175 mm thick	-	0.31	2.48	-	m²	2.48
average 200 mm thick	-	0.32	2.56	-	m²	2.56
Imported vegetable soil						
Grading; prepared for turfing or seeding; to general surfaces						
average 75 mm thick	-	0.19	1.52	0.75	m²	2.27
average 100 mm thick	-	0.20	1.60	0.97	m²	2.57
average 125 mm thick	-	0.22	1.76	1.42	m²	3.18
average 150 mm thick	-	0.23	1.84	1.87	m²	3.71
average 175 mm thick	-	0.25	2.00	2.10	m²	4.10
average 200 mm thick	-	0.26	2.08	2.32	m²	4.40
Grading; preparing for turfing or seeding; to cuttings or embankments						
average 75 mm thick	-	0.21	1.68	0.75	m²	2.43
average 100 mm thick	-	0.23	1.84	0.97	m²	2.81
average 125 mm thick	-	0.25	2.00	1.42	m²	3.42
average 150 mm thick	-	0.26	2.08	1.87	m²	3.95
average 175 mm thick	-	0.27	2.16	2.10	m²	4.26
average 200 mm thick	-	0.29	2.32	2.32	m²	4.64

Q PAVING/PLANTING/FENCING/SITE FURNITURE

	PC £	Labour hours	Labour £	Material £	Unit	Total rate £
Fertilizer; PC £7.80/25 kg						
Fertilizer 0.07 kg/m²; raking in						
general surfaces	-	0.03	0.24	0.02	m²	0.26
Selected grass seed; PC £63.40/25kg						
Grass seed; sowing at a rate of 0.042 kg/m²						
two applications; raking in						
general surfaces	-	0.06	0.48	0.23	m²	0.71
cuttings or embankments	-	0.07	0.56	0.23	m²	0.79
Preserved turf from stack on site						
Turfing						
general surfaces	-	0.19	1.52	-	m²	1.52
cuttings or embankments; shallow	-	0.20	1.60	-	m²	1.60
cuttings or embankments; steep; pegged	-	0.28	2.24	-	m²	2.24
Imported turf; selected meadow turf						
Turfing						
general surfaces	1.11	0.19	1.52	1.43	m²	2.95
cuttings or embankments; shallow	1.11	0.20	1.60	1.43	m²	3.03
cuttings or embankments; steep; pegged	1.11	0.28	2.24	1.43	m²	3.67
Q31 PLANTING						
Planting only						
Hedge plants						
height not exceeding 750 mm	-	0.23	1.84	-	nr	1.84
height 750 mm - 1.50 m	-	0.56	4.48	-	nr	4.48
Saplings						
height not exceeding 3.00 m	-	1.57	12.56	-	nr	12.56
Q40 FENCING						
NOTE: The prices for all fencing include for setting posts in position, to a depth of 0.60 m for fences not exceeding 1.40 m high and of 0.76 m for fences over 1.40 m high. The prices allow for excavating post holes; filling to within 150 mm of ground level with concrete and all backfilling.						
Strained wire fencing; BS 1722 Part 3; 4 mm diameter galvanized mild steel plain wire threaded through posts and strained with eye bolts						
Fencing; height 900 mm; three line; concrete posts at 2750 mm centres	-	-	-	-	m	7.94
Extra for						
end concrete straining post; one strut	-	-	-	-	nr	39.70
angle concrete straining post; two struts	-	-	-	-	nr	58.22
Fencing; height 1.07 m; five line; concrete posts at 2750 mm centres	-	-	-	-	m	11.44
Extra for						
end concrete straining post; one strut	-	-	-	-	nr	58.22
angle concrete straining post; two struts	-	-	-	-	nr	74.95
Fencing; height 1.20 m; six line; concrete posts at 2750 mm centres	-	-	-	-	m	11.74
Extra for						
end concrete straining post; one strut	-	-	-	-	nr	61.62
angle concrete straining post; two struts	-	-	-	-	nr	83.94
Fencing; height 1.40 m; seven line; concrete posts at 2750 mm centres	-	-	-	-	m	12.59
Extra for						
end concrete straining post; one strut	-	-	-	-	nr	65.87

Q PAVING/PLANTING/FENCING/SITE FURNITURE

	PC £	Labour hours	Labour £	Material £	Unit	Total rate £
Q40 FENCING - cont'd						
Chain link fencing; BS 1722 Part 1; 3 mm diameter galvanized mild steel wire; 50 mm mesh; galvanized mild steel tying and line wire; three line wires threaded through posts and strained with eye bolts and winding brackets						
Fencing; height 900 mm; galvanized mild steel angle posts at 3.00 m centres	-	-	-	-	m	10.89
Extra for						
end steel straining post; one strut	-	-	-	-	nr	50.69
angle steel straining post; two struts	-	-	-	-	nr	68.16
Fencing; height 900 mm; concrete posts at 3.00 m centres	-	-	-	-	m	10.59
Extra for						
end concrete straining post; one strut	-	-	-	-	nr	46.39
angle concrete straining post; two struts	-	-	-	-	nr	63.77
Fencing; height 1.20 m; galvanized mild steel angle posts at 3.00 m centres	-	-	-	-	m	14.18
Extra for						
end steel straining post; one strut	-	-	-	-	nr	61.62
angle steel straining post; two struts	-	-	-	-	nr	81.50
Fencing; height 1.20 m; concrete posts at 3.00 m centres	-	-	-	-	m	12.84
Extra for						
end concrete straining post; one strut	-	-	-	-	nr	57.28
angle concrete straining post; two struts	-	-	-	-	nr	74.85
Fencing; height 1.80 m; galvanized mild steel angle posts at 3.00 m centres	-	-	-	-	m	17.97
Extra for						
end steel straining post; one strut	-	-	-	-	nr	81.50
angle steel straining post; two struts	-	-	-	-	nr	111.16
Fencing; height 1.80 m; concrete posts at 3.00 m centres	-	-	-	-	m	16.18
Extra for						
end concrete straining post; one strut	-	-	-	-	nr	74.85
angle concrete straining post; two struts	-	-	-	-	nr	103.76
Pair of gates and gate posts; gates to match galvanized chain link fencing, with angle framing, braces, etc., complete with hinges, locking bar, lock and bolts; two 100 mm x 100 mm angle section gate posts; each with one strut						
2.44 m x 0.90 m	-	-	-	-	nr	689.09
2.44 m x 1.20 m	-	-	-	-	nr	714.06
2.44 m x 1.80 m	-	-	-	-	nr	813.91
Chain link fencing; BS 1722 Part 1; 3 mm diameter plastic coated mild steel wire; 50 mm mesh; plastic coated mild steel tying and line wire; three line wires threaded through posts and strained with eye bolts and winding brackets						
Fencing; height 900 mm; galvanized mild steel angle posts at 3.00 m centres	-	-	-	-	m	11.44
Extra for						
end steel straining post; one strut	-	-	-	-	nr	50.69
angle steel straining post; two struts	-	-	-	-	nr	68.16
Fencing; height 900 mm; concrete posts at 3.00 m centres	-	-	-	-	m	11.14
Extra for						
end concrete straining post; one strut	-	-	-	-	nr	46.54
angle concrete straining post; two struts	-	-	-	-	nr	64.91

Q PAVING/PLANTING/FENCING/SITE FURNITURE

	PC £	Labour hours	Labour £	Material £	Unit	Total rate £
Fencing; height 1.20 m; galvanized mild steel angle posts at 3.00 m centres	-	-	-	-	m	16.97
Extra for						
end steel straining post; one strut	-	-	-	-	nr	61.62
angle steel straining post; two struts	-	-	-	-	nr	81.89
Fencing; height 1.20 m; concrete posts at 3.00 m centres	-	-	-	-	m	13.33
Extra for						
end concrete straining post; one strut	-	-	-	-	nr	57.18
angle concrete straining post; two struts	-	-	-	-	nr	75.15
Fencing; height 1.80 m; galvanized mild steel angle posts at 3.00 m centres	-	-	-	-	m	19.58
Extra for						
end steel straining post; one strut	-	-	-	-	nr	81.89
angle steel straining post; two struts	-	-	-	-	nr	114.84
Fencing; height 1.80 m; concrete posts at 3.00 m centres	-	-	-	-	m	19.07
Extra for						
end concrete straining post; one strut	-	-	-	-	nr	74.10
angle concrete straining post; two struts	-	-	-	-	nr	105.86
Pair of gates and gate posts; gates to match plastic chain link fencing; with angle framing, braces, etc. complete with hinges, locking bar, lock and bolts; two 100 mm x 100 mm angle section gate posts; each with one strut						
2.44 m x 0.90 m	-	-	-	-	nr	699.07
2.44 m x 1.20 m	-	-	-	-	nr	724.03
2.44 m x 1.80 m	-	-	-	-	nr	823.90
Chain link fencing for tennis courts; BS 1722 Part 13; 2.5 diameter galvanised mild wire; 45 mm mesh; line and tying wires threaded through 45 mm x 45 mm x 5 mm galvanised mild steel angle standards, posts and struts; 60 mm x 60 mm x 6 mm straining posts and gate posts; straining posts and struts strained with eye bolts and winding brackets						
Fencing to tennis court 36.00 m x 18.00 m; including gate 1.07 m x 1.98 m; complete with hinges, locking bar, lock and bolts						
height 2745 mm fencing; standards at 3.00 m centres	-	-	-	-	nr	2197.09
height 3660 mm fencing; standards at 2.50 m centres	-	-	-	-	nr	2946.09
Cleft chestnut pale fencing; BS 1722 Part 4; pales spaced 51 mm apart; on two lines of galvanized wire; 64 mm diameter posts; 76 mm x 51 mm struts						
Fencing; height 900 mm; posts at 2.50 m centres	-	-	-	-	m	6.84
Extra for						
straining post; one strut	-	-	-	-	nr	12.44
corner straining post; two struts	-	-	-	-	nr	16.58
Fencing; height 1.05 m; posts at 2.50 m centres	-	-	-	-	m	7.54
Extra for						
straining post; one strut	-	-	-	-	nr	13.63
corner straining post; two struts	-	-	-	-	nr	18.37
Fencing; height 1.20 m; posts at 2.25 m centres	-	-	-	-	m	7.89
Extra for						
straining post; one strut	-	-	-	-	nr	15.18
corner straining post; two struts	-	-	-	-	nr	19.92
Fencing; height 1.35 m; posts at 2.25 m centres	-	-	-	-	m	8.54
Extra for						
straining post; one strut	-	-	-	-	nr	16.78
corner straining post; two struts	-	-	-	-	nr	22.52

Q PAVING/PLANTING/FENCING/SITE FURNITURE

	PC £	Labour hours	Labour £	Material £	Unit	Total rate £
Q40 FENCING - cont'd						
Close boarded fencing; BS 1722 Part 5; 76 mm x 38 mm softwood rails; 89 mm x 19 mm softwood pales lapped 13 mm; 152 mm x 25 mm softwood gravel boards; all softwood "treated"; posts at 3.00 m centres						
Fencing; two rail; concrete posts						
height 1.00 m	-	-	-	-	m	28.32
height 1.20 m	-	-	-	-	m	30.96
Fencing; three rail; concrete posts						
height 1.40 m	-	-	-	-	m	33.55
height 1.60 m	-	-	-	-	m	36.45
height 1.80 m	-	-	-	-	m	39.85
Fencing; two rail; oak posts						
height 1.00 m	-	-	-	-	m	20.97
height 1.20 m	-	-	-	-	m	23.62
Fencing; three rail; oak posts						
height 1.40 m	-	-	-	-	m	26.22
height 1.60 m	-	-	-	-	m	30.16
height 1.80 m	-	-	-	-	m	33.55
Precast concrete slab fencing; 305 mm x 38 mm x 1753 mm slabs; fitted into twice grooved concrete posts at 1830 mm centres						
Fencing						
height 1.20 m	-	-	-	-	m	41.95
height 1.50 m	-	-	-	-	m	52.43
height 1.80 m	-	-	-	-	m	62.92
Mild steel unclimbable fencing; in rivetted panels 2440 mm long; 44 mm x 13 mm flat section top and bottom rails; two 44 mm x 19 mm flat section standards; one with foot plate; and 38 mm x 13 mm raking stay with foot plate; 20 mm diameter pointed verticals at 120 mm centres; two 44 mm x 19 mm supports 760 mm long with ragged ends to bottom rail; the whole bolted together; coated with red oxide primer; setting standards and stays in ground at 2440 mm centres and supports at 815 mm centres						
Fencing						
height 1.67 m	-	-	-	-	m	104.87
height 2.13 m	-	-	-	-	m	120.59
Pair of gates and gate posts, to match mild steel unclimbable fencing; with flat section framing, braces, etc., complete with locking bar, lock, handles, drop bolt, gate stop and holding back catches; two 102 mm x 102 mm hollow section gate posts with cap and foot plates						
2.44 m x 1.67 m	-	-	-	-	nr	908.79
2.44 m x 2.13 m	-	-	-	-	nr	1048.61
4.88 m x 1.67 m	-	-	-	-	nr	1423.12
4.88 m x 2.13 m	-	-	-	-	nr	1782.64
PVC coated, galvanised mild steel high security fencing; "Sentinal Sterling" fencing or other equal and approved; Twil Wire Products Ltd; 50 mm x 50 mm mesh; 3/3.50 mm gauge wire; barbed edge - 1; "Sentinal Bi-steel" colour coated posts or other equal and approved at 2440 mm centres						
Fencing						
1.80 m	-	0.93	7.44	29.69	m	37.13
2.10 m	-	1.16	9.28	33.43	m	42.71

R DISPOSAL SYSTEMS

R10 RAINWATER PIPEWORK/GUTTERS

	PC £	Labour hours	Labour £	Material £	Unit	Total rate £
Aluminium pipes and fittings; BS 2997; ears cast on; powder coated finish						
63.50 mm diameter pipes; plugged and screwed	8.43	0.34	4.28	10.09	m	**14.37**
Extra for						
fittings with one end	-	0.20	2.52	5.23	nr	**7.75**
fittings with two ends	-	0.39	4.91	5.39	nr	**10.30**
fittings with three ends	-	0.56	7.06	7.16	nr	**14.22**
shoe	5.16	0.20	2.52	5.23	nr	**7.75**
bend	5.50	0.39	4.91	5.39	nr	**10.30**
single branch	7.17	0.56	7.06	7.16	nr	**14.22**
offset 228 projection	12.69	0.39	5.46	12.67	nr	**18.13**
offset 304 projection	14.15	0.39	4.91	14.28	nr	**19.19**
access pipe	15.67	-	-	14.93	nr	**14.93**
connection to clay pipes; cement and sand (1:2) joint	-	0.14	1.76	0.09	nr	**1.85**
76.50 mm diameter pipes; plugged and screwed	9.81	0.37	4.66	11.66	m	**16.32**
Extra for						
shoe	7.08	0.23	2.90	7.23	nr	**10.13**
bend	6.95	0.42	5.29	6.88	nr	**12.17**
single branch	8.64	0.60	7.56	8.62	nr	**16.18**
offset 228 projection	14.02	0.42	5.29	13.88	nr	**19.17**
offset 304 projection	15.52	0.42	5.29	15.54	nr	**20.83**
access pipe	17.12	-	-	16.15	nr	**16.15**
connection to clay pipes; cement and sand (1:2) joint	-	0.16	2.02	0.09	nr	**2.11**
100 mm diameter pipes; plugged and screwed	16.72	0.42	5.29	19.47	m	**24.76**
Extra for						
shoe	8.53	0.26	3.28	8.23	nr	**11.51**
bend	9.68	0.46	5.80	9.30	nr	**15.10**
single branch	11.56	0.69	8.69	10.95	nr	**19.64**
offset 228 projection	16.22	0.46	5.80	14.74	nr	**20.54**
offset 304 projection	18.02	0.46	5.80	16.72	nr	**22.52**
access pipe	20.29	-	-	17.71	nr	**17.71**
connection to clay pipes; cement and sand (1:2) joint	-	0.19	2.39	0.09	nr	**2.48**
Roof outlets; circular aluminium; with flat or domed grating; joint to pipe						
50 mm diameter	45.37	0.56	9.90	48.77	nr	**58.67**
75 mm diameter	60.20	0.60	10.60	64.72	nr	**75.32**
100 mm diameter	84.43	0.65	11.49	90.76	nr	**102.25**
150 mm diameter	105.03	0.69	12.19	112.91	nr	**125.10**
Roof outlets; d-shaped; balcony; with flat or domed grating; joint to pipe						
50 mm diameter	53.74	0.56	9.90	57.77	nr	**67.67**
75 mm diameter	61.77	0.60	10.60	66.40	nr	**77.00**
100 mm diameter	75.86	0.65	11.49	81.55	nr	**93.04**
Galvanized wire balloon grating; BS 416 for pipes or outlets						
50 mm diameter	1.28	0.06	1.06	1.38	nr	**2.44**
63 mm diameter	1.30	0.06	1.06	1.40	nr	**2.46**
75 mm diameter	1.38	0.06	1.06	1.48	nr	**2.54**
100 mm diameter	1.52	0.07	1.24	1.63	nr	**2.87**

R DISPOSAL SYSTEMS

	PC £	Labour hours	Labour £	Material £	Unit	Total rate £
R10 RAINWATER PIPEWORK/GUTTERS - cont'd						
Aluminium gutters and fittings; BS 2997; powder coated finish						
100 mm half round gutters; on brackets; screwed to timber	7.73	0.32	4.48	11.05	m	15.53
Extra for						
stop end	2.09	0.15	2.10	4.50	nr	6.60
running outlet	4.64	0.31	4.34	5.11	nr	9.45
stop end outlet	4.13	0.15	2.10	6.05	nr	8.15
angle	4.29	0.31	4.34	3.98	nr	8.32
113 mm half round gutters; on brackets; screwed to timber	8.10	0.32	4.48	11.48	m	15.96
Extra for						
stop end	2.20	0.15	2.10	4.63	nr	6.73
running outlet	5.06	0.31	4.34	5.54	nr	9.88
stop end outlet	4.73	0.15	2.10	6.68	nr	8.78
angle	4.84	0.31	4.34	4.48	nr	8.82
125 mm half round gutters; on brackets; screwed to timber	9.10	0.37	5.18	13.76	m	18.94
Extra for						
stop end	2.68	0.17	2.38	6.21	nr	8.59
running outlet	5.48	0.32	4.48	5.92	nr	10.40
stop end outlet	5.03	0.17	2.38	7.98	nr	10.36
angle	5.37	0.32	4.48	5.90	nr	10.38
100 mm ogee gutters; on brackets; screwed to timber	9.64	0.34	4.76	14.09	m	18.85
Extra for						
stop end	2.21	0.16	2.24	2.96	nr	5.20
running outlet	5.44	0.32	4.48	5.38	nr	9.86
stop end outlet	4.22	0.16	2.24	6.73	nr	8.97
angle	4.59	0.32	4.48	3.50	nr	7.98
112 mm ogee gutters; on brackets; screwed to timber	10.72	0.39	5.46	15.49	m	20.95
Extra for						
stop end	2.36	0.16	2.24	3.13	nr	5.37
running outlet	5.51	0.32	4.48	5.36	nr	9.84
stop end outlet	4.73	0.16	2.24	7.36	nr	9.60
angle	5.46	0.32	4.48	4.26	nr	8.74
125 mm ogee gutters; on brackets; screwed to timber	11.84	0.39	5.46	17.02	m	22.48
Extra for						
stop end	2.58	0.18	2.52	3.37	nr	5.89
running outlet	6.03	0.34	4.76	5.84	nr	10.60
stop end outlet	5.37	0.18	2.52	8.22	nr	10.74
angle	6.38	0.34	4.76	5.07	nr	9.83
Cast iron pipes and fittings; EN 1462; ears cast on; joints						
65 mm pipes; primed; nailed to masonry	14.52	0.48	6.05	16.70	m	22.75
Extra for						
shoe	12.62	0.30	3.78	12.86	nr	16.64
bend	7.73	0.53	6.68	7.47	nr	14.15
single branch	14.90	0.67	8.44	14.86	nr	23.30
offset 225 mm projection	13.77	0.53	6.68	13.14	nr	19.82
offset 305 mm projection	16.12	0.53	6.68	15.40	nr	22.08
connection to clay pipes; cement and sand (1:2) joint	-	0.14	1.76	0.10	nr	1.86

R DISPOSAL SYSTEMS

	PC £	Labour hours	Labour £	Material £	Unit	Total rate £
75 mm pipes; primed; nailed to masonry	14.52	0.51	6.43	16.87	m	23.30
Extra for						
shoe	12.62	0.32	4.03	12.91	nr	16.94
bend	9.38	1.11	14.82	9.34	nr	24.16
single branch	16.42	0.69	8.69	16.65	nr	25.34
offset 225 mm projection	13.77	0.56	7.06	13.20	nr	20.26
offset 305 mm projection	16.92	0.56	7.06	16.34	nr	23.40
connection to clay pipes; cement and sand (1:2) joint	-	0.16	2.02	0.10	nr	2.12
100 mm pipes; primed; nailed to masonry	19.50	0.56	7.06	22.68	m	29.74
Extra for						
shoe	16.45	0.37	4.66	16.80	nr	21.46
bend	13.26	0.60	7.56	13.28	nr	20.84
single branch	19.51	0.74	9.32	19.61	nr	28.93
offset 225 mm projection	27.02	0.60	7.56	27.13	nr	34.69
offset 305 mm projection	27.02	0.60	7.56	26.69	nr	34.25
connection to clay pipes; cement and sand (1:2) joint	-	0.19	2.39	0.09	nr	2.48
100 mm x 75 mm rectangular pipes; primed; nailing to masonry	55.79	0.56	7.06	63.65	m	70.71
Extra for						
shoe	47.27	0.37	4.66	47.48	nr	52.14
bend	45.01	0.60	7.56	44.99	nr	52.55
offset 225 mm projection	63.37	0.37	4.66	61.44	nr	66.10
offset 305 mm projection	67.74	0.37	4.66	65.00	nr	69.66
connection to clay pipes; cement and sand (1:2) joint	-	0.19	2.39	0.09	nr	2.48
Rainwater head; rectangular; for pipes						
65 mm diameter	17.23	0.53	6.68	19.25	nr	25.93
75 mm diameter	17.23	0.56	7.06	19.31	nr	26.37
100 mm diameter	23.78	0.60	7.56	26.64	nr	34.20
Rainwater head; octagonal; for pipes						
65 mm diameter	9.86	0.53	6.68	11.13	nr	17.81
75 mm diameter	11.20	0.56	7.06	12.66	nr	19.72
100 mm diameter	24.81	0.60	7.56	27.77	nr	35.33
Copper wire balloon grating; BS 416 for pipes or outlets						
50 mm diameter	1.56	0.06	0.76	1.68	nr	2.44
63 mm diameter	1.58	0.06	0.76	1.70	nr	2.46
75 mm diameter	1.77	0.06	0.76	1.90	nr	2.66
100 mm diameter	2.02	0.07	0.88	2.17	nr	3.05
Cast iron gutters and fittings; EN 1462						
100 mm half round gutters; primed; on brackets; screwed to timber	8.28	0.37	5.18	11.80	m	16.98
Extra for						
stop end	2.09	0.16	2.24	3.42	nr	5.66
running outlet	6.05	0.32	4.48	5.89	nr	10.37
angle	6.21	0.32	4.48	7.21	nr	11.69
115 mm half round gutters; primed; on brackets; screwed to timber	8.62	0.37	5.18	12.22	m	17.40
Extra for						
stop end	2.70	0.16	2.24	4.08	nr	6.32
running outlet	6.60	0.32	4.48	6.45	nr	10.93
angle	6.39	0.32	4.48	7.34	nr	11.82
125 mm half round gutters; primed; on brackets; screwed to timber	10.10	0.42	5.88	13.88	m	19.76
Extra for						
stop end	2.70	0.19	2.66	4.09	nr	6.75
running outlet	7.54	0.37	5.18	7.33	nr	12.51
angle	7.54	0.37	5.18	8.30	nr	13.48

Prices for Measured Works - Major Works
R DISPOSAL SYSTEMS

	PC £	Labour hours	Labour £	Material £	Unit	Total rate £
R10 RAINWATER PIPEWORK/GUTTERS - cont'd						
Cast iron gutters and fittings; EN 1462						
150 mm half round gutters; primed; on brackets; screwed to timber	17.25	0.46	6.44	21.62	m	28.06
Extra for						
stop end	3.60	0.20	2.80	6.51	nr	9.31
running outlet	13.04	0.42	5.88	12.67	nr	18.55
angle	13.77	0.42	5.88	14.29	nr	20.17
100 mm ogee gutters; primed; on brackets; screwed to timber	9.23	0.39	5.46	13.13	m	18.59
Extra for						
stop end	1.93	0.17	2.38	4.45	nr	6.83
running outlet	6.61	0.34	4.76	6.41	nr	11.17
angle	6.48	0.34	4.76	7.58	nr	12.34
115 mm ogee gutters; primed; on brackets; screwed to timber	10.15	0.39	5.46	14.20	m	19.66
Extra for						
stop end	2.55	0.17	2.38	5.12	nr	7.50
running outlet	7.02	0.34	4.76	6.77	nr	11.53
angle	7.02	0.34	4.76	7.98	nr	12.74
125 mm ogee gutters; primed; on brackets; screwed to timber	10.65	0.43	6.02	15.09	m	21.11
Extra for						
stop end	2.55	0.19	2.66	5.42	nr	8.08
running outlet	7.66	0.39	5.46	7.42	nr	12.88
angle	7.66	0.39	5.46	8.87	nr	14.33
3 mm thick galvanised heavy pressed steel gutters and fittings; joggle joints; BS 1091						
200 mm x 100 mm (400 mm girth) box gutter; screwed to timber	-	0.60	7.56	15.13	m	22.69
Extra for						
stop end	-	0.32	4.03	8.41	nr	12.44
running outlet	-	0.65	8.19	13.93	nr	22.12
stop end outlet	-	0.32	4.03	19.46	nr	23.49
angle	-	0.65	8.19	15.48	nr	23.67
381 mm boundary wall gutters (900 mm girth); bent twice; screwed to timber	-	0.60	7.56	24.85	m	32.41
Extra for						
stop end	-	0.37	4.66	14.33	nr	18.99
running outlet	-	0.65	8.19	19.16	nr	27.35
stop end outlet	-	0.32	4.03	27.08	nr	31.11
angle	-	0.65	8.19	22.66	nr	30.85
457 mm boundary wall gutters (1200 mm girth); bent twice; screwed to timber	-	0.69	8.69	33.13	m	41.82
Extra for						
stop end	-	0.37	4.66	18.39	nr	23.05
running outlet	-	0.74	9.32	27.71	nr	37.03
stop end outlet	-	0.37	4.66	29.42	nr	34.08
angle	-	0.74	9.32	30.56	nr	39.88
uPVC external rainwater pipes and fittings; BS 4576; slip-in joints						
50 mm pipes; fixing with pipe or socket brackets; plugged and screwed	2.10	0.28	3.53	3.20	m	6.73
Extra for						
shoe	1.24	0.19	2.39	1.70	nr	4.09
bend	1.45	0.28	3.53	1.94	nr	5.47
two bends to form offset 229 mm projection	1.45	0.28	3.53	3.13	nr	6.66
connection to clay pipes; cement and sand (1:2) joint	-	0.12	1.51	0.10	nr	1.61

R DISPOSAL SYSTEMS

	PC £	Labour hours	Labour £	Material £	Unit	Total rate £
68 mm pipes; fixing with pipe or socket brackets; plugged and screwed	1.69	0.31	3.91	2.92	m	6.83
Extra for						
shoe	1.24	0.20	2.52	1.89	nr	4.41
bend	2.19	0.31	3.91	2.93	nr	6.84
single branch	3.82	0.41	5.17	4.73	nr	9.90
two bends to form offset 229 mm projection	2.19	0.31	3.91	5.02	nr	8.93
loose drain connector; cement and sand (1:2) joint	-	0.14	1.76	1.78	nr	3.54
110 mm pipes; fixing with pipe or socket brackets; plugged and screwed	3.63	0.33	4.16	6.02	m	10.18
Extra for						
shoe	3.97	0.22	2.77	4.92	nr	7.69
bend	5.60	0.33	4.16	6.71	nr	10.87
single branch	8.28	0.44	5.54	9.67	nr	15.21
two bends to form offset 229 mm projection	5.60	0.33	4.16	12.31	nr	16.47
loose drain connector; cement and sand (1:2) joint	-	0.32	4.03	6.93	nr	10.96
65 mm square pipes; fixing with pipe or socket brackets; plugged and screwed	1.86	0.31	3.91	3.17	m	7.08
Extra for						
shoe	1.36	0.20	2.52	2.02	nr	4.54
bend	1.49	0.31	3.91	2.16	nr	6.07
single branch	4.18	0.41	5.17	5.13	nr	10.30
two bends to form offset 229 mm projection	1.49	0.31	3.91	3.55	nr	7.46
drain connector; square to round; cement and sand (1:2) joint	-	0.32	4.03	2.32	nr	6.35
Rainwater head; rectangular; for pipes						
50 mm diameter	6.76	0.42	5.29	8.19	nr	13.48
68 mm diameter	5.46	0.43	5.42	7.12	nr	12.54
110 mm diameter	11.40	0.51	6.43	13.71	nr	20.14
65 mm square	5.46	0.43	5.42	7.12	nr	12.54
uPVC gutters and fittings; BS 4576						
76 mm half round gutters; on brackets screwed to timber	1.69	0.28	3.92	2.75	m	6.67
Extra for						
stop end	0.57	0.12	1.68	0.85	nr	2.53
running outlet	1.61	0.23	3.22	1.62	nr	4.84
stop end outlet	1.60	0.12	1.68	1.83	nr	3.51
angle	1.61	0.23	3.22	2.01	nr	5.23
112 mm half round gutters; on brackets screwed to timber	1.73	0.31	4.34	3.35	m	7.69
Extra for						
stop end	0.89	0.12	1.68	1.30	nr	2.98
running outlet	1.75	0.26	3.64	1.77	nr	5.41
stop end outlet	1.75	0.12	1.68	2.09	nr	3.77
angle	1.96	0.26	3.64	2.65	nr	6.29
170 mm half round gutters; on brackets; screwed to timber	3.36	0.31	4.34	6.07	m	10.41
Extra for						
stop end	1.51	0.15	2.10	2.26	nr	4.36
running outlet	3.36	0.29	4.06	3.40	nr	7.46
stop end outlet	3.20	0.15	2.10	3.82	nr	5.92
angle	4.39	0.29	4.06	5.73	nr	9.79
114 mm rectangular gutters; on brackets; screwed to timber	1.90	0.31	4.34	3.79	m	8.13
Extra for						
stop end	0.99	0.12	1.68	1.43	nr	3.11
running outlet	1.92	0.29	4.06	1.94	nr	6.00
stop end outlet	1.92	0.12	1.68	2.29	nr	3.97
angle	2.16	0.26	3.64	2.90	nr	6.54

Prices for Measured Works - Major Works
R DISPOSAL SYSTEMS

	PC £	Labour hours	Labour £	Material £	Unit	Total rate £
R11 FOUL DRAINAGE ABOVE GROUND						
Cast iron "Timesaver" pipes and fittings or other equal and approved; BS 416						
50 mm pipes; primed; 2 m lengths; fixing with expanding bolts; to masonry	11.42	0.51	6.43	19.41	m	25.84
Extra for						
fittings with two ends	-	0.51	6.44	13.66	nr	20.10
fittings with three ends	-	0.69	8.69	23.16	nr	31.85
bends; short radius	9.03	0.51	6.43	13.66	nr	20.09
access bends; short radius	22.25	0.51	6.43	28.22	nr	34.65
boss; 38 BSP	18.70	0.51	6.43	23.99	nr	30.42
single branch	13.59	0.69	8.69	23.74	nr	32.43
access pipe	21.72	0.51	6.43	26.22	nr	32.65
roof connector; for asphalt	24.51	0.51	6.43	30.91	nr	37.34
isolated "Timesaver" coupling joint	5.12	0.28	3.53	5.64	nr	9.17
connection to clay pipes; cement and sand (1:2) joint	-	0.12	1.51	0.09	nr	1.60
75 mm pipes; primed; 3 m lengths; fixing with standard brackets; plugged and screwed to masonry	11.29	0.51	6.43	19.71	m	26.14
75 mm pipes; primed; 2 m lengths; fixing with standard brackets; plugged and screwed to masonry	11.42	0.55	6.93	19.86	m	26.79
Extra for						
bends; short radius	9.03	0.55	6.93	14.25	nr	21.18
access bends; short radius	22.25	0.51	6.43	28.82	nr	35.25
boss; 38 BSP	18.70	0.55	6.93	24.91	nr	31.84
single branch	13.59	0.79	9.95	24.80	nr	34.75
double branch	22.84	1.02	12.85	41.23	nr	54.08
offset 115 mm projection	11.13	0.55	6.93	15.08	nr	22.01
offset 150 mm projection	11.13	0.55	6.93	14.70	nr	21.63
access pipe	21.72	0.55	6.93	26.56	nr	33.49
roof connector; for asphalt	28.21	0.55	6.93	35.26	nr	42.19
isolated "Timesaver" coupling joint	5.66	0.32	4.03	6.24	nr	10.27
connection to clay pipes; cement and sand (1:2) joint	-	0.14	1.76	0.09	nr	1.85
100 mm pipes; primed; 3 m lengths; fixing with standard brackets; plugged and screwed to masonry	13.65	0.55	6.93	27.86	m	34.79
100 mm pipes; primed; 2 m lengths; fixing with standard brackets; plugged and screwed to masonry	13.78	0.62	7.81	28.01	m	35.82
Extra for						
WC bent connector; 450 mm long tail	13.86	0.55	6.93	17.89	nr	24.82
bends; short radius	12.50	0.62	7.81	19.58	nr	27.39
access bends; short radius	26.44	0.62	7.81	34.94	nr	42.75
boss; 38 BSP	22.33	0.62	7.81	30.41	nr	38.22
single branch	19.32	0.93	11.72	33.92	nr	45.64
double branch	23.90	1.20	15.12	47.11	nr	62.23
offset 225 mm projection	17.96	0.62	7.81	23.58	nr	31.39
offset 300 mm projection	20.24	0.62	7.81	25.62	nr	33.43
access pipe	22.84	0.62	7.81	28.64	nr	36.45
roof connector; for asphalt	21.59	0.62	7.81	29.05	nr	36.86
roof connector; for roofing felt	69.20	0.62	7.81	81.51	nr	89.32
isolated "Timesaver" coupling joint	7.39	0.39	4.91	8.14	nr	13.05
transitional clayware socket; cement and sand (1:2) joint	14.70	0.37	4.66	24.43	nr	29.09
150 mm pipes; primed; 3 m lengths; fixing with standard brackets; plugged and screwed to masonry	28.50	0.69	8.69	56.38	m	65.07

R DISPOSAL SYSTEMS

	PC £	Labour hours	Labour £	Material £	Unit	Total rate £
150 mm pipes; primed; 2 m lengths; fixing with standard brackets; plugged and screwed to masonry	28.28	0.77	9.70	56.13	m	**65.83**
Extra for						
bends; short radius	22.33	0.77	9.70	36.08	nr	**45.78**
access bends; short radius	37.55	0.77	9.70	52.85	nr	**62.55**
boss; 38 BSP	36.43	0.77	9.70	50.82	nr	**60.52**
single branch	47.91	1.11	13.98	76.70	nr	**90.68**
double branch	67.31	1.48	18.65	112.91	nr	**131.56**
access pipe	37.99	0.77	9.70	46.83	nr	**56.53**
roof connector; for asphalt	54.91	0.77	9.70	69.43	nr	**79.13**
isolated "Timesaver" coupling joint	-	0.46	5.80	16.26	nr	**22.06**
transitional clayware socket; cement and sand (1:2) joint	25.75	0.48	6.05	44.72	nr	**50.77**
Cast iron "Ensign" lightweight pipes and fittings or other equal and approved; EN 877						
50 mm pipes; primed; 3 m lengths; fixing with standard brackets; plugged and screwed to masonry	-	0.31	3.33	12.17	m	**15.50**
Extra for						
bends; short radius	-	0.27	2.91	9.72	nr	**12.63**
single branch	-	0.33	3.54	17.00	nr	**20.54**
access pipe	-	0.27	2.81	19.95	nr	**22.76**
70 mm pipes; primed; 3 m lengths; fixing with standard brackets; plugged and screwed to masonry	-	0.34	3.67	13.61	m	**17.28**
Extra for						
bends; short radius	-	0.30	3.21	10.85	nr	**14.06**
single branch	-	0.37	3.97	18.25	nr	**22.22**
access pipe	-	0.30	3.21	21.27	nr	**24.48**
100 mm pipes; primed; 3 m lengths; fixing with standard brackets; plugged and screwed to masonry	-	0.37	3.97	16.06	m	**20.03**
Extra for						
bends; short radius	-	0.32	3.46	13.32	nr	**16.78**
single branch	-	0.39	4.18	23.50	nr	**27.68**
double branch	-	0.46	4.94	34.42	nr	**39.36**
access pipe	-	0.32	3.46	24.19	nr	**27.65**
connector	-	0.21	2.24	28.23	nr	**30.47**
reducer	-	0.32	3.46	16.24	nr	**19.70**
Polypropylene (PP) waste pipes and fittings; BS 5254; push fit "O" - ring joints						
32 mm pipes; fixing with pipe clips; plugged and screwed	0.41	0.20	2.52	0.79	m	**3.31**
Extra for						
fittings with one end	-	0.15	1.89	0.41	nr	**2.30**
fittings with two ends	-	0.20	2.52	0.41	nr	**2.93**
fittings with three ends	-	0.28	3.53	0.41	nr	**3.94**
access plug	0.37	0.15	1.89	0.41	nr	**2.30**
double socket	0.37	0.14	1.76	0.41	nr	**2.17**
male iron to PP coupling	0.97	0.26	3.28	1.07	nr	**4.35**
sweep bend	0.37	0.20	2.52	0.41	nr	**2.93**
spigot bend	0.37	0.23	2.90	0.41	nr	**3.31**
40 mm pipes; fixing with pipe clips; plugged and screwed	0.50	0.20	2.52	0.89	m	**3.41**
Extra for						
fittings with one end	-	0.18	2.27	0.41	nr	**2.68**
fittings with two ends	-	0.28	3.53	0.41	nr	**3.94**
fittings with three ends	-	0.37	4.66	0.41	nr	**5.07**
access plug	0.37	0.18	2.27	0.41	nr	**2.68**
double socket	0.37	0.19	2.39	0.41	nr	**2.80**
universal connector	0.37	0.23	2.90	0.41	nr	**3.31**
sweep bend	0.37	0.28	3.53	0.41	nr	**3.94**
spigot bend	0.37	0.28	3.53	0.41	nr	**3.94**
reducer 40 mm - 32 mm	0.37	0.28	3.53	0.41	nr	**3.94**

R11 FOUL DRAINAGE ABOVE GROUND - cont'd

Polypropylene (PP) waste pipes and fittings;
BS 5254; push fit "O" - ring joints - cont'd

	PC £	Labour hours	Labour £	Material £	Unit	Total rate £
50 mm pipes; fixing with pipe clips; plugged and screwed	0.83	0.32	4.03	1.47	m	5.50
Extra for						
fittings with one end	-	0.19	2.39	0.73	nr	3.12
fittings with two ends	-	0.32	4.03	0.73	nr	4.76
fittings with three ends	-	0.43	5.42	0.73	nr	6.15
access plug	0.66	0.19	2.39	0.73	nr	3.12
double socket	0.66	0.21	2.65	0.73	nr	3.38
sweep bend	0.66	0.32	4.03	0.73	nr	4.76
spigot bend	0.66	0.32	4.03	0.73	nr	4.76
reducer 50 mm - 40 mm	0.66	0.32	4.03	0.73	nr	4.76
muPVC waste pipes and fittings; BS 5255; solvent welded joints						
32 mm pipes; fixing with pipe clips; plugged and screwed	0.89	0.23	2.90	1.45	m	4.35
Extra for						
fittings with one end	-	0.16	2.02	1.04	nr	3.06
fittings with two ends	-	0.23	2.90	1.08	nr	3.98
fittings with three ends	-	0.31	3.91	1.50	nr	5.41
access plug	0.81	0.16	2.02	1.04	nr	3.06
straight coupling	0.54	0.16	2.02	0.75	nr	2.77
expansion coupling	0.94	0.23	2.90	1.19	nr	4.09
male iron to muPVC coupling	0.81	0.35	4.41	0.97	nr	5.38
union coupling	2.25	0.23	2.90	2.63	nr	5.53
spigot/socket bend	-	0.23	2.90	1.21	nr	4.11
sweep tee	1.19	0.31	3.91	1.50	nr	5.41
40 mm pipes; fixing with pipe clips; plugged and screwed	1.10	0.28	3.53	1.78	m	5.31
Extra for						
fittings with one end	-	0.18	2.27	1.11	nr	3.38
fittings with two ends	-	0.28	3.53	1.19	nr	4.72
fittings with three ends	-	0.37	4.66	1.84	nr	6.50
fittings with four ends	4.24	0.49	6.17	4.94	nr	11.11
access plug	0.87	0.18	2.27	1.11	nr	3.38
straight coupling	0.63	0.19	2.39	0.85	nr	3.24
expansion coupling	1.14	0.28	3.53	1.41	nr	4.94
male iron to muPVC coupling	0.96	0.35	4.41	1.13	nr	5.54
union coupling	2.96	0.28	3.53	3.41	nr	6.94
level invert taper	0.92	0.28	3.53	1.17	nr	4.70
sweep bend	0.94	0.28	3.53	1.19	nr	4.72
spigot/socket bend	1.06	0.28	3.53	1.32	nr	4.85
sweep tee	1.50	0.37	4.66	1.84	nr	6.50
sweep cross	4.24	0.49	6.17	4.94	nr	11.11
50 mm pipes; fixing with pipe clips; plugged and screwed	1.66	0.32	4.03	2.72	m	6.75
Extra for						
fittings with one end	-	0.19	2.39	1.72	nr	4.11
fittings with two ends	-	0.32	4.03	1.87	nr	5.90
fittings with three ends	-	0.43	5.42	3.20	nr	8.62
fittings with four ends	-	0.57	7.18	5.65	nr	12.83
access plug	1.42	0.19	2.39	1.72	nr	4.11
straight coupling	0.99	0.21	2.65	1.24	nr	3.89
expansion coupling	1.55	0.32	4.03	1.86	nr	5.89
male iron to muPVC coupling	1.38	0.42	5.29	1.60	nr	6.89
union coupling	4.62	0.32	4.03	5.24	nr	9.27
level invert taper	1.28	0.32	4.03	1.56	nr	5.59
sweep bend	1.56	0.32	4.03	1.87	nr	5.90
spigot/socket bend	2.52	0.32	4.03	2.93	nr	6.96
sweep tee	1.50	0.37	4.66	1.84	nr	6.50
sweep cross	4.89	0.57	7.18	5.65	nr	12.83

R DISPOSAL SYSTEMS

	PC £	Labour hours	Labour £	Material £	Unit	Total rate £
uPVC overflow pipes and fittings; solvent welded joints						
19 mm pipes; fixing with pipe clips; plugged and screwed	0.47	0.20	2.52	0.91	m	3.43
Extra for						
splay cut end	-	0.01	0.13	-	nr	0.13
fittings with one end	-	0.16	2.02	0.62	nr	2.64
fittings with two ends	-	0.16	2.02	0.73	nr	2.75
fittings with three ends	-	0.20	2.52	0.82	nr	3.34
straight connector	0.51	0.16	2.02	0.62	nr	2.64
female iron to uPVC coupling	-	0.19	2.39	0.98	nr	3.37
bend	0.61	0.16	2.02	0.73	nr	2.75
bent tank connector	0.94	0.19	2.39	1.05	nr	3.44
uPVC pipes and fittings; BS 4514; with solvent welded joints (unless otherwise described)						
82 mm pipes; fixing with holderbats; plugged and screwed	3.24	0.37	4.66	5.25	m	9.91
Extra for						
socket plug	2.57	0.19	2.39	3.22	nr	5.61
slip coupling; push fit	5.60	0.34	4.28	6.17	nr	10.45
expansion coupling	2.70	0.37	4.66	3.36	nr	8.02
sweep bend	4.52	0.37	4.66	5.37	nr	10.03
boss connector	2.47	0.25	3.15	3.11	nr	6.26
single branch	6.32	0.49	6.17	7.59	nr	13.76
access door	6.02	0.56	7.06	6.83	nr	13.89
connection to clay pipes; caulking ring and cement and sand (1:2) joint	4.24	0.34	4.28	4.95	nr	9.23
110 mm pipes; fixing with holderbats; plugged and screwed	3.30	0.41	5.17	5.48	m	10.65
Extra for						
socket plug	3.11	0.20	2.52	3.93	nr	6.45
slip coupling; push fit	7.01	0.37	4.66	7.72	nr	12.38
expansion coupling	2.76	0.41	5.17	3.55	nr	8.72
W.C. connector	5.01	0.27	3.40	5.79	nr	9.19
sweep bend	5.29	0.41	5.17	6.33	nr	11.50
W.C. connecting bend	8.22	0.27	3.40	9.33	nr	12.73
access bend	14.69	0.43	5.42	16.69	nr	22.11
boss connector	2.47	0.27	3.40	3.23	nr	6.63
single branch	8.50	0.54	6.80	10.14	nr	16.94
single branch with access	11.99	0.56	7.06	13.99	nr	21.05
double branch	18.09	0.68	8.57	20.98	nr	29.55
W.C. manifold	31.89	0.27	3.40	35.92	nr	39.32
access door	-	0.56	7.06	6.83	nr	13.89
access pipe connector	11.25	0.46	5.80	12.90	nr	18.70
connection to clay pipes; caulking ring and cement and sand (1:2) joint	-	0.39	4.91	1.81	nr	6.72
160 mm pipes; fixing with holderbats; plugged and screwed	8.57	0.46	5.80	14.09	m	19.89
Extra for						
socket plug	5.72	0.23	2.90	7.43	nr	10.33
slip coupling; push fit	17.95	0.42	5.29	19.78	nr	25.07
expansion coupling	8.30	0.46	5.80	10.27	nr	16.07
sweep bend	13.19	0.46	5.80	15.66	nr	21.46
boss connector	3.50	0.31	3.91	4.98	nr	8.89
single branch	31.07	0.61	7.69	35.90	nr	43.59
double branch	72.64	0.77	9.70	82.25	nr	91.95
access door	10.76	0.56	7.06	12.05	nr	19.11
access pipe connector	11.25	0.46	5.80	12.90	nr	18.70
connection to clay pipes; caulking ring and cement and sand (1:2) joint	-	0.46	5.80	2.99	nr	8.79

Prices for Measured Works - Major Works
R DISPOSAL SYSTEMS

	PC £	Labour hours	Labour £	Material £	Unit	Total rate £
R11 FOUL DRAINAGE ABOVE GROUND - cont'd						
uPVC pipes and fittings; BS 4514; with solvent welded joints (unless otherwise described) - cont'd						
Weathering apron; for pipe						
82 mm diameter	1.28	0.31	3.91	1.60	nr	5.51
110 mm diameter	1.46	0.35	4.41	1.88	nr	6.29
160 mm diameter	4.40	0.39	4.91	5.39	nr	10.30
Weathering slate; for pipe						
110 mm diameter	21.01	0.83	10.46	23.42	nr	33.88
Vent cowl; for pipe						
82 mm diameter	1.28	0.31	3.91	1.60	nr	5.51
110 mm diameter	1.29	0.31	3.91	1.69	nr	5.60
160 mm diameter	3.37	0.31	3.91	4.26	nr	8.17
Polypropylene ancillaries; screwed joint to waste fitting						
Tubular "S" trap; bath; shallow seal						
40 mm diameter	3.07	0.51	6.43	3.38	nr	9.81
Trap; "P"; two piece; 76 mm seal						
32 mm diameter	2.07	0.35	4.41	2.28	nr	6.69
40 mm diameter	2.39	0.42	5.29	2.63	nr	7.92
Trap; "S"; two piece; 76 mm seal						
32 mm diameter	2.63	0.35	4.41	2.90	nr	7.31
40 mm diameter	3.07	0.42	5.29	3.38	nr	8.67
Bottle trap; "P"; 76 mm seal						
32 diameter	2.31	0.35	4.41	2.55	nr	6.96
40 diameter	2.76	0.42	5.29	3.04	nr	8.33
Bottle trap; "S"; 76 mm seal						
32 diameter	2.78	0.35	4.41	3.06	nr	7.47
40 diameter	3.38	0.42	5.29	3.72	nr	9.01
R12 DRAINAGE BELOW GROUND						
NOTE: Prices for drain trenches are for excavation in "firm" soil and it has been assumed that earthwork support will only be required for trenches 1.00 m or more in depth.						
Excavating trenches; by machine; grading bottoms; earthwork support; filling with excavated material and compacting; disposal of surplus soil; spreading on site average 50 m						
Pipes not exceeding 200 mm nominal size						
average depth of trench 0.50 m	-	0.28	2.24	1.60	m	3.84
average depth of trench 0.75 m	-	0.37	2.96	2.39	m	5.35
average depth of trench 1.00 m	-	0.79	6.32	4.73	m	11.05
average depth of trench 1.25 m	-	1.16	9.28	5.42	m	14.70
average depth of trench 1.50 m	-	1.48	11.84	6.16	m	18.00
average depth of trench 1.75 m	-	1.85	14.80	6.86	m	21.66
average depth of trench 2.00 m	-	2.13	17.04	7.79	m	24.83
average depth of trench 2.25 m	-	2.64	21.11	9.79	m	30.90
average depth of trench 2.50 m	-	3.10	24.79	11.45	m	36.24
average depth of trench 2.75 m	-	3.42	27.35	12.75	m	40.10
average depth of trench 3.00 m	-	3.75	29.99	14.04	m	44.03
average depth of trench 3.25 m	-	4.07	32.55	14.94	m	47.49
average depth of trench 3.50 m	-	4.35	34.79	15.84	m	50.63

R DISPOSAL SYSTEMS

	PC £	Labour hours	Labour £	Material £	Unit	Total rate £
Pipes exceeding 200 mm nominal size; 225 mm nominal size						
average depth of trench 0.50 m	-	0.28	2.24	1.60	m	3.84
average depth of trench 0.75 m	-	0.37	2.96	2.39	m	5.35
average depth of trench 1.00 m	-	0.79	6.32	4.73	m	11.05
average depth of trench 1.25 m	-	1.16	9.28	5.42	m	14.70
average depth of trench 1.50 m	-	1.48	11.84	6.16	m	18.00
average depth of trench 1.75 m	-	1.85	14.80	6.86	m	21.66
average depth of trench 2.00 m	-	2.13	17.04	7.79	m	24.83
average depth of trench 2.25 m	-	2.64	21.11	9.79	m	30.90
average depth of trench 2.50 m	-	3.10	24.79	11.45	m	36.24
average depth of trench 2.75 m	-	3.42	27.35	12.75	m	40.10
average depth of trench 3.00 m	-	3.75	29.99	14.04	m	44.03
average depth of trench 3.25 m	-	4.07	32.55	14.94	m	47.49
average depth of trench 3.50 m	-	4.35	34.79	15.84	m	50.63
Pipes exceeding 200 mm nominal size; 300 mm nominal size						
average depth of trencg 0.75 m	-	0.44	3.52	2.99	m	6.51
average depth of trench 1.00 m	-	0.93	7.44	4.73	m	12.17
average depth of trench 1.25 m	-	1.25	10.00	5.62	m	15.62
average depth of trench 1.50 m	-	1.62	12.96	6.36	m	19.32
average depth of trench 1.75 m	-	1.85	14.80	7.05	m	21.85
average depth of trench 2.00 m	-	2.13	17.04	8.39	m	25.43
average depth of trench 2.25 m	-	2.64	21.11	10.18	m	31.29
average depth of trench 2.50 m	-	3.10	24.79	11.65	m	36.44
average depth of trench 2.75 m	-	3.42	27.35	12.95	m	40.30
average depth of trench 3.00 m	-	3.75	29.99	14.24	m	44.23
average depth of trench 3.25 m	-	4.07	32.55	15.54	m	48.09
average depth of trench 3.50 m	-	4.35	34.79	16.24	m	51.03
Pipes exceeding 200 mm nominal size; 375 mm nominal size						
average depth of trench 0.75 m	-	0.46	3.68	3.59	m	7.27
average depth of trench 1.00 m	-	0.97	7.76	5.32	m	13.08
average depth of trench 1.25 m	-	1.34	10.72	6.62	m	17.34
average depth of trench 1.50 m	-	1.71	13.68	7.15	m	20.83
average depth of trench 1.75 m	-	1.99	15.92	8.05	m	23.97
average depth of trench 2.00 m	-	2.27	18.16	8.58	m	26.74
average depth of trench 2.25 m	-	2.82	22.55	10.78	m	33.33
average depth of trench 2.50 m	-	3.38	27.03	12.45	m	39.48
average depth of trench 2.75 m	-	3.70	29.59	13.54	m	43.13
average depth of trench 3.00 m	-	4.02	32.15	14.64	m	46.79
average depth of trench 3.25 m	-	4.35	34.79	15.94	m	50.73
average depth of trench 3.50 m	-	4.67	37.35	17.04	m	54.39
Pipes exceeding 200 mm nominal size; 450 mm nominal size						
average depth of trench 0.75 m	-	0.51	4.08	3.59	m	7.67
average depth of trench 1.00 m	-	1.02	8.16	5.72	m	13.88
average depth of trench 1.25 m	-	1.48	11.84	7.02	m	18.86
average depth of trench 1.50 m	-	1.85	14.80	7.75	m	22.55
average depth of trench 1.75 m	-	2.13	17.04	8.45	m	25.49
average depth of trench 2.00 m	-	2.45	19.60	9.18	m	28.78
average depth of trench 2.25 m	-	3.05	24.39	11.18	m	35.57
average depth of trench 2.50 m	-	3.61	28.87	13.04	m	41.91
average depth of trench 2.75 m	-	3.98	31.83	14.34	m	46.17
average depth of trench 3.00 m	-	4.26	34.07	15.64	m	49.71
average depth of trench 3.25 m	-	4.63	37.03	17.14	m	54.17
average depth of trench 3.50 m	-	5.00	39.99	18.64	m	58.63
Pipes exceeding 200 mm nominal size; 600 mm nominal size						
average depth of trench 1.00 m	-	1.11	8.88	6.12	m	15.00
average depth of trench 1.25 m	-	1.57	12.56	7.42	m	19.98

R DISPOSAL SYSTEMS

	PC £	Labour hours	Labour £	Material £	Unit	Total rate £
R12 DRAINAGE BELOW GROUND - cont'd						
Excavating trenches; by machine; grading bottoms; earthwork support; filling with excavated material and compacting; disposal of surplus soil; spreading on site average 50 m - cont'd						
Pipes exceeding 200 mm nominal size; 600 mm nominal size - cont'd						
average depth of trench 1.50 m	-	2.04	16.32	8.55	m	24.87
average depth of trench 1.75 m	-	2.31	18.48	9.05	m	27.53
average depth of trench 2.00 m	-	2.73	21.83	9.98	m	31.81
average depth of trench 2.25 m	-	3.28	26.23	12.38	m	38.61
average depth of trench 2.50 m	-	3.89	31.11	14.44	m	45.55
average depth of trench 2.75 m	-	4.30	34.39	16.14	m	50.53
average depth of trench 3.00 m	-	4.72	37.75	17.64	m	55.39
average depth of trench 3.25 m	-	5.09	40.71	18.93	m	59.64
average depth of trench 3.50 m	-	5.46	43.67	20.03	m	63.70
Pipes exceeding 200 mm nominal size; 900 mm nominal size						
average depth of trench 1.25 m	-	1.90	15.20	8.62	m	23.82
average depth of trench 1.50 m	-	2.41	19.28	9.75	m	29.03
average depth of trench 1.75 m	-	2.78	22.23	10.45	m	32.68
average depth of trench 2.00 m	-	3.10	24.79	11.98	m	36.77
average depth of trench 2.25 m	-	3.84	30.71	14.57	m	45.28
average depth of trench 2.50 m	-	4.53	36.23	16.84	m	53.07
average depth of trench 2.75 m	-	5.00	39.99	18.53	m	58.52
average depth of trench 3.00 m	-	5.46	43.67	20.23	m	63.90
average depth of trench 3.25 m	-	5.92	47.35	21.93	m	69.28
average depth of trench 3.50 m	-	6.38	51.03	23.42	m	74.45
Pipes exceeding 200 mm nominal size; 1200 mm nominal size						
average depth of trench 1.50 m	-	2.73	21.83	10.35	m	32.18
average depth of trench 1.75 m	-	3.19	25.51	12.04	m	37.55
average depth of trench 2.00 m	-	3.56	28.47	13.77	m	42.24
average depth of trench 2.25 m	-	4.35	34.79	16.77	m	51.56
average depth of trench 2.50 m	-	5.18	41.43	19.23	m	60.66
average depth of trench 2.75 m	-	5.69	45.51	21.33	m	66.84
average depth of trench 3.00 m	-	6.20	49.59	23.22	m	72.81
average depth of trench 3.25 m	-	6.75	53.99	25.12	m	79.11
average depth of trench 3.50 m	-	7.26	58.07	27.02	m	85.09
Extra over excavating trenches; irrespective of depth; breaking out existing materials						
brick	-	1.80	14.40	6.93	m³	21.33
concrete	-	2.54	20.31	9.58	m³	29.89
reinforced concrete	-	3.61	28.87	13.86	m³	42.73
Extra over excavating trenches; irrespective of depth; breaking out existing hard pavings; 75 mm thick						
tarmacadam	-	0.19	1.52	0.73	m²	2.25
Extra over excavating trenches; irrsepective of depth; breaking out existing hard pavings; 150 mm thick						
concrete	-	0.37	2.96	1.63	m²	4.59
tarmacadam and hardcore	-	0.28	2.24	0.84	m²	3.08
Excavating trenches; by hand; grading bottoms; earthwork support; filling with excavated material and compacting; disposal of surplus soil on site; spreading on site average 50 m						
Pipes not exceeding 200 mm nominal size						
average depth of trench 0.50 m	-	0.93	7.44	-	m	7.44
average depth of trench 0.75 m	-	1.39	11.12	-	m	11.12
average depth of trench 1.00 m	-	2.04	16.32	1.33	m	17.65

R DISPOSAL SYSTEMS

	PC £	Labour hours	Labour £	Material £	Unit	Total rate £
Pipes not exceeding 200 mm nominal size - cont'd						
average depth of trench 1.25 m	-	2.87	22.95	1.83	m	24.78
average depth of trench 1.50 m	-	3.93	31.43	2.23	m	33.66
average depth of trench 1.75 m	-	5.18	41.43	2.67	m	44.10
average depth of trench 2.00 m	-	5.92	47.35	3.00	m	50.35
average depth of trench 2.25 m	-	7.40	59.19	4.00	m	63.19
average depth of trench 2.50 m	-	8.88	71.02	4.66	m	75.68
average depth of trench 2.75 m	-	9.76	78.06	5.16	m	83.22
average depth of trench 3.00 m	-	10.64	85.10	5.66	m	90.76
average depth of trench 3.25 m	-	11.52	92.14	6.16	m	98.30
average depth of trench 3.50 m	-	12.40	99.18	6.66	m	105.84
Pipes exceeding 200 mm nominal size; 225 mm nominal size						
average depth of trench 0.50 m	-	0.93	7.44	-	m	7.44
average depth of trench 0.75 m	-	1.39	11.12	-	m	11.12
average depth of trench 1.00 m	-	2.04	16.32	1.33	m	17.65
average depth of trench 1.25 m	-	2.87	22.95	1.83	m	24.78
average depth of trench 1.50 m	-	3.93	31.43	2.23	m	33.66
average depth of trench 1.75 m	-	5.18	41.43	2.67	m	44.10
average depth of trench 2.00 m	-	5.92	47.35	3.00	m	50.35
average depth of trench 2.25 m	-	7.40	59.19	4.00	m	63.19
average depth of trench 2.50 m	-	8.88	71.02	4.66	m	75.68
average depth of trench 2.75 m	-	9.76	78.06	5.16	m	83.22
average depth of trench 3.00 m	-	10.64	85.10	5.66	m	90.76
average depth of trench 3.25 m	-	11.52	92.14	6.16	m	98.30
average depth of trench 3.50 m	-	12.40	99.18	6.66	m	105.84
Pipes exceeding 200 mm nominal size; 300 mm nominal size						
average depth of trench 0.75 m	-	1.62	12.96	-	m	12.96
average depth of trench 1.00 m	-	2.36	18.88	1.33	m	20.21
average depth of trench 1.25 m	-	3.33	26.63	1.83	m	28.46
average depth of trench 1.50 m	-	4.44	35.51	2.23	m	37.74
average depth of trench 1.75 m	-	5.18	41.43	2.67	m	44.10
average depth of trench 2.00 m	-	5.92	47.35	3.00	m	50.35
average depth of trench 2.25 m	-	7.40	59.19	4.00	m	63.19
average depth of trench 2.50 m	-	8.88	71.02	4.66	m	75.68
average depth of trench 2.75 m	-	9.76	78.06	5.16	m	83.22
average depth of trench 3.00 m	-	10.64	85.10	5.66	m	90.76
average depth of trench 3.25 m	-	11.52	92.14	6.16	m	98.30
average depth of trench 3.50 m	-	12.40	99.18	6.66	m	105.84
Pipes exceeding 200 mm nominal size; 375 mm nominal size						
average depth of trench 0.75 m	-	1.80	14.40	-	m	14.40
average depth of trench 1.00 m	-	2.64	21.11	1.33	m	22.44
average depth of trench 1.25 m	-	3.70	29.59	1.83	m	31.42
average depth of trench 1.50 m	-	4.93	39.43	2.23	m	41.66
average depth of trench 1.75 m	-	5.74	45.91	2.67	m	48.58
average depth of trench 2.00 m	-	6.57	52.55	3.00	m	55.55
average depth of trench 2.25 m	-	8.23	65.82	4.00	m	69.82
average depth of trench 2.50 m	-	9.90	79.18	4.66	m	83.84
average depth of trench 2.75 m	-	10.87	86.94	5.16	m	92.10
average depth of trench 3.00 m	-	11.84	94.70	5.66	m	100.36
average depth of trench 3.25 m	-	12.86	102.85	6.16	m	109.01
average depth of trench 3.50 m	-	13.88	111.01	6.66	m	117.67
Pipes exceeding 200 mm nominal size; 450 mm nominal size						
average depth of trench 0.75 m	-	2.04	16.32	-	m	16.32
average depth of trench 1.00 m	-	2.94	23.51	1.33	m	24.84
average depth of trench 1.25 m	-	4.13	33.03	1.83	m	34.86
average depth of trench 1.50 m	-	5.41	43.27	2.23	m	45.50
average depth of trench 1.75 m	-	6.31	50.47	2.67	m	53.14

R12 DRAINAGE BELOW GROUND - cont'd

	PC £	Labour hours	Labour £	Material £	Unit	Total rate £
Excavating trenches; by hand; grading bottoms; earthwork support; filling with excavated material and compacting; disposal of surplus soil on site; spreading on site average 50 m - cont'd						
Pipes exceeding 200 mm nominal size; 450 mm nominal size - cont'd						
average depth of trench 2.00 m	-	7.22	57.75	3.00	m	60.75
average depth of trench 2.25 m	-	9.05	72.38	4.00	m	76.38
average depth of trench 2.50 m	-	10.87	86.94	4.66	m	91.60
average depth of trench 2.75 m	-	11.96	95.66	5.16	m	100.82
average depth of trench 3.00 m	-	13.04	104.29	5.66	m	109.95
average depth of trench 3.25 m	-	14.11	112.85	6.16	m	119.01
average depth of trench 3.50 m	-	15.17	121.33	6.66	m	127.99
Pipes exceeding 200 mm nominal size; 600 mm nominal size						
average depth of trench 1.00 m	-	3.24	25.91	1.33	m	27.24
average depth of trench 1.25 m	-	4.63	37.03	1.83	m	38.86
average depth of trench 1.50 m	-	6.20	49.59	2.23	m	51.82
average depth of trench 1.75 m	-	7.17	57.35	2.67	m	60.02
average depth of trench 2.00 m	-	8.19	65.50	3.00	m	68.50
average depth of trench 2.25 m	-	9.20	73.58	4.00	m	77.58
average depth of trench 2.50 m	-	11.56	92.46	4.66	m	97.12
average depth of trench 2.75 m	-	12.35	98.78	5.16	m	103.94
average depth of trench 3.00 m	-	14.80	118.37	5.66	m	124.03
average depth of trench 3.25 m	-	16.03	128.21	6.16	m	134.37
average depth of trench 3.50 m	-	17.25	137.97	6.66	m	144.63
Pipes exceeding 200 mm nominal size; 900 mm nominal size						
average depth of trench 1.25 m	-	5.78	46.23	1.83	m	48.06
average depth of trench 1.50 m	-	7.63	61.02	2.23	m	63.25
average depth of trench 1.75 m	-	8.88	71.02	2.67	m	73.69
average depth of trench 2.00 m	-	10.13	81.02	3.00	m	84.02
average depth of trench 2.25 m	-	12.72	101.73	4.00	m	105.73
average depth of trench 2.50 m	-	15.31	122.45	4.66	m	127.11
average depth of trench 2.75 m	-	16.84	134.69	5.16	m	139.85
average depth of trench 3.00 m	-	18.32	146.52	5.66	m	152.18
average depth of trench 3.25 m	-	19.84	158.68	6.16	m	164.84
average depth of trench 3.50 m	-	21.37	170.92	6.66	m	177.58
Pipes exceeding 200 mm nominal size; 1200 mm nominal size						
average depth of trench 1.50 m	-	9.11	72.86	2.23	m	75.09
average depth of trench 1.75 m	-	10.59	84.70	2.67	m	87.37
average depth of trench 2.00 m	-	12.12	96.94	3.00	m	99.94
average depth of trench 2.25 m	-	15.20	121.57	4.00	m	125.57
average depth of trench 2.50 m	-	18.27	146.12	4.66	m	150.78
average depth of trench 2.75 m	-	20.07	160.52	5.16	m	165.68
average depth of trench 3.00 m	-	21.88	175.00	5.66	m	180.66
average depth of trench 3.25 m	-	23.66	189.23	6.16	m	195.39
average depth of trench 3.50 m	-	25.44	203.47	6.66	m	210.13
Extra over excavating trenches irrespective of depth; breaking out existing materials						
brick	-	2.78	22.23	4.97	m³	27.20
concrete	-	4.16	33.27	8.28	m³	41.55
reinforced concrete	-	5.55	44.39	11.59	m³	55.98
concrete; 150 mm thick	-	0.65	5.20	1.16	m²	6.36
tarmacadam and hardcore; 150 mm thick	-	0.46	3.68	0.83	m²	4.51
Extra over excavating trenches irrespective of depth; breaking out existing hard pavings, 75 mm thick						
tarmacadam	-	0.37	2.96	0.66	m²	3.62

R DISPOSAL SYSTEMS

	PC £	Labour hours	Labour £	Material £	Unit	Total rate £
Sand filling						
Beds; to receive pitch fibre pipes						
600 mm x 50 mm thick	-	0.07	0.56	0.61	m	1.17
700 mm x 50 mm thick	-	0.09	0.72	0.72	m	1.44
800 mm x 50 mm thick	-	0.11	0.88	0.82	m	1.70
Granular (shingle) filling						
Beds; 100 mm thick; to pipes						
100 mm nominal size	-	0.09	0.72	1.35	m	2.07
150 mm nominal size	-	0.09	0.72	1.58	m	2.30
225 mm nominal size	-	0.11	0.88	1.80	m	2.68
300 mm nominal size	-	0.13	1.04	2.03	m	3.07
375 mm nominal size	-	0.15	1.20	2.25	m	3.45
450 mm nominal size	-	0.17	1.36	2.48	m	3.84
600 mm nominal size	-	0.19	1.52	2.70	m	4.22
Beds; 150 mm thick; to pipes						
100 mm nominal size	-	0.13	1.04	2.03	m	3.07
150 mm nominal size	-	0.15	1.20	2.25	m	3.45
225 mm nominal size	-	0.17	1.36	2.48	m	3.84
300 mm nominal size	-	0.19	1.52	2.70	m	4.22
375 mm nominal size	-	0.22	1.76	3.38	m	5.14
450 mm nominal size	-	0.24	1.92	3.60	m	5.52
600 mm nominal size	-	0.28	2.24	4.28	m	6.52
Beds and benchings; beds 100 mm thick; to pipes						
100 nominal size	-	0.21	1.68	2.48	m	4.16
150 nominal size	-	0.23	1.84	2.48	m	4.32
225 nominal size	-	0.28	2.24	3.38	m	5.62
300 nominal size	-	0.32	2.56	3.83	m	6.39
375 nominal size	-	0.42	3.36	5.18	m	8.54
450 nominal size	-	0.48	3.84	5.86	m	9.70
600 nominal size	-	0.62	4.96	7.66	m	12.62
Beds and benchings; beds 150 mm thick; to pipes						
100 nominal size	-	0.23	1.84	2.70	m	4.54
150 nominal size	-	0.26	2.08	2.93	m²	5.01
225 nominal size	-	0.32	2.56	4.05	m	6.61
300 nominal size	-	0.42	3.36	4.96	m	8.32
375 nominal size	-	0.48	3.84	5.86	m	9.70
450 nominal size	-	0.57	4.56	6.98	m	11.54
600 nominal size	-	0.68	5.44	9.01	m	14.45
Beds and coverings; 100 mm thick; to pipes						
100 nominal size	-	0.33	2.64	3.38	m	6.02
150 nominal size	-	0.42	3.36	4.05	m	7.41
225 nominal size	-	0.56	4.48	5.63	m	10.11
300 nominal size	-	0.67	5.36	6.76	m	12.12
375 nominal size	-	0.80	6.40	8.11	m	14.51
450 nominal size	-	0.94	7.52	9.69	m	17.21
600 nominal size	-	1.22	9.76	12.39	m	22.15
Beds and coverings; 150 mm thick; to pipes						
100 nominal size	-	0.50	4.00	4.96	m	8.96
150 nominal size	-	0.56	4.48	5.63	m	10.11
225 nominal size	-	0.72	5.76	7.21	m	12.97
300 nominal size	-	0.86	6.88	8.56	m	15.44
375 nominal size	-	1.00	8.00	10.14	m	18.14
450 nominal size	-	1.19	9.52	12.16	m	21.68
600 nominal size	-	1.44	11.52	14.64	m	26.16
Plain in situ ready mixed designated concrete; C10 - 40 mm aggregate						
Beds; 100 mm thick; to pipes						
100 mm nominal size	-	0.17	1.50	3.16	m	4.66
150 mm nominal size	-	0.17	1.50	3.16	m	4.66

Prices for Measured Works - Major Works
R DISPOSAL SYSTEMS

	PC £	Labour hours	Labour £	Material £	Unit	Total rate £
R12 DRAINAGE BELOW GROUND - cont'd						
Plain in situ ready mixed designated concrete; C10 - 40 mm aggregate - cont'd						
Beds; 100 mm thick; to pipes - cont'd						
225 mm nominal size	-	0.20	1.77	3.78	m	5.55
300 mm nominal size	-	0.23	2.03	4.42	m	6.45
375 mm nominal size	-	0.27	2.39	5.04	m	7.43
450 mm nominal size	-	0.30	2.65	5.68	m	8.33
600 mm nominal size	-	0.33	2.92	6.31	m	9.23
900 mm nominal size	-	0.40	3.53	7.57	m	11.10
1200 mm nominal size	-	0.54	4.77	10.09	m	14.86
Beds; 150 mm thick; to pipes						
100 mm nominal size	-	0.23	2.03	4.42	m	6.45
150 mm nominal size	-	0.27	2.39	5.04	m	7.43
225 mm nominal size	-	0.30	2.65	5.68	m	8.33
300 mm nominal size	-	0.33	2.92	6.31	m	9.23
375 mm nominal size	-	0.40	3.53	7.57	m	11.10
450 mm nominal size	-	0.43	3.80	8.20	m	12.00
600 mm nominal size	-	0.50	4.42	9.46	m	13.88
900 mm nominal size	-	0.63	5.57	11.98	m	17.55
1200 mm nominal size	-	0.77	6.80	14.50	m	21.30
Beds and benchings; beds 100 mm thick; to pipes						
100 mm nominal size	-	0.33	2.92	5.68	m	8.60
150 mm nominal size	-	0.38	3.36	6.31	m	9.67
225 mm nominal size	-	0.45	3.98	7.57	m	11.55
300 mm nominal size	-	0.53	4.68	8.83	m	13.51
375 mm nominal size	-	0.68	6.01	11.35	m	17.36
450 mm nominal size	-	0.80	7.07	13.24	m	20.31
600 mm nominal size	-	1.02	9.01	17.03	m	26.04
900 mm nominal size	-	1.65	14.58	27.74	m	42.32
1200 mm nominal size	-	2.44	21.56	40.98	m	62.54
Beds and benchings; beds 150 mm thick; to pipes						
100 mm nominal size	-	0.38	3.36	6.31	m	9.67
150 mm nominal size	-	0.42	3.71	6.93	m	10.64
225 mm nominal size	-	0.53	4.68	8.83	m	13.51
300 mm nominal size	-	0.68	6.01	11.35	m	17.36
375 mm nominal size	-	0.80	7.07	13.24	m	20.31
450 mm nominal size	-	0.94	8.31	15.76	m	24.07
600 mm nominal size	-	1.20	10.60	20.18	m	30.78
900 mm nominal size	-	1.91	16.88	32.16	m	49.04
1200 mm nominal size	-	2.70	23.86	45.40	m	69.26
Beds and coverings; 100 mm thick; to pipes						
100 mm nominal size	-	0.50	4.42	7.57	m	11.99
150 mm nominal size	-	0.58	5.13	8.83	m	13.96
225 mm nominal size	-	0.83	7.33	12.61	m	19.94
300 mm nominal size	-	1.00	8.84	15.13	m	23.97
375 mm nominal size	-	1.21	10.69	18.28	m	28.97
450 mm nominal size	-	1.42	12.55	21.44	m	33.99
600 mm nominal size	-	1.83	16.17	27.74	m	43.91
900 mm nominal size	-	2.79	24.65	42.25	m	66.90
1200 mm nominal size	-	3.83	33.84	58.01	m	91.85
Beds and coverings; 150 mm thick; to pipes						
100 mm nominal size	-	0.75	6.63	11.35	m	17.98
150 mm nominal size	-	0.83	7.33	12.61	m	19.94
225 mm nominal size	-	1.08	9.54	16.39	m	25.93
300 mm nominal size	-	1.30	11.49	19.54	m	31.03
375 mm nominal size	-	1.50	13.25	22.70	m	35.95
450 mm nominal size	-	1.79	15.82	27.11	m	42.93
600 mm nominal size	-	2.16	19.09	32.79	m	51.88
900 mm nominal size	-	3.54	31.28	53.59	m	84.87

R DISPOSAL SYSTEMS

	PC £	Labour hours	Labour £	Material £	Unit	Total rate £
Plain in situ ready mixed designated concrete; C20 - 40 mm aggregate						
Beds; 100 mm thick; to pipes						
100 mm nominal size	-	0.17	1.50	3.26	m	**4.76**
150 mm nominal size	-	0.17	1.50	3.26	m	**4.76**
225 mm nominal size	-	0.20	1.77	3.91	m	**5.68**
300 mm nominal size	-	0.23	2.03	4.57	m	**6.60**
375 mm nominal size	-	0.27	2.39	5.22	m	**7.61**
450 mm nominal size	-	0.30	2.65	5.87	m	**8.52**
600 mm nominal size	-	0.33	2.92	6.52	m	**9.44**
900 mm nominal size	-	0.40	3.53	7.83	m	**11.36**
1200 mm nominal size	-	0.54	4.77	10.44	m	**15.21**
Beds; 150 mm thick; to pipes						
100 mm nominal size	-	0.23	2.03	4.57	m	**6.60**
150 mm nominal size	-	0.27	2.39	5.22	m	**7.61**
225 mm nominal size	-	0.30	2.65	5.87	m	**8.52**
300 mm nominal size	-	0.33	2.92	6.52	m	**9.44**
375 mm nominal size	-	0.40	3.53	7.83	m	**11.36**
450 mm nominal size	-	0.43	3.80	8.48	m	**12.28**
600 mm nominal size	-	0.50	4.42	9.79	m	**14.21**
900 mm nominal size	-	0.63	5.57	12.40	m	**17.97**
1200 mm nominal size	-	0.77	6.80	15.00	m	**21.80**
Beds and benchings; beds 100 mm thick; to pipes						
100 mm nominal size	-	0.33	2.92	5.87	m	**8.79**
150 mm nominal size	-	0.38	3.36	6.52	m	**9.88**
225 mm nominal size	-	0.45	3.98	7.83	m	**11.81**
300 mm nominal size	-	0.53	4.68	9.13	m	**13.81**
375 mm nominal size	-	0.68	6.01	11.74	m	**17.75**
450 mm nominal size	-	0.80	7.07	13.69	m	**20.76**
600 mm nominal size	-	1.02	9.01	17.61	m	**26.62**
900 mm nominal size	-	1.65	14.58	28.70	m	**43.28**
1200 mm nominal size	-	2.44	21.56	42.39	m	**63.95**
Beds and benchings; beds 150 mm thick; to pipes						
100 mm nominal size	-	0.38	3.36	6.52	m	**9.88**
150 mm nominal size	-	0.42	3.71	7.17	m	**10.88**
225 mm nominal size	-	0.53	4.68	9.13	m	**13.81**
300 mm nominal size	-	0.68	6.01	11.74	m	**17.75**
375 mm nominal size	-	0.80	7.07	13.69	m	**20.76**
450 mm nominal size	-	0.94	8.31	16.30	m	**24.61**
600 mm nominal size	-	1.20	10.60	20.87	m	**31.47**
900 mm nominal size	-	1.91	16.88	33.27	m	**50.15**
1200 mm nominal size	-	2.70	23.86	46.96	m	**70.82**
Beds and coverings; 100 mm thick; to pipes						
100 mm nominal size	-	0.50	4.42	7.83	m	**12.25**
150 mm nominal size	-	0.58	5.13	9.13	m	**14.26**
225 mm nominal size	-	0.83	7.33	13.04	m	**20.37**
300 mm nominal size	-	1.00	8.84	15.65	m	**24.49**
375 mm nominal size	-	1.21	10.69	18.91	m	**29.60**
450 mm nominal size	-	1.42	12.55	22.18	m	**34.73**
600 mm nominal size	-	1.83	16.17	28.70	m	**44.87**
900 mm nominal size	-	2.79	24.65	43.70	m	**68.35**
1200 mm nominal size	-	3.83	33.84	60.01	m	**93.85**
Beds and coverings; 150 mm thick; to pipes						
100 mm nominal size	-	0.75	6.63	11.74	m	**18.37**
150 mm nominal size	-	0.83	7.33	13.04	m	**20.37**
225 mm nominal size	-	1.08	9.54	16.96	m	**26.50**
300 mm nominal size	-	1.30	11.49	20.22	m	**31.71**
375 mm nominal size	-	1.50	13.25	23.48	m	**36.73**
450 mm nominal size	-	1.79	15.82	28.04	m	**43.86**
600 mm nominal size	-	2.16	19.09	33.92	m	**53.01**
900 mm nominal size	-	3.54	31.28	55.44	m	**86.72**

Prices for Measured Works - Major Works
R DISPOSAL SYSTEMS

	PC £	Labour hours	Labour £	Material £	Unit	Total rate £
R12 DRAINAGE BELOW GROUND - cont'd						
NOTE: The following items unless otherwise described include for all appropriate joints/couplings in the running length. The prices for gullies and rainwater shoes, etc. include for appropriate joints to pipes and for setting on and surrounding accessory with site mixed in situ concrete 10.00 N/mm² - 40 mm aggregate (1:3:6).						
Cast iron "Timesaver" drain pipes and fittings or other equal and approved; BS 437; coated; with mechanical coupling joints						
75 mm pipes; laid straight	17.15	0.42	3.29	22.91	m	26.20
75 mm pipes; in runs not exceeding 3 m long	16.66	0.56	4.39	29.25	m	33.64
Extra for						
bend; medium radius	17.01	0.46	3.60	26.28	nr	29.88
single branch	23.64	0.65	5.09	42.10	nr	47.19
isolated "Timesaver" joint	9.48	0.28	2.19	10.45	nr	12.64
100 mm pipes; laid straight	18.12	0.46	3.60	24.73	m	28.33
100 mm pipes; in runs not exceeding 3 m long	17.19	0.63	4.94	31.96	m	36.90
Extra for						
bend; medium radius	21.31	0.56	4.39	32.97	nr	37.36
bend; medium radius with access	55.97	0.56	4.39	71.16	nr	75.55
bend; long radius	31.77	0.56	4.39	43.48	nr	47.87
rest bend	24.44	0.56	4.39	35.40	nr	39.79
diminishing pipe	17.82	0.56	4.39	28.11	nr	32.50
single branch	28.28	0.69	5.41	51.27	nr	56.68
single branch; with access	65.21	0.79	6.19	91.96	nr	98.15
double branch	45.44	0.88	6.90	81.11	nr	88.01
double branch; with access	77.39	0.88	6.90	116.31	nr	123.21
isolated "Timesaver" joint	11.40	0.32	2.51	12.56	nr	15.07
transitional pipe; for WC	16.22	0.46	3.60	30.43	nr	34.03
150 mm pipes; laid straight	33.55	0.56	4.39	43.04	m	47.43
150 mm pipes; in runs not exceeding 3 m long	32.03	0.76	5.96	51.35	m	57.31
Extra for						
bend; medium radius	49.02	0.65	5.09	63.53	nr	68.62
bend; medium radius with access	103.96	0.65	5.09	124.07	nr	129.16
bend; long radius	65.66	0.65	5.09	79.97	nr	85.06
diminishing pipe	27.78	0.65	5.09	38.23	nr	43.32
single branch	61.05	0.79	6.19	68.45	nr	74.64
isolated "Timesaver" joint	13.79	0.39	3.06	15.19	nr	18.25
Accessories in "Timesaver" cast iron or other equal and approved; with mechanical coupling joints						
Gully fittings; comprising low invert gully trap and round hopper						
75 mm outlet	17.82	0.83	6.50	33.35	nr	39.85
100 mm outlet	28.28	0.88	6.90	46.99	nr	53.89
150 mm outlet	70.35	1.20	9.40	96.63	nr	106.03
Add to above for bellmouth 300 mm high; circular plain grating						
100 mm nominal size; 200 mm grating	29.45	0.42	3.29	48.67	nr	51.96
100 mm nominal size; 100 mm horizontal inlet; 200 mm grating	36.00	0.42	3.29	55.89	nr	59.18
100 mm nominal size; 100 mm horizontal inlet; 200 mm grating	36.92	0.42	3.29	56.90	nr	60.19
Yard gully (Deans); trapped; galvanized sediment pan; 267 mm round heavy grating						
100 mm outlet	191.08	2.68	21.00	242.22	nr	263.22

R DISPOSAL SYSTEMS

	PC £	Labour hours	Labour £	Material £	Unit	Total rate £
Yard gully (garage); trapless; galvanized sediment pan; 267 mm round heavy grating						
100 mm outlet	186.62	2.50	19.59	222.11	nr	241.70
Yard gully (garage); trapped; with rodding eye, galvanised perforated sediment pan; stopper; 267 mm round heavy grating						
100 mm outlet	362.48	2.50	19.59	454.65	nr	474.24
Grease trap; internal access; galvanized perforated bucket; lid and frame						
100 mm outlet; 20 gallon capacity	332.10	3.70	29.00	400.84	nr	429.84
Cast iron "Ensign" lightweight drain pipes and fittings or other equal and approved; EN 877; ductile iron couplings						
100 mm pipes; laid straight	-	0.19	2.03	17.34	m	19.37
Extra for						
bend; long radius	-	0.19	2.03	29.57	nr	31.60
single branch	-	0.23	2.49	29.87	nr	32.36
150 mm pipes; laid straight	-	0.22	2.36	34.67	m	37.03
Extra for						
bend; long radius	-	0.22	2.36	78.11	nr	80.47
single branch	-	0.28	3.00	65.76	nr	68.76
Extra strength vitrified clay pipes and fittings; Hepworth "Supersleve" or other equal and approved; plain ends with push fit polypropylene flexible couplings						
100 mm pipes; laid straight	4.20	0.19	1.49	4.74	m	6.23
Extra for						
bend	3.88	0.19	1.49	8.19	nr	9.68
access bend	25.54	0.19	1.49	32.64	nr	34.13
rest bend	8.89	0.19	1.49	13.85	nr	15.34
access pipe	22.19	0.19	1.49	28.54	nr	30.03
socket adaptor	4.11	0.16	1.25	6.71	nr	7.96
adaptor to "HepSeal" pipe	3.35	0.16	1.25	5.85	nr	7.10
saddle	8.23	0.69	5.41	11.68	nr	17.09
single junction	8.38	0.23	1.80	15.34	nr	17.14
single access junction	29.54	0.23	1.80	39.23	nr	41.03
150 mm pipes; laid straight	8.01	0.23	1.80	9.04	m	10.84
Extra for						
bend	7.99	0.22	1.72	15.74	nr	17.46
access bend	2.12	0.22	1.72	40.89	nr	42.61
rest bend	10.27	0.22	1.72	18.32	nr	20.04
taper pipe	11.84	0.22	1.72	18.14	nr	19.86
access pipe	28.82	0.22	1.72	38.60	nr	40.32
socket adaptor	7.75	0.19	1.49	12.44	nr	13.93
adaptor to "HepSeal" pipe	5.50	0.19	1.49	9.90	nr	11.39
saddle	11.72	0.83	6.50	17.57	nr	24.07
single junction	11.74	0.28	2.19	23.67	nr	25.86
single access junction	41.97	0.28	2.19	57.79	nr	59.98
Extra strength vitrified clay pipes and fittings; Hepworth "HepSeal" or equivalent; socketted; with push-fit flexible joints						
100 mm pipes; laid straight	7.68	0.25	1.96	8.67	m	10.63
Extra for						
bend	11.10	0.20	1.57	9.93	nr	11.50
rest bend	13.19	0.20	1.57	12.29	nr	13.86
stopper	4.18	0.13	1.02	4.72	nr	5.74
access pipe	23.38	0.22	1.72	22.92	nr	24.64
single junction	15.41	0.25	1.96	13.93	nr	15.89

R DISPOSAL SYSTEMS

	PC £	Labour hours	Labour £	Material £	Unit	Total rate £
R12 DRAINAGE BELOW GROUND - cont'd						
Extra strength vitrified clay pipes and fittings; Hepworth "HepSeal" or equivalent; socketted; with push-fit flexible joints - cont'd						
150 mm pipes; laid straight	9.97	0.30	2.35	11.25	m	13.60
Extra for						
bend	18.30	0.23	1.80	17.28	nr	19.08
rest bend	21.85	0.20	1.57	21.29	nr	22.86
stopper	6.24	0.15	1.18	7.04	nr	8.22
taper reducer	27.43	0.23	1.80	27.59	nr	29.39
access pipe	37.44	0.23	1.80	37.76	nr	39.56
saddle	9.00	0.75	5.88	10.16	nr	16.04
single junction	9.97	0.30	2.35	22.49	nr	24.84
single access junction	45.73	0.30	2.35	48.15	nr	50.50
double junction	46.12	0.44	3.45	46.43	nr	49.88
double collar	16.29	0.19	1.49	18.39	nr	19.88
225 mm pipes; laid straight	19.73	0.38	2.98	22.27	m	25.25
Extra for						
bend	40.92	0.30	2.35	39.51	nr	41.86
rest bend	49.65	0.30	2.35	49.36	nr	51.71
stopper	12.83	0.19	1.49	14.48	nr	15.97
taper reducer	38.08	0.30	2.35	36.30	nr	38.65
access pipe	94.70	0.30	2.35	97.98	nr	100.33
saddle	17.42	1.00	7.84	19.66	nr	27.50
single junction	72.69	0.38	2.98	73.14	nr	76.12
single access junction	103.64	0.38	2.98	108.08	nr	111.06
double junction	103.69	0.56	4.39	105.90	nr	110.29
double collar	35.74	0.25	1.96	40.34	nr	42.30
300 mm pipes; laid straight	30.25	0.50	3.92	34.14	m	38.06
Extra for						
bend	77.72	0.40	3.13	77.48	nr	80.61
rest bend	110.75	0.40	3.13	114.77	nr	117.90
stopper	29.76	0.25	1.96	33.59	nr	35.55
taper reducer	84.31	0.40	3.13	84.92	nr	88.05
saddle	88.65	1.33	10.42	100.06	nr	110.48
single junction	152.19	0.50	3.92	158.13	nr	162.05
double junction	220.25	0.75	5.88	231.53	nr	237.41
double collar	58.09	0.33	2.59	65.57	nr	68.16
400 mm pipes; laid straight	61.87	0.67	5.25	69.84	m	75.09
Extra for						
bend	232.49	0.54	4.23	241.47	nr	245.70
single unequal junction	217.83	0.67	5.25	217.94	nr	223.19
450 mm pipes; laid straight	80.36	0.83	6.50	90.71	m	97.21
Extra for						
bend	306.13	0.67	5.25	318.33	nr	323.58
single unequal junction	260.58	0.83	6.50	257.85	nr	264.35
British Standard quality vitrified clay pipes and fittings; socketted; cement and sand (1:2) joints						
100 mm pipes; laid straight	5.37	0.37	2.90	6.15	m	9.05
Extra for						
bend (short/medium/knuckle)	10.75	0.30	2.35	4.33	nr	6.68
bend (long/rest/elbow)	8.83	0.30	2.35	8.24	nr	10.59
single junction	9.87	0.37	2.90	8.83	nr	11.73
double junction	16.42	0.56	4.39	15.63	nr	20.02
double collar	6.48	0.25	1.96	7.40	nr	9.36

R DISPOSAL SYSTEMS

	PC £	Labour hours	Labour £	Material £	Unit	Total rate £
150 mm pipes; laid straight	8.27	0.42	3.29	9.42	m	**12.71**
Extra for						
bend (short/medium/knuckle)	8.27	0.33	2.59	6.62	nr	**9.21**
bend (long/rest/elbow)	14.93	0.33	2.59	14.14	nr	**16.73**
taper	19.50	0.33	2.59	19.02	nr	**21.61**
single junction	16.34	0.42	3.29	14.82	nr	**18.11**
double junction	39.08	0.63	4.94	39.57	nr	**44.51**
double collar	10.79	0.28	2.19	12.27	nr	**14.46**
225 mm pipes; laid straight	16.38	0.51	4.00	18.70	m	**22.70**
Extra for						
bend (short/medium/knuckle)	25.89	0.41	3.21	23.79	nr	**27.00**
taper	42.36	0.33	2.59	41.82	nr	**44.41**
double collar	25.25	0.33	2.59	28.59	nr	**31.18**
300 mm pipes; laid straight	27.60	0.69	5.41	31.36	m	**36.77**
Extra for						
bend (short/medium/knuckle)	45.04	0.56	4.39	41.60	nr	**45.99**
double collar	51.23	0.37	2.90	57.98	nr	**60.88**
400 mm pipes; laid straight	50.55	0.93	7.29	57.36	m	**64.65**
450 mm pipes; laid straight	65.33	1.16	9.09	74.04	m	**83.13**
500 mm pipes; laid straight	81.90	1.34	10.50	92.76	m	**103.26**
Accessories in vitrified clay; set in concrete; with polypropylene coupling joints to pipes						
Rodding point; with oval aluminium plate						
100 mm nominal size	19.95	0.46	3.60	26.17	nr	**29.77**
Gully fittings; comprising low back trap and square hopper; 150 mm x 150 mm square gully grid						
100 mm nominal size	21.45	0.79	6.19	30.26	nr	**36.45**
Access gully; trapped with rodding eye and integral vertical back inlet; stopper; 150 mm x 150 mm square gully grid						
100 mm nominal size	27.79	0.60	4.70	35.02	nr	**39.72**
Inspection chamber; comprising base; 300 mm or 450 mm raising piece; integral alloy cover and frame; 100 mm inlets						
straight through; 2 nr inlets	81.95	1.85	14.50	97.34	nr	**111.84**
single junction; 3 nr inlets	88.35	2.04	15.99	106.96	nr	**122.95**
double junction; 4 nr inlets	95.78	2.22	17.40	117.74	nr	**135.14**
Accessories in polypropylene; cover set in concrete; with coupling joints to pipes						
Inspection chamber; 5 nr 100 mm inlets; cast iron cover and frame						
475 mm diameter x 585 mm deep	136.98	2.13	16.69	160.29	nr	**176.98**
475 mm diameter x 930 mm deep	167.60	2.31	18.10	194.84	nr	**212.94**
Accessories in vitrified clay; set in concrete; with cement and sand (1:2) joints to pipes						
Gully fittings; comprising low back trap and square hopper; square gully grid						
100 mm outlet; 150 mm x 150 mm grid	29.36	0.93	7.29	33.45	nr	**40.74**
150 mm outlet; 225 mm x 225 mm grid	51.85	0.93	7.29	58.83	nr	**66.12**
Yard gully (mud); trapped with rodding eye; galvanized square bucket; stopper; square hinged grate and frame						
100 mm outlet; 225 mm x 225 mm grid	73.07	2.78	21.79	82.87	nr	**104.66**
150 mm outlet; 300 mm x 300 mm grid	130.14	3.70	29.00	147.60	nr	**176.60**
Yard gully (garage); trapped with rodding eye; galvanized perforated round bucket; stopper; round hinged grate and frame						
100 mm outlet; 273 mm grid	72.82	2.78	21.79	104.39	nr	**126.18**
150 mm outlet; 368 mm grid	133.60	3.70	29.00	151.18	nr	**180.18**

R12 DRAINAGE BELOW GROUND - cont'd

	PC £	Labour hours	Labour £	Material £	Unit	Total rate £
Accessories in vitrified clay; set in concrete; with cement and sand (1:2) joints to pipes - cont'd						
Road gully; trapped with rodding eye and stopper (grate not included)						
300 mm x 600 mm x 100 mm outlet	50.86	3.05	23.90	71.99	nr	95.89
300 mm x 600 mm x 150 mm outlet	52.08	3.05	23.90	73.37	nr	97.27
400 mm x 750 mm x 150 mm outlet	60.40	3.70	29.00	90.33	nr	119.33
450 mm x 900 mm x 150 mm outlet	81.73	4.65	36.44	118.82	nr	155.26
Grease trap; with internal access; galvanized perforated bucket; lid and frame						
450 mm x 300 mm x 525 mm deep; 100 mm outlet	361.50	3.24	25.39	424.11	nr	449.50
600 mm x 450 mm x 600 mm deep; 100 mm outlet	456.30	3.89	30.49	536.83	nr	567.32
Interceptor; trapped with inspection arm; lever locking stopper; chain and staple; cement and sand (1:2) joints to pipes; building in, and cutting and fitting brickwork around						
100 mm outlet; 100 mm inlet	65.54	3.70	29.00	74.34	nr	103.34
150 mm outlet; 150 mm inlet	93.00	4.16	32.60	105.35	nr	137.95
225 mm outlet; 225 mm inlet	253.15	4.63	36.28	286.16	nr	322.44
Accessories; grates and covers						
Aluminium alloy gully grids; set in position						
120 mm x 120 mm	2.21	0.09	0.71	2.49	nr	3.20
150 mm x 150 mm	2.21	0.09	0.71	2.49	nr	3.20
225 mm x 225 mm	6.56	0.09	0.71	7.40	nr	8.11
100 mm diameter	2.21	0.09	0.71	2.49	nr	3.20
150 mm diameter	3.37	0.09	0.71	3.80	nr	4.51
225 mm diameter	7.34	0.09	0.71	8.29	nr	9.00
Aluminium alloy sealing plates and frames; set in cement and sand (1:3)						
150 mm x 150 mm	8.46	0.23	1.80	9.64	nr	11.44
225 mm x 225 mm	15.50	0.23	1.80	17.58	nr	19.38
140 mm diameter (for 100 mm)	6.90	0.23	1.80	7.88	nr	9.68
197 mm diameter (for 150 mm)	9.92	0.23	1.80	11.28	nr	13.08
273 mm diameter (for 225 mm)	15.88	0.23	1.80	18.01	nr	19.81
Coated cast iron heavy duty road gratings and frame; BS 497 Tables 6 and 7; bedding and pointing in cement and sand (1:3); one course half brick thick wall in semi-engineering bricks in cement mortar (1:3)						
445 mm x 400 mm; Grade A1, ref GA1-450 (90 kg)	95.34	2.31	18.10	107.21	nr	125.31
400 mm x 310 mm; Grade A2, ref GA2-325 (35 kg)	74.61	2.31	18.10	84.38	nr	102.48
500 mm x 310 mm; Grade A2, ref GA2-325 (65 kg)	103.63	2.31	18.10	116.36	nr	134.46
Vibrated concrete pipes and fittings; with flexible joints; BS 5911 Part 1						
300 mm pipes Class M; laid straight	11.57	0.65	5.09	13.06	m	18.15
Extra for						
bend; 22.5 degree	-	0.65	5.09	48.97	nr	54.06
bend; 45 degree	-	0.65	5.09	81.62	nr	86.71
junction; 300 mm x 100 mm	23.84	0.46	3.60	25.63	nr	29.23
450 mm pipes Class H; laid straight	24.93	1.02	7.99	28.14	m	36.13
Extra for						
bend; 22.5 degree	-	1.02	7.99	105.52	nr	113.51
bend; 45 degree	-	1.02	7.99	175.87	nr	183.86
junction; 450 mm x 150 mm	26.33	0.65	5.09	28.30	nr	33.39

Prices for Measured Works - Major Works
R DISPOSAL SYSTEMS

	PC £	Labour hours	Labour £	Material £	Unit	Total rate £
600 mm pipes Class H; laid straight	33.65	1.48	11.60	37.98	m	49.58
Extra for						
bend; 22.5 degree	-	1.48	11.60	142.43	nr	154.03
bend; 45 degree	-	1.48	11.60	237.39	nr	248.99
junction; 600 mm x 150 mm	26.95	0.83	6.50	28.97	nr	35.47
900 mm pipes Class H; laid straight	85.69	2.59	20.30	96.72	m	117.02
Extra for						
bend; 22.5 degree	-	2.59	20.30	362.71	nr	383.01
bend; 45 degree	-	2.59	20.30	604.52	nr	624.82
junction; 900 mm x 225 mm	45.96	1.02	7.99	49.41	nr	57.40
1200 mm pipes Class H; laid straight	109.06	3.70	29.00	123.10	m	152.10
Extra for						
bend; 22.5 degree	-	3.70	29.00	461.63	nr	490.63
bend; 45 degree	-	3.70	29.00	769.38	nr	798.38
junction; 300 mm x 100 mm	56.09	1.48	11.60	60.30	nr	71.90
Accessories in precast concrete; top set in with rodding eye and stopper; cement and sand (1:2) joint to pipe						
Concrete road gully; BS 5911; trapped with rodding eye and stopper; cement and sand (1:2) joint to pipe						
450 mm diameter x 1050 mm deep; 100 mm or 150 mm outlet	26.49	4.39	34.40	45.31	nr	79.71
"Osmadrain" uPVC pipes and fittings or other equal and approved; BS 4660; with ring seal joints						
82 mm pipes; laid straight	3.35	0.15	1.18	3.78	m	4.96
Extra for						
bend; short radius	5.89	0.13	1.02	6.49	nr	7.51
spigot/socket bend	4.95	0.13	1.02	5.45	nr	6.47
adaptor	2.58	0.07	0.55	2.84	nr	3.39
single junction	7.65	0.18	1.41	8.43	nr	9.84
slip coupler	3.63	0.07	0.55	4.00	nr	4.55
100 mm pipes; laid straight	2.96	0.17	1.33	3.73	m	5.06
Extra for						
bend; short radius	7.85	0.15	1.18	8.45	nr	9.63
bend; long radius	12.71	0.15	1.18	13.00	nr	14.18
spigot/socket bend	6.63	0.15	1.18	9.15	nr	10.33
socket plug	3.18	0.04	0.31	3.50	nr	3.81
adjustable double socket bend	8.68	0.15	1.18	11.27	nr	12.45
adaptor to clay	6.78	0.09	0.71	7.36	nr	8.07
single junction	9.36	0.21	1.65	9.31	nr	10.96
sealed access junction	17.15	0.19	1.49	17.89	nr	19.38
slip coupler	3.63	0.09	0.71	4.00	nr	4.71
160 mm pipes; laid straight	6.74	0.21	1.65	8.33	m	9.98
Extra for						
bend; short radius	17.26	0.18	1.41	18.56	nr	19.97
spigot/socket bend	15.65	0.18	1.41	20.55	nr	21.96
socket plug	5.65	0.07	0.55	6.23	nr	6.78
adaptor to clay	13.63	0.12	0.94	14.65	nr	15.59
level invert taper	23.39	0.18	1.41	28.62	nr	30.03
single junction	28.26	0.24	1.88	31.14	nr	33.02
sealed access junction	47.13	0.22	1.72	51.93	nr	53.65
slip coupler	10.19	0.11	0.86	11.23	nr	12.09

R DISPOSAL SYSTEMS

	PC £	Labour hours	Labour £	Material £	Unit	Total rate £
R12 DRAINAGE BELOW GROUND - cont'd						
uPVC Osma "Ultra-Rib" ribbed pipes and fittings or other equal and approved; WIS approval; with sealed ring push-fit joints						
150 mm pipes; laid straight	-	0.19	1.49	3.32	m	4.81
Extra for						
bend; short radius	9.03	0.17	1.33	9.75	nr	11.08
adaptor to 160 mm diameter upvc	10.57	0.10	0.78	11.25	nr	12.03
adaptor to clay	21.70	0.10	0.78	23.71	nr	24.49
level invert taper	3.94	0.18	1.41	3.74	nr	5.15
single junction	16.24	0.22	1.72	16.90	nr	18.62
225 mm pipes; laid straight	6.75	0.22	1.72	7.62	m	9.34
Extra for						
bend; short radius	32.54	0.20	1.57	35.40	nr	36.97
adaptor to clay	27.04	0.13	1.02	28.88	nr	29.90
level invert taper	5.66	0.20	1.57	4.86	nr	6.43
single junction	48.30	0.27	2.12	50.93	nr	53.05
300 mm pipes; laid straight	10.13	0.32	2.51	11.43	m	13.94
Extra for						
bend; short radius	51.25	0.29	2.27	55.79	nr	58.06
adaptor to clay	71.11	0.14	1.10	76.98	nr	78.08
level invert taper	16.97	0.29	2.27	16.64	nr	18.91
single junction	103.02	0.37	2.90	110.09	nr	112.99
Interconnecting drainage channel 100 wide; ACO Polymer Products Ltd or other equal and approved; reinforced slotted galvanised steel grating ref 423/4; bedding and haunching in in situ concrete (not included)						
100 mm wide						
laid level or to falls	-	0.46	3.60	64.37	m	67.97
extra for sump unit	-	1.39	10.89	110.88	nr	121.77
extra for end caps	-	0.09	0.71	6.32	nr	7.03
Interconnecting drainage channel; "Birco-lite" ref 8012 or other equal and approved; Marshalls Plc; galvanised steel grating ref 8041; bedding and haunching in in situ concrete (not included)						
100 mm wide						
laid level or to falls	-	0.46	3.60	31.41	m	35.01
extra for 100 mm diameter trapped outlet unit	-	1.39	10.89	54.24	nr	65.13
extra for end caps	-	0.09	0.71	3.92	nr	4.63
Accessories in uPVC; with ring seal joints to pipes (unless otherwise described)						
Cast iron access point						
110 mm diameter	16.96	0.75	5.88	18.69	nr	24.57
Rodding eye						
110 mm diameter	23.88	0.43	3.37	29.47	nr	32.84
Universal gulley fitting; comprising gulley trap, plain hopper						
150 mm x 150 mm grate	13.97	0.93	7.29	19.80	nr	27.09
Bottle gulley; comprising gulley with bosses closed; sealed access covers						
217 mm x 217 mm grate	25.95	0.78	6.11	33.01	nr	39.12
Shallow access pipe; light duty screw down access door assembly						
110 mm diameter	20.16	0.78	6.11	26.63	nr	32.74
Shallow access junction; 3 nr 110 mm inlets; light duty screw down access door assembly						
110 mm diameter	36.20	1.11	8.70	53.18	nr	61.88

Prices for Measured Works - Major Works
R DISPOSAL SYSTEMS

	PC £	Labour hours	Labour £	Material £	Unit	Total rate £
Shallow inspection chamber; 250 mm diameter; 600 mm deep; sealed cover and frame						
4 nr 110 mm outlets/inlets	50.36	1.28	10.03	71.25	nr	81.28
Universal inspection chamber; 450 mm diameter; single seal cast iron cover and frame; 4 nr 110 mmoutlets/inlets						
500 mm deep	95.57	1.35	10.58	121.07	nr	131.65
730 mm deep	95.57	1.60	12.54	137.27	nr	149.81
960 mm deep	95.57	1.85	14.50	153.46	nr	167.96
Equal manhole base; 750 mm diameter						
6 nr 160 mm outlets/inlets	102.38	1.21	9.48	122.27	nr	131.75
Unequal manhole base; 750 mm diameter						
2 nr 160 mm, 4nr 110 mm outlets/inlets	95.25	1.21	9.48	114.41	nr	123.89
Kerb to gullies; class B engineering bricks on edge to three sides in cement mortar (1:3) rendering in cement mortar (1:3) to top and two sides and skirting to brickwork 230 mm high; dishing in cement mortar (1:3) to gully; steel trowelled						
230 mm x 230 mm internally	-	1.39	10.89	1.18	nr	12.07
Excavating; by machine						
Manholes						
maximum depth not exceeding 1.00 m	-	0.19	1.52	4.19	m^3	5.71
maximum depth not exceeding 2.00 m	-	0.21	1.68	4.59	m^3	6.27
maximum depth not exceeding 4.00 m	-	0.25	2.00	5.39	m^3	7.39
Excavating; by hand						
Manholes						
maximum depth not exceeding 1.00 m	-	3.05	24.39	-	m^3	24.39
maximum depth not exceeding 2.00 m	-	3.61	28.87	-	m^3	28.87
maximum depth not exceeding 4.00 m	-	4.63	37.03	-	m^3	37.03
Earthwork support (average "risk" prices)						
Maximum depth not exceeding 1.00 m						
distance between opposing faces not exceeding 2.00 m	-	0.14	1.12	3.42	m^2	4.54
Maximum depth not exceeding 2.00 m						
distance between opposing faces not exceeding 2.00 m	-	0.18	1.44	6.50	m^2	7.94
Maximum depth not exceeding 4.00 m						
distance between opposing faces not exceeding 2.00 m	-	0.22	1.76	9.59	m^2	11.35
Disposal; by machine						
Excavated material						
off site; to tip not exceeding 13 km (using lorries) including Landfill Tax based on inactive waste	-	-	-	14.48	m^3	14.48
on site; depositing on site in spoil heaps; average 50 m distance	-	0.14	1.12	2.71	m^3	3.83
Disposal; by hand						
Excavated material						
off site; to tip not exceeding 13 km (using lorries) including Landfill Tax based on inactive waste	-	0.74	5.92	23.47	m^3	29.39
on site; depositing on site in spoil heaps; average 50 m distance	-	1.20	9.60	-	m^3	9.60

R DISPOSAL SYSTEMS

	PC £	Labour hours	Labour £	Material £	Unit	Total rate £
R12 DRAINAGE BELOW GROUND - cont'd						
Filling to excavations; by machine						
Average thickness not exceeding 0.25 m						
arising excavations	-	0.14	1.12	1.99	m^3	3.11
Filling to excavations; by hand						
Average thickness not exceeding 0.25 m						
arising from excavations	-	0.93	7.44	-	m^3	7.44
Plain in situ ready mixed designated concrete; C10 - 40 mm aggregate						
Beds						
thickness not exceeding 150 mm	54.56	2.78	24.57	66.21	m^3	90.78
thickness 150 mm - 450 mm	54.56	2.08	18.38	66.21	m^3	84.59
thickness exceeding 450 mm	54.56	1.76	15.55	66.21	m^3	81.76
Plain in situ ready mixed designated concrete; C20 - 20 mm aggregate						
Beds						
thickness not exceeding 150 mm	56.44	2.78	24.57	68.49	m^3	93.06
thickness 150 mm - 450 mm	56.44	2.08	18.38	68.49	m^3	86.87
thickness exceeding 450 mm	56.44	1.76	15.55	68.49	m^3	84.04
Plain in situ ready mixed designated concrete; C25 - 20 mm aggregate; (small quantities)						
Benching in bottoms						
150 mm - 450 mm average thickness	56.87	8.33	90.89	65.72	m^3	156.61
Reinforced in situ ready mixed designated concrete; C20 - 20 mm aggregate; (small quantities)						
Isolated cover slabs						
thickness not exceeding 150 mm	56.44	6.48	57.26	65.22	m^3	122.48
Reinforcement; fabric to BS 4483; lapped; in beds or suspended slabs						
Ref A98 (1.54 kg/m^2)						
400 mm minimum laps	0.85	0.11	1.22	1.01	m^2	2.23
Ref A142 (2.22 kg/m^2)						
400 mm minimum laps	0.88	0.11	1.22	1.04	m^2	2.26
Ref A193 (3.02 kg/m^2)						
400 mm minimum laps	1.21	0.11	1.22	1.43	m^2	2.65
Formwork; basic finish						
Soffits of isolated cover slabs						
horizontal	-	2.64	36.50	7.39	m^2	43.89
Edges of isolated cover slabs						
height not exceeding 250 mm	-	0.78	10.78	2.08	m	12.86
Precast concrete rectangular access and inspection chambers; "Hepworth" chambers or other equal and approved; comprising cover frame to receive manhole cover (not included) intermediate wall sections and base section with cut outs; bedding; jointing and pointing in cement mortar (1:3) on prepared bed						
Drainage chamber; size 600 mm x 450 mm internally; depth to invert						
600 mm deep	-	4.16	32.60	88.04	nr	120.64
900 mm deep	-	5.55	43.49	110.63	nr	154.12

R DISPOSAL SYSTEMS

	PC £	Labour hours	Labour £	Material £	Unit	Total rate £
Drainage chamber; 1200 mm x 750 mm reducing to 600 mm x 600 mm; no base unit; depth of invert						
1050 mm deep	-	6.94	54.39	164.98	nr	219.37
1650 mm deep	-	8.33	65.28	279.05	nr	344.33
2250 mm deep	-	10.18	79.78	392.47	nr	472.25
Precast concrete circular manhole rings; BS5911 Part 1; bedding, jointing and pointing in cement mortar (1:3) on prepared bed						
Chamber or shaft rings; plain						
900 mm diameter	35.05	5.09	39.89	38.33	m	78.22
1050 mm diameter	38.95	6.01	47.10	43.18	m	90.28
1200 mm diameter	50.63	6.94	54.39	56.40	m	110.79
Chamber or shaft rings; reinforced						
1350 mm diameter	74.00	7.86	61.60	82.17	m	143.77
1500 mm diameter	85.69	8.79	68.89	96.05	m	164.94
1800 mm diameter	109.06	11.10	86.99	123.14	m	210.13
2100 mm diameter	151.91	13.88	108.77	171.17	m	279.94
extra for step irons built in	2.92	0.14	1.10	3.14	nr	4.24
Reducing slabs						
1200 mm diameter	58.42	5.55	43.49	64.11	nr	107.60
1350 mm diameter	77.90	8.79	68.89	86.37	nr	155.26
1500 mm diameter	93.48	10.18	79.78	103.77	nr	183.55
1800 mm diameter	132.43	12.95	101.49	147.61	nr	249.10
Heavy duty cover slabs; to suit rings						
900 mm diameter	31.16	2.78	21.79	34.15	nr	55.94
1050 mm diameter	38.95	3.24	25.39	42.66	nr	68.05
1200 mm diameter	50.63	3.70	29.00	55.74	nr	84.74
1350 mm diameter	74.00	4.16	32.60	81.52	nr	114.12
1500 mm diameter	85.69	4.63	36.28	94.74	nr	131.02
1800 mm diameter	120.75	5.55	43.49	133.45	nr	176.94
2100 mm diameter	245.38	6.48	50.78	268.70	nr	319.48
Common bricks; in cement mortar (1:3)						
Walls to manholes						
one brick thick	0.16	2.22	38.80	31.15	m²	69.95
one and a half brick thick	0.16	3.24	56.63	46.72	m²	103.35
Projections of footings						
two brick thick	0.16	4.53	79.18	62.30	m²	141.48
Class A engineering bricks; in cement mortar (1:3)						
Walls to manholes						
one brick thick	0.24	2.50	43.70	43.92	m²	87.62
one and a half brick thick	0.24	3.61	63.10	45.89	m²	108.99
Projections of footings						
two brick thick	0.24	5.09	88.97	87.84	m³	176.81
Class B engineering bricks; in cement mortar (1:3)						
Walls to manholes						
one brick thick	0.17	2.50	43.70	32.42	m²	76.12
one and a half brick thick	0.17	3.61	63.10	48.63	m²	111.73
Projections of footings						
two brick thick	0.17	5.09	88.97	64.84	m²	153.81
Brickwork sundries						
Extra over for fair face; flush smooth pointing						
manhole walls	-	0.19	3.32	-	m²	3.32
Building ends of pipes into brickwork; making good fair face or rendering						
not exceeding 55 mm nominal size	-	0.09	1.57	-	nr	1.57
55 mm - 110 mm nominal size	-	0.14	2.45	-	nr	2.45

R DISPOSAL SYSTEMS

	PC £	Labour hours	Labour £	Material £	Unit	Total rate £
R12 DRAINAGE BELOW GROUND - cont'd						
Brickwork sundries - cont'd						
Step irons; BS 1247; malleable; galvanized; building into joints						
general purpose pattern	-	0.14	2.45	4.49	nr	6.94
Cement and sand (1:3) in situ finishings; steel trowelled						
13 mm work to manhole walls; one coat; to						
brickwork base over 300 wide	-	0.65	11.36	1.31	m²	12.67
Cast iron inspection chambers; with bolted flat covers; BS 437; bedded in cement mortar (1:3); with mechanical coupling joints						
100 mm x 100 mm						
one branch	88.73	0.97	7.60	110.99	nr	118.59
one branch either side	111.02	1.43	11.21	148.11	nr	159.32
150 mm x 100 mm						
one branch	141.21	1.16	9.09	156.25	nr	165.34
one branch either side	154.26	1.67	13.09	183.85	nr	196.94
150 mm x 150 mm						
one branch	162.33	1.25	9.80	195.36	nr	205.16
one branch either side	177.59	1.76	13.79	227.37	nr	241.16
Access covers and frames; Drainage Systems or other equal and approved; coated; bedding frame in cement and sand (1:3); cover in grease and sand						
Grade A; light duty; rectangular single seal solid top						
450 mm x 450 mm; ref MC1-45/45	39.33	1.39	10.89	44.81	nr	55.70
600 mm x 450 mm; ref MC1-60/45	39.55	1.39	10.89	45.18	nr	56.07
600 mm x 600 mm; ref MC1-60/60	84.13	1.39	10.89	94.44	nr	105.33
Grade A; light duty; rectangular single seal recessed						
600 mm x 450 mm; ref MC1R-60/45	80.66	1.39	10.89	90.48	nr	101.37
600 mm x 600 mm; ref MC1R-60/60	111.43	1.39	10.89	124.52	nr	135.41
Grade A; light duty; rectangular double seal solid top						
450 mm x 450 mm; ref MC2-45/45	60.51	1.39	10.89	68.14	nr	79.03
600 mm x 450 mm; ref MC2-60/45	79.28	1.39	10.89	88.96	nr	99.85
600 mm x 600 mm; ref MC2-60/60	111.79	1.39	10.89	124.92	nr	135.81
Grade A; light duty; rectangular double seal recessed						
450 mm x 450 mm; ref MC2R-45/45	100.53	1.39	10.89	112.24	nr	123.13
600 mm x 450 mm; ref MC2R-60/45	119.81	1.39	10.89	133.62	nr	144.51
600 mm x 600 mm; ref MC2R-60/60	134.68	1.39	10.89	150.14	nr	161.03
Grade B; medium duty; circular single seal solid top						
300 mm diameter; ref MB2-50	116.46	1.85	14.50	129.77	nr	144.27
550 mm diameter; ref MB2-55	88.94	1.85	14.50	99.45	nr	113.95
600 mm diameter; ref MB2-60	77.71	1.85	14.50	87.07	nr	101.57
Grade B; medium duty; rectangular single seal solid top						
600 mm x 450 mm; ref MB2-60/45	69.77	1.85	14.50	78.48	nr	92.98
600 mm x 600 mm; ref MB2-60/60	89.38	1.85	14.50	100.23	nr	114.73
Grade B; medium duty; rectangular singular seal recessed						
600 mm x 450 mm; ref MB2R-60/45	119.01	1.85	14.50	132.74	nr	147.24
600 mm x 600 mm; ref MB2R-60/60	135.41	1.85	14.50	150.94	nr	165.44
Grade B; "Chevron"; medium duty; double triangular solid top						
600 mm x 600 mm; ref MB1-60/60	85.51	1.85	14.50	95.96	nr	110.46
Grade C; "Vulcan" heavy duty; single triangular solid top						
550 mm x 495 mm; ref MA-T	136.65	2.31	18.10	152.17	nr	170.27
Grade C; "Chevron"; heavy duty double triangular solid top						
550 mm x 550 mm; ref MA-55	123.16	2.78	21.79	137.36	nr	159.15
600 mm x 600 mm; ref MA-60	140.62	2.78	21.79	156.69	nr	178.48

R DISPOSAL SYSTEMS

	PC £	Labour hours	Labour £	Material £	Unit	Total rate £
British Standard best quality vitrified clay channels; bedding and jointing in cement and sand (1:2)						
Half section straight						
100 mm diameter x 1 m long	4.06	0.74	5.80	4.58	nr	10.38
150 mm diameter x 1 m long	6.75	0.93	7.29	7.62	nr	14.91
225 mm diameter x 1 m long	15.18	1.20	9.40	17.13	nr	26.53
300 mm diameter x 1 m long	31.15	1.48	11.60	35.16	nr	46.76
Half section bend						
100 mm diameter	4.37	0.56	4.39	4.93	nr	9.32
150 mm diameter	7.21	0.69	5.41	8.14	nr	13.55
225 mm diameter	24.04	0.93	7.29	27.14	nr	34.43
300 mm diameter	49.02	1.11	8.70	55.33	nr	64.03
Half section taper straight						
150 mm - 100 mm diameter	18.17	0.65	5.09	20.51	nr	25.60
225 mm - 150 mm diameter	40.55	0.83	6.50	45.77	nr	52.27
300 mm - 225 mm diameter	160.85	1.02	7.99	181.56	nr	189.55
Half section taper bend						
150 mm - 100 mm diameter	27.66	0.83	6.50	31.22	nr	37.72
225 mm - 150 mm diameter	79.24	1.06	8.31	89.44	nr	97.75
300 mm - 225 mm diameter	160.85	1.30	10.19	181.56	nr	191.75
Three quarter section branch bend						
100 mm diameter	9.85	0.46	3.60	11.12	nr	14.72
150 mm diameter	17.13	0.69	5.41	19.34	nr	24.75
225 mm diameter	60.31	0.93	7.29	68.08	nr	75.37
300 mm diameter	126.22	1.23	9.64	142.47	nr	152.11
uPVC channels; with solvent weld or lip seal coupling joints; bedding in cement and sand						
Half section cut away straight; with coupling either end						
110 mm diameter	18.68	0.28	2.19	25.55	nr	27.74
160 mm diameter	32.21	0.37	2.90	47.60	nr	50.50
Half section cut away long radius bend; with coupling either end						
110 mm diameter	19.12	0.28	2.19	26.05	nr	28.24
160 mm diameter	47.28	0.37	2.90	64.61	nr	67.51
Channel adaptor to clay; with one coupling						
110 mm diameter	5.35	0.23	1.80	8.27	nr	10.07
160 mm diameter	11.54	0.31	2.43	18.65	nr	21.08
Half section bend						
110 mm diameter	6.95	0.31	2.43	8.08	nr	10.51
160 mm diameter	10.36	0.46	3.60	12.24	nr	15.84
Half section channel connector						
110 mm diameter	2.75	0.07	0.55	3.57	nr	4.12
160 mm diameter	6.84	0.09	0.71	8.81	nr	9.52
Half section channel junction						
110 mm diameter	7.79	0.46	3.60	9.03	nr	12.63
160 mm diameter	23.49	0.56	4.39	27.06	nr	31.45
Polypropylene slipper bend						
110 mm diameter	10.46	0.37	2.90	12.04	nr	14.94
Glass fibre septic tank; "Klargester" or other equal and approved; fixing lockable manhole cover and frame; placing in position						
3750 litre capacity; 2000 mm diameter; depth to invert						
1000 mm deep; standard grade	696.75	2.27	17.79	823.90	nr	841.69
1500 mm deep; heavy duty grade	863.97	2.54	19.91	1003.66	nr	1023.57
6000 litre capacity; 2300 mm diameter; depth to invert						
1000 mm deep; standard grade	1114.80	2.45	19.20	1293.28	nr	1312.48
1500 mm deep; heavy duty grade	1416.72	2.73	21.39	1617.84	nr	1639.23

R DISPOSAL SYSTEMS

	PC £	Labour hours	Labour £	Material £	Unit	Total rate £
R12 DRAINAGE BELOW GROUND - cont'd						
Glass fibre septic tank; "Klargester" or other equal and approved; fixing lockable manhole cover and frame; placing in position - cont'd						
9000 litre capacity; 2660 mm diameter; depth to invert						
1000 mm deep; standard grade	1611.82	2.64	20.69	1827.58	nr	1848.27
1500 mm deep; heavy duty grade	2043.80	2.91	22.81	2291.95	nr	2314.76
Glass fibre petrol interceptors; "Klargester" or other equal and approved; placing in position						
2000 litre capacity; 2370 mm x 1300 mm diameter; depth to invert						
1000 mm deep	789.65	2.50	19.59	848.87	nr	868.46
4000 litre capacity; 4370 mm x 1300 mm diameter; depth to invert						
1000 mm deep	1314.54	2.68	21.00	1413.13	nr	1434.13
R13 LAND DRAINAGE						
Excavating; by hand; grading bottoms; earthwork support; filling to within 150 mm of surface with gravel rejects; remainder filled with excavated material and compacting; disposal of surplus soil on site; spreading on site average 50 m						
Pipes not exceeding 200 nominal size						
average depth of trench 0.75 m	-	1.57	12.56	7.23	m	19.79
average depth of trench 1.00 m	-	2.08	16.64	11.33	m	27.97
average depth of trench 1.25 m	-	2.91	23.27	14.18	m	37.45
average depth of trench 1.50 m	-	5.00	39.99	17.23	m	57.22
average depth of trench 1.75 m	-	5.92	47.35	20.07	m	67.42
average depth of trench 2.00 m	-	6.85	54.79	23.13	m	77.92
Disposal; by machine						
Excavated material						
off site; to tip not exceeding 13 km (using lorries);						
including Landfill Tax based on inactive waste	-	-	-	14.48	m³	14.48
hand loaded	-	-	-	23.47	m³	23.47
Disposal; by hand						
Excavated material						
off site; to tip not exceeding 13 km (using lorries); including Landfill Tax based on inactive waste	-	0.74	5.92	23.47	m³	29.39
Vitrified clay perforated sub-soil pipes; BS 65; Hepworth "Hepline" or other equal and approved						
Pipes; laid straight						
100 mm diameter	4.82	0.20	1.57	5.44	m	7.01
150 mm diameter	8.77	0.25	1.96	9.90	m	11.86
225 mm diameter	16.12	0.33	2.59	18.20	m	20.79

S PIPED SUPPLY SYSTEMS

	PC £	Labour hours	Labour £	Material £	Unit	Total rate £
S PIPED SUPPLY SYSTEMS						
S10/S11 HOT AND COLD WATER						
Copper pipes; EN1057:1996; capillary fittings						
15 mm pipes; fixing with pipe clips and screwed	0.86	0.34	4.28	1.04	m	5.32
Extra for						
made bend	-	0.14	1.76	-	nr	1.76
stop end	0.68	0.10	1.26	0.75	nr	2.01
straight coupling	0.12	0.16	2.02	0.13	nr	2.15
union coupling	3.67	0.16	2.02	4.04	nr	6.06
reducing coupling	1.25	0.16	2.02	1.38	nr	3.40
copper to lead connector	1.17	0.20	2.52	1.29	nr	3.81
imperial to metric adaptor	1.50	0.20	2.52	1.65	nr	4.17
elbow	0.23	0.16	2.02	0.25	nr	2.27
return bend	4.13	0.16	2.02	4.55	nr	6.57
tee; equal	0.41	0.23	2.90	0.45	nr	3.35
tee; reducing	3.00	0.23	2.90	3.31	nr	6.21
straight tap connector	1.06	0.47	5.92	1.17	nr	7.09
bent tap connector	1.06	0.63	7.94	1.17	nr	9.11
tank connector	3.28	0.23	2.90	3.61	nr	6.51
overflow bend	6.97	0.20	2.52	7.68	nr	10.20
22 mm pipes; fixing with pipe clips and screwed	1.72	0.40	5.04	2.02	m	7.06
Extra for						
made bend	-	0.19	2.39	-	nr	2.39
stop end	1.29	0.12	1.51	1.42	nr	2.93
straight coupling	0.32	0.20	2.52	0.35	nr	2.87
union coupling	5.88	0.20	2.52	6.48	nr	9.00
reducing coupling	1.25	0.20	2.52	1.38	nr	3.90
copper to lead connector	1.76	0.29	3.65	1.94	nr	5.59
elbow	0.56	0.20	2.52	0.62	nr	3.14
backplate elbow	5.92	0.41	5.17	6.52	nr	11.69
return bend	8.11	0.20	2.52	8.94	nr	11.46
tee; equal	1.32	0.31	3.91	1.45	nr	5.36
tee; reducing	1.05	0.31	3.91	1.16	nr	5.07
straight tap connector	1.07	0.16	2.02	1.18	nr	3.20
28 mm pipes; fixing with pipe clips and screwed	2.23	0.43	5.42	2.58	m	8.00
Extra for						
made bend	-	0.23	2.90	-	nr	2.90
stop end	2.26	0.14	1.76	2.49	nr	4.25
straight coupling	0.64	0.26	3.28	0.71	nr	3.99
reducing coupling	1.72	0.26	3.28	1.90	nr	5.18
union coupling	5.88	0.26	3.28	6.48	nr	9.76
copper to lead connector	2.38	0.36	4.54	2.62	nr	7.16
imperial to metric adaptor	2.59	0.36	4.54	2.85	nr	7.39
elbow	1.02	0.26	3.28	1.12	nr	4.40
return bend	10.37	0.26	3.28	11.43	nr	14.71
tee; equal	2.84	0.38	4.79	3.13	nr	7.92
tank connector	6.57	0.38	4.79	7.24	nr	12.03
35 mm pipes; fixing with pipe clips and screwed	5.59	0.50	6.30	6.38	m	12.68
Extra for						
made bend	-	0.28	3.53	-	nr	3.53
stop end	4.99	0.16	2.02	5.50	nr	7.52
straight coupling	2.08	0.31	3.91	2.29	nr	6.20
reducing coupling	4.04	0.31	3.91	4.45	nr	8.36
union coupling	11.24	0.31	3.91	12.39	nr	16.30
flanged connector	31.00	0.41	5.17	34.16	nr	39.33
elbow	4.46	0.31	3.91	4.91	nr	8.82
obtuse elbow	6.59	0.31	3.91	7.26	nr	11.17
tee; equal	7.11	0.43	5.42	7.83	nr	13.25
tank connector	8.43	0.43	5.42	9.29	nr	14.71

S PIPED SUPPLY SYSTEMS

	PC £	Labour hours	Labour £	Material £	Unit	Total rate £
S10/S11 HOT AND COLD WATER - cont'd						
Copper pipes; EN1057:1996; capillary fittings - cont'd						
42 mm pipes; fixing with pipe clips; plugged and screwed	6.86	0.56	7.06	7.81	m	14.87
Extra for						
made bend	-	0.37	4.66	-	nr	4.66
stop end	8.59	0.18	2.27	9.47	nr	11.74
straight coupling	3.40	0.36	4.54	3.75	nr	8.29
reducing coupling	6.76	0.36	4.54	7.45	nr	11.99
union coupling	16.43	0.36	4.54	18.10	nr	22.64
flanged connector	37.05	0.46	5.80	40.82	nr	46.62
elbow	7.36	0.36	4.54	8.11	nr	12.65
obtuse elbow	11.73	0.36	4.54	12.93	nr	17.47
tee; equal	11.41	0.48	6.05	12.57	nr	18.62
tank connector	11.05	0.48	6.05	12.18	nr	18.23
54 mm pipes; fixing with pipe clips; plugged and screwed	8.81	0.62	7.81	10.01	m	17.82
Extra for						
made bend	-	0.51	6.43	-	nr	6.43
stop end	11.99	0.19	2.39	13.21	nr	15.60
straight coupling	6.27	0.41	5.17	6.91	nr	12.08
reducing coupling	11.35	0.41	5.17	12.51	nr	17.68
union coupling	31.25	0.41	5.17	34.43	nr	39.60
flanged connector	56.01	0.46	5.80	61.72	nr	67.52
elbow	15.19	0.41	5.17	16.74	nr	21.91
obtuse elbow	21.22	0.41	5.17	23.38	nr	28.55
tee; equal	23.00	0.53	6.68	25.34	nr	32.02
tank connector	16.88	0.53	6.68	18.60	nr	25.28
Copper pipes; EN1057:1996; compression fittings						
15 mm pipes; fixing with pipe clips; plugged and screwed	0.86	0.39	4.91	1.04	m	5.95
Extra for						
made bend	-	0.14	1.76	-	nr	1.76
stop end	1.40	0.09	1.13	1.54	nr	2.67
straight coupling	1.13	0.14	1.76	1.25	nr	3.01
reducing set	1.20	0.16	2.02	1.32	nr	3.34
male coupling	1.00	0.19	2.39	1.10	nr	3.49
female coupling	1.20	0.19	2.39	1.32	nr	3.71
90 degree bend	1.36	0.14	1.76	1.50	nr	3.26
90 degree backplate bend	3.35	0.28	3.53	3.69	nr	7.22
tee; equal	1.91	0.20	2.52	2.10	nr	4.62
tee; backplate	5.62	0.20	2.52	6.19	nr	8.71
tank coupling	3.31	0.20	2.52	3.65	nr	6.17
22 mm pipes; fixing with pipe clips; plugged and screwed	1.72	0.44	5.54	2.02	m	7.56
Extra for						
made bend	-	0.19	2.39	-	nr	2.39
stop end	2.03	0.11	1.39	2.24	nr	3.63
straight coupling	1.84	0.19	2.39	2.03	nr	4.42
reducing set	1.29	0.05	0.63	1.42	nr	2.05
male coupling	2.16	0.26	3.28	2.38	nr	5.66
female coupling	1.77	0.26	3.28	1.95	nr	5.23
90 degree bend	2.17	0.19	2.39	2.39	nr	4.78
tee; equal	3.16	0.28	3.53	3.48	nr	7.01
tee; reducing	4.59	0.28	3.53	5.06	nr	8.59
tank coupling	3.54	0.28	3.53	3.90	nr	7.43

S PIPED SUPPLY SYSTEMS

	PC £	Labour hours	Labour £	Material £	Unit	Total rate £
28 mm pipes; fixing with pipe clips; plugged and screwed	2.23	0.48	6.05	2.58	m	8.63
Extra for						
made bend	-	0.23	2.90	-	nr	2.90
stop end	3.98	0.13	1.64	4.39	nr	6.03
straight coupling	3.81	0.23	2.90	4.20	nr	7.10
male coupling	2.70	0.32	4.03	2.98	nr	7.01
female coupling	3.50	0.32	4.03	3.86	nr	7.89
90 degree bend	4.92	0.23	2.90	5.42	nr	8.32
tee; equal	7.84	0.34	4.28	8.64	nr	12.92
tee; reducing	7.57	0.34	4.28	8.34	nr	12.62
tank coupling	6.09	0.34	4.28	6.71	nr	10.99
35 mm pipes; fixing with pipe clips; plugged and screwed	5.59	0.55	6.93	6.38	m	13.31
Extra for						
made bend	-	0.28	3.53	-	nr	3.53
stop end	6.13	0.15	1.89	6.75	nr	8.64
straight coupling	7.91	0.28	3.53	8.72	nr	12.25
male coupling	6.01	0.37	4.66	6.62	nr	11.28
female coupling	7.22	0.37	4.66	7.96	nr	12.62
tee; equal	13.90	0.39	4.91	15.32	nr	20.23
tee; reducing	13.57	0.39	4.91	14.95	nr	19.86
tank coupling	7.24	0.39	4.91	7.98	nr	12.89
42 mm pipes; fixing with pipe clips; plugged and screwed	6.86	0.61	7.69	7.81	m	15.50
Extra for						
made bend	-	0.37	4.66	-	nr	4.66
stop end	10.19	0.17	2.14	11.23	nr	13.37
straight coupling	10.40	0.32	4.03	11.46	nr	15.49
male coupling	9.02	0.42	5.29	9.94	nr	15.23
female coupling	9.71	0.42	5.29	10.70	nr	15.99
tee; equal	21.84	0.43	5.42	24.06	nr	29.48
tee; reducing	20.97	0.43	5.42	23.11	nr	28.53
54 mm pipes; fixing with pipe clips; plugged and screwed	8.81	0.67	8.44	10.01	m	18.45
Extra for						
made bend	-	0.51	6.43	-	nr	6.43
straight coupling	15.56	0.37	4.66	17.15	nr	21.81
male coupling	13.32	0.46	5.80	14.68	nr	20.48
female coupling	14.24	0.46	5.80	15.69	nr	21.49
tee; equal	35.08	0.48	6.05	38.65	nr	44.70
tee; reducing	35.08	0.48	6.05	38.65	nr	44.70
Copper, brass and gunmetal ancillaries; screwed joints to fittings						
Stopcock; brass/gunmetal capillary joints to copper						
15 mm nominal size	2.59	0.19	2.39	2.85	nr	5.24
22 mm nominal size	4.83	0.25	3.15	5.32	nr	8.47
28 mm nominal size	13.75	0.31	3.91	15.15	nr	19.06
Stopcock; brass/gunmetal compression joints to copper						
15 mm nominal size	3.55	0.17	2.14	3.91	nr	6.05
22 mm nominal size	6.25	0.22	2.77	6.89	nr	9.66
28 mm nominal size	16.29	0.28	3.53	17.95	nr	21.48
Stopcock; brass/gunmetal compression joints to polyethylene						
15 mm nominal size	9.46	0.24	3.02	10.42	nr	13.44
22 mm nominal size	9.74	0.31	3.91	10.73	nr	14.64
28 mm nominal size	14.63	0.37	4.66	16.12	nr	20.78

S PIPED SUPPLY SYSTEMS

	PC £	Labour hours	Labour £	Material £	Unit	Total rate £
S10/S11 HOT AND COLD WATER - cont'd						
Copper, brass and gunmetal ancillaries; screwed joints to fittings - cont'd						
Gunmetal "Fullway" gate valve; capillary joints to copper						
15 mm nominal size	8.07	0.19	2.39	8.89	nr	11.28
22 mm nominal size	9.35	0.25	3.15	10.30	nr	13.45
28 mm nominal size	13.01	0.31	3.91	14.34	nr	18.25
35 mm nominal size	29.04	0.38	4.79	32.00	nr	36.79
42 mm nominal size	36.31	0.43	5.42	40.01	nr	45.43
54 mm nominal size	52.67	0.49	6.17	58.04	nr	64.21
Brass gate valve; compression joints to copper						
15 mm nominal size	5.77	0.28	3.53	6.36	nr	9.89
22 mm nominal size	6.72	0.37	4.66	7.40	nr	12.06
28 mm nominal size	11.75	0.46	5.80	12.95	nr	18.75
Chromium plated; lockshield radiator valve; union outlet						
15 mm nominal size	5.10	0.20	2.52	5.62	nr	8.14
Water tanks/cisterns						
Polyethylene cold water feed and expansion cistern; BS 4213; with covers						
ref SC15; 68 litres	29.69	1.16	14.61	31.92	nr	46.53
ref SC25; 114 litres	34.88	1.34	16.88	37.50	nr	54.38
ref SC40; 182 litres	40.80	1.34	16.88	43.85	nr	60.73
ref SC50; 227 litres	56.79	1.80	22.68	61.06	nr	83.74
GRP cold water storage cistern; with covers						
ref 899.10; 30 litres	60.69	1.02	12.85	65.24	nr	78.09
ref 899.25; 68 litres	75.95	1.16	14.61	81.64	nr	96.25
ref 899.40; 114 litres	95.57	1.34	16.88	102.74	nr	119.62
ref 899.70; 227 litres	118.69	1.80	22.68	127.59	nr	150.27
Storage cylinders/calorifiers						
Copper cylinders; single feed coil indirect; BS 1566 Part 2; grade 3						
ref 2; 96 litres	133.25	1.85	23.31	143.24	nr	166.55
ref 3; 114 litres	84.78	2.08	26.21	91.14	nr	117.35
ref 7; 117 litres	83.66	2.31	29.10	89.93	nr	119.03
ref 8; 140 litres	94.72	2.78	35.03	101.82	nr	136.85
ref 9; 162 litres	120.98	3.24	40.82	130.05	nr	170.87
Combination copper hot water storage units; coil direct; BS 3198; (hot/cold)						
400 mm x 900 mm; 65/20 litres	98.47	2.59	32.63	105.86	nr	138.49
450 mm x 900 mm; 85/25 litres	101.41	3.61	45.48	109.02	nr	154.50
450 mm x 1075 mm; 115/25 litres	111.55	4.53	57.07	119.92	nr	176.99
450 mm x 1200 mm; 115/45 litres	118.75	5.09	64.13	127.66	nr	191.79
Combination copper hot water storage						
450 mm x 900 mm; 85/25 litres	127.17	4.07	51.28	136.71	nr	187.99
450 mm x 1200 mm; 115/45 litres	145.42	5.55	69.92	156.33	nr	226.25
Thermal insulation						
19 mm thick rigid mineral glass fibre sectional pipe lagging; plain finish; fixed with aluminium bands to steel or copper pipework; including working over pipe fittings						
around 15/15 pipes	4.24	0.06	0.76	4.79	m	5.55
around 20/22 pipes	4.87	0.09	1.13	5.50	m	6.63
around 25/28 pipes	5.27	0.10	1.26	5.95	m	7.21
around 32/35 pipes	5.85	0.11	1.39	6.60	m	7.99
around 40/42 pipes	6.20	0.12	1.51	7.00	m	8.51
around 50/54 pipes	7.18	0.14	1.76	8.10	m	9.86

S PIPED SUPPLY SYSTEMS

	PC £	Labour hours	Labour £	Material £	Unit	Total rate £
19 mm thick rigid mineral glass fibre sectional pipe lagging; canvas or class O lacquered aluminium finish; fixed with aluminium bands to steel or copper pipework; including working over pipe fittings						
around 15/15 pipes	6.17	0.06	0.76	6.96	m	7.72
around 20/22 pipes	6.72	0.09	1.13	7.59	m	8.72
around 25/28 pipes	7.37	0.10	1.26	8.32	m	9.58
around 32/35 pipes	8.02	0.11	1.39	9.05	m	10.44
around 40/42 pipes	8.63	0.12	1.51	9.74	m	11.25
around 50/54 pipes	10.01	0.14	1.76	11.30	m	13.06
60 mm thick glass-fibre filled polyethylene insulating jackets for GRP or polyethylene cold water cisterns; complete with fixing bands; for cisterns size (supply not included)						
450 mm x 300 mm x 300 mm (45 litres)	-	0.37	4.66	-	nr	4.66
650 mm x 500 mm x 400 mm (91 litres)	-	0.56	7.06	-	nr	7.06
675 mm x 525 mm x 500 mm (136 litres)	-	0.65	8.19	-	nr	8.19
675 mm x 575 mm x 525 mm (182 litres)	-	0.74	9.32	-	nr	9.32
1000 mm x 625 mm x 525 mm (273 litres)	-	0.79	9.95	-	nr	9.95
1125 mm x 650 mm x 575 mm (341 litres)	-	0.79	9.95	-	nr	9.95
80 mm thick glass-fibre filled insulating jackets in flame retardant PVC to BS 1763; type 1B; segmental type for hot water cylinders; complete with fixing bands; for cylinders size (supply not included)						
400 mm x 900 mm; ref 2	-	0.31	3.91	-	nr	3.91
450 mm x 900 mm; ref 7	-	0.31	3.91	-	nr	3.91
450 mm x 1050 mm; ref 8	-	0.37	4.66	-	nr	4.66
450 mm x 1200 mm	-	0.46	5.80	-	nr	5.80

S13 PRESSURISED WATER

Blue MDPE pipes; BS 6572; mains pipework; no joints in the running length; laid in trenches

	PC £	Labour hours	Labour £	Material £	Unit	Total rate £
Pipes						
20 mm nominal size	0.34	0.10	1.26	0.38	m	1.64
25 mm nominal size	0.43	0.11	1.39	0.48	m	1.87
32 mm nominal size	0.73	0.12	1.51	0.81	m	2.32
50 mm nominal size	1.73	0.14	1.76	1.92	m	3.68
63 mm nominal size	2.77	0.15	1.89	3.07	m	4.96
Ductile iron bitumen coated pipes and fittings; BS 4772; class K9; Stanton's "Tyton" water main pipes or other equal and approved; flexible joints						
100 mm pipes; laid straight	13.45	0.56	4.39	20.17	m	24.56
Extra for						
bend; 45 degrees	30.90	0.56	4.39	44.86	nr	49.25
branch; 45 degrees; socketed	222.12	0.83	6.50	265.69	nr	272.19
tee	48.14	0.83	6.50	69.31	nr	75.81
flanged spigot	32.80	0.56	4.39	42.01	nr	46.40
flanged socket	32.11	0.56	4.39	41.23	nr	45.62
150 mm pipes; laid straight	20.25	0.65	5.09	28.29	m	33.38
Extra for						
bend; 45 degrees	52.78	0.65	5.09	70.43	nr	75.52
branch; 45 degrees; socketed	283.50	0.97	7.60	336.29	nr	343.89
tee	100.03	0.97	7.60	129.20	nr	136.80
flanged spigot	39.13	0.65	5.09	49.60	nr	54.69
flanged socket	51.10	0.65	5.09	63.11	nr	68.20

Prices for Measured Works - Major Works
S PIPED SUPPLY SYSTEMS

	PC £	Labour hours	Labour £	Material £	Unit	Total rate £
S13 PRESSURISED WATER - cont'd						
Ductile iron bitumen coated pipes and fittings; BS 4772; class K9; Stanton's "Tyton" water main pipes or other equal and approved; flexible joints - cont'd						
200 mm pipes; laid straight	27.37	0.93	7.29	38.50	m	45.79
Extra for						
bend; 45 degrees	95.24	0.93	7.29	122.72	nr	130.01
branch; 45 degrees; socketed	321.96	1.39	10.89	386.24	nr	397.13
tee	137.38	1.39	10.89	177.89	nr	188.78
flanged spigot	85.32	0.93	7.29	103.91	nr	111.20
flanged socket	80.83	0.93	7.29	98.84	nr	106.13
S32 NATURAL GAS						
Ductile iron bitumen coated pipes and fittings; BS 4772; class K9; Stanton's "Stanlock" gas main pipes or other equal and approved; bolted gland joints						
100 mm pipes; laid straight	26.97	0.65	5.09	42.67	m	47.76
Extra for						
bend; 45 degrees	40.75	0.65	5.09	64.35	nr	69.44
tee	62.18	0.97	7.60	100.77	nr	108.37
flanged spigot	33.75	0.65	5.09	50.33	nr	55.42
flanged socket	33.31	0.65	5.09	49.83	nr	54.92
isolated "Stanlock" joint	10.84	0.32	2.51	12.24	nr	14.75
150 mm pipes; laid straight	41.07	0.83	6.50	63.81	m	70.31
Extra for						
bend; 45 degrees	58.84	0.83	6.50	92.59	nr	99.09
tee	104.10	1.25	9.80	161.13	nr	170.93
flanged spigot	39.13	0.83	6.50	61.62	nr	68.12
flanged socket	49.11	0.83	6.50	72.88	nr	79.38
isolated "Stanlock" joint	15.46	0.42	3.29	17.45	nr	20.74
200 mm pipes; laid straight	54.67	1.20	9.40	85.07	m	94.47
Extra for						
bend; 45 degrees	95.83	1.20	9.40	131.53	nr	140.93
tee	144.63	1.80	14.11	233.35	nr	247.46
flanged spigot	85.32	1.20	9.40	119.67	nr	129.07
flanged socket	75.98	1.20	9.40	109.13	nr	118.53
isolated "Stanlock" joint	20.70	0.60	4.70	23.37	nr	28.07

T MECHANICAL HEATING/COOLING ETC. SYSTEMS

	PC £	Labour hours	Labour £	Material £	Unit	Total rate £
T MECHANICAL HEATING/COOLING ETC SYSTEMS						
T10 GAS/OIL FIRED BOILERS						
Boilers						
Gas fired floor standing domestic boilers; cream or white; enamelled casing; 32 mm diameter BSPT female flow and return tappings; 102 mm diameter flue socket 13 mm diameter BSPT male draw-off outlet						
13.19 kW output (45,000 Btu/Hr)	-	4.63	56.59	444.62	nr	**501.21**
23.45 kW output (80,000 Btu/Hr)	-	4.63	56.59	571.65	nr	**628.24**
T31 LOW TEMPERATURE HOT WATER HEATING						
NOTE: The reader is referred to section "S10/S11 Hot and Cold Water" for rates for copper pipework which will equally apply to this section of work. For further and more detailed information the reader is advised to consult *Spon's Mechanical and Electrical Services Price Book*.						
Radiators; Myson Heat Emitters or other equal and approved						
"Premier HE" single panel type; steel; 690 mm high; wheelhead and lockshield valves						
540 mm long	20.34	1.85	22.61	33.24	nr	**55.85**
1149 mm long	41.04	2.08	25.42	55.50	nr	**80.92**
2165 mm long	79.32	2.31	28.23	96.65	nr	**124.88**
2978 mm long	121.23	2.54	31.05	141.77	nr	**172.82**

V ELECTRICAL SYSTEMS

	PC £	Labour hours	Labour £	Material £	Unit	Total rate £
V ELECTRICAL SYSTEMS						
V21/V22 GENERAL LIGHTING AND LV POWER						
NOTE: The following items indicate approximate prices for wiring of lighting and power points complete, including accessories and socket outlets, but excluding lighting fittings. Consumer control units are shown separately. For a more detailed breakdown of these costs and specialist costs for a complete range of electrical items, reference should be made to *Spon's Mechanical and Electrical Services Price Book.*						
Consumer control units						
8-way 60 amp SP&N surface mounted insulated consumer control units fitted with miniature circuit breakers including 2.00 m long 32 mm screwed welded conduit with three runs of 16 mm² PVC cables ready for final connections	-	-	-	-	nr	154.80
extra for current operated ELCB of 30 mA tripping current	-	-	-	-	nr	64.91
As above but 100 amp metal cased consumer unit and 25 mm² PVC cables	-	-	-	-	nr	169.77
extra for current operated ELCB of 30 mA tripping current	-	-	-	-	nr	144.81
Final circuits						
Lighting points						
wired in PVC insulated and PVC sheathed cable in flats and houses; insulated in cavities and roof space; protected where buried by heavy gauge PVC conduit	-	-	-	-	nr	54.92
as above but in commercial property	-	-	-	-	nr	69.91
wired in PVC insulated cable in screwed welded conduit in flats and houses	-	-	-	-	nr	114.84
as above but in commercial property	-	-	-	-	nr	144.81
as above but in industrial property	-	-	-	-	nr	164.78
wired in MICC cable in flats and houses	-	-	-	-	nr	99.87
as above but in commercial property	-	-	-	-	nr	114.84
as above but in industrial property with PVC sheathed cable	-	-	-	-	nr	134.83
Single 13 amp switched socket outlet points						
wired in PVC insulated and PVC sheathed cable in flats and houses on a ring main circuit; protected where buried by heavy gauge PVC conduit	-	-	-	-	nr	49.93
as above but in commercial property	-	-	-	-	nr	64.91
wired in PVC insulated cable in screwed welded conduit throughout on a ring main circuit in flats and houses	-	-	-	-	nr	89.88
as above but in commercial property	-	-	-	-	nr	99.87
as above but in industrial property	-	-	-	-	nr	114.84
wired in MICC cable on a ring main circuit in flats and houses	-	-	-	-	nr	89.88
as above but in commercial property	-	-	-	-	nr	104.87
as above but in industrial property with PVC sheathed cable	-	-	-	-	nr	134.83

V ELECTRICAL SYSTEMS

	PC £	Labour hours	Labour £	Material £	Unit	Total rate £
Cooker control units						
45 amp circuit including unit wired in PVC insulated and PVC sheathed cable; protected where buried by heavy gauge PVC conduit	-	-	-	-	nr	**119.84**
as above but wired in PVC insulated cable in screwed welded conduit	-	-	-	-	nr	**174.76**
as above but wired in MICC cable	-	-	-	-	nr	**194.75**

W SECURITY SYSTEMS

	PC £	Labour hours	Labour £	Material £	Unit	Total rate £
W SECURITY SYSTEMS						
W20 LIGHTNING PROTECTION						
Lightning protection equipment						
Copper strip roof or down conductors fixed with bracket or saddle clips						
20 mm x 3 mm flat section	-	-	-	-	m	14.99
25 mm x 3 mm flat section	-	-	-	-	m	17.48
Aluminium strip roof or down conductors fixed with bracket or saddle clips						
20 mm x 3 mm flat section	-	-	-	-	m	10.99
25 mm x 3 mm flat section	-	-	-	-	m	11.99
Joints in tapes	-	-	-	-	nr	8.49
Bonding connections to roof and structural metalwork	-	-	-	-	nr	49.93
Testing points	-	-	-	-	nr	24.97
Earth electrodes						
16 mm diameter driven copper electrodes in 1220 mm long sectional lengths (minimum 2440 mm long overall)	-	-	-	-	nr	129.83
first 2440 mm length driven and tested 25 mm x 3 mm copper strip electrode in 457 mm deep prepared trench	-	-	-	-	m	9.99

DAVIS LANGDON & EVEREST
Authors of Spon's Price Books

www.davislangdon.com

Davis Langdon & Everest is an independent practice of Chartered Quantity Surveyors, with some 1000 staff in 20 UK offices and, through Davis Langdon & Seah International, some 2,500 staff in 85 offices worldwide.

DLE manages client requirements, controls risk, manages cost and maximises value for money, throughout the course of construction projects, always aiming to be - and to deliver - the best.

TYPICAL PROJECT STAGES, DLE INTEGRATED SERVICES AND THEIR EFFECT:

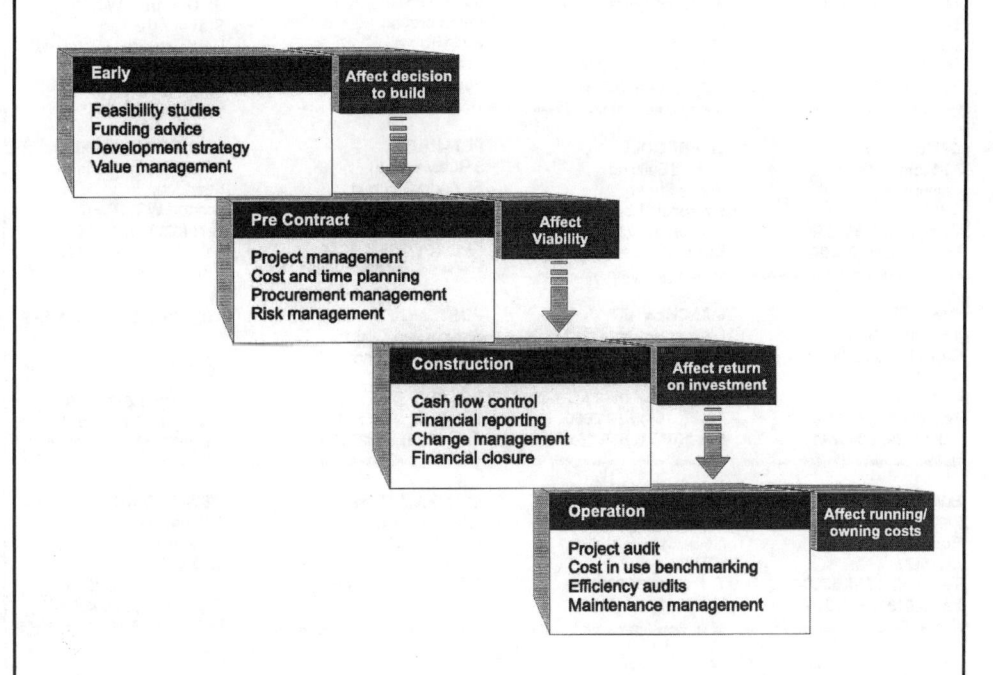

Early
- Feasibility studies
- Funding advice
- Development strategy
- Value management

→ Affect decision to build

Pre Contract
- Project management
- Cost and time planning
- Procurement management
- Risk management

→ Affect Viability

Construction
- Cash flow control
- Financial reporting
- Change management
- Financial closure

→ Affect return on investment

Operation
- Project audit
- Cost in use benchmarking
- Efficiency audits
- Maintenance management

→ Affect running/owning costs

EUROPE ⇨ ASIA - AUSTRALIA - AFRICA - AMERICA
GLOBAL REACH - LOCAL DELIVERY

DAVIS LANGDON & EVEREST

LONDON
Princes House
39 Kingsway
London
WC2B 6TP
Tel : (020) 7497 9000
Fax : (020) 7497 8858
Email:rob.smith@davislangdon-uk.com

GATESHEAD
11 Regent Terrace
Gateshead
Tyne and Wear
NE8 1LU
Tel : (0191) 477 3844
Fax : (0191) 490 1742
Email:gary.lockey@davislangdon-uk.com

MILTON KEYNES
Everest House
Rockingham Drive
Linford Wood
Milton Keynes MK14 6LY
Tel : (01908) 304700
Fax : (01908) 660059
Email:kevin.sims@davislangdon-uk.com

DAVIS LANGDON CONSULTANCY
Princes House
39 Kingsway
London WC2B 6TP
Tel : (020) 7379 3322
Fax : (020) 7379 3030
Email:jim.meikle@davislangdon-uk.com

BIRMINGHAM
29 Woodbourne Road
Harborne
Birmingham
B17 8BY
Tel : (0121) 4299511
Fax : (0121) 4292544
Email:richard.d.taylor@davislangdon-uk.com

GLASGOW
Cumbrae House
15 Carlton Court
Glasgow
G5 9JP
Tel : (0141) 429 6677
Fax : (0141) 429 2255
Email:hugh.fisher@davislangdon-uk.com

OXFORD
Avalon House
Marcham Road
Abingdon
Oxford OX14 1TZ
Tel : (01235) 555025
Fax : (01235) 554909
Email:paul.coomber@davislangdon-uk.com

MOTT GREEN & WALL
Africa House
64-78 Kingsway
London
WC2B 6NN
Tel : (020) 7836 0836
Fax : (020) 7242 0394
Email:barry.nugent@mottgreenwall.co.uk

BRISTOL
St Lawrence House
29/31 Broad Street
Bristol
BS1 2HF
Tel : (0117) 9277832
Fax : (0117) 9251350
Email:alan.trolley@davislangdon-uk.com

LEEDS
Duncan House
14 Duncan Street
Leeds
LS1 6DL
Tel : (0113) 2432481
Fax : (0113) 2424601
Email:tony.brennan@davislangdon-uk.com

PETERBOROUGH
Charterhouse
66 Broadway
Peterborough
PE1 1SU
Tel: (01733) 343625
Fax: (01733) 349177
Email:colin.harrison@davislangdon-uk.com

SCHUMANN SMITH
14th Flr, Southgate House
St Georges Way
Stevenage
Hertfordshire SG1 1HG
Tel : (01438) 742642
Fax : (01438) 742632
Email:nschumann@schumannsmith.com

CAMBRIDGE
36 Storey's Way
Cambridge
CB3 0DT
Tel : (01223) 351258
Fax : (01223) 321002
Email:stephen.bugg@davislangdon-uk.com

LIVERPOOL
Cunard Building
Water Street
Liverpool L3 1JR
Tel : (0151) 2361992
Fax : (0151) 2275401
Email:john.davenport@davislangdon-uk.com

PLYMOUTH
3 Russell Court
St Andrew Street
Plymouth PL1 2AX
Tel : (01752) 668372
Fax : (01752) 221219
Email:gareth.steventon@davislangdon-uk.com

NBW CROSHER & JAMES
1-5 Exchange Court
Strand
London WC2R 0PQ
Tel : (020) 7845 0600
Fax : (020) 7845 0601
Email:sanders@nbwcrosherjames.com

CARDIFF
4 Pierhead Street
Capital Waterside
Cardiff
CF10 4QP
Tel : (029) 20497497
Fax : (029) 20497111
Email:paul.edwards@davislangdon-uk.com

MANCHESTER
Cloister House
Riverside
New Bailey Street
Manchester M3 5AG
Tel : (0161) 819 7600
Fax : (0161) 819 1818
Email:paul.stanion@davislangdon-uk.com

PORTSMOUTH
St Andrews Court
St Michaels Road
Portsmouth
PO1 2PR
Tel : (023) 92815218
Fax : (023) 92827156
Email:brian.bartholomew@davislangdon-uk.com

NBW CROSHER & JAMES
102 New Street
Birmingham
B2 4HQ
Tel : (0121) 632 3600
Fax : (0121) 632 3601
Email:whittakerr@bir.nbwcrosherjames.com

EDINBURGH
74 Great King Street
Edinburgh
EH3 6QU
Tel : (0131) 557 5306
Fax : (0131) 557 5704
Email:ian.mcandie@davislangdon-uk.com

NORWICH
63 Thorpe Road
Norwich
NR1 1UD
Tel : (01603) 628194
Fax : (01603) 615928
Email:michael.ladbrook@davislangdon-uk.com

SOUTHAMPTON
Brunswick House
Brunswick Place
Southampton SO15 2AP
Tel : (023) 80333438
Fax : (023) 80226099
Email:richard.pitman@davislangdon-uk.com

NBW CROSHER & JAMES
5 Coates Crescent
Edinburgh
EH3 9AL
Tel : (0131) 220 4225
Fax : (0131) 220 4226
Email:mcfarlanel@nbwcrosherjames.com

with offices throughout Europe, the Middle East, Asia, Australia, Africa and the USA which together form

DAVIS LANGDON & SEAH INTERNATIONAL

PRICES FOR MEASURED WORK - MINOR WORKS

INTRODUCTION

The "Prices for Measured Work - Minor Works" are intended to apply to a small project in the outer London area costing about £65,000 (excluding Preliminaries).

The format of this section follows that of the "Major Works" section with minor variations because of the different nature of the work, and reference should be made to the "Introduction" to that section on page 9

It has been assumed that reasonable quantities of work are involved, equivalent to quantities for two houses, although clearly this would not apply to all trades and descriptions of work in a project of this value. Where smaller quantities of work are involved it will be necessary to adjust the prices accordingly.

For section "C Demolition/Alteration/Renovation" even smaller quantities have been assumed as can be seen from the stated "PC" of the materials involved.

Where work in an existing building is concerned it has been assumed that the building is vacated and that in all cases there is reasonable access and adequate storage space. Should this not be the case, and if any abnormal circumstances have to be taken into account, an allowance can be made either by a lump sum addition or by suitably modifying the percentage factor for overheads and profit. As for the major works section all rates exclude overheads and profit.

Labour rates are based upon typical gang costs divided by the number of primary working operatives for the trade concerned; and for general building work include an allowance for trade supervision, but exclude overheads and profit. The "Labour hours" column gives the total hours allocated to a particular item and the "Labour £" the consolidated cost of such labour. "Labour hours" have not always been given for "spot" items because of the inclusion of Sub-Contractor's labour.

The "Material Plant £" column includes the cost of removal of debris by skips or lorries. Alternative materials prices tables can be found in the appropriate "Prices for Measured Works - Major Works" section. As stated earlier, these prices are "list" prices before deduction of quantity discounts, and therefore can be substituted directly in place of "PC" figures given. The reader should bear in mind that although large orders are delivered free of charge, smaller orders generally attract a delivery or part load charge and this should be added to the alternative material price prior to substitution in a rate.

No allowance has been made for any Value Added Tax which will probably be payable on the majority of work of this nature.

A PRELIMINARIES/CONTRACT CONDITIONS FOR MINOR WORKS

When pricing Preliminaries all factors affecting the execution of the works must be considered; some of the more obvious have already been mentioned above.

As mentioned in "A Preliminaries" in the "Prices for Measured Work - Major Works" section (page 119), the current trend is for Preliminaries to be priced at approximately 11 to 13%, but for alterations and additions work in particular, care must be exercised in ensuring that all adverse factors are covered. The reader is advised to identify systematically and separately price all preliminary items with cost/time implications in order to reflect as accurately as possible preliminary costs likely to stem from any particular scheme.

Where the Standard Form of Contract applies two clauses which will affect the pricing of Preliminaries should be noted.

(a) Insurance of the works against Clause 22 Perils
    Clause 22C will apply whereby the Employer and not the Contractor effects the insurance.
(b) Fluctuations
    An allowance for any shortfall in recovery of increased costs under whichever clause is contained in the Contract may be covered by the inclusion of a lump sum in the Preliminaries or by increasing the prices by a suitable percentage

ADDITIONS AND NEW WORKS WITHIN EXISTING BUILDINGS

Depending upon the contract size either the prices in "Prices for Measured Work - Major Works" or those prices in "Prices for Measured Work - Minor Works" will best apply.

It is likely, however, that conditions affecting the excavations for foundations might preclude the use of mechanical plant, and that it will be necessary to restrict prices to those applicable to hand excavation.

If, in any circumstances, less than what might be termed "normal quantities" are likely to be involved it is stressed that actual quotations should be invited from specialist Sub-contractors for these works.

JOBBING WORK

Jobbing work is outside the scope of this section and no attempt has been made to include prices for such work.

NEW FROM SPON PRESS

The Architectural Expression of Environmental Control Systems

George Baird, Victoria University of Wellington, New Zealand

The Architectural Expression of Environmental Control Systems examines the way project teams can approach the design and expression of both active and passive thermal environmental control systems in a more creative way. Using seminal case studies from around the world and interviews with the architects and environmental engineers involved, the book illustrates innovative responses to client, site and user requirements, focusing upon elegant design solutions to a perennial problem.

This book will inspire architects, building scientists and building services engineers to take a more creative approach to the design and expression of environmental control systems - whether active or passive, whether they influence overall building form or design detail.

March 2001: 276x219: 304pp
135 b+w photos, 40 colour, 90 line illustrations
Hb: 0-419-24430-1: £49.95

To Order: Tel: +44 (0) 8700 768853, or +44 (0) 1264 343071 Fax: +44 (0) 1264 343005, or
Post: Spon Press Customer Services, ITPS Andover, Hants, SP10 5BE, UK Email: book.orders@tandf.co.uk.

Postage & Packing: UK: 5% of order value (min. charge £1, max. charge £10) for 3-5 days delivery. Option of next day delivery at an additional £6.50. Europe: 10% of order value (min. charge £2.95, max. charge £20) for delivery surface post. Option of airmail at an additional £6.50. ROW: 15% of order value (min.charge£6.50, max. charge £30) for airmail delivery.

For a complete listing of all our titles visit: www.sponpress.com

DAVIS LANGDON & EVEREST
Authors of Spon's Price Books

www.davislangdon.com

Davis Langdon & Everest is an independent practice of Chartered Quantity Surveyors, with some 1000 staff in 20 UK offices and, through Davis Langdon & Seah International, some 2,500 staff in 85 offices worldwide.

DLE manages client requirements, controls risk, manages cost and maximises value for money, throughout the course of construction projects, always aiming to be - and to deliver - the best.

TYPICAL PROJECT STAGES, DLE INTEGRATED SERVICES AND THEIR EFFECT:

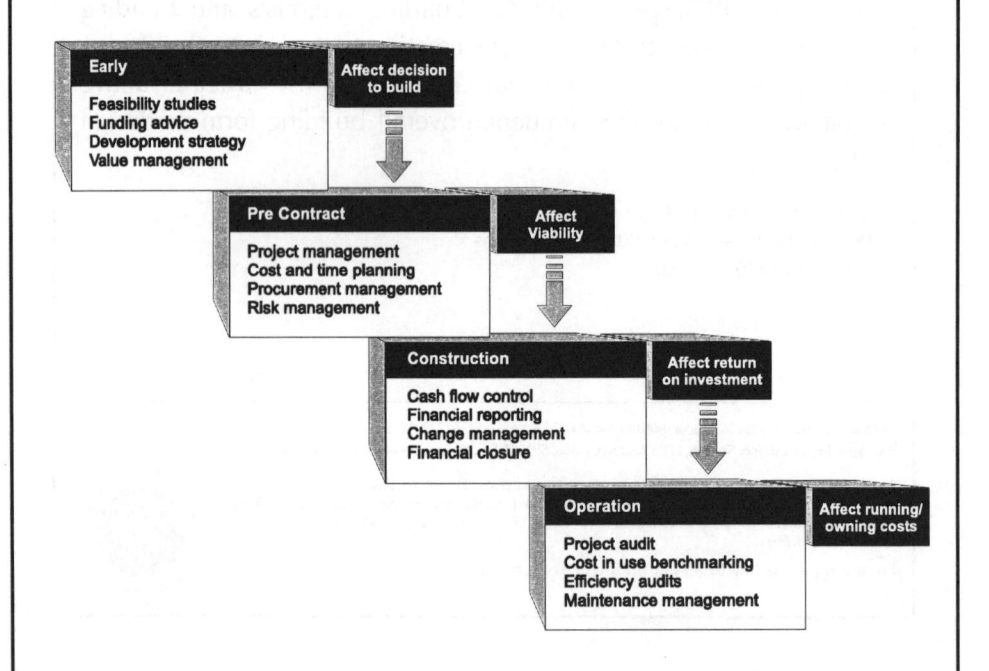

Early
- Feasibility studies
- Funding advice
- Development strategy
- Value management

Affect decision to build

Pre Contract
- Project management
- Cost and time planning
- Procurement management
- Risk management

Affect Viability

Construction
- Cash flow control
- Financial reporting
- Change management
- Financial closure

Affect return on investment

Operation
- Project audit
- Cost in use benchmarking
- Efficiency audits
- Maintenance management

Affect running/ owning costs

EUROPE ⇨ ASIA - AUSTRALIA - AFRICA - AMERICA
GLOBAL REACH - LOCAL DELIVERY

DAVIS LANGDON & EVEREST

LONDON
Princes House
39 Kingsway
London
WC2B 6TP
Tel : (020) 7497 9000
Fax : (020) 7497 8858
Email:rob.smith@davislangdon-uk.com

BIRMINGHAM
29 Woodbourne Road
Harborne
Birmingham
B17 8BY
Tel : (0121) 4299511
Fax : (0121) 4292544
Email:richard.d.taylor@davislangdon-uk.com

BRISTOL
St Lawrence House
29/31 Broad Street
Bristol
BS1 2HF
Tel : (0117) 9277832
Fax : (0117) 9251350
Email:alan.trolley@davislangdon-uk.com

CAMBRIDGE
36 Storey's Way
Cambridge
CB3 0DT
Tel : (01223) 351258
Fax : (01223) 321002
Email:stephen.bugg@davislangdon-uk.com

CARDIFF
4 Pierhead Street
Capital Waterside
Cardiff
CF10 4QP
Tel : (029) 20497497
Fax : (029) 20497111
Email:paul.edwards@davislangdon-uk.com

EDINBURGH
74 Great King Street
Edinburgh
EH3 6QU
Tel : (0131) 557 5306
Fax : (0131) 557 5704
Email:ian.mcandie@davislangdon-uk.com

GATESHEAD
11 Regent Terrace
Gateshead
Tyne and Wear
NE8 1LU
Tel : (0191) 477 3844
Fax : (0191) 490 1742
Email:gary.lockey@davislangdon-uk.com

GLASGOW
Cumbrae House
15 Carlton Court
Glasgow
G5 9JP
Tel : (0141) 429 6677
Fax : (0141) 429 2255
Email:hugh.fisher@davislangdon-uk.com

LEEDS
Duncan House
14 Duncan Street
Leeds
LS1 6DL
Tel : (0113) 2432481
Fax : (0113) 2424601
Email:tony.brennan@davislangdon-uk.com

LIVERPOOL
Cunard Building
Water Street
Liverpool L3 1JR
Tel : (0151) 2361992
Fax : (0151) 2275401
Email:john.davenport@davislangdon-uk.com

MANCHESTER
Cloister House
Riverside
New Bailey Street
Manchester M3 5AG
Tel : (0161) 819 7600
Fax : (0161) 819 1818
Email:paul.stanion@davislangdon-uk.com

NORWICH
63 Thorpe Road
Norwich
NR1 1UD
Tel : (01603) 628194
Fax : (01603) 615928
Email:michael.ladbrook@davislangdon-uk.com

MILTON KEYNES
Everest House
Rockingham Drive
Linford Wood
Milton Keynes MK14 6LY
Tel : (01908) 304700
Fax : (01908) 660059
Email:kevin.sims@davislangdon-uk.com

OXFORD
Avalon House
Marcham Road
Abingdon
Oxford OX14 1TZ
Tel : (01235) 555025
Fax : (01235) 554909
Email:paul.coomber@davislangdon-uk.com

PETERBOROUGH
Charterhouse
66 Broadway
Peterborough
PE1 1SU
Tel: (01733) 343625
Fax: (01733) 349177
Email:colin.harrison@davislangdon-uk.com

PLYMOUTH
3 Russell Court
St Andrew Street
Plymouth PL1 2AX
Tel : (01752) 668372
Fax : (01752) 221219
Email:gareth.steventon@davislangdon-uk.com

PORTSMOUTH
St Andrews Court
St Michaels Road
Portsmouth
PO1 2PR
Tel : (023) 92815218
Fax : (023) 92827156
Email:brian.bartholomew@davislangdon-uk.com

SOUTHAMPTON
Brunswick House
Brunswick Place
Southampton SO15 2AP
Tel : (023) 80333438
Fax : (023) 80226099
Email:richard.pitman@davislangdon-uk.com

DAVIS LANGDON CONSULTANCY
Princes House
39 Kingsway
London WC2B 6TP
Tel : (020) 7379 3322
Fax : (020) 7379 3030
Email:jim.meikle@davislangdon-uk.com

MOTT GREEN & WALL
Africa House
64-78 Kingsway
London
WC2B 6NN
Tel : (020) 7836 0836
Fax : (020) 7242 0394
Email:barry.nugent@mottgreenwall.co.uk

SCHUMANN SMITH
14th Flr, Southgate House
St Georges Way
Stevenage
Hertfordshire SG1 1HG
Tel : (01438) 742642
Fax : (01438) 742632
Email:nschumann@schumannsmith.com

NBW CROSHER & JAMES
1-5 Exchange Court
Strand
London WC2R 0PQ
Tel : (020) 7845 0600
Fax : (020) 7845 0601
Email:sandersr@nbwcrosherjames.com

NBW CROSHER & JAMES
102 New Street
Birmingham
B2 4HQ
Tel : (0121) 632 3600
Fax : (0121) 632 3601
Email:whittakerr@bir.nbwcrosherjames.com

NBW CROSHER & JAMES
5 Coates Crescent
Edinburgh
EH3 9AL
Tel : (0131) 220 4225
Fax : (0131) 220 4226
Email:mcfarlanel@nbwcrosherjames.com

with offices throughout Europe, the Middle East, Asia, Australia, Africa and the USA which together form

DAVIS LANGDON & SEAH INTERNATIONAL

NEW FROM SPON PRESS

Dictionary of Property and Construction Law

Edited by J. Rostron, School of the Built Environment, Liverpool John Moores University, UK

This is a new dictionary containing over 6,000 entries. It provides clear and concise explanations of the terms used in land, property and construction law and management. The four key areas of coverage are: planning/construction law, land law, equity/trusts and finance and administration. It will be a useful reference for property and building professionals and for students of property and construction law. Jack Rostron is an experienced author and editor whose 1997 Spon title *Sick Building Syndrome* has been well received and widely reviewed.

August 2001: 234x156: 160pp:
Pb: 0-419-26110-9: £19.99
Hb: 0-419-26100-1: £50.00

To Order: Tel: +44 (0) 8700 768853, or +44 (0) 1264 343071 Fax: +44 (0) 1264 343005, or
Post: Spon Press Customer Services, ITPS Andover, Hants, SP10 5BE, UK Email: book.orders@tandf.co.uk.

Postage & Packing: UK: 5% of order value (min. charge £1, max. charge £10) for 3-5 days delivery. Option of next day delivery at an additional £6.50. Europe: 10% of order value (min. charge £2.95, max. charge £20) for delivery surface post. Option of airmail at an additional £6.50. ROW: 15% of order value (min.charge£6.50, max. charge £30) for airmail delivery.

For a complete listing of all our titles visit: www.sponpress.com

SPON PRESS
Taylor & Francis Group

C DEMOLITION/ALTERATION/RENOVATION

C20 DEMOLITION

NOTE: Demolition rates vary considerably from one scheme to another, depending upon access, the type of construction, the method of demolition, whether there are any redundant materials etc. Therefore, it is advisable to obtain specific quotations for each scheme under consideration, however, the following rates (excluding scaffolding costs) for simple demolitions may be of some assistance for comparative purposes.

	PC £	Labour hours	Labour £	Material £	Unit	Total rate £
Demolishing all structures						
Demolishing to ground level; single storey brick out-building; timber flat roofs; volume						
50 m³	-	-	-	-	m³	7.74
200 m³	-	-	-	-	m³	5.70
500 m³	-	-	-	-	m³	2.89
Demolishing to ground level; two storey brick out-building; timber joisted suspended floor and timber flat roofs; volume						
200 m³	-	-	-	-	m³	4.34
Demolishing parts of structures						
Breaking up concrete bed						
100 mm thick	-	0.50	4.11	1.86	m²	5.97
150 mm thick	-	0.74	6.08	3.81	m²	9.89
200 mm thick	-	1.00	8.22	3.73	m²	11.95
300 mm thick	-	1.48	12.17	5.47	m²	17.64
Breaking up reinforced concrete bed						
100 mm thick	-	0.56	4.60	2.12	m²	6.72
150 mm thick	-	0.83	6.82	3.14	m²	9.96
200 mm thick	-	1.11	9.12	4.25	m²	13.37
300 mm thick	-	1.67	13.73	6.37	m²	20.10
Demolishing reinforced concrete column or cutting away casing to steel column	-	11.10	91.24	29.00	m³	120.24
Demolishing reinforced concrete beam or cutting away casing to steel beam	-	12.75	104.81	31.82	m³	136.63
Demolishing reinforced concrete wall						
100 mm thick	-	1.11	9.12	2.88	m²	12.00
150 mm thick	-	1.67	13.73	4.31	m²	18.04
225 mm thick	-	2.50	20.55	6.49	m²	27.04
300 mm thick	-	3.33	27.37	8.69	m²	36.06
Demolishing reinforced concrete suspended slabs						
100 mm thick	-	0.93	7.64	2.67	m²	10.31
150 mm thick	-	1.39	11.43	3.90	m²	15.33
225 mm thick	-	2.08	17.10	5.84	m²	22.94
300 mm thick	-	2.78	22.85	7.89	m²	30.74
Breaking up concrete plinth; making good structures	-	4.26	35.02	18.09	m³	53.11
Breaking up precast concrete kerb	-	0.46	3.78	0.59	m	4.37
Removing precast concrete window sill; materials for re-use	-	1.48	12.17	-	m	12.17
Breaking up concrete hearth	-	1.67	13.73	0.98	nr	14.71

C DEMOLITION/ALTERATION/RENOVATION

Prices for Measured Works - Minor Works

	PC £	Labour hours	Labour £	Material £	Unit	Total rate £
C20 DEMOLITION - cont'd						
Demolishing parts of structures - cont'd						
Demolishing external brick walls; in gauged mortar						
half brick thick	-	0.65	5.34	1.47	m²	6.81
two half brick thick skins	-	1.11	9.12	3.15	m²	12.27
one brick thick	-	1.11	9.12	3.15	m²	12.27
one and a half brick thick	-	1.57	12.91	4.92	m²	17.83
two brick thick	-	2.04	16.77	6.29	m²	23.06
add for plaster, render or pebbledash per side	-	0.09	0.74	0.29	m²	1.03
Demolishing external brick walls; in cement mortar						
half brick thick	-	0.97	7.97	1.47	m²	9.44
two half brick thick skins	-	1.62	13.32	3.15	m²	16.47
one brick thick	-	1.67	13.73	3.15	m²	16.88
one and a half brick thick	-	2.27	18.66	4.92	m²	23.58
two brick thick	-	2.91	23.92	6.29	m²	30.21
add for plaster, render or pebbledash per side	-	0.09	0.74	0.29	m²	1.03
Demolishing internal partitions; gauged mortar						
half brick thick	-	0.97	7.97	1.47	m²	9.44
one brick thick	-	1.67	13.73	3.15	m²	16.88
one and a half brick thick	-	2.36	19.40	4.92	m²	24.32
75 mm blockwork	-	0.65	5.34	1.08	m²	6.42
90 mm blockwork	-	0.69	5.67	1.28	m²	6.95
100 mm blockwork	-	0.74	6.08	1.47	m²	7.55
115 mm blockwork	-	0.79	6.49	1.47	m²	7.96
125 mm blockwork	-	0.83	6.82	1.57	m²	8.39
140 mm blockwork	-	0.88	7.23	1.67	m²	8.90
150 mm blockwork	-	0.93	7.64	1.87	m²	9.51
190 mm blockwork	-	1.09	8.96	2.36	m²	11.32
215 mm blockwork	-	1.20	9.86	2.56	m²	12.42
255 mm blockwork	-	1.39	11.43	3.05	m²	14.48
add for plaster per side	-	0.09	0.74	0.29	m²	1.03
Demolishing internal partitions; cement mortar						
half brick thick	-	1.48	12.17	1.47	m²	13.64
one brick thick	-	2.45	20.14	3.15	m²	23.29
one and a half brick thick	-	3.42	28.11	4.92	m²	33.03
add for plaster per side	-	0.09	0.74	0.29	m²	1.03
Breaking up brick plinths	-	3.70	30.41	9.83	m³	40.24
Demolishing bund walls or piers in cement mortar						
one brick thick	-	1.30	10.69	3.15	m²	13.84
Demolishing walls to roof ventilator housing						
one brick thick	-	1.48	12.17	3.15	m²	15.32
Demolishing brick chimney to 300 mm below roof level; sealing off flues with slates						
680 mm x 680 mm x 900 mm high above roof	-	11.56	101.05	16.40	nr	117.45
add for each additional 300 height	-	2.31	20.19	3.05	nr	23.24
680 mm x 1030 mm x 900 mm high above roof	-	17.40	152.11	23.86	nr	175.97
add for each additional 300 height	-	3.46	30.21	4.90	nr	35.11
1030 mm x 1030 mm x 900 mm high above roof	-	26.69	233.46	36.77	nr	270.23
add for each additional 300 height	-	5.23	45.80	7.81	nr	53.61
Demolishing brick chimneys to 300 mm below roof level; sealing off flues with slates; piecing in "treated" sawn softwood rafters and making good roof coverings over to match existing (scaffolding excluded)						
680 mm x 680 mm x 900 mm high above roof	-	-	-	-	nr	159.79
add for each additional 300 mm height	-	-	-	-	nr	26.47
680 mm x 1030 mm x 900 mm high above roof	-	-	-	-	nr	234.68
add for each additional 300 mm height	-	-	-	-	nr	47.44
1030 mm x 1030 mm x 900 mm high above roof	-	-	-	-	nr	349.54
add for each additional 300 mm height	-	-	-	-	nr	114.84

C DEMOLITION/ALTERATION/RENOVATION

	PC £	Labour hours	Labour £	Material £	Unit	Total rate £
Removing existing chimney pots; materials for re-use; demolishing defective chimney stack to roof level; re-building using 25% new facing bricks to match existing; providing new lead flashings; parge and core flues, resetting chimney pots including flaunching in cement:mortar (scaffolding excluded)						
680 mm x 680 mm x 900 mm high above roof	-	-	-	-	nr	359.52
add for each additional 300 mm height	-	-	-	-	nr	59.92
680 mm x 1030 mm x 900 mm high above roof	-	-	-	-	nr	544.28
add for each additional 300 mm height	-	-	-	-	nr	79.89
1030 mm x 1030 mm x 900 mm high above roof	-	-	-	-	nr	798.94
add for each additional 300 mm height	-	-	-	-	nr	119.84
Removing fireplace surround and hearth						
interior tiled	-	1.71	14.06	2.26	nr	16.32
cast iron; materials for re-use	-	2.87	23.59	-	nr	23.59
stone iron; materials for re-use	-	7.49	61.57	-	nr	61.57
Removing fireplace; filling in opening; plastering and extending skirtings; fixing air brick; breaking up hearth and re-screeding						
tiled	-	-	-	-	nr	134.83
cast iron; set aside	-	-	-	-	nr	124.84
stone; set aside	-	-	-	-	nr	204.72
Removing brick-on-edge coping; prepare walls for raising						
one brick thick	-	0.42	6.83	0.20	m	7.03
one and a half brick thick	-	0.56	9.11	0.29	m	9.40
Demolishing external stone walls in lime mortar						
300 mm thick	-	1.11	9.12	2.95	m²	12.07
400 mm thick	-	1.48	12.17	3.93	m²	16.10
600 mm thick	-	2.22	18.25	5.90	m²	24.15
Demolishing stone walls in lime mortar; clean off; set aside for re-use						
300 mm thick	-	1.67	13.73	0.98	m²	14.71
400 mm thick	-	2.22	18.25	1.28	m²	19.53
600 mm thick	-	3.33	27.37	1.97	m²	29.34
Demolishing metal partitions						
corrugated metal partition	-	0.32	2.63	0.29	m²	2.92
lightweight steel mesh security screen	-	0.46	3.78	0.49	m²	4.27
solid steel demountable partition	-	0.69	5.67	0.69	m²	6.36
glazed sheet demountable partition; including removal of glass	-	0.93	7.64	0.98	m²	8.62
Removing metal shutter door and track						
6.20 m x 4.60 m (12.60 m long track)	-	11.10	91.24	14.75	nr	105.99
12.40 m x 4.60 m (16.40 m long track)	-	13.88	114.09	29.49	nr	143.58
Removing roof timbers complete; including rafters, purlins, ceiling joists, plates, etc., (measured flat on plan)	-	0.31	2.93	1.08	m²	4.01
Removing softwood floor construction						
100 mm deep joists at ground level	-	0.23	1.89	0.20	m²	2.09
175 mm deep joists at first floor level	-	0.46	3.78	0.39	m²	4.17
125 mm deep joists at roof level	-	0.65	5.34	0.29	m²	5.63
Removing individual floor or roof members	-	0.25	2.39	0.20	m	2.59
Removing infected or decayed floor plates	-	0.34	3.27	0.20	m	3.47
Removing boarding; withdrawing nails						
25 mm thick softwood flooring; at ground floor level	-	0.34	3.13	0.29	m²	3.42
25 mm thick softwood flooring; at first floor level	-	0.58	5.49	0.29	m²	5.78
25 mm thick softwood roof boarding	-	0.68	6.45	0.29	m²	6.74
25 mm thick softwood gutter boarding	-	0.74	7.04	0.29	m²	7.33
22 mm thick chipboard flooring; at first floor level	-	0.34	3.13	0.29	m²	3.42

Prices for Measured Works - Minor Works
C DEMOLITION/ALTERATION/RENOVATION

	PC £	Labour hours	Labour £	Material £	Unit	Total rate £
C20 DEMOLITION - cont'd						
Demolishing parts of structures - cont'd						
Removing tilting fillet or roll	-	0.14	1.34	0.10	m	1.44
Removing fascia or barge boards	-	0.56	5.32	0.10	m	5.42
Demolishing softwood stud partitions; including finishings both sides etc						
solid	-	0.42	3.45	0.98	m^2	4.43
glazed; including removal of glass	-	0.56	4.60	0.98	m^2	5.58
Removing windows and doors; and set aside or clear away						
single door	-	0.37	6.08	0.29	nr	6.37
single door and frame or lining	-	0.74	12.17	0.49	nr	12.66
pair of doors	-	0.65	10.69	0.59	nr	11.28
pair of doors and frame or lining	-	1.11	18.25	0.98	nr	19.23
extra for taking out floor spring box	-	0.70	11.51	0.20	nr	11.71
casement window and frame	-	1.11	18.25	0.49	nr	18.74
double hung sash window and frame	-	1.57	25.81	0.98	nr	26.79
pair of french windows and frame	-	3.70	60.83	1.47	nr	62.30
Removing double hung sash window and frame; remove and store for re-use elsewhere	-	2.22	36.50	-	nr	36.50
Demolishing staircase; including balustrades						
single straight flight	-	3.24	53.27	9.83	m	63.10
dogleg flight	-	4.63	76.12	14.75	m	90.87
C30 SHORING/FACADE RETENTION						
NOTE: The requirements for shoring and strutting for the formation of large openings are dependant upon a number of factors, for example, the weight of the superimposed structure to be supported, the number (if any) of windows above, the number of floors and the type of roof to be strutted, whether raking shores are required, the depth to a load-bearing surface, and the duration the support is to be in place. Prices, would therefore, be best built-up by assessing the use and waste of materials and the labour involved, including getting timber from and returning to a yard, cutting away and making good, overhead and profit. This method is considered a more practical way of pricing than endeavouring to price the work on a cubic metre basis of timber used, and has been adopted in preparing the prices of the examples which follow.						
Support of structures not to be demolished						
Strutting to window openings over proposed new openings	-	0.56	7.29	5.15	nr	12.44
Plates, struts, braces and hardwood wedges in supports to floors and roof of opening	-	1.11	14.45	15.10	nr	29.55
Dead shore and needle using die square timber with sole plates, braces, hardwood wedges and steel dogs	-	27.75	361.31	62.75	nr	424.06
Set of two raking shores using die square timber with 50 mm thick wall piece; hardwood wedges and steel dogs; including forming holes for needles and making good	-	33.30	433.57	63.39	nr	496.96
Cut holes through one brick wall for die square needle and make good; including facings externally and plaster internally	-	5.56	90.44	1.08	nr	91.52

	PC £	Labour hours	Labour £	Material £	Unit	Total rate £
C41 REPAIRING/RENOVATING/CONSERVING MASONRY						
Repairing/renovating plain/reinforced concrete work						
Reinstating plain concrete bed with site mixed in situ concrete; mix 20.00 N/mm² - 20 mm aggregate (1:2:4), where opening no longer required						
100 mm thick	-	0.44	4.45	7.45	m²	11.90
150 mm thick	-	0.72	6.75	11.17	m²	17.92
Reinstating reinforced concrete bed with site mixed in situ concrete; mix 20.00 N/mm² - 20 mm aggregate (1:2:4); including mesh reinforcement; where opening no longer required						
100 mm thick	-	0.66	6.26	9.53	m²	15.79
150 mm thick	-	0.91	8.32	13.26	m²	21.58
Reinstating reinforced concrete suspended floor with site mixed in situ concrete; mix 25.00 N/mm² - 20 mm aggregate (1:1.5:3); including mesh reinforcement and formwork; where opening no longer required						
150 mm thick	-	2.96	29.44	16.48	m²	45.92
225 mm thick	-	3.47	28.52	22.26	m²	50.78
300 mm thick	-	3.84	31.56	28.86	m²	60.42
Reinstating 150 mm x 150 mm x 150 mm perforation through concrete suspended slab; with site mixed in situ concrete; mix 20.00 N/mm² - 20 mm aggregate (1:2:4); including formwork; where opening no longer required	-	0.85	6.99	0.11	nr	7.10
Cleaning surfaces of concrete to receive new damp proof membrane	-	0.14	1.15	-	m²	1.15
Cleaning out existing minor crack and fill in with cement mortar mixed with bonding agent	-	0.31	2.55	0.61	m	3.16
Cleaning out existing crack to form 20 mm x 20 mm groove and fill in with fine cement mixed with bonding agent	-	0.61	5.01	2.84	m	7.85
Making good hole where existing pipe removed; 150 mm deep						
50 mm diameter	-	0.39	3.21	0.40	nr	3.61
100 mm diameter	-	0.51	4.19	0.49	nr	4.68
150 mm diameter	-	0.65	5.34	0.65	nr	5.99
Add for each additional 25 mm thick up to 300 mm thick						
50 mm diameter	-	0.08	0.66	0.07	nr	0.73
100 mm diameter	-	0.11	0.90	0.10	nr	1.00
150 mm diameter	-	0.14	1.15	0.13	nr	1.28
Repairing/renovating brick/blockwork						
Cutting out decayed, defective or cracked work and replacing with new common bricks PC £189.70/1000; in gauged mortar (1:1:6)						
half brick thick	-	4.56	74.15	15.08	m²	89.23
one brick thick	0.18	8.88	144.39	31.34	m²	175.73
one and a half brick thick	0.18	12.58	204.55	47.60	m²	252.15
two brick thick	0.18	16.10	261.79	63.86	m²	325.65
individual bricks; half brick thick	-	0.28	4.55	0.26	nr	4.81

Prices for Measured Works - Minor Works
C DEMOLITION/ALTERATION/RENOVATION

	PC £	Labour hours	Labour £	Material £	Unit	Total rate £
C41 REPAIRING/RENOVATING/CONSERVING MASONRY - cont'd						
Repairing/renovating brick/blockwork - cont'd						
Cutting out decayed, defective or cracked work and replacing with new facing brickwork in gauged mortar (1:1:6); half brick thick; facing and pointing one side						
small areas; machine made facings PC £330.00/1000	0.33	6.75	109.75	25.20	m²	134.95
small areas; hand made facings PC £540.00/1000	0.54	6.75	109.75	39.42	m²	149.17
individual bricks; machine made facings PC £330.00/1000	0.33	0.42	6.83	0.41	nr	7.24
individual bricks; hand made facings PC £540.00/1000	0.54	0.42	6.83	0.64	nr	7.47
ADD or DEDUCT for variation of £10.00/1000 in PC of facing bricks; in flemish bond						
half brick thick	-	-	-	0.65	m²	0.65
Cutting out decayed, defective or cracked soldier arch and replacing with new; repointing to match existing						
machine made facings PC £330.00/1000	0.33	1.80	29.27	6.13	m	35.40
hand made facings PC £540.00/1000	0.54	1.80	29.27	9.74	m	39.01
Cutting out decayed, defective or cracked work in uncoursed stonework; replacing with cement:mortar to match existing						
small areas; 300 mm thick wall	-	5.18	84.23	11.15	m²	95.38
small areas; 400 mm thick wall	-	6.48	105.36	15.09	m²	120.45
small areas; 600 mm thick wall	-	9.25	150.41	22.96	m²	173.37
Cutting out staggered cracks and repointing to match existing along brick joints	-	0.37	6.02	-	m	6.02
Cutting out raking cracks in brickwork; stitching in new common bricks and repointing to match existing						
half brick thick	0.18	2.96	48.13	6.95	m²	55.08
one brick thick	0.18	5.41	87.97	14.45	m²	102.42
one and a half brick thick	0.18	8.09	131.54	21.40	m²	152.94
Cutting out raking cracks in brickwork; stitching in new facing bricks; half brick thick; facing and pointing one side to match existing						
machine made facings PC £330.00/1000	0.33	4.44	72.19	10.95	m²	83.14
hand made facings PC £540.00/1000	0.54	4.44	72.19	17.05	m²	89.24
Cutting out raking cracks in cavity brickwork; stitching in new common bricks PC £175.00/1000 one side; facing bricks the other side; both skins half brick thick; facing and pointing one side to match existing						
machine made facings PC £330.00/1000	0.33	7.59	123.41	17.84	m²	141.25
hand made facings PC £540.00/1000	0.54	7.59	123.41	23.93	m²	147.34
Cutting away and replacing with new cement mortar (1:3); angle fillets; 50 mm face width	-	0.23	3.74	1.76	m	5.50

C DEMOLITION/ALTERATION/RENOVATION

	PC £	Labour hours	Labour £	Material £	Unit	Total rate £
Cutting out ends of joists and plates from walls; making good in common bricks PC £189.70/1000; in cement mortar (1:3)						
175 mm deep joists; 400 mm centres	-	0.60	9.76	5.14	m	14.90
225 mm deep joists; 400 mm centres	-	0.74	12.03	5.96	m	17.99
Cutting and pinning to existing brickwork ends of joists	-	0.37	6.02	-	nr	6.02
Making good adjacent work; where intersecting wall removed						
half brick thick	-	0.28	4.55	0.59	m	5.14
one brick thick	-	0.37	6.02	1.18	m	7.20
100 blockwork	-	0.23	3.74	0.59	m	4.33
150 blockwork	-	0.27	4.39	0.59	m	4.98
215 blockwork	-	0.32	5.20	1.18	m	6.38
255 blockwork	-	0.36	5.85	1.18	m	7.03
Removing defective parapet wall; 600 mm high; with two courses of tiles and brick coping over; re-building in new facing bricks, tiles and coping stones						
one brick thick	-	6.16	100.16	44.65	m	144.81
Removing defective capping stones and haunching; replacing stones and re-haunching in cement:mortar to match existing						
300 mm thick wall	-	1.25	20.32	3.28	m²	23.60
400 mm thick wall	-	1.39	22.60	3.94	m²	26.54
600 mm thick wall	-	1.62	26.34	5.90	m²	32.24
Cleaning surfaces; moss and lichen from walls	-	0.28	2.30	-	m²	2.30
Cleaning surfaces; lime mortar off brickwork; sort and stack for re-use	-	9.25	76.03	-	t	76.03
Repointing in cement mortar (1:1:6); to match existing						
raking out existing decayed joints in brickwork walls	-	0.69	11.22	0.59	m²	11.81
raking out existing decayed joints in chimney stacks	-	1.11	18.05	0.59	m²	18.64
raking out existing decayed joints in brickwork; re-wedging horizontal flashing	-	0.23	3.74	0.29	m	4.03
raking out existing decayed joints in brickwork; re-wedging stepped flashing	-	0.34	5.53	0.29	m	5.82
Repointing in cement:mortar (1:3); to match existing						
raking out existing decayed joints in uncoursed stonework	-	1.11	18.05	0.66	m²	18.71
Making good hole where small pipe removed						
102 mm brickwork	-	0.10	1.56	0.07	nr	1.63
215 mm brickwork	-	0.19	1.56	0.07	nr	1.63
327 mm brickwork	-	0.19	1.56	0.07	nr	1.63
440 mm brickwork	-	0.19	1.56	0.07	nr	1.63
100 mm blockwork	-	0.19	1.56	0.07	nr	1.63
150 mm blockwork	-	0.19	1.56	0.07	nr	1.63
215 mm blockwork	-	0.19	1.56	0.07	nr	1.63
255 mm blockwork	-	0.19	1.56	0.07	nr	1.63
Making good hole and facings one side where small pipe removed						
102 mm brickwork	-	0.19	3.09	0.66	nr	3.75
215 mm brickwork	-	0.19	3.09	0.66	nr	3.75
327 mm brickwork	-	0.19	3.09	0.66	nr	3.75
440 mm brickwork	-	0.19	3.09	0.66	nr	3.75

Prices for Measured Works - Minor Works
C DEMOLITION/ALTERATION/RENOVATION

	PC £	Labour hours	Labour £	Material £	Unit	Total rate £
C41 REPAIRING/RENOVATING/CONSERVING MASONRY - cont'd						
Repairing/renovating brick/blockwork - cont'd						
Making good hole where large pipe removed						
102 mm brickwork	-	0.28	2.30	0.07	nr	2.37
215 mm brickwork	-	0.42	3.45	0.23	nr	3.68
327 mm brickwork	-	0.56	4.60	0.39	nr	4.99
440 mm brickwork	-	0.69	5.67	0.49	nr	6.16
100 mm blockwork	-	0.28	2.30	0.07	nr	2.37
150 mm blockwork	-	0.32	2.63	0.13	nr	2.76
215 mm blockwork	-	0.37	3.04	0.16	nr	3.20
255 mm blockwork	-	0.42	3.45	0.23	nr	3.68
Making good hole and facings one side where large pipe removed						
half brick thick	-	0.25	4.07	0.66	nr	4.73
one brick thick	-	0.33	5.37	0.72	nr	6.09
one and a half brick thick	-	0.42	6.83	0.79	nr	7.62
two brick thick	-	0.50	8.13	1.05	nr	9.18
Making good hole where extra large pipe removed						
half brick thick	-	0.37	3.04	0.26	nr	3.30
one brick thick	-	0.56	4.60	0.56	nr	5.16
one and a half brick thick	-	0.74	6.08	0.98	nr	7.06
two brick thick	-	0.93	7.64	1.18	nr	8.82
100 mm blockwork	-	0.37	3.04	0.26	nr	3.30
150 mm blockwork	-	0.43	3.53	0.39	nr	3.92
215 mm blockwork	-	0.46	3.78	0.56	nr	4.34
255 mm blockwork	-	0.51	4.19	0.68	nr	4.87
Making good hole and facings one side where extra large pipe removed						
half brick thick	-	0.33	5.37	1.02	nr	6.39
one brick thick	-	0.44	7.15	1.25	nr	8.40
one and a half brick thick	-	0.56	9.11	1.57	nr	10.68
two brick thick	-	0.67	10.89	2.13	nr	13.02
C50 REPAIRING/RENOVATING/CONSERVING METAL						
Repairing metal						
Overhauling and repairing metal casement windows; adjusting and oiling ironmongery; bringing forward affected parts for redecoration	-	1.39	11.43	2.65	nr	14.08
C51 REPAIRING/RENOVATING/CONSERVING TIMBER						
Repairing timber						
Removing or punching in projecting nails; re-fixing softwood or hardwood flooring						
loose boards	-	0.14	1.82	-	m^2	1.82
floorboards previously set aside	-	0.74	9.63	0.36	m^2	9.99
Removing damaged softwood flooring; providing and fixing new 25 mm thick plain edge softwood boarding						
small areas	-	1.06	13.80	15.99	m^2	29.79
individual boards 150 mm wide	-	0.28	3.65	1.59	m	5.24
Sanding down and resurfacing existing flooring; preparing, bodying in with shellac and wax polish						
softwood	-	-	-	-	m^2	9.14
hardwood	-	-	-	-	m^2	8.59

C DEMOLITION/ALTERATION/RENOVATION

	PC £	Labour hours	Labour £	Material £	Unit	Total rate £
Fitting existing softwood skirting to new frames or architraves						
75 mm high	-	0.09	1.17	-	m	**1.17**
150 mm high	-	0.12	1.56	-	m	**1.56**
225 mm high	-	0.15	1.95	-	m	**1.95**
Piecing in new 25 mm x 150 mm moulded softwood skirtings to match existing where old removed; bringing forward for redecoration	-	0.35	3.90	3.85	m	**7.75**
Piecing in new 25 mm x 150 mm moulded softwood skirtings to match existing where socket outlet removed; bringing forward for redecoration	-	0.20	2.22	2.14	nr	**4.36**
Easing and adjusting softwood doors, oiling ironmongery; bringing forward affected parts for redecoration	-	0.71	8.78	0.59	nr	**9.37**
Removing softwood doors, easing and adjusting; re-hanging; oiling ironmongery; bringing forward affected parts for redecoration	-	1.11	13.87	0.79	nr	**14.66**
Removing mortice lock, piecing in softwood doors; bringing forward affected parts for redecoration	-	1.02	12.89	0.49	nr	**13.38**
Fixing only salvaged softwood door	-	1.42	18.49	-	nr	**18.49**
Removing softwood doors; planing 12 mm from bottom edge; re-hanging	-	1.11	14.45	-	nr	**14.45**
Removing softwood doors; altering ironmongery; piecing in and rebating frame and door; re-hanging on opposite stile; bringing forward affected parts for redecoration	-	2.45	30.93	0.99	nr	**31.92**
Removing softwood doors to prepare for fire upgrading; removing ironmongery; replacing existing beads with 25 mm x 38 mm hardwood screwed beads; repairing minor damaged areas; re-hanging on wider butt hinges; adjusting all ironmongery; sealing around frame in cement mortar; bringing forward affected parts for redecoration (replacing glass panes not included)	-	4.85	61.79	20.19	nr	**81.98**
Upgrading and facing up one side of flush doors with 9 mm thick "Supalux"; screwing	-	1.16	15.10	34.46	nr	**49.56**
Upgrading and facing up one side of softwood panelled doors with 9 mm thick "Supalux"; screwing; plasterboard infilling to recesses	-	2.50	32.55	36.59	nr	**69.14**
Taking off existing softwood doorstops; providing and screwing on new 25 mm x 38 mm doorstop; bringing forward for redecoration	-	0.20	2.22	1.82	nr	**4.04**
Cutting away defective 75 mm x 100 mm softwood external door frames; providing and splicing in new piece 300 mm long; bedding in cement mortar (1:3); pointing one side; bringing forward for redecoration	-	1.30	16.15	6.82	nr	**22.97**
Sealing roof trap flush with ceiling	-	0.56	7.29	2.88	nr	**10.17**
Forming opening 762 mm x 762 mm in existing ceiling for new standard roof trap comprising softwood linings, architraves and 6 mm thick plywood trap doors; trimming ceiling joists (making good to ceiling plaster not included)	-	2.50	32.55	47.81	nr	**80.36**
Easing and adjusting softwood casement windows, oiling ironmongery; bringing forward affected parts for redecoration	-	0.48	5.79	0.40	nr	**6.19**

Prices for Measured Works - Minor Works
C DEMOLITION/ALTERATION/RENOVATION

	PC £	Labour hours	Labour £	Material £	Unit	Total rate £
C51 REPAIRING/RENOVATING/CONSERVING TIMBER - cont'd						
Repairing timber - cont'd						
Removing softwood casement windows; easing and adjusting; re-hanging; oiling ironmongery; bringing forward affected parts for redecoration	-	0.71	8.78	0.40	nr	9.18
Renewing solid mullion jambs or transoms of softwood casement windows to match existing; bringing forward affected parts for redecoration (taking off and re-hanging adjoining casements not included)	-	2.59	32.56	13.48	nr	46.04
Temporary linings 6 mm thick plywood infill to window while casement under repair	-	0.74	9.63	5.29	nr	14.92
Overhauling softwood double hung sash windows; easing, adjusting and oiling pulley wheels; re-hanging sashes on new hemp sash lines; re-assembling; bringing forward affected parts for redecoration	-	2.45	31.32	4.01	nr	35.33
Cutting away defective parts of softwood window sills; providing and splicing in new 75 mm x 100 mm weathered and throated pieces 300 mm long; bringing forward affected parts for redecoration	-	1.90	24.16	7.43	nr	31.59
Renewing broken stair nosings to treads or landings	-	1.67	21.74	2.87	nr	24.61
Cutting out infected or decayed structural members; shoring up adjacent work; providing and fixing new "treated" sawn softwood members pieced in						
Floors or flat roofs						
50 mm x 125 mm	-	0.37	4.82	2.15	m	6.97
50 mm x 150 mm	-	0.41	5.34	2.63	m	7.97
50 mm x 175 mm	-	0.44	5.73	3.04	m	8.77
Pitched roofs						
38 mm x 100 mm	-	0.33	4.30	1.31	m	5.61
50 mm x 100 mm	-	0.42	5.47	1.68	m	7.15
50 mm x 125 mm	-	0.46	5.99	2.06	m	8.05
50 mm x 150 mm	-	0.51	6.64	2.50	m	9.14
Kerbs bearers and the like						
50 mm x 75 mm	-	0.42	5.47	1.30	m	6.77
50 mm x 100 mm	-	0.52	6.77	1.68	m	8.45
75 mm x 100 mm	-	0.63	8.20	2.36	m	10.56
Scarfed joint; new to existing; over 450 mm^2	-	0.93	12.11	-	nr	12.11
Scarfed and bolted joint; new to existing; including bolt let in flush; over 450 mm^2	-	1.34	17.45	1.21	nr	18.66
C52 FUNGUS/BEETLE ERADICATION						
Treating existing timber						
Removing cobwebs, dust and roof insulation; de-frass; treat exposed joists/rafters with two coats of proprietary insecticide and fungicide; by spray application	-	-	-	-	m^2	7.44
Treating boarding with two coats of proprietary insecticide and fungicide; by spray application	-	-	-	-	m^2	3.25
Treating individual timbers with two coats proprietary insecticide and fungicide; by brush application						
boarding	-	-	-	-	m^2	3.59
structural members	-	-	-	-	m^2	3.59
skirtings	-	-	-	-	m	3.59

Prices for Measured Works - Minor Works
C DEMOLITION/ALTERATION/RENOVATION

	PC £	Labour hours	Labour £	Material £	Unit	Total rate £
Lifting necessary floorboards; treating floors with two coats proprietary insecticide and fungicide; by spray application; re-fixing boards	-	-	-	-	m²	5.94
Treating surfaces of adjoining concrete or brickwork with two coats of dry rot fluid; by spray application	-	-	-	-	m²	3.15

C90 ALTERATIONS - SPOT ITEMS

Composite "spot" items

NOTE: Few exactly similar composite items of alteration works are encountered on different schemes; for this reason it is considered more accurate for the reader to build up the value of such items from individual prices in the following section. However, for estimating purposes, the following "spot" items have been prepared. Prices include for removal of debris from site but do not include for shoring, scaffolding or re-decoration, except where stated.

Removing fittings and fixtures						
Removing shelves, window boards and the like	-	0.31	2.55	0.10	m	2.65
Removing handrails and balustrades						
tubular handrailing and brackets	-	0.28	2.30	0.10	m	2.40
metal balustrades	-	0.46	3.78	0.29	m	4.07
Removing handrails and brackets	-	0.09	1.48	0.29	m	1.77
Removing sloping timber ramps in corridors; at changes of levels	-	1.85	30.41	1.47	nr	31.88
Removing bath panels and bearers	-	0.37	6.08	0.49	nr	6.57
Removing kitchen fittings						
wall units	-	0.42	6.90	1.47	nr	8.37
floor units	-	0.28	4.60	2.16	nr	6.76
larder units	-	0.37	6.08	4.92	nr	11.00
built-in cupboards	-	1.39	22.85	9.83	nr	32.68
Removing bathroom fittings; making good works disturbed						
toilet roll holder or soap dispenser	-	0.28	2.30	-	nr	2.30
towel holder	-	0.56	4.60	-	nr	4.60
mirror	-	0.60	4.93	-	nr	4.93
Removing pipe casings	-	0.28	4.60	0.39	m	4.99
Removing ironmongery; in preparation for re-decoration; and subsequently re-fixing; including providing any new screws necessary	-	0.23	3.78	0.10	nr	3.88
Removing, withdrawing nails, etc; making good holes						
carpet fixing strip from floors	-	0.04	0.33	-	m	0.33
curtain track from head of window	-	0.23	1.89	-	m	1.89
nameplates or numerals from face of door	-	0.46	3.78	-	nr	3.78
fly screen and frame from window	-	0.83	6.82	-	nr	6.82
small notice board and frame from walls	-	0.82	6.74	-	nr	6.74
fire extinguisher and bracket from walls	-	1.16	9.54	-	nr	9.54
Removing plumbing and engineering installations						
Removing sanitary fittings and supports; temporarily capping off services; to receive new (not included)						
sink or lavatory basin	-	0.93	9.39	5.90	nr	15.29
bath	-	1.85	18.71	8.85	nr	27.56
WC suite	-	1.39	14.05	5.90	nr	19.95

Prices for Measured Works - Minor Works
C DEMOLITION/ALTERATION/RENOVATION

	PC £	Labour hours	Labour £	Material £	Unit	Total rate £
C90 ALTERATIONS - SPOT ITEMS - cont'd						
Removing plumbing and engineering installations - cont'd						
Removing sanitary fittings and supports, complete with associated services, overflows and waste pipes; making good all holes and other works disturbed; bringing forward all surfaces ready for re-decoration						
sink or lavatory basin	-	3.70	37.41	8.15	nr	45.56
range of three lavatory basins	-	7.40	74.83	14.50	nr	89.33
bath	-	5.55	56.12	11.00	nr	67.12
WC suite	-	7.40	74.83	17.18	nr	92.01
2 stall urinal	-	14.80	149.66	15.63	nr	165.29
3 stall urinal	-	22.20	224.48	30.56	nr	255.04
4 stall urinal	-	29.60	299.31	45.24	nr	344.55
Removing taps	-	0.09	1.05	-	nr	1.05
Clearing blocked wastes without dismantling						
sinks	-	0.46	5.23	-	nr	5.23
WC traps	-	0.56	6.37	-	nr	6.37
Removing gutterwork and supports						
uPVC or asbestos	-	0.28	2.30	0.10	m	2.40
cast iron	-	0.32	2.63	0.20	m	2.83
Overhauling sections of rainwater gutterings; cutting out existing joints; adjusting brackets to correct falls; re-making joints						
100 mm diameter uPVC	-	0.23	2.99	0.03	m	3.02
100 mm diameter cast iron including bolt	-	0.83	10.81	0.08	m	10.89
Removing rainwater heads and supports						
uPVC or asbestos	-	0.27	2.22	0.10	nr	2.32
cast iron	-	0.37	3.04	0.20	nr	3.24
Removing pipework and supports						
uPVC or asbestos rainwater stack	-	0.28	2.30	0.10	m	2.40
cast iron rainwater stack	-	0.32	2.63	0.20	m	2.83
cast iron jointed soil stack	-	0.56	4.60	0.20	m	4.80
copper or steel water or gas pipework	-	0.14	1.15	0.10	m	1.25
cast iron rainwater shoe	-	0.07	0.58	0.10	m	0.68
Overhauling and re-making leaking joints in pipework						
100 mm diameter upvc	-	0.19	1.56	0.05	nr	1.61
100 mm diameter cast iron including bolt	-	0.74	6.08	0.16	nr	6.24
Cleaning out existing rainwater installations						
rainwater gutters	-	0.07	0.58	-	m	0.58
rainwater gully	-	0.19	1.56	-	nr	1.56
rainwater stack; including head, swan-neck and shoe (not exceeding 10 m long)		0.69	5.67	-	nr	5.67
Removing the following equipment and ancillaries; capping off services; making good works disturbed (excluding any draining down of system)						
expansion tank; 900 mm x 450 mm x 900 mm	PC	1.67	16.88	3.93	nr	20.81
hot water cylinder; 450 mm diameter x 1050 mm high		1.11	11.22	1.67	nr	12.89
cold water tank; 1540 mm x 900 mm x 900 mm	-	2.22	22.45	12.09	nr	34.54
cast iron radiator	-	1.85	18.71	3.15	nr	21.86
gas water heater	-	3.70	37.41	1.97	nr	39.38
gas fire	-	1.85	18.71	2.56	nr	21.27
Removing cold water tanks and housing on roof; stripping out and capping off all associated piping; making good works disturbed and roof finishings						
1540 mm x 900 mm x 900 mm	-	11.10	112.24	15.04	nr	127.28

C DEMOLITION/ALTERATION/RENOVATION

	PC £	Labour hours	Labour £	Material £	Unit	Total rate £
Turning off supplies; dismantling the following fittings; replacing washers; re-assembling and testing						
15 mm diameter tap	-	0.23	2.62	-	nr	2.62
15 mm diameter ball valve	-	0.32	3.64	-	nr	3.64
Turning off supplies; removing the following fittings; testing and replacing						
15 mm diameter ball valve	-	0.46	5.23	5.99	nr	11.22
Removing lagging from pipes						
up to 42 mm diameter	-	0.09	0.74	0.10	nr	0.84
Removing finishings						
Removing plasterboard wall finishings	-	0.37	3.04	-	m^2	3.04
Removing wall finishings; cutting out and making good cracks						
plasterboard wall finishing	-	0.37	3.04	-	m^2	3.04
decorative wallpaper and lining	-	0.19	1.56	0.16	m^2	1.72
heavy wallpaper and lining	-	0.32	2.63	0.16	m^2	2.79
Hacking off wall finishings						
plaster	-	0.19	1.56	0.49	m^2	2.05
cement rendering or pebbledash	-	0.37	3.04	0.49	m^2	3.53
wall tiling and screed	-	0.46	3.78	0.79	m^2	4.57
Removing wall linings; including battening behind						
plain sheeting	-	0.28	2.30	0.39	m^2	2.69
matchboarding	-	0.37	3.04	0.59	m^2	3.63
Removing oak dado wall panel finishings; cleaning off and setting aside for re-use	-	0.60	9.86	-	m^2	9.86
Removing defective or damaged plaster wall finishings; re-plastering walls with two coats of gypsum plaster; including dubbing out; jointing new to existing						
small areas	-	1.48	18.94	4.80	m^2	23.74
isolated areas not exceeding 0.50 m^2	-	1.06	13.57	2.40	nr	15.97
Making good plaster wall finishings with two coats of gypsum plaster where wall or partition removed; dubbing out; trimming back existing and fair jointing to new work						
150 mm wide	-	0.60	7.68	0.72	m	8.40
225 mm wide	-	0.74	9.47	1.08	m	10.55
300 mm wide	-	0.88	11.26	1.44	m	12.70
Removing defective or damaged damp plaster wall finishings, investigating and treating wall; re-plastering walls with two coats of "Thistle Renovating" plaster; including dubbing out; fair jointing to existing work						
small areas	-	1.53	19.58	4.24	m^2	23.82
isolated areas not exceeding 0.50 m^2	-	1.13	14.46	2.12	m^2	16.58
Dubbing out in cement and sand; average 13 mm thick						
over 300 mm wide	-	0.46	5.89	0.85	m^2	6.74
Making good plaster wall finishings with plasterboard and skim where wall or partition removed; trimming back existing and fair joint to new work						
150 mm wide	-	0.69	8.83	1.50	m	10.33
225 mm wide	-	0.83	10.62	1.73	m	12.35
300 mm wide	-	0.93	11.90	1.97	m	13.87
Cutting out; making good cracks in plaster wall finishings						
walls	-	0.23	2.94	0.16	m	3.10
ceilings	-	0.31	3.97	0.16	m	4.13

C DEMOLITION/ALTERATION/RENOVATION

	PC £	Labour hours	Labour £	Material £	Unit	Total rate £
C90 ALTERATIONS - SPOT ITEMS - cont'd						
Removing finishings - cont'd						
Making good plaster wall finishings where items removed or holes left						
small pipe or conduit	-	0.06	0.77	0.08	nr	0.85
large pipe	-	0.09	1.15	0.16	nr	1.31
extra large pipe	-	0.14	1.79	0.71	nr	2.50
small recess; eg. electrical switch point	-	0.07	0.90	0.15	nr	1.05
Making good plasterboard and skim wall finishings where items removed or holes left						
small pipe or conduit	-	0.06	0.77	0.08	nr	0.85
large pipe	-	0.21	2.69	0.54	nr	3.23
extra large pipe	-	0.28	3.58	0.69	nr	4.27
Removing floor finishings						
carpet and underfelt	-	0.11	0.90	-	m²	0.90
linoleum sheet flooring	-	0.09	0.74	-	m²	0.74
carpet gripper	-	0.02	0.16	-	m	0.16
Removing floor finishings; preparing screed to receive new						
carpet and underfelt	-	0.61	5.01	-	m²	5.01
vinyl or thermoplastic tiles	-	0.79	6.49	-	m²	6.49
Removing woodblock floor finishings; cleaning off and setting aside for re-use	-	0.69	5.67	-	m²	5.67
Breaking up floor finishings						
floor screed	-	0.60	4.93	-	m²	4.93
granolithic flooring and screed	-	0.79	6.49	-	m²	6.49
terrazzo or ceramic floor tiles and screed	-	0.97	7.97	-	m²	7.97
Levelling and repairing floor finishings screed; 5 mm thick						
screed; 5 mm thick; in small areas	-	0.46	5.89	6.99	m²	12.88
screed; 5 mm thick; in isolated areas not exceeding 0.50 m²	-	0.32	4.10	3.49	m²	7.59
Removing softwood skirtings, picture rails, dado rails, architraves and the like	-	0.09	0.74	-	m	0.74
Removing softwood skirtings; cleaning off and setting aside for re-use in making good	-	0.23	1.89	-	m	1.89
Breaking up paving						
asphalt	-	0.56	4.60	-	m²	4.60
Removing ceiling finishings						
plasterboard and skim; withdrawing nails	-	0.28	2.30	0.29	m²	2.59
wood lath and plaster; withdrawing nails	-	0.46	3.78	0.49	m²	4.27
suspended ceilings	-	0.69	5.67	0.49	m²	6.16
plaster moulded cornice; 25 mm girth	-	0.14	1.15	0.10	m	1.25
Removing part of plasterboard ceiling finishings to facilitate insertion of new steel beam	-	1.02	8.38	0.59	m	8.97
Removing ceiling linings; including battening behind						
plain sheeting	-	0.42	3.45	0.39	m²	3.84
matchboarding	-	0.56	4.60	0.59	m²	5.19
Removing defective or damaged ceiling plaster finishings; removing laths or cutting back boarding; preparing and fixing new plasterboard; applying one skim coat of gypsum plaster; fair jointing new to existing						
small areas	-	1.57	20.10	4.26	m²	24.36
isolated areas not exceeding 0.50m²	-	1.13	14.46	2.36	m²	16.82

Prices for Measured Works - Minor Works
C DEMOLITION/ALTERATION/RENOVATION

	PC £	Labour hours	Labour £	Material £	Unit	Total rate £
Removing coverings						
Removing roof coverings						
slates	-	0.46	3.78	0.20	m²	3.98
slates; set aside for re-use	-	0.56	4.60	-	m²	4.60
nibbed tiles	-	0.37	3.04	0.20	m²	3.24
nibbed tiles; set aside for re-use	-	0.46	3.78	-	m²	3.78
corrugated asbestos sheeting	-	0.37	3.04	0.20	m²	3.24
corrugated metal sheeting	-	0.37	3.04	0.20	m²	3.24
underfelt and nails	-	0.04	0.33	0.10	m²	0.43
three layer felt roofing; cleaning base off for new coverings	-	0.23	1.89	0.20	m²	2.09
sheet metal coverings	-	0.46	3.78	0.20	m²	3.98
Removing roof coverings; selecting and re-fixing; including providing 25% new; including nails, etc.						
asbestos-free artificial blue/black slates; 500 mm x 250 mm; PC £986.70/1000	-	1.02	16.77	7.05	m²	23.82
asbestos-free artificial blue/black slates; 600 mm x 300 mm; PC £1311.30/1000	-	0.93	15.29	5.94	m²	21.23
natural slates; Welsh blue 510 mm x 255 mm; PC £2197.10/1000	-	1.11	18.25	11.62	m²	29.87
natural slates; Welsh blue 600 mm x 300 mm; PC £4294.30/1000	-	0.97	15.95	14.77	m²	30.72
clay plain tiles "Dreadnought" machine made; 265 mm x 165 mm; PC £229.70/1000	-	1.02	16.77	4.85	m²	21.62
concrete interlocking tiles; Marley "Ludlow Major" or other equal and approved; 413 mm x 330 mm; PC £723.00/1000	-	0.65	10.69	2.32	m²	13.01
concrete interlocking tiles; Redland "Renown" or other equal and approved; 417 mm x 330 mm; PC £623.20/1000	-	0.65	10.69	2.09	m²	12.78
Removing damaged roof coverings in area less than 10 m²; providing and fixing new; including nails, etc.						
asbestos-free artificial blue/black slates; 500 mm x 250 mm; PC £986.70/1000	-	1.25	20.55	21.76	m²	42.31
asbestos-free artificial blue/black slates; 600 mm x 300 mm; PC £1311.30/1000	-	1.16	19.07	19.12	m²	38.19
natural slates; Welsh blue 510 mm x 255 mm; PC £2197.10/1000	-	1.34	22.03	42.86	m²	64.89
natural slates; Welsh blue 600 mm x 300 mm; PC £4294.30/1000	-	1.20	19.73	56.26	m²	75.99
clay plain tiles "Dreadnought" machine made or other equal and approved; 265 mm x 165 mm; PC £229.70/1000	-	1.25	20.55	15.81	m²	36.36
concrete interlocking tiles; Marley "Ludlow Major" or other equal and approved; 413 mm x 330 mm; PC £723.00/1000	-	0.83	13.65	7.84	m²	21.49
concrete interlocking tiles; Redland "Renown" or other equal and approved; 417 mm x 330 mm; PC £623.20/1000	-	0.83	13.65	6.90	m²	20.55
Removing individual damaged roof coverings; providing and fixing new; including nails, etc.						
asbestos-free artificial blue/black slates; 500 mm x 250 mm; PC £986.70/1000	-	0.23	3.78	1.40	nr	5.18
asbestos-free artificial blue/black slates; 600 mm x 300 mm; PC £1311.30/1000	-	0.23	3.78	1.43	nr	5.21
natural slates; Welsh blue 510 mm x 255 mm; PC £2197.10/1000	-	0.28	4.60	2.36	nr	6.96
natural slates; Welsh blue 600 mm x 300 mm; PC £4294.30/1000	-	0.28	4.60	4.54	nr	9.14

Prices for Measured Works - Minor Works
C DEMOLITION/ALTERATION/RENOVATION

	PC £	Labour hours	Labour £	Material £	Unit	Total rate £
C90 ALTERATIONS - SPOT ITEMS - cont'd						
Removing coverings - cont'd						
Removing individual damaged roof coverings; providing and fixing new; including nails, etc. - cont'd						
clay plain tiles "Dreadnought" machine made or other equal and approved; 265 mm x 165 mm; PC £229.70/1000	-	0.14	2.30	0.26	nr	2.56
concrete interlocking tiles; Marley "Ludlow Major" or other equal and approved; 413 mm x 330 mm; PC £723.00/1000	-	0.19	3.12	0.78	nr	3.90
concrete interlocking tiles; Redland "Renown" or other equal and approved; 417 mm x 330 mm; PC £623.20/1000	-	0.19	3.12	0.67	nr	3.79
Breaking up roof coverings						
asphalt	-	0.93	7.64	-	m²	7.64
Removing half round ridge or hip tile 300 mm long; providing and fixing new	-	0.46	7.56	2.67	nr	10.23
Removing defective metal flashings						
horizontal	-	0.19	1.56	0.20	m	1.76
stepped	-	0.23	1.89	0.10	m	1.99
Turning back bitumen felt and later dressing up face of new brickwork as skirtings; not exceeding 150 mm girth	-	0.93	11.18	0.11	m	11.29
Cutting out crack in asphalt roof coverings; making good to match existing						
20 mm thick two coat	-	1.53	14.00	-	m	14.00
Removing bitumen felt roof coverings and boarding to allow access for work to top of walls or beams beneath	-	0.74	6.08	-	m	6.08
Removing tiling battens; withdrawing nails	-	0.07	0.58	0.10	m²	0.68
Examining roof battens; re-nailing where loose; providing and fixing 25% new						
25 mm x 50 mm slating battens at 262 mm centres	-	0.07	1.15	0.61	m²	1.76
25 mm x 38 mm tiling battens at 100 mm centres	-	0.19	3.12	1.56	m²	4.68
Removing roof battens and nails; providing and fixing new "treated" softwood battens throughout						
25 mm x 50 mm slating battens at 262 mm centres	-	0.11	1.81	2.14	m²	3.95
25 mm x 38 mm tiling battens at 100 mm centres	-	0.23	3.78	4.77	m²	8.55
Removing underfelt and nails; providing and fixing new						
unreinforced felt	0.58	0.09	1.48	0.88	m²	2.36
reinforced felt	1.08	0.09	1.48	1.44	m²	2.92
Cutting openings or recesses						
Cutting openings or recesses through reinforced concrete walls						
150 mm thick	-	5.18	53.25	8.07	m²	61.32
225 mm thick	-	7.08	71.19	12.05	m²	83.24
300 mm thick	-	9.02	89.22	16.23	m²	105.45
Cutting openings or recesses through reinforced concrete suspended slabs						
150 mm thick	-	3.93	39.26	14.76	m²	54.02
225 mm thick	-	5.83	58.36	14.10	m²	72.46
300 mm thick	-	7.26	71.51	17.22	m²	88.73

C DEMOLITION/ALTERATION/RENOVATION

	PC £	Labour hours	Labour £	Material £	Unit	Total rate £
Cutting openings or recesses through slated, boarded and timbered roof; 700 mm x 1100 mm; for new rooflight; including cutting structure and finishings; trimming timbers in rafters and making good roof coverings (kerb and rooflight not included)	-	-	-	-	nr	299.60
Cutting openings or recesses through brick or block walls or partitions; for lintels or beams above openings; in gauged mortar						
half brick thick	-	2.45	39.84	1.67	m²	41.51
one brick thick	-	4.07	66.18	3.34	m²	69.52
one and a half brick thick	-	5.69	92.52	5.01	m²	97.53
two brick thick	-	7.31	118.86	6.68	m²	125.54
75 mm blockwork	-	1.48	24.06	1.08	m²	25.14
90 mm blockwork	-	1.67	27.15	1.28	m²	28.43
100 mm blockwork	-	1.80	29.27	1.47	m²	30.74
115 mm blockwork	-	1.84	29.92	1.67	m²	31.59
125 mm blockwork	-	2.04	33.17	1.87	m²	35.04
140 mm blockwork	-	2.17	35.28	2.06	m²	37.34
150 mm blockwork	-	2.27	36.91	2.26	m²	39.17
190 mm blockwork	-	2.53	41.14	2.85	m²	43.99
215 mm blockwork	-	2.68	43.58	3.15	m²	46.73
255 mm blockwork	-	2.94	47.80	3.74	m²	51.54
Cutting openings or recesses through brick walls or partitions; for lintels or beams above openings; in cement mortar						
half brick thick	-	3.52	57.24	1.67	m²	58.91
one brick thick	-	5.83	94.80	3.34	m²	98.14
one and a half brick thick	-	8.14	132.36	5.01	m²	137.37
two brick thick	-	10.45	169.92	6.68	m²	176.60
Cutting openings or recesses through brick or block walls or partitions; for door or window openings; in gauged mortar						
half brick thick	-	1.25	20.32	1.67	m²	21.99
one brick thick	-	2.04	33.17	3.34	m²	36.51
one and a half brick thick	-	2.82	45.85	5.01	m²	50.86
two brick thick	-	3.65	59.35	6.68	m²	66.03
75 mm blockwork	-	0.74	12.03	1.08	m²	13.11
90 mm blockwork	-	0.85	13.82	1.28	m²	15.10
100 mm blockwork	-	0.93	15.12	1.47	m²	16.59
115 mm blockwork	-	0.98	15.93	1.67	m²	17.60
125 mm blockwork	-	1.02	16.59	1.87	m²	18.46
140 mm blockwork	-	1.07	17.40	2.06	m²	19.46
150 mm blockwork	-	1.11	18.05	2.26	m²	20.31
190 mm blockwork	-	1.22	19.84	2.85	m²	22.69
215 mm blockwork	-	1.34	21.79	3.15	m²	24.94
255 mm blockwork	-	1.48	24.06	3.74	m²	27.80
Cutting openings or recesses through brick or block walls or partitions; for door or window openings; in cement mortar						
half brick thick	-	1.76	28.62	1.67	m²	30.29
one brick thick	-	2.91	47.32	3.34	m²	50.66
one and a half brick thick	-	4.02	65.37	5.01	m²	70.38
two brick thick	-	5.23	85.04	6.68	m²	91.72

	PC £	Labour hours	Labour £	Material £	Unit	Total rate £
C90 ALTERATIONS - SPOT ITEMS - cont'd						
Cutting openings or recesses - cont'd						
Cutting openings or recesses through faced wall 1200 mm x 1200 mm (1.44 m²) for new window; including cutting structure, quoining up jambs, cutting and pinning in suitable precast concrete boot lintel with galvanised steel angle bolted on to support, outer brick soldier course in facing bricks to match existing (new window and frame not included)						
one brick thick wall or two half brick thick skins	-	-	-	-	nr	404.47
one and a half brick thick wall	-	-	-	-	nr	424.43
two brick thick wall	-	-	-	-	nr	459.39
Cutting openings or recesses through 100 mm thick softwood stud partition including framing studwork around, making good boarding and any plaster either side and extending floor finish through opening (new door and frame not included)						
single door and frame	-	-	-	-	nr	209.72
pair of doors and frame	-	-	-	-	nr	274.63
Cutting openings or recesses through internal plastered wall for single door and frame; including cutting structure, quoining or making good jambs, cutting and pinning in suitable precast concrete plate lintel(s), making good plasterwork up to new frame both sides and extending floor finish through new opening (new door and frame not included)						
150 mm reinforced concrete wall	-	-	-	-	nr	239.68
225 mm reinforced concrete wall	-	-	-	-	nr	329.56
half brick thick wall	-	-	-	-	nr	224.71
one brick thick wall or two half brick thick skins	-	-	-	-	nr	294.61
one and a half brick thick wall	-	-	-	-	nr	359.52
two brick thick wall	-	-	-	-	nr	434.43
100 mm block wall	-	-	-	-	nr	209.72
215 mm block wall	-	-	-	-	nr	274.63
Cutting openings or recesses through internal plastered wall for pair of doors and frame; including cutting structure, quoining or making good jambs, cutting and pinning in suitable precast concrete plate lintel(s), making good plasterwork up to new frame both sides and extending floor finish through new opening (new door and frame not included)						
150 mm reinforced concrete wall	-	-	-	-	nr	344.54
225 mm reinforced concrete wall	-	-	-	-	nr	439.42
half brick thick wall	-	-	-	-	nr	264.65
one brick thick wall or two half brick thick skins	-	-	-	-	nr	364.51
one and a half brick thick wall	-	-	-	-	nr	469.38
two brick thick wall	-	-	-	-	nr	559.26
100 mm block wall	-	-	-	-	nr	249.67
215 mm block wall	-	-	-	-	nr	339.55
Cutting back projections						
Cutting back brick projections flush with adjacent wall						
225 mm x 112 mm	-	0.28	4.55	0.10	m	4.65
225 mm x 225 mm	-	0.46	7.48	0.20	m	7.68
337 mm x 112 mm	-	0.65	10.57	0.29	m	10.86
450 mm x 225 mm	-	0.83	13.50	0.39	m	13.89

C DEMOLITION/ALTERATION/RENOVATION

	PC £	Labour hours	Labour £	Material £	Unit	Total rate £
Cutting back chimney breasts flush with adjacent wall						
half brick thick	-	1.62	26.34	3.15	m²	29.49
one brick thick	-	2.17	35.28	4.92	m²	40.20
Filling in openings						
Removing doors and frames; making good plaster and skirtings across reveals and heads; leaving as blank openings						
single doors	-	-	-	-	nr	89.88
pair of doors	-	-	-	-	nr	104.87
Removing doors and frames in 100 mm thick softwood partitions; filling in openings with timber covered on both sides with boarding or lining to match existing; extending skirtings both sides						
single doors	-	-	-	-	nr	139.81
pair of doors	-	-	-	-	nr	184.76
Removing single doors and frames in internal walls; filling in openings with brickwork or blockwork; plastering walls and extending skirtings both sides						
half brick thick	-	-	-	-	nr	154.80
one brick thick	-	-	-	-	nr	214.72
one and a half brick thick	-	-	-	-	nr	269.64
two brick thick	-	-	-	-	nr	339.55
100 mm blockwork	-	-	-	-	nr	124.84
215 mm blockwork	-	-	-	-	nr	184.76
Removing pairs of doors and frames in internal walls; filling in openings with brickwork or blockwork; plastering walls and extend skirtings both sides						
half brick thick	-	-	-	-	nr	249.67
one brick thick	-	-	-	-	nr	349.54
one and a half brick thick	-	-	-	-	nr	449.40
two brick thick	-	-	-	-	nr	544.28
100 mm blockwork	-	-	-	-	nr	214.72
215 mm blockwork	-	-	-	-	nr	284.62
Removing 825 mm x 1046 mm (1.16 m²) sliding sash windows and frames in external faced walls; filling in openings with facing brickwork on outside to match existing and common brickwork on inside; plastering internally						
one brick thick or two half brick thick skins	-	-	-	-	nr	204.72
one and a half brick thick	-	-	-	-	nr	229.70
two brick thick	-	-	-	-	nr	264.65
Removing 825 mm x 1406 mm (1.16 m²) curved headed sliding sashed windows in external stuccoed walls; filling in openings with common bricks; stucco on outside and plastering internally						
one brick thick or two half brick thick skins	-	-	-	-	nr	229.7
one and a half brick thick	-	-	-	-	nr	264.65
two brick thick	-	-	-	-	nr	329.56
Removing 825 mm x 1406 mm (1.16 m²) curved headed sliding sash windows in external masonry faced brick walls; filling in openings with facing brickwork on outside and common brickwork on inside; plastering internally						
350 mm wall	-	-	-	-	nr	599.21
500 mm wall	-	-	-	-	nr	654.14
600 mm wall	-	-	-	-	nr	719.05

Prices for Measured Works - Minor Works
C DEMOLITION/ALTERATION/RENOVATION

	PC £	Labour hours	Labour £	Material £	Unit	Total rate £
C90 ALTERATIONS - SPOT ITEMS - cont'd						
Filling in openings - cont'd						
Quoining up jambs in common bricks; PC £175.00/1000; in gauged mortar (1:1:6); as the work proceeds						
half brick thick or skin of hollow wall	-	0.93	15.12	3.44	m	18.56
one brick thick	-	1.39	22.60	6.89	m	29.49
one and a half brick thick	-	1.80	29.27	10.33	m	39.60
two brick thick	-	2.22	36.10	12.60	m	48.70
75 mm blockwork	-	0.58	9.43	3.58	m	13.01
90 mm blockwork	-	0.62	10.08	4.29	m	14.37
100 mm blockwork	-	0.65	10.57	4.87	m	15.44
115 mm blockwork	-	0.70	11.38	5.52	m	16.90
125 mm blockwork	-	0.74	12.03	6.26	m	18.29
140 mm blockwork	-	0.80	13.01	7.19	m	20.20
150 mm blockwork	-	0.83	13.50	7.91	m	21.41
190 mm blockwork	-	0.93	15.12	9.63	m	24.75
215 mm blockwork	-	1.00	16.26	10.70	m	26.96
225 mm blockwork	-	1.10	17.89	13.14	m	31.03
Closing at jambs with common brickwork half brick thick 50 mm cavity; including lead-lined hessian based vertical damp proof course	-	0.37	6.02	6.48	m	12.50
Quoining up jambs in sand faced facings; PC £174.50/1000; in gauged mortar (1:1:6); facing and pointing one side to match existing						
half brick thick or skin of hollow wall	-	1.16	18.86	4.42	m	23.28
one brick thick	-	1.39	22.60	9.04	m	31.64
one and a half brick thick	-	2.13	34.63	13.42	m	48.05
two brick thick	-	2.59	42.11	17.89	m	60.00
Quoining up jambs in machine made facings; PC £330.00/1000; in gauged mortar (1:1:6); facing and pointing one side to match existing						
half brick thick or skin of hollow wall	-	1.16	18.86	7.09	m	25.95
one brick thick	-	1.39	22.60	13.00	m	35.60
one and a half brick thick	-	2.13	34.63	19.34	m	53.97
two brick thick	-	2.59	42.11	25.55	m	67.66
Quoining up jambs in hand made facings; PC £540.00/1000; in gauged mortar (1:1:6); facing and pointing one side to match existing						
half brick thick or skin of hollow wall	-	1.16	18.86	10.48	m	29.34
one brick thick	-	1.39	22.60	19.78	m	42.38
one and a half brick thick	-	2.13	34.63	29.50	m	64.13
two brick thick	-	2.59	42.11	39.10	m	81.21
Filling existing openings with common brickwork or blockwork in gauged mortar (1:1:6) (cutting and bonding not included)						
half brick thick	-	1.71	27.80	13.41	m^2	41.21
one brick thick	-	2.82	45.85	28.00	m^2	73.85
one and a half brick thick	-	3.89	63.25	42.00	m^2	105.25
two brick thick	-	4.86	79.02	55.99	m^2	135.01
75 mm blockwork	5.26	0.85	13.82	7.65	m^2	21.47
90 mm blockwork	6.31	0.93	15.12	8.95	m^2	24.07
100 mm blockwork	7.26	0.97	15.77	10.13	m^2	25.90
115 mm blockwork	8.35	1.05	17.07	11.48	m^2	28.55
125 mm blockwork	9.09	1.09	17.72	12.40	m^2	30.12
140 mm blockwork	10.17	1.16	18.86	13.74	m^2	32.60
150 mm blockwork	10.90	1.20	19.51	14.65	m^2	34.16
190 mm blockwork	13.80	1.37	22.28	18.84	m^2	41.12
215 mm blockwork	15.62	1.48	24.06	21.10	m^2	45.16
255 mm blockwork	19.25	1.65	26.83	25.61	m^2	52.44

Prices for Measured Works - Minor Works
C DEMOLITION/ALTERATION/RENOVATION

	PC £	Labour hours	Labour £	Material £	Unit	Total rate £
Cutting and bonding ends to existing						
half brick thick	-	0.37	6.02	1.11	m	7.13
one brick thick	-	0.54	8.78	2.16	m	10.94
one and a half brick thick	-	0.80	13.01	2.63	m	15.64
two brick thick	-	1.16	18.86	4.03	m	22.89
75 mm blockwork	-	0.16	2.60	0.56	m	3.16
90 mm blockwork	-	0.19	3.09	0.69	m	3.78
100 mm blockwork	-	0.21	3.41	0.77	m	4.18
115 mm blockwork	-	0.23	3.74	0.90	m	4.64
125 mm blockwork	-	0.24	3.90	0.97	m	4.87
140 mm blockwork	-	0.26	4.23	1.07	m	5.30
150 mm blockwork	10.90	0.27	4.39	1.14	m	5.53
190 mm blockwork	13.80	0.33	5.37	1.51	m	6.88
215 mm blockwork	15.62	0.38	6.18	1.68	m	7.86
255 mm blockwork	19.25	0.44	7.15	2.03	m	9.18
half brick thick in facings; to match existing	0.33	0.56	9.11	3.19	m^2	12.30
Extra over common brickwork for fair face; flush pointing						
walls and the like	-	0.19	3.09	-	m^2	3.09
Extra over common bricks; PC £189.70/1000 for facing bricks in flemish bond; facing and pointing one side						
machine made facings; PC £330.00/1000	0.33	0.97	15.77	11.31	m^2	27.08
hand made facings; PC £540.00/1000	0.54	0.97	15.77	28.24	m^2	44.01
ADD or DEDUCT for variation of £10.00/1000 in PC for facing bricks; in flemish bond						
half brick thick	-	-	-	0.65	m^2	0.65
Filling in openings to hollow walls with inner skin of common bricks PC £189.70/1000; 50 mm cavity and galvanised steel butterfly ties; outer skin of facings; all in gauged mortar (1:1:6); facing and pointing one side						
two half brick thick skins; outer skin sand faced facings	-	4.26	69.27	30.50	m^2	99.77
two half brick thick skins; outer skin machine made facings	-	4.26	69.27	38.27	m^2	107.54
two half brick thick skins; outer skin hand made facings	-	4.26	69.27	52.49	m^2	121.76
Temporary screens						
Providing and erecting; maintaining; temporary dust proof screens; with 50 mm x 75 mm sawn softwood framing; covering one side with 12 mm thick plywood						
over 300 mm wide	-	0.74	9.63	8.88	m^2	18.51
Providing and erecting; maintaining; temporary screen; with 50 mm x 100 mm sawn softwood framing; covering one side with 13 mm thick insulating board and other side with single layer of polythene sheet						
over 300 mm wide	-	0.93	12.11	14.81	m^2	26.92
Providing and erecting; maintaining; temporary screen; with 50 mm x 100 mm sawn softwood framing; covering one side with 19 mm thick exterior quality plywood; softwood cappings; including three coats of gloss paint; clearing away						
over 300 mm wide	-	1.85	20.80	20.45	m^2	41.25

… Prices for Measured Works - Minor Works
D GROUNDWORK

	PC £	Labour hours	Labour £	Material £	Unit	Total rate £
D GROUNDWORK						
D20 EXCAVATING AND FILLING						
NOTE: Prices are applicable to excavation in firm soil. Multiplying factors for other soils are as follows:						
	Mechanical	Hand				
Clay	x 2.00	x 1.20				
Compact gravel	x 3.00	x 1.50				
Soft chalk	x 4.00	x 2.00				
Hard rock	x 5.00	x 6.00				
Running sand or silt	x 6.00	x 2.00				
Site preparation						
Removing trees						
girth 600 mm - 1.50 m	-	20.35	151.40	-	nr	151.40
girth 1.50 - 3.00 m	-	35.61	264.94	-	nr	264.94
girth exceeding 3.00 m	-	50.88	378.55	-	nr	378.55
Removing tree stumps						
girth 600 mm - 1.50 m	-	1.02	7.59	41.97	nr	49.56
girth 1.50 m - 3.00 m	-	1.02	7.59	61.16	nr	68.75
girth exceeding 3.00 m	-	1.02	7.59	83.93	nr	91.52
Clearing site vegetation						
bushes, scrub, undergrowth, hedges and trees and tree stumps not exceeding 600 mm girth	-	0.03	0.22	-	m²	0.22
Lifting turf for preservation						
stacking	-	0.36	2.68	-	m²	2.68
Excavating; by machine						
Topsoil for preservation						
average depth 150 mm	-	0.02	0.15	1.21	m²	1.36
add or deduct for each 25 mm variation in average depth	-	0.01	0.07	0.30	m²	0.37
To reduce levels						
maximum depth not exceeding 0.25 m	-	0.06	0.45	1.51	m³	1.96
maximum depth not exceeding 1.00 m	-	0.04	0.30	1.06	m³	1.36
maximum depth not exceeding 2.00 m	-	0.06	0.45	1.51	m³	1.96
maximum depth not exceeding 4.00 m	-	0.08	0.60	1.90	m³	2.50
Basements and the like; commencing level exceeding 0.25 m below exitsing ground level						
maximum depth not exceeding 1.00 m	-	0.09	0.67	1.43	m³	2.10
maximum depth not exceeding 2.00 m	-	0.06	0.45	1.07	m³	1.52
maximum depth not exceeding 4.00 m	-	0.09	0.67	1.43	m³	2.10
maximum depth not exceeding 6.00 m	-	0.10	0.74	1.79	m³	2.53
maximum depth not exceeding 8.00 m	-	0.13	0.97	2.14	m³	3.11
Pits						
maximum depth not exceeding 0.25m	-	0.36	2.68	5.71	m³	8.39
maximum depth not exceeding 1.00 m	-	0.38	2.83	5.00	m³	7.83
maximum depth not exceeding 2.00 m	-	0.44	3.27	5.71	m³	8.98
maximum depth not exceeding 4.00 m	-	0.53	3.94	6.43	m³	10.37
maximum depth not exceeding 6.00 m	-	0.55	4.09	6.78	m³	10.87
Extra over pit excavating for commencing level exceeding 0.25 m below existing ground level						
1.00 m below	-	0.03	0.22	0.71	m³	0.93
2.00 m below	-	0.06	0.45	1.07	m³	1.52
3.00 m below	-	0.07	0.52	1.43	m³	1.95
4.00 m below	-	0.10	0.74	1.79	m³	2.53

D GROUNDWORK

	PC £	Labour hours	Labour £	Material £	Unit	Total rate £
Trenches, width not exceeding 0.30 m						
maximum depth not exceeding 0.25 m	-	0.30	2.23	4.64	m^3	6.87
maximum depth not exceeding 1.00 m	-	0.32	2.38	3.93	m^3	6.31
maximum depth not exceeding 2.00 m	-	0.37	2.75	4.64	m^3	7.39
maximum depth not exceeding 4.00 m	-	0.45	3.35	5.71	m^3	9.06
maximum depth not exceeding 6.00 m	-	0.52	3.87	6.78	m^3	10.65
Trenches, width exceeding 0.30 m						
maximum depth 0.25 m	-	0.27	2.01	4.28	m^3	6.29
maximum depth 1.00 m	-	0.28	2.08	3.57	m^3	5.65
maximum depth 2.00 m	-	0.34	2.53	4.28	m^3	6.81
maximum depth 4.00 m	-	0.40	2.98	5.00	m^3	7.98
maximum depth 6.00 m	-	0.49	3.65	6.43	m^3	10.08
Extra over trench excavating for commencing level exceeding 0.25 m below existing ground level						
1.00 m below		0.03	0.22	0.71	m^3	0.93
2.00 m below		0.06	0.45	1.07	m^3	1.52
3.00 m below		0.07	0.52	1.43	m^3	1.95
4.00 m below		0.10	0.74	1.79	m^3	2.53
For pile caps and ground beams between piles						
maximum depth not exceeding 0.25 m		0.45	3.35	7.50	m^3	10.85
maximum depth not exceeding 1.00 m		0.40	2.98	6.78	m^3	9.76
maximum depth not exceeding 2.00 m		0.45	3.35	7.50	m^3	10.85
To bench sloping ground to receive filling						
maximum depth not exceeding 0.25 m	-	0.10	0.74	1.79	m^3	2.53
maximum depth not exceeding 1.00 m	-	0.07	0.52	2.14	m^3	2.66
maximum depth not exceeding 2.00 m	-	0.10	0.74	1.79	m^3	2.53
Extra over any types of excavating irrespective of depth						
excavating below ground water level	-	0.16	1.19	2.50	m^3	3.69
next existing services	-	2.90	21.58	1.43	m^3	23.01
around existing services crossing excavation	-	6.70	49.85	3.93	m^3	53.78
Extra over any types of excavating irrespective of depth for breaking out existing materials						
rock	-	3.42	25.44	14.89	m^3	40.33
concrete	-	2.90	21.58	10.73	m^3	32.31
reinforced concrete	-	4.17	31.02	15.98	m^3	47.00
brickwork; blockwork or stonework	-	2.08	15.48	8.14	m^3	23.62
Extra over any types of excavating irrespective of depth for breaking out existing hard pavings, 75 mm thick						
coated macadam or asphalt	-	0.22	1.64	0.58	m^2	2.22
Extra over any types of excavating irrespective of depth for breaking out existing hard pavings; 150 mm thick						
concrete	-	0.45	3.35	1.75	m^2	5.10
reinforced concrete	-	0.68	5.06	2.27	m^2	7.33
coated macadam or asphalt and hardcore	-	0.30	2.23	0.63	m^2	2.86
Working space allowance to excavations						
reduce levels; basements and the like		0.09	0.67	1.43	m^2	2.10
pits		0.22	1.64	3.93	m^2	5.57
trenches		0.20	1.49	3.57	m^2	5.06
pile caps and ground beams between piles		0.23	1.71	3.93	m^2	5.64
Extra over excavating for working space for backfilling with special materials						
hardcore	-	0.16	1.19	14.30	m^2	15.49
sand	-	0.16	1.19	21.68	m^2	22.87
40 mm - 20 mm gravel	-	0.16	1.19	22.11	m^2	23.30
plain in situ ready mixed designated concrete C7.5 - 40 mm aggregate	-	1.07	8.80	45.86	m^2	54.66

D GROUNDWORK

	PC £	Labour hours	Labour £	Material £	Unit	Total rate £
D20 EXCAVATING AND FILLING - cont'd						
Excavating; by hand						
Topsoil for preservation						
average depth 150 mm	-	0.26	1.93	-	m²	1.93
add or deduct for each 25 mm variation in						
average depth	-	0.03	0.22	-	m²	0.22
To reduce levels						
maximum depth not exceeding 0.25 m	-	1.59	11.83	-	m³	11.83
maximum depth not exceeding 1.00 m	-	1.80	13.39	-	m³	13.39
maximum depth not exceeding 2.00 m	-	1.99	14.81	-	m³	14.81
maximum depth not exceeding 4.00 m	-	2.19	16.29	-	m³	16.29
Basements and the like; commencing level						
exceeding 0.25 m below existing ground level						
maximum depth not exceeding 1.00 m	-	2.09	15.55	-	m³	15.55
maximum depth not exceeding 2.00 m	-	2.24	16.67	-	m³	16.67
maximum depth not exceeding 4.00 m	-	3.01	22.39	-	m³	22.39
maximum depth not exceeding 6.00 m	-	3.66	27.23	-	m³	27.23
maximum depth not exceeding 8.00 m	-	4.35	32.36	-	m³	32.36
Pits						
maximum depth not exceeding 0.25 m	-	2.34	17.41	-	m³	17.41
maximum depth not exceeding 1.00 m	-	2.90	21.58	-	m³	21.58
maximum depth not exceeding 2.00 m	-	3.60	26.78	-	m³	26.78
maximum depth not exceeding 4.00 m	-	4.56	33.93	-	m³	33.93
maximum depth not exceeding 6.00 m	-	5.64	41.96	-	m³	41.96
Extra over pit excavating for commencing level						
exceeding 0.25 m below existing ground level						
1.00 m below	-	0.45	3.35	-	m³	3.35
2.00 m below	-	0.97	7.22	-	m³	7.22
3.00 m below	-	1.43	10.64	-	m³	10.64
4.00 m below	-	1.88	13.99	-	m³	13.99
Trenches, width not exceeding 0.30 m						
maximum depth not exceeding 0.25 m	-	2.03	15.10	-	m³	15.10
maximum depth not exceeding 1.00 m	-	2.99	22.25	-	m³	22.25
maximum depth not exceeding 2.00 m	-	3.51	26.11	-	m³	26.11
maximum depth not exceeding 4.00 m	-	4.29	31.92	-	m³	31.92
maximum depth not exceeding 6.00 m	-	5.53	41.14	-	m³	41.14
Trenches, width exceeding 0.30 m						
maximum depth not exceeding 0.25 m	-	1.99	14.81	-	m³	14.81
maximum depth not exceeding 1.00 m	-	2.67	19.86	-	m³	19.86
maximum depth not exceeding 2.00 m	-	3.12	23.21	-	m³	23.21
maximum depth not exceeding 4.00 m	-	3.97	29.54	-	m³	29.54
maximum depth not exceeding 6.00 m	-	5.07	37.72	-	m³	37.72
Extra over trench excavating for commencing level						
exceeding 0.25 m below existing ground level						
1.00 m below	-	0.45	3.35	-	m³	3.35
2.00 m below	-	0.97	7.22	-	m³	7.22
3.00 m below	-	1.46	10.86	-	m³	10.86
4.00 m below	-	1.88	13.99	-	m³	13.99
For pile caps and ground beams between piles						
maximum depth not exceeding 0.25 m	-	3.05	22.69	-	m³	22.69
maximum depth not exceeding 1.00 m	-	3.26	24.25	-	m³	24.25
maximum depth not exceeding 2.00 m	-	3.87	28.79	-	m³	28.79
To bench sloping ground to receive filling						
maximum depth not exceeding 0.25 m	-	1.43	10.64	-	m³	10.64
maximum depth not exceeding 1.00 m	-	1.63	12.13	-	m³	12.13
maximum depth not exceeding 2.00 m	-	1.83	13.62	-	m³	13.62
Extra over any types of excavating irrespective of depth						
excavating below ground water level	-	0.36	2.68	-	m³	2.68
next existing services	-	1.02	7.59	-	m³	7.59
around existing services crossing excavation	-	2.04	15.18	-	m³	15.18

D GROUNDWORK

	PC £	Labour hours	Labour £	Material £	Unit	Total rate £
Extra over any types of excavating irrespective of depth for breaking out existing materials						
rock	-	5.09	37.87	6.50	m³	**44.37**
concrete	-	4.58	34.08	5.42	m³	**39.50**
reinforced concrete	-	6.11	45.46	7.58	m³	**53.04**
brickwork; blockwork or stonework	-	3.05	22.69	3.25	m³	**25.94**
Extra over any types of excavating irrespective of depth for breaking out existing hard pavings, 60 mm thick						
precast concrete paving slabs	-	0.31	2.31	-	m²	**2.31**
Extra over any types of excavating irrespective of depth for breaking out existing hard pavings, 75 mm thick						
coated macadam or asphalt	-	0.44	3.27	0.89	m²	**4.16**
Extra over any types of excavating irrespective of depth for breaking out existing hard pavings, 150 mm thick						
concrete	-	0.71	5.28	0.76	m²	**6.04**
reinforced concrete	-	0.92	6.84	1.08	m²	**7.92**
coated macadam or asphalt and hardcore	-	0.51	3.79	0.54	m²	**4.33**
Working space allowance to excavations						
reduce levels; basements and the like	-	2.34	17.41	-	m²	**17.41**
pits	-	2.44	18.15	-	m²	**18.15**
trenches	-	2.14	15.92	-	m²	**15.92**
pile caps and ground beams between piles	-	2.54	18.90	-	m²	**18.90**
Extra over excavation for working space for backfilling with special materials						
hardcore	-	0.81	6.03	12.19	m²	**18.22**
sand	-	0.81	6.03	19.57	m²	**25.60**
40 mm - 20 mm gravel	-	0.81	6.03	20.00	m²	**26.03**
plain in situ ready mixed designated concrete; C7.5 - 40 mm aggregate		1.12	9.21	43.75	m²	**52.96**
Excavating; by hand; inside existing buildings						
Basements and the like; commencing level exceeding 0.25 m below existing ground level						
maximum depth not exceeding 1.00 m	-	3.14	23.36	-	m³	**23.36**
maximum depth not exceeding 2.00 m	-	3.36	25.00	-	m³	**25.00**
maximum depth not exceeding 4.00 m	-	4.51	33.55	-	m³	**33.55**
maximum depth not exceeding 6.00m	-	5.50	40.92	-	m³	**40.92**
maximum depth not exceeding 8.00 m	-	6.65	49.48	-	m³	**49.48**
Pits						
maximum depth not exceeding 0.25 m	-	3.51	26.11	-	m³	**26.11**
maximum depth not exceeding 1.00 m	-	3.82	28.42	-	m³	**20.42**
maximum depth not exceeding 2.00 m	-	4.58	34.08	-	m³	**34.08**
maximum depth not exceeding 4.00 m	-	5.80	43.15	-	m³	**43.15**
maximum depth not exceeding 6.00 m	-	7.17	53.34	-	m³	**53.34**
Extra over pit excavating for commencing level exceeding 0.25 m below existing ground level						
1.00 m below	-	0.68	5.06	-	m³	**5.06**
2.00 m below	-	1.45	10.79	-	m³	**10.79**
3.00 m below	-	2.14	15.92	-	m³	**15.92**
4.00 m below	-	2.81	20.91	-	m³	**20.91**
Trenches, width not exceeding 0.30 m						
maximum depth not exceeding 0.25 m	-	3.05	22.69	-	m³	**22.69**
maximum depth not exceeding 1.00 m	-	3.51	26.11	-	m³	**26.11**
maximum depth not exceeding 2.00 m	-	4.13	30.73	-	m³	**30.73**
maximum depth not exceeding 4.00 m	-	5.04	37.50	-	m³	**37.50**
maximum depth not exceeding 6.00 m	-	6.49	48.29	-	m³	**48.29**

D GROUNDWORK

	PC £	Labour hours	Labour £	Material £	Unit	Total rate £
D20 EXCAVATING AND FILLING - cont'd						
Excavating; by hand; inside existing buildings - cont'd						
Trenches, width exceeding 0.30 m						
maximum depth not exceeding 0.25 m	-	2.99	22.25	-	m^3	22.25
maximum depth not exceeding 1.00 m	-	3.14	23.36	-	m^3	23.36
maximum depth not exceeding 2.00 m	-	3.66	27.23	-	m^3	27.23
maximum depth not exceeding 4.00 m	-	4.66	34.67	-	m^3	34.67
maximum depth not exceeding 6.00 m	-	5.96	44.34	-	m^3	44.34
Extra over trench excavating for commencing level exceeding 0.25 m below existing ground level						
1.00 m below	-	0.68	5.06	-	m^3	5.06
2.00 m below	-	1.45	10.79	-	m^3	10.79
3.00 m below	-	2.14	15.92	-	m^3	15.92
4.00 m below	-	2.81	20.91	-	m^3	20.91
Extra over any types of excavating irrespective of depth						
excavating below ground water level	-	0.55	4.09	-	m^3	4.09
Extra over any types of excavating irrespective of depth for breaking out existing materials						
concrete	-	6.86	51.04	5.42	m^3	56.46
reinforced concrete	-	9.16	68.15	7.58	m^3	75.73
brickwork; blockwork or stonework	-	4.58	34.08	3.25	m^3	37.33
Extra over any types of excavating irrespective of depth for breaking out existing hard pavings, 150 mm thick						
concrete	-	1.07	7.96	0.76	m^2	8.72
reinforced concrete	-	1.38	10.27	1.08	m^2	11.35
Working space allowance to excavations						
pits	-	3.60	26.78	-	m^2	26.78
trenches	-	3.21	23.88	-	m^2	23.88
Earthwork support (average "risk" prices)						
Maximum depth not exceeding 1.00 m						
distance between opposing faces not exceeding 2.00 m	-	0.11	0.82	0.33	m^2	1.15
distance between opposing faces 2.00 - 4.00 m	-	0.12	0.89	0.39	m^2	1.28
distance between opposing faces exceeding 4.00 m	-	0.13	0.97	0.49	m^2	1.46
Maximum depth not exceeding 2.00 m						
distance between opposing faces not exceeding 2.00 m	-	0.13	0.97	0.39	m^2	1.36
distance between opposing faces 2.00 - 4.00 m	-	0.14	1.04	0.49	m^2	1.53
distance between opposing faces exceeding 4.00 m	-	0.16	1.19	0.62	m^2	1.81
Maximum depth not exceeding 4.00 m						
distance between opposing faces not exceeding 2.00 m	-	0.16	1.19	0.49	m^2	1.68
distance between opposing faces 2.00 - 4.00 m	-	0.18	1.34	0.62	m^2	1.96
distance between opposing faces exceeding 4.00 m	-	0.19	1.41	0.78	m^2	2.19
Maximum depth not exceeding 6.00 m						
distance between opposing faces not exceeding 2.00 m	-	0.19	1.41	0.59	m^2	2.00
distance between opposing faces 2.00 - 4.00 m	-	0.21	1.56	0.78	m^2	2.34
distance between opposing faces exceeding 4.00 m	-	0.24	1.79	0.98	m^2	2.77
Maximum depth not exceeding 8.00 m						
distance between opposing faces not exceeding 2.00 m	-	0.26	1.93	0.78	m^2	2.71
distance between opposing faces 2.00 - 4.00 m	-	0.31	2.31	0.98	m^2	3.29
distance between opposing faces exceeding 4.00 m	-	0.37	2.75	1.17	m^2	3.92

D GROUNDWORK

	PC £	Labour hours	Labour £	Material £	Unit	Total rate £
Earthwork support (open boarded)						
Maximum depth not exceeding 1.00 m						
distance between opposing faces not exceeding 2.00 m	-	0.31	2.31	0.69	m²	**3.00**
distance between opposing faces 2.00 - 4.00 m	-	0.34	2.53	0.78	m²	**3.31**
distance between opposing faces exceeding 4.00 m	-	0.39	2.90	0.98	m²	**3.88**
Maximum depth not exceeding 2.00 m						
distance between opposing faces not exceeding 2.00 m	-	0.39	2.90	0.78	m²	**3.68**
distance between opposing faces 2.00 - 4.00 m	-	0.43	3.20	0.94	m²	**4.14**
distance between opposing faces exceeding 4.00 m	-	0.49	3.65	1.17	m²	**4.82**
Maximum depth not exceeding 4.00 m						
distance between opposing faces not exceeding 2.00 m	-	0.49	3.65	0.89	m²	**4.54**
distance between opposing faces 2.00 - 4.00 m	-	0.55	4.09	1.09	m²	**5.18**
distance between opposing faces exceeding 4.00 m	-	0.61	4.54	1.37	m²	**5.91**
Maximum depth not exceeding 6.00 m						
distance between opposing faces not exceeding 2.00 m	-	0.61	4.54	0.98	m²	**5.52**
distance between opposing faces 2.00 - 4.00 m	-	0.68	5.06	1.23	m²	**6.29**
distance between opposing faces exceeding 4.00 m	-	0.78	5.80	1.56	m²	**7.36**
Maximum depth not exceeding 8.00 m						
distance between opposing faces not exceeding 2.00 m	-	0.81	6.03	1.28	m²	**7.31**
distance between opposing faces 2.00 - 4.00 m	-	0.92	6.84	1.47	m²	**8.31**
distance between opposing faces exceeding 4.00 m	-	1.06	7.89	1.96	m²	**9.85**
Earthwork support (close boarded)						
Maximum depth not exceeding 1.00 m						
distance between opposing faces not exceeding 2.00 m	-	0.81	6.03	1.37	m²	**7.40**
distance between opposing faces 2.00 - 4.00 m	-	0.90	6.70	1.56	m²	**8.26**
distance between opposing faces exceeding 4.00 m	-	0.99	7.37	1.96	m²	**9.33**
Maximum depth not exceeding 2.00 m						
distance between opposing faces not exceeding 2.00 m	-	1.02	7.59	1.56	m²	**9.15**
distance between opposing faces 2.00 - 4.00 m	-	1.12	8.33	1.87	m²	**10.20**
distance between opposing faces exceeding 4.00 m	-	1.22	9.08	2.35	m²	**11.43**
Maximum depth not exceeding 4.00 m						
distance between opposing faces not exceeding 2.00 m	-	1.28	9.52	1.76	m²	**11.28**
distance between opposing faces 2.00 - 4.00 m	-	1.43	10.64	2.19	m²	**12.83**
distance between opposing faces exceeding 4.00 m	-	1.58	11.76	2.74	m²	**14.50**
Maximum depth not exceeding 6.00 m						
distance between opposing faces not exceeding 2.00 m	-	1.59	11.83	1.96	m²	**13.79**
distance between opposing faces 2.00 - 4.00 m	-	1.73	12.87	2.46	m²	**15.33**
distance between opposing faces exceeding 4.00 m	-	1.93	14.36	3.13	m²	**17.49**

Prices for Measured Works - Minor Works
D GROUNDWORK

	PC £	Labour hours	Labour £	Material £	Unit	Total rate £
D20 EXCAVATING AND FILLING - cont'd						
Earthwork support (close boarded) - cont'd						
Maximum depth not exceeding 8.00 m						
distance between opposing faces not exceeding 2.00 m	-	1.93	14.36	2.54	m²	16.90
distance between opposing faces 2.00 - 4.00 m	-	2.14	15.92	2.93	m²	18.85
distance between opposing faces exceeding 4.00 m	-	2.44	18.15	3.52	m²	21.67
Extra over earthwork support for						
Curved	-	0.02	0.15	0.33	m²	0.48
Below ground water level	-	0.31	2.31	0.30	m²	2.61
Unstable ground	-	0.51	3.79	0.59	m²	4.38
Next to roadways	-	0.41	3.05	0.49	m²	3.54
Left in	-	0.67	4.98	13.69	m²	18.67
Earthwork support (average "risk" prices - inside existing existing buildings)						
Maximum depth not exceeding 1.00 m						
distance between opposing faces not exceeding 2.00 m	-	0.19	1.41	0.49	m²	1.90
distance between opposing faces 2.00 - 4.00 m	-	0.21	1.56	0.56	m²	2.12
distance between opposing faces exceeding 4.00 m	-	0.24	1.79	0.69	m²	2.48
Maximum depth not exceeding 2.00 m						
distance between opposing faces not exceeding 2.00 m	-	0.24	1.79	0.56	m²	2.35
distance between opposing faces 2.00 - 4.00 m	-	0.27	2.01	0.75	m²	2.76
distance between opposing faces exceeding 4.00 m	-	0.32	2.38	0.85	m²	3.23
Maximum depth not exceeding 4.00 m						
distance between opposing faces not exceeding 2.00 m	-	0.31	2.31	0.75	m²	3.06
distance between opposing faces 2.00 - 4.00 m	-	0.34	2.53	0.89	m²	3.42
distance between opposing faces exceeding 4.00 m	-	0.38	2.83	1.04	m²	3.87
Maximum depth not exceeding 6.00 m						
distance between opposing faces not exceeding 2.00 m	-	0.38	2.83	0.84	m²	3.67
distance between opposing faces 2.00 - 4.00 m	-	0.42	3.12	1.04	m²	4.16
distance between opposing faces exceeding 6.00 m	-	0.48	3.57	1.23	m²	4.80
Disposal; by machine						
Excavated material						
off site; to tip not exceeding 13 km (using lorries); including Landfill Tax based on inactive waste	-	-	-	16.97	m³	16.97
on site; depositing in spoil heaps; average 25 m distance	-	-	-	0.91	m³	0.91
on site; spreading; average 25 m distance	-	0.23	1.71	0.60	m³	2.31
on site; depositing in spoil heaps; average 50 m distance	-	-	-	1.51	m³	1.51
on site; spreading; average 50 m distance	-	0.23	1.71	1.21	m³	2.92
on site; depositing in spoil heaps; average 100 m distance	-	-	-	2.72	m³	2.72
on site; spreading; average 100 m distance	-	0.23	1.71	1.81	m³	3.52
on site; depositing in spoil heaps; average 200 m distance	-	-	-	3.92	m³	3.92
on site; spreading; average 200 m distance	-	0.23	1.71	2.41	m³	4.12

D GROUNDWORK

	PC £	Labour hours	Labour £	Material £	Unit	Total rate £
Disposal; by hand						
Excavated material						
off site; to tip not exceeding 13 km (using lorries); including Landfill Tax based on inactive waste	-	0.81	6.03	25.96	m³	31.99
on site; depositing in spoil heaps; average 25 m distance	-	1.12	8.33	-	m³	8.33
on site; spreading; average 25 m distance	-	1.48	11.01	-	m³	11.01
on site; depositing in spoil heaps; average 50 m distance	-	1.48	11.01	-	m³	11.01
on site; spreading; average 50 m distance	-	1.79	13.32	-	m³	13.32
on site; depositing in spoil heaps; average 100 m distance	-	2.14	15.92	-	m³	15.92
on site; spreading; average 100 m distance	-	2.44	18.15	-	m³	18.15
on site; depositing in spoil heaps; average 200 m distance	-	3.15	23.44	-	m³	23.44
on site; spreading; average 200 m distance	-	3.46	25.74	-	m³	25.74
Filling to excavations; by machine						
Average thickness not exceeding 0.25 m						
arising from the excavations	-	0.19	1.41	2.41	m³	3.82
obtained off site; hardcore	9.38	0.21	1.56	26.70	m³	28.26
obtained off site; granular fill type one	13.33	0.21	1.56	38.42	m³	39.98
obtained off site; granular fill type two	12.82	0.21	1.56	37.12	m³	38.68
Average thickness exceeding 0.25 m						
arising from the excavations	-	0.16	1.19	1.81	m³	3.00
obtained off site; hardcore	9.38	0.18	1.34	22.62	m³	23.96
obtained off site; granular fill type one	13.33	0.18	1.34	37.51	m³	38.85
obtained off site; granular fill type two	12.82	0.18	1.34	36.21	m³	37.55
Filling to make up levels; by machine						
Average thickness not exceeding 0.25 m						
arising from the excavations	-	0.27	2.01	2.70	m³	4.71
obtained off site; imported topsoil	9.71	0.27	2.01	13.74	m³	15.75
obtained off site; hardcore	9.38	0.31	2.31	26.81	m³	29.12
obatined off site; granular fill type one	13.33	0.31	2.31	38.53	m³	40.84
obtained off site; granular fill type two	12.82	0.31	2.31	37.23	m³	39.54
obtained off site; sand	13.33	0.31	2.31	38.53	m³	40.84
Average thickness exceeding 0.25 m						
arising from the excavations	-	0.22	1.64	1.94	m³	3.58
obtained off site; imported topsoil	9.71	0.22	1.64	12.99	m³	14.63
obtained off site; hardcore	9.38	0.27	2.01	22.61	m³	24.62
obatined off site; granular fill type one	13.33	0.27	2.01	37.50	m³	39.51
obtained off site; granular fill type two	12.82	0.27	2.01	36.20	m³	38.21
obtained off site; sand	13.33	0.27	2.01	37.50	m³	39.51
Filling to excavations; by hand						
Average thickness not exceeding 0.25 m						
arising from the excavations	-	1.25	9.30	-	m³	9.30
obtained off site; hardcore	9.38	1.35	10.04	24.28	m³	34.32
obtained off site; granular fill type one	13.33	1.60	11.90	33.89	m³	45.79
obtained off site; granular fill; type two	12.82	1.60	11.90	32.59	m³	44.49
obtained off site; sand	13.33	1.60	11.90	33.89	m³	45.79

D GROUNDWORK

	PC £	Labour hours	Labour £	Material £	Unit	Total rate £
D20 EXCAVATING AND FILLING - cont'd						
Filling to excavations; by hand - cont'd						
Average thickness exceeding 0.25 m						
arising from the excavations	-	1.02	7.59	-	m³	**7.59**
obtained off site; hardcore	9.38	1.19	8.85	20.81	m³	**29.66**
obtained off site; granular fill type one	13.33	1.32	9.82	33.89	m³	**43.71**
obtained off site; granular fill; type two	12.82	1.32	9.82	32.59	m³	**42.41**
obtained off site; sand	13.33	1.32	9.82	33.89	m³	**43.71**
Filling to make up levels; by hand						
Average thickness not exceeding 0.25 m						
arising from the excavations	-	1.38	10.27	4.58	m³	**14.85**
obtained off site; imported soil	9.71	1.38	10.27	15.02	m³	**25.29**
obtained off site; hardcore	9.38	1.71	12.72	29.97	m³	**42.69**
obtained off site; granular fill type one	13.33	1.82	13.54	39.95	m³	**53.49**
obtained off site; granular fill type two	12.82	1.82	13.54	38.65	m³	**52.19**
obtained off site; sand	13.33	1.82	13.54	39.95	m³	**53.49**
Average thickness exceeding 0.25 m						
arising from the excavations	-	1.19	8.85	3.72	m³	**12.57**
arising from on site spoil heaps; average 25 m distance; multiple handling	-	2.44	18.15	8.12	m³	**26.27**
obtained off site; imported soil	9.71	1.19	8.85	14.16	m³	**23.01**
obtained off site; hardcore	9.38	1.57	11.68	26.04	m³	**37.72**
obtained off site; granular fill type one	13.33	1.68	12.50	39.49	m³	**51.99**
obtained off site; granular fill type two	12.82	1.68	12.50	38.19	m³	**50.69**
obtained off site; sand	13.33	1.68	12.50	39.49	m³	**51.99**
Surface packing to filling						
To vertical or battered faces	-	0.19	1.41	0.21	m²	**1.62**
Surface treatments						
Compacting						
filling; blinding with sand	-	0.05	0.37	1.86	m²	**2.23**
bottoms of excavations	-	0.05	0.37	0.03	m²	**0.40**
Trimming						
sloping surfaces	-	0.19	1.41	-	m²	**1.41**
sloping surfaces; in rock	-	1.02	7.44	1.52	m²	**8.96**
Filter membrane; one layer; laid on earth to receive granular material						
"Terram 500" filter membrane; one layer; laid on earth	-	0.05	0.37	0.50	m²	**0.87**
"Terram 700" filter membrane; one layer; laid on earth	-	0.05	0.37	0.53	m²	**0.90**
"Terram 1000" filter membrane; one layer; laid on earth	-	0.05	0.37	0.56	m²	**0.93**
"Terram 2000" filter membrane; one layer; laid on earth	-	0.05	0.37	1.19	m²	**1.56**
D50 UNDERPINNING						
Excavating; by machine						
Preliminary trenches						
maximum depth not exceeding 1.00 m	-	0.23	1.71	6.82	m³	**8.53**
maximum depth not exceeding 2.00 m	-	0.28	2.08	8.18	m³	**10.26**
maximum depth not exceeding 4.00 m	-	0.32	2.38	9.55	m³	**11.93**
Extra over preliminary trench excavating for breaking out existing hard pavings, 150 mm thick						
concrete	-	0.65	4.84	0.76	m²	**5.60**

D GROUNDWORK

	PC £	Labour hours	Labour £	Material £	Unit	Total rate £
Excavating; by hand						
Preliminary trenches						
maximum depth not exceeding 1.00 m	-	2.68	19.94	-	m³	19.94
maximum depth not exceeding 2.00 m	-	3.05	22.69	-	m³	22.69
maximum depth not exceeding 4.00 m	-	3.93	29.24	-	m³	29.24
Extra over preliminary trench excavating for breakig out existing hard pavings, 150 mm thick concrete	-	0.28	2.08	1.91	m²	3.99
Underpinning pits; commencing from 1.00 m below existing ground level						
maximum depth not exceeding 0.25 m	-	4.07	30.28	-	m³	30.28
maximum depth not exceeding 1.00 m	-	4.44	33.03	-	m³	33.03
maximum depth not exceeding 2.00 m	-	5.32	39.58	-	m³	39.58
Underpinning pits; commencing from 2.00 m below existing ground level						
maximum depth not exceeding 0.25 m	-	5.00	37.20	-	m³	37.20
maximum depth not exceeding 1.00 m	-	5.37	39.95	-	m³	39.95
maximum depth not exceeding 2.00 m	-	6.24	46.43	-	m³	46.43
Underpinning pits; commencing from 4.00 m below existing ground level						
maximum depth not exceeding 0.25 m	-	5.92	44.04	-	m³	44.04
maximum depth not exceeding 1.00 m	-	6.29	46.80	-	m³	46.80
maximum depth not exceeding 2.00 m	-	7.17	53.34	-	m³	53.34
Extra over any types of excavating irrespective of depth excavating below ground water level	-	0.32	2.38	-	m³	2.38
Earthwork support to preliminary trenches (open boarded - in 3.00 m lengths)						
Maximum depth not exceeding 1.00 m distance between opposing faces not exceeding 2.00 m	-	0.37	2.75	1.28	m²	4.03
Maximum depth not exceeding 2.00 m distance between opposing faces not exceeding 2.00 m	-	0.46	3.42	1.56	m²	4.98
Maximum depth not exceeding 4.00 m distance between opposing faces not exceeding 2.00 m	-	0.59	4.39	1.96	m²	6.35
Earthwork support to underpinning pits (open boarded - in 3.00 m lengths)						
Maximum depth not exceeding 1.00 m distance between opposing faces not exceeding 2.00 m	-	0.41	3.05	1.37	m²	4.42
Maximum depth not exceeding 2.00 m distance between opposing faces not exceeding 2.00 m	-	0.51	3.79	1.76	m²	5.55
Maximum depth not exceeding 4.00 m distance between opposing faces not exceeding 2.00 m	-	0.65	4.84	2.15	m²	6.99
Earthwork support to preliminary trenches (closed boarded - in 3.00 m lengths)						
Maximum depth not exceeding 1.00 m distance between opposing faces not exceeding 2.00 m	-	0.93	6.92	2.15	m²	9.07
Maximum depth not exceeding 2.00 m distance between opposing faces not exceeding 2.00 m	-	1.16	8.63	2.74	m²	11.37
Maximum depth not exceeding 4.00 m distance between opposing faces not exceeding 2.00 m	-	1.43	10.64	3.32	m²	13.96

Prices for Measured Works - Minor Works
D GROUNDWORK

	PC £	Labour hours	Labour £	Material £	Unit	Total rate £
D50 UNDERPINNING - cont'd						
Earthwork support to underpinning pits (closed boarded - in 3.00 m lengths)						
Maximum depth not exceeding 1.00 m						
distance between opposing faces not exceeding 2.00 m	-	1.02	7.59	2.35	m²	9.94
Maximum depth not exceeding 2.00 m						
distance between opposing faces not exceeding 2.00 m	-	1.28	9.52	2.33	m²	12.45
Maximum depth not exceeding 4.00 m						
distance between opposing faces not exceeding 2.00 m	-	1.57	11.68	3.72	m²	15.40
Extra over earthwork support for						
Left in	-	0.69	5.13	13.69	m²	18.82
Cutting away existing projecting foundations						
Concrete						
maximum width 150 mm; maximum depth 150 mm	-	0.15	1.12	0.15	m	1.27
maximum width 150 mm; maximum depth 225 mm	-	0.22	1.64	0.22	m	1.86
maximum width 150 mm; maximum depth 300 mm	-	0.30	2.23	0.30	m	2.53
maximum width 300 mm; maximum depth 300 mm	-	0.58	4.32	0.58	m	4.90
Masonry						
maximum width one brick thick; maximum depth one course high	-	0.04	0.30	0.05	m	0.35
maximum width one brick thick; maximum depth two courses high	-	0.13	0.97	0.13	m	1.10
maximum wodth one brick thick; maximum depth three courses high	-	0.25	1.86	0.25	m	2.11
maximum width one brick thick; maximum depth four courses high	-	0.42	3.12	0.42	m	3.54
Preparing the underside of the existing work to receive the pinning up of the new work						
Width of existing work						
380 mm	-	0.56	4.17	-	m	4.17
600 mm	-	0.74	5.51	-	m	5.51
900 mm	-	0.93	6.92	-	m	6.92
1200 mm	-	1.11	8.26	-	m	8.26
Disposal; by hand						
Excavated material						
off site; to tip not exceeding 13 km (using lorries); including Landfill Tax based on inactive waste	-	0.74	5.51	32.45	m³	37.96
Filling to excavations; by hand						
Average thickness exceeding 0.25 m						
arising from the excavations	-	0.93	6.92	-	m³	6.92
Surface treatments						
Compacting						
bottoms of excavations	-	0.05	0.37	0.03	m²	0.40
Plain in situ ready mixed designated concrete C10 - 40 mm aggregate; poured against faces of excavation						
Underpinning						
thickness not exceeding 150 mm	-	3.42	28.11	80.73	m³	108.84
thickness 150 - 450 mm	-	2.87	23.59	80.73	m³	104.32
thickness exceeding 450 mm	-	2.50	20.55	80.73	m³	101.28

D GROUNDWORK

	PC £	Labour hours	Labour £	Material £	Unit	Total rate £
Plain in situ ready mixed designated concrete C20 - 20 mm aggregate; poured against faces of excavation						
Underpinning						
thickness not exceeding 150 mm	-	3.42	28.11	83.50	m³	111.61
thickness 150 - 450 mm	-	2.87	23.59	83.50	m³	107.09
thickness exceeding 450 mm	-	2.50	20.55	83.50	m³	104.05
Extra for working around reinforcement	-	0.28	2.30	-	m³	2.30
Sawn formwork; sides of foundations in underpinning						
Plain vertical						
height exceeding 1.00 m	-	1.48	19.03	4.89	m²	23.92
height not exceeding 250 mm	-	0.51	6.56	1.40	m²	7.96
height 250 - 500 mm	-	0.79	10.16	2.61	m²	12.77
height 500 mm - 1.00 m	-	1.20	15.43	4.89	m²	20.32
Reinforcement bar; BS4449; hot rolled plain round mild steel bars						
20 mm diameter nominal size						
bent	284.09	28.80	289.52	351.90	t	641.42
16 mm diameter nominal size						
bent	284.09	30.70	309.07	355.80	t	664.87
12 mm diameter nominal size						
bent	287.15	32.65	329.14	363.16	t	692.30
10 mm diameter nominal size						
bent	292.26	34.60	349.20	375.58	t	724.78
8 mm diameter nominal size						
bent	292.26	36.50	366.48	379.48	t	745.96
Reinforcement bar; BS4461; cold worked deformed high yield steel bars						
20 mm diameter nominal size						
bent	284.09	28.80	289.52	351.90	t	641.42
16 mm diameter nominal size						
bent	284.09	30.70	309.07	355.80	t	664.87
12 mm diameter nominal size						
bent	287.15	32.65	329.14	363.16	t	692.30
10 mm diameter nominal size						
bent	292.26	34.60	349.20	375.58	t	724.78
8 mm diameter nominal size						
bent	292.26	36.50	366.48	379.48	t	745.96
Common bricks; PC £180.00/1000; in cement mortar (1:3)						
Walls in underpinning						
one brick thick	-	2.22	36.10	27.88	m²	63.98
one and a half brick thick	-	3.05	49.59	41.50	m²	91.09
two brick thick	-	3.79	61.63	57.80	m²	119.43
Class A engineering bricks; PC £354.60/1000; in cement mortar (1:3)						
Walls in underpinning						
one brick thick	-	2.22	36.10	38.43	m²	74.53
one and a half brick thick	-	3.05	49.59	57.31	m²	106.90
two brick thick	-	3.79	61.63	78.89	m²	140.52

D GROUNDWORK

	PC £	Labour hours	Labour £	Material £	Unit	Total rate £
D50 UNDERPINNING - cont'd						
Class B engineering bricks; PC £251.70/1000; in cement mortar (1:3)						
Walls in underpinning						
one brick thick	-	2.22	36.10	28.27	m²	64.37
one and a half brick thick	-	3.05	49.59	42.08	m²	91.67
two brick thick	-	3.79	61.63	58.58	m²	120.21
Add or deduct for variation of £10.00/1000 in PC of bricks						
one brick thick	-	-	-	1.29	m²	1.29
one and a half brick thick	-	-	-	1.94	m²	1.94
two brick thick	-	-	-	2.58	m²	2.58
"Pluvex" (hessian based) damp proof course or equal and approved; 200 mm laps; in cement mortar (1:3)						
Horizontal						
width exceeding 225 mm	6.93	0.23	3.74	8.61	m²	12.35
width not exceeding 225 mm	6.93	0.45	7.32	8.81	m²	16.13
"Hyload" (pitch polymer) damp proof course or equal and approved; 150 mm laps; in cement mortar (1:3)						
Horizontal						
width exceeding 225 mm	6.11	0.23	3.74	7.59	m²	11.33
width not exceeding 225 mm	6.11	0.46	7.48	7.77	m²	15.25
"Ledkore" grade A (bitumen based lead cored) damp proof course or other equal and approved; 200 mm laps; in cement mortar (1:3)						
Horizontal						
width exceeding 225 mm	18.79	0.31	5.04	23.34	m²	28.38
width not exceeding 225 mm	18.79	0.61	9.92	23.89	m²	33.81
Two courses of slates in cement mortar (1:3)						
Horizontal						
width exceeding 225 mm	-	1.39	22.60	25.53	m²	48.13
width not exceeding 225 mm	-	2.31	37.56	26.11	m²	63.67
Wedging and pinning						
To underside of existing construction with slates in cement mortar (1:3)						
width of wall - half brick thick	-	1.02	16.59	5.93	m	22.52
width of wall - one brick thick	-	1.20	19.51	11.85	m	31.36
width of wall - one and a half brick thick	-	1.39	22.60	17.78	m	40.38

E IN SITU CONCRETE/LARGE PRECAST CONCRETE

	PC £	Labour hours	Labour £	Material £	Unit	Total rate £
E IN SITU CONCRETE/LARGE PRECAST						
E05 IN SITU CONCRETE CONSTRUCTING GENERALLY						
Plain in situ ready mixed designated concrete; C10 - 40 mm aggregate						
Foundations	-	1.39	11.43	75.23	m³	86.66
Isolated foundations	-	1.62	13.32	75.23	m³	88.55
Beds						
thickness not exceeding 150 mm	-	1.90	15.62	75.23	m³	90.85
thickness 150 - 450 mm	-	1.30	10.69	75.23	m³	85.92
thickness exceeding 450 mm	-	1.06	8.71	75.23	m³	83.94
Filling hollow walls						
thickness not exceeding 150 mm	-	3.61	29.67	75.23	m³	104.90
Plain in situ ready mixed designated concrete; C10 - 40 mm aggregate; poured on or against earth or unblinded hardcore						
Foundations	-	1.43	11.75	77.07	m³	88.82
Isolated foundations	-	1.71	14.06	77.07	m³	91.13
Beds						
thickness not exceeding 150 mm	-	1.99	16.36	77.07	m³	93.43
thickness 150 - 450 mm	-	1.43	11.75	77.07	m³	88.82
thickness exceeding 450 mm	-	1.11	9.12	77.07	m³	86.19
Plain in situ ready mixed designated concrete; C20 - 20 mm aggregate						
Foundations	-	1.39	11.43	77.81	m³	89.24
Isolated foundations	-	1.62	13.32	77.81	m³	91.13
Beds						
thickness not exceeding 150 mm	-	2.04	16.77	77.81	m³	94.58
thickness 150 - 450 mm	-	1.39	11.43	77.81	m³	89.24
thickness exceeding 450 mm	-	1.06	8.71	77.81	m³	86.52
Filling hollow walls						
thickness not exceeding 150 mm	-	3.61	29.67	77.81	m³	107.48
Plain in situ ready mixed concrete; C20 - 20 mm aggregate; poured on or against earth or unblinded hardcore						
Foundations	-	1.43	11.75	79.71	m³	91.46
Isolated foundations	-	1.71	14.06	79.71	m³	93.77
Beds						
thickness not exceeding 150 mm	-	2.13	17.51	79.71	m³	97.22
thickness 150 - 450 mm	-	1.48	12.17	79.71	m³	91.88
thickness exceeding 450 mm	-	1.11	9.12	79.71	m³	88.83
Reinforced in situ ready mixed designated concrete; C20 - 20 mm aggregate						
Foundations	-	1.48	12.17	77.81	m³	89.98
Ground beams	-	2.96	24.33	77.81	m³	102.14
Isolated foundations	-	1.80	14.80	77.81	m³	92.61
Beds						
thickness not exceeding 150 mm	-	2.36	19.40	77.81	m³	97.21
thickness 150 - 450 mm	-	1.71	14.06	77.81	m³	91.87
thickness exceeding 450 mm	-	1.39	11.43	77.81	m³	89.24
Slabs						
thickness not exceeding 150 mm	-	3.75	30.82	77.81	m³	108.63
thickness 150 - 450 mm	-	2.96	24.33	77.81	m³	102.14
thickness exceeding 450 mm	-	2.68	22.03	77.81	m³	99.84

Prices for Measured Works - Minor Works
E IN SITU CONCRETE/LARGE PRECAST CONCRETE

	PC £	Labour hours	Labour £	Material £	Unit	Total rate £
E05 IN SITU CONCRETE CONSTRUCTING GENERALLY - cont'd						
Reinforced in situ ready mixed designated concrete; C20 - 20 mm aggregate - cont'd						
Coffered and troughed slabs						
thickness 150 - 450 mm	-	3.42	28.11	77.81	m³	105.92
thickness exceeding 450 mm	-	2.96	24.33	77.81	m³	102.14
Extra over for sloping						
not exceeding 15 degrees	-	0.28	2.30	-	m³	2.30
over 15 degrees	-	0.56	4.60	-	m³	4.60
Walls						
thickness not exceeding 150 mm	-	3.93	32.30	77.81	m³	110.11
thickness 150 - 450 mm	-	3.15	25.89	77.81	m³	103.70
thickness exceeding 450 mm	-	2.75	22.61	77.81	m³	100.42
Beams						
isolated	-	4.26	35.02	77.81	m³	112.83
isolated deep	-	4.67	38.39	77.81	m³	116.20
attached deep	-	4.26	35.02	77.81	m³	112.83
Beam casings						
isolated	-	4.67	38.39	77.81	m³	116.20
isolated deep	-	5.09	41.84	77.81	m³	119.65
attached deep	-	4.67	38.39	77.81	m³	116.20
Columns	-	5.09	41.84	77.81	m³	119.65
Column casings	-	5.64	46.36	77.81	m³	124.17
Staircases	-	6.38	52.44	77.81	m³	130.25
Upstands	-	4.12	33.87	77.81	m³	111.68
Reinforced in situ ready mixed designated concrete; C25 - 20 mm aggregate						
Foundations	-	1.48	12.17	78.40	m³	90.57
Ground beams	-	2.96	24.33	78.40	m³	102.73
Isolated foundations	-	1.80	14.80	78.40	m³	93.20
Beds						
thickness not exceeding 150 mm	-	2.17	17.84	78.40	m³	96.24
thickness 150 - 450 mm	-	1.62	13.32	78.40	m³	91.72
thickness exceeding 450 mm	-	1.39	11.43	78.40	m³	89.83
Slabs						
thickness not exceeding 150 mm	-	3.56	29.26	78.40	m³	107.66
thickness 150 - 450 mm	-	2.91	23.92	78.40	m³	102.32
thickness exceeding 450 mm	-	2.68	22.03	78.40	m³	100.43
Coffered and troughed slabs						
thickness 150 - 450 mm	-	3.33	27.37	78.40	m³	105.77
thickness exceeding 450 mm	-	2.96	24.33	78.40	m³	102.73
Extra over for sloping						
not exceeding 15 degrees	-	0.28	2.30	-	m³	2.30
over 15 degrees	-	0.56	4.60	-	m³	4.60
Walls						
thickness not exceeding 150 mm	-	3.84	31.56	78.40	m³	109.96
thickness 150 - 450 mm	-	3.10	25.48	78.40	m³	103.88
thickness exceeding 450 mm	-	2.78	22.85	78.40	m³	101.25
Beams						
isolated	-	4.26	35.02	78.40	m³	113.42
isolated deep	-	4.67	38.39	78.40	m³	116.79
attached deep	-	4.26	35.02	78.40	m³	113.42
Beam casings						
isolated	-	4.67	38.39	78.40	m³	116.79
isolated deep	-	5.09	41.84	78.40	m³	120.24
attached deep	-	4.67	38.39	78.40	m³	116.79
Columns	-	5.09	41.84	78.40	m³	120.24
Column casings	-	5.64	46.36	78.40	m³	124.76

Prices for Measured Works - Minor Works

E IN SITU CONCRETE/LARGE PRECAST CONCRETE

	PC £	Labour hours	Labour £	Material £	Unit	Total rate £
Staircases	-	6.38	52.44	78.40	m³	130.84
Upstands	-	4.12	33.87	78.40	m³	112.27
Reinforced in situ ready mixed designated concrete; C30 - 20 mm aggregate						
Foundations	-	1.48	12.17	79.00	m³	91.17
Ground beams	-	2.96	24.33	79.00	m³	103.33
Isolated foundations	-	1.80	14.80	79.00	m³	93.80
Beds						
thickness not exceeding 150 mm	-	2.36	19.40	79.00	m³	98.40
thickness 150 - 450 mm	-	1.71	14.06	79.00	m³	93.06
thickness exceeding 450 mm	-	1.39	11.43	79.00	m³	90.43
Slabs						
thickness not exceeding 150 mm	-	3.75	30.82	79.00	m³	109.82
thickness 150 - 450 mm	-	2.96	24.33	79.00	m³	103.33
thickness exceeding 450 mm	-	2.68	22.03	79.00	m³	101.03
Coffered and troughed slabs						
thickness 150 - 450 mm	-	3.42	28.11	79.00	m³	107.11
thickness exceeding 450 mm	-	2.96	24.33	79.00	m³	103.33
Extra over for sloping						
not exceeding 15 degrees	-	0.28	2.30	-	m³	2.30
over 15 degrees	-	0.56	4.60	-	m³	4.60
Walls						
thickness not exceeding 150 mm	-	3.93	32.30	79.00	m³	111.30
thickness 150 - 450 mm	-	3.15	25.89	79.00	m³	104.89
thickness exceeding 450 mm	-	2.78	22.85	79.00	m³	101.85
Beams						
isolated	-	4.26	35.02	79.00	m³	114.02
isolated deep	-	4.67	38.39	79.00	m³	117.39
attached deep	-	4.26	35.02	79.00	m³	114.02
Beam casings						
isolated	-	4.67	38.39	79.00	m³	117.39
isolated deep	-	5.09	41.84	79.00	m³	120.84
attached deep	-	4.67	38.39	79.00	m³	117.39
Columns	-	5.09	41.84	79.00	m³	120.84
Column casings	-	5.64	46.36	79.00	m³	125.36
Staircases	-	6.38	52.44	79.00	m³	131.44
Upstands	-	4.12	33.87	79.00	m³	112.87
Extra over vibrated concrete for						
Reinforcement content over 5%	-	0.58	4.77	-	m³	4.77
Grouting with cement mortar (1:1)						
Stanchion bases						
10 mm thick	-	1.06	8.71	0.12	nr	8.83
25 mm thick	-	1.33	10.93	0.32	nr	11.25
Grouting with epoxy resin						
Stanchion bases						
10 mm thick	-	1.33	10.93	9.39	nr	20.32
25 mm thick	-	1.60	13.15	23.47	nr	36.62
Grouting with "Conbextra GP" cementitious grout or other equal and approved						
Stanchion bases						
10 mm thick	-	1.33	10.93	1.20	nr	12.13
25 mm thick	-	1.60	13.15	3.20	nr	16.35

Prices for Measured Works - Minor Works
E IN SITU CONCRETE/LARGE PRECAST CONCRETE

	PC £	Labour hours	Labour £	Material £	Unit	Total rate £
E05 IN SITU CONCRETE CONSTRUCTING GENERALLY - cont'd						
Filling; plain ready mixed designated concrete; C20 - 20 mm aggregate						
Mortices	-	0.11	0.90	0.43	nr	1.33
Holes	-	0.27	2.22	89.96	m³	92.18
Chases exceeding 0.01 m²	-	0.21	1.73	89.96	m³	91.69
Chases not exceeding 0.01 m²	-	0.16	1.32	0.90	m	2.22
Sheeting to prevent moisture loss						
Building paper; lapped joints						
subsoil grade; horizontal on foundations	-	0.02	0.16	0.64	m²	0.80
standard grade; horizontal on slabs	-	0.05	0.41	0.90	m²	1.31
Polythene sheeting; lapped joints; horizontal on slabs						
250 microns; 0.25 mm thick	-	0.05	0.41	0.35	m²	0.76
"Visqueen" sheeting or other equal and approved; lapped joints; horizontal on slabs						
250 microns; 0.25 mm thick	-	0.05	0.41	0.29	m²	0.70
300 microns; 0.30 mm thick	-	0.06	0.49	0.40	m²	0.89
E20 FORMWORK FOR IN SITU CONCRETE						
NOTE: Generally all formwork based on four uses unless otherwise stated.						
Sides of foundations; basic finish						
Plain vertical						
height exceeding 1.00 m	-	1.70	21.86	8.29	m²	30.15
height exceeding 1.00 m; left in	-	1.49	19.16	19.94	m²	39.10
height not exceeding 250 mm	-	0.48	6.17	3.28	m	9.45
height not exceeding 250 mm; left in	-	0.48	6.17	5.79	m	11.96
height 250 - 500 mm	-	0.91	11.70	6.46	m	18.16
height 250 - 500 mm; left in	-	0.80	10.29	12.98	m	23.27
height 500 mm - 1.00 m	-	1.28	16.46	8.29	m	24.75
height 500 mm - 1.00 m ; left in	-	1.22	15.69	19.94	m	35.63
Sides of foundations; polystyrene sheet formwork; Cordek "Claymaster" or other equal and approved; 50 mm thick						
Plain vertical						
height exceeding 1.00 m; left in	-	0.34	4.37	8.25	m²	12.62
height not exceeding 250 mm; left in	-	0.11	1.41	2.06	m	3.47
height 250 - 500 mm; left in	-	0.19	2.44	4.13	m	6.57
height 500 mm - 1.00 m; left in	-	0.28	3.60	8.25	m	11.85
Sides of foundation; polystyrene sheet formwork; Cordek "Claymaster" or other equal and approved; 100 mm thick						
Plain vertical						
height exceeding 1.00 m; left in	-	0.37	4.76	16.48	m²	21.24
height not exceeding 250 mm; left in	-	0.12	1.54	4.12	m	5.66
height 250 - 500 mm; left in	-	0.20	2.57	8.24	m	10.81
height 500 mm - 1.00 m; left in	-	0.31	3.99	16.48	m	20.47
Sides of ground beams and edges of beds; basic finish						
Plain vertical						
height exceeding 1.00 m	-	1.76	22.63	8.23	m²	30.86
height not exceeding 250 mm	-	0.53	6.82	3.23	m	10.05
height 250 - 500 mm	-	0.95	12.22	6.40	m	18.62
height 500 mm - 1.00 m	-	1.33	17.10	8.23	m	25.33

E IN SITU CONCRETE/LARGE PRECAST CONCRETE

	PC £	Labour hours	Labour £	Material £	Unit	Total rate £
Edges of suspended slabs; basic finish						
Plain vertical						
height not exceeding 250 mm	-	0.80	10.29	3.34	m	13.63
height 250 - 500 mm	-	1.17	15.05	5.38	m	20.43
height 500 mm - 1.00 m	-	1.86	23.92	8.34	m	32.26
Sides of upstands; basic finish						
Plain vertical						
height exceeding 1.00 m	-	2.13	27.39	9.95	m^2	37.34
height not exceeding 250 mm	-	0.67	8.62	3.44	m	12.06
height 250 - 500 mm	-	1.06	13.63	6.62	m	20.25
height 500 mm - 1.00 m	-	1.86	23.92	9.95	m	33.87
Steps in top surfaces; basic finish						
Plain vertical						
height not exceeding 250 mm	-	0.53	6.82	3.50	m	10.32
height 250 - 500 mm	-	0.85	10.93	6.67	m	17.60
Steps in soffits; basic finish						
Plain vertical						
height not exceeding 250 mm	-	0.58	7.46	2.74	m	10.20
height 250 - 500 mm	-	0.93	11.96	4.95	m	16.91
Machine bases and plinths; basic finish						
Plain vertical						
height exceeding 1.00 m	-	1.70	21.86	8.23	m^2	30.09
height not exceeding 250 mm	-	0.53	6.82	3.23	m	10.05
height 250 - 500 mm	-	0.91	11.70	6.40	m	18.10
height 500 mm - 1.00 m	-	1.33	17.10	8.23	m	25.33
Soffits of slabs; basic finish						
Slab thickness not exceeding 200 mm						
horizontal; height to soffit not exceeding 1.50 m	-	1.92	24.69	7.64	m^2	32.33
horizontal; height to soffit 1.50 - 3.00 m	-	1.86	23.92	7.75	m^2	31.67
horizontal; height to soffit 1.50 - 3.00 m (based on 5 uses)	-	1.76	22.63	6.42	m^2	29.05
horizontal; height to soffit 1.50 - 3.00 m (based on 6 uses)	-	1.70	21.86	5.53	m^2	27.39
horizontal; height to soffit 3.00 - 4.50 m	-	1.81	23.28	8.02	m^2	31.30
horizontal; height to soffit 4.50 - 6.00 m	-	1.92	24.69	8.29	m^2	32.98
Slab thickness 200 - 300 mm						
horizontal; height to soffit 1.50 - 3.00 m	-	1.92	24.69	9.79	m^2	34.48
Slab thickness 300 - 400 mm						
horizontal; height to soffit 1.50 - 3.00 m	-	1.97	25.33	10.81	m^2	36.14
Slab thickness 400 - 500 mm						
horizontal; height to soffit 1.50 - 3.00 m	-	2.07	26.62	11.84	m^2	38.46
Slab thickness 500 - 600 mm						
horizontal; height to soffit 1.50 - 3.00 m	-	2.23	28.68	11.84	m^2	40.52
Extra over soffits of slabs for						
sloping not exceeding 15 degrees	-	0.21	2.70	-	m^2	2.70
sloping exceeding 15 degrees	-	0.43	5.53	-	m^2	5.53
Soffits of slabs; Expamet "Hy-rib" permanent shuttering and reinforcement or other equal and approved; ref. 2411						
Slab thickness not exceeding 200 mm						
horizontal; height to soffit 1.50 - 3.00 m	-	1.60	20.58	19.32	m^2	39.90

Prices for Measured Works - Minor Works
E IN SITU CONCRETE/LARGE PRECAST CONCRETE

	PC £	Labour hours	Labour £	Material £	Unit	Total rate £
E20 FORMWORK FOR IN SITU CONCRETE - cont'd						
Soffits of slabs; Richard Lees "Ribdeck AL" steel deck permanent shuttering or other equal and approved; 0.90 mm gauge; shot-fired to frame (not included)						
Slab thickness not exceeding 200 mm						
horizontal; height to soffit 1.50 - 3.00 m	-	0.74	9.52	17.14	m²	**26.66**
horizontal; height to soffit 3.00 - 4.50 m	-	0.85	10.93	17.21	m²	**28.14**
horizontal; height to soffit 4.50 - 6.00 m	-	0.95	12.22	17.28	m²	**29.50**
Soffits of slabs; Richard Lees "Ribdeck AL" steel deck permanent shuttering or other equal and approved; 1.20 mm gauge; shot-fired to frame (not included)						
Slab thickness not exceeding 200 mm						
horizontal; height to soffit 1.50 - 3.00 m	-	0.85	10.93	20.25	m²	**31.18**
horizontal; height to soffit 3.00 - 4.50 m	-	0.95	12.22	20.02	m²	**32.24**
horizontal; height to soffit 4.50 - 6.00 m	-	1.12	14.40	20.12	m²	**34.52**
Soffits of slabs; Richard Lees "Super Holorib" steel deck permanent shuttering or other equal and approved; 0.90 mm gauge						
Slab thickness not exceeding 200 mm						
horizontal; height to soffit 1.50 - 3.00 m	-	0.74	9.52	24.22	m²	**33.74**
horizontal; height to soffit 3.00 - 4.50 m	-	0.82	10.55	24.76	m²	**35.31**
horizontal; height to soffit 4.50 - 6.00 m	-	0.95	12.22	25.30	m²	**37.52**
Soffits of slabs; Richard Lees "Super Holorib" steel deck permanent shuttering or other equal and approved; 1.20 mm gauge						
Slab thickness not exceeding 200 mm						
horizontal; height to soffit 1.50 - 3.00 m	-	0.85	10.93	28.85	m²	**39.78**
horizontal; height to soffit 3.00 - 4.50 m	-	0.85	10.93	29.65	m²	**40.58**
horizontal; height to soffit 4.50 - 6.00 m	-	1.12	14.40	30.19	m²	**44.59**
Soffits of landings; basic finish						
Slab thickness not exceeding 200 mm						
horizontal; height to soffit 1.50 - 3.00 m	-	1.92	24.69	8.16	m²	**32.85**
Slab thickness 200 - 300 mm						
horizontal; height to soffit 1.50 - 3.00 m	-	2.02	25.98	10.44	m²	**36.42**
Slab thickness 300 - 400 mm						
horizontal; height to soffit 1.50 - 3.00 m	-	2.07	26.62	11.53	m²	**38.15**
Slab thickness 400 - 500 mm						
horizontal; height to soffit 1.50 - 3.00 m	-	2.18	28.03	12.65	m²	**40.68**
Slab thickness 500 - 600 mm						
horizontal; height to soffit 1.50 - 3.00 m	-	2.34	30.09	12.65	m²	**42.74**
Extra over soffits of landings for						
sloping not exceeding 15 degrees	-	0.21	2.70	-	m²	**2.70**
sloping exceeding 15 degrees	-	0.43	5.53	-	m²	**5.53**
Soffits of coffered or troughed slabs; basic finish Cordek "Correx" trough mould or other equal and approved; 300 mm deep; ribs of mould at 600 mm centres and cross ribs at centres of bay; slab thickness 300 - 400 mm						
horizontal; height to soffit 1.50 - 3.00 m	-	2.66	34.21	12.78	m²	**46.99**
horizontal; height to soffit 3.00 - 4.50 m	-	2.77	35.62	13.04	m²	**48.66**
horizontal; height to soffit 4.50 - 6.00 m	-	2.88	37.04	13.21	m²	**50.25**

Prices for Measured Works - Minor Works

E IN SITU CONCRETE/LARGE PRECAST CONCRETE

	PC £	Labour hours	Labour £	Material £	Unit	Total rate £
Top formwork; basic finish						
Sloping exceeding 15 degrees	-	1.60	20.58	5.98	m²	26.56
Walls; basic finish						
Vertical	-	1.92	24.69	9.79	m²	34.48
Vertical; height exceeding 3.00 m above floor level	-	2.34	30.09	10.06	m²	40.15
Vertical; interrupted	-	2.23	28.68	10.06	m²	38.74
Vertical; to one side only	-	3.73	47.97	12.37	m²	60.34
Vertical; exceeding 3.00 m high; inside stairwell	-	2.34	30.09	10.06	m²	40.15
Battered	-	2.98	38.32	10.55	m²	48.87
Beams; basic finish						
Attached to slabs						
regular shaped; square or rectangular; height to soffit 1.50 - 3.00 m	-	2.34	30.09	9.47	m²	39.56
regular shaped; square or rectangular; height to soffit 3.00 - 4.50 m	-	2.44	31.38	9.79	m²	41.17
regular shaped; spaure or rectangular; height to soffit 4.50 - 6.00 m	-	2.55	32.79	10.06	m²	42.85
Attached to walls						
regular shaped; square or rectangular; height to soffit 1.50 - 3.00 m	-	2.44	31.38	9.47	m²	40.85
Isolated						
regular shaped; square or rectangular; height to soffit 1.50 - 3.00 m	-	2.55	32.79	9.47	m²	42.26
regular shaped; square or rectangular; height to soffit 3.00 - 4.50 m	-	2.66	34.21	9.79	m²	44.00
regular shaped; square or rectangular; height to soffit 4.50 - 6.00 m	-	2.77	35.62	10.06	m²	45.68
Extra over beams for						
regular shaped; sloping not exceeding 15 degrees	-	0.32	4.12	0.97	m²	5.09
regular shaped; sloping exceeding 15 degrees	-	0.64	8.23	1.93	m²	10.16
Beam casings; basic finish						
Attached to slabs						
regular shaped; square or rectangular; height to soffit 1.50 - 3.00 m	-	2.44	31.38	9.47	m²	40.85
regular shaped; square or rectangular; height to soffit 3.00 - 4.50 m	-	2.55	32.79	9.79	m²	42.58
Attached to walls						
regular shaped; square or rectangular; height to soffit 1.50 - 3.00 m	-	2.55	32.79	9.47	m²	42.26
Isolated						
regular shaped; square or rectangular; height to soffit 1.50 - 3.00 m	-	2.66	34.21	9.47	m²	43.68
regular shaped; square or rectangular; height to soffit 3.00 - 4.50 m	-	2.77	35.62	9.79	m²	45.41
Extra over beam casings for						
regular shaped; sloping not exceeding 15 degrees	-	0.32	4.12	0.97	m²	5.09
regular shaped; sloping exceeding 15 degrees	-	0.64	8.23	1.93	m²	10.16
Columns; basic finish						
Attached to walls						
regular shaped; square or rectangular; height to soffit 1.50 - 3.00 m	-	2.34	30.09	8.29	m²	38.38
Isolated						
regular shaped; square or rectangular; height to soffit 1.50 - 3.00 m	-	2.44	31.38	8.29	m²	39.67
regular shaped; circular; not exceeding 300 mm diameter; height to soffit 1.50 - 3.00 m	-	4.26	54.78	13.34	m²	68.12

E IN SITU CONCRETE/LARGE PRECAST CONCRETE

	PC £	Labour hours	Labour £	Material £	Unit	Total rate £
E20 FORMWORK FOR IN SITU CONCRETE - cont'd						
Columns; basic finish - cont'd						
Isolated - cont'd						
regular shaped; circular; 300 - 600 mm diameter; height to soffit 1.50 - 3.00 m	-	3.99	51.31	11.84	m²	**63.15**
regular shaped; circular; 600 - 900 mm diameter; height to soffit 1.50 - 3.00 m	-	3.73	47.97	11.57	m²	**59.54**
Column casings; basic finish						
Attached to walls						
regular shaped; square or rectangular; height to soffit; 1.50 - 3.00 m	-	2.44	31.38	8.29	m²	**39.67**
Isolated						
regular shaped; square or rectangular; height to soffit 1.50 - 3.00 m	-	2.55	32.79	8.29	m²	**41.08**
Recesses or rebates						
12 x 12 mm	-	0.07	0.90	0.08	m	**0.98**
25 x 25 mm	-	0.07	0.90	0.13	m	**1.03**
25 x 50 mm	-	0.07	0.90	0.17	m	**1.07**
50 x 50 mm	-	0.07	0.90	0.28	m	**1.18**
Nibs						
50 x 50 mm	-	0.58	7.46	1.00	m	**8.46**
100 x 100 mm	-	0.83	10.67	1.33	m	**12.00**
100 x 200 mm	-	1.11	14.27	2.43	m	**16.70**
Extra over a basic finish for fine formed finishes						
Slabs	-	0.35	4.50	-	m²	**4.50**
Walls	-	0.35	4.50	-	m²	**4.50**
Beams	-	0.35	4.50	-	m²	**4.50**
Columns	-	0.35	4.50	-	m²	**4.50**
Add to prices for basic formwork for						
Curved radius 6.00 m - 50%						
Curved radius 2.00 m - 100%						
Coating with retardant agent	-	0.01	0.13	0.50	m²	**0.63**
Wall kickers; basic finish						
Height 150 mm	-	0.53	6.82	2.39	m	**9.21**
Height 225 mm	-	0.69	8.87	2.94	m	**11.81**
Suspended wall kickers; basic finish						
Height 150 mm	-	0.67	8.62	1.95	m	**10.57**
Wall ends, soffits and steps in walls; basic finish						
Plain						
width exceeding 1.00 m	-	2.02	25.98	9.79	m²	**35.77**
width not exceeding 250 mm	-	0.64	8.23	2.48	m	**10.71**
width 250 - 500 mm	-	1.01	12.99	5.43	m	**18.42**
width 500 mm - 1.00 m	-	1.60	20.58	9.79	m	**30.37**
Openings in walls						
Plain						
width exceeding 1.00 m	-	2.23	28.68	9.79	m²	**38.47**
width not exceeding 250 mm	-	0.69	8.87	2.48	m	**11.35**
width 250 - 500 mm	-	1.17	15.05	5.43	m	**20.48**
width 500 mm - 1.00 m	-	1.81	23.28	9.79	m	**33.07**

E IN SITU CONCRETE/LARGE PRECAST CONCRETE

	PC £	Labour hours	Labour £	Material £	Unit	Total rate £
Stairflights						
Width 1.00 m; 150 mm waist; 150 mm undercut risers string, width 300 mm	-	5.32	68.42	21.48	m	89.90
Width 2.00 m; 300 mm waist; 150 mm undercut risers string, width 350 mm	-	9.57	123.07	37.59	m	160.66
Mortices						
Girth not exceeding 500 mm						
depth not exceeding 250 mm; circular	-	0.16	2.06	0.82	nr	2.88
Holes						
Girth not exceeding 500 mm						
depth not exceeding 250 mm; circular	-	0.21	2.70	1.07	nr	3.77
depth 250 - 500 mm; circular	-	0.33	4.24	2.60	nr	6.84
Girth 500 mm - 1.00 m						
depth not exceeding 250 mm; circular	-	0.27	3.47	1.80	nr	5.27
depth 250 - 500 mm; circular	-	0.41	5.27	4.82	nr	10.09
Girth 1.00 - 2.00 m						
depth not exceeding 250 mm; circular	-	0.48	6.17	4.82	nr	10.99
depth 250 - 500 mm; circular	-	0.71	9.13	10.09	nr	19.22
Girth 2.00 - 3.00 m						
depth not exceeding 250 mm; circular	-	0.64	8.23	9.70	nr	17.93
depth 250 - 500 mm; circular	-	0.95	12.22	17.27	nr	29.49
E30 REINFORCEMENT FOR IN SITU CONCRETE						
Bar; BS 4449; hot rolled plain round mild steel bars						
40 mm diameter nominal size						
straight or bent	319.85	22.00	220.17	388.36	t	608.53
curved	348.93	22.00	220.17	421.18	t	641.35
32 mm diameter nominal size						
straight or bent	295.33	23.00	230.46	361.83	t	592.29
curved	322.18	23.00	230.46	392.14	t	622.60
25 mm diameter nominal size						
straight or bent	290.22	25.00	251.04	356.04	t	607.08
curved	316.60	25.00	251.04	385.81	t	636.85
20 mm diameter nominal size						
straight or bent	287.15	27.00	271.62	355.35	t	626.97
curved	313.26	27.00	271.62	384.82	t	656.44
16 mm diameter nominal size						
straight or bent	287.15	29.00	292.20	359.25	t	651.45
curved	313.26	29.00	292.20	388.73	t	680.93
12 mm diameter nominal size						
straight or bent	292.26	31.00	312.78	368.93	t	681.71
curved	318.83	31.00	312.78	398.92	t	711.70
10 mm diameter nominal size						
straight or bent	298.39	33.00	333.36	382.50	t	715.86
curved	325.52	33.00	333.36	413.12	t	746.48
8 mm diameter nominal size						
straight or bent	308.61	35.00	351.87	397.94	t	749.81
links	308.61	38.00	382.74	400.24	t	782.98
curved	336.67	35.00	351.87	429.61	t	781.48
6 mm diameter nominal size						
straight or bent	336.21	40.00	403.32	429.09	t	832.41
links	336.21	44.00	444.48	429.09	t	873.57
curved	366.77	40.00	403.32	463.59	t	866.91

Prices for Measured Works - Minor Works
E IN SITU CONCRETE/LARGE PRECAST CONCRETE

	PC £	Labour hours	Labour £	Material £	Unit	Total rate £
E30 REINFORCEMENT FOR IN SITU CONCRETE						
- cont'd						
Bar; BS 4449; hot rolled deformed high steel bars grade 460						
40 mm diameter nominal size						
straight or bent	292.26	22.00	220.17	357.21	t	577.38
curved	318.83	22.00	220.17	387.21	t	607.38
32 mm diameter nominal size						
straight or bent	287.15	23.00	230.46	352.60	t	583.06
curved	313.26	23.00	230.46	382.07	t	612.53
25 mm diameter nominal size						
straight or bent	282.04	25.00	251.04	346.81	t	597.85
curved	307.68	25.00	251.04	375.75	t	626.79
20 mm diameter nominal size						
straight or bent	284.09	27.00	271.62	351.90	t	623.52
curved	309.91	27.00	271.62	381.04	t	652.66
16 mm diameter nominal size						
straight or bent	284.09	29.00	292.20	355.80	t	648.00
curved	309.91	29.00	292.20	384.94	t	677.14
12 mm diameter nominal size						
straight or bent	287.15	31.00	312.78	363.16	t	675.94
curved	313.26	31.00	312.78	392.63	t	705.41
10 mm diameter nominal size						
straight or bent	292.26	33.00	333.36	375.58	t	708.94
curved	318.83	33.00	333.36	405.57	t	738.93
8 mm diameter nominal size						
straight or bent	303.50	35.00	351.87	392.17	t	744.04
links	303.50	38.00	382.74	394.48	t	777.22
curved	331.10	35.00	351.87	423.32	t	775.19
Bar; stainless steel						
32 mm diameter nominal size						
straight or bent	2454.65	23.00	230.46	2667.23	t	2897.69
curved	2548.59	23.00	230.46	2768.21	t	2998.67
25 mm diameter nominal size						
straight or bent	2454.65	25.00	251.04	2667.20	t	2918.24
curved	2548.59	25.00	251.04	2768.19	t	3019.23
20 mm diameter nominal size						
straight or bent	2454.65	27.00	271.62	2669.98	t	2941.60
curved	2548.59	27.00	271.62	2770.96	t	3042.58
16 mm diameter nominal size						
straight or bent	2003.74	29.00	292.20	2189.15	t	2481.35
curved	2097.68	29.00	292.20	2290.14	t	2582.34
12 mm diameter nominal size						
straight or bent	2003.74	31.00	312.78	2193.06	t	2505.84
curved	2128.68	31.00	312.78	2327.37	t	2640.15
10 mm diameter nominal size						
straight or bent	2003.74	33.00	333.36	2199.71	t	2533.07
curved	2144.65	33.00	333.36	2351.19	t	2684.55
8 mm diameter nominal size						
straight or bent	2003.74	35.00	351.87	2203.62	t	2555.49
curved	2175.65	35.00	351.87	2388.42	t	2740.29
Fabric; BS 4483						
Ref A98 (1.54 kg/m^2)						
400 mm minimum laps	0.75	0.13	1.34	0.89	m^2	2.23
strips in one width; 600 mm width	0.75	0.16	1.65	0.89	m^2	2.54
strips in one width; 900 mm width	0.75	0.15	1.54	0.89	m^2	2.43
strips in one width; 1200 mm width	0.75	0.14	1.44	0.89	m^2	2.33

E IN SITU CONCRETE/LARGE PRECAST CONCRETE

	PC £	Labour hours	Labour £	Material £	Unit	Total rate £
Ref A142 (2.22 kg/m^2)						
400 mm minimum laps	0.78	0.13	1.34	0.92	m^2	2.26
strips in one width; 600 mm width	0.78	0.16	1.65	0.92	m^2	2.57
strips in one width; 900 mm width	0.78	0.15	1.54	0.92	m^2	2.46
strips in one width; 1200 mm width	0.78	0.14	1.44	0.92	m^2	2.36
Ref A193 (3.02 kg/m^2)						
400 mm minimum laps	1.06	0.13	1.34	1.25	m^2	2.59
strips in one width; 600 mm width	1.06	0.16	1.65	1.25	m^2	2.90
strips in one width; 900 mm width	1.06	0.15	1.54	1.25	m^2	2.79
strips in one width; 1200 mm width	1.06	0.14	1.44	1.25	m^2	2.69
Ref A252 (3.95 kg/m^2)						
400 mm minimum laps	1.39	0.14	1.44	1.64	m^2	3.08
strips in one width; 600 mm width	1.39	0.17	1.75	1.64	m^2	3.39
strips in one width; 900 mm width	1.39	0.16	1.65	1.64	m^2	3.29
strips in one width; 1200 mm width	1.39	0.15	1.54	1.64	m^2	3.18
Ref A393 (6.16 kg/m^2)						
400 mm minimum laps	2.13	0.16	1.65	2.52	m^2	4.17
strips in one width; 600 mm width	2.13	0.19	1.96	2.52	m^2	4.48
strips in one width; 900 mm width	2.13	0.18	1.85	2.52	m^2	4.37
strips in one width; 1200 mm width	2.13	0.17	1.75	2.52	m^2	4.27
Ref B196 (3.05 kg/m^2)						
400 mm minimum laps	1.10	0.13	1.34	1.30	m^2	2.64
strips in one width; 600 mm width	1.10	0.16	1.65	1.30	m^2	2.95
strips in one width; 900 mm width	1.10	0.15	1.54	1.30	m^2	2.84
strips in one width; 1200 mm width	1.10	0.14	1.44	1.30	m^2	2.74
Ref B283 (3.73 kg/m^2)						
400 mm minimum laps	1.33	0.13	1.34	1.57	m^2	2.91
strips in one width; 600 mm width	1.33	0.16	1.65	1.57	m^2	3.22
strips in one width; 900 mm width	1.33	0.15	1.54	1.57	m^2	3.11
strips in one width; 1200 mm width	1.33	0.14	1.44	1.57	m^2	3.01
Ref B385 (4.53 kg/m^2)						
400 mm minimum laps	1.62	0.14	1.44	1.92	m^2	3.36
strips in one width; 600 mm width	1.62	0.17	1.75	1.92	m^2	3.67
strips in one width; 900 mm width	1.62	0.16	1.65	1.92	m^2	3.57
strips in one width; 1200 mm width	1.62	0.15	1.54	1.92	m^2	3.46
Ref B503 (5.93 kg/m^2)						
400 mm minimum laps	2.13	0.16	1.65	2.52	m^2	4.17
strips in one width; 600 mm width	2.13	0.19	1.96	2.52	m^2	4.48
strips in one width; 900 mm width	2.13	0.18	1.85	2.52	m^2	4.37
strips in one width; 1200 mm width	2.13	0.17	1.75	2.52	m^2	4.27
Ref B785 (8.14 kg/m^2)						
400 mm minimum laps	2.91	0.19	1.96	3.44	m^2	5.40
strips in one width; 600 mm width	2.91	0.22	2.26	3.44	m^2	5.70
strips in one width; 900 mm width	2.91	0.21	2.16	3.44	m^2	5.60
strips in one width; 1200 mm width	2.91	0.20	2.06	3.44	m^2	5.50
Ref B1131 (10.90 kg/m^2)						
400 mm minimum laps	3.79	0.20	2.06	4.48	m^2	6.54
strips in one width; 600 mm width	3.79	0.26	2.68	4.48	m^2	7.16
strips in one width; 900 mm width	3.79	0.24	2.47	4.48	m^2	6.95
strips in one width; 1200 mm width	3.79	0.22	2.26	4.48	m^2	6.74
Ref C385 (3.41 kg/m^2)						
400 mm minimum laps	1.26	0.13	1.34	1.49	m^2	2.83
strips in one width; 600 mm width	1.26	0.16	1.65	1.49	m^2	3.14
strips in one width; 900 mm width	1.26	0.15	1.54	1.49	m^2	3.03
strips in one width; 1200 mm width	1.26	0.14	1.44	1.49	m^2	2.93
Ref C503 (4.34 kg/m^2)						
400 mm minimum laps	1.60	0.14	1.44	1.89	m^2	3.33
strips in one width; 600 mm width	1.60	0.17	1.75	1.89	m^2	3.64
strips in one width; 900 mm width	1.60	0.16	1.65	1.89	m^2	3.54
strips in one width; 1200 mm width	1.60	0.15	1.54	1.89	m^2	3.43

E IN SITU CONCRETE/LARGE PRECAST CONCRETE

	PC £	Labour hours	Labour £	Material £	Unit	Total rate £
E30 REINFORCEMENT FOR IN SITU CONCRETE - cont'd						
Fabric; BS 4483 - cont'd						
Ref C636 (5.55 kg/m^2)						
400 mm minimum laps	2.04	0.15	1.54	2.41	m^2	3.95
strips in one width; 600 mm width	2.04	0.18	1.85	2.41	m^2	4.26
strips in one width; 900 mm width	2.04	0.17	1.75	2.41	m^2	4.16
strips in one width; 1200 mm width	2.04	0.16	1.65	2.41	m^2	4.06
Ref C785 (6.72 kg/m^2)						
400 mm minimum laps	2.48	0.15	1.54	2.93	m^2	4.47
strips in one width; 600 mm width	2.48	0.18	1.85	2.93	m^2	4.78
strips in one width; 900 mm width	2.48	0.17	1.75	2.93	m^2	4.68
strips in one width; 1200 mm width	2.48	0.16	1.65	2.93	m^2	4.58
Ref D49 (0.77 kg/m^2)						
100 mm minimum laps; bent	0.87	0.26	2.68	1.03	m^2	3.71
Ref D98 (1.54 kg/m^2)						
200 mm minimum laps; bent	0.65	0.26	2.68	0.77	m^2	3.45
E40 DESIGNED JOINTS IN IN SITU CONCRETE						
Formed; Fosroc Expandite "Flexcell" impregnated fibreboard joint filler or other equal and approved						
Width not exceeding 150 mm						
12.50 mm thick	-	0.16	2.06	1.36	m	3.42
20 mm thick	-	0.20	2.57	1.97	m	4.54
25 mm thick	-	0.20	2.57	2.24	m	4.81
Width 150 - 300 mm						
12.50 mm thick	-	0.20	2.57	2.08	m	4.65
20 mm thick	-	0.26	3.34	3.03	m	6.37
25 mm thick	-	0.26	3.34	3.50	m	6.84
Width 300 - 450 mm						
12.50 mm thick	-	0.26	3.34	3.11	m	6.45
20 mm thick	-	0.31	3.99	4.54	m	8.53
25 mm thick	-	0.31	3.99	5.26	m	9.25
Formed; Grace Servicised "Kork-pak" waterproof bonded cork joint filler board or other equal and approved						
Width not exceeding 150 mm						
10 mm thick	-	0.16	2.06	2.92	m	4.98
13 mm thick	-	0.16	2.06	2.96	m	5.02
19 mm thick	-	0.16	2.06	3.84	m	5.90
25 mm thick	-	0.16	2.06	4.37	m	6.43
Width 150 - 300 mm						
10 mm thick	-	0.20	2.57	5.47	m	8.04
13 mm thick	-	0.20	2.57	5.55	m	8.12
19 mm thick	-	0.20	2.57	7.31	m	9.88
25 mm thick	-	0.20	2.57	8.37	m	10.94
Width 300 - 450 mm						
10 mm thick	-	0.26	3.34	8.31	m	11.65
13 mm thick	-	0.26	3.34	8.43	m	11.77
19 mm thick	-	0.26	3.34	11.07	m	14.41
25 mm thick	-	0.26	3.34	12.66	m	16.00
Sealants; Fosroc Expandite "Pliastic 77" hot poured rubberized bituminous compound or other equal and approved						
Width 10 mm						
25 mm depth	-	0.19	2.44	0.86	m	3.30

E IN SITU CONCRETE/LARGE PRECAST CONCRETE

	PC £	Labour hours	Labour £	Material £	Unit	Total rate £
Width 12.50 mm						
25 mm depth	-	0.20	2.57	1.06	m	3.63
Width 20 mm						
25 mm depth	-	0.21	2.70	1.73	m	4.43
Width 25 mm						
25 mm depth	-	0.22	2.83	2.12	m	4.95
Sealants; Fosroc Expandite "Thioflex 600" gun grade two part polysulphide or other equal and approved						
Width 10 mm						
25 mm depth	-	0.06	0.77	3.29	m	4.06
Width 12.50 mm						
25 mm depth	-	0.07	0.90	4.11	m	5.01
Width 20 mm						
25 mm depth	-	0.08	1.03	6.57	m	7.60
Width 25 mm						
25 mm depth	-	0.09	1.16	8.22	m	9.38
Sealants; Grace Servicised "Paraseal" polysulphide compound or other equal and approved; priming with Grace Servicised "Servicised P" or other equal and approved						
Width 10 mm						
25 mm depth	-	0.20	1.64	2.10	m	3.74
Width 13 mm						
25 mm depth	-	0.20	1.64	2.68	m	4.32
Width 19 mm						
25 mm depth	-	0.26	2.14	3.86	m	6.00
Width 25 mm						
25 mm depth	-	0.26	2.14	5.03	m	7.17
Waterstops						
PVC water stop; flat dumbbell type; heat welded joints; cast into concrete						
100 mm wide	1.89	0.20	2.06	2.25	m	4.31
flat angle	7.88	0.26	2.68	7.30	nr	9.98
vertical angle	11.85	0.26	2.68	11.67	nr	14.35
flat three way intersection	10.19	0.36	3.70	10.20	nr	13.90
vertical three way intersection	14.00	0.36	3.70	14.40	nr	18.10
four way intersection	10.54	0.43	4.42	10.93	nr	15.35
170 mm wide	2.84	0.26	2.68	3.32	m	6.00
flat angle	9.04	0.31	3.19	8.17	nr	11.36
vertical angle	12.99	0.31	3.19	12.53	nr	15.72
flat three way intersection	11.52	0.41	4.22	11.59	nr	15.81
vertical three way intersection	16.28	0.41	4.22	16.83	nr	21.05
four way intersection	12.58	0.51	5.25	13.42	nr	18.67
210 mm wide	3.70	0.26	2.68	4.29	m	6.97
flat angle	11.21	0.31	3.19	10.42	nr	13.61
vertical angle	12.36	0.31	3.19	11.70	nr	14.89
flat three way intersection	14.63	0.41	4.22	15.27	nr	19.49
vertical three way intersection	11.82	0.41	4.22	12.17	nr	16.39
four way intersection	16.02	0.51	5.25	17.87	nr	23.12
250 mm wide	4.90	0.31	3.19	5.65	m	8.84
flat angle	13.48	0.36	3.70	12.83	nr	16.53
vertical angle	13.90	0.36	3.70	13.29	nr	16.99
flat three way intersection	17.43	0.45	4.63	18.88	nr	23.51
vertical three way intersection	13.27	0.45	4.63	14.29	nr	18.92
four way intersection	19.34	0.56	5.76	22.66	nr	28.42

Prices for Measured Works - Minor Works
E IN SITU CONCRETE/LARGE PRECAST CONCRETE

	PC £	Labour hours	Labour £	Material £	Unit	Total rate £
E40 DESIGNED JOINTS IN IN SITU CONCRETE - cont'd						
Waterstops - cont'd						
PVC water stop; centre bulb type; heat welded joints; cast into concrete						
160 mm wide	3.07	0.26	2.68	3.58	m	6.26
flat angle	9.31	0.31	3.19	8.32	nr	11.51
vertical angle	13.46	0.31	3.19	12.89	nr	16.08
flat three way intersection	11.80	0.41	4.22	11.80	nr	16.02
vertical three way intersection	12.90	0.41	4.22	13.00	nr	17.22
four way intersection	12.90	0.51	5.25	13.72	nr	18.97
210 mm wide	4.39	0.26	2.68	5.07	m	7.75
flat angle	11.83	0.31	3.19	10.73	nr	13.92
vertical angle	14.97	0.31	3.19	14.19	nr	17.38
flat three way intersection	15.16	0.41	4.22	15.68	nr	19.90
vertical three way intersection	14.32	0.41	4.22	14.75	nr	18.97
four way intersection	16.56	0.51	5.25	18.48	nr	23.73
260 mm wide	5.13	0.31	3.19	5.90	m	9.09
flat angle	13.78	0.36	3.70	13.07	nr	16.77
vertical angle	14.09	0.36	3.70	13.40	nr	17.10
flat three way intersection	17.77	0.45	4.63	19.23	nr	23.86
vertical three way intersection	13.54	0.45	4.63	14.57	nr	19.20
four way intersection	19.68	0.56	5.76	23.09	nr	28.85
325 mm wide	11.39	0.36	3.70	12.97	m	16.67
flat angle	28.17	0.41	4.22	28.83	nr	33.05
vertical angle	20.90	0.41	4.22	20.82	nr	25.04
flat three way intersection	31.81	0.51	5.25	38.03	nr	43.28
vertical three way intersection	20.27	0.51	5.25	25.30	nr	30.55
four way intersection	35.13	0.61	6.28	46.84	nr	53.12
E41 WORKED FINISHES/CUTTING TO IN SITU CONCRETE						
Worked finishes						
Tamping by mechanical means	-	0.02	0.16	0.10	m^2	0.26
Power floating	-	0.18	1.48	0.34	m^2	1.82
Trowelling	-	0.33	2.71	-	m^2	2.71
Hacking						
by mechanical means	25.85	0.33	2.71	0.31	m^2	3.02
by hand	-	0.71	5.84	-	m^2	5.84
Lightly shot blasting surface of concrete	-	0.41	3.37	-	m^2	3.37
Blasting surface of concrete						
to produce textured finish	-	0.71	5.84	0.65	m^2	6.49
Wood float finish	-	0.13	1.07	-	m^2	1.07
Tamped finish						
level or to falls	-	0.07	0.58	-	m^2	0.58
to falls	-	0.10	0.82	-	m^2	0.82
Spade finish	-	0.16	1.32	-	m^2	1.32
Cutting chases						
Depth not exceeding 50 mm						
width 10 mm	-	0.33	2.71	0.85	m	3.56
width 50 mm	-	0.51	4.19	1.00	m	5.19
width 75 mm	-	0.68	5.59	1.15	m	6.74
Depth 50 - 100 mm						
width 75 mm	-	0.92	7.56	1.82	m	9.38
width 100 mm	-	1.02	8.38	1.91	m	10.29
width 100 mm; in reinforced concrete	-	1.53	12.58	2.96	m	15.54

E IN SITU CONCRETE/LARGE PRECAST CONCRETE

	PC £	Labour hours	Labour £	Material £	Unit	Total rate £
Depth 100 - 150 mm						
width 100 mm	-	1.32	10.85	2.30	m	**13.15**
width 100 mm; in reinforced concrete	-	2.04	16.77	3.76	m	**20.53**
width 150 mm	-	1.63	13.40	2.58	m	**15.98**
width 150 mm; in reinforced concrete	-	2.44	20.06	4.19	m	**24.25**
Cutting rebates						
Depth not exceeding 50 mm						
width 50 mm	-	0.51	4.19	1.00	m	**5.19**
Depth 50 - 100 mm						
width 100 mm	-	1.02	8.38	1.91	m	**10.29**

NOTE: The following rates for cutting mortices and holes in reinforced concrete allow for diamond drilling.

	PC £	Labour hours	Labour £	Material £	Unit	Total rate £
Cutting mortices						
Depth not exceeding 100 mm						
cross sectional size 20 mm diameter; making good	-	0.16	1.32	0.10	nr	**1.42**
cross sectional size 50 mm diameter; making good	-	0.18	1.48	0.13	nr	**1.61**
cross sectional size 150 mm x 150 mm; making good	-	0.36	2.96	0.33	nr	**3.29**
cross sectional size 300 mm x 300 mm; making good	-	0.71	5.84	0.71	nr	**6.55**
cross sectional size 50 mm diameter; in reinforced concrete; making good	-	-	-	-	nr	**8.8**
cross sectional size 50 mm - 75 mm diameter; in reinforced concrete; making good	-	-	-	-	nr	**10.24**
cross sectional size 75 mm - 100 mm diameter; in reinforced concrete; making good	-	-	-	-	nr	**10.67**
cross sectional size 100 mm - 125 mm diameter; in reinforced concrete; making good	-	-	-	-	nr	**11.73**
cross sectional size 125 mm - 150 mm diameter; in reinforced concrete; making good	-	-	-	-	nr	**13.21**
cross sectional size 150 mm x 150 mm; in reinforced concrete; making good	-	0.56	4.60	0.52	nr	**5.12**
cross sectional size 300 mm x 300 mm; in reinforced concrete; making good	-	1.06	8.71	1.00	nr	**9.71**
Depth 100 - 200 mm						
cross sectional size 50 mm diameter; in reinforced concrete; making good	-	-	-	-	nr	**12.29**
cross sectional size 50 mm - 75 mm diameter; in reinforced concrete; making good	-	-	-	-	nr	**15.19**
cross sectional size 75 mm - 100 mm diameter; in reinforced concrete; making good	-	-	-	-	nr	**15.4**
cross sectional size 100 mm - 125 mm diameter; in reinforced concrete; making good	-	-	-	-	nr	**17.62**
cross sectional size 125 mm - 150 mm diameter; in reinforced concrete; making good	-	-	-	-	nr	**19.81**
Depth 200 - 300 mm						
cross sectional size 50 mm diameter; in reinforced concrete; making good	-	-	-	-	nr	**24.61**
cross sectional size 50 mm - 75 mm diameter; in reinforced concrete; making good	-	-	-	-	nr	**28.19**
cross sectional size 75 mm - 100 mm diameter; in reinforced concrete; making good	-	-	-	-	nr	**30.83**
cross sectional size; 100 mm - 125 mm diameter; in reinforced concrete; making good	-	-	-	-	nr	**35.18**
cross sectional size; 125 mm - 150 mm diameter; in reinforced concrete; making good	-	-	-	-	nr	**39.59**

	PC £	Labour hours	Labour £	Material £	Unit	Total rate £
E41 WORKED FINISHES/CUTTING TO IN SITU CONCRETE - cont'd						
Cutting mortices - cont'd						
Depth exceeding 300 mm; 400 mm depth						
cross sectional size 50 mm diameter; in reinforced concrete; making good	-	-	-	-	nr	32.81
cross sectional size 50 mm - 75 mm diameter; in reinforced concrete; making good	-	-	-	-	nr	37.54
cross sectional size 75 mm - 100 mm diameter; in reinforced concrete; making good	-	-	-	-	nr	41.01
cross sectional size 100 mm - 125 mm diameter; in reinforced concrete; making good	-	-	-	-	nr	46.91
cross sectional size 125 mm - 150 mm diameter; in reinforced concrete; making good	-	-	-	-	nr	52.8
Depth exceeding 300 mm; 500 mm depth						
cross sectional size 50 mm diameter; in reinforced concrete; making good	-	-	-	-	nr	41.01
cross sectional size 50 mm - 75 mm diameter; in reinforced concrete; making good	-	-	-	-	nr	46.91
cross sectional size 75 mm - 100 mm diameter; in reinforced concrete; making good	-	-	-	-	nr	51.31
cross sectional size 100 mm - 125 mm diameter; in reinforced concrete; making good	-	-	-	-	nr	58.63
cross sectional size 125 mm - 150 mm diameter; in reinforced concrete; making good	-	-	-	-	nr	65.94
Depth exceeding 300 mm; 600 mm depth						
cross sectional size 50 mm diameter; in reinforced concrete; making good	-	-	-	-	nr	49.28
cross sectional size 50 mm - 75 mm diameter; in reinforced concrete; making good	-	-	-	-	nr	56.27
cross sectional size 75 mm - 100 mm diameter; in reinforced concrete; making good	-	-	-	-	nr	61.55
cross sectional size 100 mm - 125 mm diameter; in reinforced concrete; making good	-	-	-	-	nr	70.35
cross sectional size 125 mm - 150 mm diameter; in reinforced concrete; making good	-	-	-	-	nr	79.17
Cutting holes						
Depth not exceeding 100 mm						
cross sectional size 50 mm diameter; making good	-	0.36	2.96	0.34	nr	3.30
cross sectional size 50 mm diameter; in reinforced concrete; making good	-	0.56	4.60	0.54	nr	5.14
cross sectional size 100 mm diameter; making good	-	0.41	3.37	0.39	nr	3.76
cross sectional size 100 mm diameter; in reinforced concrete; making good	-	0.61	5.01	0.59	nr	5.60
cross sectional size 150 mm x 150 mm; making good	-	0.45	3.70	0.44	nr	4.14
cross sectional size 150 mm x 150 mm; in reinforced concrete; making good	-	0.71	5.84	0.68	nr	6.52
cross sectional size 300 mm x 300 mm; making good	-	0.56	4.60	0.54	nr	5.14
cross sectional size 300 mm x 300 mm; in reinforced concrete; making good	-	0.87	7.15	0.83	nr	7.98
Depth 100 - 200 mm						
cross sectional size 50 mm diameter; making good	-	0.51	4.19	0.49	nr	4.68
cross sectional size 50 mm diameter; in reinforced concrete; making good	-	0.77	6.33	0.73	nr	7.06
cross sectional size 100 mm diameter; making good	-	0.61	5.01	0.59	nr	5.60
cross sectional size 100 mm diameter; in reinforced concrete; making good	-	0.92	7.56	0.88	nr	8.44
cross sectional size 150 mm x 150 mm; making good	-	0.77	6.33	0.73	nr	7.06

E IN SITU CONCRETE/LARGE PRECAST CONCRETE

	PC £	Labour hours	Labour £	Material £	Unit	Total rate £
Depth 100 - 200 mm - cont'd						
cross sectional size 150 mm x 150 mm; in						
reinforced concrete; making good	-	1.17	9.62	1.13	nr	10.75
cross sectional size 300 mm x 300 mm; making good	-	0.97	7.97	0.93	nr	8.90
cross sectional size 300 mm x 300 mm; in						
reinforced concrete; making good	-	1.48	12.17	1.42	nr	13.59
cross sectional size not exceeding 0.10 m²; in						
reinforced concrete; making good	-	-	-	-	nr	102.61
cross sectional size 0.10 m² - 0.20 m²; in						
reinforced concrete; making good	-	-	-	-	nr	205.17
cross sectional size 0.20 m² - 0.30 m²; in						
reinforced concrete; making good	-	-	-	-	nr	231.54
cross sectional size 0.30 m² - 0.40 m²; in						
reinforced concrete; making good	-	-	-	-	nr	269.71
cross sectional size 0.40 m² - 0.50 m²; in						
reinforced concrete; making good	-	-	-	-	nr	309.22
cross sectional size 0.50 m² - 0.60 m²; in						
reinforced concrete; making good	-	-	-	-	nr	360.53
cross sectional size 0.60 m² - 0.70 m²; in						
reinforced concrete; making good	-	-	-	-	nr	412.9
cross sectional size 0.70 m² - 0.80 m²; in						
reinforced concrete; making good	-	-	-	-	nr	473.44
Depth 200 - 300 mm						
cross sectional size 50 mm diameter; making good	-	0.77	6.33	0.73	nr	7.06
cross sectional size 50 mm diameter; in						
reinforced concrete; making good	-	1.17	9.62	1.13	nr	10.75
cross sectional size 100 mm diameter; making good	-	0.92	7.56	0.88	nr	8.44
cross sectional size 100 mm diameter; in						
reinforced concrete; making good	-	1.38	11.34	1.32	nr	12.66
cross sectional size 150 mm x 150 mm; making good	-	1.12	9.21	1.08	nr	10.29
cross sectional size 150 mm x 150 mm; in						
reinforced concrete; making good	-	1.68	13.81	1.61	nr	15.42
cross sectional size 300 mm x 300 mm; making good	-	1.43	11.75	1.37	nr	13.12
cross sectional size 300 mm x 300 mm; in						
reinforced concrete; making good	-	2.14	17.59	2.05	nr	19.64
E42 ACCESSORIES CAST INTO IN SITU CONCRETE						
Foundation bolt boxes						
Temporary plywood; for group of 4 nr bolts						
75 mm x 75 mm x 150 mm	-	0.45	5.79	1.01	nr	6.80
75 mm x 75 mm x 250 mm	-	0.45	5.79	1.25	nr	7.04
Expanded metal, Expamet Building Products						
Ltd or other equal and approved						
75 mm diameter x 150 mm long	-	0.31	3.99	1.27	nr	5.26
75 mm diameter x 300 mm long	-	0.31	3.99	1.74	nr	5.73
100 mm diameter x 450 mm long	-	0.31	3.99	3.09	nr	7.08
Foundation bolts and nuts						
Black hexagon						
10 mm diameter x 100 mm long	-	0.26	3.34	0.52	nr	3.86
12 mm diameter x 120 mm long	-	0.26	3.34	0.79	nr	4.13
16 mm diameter x 160 mm long	-	0.31	3.99	2.18	nr	6.17
20 mm diameter x 180 mm long	-	0.31	3.99	2.54	nr	6.53
Masonry slots						
Galvanised steel; dovetail slots; 1.20 mm thick; 18G						
75 mm long	-	0.08	1.03	0.19	nr	1.22
100 mm long	-	0.08	1.03	0.21	nr	1.24
150 mm long	-	0.09	1.16	0.26	nr	1.42
225 mm long	-	0.10	1.29	0.36	nr	1.65

Prices for Measured Works - Minor Works
E IN SITU CONCRETE/LARGE PRECAST CONCRETE

	PC £	Labour hours	Labour £	Material £	Unit	Total rate £
E42 ACCESSORIES CAST INTO IN SITU CONCRETE - cont'd						
Masonry slots - cont'd						
Galvanised steel; metal insert slots; Halfen Ltd or other equal and approved; 2.50 mm thick; end caps and foam filling						
41 mm x 41 mm; ref P3270	-	0.41	5.27	8.92	m	14.19
41 mm x 41 mm x 75 mm; ref P3249	-	0.10	1.29	1.78	nr	3.07
41 mm x 41 mm x 100 mm; ref P3250	-	0.10	1.29	2.21	nr	3.50
41 mm x 41 mm x 150 mm; ref P3251	-	0.10	1.29	2.67	nr	3.96
Cramps						
Mild steel; once bent; one end shot fired into concrete; other end flanged and built into brickwork joint						
200 mm girth	-	0.16	2.08	0.49	nr	2.57
Column guards						
White nylon coated steel; "Rigifix"; Huntley and Sparks Ltd or other equal and approved; plugging; screwing to concrete; 1.50 mm thick						
75 mm x 75 mm x 1000 mm	-	0.81	10.42	17.23	nr	27.65
Galvanised steel; "Rigifix"; Huntley and Sparks Ltd or other equal and approved; 3 mm thick						
75 mm x 75 mm x 1000 mm	-	0.61	7.84	12.13	nr	19.97
Galvanised steel; "Rigifix"; Huntley and Sparks Ltd or other equal and approved; 4.50 mm thick						
75 mm x 75 mm x 1000 mm	-	0.61	7.84	16.31	nr	24.15
Stainless steel; "HKW"; Halfen Ltd or other equal and approved; 5 mm thick						
50 mm x 50 mm x 1500 mm	-	1.02	13.12	47.27	nr	60.39
50 mm x 50 mm x 2000 mm	-	1.22	15.69	68.15	nr	83.84
E60 PRECAST/COMPOSITE CONCRETE DECKING						
Prestressed precast flooring planks; Bison "Drycast" or other equal and approved; cement:sand (1:3) grout between planks and on prepared bearings						
100 mm thick suspended slabs; horizontal						
600 mm wide planks	-	-	-	-	m^2	38.45
1200 mm wide planks	-	-	-	-	m^2	36.47
150 mm thick suspended slabs; horizontal						
1200 mm wide planks	-	-	-	-	m^2	38.03

F MASONRY

F10 BRICK/BLOCK WALLING

	PC £	Labour hours	Labour £	Material £	Unit	Total rate £
Common bricks; PC £189.70/1000; in cement mortar (1:3)						
Walls						
half brick thick	-	1.06	17.24	14.20	m²	31.44
half brick thick; building against other work; concrete	-	1.16	18.86	15.52	m²	34.38
half brick thick; building overhand	-	1.34	21.79	14.20	m²	35.99
half brick thick; curved; 6.00 m radii	-	1.39	22.60	14.20	m²	36.80
half brick thick; curved; 1.50 m radii	-	1.80	29.27	16.24	m²	45.51
one brick thick	-	1.80	29.27	28.41	m²	57.68
one brick thick; curved; 6.00 m radii	-	2.36	38.37	30.45	m²	68.82
one brick thick; curved; 1.50 m radii	-	2.91	47.32	31.10	m²	78.42
one and a half brick thick	-	2.45	39.84	42.61	m²	82.45
one and a half brick thick; battering	-	2.82	45.85	42.61	m²	88.46
two brick thick	-	2.96	48.13	56.81	m²	104.94
two brick thick; battering	-	3.52	57.24	56.81	m²	114.05
337 mm average thick; tapering, one side	-	3.10	50.41	42.61	m²	93.02
450 mm average thick; tapering, one side	-	4.30	69.92	8.68	m²	78.60
337 mm average thick; tapering, both sides	-	3.56	57.89	42.61	m²	100.50
450 mm average thick; tapering, both sides	-	4.49	73.01	57.47	m²	130.48
facework one side, half brick thick	-	1.16	18.86	14.20	m²	33.06
facework one side, one brick thick	-	1.90	30.89	28.41	m²	59.30
facework one side, one and a half brick thick	-	2.54	41.30	42.61	m²	83.91
facework one side, two brick thick	-	3.10	50.41	56.81	m²	107.22
facework both sides, half brick thick	-	1.30	21.14	14.20	m²	35.34
facework both sides, one brick thick	-	2.04	33.17	28.41	m²	61.58
facework both sides, one and a half brick thick	-	2.68	43.58	42.61	m²	86.19
facework both sides, two brick thick	-	3.19	51.87	56.81	m²	108.68
Isolated piers						
one brick thick	-	2.73	44.39	28.41	m²	72.80
two brick thick	-	4.26	69.27	57.47	m²	126.74
three brick thick	-	5.37	87.32	86.53	m²	173.85
Isolated casings						
half brick thick	-	1.39	22.60	14.20	m²	36.80
one brick thick	-	2.36	38.37	28.41	m²	66.78
Chimney stacks						
one brick thick	-	2.73	44.39	28.41	m²	72.80
two brick thick	-	4.26	69.27	57.47	m²	126.74
three brick thick	-	5.37	87.32	86.53	m²	173.85
Projections						
225 mm width; 112 mm depth; vertical	-	0.32	5.20	2.95	m	8.15
225 mm width; 225 mm depth; vertical	-	0.65	10.57	5.90	m	16.47
327 mm width; 225 mm depth; vertical	-	0.97	15.77	8.85	m	24.62
440 mm width; 225 mm depth; vertical	-	1.06	17.24	11.79	m	29.03
Closing cavities						
with of 50 mm, closing with common brickwork half brick thick; vertical	-	0.32	5.20	0.72	m	5.92
with of 50 mm, closing with common brickwork half brick thick; horizontal	-	0.32	5.20	2.18	m	7.38
with of 50 mm, closing with common brickwork half brick thick; including damp proof course; vertical	-	0.43	6.99	1.60	m	8.59
with of 50 mm, closing with common brickwork half brick thick; including damp proof course; horizontal	-	0.37	6.02	2.82	m	8.84
with of 75 mm, closing with common brickwork half brick thick; vertical	-	0.32	5.20	1.06	m	6.26

Prices for Measured Works - Minor Works
F MASONRY

	PC £	Labour hours	Labour £	Material £	Unit	Total rate £
F10 BRICK/BLOCK WALLING - cont'd						
Common bricks; PC £189.70/1000; in cement mortar (1:3) - cont'd						
Closing cavities - cont'd						
with of 75 mm, closing with common brickwork half brick thick; horizontal	-	0.32	5.20	3.20	m	8.40
with of 75 mm, closing with common brickwork half brick thick; including damp proof course; vertical	-	0.43	6.99	1.94	m	8.93
with of 50 mm, closing with common brickwork half brick thick; including damp proof course; horizontal	-	0.37	6.02	3.84	m	9.86
Bonding to existing						
half brick thick	-	0.32	5.20	0.78	m	5.98
one brick thick	-	0.46	7.48	1.56	m	9.04
one and a half brick thick	-	0.74	12.03	2.33	m	14.36
two brick thick	-	1.02	16.59	3.11	m	19.70
ADD or DEDUCT to walls for variation of £10.00/1000 in PC of common bricks						
half brick thick	-	-	-	0.65	m²	0.65
one brick thick	-	-	-	1.29	m²	1.29
one and a half brick thick	-	-	-	1.94	m²	1.94
two brick thick	-	-	-	2.58	m²	2.58
Common bricks; PC £189.70/1000; in gauged mortar (1:1:6)						
Walls						
half brick thick	-	1.06	17.24	14.00	m²	31.24
half brick thick; building against other work; concrete	-	1.16	18.86	15.17	m²	34.03
half brick thick; building overhand	-	1.34	21.79	14.00	m²	35.79
half brick thick; curved; 6.00 m radii	-	1.39	22.60	14.00	m²	36.60
half brick thick; curved; 1.50 m radii	-	1.80	29.27	16.04	m²	45.31
one brick thick	-	1.80	29.27	28.00	m²	57.27
one brick thick; curved; 6.00 m radii	-	2.36	38.37	30.04	m²	68.41
one brick thick; curved; 1.50 m radii	-	2.91	47.32	30.62	m²	77.94
one and a half brick thick	-	2.45	39.84	42.00	m²	81.84
one and a half brick thick; battering	-	2.82	45.85	42.00	m²	87.85
two brick thick	-	2.96	48.13	55.99	m²	104.12
two brick thick; battering	-	3.52	57.24	55.99	m²	113.23
337 mm average thick; tapering, one side	-	3.10	50.41	42.00	m²	92.41
450 mm average thick; tapering, one side	-	3.98	64.71	55.99	m²	120.70
337 mm average thick; tapering both sides	-	3.56	57.89	42.00	m²	99.89
450 mm average thick; tapering both sides	-	4.49	73.01	56.58	m²	129.59
facework one side, half brick thick	-	1.16	18.86	14.00	m²	32.86
facework one side, one brick thick	-	1.90	30.89	28.00	m²	58.89
facework one side, one and a half brick thick	-	2.54	41.30	42.00	m²	83.30
facework one side, two brick thick	-	3.10	50.41	55.99	m²	106.40
facework both sides, half brick thick	PC	1.30	21.14	14.00	m²	35.14
facework both sides, one brick thick	-	2.04	33.17	28.00	m²	61.17
facework both sides, one and a half brick thick	-	2.68	43.58	42.00	m²	85.58
facework both sides, two brick thick	-	3.19	51.87	55.99	m²	107.86
Isolated piers						
one brick thick	-	2.73	44.39	28.00	m²	72.39
two brick thick	-	4.26	69.27	56.58	m²	125.85
three brick thick	-	5.37	87.32	85.17	m²	172.49
Isolated casings						
half brick thick	-	1.39	22.60	14.00	m²	36.60
one brick thick	-	2.36	38.37	28.00	m²	66.37

F MASONRY

	PC £	Labour hours	Labour £	Material £	Unit	Total rate £
Chimney stacks						
one brick thick	-	2.73	44.39	28.00	m²	72.39
two brick thick	-	4.26	69.27	56.58	m²	125.85
three brick thick	-	5.37	87.32	85.17	m²	172.49
Projections						
225 mm width; 112 mm depth; vertical	-	0.32	5.20	2.92	m	8.12
225 mm width; 225 mm depth; vertical	-	0.65	10.57	5.85	m	16.42
337 mm width; 225 mm depth; vertical	-	0.97	15.77	8.77	m	24.54
440 mm width; 225 mm depth; vertical	-	1.06	17.24	11.70	m	28.94
Closing cavities						
with of 50 mm, closing with common brickwork half brick thick; vertical	-	0.32	5.20	0.71	m	5.91
with of 50 mm, closing with common brickwork half brick thick; horizontal	-	0.32	5.20	2.16	m	7.36
with of 50 mm, closing with common brickwork half brick thick; including damp proof course; vertical	-	0.43	6.99	1.59	m	8.58
with of 50 mm, closing with common brickwork half brick thick; including damp proof course; horizontal	-	0.37	6.02	2.80	m	8.82
with of 75 mm, closing with common brickwork half brick thick; vertical	-	0.32	5.20	1.04	m	6.24
with of 75 mm, closing with common brickwork half brick thick; horizontal	-	0.32	5.20	3.18	m	8.38
with of 75 mm, closing with common brickwork half brick thick; including damp proof course; vertical	-	0.43	6.99	1.92	m	8.91
with of 50 mm, closing with common brickwork half brick thick; including damp proof course; horizontal	-	0.37	6.02	3.82	m	9.84
Bonding to existing						
half brick thick	-	0.32	5.20	0.77	m	5.97
one brick thick	-	0.46	7.48	1.54	m	9.02
one and a half brick thick	-	0.74	12.03	2.30	m	14.33
two brick thick	-	1.02	16.59	3.07	m	19.66
Arches						
height on face 102 mm, width of exposed soffit 102 mm, shape of arch - segmental, one ring	-	1.80	22.87	7.89	m	30.76
height on face 102 mm, width of exposed soffit 215 mm, shape of arch - segmental, segmental, one ring	-	2.36	32.26	9.59	m	41.85
height on face 102 mm, width of exposed soffit 102 mm, shape of arch - semi-circular, one ring	-	2.31	31.45	7.89	m	39.34
height on face 102 mm, width of exposed soffit 215 mm, shape of arch - semi-circular, one ring	-	2.87	40.84	9.59	m	50.43
height on face 215 mm, width of exposed soffit 102 mm, shape of arch - segmental, two ring	-	2.31	31.45	9.37	m	40.82
height on face 215 mm, width of exposed soffit 215 mm, shape of arch - segmental, two ring	-	2.82	40.02	12.54	m	52.56
height on face 215 mm, width of exposed soffit 102 mm, shape of arch semi-circular, two ring	-	3.10	44.72	9.37	m	54.09
height on face 215 mm, width of exposed soffit 215 mm, shape of arch - semi-circular, two ring	-	3.52	51.76	12.54	m	64.30
ADD or DEDUCT to walls for variation of £10.00/1000 in PC of common bricks						
half brick thick	-	-	-	0.65	m²	0.65
one brick thick	-	-	-	1.29	m²	1.29
one and a half brick thick	-	-	-	1.94	m²	1.94
two brick thick	-	-	-	2.58	m²	2.58

Prices for Measured Works - Minor Works
F MASONRY

	PC £	Labour hours	Labour £	Material £	Unit	Total rate £
F10 BRICK/BLOCK WALLING - cont'd						
Class A engineering bricks; PC £277.90/1000; in cement mortar (1:3) - cont'd						
Walls						
half brick thick	-	1.16	18.86	19.48	m²	38.34
one brick thick	-	1.90	30.89	38.95	m²	69.84
one brick thick; building against other work	-	2.27	36.91	40.92	m²	77.83
one brick thick; curved; 6.00 m radii	-	2.54	41.30	38.95	m²	80.25
one and a half brick thick	-	2.54	41.30	58.43	m²	99.73
one and a half brick thick; building against other work	-	3.10	50.41	58.43	m²	108.84
two brick thick	-	3.19	51.87	77.90	m²	129.77
337 mm average thick; tapering, one side	-	3.28	53.33	58.43	m²	111.76
450 mm average thick; tapering, one side	-	4.26	69.27	77.90	m²	147.17
337 mm average thick; tapering, both sides	-	3.84	62.44	58.43	m²	120.87
450 mm average thick; tapering, both sides	-	4.86	79.02	78.56	m²	157.58
facework one side, half brick thick	-	1.30	21.14	19.48	m²	40.62
facework one side, one brick thick	-	2.04	33.17	38.95	m²	72.12
facework one side, one and a half brick thick	-	2.68	43.58	58.43	m²	102.01
facework one side, two brick thick	-	3.28	53.33	77.90	m²	131.23
facework both sides, half brick thick	-	1.39	22.60	19.48	m²	42.08
facework both sides, one brick thick	-	2.13	34.63	38.95	m²	73.58
facework both sides, one and a half brick thick	-	2.78	45.20	58.43	m²	103.63
facework both sides, two brick thick	-	3.42	55.61	77.90	m²	133.51
Isolated piers						
one brick thick	-	2.96	48.13	38.95	m²	87.08
two brick thick	-	4.67	75.93	78.56	m²	154.49
three brick thick	-	5.73	93.17	118.17	m²	211.34
Isolated casings						
half brick thick	-	1.48	24.06	19.48	m²	43.54
one brick thick	-	2.54	41.30	38.95	m²	80.25
Projections						
225 mm width; 112 mm depth; vertical	-	0.37	6.02	4.12	m	10.14
225 mm width; 225 mm depth; vertical	-	0.69	11.22	8.24	m	19.46
337 mm width; 225 mm depth; vertical	-	1.02	16.59	12.36	m	28.95
440 mm width; 225 mm depth; vertical	-	1.16	18.86	16.48	m	35.34
Bonding to existing						
half brick thick	-	0.37	6.02	1.07	m	7.09
one brick thick	-	0.56	9.11	2.14	m	11.25
one and a half brick thick	-	0.74	12.03	3.21	m	15.24
two brick thick	-	1.11	18.05	4.28	m	22.33
ADD or DEDUCT to walls for variation of £10.00/1000 in PC of bricks						
half brick thick	-	-	-	0.65	m²	0.65
one brick thick	-	-	-	1.29	m²	1.29
one and a half brick thick	-	-	-	1.94	m²	1.94
two brick thick	-	-	-	2.58	m²	2.58
Class B engineering bricks; PC £197.30/1000; in cement mortar (1:3)						
Walls						
half brick thick	-	1.16	18.86	14.40	m²	33.26
one brick thick	-	1.90	30.89	28.80	m²	59.69
one brick thick; building against other work	-	2.27	36.91	30.76	m²	67.67
one brick thick; curved; 6.00 m radii	-	2.54	41.30	28.80	m²	70.10
one and a half brick thick	-	2.54	41.30	43.19	m²	84.49
one and a half brick thick; building against other work	-	3.10	50.41	43.19	m²	93.60
two brick thick	-	3.27	53.17	57.59	m²	110.76
337 mm average thick; tapering, one side	-	3.28	53.33	43.19	m²	96.52

F MASONRY

	PC £	Labour hours	Labour £	Material £	Unit	Total rate £
Walls - cont'd						
450 mm average thick; tapering, one side	-	4.26	69.27	57.59	m²	126.86
337 mm average thick; tapering, both sides	-	3.84	62.44	43.19	m²	105.63
450 mm average thick; tapering, both sides	-	4.86	79.02	58.25	m²	137.27
facework one side, half brick thick	-	1.30	21.14	14.40	m²	35.54
facework one side, one brick thick	-	2.04	33.17	28.80	m²	61.97
facework one side, one and a half brick thick	-	2.68	43.58	43.19	m²	86.77
facework one side, two brick thick	-	3.28	53.33	57.59	m²	110.92
facework both sides, half brick thick	-	1.39	22.60	14.40	m²	37.00
facework both sides, one brick thick	-	2.13	34.63	28.80	m²	63.43
facework both sides, one and a half brick thick	-	2.78	45.20	43.19	m²	88.39
facework both sides, two brick thick	-	3.42	55.61	57.59	m²	113.20
Isolated piers						
one brick thick	-	2.96	48.13	28.80	m²	76.93
two brick thick	-	4.67	75.93	58.25	m²	134.18
three brick thick	-	5.74	93.33	87.70	m²	181.03
Isolated casings						
half brick thick	-	1.48	24.06	14.40	m²	38.46
one brick thick	-	2.54	41.30	28.80	m²	70.10
Projections						
225 mm width; 112 mm depth; vertical	-	0.37	6.02	2.99	m	9.01
225 mm width; 225 mm depth; vertical	-	0.69	11.22	5.98	m	17.20
337 mm width; 225 mm depth; vertical	-	1.02	16.59	8.98	m	25.57
440 mm width; 225 mm depth; vertical	-	1.16	18.86	11.97	m	30.83
Bonding to existing						
half brick thick	-	0.37	6.02	0.79	m	6.81
one brick thick	-	0.56	9.11	1.58	m	10.69
one and a half brick thick	-	0.74	12.03	2.37	m	14.40
two brick thick	-	1.11	18.05	3.16	m	21.21
ADD or DEDUCT to walls for variation of £10.00/1000 in PC of bricks						
half brick thick	-	-	-	0.65	m²	0.65
one brick thick	-	-	-	1.29	m²	1.29
one and a half brick thick	-	-	-	1.94	m²	1.94
two brick thick	-	-	-	2.58	m²	2.58
Facing bricks; sand faced; PC £155.00/1000 (unless otherwise stated); in gauged mortar (1:1:6)						
Walls						
facework one side, half brick thick; stretcher bond	-	1.39	22.60	11.12	m²	33.72
facework one side, half brick thick; flemish bond with snapped headers	-	1.62	26.34	11.12	m²	37.46
facework one side, half brick thick; stretcher bond; building against other work; concrete	-	1.48	24.06	12.29	m²	36.35
facework one side, half brick thick; flemish bond with snapped headers; building against other work; concrete	-	1.71	27.80	12.29	m²	40.09
facework one side, half brick thick; stretcher bond; building overhand	-	1.71	27.80	11.12	m²	38.92
facework one side, half brick thick; flemish bond with snapped headers; building overhand	-	1.90	30.89	11.12	m²	42.01
facework one side, half brick thick; stretcher bond; curved; 6.00 m radii	-	2.04	33.17	11.12	m²	44.29
facework one side, half brick thick; flemish bond with snapped headers; curved; 6.00 m	-	2.27	36.91	11.12	m²	48.03
facework one side, half brick thick; stretcher bond; curved; 1.50 m radii	-	2.54	41.30	11.12	m²	52.42
facework one side, half brick thick; flemish bond with snapped headers; curved; 1.50 m radii	-	2.96	48.13	11.12	m²	59.25

Prices for Measured Works - Minor Works
F MASONRY

	PC £	Labour hours	Labour £	Material £	Unit	Total rate £
F10 BRICK/BLOCK WALLING - cont'd						
Facing bricks; sand faced; PC £155.00/1000 (unless otherwise stated); in gauged mortar (1:1:6) - cont'd						
Walls - cont'd						
facework both sides, one brick thick; two stretcher skins tied together	-	2.41	39.19	23.54	m²	62.73
facework both sides, one brick thick; flemish bond	-	2.45	39.84	22.23	m²	62.07
facework both sides, one brick thick; two stretcher skins tied together; curved; 6.00 m radii	-	3.28	53.33	25.10	m²	78.43
facework both sides, one brick thick; flemish bond; curved; 6.00 m radii	-	3.42	55.61	23.79	m²	79.40
facework both sides, one brick thick; two stretcher skins tied together; curved; 1.50 m radii	-	4.12	66.99	27.25	m²	94.24
facework both sides, one brick thick; flemish bond; curved; 1.50 m radii	-	4.26	69.27	25.94	m²	95.21
Isolated piers						
facework both sides, one brick thick; two stretcher skins tied together	-	2.82	45.85	24.29	m²	70.14
facework both sides, one brick thick; flemish bond	-	2.87	46.67	24.29	m²	70.96
Isolated casings						
isolated casings; facework one side, half brick thick; stretcher bond	-	2.13	34.63	11.12	m²	45.75
isolated casings; facework one side, half brick thick; flemish bond with snapped headers	-	2.36	38.37	11.12	m²	49.49
Projections						
225 mm width; 112 mm depth; stretcher bond; vertical	-	0.32	5.20	2.28	m	7.48
225 mm width; 112 mm depth; flemish bond with snapped headers; vertical	-	0.42	6.83	2.28	m	9.11
225 mm width; 225 mm depth; flemish bond; vertical	-	0.69	11.22	4.57	m	15.79
328 mm width; 112 mm depth; stretcher bond; vertical	-	0.65	10.57	3.43	m	14.00
328 mm width; 112 mm depth; flemish bond with snapped headers; vertical	-	0.74	12.03	3.43	m	15.46
328 mm width; 225 mm depth; flemish bond; vertical	-	1.30	21.14	6.82	m	27.96
440 mm width; 112 mm depth; stretcher bond; vertical	-	0.97	15.77	4.57	m	20.34
440 mm width; 112 mm depth; flemish bond with snapped headers; vertical	-	1.02	16.59	4.57	m	21.16
440 mm width; 225 mm depth; flemish bond; vertical	-	1.85	30.08	9.14	m	39.22
Arches						
height on face 215 mm, width of exposed soffit 102 mm, shape of arch - flat	-	1.06	14.68	3.89	m	18.57
height on face 215 mm, width of exposed soffit 215 mm, shape of arch - flat	-	1.57	23.25	6.28	m	29.53
height one face 215 mm, width of exposed soffit 102 mm, shape of arch - segmental, one ring	-	2.06	26.31	8.72	m	35.03
height on face 215 mm, width of exposed soffit 215 mm, shape of arch - segmental, one ring	-	2.45	33.08	11.03	m	44.11
height on face 215 mm, width of exposed soffit 102 mm, shape of arch - semi-circular, one ring	-	3.10	44.01	8.72	m	52.73
height on face 215 mm, width of exposed soffit 215 mm, shape of arch - semi-circular, one ring	-	4.16	61.88	11.03	m	72.91

F MASONRY

	PC £	Labour hours	Labour £	Material £	Unit	Total rate £
Arches - cont'd						
height on face 215 mm, width of exposed soffit 102 mm, shape of arch - segmental, two ring	-	2.50	33.90	8.72	m	42.62
height on face 215 mm, width of exposed soffit 215 mm, shape of arch - segmental, two ring	-	3.24	46.35	11.03	m	57.38
height on face 215 mm, width of exposed soffit 102 mm, shape of arch - semi-circular, two ring	-	4.16	61.88	8.72	m	70.60
height on face 215 mm, width of exposed soffit 215 mm, shape of arch - semi-circular, two ring	-	5.73	88.33	11.03	m	99.36
Arches; cut voussoirs; PC £198.00/100						
height on face 215 mm, width of exposed soffit 102 mm, shape of arch - segmental, one ring	-	2.08	26.85	35.17	m	62.02
height on face 215 mm, width of exposed soffit 215 mm, shape of arch - segmental, one ring	-	2.59	35.43	63.92	m	99.35
height on face 215 mm, width of exposed soffit 102 mm, shape of arch - semi-circular, one ring	-	2.12	29.35	35.17	m	64.52
height on face 215 mm, width of exposed soffit 215 mm, shape of arch - semi-circular, one ring	-	3.20	43.86	63.92	m	107.78
height on face 320 mm, width of exposed soffit 102 mm, shape of arch - segmental, one and a half ring	-	2.78	38.59	63.81	m	102.40
height on face 320 mm, width of exposed soffit 215 mm, shape of arch - segmental, one and a half ring	-	3.61	52.58	127.79	m	180.37
Arches; bullnosed specials; PC £1265.00/1000						
height on face 215 mm, width of exposed soffit 102 mm, shape of arch - flat	-	1.11	15.49	20.70	m	36.19
height on face 215 mm, width of exposed soffit 215 mm, shape of arch - flat	-	1.62	24.07	40.34	m	64.41
Bullseye windows; 600 mm diameter						
height on face 215 mm, width of exposed soffit 102 mm, two rings	-	5.32	81.38	7.29	nr	88.67
height on face 215 mm, width of exposed soffit 215 mm, two rings	-	7.55	118.14	12.39	nr	130.53
Bullseye windows; 600 mm diameter; cut voussoirs; PC £198.00/100						
height on face 215 mm, width of exposed soffit 102 mm, one ring	-	4.49	67.39	77.81	nr	145.20
height on face 215 mm, width of exposed soffit 215 mm, one ring	-	6.15	95.38	153.41	nr	248.79
Bullseye windows; 1200 mm diameter						
height on face 215 mm, width of exposed soffit 102 mm, two rings	-	8.38	132.13	18.44	nr	150.57
height on face 215 mm, width of exposed soffit 215 mm, two rings	-	11.93	192.70	29.16	nr	221.86
Bullseye windows; 1200 mm diameter; cut voussoirs; PC £198.00/1000						
height on face 215 mm, width of exposed soffit 102 mm, one ring	-	7.03	110.18	138.66	nr	248.84
height on face 215 mm, width of exposed soffit 215 mm, one ring	-	9.99	160.02	268.43	nr	428.45
ADD or DEDUCT for variation of £10.00/1000 in PC of facing bricks in 102 mm high arches with 215 mm soffit	-	-	-	0.29	m	0.29
Facework sills						
150 mm x 102 mm; headers on edge; pointing top and one side; set weathering; horizontal	-	0.60	9.76	2.28	m	12.04
150 mm x 102 mm; cant headers on edge; PC £1265.00/1000; pointing top and one side; set weathering; horizontal	-	0.65	10.57	19.10	m	29.67

Prices for Measured Works - Minor Works
F MASONRY

	PC £	Labour hours	Labour £	Material £	Unit	Total rate £
F10 BRICK/BLOCK WALLING - cont'd						
Facing bricks; sand faced; PC £155.00/1000 (unless otherwise stated); in gauged mortar (1:1:6) - cont'd						
Facework sills - cont'd						
150 mm x 102 mm; bullnosed specials; PC £1265.00/1000; headers on flat; pointing top and one side; horizontal	-	0.56	9.11	19.10	m	28.21
Facework copings						
215 mm x 102 mm; headers on edge; pointing top and both sides; horizontal	-	0.46	7.48	2.37	m	9.85
260 mm x 102 mm; headers on edge; pointing top and both sides; horizontal	-	0.74	12.03	3.49	m	15.52
215 mm x 102 mm; double bullnose specials; PC £1265.00/1000; headers on edge; pointing top and both sides; horizontal	-	0.56	9.11	19.19	m	28.30
260 mm x 102 mm; single bullnose specials; PC £1265.00/1000; headers on edge; pointing top and both sides; horizontal	-	0.74	12.03	38.16	m	50.19
ADD or DEDUCT for variation of £10.00/1000 in PC of facing bricks in copings 215 mm wide, 102 mm high	-	-	-	0.14	m	0.14
Extra over facing bricks for; facework and ornamental bands and the like, plain bands; PC £165/1000						
flush; horizontal; 225 mm width; entirely of stretchers	-	0.23	3.74	0.29	m	4.03
Extra over facing bricks for; facework quoins; PC £165.00/1000						
flush; mean girth 320 mm	-	0.32	5.20	0.29	m	5.49
Bonding to existing						
facework one side, half brick thick; stretcher bond	-	0.56	9.11	0.61	m	9.72
facework one side, half brick thick, flemish bond with snapped headers	-	0.56	9.11	0.61	m	9.72
facework both sides, one brick thick; two stretcher skins tied together	-	0.74	12.03	1.22	m	13.25
facework both sides, one brick thick; flemish bond	-	0.74	12.03	1.22	m	13.25
ADD or DEDUCT for variation of £10.00/1000 in PC of facing bricks; in walls built entirely of facings; in stretcher or flemish bond						
half brick thick	-	-	-	0.65	m²	0.65
one brick thick	-	-	-	1.29	m²	1.29
Facing bricks; white sandlime; PC £180.00/1000 in gauged mortar (1:1:6)						
Walls						
facework one side, half brick thick; stretcher bond	-	1.39	22.60	13.37	m²	35.97
facework both sides, one brick thick; flemish bond	-	2.25	36.59	26.75	m²	63.34
ADD or DEDUCT for variation of £10.00/1000 in PC of facing bricks; in walls built entirely of facings; in stretcher or flemish bond						
half brick thick	-	-	-	0.65	m²	0.65
one brick thick	-	-	-	1.29	m²	1.29

F MASONRY

	PC £	Labour hours	Labour £	Material £	Unit	Total rate £
Facing bricks; machine made facings; PC £330.00/1000 (unless otherwise stated); in gauged mortar (1:1:6)						
Walls						
facework one side, half brick thick; stretcher bond	-	1.39	22.60	23.05	m²	45.65
facework one side, half brick thick, flemish bond with snapped headers	-	1.62	26.34	23.05	m²	49.39
facework one side, half brick thick; stretcher bond; building against other work; concrete	-	1.48	24.06	24.22	m²	48.28
facework one side, half brick thick; flemish bond with snapped headers; building against other work; concrete	-	1.71	27.80	24.22	m²	52.02
facework one side, half brick thick, stretcher bond; building overhand	-	1.71	27.80	23.05	m²	50.85
facework one side, half brick thick; flemish bond with snapped headers; building overhand	-	1.90	30.89	23.05	m²	53.94
facework one side, half brick thick; stretcher bond; curved; 6.00 m radii	-	2.04	33.17	23.05	m²	56.22
facework one side, half brick thick; flemish bond with snapped headers; curved; 6.00 m radii	-	2.27	36.91	23.05	m²	59.96
facework one side, half brick thick; stretcher bond; curved; 1.50 m radii	-	2.54	41.30	26.60	m²	67.90
facework one side; half brick thick; stretcher bond; curved; 1.50 m radii	-	2.96	48.13	26.60	m²	74.73
facework both sides, one brick thick; two stretcher skins tied together	-	2.41	39.19	47.41	m²	86.60
facework both sides, one brick thick; flemish bond	-	3.28	53.33	4o 10	m²	99.43
facework both sides, one brick thick; two stretcher skins tied together; curved; 6.00 m radii	-	3.42	55.61	50.95	m²	106.56
facework both sides, one brick thick; flemish bond; curved; 6.00 m radii	-	3.42	55.61	49.64	m²	105.25
facework both sides, one brick thick; two stretcher skins tied together; curved; 1.50 m radii	-	4.12	66.99	55.09	m²	122.08
facework both sides, one brick thick; flemish bond; curved; 1.50 m radii	-	2.26	36.75	53.78	m²	90.53
Isolated piers						
facework both sides, one brick thick; two ctrotcher skins tied together	-	2.82	45.85	48.16	m²	94.01
facework both sides, one brick thick; flemish bond	-	2.87	46.67	48.16	m²	94.83
Isolated casings						
facework one side, half brick thick; stretcher bond	-	2.13	34.63	23.05	m²	57.68
facework one side, half brick thick; flemish bond with snapped headers	-	2.36	38.37	23.05	m²	61.42
Projections						
225 mm width; 112 mm depth; stretcher bond; vertical	-	0.32	5.20	4.94	m	10.14
225 mm width; 112 mm depth; flemish bond with snapped headers; vertical	-	0.42	6.83	4.94	m	11.77
225 mm width; 225 mm depth; flemish bond; vertical	-	0.69	11.22	9.87	m	21.09
328 mm width; 112 mm depth; stretcher bond; vertical	-	0.65	10.57	7.41	m	17.98
328 mm width; 112 mm depth; flemish bond with snapped headers; vertical	-	0.74	12.03	7.41	m	19.44

F MASONRY

	PC £	Labour hours	Labour £	Material £	Unit	Total rate £
F10 BRICK/BLOCK WALLING - cont'd						
Facing bricks; machine made facings; PC £330.00/1000 (unless otherwise stated); in gauged mortar (1:1:6) - cont'd						
Projections - cont'd						
328 mm width; 225 mm depth; flemish bond; vertical	-	1.30	21.14	14.78	m	35.92
440 mm width; 112 mm depth; stretcher bond; vertical	-	0.97	15.77	9.87	m	25.64
440 mm width; 112 mm depth; flemish bond with snapped headers; vertical	-	1.02	16.59	9.87	m	26.46
440 mm width; 225 mm depth; flemish bond; vertical	-	1.85	30.08	19.74	m	49.82
Arches						
height on face 215 mm, width of exposed soffit 102 mm, shape of arch - flat	-	1.06	14.68	6.54	m	21.22
height on face 215 mm, width of exposed soffit 215 mm, shape of arch - flat	-	1.57	23.25	11.65	m	34.90
height on face 215 mm, width of exposed soffit 102 mm, shape of arch - segmental, one ring	-	2.04	26.13	11.38	m	37.51
height on face 215 mm, width of exposed soffit 215 mm, shape of arch - segmental, one ring	-	2.45	33.08	16.33	m	49.41
height on face 215 mm, width of exposed soffit 102 mm, shape of arch - semi-circular, one ring	-	3.10	44.01	11.38	m	55.39
height on face 215 mm, width of exposed soffit 215 mm, shape of arch - semi-circular, one ring	-	4.16	61.88	16.33	m	78.21
height on face 215 mm, width of exposed soffit 102 mm, shape of arch - segmental, two ring	-	2.50	33.90	11.38	m	45.28
height on face 215 mm, width of exposed soffit 215 mm, shape of arch - segmental, two ring	-	3.24	46.35	16.33	m	62.68
height on face 215 mm, width of exposed soffit 102 mm, shape of arch - semi-circular, two ring	-	4.16	61.88	11.38	m	73.26
height on face 215 mm, width of exposed soffit 215 mm, shape of arch - semi-circular, two ring	-	5.74	88.43	16.33	m	104.76
Arches; cut voussoirs; PC £2570.00/1000						
height on face 215 mm, width of exposed soffit 102 mm, shape of arch - segmental, one ring	-	2.08	26.85	43.48	m	70.33
height on face 215 mm, width of exposed soffit 215 mm, shape of arch - segmental, one ring	-	2.49	34.52	80.54	m	115.06
height on face 215 mm, width of exposed soffit 102 mm, shape of arch - semi-circular, one ring	-	2.36	31.55	43.48	m	75.03
height on face 215 mm, width of exposed soffit 215 mm, shape of arch - semi-circular, one ring	-	2.96	41.66	80.54	m	122.20
height on face 320 mm, width of exposed soffit 102 mm, shape of arch - segmental, one and a half ring	-	2.78	38.59	80.44	m	119.03
height on face 320 mm, width of exposed soffit 215 mm, shape of arch - segmental, one and a half ring	-	3.61	52.58	161.03	m	213.61
Arches; bullnosed specials; PC £1186.00/1000						
height on face 215 mm, width of exposed soffit 102 mm, shape of arch - flat	-	1.11	15.49	18.81	m	34.30
height on face 215 mm, width of exposed soffit 215 mm, shape of arch - flat	-	1.62	24.07	36.50	m	60.57
Bullseye windows; 600 mm diameter						
height on face 215 mm, width of exposed soffit 102 mm, two rings	-	5.32	81.38	12.86	nr	94.24
height on face 215 mm, width of exposed soffit 215 mm, two rings	-	7.55	118.14	23.52	nr	141.66

F MASONRY

	PC £	Labour hours	Labour £	Material £	Unit	Total rate £
Bullseye windows; 600 mm diameter; cut voussoirs; PC £2570.00/1000						
height on face 215 mm, width of exposed soffit 102 mm, one ring	-	4.49	67.39	99.63	nr	167.02
height on face 215 mm, width of exposed soffit, 215 mm, one ring	-	6.15	95.38	197.05	nr	292.43
Bullseye windows; 1200 mm diameter						
height on face 215 mm, width of exposed soffit 102 mm, two rings	-	8.28	131.22	29.58	nr	160.80
height on face 215 mm, width of exposed soffit 215 mm, two rings	-	11.93	192.70	51.43	nr	244.13
Bullseye windows; 1200 mm diameter; cut voussoirs; PC £2570.00/1000						
height on face 215 mm, width of exposed soffit 102 mm, one ring	-	7.03	110.18	176.07	nr	286.25
height on face 215 mm, width of exposed soffit 215 mm, one ring	-	9.99	160.02	343.23	nr	503.25
ADD or DEDUCT for variation of £10.00/1000 in PC of facing bricks in 102 mm high arches with 215 mm soffit	-	-	-	0.29	m	0.29
Facework sills						
150 mm x 102 mm; headers on edge; pointing top and one side; set weathering; horizontal	-	0.60	9.76	4.94	m	14.70
150 mm x 102 mm; cant headers on edge; PC £1186.00/1000; pointing top and one side; set weathering; horizontal	-	0.65	10.57	17.21	m	27.78
150 mm x 102 mm; bullnosed specials; PC £1186.00/1000; headers on flat; pointing top and one side; horizontal	-	0.56	9.11	17.21	m	26.32
Facework copings						
215 mm x 102 mm; headers on edge; pointing top and both sides; horizontal	-	0.46	7.48	5.02	m	12.50
260 mm x 102 mm; headers on edge; pointing top and both sides; horizontal	-	0.74	12.03	7.47	m	19.50
215 mm x 102 mm; double bullnose specials; PC £1186.00/1000; headers on edge; pointing top and both sides; horizontal	-	0.56	9.11	17.29	m	26.40
260 mm x 102 mm; single bullnose specials; PC £1681.90/1000; headers on edge; pointing top and both sides; horizontal	-	0.74	12.03	41.47	m	53.50
ADD or DEDUCT for variation of £10.00/1000 in PC of facing bricks in copings 215 mm wide, 102 mm high	-	-	-	0.14	m	0.14
Extra over facing bricks for; facework ornamental bands and the like, plain bands; PC £360.00/1000						
flush; horizontal; 225 mm width; entirely of stretchers	-	0.23	3.74	0.44	m	4.18
Extra over facing bricks for; facework quoins; PC £360.00/1000						
flush; mean girth 320 mm	-	0.32	5.20	0.43	m	5.63
Bonding to existing						
facework one side, half brick thick; stretcher bond	-	0.56	9.11	1.27	m	10.38
facework one side, half brick thick; flemish bond with snapped headers	-	0.56	9.11	1.27	m	10.38
facework both sides, one brick thick; two stretcher skins tied together	-	0.74	12.03	2.54	m	14.57
facework both sides, one brick thick; flemish bond	-	0.74	12.03	2.54	m	14.57

Prices for Measured Works - Minor Works
F MASONRY

	PC £	Labour hours	Labour £	Material £	Unit	Total rate £
F10 BRICK/BLOCK WALLING - cont'd						
Facing bricks; machine made facings; PC £330.00/1000 (unless otherwise stated); in gauged mortar (1:1:6) - cont'd						
ADD or DEDUCT for variation of ú10.00/1000 in PC of facing bricks; in walls built entirely of facings; in stretcher or flemish bond						
half brick thick	-	-	-	0.65	m²	0.65
one brick thick	-	-	-	1.29	m²	1.29
Facing bricks; hand made; PC £540.00/1000 (unless otherwise stated); in gauged mortar (1:1:6)						
Walls						
facework one side, half brick thick; stretcher bond	-	1.39	22.60	36.59	m²	59.19
facework one side, half brick thick; flemish bond with snapped headers	-	1.62	26.34	36.59	m²	62.93
facework one side, half brick thick; stretcher bond; building against other work; concrete	-	1.48	24.06	37.77	m²	61.83
facework one side, half brick thick; flemish bond with snapped headers; building against other work; concrete	-	1.71	27.80	37.77	m²	65.57
facework one side, half brick thick; stretcher bond; building overhand	-	1.71	27.80	36.59	m²	64.39
facework one side, half brick thick; flemish bond with snapped headers; building overhand	-	1.90	30.89	36.59	m²	67.48
facework one side, half brick thick; stretcher bond; curved; 6.00 m radii	-	2.04	33.17	36.59	m²	69.76
facework one side, half brick thick; flemish bond with snapped headers; curved; 6.00 m radii	-	2.27	36.91	40.95	m²	77.86
facework one side; half brick thick; stretcher bond; curved; 1.50 m radii	-	2.54	41.30	36.59	m²	77.89
facework one side; half brick thick; flemish bond with snapped headers; curved; 1.50 m radii	-	2.96	48.13	43.85	m²	91.98
facework both sides, one brick thick; two stretcher skins tied together	-	2.41	39.19	74.50	m²	113.69
facework both sides; one brick thick; flemish bond	-	2.45	39.84	73.19	m²	113.03
facework both sides; one brick thick; two stretcher skins tied together; curved; 6.00 m radii	-	3.28	53.33	80.30	m²	133.63
facework both sides; one brick thick; flemish bond; curved; 6.00 m radii	-	3.42	55.61	78.99	m²	134.60
facework both sides; one brick thick; two stretcher skins tied together; curved; 1.50 m radii	-	4.12	66.99	86.69	m²	153.68
facework both sides; one brick thick; flemish bond; curved; 1.50 m radii	-	4.26	69.27	85.38	m²	154.65
Isolated piers						
facework both sides, one brick thick; two stretcher skins tied together	-	2.82	45.85	75.25	m²	121.10
facework both sides, one brick thick; flemish bond	-	2.87	46.67	75.25	m²	121.92
Isolated casings						
facework one side, half brick thick; stretcher bond	-	2.13	34.63	36.59	m²	71.22
facework one side, half brick thick; flemish bond with snapped headers	-	2.36	38.37	36.59	m²	74.96

F MASONRY

	PC £	Labour hours	Labour £	Material £	Unit	Total rate £
Projections						
225 mm width; 112 mm depth; stretcher bond; vertical	-	0.32	5.20	7.95	m	13.15
225 mm width; 112 mm depth; flemish bond with snapped headers; vertical	-	0.42	6.83	7.95	m	14.78
225 mm width; 225 mm depth; flemish bond; vertical	-	0.69	11.22	15.89	m	27.11
328 mm width; 112 mm depth; stretcher bond; vertical	-	0.65	10.57	11.93	m	22.50
328 mm width; 112 mm depth; flemish bond with snapped headers; vertical	-	0.74	12.03	11.93	m	23.96
328 mm width; 225 mm depth; flemish bond; vertical	-	1.30	21.14	23.81	m	44.95
440 mm width; 112 mm depth; stretcher bond; vertical	-	0.97	15.77	15.89	m	31.66
440 mm width; 112 mm depth; flemish bond with snapped headers; vertical	-	1.02	16.59	15.89	m	32.48
440 mm width; 225 mm depth; flemish bond; vertical	-	1.85	30.08	31.78	m	61.86
Arches						
height on face 215 mm, width of exposed soffit 102 mm, shape of arch - flat	-	1.06	14.68	9.55	m	24.23
height on face 215 mm, width of exposed soffit 215 mm, shape of arch - flat	-	1.57	23.25	17.75	m	41.00
height on face 215 mm, width of exposed soffit 102 mm, shape of arch - segmental, one ring	-	2.04	26.13	14.39	m	40.52
height on face 215 mm, width of exposed soffit 215 mm, shape of arch - segmental, one ring	-	2.45	33.08	22.35	m	55.43
height on face 215 mm, width of exposed soffit 102 mm, shape of arch - semi-circular, one ring	-	2.50	33.90	14.39	m	48.29
height on face 215 mm, width of exposed soffit 215 mm, shape of arch - semi-circular, one ring	-	3.24	46.35	22.35	m	68.70
height on face 215 mm, width of exposed soffit 102 mm, shape of arch - segmental, two ring	-	2.50	33.90	14.39	m	48.29
height on face 215 mm, width of exposed soffit 215 mm, shape of arch - segmental, two ring	-	3.24	46.35	22.35	m	68.70
height on face 215 mm, width of exposed soffit 102 mm, shape of arch - semi-circular, two ring	-	4.16	61.88	14.39	m	76.27
height on face 215 mm, width of exposed soffit 215 mm, shape of arch - semi-circular, two ring	-	5.74	88.43	22.35	m	110.78
Arches; cut voussoirs; PC £2520.00/1000						
height on face 215 mm, width of exposed soffit 102 mm, shape of arch - segmental, one ring	-	2.08	26.85	42.77	m	69.62
height on face 215 mm, width of exposed soffit 215 mm, shape of arch - segmental, one ring	-	2.59	35.43	79.11	m	114.54
height on face 215 mm, width of exposed soffit 102 mm, shape of arch - semi-circular, one ring	-	2.36	31.55	42.77	m	74.32
height on face 215 mm, width of exposed soffit 215 mm, shape of arch - semi-circular, one ring	-	2.96	41.66	79.11	m	120.77
height on face 320 mm, width of exposed soffit 102 mm, shape of arch - segmental, one and a half ring	-	2.78	38.59	79.00	m	117.59
height on face 320 mm, width of exposed soffit 215 mm, shape of arch - segmental, one and a half ring	-	3.61	52.58	158.17	m	210.75
Arches; bullnosed specials; PC £1160.00/1000						
height on face 215 mm, width of exposed soffit 102 mm, shape of arch - flat	-	1.11	15.49	18.44	m	33.93
height on face 215 mm, width of exposed soffit 215 mm, shape of arch - flat	-	1.62	24.07	35.74	m	59.81

Prices for Measured Works - Minor Works
F MASONRY

	PC £	Labour hours	Labour £	Material £	Unit	Total rate £
F10 BRICK/BLOCK WALLING - cont'd						
Facing bricks; hand made; PC £540.00/1000 (unless otherwise stated); in gauged mortar (1:1:6) - cont'd						
Bullseye windows; 600 mm diameter						
height on face 215 mm, width of exposed soffit 102 mm, two ring	-	5.32	81.38	19.18	nr	100.56
height on face 215 mm, width of exposed soffit 215 mm, two ring	-	7.45	117.23	46.74	nr	163.97
Bullseye windows; 1200 mm diameter						
height on face 215 mm, width of exposed soffit 102 mm, two ring	-	8.28	131.22	42.22	nr	173.44
height on face 215 mm, width of exposed soffit 215 mm, two ring	-	11.93	192.70	76.72	nr	269.42
Bullseye windows; 600 mm diameter; cut voussoirs; PC £2520.00/1000						
height on face 215 mm, width of exposed soffit 102 mm, one ring	-	4.49	67.39	97.74	nr	165.13
height on face 215 mm, width of exposed soffit 215 mm, one ring	-	6.15	95.38	193.29	nr	288.67
Bullseye windows; 1200 mm diameter; cut voussoirs; PC £2520.00/1000						
height on face 215 mm, width of exposed soffit 102 mm, one ring	-	7.03	110.18	172.84	nr	283.02
height on face 215 mm, width of exposed soffit 215 mm, one ring	-	9.99	160.02	336.78	nr	496.80
ADD or DEDUCT for variation of £10.00/1000 in PC of facing bricks in 102 high arches with 215 mm soffit	-	-	-	0.29	m	0.29
Facework sills						
150 mm x 102 mm; headers on edge; pointing top and one side; set weathering; horizontal	-	0.60	9.76	7.95	m	17.71
150 mm x 102 mm; cant headers on edge; PC £1160.00/1000; pointing top and one side; set weathering; horizontal	-	0.65	10.57	16.83	m	27.40
150 mm x 102 mm; bullnosed specials; PC £1160.00/1000; headers on flat; pointing top and one side; horizontal	-	0.56	9.11	16.83	m	25.94
Facework copings						
215 mm x 102 mm; headers on edge; pointing top and both sides; horizontal	-	0.46	7.48	8.03	m	15.51
260 mm x 102 mm; headers on edge; pointing top and both sides; horizontal	-	0.74	12.03	11.98	m	24.01
215 mm x 102 mm; double bullnose specials; PC £1160.00/1000; headers on edge; pointing top and both sides	-	0.56	9.11	16.92	m	26.03
260 mm x 102 mm; single bullnose specials; PC £1160.00/1000; headers on edge; pointing top and both sides	-	0.74	12.03	33.62	m	45.65
ADD or DEDUCT for variation of £10.00/1000 in PC of facing bricks in copings 215 mm wide, 102 mm high	-	-	-	0.14	m	0.14
Extra over facing bricks for; facework ornamental bands and the like, plain bands; PC £580.00/1000						
flush; horizontal; 225 mm width; entirely of stretchers	-	0.23	3.74	0.57	m	4.31

Prices for Measured Works - Minor Works

F MASONRY

	PC £	Labour hours	Labour £	Material £	Unit	Total rate £
Extra over facing bricks for; facework quoins; PC £580.00/1000						
flush; mean girth 320 mm	-	0.32	5.20	0.57	m	5.77
Bonding ends to existing						
facework one side, half brick thick; stretcher bond	-	0.56	9.11	2.02	m	11.13
facework one side, half brick thick; flemish bond with snapped headers	-	0.56	9.11	2.02	m	11.13
facework both sides, one brick thick; two stretcher skins tied together	-	0.74	12.03	4.05	m	16.08
facework both sides, one brick thick; flemish bond	-	0.74	12.03	4.05	m	16.08
ADD or DEDUCT for variation of £10.00/1000 in PC of facing bricks; in walls built entirely of facings; in stretcher or flemish bond						
half brick thick	-	-	-	0.65	m²	0.65
one brick thick	-	-	-	1.29	m²	1.29
Facing bricks slips 50 mm thick; PC £1050.00/100; in gauged mortar (1:1:6) built up against concrete including flushing up at back (ties not included)						
Walls	2.13	34.63	69.44		m²	104.07
Edges of suspended slabs; 200 mm wide	-	0.65	10.57	13.89	m	24.46
Columns; 400 mm wide	-	1.30	21.14	27.78	m	48.92
Engineering bricks; PC £277.90/1000; and specials at PC £1186.00/100; in cement mortar (1:3)						
Facework steps						
215 mm x 102 mm; all headers-on-edge; edges set with bullnosed specials; pointing top and one side; set weathering; horizontal	-	0.60	9.76	17.23	m	26.99
returned ends pointed	-	0.14	2.28	3.83	nr	6.11
430 mm x 102 mm; all headers-on-edge; edges set with bullnosed specials; pointing top and one side; set weathering; horizontal	-	0.83	13.50	21.35	m	34.85
returned ends pointed	-	0.23	3.74	4.92	nr	8.66
Lightweight aerated concrete blocks; Thermalite "Turbo" blocks or other equal and approved; in gauged mortar (1:2:9)						
Walls						
100 mm thick	7.26	0.56	9.11	8.97	m²	18.08
115 mm thick	8.35	0.56	9.11	10.31	m²	19.42
125 mm thick	9.09	0.56	9.11	11.22	m²	20.33
130 mm thick	9.45	0.56	9.11	11.66	m²	20.77
140 mm thick	10.17	0.60	9.76	12.56	m²	22.32
150 mm thick	10.90	0.60	9.76	13.46	m²	23.22
190 mm thick	13.80	0.65	10.57	17.04	m²	27.61
200 mm thick	14.53	0.65	10.57	17.94	m²	28.51
215 mm thick	15.62	0.65	10.57	19.28	m²	29.85
Isolated piers or chimney stacks						
190 mm thick	-	0.97	15.77	17.04	m²	32.81
215 mm thick	-	0.97	15.77	19.28	m²	35.05
Isolated casings						
100 mm thick	-	0.60	9.76	8.97	m²	18.73
115 mm thick	-	0.60	9.76	10.31	m²	20.07
125 mm thick	-	0.60	9.76	11.22	m²	20.98
140 mm thick	-	0.65	10.57	12.56	m²	23.13
Extra over for fair face; flush pointing						
walls; one side	-	0.04	0.65	-	m²	0.65
walls; both sides	-	0.09	1.46	-	m²	1.46

F10 BRICK/BLOCK WALLING - cont'd

	PC £	Labour hours	Labour £	Material £	Unit	Total rate £
Lightweight aerated concrete blocks; Thermalite "Turbo" blocks or other equal and approved; in gauged mortar (1:2:9) - cont'd						
Closing cavities						
width of cavity 50 mm, closing with lightweight blockwork 100 mm; thick	-	0.27	4.39	0.50	m	4.89
width of cavity 50 mm, closing with lightweight blockwork 100 mm; thick; including damp proof course; vertical	-	0.32	5.20	1.38	m	6.58
width of cavity 75 mm, closing with lightweight blockwork 100 mm; thick	-	0.27	4.39	0.72	m	5.11
width of cavity 75 mm, closing with lightweight blockwork 100 mm; thick; including damp proof course; vertical	-	0.32	5.20	1.60	m	6.80
Bonding ends to common brickwork						
100 mm thick	-	0.14	2.28	1.03	m	3.31
115 mm thick	-	0.14	2.28	1.18	m	3.46
125 mm thick	-	0.28	4.55	1.29	m	5.84
130 mm thick	-	0.28	4.55	1.34	m	5.89
140 mm thick	-	0.28	4.55	1.45	m	6.00
150 mm thick	-	0.28	4.55	1.55	m	6.10
190 mm thick	-	0.32	5.20	1.96	m	7.16
200 mm thick	-	0.32	5.20	2.06	m	7.26
215 mm thick	-	0.37	6.02	2.22	m	8.24
Lightweight aerated concrete blocks; Thermalite "Shield 2000" blocks or other equal and approved; in gauged mortar (1:2:9)						
Walls						
75 mm thick	5.26	0.46	7.48	6.50	m²	13.98
90 mm thick	6.31	0.46	7.48	7.80	m²	15.28
100 mm thick	7.00	0.56	9.11	7.91	m²	17.02
140 mm thick	9.81	0.60	9.76	11.07	m²	20.83
150 mm thick	10.51	0.60	9.76	11.87	m²	21.63
190 mm thick	13.31	0.60	9.76	15.02	m²	24.78
200 mm thick	14.01	0.65	10.57	15.82	m²	26.39
Isolated piers or chimney stacks						
190 mm thick	-	0.97	15.77	15.02	m²	30.79
Isolated casings						
75 mm thick	-	0.60	9.76	6.50	m²	16.26
90 mm thick	-	0.60	9.76	7.80	m²	17.56
100 mm thick	-	0.60	9.76	7.91	m²	17.67
140 mm thick	-	0.65	10.57	11.07	m²	21.64
Extra over for fair face; flush pointing						
walls; one side	-	0.04	0.65	-	m²	0.65
walls; both sides	-	0.09	1.46	-	m²	1.46
Closing cavities						
width of cavity 50 mm, closing with lightweight blockwork 100 mm; thick	-	0.27	4.39	0.44	m	4.83
width of cavity 50 mm, closing with lightweight blockwork 100 mm; thick; including damp proof course; vertical	-	0.32	5.20	1.33	m	6.53
width of cavity 75 mm, closing with lightweight blockwork 100 mm; thick	-	0.27	4.39	0.64	m	5.03
width of cavity 75 mm, closing with lightweight blockwork 100 mm; thick; including damp proof course; vertical	-	0.32	5.20	1.52	m	6.72

F MASONRY

	PC £	Labour hours	Labour £	Material £	Unit	Total rate £
Bonding ends to common brickwork						
75 mm thick	-	0.09	1.46	0.75	m	2.21
90 mm thick	-	0.09	1.46	0.90	m	2.36
100 mm thick	-	0.14	2.28	0.91	m	3.19
140 mm thick	-	0.28	4.55	1.29	m	5.84
150 mm thick	-	0.28	4.55	1.37	m	5.92
190 mm thick	-	0.32	5.20	1.73	m	6.93
200 mm thick	-	0.32	5.20	1.83	m	7.03
Lightweight smooth face aerated concrete blocks; Thermalite "Smooth Face" blocks or other equal and approved; in gauged mortar (1:2:9); flush pointing one side						
Walls						
100 mm thick	12.04	0.65	10.57	14.62	m²	25.19
140 mm thick	16.85	0.74	12.03	20.45	m²	32.48
150 mm thick	18.06	0.74	12.03	21.92	m²	33.95
190 mm thick	22.87	0.83	13.50	27.76	m²	41.26
200 mm thick	24.08	0.83	13.50	29.23	m²	42.73
215 mm thick	25.88	0.83	13.50	31.42	m²	44.92
Isolated piers or chimney stacks						
190 mm thick	-	1.06	17.24	27.76	m²	45.00
200 mm thick	-	1.06	17.24	29.23	m²	46.47
215 mm thick	-	1.06	17.24	31.42	m²	48.66
Isolated casings						
100 mm thick	-	0.79	12.85	14.62	m²	27.47
140 mm thick	-	0.83	13.50	20.45	m²	33.95
Extra over for flush pointing						
walls; both sides	-	0.04	0.65	-	m²	0.65
Bonding ends to common brickwork						
100 mm thick	-	0.28	4.55	1.66	m	6.21
140 mm thick	-	0.28	4.55	2.33	m	6.88
150 mm thick	-	0.32	5.20	2.49	m	7.69
190 mm thick	-	0.37	6.02	3.15	m	9.17
200 mm thick	-	0.37	6.02	3.32	m	9.34
215 mm thick	-	0.37	6.02	3.57	m	9.59
Lightweight smooth face aerated concrete blocks; Thermalite "Party Wall" blocks (650 kg/m³) or other equal and approved; in gauged mortar (1:2:9); flush pointing one side						
Walls						
100 mm thick	7.12	0.65	10.57	8.04	m²	18.61
215 mm thick	15.30	0.83	13.50	17.26	m²	30.76
Isolated piers or chimney stacks						
215 mm thick	-	1.06	17.24	17.26	m²	34.50
Isolated casings						
100 mm thick	-	0.79	12.85	8.04	m²	20.89
Extra over for flush pointing						
walls; both sides	-	0.04	0.65	-	m²	0.65
Bonding ends to common brickwork						
100 mm thick	-	0.28	4.55	0.92	m	5.47
215 mm thick	-	0.37	6.02	2.00	m	8.02
Lightweight smooth face aerated concrete blocks; Thermalite "Party Wall" blocks (880 kg/m³) or other equal and approved; in gauged mortar (1:2:9); flush pointing one side						
Walls						
255 mm thick	18.14	1.02	16.59	20.31	m²	36.90

F MASONRY

	PC £	Labour hours	Labour £	Material £	Unit	Total rate £
F10 BRICK/BLOCK WALLING - cont'd						
Lightweight smooth face aerated concrete blocks; Thermalite "Party Wall" blocks (880 kg/m³) or other equal and approved; in gauged mortar (1:2:9); flush pointing one side - cont'd						
Isolated piers or chimney stacks						
215 mm thick	-	1.34	21.79	23.33	m²	45.12
Extra over for flush pointing						
walls; both sides	-	0.04	0.65	-	m²	0.65
Bonding ends to common brickwork						
215 mm thick	-	0.46	7.48	4.50	m	11.98
Lightweight aerated high strength concrete blocks (7.00 N/mm²); Thermalite "High Strength" blocks or other equal and approved; in cement mortar (1:3)						
Walls						
100 mm thick	8.81	0.56	9.11	10.86	m²	19.97
140 mm thick	12.33	0.60	9.76	15.19	m²	24.95
150 mm thick	13.21	0.60	9.76	16.28	m²	26.04
190 mm thick	16.73	0.65	10.57	20.61	m²	31.18
200 mm thick	17.61	0.65	10.57	21.70	m²	32.27
215 mm thick	18.93	0.65	10.57	23.32	m²	33.89
Isolated piers or chimney stacks						
190 mm thick	-	0.97	15.77	20.61	m²	36.38
200 mm thick	-	0.97	15.77	21.70	m²	37.47
215 mm thick	-	0.97	15.77	23.32	m²	39.09
Isolated casings						
100 mm thick	-	0.60	9.76	10.86	m²	20.62
140 mm thick	-	0.65	10.57	15.19	m²	25.76
150 mm thick	-	0.65	10.57	16.28	m²	26.85
190 mm thick	-	0.79	12.85	20.61	m²	33.46
200 mm thick	-	0.79	12.85	21.70	m²	34.55
215 mm thick	-	0.79	12.85	23.32	m²	36.17
Extra over for flush pointing						
walls; one side	-	0.04	0.65	-	m²	0.65
walls; both sides	-	0.09	1.46	-	m²	1.46
Bonding ends to common brickwork						
100 mm thick	-	0.28	4.55	1.24	m	5.79
140 mm thick	-	0.28	4.55	1.75	m	6.30
150 mm thick	-	0.32	5.20	1.87	m	7.07
190 mm thick	-	0.37	6.02	2.36	m	8.38
200 mm thick	-	0.37	6.02	2.49	m	8.51
215 mm thick	-	0.37	6.02	2.68	m	8.70
Lightweight aerated high strength concrete blocks; Thermalite "Trenchblock" blocks or other equal and approved; in cement mortar (1:4)						
Walls						
255 mm thick	18.14	0.83	13.50	20.35	m²	33.85
275 mm thick	19.57	0.88	14.31	22.12	m²	36.43
305 mm thick	21.70	0.93	15.12	24.53	m²	39.65
355 mm thick	25.26	1.02	16.59	28.51	m²	45.10

F MASONRY

	PC £	Labour hours	Labour £	Material £	Unit	Total rate £
Dense aggregate concrete blocks; Hanson "Conbloc" or other equal and approved; in gauged mortar (1:2:9)						
Walls						
75 mm thick; solid	5.56	0.65	10.56	6.86	m²	17.42
100 mm thick; solid	6.15	0.79	12.83	7.65	m²	20.48
140 mm thick; solid	12.10	0.97	15.75	14.84	m²	30.59
140 mm thick; hollow	11.61	0.83	13.48	14.26	m²	27.74
190 mm thick; hollow	13.71	1.06	17.21	15.93	m²	34.14
215 mm thick; hollow	14.29	1.16	18.84	17.71	m²	36.55
Isolated piers or chimney stacks						
140 mm thick; hollow	-	1.16	18.84	14.26	m²	33.10
190 mm thick; hollow	-	1.53	24.85	16.93	m²	41.78
215 mm thick; hollow	-	1.76	28.58	17.71	m²	46.29
Isolated casings						
75 mm thick; solid	-	0.79	12.83	6.86	m²	19.69
100 mm thick; solid	-	0.83	13.48	7.65	m²	21.13
140 mm thick; solid	-	1.06	17.21	14.84	m²	32.05
Extra over for fair face; flush pointing						
walls; one side	-	0.09	1.46	-	m²	1.46
walls; both sides	-	0.14	2.27	-	m²	2.27
Bonding ends to common brickwork						
75 mm thick solid	-	0.14	2.27	0.79	m	3.06
100 mm thick solid	-	0.28	4.55	0.88	m	5.43
140 mm thick solid	-	0.32	5.20	1.70	m	6.90
140 mm thick hollow	-	0.32	5.20	1.64	m	6.84
190 mm thick hollow	-	0.37	6.01	1.94	m	7.95
215 mm thick hollow	-	0.42	6.82	2.05	m	8.87
Dense aggregate concrete blocks; (7.00 N/mm²) Forticrete "Shepton Mallet Common" blocks or other equal and approved; in cement mortar (1:3)						
Walls						
75 mm thick; solid	6.10	0.65	10.56	7.54	m²	18.10
100 mm thick; hollow	7.13	0.79	12.83	8.87	m²	21.70
100 mm thick; solid	4.60	0.79	12.83	5.88	m²	18.71
140 mm thick; hollow	9.84	0.83	13.48	12.25	m²	25.73
140 mm thick; solid	7.05	0.97	15.75	8.95	m²	24.70
190 mm thick; hollow	13.06	1.06	17.21	16.27	m²	33.48
190 mm thick; solid	9.63	1.16	18.84	12.21	m²	31.05
215 mm thick; hollow	10.39	1.16	18.84	13.22	m²	32.06
215 mm thick; solid	10.27	1.34	21.76	13.08	m²	34.84
Dwarf support wall						
140 mm thick; solid	-	1.34	21.76	8.95	m²	30.71
190 mm thick; solid	-	1.53	24.85	12.21	m²	37.06
215 mm thick; solid	-	1.76	28.58	13.08	m²	41.66
Isolated piers or chimney stacks						
140 mm thick; hollow	-	1.16	18.84	12.25	m²	31.09
190 mm thick; hollow	-	1.53	24.85	16.27	m²	41.12
215 mm thick; hollow	-	1.76	28.58	13.22	m²	41.80
Isolated casings						
75 mm thick; solid	-	0.79	12.83	7.54	m²	20.37
100 mm thick; solid	-	0.83	13.48	5.88	m²	19.36
140 mm thick; solid	-	1.06	17.21	8.95	m²	26.16
Extra over for fair face; flush pointing						
walls; one side	-	0.09	1.46	-	m²	1.46
walls; both sides	-	0.14	2.27	-	m²	2.27

F MASONRY

	PC £	Labour hours	Labour £	Material £	Unit	Total rate £
F10 BRICK/BLOCK WALLING - cont'd						
Dense aggregate concrete blocks; (7.00 N/mm^2) Forticrete "Shepton Mallet Common" blocks or other equal and approved; in cement mortar (1:3) - cont'd						
Bonding ends to common brickwork						
75 mm thick solid	-	0.14	2.27	0.86	m	3.13
100 mm thick solid	-	0.28	4.55	0.68	m	5.23
140 mm thick solid	-	0.32	5.20	1.04	m	6.24
190 mm thick solid	-	0.37	6.01	1.41	m	7.42
215 mm thick solid	-	0.42	6.82	1.52	m	8.34
Dense aggregate coloured concrete blocks; Forticrete "Shepton Mallet Bathstone" or other equal and approved; in coloured gauged mortar (1:1:6); flush pointing one side						
Walls						
100 mm thick hollow	15.90	0.83	13.48	18.44	m^2	31.92
100 mm thick solid	19.27	0.83	13.48	22.24	m^2	35.72
140 mm thick hollow	20.93	0.97	15.75	24.30	m^2	40.05
140 mm thick solid	28.07	1.06	17.21	32.36	m^2	49.57
190 mm thick hollow	26.61	1.16	18.84	30.96	m^2	49.80
190 mm thick solid	38.54	1.34	21.76	44.42	m^2	66.18
215 mm thick hollow	28.08	1.34	21.76	32.74	m^2	54.50
215 mm thick solid	40.85	1.39	22.57	47.15	m^2	69.72
Isolated piers or chimney stacks						
140 mm thick solid	-	1.43	23.22	32.36	m^2	55.58
190 mm thick solid	-	1.62	26.31	44.42	m^2	70.73
215 mm thick solid	-	1.80	29.23	47.15	m^2	76.38
Extra over blocks for						
100 mm thick half lintel blocks ref D14	-	0.28	4.55	17.02	m	21.57
140 mm thick half lintel blocks ref H14	-	0.32	5.20	21.60	m	26.80
140 mm thick quoin blocks ref H16	-	0.37	6.01	24.74	m	30.75
140 mm thick cavity closer blocks ref H17	-	0.37	6.01	29.68	m	35.69
140 mm thick cill blocks ref H21	-	0.32	5.20	17.61	m	22.81
190 mm thick half lintel blocks ref A14	-	0.37	6.01	24.77	m	30.78
190 mm thick cill blocks ref A21	-	0.37	6.01	18.33	m	24.34
Astra-Glaze satin-gloss glazed finish blocks or other equal and approved; Aldwick Design Ltd; standard colours; in gauged mortar (1:1:6); joints raked out; gun applied latex grout to joints						
Walls						
100 mm thick; glazed one side	71.53	1.06	17.21	81.13	m^2	98.34
extra; glazed square end return	9.71	0.42	6.82	27.29	m	34.11
150 mm thick; glazed one side	81.75	1.16	18.84	92.86	m^2	111.70
extra; glazed square end return	15.53	0.56	9.09	55.09	m	64.18
200 mm thick; glazed one side	91.97	1.34	21.76	104.59	m^2	126.35
extra; glazed square end return	-	0.60	9.74	77.83	m	87.57
100 mm thick; glazed both sides	102.19	1.30	21.11	115.74	m^2	136.85
150 mm thick; glazed both sides	112.41	1.39	22.57	127.47	m^2	150.04
100 mm thick lintel 200 mm high; glazed one side	-	0.97	12.07	22.17	m	34.24
150 mm thick lintel 200 mm high; glazed one side	-	1.17	14.41	26.76	m	41.17
200 mm thick lintel 200 mm high; glazed one side	-	1.42	17.96	24.97	m	42.93

F MASONRY

	PC £	Labour hours	Labour £	Material £	Unit	Total rate £
F20 NATURAL STONE RUBBLE WALLING						
Cotswold Guiting limestone or other equal and approved; laid dry						
Uncoursed random rubble walling						
275 mm thick	-	2.36	36.25	33.60	m²	69.85
350 mm thick	-	2.82	42.91	42.76	m²	85.67
425 mm thick	-	3.24	48.83	51.92	m²	100.75
500 mm thick	-	3.61	54.33	61.09	m²	115.42
Cotswold Guiting limestone or other equal and approved; bedded; jointed and pointed in cement; lime mortar (1:2:9)						
Uncoursed random rubble walling; faced and pointed; both sides						
275 mm thick	-	2.27	34.69	37.04	m²	71.73
350 mm thick	-	2.50	37.40	47.14	m²	84.54
425 mm thick	-	2.78	40.94	57.24	m²	98.18
500 mm thick	-	2.96	43.24	67.35	m²	110.59
Coursed random rubble walling; rough dressed; faced and pointed one side						
114 mm thick	-	1.71	27.72	46.57	m²	74.29
150 mm thick	-	2.16	34.20	51.43	m²	85.63
Fair returns on walling						
114 mm wide	-	0.02	0.33	-	m	0.33
150 mm wide	-	0.04	0.65	-	m	0.65
275 mm wide	-	0.07	1.14	-	m	1.14
350 mm wide	-	0.09	1.46	-	m	1.46
425 mm wide	-	0.12	1.95	-	m	1.95
500 mm wide	-	0.14	2.28	-	m	2.28
Fair raking cutting on walling						
114 mm wide	-	0.23	3.74	6.77	m	10.51
150 mm wide	-	0.29	4.72	7.43	m	12.15
Level uncoursed rubble walling for damp proof courses and the like						
275 mm wide	-	0.22	3.62	2.40	m	6.02
350 mm wide	-	0.23	3.78	3.01	m	6.79
425 mm wide	-	0.24	3.95	3.67	m	7.62
500 mm wide	-	0.26	4.27	4.33	m	8.60
Copings formed of rough stones; faced and pointed all round						
275 mm x 200 mm (average) high	-	0.64	10.11	8.69	m	18.80
350 mm x 250 mm (average) high	-	0.86	13.40	12.09	m	25.49
425 mm x 300 mm (average) high	-	1.12	17.26	16.73	m	33.99
500 mm x 300 mm (average) high	-	1.42	21.62	22.58	m	44.20
F22 CAST STONE ASHLAR WALLING DRESSINGS						
Reconstructed limestone walling; "Bradstone 100 bed Weathered Cotswold" or "North Cerney" masonry blocks or other equal and approved; laid to pattern or course recommended; bedded; jointed and pointed in approved coloured cement:lime mortar (1:2:9)						
Walls; facing and pointing one side						
masonry blocks; random uncoursed	-	1.19	19.35	23.28	m²	42.63
extra; returned ends	-	0.43	6.99	24.90	m	31.89
extra; plain L shaped quoins	-	0.14	2.28	8.48	m	10.76
traditional walling; coursed squared	-	1.49	24.23	23.23	m²	47.46

F MASONRY

	PC £	Labour hours	Labour £	Material £	Unit	Total rate £
F22 CAST STONE ASHLAR WALLING DRESSINGS - cont'd						
Reconstructed limestone walling; "Bradstone 100 bed Weathered Cotswold" or "North Cerney" masonry blocks or other equal and approved; laid to pattern or course recommended; bedded; jointed and pointed in approved coloured cement:lime mortar (1:2:9) - cont'd						
Walls; facing and pointing one side - cont'd						
squared random rubble	-	1.49	24.23	25.74	m²	**49.97**
squared coursed rubble (large module)	-	1.38	22.44	25.60	m²	**48.04**
squared coursed rubble (small module)	-	1.43	23.25	25.89	m²	**49.14**
squared and pitched rock faced walling; coursed	-	1.54	25.04	24.12	m²	**49.16**
rough hewn rock faced walling; random	-	1.60	26.02	23.97	m²	**49.99**
extra; returned ends	-	0.17	2.76	-	m	**2.76**
Isolated piers or chimney stacks; facing and pointing one side						
masonry blocks; random uncoursed	-	1.65	26.83	23.28	m²	**50.11**
traditional walling; coursed squared	-	2.07	33.66	23.23	m²	**56.89**
squared random rubble	-	2.07	33.66	25.74	m²	**59.40**
squared coursed rubble (large module)	-	1.91	31.06	25.60	m²	**56.66**
squared coursed rubble (small module)	-	2.02	32.85	25.39	m²	**58.74**
squared and pitched rock faced walling; coursed	-	2.18	35.45	24.12	m²	**59.57**
rough hewn rock faced walling; random	-	2.23	36.26	23.97	m²	**60.23**
Isolated casings; facing and pointing one side						
masonry blocks; random uncoursed	-	1.43	23.25	23.28	m²	**46.53**
traditional walling; coursed squared	-	1.81	29.43	23.23	m²	**52.66**
squared random rubble	-	1.81	29.43	25.74	m²	**55.17**
squared coursed rubble (large module)	-	1.65	26.83	25.60	m²	**52.43**
squared coursed rubble (small module)	-	1.76	28.62	25.89	m²	**54.51**
squared and pitched rock faced walling; coursed	-	1.86	30.24	24.12	m²	**54.36**
rough hewn rock faced walling; random	-	1.91	31.06	23.97	m²	**55.03**
Fair returns 100 mm wide						
masonry blocks; random uncoursed		0.13	2.11	-	m²	**2.11**
traditional walling; coursed squared		0.16	2.60	-	m²	**2.60**
squared random rubble		0.16	2.60	-	m²	**2.60**
squared coursed rubble (large module)		0.15	2.44	-	m²	**2.44**
squared coursed rubble (small module)		0.15	2.44	-	m²	**2.44**
squared and pitched rock faced walling; coursed	-	0.16	2.60	-	m²	**2.60**
rough hewn rock faced walling; random	-	0.17	2.76	-	m²	**2.76**
Fair raking cutting on masonry blocks						
100 mm wide		0.19	3.09	-	m	**3.09**
Reconstructed limestone dressings; "Bradstone Architectural" dressings in weathered "Cotswold" or "North Cerney" shades or other equal and approved; bedded, jointed and pointed in approved coloured cement:lime mortar (1:2:9)						
Copings; twice weathered and throated						
152 mm x 76 mm; type A	-	0.35	5.69	10.77	m	**16.46**
178 mm x 64 mm; type B	-	0.35	5.69	10.46	m	**16.15**
305 mm x 76 mm; type A	-	0.43	6.99	20.42	m	**27.41**
Extra for						
fair end	-	-	-	4.44	nr	**4.44**
returned mitred fair end	-	-	-	4.44	nr	**4.44**

F MASONRY

	PC £	Labour hours	Labour £	Material £	Unit	Total rate £
Copings; once weathered and throated						
191 mm x 76 mm	-	0.35	5.69	11.61	m	17.30
305 mm x 76 mm	-	0.43	6.99	20.11	m	27.10
365 mm x 76 mm	-	0.43	6.99	21.53	m	28.52
Extra for						
fair end	-	-	-	4.44	nr	4.44
returned mitred fair end	-	-	-	4.44	nr	4.44
Chimney caps; four times weathered and throated; once holed						
553 mm x 553 mm x 76 mm	-	0.43	6.99	25.08	nr	32.07
686 mm x 686 mm x 76 mm	-	0.43	6.99	41.08	nr	48.07
Pier caps; four times weathered and throated						
305 mm x 305 mm	-	0.27	4.39	9.03	nr	13.42
381 mm x 381 mm	-	0.27	4.39	12.52	nr	16.91
457 mm x 457 mm	-	0.32	5.20	16.96	nr	22.16
533 mm x 533 mm	-	0.32	5.20	23.96	nr	29.16
Splayed corbels						
479 mm x 100 mm x 215 mm	-	0.16	2.60	11.63	nr	14.23
665 mm x 100 mm x 215 mm	-	0.21	3.41	18.69	nr	22.10
Air bricks						
300 mm x 140 mm x 76 mm	-	0.08	1.30	6.62	nr	7.92
100 mm x 140 mm lintels; rectangular; reinforced with mild steel bars						
not exceeding 1.22 m long	-	0.26	4.23	18.33	m	22.56
1.37 m - 1.67 m long	-	0.28	4.55	18.33	m	22.88
1.83 m - 1.98 m long	-	0.30	4.88	18.33	m	23.21
100 mm x 215 mm lintels; rectangular; reinforced with mild steel bars						
not exceeding 1.67 m long	-	0.28	4.55	24.67	m	29.22
1.83 m - 1.98 m long	-	0.30	4.88	24.67	m	29.55
2.13 m - 2.44 m long	-	0.32	5.20	24.67	m	29.87
2.59 m - 2.90 m long	-	0.34	5.53	24.67	m	30.20
197 mm x 67 mm sills to suit standard softwood windows; stooled at ends						
not exceeding 2.54 m long	-	0.32	5.20	28.73	m	33.93
Window surround; traditional with label moulding; for single light; sill 146 mm x 133 mm; jambs 146 mm x 146 mm; head 146 mm x105 mm; including all dowels and anchors						
overall size 508 mm x 1479 mm	134.71	0.95	15.45	146.29	nr	161.74
Window surround; traditional with label moulding; three light; for windows 508 mm x 1219 mm; sill 146 mm x 133 mm; jambs 146 mm x 146 mm; head 146 mm x 103 mm; mullions 146 mm x 108 mm; including all dowels and anchors						
overall size 1975 mm x 1479 mm	287.99	2.50	40.65	313.27	nr	353.92
Door surround; moulded continuous jambs and head with label moulding; including all dowels and anchors						
door 839 mm x 1981 mm in 102 mm x 64 mm frame	246.19	1.76	28.62	265.39	nr	294.01

F30 ACCESSORIES/SUNDRY ITEMS FOR BRICK/BLOCK/STONE WALLING

	PC £	Labour hours	Labour £	Material £	Unit	Total rate £
Forming cavities						
In hollow walls						
width of cavity 50 mm; galvanised steel butterfly wall ties; three wall ties per m^2	-	0.06	0.98	0.17	m^2	1.15
width of cavity 50 mm; galvanised steel twisted wall ties; three wall ties per m^2	-	0.06	0.98	0.33	m^2	1.31

Prices for Measured Works - Minor Works
F MASONRY

	PC £	Labour hours	Labour £	Material £	Unit	Total rate £
F30 ACCESSORIES/SUNDRY ITEMS FOR BRICK/BLOCK/STONE WALLING - cont'd						
Forming cavities - cont'd						
In hollow walls - cont'd						
width of cavity 50 mm; stainless steel butterfly wall ties; three wall ties per m^2	-	0.06	0.98	0.35	m^2	1.33
width of cavity 50 mm; stainless steel twisted wall ties; three wall ties per m^2	-	0.06	0.98	0.69	m^2	1.67
width of cavity 75 mm; galvanised steel butterfly wall ties; three wall ties per m^2	-	0.06	0.98	0.21	m^2	1.19
width of cavity 75 mm; galvanised steel twisted wall ties; three wall ties per m^2	-	0.06	0.98	0.37	m^2	1.35
width of cavity 75 mm; stainless steel butterfly wall ties; three wall ties per m^2	-	0.06	0.98	0.50	m^2	1.48
width of cavity 75 mm; stainless steel twisted wall ties; three wall ties per m^2	-	0.06	0.98	0.75	m^2	1.73
Damp proof courses						
"Pluvex No 1" (hessian based) damp proof course or other equal and approved; 200 mm laps; in gauged mortar (1:1:6)						
width exceeding 225 mm; horizontal	6.93	0.27	4.39	8.01	m^2	12.40
width exceeding 225 mm; forming cavity gutters in hollow walls	-	0.43	6.99	8.01	m^2	15.00
width not exceeding 225 mm; horizontal	-	0.53	8.62	8.01	m^2	16.63
width not exceeding 225 mm; horizontal	-	0.80	13.01	8.01	m^2	21.02
"Pluvex No 2" (fibre based) damp proof course or other equal and approved; 200 mm laps; in gauged mortar (1:1:6)						
width exceeding 225 mm; horizontal	4.65	0.27	4.39	5.37	m^2	9.76
width exceeding 225 mm; forming cavity gutters in hollow walls	-	0.43	6.99	5.37	m^2	12.36
width not exceeding 225 mm; horizontal	-	0.53	8.62	5.37	m^2	13.99
width not exceeding 225 mm; vertical	-	0.80	13.01	5.37	m^2	18.38
"Rubberthene" polythylene damp proof course or other equal and approved; 200 mm laps; in gauged mortar (1:1:6)						
width exceeding 225 mm; horizontal	1.08	0.27	4.39	1.25	m^2	5.64
width exceeding 225 mm; forming cavity gutters in hollow walls	-	0.43	6.99	1.25	m^2	8.24
width not exceeding 225 mm; horizontal	-	0.53	8.62	1.25	m^2	9.87
width not exceeding 225 mm; vertical	-	0.80	13.01	1.25	m^2	14.26
"Permabit" bitumen polymer damp proof course or other equal and approved; 150 mm laps; in gauged mortar (1:1:6)						
width exceeding over 225 mm	6.11	0.27	4.39	7.06	m^2	11.45
width exceeding 225 mm; forming cavity gutters in hollow walls	-	0.43	6.99	7.06	m^2	14.05
width not exceeding 225 mm; horizontal	-	0.53	8.62	7.06	m^2	15.68
width not exceeding 225 mm; vertical	-	0.80	13.01	7.06	m^2	20.07
"Hyload" (pitch polymer) damp proof course or other equal and approved; 150 mm laps; in gauged mortar (1:1:6)						
width exceeding 225 mm; horizontal	6.11	0.27	4.39	7.06	m^2	11.45
width exceeding 225 mm; forming cavity gutters in hollow walls	-	0.43	6.99	7.06	m^2	14.05
width not exceeding 225 mm; horizontal	-	0.53	8.62	7.06	m^2	15.68
width not exceeding 225 mm; vertical	-	0.80	13.01	7.06	m^2	20.07

F MASONRY

	PC £	Labour hours	Labour £	Material £	Unit	Total rate £
"Ledkore" grade A (bitumen based lead cored) damp proof course or other equal and approved; 200 mm laps; in gauged mortar (1:1:6)						
width exceeding 225 mm; horizontal	18.79	0.35	5.69	21.71	m²	27.40
width exceeding 225 mm wide; forming cavity gutters in hollow walls; horizontal	-	0.56	9.11	21.71	m²	30.82
width not exceeding 225 mm; horizontal	-	0.68	11.06	21.71	m²	32.77
width not exceeding 225 mm; vertical	-	0.80	13.01	21.71	m²	34.72
Milled lead damp proof course; BVS 1178; 1.80 mm thick (code 4), 175 mm laps, in cement:lime mortar (1:2:9)						
width exceeding 225 mm; horizontal	-	2.13	34.63	15.59	m²	50.22
width not exceeding 225 mm; horizontal	-	3.19	51.87	15.59	m²	67.46
Two courses slates in cement mortar (1:3)						
width exceeding 225 mm; horizontal	-	1.60	26.02	11.20	m²	37.22
width exceeding 225 mm; vertical	-	2.40	39.02	11.20	m²	50.22
"Peter Cox" chemical transfusion damp proof course system or other equal and approved						
half brick thick; horizontal	-	-	-	-	m	13.68
one brick thick; horizontal	-	-	-	-	m	27.26
one and a half brick thick; horizontal	-	-	-	-	m	40.89
Silicone injection damp-proofing; 450 mm centres; making good brickwork						
half brick thick; horizontal	-	-	-	-	m	7.29
one brick thick; horizontal	-	-	-	-	m	13.48
one and a half brick thick; horizontal	-	-	-	-	m	17.28
"Synthprufe" damp proof membrane or other equal and approved; PC £65.33/25 litres; three coats brushed on						
width not exceeding 150 mm; vertical	-	0.36	2.62	7.02	m²	9.64
width 150 mm - 225 mm; vertical	-	0.34	2.48	7.02	m²	9.50
width 225 mm - 300 mm; vertical	-	0.32	2.33	7.02	m²	9.35
width exceeding 300 mm; vertical	-	0.30	2.19	7.02	m²	9.21
Joint reinforcement						
"Brickforce" galvanised steel joint reinforcement or other equal and approved						
width 60 mm; ref GBF35W60B	-	0.02	0.33	0.42	m	0.75
width 100 mm; ref GBF35W100B	-	0.04	0.65	0.50	m	1.15
width 160 mm; ref GBF35W160B	-	0.05	0.81	0.62	m	1.43
width 175 mm; ref GBF35W175B	-	0.06	0.98	0.73	m	1.71
"Brickforce" stainless steel joint reinforcement or other equal and approved						
width 60 mm; ref SBF35W60BSC	-	0.02	0.33	1.13	m	1.46
width 100 mm; ref SBF35W100BSC	-	0.04	0.65	1.18	m	1.83
width 160 mm; ref SBF35W160BSC	-	0.05	0.81	1.21	m	2.02
width 175 mm; ref SBF35W175BSC	-	0.06	0.98	1.29	m	2.27
"Wallforce" stainless steel joint reinforcement or other equal and approved						
width 210 mm; ref SWF40BSC	-	0.06	0.98	3.20	m	4.18
width 235 mm; ref SWF60BSC	-	0.07	1.14	3.28	m	4.42
width 250 mm; ref SWF100BSC	-	0.08	1.30	3.40	m	4.70
width 275 mm; ref SWF160BSC	-	0.09	1.46	3.48	m	4.94
Weather fillets						
Weather fillets in cement mortar (1:3)						
50 mm face width	-	0.13	2.11	0.04	m	2.15
100 mm face width	-	0.21	3.41	0.16	m	3.57
Bedding wall plates or similar in cement mortar (1:3)						
100 mm wide		0.06	0.98	0.07	m	1.05

F MASONRY

	PC £	Labour hours	Labour £	Material £	Unit	Total rate £
F30 ACCESSORIES/SUNDRY ITEMS FOR BRICK/BLOCK/STONE WALLING - cont'd						
Weather fillets - cont'd						
Bedding wood frames in cement mortar (1:3) and point						
one side	-	0.08	1.30	0.07	m	**1.37**
both sides	-	0.11	1.79	0.09	m	**1.88**
one side in mortar; other side in mastic	-	0.22	2.87	0.54	m	**3.41**
Angle fillets						
Angle fillets in cement mortar (1:3)						
50 mm face width	-	0.13	2.11	0.04	m	**2.15**
100 mm face width	-	0.21	3.41	0.16	m	**3.57**
Pointing in						
Pointing with mastic						
wood frames or sills	-	0.11	1.01	0.48	m	**1.49**
Pointing with polysulphide sealant						
wood frames or sills	-	0.11	1.01	1.44	m	**2.45**
Wedging and pinning						
To underside of existing construction with slates in cement mortar (1:3)						
width of wall - one brick thick	-	0.85	13.82	2.44	m	**16.26**
width of wall - one and a half brick thick	-	1.06	17.24	4.88	m	**22.12**
width of wall - two brick thick	-	1.28	20.81	7.31	m	**28.12**
Joints						
Hacking joints and faces of brickwork or blockwork						
to form key for plaster	-	0.28	2.04	-	m²	**2.04**
Raking out joint in brickwork or blockwork for turned-in edge of flashing						
horizontal	-	0.16	2.60	-	m	**2.60**
stepped	-	0.21	3.41	-	m	**3.41**
Raking out and enlarging joint in brickwork or blockwork for nib of asphalt						
horizontal	-	0.21	3.41	-	m	**3.41**
Cutting grooves in brickwork or blockwork						
for water bars and the like	-	0.27	1.97	0.54	m	**2.51**
for nib of asphalt; horizontal	-	0.27	1.97	0.54	m	**2.51**
Preparing to receive new walls						
top existing 215 mm wall	-	0.21	3.41	-	m	**3.41**
Cleaning and priming both faces; filling with pre-formed closed cell joint filler and pointing one side with polysulphide sealant; 12 mm deep						
expansion joints; 12 mm wide	-	0.27	3.68	2.87	m	**6.55**
expansion joints; 20 mm wide	-	0.31	4.05	4.20	m	**8.25**
expansion joints; 25 mm wide	-	0.37	4.59	5.07	m	**9.66**
Fire resisting horizontal expansion joints; filling with joint filler; fixed with high temperature slip adhesive; between top of wall and soffit						
wall not exceeding 215 mm wide; 10 mm wide joint with 30 mm deep filler (one hour fire seal)	-	0.27	4.39	6.15	m	**10.54**
wall not exceeding 215 mm wide; 10 mm wide joint with 30 mm deep filler (two hour fire seal)	-	0.27	4.39	6.15	m	**10.54**
wall not exceeding 215 mm wide; 20 mm wide joint with 45 mm deep filler (two hour fire seal)	-	0.32	5.20	9.40	m	**14.60**
wall not exceeding 215 mm wide; 30 mm wide joint with 75 mm deep filler (three hour fire seal)	-	0.37	6.02	23.18	m	**29.20**

F MASONRY

	PC £	Labour hours	Labour £	Material £	Unit	Total rate £
Fire resisting vertical expansion joints; filling with joint filler; fixed with high temperature slip adhesive; with polysulphide sealant one side; between end of wall of wall and concrete						
wall not exceeding 215 mm wide; 20 mm wide joint with 45 mm deep filler (two hour fire seal)	-	0.43	6.42	12.27	m	**18.69**
Slates and tile sills						
Sills; two courses of machine made plain roofing tiles						
set weathering; bedded and pointed	-	0.64	10.41	4.02	m	**14.43**
Flue linings						
Flue linings; Marflex "ML 200" square refractory concrete flue linings or other equal and approved; rebated joints in flue joint mortar mix						
linings; ref. MLS1U	-	0.32	5.20	10.37	m	**15.57**
bottom swivel unit; ref MLS2U	-	0.11	1.79	8.52	nr	**10.31**
45 degree bend; ref MLS4U	-	0.11	1.79	6.22	nr	**8.01**
90 mm offset liner; ref MLS5U	-	0.11	1.79	4.12	nr	**5.91**
70 mm offset liner; ref MLS6U	-	0.11	1.79	3.72	nr	**5.51**
57 mm offset liner; ref MLS7U	-	0.11	1.79	9.04	nr	**10.83**
pot; ref MLSP83R	-	0.11	1.79	16.05	nr	**17.84**
single cap unit; ref MLS10U	-	0.27	4.39	20.43	nr	**24.82**
double cap unit; ref MLS11U	-	0.32	5.20	18.78	nr	**23.98**
combined gather lintel; ref MLSLUL4	-	0.32	5.20	19.09	nr	**24.29**
Air bricks						
Air bricks; red terracotta; building into prepared openings						
215 mm x 65 mm	-	0.08	1.30	1.78	nr	**3.08**
215 mm x 140 mm	-	0.08	1.30	2.47	nr	**3.77**
215 mm x 215 mm	-	0.08	1.30	6.72	nr	**8.02**
Gas flue blocks						
Gas flue system; Marflex "HP" or other equal and approved; concrete blocks built in; in flue joint mortar mix; cutting brickwork or blockwork around						
recess; ref HP1	-	0.11	1.79	1.71	nr	**3.50**
cover; ref HP2	-	0.11	1.79	3.82	nr	**5.61**
222 mm standard block with nib; ref HP3	-	0.11	1.79	2.69	nr	**4.48**
112 mm standard block with nib; ref HP3112	-	0.11	1.79	2.12	nr	**3.91**
72 mm standard block with nib; ref HP372	-	0.11	1.79	2.12	nr	**3.91**
vent block; ref HP3BH	-	0.11	1.79	2.08	nr	**3.87**
222 mm standard block without nib; ref HP4	-	0.11	1.79	2.64	nr	**4.43**
112 mm standard block without nib; ref HP4112	-	0.11	1.79	2.12	nr	**3.91**
72 mm standard block without nib; ref HP472	-	0.11	1.79	2.12	nr	**3.91**
120 mm side offset block; ref HP5	-	0.11	1.79	2.85	nr	**4.64**
70 mm back offset block; ref HP6	-	0.11	1.79	8.32	nr	**10.11**
vertical exit block; ref HP7	-	0.11	1.79	5.45	nr	**7.24**
angled entry/exit block; ref HP8	-	0.11	1.79	5.45	nr	**7.24**
reverse rebate block; ref HP9	-	0.11	1.79	3.92	nr	**5.71**
corbel block; ref HP10	-	0.11	1.79	5.25	nr	**7.04**
lintel unit; ref HP11	-	0.11	1.79	4.94	nr	**6.73**

Prices for Measured Works - Minor Works
F MASONRY

	PC £	Labour hours	Labour £	Material £	Unit	Total rate £
F30 ACCESSORIES/SUNDRY ITEMS FOR BRICK/BLOCK/STONE WALLING - cont'd						
Proprietary items						
"Thermabate" cavity closers or other equal and approved; RMC Panel Products Ltd						
closing cavities; width of cavity 50 mm	-	0.16	2.60	5.29	m	7.89
closing cavities; width of cavity 75 mm	-	0.16	2.60	5.67	m	8.27
"Westbrick" cavity closers or other equal and approved; Manthorpe Building Products Ltd						
closing cavities; width of cavity 50 mm	-	0.16	2.60	5.23	m	7.83
"Type H cavicloser" or other equal and approved; uPVC universal cavity closer, insulator and damp-proof course by Cavity Trays Ltd; built into cavity wall as work proceeds, complete with face closer and ties						
closing cavities; width of cavity 50 mm - 100 mm	-	0.08	1.30	5.46	m	6.76
"Type L" durropolythelene lintel stop ends or other equal and approved; Cavity Trays Ltd; fixing to lintel with butyl anchoring strip; building in as the work proceeds						
adjusted to lintel as required	-	0.05	0.81	0.49	nr	1.30
"Type W" polypropylene weeps/vents or other equal and approved; Cavity Trays Ltd; built into cavity wall as work proceeds						
100 mm/115 mm x 65 mm x 10 mm including lock fit wedges	-	0.05	0.81	0.42	nr	1.23
extra; extension duct 200 mm/225 mm x 65 mm x 10 mm	-	0.08	1.30	0.71	nr	2.01
"Type X" polypropylene abutment cavity tray or other equal and approved; Cavity Trays; built into facing brickwork as the work proceeds; complete with Code 4 lead flashing; intermediate/catchment tray with short leads (requiring soakers); to suit roof of						
17 - 20 degree pitch	-	0.06	0.98	4.64	nr	5.62
21 - 25 degree pitch	-	0.06	0.98	4.30	nr	5.28
26 - 45 degree pitch	-	0.06	0.98	4.11	nr	5.09
"Type X" polypropylene abutment cavity tray or other equal and approved; Cavity Trays; built into facing brickwork as the work proceeds; complete with Code 4 lead flashing; intermediate/catchment tray with long leads (suitable only for corrugated roof tiles); to suit roof of						
17 - 20 degree pitch	-	0.06	0.98	6.26	nr	7.24
21 - 25 degree pitch	-	0.06	0.98	5.74	nr	6.72
26 - 45 degree pitch	-	0.06	0.98	5.30	nr	6.28
"Type X" polypropylene abutment cavity tray or other equal and approved; Cavity Trays; built into facing brickwork as the work proceeds; complete with Code 4 lead flashing; ridge tray with short/long leads; to suit roof of						
17 - 20 degree pitch	-	0.06	0.98	10.52	nr	11.50
21 - 25 degree pitch	-	0.06	0.98	9.76	nr	10.74
26 - 45 degree pitch	-	0.06	0.98	8.68	nr	9.66
Servicised "Bit-uthene" self-adhesive cavity flashing or other equal and approved; type "CA"; well lapped at joints; in gauged mortar (1:1:6)						
width exceeding 225 mm; horizontal	-	0.91	14.80	20.61	m^2	35.41

F MASONRY

	PC £	Labour hours	Labour £	Material £	Unit	Total rate £
"Expamet" stainless steel wall starters or other equal and approved; plugged and screwed						
to suit walls 60 mm - 75 mm thick	-	0.27	1.97	7.49	m	9.46
to suit walls 100 mm - 115 mm thick	-	0.27	1.97	8.24	m	10.21
to suit walls 125 mm - 180 mm thick	-	0.43	3.13	11.03	m	14.16
to suit walls 190 mm - 260 mm thick	-	0.53	3.86	14.08	m	17.94
Ties in walls; 200 mm long butterfly type; building into joints of brickwork or blockwork						
galvanised steel or polypropylene	-	0.02	0.33	0.10	nr	0.43
stainless steel	-	0.02	0.33	0.14	nr	0.47
Ties in walls; 20 mm x 3 mm x 200 mm long twisted wall type; building into joints of brickwork or blockwork						
galvanised steel	-	0.02	0.33	0.14	nr	0.47
stainless steel	-	0.02	0.33	0.28	nr	0.61
Anchors in walls; 25 mm x 3 mm x 100 mm long; one end dovetailed; other end building into joints of brickwork or blockwork						
galvanised steel	-	0.06	0.98	0.19	nr	1.17
stainless steel	-	0.06	0.98	0.29	nr	1.27
Fixing cramp 25 mm x 3 mm x 250 mm long; once bent; fixed to back of frame; other end building into joints of brickwork or blockwork						
galvanised steel	-	0.06	0.98	0.15	nr	1.13
Chimney pots; red terracotta; plain or cannon-head; setting and flaunching in cement mortar (1:3)						
185 mm diameter x 300 mm long	18.89	1.91	31.06	21.62	nr	52.68
185 mm diameter x 600 mm long	33.64	2.13	34.63	37.47	nr	72.10
185 mm diameter x 900 mm long	61.31	2.13	34.63	67.22	nr	101.85
Galvanised steel lintels; "Catnic" or other equal and approved; built into brickwork or blockwork; "CN7" combined lintel; 143 mm high; for standard cavity walls						
750 mm long	20.30	0.27	4.39	21.86	nr	26.25
900 mm long	24.59	0.32	5.20	26.47	nr	31.67
1200 mm long	31.86	0.37	6.02	34.28	nr	40.30
1500 mm long	36.43	0.43	6.99	39.20	nr	46.19
1800 mm long	51.20	0.48	7.80	55.08	nr	62.88
2100 mm long	58.77	0.53	8.62	63.21	nr	71.83
Galvanised steel lintels; "Catnic" or other equal and approved; built into brickwork or blockwork; "CN8" combined lintel; 219 mm high; for standard cavity walls						
2400 mm long	71.97	0.64	10.41	77.40	nr	87.81
2700 mm long	84.85	0.74	12.03	91.25	nr	103.28
3000 mm long	113.12	0.85	13.82	121.64	nr	135.46
3300 mm long	126.27	0.95	15.45	135.78	nr	151.23
3600 mm long	138.56	1.06	17.24	148.99	nr	166.23
3900 mm long	180.27	1.17	19.02	193.83	nr	212.85
4200 mm long	197.69	0.53	8.62	212.55	nr	221.17
Galvanised steel lintels; "Catnic" or other equal and approved; built into brickwork or blockwork; "CN92" single lintel; for 75 mm internal walls						
900 mm long	3.04	0.32	5.20	3.29	nr	8.49
1050 mm long	3.65	0.32	5.20	3.94	nr	9.14
1200 mm long	4.14	0.37	6.02	4.47	nr	10.49
Galvanised steel lintels; "Catnic" or other equal and approved; builtin to brickwork or blockwork; "CN102" single lintel; for 100 mm internal walls						
900 mm long	3.74	0.32	5.20	4.04	nr	9.24
1050 mm long	4.63	0.32	5.20	4.99	nr	10.19
1200 mm long	5.11	0.37	6.02	5.51	nr	11.53

Prices for Measured Works - Minor Works
F MASONRY

	PC £	Labour hours	Labour £	Material £	Unit	Total rate £
F31 PRECAST CONCRETE SILLS/LINTELS/ COPINGS/FEATURES						
Mix 20.00 N/mm² - 20 mm aggregate (1:2:4)						
Lintels; plate; prestressed bedded						
100 mm x 70 mm x 750 mm long	3.01	0.43	6.99	3.26	nr	10.25
100 mm x 70 mm x 900 mm long	3.58	0.43	6.99	3.87	nr	10.86
100 mm x 70 mm x 1050 mm long	4.24	0.43	6.99	4.58	nr	11.57
100 mm x 70 mm x 1200 mm long	4.80	0.43	6.99	5.18	nr	12.17
150 mm x 70 mm x 900 mm long	4.91	0.53	8.62	5.31	nr	13.93
150 mm x 70 mm x 1050 mm long	5.72	0.53	8.62	6.18	nr	14.80
150 mm x 70 mm x 1200 mm long	6.54	0.53	8.62	7.06	nr	15.68
220 mm x 70 mm x 900 mm long	7.56	0.64	10.41	8.17	nr	18.58
220 mm x 70 mm x 1200 mm long	10.07	0.64	10.41	10.87	nr	21.28
220 mm x 70 mm x 1500 mm long	12.62	0.74	12.03	13.61	nr	25.64
265 mm x 70 mm x 900 mm long	8.07	0.64	10.41	8.73	nr	19.14
265 mm x 70 mm x 1200 mm long	10.78	0.64	10.41	11.64	nr	22.05
265 mm x 70 mm x 1500 mm long	13.44	0.74	12.03	14.50	nr	26.53
265 mm x 70 mm x 1800 mm long	16.09	0.85	13.82	17.35	nr	31.17
Lintels; rectangular; reinforced with mild steel bars; bedded						
100 mm x 145 mm x 900 mm long	7.31	0.64	10.41	7.38	nr	18.29
100 mm x 145 mm x 1050 mm long	8.58	0.64	10.41	9.24	nr	19.65
100 mm x 145 mm x 1200 mm long	9.35	0.64	10.41	10.07	nr	20.48
225 mm x 145 mm x 1200 mm long	12.31	0.85	13.82	13.28	nr	27.10
225 mm x 225 mm x 1800 mm long	23.86	1.60	26.02	25.70	nr	51.72
Lintels; boot; reinforced with mild steel bars; bedded						
250 mm x 225 mm x 1200 mm long	19.77	1.28	20.81	21.30	nr	42.11
275 mm x 225 mm x 1800 mm long	26.52	1.91	31.06	28.55	nr	59.61
Padstones						
300 mm x 100 mm x 75 mm	4.04	0.32	5.20	4.36	nr	9.56
225 mm x 225 mm x 150 mm	6.13	0.43	6.99	6.62	nr	13.61
450 mm x 450 mm x 150 mm	16.20	0.64	10.41	17.55	nr	27.96
Mix 30.00 N/mm² - 20 mm aggregate (1:1:2)						
Copings; once weathered; once throated; bedded and pointed						
152 mm x 76 mm	4.91	0.74	12.03	5.61	m	17.64
178 mm x 64 mm	5.42	0.74	12.03	6.18	m	18.21
305 mm x 76 mm	9.15	0.85	13.82	10.48	m	24.30
extra for fair ends	-	-	-	4.50	nr	4.50
extra for angles	-	-	-	5.11	nr	5.11
Copings; twice weathered; twice throated; bedded and pointed						
152 mm x 76 mm	4.91	0.74	12.03	5.61	m	17.64
178 mm x 64 mm	5.36	0.74	12.03	6.12	m	18.15
305 mm x 76 mm	9.15	0.85	13.82	10.48	m	24.30
extra for fair ends	-	-	-	4.50	nr	4.50
extra for angles	-	-	-	5.11	nr	5.11

G STRUCTURAL/CARCASSING METAL/TIMBER

G10 STRUCTURAL STEEL FRAMING

Framing, fabrication; weldable steel; BS EN 10025: 1993 Grade S275; hot rolled structural steel sections; welded fabrication

	PC £	Labour hours	Labour £	Material £	Unit	Total rate £
Columns						
weight not exceeding 40 kg/m	-	-	-	-	t	918.78
weight not exceeding 40 kg/m; castellated	-	-	-	-	t	1038.62
weight not exceeding 40 kg/m; curved	-	-	-	-	t	1263.32
weight not exceeding 40 kg/m; square hollow section	-	-	-	-	t	1383.16
weight not exceeding 40 kg/m; circular hollow section	-	-	-	-	t	1532.97
weight 40 - 100 kg/m	-	-	-	-	t	898.81
weight 40 - 100 kg/m; castellated	-	-	-	-	t	1013.66
weight 40 - 100 kg/m; curved	-	-	-	-	t	1243.36
weight 40 - 100 kg/m; square hollow section	-	-	-	-	t	1353.21
weight 40 - 100 kg/m; circular hollow section	-	-	-	-	t	1498.01
weight exceeding 100 kg/m	-	-	-	-	t	883.82
weight exceeding 100 kg/m; castellated	-	-	-	-	t	998.67
weight exceeding 100 kg/m; curved	-	-	-	-	t	1218.38
weight exceeding 100 kg/m; square hollow section	-	-	-	-	t	1328.24
weight exceeding 100 kg/m; circular hollow section	-	-	-	-	t	1478.04
Beams						
weight not exceeding 40 kg/m	-	-	-	-	t	963.73
weight not exceeding 40 kg/m; castellated	-	-	-	-	t	1098.54
weight not exceeding 40 kg/m; curved	-	-	-	-	t	1413.13
weight not exceeding 40 kg/m; square hollow section	-	-	-	-	t	1498.01
weight not exceeding 40 kg/m; circular hollow section	-	-	-	-	t	1722.72
weight 40 - 100 kg/m	-	-	-	-	t	938.75
weight 40 - 100 kg/m; castellated	-	-	-	-	t	1063.58
weight 40 - 100 kg/m; curved	-	-	-	-	t	1388.16
weight 40 - 100 kg/m; square hollow section	-	-	-	-	t	1468.05
weight 40 - 100 kg/m; circular hollow section	-	-	-	-	t	1687.76
weight exceeding 100 kg/m	-	-	-	-	t	928.77
weight exceeding 100 kg/m; castellated	-	-	-	-	t	1048.61
weight exceeding 100 kg/m; curved	-	-	-	-	t	1363.20
weight exceeding 100 kg/m; square hollow section	-	-	-	-	t	1453.07
weight exceeding 100 kg/m; circular hollow section	-	-	-	-	t	1667.79
Bracings						
weight not exceeding 40 kg/m	-	-	-	-	t	1238.36
weight not exceeding 40 kg/m; square hollow section	-	-	-	-	t	1722.72
weight not exceeding 40 kg/m; circular hollow section	-	-	-	-	t	1927.44
weight 40 - 100 kg/m	-	-	-	-	t	1208.40
weight 40 - 100 kg/m; square hollow section	-	-	-	-	t	1697.75
weight 40 - 100 kg/m; circular hollow section	-	-	-	-	t	1902.48
weight exceeding 100 kg/m	-	-	-	-	t	1168.45
weight exceeding 100 kg/m; square hollow section	-	-	-	-	t	1672.79
weight exceeding 100 kg/m; circular hollow section	-	-	-	-	t	1867.52

Prices for Measured Works - Minor Works
G STRUCTURAL/CARCASSING METAL/TIMBER

	PC £	Labour hours	Labour £	Material £	Unit	Total rate £
G10 STRUCTURAL STEEL FRAMING - cont'd						
Framing, fabrication; weldable steel; BS EN 10025: 1993 Grade S275; hot rolled structural steel sections; welded fabrication - cont'd						
Purlins and cladding rails						
weight not exceeding 40 kg/m	-	-	-	-	t	1088.56
weight not exceeding 40 kg/m; square hollow section	-	-	-	-	t	1552.94
weight not exceeding 40 kg/m; circular hollow section	-	-	-	-	t	1722.72
weight 40 - 100 kg/m	-	-	-	-	t	1063.58
weight 40 - 100 kg/m; square hollow section	-	-	-	-	t	1527.97
weight 40 - 100 kg/m; circular hollow section	-	-	-	-	t	1697.75
weight exceeding 100 kg/m	-	-	-	-	t	1038.62
weight exceeding 100 kg/m; square hollow section	-	-	-	-	t	1498.01
weight exceeding 100 kg/m; circular hollow section	-	-	-	-	t	1672.79
Grillages						
weight not exceeding 40 kg/m	-	-	-	-	t	749.01
weight 40 - 100 kg/m	-	-	-	-	t	788.95
weight exceeding 100 kg/m	-	-	-	-	t	749.01
Trestles, towers and built up columns						
straight	-	-	-	-	t	1438.09
Trusses and built up girders						
straight	-	-	-	-	t	1438.09
curved	-	-	-	-	t	1837.56
Fittings	-	-	-	-	t	1038.62
Framing, erection						
Trial erection	-	-	-	-	t	254.67
Permanent erection on site	-	-	-	-	t	209.72
Surface preparation						
At works						
blast cleaning	-	-	-	-	m²	2.00
Surface treatment						
At works						
galvanising	-	-	-	-	m²	12.99
grit blast and one coat zinc chromate primer	-	-	-	-	m²	3.49
touch up primer and one coat of two pack epoxy zinc phosphate primer	-	-	-	-	m²	4.15
intumescent paint fire protection, 30 minutes; spray applied	-	-	-	-	m²	12.00
intumescent paint fire protection, 60 minutes; spray applied	-	-	-	-	m²	18.00
On site						
intumescent paint fire protection, 30 minutes; spray applied	-	-	-	-	m²	9.00
intumescent paint fire protection, 60 minutes; spray applied	-	-	-	-	m²	11.00
G12 ISOLATED STRUCTURAL METAL MEMBERS						
Isolated structural member; weldable steel; BS EN 10025: 1993 Grade S275; hot rolled structural steel sections						
Plain member; beams						
weight not exceeding 40 kg/m	-	-	-	-	t	674.11
weight 40 - 100 kg/m	-	-	-	-	t	644.15
weight exceeding 100 kg/m	-	-	-	-	t	629.17

G STRUCTURAL/CARCASSING METAL/TIMBER

	PC £	Labour hours	Labour £	Material £	Unit	Total rate £
Metsec open web steel lattice beams; in single members; raised 3.50 m above ground; ends built in						
Beams; one coat zinc phosphate primer at works						
200 mm deep; to span 5.00 m (7.69 kg/m); ref B22	-	0.20	3.29	30.64	m	33.93
200 mm deep; to span 5.00 m (7.64 kg/m); ref B27	-	0.26	4.27	27.61	m	31.88
300 mm deep; to span 7.00 m (10.26 kg/m); ref B30	-	0.26	4.27	33.67	m	37.94
350 mm deep; to span 8.50 m (10.60 kg/m); ref B35	-	0.26	4.27	36.72	m	40.99
350 mm deep; to span 10.00 m (12.76 kg/m); ref D35	-	0.31	5.10	51.87	m	56.97
350 mm deep; to span 11.50 m (17.08 kg/m); ref D50	-	0.36	5.92	57.94	m	63.86
450 mm deep; to span 13.00 m (25.44 kg/m); ref G50	-	0.51	8.38	84.60	m	92.98
Beams; galvanised						
200 mm deep; to span 5.00 m (7.69 kg/m); ref B22	-	0.20	3.29	32.20	m	35.49
200 mm deep; to span 5.00 m (7.64 kg/m); ref B27	-	0.26	4.27	29.02	m	33.29
300 mm deep; to span 7.00 m (10.26 kg/m); ref B30	-	0.26	4.27	35.39	m	39.66
350 mm deep; to span 8.50 m (10.60 kg/m); ref B35	-	0.26	4.27	38.58	m	42.85
350 mm deep; to span 10.00 m (12.76 kg/m); ref D35	-	0.31	5.10	54.52	m	59.62
350 mm deep; to span 11.50 m (17.08 kg/m); ref D50	-	0.36	5.92	60.89	m	66.81
450 mm deep; to span 13.00 m (25.44 kg/m); ref G50	-	0.51	8.38	88.93	m	97.31
G20 CARPENTRY/TIMBER FRAMING/FIRST FIXING						
Sawn softwood; untreated						
Floor members						
38 mm x 100 mm	-	0.12	1.56	1.14	m	2.70
38 mm x 150 mm	-	0.14	1.82	1.66	m	3.48
50 mm x 75 mm	-	0.12	1.56	1.13	m	2.69
50 mm x 100 mm	-	0.14	1.82	1.44	m	3.26
50 mm x 125 mm	-	0.14	1.82	1.75	m	3.57
50 mm x 150 mm	-	0.16	2.08	2.12	m	4.20
50 mm x 175 mm	-	0.16	2.08	2.46	m	4.54
50 mm x 200 mm	-	0.17	2.21	2.89	m	5.10
50 mm x 225 mm	-	0.17	2.21	3.35	m	5.56
50 mm x 250 mm	-	0.18	2.34	3.80	m	6.14
75 mm x 125 mm	-	0.17	2.21	2.64	m	4.85
75 mm x 150 mm	-	0.17	2.21	3.16	m	5.37
75 mm x 175 mm	-	0.17	2.21	3.72	m	5.93
75 mm x 200 mm	-	0.18	2.34	4.34	m	6.68
75 mm x 225 mm	-	0.18	2.34	5.01	m	7.35
75 mm x 250 mm	-	0.19	2.47	5.70	m	8.17
100 mm x 160 mm	-	0.22	2.86	4.34	m	7.20
100 mm x 200 mm	-	0.23	2.99	5.93	m	8.92
100 mm x 250 mm	-	0.26	3.39	8.22	m	11.61
100 mm x 300 mm	-	0.28	3.65	10.15	m	13.80
Wall or partition members						
25 mm x 25 mm	-	0.07	0.91	0.33	m	1.24
25 mm x 38 mm	-	0.07	0.91	0.42	m	1.33
25 mm x 75 mm	-	0.09	1.17	0.57	m	1.74
38 mm x 38 mm	-	0.09	1.17	0.49	m	1.66
38 mm x 50 mm	-	0.09	1.17	0.56	m	1.73
38 mm x 75 mm	-	0.12	1.56	0.85	m	2.41
38 mm x 100 mm	-	0.16	2.08	1.14	m	3.22
50 mm x 50 mm	-	0.12	1.56	0.79	m	2.35
50 mm x 75 mm	-	0.16	2.08	1.15	m	3.23
50 mm x 100 mm	-	0.19	2.47	1.46	m	3.93
50 mm x 125 mm	-	0.19	2.47	1.77	m	4.24
75 mm x 75 mm	-	0.19	2.47	1.94	m	4.41
75 mm x 100 mm	-	0.21	2.73	2.16	m	4.89
100 mm x 100 mm	-	0.21	2.73	2.87	m	5.60

Prices for Measured Works - Minor Works
G STRUCTURAL/CARCASSING METAL/TIMBER

	PC £	Labour hours	Labour £	Material £	Unit	Total rate £
G20 CARPENTRY/TIMBER FRAMING/FIRST FIXING - cont'd						
Sawn softwood; untreated - cont'd						
Roof members; flat						
38 mm x 75 mm	-	0.14	1.82	0.85	m	2.67
38 mm x 100 mm	-	0.14	1.82	1.14	m	2.96
38 mm x 125 mm	-	0.14	1.82	1.42	m	3.24
38 mm x 150 mm	-	0.14	1.82	1.66	m	3.48
50 mm x 100 mm	-	0.14	1.82	1.44	m	3.26
50 mm x 125 mm	-	0.14	1.82	1.75	m	3.57
50 mm x 150 mm	-	0.16	2.08	2.12	m	4.20
50 mm x 175 mm	-	0.16	2.08	2.46	m	4.54
50 mm x 200 mm	-	0.17	2.21	2.89	m	5.10
50 mm x 225 mm	-	0.17	2.21	3.35	m	5.56
50 mm x 250 mm	-	0.18	2.34	3.80	m	6.14
75 mm x 150 mm	-	0.17	2.21	3.16	m	5.37
75 mm x 175 mm	-	0.17	2.21	3.72	m	5.93
75 mm x 200 mm	-	0.18	2.34	4.34	m	6.68
75 mm x 225 mm	-	0.18	2.34	5.01	m	7.35
75 mm x 250 mm	-	0.19	2.47	5.70	m	8.17
Roof members; pitched						
25 mm x 100 mm	-	0.12	1.56	0.78	m	2.34
25 mm x 125 mm	-	0.12	1.56	0.96	m	2.52
25 mm x 150 mm	-	0.16	2.08	1.21	m	3.29
25 mm x 175 mm	-	0.18	2.34	1.59	m	3.93
25 mm x 200 mm	-	0.19	2.47	1.85	m	4.32
38 mm x 100 mm	-	0.16	2.08	1.14	m	3.22
38 mm x 125 mm	-	0.16	2.08	1.42	m	3.50
38 mm x 150 mm	-	0.16	2.08	1.66	m	3.74
50 mm x 50 mm	-	0.12	1.56	0.77	m	2.33
50 mm x 75 mm	-	0.16	2.08	1.13	m	3.21
50 mm x 100 mm	-	0.19	2.47	1.44	m	3.91
50 mm x 125 mm	-	0.19	2.47	1.75	m	4.22
50 mm x 150 mm	-	0.21	2.73	2.12	m	4.85
50 mm x 175 mm	-	0.21	2.73	2.46	m	5.19
50 mm x 200 mm	-	0.21	2.73	2.89	m	5.62
50 mm x 225 mm	-	0.21	2.73	3.35	m	6.08
75 mm x 100 mm	-	0.26	3.39	2.12	m	5.51
75 mm x 125 mm	-	0.26	3.39	2.64	m	6.03
75 mm x 150 mm	-	0.26	3.39	3.16	m	6.55
100 mm x 150 mm	-	0.31	4.04	4.36	m	8.40
100 mm x 175 mm	-	0.31	4.04	5.05	m	9.09
100 mm x 200 mm	-	0.31	4.04	5.93	m	9.97
100 mm x 225 mm	-	0.33	4.30	7.24	m	11.54
100 mm x 250 mm	-	0.33	4.30	8.22	m	12.52
Plates						
38 mm x 75 mm	-	0.12	1.56	0.85	m	2.41
38 mm x 100 mm	-	0.16	2.08	1.14	m	3.22
50 mm x 75 mm	-	0.16	2.08	1.15	m	3.23
50 mm x 100 mm	-	0.19	2.47	1.44	m	3.91
75 mm x 100 mm	-	0.21	2.73	2.16	m	4.89
75 mm x 125 mm	-	0.24	3.12	2.62	m	5.74
75 mm x 150 mm	-	0.27	3.52	3.14	m	6.66

Prices for Measured Works - Minor Works
G STRUCTURAL/CARCASSING METAL/TIMBER

	PC £	Labour hours	Labour £	Material £	Unit	Total rate £
Plates; fixing by bolting						
38 mm x 75 mm	-	0.21	2.73	0.85	m	3.58
38 mm x 100 mm	-	0.25	3.25	1.14	m	4.39
50 mm x 75 mm	-	0.25	3.25	1.15	m	4.40
50 mm x 100 mm	-	0.28	3.65	1.44	m	5.09
75 mm x 100 mm	-	0.31	4.04	2.16	m	6.20
75 mm x 125 mm	-	0.33	4.30	2.62	m	6.92
75 mm x 150 mm	-	0.36	4.69	3.14	m	7.83
Joist strutting; herringbone strutting						
50 mm x 50 mm; depth of joist 150 mm	-	0.51	6.64	1.81	m	8.45
50 mm x 50 mm; depth of joist 175 mm	-	0.51	6.64	1.85	m	8.49
50 mm x 50 mm; depth of joist 200 mm	-	0.51	6.64	1.88	m	8.52
50 mm x 50 mm; depth of joist 225 mm	-	0.51	6.64	1.92	m	8.56
50 mm x 50 mm; depth of joist 250 mm	-	0.51	6.64	1.96	m	8.60
Joist strutting; block						
50 mm x 150 mm; depth of joist 150 mm	-	0.31	4.04	2.40	m	6.44
50 mm x 175 mm; depth of joist 175 mm	-	0.31	4.04	2.73	m	6.77
50 mm x 200 mm; depth of joist 200 mm	-	0.31	4.04	3.16	m	7.20
50 mm x 225 mm; depth of joist 225 mm	-	0.31	4.04	3.62	m	7.66
50 mm x 250 mm; depth of joist 250 mm	-	0.31	4.04	4.07	m	8.11
Cleats						
225 mm x 100 mm x 75 mm	-	0.20	2.60	0.56	nr	3.16
Extra for stress grading to above timbers						
general structural (GS) grade	-	-	-	16.44	m³	16.44
special structural (SS) grade	-	-	-	31.63	m³	31.63
Extra for protecting and flameproofing timber with "Celgard CF" protection						
small sections	-	-	-	120.58	m³	120.58
large sections	-	-	-	115.35	m³	115.35
Wrot surfaces						
plain; 50 mm wide	-	0.02	0.26	-	m	0.26
plain; 100 mm wide	-	0.03	0.39	-	m	0.39
plain; 150 mm wide	-	0.05	0.65	-	m	0.65
Sawn softwood; "Tanalised"						
Floor members						
38 mm x 75 mm	-	0.12	1.56	0.96	m	2.52
38 mm x 100 mm	-	0.12	1.56	1.29	m	2.85
38 mm x 150 mm	-	0.14	1.82	1.88	m	3.70
50 mm x 75 mm	-	0.12	1.56	1.27	m	2.83
50 mm x 100 mm	-	0.14	1.82	1.63	m	3.45
50 mm x 125 mm	-	0.14	1.82	1.99	m	3.81
50 mm x 150 mm	-	0.16	2.08	2.41	m	4.49
50 mm x 175 mm	-	0.16	2.08	2.80	m	4.00
50 mm x 200 mm	-	0.17	2.21	3.27	m	5.48
50 mm x 225 mm	-	0.17	2.21	3.78	m	5.99
50 mm x 250 mm	-	0.18	2.34	4.27	m	6.61
75 mm x 125 mm	-	0.17	2.21	3.00	m	5.21
75 mm x 150 mm	-	0.17	2.21	3.59	m	5.80
75 mm x 175 mm	-	0.17	2.21	4.22	m	6.43
75 mm x 200 mm	-	0.18	2.34	4.91	m	7.25
75 mm x 225 mm	-	0.18	2.34	5.66	m	8.00
75 mm x 250 mm	-	0.19	2.47	6.42	m	8.89
100 mm x 150 mm	-	0.22	2.86	4.91	m	7.77
100 mm x 200 mm	-	0.23	2.99	6.70	m	9.69
100 mm x 250 mm	-	0.26	3.39	9.18	m	12.57
100 mm x 300 mm	-	0.28	3.65	11.30	m	14.95

Prices for Measured Works - Minor Works
G STRUCTURAL/CARCASSING METAL/TIMBER

	PC £	Labour hours	Labour £	Material £	Unit	Total rate £
G20 CARPENTRY/TIMBER FRAMING/FIRST FIXING - cont'd						
Sawn softwood; "Tanalised" - cont'd						
Wall or partition members						
25 mm x 25 mm	-	0.07	0.91	0.35	m	1.26
25 mm x 38 mm	-	0.07	0.91	0.46	m	1.37
25 mm x 75 mm	-	0.09	1.17	0.65	m	1.82
38 mm x 38 mm	-	0.09	1.17	0.55	m	1.72
38 mm x 50 mm	-	0.09	1.17	0.63	m	1.80
38 mm x 75 mm	-	0.12	1.56	0.96	m	2.52
38 mm x 100 mm	-	0.16	2.08	1.29	m	3.37
50 mm x 50 mm	-	0.12	1.56	0.89	m	2.45
50 mm x 75 mm	-	0.16	2.08	1.30	m	3.38
50 mm x 100 mm	-	0.19	2.47	1.65	m	4.12
50 mm x 125 mm	-	0.19	2.47	2.02	m	4.49
75 mm x 75 mm	-	0.19	2.47	2.15	m	4.62
75 mm x 100 mm	-	0.21	2.73	2.45	m	5.18
100 mm x 100 mm	-	0.21	2.73	3.25	m	5.98
Roof members; flat						
38 mm x 75 mm	-	0.14	1.82	0.96	m	2.78
38 mm x 100 mm	-	0.14	1.82	1.29	m	3.11
38 mm x 125 mm	-	0.14	1.82	1.60	m	3.42
38 mm x 150 mm	-	0.14	1.82	1.88	m	3.70
50 mm x 100 mm	-	0.14	1.82	1.63	m	3.45
50 mm x 125 mm	-	0.14	1.82	1.99	m	3.81
50 mm x 150 mm	-	0.16	2.08	2.41	m	4.49
50 mm x 175 mm	-	0.16	2.08	2.80	m	4.88
50 mm x 200 mm	-	0.17	2.21	3.27	m	5.48
50 mm x 225 mm	-	0.17	2.21	3.78	m	5.99
50 mm x 250 mm	-	0.18	2.34	4.27	m	6.61
75 mm x 150 mm	-	0.17	2.21	3.59	m	5.80
75 mm x 175 mm	-	0.17	2.21	4.22	m	6.43
75 mm x 200 mm	-	0.18	2.34	4.91	m	7.25
75 mm x 225 mm	-	0.18	2.34	5.66	m	8.00
75 mm x 250 mm	-	0.19	2.47	6.42	m	8.89
Roof members; pitched						
25 mm x 100 mm	-	0.12	1.56	0.88	m	2.44
25 mm x 125 mm	-	0.12	1.56	1.07	m	2.63
25 mm x 150 mm	-	0.16	2.08	1.36	m	3.44
25 mm x 175 mm	-	0.18	2.34	1.76	m	4.10
25 mm x 200 mm	-	0.19	2.47	2.04	m	4.51
38 mm x 100 mm	-	0.16	2.08	1.29	m	3.37
38 mm x 125 mm	-	0.16	2.08	1.60	m	3.68
38 mm x 150 mm	-	0.16	2.08	1.88	m	3.96
50 mm x 50 mm	-	0.12	1.56	0.87	m	2.43
50 mm x 75 mm	-	0.16	2.08	1.27	m	3.35
50 mm x 100 mm	-	0.19	2.47	1.63	m	4.10
50 mm x 125 mm	-	0.19	2.47	1.99	m	4.46
50 mm x 150 mm	-	0.21	2.73	2.41	m	5.14
50 mm x 175 mm	-	0.21	2.73	2.80	m	5.53
50 mm x 200 mm	-	0.21	2.73	3.27	m	6.00
50 mm x 225 mm	-	0.21	2.73	3.78	m	6.51
75 mm x 100 mm	-	0.26	3.39	2.41	m	5.80
75 mm x 125 mm	-	0.26	3.39	3.00	m	6.39
75 mm x 150 mm	-	0.26	3.39	3.59	m	6.98
100 mm x 150 mm	-	0.31	4.04	4.93	m	8.97
100 mm x 175 mm	-	0.31	4.04	5.72	m	9.76
100 mm x 200 mm	-	0.31	4.04	6.70	m	10.74
100 mm x 225 mm	-	0.33	4.30	8.10	m	12.40
100 mm x 250 mm	-	0.33	4.30	9.18	m	13.48

G STRUCTURAL/CARCASSING METAL/TIMBER

	PC £	Labour hours	Labour £	Material £	Unit	Total rate £
Plates						
38 mm x 75 mm	-	0.12	1.56	0.96	m	2.52
38 mm x 100 mm	-	0.16	2.08	1.29	m	3.37
50 mm x 75 mm	-	0.16	2.08	1.30	m	3.38
50 mm x 100 mm	-	0.19	2.47	1.65	m	4.12
75 mm x 100 mm	-	0.21	2.73	2.45	m	5.18
75 mm x 125 mm	-	0.24	3.12	2.98	m	6.10
75 mm x 150 mm	-	0.27	3.52	3.57	m	7.09
Plates; fixing by bolting						
38 mm x 75 mm	-	0.21	2.73	0.96	m	3.69
38 mm x 100 mm	-	0.25	3.25	1.29	m	4.54
50 mm x 75 mm	-	0.25	3.25	1.30	m	4.55
50 mm x 100 mm	-	0.28	3.65	1.65	m	5.30
75 mm x 100 mm	-	0.31	4.04	2.45	m	6.49
75 mm x 125 mm	-	0.33	4.30	2.98	m	7.28
75 mm x 150 mm	-	0.36	4.69	3.57	m	8.26
Joist strutting; herringbone						
50 mm x 50 mm; depth of joist 150 mm	-	0.51	6.64	2.01	m	8.65
50 mm x 50 mm; depth of joist 175 mm	-	0.51	6.64	2.05	m	8.69
50 mm x 50 mm; depth of joist 200 mm	-	0.51	6.64	2.09	m	8.73
50 mm x 50 mm; depth of joist 225 mm	-	0.51	6.64	2.13	m	8.77
50 mm x 50 mm; depth of joist 250 mm	-	0.51	6.64	2.17	m	8.81
Joist strutting; block						
50 mm x 150 mm; depth of joist 150 mm	-	0.31	4.04	2.69	m	6.73
50 mm x 175 mm; depth of joist 175 mm	-	0.31	4.04	3.07	m	7.11
50 mm x 200 mm; depth of joist 200 mm	-	0.31	4.04	3.54	m	7.58
50 mm x 225 mm; depth of joist 225 mm	-	0.31	4.04	4.05	m	8.09
50 mm x 250 mm; depth of joist 250 mm	-	0.31	4.04	4.55	m	8.59
Cleats						
225 mm x 100 mm x 75 mm	-	0.20	2.60	0.62	nr	3.22
Extra for stress grading to above timbers						
general structural (GS) grade	-	-	-	16.44	m³	16.44
special structural (SS) grade	-	-	-	31.63	m³	31.63
Extra for protecting and flameproofing timber with "Celgard CF" protection or other equal or approved						
small sections	-	-	-	120.58	m³	120.58
large sections	-	-	-	115.35	m³	115.35
Wrot surfaces						
plain; 50 mm wide	-	0.02	0.26	-	m	0.26
plain; 100 mm wide	-	0.03	0.39	-	m	0.39
plain; 150 mm wide	-	0.05	0.65	-	m	0.65
Trussed rafters, stress graded sawn softwood pressure impregnated; raised through two storeys and fixed in position						
"W" type truss (Fink); 22.50 degree pitch; 450 mm eaves overhang						
5.00 m span	-	1.63	21.22	27.59	nr	48.81
7.60 m span	-	1.79	23.31	34.79	nr	58.10
10.00 m span	-	2.04	26.56	45.30	nr	71.86
"W" type truss (Fink); 30 degree pitch; 450 mm eaves overhang						
5.00 m span	-	1.63	21.22	27.38	nr	48.60
7.60 m span	-	1.79	23.31	35.73	nr	59.04
10.00 m span	-	2.04	26.56	51.26	nr	77.82
"W" type truss (Fink); 45 degree pitch; 450 mm eaves overhang						
4.60 m span	-	1.63	21.22	29.41	nr	50.63
7.00 m span	-	1.79	23.31	39.61	nr	62.92

Prices for Measured Works - Minor Works
G STRUCTURAL/CARCASSING METAL/TIMBER

	PC £	Labour hours	Labour £	Material £	Unit	Total rate £
G20 CARPENTRY/TIMBER FRAMING/FIRST FIXING - cont'd						
Trussed rafters, stress graded sawn softwood pressure impregnated; raised through two storeys and fixed in position - cont'd						
"Mono" type truss; 17.50 degree pitch; 450 mm eaves overhang						
3.30 m span	-	1.43	18.62	22.25	nr	40.87
5.60 m span	-	1.63	21.22	31.52	nr	52.74
7.00 m span	-	1.88	24.48	39.61	nr	64.09
"Mono" type truss; 30 degree pitch; 450 mm eaves overhang						
3.30 m span	-	1.43	18.62	23.91	nr	42.53
5.60 m span	-	1.63	21.22	34.13	nr	55.35
7.00 m span	-	1.88	24.48	44.99	nr	69.47
"Attic" type truss; 45 degree pitch; 450 mm eaves overhang						
5.00 m span	-	3.21	41.79	59.64	nr	101.43
7.60 m span	-	3.36	43.75	96.80	nr	140.55
9.00 m span	-	3.56	46.35	123.97	nr	170.32
"Moelven Toreboda" glulam timber beams or other equal and approved; Moelven Laminated Timber Structures; LB grade whitewood; pressure impregnated; phenbol resorcinal adhesive; clean planed finish; fixed						
Laminated roof beams						
56 mm x 225 mm	-	0.56	7.29	5.39	m	12.68
66 mm x 315 mm	-	0.71	9.24	10.53	m	19.77
90 mm x 315 mm	-	0.92	11.98	14.36	m	26.34
90 mm x 405 mm	-	1.17	15.23	18.47	m	33.70
115 mm x 405 mm	-	1.48	19.27	23.59	m	42.86
115 mm x 495 mm	-	1.83	23.83	28.84	m	52.67
115 mm x 630 mm	-	2.24	29.16	36.71	m	65.87
"Masterboard" or other equal and approved; 6 mm thick						
Eaves, verge soffit boards, fascia boards and the like						
over 300 mm wide	7.92	0.71	9.24	9.58	m²	18.82
75 mm wide	-	0.21	2.73	0.73	m	3.46
150 mm wide	-	0.24	3.12	1.44	m	4.56
225 mm wide	-	0.29	3.78	2.14	m	5.92
300 mm wide	-	0.31	4.04	2.85	m	6.89
Plywood; external quality; 15 mm thick						
Eaves, verge soffit boards, fascia boards and the like						
over 300 mm wide	10.64	0.83	10.81	12.72	m²	23.53
75 mm wide	-	0.26	3.39	0.96	m	4.35
150 mm wide	-	0.30	3.91	1.91	m	5.82
225 mm wide	-	0.33	4.30	2.85	m	7.15
300 mm wide	-	0.38	4.95	3.79	m	8.74

G STRUCTURAL/CARCASSING METAL/TIMBER

	PC £	Labour hours	Labour £	Material £	Unit	Total rate £
Plywood; external quality; 18 mm thick						
Eaves, verge soffit boards, fascia boards and the like						
over 300 mm wide	12.71	0.83	10.81	15.11	m²	25.92
75 mm wide	-	0.26	3.39	1.14	m	4.53
150 mm wide	-	0.30	3.91	2.27	m	6.18
225 mm wide	-	0.33	4.30	3.39	m	7.69
300 mm wide	-	0.38	4.95	4.51	m	9.46
Plywood; marine quality; 18 mm thick						
Gutter boards; butt joints						
over 300 mm wide	9.47	0.94	12.24	11.36	m²	23.60
150 mm wide	-	0.33	4.30	1.71	m	6.01
225 mm wide	-	0.38	4.95	2.57	m	7.52
300 mm wide	-	0.42	5.47	3.41	m	8.88
Eaves, verge soffit boards, fascia boards and the like						
over 300 mm wide	-	0.83	10.81	11.36	m²	22.17
75 mm wide	-	0.26	3.39	0.86	m	4.25
150 mm wide	-	0.30	3.91	1.71	m	5.62
225 mm wide	-	0.33	4.30	2.55	m	6.85
300 mm wide	-	0.38	4.95	3.39	m	8.34
Plywood; marine quality; 25 mm thick						
Gutter boards; butt joints						
over 300 mm wide	13.15	1.02	13.28	15.62	m²	28.90
150 mm wide	-	0.36	4.69	2.34	m	7.03
225 mm wide	-	0.41	5.34	3.53	m	8.87
300 mm wide	-	0.45	5.86	4.69	m	10.55
Eaves, verge soffit boards, fascia boards and the like						
over 300 mm wide	-	0.90	11.72	15.62	m²	27.34
75 mm wide	-	0.27	3.52	1.18	m	4.70
150 mm wide	-	0.31	4.04	2.34	m	6.38
225 mm wide	-	0.31	4.04	3.50	m	7.54
300 mm wide	-	0.41	5.34	4.66	m	10.00
Sawn softwood; untreated						
Gutter boards; butt joints						
19 mm thick; sloping	-	1.28	16.67	6.67	m²	23.34
19 mm thick; 75 mm wide	-	0.36	4.69	0.50	m	5.19
19 mm thick; 150 mm wide	-	0.41	5.34	0.97	m	6.31
19 mm thick; 225 mm wide	-	0.45	5.86	1.68	m	7.54
25 mm thick; sloping	-	1.28	16.67	8.52	m²	25.19
26 mm thick; 75 mm wide	-	0.36	4.69	0.60	m	5.29
25 mm thick; 150 mm wide	-	0.41	5.34	1.25	m	6.59
25 mm thick; 225 mm wide	-	0.45	5.86	2.23	m	8.09
Cesspools with 25 mm thick sides and bottom						
225 mm x 225 mm x 150 mm	-	1.22	15.88	1.71	nr	17.59
300 mm x 300 mm x 150 mm	-	1.42	18.49	2.16	nr	20.65
Individual supports; firrings						
50 mm wide x 36 mm average depth	-	0.16	2.08	1.07	m	3.15
50 mm wide x 50 mm average depth	-	0.16	2.08	1.38	m	3.46
50 mm wide x 75 mm average depth	-	0.16	2.08	1.78	m	3.86
Individual supports; bearers						
25 mm x 50 mm	-	0.10	1.30	0.49	m	1.79
38 mm x 50 mm	-	0.10	1.30	0.60	m	1.90
50 mm x 50 mm	-	0.10	1.30	0.81	m	2.11
50 mm x 75 mm	-	0.10	1.30	1.17	m	2.47

Prices for Measured Works - Minor Works
G STRUCTURAL/CARCASSING METAL/TIMBER

	PC £	Labour hours	Labour £	Material £	Unit	Total rate £
G20 CARPENTRY/TIMBER FRAMING/FIRST FIXING - cont'd						
Sawn softwood; untreated - cont'd						
Individual supports; angle fillets						
38 mm x 38 mm	-	0.10	1.30	0.54	m	1.84
50 mm x 50 mm	-	0.10	1.30	0.81	m	2.11
75 mm x 75 mm	-	0.12	1.56	1.91	m	3.47
Individual supports; tilting fillets						
19 mm x 38 mm	-	0.10	1.30	0.25	m	1.55
25 mm x 50 mm	-	0.10	1.30	0.36	m	1.66
38 mm x 75 mm	-	0.10	1.30	0.59	m	1.89
50 mm x 75 mm	-	0.10	1.30	0.76	m	2.06
75 mm x 100 mm	-	0.16	2.08	1.41	m	3.49
Individual supports; grounds or battens						
13 mm x 19 mm	-	0.05	0.65	0.31	m	0.96
13 mm x 32 mm	-	0.05	0.65	0.33	m	0.98
25 mm x 50 mm	-	0.05	0.65	0.45	m	1.10
Individial supports; grounds or battens; plugged and screwed						
13 mm x 19 mm	-	0.16	2.08	0.34	m	2.42
13 mm x 32 mm	-	0.16	2.08	0.37	m	2.45
25 mm x 50 mm	-	0.16	2.08	0.48	m	2.56
Framed supports; open-spaced grounds or battens; at 300 mm centres one way						
25 mm x 50 mm	-	0.16	2.08	1.48	m^2	3.56
25 mm x 50 mm; plugged and screwed	-	0.45	5.86	1.61	m^2	7.47
Framed supports; at 300 mm centres one way and 600 mm centres the other way						
25 mm x 50 mm	-	0.77	10.03	2.21	m^2	12.24
38 mm x 50 mm	-	0.77	10.03	2.79	m^2	12.82
50 mm x 50 mm	-	0.77	10.03	3.83	m^2	13.86
50 mm x 75 mm	-	0.77	10.03	5.62	m^2	15.65
75 mm x 75 mm	-	0.77	10.03	9.55	m^2	19.58
Framed supports; at 300 mm centres one way and 600 mm centres the other way; plugged and screwed						
25 mm x 50 mm	-	1.28	16.67	2.58	m^2	19.25
38 mm x 50 mm	-	1.28	16.67	3.16	m^2	19.83
50 mm x 50 mm	-	1.28	16.67	4.20	m^2	20.87
50 mm x 75 mm	-	1.28	16.67	5.99	m^2	22.66
75 mm x 75 mm	-	1.28	16.67	9.92	m^2	26.59
Framed supports; at 500 mm centres both ways						
25 mm x 50 mm; to bath panels	-	0.92	11.98	2.88	m^2	14.86
Framed supports; as bracketing and cradling around steelwork						
25 mm x 50 mm	-	1.42	18.49	3.13	m^2	21.62
50 mm x 50 mm	-	1.53	19.92	5.39	m^2	25.31
50 mm x 75 mm	-	1.63	21.22	7.90	m^2	29.12
Sawn softwood; "Tanalised"						
Gutter boards; butt joints						
19 mm thick; sloping	-	1.28	16.67	7.40	m^2	24.07
19 mm thick; 75 mm wide	-	0.36	4.69	0.56	m	5.25
19 mm thick; 150 mm wide	-	0.41	5.34	1.08	m	6.42
19 mm thick; 225 mm wide	-	0.45	5.86	1.84	m	7.70
25 mm thick; sloping	-	1.28	16.67	9.48	m^2	26.15
25 mm thick; 75 mm wide	-	0.36	4.69	0.67	m	5.36
25 mm thick; 150 mm wide	-	0.41	5.34	1.40	m	6.74
25 mm thick; 225 mm wide	-	0.45	5.86	2.45	m	8.31

G STRUCTURAL/CARCASSING METAL/TIMBER

	PC £	Labour hours	Labour £	Material £	Unit	Total rate £
Cesspools with 25 mm thick sides and bottom						
225 mm x 225 mm x 150 mm	-	1.22	15.88	1.90	nr	17.78
300 mm x 300 mm x 150 mm	-	1.42	18.49	2.42	nr	20.91
Individual supports; firrings						
50 mm wide x 36 mm average depth	-	0.16	2.08	1.14	m	3.22
50 mm wide x 50 mm average depth	-	0.16	2.08	1.47	m	3.55
50 mm wide x 75 mm average depth	-	0.16	2.08	1.93	m	4.01
Individual suports; bearers						
25 mm x 50 mm	-	0.10	1.30	0.54	m	1.84
38 mm x 50 mm	-	0.10	1.30	0.68	m	1.98
50 mm x 50 mm	-	0.10	1.30	0.91	m	2.21
50 mm x 75 mm	-	0.10	1.30	1.32	m	2.62
Individual supports; angle fillets						
38 mm x 38 mm	-	0.10	1.30	0.56	m	1.86
50 mm x 50 mm	-	0.10	1.30	0.86	m	2.16
75 mm x 75 mm	-	0.12	1.56	2.02	m	3.58
Individual supports; tilting fillets						
19 mm x 38 mm	-	0.10	1.30	0.26	m	1.56
25 mm x 50 mm	-	0.10	1.30	0.38	m	1.68
38 mm x 75 mm	-	0.10	1.30	0.65	m	1.95
50 mm x 75 mm	-	0.10	1.30	0.83	m	2.13
75 mm x 100 mm	-	0.16	2.08	1.56	m	3.64
Individual supports; grounds or battens						
13 mm x 19 mm	-	0.05	0.65	0.32	m	0.97
13 mm x 32 mm	-	0.05	0.65	0.35	m	1.00
25 mm x 50 mm	-	0.05	0.65	0.50	m	1.15
Individual supports; grounds or battens; plugged and screwed						
13 mm x 19 mm	-	0.16	2.08	0.35	m	2.43
13 mm x 32 mm	-	0.16	2.08	0.38	m	2.46
25 mm x 50 mm	-	0.16	2.08	0.53	m	2.61
Framed supports; open-spaced grounds or battens; at 300 mm centres one way						
25 mm x 50 mm	-	0.16	2.08	1.64	m²	3.72
25 mm x 50 mm; plugged and screwed	-	0.45	5.86	1.77	m²	7.63
Framed supports; at 300 mm centres one way and 600 mm centres the other way						
25 mm x 50 mm	-	0.77	10.03	2.45	m²	12.48
38 mm x 50 mm	-	0.77	10.03	3.15	m²	13.18
50 mm x 50 mm	-	0.77	10.03	4.31	m²	14.34
50 mm x 75 mm	-	0.77	10.03	6.34	m²	16.37
75 mm x 75 mm	-	0.77	10.03	10.62	m²	20.65
Framed supports; at 300 mm centres one way and 600 mm centres the other way; plugged and screwed						
25 mm x 50 mm	-	1.28	16.67	2.82	m²	19.49
38 mm x 50 mm	-	1.28	16.67	3.52	m²	20.19
50 mm x 50 mm	-	1.28	16.67	4.68	m²	21.35
50 mm x 75 mm	-	1.28	16.67	6.71	m²	23.38
75 mm x 75 mm	PC	1.28	16.67	10.99	m²	27.66
Framed supports; at 500 mm centres both ways						
25 mm x 50 mm; to bath panels	-	0.92	11.98	3.19	m²	15.17
Framed supports; as bracketing and cradling around steelwork						
25 mm x 50 mm	-	1.42	18.49	3.47	m²	21.96
50 mm x 50 mm	-	1.53	19.92	6.06	m²	25.98
50 mm x 75 mm	-	1.63	21.22	8.91	m²	30.13

Prices for Measured Works - Minor Works
G STRUCTURAL/CARCASSING METAL/TIMBER

	PC £	Labour hours	Labour £	Material £	Unit	Total rate £
G20 CARPENTRY/TIMBER FRAMING/FIRST FIXING - cont'd						
Wrought softwood						
Gutter boards; tongued and grooved joints						
19 mm thick; sloping	-	1.53	19.92	8.60	m²	28.52
19 mm thick; 75 mm wide	-	0.41	5.34	1.02	m	6.36
19 mm thick; 150 mm wide	-	0.45	5.86	1.26	m	7.12
19 mm thick; 225 mm wide	-	0.51	6.64	2.24	m	8.88
25 mm thick; sloping	-	1.53	19.92	10.51	m²	30.43
25 mm thick; 75 mm wide	-	0.41	5.34	1.42	m	6.76
25 mm thick; 150 mm wide	-	0.45	5.86	1.50	m	7.36
25 mm thick; 225 mm wide	-	0.51	6.64	2.85	m	9.49
Eaves, verge soffit boards, fascia boards and the like						
19 mm thick; over 300 mm wide	-	1.26	16.41	8.60	m²	25.01
19 mm thick; 150 mm wide; once grooved	-	0.20	2.60	1.54	m	4.14
25 mm thick; 150 mm wide; once grooved	-	0.20	2.60	1.88	m	4.48
25 mm thick; 175 mm wide; once grooved	-	0.22	2.86	2.15	m	5.01
32 mm thick; 225 mm wide; once grooved	-	0.26	3.39	3.78	m	7.17
Wrought softwood; "Tanalised"						
Gutter boards; tongued and grooved joints						
19 mm thick; sloping	-	1.53	19.92	9.32	m²	29.24
19 mm thick; 75 mm wide	-	0.41	5.34	1.08	m	6.42
19 mm thick; 150 mm wide	-	0.45	5.86	1.37	m	7.23
19 mm thick; 225 mm wide	-	0.51	6.64	2.41	m	9.05
25 mm thick; sloping	-	1.53	19.92	11.46	m²	31.38
25 mm thick; 75 mm wide	-	0.41	5.34	1.49	m	6.83
25 mm thick; 150 mm wide	-	0.45	5.86	1.64	m	7.50
25 mm thick; 225 mm wide	-	0.51	6.64	3.06	m	9.70
Eaves, verge soffit boards, fascia boards and the like						
19 mm thick; over 300 mm wide	-	1.26	16.41	9.32	m²	25.73
19 mm thick; 150 mm wide; once grooved	-	0.20	2.60	1.65	m	4.25
25 mm thick; 150 mm wide; once grooved	-	0.20	2.60	2.02	m	4.62
25 mm thick; 175 mm wide; once grooved	-	0.22	2.86	2.32	m	5.18
32 mm thick; 225 mm wide; once grooved	-	0.26	3.39	4.05	m	7.44
Straps; mild steel; galvanised						
Straps; standard twisted vertical restraint; fixing to softwood and brick or blockwork						
30 mm x 2.50 mm x 400 mm girth	-	0.26	3.39	2.74	nr	6.13
30 mm x 2.50 mm x 600 mm girth	-	0.27	3.52	4.08	nr	7.60
30 mm x 2.50 mm x 800 mm girth	-	0.28	3.65	5.63	nr	9.28
30 mm x 2.50 mm x 1000 mm girth	-	0.31	4.04	7.29	nr	11.33
30 mm x 2.50 mm x 1200 mm girth	-	0.31	4.04	8.63	nr	12.67
Hangers; mild steel; galvanised						
Joist hangers 0.90 mm thick; The Expanded Metal Company Ltd "Speedy" or other equal and approved; for fixing to softwood; joist sizes						
50 mm wide; all sizes to 225 mm deep	0.83	0.12	1.56	1.03	nr	2.59
75 mm wide; all sizes to 225 mm deep	0.86	0.16	2.08	1.12	nr	3.20
100 mm wide; all sizes to 225 mm deep	0.93	0.19	2.47	1.25	nr	3.72

Prices for Measured Works - Minor Works
G STRUCTURAL/CARCASSING METAL/TIMBER

	PC £	Labour hours	Labour £	Material £	Unit	Total rate £
Joist hangers 2.50 mm thick; for building in; joist sizes						
50 mm x 100 mm	1.60	0.08	1.11	1.87	nr	2.98
50 mm x 125 mm	1.61	0.08	1.11	1.88	nr	2.99
50 mm x 150 mm	1.50	0.10	1.37	1.79	nr	3.16
50 mm x 175 mm	1.59	0.10	1.37	1.89	nr	3.26
50 mm x 200 mm	1.76	0.12	1.63	2.12	nr	3.75
50 mm x 225 mm	1.87	0.12	1.63	2.24	nr	3.87
75 mm x 150 mm	2.32	0.10	1.37	2.72	nr	4.09
75 mm x 175 mm	2.19	0.10	1.37	2.57	nr	3.94
75 mm x 200 mm	2.32	0.12	1.63	2.75	nr	4.38
75 mm x 225 mm	2.49	0.12	1.63	2.94	nr	4.57
75 mm x 250 mm	2.64	0.14	1.89	3.14	nr	5.03
100 mm x 200 mm	2.90	0.12	1.63	3.40	nr	5.03
Metal connectors; mild steel; galvanised						
Round toothed plate; for 10 mm or 12 mm diameter bolts						
38 mm diameter; single sided	-	0.02	0.26	0.25	nr	0.51
38 mm diameter; double sided	-	0.02	0.26	0.28	nr	0.54
50 mm diameter; single sided	-	0.02	0.26	0.27	nr	0.53
50 mm diameter; double sided	-	0.02	0.26	0.30	nr	0.56
63 mm diameter; single sided	-	0.02	0.26	0.39	nr	0.65
63 mm diameter; double sided	-	0.02	0.26	0.43	nr	0.69
75 mm diameter; single sided	-	0.02	0.26	0.58	nr	0.84
75 mm diameter; double sided	-	0.02	0.26	0.60	nr	0.86
framing anchor	-	0.16	2.08	0.45	nr	2.53
Bolts; mild steel; galvanised						
Fixing only bolts; 50 mm - 200 mm long						
6 mm diameter	-	0.03	0.39	-	nr	0.39
8 mm diameter	-	0.03	0.39	-	nr	0.39
10 mm diameter	-	0.05	0.65	-	nr	0.65
12 mm diameter	-	0.05	0.65	-	nr	0.65
16 mm diameter	-	0.06	0.78	-	nr	0.78
20 mm diameter	-	0.06	0.78	-	nr	0.78
Bolts						
Expanding bolts; "Rawlbolt" projecting type or other equal and approved; The Rawlplug Company; plated; one nut; one washer						
6 mm diameter; ref M6 10P	-	0.10	1.30	0.77	nr	2.07
6 mm diameter; ref M6 25P	-	0.10	1.30	0.86	nr	2.16
6 mm diameter; ref M6 60P	-	0.10	1.30	0.90	nr	2.20
8 mm diameter; ref M8 25P	-	0.10	1.30	1.02	nr	2.32
8 mm diameter; ref M8 60P	-	0.10	1.30	1.09	nr	2.39
10 mm diameter; ref M10 15P	-	0.10	1.30	1.33	nr	2.63
10 mm diameter; ref M10 30P	-	0.10	1.30	1.39	nr	2.69
10 mm diameter; ref M10 60P	-	0.10	1.30	1.45	nr	2.75
12 mm diameter; ref M12 15P	-	0.10	1.30	2.11	nr	3.41
12 mm diameter; ref M12 30P	-	0.10	1.30	2.27	nr	3.57
12 mm diameter; ref M12 75P	-	0.10	1.30	2.83	nr	4.13
16 mm diameter; ref M16 35P	-	0.10	1.30	5.40	nr	6.70
16 mm diameter; ref M16 75P	-	0.10	1.30	5.67	nr	6.97

Prices for Measured Works - Minor Works
G STRUCTURAL/CARCASSING METAL/TIMBER

	PC £	Labour hours	Labour £	Material £	Unit	Total rate £
G20 CARPENTRY/TIMBER FRAMING/FIRST FIXING - cont'd						
Bolts - cont'd						
Expanding bolts; "Rawlbolt" loose bolt type or other equal and approved; The Rawlplug Company; plated; one bolt; one washer						
6 mm diameter; ref M6 10L	-	0.10	1.30	0.77	nr	2.07
6 mm diameter; ref M6 25L	-	0.10	1.30	0.82	nr	2.12
6 mm diameter; ref M6 40L	-	0.10	1.30	0.82	nr	2.12
8 mm diameter; ref M8 25L	-	0.10	1.30	1.00	nr	2.30
8 mm diameter; ref M8 40L	-	0.10	1.30	1.06	nr	2.36
10 mm diameter; ref M10 10L	-	0.10	1.30	1.28	nr	2.58
10 mm diameter; ref M10 25L	-	0.10	1.30	1.32	nr	2.62
10 mm diameter; ref M10 50L	-	0.10	1.30	1.39	nr	2.69
10 mm diameter; ref M10 75L	-	0.10	1.30	1.44	nr	2.74
12 mm diameter; ref M12 10L	-	0.10	1.30	1.92	nr	3.22
12 mm diameter; ref M12 25L	-	0.10	1.30	2.11	nr	3.41
12 mm diameter; ref M12 40L	-	0.10	1.30	2.21	nr	3.51
12 mm diameter; ref M12 60L	-	0.10	1.30	2.32	nr	3.62
16 mm diameter; ref M16 30L	-	0.10	1.30	5.14	nr	6.44
16 mm diameter; ref M16 60L	-	0.10	1.30	5.56	nr	6.86
Truss clips; mild steel galvanised						
Truss clips; fixing to softwood; joist size						
38 mm wide	0.28	0.16	2.08	0.51	nr	2.59
50 mm wide	0.31	0.16	2.08	0.54	nr	2.62
Sole plate angles; mild steel; galvanised						
Sole plate angles; fixing to softwood and concrete						
112 mm x 40 mm x 76 mm	0.37	0.20	2.60	1.25	nr	3.85
Chemical anchors						
"Kemfix" capsules and standard studs or other equal and approved; The Rawlplug Company; with nuts and washers; drilling masonry						
capsule ref 60-408; stud ref 60-448	-	0.28	3.65	1.45	nr	5.10
capsule ref 60-410; stud ref 60-454	-	0.31	4.04	1.63	nr	5.67
capsule ref 60-412; stud ref 60-460	-	0.34	4.43	1.99	nr	6.42
capsule ref 60-416; stud ref 60-472	-	0.38	4.95	3.13	nr	8.08
capsule ref 60-420; stud ref 60-478	-	0.40	5.21	4.77	nr	9.98
capsule ref 60-424; stud ref 60-484	-	0.43	5.60	7.10	nr	12.70
"Kemfix" capsules and stainless steel studs or other equal and approved; The Rawlplug Company; with nuts and washers; drilling masonry						
capsule ref 60-408; stud ref 60-905	-	0.28	3.65	2.57	nr	6.22
capsule ref 60-410; stud ref 60-910	-	0.31	4.04	3.43	nr	7.47
capsule ref 60-412; stud ref 60-915	-	0.34	4.43	4.71	nr	9.14
capsule ref 60-416; stud ref 60-920	-	0.38	4.95	8.02	nr	12.97
capsule ref 60-420; stud ref 60-925	-	0.40	5.21	12.38	nr	17.59
capsule ref 60-424; stud ref 60-930	-	0.43	5.60	20.02	nr	25.62
"Kemfix" capsules and standard internal threaded sockets or other equal and approved; The Rawlplug Company; drilling masonry						
capsule ref 60-408; socket ref 60-650	-	0.28	3.65	1.50	nr	5.15
capsule ref 60-410; socket ref 60-656	-	0.31	4.04	1.82	nr	5.86
capsule ref 60-412; socket ref 60-662	-	0.34	4.43	2.27	nr	6.70
capsule ref 60-416; socket ref 60-668	-	0.38	4.95	3.12	nr	8.07
capsule ref 60-420; socket ref 60-674	-	0.40	5.21	3.75	nr	8.96
capsule ref 60-424; socket ref 60-676	-	0.43	5.60	6.09	nr	11.69

Prices for Measured Works - Minor Works
G STRUCTURAL/CARCASSING METAL/TIMBER

	PC £	Labour hours	Labour £	Material £	Unit	Total rate £
"Kemfix" capsules and stainless steel internal threaded sockets or other equal and approved; The Rawlplug Company; drilling masonry						
capsule ref 60-408; socket ref 60-943	-	0.28	3.65	2.85	nr	**6.50**
capsule ref 60-410; socket ref 60-945	-	0.31	4.04	3.21	nr	**7.25**
capsule ref 60-412; socket ref 60-947	-	0.34	4.43	3.71	nr	**8.14**
capsule ref 60-416; socket ref 60-949	-	0.38	4.95	5.27	nr	**10.22**
capsule ref 60-420; socket ref 60-951	-	0.40	5.21	6.36	nr	**11.57**
capsule ref 60-424; socket ref 60-955	-	0.43	5.60	11.63	nr	**17.23**
"Kemfix" capsules, perforated sleeves and standard studs or other equal and approved; The Rawlplug Company; in low density material; with nuts and washers; drilling masonry						
capsule ref 60-408; sleeve ref 60-538 stud ref 60-448	-	0.28	3.65	2.14	nr	**5.79**
capsule ref 60-410; sleeve ref 60-544; stud ref 60-454	-	0.31	4.04	2.35	nr	**6.39**
capsule ref 60-412; sleeve ref 60-550; stud ref 60-460	-	0.34	4.43	2.82	nr	**7.25**
capsule ref 60-416; sleeve ref 60-562; stud ref 60-472	-	0.38	4.95	4.15	nr	**9.10**
"Kemfix" capsules, perforated sleeves and stainless steel studs or other equal and approved; The Rawlplug Company; in low density material; with nuts and washers; drilling masonry						
capsule ref 60-408; sleeve ref 60-538; stud ref 60-905	-	0.28	3.65	3.25	nr	**6.90**
capsule ref 60-410; sleeve ref 60-544; stud ref 60-910	-	0.31	4.04	4.15	nr	**8.19**
capsule ref 60-412; sleeve ref 60-550; stud ref 60-915	-	0.34	4.43	5.54	nr	**9.97**
capsule ref 60-416; sleeve ref 60-562; stud ref 60-920	-	0.38	4.95	9.04	nr	**13.99**
"Kemfix" capsules, perforated sleeves and standard internal threaded sockets or other equal and approved; The Rawlplug Company; in low density material; with nuts and washers; drilling masonry						
capsule ref 60-408; sleeve ref 60-538; socket ref 60-650	-	0.28	3.65	2.18	nr	**5.83**
capsule ref 60-410; sleeve ref 60-544; socket ref 60-656	-	0.31	4.04	2.54	nr	**6.58**
capsule ref 60-412; sleeve ref 60-550; socket ref 60-662	-	0.34	4.43	3.10	nr	**7.53**
capsule ref 60-416; sleeve ref 60-562; socket ref 60-668	-	0.38	4.95	4.13	nr	**9.08**
"Kemfix" capsules, perforated sleeves and stainless steel internal threaded sockets or other equal and approved; The Rawlplug Company; in low density material; drilling masonry						
capsule ref 60-408; sleeve ref 60-538; socket ref 60-943	-	0.28	3.65	3.54	nr	**7.19**
capsule ref 60-410; sleeve ref 60-544; socket ref 60-945	-	0.31	4.04	3.93	nr	**7.97**
capsule ref 60-412; sleeve ref 60-550; socket ref 60-947	-	0.34	4.43	4.54	nr	**8.97**
capsule ref 60-416; sleeve ref 60-562; socket ref 60-949	-	0.38	4.95	6.29	nr	**11.24**

G STRUCTURAL/CARCASSING METAL/TIMBER

	PC £	Labour hours	Labour £	Material £	Unit	Total rate £
G32 EDGE SUPPORTED/REINFORCED WOODWOOL SLAB DECKING						
Woodwool interlocking reinforced slabs; Torvale "Woodcelip" or other equal and approved; natural finish; fixing to timber or steel with galvanized nails or clips; flat or sloping						
50 mm thick slabs; type 503; maximum span 2100 mm						
1800 mm - 2100 mm lengths	15.23	0.51	6.64	17.42	m²	24.06
2400 mm lengths	15.96	0.51	6.64	18.25	m²	24.89
2700 mm - 3000 mm lengths	16.22	0.51	6.64	18.54	m²	25.18
75 mm thick slabs; type 751; maximum span 2100 mm						
1800 mm - 2400 mm lengths	22.83	0.56	7.29	26.04	m²	33.33
2700 mm - 3000 mm lengths	22.94	0.56	7.29	26.17	m²	33.46
75 mm thick slabs; type 752; maximum span 2100 mm						
1800 mm - 2400 mm lengths	22.76	0.56	7.29	25.97	m²	33.26
2700 mm - 3000 mm lengths	22.81	0.56	7.29	26.02	m²	33.31
75 mm thick slabs; type 753; maximum span 3600 mm						
2400 mm lengths	22.08	0.56	7.29	25.20	m²	32.49
2700 mm - 3000 mm lengths	23.04	0.56	7.29	26.28	m²	33.57
3300 mm - 3900 mm lengths	27.77	0.56	7.29	31.62	m²	38.91
extra for holes for pipes and the like	-	0.12	1.56	-	nr	1.56
100 mm thick slabs; type 1001; maximum span 3600 mm						
3000 mm lengths	30.40	0.61	7.94	34.63	m²	42.57
3300 mm - 3600 mm lengths	32.90	0.61	7.94	37.45	m²	45.39
100 mm thick slabs; type 1002; maximum span 3600 mm						
3000 mm lengths	29.67	0.61	7.94	33.81	m²	41.75
3300 mm - 3600 mm lengths	31.51	0.61	7.94	35.88	m²	43.82
100 mm thick slabs; type 1003; maximum span 4000 mm						
3000 mm - 3600 mm lengths	28.19	0.61	7.94	32.14	m²	40.08
3900 mm - 4000 mm lengths	28.19	0.61	7.94	32.14	m²	40.08
125 mm thick slabs; type 1252; maximum span 3000 mm						
2400 mm - 3000 mm lengths	32.07	0.61	7.94	36.52	m²	44.46
Extra over slabs for						
pre-screeded deck	-	-	-	1.14	m²	1.14
pre-screeded soffit	-	-	-	2.93	m²	2.93
pre-screeded deck and soffit	-	-	-	3.83	m²	3.83
pre-screeded and proofed deck	-	-	-	2.36	m²	2.36
pre-screeded and proofed deck plus pre-screeded soffit	-	-	-	5.49	m²	5.49
pre-felted deck (glass fibre)	-	-	-	3.23	m²	3.23
pre-felted deck plus pre-screeded soffit	-	-	-	6.16	m²	6.16

H CLADDING/COVERING

	PC £	Labour hours	Labour £	Material £	Unit	Total rate £

H10 PATENT GLAZING

Patent glazing; aluminium alloy bars 2.55 m long at 622 mm centres; fixed to supports

	PC £	Labour hours	Labour £	Material £	Unit	Total rate £
Roof cladding						
glazing with 7 mm thick Georgian wired cast glass	-	-	-	-	m²	129.83
Associated code 4 lead flashings						
top flashing; 210 mm girth	-	-	-	-	m	44.94
bottom flashing; 240 mm girth	-	-	-	-	m	33.45
end flashing; 300 mm girth	-	-	-	-	m	74.90
Wall cladding						
glazing with 7 mm thick Georgian wired cast glass	-	-	-	-	m²	134.83
glazing with 6 mm thick plate glass	-	-	-	.	m²	124.84
Extra for aluminium alloy members						
38 mm x 38 mm x 3 mm angle jamb	-	-	-	-	m	19.97
pressed cill member	-	-	-	..	m	39.95
pressed channel head and PVC case	-	-	-	-	m	32.45

"Kawneer" window frame system or other equal and approved; polyester powder coated glazing bars; glazed with double hermetically sealed units in toughened safety glass; one 6 mm thick air space; overall 18 mm thick

	PC £	Labour hours	Labour £	Material £	Unit	Total rate £
Vertical surfaces						
single tier; aluminium glazing bars at 850 mm centres x 890 mm long; timber supports at 890 mm centres	-	-	-	-	m²	237.19

"Kawneer" window frame system or other equal and approved; polyester powder coated glazing bars; glazed with clear toughened safety glass; 10.70 mm thick

	PC £	Labour hours	Labour £	Material £	Unit	Total rate £
Vertical surfaces						
single tier; aluminium glazing bars at 850 mm centres x 890 mm long; timber supports at 890 mm centres	-	-	-	-	m²	224.71

H20 RIGID SHEET CLADDING

"Resoplan" sheet or other equal and approved; Eternit UK Ltd; flexible neoprene gasket joints; fixing with stainless steel screws and coloured caps

	PC £	Labour hours	Labour £	Material £	Unit	Total rate £
6 mm thick cladding to walls						
over 300 mm wide	-	2.23	29.03	54.90	m²	83.93
not exceeding 300 mm wide	-	0.74	9.63	19.21	m	28.84

Eternit 2000 "Glasal" sheet or other equal and approved; Eternit UK Ltd; flexible neoprene gasket joints; fixing with stainless steel screws and coloured caps

	PC £	Labour hours	Labour £	Material £	Unit	Total rate £
7.50 mm thick cladding to walls						
over 300 mm wide	-	2.23	29.03	50.17	m²	79.20
not exceeding 300 mm wide	-	0.74	9.63	17.79	m	27.42
external angle trim	-	0.11	1.43	9.19	m	10.62
7.50 mm thick cladding to eaves; verges fascias or the like						
100 mm wide	-	0.53	6.90	8.61	m	15.51
200 mm wide	-	0.64	8.33	13.20	m	21.53
300 mm wide	-	0.74	9.63	17.79	m	27.42

Prices for Measured Works - Minor Works
H CLADDING/COVERING

	PC £	Labour hours	Labour £	Material £	Unit	Total rate £
H30 FIBRE CEMENT PROFILED SHEET CLADDING/COVERING/SIDING						
Asbestos-free corrugated sheets; Eternit "2000" or other equal and approved						
Roof cladding; sloping not exceeding 50 degrees; fixing to timber purlins with drive screws						
"Profile 3"; natural grey	10.47	0.21	3.45	13.44	m²	16.89
"Profile 3"; coloured	12.56	0.21	3.45	15.69	m²	19.14
"Profile 6"; natural grey	9.23	0.27	4.44	12.04	m²	16.48
"Profile 6"; coloured	11.07	0.27	4.44	14.01	m²	18.45
"Profile 6"; natural grey; insulated 60 mm glass fibre infill; lining panel	-	0.48	7.89	29.77	m²	37.66
Roof cladding; sloping not exceeding 50 degrees; fixing to steel purlins with hook bolts						
"Profile 3"; natural grey	-	0.27	4.44	14.68	m²	19.12
"Profile 3"; coloured	-	0.27	4.44	16.93	m²	21.37
"Profile 6"; natural grey	-	0.32	5.26	13.27	m²	18.53
"Profile 6"; coloured	-	0.32	5.26	15.25	m²	20.51
"Profile 6"; natural grey; insulated 60 mm glass fibre infill; lining panel	-	0.53	8.71	27.47	m²	36.18
Wall cladding; vertical; fixing to steel rails with hook bolts						
"Profile 3"; natural grey	-	0.32	5.26	14.68	m²	19.94
"Profile 3"; coloured	-	0.32	5.26	16.93	m²	22.19
"Profile 6"; natural grey	-	0.37	6.08	13.27	m²	19.35
"Profile 6"; coloured	-	0.37	6.08	15.25	m²	21.33
"Profile 6"; natural grey; insulated 60 mm glass fibre infill; lining panel	-	0.58	9.54	27.47	m²	37.01
raking cutting	-	0.16	2.63	-	m	2.63
holes for pipes and the like	-	0.16	2.63	-	nr	2.63
Accessories; to "Profile 3" cladding; natural grey						
eaves filler	-	0.11	1.81	10.30	m	12.11
vertical corrugation closer	-	0.13	2.14	10.30	m	12.44
apron flashing	-	0.13	2.14	10.36	m	12.50
plain wing or close fitting two piece adjustable capping to ridge	-	0.19	3.12	9.69	m	12.81
ventilating two piece adjustable capping to ridge	-	0.19	3.12	14.62	m	17.74
Accessories; to "Profile 3" cladding; coloured						
eaves filler	-	0.11	1.81	12.90	m	14.71
vertical corrugation closer	-	0.13	2.14	12.90	m	15.04
apron flashing	-	0.13	2.14	12.93	m	15.07
plain wing or close fitting two piece adjustable capping to ridge	-	0.19	3.12	11.67	m	14.79
ventilating two piece adjustable capping to ridge	-	0.19	3.12	18.25	m	21.37
Accessories; to "Profile 6" cladding; natural grey						
eaves filler	-	0.11	1.81	7.64	m	9.45
vertical corrugation closer	-	0.13	2.14	7.64	m	9.78
apron flashing	-	0.13	2.14	8.22	m	10.36
underglazing flashing	-	0.13	2.14	7.86	m	10.00
plain cranked crown to ridge	-	0.19	3.12	15.43	m	18.55
plain wing or close fitting two piece adjustable capping to ridge	-	0.19	3.12	15.08	m	18.20
ventilating two piece adjustable capping to ridge	-	0.19	3.12	19.24	m	22.36

H CLADDING/COVERING

	PC £	Labour hours	Labour £	Material £	Unit	Total rate £
Accessories; to "Profile 6" cladding; coloured						
eaves filler	-	0.11	1.81	9.55	m	11.36
vertical corrugation closer	-	0.13	2.14	9.55	m	11.69
apron flashing	-	0.13	2.14	10.30	m	12.44
underglazing flashing	-	0.13	2.14	9.84	m	11.98
plain cranked crown to ridge	-	0.19	3.12	19.30	m	22.42
plain wing or close fitting two piece adjustable capping to ridge	-	0.19	3.12	18.84	m	21.96
ventilating two piece adjustable capping to ridge	-	0.19	3.12	24.09	m	27.21

H31 METAL PROFILED/FLAT SHEET CLADDING/COVERING/SIDING

Lightweight galvanised steel roof tiles; Decra Roof Systems (UK) Ltd or other equal and approved; coated finish

	PC £	Labour hours	Labour £	Material £	Unit	Total rate £
Roof coverings	-	0.26	4.27	14.95	m²	19.22
Accessories for roof cladding						
pitched "D" ridge	-	0.10	1.64	8.03	m	9.67
barge cover (handed)	-	0.10	1.64	8.61	m	10.25
in line air vent	-	0.10	1.64	39.55	nr	41.19
in line soil vent	-	0.10	1.64	61.98	nr	63.62
gas flue terminal	-	0.20	3.29	81.80	nr	85.09

H32 PLASTICS PROFILED SHEET CLADDING/COVERING/SIDING

Extended, hard skinned, foamed PVC-UE profiled sections; Swish Celuka or other equal and approved; Class 1 fire rated to BS 476; Part 7; 1987; in white finish

	PC £	Labour hours	Labour £	Material £	Unit	Total rate £
Wall cladding; vertical; fixing to timber						
100 mm shiplap profiles Code 001	-	0.39	5.08	41.47	m²	46.55
150 mm shiplap profiles Code 002	-	0.36	4.69	36.85	m²	41.54
125 mm feather-edged profiles Code C208	-	0.38	4.95	40.14	m²	45.09
Vertical angles	-	0.20	2.60	4.41	m	7.01
Raking cutting	-	0.16	2.08	-	m	2.08
Holes for pipes and the like	-	0.03	0.39	-	nr	0.39

H41 GLASS REINFORCED PLASTICS CLADDING/FEATURES

Glass fibre translucent sheeting grade AB class 3

	PC £	Labour hours	Labour £	Material £	Unit	Total rate £
Roof cladding; sloping not exceeding 50 degrees; fixing to timber purlins with drive screws; to suit						
"Profile 3" or other equal and approved	12.44	0.21	3.45	15.69	m²	19.14
"Profile 6" or other equal and approved	11.61	0.27	4.44	14.75	m²	19.19
Roof cladding; sloping not exceeding 50 degrees; fixing to timber purlins with hook bolts; to suit						
"Profile 3" or other equal and approved	12.44	0.27	4.44	16.92	m²	21.36
Profile 6" or other equal and approved	11.61	0.32	5.26	15.98	m²	21.24
"Longrib 1000" or other equal and approved	11.40	0.32	5.26	15.75	m²	21.01

Prices for Measured Works - Minor Works
H CLADDING/COVERING

	PC £	Labour hours	Labour £	Material £	Unit	Total rate £
H60 PLAIN ROOF TILING						
NOTE: The following items of tile roofing unless otherwise described, include for conventional fixing assuming normal exposure with appropriate nails and/or rivets or clips to pressure impregnated softwood battens fixed with galvanised nails; prices also include for all bedding and pointing at verges, beneath ridge tiles etc.						
Clay interlocking pantiles; Sandtoft Goxhill "Tudor" red sand faced or other equal and approved; PC £1113.50/1000; 470 mm x 285 mm; to 100 mm lap; on 25 mm x 38 mm battens and type 1F reinforced underlay						
Roof coverings	-	0.43	7.07	14.69	m²	21.76
Extra over coverings for						
fixing every tile	-	0.08	1.32	1.37	m²	2.69
eaves course with plastic filler	-	0.32	5.26	6.13	m	11.39
verges; extra single undercloak course of plain tiles	-	0.32	5.26	8.49	m	13.75
open valleys; cutting both sides	-	0.19	3.12	4.68	m	7.80
ridge tiles	-	0.64	10.52	12.27	m	22.79
hips; cutting both sides	-	0.80	13.15	16.95	m	30.10
holes for pipes and the like	-	0.21	3.45	-	nr	3.45
Clay pantiles; Sandtoft Goxhill "Old English"; red sand faced or other equal and approved; PC £749.00/1000; 342 mm x 241 mm; to 75 mm lap; on 25 mm x 38 mm battens and type 1F reinforced underlay						
Roof coverings	-	0.48	7.89	16.76	m²	24.65
Extra over coverings for						
fixing every tile	-	0.02	0.33	3.42	m²	3.75
other colours	-	-	-	0.59	m²	0.59
double course at eaves	-	0.35	5.75	4.28	m	10.03
verges; extra single undercloak course of plain tiles	-	0.32	5.26	8.98	m	14.24
open valleys; cutting both sides	-	0.19	3.12	3.15	m	6.27
ridge tiles; tile slips	-	0.64	10.52	18.56	m	29.08
hips; cutting both sides	-	0.80	13.15	21.71	m	34.86
holes for pipes and the like	-	0.21	3.45	-	nr	3.45
Clay pantiles; William Blyth's "Lincoln" natural or other equal and approved; 343 mm x 280 mm; to 75 mm lap; PC £898.80/1000; on 19 mm x 38 mm battens and type 1F reinforced underlay						
Roof coverings	-	0.48	7.89	18.80	m²	26.69
Extra over coverings for						
fixing every tile	-	0.02	0.33	3.42	m²	3.75
other colours	-	-	-	1.63	m²	1.63
double course at eaves	-	0.35	5.75	4.91	m	10.66
verges; extra single undercloak course of plain tiles	-	0.32	5.26	11.80	m	17.06
open valleys; cutting both sides	-	0.19	3.12	3.77	m	6.89
ridge tiles; tile slips	-	0.64	10.52	19.54	m	30.06
hips; cutting both sides	-	0.80	13.15	23.32	m	36.47
holes for pipes and the like	-	0.21	3.45	-	nr	3.45

H CLADDING/COVERING

	PC £	Labour hours	Labour £	Material £	Unit	Total rate £
Clay plain tiles; Hinton, Perry and Davenhill "Dreadnought" smooth red machine-made or other equal and approved; PC £229.70/1000; 265 mm x 165 mm; on 19 mm x 38 mm battens and type 1F reinforced underlay						
Roof coverings; to 64 mm lap	-	1.12	18.41	21.60	m^2	40.01
Wall coverings; to 38 mm lap	-	1.33	21.87	18.38	m^2	40.25
Extra over coverings for						
other colours	-	-	-	7.92	m^2	7.92
ornamental tiles	-	-	-	11.89	m^2	11.89
double course at eaves	-	0.27	4.44	2.58	m	7.02
verges	-	0.32	5.26	1.34	m	6.60
swept valleys; cutting both sides	-	0.69	11.34	30.00	m	41.34
bonnet hips; cutting both sides	-	0.85	13.97	30.00	m	43.97
external vertical angle tiles; supplementary nail fixings	-	0.43	7.07	33.58	m	40.65
half round ridge tiles	-	0.64	10.52	9.74	m	20.26
holes for pipes and the like	-	0.21	3.45	-	nr	3.45
Clay plain tiles; Keymer best hand-made sand-faced tiles or other equal or approved; PC £594.20/1000; 265 mm x 165 mm; on 25 mm x 38 mm battens and type 1F reinforced underlay						
Roof coverings; to 64 mm lap	-	1.12	18.41	44.56	m^2	62.97
Wall coverings; to 38 mm lap	-	1.33	21.87	38.66	m^2	60.53
Extra over coverings for						
ornamental tiles	-	-	-	8.18	m^2	8.18
double course at eaves	-	0.27	4.44	4.88	m	9.32
verges	-	0.32	5.26	2.06	m	7.32
swept valleys; cutting both sides	-	0.69	11.34	35.48	m	46.82
bonnet hips; cutting both sides	-	0.85	13.97	35.48	m	49.45
external vertical angle tiles; supplementary nail fixings	-	0.43	7.07	48.49	m	55.56
half round ridge tiles	-	0.64	10.52	10.58	m	21.10
holes for pipes and the like	-	0.21	3.45	-	nr	3.45
Concrete interlocking tiles; Marley "Bold Roll" granule finish tiles or other equal and aqpproved; PC £763.00/1000; 420 mm x 330 mm; to 75 mm lap; on 25 mm x 38 mm battens and type 1F reinforced underlay						
Roof ooverings	-	0.37	6.08	10.69	m^2	16.77
Extra over coverings for						
fixing every tile	-	0.02	0.33	0.57	m^2	0.90
eaves; eaves filler	-	0.05	0.82	1.04	m	1.86
verges; 150 mm wide asbestos free strip undercloak	-	0.24	3.95	1.77	m	5.72
valley trough tiles; cutting both sides	-	0.58	9.54	19.61	m	29.15
segmental ridge tiles; tile slips	-	0.58	9.54	10.65	m	20.19
segmental hip tiles; tile slips; cutting both sides	-	0.74	12.17	13.05	m	25.22
dry ridge tiles; segmental including batten sections; unions and filler pieces	-	0.32	5.26	15.74	m	21.00
segmental mono-ridge tiles	-	0.58	9.54	15.02	m	24.56
gas ridge terminal	-	0.53	8.71	55.58	nr	64.29
holes for pipes and the like	-	0.21	3.45	-	nr	3.45

H CLADDING/COVERING

	PC £	Labour hours	Labour £	Material £	Unit	Total rate £
H60 PLAIN ROOF TILING - cont'd						
Concrete interlocking tiles; Marley "Ludlow Major" granule finish tiles or other equal and approved; PC £723.00/1000; 420 mm x 330 mm; to 75 mm lap; on 25 mm x 38 mm battens and type 1F reinforced underlay						
Roof coverings	-	0.37	6.08	10.36	m²	16.44
Extra over coverings for						
fixing every tile	-	0.02	0.33	0.57	m²	0.90
eaves; eaves filler	-	0.05	0.82	0.43	m	1.25
verges; 150 mm wide asbestos free strip undercloak	-	0.24	3.95	1.77	m	5.72
dry verge system; extruded white pvc	-	0.16	2.63	8.01	m	10.64
segmental ridge cap to dry verge	-	0.02	0.33	2.93	m	3.26
valley trough tiles; cutting both sides	-	0.58	9.54	19.52	m	29.06
segmental ridge tiles	-	0.53	8.71	7.37	m	16.08
segmental hip tiles; cutting both sides	-	0.69	11.34	9.65	m	20.99
dry ridge tiles; segmental including batten sections; unions and filler pieces	-	0.32	5.26	15.74	m	21.00
segmental mono-ridge tiles	-	0.53	8.71	13.11	m	21.82
gas ridge terminal	-	0.53	8.71	55.58	nr	64.29
holes for pipes and the like	-	0.21	3.45	-	nr	3.45
Concrete interlocking tiles; Marley "Mendip" granule finish double pantiles or other equal and approved; PC £791.37/1000; 420 mm x 330 mm; to 75 mm lap; on 22 mm x 38 mm battens and type 1F reinforced underlay						
Roof coverings	-	0.37	6.08	10.84	m²	16.92
Extra over coverings for						
fixing every tile	-	0.02	0.33	0.57	m²	0.90
eaves; eaves filler	-	0.02	0.33	0.70	m	1.03
verges; 150 mm wide asbestos free strip undercloak	-	0.24	3.95	1.77	m	5.72
dry verge system; extruded white pvc	-	0.16	2.63	8.01	m	10.64
segmental ridge cap to dry verge	-	0.02	0.33	2.93	m	3.26
valley trough tiles; cutting both sides	-	0.58	9.54	19.04	m	29.18
segmental ridge tiles	-	0.58	9.54	10.65	m	20.19
segmental hip tiles; cutting both sides	-	0.74	12.17	13.09	m	25.26
dry ridge tiles; segmental including batten sections; unions and filler pieces	-	0.32	5.26	15.74	m	21.00
segmental mono-ridge tiles	-	0.53	8.71	14.74	m	23.45
gas ridge terminal	-	0.53	8.71	55.58	nr	64.29
holes for pipes and the like	-	0.21	3.45	-	nr	3.45
Concrete interlocking tiles; Marley "Modern" smooth finish tiles or other equal and approved; PC £723.00/1000; 420 mm x 220 mm; to 75 mm lap; on 25 mm x 38 mm battens and type 1F reinforced underlay						
Roof coverings	-	0.37	6.08	10.57	m²	16.65
Extra over coverings for						
fixing every tile	-	0.02	0.33	0.57	m²	0.90
verges; 150 mm wide asbestos free strip undercloak	-	0.24	3.95	1.77	m	5.72
dry verge system; extruded white pvc	-	0.21	3.45	8.01	m	11.46
"Modern" ridge cap to dry verge	-	0.02	0.33	2.93	m	3.26
valley trough tiles; cutting both sides	-	0.58	9.54	19.52	m	29.06
"Modern" ridge tiles	-	0.53	8.71	7.75	m	16.46
"Modern" hip tiles; cutting both sides	-	0.69	11.34	10.02	m	21.36
dry ridge tiles; "Modern"; including batten sections; unions and filler pieces	-	0.32	5.26	16.12	m	21.38
"Modern" mono-ridge tiles	-	0.53	8.71	13.11	m	21.82
gas ridge terminal	-	0.53	8.71	55.58	nr	64.29
holes for pipes and the like	-	0.21	3.45	-	nr	3.45

Prices for Measured Works - Minor Works
H CLADDING/COVERING

	PC £	Labour hours	Labour £	Material £	Unit	Total rate £
Concrete interlocking tiles; Marley "Wessex" smooth finish tiles or other equal and approved; PC £1196.40/1000; 413 mm x 330 mm; to 75 mm lap; on 25 mm x 38 mm battens and type 1F reinforced underlay						
Roof coverings	-	0.37	6.08	15.50	m²	21.58
Extra over coverings for						
fixing every tile	-	0.02	0.33	0.57	m²	0.90
verges; 150 mm wide asbestos free strip undercloak	-	0.24	3.95	1.77	m	5.72
dry verge system; extruded white pvc	-	0.21	3.45	8.01	m	11.46
"Modern" ridge cap to dry verge	-	0.02	0.33	2.93	m	3.26
valley trough tiles; cutting both sides	-	0.58	9.54	20.52	m	30.06
"Modern" ridge tiles	-	0.53	8.71	7.75	m	16.46
"Modern" hip tiles; cutting both sides	-	0.69	11.34	11.52	m	22.86
dry ridge tiles; "Modern"; including batten sections; unions and filler pieces	-	0.32	5.26	16.12	m	21.38
"Modern" mono-ridge tiles	-	0.53	8.71	13.11	m	21.82
gas ridge terminal	-	0.53	8.71	55.58	nr	64.29
holes for pipes and the like	-	0.21	3.45	-	nr	3.45
Concrete interlocking slates; Redland "Richmond" smooth finish tiles or other equal and approved; PC £898.80/1000; 430 mm x 380 mm; to 75 mm lap; on 25 mm x 38 mm battens and type 1F reinforced underlay						
Roof coverings	-	0.37	6.08	10.68	m²	16.76
Extra over coverings for						
fixing every tile	-	0.02	0.33	0.57	m²	0.90
eaves; eaves filler	-	0.02	0.33	0.60	m	0.93
verges; extra single undercloak course of plain tiles	-	0.27	4.44	3.07	m	7.51
ambi-dry verge system	-	0.21	3.45	8.78	m	12.23
ambi-dry verge eave/ridge end piece	-	0.02	0.33	3.15	m	3.48
valley trough tiles; cutting both sides	-	0.64	10.52	25.00	m	35.52
"Delta" ridge tiles	-	0.53	8.71	9.17	m	17.88
"Delta" hip tiles; cutting both sides	-	0.69	11.34	12.01	m	23.35
"Delta" mono-ridge tiles	-	0.53	8.71	18.21	m	26.92
gas ridge terminal	-	0.53	8.71	73.40	nr	82.11
ridge vent with 110 mm diameter flexible adaptor	-	0.53	8.71	64.49	nr	73.20
holes for pipes and the like	-	0.21	3.45	-	nr	3.45
Concrete interlocking tiles; Redland "Norfolk" smooth finish pantiles or other equal and approved; PC £457.40/1000; 381 mm x 229 mm; to 75 mm lap; on 25 mm x 38 mm battens and type 1F reinforced underlay						
Roof coverings	-	0.48	7.89	11.90	m²	19.79
Extra over coverings for						
fixing every tile	-	0.05	0.82	0.25	m²	1.07
eaves; eaves filler	-	0.05	0.82	0.94	m	1.76
verges; extra single undercloak course of plain tiles	-	0.32	5.26	5.46	m	10.72
valley trough tiles; cutting both sides	-	0.64	10.52	23.62	m	34.14
universal ridge tiles	-	0.53	8.71	9.79	m	18.50
universal hip tiles; cutting both sides	-	0.69	11.34	12.19	m	23.53
universal gas flue ridge tile	-	0.53	8.71	74.28	nr	82.99
universal ridge vent tile with 110 mm diameter adaptor	-	0.53	8.71	67.14	nr	75.85
holes for pipes and the like	-	0.21	3.45	-	nr	3.45

Prices for Measured Works - Minor Works
H CLADDING/COVERING

	PC £	Labour hours	Labour £	Material £	Unit	Total rate £
H60 PLAIN ROOF TILING - cont'd						
Concrete interlocking tiles; Redland "Regent" granule finish bold roll tiles or other equal and approved; PC £655.10/1000; 418 mm x 332 mm; to 75 mm lap; on 25 mm x 38 mm battens and type 1F reinforced underlay						
Roof coverings	-	0.37	6.08	9.80	m²	15.88
Extra over coverings for						
fixing every tile	-	0.04	0.66	0.61	m²	1.27
eaves; eaves filler	-	0.05	0.82	0.73	m	1.55
verges; extra single undercloak course of plain tiles	-	0.27	4.44	2.52	m	6.96
cloaked verge system	-	0.16	2.63	5.65	m	8.28
valley trough tiles; cutting both sides	-	0.58	9.54	23.28	m	32.82
universal ridge tiles	-	0.53	8.71	9.79	m	18.50
universal hip tiles; cutting both sides	-	0.69	11.34	11.85	m	23.19
dry ridge system; universal ridge tiles	-	0.27	4.44	31.96	m	36.40
universal half round mono-pitch ridge tiles	-	0.58	9.54	20.80	m	30.34
universal gas flue ridge tile	-	0.53	8.71	74.28	nr	82.99
universal ridge vent tile with 110 mm diameter adaptor	-	0.53	8.71	67.14	nr	75.85
holes for pipes and the like	-	0.21	3.45	-	nr	3.45
Concrete interlocking tiles; Redland "Renown" granule finish tiles or other equal and approved; PC £623.20/1000; 418 mm x 330 mm; to 75 mm lap; on 25 mm x 38 mm battens and type 1F reinforced underlay						
Roof coverings	-	0.37	6.08	9.48	m²	15.56
Extra over coverings for						
fixing every tile	-	0.02	0.33	0.30	m²	0.63
verges; extra single undercloak course of plain tiles	-	0.27	4.44	3.07	m	7.51
cloaked verge system	-	0.16	2.63	5.74	m	8.37
valley trough tiles; cutting both sides	-	0.58	9.54	23.18	m	32.72
universal ridge tiles	-	0.53	8.71	9.79	m	18.50
universal hip tiles; cutting both sides	-	0.69	11.34	11.75	m	23.09
dry ridge system; universal ridge tiles	-	0.27	4.44	31.96	m	36.40
universal half round mono-pitch ridge tiles	-	0.58	9.54	20.80	m	30.34
universal gas flue ridge tile	-	0.53	8.71	74.28	nr	82.99
universal ridge vent tile with 110 mm diameter adaptor	-	0.53	8.71	67.14	nr	75.85
holes for pipes and the like	-	0.21	3.45	-	nr	3.45
Concrete interlocking tiles; Redland "Stonewold II" smooth finish tiles or other equal and approved; PC £1269.40/1000; 430 mm x 380 mm; to 75 mm lap; on 25 mm x 38 mm battens and type 1F reinforced underlay						
Roof coverings	-	0.37	6.08	16.18	m²	22.26
Extra over coverings for						
fixing every tile	-	0.02	0.33	0.94	m²	1.27
verges; extra single undercloak course of plain tiles	-	0.31	5.10	3.07	m	8.17
ambi-dry verge system	-	0.21	3.45	8.78	m	12.23
ambi-dry verge eave/ridge end piece	-	0.02	0.33	3.15	m	3.48
valley trough tiles; cutting both sides	-	0.58	9.54	25.22	m	34.76
universal angle ridge tiles	-	0.53	8.71	7.23	m	15.94
universal hip tiles; cutting both sides	-	0.69	11.34	11.23	m	22.57
dry ridge system; universal angle ridge tiles	-	0.27	4.44	19.39	m	23.83
universal monopitch angle ridge tiles	-	0.58	9.54	14.15	m	23.69
universal gas flue angle ridge tile	-	0.53	8.71	74.28	nr	82.99
universal angle ridge vent tile with 110 mm diameter adaptor	-	0.53	8.71	67.14	nr	75.85
holes for pipes and the like	-	0.21	3.45	-	nr	3.45

H CLADDING/COVERING

	PC £	Labour hours	Labour £	Material £	Unit	Total rate £
Concrete plain tiles; EN 490 and 491 group A; PC £272.60/1000; 267 mm x 165 mm; on 25 mm x 38 mm battens and type 1F reinforced underlay						
Roof coverings; to 64 mm lap	-	1.12	18.41	24.30	m^2	42.71
Wall coverings; to 38 mm lap	-	1.33	21.87	20.76	m^2	42.63
Extra over coverings for						
ornamental tiles	-	-	-	14.53	m^2	14.53
double course at eaves	-	0.27	4.44	2.85	m	7.29
verges	-	0.35	5.75	1.06	m	6.81
swept valleys; cutting both sides	-	0.69	11.34	24.47	m	35.81
bonnet hips; cutting both sides	-	0.85	13.97	24.47	m	38.44
external vertical angle tiles; supplementary nail fixings	-	0.43	7.07	17.23	m	24.30
half round ridge tiles	-	0.53	8.71	5.94	m	14.65
third round hip tiles; cutting both sides	-	0.53	8.71	7.66	m	16.37
holes for pipes and the like	-	0.21	3.45	-	nr	3.45
Sundries						
Hip irons						
galvanised mild steel; fixing with screws	-	0.11	1.81	2.76	nr	4.57
"Rytons Clip strip" or other equal and approved; continuous soffit ventilator						
51 mm wide; plastic; code CS351	-	0.32	5.26	1.08	m	6.34
"Rytons over fascia ventilator" or other equal and approved; continuous eaves ventilator						
40 mm wide; plastic; code OFV890	-	0.11	1.81	1.81	m	3.62
"Rytons roof ventilator" or other equal and approved; to suit rafters at 600 centres						
250 mm deep x 43 mm high; plastic; code TV600	-	0.11	1.81	1.72	m	3.53
"Rytons push and lock ventilators" or other equal and approved; circular						
83 mm diameter; plastic; code PL235	-	0.05	0.65	0.29	nr	0.94
Fixing only						
lead soakers (supply cost given elsewhere)	-	0.08	0.91	-	nr	0.91
Pressure impregnated softwood counter battens; 25 mm x 50 mm						
450 mm centres	-	0.07	1.15	1.18	m^2	2.33
600 mm centres	-	0.05	0.82	0.89	m^2	1.71
Underlay; BS 747 type 1B; bitumen felt weighing 14kg/10m^2; 75 mm laps						
To sloping or vertical surfaces	0.58	0.03	0.49	1.15	m^2	1.64
Underlay; BS 747 type 1F; reinforced bitumen felt weighing 22.5kg/10 m^2; 75 mm laps						
To sloping or vertical surfaces	1.08	0.03	0.49	1.72	m^2	2.21
Underlay; Visqueen "Tilene 200P" or other equal and approved; micro-perforated sheet; 75 mm laps						
To sloping or vertical surfaces	0.50	0.03	0.49	1.09	m^2	1.58
Underlay; "Powerlon 250 BM" or other equal and approved; reinforced breather membrane; 75 mm laps						
To sloping or vertical surfaces	0.61	0.03	0.49	1.22	m^2	1.71
Underlay; "Anticon" sarking membrane or other equal and approved; Euroroof Ltd; polyethylene; 75 mm laps						
To sloping or vertical surfaces	0.89	0.03	0.49	1.50	m^2	1.99

Prices for Measured Works - Minor Works
H CLADDING/COVERING

	PC £	Labour hours	Labour £	Material £	Unit	Total rate £
H61 FIBRE CEMENT SLATING						
Asbestos-free artificial slates; Eternit "2000" or other equal and approved; to 75 mm lap; on 19 mm x 50 mm battens and type 1F reinforced underlay						
Coverings; 400 mm x 200 mm slates						
roof coverings	-	0.85	13.97	27.79	m²	41.76
wall coverings	-	1.12	18.41	27.79	m²	46.20
Coverings; 500 mm x 250 mm slates						
roof coverings	-	0.69	11.34	25.18	m²	36.52
wall coverings	-	0.85	13.97	25.18	m²	39.15
Coverings; 600 mm x 300 mm slates						
roof coverings	-	0.53	8.71	22.12	m²	30.83
wall coverings	-	0.69	11.34	22.12	m²	33.46
Extra over slate coverings for						
double course at eaves	-	0.27	4.44	5.49	m	9.93
verges; extra single undercloak course	-	0.35	5.75	1.31	m	7.06
open valleys; cutting both sides	-	0.21	3.45	5.23	m	8.68
valley gutters; cutting both sides	-	0.58	9.54	22.36	m	31.90
half round ridge tiles	-	0.53	8.71	19.89	m	28.60
stop end	-	0.11	1.81	6.79	nr	8.60
roll top ridge tiles	-	0.64	10.52	21.64	m	32.16
stop end	-	0.11	1.81	10.12	nr	11.93
mono-pitch ridge tiles	-	0.53	8.71	25.39	m	34.10
stop end	-	0.11	1.81	30.56	nr	32.37
duo-pitch ridge tiles	-	0.53	8.71	20.53	m	29.24
stop end	-	0.11	1.81	22.45	nr	24.26
mitred hips; cutting both sides	-	0.21	3.45	5.23	m	8.68
half round hip tiles; cutting both sides	-	0.21	3.45	25.12	m	28.57
holes for pipes and the like	-	0.21	3.45	-	nr	3.45
H62 NATURAL SLATING						
NOTE: The following items of slate roofing unless otherwise described, include for conventional fixing assuming "normal exposure" with appropriate nails and/or rivets or clips to pressure impregnated softwood battens fixed with galvanised nails; prices also include for all bedding and pointing at verges, beneath verge tiles etc.						
Natural slates; BS 680 Part 2; Welsh blue; uniform size; to 75 mm lap; on 25 mm x 50 mm battens and type 1F reinforced underlay						
Coverings; 405 mm x 255 mm slates						
roof coverings	-	0.95	15.62	38.43	m²	54.05
wall coverings	-	1.22	20.06	38.43	m²	58.49
Coverings; 510 mm x 255 mm slates						
roof coverings	-	0.80	13.15	46.41	m²	59.56
wall coverings	-	0.95	15.62	46.41	m²	62.03
Coverings; 610 mm x 305 mm slates						
roof coverings	-	0.64	10.52	59.55	m²	70.07
wall coverings	-	0.80	13.15	59.55	m²	72.70
Extra over coverings for						
double course at eaves	-	0.32	5.26	15.64	m	20.90
verges; extra single undercloak course	-	0.44	7.23	9.13	m	16.36
open valleys; cutting both sides	-	0.23	3.78	36.07	m	39.85
blue/black glazed ware 152 mm half round ridge tiles	-	0.53	8.71	7.70	m	16.41

H CLADDING/COVERING

	PC £	Labour hours	Labour £	Material £	Unit	Total rate £
Extra over coverings for - cont'd						
blue/black glazed ware 125 mm x 125 mm						
plain angle ridge tiles	-	0.53	8.71	7.70	m	16.41
mitred hips; cutting both sides	-	0.23	3.78	36.07	m	39.85
blue/black glazed ware 152 mm half round hip						
tiles; cutting both sides	-	0.74	12.17	43.77	m	55.94
blue/black glazed ware 125 mm x 125 mm						
plain angle hip tiles; cutting both sides	-	0.74	12.17	43.77	m	55.94
holes for pipes and the like	-	0.21	3.45	-	nr	3.45
Natural slates; Westmoreland green; PC £1700.75/t; random lengths; 457 mm - 229 mm proportionate widths to 75 mm lap; in diminishing courses; on 25 mm x 50 mm battens and type 1F underlay						
Roof coverings	-	1.22	20.06	101.87	m²	121.93
Wall coverings	-	1.54	25.32	101.87	m²	127.19
Extra over coverings for						
double course at eaves	-	0.70	11.51	18.78	m	30.29
verges; extra single undercloak course slates						
152 mm wide	-	0.80	13.15	16.16	m	29.31
holes for pipes and the like	-	0.32	5.26	-	nr	5.26
H63 RECONSTRUCTED STONE SLATING/TILING						
Reconstructed stone slates; "Hardrow Slates" or other equal and approved; standard colours; or similar; PC £924.00/1000 75 lap; on 25 mm x 50 mm battens and type 1F reinforced underlay						
Coverings; 457 mm x 305 mm slates						
roof coverings	-	0.85	13.97	22.27	m²	36.24
wall coverings	-	1.06	17.43	22.27	m²	39.70
Coverings; 457 mm x 457 mm slates						
roof coverings	-	0.69	11.34	21.96	m²	33.30
wall coverings	-	0.91	14.96	21.96	m²	36.92
Extra over 457 mm x 305 mm coverings for						
double course at eaves	-	0.32	5.26	4.02	m	9.28
verges; pointed	-	0.44	7.23	0.06	m	7.29
open valleys; cutting both sides	-	0.23	3.78	9.70	m	13.48
ridge tiles	-	0.53	8.71	25.36	m	34.07
hip tiles; cutting both sides	-	0.74	12.17	21.04	m	33.21
holes for pipes and the like	-	0.21	3.45	-	nr	3.45
Reconstructed stone slates; "Hardrow Slates" green/ oldstone colours or other equal and approved; PC £924.00/1000 75 mm lap; on 25 mm x 50 mm battens and type 1F reinforced underlay						
Coverings; 457 mm x 305 mm slates						
roof coverings	-	0.85	13.97	22.27	m²	36.24
wall coverings	-	1.06	17.43	22.27	m²	39.70
Coverings; 457 mm x 457 mm slates						
roof coverings	-	0.69	11.34	21.96	m²	33.30
wall coverings	-	0.91	14.96	21.96	m²	36.92
Extra over 457 mm x 305 mm coverings for						
double course at eaves	-	0.32	5.26	4.02	m	9.28
verges; pointed	-	0.44	7.23	0.06	m	7.29
open valleys; cutting both sides	-	0.23	3.78	9.70	m	13.48
ridge tiles	-	0.53	8.71	25.36	m	34.07
hip tiles; cutting both sides	-	0.74	12.17	21.04	m	33.21
holes for pipes and the like	-	0.21	3.45	-	nr	3.45

Prices for Measured Works - Minor Works
H CLADDING/COVERING

	PC £	Labour hours	Labour £	Material £	Unit	Total rate £
H63 RECONSTRUCTED STONE SLATING/TILING - cont'd						
Reconstructed stone slates; Bradstone "Cotswold" style or other equal and approved; PC £22.27/m²; random lengths 550 mm - 300 mm; proportional widths; to 80 mm lap; in diminishing courses; on 25 mm x 50 mm battens and type 1F reinforced underlay						
Roof coverings (all-in rate inclusive of eaves and verges)	-	1.12	18.41	28.97	m²	**47.38**
Extra over coverings for						
open valleys/mitred hips; cutting both sides	-	0.48	7.89	10.52	m²	**18.41**
ridge tiles	-	0.70	11.51	14.03	m	**25.54**
hip tiles; cutting both sides	-	1.12	18.41	23.81	m	**42.22**
holes for pipes and the like	-	0.32	5.26	-	nr	**5.26**
Reconstructed stone slates; Bradstone "Moordale" style or other equal and approved; PC £22.02/m²; random lengths 550 mm - 450 mm; proportional widths; to 80 lap; in diminishing course; on 25 mm x 50 mm battens and type 1F reinforced underlay						
Roof coverings (all-in rate inclusive of eaves and verges)	-	1.12	18.41	27.61	m²	**46.02**
Extra over coverings for						
open valleys/mitred hips; cutting both sides	-	0.48	7.89	9.91	m²	**17.80**
ridge tiles	-	0.70	11.51	14.03	m	**25.54**
holes for pipes and the like	-	0.32	5.26	-	nr	**5.26**
H64 TIMBER SHINGLING						
Red cedar sawn shingles preservative treated; PC £34.96/bundle; uniform length 400 mm; to 125 mm gauge; on 25 mm x 38 mm battens and type 1F reinforced underlay						
Roof coverings; 125 mm gauge, 2.28 m²/bundle	-	1.12	18.41	21.80	m²	**40.21**
Wall coverings; 190 mm gauge, 3.47 m²/bundle	-	0.85	13.97	14.80	m²	**28.77**
Extra over coverings for						
double course at eaves	32.52	0.21	3.45	2.08	m	**5.53**
open valleys; cutting both sides	32.52	0.21	3.45	3.94	m	**7.39**
pre-formed ridge capping	-	0.32	5.26	7.00	m	**12.26**
pre-formed hip capping; cutting both sides	-	0.53	8.71	10.95	m	**19.66**
double starter course to cappings	-	0.11	1.81	0.73	m	**2.54**
holes for pipes and the like	-	0.16	2.63	-	nr	**2.63**
H71 LEAD SHEET COVERINGS/FLASHINGS						
Milled lead; BS 1178; PC £732.50/t						
1.32 mm thick (code 3) roof coverings						
flat	-	2.88	32.75	11.43	m²	**44.18**
sloping 10 - 50 degrees	-	3.19	36.27	11.43	m²	**47.70**
vertical or sloping over 50 degrees	-	3.51	39.91	11.43	m²	**51.34**
1.80 mm thick (code 4) roof coverings						
flat	-	3.08	35.02	15.59	m²	**50.61**
sloping 10 - 50 degrees	-	3.40	38.66	15.59	m²	**54.25**
vertical or sloping over 50 degrees	-	3.73	42.41	15.59	m²	**58.00**

H CLADDING/COVERING

	PC £	Labour hours	Labour £	Material £	Unit	Total rate £
1.80 mm thick (code 4) dormer coverings						
flat	-	3.62	41.16	15.59	m²	56.75
sloping 10 - 50 degrees	-	4.15	47.19	15.59	m²	62.78
vertical or sloping over 50 degrees	-	4.47	50.82	15.59	m²	66.41
2.24 mm thick (code 5) roof coverings						
flat	-	3.29	37.41	20.78	m²	58.19
sloping 10 - 50 degrees	-	3.62	41.16	20.78	m²	61.94
vertical or sloping over 50 degrees	-	3.93	44.68	20.78	m²	65.46
2.24 mm thick (code 5) dormer coverings						
flat	-	3.93	44.68	20.78	m²	65.46
sloping 10 - 50 degrees	-	4.36	49.57	20.78	m²	70.35
vertical or sloping over 50 degrees	-	4.78	54.35	20.78	m²	75.13
2.65 mm thick (code 6) roof coverings						
flat	-	3.51	39.91	22.95	m²	62.86
sloping 10 - 50 degrees	-	3.83	43.55	22.95	m²	66.50
vertical or sloping over 50 degrees	-	4.15	47.19	22.95	m²	70.14
2.65 mm thick (code 6) dormer coverings						
flat	-	4.26	48.44	22.95	m²	71.39
sloping 10 - 50 degrees	-	4.57	51.96	22.95	m²	74.91
vertical or sloping over 50 degrees	-	5.00	56.85	22.95	m²	79.80
3.15 mm thick (code 7) roof coverings						
flat	-	3.73	42.41	27.28	m²	69.69
sloping 10 - 50 degrees	-	4.04	45.93	27.28	m²	73.21
vertical or sloping over 50 degrees	-	4.36	49.57	27.28	m²	76.85
3.15 mm thick (code 7) dormer coverings						
flat	-	4.47	50.82	27.28	m²	78.10
sloping 10 - 50 degrees	-	4.78	54.35	27.28	m²	81.63
vertical or sloping over 50 degrees	-	5.22	59.35	27.28	m²	86.63
3.55 mm thick (code 8) roof coverings						
flat	-	3.93	44.68	30.74	m²	75.42
sloping 10 - 50 degrees	-	4.26	48.44	30.74	m²	79.18
vertical or sloping over 50 degrees	-	4.57	51.96	30.74	m²	82.70
3.55 mm thick (code 8) dormer coverings						
flat	-	4.78	54.35	30.74	m²	85.09
sloping 10 - 50 degrees	-	5.11	58.10	30.74	m²	88.84
vertical or sloping over 50 degrees	-	5.53	62.88	30.74	m²	93.62
Dressing over glazing bars and glass	-	0.35	3.98	-	m	3.98
Soldered dot	-	0.16	1.82	-	nr	1.82
Copper nailing 75 mm centres	-	0.21	2.39	0.34	m	2.73
1.80 mm thick (code 4) lead flashings, etc.						
Flashings; wedging into grooves						
150 mm girth	-	0.85	9.66	2.43	m	12.09
240 mm girth	-	0.95	10.80	3.87	m	14.67
Stepped flashings; wedging into grooves						
180 mm girth	-	0.95	10.80	2.89	m	13.69
270 mm girth	-	1.06	12.05	4.33	m	16.38
Linings to sloping gutters						
390 mm girth	-	1.28	14.55	6.30	m	20.85
450 mm girth	-	1.38	15.69	7.21	m	22.90
750 mm girth	-	1.28	14.55	11.69	m	26.24
Cappings to hips or ridges						
450 mm girth	-	1.60	18.19	7.21	m	25.40
600 mm girth	-	1.70	19.33	9.65	m	28.98
Soakers						
200 mm x 200 mm	-	0.16	1.82	0.62	nr	2.44
300 mm x 300 mm	-	0.21	2.39	1.40	nr	3.79
Saddle flashings; at intersections of hips and ridges; dressing and bossing						
450 mm x 600 mm	-	1.91	21.72	4.33	nr	26.05

H CLADDING/COVERING

	PC £	Labour hours	Labour £	Material £	Unit	Total rate £
H71 LEAD SHEET COVERINGS/FLASHINGS - cont'd						
1.80 mm thick (code 4) lead flashings, etc. - cont'd						
Slates; with 150 mm high collar						
450 mm x 450 mm; to suit 50 mm diameter pipe	-	1.81	20.58	3.73	nr	24.31
450 mm x 450 mm; to suit 100 mm diameter pipe	-	2.13	24.22	4.09	nr	28.31
2.24 mm (code 5) lead flashings, etc.						
Flashings; wedging into grooves						
150 mm girth	-	0.85	9.66	3.02	m	12.68
240 mm girth	-	0.95	10.80	4.78	m	15.58
Stepped flashings; wedging into grooves						
180 mm girth	-	0.95	10.80	3.62	m	14.42
270 mm girth	-	1.06	12.05	5.39	m	17.44
Linings to sloping gutters						
390 mm girth	-	1.28	14.55	7.81	m	22.36
450 mm girth	-	1.38	15.69	8.99	m	24.68
Cappings to hips or ridges						
450 mm girth	-	1.60	18.19	8.99	m	27.18
600 mm girth	-	1.70	19.33	12.01	m	31.34
Soakers						
200 mm x 200 mm	-	0.16	1.82	0.83	nr	2.65
300 mm x 300 mm	-	0.21	2.39	1.87	nr	4.26
Saddle flashings; at intersections of hips and ridges; dressing and bossing						
450 mm x 600 mm		1.91	21.72	5.39	nr	27.11
Slates; with 150 mm high collar						
450 mm x 450 mm; to suit 50 mm diameter pipe	-	1.81	20.58	4.65	nr	25.23
450 mm x 450 mm; to suit 100 mm diameter pipe	-	2.13	24.22	5.10	nr	29.32
H72 ALUMINIUM SHEET COVERINGS/FLASHINGS						
Aluminium roofing; commercial grade						
0.90 mm thick roof coverings						
flat	-	3.19	36.27	6.68	m²	42.95
sloping 10 - 50 degrees	-	3.52	40.02	6.68	m²	46.70
vertical or sloping over 50 degrees	-	3.84	43.66	6.68	m²	50.34
0.90 mm thick dormer coverings						
flat	-	3.84	43.66	6.68	m²	50.34
sloping 10 - 50 degrees	-	4.26	48.44	6.68	m²	55.12
vertical or sloping over 50 degrees	-	4.67	53.10	6.68	m²	59.78
Aluminium nailing; 75 mm spacing	-	0.21	2.39	0.25	m	2.64
0.90 mm commercial grade aluminium flashings, etc.						
Flashings; wedging into grooves						
150 mm girth	-	0.85	9.66	1.04	m	10.70
240 mm girth	-	0.97	11.03	1.68	m	12.71
300 mm girth	-	1.11	12.62	2.05	m	14.67
Stepped flashings; wedging into grooves						
180 mm girth	-	0.97	11.03	1.23	m	12.26
270 mm girth	-	1.06	12.05	1.92	m	13.97
1.20 mm commercial grade aluminium flashings; polyester powder coated						
Flashings; fixed with self tapping screws						
170 mm girth	-	0.97	11.03	1.43	m	12.46
200 mm girth	-	1.02	11.60	1.90	m	13.50
280 mm girth	-	1.16	13.19	2.48	m	15.67

H CLADDING/COVERING

	PC £	Labour hours	Labour £	Material £	Unit	Total rate £
H73 COPPER STRIP/SHEET COVERINGS/ FLASHINGS						
Copper roofing; BS 2870						
0.56 mm thick (24 swg) roof coverings						
flat	18.65	3.42	38.89	21.05	m²	59.94
sloping 10 - 50 degrees	-	3.75	42.64	21.05	m²	63.69
vertical or sloping over 50 degrees	-	4.02	45.71	21.05	m²	66.76
0.56 mm thick (24 swg) dormer coverings						
flat	18.65	4.02	45.71	21.05	m²	66.76
sloping 10 - 50 degrees	-	4.49	51.05	21.05	m²	72.10
vertical or sloping over 50 degrees	-	4.90	55.71	21.05	m²	76.76
0.61 mm thick (23 swg) roof coverings						
flat	-	3.42	38.89	23.02	m²	61.91
sloping 10 - 50 degrees	-	3.75	42.64	23.02	m²	65.66
vertical or sloping over 50 degrees	-	4.02	45.71	23.02	m²	68.73
0.61 mm thick (23 swg) dormer coverings						
flat	-	4.02	45.71	23.02	m²	68.73
sloping 10 - 50 degrees	-	4.49	51.05	23.02	m²	74.07
vertical or sloping over 50 degrees	-	4.90	55.71	23.02	m²	78.73
Copper nailing; 75 mm spacing	-	0.21	2.39	0.34	m	2.73
0.56 mm thick copper flashings, etc.						
Flashings; wedging into grooves						
150 mm girth	-	0.85	9.66	3.31	m	12.97
240 mm girth	-	0.97	11.03	5.31	m	16.34
300 mm girth	-	1.11	12.62	6.66	m	19.28
Stepped flashings; wedging into grooves						
180 mm girth	-	0.97	11.03	4.01	m	15.04
270 mm girth	-	1.06	12.05	5.97	m	18.02
0.61 mm thick copper flashings, etc.						
Flashings; wedging into grooves						
150 mm girth	-	0.85	9.66	3.60	m	13.26
240 mm girth	-	0.97	11.03	5.79	m	16.82
300 mm girth	-	1.11	12.62	7.24	m	19.86
Stepped flashings; wedging into grooves						
180 mm girth	-	0.97	11.03	4.33	m	15.36
270 mm girth	-	1.06	12.05	6.50	m	18.55
H74 ZINC STRIP/SHEET COVERINGS/FLASHINGS						
Zinc, BS 849						
0.80 mm roof coverings						
flat	-	3.42	38.89	15.17	m²	54.06
sloping 10 - 50 degrees	-	3.75	42.64	15.17	m²	57.81
vertical or sloping over 50 degrees	-	4.02	45.71	15.17	m²	60.88
0.61 mm thick (23 swg) dormer coverings						
flat	-	4.02	45.71	15.17	m²	60.88
sloping 10 - 50 degrees	-	4.49	51.05	15.17	m²	66.22
vertical or sloping over 50 degrees	-	4.90	55.71	15.17	m²	70.88
0.80 mm thick zinc flashings, etc.						
Flashings; wedging into grooves						
150 mm girth	-	0.85	9.66	2.35	m	12.01
240 mm girth	-	0.97	11.03	3.80	m	14.83
300 mm girth	-	1.11	12.62	4.79	m	17.41
Stepped flashings; wedging into grooves						
180 mm girth	-	0.97	11.03	2.87	m	13.90
270 mm girth	-	1.06	12.05	4.29	m	16.34

Prices for Measured Works - Minor Works
H CLADDING/COVERING

	PC £	Labour hours	Labour £	Material £	Unit	Total rate £
H75 STAINLESS STEEL SHEET COVERINGS/FLASHINGS						
Terne coated stainless steel roofing; Eurocom Enterprise Ltd or other equal and approved						
Roof coverings; "Uginox AME"; 0.40 mm thick						
flat; fixing with stainless steel screws to timber	13.66	1.02	16.77	20.22	m^2	36.99
flat; including batten rolls; fixing with stainless steel screws to timber	-	1.02	16.77	21.70	m^2	38.47
Roof coverings; "Uginox AE"; 0.40 mm thick						
flat; fixing with stainless steel screws to timber	11.46	1.02	16.77	17.33	m^2	34.10
flat; including batten rolls; fixing with stainless steel screws to timber	-	1.02	16.77	18.77	m^2	35.54
Wall coverings; "Uginox AME"; 0.40 mm thick						
vertical; coulisseau joints; fixing with stainless steel screws to timber	-	0.92	15.12	19.07	m^2	34.19
Wall coverings; "Uginox AE"; 0.40 mm thick						
vertical; coulisseau joints; fixing with stainless steel screws to timber	-	0.92	15.12	16.25	m^2	31.37
Flashings						
head abutments; swiss fold and cover flashings 150 mm girth	-	0.41	6.74	4.01	m	10.75
head abutments; manchester fold and cover flashings 150 mm girth	-	0.51	8.38	4.01	m	12.39
head abutments; soldered saddle flashings 150 mm girth	-	0.51	8.38	4.54	m	12.92
head abutments; into brick wall with timber infill; cover flashings 150 mm girth	-	0.81	13.32	6.28	m	19.60
head abutments; into brick wall with soldered apron; cover flashings 150 mm girth	-	0.97	15.95	7.03	m	22.98
side abutments; "Uginox" stepped flashings 170 mm girth	-	1.63	26.80	5.05	m	31.85
eaves flashings 140 mm girth; approned with single lock welts	-	0.36	5.92	7.21	m	13.13
eaves flashings 190 mm girth; approned with flashings and soldered strips	-	0.51	8.38	8.84	m	17.22
verge flashings 150 mm girth	-	0.26	4.27	7.42	m	11.69
Aprons						
fan aprons 250 mm girth	-	0.28	4.60	6.22	m	10.82
Ridges						
ridge with coulisseau closures	-	0.61	10.03	2.90	m	12.93
monoridge flanged apex (AME)	-	0.31	5.10	4.30	m	9.40
ridge with tapered timber infill and flat saddles	-	1.02	16.77	5.88	m	22.65
monoridge with timber infill and flat saddles	-	0.56	9.21	9.21	m	18.42
ventilated ridge with timber kerb and flat saddles	-	1.22	20.06	15.68	m	35.74
ventilated ridge with stainless steel brackets	-	0.92	15.12	27.17	m	42.29
Edges						
downstand	-	0.31	5.10	5.77	m	10.87
flanged	-	0.26	4.27	6.22	m	10.49
Gutters						
double lock welted valley gutters	-	1.07	17.59	9.56	m	27.15
eaves gutters; against masonry work	-	1.22	20.06	13.00	m	33.06
recessed valley gutters	-	0.92	15.12	18.37	m	33.49
recessed valley gutters with laplocks	-	1.12	18.41	21.64	m	40.05
secret	-	1.02	16.77	22.59	m	39.36
saddle expansion joints in gutters	-	1.02	16.77	7.77	-	24.54

H CLADDING/COVERING

	PC £	Labour hours	Labour £	Material £	Unit	Total rate £
H76 FIBRE BITUMEN THERMOPLASTIC SHEET COVERINGS/FLASHINGS						
Glass fibre reinforced bitumen strip slates; "Ruberglas 105" or other equal and approved; 1000 mm x 336 mm mineral finish; to external quality plywood boarding (boarding not included)						
Roof coverings	-	0.27	4.44	13.40	m²	17.84
Wall coverings	-	0.43	7.07	13.40	m²	20.47
Extra over coverings for						
double course at eaves; felt soaker	-	0.21	3.45	9.05	m	12.50
verges; felt soaker	-	0.16	2.63	7.50	m	10.13
valley slate; cut to shape; felt soaker and cutting both sides	-	0.48	7.89	11.86	m	19.75
ridge slate; cut to shape	-	0.32	5.26	7.50	m	12.76
hip slate; cut to shape; felt soaker and cutting both sides	-	0.48	7.89	11.76	m	19.65
holes for pipes and the like	-	0.56	9.21	-	nr	9.21
Evode "Flashband Original" sealing strips and flashings or other equal and approved; special grey finish						
Flashings; wedging at top if required; pressure bonded; flashband primer before application; to walls						
75 mm girth	-	0.18	2.05	0.89	m	2.94
100 mm girth	-	0.27	3.07	1.22	m	4.29
150 mm girth	-	0.35	3.98	1.82	m	5.80
225 mm girth	-	0.43	4.89	2.86	m	7.75
300 mm girth	-	0.48	5.46	3.59	m	9.05

J WATERPROOFING

	PC £	Labour hours	Labour £	Material £	Unit	Total rate £
J WATERPROOFING						
J10 SPECIALIST WATERPROOF RENDERING						
"Sika" waterproof rendering or other equal and approved; steel trowelled						
20 mm work to walls; three coat; to concrete base						
over 300 mm wide	-	-	-	-	m²	36.45
not exceeding 300 mm wide	-	-	-	-	m²	56.92
25 mm work to walls; three coat; to concrete base						
over 300 mm wide	-	-	-	-	m²	41.44
not exceeding 300 mm wide	-	-	-	-	m²	64.91
40 mm work to walls; four coat; to concrete base						
over 300 mm wide	-	-	-	-	m²	61.92
not exceeding 300 mm wide	-	-	-	-	m²	99.87
J20 MASTIC ASPHALT TANKING/DAMP PROOF MEMBRANES						
Mastic asphalt to BS 6925 Type T 1097						
13 mm thick one coat coverings to concrete base; flat; subsequently covered						
over 300 mm wide	-	-	-	-	m²	11.46
225 mm - 300 mm wide	-	-	-	-	m²	27.19
150 mm - 225 mm wide	-	-	-	-	m²	29.89
not exceeding 150 mm wide	-	-	-	-	m²	37.30
20 mm thick two coat coverings to concrete base; flat; subsequently covered						
over 300 mm wide	-	-	-	-	m²	16.55
225 mm - 300 mm wide	-	-	-	-	m²	30.26
150 mm - 225 mm wide	-	-	-	-	m²	33.03
not exceeding 150 mm wide	-	-	-	-	m²	40.37
30 mm thick three coat coverings to concrete base; flat; subsequently covered						
over 300 mm wide	-	-	-	-	m²	23.67
225 mm - 300 mm wide	-	-	-	-	m²	34.38
150 mm - 225 mm wide	-	-	-	-	m²	37.30
not exceeding 150 mm wide	-	-	-	-	m²	44.49
13 mm thick two coat coverings to brickwork base; vertical; subsequently covered						
over 300 mm wide	-	-	-	-	m²	45.99
225 mm - 300 mm wide	-	-	-	-	m²	56.10
150 mm - 225 mm wide	-	-	-	-	m²	66.28
not exceeding 150 mm wide	-	-	-	-	m²	85.91
20 mm thick three coat coverings to brickwork base; vertical; subsequently covered						
over 300 mm wide	-	-	-	-	m²	71.98
225 mm - 300 mm wide	-	-	-	-	m²	77.44
150 mm - 225 mm wide	-	-	-	-	m²	82.24
not exceeding 150 mm wide	-	-	-	-	m²	106.36
Turning into groove 20 mm deep	-	-	-	-	m	0.90
Internal angle fillets; subsequently covered	-	-	-	-	m	5.32

Prices for Measured Works - Minor Works
J WATERPROOFING

	PC £	Labour hours	Labour £	Material £	Unit	Total rate £
J21 MASTIC ASPHALT ROOFING/ INSULATION/FINISHES						
Mastic asphalt to BS 6925 Type R 988						
20 mm thick two coat coverings; felt isolating membrane; to concrete (or timber) base; flat or to falls or slopes not exceeding 10 degrees from horizontal						
over 300 mm wide	-	-	-	-	m²	16.33
225 mm - 300 mm wide	-	-	-	-	m²	29.96
150 mm - 225 mm wide	-	-	-	-	m²	33.19
not exceeding 150 mm wide	-	-	-	-	m²	40.08
Add to the above for covering with:						
10 mm thick limestone chippings in hot bitumen	-	-	-	-	m²	3.23
coverings with solar reflective paint	-	-	-	-	m²	2.92
300 mm x 300 mm x 8 mm g.r.p. tiles in hot bitumen	-	-	-	-	m²	42.76
Cutting to line; jointing to old asphalt	-	-	-	-	m	7.19
13 mm thick two coat skirtings to brickwork base						
not exceeding 150 mm girth	-	-	-	-	m	12.80
150 mm - 225 mm girth	-	-	-	-	m	14.76
225 mm - 300 mm girth	-	-	-	-	m	16.55
13 mm thick three coat skirtings; expanded metal lathing reinforcement nailed to timber base						
not exceeding 150 mm girth	-	-	-	-	m	20.30
150 mm - 225 mm girth	-	-	-	-	m	23.97
225 mm - 300 mm girth	-	-	-	-	m	27.94
13 mm thick two coat fascias to concrete base						
not exceeding 150 mm girth	-	-	-	-	m	12.80
150 mm - 225 mm girth	-	-	-	-	m	14.76
20 mm thick two coat linings to channels to concrete base						
not exceeding 150 mm girth	-	-	-	-	m	31.30
150 mm - 225 mm girth	-	-	-	-	m	33.19
225 mm - 300 mm girth	-	-	-	-	m	37.45
20 mm thick two coat lining to cesspools						
250 mm x 150 mm x 150 mm deep	-	-	-	-	nr	30.41
Collars around pipes, standards and like members	-	-	-	-	nr	14.76
Accessories						
Eaves trim; extruded aluminium alloy; working asphalt into trim						
"Alutrim"; type A roof edging or other equal and approved	-	-	-	-	m	11.86
extra; angle	-	-	-	-	nr	6.61
Roof screed ventilator - aluminium alloy						
"Extr-aqua-vent" or other equal and approved; set on screed over and including dished sinking; working collar around ventilator	-	-	-	-	nr	22.97
J30 LIQUID APPLIED TANKING/DAMP PROOF MEMBRANES						
Tanking and damp proofing						
"Synthaprufe" or other equal and approved; blinding with sand; horizontal on slabs						
two coats	-	0.20	1.64	3.25	m²	4.89
three coats	-	0.25	2.38	4.81	m²	7.19

J WATERPROOFING

	PC £	Labour hours	Labour £	Material £	Unit	Total rate £
J30 LIQUID APPLIED TANKING/DAMP PROOF MEMBRANES - cont'd						
Tanking and damp proofing - cont'd						
"Tretolastex 202T" or other equal and approved; on vertical surfaces of concrete						
two coats	-	0.20	1.64	0.54	m²	2.18
three coats	-	0.29	2.38	0.82	m²	3.20
One coat Vandex "Super" 0.75 kg/m² slurry or other equal and approved; one consolidating coat of Vandex "Premix" 1 kg/m² slurry or simiar; horizontal on beds						
over 225 mm wide	-	0.36	2.96	2.99	m²	5.95
J40 FLEXIBLE SHEET TANKING/DAMP PROOF MEMBRANES						
Tanking and damp proofing						
"Bituthene" sheeting or other equal and approved; lapped joints; horizontal on slabs						
500 grade	-	0.10	0.82	2.80	m²	3.62
1000 grade	-	0.11	0.90	3.82	m²	4.72
heavy duty grade	-	0.13	1.07	5.54	m²	6.61
"Bituthene" sheeting or other equal and approved; lapped joints; dressed up vertical face of concrete						
1000 grade	-	0.19	1.56	3.82	m²	5.38
"Servi-pak" protection board or other equal and approved; butt jointed; taped joints; to horizontal surfaces;						
3 mm thick	-	0.16	1.32	5.27	m²	6.59
6 mm thick	-	0.16	1.32	7.29	m²	8.61
12 mm thick	-	0.20	1.64	11.40	m²	13.04
"Servi-pak" protection board or other equal and approved; butt jointed; taped joints; to vertical surfaces						
3 mm thick	-	0.20	1.64	5.27	m²	6.91
6 mm thick	-	0.20	1.64	7.29	m²	8.93
12 mm thick	-	0.26	2.14	11.40	m²	13.54
"Bituthene" fillet or other equal and approved						
40 mm x 40 mm	-	0.10	0.82	5.02	m	5.84
"Bituthene" reinforcing strip or other equal and approved; 300 mm wide						
1000 grade	-	0.10	0.82	1.15	m	1.97
Expandite "Famflex" hot bitumen bonded waterproof tanking or other equal and approved; 150 mm laps						
horizontal; over 300 mm wide	-	0.41	3.37	12.83	m²	16.20
vertical; over 300 mm wide	-	0.67	5.51	12.83	m²	18.34
J41 BUILT UP FELT ROOF COVERINGS						
NOTE: The following items of felt roofing, unless otherwise described, include for conventional lapping, laying and bonding between layers and to base; and laying flat or to falls, to crossfalls or to slopes not exceeding 10 degrees - but exclude any insulation etc.						
Felt roofing; BS 747; suitable for flat roofs						
Three layer coverings first layer type 3G; subsequent layers type 3B bitumen glass fibre based felt	-	-	-	-	m²	7.74

J WATERPROOFING

	PC £	Labour hours	Labour £	Material £	Unit	Total rate £
Extra over felt for covering with and bedding in hot bitumen						
13 mm thick stone chippings	-	-	-	-	m²	2.55
300 mm x 300 mm x 8 mm g.r.p. tiles	-	-	-	-	m²	42.8
working into outlet pipes and the like	-	-	-	-	m²	10.54
Skirtings; three layer; top layer mineral surfaced; dressed over tilting fillet; turned into groove						
not exceeding 200 mm girth	-	-	-	-	m	7.79
200 mm - 400 mm girth	-	-	-	-	m	9.34
Coverings to kerbs; three layer						
400 mm - 600 mm girth	-	-	-	-	m	19.67
Linings to gutters; three layer						
400 mm - 600 mm girth	-	-	-	-	m	13.33
Collars around pipes and the like; three layer mineral surface; 150 mm high						
not exceeding 55 mm nominal size	-	-	-	-	nr	8.89
55 mm - 110 mm nominal size	-	-	-	-	nr	8.89
Three layer coverings; two base layers type 5U bitumen polyester based felt; top layer type 5B polyester based mineral surfaced felt; 10 mm stone chipping covering; bitumen bonded	-	-	-	-	m²	16.38
Coverings to kerbs						
not exceeding 200 mm girth	-	-	-	-	m	10.29
200 mm - 400 mm girth	-	-	-	-	m	12.93
Outlets and dishing to gullies						
300 mm diameter	-	-	-	-	nr	10.54
"Andersons" high performance polyester-based roofing system or other equal and approved						
Two layer coverings; first layer HT 125 underlay; second layer HT 350; fully bonded to wood; fibre or cork base	-	-	-	-	m²	11.69
Extra over for						
top layer mineral surfaced	-	-	-	-	m²	2.25
13 mm thick stone chippings	-	-	-	-	m²	2.55
third layer of type 3B as underlay for concrete or screeded base	-	-	-	-	m²	2.25
working into outlet pipes and the like	-	-	-	-	nr	10.54
Skirtings; two layer; top layer mineral surfaced; dressed over tilting fillet; turned into groove						
not exceeding 200 mm girth	-	-	-	-	m	8.29
200 mm - 400 mm girth	-	-	-	-	m	9.99
Coverings to kerbs; two layer						
400 mm - 600 mm girth	-	-	-	-	m	23.22
Linings to gutters; three layer						
400 mm - 600 mm girth	-	-	-	-	m	17.03
Collars around pipes and the like; two layer; 150 mm high						
not exceeding 55 mm nominal size	-	-	-	-	nr	8.89
55 mm - 110 mm nominal size	-	-	-	-	nr	8.89
"Ruberglas 120 GP" high performance roofing or other equal and approved						
Two layer coverings; first and second layers "Ruberglas 120 GP"; fully bonded to wood, fibre or cork base	-	-	-	-	m²	7.25
Extra over for						
top layer mineral surfaced	-	-	-	-	m²	2.25
13 mm thick stone chippings	-	-	-	-	m²	2.55
third layer of "Rubervent 3G" as underlay for concrete or screeded base	-	-	-	-	m²	2.25
working into outlet pipes and the like	-	-	-	-	nr	10.54

J WATERPROOFING

	PC £	Labour hours	Labour £	Material £	Unit	Total rate £
J41 BUILT UP FELT ROOF COVERINGS - cont'd						
"Ruberglas 120 GP" high performance roofing or other equal and approved - cont'd						
Skirtings; two layer; top layer mineral surfaced; dressed over tilting fillet; turned into groove						
not exceeding 200 mm girth	-	-	-	-	m	7.25
200 mm - 400 mm girth	-	-	-	-	m	8.38
Coverings to kerbs; two layer						
400 mm - 600 mm girth	-	-	-	-	m	14.48
Linings to gutters; three layer						
400 mm - 600 mm girth	-	-	-	-	m	11.99
Collars around pipes and the like; two layer, 150 mm high						
not exceeding 55 mm nominal size	-	-	-	-	nr	8.89
55 mm - 110 mm nominal size	-	-	-	-	nr	8.89
"Ruberfort HP 350" high performance roofing or other equal and approved						
Two layer coverings; first layer "Ruberfort HP 180"; second layer "Ruberfort HP 350"; fully bonded; to wood; fibre or cork base	-	-	-	-	m^2	11.18
Extra over for						
top layer mineral surfaced	-	-	-	-	m^2	2.25
13 mm thick stone chippings	-	-	-	-	m^2	2.55
third layer of "Rubervent 3G"; as underlay for concrete or screeded base	-	-	-	-	m^2	2.34
working into outlet pipes and the like	-	-	-	-	nr	10.54
Skirtings; two layer; top layer mineral surface; dressed over tilting fillet; turned into groove						
not exceeding 200 mm girth	-	-	-	-	m	7.99
200 mm - 400 mm girth	-	-	-	-	m	9.44
Coverings to kerbs; two layer						
400 mm - 600 mm girth	-	-	-	-	m	22.42
Linings to gutters; three layer						
400 mm - 600 mm girth	-	-	-	-	m	16.13
Collars around pipes and the like; two layer; 150 mm high						
not exceeding 55 mm nominal size	-	-	-	-	nr	8.89
55 mm - 110 mm nominal size	-	-	-	-	nr	8.89
"Polybit 350" elastomeric roofing or other equal or applied						
Two layer coverings; first layer "Polybit 180"; second layer "Polybit 350"; fully bonded to wood; fibre or cork base	-	-	-	-	m^2	11.44
Extra over for						
top layer mineral surfaced	-	-	-	-	m^2	2.25
13 mm thick stone chippings	-	-	-	-	m^2	2.55
third layer of "Rubervent 3G" as underlay for concrete or screeded base	-	-	-	-	m^2	2.25
working into outlet pipes and the like	-	-	-	-	nr	10.54
Skirtings; two layer; top layer mineral surfaced; dressed over tilting fillet; turned into groove						
not exceeding 200 mm girth	-	-	-	-	m	8.19
200 mm - 400 mm girth	-	-	-	-	m	9.78
Coverings to kerbs; two layer						
400 mm - 600 mm girth	-	-	-	-	m	22.91
Linings to gutters; three layer						
400 mm - 600 mm girth	-	-	-	-	m	16.63
Collars around pipes and the like; two layer; 150 mm high						
not exceeding 55 mm nominal size	-	-	-	-	nr	8.89
55 mm - 110 mm nominal size	-	-	-	-	nr	8.89

J WATERPROOFING

	PC £	Labour hours	Labour £	Material £	Unit	Total rate £
"Hyload 150 E" elastomeric roofing or other equal and approved						
Two layer coverings; first layer "Ruberglas 120 GHP"; second layer "Hyload 150 E" fully bonded to wood; fibre or cork base	-	-	-	-	m^2	23.37
Extra over for						
13 mm thick stone chippings	-	-	-	-	m^2	2.55
third layer of "Rubervent 3G" as underlay for concrete or screeded base	-	-	-	-	m^2	2.25
working into outlet pipes and the like	-	-	-	-	nr	10.48
Skirtings; two layer; dressed over tilting fillet; turned into groove						
not exceeding 200 mm girth	-	-	-	-	m	13.14
200 mm - 400 mm girth	-	-	-	-	m	21.37
Coverings to kerbs; two layer						
400 mm - 600 mm girth	-	-	-	-	m	43.59
Linings to gutters; three layer						
400 mm - 600 mm girth	-	-	-	-	m	28.91
Collars around pipes and the like; two layer; 150 mm high						
not exceeding 55 mm nominal size	-	-	-	-	nr	8.89
55 mm - 110 mm nominal size	-	-	-	-	nr	8.89
Accessories						
Eaves trim; extruded aluminium alloy; working felt into trim "Alutrim"; type F roof edging or other equal and approved	-	-	-	-	m	7.79
extra over for external angle	-	-	-	-	nr	10.54
Roof screed ventilator - aluminium alloy "Extr-aqua-vent" or other equal and approved - set on screed over and including dished sinking and collar	-	-	-	-	nr	23.22
Insulation board underlays						
Vapour barrier						
reinforced; metal lined	-	-	-	-	m^2	11.18
Cork boards; density 112 - 125 kg/m³						
60 mm thick	-	-	-	-	m^2	14.68
Rockwool; slab RW4 or other equal and approved						
60 mm thick	-	-	-	-	m^2	17.78
Perlite boards or other equal and approved; density 170 - 180 kg/m³						
60 mm thick	-	-	-	-	m^2	23.22
Polyurethane boards; density 32 kg/m³						
30 mm thick	-	-	-	-	m^2	9.19
35 mm thick	-	-	-	-	m^2	9.84
50 mm thick	-	-	-	-	m^2	10.78
Wood fibre boards; impregnated; density 220 - 350 kg/m³						
12.70 mm thick	-	-	-	-	m^2	4.04
Insulation board overlays						
Dow "Roofmate SL" extruded polystyrene foam boards or other equal and approved						
50 mm thick	-	0.32	5.26	14.26	m^2	19.52
75 mm thick	-	0.32	5.26	19.48	m^2	24.74
Dow "Roofmate LG" extruded polystyrene foam boards or other equal and approved						
60 mm thick	-	0.32	5.26	28.12	m^2	33.38
90 mm thick	-	0.32	5.26	31.92	m^2	37.18
110 mm thick	-	0.32	5.26	35.27	m^2	40.53

Prices for Measured Works - Minor Works
J WATERPROOFING

	PC £	Labour hours	Labour £	Material £	Unit	Total rate £
J41 BUILT UP FELT ROOF COVERINGS - cont'd						
Insulation board overlays - cont'd						
Dow "Roofmate PR" extruded polystyrene foam boards or other equal and approved						
90 mm thick	-	0.32	5.26	21.94	m^2	**27.20**
120 mm thick	-	0.32	5.26	29.23	m^2	**34.49**
J42 SINGLE LAYER PLASTICS ROOF COVERINGS						
"Trocal S" PVC roofing or other equal and approved						
Coverings	-	-	-	-	m^2	**17.23**
Skirtings; dressed over metal upstands						
not exceeding 200 mm girth	-	-	-	-	m	**13.38**
200 mm - 400 mm girth	-	-	-	-	m	**16.43**
Coverings to kerbs						
400 mm - 600 mm girth	-	-	-	-	m	**30.04**
Collars around pipes and the like; 150 mm high						
not exceeding 55 mm nominal size	-	-	-	-	nr	**9.19**
55 mm - 110 mm nominal size	-	-	-	-	nr	**9.19**
Accessories						
"Trocal" metal upstands or other equal and approved						
not exceeding 200 mm girth	-	-	-	-	m	**9.76**
200 mm - 400 mm girth	-	-	-	-	m	**12.63**

Prices for Measured Works - Minor Works
K LININGS/SHEATHING/DRY PARTITIONING

	PC £	Labour hours	Labour £	Material £	Unit	Total rate £
K LININGS/SHEATHING/DRY PARTITIONING						
K10 PLASTERBOARD DRY LINING/ PARTITIONS/CEILINGS						
"Gyproc" laminated proprietary partitions or other equal and approved; two skins of gypsum plasterboard bonded to a centre core of plasterboard square edge plank 19 mm thick; fixing with nails to softwood perimeter battens (not included); joints filled with filler and joint tape; to receive direct decoration						
50 mm partition; two outer skins of 12.50 mm thick tapered edge wallboard						
height 2.10 m - 2.40 m	-	2.04	26.36	20.37	m	46.73
height 2.40 m - 2.70 m	-	2.64	34.09	23.26	m	57.35
height 2.70 m - 3.00 m	-	2.96	38.20	25.29	m	63.49
height 3.00 m - 3.30 m	-	3.24	41.80	27.33	m	69.13
height 3.30 m - 3.60 m	-	3.56	45.91	29.45	m	75.36
65 mm partition; two outer skins of 19 mm thick tapered edge planks						
height 2.10 m - 2.40 m	-	2.27	29.36	25.36	m	54.72
height 2.40 m - 2.70 m	-	2.82	36.43	28.57	m	65.00
height 2.70 m - 3.00 m	-	3.15	40.67	30.91	m	71.58
height 3.00 m - 3.30 m	-	3.52	45.45	33.26	m	78.71
height 3.30 m - 3.60 m	-	3.84	49.56	35.70	m	85.26
Labours and associated additional wrought softwood studwork						
Floor, wall or ceiling battens						
25 mm x 38 mm	-	0.12	1.56	0.40	m	1.96
Forming openings in 2400 mm high partition; 25 mm x 38 mm softwood framing						
900 mm x 2100 mm	-	0.51	6.64	2.24	nr	8.88
fair ends	-	0.20	2.58	0.44	m	3.02
angle	-	0.31	3.99	0.82	m	4.81
Cutting and fitting around steel joists, angles, trunking, ducting, ventilators, pipes, tubes, etc.						
over 2.00 m girth	-	0.09	1.17	-	nr	1.17
not exceeding 0.30 m girth	-	0.05	0.65	-	nr	0.65
0.30 m - 1.00 m girth	-	0.07	0.91	-	nr	0.91
1.00 m - 2.00 m girth	-	0.11	1.43	-	nr	1.43
"Gyproc" metal stud proprietary partitions or other equal and approved; comprising 48 mm wide metal stud frame; 50 mm wide floor channel plugged and screwed to concrete through 38 mm x 48 mm tanalised softwood sole plate Tapered edge panels; joints filled with joint filler and joint tape to receive direct decoration; 80 mm thick partition; one hour; one layer of 15 mm thick "Fireline" board each side						
height 2.10 m - 2.40 m	-	4.30	55.79	23.78	m	79.57
height 2.40 m - 2.70 m	-	4.90	63.51	27.43	m	90.94
height 2.70 m - 3.00 m	-	5.50	71.27	30.19	m	101.46
height 3.00 m - 3.30 m	-	6.34	82.16	33.22	m	115.38
height 3.30 m - 3.60 m	-	6.99	90.44	35.82	m	126.26
height 3.60 m - 3.90 m	-	8.33	107.97	38.65	m	146.62
height 3.90 m - 4.20 m	-	8.93	115.74	41.49	m	157.23
angles	-	0.20	2.58	1.45	m	4.03
T-junctions	-	0.10	1.30	-	m	1.30
fair ends	-	0.20	2.58	0.52	m	3.10

K LININGS/SHEATHING/DRY PARTITIONING

	PC £	Labour hours	Labour £	Material £	Unit	Total rate £
K10 PLASTERBOARD DRY LINING/ PARTITIONS/CEILINGS						
"Gyproc" metal stud proprietary partitions or other equal and approved; comprising 48 mm wide metal stud frame; 50 mm wide floor channel plugged and screwed to concrete through 38 mm x 48 mm tanalised softwood sole plate - cont'd Tapered edge panels; joints filled with joint filler and joint tape to receive direct decoration; 100 mm thick partition; two hour; two layers of 12.50 mm thick "Fireline" board both sides						
height 2.10 m - 2.40 m	-	5.32	69.07	33.12	m	102.19
height 2.40 m - 2.70 m	-	6.06	78.62	37.96	m	116.58
height 2.70 m - 3.00 m	-	6.80	88.20	41.88	m	130.08
height 3.00 m - 3.30 m	-	6.75	87.50	46.08	m	133.58
height 3.30 m - 3.60 m	-	8.51	110.36	49.84	m	160.20
height 3.60 m - 3.90 m	-	8.33	107.97	53.90	m	161.87
height 3.90 m - 4.20 m	-	10.73	139.18	57.83	m	197.01
angles	-	0.31	4.01	1.58	m	5.59
T-junctions	-	0.10	1.30	-	m	1.30
fair ends	-	0.31	4.01	0.64	m	4.65
Gypsum plasterboard; BS 1230; plain grade tapered edge wallboard; fixing with nails; joints left open to receive "Artex" finish or other equal and approved; to softwood base						
9.50 mm board to ceilings						
over 300 mm wide	-	0.26	4.41	1.99	m²	6.40
9.50 mm board to beams						
girth not exceeding 600 mm	-	0.31	5.26	1.21	m²	6.47
girth 600 mm - 1200 mm	-	0.41	6.96	2.41	m²	9.37
12.50 mm board to ceilings						
over 300 mm wide	-	0.33	5.60	2.33	m²	7.93
12.50 mm board to beams						
girth not exceeding 600 mm	-	0.31	5.26	1.43	m²	6.69
girth 600 mm - 1200 mm	-	0.41	6.96	2.80	m²	9.76
Gypsum plasterboard to BS 1230; fixing with nails; joints filled with joint filler and joint tape to receive direct decoration; to softwood base						
Plain grade tapered edge wallboard						
9.50 mm board to walls						
wall height 2.40 m - 2.70 m	-	1.02	13.14	6.61	m	19.75
wall height 2.70 m - 3.00 m	-	1.16	14.94	7.36	m	22.30
wall height 3.00 m - 3.30 m	-	1.34	17.25	8.10	m	25.35
wall height 3.30 m - 3.60 m	-	1.53	19.70	8.88	m	28.58
9.50 mm board to reveals and soffits of openings and recesses						
not exceeding 300 mm wide	-	0.20	2.58	1.32	m	3.90
300 mm - 600 mm wide	-	0.41	5.29	1.90	m	7.19
9.50 mm board to faces of columns - 4 nr						
not exceeding 600 mm total girth	-	0.52	6.71	2.68	m	9.39
600 mm - 1200 mm total girth	-	1.02	13.15	3.85	m	17.00
1200 mm - 1800 mm total girth	-	1.32	16.99	5.02	m	22.01
9.50 mm board to ceilings						
over 300 mm wide	-	0.43	5.55	2.45	m²	8.00

K LININGS/SHEATHING/DRY PARTITIONING

	PC £	Labour hours	Labour £	Material £	Unit	Total rate £
9.50 mm board to faces of beams - 3 nr						
not exceeding 600 mm total girth	-	0.61	7.87	2.64	m	10.51
600 mm - 1200 mm total girth	-	1.12	14.44	3.81	m	18.25
1200 mm - 1800 mm total girth	-	1.43	18.41	4.98	m	23.39
add for "Duplex" insulating grade	-	-	-	0.60	m^2	0.60
12.50 mm board to walls						
wall height 2.40 m - 2.70 m	-	1.20	15.46	7.45	m	22.91
wall height 2.70 m - 3.00 m	-	1.34	17.26	8.28	m	25.54
wall height 3.00 m - 3.30 m	-	1.48	19.06	9.11	m	28.17
wall height 3.30 m - 3.60 m	-	1.76	22.67	9.98	m	32.65
12.50 mm board to reveals and soffits of openings and recesses						
not exceeding 300 mm wide	-	0.20	2.58	1.44	m	4.02
300 mm - 600 mm wide	-	0.41	5.29	2.10	m	7.39
12.50 mm board to faces of columns - 4 nr						
not exceeding 600 mm total girth	-	0.52	6.71	2.92	m	9.63
600 mm - 1200 mm total girth	-	1.02	13.15	4.27	m	17.42
1200 mm - 1800 mm total girth	-	1.32	16.99	5.61	m	22.60
12.50 mm board to ceilings						
over 300 mm wide	-	0.44	5.68	2.76	m^2	8.44
12.50 mm board to faces of beams - 3 nr						
not exceeding 600 mm total girth	-	0.61	7.87	2.86	m	10.73
600 mm - 1200 mm total girth	-	1.12	14.44	4.20	m	18.64
1200 mm - 1800 mm total girth	-	1.43	18.41	5.54	m	23.95
external angle; with joint tape bedded and covered with "Jointex" or other equal and approved	-	0.12	1.54	0.34	m	1.88
add for "Duplex" insulating grade or other equal and approved	-	-	-	0.60	m^2	0.60
Tapered edge plank						
19 mm plank to walls						
wall height 2.40 m - 2.70 m	-	1.11	14.31	10.26	m	24.57
wall height 2.70 m - 3.00 m	-	1.30	16.76	11.41	m	28.17
wall height 3.00 m - 3.30 m	-	1.43	18.42	12.56	m	30.98
wall height 3.30 m - 3.60 m	-	1.71	22.04	13.74	m	35.78
19 mm plank to reveals and soffits of openings and recesses						
not exceeding 300 mm wide	-	0.22	2.84	1.75	m	4.59
300 mm - 600 mm wide	-	0.46	5.94	2.73	m	8.67
19 mm plank to faces of columns - 4 nr						
not exceeding 600 mm total girth	-	0.56	7.23	3.55	m	10.78
600 mm - 1200 mm total girth	-	1.07	13.80	5.52	m	19.32
1200 mm - 1800 mm total girth	-	1.38	17.77	7.49	m	25.26
19 mm plank to ceilings						
over 300 mm wide	-	0.47	6.07	3.80	m^2	9.87
19 mm plank to faces of beams - 3 nr						
not exceeding 600 mm total girth	-	0.67	8.65	3.48	m	12.13
600 mm - 1200 mm total girth	-	1.17	15.09	5.45	m	20.54
1200 mm - 1800 mm total girth	-	1.48	19.06	7.42	m	26.48
Thermal Board						
27 mm board to walls						
wall height 2.40 m - 2.70 m	-	1.16	14.96	18.83	m	33.79
wall height 2.70 m - 3.00 m	-	1.34	17.28	20.92	m	38.20
wall height 3.00 m - 3.30 m	-	1.48	19.08	23.02	m	42.10
wall height 3.30 m - 3.60 m	-	1.80	23.22	25.16	m	48.38
27 mm board to reveals and soffits of openings and recesses						
not exceeding 300 mm wide	-	0.23	2.97	2.70	m	5.67
300 mm - 600 mm wide	-	0.47	6.07	4.03	m	10.70

Prices for Measured Works - Minor Works
K LININGS/SHEATHING/DRY PARTITIONING

	PC £	Labour hours	Labour £	Material £	Unit	Total rate £
K10 PLASTERBOARD DRY LINING/ PARTITIONS/CEILINGS - cont'd						
Thermal Board - cont'd						
27 mm board to faces of columns - 4 nr						
not exceeding 600 mm total girth	-	0.57	7.36	5.45	m	12.81
600 mm - 1200 mm total girth	-	1.13	14.58	9.32	m	23.90
1200 mm - 1800 mm total girth	-	1.42	18.29	13.20	m	31.49
27 mm board to ceilings						
over 300 mm wide	-	0.51	6.59	6.97	m²	13.56
27 mm board to faces of beams - 3 nr						
not exceeding 600 mm total girth	-	0.62	8.01	5.38	m	13.39
600 mm - 1200 mm total girth	-	1.17	15.10	9.26	m	24.36
1200 mm - 1800 mm total girth	-	1.58	20.36	13.13	m	33.49
50 mm board to walls						
wall height 2.40 m - 2.70 m	-	1.16	14.96	18.74	m	33.70
wall height 2.70 m - 3.00 m	-	1.39	17.93	20.85	m	38.78
wall height 3.00 m - 3.30 m	-	1.62	20.91	22.94	m	43.85
wall height 3.30 m - 3.60 m	-	1.90	24.52	25.07	m	49.59
50 mm board to reveals and soffits of openings and recesses						
not exceeding 300 mm wide	-	0.26	3.36	2.70	m	6.06
300 mm - 600 mm wide	-	0.51	6.59	4.62	m	11.21
50 mm board to faces of columns - 4 nr						
not exceeding 600 mm total girth	-	0.62	8.01	5.49	m	13.50
600 mm - 1200 mm total girth	-	1.23	15.88	9.34	m	25.22
1200 mm - 1800 mm total girth	-	1.58	20.38	13.18	m	33.56
50 mm board to ceilings						
over 300 mm wide	-	0.54	6.98	6.95	m²	13.93
50 mm board to faces of beams - 3 nr						
not exceeding 600 mm total girth	-	0.65	8.40	5.54	m	13.94
600 mm - 1200 mm total girth	-	1.30	16.79	9.43	m	26.22
1200 mm - 1800 mm total girth	-	1.74	22.45	13.32	m	35.77
White plastic faced gypsum plasterboard to BS 1230; industrial grade square edge wallboard; fixing with screws; butt joints; to softwood base						
9.50 mm board to walls						
wall height 2.40 m - 2.70 m	-	0.69	8.98	10.25	m	19.23
wall height 2.70 m - 3.00 m	-	0.88	11.46	11.37	m	22.83
wall height 3.00 m - 3.30 m	-	1.02	13.28	12.50	m	25.78
wall height 3.30 m - 3.60 m	-	1.16	15.10	13.62	m	28.72
9.50 mm board to reveals and soffits of openings and recesses						
not exceeding 300 mm wide	-	0.16	2.08	1.17	m	3.25
300 mm - 600 mm wide	-	0.31	4.04	2.28	m	6.32
9.50 mm board to faces of columns - 4 nr						
not exceeding 600 mm total girth	-	0.41	5.34	2.47	m	7.81
600 mm - 1200 mm total girth	-	0.81	10.55	4.82	m	15.37
1200 mm - 1800 mm total girth	-	1.06	13.80	7.11	m	20.91
12.50 mm board to walls						
wall height 2.40 m - 2.70 m	-	0.79	10.29	10.89	m	21.18
wall height 2.70 m - 3.00 m	-	0.93	12.11	12.09	m	24.20
wall height 3.00 m - 3.30 m	-	1.06	13.80	13.29	m	27.09
wall height 3.30 m - 3.60 m	-	1.20	15.62	14.48	m	30.10
12.50 mm board to reveals and soffits of openings and recesses						
not exceeding 300 mm wide	-	0.17	2.21	1.24	m	3.45
300 mm - 600 mm wide	-	0.32	4.17	2.42	m	6.59

Prices for Measured Works - Minor Works
K LININGS/SHEATHING/DRY PARTITIONING

	PC £	Labour hours	Labour £	Material £	Unit	Total rate £
12.50 mm board to faces of columns - 4 nr						
not exceeding 600 mm total girth	-	0.43	5.60	2.62	m	8.22
600 mm - 1200 mm total girth	-	0.85	11.07	5.13	m	16.20
1200 mm - 1800 mm total girth	-	1.12	14.58	7.56	m	22.14
Plasterboard jointing system; filling joint with jointing compounds						
To ceilings						
to suit 9.50 mm or 12.50 mm thick boards	-	0.10	1.30	1.49	m	2.79
Angle trim; plasterboard edge support system						
To ceilings						
to suit 9.50 mm or 12.50 mm thick boards	-	0.10	1.30	1.41	m	2.71
Two layers of gypsum plasterboard to BS 1230; plain grade square and tapered edge wallboard; fixing with nails; joints filled with joint filler and joint tape; top layer to receive direct decoration; to softwood base						
19 mm two layer board to walls						
wall height 2.40 m - 2.70 m	-	1.43	18.48	12.13	m	30.61
wall height 2.70 m - 3.00 m	-	1.62	20.93	13.49	m	34.42
wall height 3.00 m - 3.30 m	-	1.85	23.89	14.85	m	38.74
wall height 3.30 m - 3.60 m	-	2.13	27.51	16.24	m	43.75
19 mm two layer board to reveals and soffits of openings and recesses						
not exceeding 300 mm wide	-	0.31	4.01	1.99	m	6.00
300 mm - 600 mm wide	-	0.61	7.89	3.16	m	11.05
19 mm two layer board to faces of columns - 4 nr						
not exceeding 600 mm total girth	-	0.77	9.96	4.08	m	14.04
600 mm - 1200 mm total girth	-	1.38	17.83	6.44	m	24.27
1200 mm - 1800 mm total girth	-	1.83	23.63	8.80	m	32.43
25 mm two layer board to walls						
wall height 2.40 m - 2.70 m	-	1.53	19.78	13.64	m	33.42
wall height 2.70 m - 3.00 m	-	1.71	22.10	15.16	m	37.26
wall height 3.00 m - 3.30 m	-	1.94	25.06	16.69	m	41.75
wall height 3.30 m - 3.60 m	-	2.27	29.34	18.24	m	47.58
25 mm two layer board to reveals and soffits of openings and recesses						
not exceeding 300 mm wide	-	0.31	4.01	2.19	m	6.20
300 mm - 600 mm wide	-	0.61	7.89	3.51	m	11.40
25 mm two layer board to faces of columns - 4 nr						
not exceeding 600 mm total girth	-	0.77	9.96	4.47	m	14.43
600 mm - 1200 mm total girth	-	1.38	17.83	7.15	m	24.98
1200 mm - 1800 mm total girth	-	1.83	23.63	9.84	m	33.47
Gyproc Dri-Wall dry lining system or other equal and approved; plain grade tapered edge wallboard; fixed to walls with adhesive; joints filled with joint filler and joint tape; to receive direct decoration						
9.50 mm board to walls						
wall height 2.40 m - 2.70 m	-	1.20	15.48	8.19	m	23.67
wall height 2.70 m - 3.00 m	-	1.39	17.93	9.08	m	27.01
wall height 3.00 m - 3.30 m	-	1.62	20.89	9.98	m	30.87
wall height 3.30 m - 3.60 m	-	1.85	23.87	10.92	m	34.79
9.50 mm board to reveals and soffits of openings and recesses						
not exceeding 300 mm wide	-	0.26	3.36	1.44	m	4.80
300 mm - 600 mm wide	-	0.51	6.59	2.20	m	8.79

K LININGS/SHEATHING/DRY PARTITIONING

	PC £	Labour hours	Labour £	Material £	Unit	Total rate £
K10 PLASTERBOARD DRY LINING/ PARTITIONS/CEILINGS - cont'd						
Gyproc Dri-Wall dry lining system or other equal and approved; plain grade tapered edge wallboard; fixed to walls with adhesive; joints filled with joint filler and joint tape; to receive direct decoration - cont'd						
9.50 mm board to faces of columns - 4 nr						
not exceeding 600 mm total girth	-	0.65	8.40	2.89	m	11.29
600 mm - 1200 mm total girth	-	1.26	16.27	4.53	m	20.80
1200 mm - 1800 mm total girth	-	1.58	20.38	5.91	m	26.29
Angle; with joint tape bedded and covered with "Jointex" or other equal and approved						
internal	-	0.06	0.77	0.34	m	1.11
external	-	0.12	1.54	0.34	m	1.88
Gyproc Dri-Wall M/F dry lining system or other equal and approved; mild steel furrings fixed to walls with adhesive; tapered edge wallboard screwed to furrings; joints filled with joint filler and joint tape						
12.50 mm board to walls						
wall height 2.40 m - 2.70 m	-	1.62	20.95	14.70	m	35.65
wall height 2.70 m - 3.00 m	-	1.85	23.92	16.32	m	40.24
wall height 3.00 m - 3.30 m	-	2.13	27.53	17.96	m	45.49
wall height 3.30 m - 3.60 m	-	2.45	31.68	19.63	m	51.31
12.50 mm board to reveals and soffits of openings and recesses						
not exceeding 300 mm wide	-	0.26	3.36	1.30	m	4.66
300 mm - 600 mm wide	-	0.51	6.59	1.91	m	8.50
add for one coat of "Dry-Wall" top coat or other equal and approved	-	0.05	0.64	0.84	m^2	1.48
Vermiculite gypsum cladding; "Vicuclad 900R" board or other equal and approved; fixed with adhesive; joints pointed in adhesive						
25 mm thick column linings, faces - 4; 2 hour fire protection rating						
not exceeding 600 mm girth	-	0.77	10.03	14.49	m	24.52
600 mm - 1200 mm girth	-	0.92	11.98	28.77	m	40.75
1200 mm - 1800 mm girth	-	1.28	16.67	43.05	m	59.72
30 mm thick beam linings, faces - 3; 2 hour fire protection rating						
not exceeding 600 mm girth	-	0.61	7.94	17.15	m	25.09
600 mm - 1200 mm girth	-	0.77	10.03	34.09	m	44.12
1200 mm - 1800 mm girth	-	0.92	11.98	51.03	m	63.01
Vermiculite gypsum cladding; "Vicuclad 1050R" board or other equal and approved; fixed with adhesive; joints pointed in adhesive						
55 mm thick column linings, faces - 4 ; 4 hour fire protection rating						
not exceeding 600 mm girth	-	0.92	11.98	38.83	m	50.81
600 mm - 1200 mm girth	-	1.12	14.58	77.45	m	92.03
1200 mm - 1800 mm girth	-	1.53	19.92	116.07	m	135.99
60 mm thick beam linings, faces - 3; 4 hour fire protection rating						
not exceeding 600 mm girth	-	0.77	10.03	42.59	m	52.62
600 mm - 1200 mm girth	-	0.92	11.98	84.97	m	96.95
1200 mm - 1800 mm girth	-	1.12	14.58	127.35	m	141.93

K LININGS/SHEATHING/DRY PARTITIONING

	PC £	Labour hours	Labour £	Material £	Unit	Total rate £
Add to the above for						
plus 3% for work 3.50 m -5.00 m high						
plus 6% for work 5.00 m - 6.50 m high						
plus 12% for work 6.50 m - 8.00 m high						
plus 18% for work over 8.00 m high						
Cutting and fitting around steel joints, angles, trunking, ducting, ventilators, pipes, tubes, etc.						
over 2.00 m girth	-	0.45	5.86	-	m	5.86
not exceeding 0.30 m girth	-	0.31	4.04	-	nr	4.04
0.30 m - 1.00 m girth	-	0.41	5.34	-	nr	5.34
1.00 m - 2.00 m girth	-	0.56	7.29	-	nr	7.29
K11 RIGID SHEET FLOORING/SHEATHING/ LININGS/CASING						
Blockboard (Birch faced)						
Lining to walls 12 mm thick						
over 300 mm wide	9.12	0.48	6.25	10.71	m²	16.96
not exceeding 300 mm wide	-	0.31	4.04	3.22	m	7.26
holes for pipes and the like	-	0.03	0.39	-	nr	0.39
Two-sided 12 mm thick pipe casing; to softwood framing (not included)						
300 mm girth	-	0.79	10.29	3.29	m	13.58
600 mm girth	-	0.93	12.11	6.45	m	18.56
Three-sided 12 mm thick pipe casing; to softwood framing (not included)						
450 mm girth	-	1.02	13.28	4.93	m	18.21
900 mm girth	-	1.20	15.62	9.70	m	25.32
extra for 400 x 400 removable access panel; brass cups and screws; additional framing	-	1.02	13.28	0.81	nr	14.09
Lining to walls 18 mm thick						
over 300 mm wide	12.75	0.51	6.64	14.90	m²	21.54
not exceeding 300 mm wide	-	0.32	4.17	4.48	m	8.65
holes for pipes and the like	-	0.05	0.65	-	nr	0.65
Lining to walls 25 mm thick						
over 300 mm wide	16.60	0.55	7.16	19.39	m²	26.55
not exceeding 300 mm wide	-	0.36	4.69	5.82	m	10.51
holes for pipes and the like	-	0.06	0.78	-	nr	0.78
Chipboard (plain)						
Lining to walls 12 mm thick						
over 300 mm wide	2.03	0.39	5.08	2.52	m²	7.60
not exceeding 300 mm wide	-	0.22	2.86	0.77	m	3.63
holes for pipes and the like	-	0.02	0.26	-	nr	0.26
Lining to walls 15 mm thick						
over 300 mm wide	2.33	0.41	5.34	2.86	m²	8.20
not exceeding 300 mm wide	-	0.24	3.12	0.87	m	3.99
holes for pipes and the like	-	0.03	0.39	-	nr	0.39
Two-sided 15 mm thick pipe casing; to softwood framing (not included)						
300 mm girth	-	0.60	7.81	0.93	m	8.74
600 mm girth	-	0.69	8.98	1.74	m	10.72
Three-sided 15 mm thick pipe casing; to softwood framing (not included)						
450 mm girth	-	1.30	16.93	1.40	m	18.33
900 mm girth	-	1.53	19.92	2.64	m	22.56
extra for 400 x 400 removable access panel; brass cups and screws; additional framing	-	1.02	13.28	0.81	nr	14.09
Lining to walls 18 mm thick						
over 300 mm wide	2.80	0.43	5.60	3.44	m²	9.04
not exceeding 300 mm wide	-	0.28	3.65	1.03	m	4.68

K LININGS/SHEATHING/DRY PARTITIONING

	PC £	Labour hours	Labour £	Material £	Unit	Total rate £
K11 RIGID SHEET FLOORING/SHEATHING/ LININGS/CASING - cont'd						
Fire-retardant chipboard; Antivlam or other equal and approved; Class 1 spread of flame						
Lining to walls 12 mm thick						
over 300 mm wide	-	0.39	5.08	8.88	m²	13.96
not exceeding 300 mm wide	-	0.22	2.86	2.68	m	5.54
holes for pipes and the like	-	0.02	0.26	-	nr	0.26
Lining to walls 18 mm thick						
over 300 mm wide	-	0.43	5.60	11.72	m²	17.32
not exceeding 300 mm wide	-	0.28	3.65	3.53	m	7.18
holes for pipes and the like	-	0.05	0.65	-	nr	0.65
Lining to walls 22 mm thick						
over 300 mm wide	-	0.44	5.73	14.88	m²	20.61
not exceeding 300 mm wide	-	0.31	4.04	4.48	m	8.52
holes for pipes and the like	-	0.06	0.78	-	nr	0.78
Chipboard Melamine faced; white matt finish; laminated masking strips						
Lining to walls 15 mm thick						
over 300 mm wide	3.76	1.06	13.80	4.69	m²	18.49
not exceeding 300 mm wide	-	0.69	8.98	1.49	m	10.47
holes for pipes and the like	-	0.07	0.91	-	nr	0.91
Chipboard boarding and flooring						
Boarding to floors; butt joints						
18 mm thick	2.73	0.31	4.04	3.37	m²	7.41
22 mm thick	3.54	0.33	4.30	4.31	m²	8.61
Boarding to floors; tongued and grooved joints						
18 mm thick	2.89	0.32	4.17	3.55	m²	7.72
22 mm thick	3.54	0.36	4.69	4.31	m²	9.00
Boarding to roofs; butt joints						
18 mm thick	3.59	0.33	4.30	4.36	m²	8.66
Durabella "Westbourne" flooring system or other equal and approved; comprising 19 mm thick tongued and grooved chipboard panels secret nailed to softwood MK 10X-profiled foam backed battens at 600 mm centres; on concrete floor						
Flooring tongued and grooved joints						
63 mm thick overall; 44 mm x 54 mm nominal size battens	-	-	-	-	m²	19.27
75 mm thick overall; 56 mm x 54 mm nominal size battens	-	-	-	-	m²	20.82
Plywood flooring						
Boarding to floors; tongued and grooved joints						
15 mm thick	16.23	0.41	5.34	18.97	m²	24.31
18 mm thick	18.85	0.44	5.73	21.99	m²	27.72
Plywood; external quality; 18 mm thick						
Boarding to roofs; butt joints						
flat to falls	12.71	0.41	5.34	14.90	m²	20.24
sloping	12.71	0.43	5.60	14.90	m²	20.50
vertical	12.71	0.58	7.55	14.90	m²	22.45

Prices for Measured Works - Minor Works
K LININGS/SHEATHING/DRY PARTITIONING

	PC £	Labour hours	Labour £	Material £	Unit	Total rate £
Plywood; external quality; 12 mm thick						
Boarding to roofs; butt joints						
flat to falls	8.47	0.41	5.34	10.00	m²	15.34
sloping	8.47	0.43	5.60	10.00	m²	15.60
vertical	8.47	0.58	7.55	10.00	m²	17.55
Glazed hardboard to BS 1142; on and including 38 mm x 38 mm sawn softwood framing						
3.20 mm thick panel						
to side of bath	-	1.82	23.70	4.61	nr	28.31
to end of bath	-	0.71	9.24	1.27	nr	10.51
Insulation board to BS 1142						
Lining to walls 12 mm thick						
over 300 mm wide	1.57	0.24	3.12	1.98	m²	5.10
not exceeding 300 mm wide	-	0.14	1.82	0.61	m	2.43
holes for pipes and the like	-	0.01	0.13	-	nr	0.13
Lining to walls 18 mm thick						
over 300 mm wide	2.47	0.27	3.52	3.02	m²	6.54
not exceeding 300 mm wide	-	0.17	2.21	0.92	m	3.13
holes for pipes and the like	-	0.01	0.13	-	nr	0.13
Laminboard (Birch Faced); 18 mm thick						
Lining to walls						
over 300 mm wide	15.91	0.54	7.03	18.56	m²	25.59
not exceeding 300 mm wide	-	0.34	4.43	5.58	m	10.01
holes for pipes and the like	-	0.06	0.78	-	nr	0.78
Non-asbestos board; "Masterboard" or other equal and approved; sanded finish						
Lining to walls 6 mm thick						
over 300 mm wide	7.92	0.33	4.30	9.28	m²	13.58
not exceeding 300 mm wide	-	0.20	2.60	2.79	m	5.39
Lining to ceilings 6 mm thick						
over 300 mm wide	7.92	0.44	5.73	9.28	m²	15.01
not exceeding 300 mm wide	-	0.28	3.65	2.79	m	6.44
holes for pipes and the like	-	0.02	0.26	-	nr	0.26
Lining to walls 9 mm thick						
over 300 mm wide	16.70	0.37	4.82	19.43	m²	24.25
not exceeding 300 mm wide	-	0.20	2.60	5.83	m	8.43
Lining to ceilings 9 mm thick						
over 300 mm wide	16.70	0.45	5.86	19.43	m²	25.29
not exceeding 300 mm wide	-	0.30	3.91	5.83	m	9.74
holes for pipes and the like	-	0.03	0.39	-	nr	0.39
Non-asbestos board; "Supalux" or other equal and approved; sanded finish						
Lining to walls 6 mm thick						
over 300 mm wide	11.82	0.33	4.30	13.79	m²	18.09
not exceeding 300 mm wide	-	0.20	2.60	4.14	m	6.74
Lining to ceilings 6 mm thick						
over 300 mm wide	11.82	0.44	5.73	13.79	m²	19.52
not exceeding 300 mm wide	-	0.28	3.65	4.14	m	7.79
holes for pipes and the like	-	0.02	0.26	-	nr	0.26
Lining to walls 9 mm thick						
over 300 mm wide	17.58	0.37	4.82	20.44	m²	25.26
not exceeding 300 mm wide	-	0.20	2.60	6.14	m	8.74
Lining to ceilings 9 mm thick						
over 300 mm wide	17.58	0.45	5.86	20.44	m²	26.30
not exceeding 300 mm wide	-	0.30	3.91	6.14	m	10.05
holes for pipes and the like	-	0.03	0.39	-	nr	0.39

K LININGS/SHEATHING/DRY PARTITIONING

	PC £	Labour hours	Labour £	Material £	Unit	Total rate £
K11 RIGID SHEET FLOORING/SHEATHING/ LININGS/CASING - cont'd						
Non-asbestos board; "Supalux" or other equal and approved; sanded finish - cont'd						
Lining to walls 12 mm thick						
over 300 mm wide	23.27	0.41	5.34	27.02	m²	32.36
not exceeding 300 mm wide	-	0.24	3.12	8.11	m	11.23
Lining to ceilings 12 mm thick						
over 300 mm wide	23.27	0.54	7.03	27.02	m²	34.05
not exceeding 300 mm wide	-	0.32	4.17	8.11	m	12.28
holes for pipes and the like	-	0.05	0.65	-	nr	0.65
Non-asbestos board; "Monolux 40" or other equal and approved; 6 mm x 50 mm "Supalux" cover fillets or other equal and approved one side						
Lining to walls 19 mm thick						
over 300 mm wide	43.26	0.71	9.24	52.02	m²	61.26
not exceeding 300 mm wide	-	0.51	6.64	17.41	m	24.05
Lining to walls 25 mm thick						
over 300 mm wide	50.86	0.77	10.03	60.80	m²	70.83
not exceeding 300 mm wide	-	0.54	7.03	20.05	m	27.08
Plywood (Far Eastern); internal quality						
Lining to walls 4 mm thick						
over 300 mm wide	1.81	0.38	4.95	2.26	m²	7.21
not exceeding 300 mm wide	-	0.24	3.12	0.69	m	3.81
Lining to ceilings 4 mm thick						
over 300 mm wide	1.81	0.51	6.64	2.26	m²	8.90
not exceeding 300 mm wide	-	0.32	4.17	0.69	m	4.86
holes for pipes and the like	-	0.02	0.26	-	nr	0.26
Lining to walls 6 mm thick						
over 300 mm wide	2.15	0.41	5.34	2.65	m²	7.99
not exceeding 300 mm wide	-	0.27	3.52	0.81	m	4.33
Lining to ceilings 6 mm thick						
over 300 mm wide	2.15	0.54	7.03	2.65	m²	9.68
not exceeding 300 mm wide	-	0.36	4.69	0.81	m	5.50
holes for pipes and the like	-	0.02	0.26	-	nr	0.26
Two-sided 6 mm thick pipe casings; to softwood framing (not included)						
300 mm girth	-	0.83	10.81	0.87	m	11.68
600 mm girth	-	1.02	13.28	1.61	m	14.89
Three-sided 6 mm thick pipe casing; to softwood framing (not included)						
450 mm girth	-	1.16	15.10	1.31	m	16.41
900 mm girth	-	1.39	18.10	2.44	m	20.54
Lining to walls 9 mm thick						
over 300 mm wide	3.14	0.43	5.60	3.79	m²	9.39
not exceeding 300 mm wide	-	0.29	3.78	1.15	m	4.93
Lining to ceilings 9 mm thick						
over 300 mm wide	3.14	0.58	7.55	3.79	m²	11.34
not exceeding 300 mm wide	-	0.38	4.95	1.15	m	6.10
holes for pipes and the like	-	0.03	0.39	-	nr	0.39
Lining to walls 12 mm thick						
over 300 mm wide	4.05	0.47	6.12	4.85	m²	10.97
not exceeding 300 mm wide	-	0.31	4.04	1.47	m	5.51
Lining to ceilings 12 mm thick						
over 300 mm wide	4.05	0.62	8.07	4.85	m²	12.92
not exceeding 300 mm wide	-	0.41	5.34	1.47	m	6.81
holes for pipes and the like	-	0.03	0.39	-	nr	0.39

K LININGS/SHEATHING/DRY PARTITIONING

	PC £	Labour hours	Labour £	Material £	Unit	Total rate £
Plywood (Far Eastern); external quality						
Lining to walls 4 mm thick						
over 300 mm wide	3.26	0.38	4.95	3.94	m²	8.89
not exceeding 300 mm wide	-	0.24	3.12	1.20	m	4.32
Lining to ceilings 4 mm thick						
over 300 mm wide	3.26	0.51	6.64	3.94	m²	10.58
not exceeding 300 mm wide	-	0.32	4.17	1.20	m	5.37
holes for pipes and the like	-	0.02	0.26	-	nr	0.26
Lining to walls 6 mm thick						
over 300 mm wide	4.30	0.41	5.34	5.14	m²	10.48
not exceeding 300 mm wide	-	0.27	3.52	1.55	m	5.07
Lining to ceilings 6 mm thick						
over 300 mm wide	4.30	0.54	7.03	5.14	m²	12.17
not exceeding 300 mm wide	-	0.36	4.69	1.55	m	6.24
holes for pipes and the like	-	0.02	0.26	-	nr	0.26
Two-sided 6 mm thick pipe casings; to softwood framing (not included)						
300 mm girth	-	0.83	10.81	1.62	m	12.43
600 mm girth	-	1.02	13.28	3.11	m	16.39
Three-sided 6 mm thick pipe casing; to softwood framing (not included)						
450 mm girth	-	1.16	15.10	2.43	m	17.53
900 mm girth	-	1.39	18.10	4.69	m	22.79
Lining to walls 9 mm thick						
over 300 mm wide	5.73	0.43	5.60	6.79	m²	12.39
not exceeding 300 mm wide	-	0.29	3.78	2.05	m	5.83
Lining to ceilings 9 mm thick						
over 300 mm wide	5.73	0.58	7.55	6.79	m²	14.34
not exceeding 300 mm wide	-	0.38	4.95	2.05	m	7.00
holes for pipes and the like	-	0.03	0.39	-	nr	0.39
Lining to walls 12 mm thick						
over 300 mm wide	7.61	0.47	6.12	8.96	m²	15.08
not exceeding 300 mm wide	-	0.31	4.04	2.70	m	6.74
Lining to ceilings 12 mm thick						
over 300 mm wide	7.61	0.62	8.07	8.96	m²	17.03
not exceeding 300 mm wide	-	0.41	5.34	2.70	m	8.04
holes for pipes and the like	-	0.03	0.39	-	nr	0.39
Extra over wall linings fixed with nails for screwing	-	-	-	-	m²	1.71
Preformed white melamine faced plywood casings; Pendock Profiles Ltd or other equal and approved; to softwood battens (not included)						
Skirting trunking profile; plain butt joints in the running length						
45 mm x 150 mm; ref TK150	-	0.12	1.56	8.10	m	9.66
extra for stop end	-	0.05	0.65	6.74	nr	7.39
extra for external corner	-	0.10	1.30	8.30	nr	9.60
extra for internal corner	-	0.10	1.30	6.20	nr	7.50
Casing profiles						
150 mm x 150 mm ref MX150/150; 5 mm thick	-	0.12	1.56	12.90	m	14.46
extra for stop end	-	0.05	0.65	5.28	nr	5.93
extra for external corner	-	0.10	1.30	19.60	nr	20.90
extra for internal corner	-	0.10	1.30	6.20	nr	7.50

K LININGS/SHEATHING/DRY PARTITIONING

	PC £	Labour hours	Labour £	Material £	Unit	Total rate £
K11 RIGID SHEET FLOORING/SHEATHING/ LININGS/CASING - cont'd						
Woodwool unreinforced slabs; Torvale "Woodcemair" or other and approved; BS 1105 type SB; natural finish; fixing to timber or steel with galvanized nails or clips; flat or sloping						
50 mm thick slabs; type 500; 600 mm maximum span						
1800 mm - 2400 mm lengths	6.22	0.41	5.34	7.25	m²	12.59
2700 mm - 3000 mm lengths	6.40	0.41	5.34	7.46	m²	12.80
75 mm thick slabs; type 750; 900 mm maximum span						
2100 mm lengths	7.41	0.45	5.86	8.64	m²	14.50
2400 mm - 2700 mm lengths	7.52	0.45	5.86	8.76	m²	14.62
3000 mm lengths	7.53	0.45	5.86	8.77	m²	14.63
100 mm thick slabs; type 1000; 1200 mm maximum span						
3000 mm - 3600 mm lengths	10.27	0.51	6.64	11.83	m²	18.47
Internal quality American Cherry veneered plywood; 6 mm thick						
Lining to walls						
over 300 mm wide	5.44	0.44	5.73	6.42	m²	12.15
not exceeding 300 mm wide	-	0.30	3.91	1.95	m	5.86
"Tacboard" or other equal and approved; Eternit UK Ltd; fire resisting boards; butt joints; to softwood base						
Lining to walls; 6 mm thick						
over 300 mm wide	-	0.33	4.30	10.06	m²	14.36
not exceeding 300 mm wide	-	0.20	2.60	3.04	m	5.64
Lining to walls; 9 mm thick						
over 300 mm wide	-	0.37	4.82	18.44	m²	23.26
not exceeding 300 mm wide	-	0.22	2.86	5.56	m	8.42
Lining to walls; 12 mm thick						
over 300 mm wide	-	0.41	5.34	23.97	m²	29.31
not exceeding 300 mm wide	-	0.24	3.12	7.22	m	10.34
"Tacfire" or other equal and approved; Eternit UK Ltd; fire resisting boards						
Lining to walls; 6 mm thick						
over 300 mm wide	-	0.33	4.30	13.49	m²	17.79
not exceeding 300 mm wide	-	0.20	2.60	4.07	m	6.67
Lining to walls; 9 mm thick						
over 300 mm wide	-	0.37	4.82	20.58	m²	25.40
not exceeding 300 mm wide	-	0.22	2.86	6.20	m	9.06
Lining to walls; 12 mm thick						
over 300 mm wide	-	0.41	5.34	27.10	m²	32.44
not exceeding 300 mm wide	-	0.24	3.12	8.16	m	11.28

K LININGS/SHEATHING/DRY PARTITIONING

	PC £	Labour hours	Labour £	Material £	Unit	Total rate £
K14 GLASS REINFORCED GYPSUM LININGS/ PANELLING/CASINGS/MOULDINGS						
Glass reinforced gypsum Glasroc Multi-board or other equal and approved; fixing with nails; joints filled with joint filler and joint tape; finishing with "Jointex" or other equal and approved to receive decoration; to softwood base						
10 mm board to walls						
wall height 2.40 m - 2.70 m	-	1.02	13.14	45.55	m	58.69
wall height 2.70 m - 3.00 m	-	1.16	14.94	50.62	m	65.56
wall height 3.00 m - 3.30 m	-	1.34	17.25	55.68	m	72.93
wall height 3.30 m - 3.60 m	-	1.53	19.70	60.79	m	80.49
12.50 mm board to walls						
wall height 2.40 m - 2.70 m	-	1.06	13.66	59.55	m	73.21
wall height 2.70 m - 3.00 m	-	1.20	15.46	66.17	m	81.63
wall height 3.00 m - 3.30 m	-	1.39	17.90	72.79	m	90.69
wall height 3.30 m - 3.60 m	-	1.57	20.22	79.45	m	99.67
K20 TIMBER BOARD FLOORING/SHEATHING/ LININGS/CASINGS						
Sawn softwood; untreated						
Boarding to roofs; 150 mm wide boards; butt joints						
19 mm thick; flat; over 300 mm wide	0.77	0.45	5.86	6.36	m²	12.22
19 mm thick; flat; not exceeding 300 mm wide	-	0.31	4.04	1.93	m	5.97
19 mm thick; sloping; over 300 mm wide	-	0.51	6.64	6.36	m²	13.00
19 mm thick; sloping; not exceeding 300 mm wide	-	0.33	4.30	1.93	m	6.23
19 mm thick; sloping; laid diagonally; over 300 mm wide	-	0.64	8.33	6.36	m²	14.69
19 mm thick; sloping; laid diagonally; not exceeding 300 mm wide	-	0.41	5.34	1.93	m	7.27
25 mm thick; flat; over 300 mm wide	-	0.45	5.86	8.20	m²	14.06
25 mm thick; flat; not exceeding 300 mm wide	-	0.31	4.04	2.48	m	6.52
25 mm thick; sloping; over 300 mm wide	-	0.51	6.64	8.20	m²	14.84
25 mm thick; sloping; not exceeding 300 mm wide	-	0.33	4.30	2.48	m	6.78
25 mm thick; sloping; laid diagonally; over 300 mm wide	-	0.64	8.33	8.20	m²	16.53
25 mm thick; sloping; laid diagonally; not exceeding 300 mm wide	-	0.41	5.34	2.48	m	7.82
Boarding to tops or cheeks of dormers; 150 mm wide boards; butt joints						
19 mm thick; laid diagonally; over 300 mm wide	-	0.81	10.55	6.36	m²	16.91
19 mm thick; laid diagonally; not exceeding 300 mm wide	-	0.51	6.64	1.93	m	8.57
19 mm thick; laid diagonally; area not exceeding 1.00 m² irrespective of width	-	1.02	13.28	6.08	nr	19.36
Sawn softwood; "Tanalised"						
Boarding to roofs; 150 mm wide boards; butt joints						
19 mm thick; flat; over 300 mm wide	-	0.45	5.86	7.08	m²	12.94
19 mm thick; flat; not exceeding 300 mm wide	-	0.31	4.04	2.14	m	6.18
19 mm thick; sloping; over 300 mm wide	-	0.51	6.64	7.08	m²	13.72
19 mm thick; sloping; not exceeding 300 mm wide	-	0.33	4.30	2.14	m	6.44
19 mm thick; sloping; laid diagonally; over 300 mm wide	-	0.64	8.33	7.08	m²	15.41
19 mm thick; sloping; laid diagonally; not exceeding 300 mm wide	-	0.41	5.34	2.14	m	7.48
25 mm thick; flat; over 300 mm wide	-	0.45	5.86	9.16	m²	15.02

Prices for Measured Works - Minor Works
K LININGS/SHEATHING/DRY PARTITIONING

	PC £	Labour hours	Labour £	Material £	Unit	Total rate £
K20 TIMBER BOARD FLOORING/SHEATHING/ LININGS/CASINGS - cont'd						
Sawn softwood; "Tanalised" - cont'd						
Boarding to roofs; 150 mm wide boards; butt joints - cont'd						
25 mm thick; flat; not exceeding 300 mm wide	-	0.31	4.04	2.77	m	6.81
25 mm thick; sloping; over 300 mm wide	-	0.51	6.64	9.16	m²	15.80
25 mm thick; sloping; not exceeding 300 mm wide	-	0.33	4.30	2.77	m	7.07
25 mm thick; sloping; laid diagonally; over 300 mm wide	-	0.64	8.33	9.16	m²	17.49
25 mm thick; sloping; laid diagonally; not exceeding 300 mm wide	-	0.41	5.34	2.77	m	8.11
Boarding to tops or cheeks of dormers; 150 mm wide boards; butt joints						
19 mm thick; laid diagonally; over 300 mm wide	-	0.81	10.55	7.08	m²	17.63
19 mm thick; laid diagonally; not exceeding 300 mm wide	-	0.51	6.64	2.14	m	8.78
19 mm thick; laid diagonally; area not exceeding 1.00 m² irrespective of width	-	1.02	13.28	6.81	nr	20.09
Wrought softwood						
Boarding to floors; butt joints						
19 mm x 75 mm boards	-	0.61	7.94	6.74	m²	14.68
19 mm x 125 mm boards	-	0.56	7.29	7.43	m²	14.72
22 mm x 150 mm boards	-	0.52	6.77	7.97	m²	14.74
25 mm x 100 mm boards	-	0.56	7.29	8.60	m²	15.89
25 mm x 150 mm boards	-	0.52	6.77	9.24	m²	16.01
25 mm boarding and bearers to floors; butt joints; in making good where partitions removed or openings formed (boards running in direction of partition)						
150 mm wide	-	0.28	3.65	1.73	m	5.38
225 mm wide	-	0.42	5.47	2.80	m	8.27
300 mm wide	-	0.56	7.29	3.41	m	10.70
25 mm boarding and bearers to floors; butt joints; in making good where partitions removed or openings formed (boards running at right angles to partition)						
150 mm wide	-	0.42	5.47	4.13	m	9.60
225 mm wide	-	0.65	8.46	5.45	m	13.91
300 mm wide	-	0.83	10.81	6.25	m	17.06
450 mm wide	-	1.25	16.27	8.19	m	24.46
Boarding to floors; tongued and grooved joints						
19 mm x 75 mm boards	-	0.71	9.24	7.37	m²	16.61
19 mm x 125 mm boards	-	0.67	8.72	8.05	m²	16.77
22 mm x 150 mm boards	-	0.62	8.07	8.60	m²	16.67
25 mm x 100 mm boards	-	0.67	8.72	9.24	m²	17.96
25 mm x 150 mm boards	-	0.62	8.07	9.86	m²	17.93
Boarding to internal walls; tongued and grooved and V-jointed						
12 mm x 100 mm boards	-	0.81	10.55	6.96	m²	17.51
16 mm x 100 mm boards	-	0.81	10.55	9.35	m²	19.90
19 mm x 100 mm boards	-	0.81	10.55	10.50	m²	21.05
19 mm x 125 mm boards	-	0.77	10.03	11.08	m²	21.11
19 mm x 125 mm boards; chevron pattern	-	1.22	15.88	11.08	m²	26.96
25 mm x 125 mm boards	-	0.77	10.03	14.30	m²	24.33
12 mm x 100 mm boards; knotty pine	-	0.81	10.55	5.72	m²	16.27

K LININGS/SHEATHING/DRY PARTITIONING

	PC £	Labour hours	Labour £	Material £	Unit	Total rate £
Boarding to internal ceilings						
12 mm x 100 mm boards	-	1.02	13.28	6.96	m²	20.24
16 mm x 100 mm boards	-	1.02	13.28	9.35	m²	22.63
19 mm x 100 mm boards	-	1.02	13.28	10.50	m²	23.78
19 mm x 125 mm boards	-	0.97	12.63	11.08	m²	23.71
19 mm x 125 mm boards; chevron pattern	-	1.42	18.49	11.08	m²	29.57
25 mm x 125 mm boards	-	0.97	12.63	14.30	m²	26.93
12 mm x 100 mm boards; knotty pine	-	1.02	13.28	5.72	m²	19.00
Boarding to roofs; tongued and grooved joints						
19 mm thick; flat to falls	-	0.56	7.29	8.28	m²	15.57
19 mm thick; sloping	-	0.61	7.94	8.28	m²	16.22
19 mm thick; sloping; laid diagonally	-	0.79	10.29	8.28	m²	18.57
25 mm thick; flat to falls	-	0.56	7.29	9.98	m²	17.27
25 mm thick; sloping	-	0.61	7.94	9.98	m²	17.92
Boarding to tops or cheeks of dormers; tongued and grooved joints						
19 mm thick; laid diagonally	-	1.02	13.28	8.28	m²	21.56
Wrought softwood; "Tanalised"						
Boarding to roofs; tongued and grooved joints						
19 mm thick; flat to falls	-	0.56	7.29	9.01	m²	16.30
19 mm thick; sloping	-	0.61	7.94	9.01	m²	16.95
19 mm thick; sloping; laid diagonally	-	0.79	10.29	9.01	m²	19.30
25 mm thick; flat to falls	-	0.56	7.29	10.93	m²	18.22
25 mm thick; sloping	-	0.61	7.94	10.93	m²	18.87
Boarding to tops or cheeks of dormers; tongued and grooved joints						
19 mm thick; laid diagonally	-	1.02	13.28	9.01	m²	22.29
Wood strip; 22 mm thick; "Junckers" pre-treated or other and approved; tongued and grooved joints; pre-finished boards; level fixing to resilient battens; to cement and sand base						
Strip flooring; over 300 mm wide						
beech; prime	-	-	-	-	m²	59.92
beech; standard	-	-	-	-	m²	52.43
beech; sylvia squash	-	-	-	-	m²	54.92
oak; quality A	-	-	-	-	m²	64.91
Wrought hardwood						
Strip flooring to floors; 25 mm thick x 75 mm wide; tongue and grooved joints; secret fixing; surface sanded after laying						
american oak	-	-	-	-	m²	44.94
canadian maple	-	-	-	-	m²	47.44
gurjun	-	-	-	-	m²	41.44
iroko	-	-	-	-	m²	39.95

Prices for Measured Works - Minor Works
L WINDOWS/DOORS/STAIRS

	PC £	Labour hours	Labour £	Material £	Unit	Total rate £
L WINDOWS/DOORS/STAIRS						
L10 WINDOWS/ROOFLIGHTS/SCREENS/ LOUVRES						
Standard windows; "treated" wrought softwood; Rugby Joinery or other equal and approved						
Side hung casement windows without glazing bars; with 140 mm wide softwood sills; opening casements and ventilators hung on rustproof hinges; fitted with aluminized lacquered finish casement stays and fasteners						
500 mm x 750 mm; ref N07V	56.91	0.69	8.98	61.40	nr	70.38
500 mm x 900 mm; ref N09V	58.82	0.83	10.81	63.45	nr	74.26
600 mm x 750 mm; ref 107V	68.63	0.83	10.81	74.00	nr	84.81
600 mm x 750 mm; ref 107C	70.61	0.83	10.81	76.13	nr	86.94
600 mm x 900 mm; ref 109V	69.43	0.93	12.11	74.86	nr	86.97
600 mm x 900 mm; ref 109C	74.91	0.83	10.81	80.75	nr	91.56
600 mm x 1050 mm; ref 109V	71.07	0.83	10.81	76.69	nr	87.50
600 mm x 1050 mm; ref 110V	75.23	1.02	13.28	81.17	nr	94.45
915 mm x 900 mm; ref 2NO9W	84.72	1.11	14.45	91.29	nr	105.74
915 mm x 1050 mm; ref 2N1OW	85.90	1.16	15.10	92.64	nr	107.74
915 mm x 1200 mm; ref 2N12W	92.68	1.20	15.62	99.92	nr	115.54
915 mm x 1350 mm; ref 2N13W	94.54	1.39	18.10	101.92	nr	120.02
915 mm x 1500 mm; ref 2N15W	95.81	1.43	18.62	103.36	nr	121.98
1200 mm x 750 mm; ref 2O7C	92.18	1.16	15.10	99.39	nr	114.49
1200 mm x 750 mm; ref 2O7CV	115.25	1.16	15.10	124.19	nr	139.29
1200 mm x 900 mm; ref 2O9C	94.31	1.20	15.62	101.68	nr	117.30
1200 mm x 900 mm; ref 2O9W	100.78	1.20	15.62	108.63	nr	124.25
1200 mm x 900 mm; ref 2O9CV	116.38	1.20	15.62	125.40	nr	141.02
1200 mm x 1050 mm; ref 210C	94.41	1.39	18.10	101.86	nr	119.96
1200 mm x 1050 mm; ref 210W	102.54	1.39	18.10	110.60	nr	128.70
1200 mm x 1050 mm; ref 210T	119.50	1.39	18.10	128.83	nr	146.93
1200 mm x 1050 mm; ref 210CV	116.52	1.39	18.10	125.63	nr	143.73
1200 mm x 1200 mm; ref 212C	98.07	1.48	19.27	105.86	nr	125.13
1200 mm x 1200 mm; ref 212W	104.49	1.48	19.27	112.77	nr	132.04
1200 mm x 1200 mm; ref 212TX	129.09	1.48	19.27	139.21	nr	158.48
1200 mm x 1200 mm; ref 212CV	120.22	1.48	19.27	129.68	nr	148.95
1200 mm x 1350 mm; ref 213W	106.61	1.57	20.44	115.05	nr	135.49
1200 mm x 1350 mm; ref 213CV	133.12	1.57	20.44	143.54	nr	163.98
1200 mm x 1500 mm; ref 215W	109.11	1.71	22.26	117.81	nr	140.07
1770 mm x 750 mm; ref 307CC	135.70	1.43	18.62	146.32	nr	164.94
1770 mm x 900 mm; ref 309CC	138.95	1.71	22.26	149.81	nr	172.07
1770 mm x 1050 mm; ref 310C	116.97	1.80	23.44	126.18	nr	149.62
1770 mm x 1050 mm; ref 310T	139.22	1.71	22.26	150.10	nr	172.36
1770 mm x 1050 mm; ref 310CC	139.49	1.43	18.62	150.39	nr	169.01
1770 mm x 1050 mm; ref 310WW	160.88	1.43	18.62	173.39	nr	192.01
1770 mm x 1200 mm; ref 312C	120.40	1.85	24.09	129.94	nr	154.03
1770 mm x 1200 mm; ref 312T	143.11	1.85	24.09	154.36	nr	178.45
1770 mm x 1200 mm; ref 312CC	144.56	1.85	24.09	155.91	nr	180.00
1770 mm x 1200 mm; ref 312WW	165.00	1.85	24.09	177.89	nr	201.98
1770 mm x 1200 mm; ref 312CVC	176.26	1.85	24.09	189.99	nr	214.08
1770 mm x 1350 mm; ref 313CC	163.06	1.94	25.26	175.80	nr	201.06
1770 mm x 1350 mm; ref 313WW	168.80	1.94	25.26	181.97	nr	207.23
1770 mm x 1350 mm; ref 313CVC	186.57	1.94	25.26	201.08	nr	226.34
1770 mm x 1500 mm; ref 315T	157.26	2.04	26.56	169.57	nr	196.13
2340 mm x 1050 mm; ref 410CWC	199.45	1.99	25.91	214.92	nr	240.83
2340 mm x 1200 mm; ref 412CWC	205.84	2.08	27.08	221.86	nr	248.94
2340 mm x 1350 mm; ref 413CWC	217.13	2.22	28.90	234.07	nr	262.97

Prices for Measured Works - Minor Works
L WINDOWS/DOORS/STAIRS

	PC £	Labour hours	Labour £	Material £	Unit	Total rate £
Top hung casement windows; with 140 mm wide softwood sills; opening casements and ventilators hung on rustproof hinges; fitted with aluminized lacquered finish casement stays						
600 mm x 750 mm; ref 107A	78.62	0.83	10.81	84.74	nr	95.55
600 mm x 900 mm; ref 109A	81.01	0.93	12.11	87.31	nr	99.42
600 mm x 1050 mm; ref 110A	84.22	1.02	13.28	90.83	nr	104.11
915 mm x 750 mm; ref 2N07A	93.76	1.06	13.80	101.01	nr	114.81
915 mm x 900 mm; ref 2N09A	100.42	1.11	14.45	108.17	nr	122.62
915 mm x 1050 mm; ref 2N10A	103.99	1.16	15.10	112.08	nr	127.18
915 mm x 1350 mm; ref 2N13AS	117.61	1.39	18.10	126.80	nr	144.90
1200 mm x 750 mm; ref 207A	105.25	1.16	15.10	113.44	nr	128.54
1200 mm x 900 mm; ref 209A	111.32	1.20	15.62	119.96	nr	135.58
1200 mm x 1050 mm; ref 210A	114.85	1.39	18.10	123.83	nr	141.93
1200 mm x 1200 mm; ref 212A	118.24	1.48	19.27	127.47	nr	146.74
1200 mm x 1350 mm; ref 213AS	129.09	1.57	20.44	139.21	nr	159.65
1200 mm x 1500 mm; ref 215AS	132.58	1.71	22.26	142.96	nr	165.22
1770 mm x 1050 mm; ref 310AE	134.87	1.71	22.26	145.42	nr	167.68
1770 mm x 1200 mm; ref 312A	138.86	1.85	24.09	149.71	nr	173.80
High performance single light with canopy sash windows; ventilators; weather stripping; opening sashes and fanlights hung on rustproof hinges; fitted with aluminized lacquered espagnolette bolts						
600 mm x 900 mm; ref AL0609	161.07	0.93	12.11	173.37	nr	185.48
900 mm x 900 mm; ref AL0909	167.09	1.11	14.45	179.84	nr	194.29
900 mm x 1050 mm; ref AL0910	167.99	1.16	15.10	180.81	nr	195.91
900 mm x 1200 mm; ref AL0912	171.31	1.30	16.93	184.45	nr	201.38
900 mm x 1500 mm; ref AL0915	180.51	1.43	18.62	194.34	nr	212.96
1200 mm x 1050 mm; ref AL1210	194.69	1.39	18.10	209.58	nr	227.68
1200 mm x 1200 mm; ref AL1212	195.94	1.48	19.27	211.00	nr	230.27
1200 mm x 1500 mm; ref AL1215	221.22	1.71	22.26	238.25	nr	260.51
1500 mm x 1050 mm; ref AL1510	195.03	1.71	22.26	210.02	nr	232.28
1500 mm x 1200 mm; ref AL1512	204.80	1.71	22.26	220.53	nr	242.79
1500 mm x 1500 mm; ref AL1515	220.55	1.85	24.09	237.60	nr	261.69
1500 mm x 1650 mm; ref AL1516	227.64	1.94	25.26	245.15	nr	270.41
High performance double hung sash windows with glazing bars; solid frames; 63 mm x 175 mm softwood sills; standard flush external linings; spiral spring balances and sash catch						
635 mm x 1050 mm; ref VS0610B	255.59	2.04	26.56	275.05	nr	301.61
635 mm x 1350 mm; ref VS0613B	276.82	2.22	28.90	297.95	nr	326.85
635 mm x 1650 mm; ref VS0616B	295.73	2.50	32.55	318.35	nr	350.90
860 mm x 1050 mm; ref VS0810B	278.82	2.36	30.73	300.02	nr	330.75
860 mm x 1350 mm; ref VS0813B	301.80	2.64	34.37	324.80	nr	359.17
860 mm x 1650 mm; ref VS0816B	324.73	3.05	39.71	349.52	nr	389.23
1085 mm x 1050 mm; ref VS1010B	303.26	2.64	34.37	326.30	nr	360.67
1085 mm x 1350 mm; ref VS1013B	332.64	3.05	39.71	357.95	nr	397.66
1085 mm x 1650 mm; ref VS1016B	351.92	3.75	48.83	378.75	nr	427.58
1725 mm x 1050 mm; ref VS1710B	555.26	3.75	48.83	597.27	nr	646.10
1725 mm x 1350 mm; ref VS1713B	605.16	4.67	60.80	650.99	nr	711.79
1725 mm x 1650 mm; ref VS1716B	649.31	4.76	61.98	698.52	nr	760.50

Prices for Measured Works - Minor Works
L WINDOWS/DOORS/STAIRS

	PC £	Labour hours	Labour £	Material £	Unit	Total rate £
L10 WINDOWS/ROOFLIGHTS/SCREENS/ LOUVRES - cont'd						
Standard windows; Premdor Crosby or other equal and approved; Meranti; factory applied preservative stain base coat						
Side hung casement windows; 45 mm x 140 mm hardwood sills; weather stripping; opening sashes on canopy hinges; fitted with fasteners; brown finish ironmongery						
630 mm x 750 mm; ref 107C	134.18	0.97	12.63	144.46	nr	157.09
630 mm x 900 mm; ref 109C	139.99	1.20	15.62	150.71	nr	166.33
630 mm x 900 mm; ref 109V	133.40	0.97	12.63	143.62	nr	156.25
630 mm x 1050 mm; ref 110V	136.79	1.34	17.45	147.27	nr	164.72
915 mm x 900 mm; ref 2N09W	166.71	1.53	19.92	179.51	nr	199.43
915 mm x 1050 mm; ref 2N10W	169.39	1.62	21.09	182.46	nr	203.55
915 mm x 1200 mm; ref 2N12W	173.64	1.71	22.26	187.03	nr	209.29
915 mm x 1350 mm; ref 2N13W	177.57	1.85	24.09	191.25	nr	215.34
915 mm x 1500 mm; ref 2N15W	181.47	1.94	25.26	195.52	nr	220.78
1200 mm x 900 mm; ref 209C	181.20	1.71	22.26	195.01	nr	217.27
1200 mm x 900 mm; ref 209W	190.35	1.71	22.26	204.85	nr	227.11
1200 mm x 1050 mm; ref 210C	187.78	1.85	24.09	202.08	nr	226.17
1200 mm x 1050 mm; ref 210W	194.01	1.85	24.09	208.78	nr	232.87
1200 mm x 1200 mm; ref 212C	195.68	1.99	25.91	210.65	nr	236.56
1200 mm x 1200 mm; ref 212W	197.90	1.99	25.91	213.04	nr	238.95
1200 mm x 1350 mm; ref 213W	202.23	2.13	27.73	217.69	nr	245.42
1200 mm x 1500 mm; ref 215W	206.08	2.22	28.90	221.83	nr	250.73
1770 mm x 1050 mm; ref 310C	220.89	2.31	30.08	237.82	nr	267.90
1770 mm x 1050 mm; ref 310CC	285.66	2.31	30.08	307.45	nr	337.53
1770 mm x 1200 mm; ref 312C	228.01	2.45	31.90	245.48	nr	277.38
1770 mm x 1200 mm; ref 312CC	296.88	2.45	31.90	319.51	nr	351.41
2339 mm x 1200 mm; ref 412CMC	370.61	2.64	34.37	398.85	nr	433.22
Top hung casement windows; 45 mm x 140 mm hardwood sills; weather stripping; opening sashes on canopy hinges; fitted with fasteners; brown finish ironmongery						
630 mm x 900 mm; ref 109A	127.05	0.97	12.63	136.80	nr	149.43
630 mm x 1050 mm; ref 110A	135.00	1.34	17.45	145.42	nr	162.87
915 mm x 900 mm; ref 2N09A	154.94	1.53	19.92	166.78	nr	186.70
915 mm x 1050 mm; ref 2N10A	162.89	1.62	21.09	175.40	nr	196.49
915 mm x 1200 mm; ref 2N12A	171.61	1.71	22.26	184.77	nr	207.03
915 mm x 1350 mm; ref 2N13D	206.75	1.85	24.09	222.62	nr	246.71
1200 mm x 900 mm; ref 209A	174.96	1.53	19.92	188.30	nr	208.22
1200 mm x 1050 mm; ref 210A	182.95	1.71	22.26	196.96	nr	219.22
1200 mm x 1200 mm; ref 212A	191.67	1.85	24.09	206.34	nr	230.43
1770 mm x 1050 mm; ref 310AE	221.66	1.94	25.26	238.65	nr	263.91
1770 mm x 1200 mm; ref 312AE	233.04	2.05	26.69	250.88	nr	277.57
Purpose made window casements; "treated" wrought softwood						
Casements; rebated; moulded - supply only						
38 mm thick	-	-	-	25.81	m^2	25.81
50 mm thick	-	-	-	30.69	m^2	30.69
Casements; rebated; moulded; in medium panes - supply only						
38 mm thick	-	-	-	37.72	m^2	37.72
50 mm thick	-	-	-	46.64	m^2	46.64
Casements; rebated; moulded; with semi-circular head - supply only						
38 mm thick	-	-	-	47.11	m^2	47.11
50 mm thick	-	-	-	58.35	m^2	58.35

L WINDOWS/DOORS/STAIRS

	PC £	Labour hours	Labour £	Material £	Unit	Total rate £
Casements; rebated; moulded; to bullseye window - supply only						
38 mm thick; 600 mm diameter	-	-	-	84.73	nr	84.73
38 mm thick; 900 mm diameter	-	-	-	125.68	nr	125.68
50 mm thick; 600 mm diameter	-	-	-	97.46	nr	97.46
50 mm thick; 900 mm diameter	-	-	-	144.48	nr	144.48
Fitting and hanging casements						
square or rectangular	-	0.51	6.64	-	nr	6.64
semi-circular	-	1.30	16.93	-	nr	16.93
bullseye	-	2.04	26.56	-	nr	26.56
Purpose made window casements; selected West African Mahogany						
Casements; rebated; moulded - supply only						
38 mm thick	-	-	-	43.46	m²	43.46
50 mm thick	-	-	-	56.65	m²	56.65
Casements; rebated; moulded; in medium panes - supply only						
38 mm thick	-	-	-	52.73	m²	52.73
50 mm thick	-	-	-	70.55	m²	70.55
Casements; rebated; moulded with semi-circular head - supply only						
38 mm thick	-	-	-	65.93	m²	65.93
50 mm thick	-	-	-	88.27	m²	88.27
Casements; rebated; moulded; to bullseye window - supply only						
38 mm thick; 600 mm diameter	-	-	-	142.70	nr	142.70
38 mm thick; 900 mm diameter	-	-	-	213.56	nr	213.56
50 mm thick; 600 mm diameter	-	-	-	164.09	nr	164.09
50 mm thick; 900 mm diameter	-	-	-	244.66	nr	244.66
Fitting and hanging casements						
square or rectangular	-	0.69	8.98	-	nr	8.98
semi-circular	-	1.71	22.26	-	nr	22.26
bullseye	-	2.73	35.54	-	nr	35.54
Purpose made window casements; American White Ash						
Casements; rebated; moulded - supply only						
38 mm thick	-	-	-	51.37	m²	51.37
50 mm thick	-	-	-	63.07	m²	63.07
Casements; rebated; moulded; in medium panes - supply only						
38 mm thick	-	-	-	66.57	m²	66.57
50 mm thick	-	-	-	80.01	m²	80.01
Casements; rebated; moulded with semi-circular head - supply only						
38 mm thick	-	-	-	83.24	m²	83.24
50 mm thick	-	-	-	100.04	m²	100.04
Casements; rebated; moulded; to bullseye window - supply only						
38 mm thick; 600 mm diameter	-	-	-	168.74	nr	168.74
38 mm thick; 900 mm diameter	-	-	-	252.12	nr	252.12
50 mm thick; 600 mm diameter	-	-	-	194.64	nr	194.64
50 mm thick; 900 mm diameter	-	-	-	290.19	nr	290.19
Fitting and hanging casements						
square or rectangular	-	0.69	8.98	-	nr	8.98
semi-circular	-	1.71	22.26	-	nr	22.26
bullseye	-	2.73	35.54	-	nr	35.54

Prices for Measured Works - Minor Works
L WINDOWS/DOORS/STAIRS

	PC £	Labour hours	Labour £	Material £	Unit	Total rate £
L10 WINDOWS/ROOFLIGHTS/SCREENS/ LOUVRES - cont'd						
Purpose made window frames; "treated" wrought softwood; prices indicated are for supply only as part of a complete window frame; the reader is referred to the previous pages for fixing costs for frames based on the overall window size						
Frames; rounded; rebated check grooved						
25 mm x 120 mm	-	-	-	7.87	m	7.87
50 mm x 75 mm	-	-	-	8.87	m	8.87
50 mm x 100 mm	-	-	-	10.87	m	10.87
50 mm x 125 mm	-	-	-	13.19	m	13.19
63 mm x 100 mm	-	-	-	13.52	m	13.52
75 mm x 150 mm	-	-	-	22.74	m	22.74
90 mm x 140 mm	-	-	-	29.33	m	29.33
Mullions and transoms; twice rounded, rebated and check grooved						
50 mm x 75 mm	-	-	-	10.90	m	10.90
50 mm x 100 mm	-	-	-	12.89	m	12.89
63 mm x 100 mm	-	-	-	15.56	m	15.56
75 mm x 150 mm	-	-	-	24.76	m	24.76
Sill; sunk weathered, rebated and grooved						
75 mm x 100 mm	-	-	-	22.79	m	22.79
75 mm x 150 mm	-	-	-	29.35	m	29.85
Add 5% to the above material prices for "selected" softwood for staining						
Purpose made window frames; selected West African Mahogany; prices given are for supply only as part of a complete window frame; the reader is referred to the previous pages for fixing costs for frames based on the overall window size						
Frames; rounded; rebated check grooved						
25 mm x 120 mm	-	-	-	13.57	m	13.57
50 mm x 75 mm	-	-	-	15.44	m	15.44
50 mm x 100 mm	-	-	-	19.11	m	19.11
50 mm x 125 mm	-	-	-	23.31	m	23.31
63 mm x 100 mm	-	-	-	23.84	m	23.84
75 mm x 150 mm	-	-	-	40.38	m	40.38
90 mm x 140 mm	-	-	-	52.36	m	52.36
Mullions and transoms; twice rounded, rebated and check grooved						
50 mm x 75 mm	-	-	-	18.46	m	18.46
50 mm x 100 mm	-	-	-	22.12	m	22.12
63 mm x 100 mm	-	-	-	26.88	m	26.88
75 mm x 150 mm	-	-	-	43.40	m	43.40
Sill; sunk weathered, rebated and grooved						
75 mm x 100 mm	-	-	-	38.38	m	38.38
75 mm x 150 mm	-	-	-	51.13	m	51.13

	PC £	Labour hours	Labour £	Material £	Unit	Total rate £
Purpose made window frames; American White Ash; prices indicated are for supply only as part of a complete window frame; the reader is referred to the previous pages for fixing costs for frames based on the overall window size						
Frames; rounded; rebated check grooved						
25 mm x 120 mm	-	-	-	15.10	m	15.10
50 mm x 75 mm	-	-	-	17.28	m	17.28
50 mm x 100 mm	-	-	-	21.56	m	21.56
50 mm x 125 mm	-	-	-	26.38	m	26.38
63 mm x 100 mm	-	-	-	26.96	m	26.96
75 mm x 150 mm	-	-	-	45.91	m	45.91
90 mm x 140 mm	-	-	-	59.79	m	59.79
Mullions and transoms; twice rounded, rebated and check grooved						
50 mm x 75 mm	-	-	-	20.29	m	20.29
50 mm x 100 mm	-	-	-	24.59	m	24.59
63 mm x 100 mm	-	-	-	29.98	m	29.98
75 mm x 150 mm	-	-	-	48.93	m	48.93
Sill; sunk weathered, rebated and grooved						
75 mm x 100 mm	-	-	-	42.08	m	42.08
75 mm x 150 mm	-	-	-	56.68	m	56.68
Purpose made double hung sash windows; "treated" wrought softwood						
Cased frames of 100 mm x 25 mm grooved inner linings; 114 mm x 25 mm grooved outer linings; 125 mm x 38 mm twice rebated head linings; 125 mm x 32 mm twice rebated grooved pulley stiles; 150 mm x 13 mm linings; 50 mm x 19 mm partings slips; 25 mm x 19 mm inside beads; 150 mm x 75 mm Oak twice sunk weathered throated sill; 50 mm thick rebated and moulded sashes; moulded horns						
over 1.25 m^2 each; both sashes in medium panes; including spiral spring balances	175.81	2.31	30.08	262.26	m^2	292.34
As above but with cased mullions	185.49	2.54	33.07	272.67	m^2	305.74
Purpose made double hung sash windows; selected West African Mahogany						
Cased frames of 100 mm x 25 mm grooved inner linings; 114 mm x 25 mm grooved outer linings; 125 mm x 38 mm twice reabated head linings; 125 mm x 32 mm twice rebated grooved pulley stiles; 150 mm x 13 mm linings; 50 mm x 19 mm parting slips; 25 mm x 19 mm inside beads; 150 mm x 75 mm Oak twice sunk weathered throated sill; 50 mm thick rebated and moulded sashes; moulded horns						
over 1.25 m^2 each; both sashes in medium panes; including spiral sash balances	280.63	3.05	39.71	374.95	m^2	414.66
As above but with cased mullions	296.05	3.38	44.01	391.52	m^2	435.53

Prices for Measured Works - Minor Works
L WINDOWS/DOORS/STAIRS

	PC £	Labour hours	Labour £	Material £	Unit	Total rate £
L10 WINDOWS/ROOFLIGHTS/SCREENS/ LOUVRES - cont'd						
Purpose made double hung sash windows; American White Ash						
Cased frames of 100 mm x 25 mm grooved inner linings; 114 mm x 25 mm grooved outer linings; 125 mm x 38 mm twice rebated head linings; 125 mm x 32 mm twice rebated grooved pulley stiles; 150 mm x 13 mm linings; 50 mm x 19 mm parting slips; 25 mm x 19 mm inside beads; 150 mm x 75 mm Oak twice sunk weathered throated sill; 50 mm thick rebated and moulded sashes; moulded horns						
over 1.25m² each; both sashes in medium panes; including spiral spring balances	304.34	3.05	39.71	400.43	m²	440.14
As above but with cased mullions	321.05	3.38	44.01	418.40	m²	462.41
Galvanised steel fixed, casement and fanlight windows; Crittal "Homelight" range or other equal and approved; site glazing not included; fixed in position including lugs plugged and screwed to brickwork or blockwork						
Basic fixed lights; including easy-glaze beads						
628 mm x 292 mm; ref ZNG5	21.14	1.20	15.62	22.87	nr	38.49
628 mm x 923 mm; ref ZNC5	29.02	1.20	15.62	31.42	nr	47.04
628 mm x 1513 mm; ref ZNDV5	41.45	1.20	15.62	44.92	nr	60.54
1237 mm x 292 mm; ref ZNG13	33.16	1.20	15.62	35.79	nr	51.41
1237 mm x 923 mm; ref ZNC13	41.45	1.80	23.44	44.78	nr	68.22
1237 mm x 1218 mm; ref ZND13	49.74	1.80	23.44	53.76	nr	77.20
1237 mm x 1513 mm; ref ZNDV13	53.88	1.80	23.44	58.29	nr	81.73
1846 mm x 292 mm; ref ZNG14	41.45	1.20	15.62	44.78	nr	60.40
1846 mm x 923 mm; ref ZNC14	49.74	1.80	23.44	53.84	nr	77.28
1846 mm x 1513 mm; ref ZNDV14	62.17	2.22	28.90	67.35	nr	96.25
Basic opening lights; including easy-glaze beads and weatherstripping						
628 mm x 292 mm; ref ZNG1	53.88	1.20	15.62	58.07	nr	73.69
1237 mm x 292 mm; ref ZNG13G	78.75	1.20	15.62	84.80	nr	100.42
1846 mm x 292 mm; ref ZNG4	132.64	1.20	15.62	142.88	nr	158.50
One piece composites; including easy-glaze beads and weatherstripping						
628 mm x 923 mm; ref ZNC5F	74.61	1.20	15.62	80.43	nr	96.05
628 mm x 1513 mm; ref ZNDV5F	87.05	1.20	15.62	93.94	nr	109.56
1237 mm x 923 mm; ref ZNC2F	140.93	1.80	23.44	151.72	nr	175.16
1237 mm x 1218 mm; ref ZND2F	165.80	1.80	23.44	178.53	nr	201.97
1237 mm x 1513 mm; ref ZNDV2V	194.81	1.80	23.44	209.79	nr	233.23
1846 mm x 923 mm; ref ZNC4F	215.54	1.80	23.44	232.07	nr	255.51
1846 mm x 1218 mm; ref ZND10F	207.25	2.22	28.90	223.23	nr	252.13
Reversible windows; including easy-glaze beads						
997 mm x 923 mm; ref NC13R	186.53	1.57	20.44	200.74	nr	221.18
997 mm x 1067 mm; ref NCO13R	194.81	1.57	20.44	209.64	nr	230.08
1237 mm x 923 mm; ref ZNC13R	203.10	2.36	30.73	218.55	nr	249.28
1237 mm x 1218 mm; ref ZND13R	219.69	2.36	30.73	236.46	nr	267.19
1237 mm x 1513 mm; ref ZNDV13RS	248.70	2.36	30.73	267.72	nr	298.45
Pressed steel sills; to suit above window widths						
628 mm long	12.85	0.37	4.82	13.81	nr	18.63
997 mm long	16.58	0.46	5.99	17.82	nr	23.81
1237 mm long	21.14	0.56	7.29	22.73	nr	30.02
1846 mm long	29.02	0.79	10.29	31.20	nr	41.49

Prices for Measured Works - Minor Works
L WINDOWS/DOORS/STAIRS

	PC £	Labour hours	Labour £	Material £	Unit	Total rate £
Factory finished steel fixed light; casement and fanlight windows; Crittall polyester powder coated "Homelight" range or other equal and approved; site glazing not included; fixed in position; including lugs plugged and screwed to brickwork or blockwork						
Basic fixed lights; including easy-glaze beads						
628 mm x 292 mm; ref ZNG5	26.94	1.20	15.62	29.11	nr	44.73
628 mm x 923 mm; ref ZNC5	37.30	1.20	15.62	40.32	nr	55.94
628 mm x 1513 mm; ref ZNDV5	53.88	1.20	15.62	58.29	nr	73.91
1237 mm x 292 mm; ref ZNG13	41.45	1.20	15.62	44.71	nr	60.33
1237 mm x 923 mm; ref ZNC13	53.88	1.80	23.44	58.14	nr	81.58
1237 mm x 1218 mm; ref ZND13	62.17	1.80	23.44	67.13	nr	90.57
1237 mm x 1513 mm; ref ZNDV13	66.32	1.80	23.44	71.66	nr	95.10
1846 mm x 292 mm; ref ZNG14	53.88	1.20	15.62	58.07	nr	73.69
1846 mm x 923 mm; ref ZNC14	66.32	1.80	23.44	71.51	nr	94.95
1846 mm x 1513 mm; ref ZNDV14	78.75	2.22	28.90	85.02	nr	113.92
Basic opening lights; including easy-glaze beads and weatherstripping						
628 mm x 292 mm; ref ZNG1	64.25	1.20	15.62	69.22	nr	84.84
1237 mm x 292 mm; ref ZNG13G	91.19	1.20	15.62	98.18	nr	113.80
1846 mm x 292 mm; ref ZNG4	157.51	1.20	15.62	169.47	nr	185.09
One piece composites; including easy-glaze beads and weatherstripping						
628 mm x 923 mm; ref ZNC5F	87.05	1.20	15.62	93.80	nr	109.42
628 mm x 1513 mm; ref ZNDV5F	107.77	1.20	15.62	116.22	nr	131.84
1237 mm x 923 mm; ref ZNC2F	174.09	1.80	23.44	187.37	nr	210.81
1237 mm x 1218 mm; ref ZND2F	198.96	1.80	23.44	214.10	nr	237.54
1237 mm x 1513 mm; ref ZNDV2V	232.12	1.80	23.44	249.90	nr	273.34
1846 mm x 923 mm; ref ZNC4F	261.13	1.80	23.44	280.93	nr	304.37
1846 mm x 1218 mm; ref ZND10F	248.70	2.22	28.90	267.72	nr	296.62
Reversible windows; including easy-glaze beads						
997 mm x 923 mm; ref NC13R	240.41	1.57	20.44	258.66	nr	279.10
997 mm x 1218 mm; ref NCO13R	252.84	1.57	20.44	272.02	nr	292.46
1237 mm x 923 mm; ref ZNC13R	261.13	2.36	30.73	280.93	nr	311.66
1237 mm x 1218 mm; ref ZND13R	286.00	2.36	30.73	307.74	nr	338.47
1237 mm x 1513 mm; ref ZNDV13RS	319.17	2.36	30.73	343.47	nr	374.20
Pressed steel sills; to suit above window widths						
628 mm long	16.58	0.37	4.82	17.82	nr	22.64
997 mm long	22.38	0.46	5.99	24.06	nr	30.05
1237 mm long	26.53	0.56	7.29	28.52	nr	35.81
1846 mm long	37.30	0.79	10.29	40.10	nr	50.39
uPVC windows to BS 2782; reinforced where appropriate with aluminium alloy; including standard ironmongery; cills and glazing; fixed in position; including lugs plugged and screwed to brickwork or blockwork						
Fixed light; including e.p.d.m. glazing gaskets and weather seals						
600 mm x 900 mm; single glazed	37.07	1.53	19.92	40.14	nr	60.06
600 mm x 900 mm; double glazed	47.80	1.53	19.92	51.68	nr	71.60
Casement/fixed light; including e.p.d.m. glazing gaskets and weather seals						
600 mm x 1200 mm; single glazed	79.99	1.80	23.44	86.36	nr	109.80
600 mm x 1200 mm; double glazed	95.59	1.80	23.44	103.13	nr	126.57
1200 mm x 1200 mm; single glazed	106.32	2.04	26.56	114.73	nr	141.29
1200 mm x 1200 mm; double glazed	149.24	2.04	26.56	160.87	nr	187.43
1800 mm x 1200 mm; single glazed	117.05	2.04	26.56	126.27	nr	152.83
1800 mm x 1200 mm; double glazed	159.97	2.04	26.56	172.41	nr	198.97

L WINDOWS/DOORS/STAIRS

	PC £	Labour hours	Labour £	Material £	Unit	Total rate £
L10 WINDOWS/ROOFLIGHTS/SCREENS/ LOUVRES - cont'd						
uPVC windows to BS 2782; reinforced where appropriate with aluminium alloy; including standard ironmongery; cills and glazing; fixed in position; including lugs plugged and screwed to brickwork or blockwork - cont'd "Tilt & Turn" light; including e.p.d.m. glazing gaskets and weather seals						
1200 mm x 1200 mm; single glazed	111.20	2.04	26.56	119.98	nr	**146.54**
1200 mm x 1200 mm; double glazed	138.51	2.04	26.56	149.34	nr	**175.90**
Rooflights, skylights, roof windows and frames; pre-glazed; "treated" Nordic Red Pine and aluminium trimmed "Velux" windows or other equal and approved; type U flashings and soakers (for tiles and pantiles), and sealed double glazing unit (trimming opening not included) Roof windows						
550 mm x 780 mm; ref GGL-3000-102	128.08	2.05	26.69	137.93	nr	**164.62**
550 mm x 980 mm; ref GGL-3000-104	141.52	2.31	30.08	152.39	nr	**182.47**
660 mm x 1180 mm; ref GGL-3000-206	166.83	2.54	33.07	179.63	nr	**212.70**
780 mm x 980 mm; ref GGL-3000-304	160.50	2.54	33.07	172.84	nr	**205.91**
780 mm x 1400 mm; ref GGL-3000-308	192.92	2.54	33.07	207.77	nr	**240.84**
940 mm x 1600 mm; ref GGL-3000-410	231.66	3.05	39.71	249.48	nr	**289.19**
1140 mm x 1180 mm; ref GGL-3000-606	213.47	3.05	39.71	229.91	nr	**269.62**
1340 mm x 980 mm; ref GGL-3000-804	220.59	3.05	39.71	237.58	nr	**277.29**
1340 mm x 1400 mm; ref GGL-3000-808	261.70	3.05	39.71	281.77	nr	**321.48**
Rooflights, skylights, roof windows and frames; galvanised steel; plugges and screwed to concrete; or screwed to timber Rooflight "Coxdome TPX" or other equal and approved; dome; galvanised steel; UV protected polycarbonate glazing; GRP splayed upstand; hit and miss vent						
600 mm x 600 mm	148.18	1.80	23.44	159.57	nr	**183.01**
900 mm x 600 mm	208.47	1.94	25.26	224.42	nr	**249.68**
900 mm x 900 mm	237.08	2.08	27.08	255.23	nr	**282.31**
1200 mm x 900 mm	276.93	2.22	28.90	298.11	nr	**327.01**
1200 mm x 1200 mm	327.01	2.41	31.38	351.99	nr	**383.37**
1800 mm x 1200 mm	431.24	2.54	33.07	464.08	nr	**497.15**
Rooflights, skylights, roof windows and frames; uPVC; plugged and screwed to concrete; or screwed to timber Rooflight; "Coxdome Universal Dome" or other equal and approved; acrylic double skin; dome						
600 mm x 600 mm	96.06	1.53	19.92	103.26	nr	**123.18**
900 mm x 600 mm	178.83	1.67	21.74	192.24	nr	**213.98**
900 mm x 900 mm	178.83	1.85	24.09	192.24	nr	**216.33**
1200 mm x 900 mm	216.64	1.99	25.91	232.89	nr	**258.80**
1200 mm x 1200 mm	273.87	2.13	27.73	294.41	nr	**322.14**
1500 mm x 1050 mm	406.72	2.22	28.90	437.22	nr	**466.12**
Rooflight; "Coxdome Universal Dome" or other equal and approved; dome; polycarbonate single skin; GRP splayed upstand; hit and miss vent						
600 mm x 600 mm	148.18	1.80	23.44	159.57	nr	**183.01**
900 mm x 600 mm	208.47	1.94	25.26	224.42	nr	**249.68**
900 mm x 900 mm	237.08	2.08	27.08	255.23	nr	**282.31**
1200 mm x 900 mm	276.93	2.22	28.90	298.11	nr	**327.01**

L WINDOWS/DOORS/STAIRS

	PC £	Labour hours	Labour £	Material £	Unit	Total rate £
Rooflight; "Coxdome Universal Dome" or other equal and approved; dome; polycarbonate single skin; GRP splayed upstand; hit and miss vent - cont'd						
1200 mm x 1200 mm	327.01	2.41	31.38	351.99	nr	383.37
1800 mm x 1200 mm	431.24	2.54	33.07	464.08	nr	497.15
Louvres and frames; polyester powder coated aluminium; fixing in position including brackets						
Louvre; Gil Airvac "Plusaire 75SP" weatherlip or other equal and approved						
1250 mm x 675 mm	-	-	-	-	nr	247.17
2000 mm x 530 mm	-	-	-	-	nr	280.12
L20 DOORS/SHUTTERS/HATCHES						
Doors; standard matchboarded; wrought softwood						
Matchboarded, ledged and braced doors; 25 mm thick ledges and braces; 19 mm thick tongued, grooved and V-jointed boarding; one side vertical boarding						
762 mm x 1981 mm	49.78	1.53	19.92	53.51	nr	73.43
838 mm x 1981 mm	49.78	1.53	19.92	53.51	nr	73.43
Matchboarded, framed, ledged and braced doors; 44 mm thick overall; 19 mm thick tongued, grooved and V-jointed boarding; one side vertical boarding						
762 mm x 1981 mm	61.00	1.85	24.09	65.58	nr	89.67
838 mm x 1981 mm	61.00	1.85	24.09	65.58	nr	89.67
Doors; standard flush; softwood composition						
Flush door; internal quality; skeleton or cellular core; hardboard faced both sides; Premdor Crosby "Primaseal" or other equal and approved						
457 mm x 1981 mm x 35 mm	19.90	1.30	16.93	21.39	nr	38.32
533 mm x 1981 mm x 35 mm	19.90	1.30	16.93	21.39	nr	38.32
610 mm x 1981 mm x 35 mm	19.90	1.30	16.93	21.39	nr	38.32
686 mm x 1981 mm x 35 mm	19.90	1.30	16.93	21.39	nr	38.32
762 mm x 1981 mm x 35 mm	19.90	1.30	16.93	21.39	nr	38.32
838 mm x 1981 mm x 35 mm	20.92	1.30	16.93	22.49	nr	39.42
526 mm x 2040 mm x 40 mm	21.54	1.30	16.93	23.16	nr	40.09
626 mm x 2040 mm x 40 mm	21.54	1.30	16.93	23.16	nr	40.09
726 mm x 2040 mm x 40 mm	21.54	1.30	16.93	23.16	nr	40.09
826 mm x 2040 mm x 40 mm	21.54	1.30	16.93	23.16	nr	40.09
Flush door; internal quality; skeleton or cellular core; chipboard veneered; faced both sides; lipped on two long edges; Premdor Crosby "Popular" or other equal and approved						
457 mm x 1981 mm x 35 mm	28.63	1.30	16.93	30.78	nr	47.71
533 mm x 1981 mm x 35 mm	28.63	1.30	16.93	30.78	nr	47.71
610 mm x 1981 mm x 35 mm	28.63	1.30	16.93	30.78	nr	47.71
686 mm x 1981 mm x 35 mm	28.63	1.30	16.93	30.78	nr	47.71
762 mm x 1981 mm x 35 mm	28.63	1.30	16.93	30.78	nr	47.71
838 mm x 1981 mm x 35 mm	29.95	1.30	16.93	32.20	nr	49.13
526 mm x 2040 mm x 40 mm	31.47	1.30	16.93	33.83	nr	50.76
626 mm x 2040 mm x 40 mm	31.47	1.30	16.93	33.83	nr	50.76
726 mm x 2040 mm x 40 mm	31.47	1.30	16.93	33.83	nr	50.76
826 mm x 2040 mm x 40 mm	31.47	1.30	16.93	33.83	nr	50.76

L20 DOORS/SHUTTERS/HATCHES - cont'd

	PC £	Labour hours	Labour £	Material £	Unit	Total rate £
Doors; standard flush; softwood composition - cont'd						
Flush door; internal quality; skeleton or cellular core; Sapele faced both sides; lipped on all four edges; Premdor Crosby "Landscape Sapele" or other equal and approved						
457 mm x 1981 mm x 35 mm	32.48	1.39	18.10	34.92	nr	**53.02**
533 mm x 1981 mm x 35 mm	32.48	1.39	18.10	34.92	nr	**53.02**
610 mm x 1981 mm x 35 mm	32.48	1.39	18.10	34.92	nr	**53.02**
686 mm x 1981 mm x 35 mm	32.48	1.39	18.10	34.92	nr	**53.02**
762 mm x 1981 mm x 35 mm	32.48	1.39	18.10	34.92	nr	**53.02**
838 mm x 1981 mm x 35 mm	33.81	1.39	18.10	36.35	nr	**54.45**
526 mm x 2040 mm x 40 mm	35.02	1.39	18.10	37.65	nr	**55.75**
626 mm x 2040 mm x 40 mm	35.02	1.39	18.10	37.65	nr	**55.75**
726 mm x 2040 mm x 40 mm	35.02	1.39	18.10	37.65	nr	**55.75**
826 mm x 2040 mm x 40 mm	35.02	1.39	18.10	37.65	nr	**55.75**
Flush door; half-hour fire check (FD20); hardboard faced both sides; Premdor Crosby "Primaseal Fireshield" or other equal and approved						
762 mm x 1981 mm x 44 mm	47.91	1.80	23.44	51.50	nr	**74.94**
838 mm x 1981 mm x 44 mm	49.66	1.80	23.44	53.38	nr	**76.82**
726 mm x 2040 mm x 44 mm	50.87	1.80	23.44	54.69	nr	**78.13**
826 mm x 2040 mm x 44 mm	50.87	1.80	23.44	54.69	nr	**78.13**
Flush door; half-hour fire check (30/20); chipboard veneered; faced both sides; lipped on all four edges; Premdor Crosby "Popular Fireshield" or other equal and approved						
457 mm x 1981 mm x 44 mm	47.13	1.80	23.44	50.66	nr	**74.10**
533 mm x 1981 mm x 44 mm	47.13	1.80	23.44	50.66	nr	**74.10**
610 mm x 1981 mm x 44 mm	47.13	1.80	23.44	50.66	nr	**74.10**
686 mm x 1981 mm x 44 mm	47.13	1.80	23.44	50.66	nr	**74.10**
762 mm x 1981 mm x 44 mm	47.13	1.80	23.44	50.66	nr	**74.10**
838 mm x 1981 mm x 44 mm	49.08	1.80	23.44	52.76	nr	**76.20**
526 mm x 2040 mm x 44 mm	51.53	1.80	23.44	55.39	nr	**78.83**
626 mm x 2040 mm x 44 mm	51.53	1.80	23.44	55.39	nr	**78.83**
726 mm x 2040 mm x 44 mm	51.53	1.80	23.44	55.39	nr	**78.83**
826 mm x 2040 mm x 44 mm	51.53	1.80	23.44	55.39	nr	**78.83**
Flush door; half hour fire resisting "Leaderflush" (type B30) or other equal and approved; chipboard for painting; hardwood lipping two long edges						
526 mm x 2040 mm x 44 mm	79.82	1.90	24.74	85.81	nr	**110.55**
626 mm x 2040 mm x 44 mm	83.87	1.90	24.74	90.16	nr	**114.90**
726 mm x 2040 mm x 44 mm	82.76	1.90	24.74	88.97	nr	**113.71**
826 mm x 2040 mm x 44 mm	83.65	1.90	24.74	89.92	nr	**114.66**
Flush door; half-hour fire check (FD20); Sapele faced both sides; lipped on all four edges; Premdor Crosby "Landscape Sapele Fireshield" or other equal and approved						
686 mm x 1981 mm x 44 mm	68.36	1.90	24.74	73.49	nr	**98.23**
762 mm x 1981 mm x 44 mm	68.36	1.90	24.74	73.49	nr	**98.23**
838 mm x 1981 mm x 44 mm	70.11	1.90	24.74	75.37	nr	**100.11**
726 mm x 2040 mm x 44 mm	71.32	1.90	24.74	76.67	nr	**101.41**
826 mm x 2040 mm x 44 mm	71.32	1.90	24.74	76.67	nr	**101.41**
Flush door; half-hour fire resisting (30/30) Sapele faced both sides; lipped on all four edges						
610 mm x 1981 mm x 44 mm	71.59	1.90	24.74	76.96	nr	**101.70**
686 mm x 1981 mm x 44 mm	71.59	1.90	24.74	76.96	nr	**101.70**
762 mm x 1981 mm x 44 mm	71.59	1.90	24.74	76.96	nr	**101.70**
838 mm x 1981 mm x 44 mm	71.59	1.90	24.74	76.96	nr	**101.70**
726 mm x 2040 mm x 44 mm	71.59	1.90	24.74	76.96	nr	**101.70**
826 mm x 2040 mm x 44 mm	71.59	1.90	24.74	76.96	nr	**101.70**

L WINDOWS/DOORS/STAIRS

	PC £	Labour hours	Labour £	Material £	Unit	Total rate £
Flush door; half hour fire resisting "Leaderflush" (type B30) or other equal and approved; American light oak veneer; hardwood lipping all edges						
526 mm x 2040 mm x 44 mm	113.49	1.90	24.74	122.00	nr	146.74
626 mm x 2040 mm x 44 mm	119.01	1.90	24.74	127.94	nr	152.68
726 mm x 2040 mm x 44 mm	119.35	1.90	24.74	128.30	nr	153.04
826 mm x 2040 mm x 44 mm	121.66	1.90	24.74	130.78	nr	155.52
Flush door; one hour fire check (60/45); plywood faced both sides; lipped on all four edges; Premdor Crosby "Popular Firemaster" or other equal and approved						
762 mm x 1981 mm x 54 mm	178.43	2.04	26.56	191.81	nr	218.37
838 mm x 1981 mm x 54 mm	187.08	2.04	26.56	201.11	nr	227.67
726 mm x 2040 mm x 54 mm	184.23	2.04	26.56	198.05	nr	224.61
826 mm x 2040 mm x 54 mm	184.23	2.04	26.56	198.05	nr	224.61
Flush door; one hour fire check (60/45); Sapele faced both sides; lipped on all four edges						
762 mm x 1981 mm x 54 mm	294.97	2.13	27.73	317.09	nr	344.82
838 mm x 1981 mm x 54 mm	308.68	2.13	27.73	331.83	nr	359.56
Flush door; one hour fire resisting "Leaderflush" (type B60) or other equal and approved; Afrormosia veneer; hardwood lipping all edges including groove and "Leaderseal" intumescent strip or similar						
457 mm x 1981 mm x 54 mm	149.76	2.13	27.73	160.99	nr	188.72
533 mm x 1981 mm x 54 mm	147.19	2.13	27.73	158.23	nr	185.96
610 mm x 1981 mm x 54 mm	166.18	2.13	27.73	178.64	nr	206.37
686 mm x 1981 mm x 54 mm	168.76	2.13	27.73	181.42	nr	209.15
762 mm x 1981 mm x 54 mm	171.32	2.13	27.73	184.17	nr	211.90
838 mm x 1981 mm x 54 mm	180.74	2.13	27.73	194.30	nr	222.03
526 mm x 2040 mm x 54 mm	146.97	2.13	27.73	157.99	nr	185.72
626 mm x 2040 mm x 54 mm	166.62	2.13	27.73	179.12	nr	206.85
726 mm x 2040 mm x 54 mm	169.87	2.13	27.73	182.61	nr	210.34
826 mm x 2040 mm x 54 mm	173.12	2.13	27.73	186.10	nr	213.83
Flush door; external quality; skeleton or cellular core; plywood faced both sides; lipped on all four edges						
762 mm x 1981 mm x 54 mm	44.93	1.80	23.44	48.30	nr	71.74
838 mm x 1981 mm x 54 mm	46.50	1.80	23.44	49.99	nr	73.43
Flush door; external quality with standard glass opening; skeleton or cellular core; plywood faced both sides; lipped on all four edges; including glazing beads						
762 mm x 1981 mm x 54 mm	56.45	1.80	23.44	60.68	nr	84.12
838 mm x 1981 mm x 54 mm	57.93	1.80	23.44	62.27	nr	85.71
Doors; purpose made panelled; wrought softwood						
Panelled doors; one open panel for glass; including glazing beads						
686 mm x 1981 mm x 44 mm	59.08	1.80	23.44	63.51	nr	86.95
762 mm x 1981 mm x 44 mm	60.90	1.80	23.44	65.47	nr	88.91
838 mm x 1981 mm x 44 mm	62.69	1.80	23.44	67.39	nr	90.83
Panelled doors; two open panel for glass; including glazing beads						
686 mm x 1981 mm x 44 mm	81.71	1.80	23.44	87.84	nr	111.28
762 mm x 1981 mm x 44 mm	86.10	1.80	23.44	92.56	nr	116.00
838 mm x 1981 mm x 44 mm	90.37	1.80	23.44	97.15	nr	120.59
Panelled doors; four 19 mm thick plywood panels; mouldings worked on solid both sides						
686 mm x 1981 mm x 44 mm	112.79	1.80	23.44	121.25	nr	144.69
762 mm x 1981 mm x 44 mm	117.49	1.80	23.44	126.30	nr	149.74
838 mm x 1981 mm x 44 mm	123.41	1.80	23.44	132.67	nr	156.11

Prices for Measured Works - Minor Works
L WINDOWS/DOORS/STAIRS

	PC £	Labour hours	Labour £	Material £	Unit	Total rate £
L20 DOORS/SHUTTERS/HATCHES - cont'd						
Doors; purpose made panelled; wrought softwood - cont'd						
Panelled doors; six 25 mm thick panels raised and fielded; mouldings worked on solid both sides						
686 mm x 1981 mm x 44 mm	257.68	2.13	27.73	277.01	nr	304.74
762 mm x 1981 mm x 44 mm	268.39	2.13	27.73	288.52	nr	316.25
838 mm x 1981 mm x 44 mm	281.82	2.13	27.73	302.96	nr	330.69
rebated edges beaded	-	-	-	1.46	m	1.46
rounded edges or heels	-	-	-	0.73	m	0.73
weatherboard fixed to bottom rail	-	0.28	3.65	2.82	m	6.47
stopped groove for weatherboard	-	-	-	4.38	m	4.38
Doors; purpose made panelled; selected West African Mahogany						
Panelled doors; one open panel for glass; including glazing beads						
686 mm x 1981 mm x 50 mm	119.47	2.54	33.07	128.43	nr	161.50
762 mm x 1981 mm x 50 mm	123.41	2.54	33.07	132.67	nr	165.74
838 mm x 1981 mm x 50 mm	127.35	2.54	33.07	136.90	nr	169.97
686 mm x 1981 mm x 63 mm	145.38	2.82	36.72	156.28	nr	193.00
762 mm x 1981 mm x 63 mm	150.31	2.82	36.72	161.58	nr	198.30
838 mm x 1981 mm x 63 mm	155.20	2.82	36.72	166.84	nr	203.56
Panelled doors; 250 mm wide cross tongued intermediate rail; two open panels for glass; mouldings worked on the solid one side; 19 mm x 13 mm beads one side; fixing with brass cups and screws						
686 mm x 1981 mm x 50 mm	163.15	2.54	33.07	175.39	nr	208.46
762 mm x 1981 mm x 50 mm	171.21	2.54	33.07	184.05	nr	217.12
838 mm x 1981 mm x 50 mm	179.17	2.54	33.07	192.61	nr	225.68
686 mm x 1981 mm x 63 mm	196.12	2.82	36.72	210.83	nr	247.55
762 mm x 1981 mm x 63 mm	205.85	2.82	36.72	221.29	nr	258.01
838 mm x 1981 mm x 63 mm	215.56	2.82	36.72	231.73	nr	268.45
Panelled doors; four panels; (19 mm thick for 50 mm doors, 25 mm thick for 63 mm doors); mouldings worked on solid both sides						
686 mm x 1981 mm x 50 mm	207.88	2.54	33.07	223.47	nr	256.54
762 mm x 1981 mm x 50 mm	216.52	2.54	33.07	232.76	nr	265.83
838 mm x 1981 mm x 50 mm	227.36	2.54	33.07	244.41	nr	277.48
686 mm x 1981 mm x 63 mm	245.47	2.82	36.72	263.88	nr	300.60
762 mm x 1981 mm x 63 mm	254.64	2.82	36.72	273.74	nr	310.46
838 mm x 1981 mm x 63 mm	268.51	2.82	36.72	288.65	nr	325.37
Panelled doors; 150 mm wide stiles in one width; 430 mm wide cross tongued bottom rail; six panels raised and fielded one side; (19 mm thick for 50 mm doors, 25 mm thick for 63 mm doors); mouldings worked on solid both sides						
686 mm x 1981 mm x 50 mm	438.58	2.54	33.07	471.47	nr	504.54
762 mm x 1981 mm x 50 mm	456.86	2.54	33.07	491.12	nr	524.19
838 mm x 1981 mm x 50 mm	479.67	2.54	33.07	515.65	nr	548.72
686 mm x 1981 mm x 63 mm	540.12	2.82	36.72	580.63	nr	617.35
762 mm x 1981 mm x 63 mm	562.66	2.82	36.72	604.86	nr	641.58
838 mm x 1981 mm x 63 mm	590.78	2.82	36.72	635.09	nr	671.81
rebated edges beaded	-	-	-	2.19	m	2.19
rounded edges or heels	-	-	-	1.10	m	1.10
weatherboard fixed to bottom rail	-	0.32	4.17	6.28	m	10.45
stopped groove for weatherboard	-	-	-	5.38	m	5.38

L WINDOWS/DOORS/STAIRS

	PC £	Labour hours	Labour £	Material £	Unit	Total rate £
Doors; purpose made panelled; American White Ash						
Panelled doors; one open panel for glass; including glazing beads						
686 mm x 1981 mm x 50 mm	133.78	2.54	33.07	143.81	nr	176.88
762 mm x 1981 mm x 50 mm	138.27	2.54	33.07	148.64	nr	181.71
838 mm x 1981 mm x 50 mm	142.75	2.54	33.07	153.46	nr	186.53
686 mm x 1981 mm x 63 mm	163.24	2.82	36.72	175.48	nr	212.20
762 mm x 1981 mm x 63 mm	168.88	2.82	36.72	181.55	nr	218.27
838 mm x 1981 mm x 63 mm	174.47	2.82	36.72	187.56	nr	224.28
Panelled doors; 250 mm wide cross tongued intermediate rail; two open panels for glass; mouldings worked on the solid one side; 19 mm x 13 mm beads one side; fixing with brass cups and screws						
686 mm x 1981 mm x 50 mm	182.05	2.54	33.07	195.70	nr	228.77
762 mm x 1981 mm x 50 mm	191.16	2.54	33.07	205.50	nr	238.57
838 mm x 1981 mm x 50 mm	200.21	2.54	33.07	215.23	nr	248.30
686 mm x 1981 mm x 63 mm	219.72	2.82	36.72	236.20	nr	272.92
762 mm x 1981 mm x 63 mm	230.76	2.82	36.72	248.07	nr	284.79
838 mm x 1981 mm x 63 mm	241.77	2.82	36.72	259.90	nr	296.62
Panelled doors; four panels; (19 mm thick for 50 mm doors, 25 mm thick for 63 mm doors); mouldings worked on solid both sides						
686 mm x 1981 mm x 50 mm	231.20	2.54	33.07	248.54	nr	281.61
762 mm x 1981 mm x 50 mm	240.80	2.54	33.07	258.86	nr	291.93
838 mm x 1981 mm x 50 mm	252.84	2.54	33.07	271.80	nr	304.87
686 mm x 1981 mm x 63 mm	274.75	2.82	36.72	295.36	nr	332.08
762 mm x 1981 mm x 63 mm	286.21	2.82	36.72	307.68	nr	344.40
838 mm x 1981 mm x 63 mm	300.52	2.82	36.72	323.06	nr	359.78
Panelled doors; 150 mm wide stiles in one width; 430 mm wide cross tongued bottom rail; six panels raised and fielded one side; (19 mm thick for 50 mm doors, 25 mm thick for 63 mm doors); mouldings worked on solid both sides						
686 mm x 1981 mm x 50 mm	476.49	2.54	33.07	512.23	nr	545.30
762 mm x 1981 mm x 50 mm	496.35	2.54	33.07	533.58	nr	566.65
838 mm x 1981 mm x 50 mm	521.17	2.54	33.07	560.26	nr	593.33
686 mm x 1981 mm x 63 mm	588.13	2.82	36.72	632.24	nr	668.96
762 mm x 1981 mm x 63 mm	612.63	2.82	36.72	658.58	nr	695.30
838 mm x 1981 mm x 63 mm	643.28	2.82	36.72	691.53	nr	728.25
rebated edges beaded	-	-	-	2.19	m	2.19
rounded edges or heels	-	-	-	1.10	m	1.10
weatherboard fixed to bottom rail	-	0.32	4.17	6.88	m	11.05
stopped groove for weatherboard	-	-	-	5.38	m	5.38
Doors; galvanised steel "up and over" type garage doors; Catnic "Horizon 90" or other equal and approved; spring counterbalanced; fixed to timber frame (not included)						
Garage door						
2135 mm x 1980 mm	203.05	4.07	52.99	218.57	nr	271.56
2135 mm x 2135 mm	228.54	4.07	52.99	245.97	nr	298.96
2400 mm x 2135 mm	276.88	4.07	52.99	297.99	nr	350.98
3965 mm x 2135 mm	728.69	6.11	79.55	784.03	nr	863.58

Prices for Measured Works - Minor Works
L WINDOWS/DOORS/STAIRS

	PC £	Labour hours	Labour £	Material £	Unit	Total rate £
L20 DOORS/SHUTTERS/HATCHES - cont'd						
Doorsets; galvanised steel IG "Weatherbeater Original" door and frame units or other equal and approved; treated softwood frame, primed hardwood sill; fixing in position; plugged and screwed to brickwork or blockwork						
Door and frame						
762 mm x 1981 mm; ref IGD1	123.56	3.05	39.71	133.12	nr	172.83
Doorsets; steel door and frame units; Jandor Architectural Ltd or other equal and approved; polyester powder coated; ironmongery						
Single action door set; "Metset MD01" doors and "Metset MF" frames						
900 mm x 2100 mm	-	-	-	-	nr	1757.67
pair 1800 mm x 2100 mm	-	-	-	-	nr	2416.79
Rolling shutters and collapsible gates; steel counter shutters; Bolton Brady Ltd or other equal and approved; push-up, self-coiling; polyester power coated; fixing by bolting						
Shutters						
3000 mm x 1000 mm	-	-	-	-	nr	948.74
4000 mm x 1000 mm; in two panels	-	-	-	-	nr	1647.81
Rolling shutters and collapsible gates; galvanised steel; Bolton Brady Type 474 or other equal and approved; one hour fibre resisting; self-coiling; activated by fusible link; fixing with bolts						
Rolling shutters and collapsible gates						
1000 mm x 2750 mm	-	-	-	-	nr	1148.48
1500 mm x 2750 mm	-	-	-	-	nr	1208.40
2400 mm x 2750 mm	-	-	-	-	nr	1398.14
Sliding/folding partitions; aluminium double glazed sliding patio doors; Crittal "Luminaire" or other equal and approved; white acrylic finish; with and including 18 thick annealed double glazing; fixed in position; including lugs plugged and screwed to brickwork or blockwork						
Patio doors						
1800 mm x 2100 mm; ref PF1821	1243.50	2.54	33.07	1337.64	nr	1370.71
2400 mm x 2100 mm; ref PF2421	1492.20	3.05	39.71	1604.99	nr	1644.70
2700 mm x 2100 mm; ref PF2721	1658.00	3.56	46.35	1783.23	nr	1829.58
Grilles; "Galaxy" nylon rolling counter grille or other equal and approved; Bolton Brady Ltd; colour, off-white; self-coiling; fixing by bolting						
Grilles						
3000 mm x 1000 mm	-	-	-	-	nr	848.87
4000 mm x 1000 mm	-	-	-	-	nr	1298.28

L WINDOWS/DOORS/STAIRS

	PC £	Labour hours	Labour £	Material £	Unit	Total rate £
Door frames and door linings, sets; standard joinery; wrought softwood						
Internal door frame or lining composite sets for 686 mm x 1981mm door; all with loose stops unless rebated; "finished sizes"						
27 mm x 94 mm lining	27.53	0.83	10.81	29.89	nr	40.70
27 mm x 107 mm lining	28.96	0.83	10.81	31.42	nr	42.23
35 mm x 107 mm rebated lining	30.06	0.83	10.81	32.61	nr	43.42
27 mm x 121 mm lining	31.32	0.83	10.81	33.96	nr	44.77
27 mm x 121 mm lining with fanlight over	44.11	0.97	12.63	47.78	nr	60.41
27 mm x 133 mm lining	33.77	0.83	10.81	36.60	nr	47.41
35 mm x 133 mm rebated lining	34.69	0.83	10.81	37.58	nr	48.39
27 mm x 133 mm lining with fanlight over	47.40	0.97	12.63	51.32	nr	63.95
33 mm x 57 mm frame	22.18	0.83	10.81	24.14	nr	34.95
33 mm x 57 mm storey height frame	27.37	0.97	12.63	29.79	nr	42.42
33 mm x 57 mm frame with fanlight over	33.18	0.97	12.63	36.03	nr	48.66
33 mm x 64 mm frame	23.91	0.83	10.81	26.00	nr	36.81
33 mm x 64 mm storey height frame	29.30	0.97	12.63	31.86	nr	44.49
33 mm x 64 mm frame with fanlight over	35.02	0.97	12.63	38.01	nr	50.64
44 mm x 94 mm frame	37.05	0.93	12.11	40.12	nr	52.23
44 mm x 94 mm storey height frame	44.08	1.11	14.45	47.75	nr	62.20
44 mm x 94 mm frame with fanlight over	51.39	1.11	14.45	55.61	nr	70.06
44 mm x 107 mm frame	42.36	1.02	13.28	45.83	nr	59.11
44 mm x 107 mm storey height frame	49.63	1.11	14.45	53.72	nr	68.17
44 mm x 107 mm frame with fanlight over	56.62	1.11	14.45	61.23	nr	75.68
Internal door frame or lining composite sets for 762 mm x 1981 mm door; all with loose stops unless rebated; "finished sizes"						
27 mm x 94 mm lining	27.53	0.83	10.81	29.89	nr	40.70
27 mm x 107 mm lining	28.96	0.83	10.81	31.42	nr	42.23
35 mm x 107 mm rebated lining	30.06	0.83	10.81	32.61	nr	43.42
27 mm x 121 mm lining	31.32	0.83	10.81	33.96	nr	44.77
27 mm x 121 mm lining with fanlight over	44.11	0.97	12.63	47.78	nr	60.41
27 mm x 133 mm lining	33.77	0.83	10.81	36.60	nr	47.41
35 mm x 133 mm rebated lining	34.69	0.83	10.81	37.58	nr	48.39
27 mm x 133 mm lining with fanlight over	47.40	0.97	12.63	51.32	nr	63.95
33 mm x 57 mm frame	22.18	0.83	10.81	24.14	nr	34.95
33 mm x 57 mm storey height frame	27.37	0.97	12.63	29.79	nr	42.42
33 mm x 57 mm frame with fanlight over	33.18	0.97	12.63	36.03	nr	48.66
33 mm x 64 mm frame	23.91	0.83	10.81	26.00	nr	36.81
33 mm x 64 mm storey height frame	29.30	0.97	12.63	31.86	nr	44.49
33 mm x 64 mm frame with fanlight over	35.02	0.97	12.63	38.01	nr	50.64
44 mm x 94 mm frame	37.05	0.93	12.11	40.12	nr	52.23
44 mm x 94 mm storey height frame	44.08	1.11	14.45	47.75	nr	62.20
44 mm x 94 mm frame with fanlight over	51.39	1.11	14.45	55.61	nr	70.06
44 mm x 107 mm frame	42.36	1.02	13.28	45.83	nr	59.11
44 mm x 107 mm storey height frame	49.63	1.11	14.45	53.72	nr	68.17
44 mm x 107 mm frame with fanlight over	56.62	1.11	14.45	61.23	nr	75.68
Internal door frame or lining composite sets for 726 mm x 2040 mm door; with loose stops						
30 mm x 94 mm lining	32.11	0.83	10.81	34.81	nr	45.62
30 mm x 94 mm lining with fanlight over	45.76	0.97	12.63	49.56	nr	62.19
30 mm x 107 mm lining	37.76	0.83	10.81	40.88	nr	51.69
30 mm x 107 mm lining with fanlight over	51.57	0.97	12.63	55.80	nr	68.43
30 mm x 133 mm lining	42.31	0.83	10.81	45.78	nr	56.59
30 mm x 133 mm lining with fanlight over	55.61	0.97	12.63	60.15	nr	72.78

Prices for Measured Works - Minor Works
L WINDOWS/DOORS/STAIRS

	PC £	Labour hours	Labour £	Material £	Unit	Total rate £
L20 DOORS/SHUTTERS/HATCHES - cont'd						
Door frames and door linings, sets; standard joinery; wrought softwood - cont'd						
Internal door frame or lining composite sets for 826 mm x 2040 mm door; with loose stops						
30 mm x 94 mm lining	32.11	0.83	10.81	34.81	nr	45.62
30 mm x 94 mm lining with fanlight over	45.76	0.97	12.63	49.56	nr	62.19
30 mm x 107 mm lining	37.76	0.83	10.81	40.88	nr	51.69
30 mm x 107 mm lining with fanlight over	51.57	0.97	12.63	55.80	nr	68.43
30 mm x 133 mm lining	42.31	0.83	10.81	45.78	nr	56.59
30 mm x 133 mm lining with fanlight over	55.61	0.97	12.63	60.15	nr	72.78
Door frames and door linings, sets; purpose made; wrought softwood						
Jambs and heads; as linings						
32 mm x 63 mm	-	0.19	2.47	4.45	m	6.92
32 mm x 100 mm	-	0.19	2.47	5.71	m	8.18
32 mm x 140 mm	-	0.19	2.47	7.31	m	9.78
Jambs and heads; as frames; rebated, rounded and grooved						
38 mm x 75 mm	-	0.19	2.47	5.66	m	8.13
38 mm x 100 mm	-	0.19	2.47	6.71	m	9.18
38 mm x 115 mm	-	0.19	2.47	7.80	m	10.27
38 mm x 140 mm	-	0.19	2.47	9.08	m	11.55
50 mm x 100 mm	-	0.19	2.47	8.45	m	10.92
50 mm x 125 mm	-	0.19	2.47	10.04	m	12.51
63 mm x 88 mm	-	0.19	2.47	9.68	m	12.15
63 mm x 100 mm	-	0.19	2.47	10.34	m	12.81
63 mm x 125 mm	-	0.19	2.47	12.32	m	14.79
75 mm x 100 mm	-	0.19	2.47	11.84	m	14.31
75 mm x 125 mm	-	0.23	2.99	14.65	m	17.64
75 mm x 150 mm	-	0.23	2.99	17.09	m	20.08
100 mm x 100 mm	-	0.28	3.65	15.40	m	19.05
100 mm x 150 mm	-	0.28	3.65	22.07	m	25.72
Mullions and transoms; in linings						
32 mm x 63 mm	-	0.14	1.82	7.28	m	9.10
32 mm x 100 mm	-	0.14	1.82	8.57	m	10.39
32 mm x 140 mm	-	0.14	1.82	10.07	m	11.89
Mullions and transoms; in frames; twice rebated, rounded and grooved						
38 mm x 75 mm	-	0.14	1.82	8.69	m	10.51
38 mm x 100 mm	-	0.14	1.82	9.73	m	11.55
38 mm x 115 mm	-	0.14	1.82	10.57	m	12.39
38 mm x 140 mm	-	0.14	1.82	11.85	m	13.67
50 mm x 100 mm	-	0.14	1.82	11.21	m	13.03
50 mm x 125 mm	-	0.14	1.82	12.96	m	14.78
63 mm x 88 mm	-	0.14	1.82	12.51	m	14.33
63 mm x 100 mm	-	0.14	1.82	13.16	m	14.98
75 mm x 100 mm	-	0.14	1.82	14.77	m	16.59
Add 5% to the above material prices for selected softwood for staining						
Door frames and door linings, sets; purpose made; selected West African Mahogany						
Jambs and heads; as linings						
32 mm x 63 mm	7.63	0.23	2.99	8.28	m	11.27
32 mm x 100 mm	9.93	0.23	2.99	10.75	m	13.74
32 mm x 140 mm	12.60	0.23	2.99	13.69	m	16.68

Prices for Measured Works - Minor Works
L WINDOWS/DOORS/STAIRS

	PC £	Labour hours	Labour £	Material £	Unit	Total rate £
Jambs and heads; as frames; rebated, rounded and grooved						
38 mm x 75 mm	9.63	0.23	2.99	10.43	m	13.42
38 mm x 100 mm	11.66	0.23	2.99	12.61	m	15.60
38 mm x 115 mm	13.41	0.23	2.99	14.56	m	17.55
38 mm x 140 mm	15.50	0.28	3.65	16.81	m	20.46
50 mm x 100 mm	14.71	0.28	3.65	15.96	m	19.61
50 mm x 125 mm	17.75	0.28	3.65	19.23	m	22.88
63 mm x 88 mm	16.49	0.28	3.65	17.80	m	21.45
63 mm x 100 mm	18.28	0.28	3.65	19.72	m	23.37
63 mm x 125 mm	21.97	0.28	3.65	23.76	m	27.41
75 mm x 100 mm	21.06	0.28	3.65	22.79	m	26.44
75 mm x 125 mm	26.20	0.32	4.17	28.31	m	32.48
75 mm x 150 mm	30.75	0.32	4.17	33.20	m	37.37
100 mm x 100 mm	27.58	0.32	4.17	29.79	m	33.96
100 mm x 150 mm	40.02	0.32	4.17	43.17	m	47.34
Mullions and transoms; in linings						
32 mm x 63 mm	11.70	0.19	2.47	12.58	m	15.05
32 mm x 100 mm	13.99	0.19	2.47	15.04	m	17.51
32 mm x 140 mm	16.67	0.19	2.47	17.92	m	20.39
Mullions and transoms; in frames; twice rebated, rounded and grooved						
38 mm x 75 mm	13.91	0.19	2.47	14.95	m	17.42
38 mm x 100 mm	15.94	0.19	2.47	17.14	m	19.61
38 mm x 115 mm	17.49	0.19	2.47	18.80	m	21.27
38 mm x 140 mm	19.56	0.19	2.47	21.03	m	23.50
50 mm x 100 mm	18.79	0.19	2.47	20.20	m	22.67
50 mm x 125 mm	22.03	0.19	2.47	23.68	m	26.15
63 mm x 88 mm	20.56	0.19	2.47	22.10	m	24.57
63 mm x 100 mm	22.34	0.19	2.47	24.02	m	26.49
75 mm x 100 mm	25.35	0.19	2.47	27.25	m	29.72
Sills; once sunk weathered; once rebated, three times grooved						
63 mm x 175 mm	41.10	0.32	4.17	44.18	m	48.35
75 mm x 125 mm	37.22	0.32	4.17	40.01	m	44.18
75 mm x 150 mm	41.84	0.32	4.17	44.98	m	49.15
Door frames and door linings, sets; purpose made; American White Ash						
Jambs and heads; as linings						
32 mm x 63 mm	8.67	0.23	2.99	9.39	m	12.38
32 mm x 100 mm	11.36	0.23	2.99	12.29	m	15.28
32 mm x 140 mm	14.67	0.23	2.99	15.92	m	18.91
Jambs and heads; as frames; rebated, rounded and grooved						
38 mm x 75 mm	10.94	0.23	2.99	11.83	m	14.82
38 mm x 100 mm	13.42	0.23	2.99	14.50	m	17.49
38 mm x 115 mm	15.50	0.23	2.99	16.81	m	19.80
38 mm x 140 mm	17.91	0.28	3.65	19.40	m	23.05
50 mm x 100 mm	17.01	0.28	3.65	18.43	m	22.08
50 mm x 125 mm	20.61	0.28	3.65	22.30	m	25.95
63 mm x 88 mm	19.07	0.28	3.65	20.57	m	24.22
63 mm x 100 mm	21.18	0.28	3.65	22.84	m	26.49
63 mm x 125 mm	25.60	0.28	3.65	27.67	m	31.32
75 mm x 100 mm	24.49	0.28	3.65	26.47	m	30.12
75 mm x 125 mm	30.50	0.32	4.17	32.93	m	37.10
75 mm x 150 mm	35.90	0.32	4.17	38.74	m	42.91
100 mm x 100 mm	32.15	0.32	4.17	34.71	m	38.88
100 mm x 150 mm	46.91	0.32	4.17	50.57	m	54.74

Prices for Measured Works - Minor Works
L WINDOWS/DOORS/STAIRS

	PC £	Labour hours	Labour £	Material £	Unit	Total rate £
L20 DOORS/SHUTTERS/HATCHES - cont'd						
Door frames and door linings, sets; purpose made; American White Ash - cont'd						
Mullions and transoms; in linings						
32 mm x 63 mm	12.72	0.19	2.47	13.67	m	16.14
32 mm x 100 mm	15.44	0.19	2.47	16.60	m	19.07
32 mm x 140 mm	18.74	0.19	2.47	20.15	m	22.62
Mullions and transoms; in frames; twice rebated, rounded and grooved						
38 mm x 75 mm	15.21	0.19	2.47	16.35	m	18.82
38 mm x 100 mm	17.71	0.19	2.47	19.04	m	21.51
38 mm x 115 mm	19.56	0.19	2.47	21.03	m	23.50
38 mm x 140 mm	21.97	0.19	2.47	23.62	m	26.09
50 mm x 100 mm	21.07	0.19	2.47	22.65	m	25.12
50 mm x 125 mm	24.90	0.19	2.47	26.77	m	29.24
63 mm x 88 mm	23.18	0.19	2.47	24.92	m	27.39
63 mm x 100 mm	25.24	0.19	2.47	27.13	m	29.60
75 mm x 100 mm	28.76	0.19	2.47	30.92	m	33.39
Sills; once sunk weathered; once rebated, three times grooved						
63 mm x 175 mm	46.14	0.32	4.17	49.60	m	53.77
75 mm x 125 mm	41.52	0.32	4.17	44.63	m	48.80
75 mm x 150 mm	47.03	0.32	4.17	50.56	m	54.73
Door frames and door linings, sets; European Oak						
Sills; once sunk weathered; once rebated, three times grooved						
63 mm x 175 mm	81.33	0.32	4.17	87.43	m	91.60
75 mm x 125 mm	72.22	0.32	4.17	77.64	m	81.81
75 mm x 150 mm	82.79	0.32	4.17	89.00	m	93.17
Bedding and pointing frames						
Pointing wood frames or sills with mastic						
one side	-	0.09	0.82	0.48	m	1.30
both sides	-	0.19	1.74	0.95	m	2.69
Pointing wood frames or sills with polysulphide sealant						
one side	-	0.09	0.82	1.44	m	2.26
both sides	-	0.19	1.74	2.88	m	4.62
Bedding wood frames in cement mortar (1:3) and point						
one side	-	0.09	1.46	0.07	m	1.53
both sides	-	0.09	1.46	0.09	m	1.55
one side in mortar; other side in mastic	-	0.19	2.45	0.54	m	2.99
L30 STAIRS/WALKWAYS/BALUSTRADES						
Standard staircases; wrought softwood (parana pine)						
Stairs; 25 mm thick treads with rounded nosings; 9 mm thick plywood risers; 32 mm thick strings; bullnose bottom tread; 50 mm x 75 mm hardwood handrail; 32 mm square plain balusters; 100 mm square plain newel posts						
straight flight; 838 mm wide; 2676 mm going; 2600 mm rise; with two newel posts	-	7.12	92.70	375.57	nr	468.27
straight flight with turn; 838 wide; 2676 going; 2600 rise; with two newel posts; three top treads winding	-	7.12	92.70	475.42	nr	568.12

	PC £	Labour hours	Labour £	Material £	Unit	Total rate £
Stairs; 25 mm thick treads with rounded nosings; 9 mm thick plywood risers; 32 mm thick strings; bullnose bottom tread; 50 mm x 75 mm hardwood handrail; 32 mm square plain balusters; 100 mm square plain newel posts - cont'd						
dogleg staircase; 838 mm wide; 2676 mm going; 2600 mm rise; with two newel posts; quarter space landing third riser from top	-	7.12	92.70	400.53	nr	493.23
dogleg staircase; 838 mm wide; 2676 mm going; 2600 mm rise; with two newel posts; half space landing third riser from top	-	8.14	105.98	425.49	nr	531.47
Standard balustrades; wrought softwood						
Landing balustrade; 50 mm x 75 mm hardwood handrail; three 32 mm x 140 mm balustrade kneerails; two 32 mm x 50 mm stiffeners; one end of handrail jointed to newel post; other end built into wall; (newel post and mortices both not included)						
3.00 m long	-	2.82	36.72	129.83	nr	166.55
Landing balustrade; 50 mm x 75 mm hardwood handrail; 32 mm square plain balusters; one end of handrail jointed to newel post; other end built into wall; balusters housed in at bottom (newel post and mortices both not included)						
3.00 m long	-	4.07	52.99	79.89	nr	132.88
Hardwood staircases; purpose made; assembled at works						
Fixing only complete staircase including landings, balustrades, etc.						
plugging and screwing to brickwork or blockwork	-	15.26	198.69	1.79	nr	200.48
The following are supply only prices for purpose made staircase components in selected West African Mahogany supplied as part of an assembled staircase and may be used to arrive at a guide price for a complete hardwood staircase						
Board landings; cross-tongued joints; 100 mm x 50 mm sawn softwood bearers						
25 mm thick	-	-	-	94.07	m²	94.07
32 mm thick	-	-	-	115.36	m²	115.36
Treads; cross-tongued joints and risers; rounded nosings; tongued, grooved, glued and blocked together; one 175 mm x 50 mm sawn softwood carriage						
25 mm treads; 19 mm risers	-	-	-	116.49	m²	116.49
ends; quadrant	-	-	-	53.91	nr	53.91
ends; housed to hardwood	-	-	-	0.99	nr	0.99
32 mm treads; 25 mm risers	-	-	-	139.80	m²	139.80
ends; quadrant	-	-	-	69.29	nr	69.29
ends; housed to hardwood	-	-	-	0.99	nr	0.99
Winders; cross-tongued joints and risers in one width; rounded nosings; tongued, grooved glued and blocked together; one 175 mm x 50 mm sawn softwood carriage						
25 mm treads; 19 mm risers	-	-	-	128.13	m²	128.13
32 mm treads; 25 mm risers	-	-	-	153.79	m²	153.79
wide ends; housed to hardwood	-	-	-	2.00	nr	2.00
narrow ends; housed to hardwood	-	-	-	1.49	nr	1.49

L30 STAIRS/WALKWAYS/BALUSTRADES - cont'd

The following are supply only prices for purpose made staircase components in selected West African Mahogany supplied as part of an assembled staircase and may be used to arrive at a guide price for a complete hardwood staircase - cont'd

	PC £	Labour hours	Labour £	Material £	Unit	Total rate £
Closed strings; in one width; 230 mm wide; rounded twice						
32 mm thick	-	-	-	24.69	m	24.69
38 mm thick	-	-	-	29.65	m	29.65
50 mm thick	-	-	-	39.17	m	39.17
Closed strings; cross-tongued joints; 280 mm wide; once rounded						
32 mm thick	-	-	-	43.24	m	43.24
extra for short ramp	-	-	-	21.61	nr	21.61
38 mm thick	-	-	-	49.05	m	49.05
extra for short ramp	-	-	-	24.56	nr	24.56
50 mm thick	-	-	-	60.73	m	60.73
extra for short ramp	-	-	-	30.42	nr	30.42
Closed strings; ramped; crossed tongued joints 280 mm wide; once rounded; fixing with screws; plugging 450 mm centres						
32 mm thick	-	-	-	47.59	m	47.59
38 mm thick	-	-	-	53.97	m	53.97
50 mm thick	-	-	-	66.80	m	66.80
Apron linings; in one width 230 mm wide						
19 mm thick	-	-	-	9.80	m	9.80
25 mm thick	-	-	-	11.87	m	11.87
Handrails; rounded						
40 mm x 50 mm	-	-	-	8.17	m	8.17
50 mm x 75 mm	-	-	-	10.20	m	10.20
63 mm x 87 mm	-	-	-	13.20	m	13.20
75 mm x 100 mm	-	-	-	16.05	m	16.05
Handrails; moulded						
40 mm x 50 mm	-	-	-	8.74	m	8.74
50 mm x 75 mm	-	-	-	10.81	m	10.81
63 mm x 87 mm	-	-	-	13.79	m	13.79
75 mm x 100 mm	-	-	-	16.64	m	16.64
Handrails; rounded; ramped						
40 mm x 50 mm	-	-	-	16.32	m	16.32
50 mm x 75 mm	-	-	-	20.43	m	20.43
63 mm x 87 mm	-	-	-	26.43	m	26.43
75 mm x 100 mm	-	-	-	26.15	m	26.15
Handrails; moulded; ramped						
40 mm x 50 mm	-	-	-	17.49	m	17.49
50 mm x 75 mm	-	-	-	21.62	m	21.62
63 mm x 87 mm	-	-	-	27.63	m	27.63
75 mm x 100 mm	-	-	-	33.26	m	33.26
Heading joints to handrail; mitred or raked						
overall size not exceeding 50 mm x 75 mm	-	-	-	27.92	nr	27.92
overall size not exceeding 75 mm x 100 mm	-	-	-	35.55	nr	35.55

L WINDOWS/DOORS/STAIRS

	PC £	Labour hours	Labour £	Material £	Unit	Total rate £
Balusters; stiffeners						
25 mm x 25 mm	-	-	-	3.97	m	3.97
32 mm x 32 mm	-	-	-	4.59	m	4.59
50 mm x 50 mm	-	-	-	5.55	m	5.55
ends; housed	-	-	-	1.26	nr	1.26
Sub rails						
32 mm x 63 mm	-	-	-	7.06	m	7.06
ends; framed joint to newel	-	-	-	5.06	nr	5.06
Knee rails						
32 mm x 140 mm	-	-	-	11.04	m	11.04
ends; framed joint to newel	-	-	-	5.06	nr	5.06
Newel posts						
50 mm x 100 mm; half newel	-	-	-	11.80	m	11.80
75 mm x 75 mm	-	-	-	12.74	m	12.74
100 mm x 100 mm	-	-	-	19.69	m	19.69
Newel caps; splayed on four sides						
62.50 mm x 125 mm x 50 mm	-	-	-	7.41	nr	7.41
100 mm x 100 mm x 50 mm	-	-	-	7.41	nr	7.41
125 mm x 125 mm x 50 mm	-	-	-	8.04	nr	8.04
The following are supply only prices for purpose made staircase components in selected Oak; supplied as part of an assembled staircase						
Board landings; cross-tongued joints; 100 mm x 50 mm sawn softwood bearers						
25 mm thick	-	-	-	197.86	m²	197.86
32 mm thick	-	-	-	243.90	m²	243.90
Treads; cross-tongued joints and risers; rounded nosings; tongued, grooved, glued and blocked together; one 175 mm x 50 mm sawn softwood carriage						
25 mm treads; 19 mm risers	-	-	-	215.46	m²	215.46
ends; quadrant	-	-	-	107.76	nr	107.76
ends; housed to hardwood	-	-	-	1.33	nr	1.33
32 mm treads; 25 mm risers	-	-	-	265.26	m²	265.26
ends; quadrant	-	-	-	132.66	nr	132.66
ends; housed to hardwood	-	-	-	1.33	nr	1.33
Winders; cross-tongued joints and risers in one width; rounded nosings; tongued, grooved glued and blocked together; one 175 mm x 50 mm sawn softwood carriage						
25 mm treads; 19 mm risers	-	-	-	237.05	m²	237.05
32 mm treads; 25 mm risers	-	-	-	291.79	m²	291.79
wide ends; housed to hardwood	-	-	-	2.67	nr	2.67
narrow ends; housed to hardwood	-	-	-	2.00	nr	2.00
Closed strings; in one width; 230 mm wide; rounded twice						
32 mm thick	-	-	-	52.07	m	52.07
38 mm thick	-	-	-	62.02	m	62.02
50 mm thick	-	-	-	81.99	m	81.99
Closed strings; cross-tongued joints; 280 mm wide; once rounded						
32 mm thick	-	-	-	83.18	m	83.18
extra for short ramp	-	-	-	41.59	nr	41.59
38 mm thick	-	-	-	95.40	m	95.40
extra for short ramp	-	-	-	47.72	nr	47.72
50 mm thick	-	-	-	120.07	m	120.07
extra for short ramp	-	-	-	60.04	nr	60.04

L30 STAIRS/WALKWAYS/BALUSTRADES - cont'd

The following are supply only prices for purpose made staircase components in selected Oak; supplied as part of an assembled staircase - cont'd

Item	PC £	Labour hours	Labour £	Material £	Unit	Total rate £
Closed strings; ramped; crossed tongued joints 280 mm wide; once rounded; fixing with screws; plugging 450 mm centres						
32 mm thick	-	-	-	91.51	m	91.51
38 mm thick	-	-	-	104.96	m	104.96
50 mm thick	-	-	-	132.07	m	132.07
Apron linings; in one width 230 mm wide						
19 mm thick	-	-	-	23.78	m	23.78
25 mm thick	-	-	-	29.89	m	29.89
Handrails; rounded						
40 mm x 50 mm	-	-	-	16.95	m	16.95
50 mm x 75 mm	-	-	-	23.07	m	23.07
63 mm x 87 mm	-	-	-	31.98	m	31.98
75 mm x 100 mm	-	-	-	40.41	m	40.41
Handrails; moulded						
40 mm x 50 mm	-	-	-	17.74	m	17.74
50 mm x 75 mm	-	-	-	23.86	m	23.86
63 mm x 87 mm	-	-	-	26.86	m	26.86
75 mm x 100 mm	-	-	-	41.19	m	41.19
Handrails; rounded; ramped						
40 mm x 50 mm	-	-	-	33.87	m	33.87
50 mm x 75 mm	-	-	-	46.12	m	46.12
63 mm x 87 mm	-	-	-	63.96	m	63.96
75 mm x 100 mm	-	-	-	80.82	m	80.82
Handrails; moulded; ramped						
40 mm x 50 mm	-	-	-	35.44	m	35.44
50 mm x 75 mm	-	-	-	47.72	m	47.72
63 mm x 87 mm	-	-	-	65.53	m	65.53
75 mm x 100 mm	-	-	-	82.40	m	82.40
Heading joints to handrail; mitred or raked						
overall size not exceeding 50 mm x 75 mm	-	-	-	42.67	nr	42.67
overall size not exceeding 75 mm x 100 mm	-	-	-	51.97	nr	51.97
Balusters; stiffeners						
25 mm x 25 mm	-	-	-	6.45	m	6.45
32 mm x 32 mm	-	-	-	8.29	m	8.29
50 mm x 50 mm	-	-	-	11.14	m	11.14
ends; housed	-	-	-	1.66	nr	1.66
Sub rails						
32 mm x 63 mm	-	-	-	14.21	m	14.21
ends; framed joint to newel	-	-	-	6.75	nr	6.75
Knee rails						
32 mm x 140 mm	-	-	-	25.95	m	25.95
ends; framed joint to newel	-	-	-	6.75	nr	6.75
Newel posts						
50 mm x 100 mm; half newel	-	-	-	28.30	m	28.30
75 mm x 75 mm	-	-	-	31.07	m	31.07
100 mm x 100 mm	-	-	-	51.74	m	51.74
Newel caps; splayed on four sides						
62.50 mm x 125 mm x 50 mm	-	-	-	10.76	nr	10.76
100 mm x 100 mm x 50 mm	-	-	-	10.76	nr	10.76
125 mm x 125 mm x 50 mm	-	-	-	12.61	nr	12.61

	PC £	Labour hours	Labour £	Material £	Unit	Total rate £
Aluminium alloy folding loft ladders; "Zig Zag" stairways, model B or other equal and approved; Light Alloy Ltd; on and including plywood backboard; fixing with screws to timber lining (not included)						
Loft ladders						
ceiling height not exceeding 2500 mm; model 888801	-	1.02	13.28	486.50	nr	**499.78**
ceiling height not exceeding 2800 mm; model 888802	-	1.02	13.28	533.44	nr	**546.72**
ceiling height not exceeding 3100 mm; model 888803	-	1.02	13.28	595.35	nr	**608.63**
Flooring, balustrades and handrails; mild steel						
Chequer plate flooring; over 300 mm wide; bolted to steel supports						
6 mm thick	-	-	-	49.44	m^2	**49.44**
8 mm thick	-	-	-	65.91	m^2	**65.91**
Open grid steel flooring; Eurogrid Ltd; "Safeway" type D38 diamond pattern flooring or other equal and approved; to steel supports at 1200 mm centres; galvanised; to BS 729 part 1						
30 mm x 5 mm sections; in one mat; fixing to 10 mm thick channels with type F5 clamps	-	-	-	126.33	m^2	**126.33**
Balustrades; welded construction; 1070 mm high; galvanized; 50 mm x 50 mm x 3.20 mm rhs top rail; 38 mm x 13 mm bottom rail, 50 mm x 50 mm x 3.20 mm rhs standards at 1830 mm centres with base plate drilled and bolted to concrete; 13 mm x 13 mm balusters at 102 mm centres	-	-	-	120.84	m	**120.84**
Balusters; isolated; one end ragged and cemented in; one 76 mm x 25 mm x 6 mm flange plate welded on; ground to a smooth finish; countersunk drilled and tap screwed to underside of handrail						
19 mm square 914 mm long bar	-	-	-	16.48	nr	**16.48**
Core-rails; joints prepared, welded and ground to a smooth finish; fixing on brackets (not included)						
38 mm x 10 mm flat bar	-	-	-	21.97	m	**21.97**
50 mm x 8 mm flat bar	-	-	-	21.15	m	**21.15**
Handrails; joints prepared, welded and ground to a smooth finish; fixing on brackets (not included)						
38 mm x 12 mm half oval bar	-	-	-	29.11	m	**29.11**
44 mm x 13 mm half oval bar	-	-	-	31.86	m	**31.86**
Handrail bracket; comprising 40 mm x 5 mm plate with mitred and welded angle; one end welded to 100 mm diameter x 5 mm backplate; three times holed and plugged and screwed to brickwork; other end scribed and welded to underside of handrail						
140 mm girth	-	-	-	17.58	nr	**17.58**

Prices for Measured Works - Minor Works
L WINDOWS/DOORS/STAIRS

	PC £	Labour hours	Labour £	Material £	Unit	Total rate £
L30 STAIRS/WALKWAYS/BALUSTRADES - cont'd						
Holes						
Holes; countersunk; for screws or bolts						
6 mm diameter; 3 mm thick	-	-	-	0.88	nr	0.88
6 mm diameter; 6 mm thick	-	-	-	1.32	nr	1.32
8 mm diameter; 6 mm thick	-	-	-	1.32	nr	1.32
10 mm diameter; 6 mm thick	-	-	-	1.38	nr	1.38
12 mm diameter; 8 mm thick	-	-	-	1.76	nr	1.76
L40 GENERAL GLAZING						
Standard plain glass; BS 952; clear float;						
panes area 0.15 m^2 - 4.00 m^2						
3 mm thick; glazed with						
putty or bradded beads	-	-	-	-	m^2	25.26
bradded beads and butyl compound	-	-	-	-	m^2	25.26
screwed beads	-	-	-	-	m^2	28.56
screwed beads and butyl compound	-	-	-	-	m^2	28.56
4 mm thick; glazed with						
putty or bradded beads	-	-	-	-	m^2	26.92
bradded beads and butyl compound	-	-	-	-	m^2	26.92
screwed beads	-	-	-	-	m^2	30.21
screwed beads and butyl compound	-	-	-	-	m^2	30.21
5 mm thick; glazed with						
putty or bradded beads	-	-	-	-	m^2	31.86
bradded beads and butyl compound	-	-	-	-	m^2	31.86
screwed beads	-	-	-	-	m^2	35.15
screwed beads and butyl compound	-	-	-	-	m^2	35.15
6 mm thick; glazed with						
putty or bradded beads	-	-	-	-	m^2	33.51
bradded beads and butyl compound	-	-	-	-	m^2	33.51
screwed beads	-	-	-	-	m^2	36.80
screwed beads and butyl compound	-	-	-	-	m^2	36.80
Standard plain glass; BS 952; white patterned;						
panes area 0.15 m^2 - 4.00 m^2						
4 mm thick; glazed with						
putty or bradded beads	-	-	-	-	m^2	30.21
bradded beads and butyl compound	-	-	-	-	m^2	30.21
screwed beads	-	-	-	-	m^2	33.51
screwed beads and butyl compound	-	-	-	-	m^2	33.51
6 mm thick; glazed with						
putty or bradded beads	-	-	-	-	m^2	36.80
bradded beads and butyl compound	-	-	-	-	m^2	36.80
screwed beads	-	-	-	-	m^2	40.10
screwed beads and butyl compound	-	-	-	-	m^2	40.10
Standard plain glass; BS 952; rough cast;						
panes area 0.15 m^2 - 4.00 m^2						
6 mm thick; glazed with						
putty or bradded beads	-	-	-	-	m^2	40.10
bradded beads and butyl compound	-	-	-	-	m^2	40.10
screwed beads	-	-	-	-	m^2	43.40
screwed beads and butyl compound	-	-	-	-	m^2	43.40

L WINDOWS/DOORS/STAIRS

	PC £	Labour hours	Labour £	Material £	Unit	Total rate £
Standard plain glass; BS 952; Georgian wired cast; panes area 0.15 m² - 4.00 m²						
7 mm thick; glazed with						
putty or bradded beads	-	-	-	-	m²	43.40
bradded beads and butyl compound	-	-	-	-	m²	43.40
screwed beads	-	-	-	-	m²	46.69
screwed beads and butyl compound	-	-	-	-	m²	46.69
Extra for lining up wired glass	-	-	-	-	m²	3.35
Standard plain glass; BS 952; Georgian wired polished; panes area 0.15 m² - 4.00 m²						
6 mm thick; glazed with						
putty or bradded beads	-	-	-	-	m²	79.65
bradded beads and butyl compound	-	-	-	-	m²	79.65
screwed beads	-	-	-	-	m²	82.39
screwed beads and butyl compound	-	-	-	-	m²	82.39
Extra for lining up wired glass	-	-	-	-	m²	3.35
Special glass; BS 952; toughened clear float; panes area 0.15 m² - 4.00 m²						
4 mm thick; glazed with						
putty or bradded beads	-	-	-	-	m²	30.21
bradded beads and butyl compound	-	-	-	-	m²	30.21
screwed beads	-	-	-	-	m²	33.51
screwed beads and butyl compound	-	-	-	-	m²	33.51
5 mm thick; glazed with						
putty or bradded beads	-	-	-	-	m²	40.10
bradded beads and butyl compound	-	-	-	-	m²	40.10
screwed beads	-	-	-	-	m²	43.40
screwed beads and butyl compound	-	-	-	-	m²	43.40
6 mm thick; glazed with						
putty or bradded beads	-	-	-	-	m²	40.10
bradded beads and butyl compound	-	-	-	-	m²	40.10
screwed beads	-	-	-	-	m²	43.40
screwed beads and butyl compound	-	-	-	-	m²	43.40
10 mm thick; glazed with						
putty or bradded beads	-	-	-	-	m²	79.65
bradded beads and butyl compound	-	-	-	-	m²	79.65
screwed beads	-	-	-	-	m²	82.39
screwed beads and butyl compound	-	-	-	-	m²	82.39
Special glass; BS 952; clear laminated safety glass; panes area 0.15 m² - 4.00 m²						
4.40 mm thick; glazed with						
putty or bradded beads	-	-	-	-	m²	65.91
bradded beads and butyl compound	-	-	-	-	m²	65.91
screwed beads	-	-	-	-	m²	67.56
screwed beads and butyl compound	-	-	-	-	m²	67.56
5.40 mm thick; glazed with						
putty or bradded beads	-	-	-	-	m²	79.65
bradded beads and butyl compound	-	-	-	-	m²	79.65
screwed beads	-	-	-	-	m²	83.48
screwed beads and butyl compound	-	-	-	-	m²	83.48
6.40 mm thick; glazed with						
putty or bradded beads	-	-	-	-	m²	87.88
bradded beads and butyl compound	-	-	-	-	m²	87.88
screwed beads	-	-	-	-	m²	96.67
screwed beads and butyl compound	-	-	-	-	m²	96.67

L WINDOWS/DOORS/STAIRS

L40 GENERAL GLAZING - cont'd

	PC £	Labour hours	Labour £	Material £	Unit	Total rate £
Special glass; BS 952; "Pyran" half-hour fire resisting glass or other equal and approved 6.50 mm thick rectangular panes; glazed with screwed hardwood beads and Sealmaster "Fireglaze" intumescent compound or other equal and approved to rebated frame						
300 mm x 400 mm pane	-	0.42	6.90	31.82	nr	38.72
400 mm x 800 mm pane	-	0.51	8.38	79.60	nr	87.98
500 mm x 1400 mm pane	-	0.83	13.65	168.72	nr	182.37
600 mm x 1800 mm pane	-	1.02	16.77	288.21	nr	304.98
Special glass; BS 952; "Pyrostop" one-hour fire resisting glass or other equal and approved 15 mm thick regular panes; glazed with screwed hardwood beads and Sealmaster "Fireglaze" intumescent liner and compound or other equal and approved both sides						
300 mm x 400 mm pane	-	1.20	19.73	72.35	nr	92.08
400 mm x 800 mm pane	-	1.53	25.15	141.30	nr	166.45
500 mm x 1400 mm pane	-	2.04	33.54	278.94	nr	312.48
600 mm x 1800 mm pane	-	2.54	41.76	406.81	nr	448.57
Special glass; BS 952; clear laminated security glass 7.50 mm thick regular panes; glazed with screwed hardwood beads and "Intergens" intumescent strip or other equal and approved						
300 mm x 400 mm pane	-	0.42	6.90	19.39	nr	26.29
400 mm x 800 mm pane	-	0.51	8.38	46.95	nr	55.33
500 mm x 1400 mm pane	-	0.83	13.65	97.44	nr	111.09
600 mm x 1800 mm pane	-	1.02	16.77	177.26	nr	194.03
Mirror panels; BS 952; silvered; insulation backing 4 mm thick float; fixing with adhesive						
1000 mm x 1000 mm	-	-	-	-	nr	161.57
1000 mm x 2000 mm	-	-	-	-	nr	302.96
1000 mm x 4000 mm	-	-	-	-	nr	585.71
Glass louvres; BS 952; with long edges ground or smooth 6 mm thick float						
150 mm wide	-	-	-	-	m	12.42
7 mm thick Georgian wired cast						
150 mm wide	-	-	-	-	m	17.17
6 mm thick Georgian wire polished						
150 mm wide	-	-	-	-	m	24.49
Factory made double hermetically sealed units; to wood or metal with screwed or clipped beads Two panes; BS 952; clear float glass; 4 mm thick; 6 mm air space						
2.00 m^2 - 4.00 m^2	-	-	-	-	m^2	55.55
1.00 m^2 - 2.00 m^2	-	-	-	-	m^2	55.55
0.75 m^2 - 1.00 m^2	-	-	-	-	m^2	55.55
0.50 m^2 - 0.75 m^2	-	-	-	-	m^2	55.55
0.35 m^2 - 0.50 m^2	-	-	-	-	m^2	55.55
0.25 m^2 - 0.35 m^2	-	-	-	-	m^2	55.55
not exceeding 0.25 m^2	-	-	-	-	nr	13.38

L WINDOWS/DOORS/STAIRS

	PC £	Labour hours	Labour £	Material £	Unit	Total rate £
Two panes; BS 952; clear float glass; 6 mm thick; 6 mm air space						
2.00 m² - 4.00 m²	-	-	-	-	m²	65.64
1.00 m² - 2.00 m²	-	-	-	-	m²	65.64
0.75 m² - 1.00 m²	-	-	-	-	m²	65.64
0.50 m² - 0.75 m²	-	-	-	-	m²	65.64
0.35 m² - 0.50 m²	-	-	-	-	m²	65.64
0.25 m² - 0.35 m²	-	-	-	-	m²	65.64
not exceeding 0.25 m²	-	-	-	-	nr	16.66

M SURFACE FINISHES

M10 CEMENT:SAND/CONCRETE SCREEDS/TOPPINGS

	PC £	Labour hours	Labour £	Material £	Unit	Total rate £
Cement:sand (1:3); steel trowelled						
Work to floors; one coat; level and to falls not exceeding 15 degrees from horizontal; to concrete base; over 300 mm wide						
32 mm thick	-	0.30	3.84	2.10	m^2	5.94
40 mm thick	-	0.31	3.97	2.62	m^2	6.59
48 mm thick	-	0.33	4.22	3.15	m^2	7.37
50 mm thick	-	0.34	4.35	3.28	m^2	7.63
60 mm thick	-	0.38	4.86	3.94	m^2	8.80
65 mm thick	-	0.41	5.25	4.26	m^2	9.51
70 mm thick	-	0.42	5.38	4.59	m^2	9.97
75 mm thick	-	0.43	5.50	4.92	m^2	10.42
Add to the above for work to falls and crossfalls and to slopes						
not exceeding 15 degrees from horizontal	-	0.02	0.26	-	m^2	0.26
over 15 degrees from horizontal	-	0.10	1.28	-	m^2	1.28
water repellent additive incorporated in the mix	-	0.02	0.26	0.60	m^2	0.86
oil repellent additive incorporated in the mix	-	0.08	1.02	4.30	m^2	5.32
Cement:sand (1:3); beds and backings						
Work to floors; one coat level; to concrete base; screeded; over 300 mm wide						
25 mm thick	-	-	-	-	m^2	8.80
50 mm thick	-	-	-	-	m^2	10.38
75 mm thick	-	-	-	-	m^2	13.54
100 mm thick	-	-	-	-	m^2	16.73
Work to floors; one coat; level; to concrete base; steel trowelled; over 300 mm wide						
25 mm thick	-	-	-	-	m^2	8.80
50 mm thick	-	-	-	-	m^2	10.38
75 mm thick	-	-	-	-	m^2	13.54
100 mm thick	-	-	-	-	m^2	16.73
Fine concrete (1:4); beds and backings						
Work to floors; one coat; level; to concrete base; steel trowelled; over 300 mm wide						
50 mm thick	-	-	-	-	m^2	10.13
75 mm thick	-	-	-	-	m^2	12.99
Granolithic paving; cement and granite chippings 5 mm to dust (1:1:2); steel trowelled						
Work to floors; one coat; level; laid on concrete while green; over 300 mm wide						
20 mm thick	-	-	-	-	m^2	15.90
25 mm thick	-	-	-	-	m^2	17.16
Work to floors; two coat; laid on hacked concrete with slurry; over 300 mm wide						
38 mm thick	-	-	-	-	m^2	20.47
50 mm thick	-	-	-	-	m^2	23.52
75 mm thick	-	-	-	-	m^2	31.79
Work to landings; one coat; level; laid on concrete while green; over 300 mm wide						
20 mm thick	-	-	-	-	m^2	17.80
25 mm thick	-	-	-	-	m^2	19.07

Prices for Measured Works - Minor Works
M SURFACE FINISHES

	PC £	Labour hours	Labour £	Material £	Unit	Total rate £
Work to landings; two coat; laid on hacked concrete with slurry; over 300 mm wide						
38 mm thick	-	-	-	-	m²	22.38
50 mm thick	-	-	-	-	m²	25.45
75 mm thick	-	-	-	-	m²	33.70
Add to the above over 300 mm wide for liquid hardening additive incorporated in the mix	-	0.05	0.64	0.70	m²	1.34
oil-repellent additive incorporated in the mix	-	0.08	1.02	4.30	m²	5.32
25 mm work to treads; one coat; to concrete base						
225 mm wide	-	0.92	15.12	8.24	m	23.36
275 mm wide	-	0.92	15.12	9.23	m	24.35
returned end	-	0.19	3.12	-	nr	3.12
13 mm skirtings; rounded top edge and coved bottom junction; to brickwork or blockwork base						
75 mm wide on face	-	0.56	9.21	0.40	m	9.61
150 mm wide on face	-	0.77	12.66	7.25	m	19.91
ends; fair	-	0.05	0.82	-	nr	0.82
angles	-	0.07	1.15	-	nr	1.15
13 mm outer margin to stairs; to follow profile of and with rounded nosing to treads and risers; fair edge and arris at bottom, to concrete base						
75 mm wide	-	0.92	15.12	3.96	m	19.08
angles	-	0.07	1.15	-	nr	1.15
13 mm wall string to stairs; fair edge and arris on top; coved bottom junction with treads and risers; to brickwork or blockwork base						
275 mm (extreme) wide	-	0.81	13.32	6.92	m	20.24
ends	-	0.05	0.82	-	nr	0.82
angles	-	0.07	1.15	-	nr	1.15
ramps	-	0.08	1.32	-	nr	1.32
ramped and wreathed corners	-	0.10	1.64	-	nr	1.64
13 mm outer string to stairs; rounded nosing on top at junction with treads and risers; fair edge and arris at bottom; to concrete base						
300 mm (extreme) wide	-	0.81	13.32	8.57	m	21.89
ends	-	0.05	0.82	-	nr	0.82
angles	-	0.07	1.15	-	nr	1.15
ramps	-	0.08	1.32	-	nr	1.32
ramps and wreathed corners	-	0.10	1.64	-	nr	1.64
19 mm thick skirtings; rounded top edge and coved bottom junction; to brickwork or blockwork base						
75 mm wide on face	-	0.56	9.21	7.25	m	16.46
150 mm wide on face	-	0.77	12.66	11.21	m	23.87
ends; fair	-	0.05	0.64	-	nr	0.64
angles	-	0.07	1.15	-	nr	1.15
19 mm risers; one rounded nosing; to concrete base						
150 mm high; plain	-	0.92	15.12	6.26	m	21.38
150 mm high; undercut	-	0.92	15.12	6.26	m	21.38
180 mm high; plain	-	0.92	15.12	8.57	m	23.69
180 mm high; undercut	-	0.92	15.12	8.57	m	23.69

Prices for Measured Works - Minor Works
M SURFACE FINISHES

	PC £	Labour hours	Labour £	Material £	Unit	Total rate £
M11 MASTIC ASPHALT FLOORING/FLOOR UNDERLAYS						
Mastic asphalt flooring to BS 6925 Type F 1076; black						
20 mm thick; one coat coverings; felt isolating membrane; to concrete base; flat						
over 300 mm wide	-	-	-	-	m^2	16.18
225 mm - 300 mm wide	-	-	-	-	m^2	30.11
150 mm - 225 mm wide	-	-	-	-	m^2	33.03
not exceeding 150 mm wide	-	-	-	-	m^2	40.44
25 mm thick; one coat coverings; felt isolating membrane; to concrete base; flat						
over 300 mm wide	-	-	-	-	m^2	18.80
225 mm - 300 mm wide	-	-	-	-	m^2	32.06
150 mm - 225 mm wide	-	-	-	-	m^2	34.98
not exceeding 150 mm wide	-	-	-	-	m^2	42.40
20 mm thick; three coat skirtings to brickwork base						
not exceeding 150 mm girth	-	-	-	-	m	16.18
150 mm - 225 mm girth	-	-	-	-	m	18.80
225 mm - 300 mm girth	-	-	-	-	m	21.27
Mastic asphalt flooring; acid-resisting; black						
20 mm thick; one coat coverings; felt isolating membrane; to concrete base flat						
over 300 mm wide	-	-	-	-	m^2	18.95
225 mm - 300 mm wide	-	-	-	-	m^2	34.68
150 mm - 225 mm wide	-	-	-	-	m^2	35.80
not exceeding 150 mm wide	-	-	-	-	m^2	43.22
25 mm thick; one coat coverings; felt isolating membrane; to concrete base; flat						
over 300 mm wide	-	-	-	-	m^2	22.39
225 mm - 300 mm wide	-	-	-	-	m^2	35.66
150 mm - 225 mm wide	-	-	-	-	m^2	38.57
not exceeding 150 mm wide	-	-	-	-	m^2	45.99
20 mm thick; three coat skirtings to brickwork base						
not exceeding 150 mm girth	-	-	-	-	m	16.71
150 mm - 225 mm girth	-	-	-	-	m	19.48
225 mm - 300 mm girth	-	-	-	-	m	22.09
Mastic asphalt flooring to BS 6925 Type F 1451; red						
20 mm thick; one coat coverings; felt isolating membrane; to concrete base; flat						
over 300 mm wide	-	-	-	-	m^2	26.51
225 mm - 300 mm wide	-	-	-	-	m^2	43.82
150 mm - 225 mm wide	-	-	-	-	m^2	47.33
not exceeding 150 mm wide	-	-	-	-	m^2	56.62
20 mm thick; three coat skirtings to brickwork base						
not exceeding 150 mm girth	-	-	-	-	m	20.82
150 mm - 225 mm girth	-	-	-	-	m	26.51

M SURFACE FINISHES

	PC £	Labour hours	Labour £	Material £	Unit	Total rate £
M12 TROWELLED BITUMEN/RESIN/RUBBER LATEX FLOORING						
Latex cement floor screeds; steel trowelled						
Work to floors; level; to concrete base; over 300 mm wide						
3 mm thick; one coat	-	-	-	-	m²	4.06
5 mm thick; two coats	-	-	-	-	m²	5.76
Isocrete K screeds or other equal and approved; steel trowelled						
Work to floors; level; to concrete base; over 300 mm wide						
35 mm thick; plus polymer bonder coat	-	-	-	-	m²	13.48
40 mm thick	-	-	-	-	m²	11.90
45 mm thick	-	-	-	-	m²	12.19
50 mm thick	-	-	-	-	m²	12.53
Work to floors; to falls or crossfalls; to concrete base; over 300 mm wide						
55 mm (average) thick	-	-	-	-	m²	13.81
60 mm (average) thick	-	-	-	-	m²	14.45
65 mm (average) thick	-	-	-	-	m²	15.07
75 mm (average) thick	-	-	-	-	m²	16.35
90 mm (average) thick	-	-	-	-	m²	18.25
Bituminous lightweight insulating roof screeds						
"Bit-Ag" or similar roof screed or other equal and approved; to falls or crossfalls; bitumen felt vapour barrier; over 300 mm wide						
75 mm (average) thick	-	-	-	-	m²	42.55
100 mm (average) thick	-	-	-	-	m²	53.93
M20 PLASTERED/RENDERED/ROUGHCAST COATINGS						
Prepare and brush down 2 coats of "Unibond" bonding agent or other equal and approved						
Brick or block walls						
over 300 mm wide	-	0.16	2.05	0.87	m²	2.92
Concrete walls or ceilings						
over 300 mm wide	-	0.12	1.54	0.69	m²	2.23
Cement.sand (1:3) beds and backings						
10 mm thick work to walls; one coat; to brickwork or blockwork base						
over 300 mm wide	-	-	-	-	m²	9.84
not exceeding 300 mm wide	-	-	-	-	m	4.89
13 mm thick; work to walls; two coats; to brickwork or blockwork base						
over 300 mm wide	-	-	-	-	m²	11.76
not exceeding 300 mm wide	-	-	-	-	m	5.88
15 mm thick work to walls; two coats; to brickwork or blockwork base						
over 300 mm wide	-	-	-	-	m²	12.70
not exceeding 300 mm wide	-	-	-	-	m	6.34
Cement:sand (1:3); steel trowelled						
13 mm thick work to walls; two coats; to brickwork or blockwork base						
over 300 mm wide	-	-	-	-	m²	11.43
not exceeding 300 mm wide	-	-	-	-	m	5.72

Prices for Measured Works - Minor Works
M SURFACE FINISHES

	PC £	Labour hours	Labour £	Material £	Unit	Total rate £
M20 PLASTERED/RENDERED/ROUGHCAST COATINGS - cont'd						
Cement:sand (1:3); steel trowelled - cont'd						
16 mm thick work to walls; two coats; to brickwork or blockwork base						
over 300 mm wide	-	-	-	-	m²	12.80
not exceeding 300 mm wide	-	-	-	-	m	6.42
19 mm thick work to walls; two coats; to brickwork or blockwork base						
over 300 mm wide	-	-	-	-	m²	14.18
not exceeding 300 mm wide	-	-	-	-	m	7.09
ADD to above						
over 300 mm wide in water repellant cement	-	-	-	-	m²	2.90
finishing coat in colour cement	-	-	-	-	m²	6.69
Cement:lime:sand (1:2:9); steel trowelled						
19 mm thick work to walls; two coats; to brickwork or blockwork base						
over 300 mm wide	-	-	-	-	m²	14.34
not exceeding 300 mm wide	-	-	-	-	m	7.17
Cement:lime:sand (1:1:6); steel trowelled						
13 mm thick work to walls; two coats; to brickwork or blockwork base						
over 300 mm wide	-	-	-	-	m²	11.61
not exceeding 300 mm wide	-	-	-	-	m	5.80
Add to the above over 300 wide for						
waterproof additive	-	-	-	-	m²	1.99
19 mm thick work to ceilings; three coats; to metal lathing base						
over 300 mm wide	-	-	-	-	m²	17.60
not exceeding 300 mm wide	-	-	-	-	m	9.07
Plaster; first and finishing coats of "Carlite" pre-mixed lightweight plaster or other equal and approved; steel trowelled						
13 mm thick work to walls; two coats; to brickwork or blockwork base (or 10 mm thick work to concrete base)						
over 300 mm wide	-	-	-	-	m²	8.34
over 300 mm wide; in staircase areas or plant rooms	-	-	-	-	m²	9.63
not exceeding 300 mm wide	-	-	-	-	m	4.16
13 mm thick work to isolated piers or columns; two coats						
over 300 mm wide	-	-	-	-	m²	10.27
not exceeding 300 mm wide	-	-	-	-	m	5.13
10 mm thick work to ceilings; two coats; to concrete base						
over 300 mm wide	-	-	-	-	m²	8.93
over 300 mm wide; 3.50 m - 5.00 m high	-	-	-	-	m²	9.57
over 300 mm wide; in staircase areas or plant rooms	-	-	-	-	m²	10.20
not exceeding 300 mm wide	-	-	-	-	m	4.46
10 mm thick work to isolated beams; two coats; to concrete base						
over 300 mm wide	-	-	-	-	m²	10.86
over 300 mm wide; 3.50 m - 5.00 m high	-	-	-	-	m²	11.49
not exceeding 300 mm wide	-	-	-	-	m	5.42

M SURFACE FINISHES

	PC £	Labour hours	Labour £	Material £	Unit	Total rate £
Plaster; first coat of "Thistle Hardwall" plaster or other equal and approved; finishing coat of "Thistle Multi Finish" plaster or other equal and approved; steel trowelled						
13 mm thick work to walls; two coats; to brickwork or blockwork base						
over 300 mm wide	-	-	-	-	m²	9.09
over 300 mm wide; in staircase areas or plant rooms	-	-	-	-	m²	10.44
not exceeding 300 mm wide	-	-	-	-	m	4.56
13 mm thick work to isolated columns; two coats						
over 300 mm wide	-	-	-	-	m²	11.10
not exceeding 300 mm wide	-	-	-	-	m	5.56
Plaster; one coat "Snowplast" plaster or other equal and approved; steel trowelled						
13 mm thick work to walls; one coat; to brickwork or blockwork base						
over 300 mm wide	-	-	-	-	m²	9.70
over 300 mm wide; in staircase areas or plant rooms	-	-	-	-	m²	11.04
not exceeding 300 mm wide	-	-	-	-	m	4.85
13 mm thick work to isolated columns; one coat						
over 300 mm wide	-	-	-	-	m²	11.71
not exceeding 300 mm wide	-	-	-	-	m	5.87
Plaster; first coat of cement:sand (1:3); finishing coat of "Thistle" class B plaster or other equal and approved; steel trowelled						
13 mm thick; work to walls; two coats; to brickwork or blockwork base						
over 300 mm wide	-	-	-	-	m²	9.29
over 300 mm wide; in staircase areas or plant rooms	-	-	-	-	m²	10.62
not exceeding 300 mm wide	-	-	-	-	m	4.65
13 mm thick work to isolated columns; two coats						
over 300 mm wide	-	-	-	-	m²	11.96
not exceeding 300 mm wide	-	-	-	-	m	5.99
Plaster; first coat of cement:lime:sand (1:1:6); finishing coat of "Multi Finish" plaster or other equal and approved; steel trowelled						
13 mm thick work to walls; two coats; to brickwork or blockwork base						
over 300 mm wide	-	-	-	-	m²	9.54
over 300 mm wide; in staircase areas or plant rooms	-	-	-	-	m²	10.88
not exceeding 300 mm wide	-	-	-	-	m	4.56
13 mm thick work to isolated columns; two coats						
over 300 mm wide	-	-	-	-	m²	11.10
not exceeding 300 mm wide	-	-	-	-	m	5.56
Plaster; first coat of "Limelite" renovating plaster or other equal and approved; finishing coat of "Limelite" renovating plaster or other equal and approved; finishing coat of "Limelite" finishing plaster or other equal and approved; steel trowelled						
13 mm thick work to walls; two coats; to brickwork or blockwork base						
over 300 mm wide	-	-	-	-	m²	13.39
over 300 mm wide; in staircase areas or plant rooms	-	-	-	-	m²	14.74
not exceeding 300 mm wide	-	-	-	-	m	6.71

M SURFACE FINISHES

	PC £	Labour hours	Labour £	Material £	Unit	Total rate £
M20 PLASTERED/RENDERED/ROUGHCAST COATINGS - cont'd						
Plaster; first coat of "Limelite" renovating plaster or other equal and approved; finishing coat of "Limelite" renovating plaster or other equal and approved; finishing coat of "Limelite" finishing plaster or other equal and approved; steel trowelled - cont'd						
Dubbing out existing walls with undercoat plaster; average 6 mm thick						
over 300 mm wide	-	-	-	-	m^2	4.02
not exceeding 300 mm wide	-	-	-	-	m	2.02
Dubbing out existing walls with undercoat plaster; average 12 mm thick						
over 300 mm wide	-	-	-	-	m^2	8.05
not exceeding 300 mm wide	-	-	-	-	m	4.02
Plaster; first coat of "Thistle X-ray" plaster or other equal and approved; finishing coat of "Thistle X-ray" finishing plaster or other equal and approved; steel trowelled						
17 mm thick work to walls; two coats; to brickwork or blockwork base						
over 300 mm wide	-	-	-	-	m^2	43.57
over 300 mm wide; in staircase areas or plant rooms	-	-	-	-	m^2	44.90
not exceeding 300 mm wide	-	-	-	-	m	17.43
17 mm thick work to isolated columns; two coats						
over 300 mm wide	-	-	-	-	m^2	47.58
not exceeding 300 mm wide	-	-	-	-	m	19.02
Plaster, one coat "Thistle" projection plaster or other equal and approved; steel trowelled						
13 mm thick work to walls; one coat; to brickwork or blockwork base						
over 300 mm wide	-	-	-	-	m^2	9.35
over 300 mm wide; in staircase areas or plant rooms	-	-	-	-	m^2	10.69
not exceeding 300 mm wide	-	-	-	-	m	4.68
10 mm thick work to isolated columns; one coat						
over 300 mm wide	-	-	-	-	m^2	11.36
not exceeding 300 mm wide	-	-	-	-	m	5.68
Plaster; first, second and finishing coats of "Carlite" pre-mixed lightweight plaster or other equal and approved; steel trowelled						
13 mm thick work to ceilings; three coats to metal lathing base						
over 300 mm wide	-	-	-	-	m^2	12.08
over 300 mm wide; in staircase areas or plant rooms	-	-	-	-	m^2	13.43
not exceeding 300 mm wide	-	-	-	-	m	6.03
13 mm thick work to swept soffit of metal lathing arch former						
not exceeding 300 mm wide	-	-	-	-	m	8.05
300 mm - 400 mm wide	-	-	-	-	m	10.74
13 mm thick work to vertical face of metal lathing arch former						
not exceeding 0.50 m^2 per side	-	-	-	-	nr	11.43
0.50 m^2 - 1.00 m^2 per side	-	-	-	-	nr	17.12

M SURFACE FINISHES

	PC £	Labour hours	Labour £	Material £	Unit	Total rate £
"Tyrolean" decorative rendering or other equal and approved; 13 mm thick first coat of cement:lime:sand (1:1:6); finishing three coats of "Cullamix" or other equal and approved applied with approved hand operated machine external						
To walls; four coats; to brickwork or blockwork base						
over 300 mm wide	-	-	-	-	m²	21.45
not exceeding 300 mm wide	-	-	-	-	m	10.72
Drydash (pebbledash) finish of Derbyshire Spar chippings or other equal and approved on and including cement:lime:sand (1:2:9) backing 18 mm thick work to walls; two coats; to brickwork or blockwork base						
over 300 mm wide	-	-	-	-	m²	19.73
not exceeding 300 mm wide	-	-	-	-	m	9.87
Plaster; one coat "Thistle" board finish or other equal and approved; steel trowelled (prices included within plasterboard rates) 3 mm thick work to walls or ceilings; one coat; to plasterboard base						
over 300 mm wide	-	-	-	-	m²	4.75
over 300 mm wide; in staircase areas or plant rooms	-	-	-	-	m²	5.43
not exceeding 300 mm wide	-	-	-	-	m	2.84
Plaster; one coat "Thistle" board finish or other equal and approved; steel trowelled 3 mm work to walls or ceilings; one coat on and including gypsum plasterboard; BS 1230; fixing with nails; 3 mm joints filled with plaster and jute scrim cloth; to softwood base; plain grade baseboard or lath with rounded edges						
9.50 mm thick boards to walls						
over 300 mm wide	-	1.07	9.27	3.12	m²	12.39
not exceeding 300 mm wide	-	0.42	3.84	0.57	m	4.41
9.50 mm thick boards to walls; in staircase areas or plant rooms						
over 300 mm wide	-	1.17	10.18	3.12	m²	13.30
not exceeding 300 mm wide	-	0.52	4.76	0.57	m	5.33
9.50 mm thick boards to isolated columns						
over 300 mm wide	-	1.17	10.18	3.12	m²	13.30
not exceeding 300 mm wide	-	0.62	5.67	0.57	m	6.24
9.50 mm thick boards to ceilings						
over 300 mm wide	-	0.99	8.54	3.12	m²	11.66
over 300 mm wide; 3.50 m - 5.00 m high	-	1.14	9.91	3.12	m²	13.03
not exceeding 300 mm wide	-	0.47	4.30	0.57	m	4.87
9.50 mm thick boards to ceilings; in staircase areas or plant rooms						
over 300 mm wide	-	1.09	9.45	3.12	m²	12.57
not exceeding 300 mm wide	-	0.52	4.76	0.57	m	5.33
9.50 mm thick boards to isolated beams						
over 300 mm wide	-	1.16	10.09	3.12	m²	13.21
not exceeding 300 mm wide	-	0.56	5.12	0.57	m	5.69
add for "Duplex" insulating grade 12.50 mm thick boards or other equal and approved to walls	-	-	-	0.60	m²	0.60
12.50 mm thick boards to walls; in staircase areas or plant rooms						
over 300 mm wide	-	1.12	9.73	3.37	m²	13.10
not exceeding 300 mm wide	-	0.53	4.85	0.65	m	5.50

M20 PLASTERED/RENDERED/ROUGHCAST COATINGS - cont'd

	PC £	Labour hours	Labour £	Material £	Unit	Total rate £

Plaster; one coat "Thistle" board finish or other equal and approved; steel trowelled 3 mm work to walls or ceilings; one coat on and including gypsum plasterboard; BS 1230; fixing with nails; 3 mm joints filled with plaster and jute scrim cloth; to softwood base; plain grade baseboard or lath with rounded edges - cont'd

	PC £	Labour hours	Labour £	Material £	Unit	Total rate £
12.50 mm thick boards to isolated columns						
over 300 mm wide	-	1.16	10.09	3.37	m²	13.46
not exceeding 300 mm wide	-	0.56	5.12	0.65	m	5.77
12.50 mm thick boards to ceilings						
over 300 mm wide	-	1.05	9.09	3.37	m²	12.46
over 300 mm wide; 3.50 m - 5.00 m high	-	1.17	10.18	3.37	m²	13.55
not exceeding 300 mm wide	-	0.50	4.58	0.65	m	5.23
12.50 mm thick boards to ceilings; in staircase areas or plant rooms						
over 300 mm wide	-	1.17	10.18	3.37	m²	13.55
not exceeding 300 mm wide	-	0.56	5.12	0.65	m	5.77
12.50 mm thick boards to isolated beams						
over 300 wide	-	1.27	11.10	3.37	m²	14.47
not exceeding 300 mm wide	-	0.62	5.67	0.65	m	6.32
add for "Duplex" insulating grade 12.50 mm thick boards or other equal and approved to walls	-	-	-	0.60	m²	0.60

Accessories
"Expamet" render beads or other equal and approved; white PVC nosings; to brickwork or blockwork base

	PC £	Labour hours	Labour £	Material £	Unit	Total rate £
external stop bead; ref 573	-	0.08	0.73	2.38	m	3.11

"Expamet" render beads or other equal and approved; stainless steel; to brickwork or blockwork base

stop bead; ref 546	-	0.08	0.73	1.95	m	2.68
stop bead; ref 547	-	0.08	0.73	1.95	m	2.68

"Expamet" plaster beads or other equal and approved; galvanised steel; to brickwork or blockwork base

angle bead; ref 550	-	0.09	0.82	0.78	m	1.60
architrave bead; ref 579	-	0.11	1.01	1.20	m	2.21
stop bead; ref 562	-	0.08	0.73	0.91	m	1.64
stop beads; ref 563	-	0.08	0.73	0.91	m	1.64
movement bead; ref 588	-	0.10	0.92	4.28	m	5.20

"Expamet" plaster beads or other equal and approved; stainless steel; to brickwork or blockwork base

angle bead; ref 545	-	0.09	0.82	2.19	m	3.01
stop bead; ref 534	-	0.08	0.73	1.95	m	2.68
stop bead; ref 533	-	0.08	0.73	1.95	m	2.68

"Expamet" thin coat plaster beads or other equal and approved; galvanised steel; to timber base

angle bead; ref 553	-	0.08	0.73	0.77	m	1.50
angle bead; ref 554	-	0.08	0.73	1.07	m	1.80
stop bead; ref 560	-	0.07	0.64	0.98	m	1.62
stop bead; ref 561	-	0.07	0.64	1.06	m	1.70

M SURFACE FINISHES

	PC £	Labour hours	Labour £	Material £	Unit	Total rate £
M22 SPRAYED MINERAL FIBRE COATINGS						
Prepare and apply by spray "Mandolite CP2" fire protection or other equal and approved on structural steel/metalwork						
16 mm thick (one hour) fire protection						
to walls and columns	-	-	-	-	m²	8.99
to ceilings and beams	-	-	-	-	m²	9.99
to isolated metalwork	-	-	-	-	m²	19.67
22 mm thick (one and a half hour) fire protection						
to walls and columns	-	-	-	-	m²	10.34
to ceilings and beams	-	-	-	-	m²	11.48
to isolated metalwork	-	-	-	-	m²	23.07
28 mm thick (two hour) fire protection						
to walls and columns	-	-	-	-	m²	12.14
to ceilings and beams	-	-	-	-	m²	13.48
to isolated metalwork	-	-	-	-	m²	26.47
52 mm thick (four hour) fire protection						
to walls and columns	-	-	-	-	m²	18.58
to ceilings and beams	-	-	-	-	m²	20.47
to isolated metalwork	-	-	-	-	m²	40.95
Prepare and apply by spray; cementitious "Pyrok WF26" render or other equal and approved; on expanded metal lathing (not included)						
15 mm thick						
to ceilings and beams	-	-	-	-	m²	28.36
M30 METAL MESH LATHING/ANCHORED REINFORCEMENT FOR PLASTERED COATING						
Accessories						
Pre-formed galvanised expanded steel arch-frames; "Simpson Strong-Tie" or other equal and approved; semi-circular; to suit walls up to 230 mm thick						
375 mm radius; for 800 mm opening; ref SC 750	21.33	0.52	4.24	23.50	nr	27.74
425 mm radius; for 850 mm opening; ref SC 850	20.37	0.52	4.24	22.45	nr	26.69
450 mm radius; for 900 mm opening; ref SC 900	21.06	0.52	4.24	23.21	nr	27.45
600 mm radius; for 1200 mm opening; ref SC 1200	24.37	0.52	4.24	26.85	nr	31.09
"Newlath" damp free lathing or other equal and approved; plugging and screwing to background at 250 mm centres each way						
Linings to walls						
over 300 mm wide	-	1.28	16.67	9.86	m²	26.53
not exceeding 300 mm wide	-	0.41	5.34	3.04	m²	8.38
Lathing; Expamet "BB" expanded metal lathing or other and approved; BS 1369; 50 mm laps						
6 mm thick mesh linings to ceilings; fixing with staples; to softwood base; over 300 mm wide						
ref BB263; 0.500 mm thick	3.70	0.61	4.97	4.18	m²	9.15
ref BB264; 0.675 mm thick	4.32	0.61	4.97	4.88	m²	9.85
6 mm thick mesh linings to ceilings; fixing with wire; to steelwork; over 300 mm wide						
ref BB263; 0.500 mm thick	-	0.65	5.30	4.18	m²	9.48
ref BB264; 0.675 mm thick	-	0.65	5.30	4.88	m²	10.18

Prices for Measured Works - Minor Works
M SURFACE FINISHES

	PC £	Labour hours	Labour £	Material £	Unit	Total rate £
M30 METAL MESH LATHING/ANCHORED REINFORCEMENT FOR PLASTERED COATING - cont'd						
Lathing; Expamet "BB" expanded metal lathing or other and approved; BS 1369; 50 mm laps - cont'd						
6 mm thick mesh linings to ceilings; fixing with wire; to steelwork; not exceeding 300 mm wide						
ref BB263; 0.500 mm thick	-	0.41	3.34	4.18	m^2	7.52
ref BB264; 0.675 mm thick	-	0.41	3.34	4.88	m^2	8.22
raking cutting	-	0.20	1.83	-	m	1.83
cutting and fitting around pipes; not exceeding 0.30 m girth	-	0.31	2.84	-	nr	2.84
Lathing; Expamet "Riblath" or "Spraylath" stiffened expanded metal lathing or other equal and approved; 50 mm laps						
10 mm thick mesh lining to walls; fixing with nails; to softwood base; over 300 mm wide						
"Riblath" ref 269; 0.30 mm thick	4.64	0.52	4.24	5.33	m^2	9.57
"Riblath" ref 271; 0.50 mm thick	5.34	0.52	4.24	6.12	m^2	10.36
"Spraylath" ref 273; 0.50 mm thick	-	0.52	4.24	7.43	m^2	11.67
10 mm thick mesh lining to walls; fixing with nails; to softwood base; not exceeding 300 mm wide						
"Riblath" ref 269; 0.30 mm thick	-	0.31	2.52	1.63	m^2	4.15
"Riblath" ref 271; 0.50 mm thick	-	0.31	2.52	1.87	m^2	4.39
"Spraylath" ref 273; 0.50 mm thick	-	0.31	2.52	2.29	m^2	4.81
10 mm thick mesh lining to walls; fixing to brick or blockwork; over 300 mm wide						
"Red-rib" ref 274; 0.50 mm thick	5.89	0.41	3.34	7.26	m^2	10.60
stainless steel "Riblath" ref 267; 0.30 mm thick	11.69	0.41	3.34	13.81	m^2	17.15
10 mm thick mesh lining to ceilings; fixing with wire; to steelwork; over 300 mm wide						
"Riblath" ref 269; 0.30 mm thick	-	0.65	5.30	5.75	m^2	11.05
"Riblath" ref 271; 0.50 mm thick	-	0.65	5.30	6.55	m^2	11.85
"Spraylath" ref 273; 0.50 mm thick	-	0.65	5.30	7.85	m^2	13.15
M31 FIBROUS PLASTER						
Fibrous plaster; fixing with screws; plugging; countersinking; stopping; filling and pointing joints with plaster						
16 mm thick plain slab coverings to ceilings						
over 300 mm wide	-	-	-	-	m^2	115.35
not exceeding 300 mm wide	-	-	-	-	m	38.45
Coves; not exceeding 150 mm girth						
per 25 mm girth	-	-	-	-	m	5.49
Coves; 150 mm - 300 mm girth						
per 25 mm girth	-	-	-	-	m	6.70
Cornices						
per 25 mm girth	-	-	-	-	m	6.82
Cornice enrichments						
per 25 mm girth; depending on degree of enrichments	-	-	-	-	m	8.07
Fibrous plaster; fixing with plaster wadding filling and pointing joints with plaster; to steel base						
16 mm thick plain slab coverings to ceilings						
over 300 mm wide	-	-	-	-	m^2	115.35
not exceeding 300 mm wide	-	-	-	-	m	38.45
16 mm thick plain casings to stanchions						
per 25 mm girth	-	-	-	-	m	3.41

M SURFACE FINISHES

	PC £	Labour hours	Labour £	Material £	Unit	Total rate £
16 mm thick plain casings to beams						
per 25 mm girth	-	-	-	-	m	3.41
Gyproc cove or other equal and approved; fixing with adhesive; filling and pointing joints with plaster						
Coves						
125 mm girth	-	0.20	1.83	0.96	m	2.79
angles	-	0.03	0.27	0.60	nr	0.87
M40 STONE/CONCRETE/QUARRY/CERAMIC TILING/MOSAIC						
Clay floor quarries; BS 6431; class 1; Daniel Platt "Crown" tiles or other equal and approved; level bedding 10 mm thick and jointing in cement and sand (1:3); butt joints; straight both ways; flush pointing with grout; to cement and sand base						
Work to floors; over 300 mm wide						
150 mm x 150 mm x 12.50 mm thick; red	-	0.81	10.37	22.58	m^2	32.95
150 mm x 150 mm x 12.50 mm thick; brown	-	0.81	10.37	27.43	m^2	37.80
200 mm x 200 mm x 19 mm thick; brown	-	0.67	8.58	40.14	m^2	48.72
Works to floors; in staircase areas or plant rooms						
150 mm x 150 mm x 12.50 mm thick; red	-	0.92	11.78	22.58	m^2	34.36
150 mm x 150 mm x 12.50 mm thick; brown	-	0.92	11.78	27.43	m^2	39.21
200 mm x 200 mm x 19 mm thick; brown	-	0.77	9.86	40.14	m^2	50.00
Work to floors; not exceeding 300 mm wide						
150 mm x 150 mm x 12.50 mm thick; red	-	0.41	5.25	5.46	m^2	10.71
150 mm x 150 mm x 12.50 mm thick; brown	-	0.41	5.25	6.92	m^2	12.17
200 mm x 200 mm x 19 mm thick; brown	-	0.33	4.22	10.73	m^2	14.95
fair square cutting against flush edges of existing finishes	-	0.12	1.04	1.28	m	2.32
raking cutting	-	0.21	1.85	1.44	m	3.29
cutting around pipes; not exceeding 0.30 m girth	-	0.16	1.46	-	nr	1.46
extra for cutting and fitting into recessed manhole cover 600 mm x 600 mm	-	1.02	9.33	-	nr	9.33
Work to sills; 150 mm wide; rounded edge tiles						
150 mm x 150 mm x 12.50 mm thick; red	-	0.33	4.22	5.00	m	9.22
150 mm x 150 mm x 12.50 mm thick; brown	-	0.33	4.22	6.46	m	10.68
fitted end	-	0.16	1.46	-	m	1.46
Coved skirtings; 138 mm high; rounded top edge						
150 mm x 138 mm x 12.50 mm thick; red	-	0.26	3.33	5.53	m	8.86
150 mm x 138 mm x 12.50 mm thick; brown	-	0.26	3.33	5.74	m	9.07
ends	-	0.05	0.46	-	nr	0.46
angles	-	0.16	1.40	2.29	nr	3.76
Glazed ceramic wall tiles; BS 6431; fixing with adhesive; butt joints; straight both ways; flush pointing with white grout; to plaster base						
Work to walls; over 300 mm wide						
152 mm x 152 mm x 5.50 mm thick; white	11.14	0.61	10.03	16.82	m^2	26.85
152 mm x 152 mm x 5.50 mm thick; light colours	13.59	0.61	10.03	19.59	m^2	29.62
152 mm x 152 mm x 5.50 mm thick; dark colours	14.85	0.61	10.03	21.01	m^2	31.04
extra for RE or REX tile	-	-	-	6.45	m^2	6.45
200 mm x 100 mm x 6.50 mm thick; white and light colours	11.14	0.61	10.03	16.82	m^2	26.85
250 mm x 200 mm x 7 mm thick; white and light colours	12.07	0.61	10.03	17.87	m^2	27.90
Work to walls; in staircase areas or plant rooms						
152 mm x 152 mm x 5.50 mm thick; white	-	0.68	11.18	16.82	m^2	28.00

M SURFACE FINISHES

	PC £	Labour hours	Labour £	Material £	Unit	Total rate £
M40 STONE/CONCRETE/QUARRY/CERAMIC TILING/MOSAIC - cont'd						
Glazed ceramic wall tiles; BS 6431; fixing with adhesive; butt joints; straight both ways; flush pointing with white grout; to plaster base - cont'd						
Work to walls; not exceeding 300 mm wide						
152 mm x 152 mm x 5.50 mm thick; white	-	0.31	5.10	5.00	m	10.10
152 mm x 152 mm x 5.50 mm thick; light colours	-	0.31	5.10	7.85	m	12.95
152 mm x 152 mm x 5.50 mm thick; dark colours	-	0.31	5.10	8.28	m	13.38
200 mm x 100 mm x 6.50 mm thick; white and light colours	-	0.31	5.10	5.00	m	10.10
250 mm x 200 mm x 7 mm thick; white and light colours	-	0.26	4.27	5.31	m	9.58
cutting around pipes; not exceeding 0.30 m girth	-	0.10	0.92	-	nr	0.92
Work to sills; 150 mm wide; rounded edge tiles						
152 mm x 152 mm x 5.50 mm thick; white	-	0.26	4.27	2.50	m	6.77
fitted end	-	0.10	0.92	-	nr	0.92
198 mm x 64.50 mm x 6 mm thick wall tiles; fixing with adhesive; butt joints; straight both ways; flush pointing with white grout; to plaster base						
Work to walls						
over 300 mm wide	24.15	1.83	30.09	31.51	m²	61.60
not exceeding 300 mm wide	-	0.71	11.67	9.40	m	21.07
20 mm x 20 mm x 5.50 mm thick glazed mosaic wall tiles; fixing with adhesive; butt joints; straight both ways; flush pointing with white grout; to plaster base						
Work to walls						
over 300 mm wide	29.64	1.93	31.73	38.46	m²	70.19
not exceeding 300 mm wide	-	0.77	12.66	13.41	m	26.07
50 mm x 50 mm x 5 mm thick slip resistant mosaic floor tiles, Series 2 or other equal and approved; fixing with adhesive; butt joints; straight both ways; flush pointing with white grout; to cement:sand base						
Work to floors						
over 300 mm wide	28.33	1.93	24.70	37.85	m²	62.55
not exceeding 300 mm wide	-	0.77	9.86	12.97	m	22.83
Riven Welsh slate floor tiles; level; bedding 10 mm thick and jointing in cement and sand (1:3); butt joints; straight both ways; flush pointing with coloured mortar; to cement:sand base						
Work to floors; over 300 mm wide						
250 mm x 250 mm x 12 mm - 15 mm thick	-	0.61	10.03	34.97	m²	45.00
Work to floors; not exceeding 300 mm wide						
250 mm x 250 mm x 12 mm - 15 mm thick	-	0.31	5.10	10.56	m	15.66

M SURFACE FINISHES

	PC £	Labour hours	Labour £	Material £	Unit	Total rate £
M41 TERRAZZO TILING/ IN SITU TERRAZZO						
Terrazzo tiles BS 4131; aggregate size random ground grouted and polished to 80's grit finish; standard colour range; 3 mm joints symmetrical layout; bedding in 42 mm cement semi-dry mix (1:4); grouting with neat matching cement						
300 mm x 300 mm x 28 mm (nominal), "Quil-Terra Terrazzo" tile units or other equal and approved, Quilligotti Contracts Ltd; hydraulically pressed, mechanically vibrated, steam cured; to floors on concrete base (not included); sealed with "Quil-Shield" penetrating case hardener or other equal and approved; 2 coats applied immediately after final polishing						
plain; laid level	-	-	-	-	m^2	33.06
plain; to slopes exceeding 15 degrees from horizontal	-	-	-	-	m^2	37.75
to small areas/toilets	-	-	-	-	m^2	73.40
Accessories						
plastic division strips 6 mm x 38 mm; set into floor tiling above crack inducing joints; to the nearest full tile module	-	-	-	-	m	2.25
M42 WOOD BLOCK/COMPOSITION BLOCK/PARQUET FLOORING						
Wood block; Vigers, Stevens & Adams "Vigerflex" or other equal and approved; 7.50 mm thick; level; fixing with adhesive; to cement:sand base						
Work to floors; over 300 mm wide						
"Maple 7"	-	0.51	6.53	18.16	m^2	24.69
Wood blocks 25 mm thick; tongued and grooved joints; herringbone pattern; level; fixing with adhesive; to cement:sand base						
Work to floors; over 300 mm wide						
iroko	-	-	-	-	m^2	40.95
"Maple 7" or other equal and approved	-	-	-	-	m^2	44.94
french oak	-	-	-	-	m^2	59.92
american oak	-	-	-	-	m^2	54.92
fair square cutting against flush edges of existing finishings	-	-	-	-	m	5.40
extra for cutting and fitting into recessed duct covers 450 mm wide; lining up with adjoining work	-	-	-	-	m	13.38
cutting around pipes; not exceeding 0.30 m girth	-	-	-	-	nr	2.30
extra for cutting and fitting into recessed manhole covers 600 mm x 600 mm; lining up with adjoining work	-	-	-	-	nr	9.78
Add to wood block flooring over 300 mm wide for						
sanding; one coat sealer; one coat wax polish	-	-	-	-	m^2	4.40
sanding; two coats sealer; buffing with steel wool	-	-	-	-	m^2	4.29
sanding; three coats polyurethane lacquer; buffing down between coats	-	-	-	-	m^2	6.64

M SURFACE FINISHES

	PC £	Labour hours	Labour £	Material £	Unit	Total rate £
M50 RUBBER/PLASTICS/CORK/LINO/CARPET TILING/SHEETING						
Linoleum sheet; BS 6826; Forbo-Nairn floors or other equal and approved; level; fixing with adhesive; butt joints; to cement:sand base						
Work to floors; over 300 mm wide						
2.50 mm thick; plain	-	0.41	5.25	12.66	m^2	**17.91**
3.20 mm thick; marbled	-	0.41	5.25	13.41	m^2	**18.66**
Linoleum sheet; "Marmoleum Real" or other equal and approved; level; with welded seams; fixing with adhesive; to cement:sand base						
Work to floors; over 300 mm wide						
2.50 mm thick	-	0.51	6.53	10.84	m^2	**17.37**
Vinyl sheet; Altro "Safety" range or other equal and approved; with welded seams; level; fixing with adhesive; to cement:sand base						
Work to floors; over 300 mm wide						
2.00 mm thick; "Marine T20"	-	0.61	7.81	15.94	m^2	**23.75**
2.50 mm thick; "Classic D25"	-	0.71	9.09	18.05	m^2	**27.14**
3.50 mm thick; "Stronghold"	-	0.81	10.37	23.69	m^2	**34.06**
Slip resistant vinyl sheet; Forbo-Nairn "Surestep" or other equal and approved; level with welded seams; fixing with adhesive; to cement:sand base						
Work to floors; over 300 mm wide						
2.00 mm thick	-	0.51	6.53	13.36	m^2	**19.89**
Vinyl sheet; heavy duty; Marley "HD" or other equal and approved; level; with welded seams; fixing with adhesive; level; to cement:sand base						
Work to floors; over 300 mm wide						
2.00 mm thick	-	0.45	5.76	10.75	m^2	**16.51**
2.50 mm thick	-	0.51	6.53	12.05	m^2	**18.58**
2.00 mm thick skirtings						
100 high	-	0.12	1.54	1.91	m	**3.45**
Vinyl sheet; "Gerflex" standard sheet; "Classic" range or other equal and approved; level; with welded seams; fixing with adhesive; to cement:sand base						
Work to floors; over 300 mm wide						
2.00 mm thick	-	0.51	6.53	7.59	m^2	**14.12**
Vinyl sheet; "Armstrong Rhino Contract" or other equal and approved; level; with welded seams; fixing with adhesive; to cement:sand base						
Work to floors; over 300 mm wide						
2.50 mm thick	-	0.51	6.53	8.29	m^2	**14.82**
Vinyl tiles; "Accoflex" or other equal and approved; level; fixing with adhesive; butt joints; straight both ways; to cement:sand base						
Work to floors; over 300 mm wide						
300 mm x 300 mm x 2.00 mm thick	-	0.26	3.33	4.95	m^2	**8.28**

M SURFACE FINISHES

	PC £	Labour hours	Labour £	Material £	Unit	Total rate £
Vinyl semi-flexible tiles; "Arlon" or other equal and approved; level; fixing with adhesive; butt joints; straight both ways; to cement:sand base						
Work to floors; over 300 mm wide						
250 mm x 250 mm x 2.00 mm thick	-	0.26	3.33	5.02	m²	8.35
Vinyl semi-flexible tiles; Marley "Marleyflex" or other equal and approved; level; fixing with adhesive; butt joints; straight both ways; to cement:sand base						
Work to floors; over 300 mm wide						
300 mm x 300 mm x 2.00 mm thick	-	0.26	3.33	6.16	m²	9.49
300 mm x 300 mm x 2.50 mm thick	-	0.26	3.33	7.42	m²	10.75
Vinyl semi-flexible tiles; "Vylon" or other equal and approved; level; fixing with adhesive; butt joints; straight both ways; to cement:sand base						
Work to floors; over 300 mm wide						
250 mm x 250 mm x 2.00 mm thick	-	0.29	3.71	6.17	m²	9.88
Vinyl tiles; anti-static; level; fixing with adhesive; butt joints; straight both ways; to cement:sand base						
Work to floors; over 300 mm wide						
457 mm x 457 mm x 2.00 mm thick	-	0.45	5.76	10.47	m²	16.23
Vinyl tiles; "Polyflex" or other equal and approved; level; fixing with adhesive; butt joints; straight both ways; to cement:sand base						
Work to floors; over 300 mm wide						
300 mm x 300 mm x 1.50 mm thick	-	0.26	3.33	4.35	m²	7.68
300 mm x 300 mm x 2.00 mm thick	-	0.26	3.33	4.77	m²	8.10
Vinyl tiles; "Polyflor XL" or other equal and approved; level; fixing with adhesive; butt joints; straight both ways; to cement:sand base						
Work to floors; over 300 mm wide						
300 mm x 300 mm x 2.00 mm thick	-	0.36	4.61	6.25	m²	10.86
Vinyl tiles; Marley "HD" or other equal and approved; level; fixing with adhesive; butt joints; straight both ways; to cement:sand base						
Work to floors; over 300 mm wide						
300 mm x 300 mm x 2.00 mm thick	-	0.36	4.61	10.75	m²	15.36
Thermoplastic tiles; Marley "Marleyflex" or other equal and approved; level; fixing with adhesive; butt joints; straight both ways; to cement:sand base						
Work to floors; over 300 mm wide						
300 mm x 300 mm x 2.00 mm thick; series 2	-	0.23	2.94	5.05	m²	7.99
300 mm x 300 mm x 2.00 mm thick; series 4	-	0.23	2.94	5.83	m²	8.77
Linoleum tiles; BS 6826; Forbo-Nairn Floors or other equal and approved; level; fixing with adhesive; butt joints; straight both ways; to cement:sand base						
Work to floors; over 300 mm wide						
2.50 mm thick (marble pattern)	-	0.31	3.97	13.31	m²	17.28

Prices for Measured Works - Minor Works
M SURFACE FINISHES

	PC £	Labour hours	Labour £	Material £	Unit	Total rate £
M50 RUBBER/PLASTICS/CORK/LINO/CARPET TILING/SHEETING - cont'd						
Cork tiles Wicanders "Cork-Master" or other equal and approved; level; fixing with adhesive; butt joints; straight both ways; to cement:sand base						
Work to floors; over 300 mm wide						
300 mm x 300 mm x 4.00 mm thick	-	0.41	5.25	20.78	m²	26.03
Rubber studded tiles; Altro "Mondopave" or other equal and approved; level; fixing with adhesive; butt joints; straight to cement:sand base						
Work to floors; over 300 mm wide						
500 mm x 500 mm x 2.50 mm thick; type MRB; black	-	0.61	7.81	26.16	m²	33.97
500 mm x 500 mm x 4.00 mm thick; type MRB; black	-	0.61	7.81	24.98	m²	32.79
Work to landings; over 300 mm wide						
500 mm x 500 mm x 4.00 mm thick; type MRB; black	-	0.81	10.37	24.98	m²	35.35
4.00 mm thick to treads						
275 mm wide	-	0.51	6.53	7.43	m	13.96
4.00 mm thick to risers						
180 mm wide	-	0.61	7.81	5.25	m	13.06
Sundry floor sheeting underlays						
For floor finishings; over 300 mm wide						
building paper to BS 1521; class A; 75 mm lap (laying only)	-	0.06	0.44	-	m²	0.44
3.20 mm thick hardboard	-	0.20	3.29	1.29	m²	4.58
6.00 mm thick plywood	-	0.31	5.10	6.57	m²	11.67
Stair nosings						
Light duty hard aluminium alloy stair tread nosings; plugged and screwed in concrete						
57 mm x 32 mm	7.52	0.26	2.38	8.40	m	10.78
84 mm x 32 mm	10.41	0.31	2.84	11.58	m	14.42
Heavy duty aluminium alloy stair tread nosings; plugged and screwed to concrete						
60 mm x 32 mm	8.95	0.31	2.84	9.97	m	12.81
92 mm x 32 mm	12.26	0.36	3.29	13.62	m	16.91
Heavy duty carpet tiles; Heuga "580 Olympic" or other and approved; to cement:sand base						
Work to floors						
over 300 mm wide	19.04	0.31	3.97	21.49	m²	25.46
Nylon needlepunch carpet; "Marleytex" or other equal and approved; fixing; with adhesive; level; to cement						
Work to floors						
over 300 mm wide	-	0.26	3.33	7.40	m²	10.73
M51 EDGE FIXED CARPETING						
Fitted carpeting; Wilton wool/nylon or other equal and approved; 80/20 velvet pile; heavy domestic plain; PC £41.31/m²						
Work to floors						
over 300 mm wide	-	0.42	3.45	44.40	m²	47.85
Work to treads and risers						
over 300 mm wide	-	0.83	6.82	44.40	m²	51.22

M SURFACE FINISHES

	PC £	Labour hours	Labour £	Material £	Unit	Total rate £
Underlay to carpeting; PC £3.93/m²						
Work to floors						
over 300 mm wide	-	0.08	0.66	4.13	m²	4.79
Sundries						
Carpet gripper fixed to floor; standard edging						
22 mm wide	-	0.05	0.36	0.21	m	0.57
M52 DECORATIVE PAPERS/FABRICS						
Lining paper; PC £2.19/roll; and hanging						
Plaster walls or columns						
over 300 mm girth	-	0.20	1.83	0.47	m²	2.30
Plaster ceilings or beams						
over 300 mm girth	-	0.24	2.20	0.47	m²	2.67
Decorative vinyl wallpaper; PC £10.99/roll; and hanging						
Plaster walls or columns						
over 300 mm girth	-	0.26	2.38	2.16	m²	4.54
M60 PAINTING/CLEAR FINISHING						
PREPARATION OF EXISTING SURFACES - INTERNALLY						
NOTE: The prices for preparation given hereunder assume that existing surfaces are in fair condition and should be increased for badly dilapidated surfaces.						
Wash down walls; cut out and make good cracks						
Emulsion painted surfaces; including bringing forward bare patches	-	0.07	0.64	-	m²	0.64
Gloss painted surfaces	-	0.05	0.46	-	m²	0.46
Wash down ceilings; cut out and make good cracks						
Distempered surfaces	-	0.09	0.82	-	m²	0.82
Emulsion painted surfaces; including bringing forward bare patches	-	0.10	0.92	-	m²	0.92
Gloss painted surfaces	-	0.09	0.82	-	m²	0.82
Wash down plaster cornices; cut out and make good cracks						
Distempered surfaces	-	0.13	1.19	-	m²	1.19
Emulsion painted surfaces; including bringing forward bare patches	-	0.16	1.46	-	m²	1.46
Wash and rub down iron and steel surfaces; bringing forward						
General surfaces						
over 300 mm girth	-	0.12	1.10	-	m²	1.10
isolated surfaces not exceeding 300 mm girth	-	0.05	0.46	-	m²	0.46
isolated areas not exceeding 0.50 m² irrespective of girth	-	0.09	0.82	-	nr	0.82
Glazed windows and screens						
panes; area not exceeding 0.10 m²	-	0.20	1.83	-	m²	1.83
panes; area 0.10 - 0.50 m²	-	0.16	1.46	-	m²	1.46
panes; area 0.50 - 1.00 m²	-	0.13	1.19	-	m²	1.19
panes; area over 1.00 m²	-	0.12	1.10	-	m²	1.10

M SURFACE FINISHES

	PC £	Labour hours	Labour £	Material £	Unit	Total rate £
M60 PAINTING/CLEAR FINISHING - cont'd						
PREPARATION OF EXISTING SURFACES - INTERNALLY - cont'd						
Wash and rub down wood surfaces; prime bare patches; bringing forward						
General surfaces						
over 300 mm girth	-	0.19	1.74	-	m^2	1.74
isolated surfaces not exceeding 300 mm girth	-	0.07	0.64	-	m^2	0.64
isolated areas not exceeding 0.50 m^2						
irrespective of girth	-	0.14	1.28	-	nr	1.28
Glazed windows and screens						
panes; area not exceeding 0.10 m^2	-	0.31	2.84	-	m^2	2.84
panes; area 0.10 - 0.50 m^2	-	0.24	2.20	-	m^2	2.20
panes; area 0.50 - 1.00 m^2	-	0.20	1.83	-	m^2	1.83
panes; area over 1.00 m^2	-	0.19	1.74	-	m^2	1.74
Wash down and remove paint with chemical stripper from iron, steel or wood surfaces						
General surfaces						
over 300 mm girth	-	0.56	5.12	-	m^2	5.12
isolated surfaces not exceeding 300 mm girth	-	0.24	2.20	-	m^2	2.20
isolated areas not exceeding 0.50 m^2						
irrespective of girth	-	0.42	3.84	-	nr	3.84
Glazed windows and screens						
panes; area not exceeding 0.10 m^2	-	1.18	10.80	-	m^2	10.80
panes; area 0.10 - 0.50 m^2	-	0.94	8.60	-	m^2	8.60
panes; area 0.50 - 1.00 m^2	-	0.81	7.41	-	m^2	7.41
panes; area over 1.00 m^2	-	0.71	6.50	-	m^2	6.50
Burn off and rub down to remove paint from iron, steel or wood surfaces						
General surfaces						
over 300 mm girth	-	0.68	6.22	-	m^2	6.22
isolated surfaces not exceeding 300 mm girth	-	0.31	2.84	-	m^2	2.84
isolated areas not exceeding 0.50 m^2						
irrespective of girth	-	0.52	4.76	-	nr	4.76
Glazed windows and screens						
panes; area not exceeding 0.10 m^2	-	1.48	13.54	-	m^2	13.54
panes; area 0.10 - 0.50 m^2	-	1.18	10.80	-	m^2	10.80
panes; area 0.50 - 1.00 m^2	-	1.02	9.33	-	m^2	9.33
panes; area over 1.00 m^2	-	0.89	8.14	-	m^2	8.14
PAINTING/CLEAR FINISHING - INTERNALLY						
NOTE: The following prices include for preparing surfaces. Painting woodwork also includes for knotting prior to applying the priming coat and for the stopping of all nail holes etc.						
One coat primer; on wood surfaces before fixing						
General surfaces						
over 300 mm girth	-	0.10	0.92	0.52	m^2	1.44
isolated surfaces not exceeding 300 mm girth	-	0.03	0.27	0.19	m	0.46
isolated areas not exceeding 0.50 m^2						
irrespective of girth	-	0.07	0.64	0.21	nr	0.85

M SURFACE FINISHES

	PC £	Labour hours	Labour £	Material £	Unit	Total rate £
One coat polyurethane sealer; on wood surfaces before fixing						
General surfaces						
over 300 mm girth	-	0.12	1.10	0.77	m²	1.87
isolated surfaces not exceeding 300 mm girth	-	0.04	0.37	0.27	m	0.64
isolated areas not exceeding 0.50 m²						
irrespective of girth	-	0.09	0.82	0.37	nr	1.19
One coat of Sikkens "Cetol HLS" stain or other equal and approved; on wood surfaces before fixing						
General surfaces						
over 300 mm girth	-	0.13	1.19	0.72	m²	1.91
isolated surfaces not exceeding 300 mm girth	-	0.04	0.37	0.28	m	0.65
isolated areas not exceeding 0.50 m²						
irrespective of girth	-	0.09	0.82	0.34	nr	1.16
One coat of Sikkens "Cetol TS" interior stain or other equal and approved; on wood surfaces before fixing						
General surfaces						
over 300 mm girth	-	0.13	1.19	1.12	m²	2.31
isolated surfaces not exceeding 300 mm girth	-	0.04	0.37	0.43	m	0.80
isolated areas not exceeding 0.50 m²						
irrespective of girth	-	0.09	0.82	0.54	nr	1.36
One coat Cuprinol clear wood preservative or other equal and approved; on wood surfaces before fixing						
General surfaces						
over 300 mm girth	-	0.09	0.82	0.48	m²	1.30
isolated surfaces not exceeding 300 mm girth	-	0.03	0.27	0.17	m	0.44
isolated areas not exceeding 0.50 m²						
irrespective of girth	-	0.06	0.55	0.22	nr	0.77
One coat HCC Protective Coatings Ltd "Permacor" urethane alkyd gloss finishing coat or other equal and approved; on previously primed steelwork						
Members of roof trusses						
over 300 mm girth	-	0.08	0.73	1.48	m²	2.21
Two coats emulsion paint						
Brick or block walls						
over 300 mm girth	-	0.25	2.29	0.81	m²	3.10
Cement render or concrete						
over 300 mm girth	-	0.24	2.20	0.78	m²	2.98
isolated surfaces not exceeding 300 mm girth	-	0.12	1.10	0.24	m	1.34
Plaster walls or plaster/plasterboard ceilings						
over 300 mm girth	-	0.22	2.01	0.75	m²	2.76
over 300 mm girth; in multi colours	-	0.30	2.75	0.91	m²	3.66
over 300 mm girth; in staircase areas	-	0.25	2.29	0.87	m²	3.16
cutting in edges on flush surfaces	-	0.09	0.82	-	m	0.82
Plaster/plasterboard ceilings						
over 300 mm girth; 3.50 m - 5.00 m high	-	0.25	2.29	0.76	m²	3.05
One mist and two coats emulsion paint						
Brick or block walls						
over 300 mm girth	-	0.23	2.10	1.15	m²	3.25
Cement render or concrete						
over 300 mm girth	-	0.23	2.10	1.06	m²	3.16

Prices for Measured Works - Minor Works
M SURFACE FINISHES

	PC £	Labour hours	Labour £	Material £	Unit	Total rate £
M60 PAINTING/CLEAR FINISHING - cont'd						
PAINTING/CLEAR FINISHING - INTERNALLY - cont'd						
One mist and two coats emulsion paint - cont'd						
Plaster walls or plaster/plasterboard ceilings						
over 300 mm girth	-	0.22	2.01	1.06	m^2	3.07
over 300 mm girth; in multi colours	-	0.31	2.84	1.08	m^2	3.92
over 300 mm girth; in staircase areas	-	0.25	2.29	1.06	m^2	3.35
cutting in edges on flush surfaces	-	0.10	0.92	-	m	0.92
Plaster/plasterboard ceilings						
over 300 mm girth; 3.50 m - 5.00 m high	-	0.21	1.92	1.06	m^2	2.98
One coat "Tretol No 10" sealer or other equal and approved; two coats Tretol sprayed "Supercover Spraytone" emulsion paint or other equal and approved						
Plaster walls or plaster/plasterboard ceilings						
over 300 mm girth	-	-	-	-	m^2	4.72
Textured plastic; "Artex" finish or other equal and approved						
Plasterboard ceilings						
over 300 mm girth	-	0.23	2.10	1.38	m^2	3.98
Concrete walls or ceilings						
over 300 mm girth	-	0.25	2.29	1.81	m^2	4.10
One coat "Portabond" or similar; one coat "Portaflek" or other equal and approved; on plaster surfaces; spray applied, masking adjacent surfaces						
General surfaces						
over 300 mm girth	-	-	-	-	m^2	9.89
extra for one coat standard "HD" glaze	-	-	-	-	m^2	1.87
not exceeding 300 mm girth	-	-	-	-	m	4.29
Touch up primer; one undercoat and one finishing coat of gloss oil paint; on wood surfaces						
General surfaces						
over 300 mm girth	-	0.27	2.47	1.39	m^2	3.86
isolated surfaces not exceeding 300 mm girth	-	0.11	1.01	0.47	m	1.48
isolated areas not exceeding 0.50 m^2						
irrespective of girth	-	0.20	1.83	0.72	nr	2.55
Glazed windows and screens						
panes; area not exceeding 0.10 m^2	-	0.44	4.03	1.02	m^2	5.05
panes; area 0.10 - 0.50 m^2	-	0.36	3.29	0.79	m^2	4.08
panes; area 0.50 - 1.00 m^2	-	0.31	2.84	0.63	m^2	3.47
panes; area over 1.00 m^2	-	0.27	2.47	0.54	m^2	3.01
Touch up primer; two undercoats and one finishing coat of gloss oil paint; on wood surfaces						
General surfaces						
over 300 mm girth	-	0.38	3.48	1.29	m^2	4.77
isolated surfaces not exceeding 300 mm girth	-	0.15	1.37	0.53	m	1.90
isolated areas not exceeding 0.50 m^2						
irrespective of girth	-	0.28	2.56	0.73	nr	3.29
Glazed windows and screens						
panes; area not exceeding 0.10 m^2	-	0.63	5.76	1.39	m^2	7.15
panes; area 0.10 - 0.50 m^2	-	0.50	4.58	1.10	m^2	5.68
panes; area 0.50 - 1.00 m^2	-	0.43	3.93	0.91	m^2	4.84
panes; area over 1.00 m^2	-	0.38	3.48	0.76	m^2	4.24

M SURFACE FINISHES

	PC £	Labour hours	Labour £	Material £	Unit	Total rate £
Knot; one coat primer; stop; one undercoat and one finishing coat of gloss oil paint; on wood surfaces						
General surfaces						
over 300 mm girth	-	0.39	3.57	1.26	m²	4.83
isolated surfaces not exceeding 300 mm girth	-	0.16	1.46	0.42	m	1.88
isolated areas not exceeding 0.50 m² irrespective of girth	-	0.30	2.75	0.82	nr	3.57
Glazed windows and screens						
panes; area not exceeding 0.10 m²	-	0.67	6.13	1.26	m²	7.39
panes; area 0.10 - 0.50 m²	-	0.54	4.94	1.06	m²	6.00
panes; area 0.50 - 1.00 m²	-	0.46	4.21	1.06	m²	5.27
panes; area over 1.00 m²	-	0.39	3.57	0.77	m²	4.34
Knot; one coat primer; stop; two undercoats and one finishing coat of gloss oil paint; on wood surfaces						
General surfaces						
over 300 mm girth	-	0.50	4.58	1.77	m²	6.35
isolated surfaces not exceeding 300 mm girth	-	0.20	1.83	0.61	m	2.44
isolated areas not exceeding 0.50 m² irrespective of girth	-	0.38	3.48	0.99	nr	4.47
Glazed windows and screens						
panes; area not exceeding 0.10 m²	-	0.83	7.59	1.67	m²	9.26
panes; area 0.10 - 0.50 m²	-	0.67	6.13	1.53	m²	7.66
panes; area 0.50 - 1.00 m²	-	0.58	5.31	1.39	m²	6.70
panes; area over 1.00 m²	-	0.50	4.58	1.03	m²	5.61
One coat primer; one undercoat and one finishing coat of gloss oil paint						
Plaster surfaces						
over 300 mm girth	-	0.35	3.20	1.61	m²	4.81
One coat primer; two undercoats and one finishing coat of gloss oil paint						
Plaster surfaces						
over 300 mm girth	-	0.46	4.21	2.08	m²	6.29
One coat primer; two undercoats and one finishing coat of eggshell paint						
Plaster surfaces						
over 300 mm girth	-	0.46	4.21	2.04	m²	6.25
Touch up primer; one undercoat and one finishing coat of gloss paint; on iron or steel surfaces						
General surfaces						
over 300 mm girth	-	0.27	2.47	0.98	m²	3.45
isolated surfaces not exceeding 300 mm girth	-	0.11	1.01	0.33	m	1.34
isolated areas not exceeding 0.50 m² irrespective of girth	-	0.20	1.83	0.54	nr	2.37
Glazed windows and screens						
panes; area not exceeding 0.10 m²	-	0.44	4.03	1.01	m²	5.04
panes; area 0.10 - 0.50 m²	-	0.36	3.29	0.80	m²	4.09
panes; area 0.50 - 1.00 m²	-	0.31	2.84	0.61	m²	3.45
panes; area over 1.00 m²	-	0.27	2.47	0.51	m²	2.98
Structural steelwork						
over 300 mm girth	-	0.30	2.75	1.02	m²	3.77
Members of roof trusses						
over 300 mm girth	-	0.40	3.66	1.15	m²	4.81

M SURFACE FINISHES

	PC £	Labour hours	Labour £	Material £	Unit	Total rate £
M60 PAINTING/CLEAR FINISHING - cont'd						
PAINTING/CLEAR FINISHING - INTERNALLY - cont'd						
Touch up primer; one undercoat and one finishing coat of gloss paint; on iron or steel surfaces - cont'd						
Ornamental railings and the like; each side measured overall						
over 300 mm girth	-	0.46	4.21	1.28	m²	5.49
Iron or steel radiators						
over 300 mm girth	-	0.27	2.47	1.06	m²	3.53
Pipes or conduits						
over 300 mm girth	-	0.40	3.66	1.11	m²	4.77
not exceeding 300 mm girth	-	0.16	1.46	0.36	m	1.82
Touch up primer; two undercoats and one finishing coat of gloss oil paint; on iron or steel surfaces						
General surfaces						
over 300 mm girth	-	0.38	3.48	1.34	m²	4.82
isolated surfaces not exceeding 300 mm girth	-	0.15	1.37	0.49	m	1.86
isolated areas not exceeding 0.50 m²						
irrespective of girth	-	0.28	2.56	0.78	nr	3.34
Glazed windows and screens						
panes; area not exceeding 0.10 m²	-	0.63	5.76	1.42	m²	7.18
panes; area 0.10 - 0.50 m²	-	0.50	4.58	1.16	m²	5.74
panes; area 0.50 - 1.00 m²	-	0.43	3.93	1.00	m²	4.93
panes; area over 1.00 m²	-	0.38	3.48	0.79	m²	4.27
Structural steelwork						
over 300 girth	-	0.43	3.93	1.36	m²	5.29
Members of roof trusses						
over 300 girth	-	0.56	5.12	1.64	m²	6.76
Ornamental railings and the like; each side measured overall						
over 300 girth	-	0.64	5.86	1.76	m²	7.62
Iron or steel radiators						
over 300 girth	-	0.38	3.48	1.50	m²	4.98
Pipes or conduits						
over 300 girth	-	0.56	5.12	1.67	m²	6.79
not exceeding 300 girth	-	0.22	2.01	0.54	m	2.55
One coat primer; one undercoat and one finishing coat of gloss oil paint; on iron or steel surfaces						
General surfaces						
over 300 mm girth	-	0.27	2.47	0.98	m²	3.45
isolated surfaces not exceeding 300 mm girth	-	0.14	1.28	0.54	m	1.82
isolated areas not exceeding 0.50 m²						
irrespective of girth	-	0.27	2.47	0.93	nr	3.40
Glazed windows and screens						
panes; area not exceeding 0.10 m²	-	0.58	5.31	1.51	m²	6.82
panes; area 0.10 - 0.50 m²	-	0.46	4.21	1.20	m²	5.41
panes; area 0.50 - 1.00 m²	-	0.40	3.66	1.02	m²	4.68
panes; area over 1.00 m²	-	0.35	3.20	0.93	m²	4.13
Structural steelwork						
over 300 mm girth	-	0.39	3.57	1.46	m²	5.03
Members of roof trusses						
over 300 mm girth	-	0.53	4.85	1.56	m²	6.41
Ornamental railings and the like; each side measured overall						
over 300 mm girth	-	0.60	5.49	1.85	m²	7.34

M SURFACE FINISHES

	PC £	Labour hours	Labour £	Material £	Unit	Total rate £
Iron or steel radiators						
over 300 mm girth	-	0.35	3.20	1.56	m²	4.76
Pipes or conduits						
over 300 mm girth	-	0.53	4.85	1.56	m²	6.41
not exceeding 300 mm girth	-	0.20	1.83	0.52	m	2.35
One coat primer; two undercoats and one finishing coat of gloss oil paint; on iron or steel surfaces						
General surfaces						
over 300 mm girth	-	0.46	4.21	1.81	m²	6.02
isolated surfaces not exceeding 300 mm girth	-	0.19	1.74	0.73	m	2.47
isolated areas not exceeding 0.50 m² irrespective of girth	-	0.33	3.02	1.03	nr	4.05
Glazed windows and screens						
panes; area not exceeding 0.10 m²	-	0.75	6.86	1.86	m²	8.72
panes; area 0.10 - 0.50 m²	-	0.63	5.76	1.04	m²	6.80
panes; area 0.50 - 1.00 m²	-	0.54	4.94	1.28	m²	6.22
panes; area over 1.00 m²	-	0.46	4.21	1.08	m²	5.29
Structural steelwork						
over 300 mm girth	-	0.52	4.76	1.86	m²	6.62
Members of roof trusses						
over 300 mm girth	-	0.68	6.22	2.22	m²	8.44
Ornamental railings and the like; each side measured overall						
over 300 mm girth	-	0.78	7.14	2.45	m²	9.59
Iron or steel radiators						
over 300 mm girth	-	0.46	4.21	2.00	m²	6.21
Pipes or conduits						
over 300 mm girth	-	0.69	6.31	2.25	m²	8.56
not exceeding 300 mm girth	-	0.28	2.56	0.72	m	3.28
Two coats of bituminous paint; on iron or steel surfaces						
General surfaces						
over 300 mm girth	-	0.27	2.47	0.48	m²	2.95
Inside of galvanized steel cistern						
over 300 mm girth	-	0.40	3.66	0.57	m²	4.23
Two coats bituminous paint; first coat blinded with clean sand prior to second coat; on concrete surfaces						
General surfaces						
over 300 mm girth	-	0.93	8.51	1.42	m²	9.93
Mordant solution; one coat HCC Protective Coatings Ltd "Permacor alkyd MIO" or other equal and approved; one coat "Permacor Epoxy Gloss" finishing coat or other equal and approved on galvanised steelwork						
Structural steelwork						
over 300 mm girth	-	0.52	4.76	2.76	m²	7.52
One coat HCC Protective Coatings Ltd "Permacor Epoxy Zinc Primer" or other equal and approved; two coats "Permacor alkyd MIO" or other equal and approved; one coat "Permatex Epoxy Gloss" finishing coat or other equal and approved on steelwork						
Structural steelwork						
over 300 mm girth	-	0.74	6.77	4.98	m²	11.75

M SURFACE FINISHES

	PC £	Labour hours	Labour £	Material £	Unit	Total rate £
M60 PAINTING/CLEAR FINISHING - cont'd						
PAINTING/CLEAR FINISHING - INTERNALLY - cont'd						
Steel protection; HCC Protective Coatings Ltd "Unitherm" or other equal and approved; two coats to steelwork						
Structural steelwork						
over 300 mm girth	-	1.16	10.61	1.70	m²	12.31
Two coats epoxy resin sealer; HCC Protective Coatings Ltd "Betonol" or other equal and approved; on concrete surfaces						
General surfaces						
over 300 mm girth	-	0.23	2.10	3.85	m²	5.95
"Nitoflor Lithurin" floor hardener and dust proofer or other equal and approved; Fosroc Expandite Ltd or other equal and approved; two coats; on concrete surfaces						
General surfaces						
over 300 mm girth	-	0.28	2.04	0.48	m²	2.52
Two coats of boiled linseed oil; on hardwood surfaces						
General surfaces						
over 300 mm girth	-	0.20	1.83	1.36	m²	3.19
isolated surfaces not exceeding 300 mm girth	-	0.08	0.73	0.44	m	1.17
isolated areas not exceeding 0.50 m²						
irrespective of girth	-	0.16	1.46	0.79	nr	2.25
Two coats polyurethane varnish; on wood surfaces						
General surfaces						
over 300 mm girth	-	0.20	1.83	1.36	m²	3.19
isolated surfaces not exceeding 300 mm girth	-	0.08	0.73	0.50	m	1.23
isolated areas not exceeding 0.50 m²						
irrespective of girth	-	0.16	1.46	0.16	nr	1.62
Three coats polyurethane varnish; on wood surfaces						
General surfaces						
over 300 mm girth	-	0.31	2.84	2.08	m²	4.92
isolated surfaces not exceeding 300 mm girth	-	0.12	1.10	0.63	m	1.73
isolated areas not exceeding 0.50 m²						
irrespective of girth	-	0.23	2.10	1.15	nr	3.25
One undercoat; and one finishing coat; of "Albi" clear flame retardant surface coating or other equal and approved; on wood surfaces						
General surfaces						
over 300 mm girth	-	0.40	3.66	2.39	m²	6.55
isolated surfaces not exceeding 300 mm girth	-	0.17	1.56	1.00	m	2.56
isolated areas not exceeding 0.50 m²						
irrespective of girth	-	0.23	2.10	2.20	nr	4.30
Two undercoats; and one finishing coat; of "Albi" clear flame retardant surface coating or other equal and approved; on wood surfaces						
General surfaces						
over 300 mm girth	-	0.47	4.30	4.07	m²	8.37
isolated surfaces not exceeding 300 mm girth	-	0.24	2.20	1.47	m	3.67
isolated areas not exceeding 0.50 m²						
irrespective of girth	-	0.39	3.57	2.24	nr	5.81

Prices for Measured Works - Minor Works
M SURFACE FINISHES

	PC £	Labour hours	Labour £	Material £	Unit	Total rate £
Seal and wax polish; dull gloss finish on wood surfaces						
General surfaces						
over 300 mm girth	-	-	-	-	m²	8.89
isolated surfaces not exceeding 300 mm girth	-	-	-	-	m	4.03
isolated areas not exceeding 0.50 m²						
irrespective of girth	-	-	-	-	nr	6.20
One coat of "Sadolin Extra" or other equal and approved; clear or pigmented; one further coat of "Holdex" clear interior silk matt lacquer or other equal and approved						
General surfaces						
over 300 mm girth	-	0.30	2.75	4.48	m²	7.23
isolated surfaces not exceeding 300 mm girth	-	0.12	1.10	2.10	m	3.20
isolated areas not exceeding 0.50 m²						
irrespective of girth	-	0.22	2.01	2.18	nr	4.19
Glazed windows and screens						
panes; area not exceeding 0.10 m²	-	0.49	4.48	2.56	m²	7.04
panes; area 0.10 - 0.50 m²	-	0.39	3.57	2.39	m²	5.96
panes; area 0.50 - 1.00 m²	-	0.34	3.11	2.21	m²	5.32
panes; area over 1.00 m²	-	0.30	2.75	2.10	m²	4.85
Two coats of "Sadolins Extra" or other equal and approved; clear or pigmented; two further coats of "Holdex" clear interior silk matt lacquer or other equal and approved						
General surfaces						
over 300 mm girth	-	0.46	4.21	8.28	m²	12.49
isolated surfaces not exceeding 300 mm girth	-	0.19	1.74	4.14	m	5.88
isolated areas not exceeding 0.50 m²						
irrespective of girth	-	0.35	3.20	4.72	nr	7.92
Glazed windows and screens						
panes; area not exceeding 0.10 m²	-	0.77	7.05	5.07	m²	12.12
panes; area 0.10 - 0.50 m²	-	0.61	5.58	4.72	m²	10.30
panes; area 0.50 - 1.00 m²	-	0.54	4.94	4.37	m²	9.31
panes; area over 1.00 m²	-	0.46	4.21	4.14	m²	8.35
Two coats of Sikkens "Cetol TS" interior stain or other equal and approved; on wood surfaces						
General surfaces						
over 300 mm girth	-	0.21	1.92	2.01	m²	3.93
isolated surfaces not exceeding 300 mm girth	-	0.09	0.82	0.71	m	1.53
isolated areas not exceeding 0.50 m²						
irrespective of girth	-	0.16	1.46	1.10	nr	2.56
Body in and wax polish; dull gloss finish; on hardwood surfaces						
General surfaces						
over 300 mm girth	-	-	-	-	m²	10.23
isolated surfaces not exceeding 300 mm girth	-	-	-	-	m	4.64
isolated areas not exceeding 0.50 m²						
irrespective of girth	-	-	-	-	nr	7.16
Stain; body in and wax polish; dull gloss finish; on hardwood surfaces						
General surfaces						
over 300 mm girth	-	-	-	-	m²	14.15
isolated surfaces not exceeding 300 mm girth	-	-	-	-	m	6.37
isolated areas not exceeding 0.50 m²						
irrespective of girth	-	-	-	-	nr	9.95

Prices for Measured Works - Minor Works
M SURFACE FINISHES

	PC £	Labour hours	Labour £	Material £	Unit	Total rate £
M60 PAINTING/CLEAR FINISHING - cont'd						
PAINTING/CLEAR FINISHING - INTERNALLY - cont'd						
Seal; two coats of synthetic resin lacquer; decorative flatted finish; wire down, wax and burnish; on wood surfaces						
General surfaces						
over 300 mm girth	-	-	-	-	m²	17.01
isolated surfaces not exceeding 300 mm girth	-	-	-	-	m	7.66
isolated areas not exceeding 0.50 m²						
irrespective of girth	-	-	-	-	nr	11.91
Stain; body in and fully French polish; full gloss finish; on hardwood surfaces						
General surfaces						
over 300 mm girth	-	-	-	-	m²	19.58
isolated surfaces not exceeding 300 mm girth	-	-	-	-	m	8.84
isolated areas not exceeding 0.50 m²						
irrespective of area	-	-	-	-	nr	13.71
Stain; fill grain and fully French polish; full gloss finish; on hardwood surfaces						
General surfaces						
over 300 mm girth	-	-	-	-	m²	26.84
isolated surfaces not exceeding 300 mm girth	-	-	-	-	m	12.08
isolated areas not exceeding 0.50 m²						
irrespective of girth	-	-	-	-	nr	18.79
Stain black; body in and fully French polish; ebonized finish; on hardwood surfaces						
General surfaces						
over 300 mm girth	-	-	-	-	m²	33.55
isolated surfaces not exceeding 300 mm girth	-	-	-	-	m	13.99
isolated areas not exceeding 0.50 m²						
irrespective of girth	-	-	-	-	nr	23.49
PREPARATION OF EXISTING SURFACES - EXTERNALLY						
Wash and rub down iron and steel surfaces; bringing forward						
General surfaces						
over 300 mm girth	-	0.15	1.37	-	m²	1.37
isolated surfaces not exceeding 300 mm girth	-	0.05	0.46	-	m	0.46
isolated areas not exceeding 0.50 m²						
irrespective of girth	-	0.11	1.01	-	nr	1.01
Glazed windows and screens						
panes; area not exceeding 0.10 m²	-	0.24	2.20	-	m²	2.20
panes; area 0.10 - 0.50 m²	-	0.20	1.83	-	m²	1.83
panes; area 0.50 - 1.00 m²	-	0.17	1.56	-	m²	1.56
panes; area over 1.00 m²	-	0.15	1.37	-	m²	1.37
Wash and rub down wood surfaces; prime bare patches; bringing forward						
General surfaces						
over 300 mm girth	-	0.24	2.20	-	m²	2.20
isolated surfaces not exceeding 300 mm girth	-	0.09	0.82	-	m	0.82
isolated areas not exceeding 0.50 m²						
irrespective of girth	-	0.18	1.65	-	nr	1.65

M SURFACE FINISHES

	PC £	Labour hours	Labour £	Material £	Unit	Total rate £
Glazed windows and screens						
panes; area not exceeding 0.10 m²	-	0.41	3.75	-	m²	3.75
panes; area 0.10 - 0.50 m²	-	0.32	2.93	-	m²	2.93
panes; area 0.50 - 1.00 m²	-	0.28	2.56	-	m²	2.56
panes; area over 1.00 m²	-	0.24	2.20	-	m²	2.20
Wash down and remove paint with chemical stripper from iron, steel or wood surfaces						
General surfaces						
over 300 mm girth	-	0.74	6.77	-	m²	6.77
isolated surfaces not exceeding 300 mm girth	-	0.33	3.02	-	m	3.02
isolated areas not exceeding 0.50 m²						
irrespective of girth	-	0.56	5.12	-	nr	5.12
Glazed windows and screens						
panes; area not exceeding 0.10 m²	-	1.58	14.46	-	m²	14.46
panes; area 0.10 - 0.50 m²	-	1.26	11.53	-	m²	11.53
panes; area 0.50 - 1.00 m²	-	1.09	9.97	-	m²	9.97
panes; area over 1.00 m²	-	0.94	8.60	-	m²	8.60
Burn off and rub down to remove paint from iron, steel or wood surfaces						
General surfaces						
over 300 mm girth	-	0.91	8.33	-	m²	8.33
isolated surfaces not exceeding 300 mm girth	-	0.42	3.84	-	m	3.84
isolated surfaces not exceeding 0.50 m²	-	0.68	6.22	-	nr	6.22
Glazed windows and screens						
panes over 1m²	-	1.18	10.80	-	m²	10.80
panes 0.50 - 1.00m²	-	1.35	12.35	-	m²	12.35
panes 0.10 - 0.50m²	-	1.57	14.37	-	m²	14.37
panes not exceeding 0.10m²	-	1.97	18.03	-	m²	18.03
PAINTING/CLEAR FINISHING - EXTERNALLY						
Two coats of cement paint, "Sandtex Matt" or other equal and approved						
Brick or block walls						
over 300 mm girth	-	0.31	2.84	1.87	m²	4.71
Cement render or concrete walls						
over 300 mm girth	-	0.27	2.47	1.23	m²	3.70
Roughcast walls						
over 300 mm girth	-	0.46	4.21	1.23	m²	5.44
One coat sealer and two coats of external grade emulsion paint, Dulux "Weathershield" or other equal and approved						
Brick or block walls						
over 300 mm girth	-	0.50	4.58	6.90	m²	11.48
Cement render or concrete walls						
over 300 mm girth	-	0.39	3.57	4.60	m²	8.17
Concrete soffits						
over 300 mm girth	-	0.46	4.21	4.60	m²	8.81
One coat sealer (applied by brush) and two coats of external grade emulsion paint, Dulux "Weathershield" or other equal and approved (spray applied)						
Roughcast						
over 300 mm girth	-	0.33	3.02	9.30	m²	12.32

M SURFACE FINISHES

	PC £	Labour hours	Labour £	Material £	Unit	Total rate £
M60 PAINTING/CLEAR FINISHING - cont'd						
PAINTING/CLEAR FINISHING - EXTERNALLY - cont'd						
Two coat sealer and one coat of Anti-Graffiti paint (spray applied)						
Roughcast						
over 300 mm girth	-	0.33	3.02	39.68	m^2	42.70
Two coats solar reflective aluminium paint; on bituminous roofing						
General surfaces						
over 300 mm girth	-	0.52	4.76	11.99	m^2	16.75
Touch up primer; one undercoat and one finishing coat of gloss oil paint; on wood surfaces						
General surfaces						
over 300 mm girth	-	0.30	2.75	0.97	m^2	3.72
isolated surfaces not exceeding 300 mm girth	-	0.13	1.19	0.30	m	1.49
isolated areas not exceeding 0.50 m^2						
irrespective of girth	-	0.22	2.01	0.59	nr	2.60
Glazed windows and screens						
panes; area not exceeding 0.10 m^2	-	0.50	4.58	1.06	m^2	5.64
panes; area 0.10 - 0.50 m^2	-	0.40	3.66	0.89	m^2	4.55
panes; area 0.50 - 1.00 m^2	-	0.35	3.20	0.71	m^2	3.91
panes; area over 1.00 m^2	-	0.30	2.75	0.59	m^2	3.34
Glazed windows and screens; multi-coloured work						
panes; area not exceeding 0.10 m^2	-	0.55	5.03	1.06	m^2	6.09
panes; area 0.10 - 0.50 m^2	-	0.44	4.03	0.91	m^2	4.94
panes; area 0.50 - 1.00 m^2	-	0.38	3.48	0.71	m^2	4.19
panes; area over 1.00 m^2	-	0.33	3.02	0.59	m^2	3.61
Touch up primer; two undercoats and one finishing coat of gloss oil paint; on wood surfaces						
General surfaces						
over 300 mm girth	-	0.42	3.84	1.27	m^2	5.11
isolated surfaces not exceeding 300 mm girth	-	0.17	1.56	0.34	m	1.90
isolated areas not exceeding 0.50 m^2						
irrespective of girth	-	0.31	2.84	0.69	nr	3.53
Glazed windows and screens						
panes; area not exceeding 0.10 m^2	-	0.69	6.31	1.12	m^2	7.43
panes; area 0.10 - 0.50 m^2	-	0.69	6.31	0.94	m^2	7.25
panes; area 0.50 - 1.00 m^2	-	0.56	5.12	0.83	m^2	5.95
panes; area over 1.00 m^2	-	0.42	3.84	0.69	m^2	4.53
Glazed windows and screens; multi- coloured work						
panes; area not exceeding 0.10 m^2	-	0.80	7.32	1.12	m^2	8.44
panes; area 0.10 - 0.50 m^2	-	0.64	5.86	0.98	m^2	6.84
panes; area 0.50 - 1.00 m^2	-	0.56	5.12	0.83	m^2	5.95
panes; area over 1.00 m^2	-	0.48	4.39	0.69	m^2	5.08
Knot; one coat primer; one undercoat and one finishing coat of gloss oil paint; on wood surfaces						
General surfaces						
over 300 mm girth	-	0.44	4.03	1.34	m^2	5.37
isolated surfaces not exceeding 300 mm girth	-	0.19	1.74	0.43	m	2.17
isolated areas not exceeding 0.50 m^2						
irrespective of girth	-	0.33	3.02	0.74	nr	3.76

M SURFACE FINISHES

	PC £	Labour hours	Labour £	Material £	Unit	Total rate £
Glazed windows and screens						
panes; area not exceeding 0.10 m²	-	0.73	6.68	1.38	m²	8.06
panes; area 0.10 - 0.50 m²	-	0.58	5.31	1.24	m²	6.55
panes; area 0.50 - 1.00 m²	-	0.52	4.76	0.97	m²	5.73
panes; area over 1.00 m²	-	0.44	4.03	0.69	m²	4.72
Glazed windows and screens; multi-coloured work						
panes; area not exceeding 0.10 m²	-	0.80	7.32	1.39	m²	8.71
panes; area 0.10 - 0.50 m²	-	0.64	5.86	1.24	m²	7.10
panes; area 0.50 - 1.00 m²	-	0.56	5.12	0.95	m²	6.07
panes; area over 1.00 m²	-	0.48	4.39	0.67	m²	5.06
Knot; one coat primer; two undercoats and one finishing coat of gloss oil paint; on wood surfaces						
General surfaces						
over 300 mm girth	-	0.55	5.03	1.60	m²	6.63
isolated surfaces not exceeding 300 mm girth	-	0.22	2.01	0.57	m	2.58
isolated areas not exceeding 0.50 m²						
irrespective of girth	-	0.42	3.84	0.96	nr	4.80
Glazed windows and screens						
panes; area not exceeding 0.10 m²	-	0.92	8.42	1.78	m²	10.20
panes; area 0.10 - 0.50 m²	-	0.73	6.68	1.59	m²	8.27
panes; area 0.50 - 1.00 m²	-	0.64	5.86	1.22	m²	7.08
panes; area over 1.00 m²	-	0.55	5.03	0.85	m²	5.88
Glazed windows and screens; multi-coloured work						
panes; area not exceeding 0.10 m²	-	1.05	9.61	1.78	m²	11.39
panes; area 0.10 - 0.50 m²	-	0.85	7.78	1.60	m²	9.38
panes; area 0.50 - 1.00 m²	-	0.75	6.86	1.22	m²	8.08
panes; area over 1.00 m²	-	0.63	5.76	0.85	m²	6.61
Touch up primer; one undercoat and one finishing coat of gloss oil paint; on iron or steel surfaces						
General surfaces						
over 300 mm girth	-	0.30	2.75	0.96	m²	3.71
isolated surfaces not exceeding 300 mm girth	-	0.13	1.19	0.55	m	1.74
isolated areas not exceeding 0.50 m²						
irrespective of girth	-	0.22	2.01	0.51	nr	2.52
Glazed windows and screens						
panes; area not exceeding 0.10 m²	-	0.50	4.58	1.02	m²	5.60
panes; area 0.10 - 0.50 m²	-	0.40	3.66	0.88	m²	4.54
panes; area 0.50 - 1.00 m²	-	0.35	3.20	0.74	m²	3.94
panes; area over 1.00 m²	-	0.30	2.75	0.55	m²	3.30
Structural steelwork						
over 300 mm girth	-	0.33	3.02	1.01	m²	4.03
Members of roof trusses						
over 300 mm girth	-	0.44	4.03	1.15	m²	5.18
Ornamental railings and the like; each side measured overall						
over 300 mm girth	-	0.50	4.58	1.20	m²	5.78
Eaves gutters						
over 300 mm girth	-	0.54	4.94	1.39	m²	6.33
not exceeding 300 mm girth	-	0.21	1.92	0.46	m	2.38
Pipes or conduits						
over 300 mm girth	-	0.44	4.03	1.39	m²	5.42
not exceeding 300 mm girth	-	0.18	1.65	0.46	m	2.11
Touch up primer; two undercoats and one finishing coat of gloss oil paint; on iron or steel surfaces						
General surfaces						
over 300 mm girth	-	0.42	3.84	1.31	m²	5.15
isolated surfaces not exceeding 300 mm girth	-	0.17	1.56	0.36	m	1.92

M SURFACE FINISHES

	PC £	Labour hours	Labour £	Material £	Unit	Total rate £
M60 PAINTING/CLEAR FINISHING - cont'd						
PAINTING/CLEAR FINISHING - EXTERNALLY - cont'd						
Touch up primer; two undercoats and one finishing coat of gloss oil paint; on iron or steel surfaces - cont'd						
General surfaces - cont'd						
isolated areas not exceeding 0.50 m²						
irrespective of girth	-	0.31	2.84	0.73	nr	**3.57**
Glazed windows and screens						
panes; area not exceeding 0.10 m²	-	0.69	6.31	1.33	m²	**7.64**
panes; area 0.10 - 0.50 m²	-	0.56	5.12	1.14	m²	**6.26**
panes; area 0.50 - 1.00 m²	-	0.48	4.39	0.96	m²	**5.35**
panes; area over 1.00 m²	-	0.42	3.84	0.78	m²	**4.62**
Structural steelwork						
over 300 mm girth	-	0.47	4.30	1.37	m²	**5.67**
Members of roof trusses						
over 300 mm girth	-	0.63	5.76	1.55	m²	**7.31**
Ornamental railings and the like; each side measured overall						
over 300 mm girth	-	0.71	6.50	1.59	m²	**8.09**
Eaves gutters						
over 300 mm girth	-	0.75	6.86	1.78	m²	**8.64**
not exceeding 300 mm girth	-	0.30	2.75	0.75	m	**3.50**
Pipes or conduits						
over 300 mm girth	-	0.63	5.76	1.78	m²	**7.54**
not exceeding 300 mm girth	-	0.25	2.29	0.61	m	**2.90**
One coat primer; one undercoat and one finishing coat of gloss oil paint; on iron or steel surfaces						
General surfaces						
over 300 mm girth	-	0.38	3.48	1.15	m²	**4.63**
isolated surfaces not exceeding 300 mm girth	-	0.16	1.46	0.30	m	**1.76**
isolated areas not exceeding 0.50 m²						
irrespective of girth	-	0.30	2.75	0.60	nr	**3.35**
Glazed windows and screens						
panes; area not exceeding 0.10 m²	-	0.64	5.86	1.06	m²	**6.92**
panes; area 0.10 - 0.50 m²	-	0.52	4.76	0.93	m²	**5.69**
panes; area 0.50 - 1.00 m²	-	0.44	4.03	0.79	m²	**4.82**
panes; area over 1.00 m²	-	0.38	3.48	0.60	m²	**4.08**
Structural steelwork						
over 300 mm girth	-	0.43	3.93	1.20	m²	**5.13**
Members of roof trusses						
over 300 mm girth	-	0.56	5.12	1.34	m²	**6.46**
Ornamental railings and the like; each side measured overall						
over 300 mm girth	-	0.65	5.95	1.34	m²	**7.29**
Eaves gutters						
over 300 mm girth	-	0.68	6.22	1.62	m²	**7.84**
not exceeding 300 mm girth	-	0.28	2.56	0.56	m	**3.12**
Pipes or conduits						
over 300 mm girth	-	0.56	5.12	1.62	m²	**6.74**
not exceeding 300 mm girth	-	0.22	2.01	0.54	m	**2.55**
One coat primer; two undercoats and one finishing coat of gloss oil paint; on iron or steel surfaces						
General surfaces						
over 300 mm girth	-	0.50	4.58	1.50	m²	**6.08**
isolated surfaces not exceeding 300 mm girth	-	0.20	1.83	0.39	m	**2.22**
isolated areas not exceeding 0.50 m²						
irrespective of girth	-	0.38	3.48	0.78	nr	**4.26**

M SURFACE FINISHES

	PC £	Labour hours	Labour £	Material £	Unit	Total rate £
Glazed windows and screens						
panes; area not exceeding 0.10 m²	-	0.83	7.59	1.37	m²	8.96
panes; area 0.10 - 0.50 m²	-	0.67	6.13	1.19	m²	7.32
panes; area 0.50 - 1.00 m²	-	0.58	5.31	1.01	m²	6.32
panes; area over 1.00 m²	-	0.50	4.58	0.78	m²	5.36
Structural steelwork						
over 300 mm girth	-	0.56	5.12	1.55	m²	6.67
Members of roof trusses						
over 300 mm girth	-	0.75	6.86	1.73	m²	8.59
Ornamental railings and the like; each side measured overall						
over 300 mm girth	-	0.85	7.78	1.73	m²	9.51
Eaves gutters						
over 300 mm girth	-	0.90	8.23	2.02	m²	10.25
not exceeding 300 mm girth	-	0.36	3.29	0.69	m	3.98
Pipes or conduits						
over 300 mm girth	-	0.75	6.86	2.02	m²	8.88
not exceeding 300 mm girth	-	0.30	2.75	0.67	m	3.42
One coat of Andrews "Hammerite" paint or other equal and approved; on iron or steel surfaces						
General surfaces						
over 300 mm girth	-	0.18	1.65	1.21	m²	2.86
isolated surfaces not exceeding 300 mm girth	-	0.09	0.82	0.38	m	1.20
isolated areas not exceeding 0.50 m² irrespective of girth	-	0.13	1.19	0.70	nr	1.89
Glazed windows and screens						
panes; area not exceeding 0.10 m²	-	0.30	2.75	0.92	m²	3.67
panes; area 0.10 - 0.50 m²	-	0.23	2.10	1.01	m²	3.11
panes; area 0.50 - 1.00 m²	-	0.21	1.92	0.91	m²	2.83
panes; area over 1.00 m²	-	0.18	1.65	0.91	m²	2.56
Structural steelwork						
over 300 mm girth	-	0.20	1.83	1.11	m²	2.94
Members of roof trusses						
over 300 mm girth	-	0.27	2.47	1.21	m²	3.68
Ornamental railings and the like; each side measured overall						
over 300 mm girth	-	0.30	2.75	1.21	m²	3.96
Eaves gutters						
over 300 mm girth	-	0.31	2.84	1.31	m²	4.15
not exceeding 300 mm girth	-	0.10	0.92	0.61	m	1.53
Pipes or conduits						
over 300 mm girth	-	0.30	2.75	1.11	m²	3.86
not exceeding 300 mm girth	-	0.10	0.92	0.51	m	1.43
Two coats of creosote; on wood surfaces						
General surfaces						
over 300 mm girth	-	0.19	1.74	0.20	m²	1.94
isolated surfaces not exceeding 300 mm girth	-	0.06	0.55	0.13	m	0.68
Two coats of "Solignum" wood preservative or other equal and approved; on wood surfaces						
General surfaces						
over 300 mm girth	-	0.17	1.56	1.10	m²	2.66
isolated surfaces not exceeding 300 mm girth	-	0.06	0.55	0.34	m	0.89

Prices for Measured Works - Minor Works
M SURFACE FINISHES

	PC £	Labour hours	Labour £	Material £	Unit	Total rate £
M60 PAINTING/CLEAR FINISHING - cont'd						
PAINTING/CLEAR FINISHING - EXTERNALLY - cont'd						
Three coats of polyurethane; on wood surfaces						
General surfaces						
over 300 mm girth	-	0.33	3.02	2.29	m^2	**5.31**
isolated surfaces not exceeding 300 mm girth	-	0.13	1.19	1.13	m	**2.32**
isolated areas not exceeding 0.50 m^2						
irrespective of girth	-	0.25	2.29	1.31	nr	**3.60**
Two coats of "New Base" primer or other and approved; and two coats of "Extra"; Sadolin Ltd or other equal and approved; pigmented; on wood surfaces						
General surfaces						
over 300 mm girth	-	0.50	4.58	4.29	m^2	**8.87**
isolated surfaces not exceeding 300 mm girth	-	0.31	2.84	1.46	m	**4.30**
Glazed windows and screens						
panes; area not exceeding 0.10 m^2	-	0.84	7.69	3.05	m^2	**10.74**
panes; area 0.10 - 0.50 m^2	-	0.68	6.22	2.88	m^2	**9.10**
panes; area 0.50 - 1.00 m^2	-	0.58	5.31	2.70	m^2	**8.01**
panes; area over 1.00 m^2	-	0.50	4.58	2.17	m^2	**6.75**
Two coats Sikkens "Cetol Filter 7" exterior stain or other equal and approved; on wood surfaces						
General surfaces						
over 300 mm girth	-	0.21	1.92	3.01	m^2	**4.93**
isolated surfaces not exceeding 300 mm girth	-	0.09	0.82	1.04	m	**1.86**
isolated areas not exceeding 0.50 m^2						
irrespective of girth	-	0.16	1.46	1.53	nr	**2.99**

N FURNITURE/EQUIPMENT

N10/11 GENERAL FIXTURES/KITCHEN FITTINGS

Fixing general fixtures

NOTE: The fixing of general fixtures will vary considerably depending upon the size of the fixture and the method of fixing employed. Prices for fixing like sized kitchen fittings may be suitable for certain fixtures, although adjustment to those rates will almost invariably be necessary and the reader is directed to section "G20" for information on bolts, plugging brickwork and blockwork etc. which should prove useful in building up a suitable rate.

The following supply only prices are for purpose made fittings components in various materials supplied as part of an assembled fitting and therefore may be used to arrive at a guide price for a complete fitting.

	PC £	Labour hours	Labour £	Material £	Unit	Total rate £
Fitting components; blockboard						
Backs, fronts, sides or divisions; over 300 mm wide						
12 mm thick	-	-	-	40.24	m^2	40.24
19 mm thick	-	-	-	51.92	m^2	51.92
25 mm thick	-	-	-	68.21	m^2	68.21
Shelves or worktops; over 300 mm wide						
19 mm thick	-	-	-	51.92	m^2	51.92
25 mm thick	-	-	-	68.21	m^2	68.21
Flush doors; lipped on four edges						
450 mm x 750 mm x 19 mm	-	-	-	28.20	nr	28.20
450 mm x 750 mm x 25 mm	-	-	-	34.54	nr	34.54
600 mm x 900 mm x 19 mm	-	-	-	41.91	nr	41.91
600 mm x 900 mm x 25 mm	-	-	-	51.18	nr	51.18
Fitting components; chipboard						
Backs, fronts, sides or divisions; over 300 mm wide						
6 mm thick	-	-	-	11.92	m^2	11.92
9 mm thick	-	-	-	16.94	m^2	16.94
12 mm thick	-	-	-	20.90	m^2	20.90
19 mm thick	-	-	-	30.13	m^2	30.13
25 mm thick	-	-	-	40.67	m^2	40.67
Shelves or worktops; over 300 mm wide						
19 mm thick	-	-	-	30.13	m^2	30.13
25 mm thick	-	-	-	40.67	m^2	40.67
Flush doors; lipped on four edges						
450 mm x 750 mm x 19 mm	-	-	-	21.52	nr	21.52
450 mm x 750 mm x 25 mm	-	-	-	26.17	nr	26.17
600 mm x 900 mm x 19 mm	-	-	-	31.54	nr	31.54
600 mm x 900 mm x 25 mm	-	-	-	38.11	nr	38.11

Prices for Measured Works - Minor Works
N FURNITURE/EQUIPMENT

	PC £	Labour hours	Labour £	Material £	Unit	Total rate £
N10/11 GENERAL FIXTURES/KITCHEN FITTINGS - cont'd						
Fitting components; Melamine faced chipboard						
Backs, fronts, sides or divisions; over 300 mm wide						
12 mm thick	-	-	-	28.70	m²	28.70
19 mm thick	-	-	-	38.91	m²	38.91
Shelves or worktops; over 300 mm wide						
19 mm thick	-	-	-	38.91	m²	38.91
Flush doors; lipped on four edges						
450 mm x 750 mm x 19 mm	-	-	-	26.72	nr	26.72
600 mm x 900 mm x 25 mm	-	-	-	39.29	nr	39.29
Fitting components; "Warerite Xcel" standard colour laminated chipboard type LD2 or other equal and approved						
Backs, fronts, sides or divisions; over 300 mm wide						
13.20 mm thick	-	-	-	77.00	m²	77.00
Shelves or worktops; over 300 mm wide						
13.20 mm thick	-	-	-	77.00	m²	77.00
Flush doors; lipped on four edges						
450 mm x 750 mm x 13.20 mm	-	-	-	37.81	nr	37.81
600 mm x 900 mm x 13.20 mm	-	-	-	56.85	nr	56.85
Fitting components; plywood						
Backs, fronts, sides or divisions; over 300 mm wide						
6 mm thick	-	-	-	19.75	m²	19.75
9 mm thick	-	-	-	26.82	m²	26.82
12 mm thick	-	-	-	33.80	m²	33.80
19 mm thick	-	-	-	49.29	m²	49.29
25 mm thick	-	-	-	66.51	m²	66.51
Shelves or worktops; over 300 mm wide						
19 mm thick	-	-	-	49.29	m²	49.29
25 mm thick	-	-	-	66.51	m²	66.51
Flush doors; lipped on four edges						
450 mm x 750 mm x 19 mm	-	-	-	26.96	nr	26.96
450 mm x 750 mm x 25 mm	-	-	-	33.53	nr	33.53
600 mm x 900 mm x 19 mm	-	-	-	40.16	nr	40.16
600 mm x 900 mm x 25 mm	-	-	-	49.74	nr	49.74
Fitting components; wrought softwood						
Backs, fronts, sides or divisions; cross-tongued joints; over 300 mm wide						
25 mm thick	-	-	-	47.41	m²	47.41
Shelves or worktops; cross-tongued joints; over 300 mm wide						
25 mm thick	-	-	-	47.41	m²	47.41
Bearers						
19 mm x 38 mm	-	-	-	2.69	m	2.69
25 mm x 50 mm	-	-	-	3.38	m	3.38
50 mm x 50 mm	-	-	-	4.91	m	4.91
50 mm x 75 mm	-	-	-	6.58	m	6.58
Bearers; framed						
19 mm x 38 mm	-	-	-	5.26	m	5.26
25 mm x 50 mm	-	-	-	5.94	m	5.94
50 mm x 50 mm	-	-	-	7.50	m	7.50
50 mm x 75 mm	-	-	-	9.16	m	9.16

Prices for Measured Works - Minor Works
N FURNITURE/EQUIPMENT

	PC £	Labour hours	Labour £	Material £	Unit	Total rate £
Framing to backs, fronts or sides						
19 mm x 38 mm	-	-	-	5.26	m	5.26
25 mm x 50 mm	-	-	-	5.94	m	5.94
50 mm x 50 mm	-	-	-	7.50	m	7.50
50 mm x 75 mm	-	-	-	9.16	m	9.16
Flush doors, softwood skeleton or cellular core; plywood facing both sides; lipped on four edges						
450 mm x 750 mm x 35 mm	-	-	-	26.63	nr	26.63
600 mm x 900 mm x 35 mm	-	-	-	42.41	nr	42.41
Add 5% to the above materials prices for selected softwood for staining						
Fitting components; selected West African Mahogany						
Bearers						
19 mm x 38 mm	-	-	-	4.49	m	4.49
25 mm x 50 mm	-	-	-	5.92	m	5.92
50 mm x 50 mm	-	-	-	9.12	m	9.12
50 mm x 75 mm	-	-	-	12.46	m	12.46
Bearers; framed						
19 mm x 38 mm	-	-	-	8.37	m	8.37
25 mm x 50 mm	-	-	-	9.81	m	9.81
50 mm x 50 mm	-	-	-	12.99	m	12.99
50 mm x 75 mm	-	-	-	16.32	m	16.32
Framing to backs, fronts or sides						
19 mm x 38 mm	-	-	-	8.37	m	8.37
25 mm x 50 mm	-	-	-	9.81	m	9.81
50 mm x 50 mm	-	-	-	12.99	m	12.99
50 mm x 75 mm	-	-	-	16.32	m	16.32
Fitting components; Iroko						
Backs, fronts, sides or divisions; cross-tongued joints; over 300 mm wide						
25 mm thick	-	-	-	106.51	m²	106.51
Shelves or worktops; cross-tongued joints; over 300 mm wide						
25 mm thick	-	-	-	106.51	m²	106.51
Draining boards; cross-tongued joints; over 300 mm wide						
25 mm thick	-	-	-	112.56	m²	112.56
stopped flutes	-	-	-	3.40	m	3.40
grooves; cross-grain	-	-	-	0.80	m	0.80
Bearers						
19 mm x 38 mm	-	-	-	5.47	m	5.47
25 mm x 50 mm	-	-	-	7.08	m	7.08
50 mm x 50 mm	-	-	-	10.80	m	10.80
50 mm x 75 mm	-	-	-	14.71	m	14.71
Bearers; framed						
19 mm x 38 mm	-	-	-	10.65	m	10.65
25 mm x 50 mm	-	-	-	12.26	m	12.26
50 mm x 50 mm	-	-	-	15.96	m	15.96
50 mm x 75 mm	-	-	-	19.88	m	19.88
Framing to backs, fronts or sides						
19 mm x 38 mm	-	-	-	10.65	m	10.65
25 mm x 50 mm	-	-	-	12.26	m	12.26
50 mm x 50 mm	-	-	-	15.96	m	15.96
50 mm x 75 mm	-	-	-	19.88	m	19.88

Prices for Measured Works - Minor Works
N FURNITURE/EQUIPMENT

	PC £	Labour hours	Labour £	Material £	Unit	Total rate £
N10/11 GENERAL FIXTURES/KITCHEN FITTINGS - cont'd						
Fitting components; Teak						
Backs, fronts, sides or divisions; cross-tongued joints; over 300 mm wide						
25 mm thick	-	-	-	216.15	m²	216.15
Shelves or worktops; cross-tongued joints; over 300 mm wide						
25 mm thick	-	-	-	216.15	m²	216.15
Draining boards; cross-tongued joints; over 300 mm wide						
25 mm thick	-	-	-	220.70	m²	220.70
stopped flutes	-	-	-	3.40	m	3.40
grooves; cross-grain	-	-	-	0.80	m	0.80
Fixing kitchen fittings						
NOTE: Kitchen fittings prices vary considerably. PC supply prices for reasonable quantities for a moderately priced range of kitchen fittings (Rugby Joinery "Lambeth" range) have been shown but not extended.						
Fixing to backgrounds requiring plugging; including any pre-assembly						
Wall units						
300 mm x 580 mm x 300 mm	45.55	1.20	10.98	0.20	nr	11.18
300 mm x 720 mm x 300 mm	42.65	1.30	11.89	0.20	nr	12.09
600 mm x 580 mm x 300 mm	48.11	1.43	13.08	0.20	nr	13.28
600 mm x 720 mm x 300 mm	53.45	1.62	14.82	0.20	nr	15.02
Floor units with drawers						
600 mm x 900 mm x 500 mm	77.81	1.30	11.89	0.20	nr	12.09
600 mm x 900 mm x 600 mm	81.14	1.43	13.08	0.20	nr	13.28
1200 mm x 900 mm x 600 mm	145.10	1.71	15.65	0.20	nr	15.85
Sink units (excluding sink top)						
1000 mm x 900 mm x 600 mm	120.91	1.62	14.82	0.20	nr	15.02
1200 mm x 900 mm x 600 mm	133.46	1.85	16.93	0.20	nr	17.13
Laminated plastics worktops; single rolled edge; prices include for fixing						
28 mm thick; 500 mm wide	-	0.41	3.75	21.92	m	25.67
38 mm thick; 600 mm wide	-	0.41	3.75	27.33	m	31.08
extra for forming hole for inset sink	-	0.77	7.05	-	nr	7.05
extra for jointing strip at corner intersection of worktops	-	0.16	1.46	11.48	nr	12.94
extra for butt and scribe joint at corner intersection of worktops	-	4.58	41.91	-	nr	41.91
Lockers; The Welconstruct Company or other equal and approved						
Standard clothes lockers; steel body and door within reinforced 19G frame, powder coated finish, cam locks						
1 compartment; placing in position						
305 mm x 305 mm x 1830 mm	-	0.23	1.68	53.90	nr	55.58
305 mm x 460 mm x 1830 mm	-	0.23	1.68	65.44	nr	67.12
460 mm x 460 mm x 1830 mm	-	0.28	2.04	82.31	nr	84.35
610 mm x 460 mm x 1830 mm	-	0.28	2.04	99.20	nr	101.24

N FURNITURE/EQUIPMENT

	PC £	Labour hours	Labour £	Material £	Unit	Total rate £
Compartment lockers; steel body and door with reinforced 19G frame, powder coated finish, cam locks						
2 compartments; placing in position						
305 mm x 305 mm x 1830 mm	-	0.23	1.68	72.87	nr	74.55
305 mm x 460 mm x 1830 mm	-	0.23	1.68	76.13	nr	77.81
460 mm x 460 mm x 1830 mm	-	0.28	2.04	90.49	nr	92.53
4 compartments; placing in position						
305 mm x 305 mm x 1830 mm	-	0.23	1.68	92.06	nr	93.74
305 mm x 460 mm x 1830 mm	-	0.23	1.68	98.36	nr	100.04
460 mm x 460 mm x 1830 mm	-	0.28	2.04	93.75	nr	95.79
Wet area lockers; galvanised steel 18/22G etched primed coating body, galvanised steel 18/20G reinforced door on non-ferrous hinges, powder coated finish, cam locks						
1 compartment; placing in position						
305 mm x 305 mm x 1830 mm	-	0.23	1.68	79.38	nr	81.06
2 compartments; placing in position						
305 mm x 305 mm x 1830 mm	-	0.23	1.68	102.87	nr	104.55
305 mm x 460 mm x 1830 mm	-	0.23	1.68	107.90	nr	109.58
4 compartments; placing in position						
305 mm x 305 mm x 1830 mm	-	0.23	1.68	130.24	nr	131.92
305 mm x 460 mm x 1830 mm	-	0.23	1.68	139.36	nr	141.04
Extra for						
coin operated lock; coin returned	-	-	-	61.97	nr	61.97
coin operated lock; coin retained	-	-	-	86.92	nr	86.92
Timber clothes lockers; veneered MDF finish, routed door, cam locks						
1 compartment; placing in position						
380 mm x 380 mm x 1830 mm	-	0.28	2.04	235.41	nr	237.45
4 compartments; placing in position						
380 mm x 380 mm x 1830 mm	-	0.28	2.04	353.38	nr	355.42
Shelving support systems; The Welconstruct Company or other equal and approved						
Rolled front shelving support systems; steel body; stove enamelled finish						
open initial bay; 5 shelves; placing in position						
915 mm x 305 mm x 1905 mm	-	0.69	5.67	76.00	nr	81.67
915 mm x 460 mm x 1905 mm	-	0.69	5.67	99.07	nr	104.74
open extension bay; 5 shelves; placing in position						
915 mm x 305 mm x 1905 mm	-	0.83	6.82	61.62	nr	68.44
915 mm x 460 mm x 1905 mm	-	0.83	6.82	79.50	nr	86.32
closed initial bay; 5 shelves; placing in position						
915 mm x 305 mm x 1905 mm	-	0.69	5.67	95.28	nr	100.95
915 mm x 460 mm x 1905 mm	-	0.69	5.67	118.24	nr	123.91
closed extension bay; 5 shelves; placing in position						
915 mm x 305 mm x 1905 mm	-	0.83	6.82	85.29	nr	92.11
915 mm x 460 mm x 1905 mm	-	0.83	6.82	102.96	nr	109.78
6 mm thick rectangular glass mirrors; silver backed; fixed with chromium plated domed headed screws; to background requiring plugging						
Mirror with polished edges						
365 mm x 254 mm	5.62	0.81	7.41	6.23	nr	13.64
400 mm x 300 mm	-	0.81	7.41	7.88	nr	15.29
560 mm x 380 mm	12.26	0.92	8.42	13.37	nr	21.79
640 mm x 460 mm	15.58	1.02	9.33	16.94	nr	26.27

Prices for Measured Works - Minor Works
N FURNITURE/EQUIPMENT

	PC £	Labour hours	Labour £	Material £	Unit	Total rate £
N10/11 GENERAL FIXTURES/KITCHEN FITTINGS - cont'd						
6 mm thick rectangular glass mirrors; silver backed; fixed with chromium plated domed headed screws; to background requiring plugging - cont'd						
Mirror with bevelled edges						
365 mm x 254 mm	9.71	0.81	7.41	10.63	nr	18.04
400 mm x 300 mm	11.24	0.81	7.41	12.28	nr	19.69
560 mm x 380 mm	18.39	0.92	8.42	19.96	nr	28.38
640 mm x 460 mm	22.99	1.02	9.33	24.91	nr	34.24
Door mats						
Entrance mats; "Tuftiguard type C" or other equal and approved; laying in position; 12 mm thick						
900 mm x 550 mm	185.95	0.51	3.72	99.95	nr	103.67
1200 mm x 750 mm	185.95	0.51	3.72	179.91	nr	183.63
2400 mm x 1200 mm	185.95	1.02	7.44	575.70	nr	583.14
Matwells						
Polished aluminium matwell; comprising 34 mm x 26 mm x 6 mm angle rim; with brazed angles and lugs brazed on; to suit mat size						
914 mm x 560 mm	34.35	1.02	7.44	36.93	nr	44.37
1067 mm x 610 mm	37.73	1.02	7.44	40.56	nr	48.00
1219 mm x 762 mm	44.60	1.02	7.44	47.95	nr	55.39
Polished brass matwell; comprising 38 mm x 38 mm x 6 mm angle rim; with brazed angles and lugs welded on; to suit mat size						
914 mm x 560 mm	101.70	1.02	7.44	109.33	nr	116.77
1067 mm x 610 mm	111.70	1.02	7.44	120.08	nr	127.52
1219 mm x 762 mm	132.04	1.02	7.44	141.94	nr	149.38
Internal blinds; Luxaflex Ltd or other equal and approved						
Roller blinds; Luxaflex "Safeweave RB"; fire resisting material; 1200 mm drop; fixing with screws						
1000 mm wide	38.83	1.02	7.44	41.74	nr	49.18
2000 mm wide	75.62	1.60	11.66	81.29	nr	92.95
3000 mm wide	114.45	2.16	15.75	123.03	nr	138.78
Roller blinds; Luxaflex "Plain RB"; plain type material; 1200 mm drop; fixing with screws						
1000 mm wide	24.53	1.02	7.44	26.37	nr	33.81
2000 mm wide	47.01	1.60	11.66	50.54	nr	62.20
3000 mm wide	69.49	2.16	15.75	74.70	nr	90.45
Roller blinds; Luxaflex "Dimout plain RB"; blackout material; 1200 mm drop; fixing with screws						
1000 mm wide	38.83	1.02	7.44	41.74	nr	49.18
2000 mm wide	75.62	1.60	11.66	81.29	nr	92.95
3000 mm wide	114.45	2.16	15.75	123.03	nr	138.78
Roller blinds; Luxaflex "Lite-master Crank Op"; 100% blackout; 1200 mm drop; fixing with screws						
1000 mm wide	176.79	2.15	15.67	190.05	nr	205.72
2000 mm wide	251.39	3.02	22.02	270.24	nr	292.26
3000 mm wide	315.77	3.89	28.36	339.45	nr	367.81
Vertical louvre blinds; 89 mm wide louvres; Luxaflex "Finessa 3430" Group 0; 1200 mm drop; fixing with screws						
1000 mm wide	55.18	0.91	6.63	59.32	nr	65.95
2000 mm wide	95.04	1.42	10.35	102.17	nr	112.52
3000 mm wide	134.89	1.94	14.14	145.01	nr	159.15

Prices for Measured Works - Minor Works
N FURNITURE/EQUIPMENT

	PC £	Labour hours	Labour £	Material £	Unit	Total rate £
Vertical louvre blinds; 127 mm wide louvres; "Finessa 3420" Group 0; 1200 mm drop; fixing with screws						
1000 mm wide	49.05	0.97	7.07	52.73	nr	59.80
2000 mm wide	81.75	1.50	10.94	87.88	nr	98.82
3000 mm wide	117.52	2.00	14.58	126.33	nr	140.91
N13 SANITARY APPLIANCES/FITTINGS						
Sinks; Armitage Shanks or equal and approved						
Sinks; white glazed fireclay; BS 1206; pointing all round with Dow Corning Hansil silicone sealant ref 785						
Belfast sink; 455 mm x 380 mm x 205 mm ref 350016S; Nimbus ½" inclined bib taps ref 6610400; ½" wall mounts ref 81460PR; 1½" slotted waste, chain and plug, screw stay ref 70668M8; aluminium alloy build-in fixing brackets ref 7931WD0	-	2.78	33.98	134.93	nr	168.91
Belfast sink; 610 mm x 455 mm x 255 mm ref 350086S; Nimbus ½" inclined bib taps ref 6610400; ½" wall mounts ref 81460PR; 1½" slotted waste, chain and plug, screw stay ref 70668M8; aluminium alloy build-in fixing brackets ref 7931VE0	-	2.78	33.98	170.02	nr	204.00
Belfast sink; 760 mm x 455 mm x 255 mm ref 3500A6S; Nimbus ½" inclined bib taps ref 6610400; ½" wall mounts ref 81460PR; 1½" slotted waste, chain and plug, screw stay ref 70668M8; aluminium alloy build-in fixing brackets ref 7931VE0	-	2.78	33.98	248.04	nr	282.02
Lavatory basins; Armitage Shanks or equal and approved						
Basins; white vitreous china; BS 5506 Part 3; pointing all round with Dow Corning Hansil silcone sealant ref 785						
Portman basin 400 mm x 365 mm ref 117913J; Nuastyle 2 ½" pillar taps with anti-vandal indices ref 6973400; 1½" bead chain waste and plug, 80 mm slotted tail, bolt stay ref 90547N1; 1½" plastics bottle trap with 75 mm seal ref 70237Q4; concealed fixing bracket ref 790002Z; Isovalve servicing valve ref 9060400; screwing	-	2.13	26.03	105.35	nr	131.38
Portman basin 500 mm x 420 mm ref 117923S; Nuastyle 2 ½" pillar taps with anti-vandal indices ref 6973400; 1½" bead chain waste and plug, 80 mm slotted tail, bolt stay ref 90547N1; 1½" plastics bottle trap with 75 mm seal ref 70237Q4; concealed fixing bracket ref 790002Z; Isovalve servicing valve ref 9060400; screwing	-	2.13	26.03	127.98	nr	154.01
Tiffany basin 560 mm x 455 mm ref 121614A; Tiffany pedestal ref 132408S; Millenia ½" monobloc mixer tap with non-return valves and 1½" pop-up waste ref 6104XXX; Universal Porcelain handwheels ref 63614XXX; Isovalve servicing valve ref 9060400; screwing	-	2.13	26.03	142.88	nr	168.91

Prices for Measured Works - Minor Works
N FURNITURE/EQUIPMENT

	PC £	Labour hours	Labour £	Material £	Unit	Total rate £
N13 SANITARY APPLIANCES/FITTINGS - cont'd						
Lavatory basins; Armitage Shanks or equal and approved - cont'd						
Basins; white vitreous china; BS 5506 Part 3; pointing all round with Dow Corning Hansil silcone sealant ref 785 - cont'd						
Montana basin 510 mm x 410 mm ref 120814E; Montana pedestal ref 132408S; Millenia ½" monobloc mixer tap with non-return valves and 1½" pop-up waste ref 6104XXX; Millenia metal handwheels ref 63664XX; Isovalve servicing valve ref 9060400; screwing	-	2.31	28.23	148.66	nr	176.89
Montana basin 580 mm x 475 mm ref 120824E; Montana pedestal ref 132408S; Millenia ½" monobloc mixer tap with non-return valves and 1½" pop-up waste ref 6104XXX; Millenia metal handwheels ref 63664XX; Isovalve servicing valve ref 9060400; screwing	-	2.31	28.23	152.30	nr	180.53
Portman basin 600 mm x 480 mm ref 117933S; Nuastyle 2 ½" pillar taps with anti-vandal indices ref 6973400; 1½" bead chain waste and plug, 80 mm slotted tail, bolt stay ref 90547N1; 1½" plastics bottle trap with 75 mm seal ref 70237Q4; concealed fixing bracket ref 790002Z; Isovalve servicing valve ref 9060400; screwing	-	2.13	26.03	167.36	nr	193.39
Cottage basin 560 mm x 430 mm ref 121013S; Cottage pedestal ref 132008S; Cottage ½" pillar taps ref 9290400; 1½" bead chain waste and plug, 80 mm slotted tail, bolt stay ref 90547N1' Isovalve servicing valve ref 9060400; screwing	-	2.31	28.23	237.03	nr	265.26
Cottage basin 625 mm x 500 mm ref 121023S; Cottage pedestal ref 132008S; Cottage ½" pillar taps ref 9290400; 1½" bead chain waste and plug, 80 mm slotted tail, bolt stay ref 90547N1' Isovalve servicing valve ref 9060400; screwing	-	2.31	28.23	280.30	nr	308.53
Cliveden basin 620 mm x 525 mm ref 121223S; Cliveden pedestal ref 132308S; Cliveden ½" mixer tap with 1½" pop-up waste ref 9456400; Isovalve servicing vale ref 9060400; screwing	-	2.31	28.23	612.01	nr	640.24
Drinking fountains; Armitage Shanks or equal and approved						
White vitreous china fountains; pointing all round with Dow Corning Hansil silicone selant ref 785						
Aqualon drinking fountain ref 200100; ½" self closing non-conclussive valve with flow control, plastics strainer waste, concealed hanger ref 0275A00; 1½" plastics bottle trap with 75 mm seal; screwing	-	2.31	28.23	198.99	nr	227.22
Stainless steel fountains; pointing all round with Dow Corning Hansil silicone selant ref 785						
Purita drinking fountain, self closing non-conclussive valve with push button operation, flow control and 1½" strainer waste ref 53400Z5; screwing	-	2.31	28.23	176.24	nr	206.47
Purita drinking fountain, self closing non-conclussive valve with push button operation, flow control and 1½" strainer waste ref 53400Z5; pedestal shroud ref 5341000; screwing	-	2.78	33.98	395.13	nr	429.11

N FURNITURE/EQUIPMENT

	PC £	Labour hours	Labour £	Material £	Unit	Total rate £
Baths; Armitage Shanks or equal and approved						
Bath; reinforced acrylic rectangular pattern; chromium plated overflow chain and plug; 40 mm diameter chromium plated waste; cast brass "P" trap with plain outlet and overflow connection; pair 20 mm diameter chromium plated easy clean pillar taps to BS 1010						
1700 mm long; white	-	3.50	42.78	212.47	nr	255.25
1700 mm long; coloured	-	3.50	42.78	235.43	nr	278.21
Bath; enamelled steel; medium gauge rectangular pattern; 40 mm diameter chromium plated overflow chain and plug; 40 mm diameter chromium plated waste; cast brass "P" trap with plain outlet and overflow connection; pair 20 mm diameter chromium plated easy clean pillar taps to BS 1010						
1700 mm long; white	-	3.50	42.78	327.31	nr	370.09
1700 mm long; coloured	-	3.50	42.78	344.54	nr	387.32
Water closets; Armitage Shanks or equal and approved						
White vitreous china pans and cisterns; pointing all round base with Dow Corning Hansil silicone sealant ref 785						
Seville close coupled washdown pan ref 147001A; Seville plastics seat and cover ref 68780B1; Panekta WC pan P trap connector ref 9013000; Seville 6 litres capacity cistern and cover, bottom supply ball valve, bottom overflow and close coupling fitment ref 17156FR; Seville modern lever ref 7959STR	-	3.05	37.28	180.22	nr	217.50
Extra over for; Panekta WC S trap connector ref 9014000	-	-	-	1.68	nr	1.68
Wentworth close coupled washdown pan ref 150601A; Orion III plastics seat and cover; Panekta WC pan P trap connector ref 9013000; Group 7½ litres capacity cistern and cover, bottom supply ballvalve, bottom overflow, close coupling fitment and lever ref 17650FB	-	3.05	37.28	212.36	nr	249.64
Tiffany back to wall washdown pan ref 154601A; Saturn plastics seat and cover ref 08900D1; Panekta WC pan P trap connector ref 9013000; Conceala 7½ litres capacity cistern and cover, side supply ball valve, side overflow, flushbend and extended lever ref 42350JE	-	3.05	37.28	230.32	nr	267.60
Extra over for; Panekta WC pan S trap connector ref 9014000	-	-	-	1.68	nr	1.68
Tiffany close coupled washdown pan ref 154301A; Saturn plastics seat and cover ref 68980B1; Panekta WC pan P trap connector ref 9013000; Tiffany 7½ litres capacity cistern and cover, bottom supply ball valve, bottom overflow, close coupling fitment and side lever ref 17751FN	-	3.05	37.28	245.53	nr	282.81
Extra over for Panekta WC pan S trap connector ref 9014000	-	-	-	1.68	nr	1.68

Prices for Measured Works - Minor Works
N FURNITURE/EQUIPMENT

	PC £	Labour hours	Labour £	Material £	Unit	Total rate £
N13 SANITARY APPLIANCES/FITTINGS - cont'd						
Water closets; Armitage Shanks or equal and approved - cont'd						
White vitreous china pans and cisterns; pointing all round base with Dow Corning Hansil silicone sealant ref 785 - cont'd						
Cameo close coupled washdown pan ref 154301A; Cameo plastics seat and cover ref 6879NB2; Panekta WC pan P trap connector ref 9013000; Cameo 7½ litres capacity cistern and cover, bottom supply ball valve, bottom overflow and close coupling fitment ref 17831KR; luxury metal lever ref 7968000	-	3.05	37.28	316.64	nr	353.92
Extra over for; Panekta WC pan S trap connector ref 9014000	-	-	-	1.68	nr	1.68
Cottage close coupled washdown pan ref 152301A; mahogany seat and cover ref 68970B2; Panekta WC pan P trap connector ref 9013000; 7½ lites capacity cistern and cover, bottom supply ball valve, bottom overflow and close coupling fitment ref 17700FR; mahogany lever assembly ref S03SN23	-	3.05	37.28	495.80	nr	533.08
Extra over for; Panekta WC pan S trap connector ref 9014000	-	-	-	1.68	nr	1.68
Cliveden close coupled washdown pan ref 153201A; mahogany seat and cover ref 68970B2; Panekta WC pan P trap connector ref 9013000; 7½ litres capacity cistern and cover, bottom supply ball valve, bottom overflow and close coupling fitment ref 17720FR; brass level assembly ref S03SN01	-	3.05	37.28	550.95	nr	588.23
Extra over for; Panekta WC pan S trap connector ref 9014000	-	-	-	1.68	nr	1.68
Concept back to wall washdown pan ref 153301A; Concept plastics seat and cover ref 6896AB2; Panekta WC pan P trap connector ref 9013000; Conceala 7½ litres capacity cistern and cover, side supply ball valve, side overflow, flushbend and extended lever ref 42350JE	-	3.05	37.28	651.59	nr	688.87
Wall urinals; Armitage Shanks or equal and approved						
White vitreous china bowls and cisterns; pointing all round with Dow Corning Hansil silicone sealant ref 785						
single Sanura 400 mm bowls ref 261119E; top inlet spreader ref 74344A1; concealed steel hangers ref 7220000; 1½" plastics domed strainer waste ref 90568N0; 1½" plastics bottle trap with 75 mm seal ref 70238Q4; Conceala 4½ litres capacity cistern and cover ref 4225100; polished stainless steel exposed flushpipes ref 74450A1PO; screwing	-	3.70	45.22	239.44	nr	284.66
single Sanura 500 mm bowls ref 261129E; top inlet spreader ref 74344A1; concealed steel hangers ref 7220000; 1½" plastics domed strainer waste ref 90568N0; 1½" plastics bottle trap with 75 mm seal ref 70238Q4; Conceala 4½ litres capacity cistern and cover ref 4225100; polished stainless steel exposed flushpipes ref 74450A1PO; screwing	-	3.70	45.22	310.24	nr	355.46

N FURNITURE/EQUIPMENT

	PC £	Labour hours	Labour £	Material £	Unit	Total rate £
White vitreous china bowls and cisterns; pointing all round with Dow Corning Hansil silicone sealant ref 785 - cont'd						
range of 2 nr Sanura 400 mm bowls ref 261119E; top inlet spreaders ref 74344A1; concealed steel hangers ref 7220000; 1½" plastics domed strainer wastes ref 90568N0; 1½" plastics bottle traps with 75 mm seal ref 70238Q4; Conceala 9 litres capacity cistern and cover ref 4225200; polished stainless steel exposed flushpipes ref 74450B1PO; screwing	-	6.94	84.83	388.73	nr	473.56
range of 2 nr Sanura 500 mm bowls ref 261129E; top inlet spreaders ref 74344A1; concealed steel hangers ref 7220000; 1½" plastics domed strainer wastes ref 90568N0; 1½" plastics bottle traps with 75 mm seal ref 70238Q4; Conceala 9 litres capacity cistern and cover ref 4225200; polished stainless steel exposed flushpipes ref 74450B1PO; screwing	-	6.94	84.83	530.32	nr	615.15
range of 3 nr Sanura 400 mm bowls ref 261119E; top inlet spreaders ref 74344A1; concealed steel hangersref 7220000; 1½" plastics domed strainer wastes ref 90568N0; 1½" plastics bottle traps with 75 mm seal ref 70238Q4; Conceala 9 litres capacity cistern and cover ref 4225200; polished stainless steel flushpipes ref 74450C1PO; screwing	-	10.18	124.43	524.80	nr	649.23
range of 3 nr Sanura 500 mm bowls ref 261129E; top inlet spreaders ref 74344A1; concealed steel hangers ref 7220000; 1½" plastics domed strainer wastes ref 90568N0; 1½" plastics bottle traps with 75 mm seal ref 70238Q4; Conceala 9 litres capacity cistern and cover ref 4225200; polished stainless steel flushpipes ref 74450C1PO; screwing	-	10.18	124.43	736.94	nr	861.37
range of 4 nr Sanura 400 mm bowls ref 261119E; top inlet spreaders ref 74344A1; concealed steel hangers ref 7220000; 1½" plastics domed strainer wastes ref 90568N0; 1½" plastics bottle traps with 75 mm seal ref 70238Q4; Conceala 13.60 litres capacity cistern and cover ref 4225300; polished stainless steel flushpipes ref 74450D1PO; screwing	-	13.41	163.91	681.40	nr	845.31
range of 4 nr Sanura 500 mm bowls ref 261129E; top inlet spreaders ref 74344A1; concealed steel hangers ref 7220000; 1½" plastics domed strainer wastes ref 90568N0; 1½" plastics bottle traps with 75 mm seal ref 70238Q4; Conceala 13.60 litres capacity cistern and cover ref 4225300; polished stainless steel flushpipes ref 74450D1PO; screwing	-	13.41	163.91	964.56	nr	1128.47
White vitreous china division panels; pointing all round with Dow Corning Hansill silicone sealant ref 785						
625 mm long ref 2605000; screwing	-	0.69	8.43	61.49	nr	69.92

Prices for Measured Works - Minor Works
N FURNITURE/EQUIPMENT

	PC £	Labour hours	Labour £	Material £	Unit	Total rate £
N13 SANITARY APPLIANCES/FITTINGS - cont'd						
Bidets; Armitage Shanks or equal and approved						
Bidet; vitreous china; chromium plated waste;						
mixer tap with hand wheels						
600 mm x 400 mm x 400 mm; white	-	3.50	42.78	212.47	nr	255.25
Shower trays and fittings; Armitage Shanks						
or equal and approved						
Shower tray; glazed fireclay with outlet and						
grated waste; chain and plug; bedding and						
pointing in waterproof cement mortar						
760 mm x 760 mm x 180 mm; white	-	3.00	36.67	132.08	nr	168.75
760 mm x 760 mm x 180 mm; coloured	-	3.00	36.67	200.97	nr	237.64
Shower fitting; riser pipe with mixing valve and						
shower rose; chromium plated; plugging and						
screwing mixing valve and pipe bracket						
15 mm diameter riser pipe; 127 mm diameter						
shower rose	-	5.00	61.11	258.42	nr	319.53
Miscellaneous fittings; Pressalit or equal						
and approved						
Raised seats						
Ergosit; 50 mm high; ref R19000	-	0.50	4.92	87.98	nr	92.90
Dania; 50 mm high; ref R23000	-	0.50	4.92	97.18	nr	102.10
Dania; 100 mm high; ref R24000	-	0.50	4.92	148.93	nr	153.85
Dania; sloping 100 mm - 50 mm high; ref R25000	-	0.50	4.92	141.45	nr	146.37
Raised seats and covers						
Ergosit; 50 mm high; ref R20000	-	0.50	4.92	120.18	nr	125.10
Dania; 50 mm high; ref R33000	-	0.50	4.92	105.23	nr	110.15
Dania; 100 mm high; ref R34000	-	0.50	4.92	173.65	nr	178.57
Dania; sloping 100 mm - 50 mm high; ref R35000	-	0.50	4.92	167.33	nr	172.25
Miscellaneous fittings; Pressalit Ltd or equal						
and approved						
Grab rails						
300 mm long ref RT100000; screwing	-	0.50	4.92	43.04	nr	47.96
450 mm long ref RT101000; screwing	-	0.50	4.92	50.34	nr	55.26
600 mm long ref RT102000; screwing	-	0.50	4.92	57.80	nr	62.72
800 mm long ref RT103000; screwing	-	0.50	4.92	64.69	nr	69.61
1000 mm long ref RT104000; screwing	-	0.50	4.92	74.45	nr	79.37
Angled grab rails						
900 mm long, angled 135 ref RT110000; screwing	-	0.50	4.92	93.40	nr	98.32
1300 mm long, angled 90 ref RT119000; screwing	-	0.75	7.38	146.42	nr	153.80
Hinged grab rails						
600 mm long ref R3016000 ; screwing	-	0.35	3.44	151.47	nr	154.91
600 mm long with spring counter balance ref						
RF016000 ; screwing	-	0.35	3.44	210.04	nr	213.48
800 mm long ref R3010000 ; screwing	-	0.35	3.44	181.33	nr	184.77
800 mm long with spring counter balance ref						
RF010000 ; screwing	-	0.35	3.44	223.24	nr	226.68

N FURNITURE/EQUIPMENT

	PC £	Labour hours	Labour £	Material £	Unit	Total rate £
N15 SIGNS/NOTICES						
Plain script; in gloss oil paint; on painted or varnished surfaces						
Capital letters; lower case letters or numerals						
per coat; per 25 mm high	-	0.10	0.92	-	nr	**0.92**
Stops						
per coat	-	0.02	0.18	-	nr	**0.18**

Prices for Measured Works - Major Works
P BUILDING FABRIC SUNDRIES

	PC £	Labour hours	Labour £	Material £	Unit	Total rate £
P BUILDING FABRIC SUNDRIES						
P10 SUNDRY INSULATION/PROOFING WORK/FIRESTOPS						
"Sisalkraft" building papers/vapour barriers or other equal and approved						
Building paper; 150 mm laps; fixed to softwood						
"Moistop" grade 728 (class A1F)	-	0.10	1.30	1.27	m^2	2.57
Vapour barrier/reflective insulation 150 mm laps; fixed to softwood						
"Insulex" grade 714; single sided	-	0.10	1.30	1.58	m^2	2.88
"Insulex" grade 714; double sided	-	0.10	1.30	2.25	m^2	3.55
Mat or quilt insulation						
Glass fibre quilt; "Isowool 1000" or other equal and approved; laid loose between members at 600 mm centres						
60 mm thick	1.90	0.21	2.73	2.04	m^2	4.77
80 mm thick	2.49	0.23	2.99	2.68	m^2	5.67
100 mm thick	2.98	0.26	3.39	3.20	m^2	6.59
150 mm thick	4.54	0.31	4.04	4.88	m^2	8.92
Mineral fibre quilt; "Isowool 1200" or other equal and approved; pinned vertically to softwood						
25 mm thick	1.61	0.14	1.82	1.73	m^2	3.55
50 mm thick	2.58	0.16	2.08	2.77	m^2	4.85
Glass fibre building roll; pinned vertically to softwood						
60 mm thick	1.88	0.16	2.08	2.12	m^2	4.20
80 mm thick	2.46	0.17	2.21	2.78	m^2	4.99
100 mm thick	2.94	0.18	2.34	3.32	m^2	5.66
Glass fibre flanged building roll; paper faces; pinned vertically or to slope between timber framing						
60 mm thick	4.13	0.19	2.47	4.66	m^2	7.13
80 mm thick	5.16	0.20	2.60	5.82	m^2	8.42
100 mm thick	6.02	0.21	2.73	6.80	m^2	9.53
Glass fibre aluminium foil faced roll; pinned to softwood						
80 mm thick	-	0.19	2.47	4.92	m^2	7.39
Board or slab insulation						
Expanded polystyrene board standard grade RD/N or other equal and approved; fixed with adhesive						
25 mm thick	-	0.43	5.60	3.06	m^2	8.66
40 mm thick	-	0.45	5.86	4.44	m^2	10.30
50 mm thick	-	0.51	6.64	5.36	m^2	12.00
Jablite expanded polystyrene board; grade EHD(N) or other equal and approved						
100 mm thick	-	0.51	6.64	1.82	m^2	8.46
"Styrofoam Floormate 350" extruded polystyrene foam or other equal and approved						
50 mm thick	-	0.51	6.64	13.16	m^2	19.80
Fire stops						
Cape "Firecheck" channel or other equal and approved; intumescent coatings on cut mitres; fixing with brass cups and screws						
19 mm x 44 mm or 19 mm x 50 mm	12.90	0.61	7.94	15.05	m	22.99

P BUILDING FABRIC SUNDRIES

	PC £	Labour hours	Labour £	Material £	Unit	Total rate £
"Sealmaster" intumescent fire and smoke seals or other equal and approved; pinned into groove in timber						
type N30; for single leaf half hour door	6.39	0.31	4.04	7.21	m	11.25
type N60; for single leaf one hour door	9.72	0.33	4.30	10.98	m	15.28
type IMN or IMP; for meeting or pivot stiles of pair of one hour doors; per stile	9.72	0.33	4.30	10.98	m	15.28
intumescent plugs in timber; including boring	-	0.10	1.30	0.34	nr	1.64
Rockwool fire stops or other equal and approved; between top of brick/block wall and concrete soffit						
25 mm deep x 112 mm wide	-	0.08	1.04	1.08	m	2.12
25 mm deep x 150 mm wide	-	0.10	1.30	1.44	m	2.74
50 mm deep x 225 mm wide	-	0.16	2.08	5.09	m	7.17
Fire protection compound						
Quelfire "QF4", fire protection compound or other equal and approved; filling around pipes, ducts and the like; including all necessary formwork						
300 mm x 300 mm x 250 mm; pipes - 2	-	1.25	13.52	39.65	nr	53.17
Fire barriers						
Rockwool fire barrier or other and approved between top of suspended ceiling and concrete soffit						
one 50 mm layer x 900 mm wide; half hour	-	0.61	7.94	7.36	m^2	15.30
two 50 mm layers x 900 mm wide; one hour	-	0.92	11.98	14.42	m^2	26.40
Lamatherm fire barrier or other equal and approved; to void below raised access floors						
75 mm thick x 300 mm wide; half hour	-	0.17	2.21	14.15	m	16.36
75 mm thick x 600 mm wide; half hour	-	0.17	2.21	31.83	m	34.04
90 mm thick x 300 mm wide; half hour	-	0.17	2.21	17.82	m	20.03
90 mm thick x 600 mm wide; half hour	-	0.17	2.21	42.46	m	44.67
Dow Chemicals "Styrofoam 1B" or other equal and approved; cold bridging insulation fixed with adhesive to brick, block or concrete base						
Insulation to walls						
25 mm thick	-	0.34	4.43	6.28	m^2	10.71
50 mm thick	-	0.37	4.82	11.52	m^2	16.34
75 mm thick	-	0.39	5.08	17.17	m^2	22.25
Insulation to isolated columns						
25 mm thick	-	0.42	5.47	6.28	m^2	11.75
50 mm thick	-	0.44	5.73	11.52	m^2	17.25
75 mm thick	-	0.48	6.25	17.17	m^2	23.42
Insulation to ceilings						
25 mm thick	-	0.37	4.82	6.28	m^2	11.10
50 mm thick	-	0.40	5.21	11.52	m^2	16.73
75 mm thick	-	0.43	5.60	17.17	m^2	22.77
Insulation to isolated beams						
25 mm thick	-	0.44	5.73	6.28	m^2	12.01
50 mm thick	-	0.48	6.25	11.52	m^2	17.77
75 mm thick	-	0.51	6.64	17.17	m^2	23.81
P11 FOAMED/FIBRE/BEAD CAVITY WALL INSULATION						
Injected insulation						
Cavity wall insulation; injecting 65 mm cavity with						
blown EPS granules	-	-	-	-	m^2	2.60
blown mineral wool	-	-	-	-	m^2	2.70

P BUILDING FABRIC SUNDRIES

	PC £	Labour hours	Labour £	Material £	Unit	Total rate £
P20 UNFRAMED ISOLATED TRIMS/ SKIRTINGS/SUNDRY ITEMS						
Blockboard (Birch faced); 18 mm thick Window boards and the like; rebated; hardwood lipped on one edge						
18 mm x 200 mm	-	0.26	3.39	4.21	m	7.60
18 mm x 250 mm	-	0.29	3.78	4.95	m	8.73
18 mm x 300 mm	-	0.31	4.04	5.69	m	9.73
18 mm x 350 mm	-	0.34	4.43	6.42	m	10.85
returned and fitted ends	-	0.22	2.86	0.43	nr	3.29
Blockboard (Sapele veneered one side); 18 mm thick Window boards and the like; rebated; hardwood lipped on one edge						
18 mm x 200 mm	-	0.28	3.65	3.76	m	7.41
18 mm x 250 mm	-	0.31	4.04	4.38	m	8.42
18 mm x 300 mm	-	0.33	4.30	5.00	m	9.30
18 mm x 350 mm	-	0.37	4.82	5.62	m	10.44
returned and fitted ends	-	0.22	2.86	0.43	nr	3.29
Blockboard (American White Ash veneered one side); 18 mm thick Window boards and the like; rebated; hardwood lipped on one edge						
18 mm x 200 mm	-	0.28	3.65	4.29	m	7.94
18 mm x 250 mm	-	0.31	4.04	5.05	m	9.09
18 mm x 300 mm	-	0.33	4.30	5.81	m	10.11
18 mm x 350 mm	-	0.37	4.82	6.57	m	11.39
returned and fitted ends	-	0.22	2.86	0.43	nr	3.29
Medium density fibreboard primed profiles; Balcas Kildare Ltd or other equal and approved Window boards; rounded and rebated						
25 mm x 220 mm	-	0.31	4.04	4.09	m	8.13
25 mm x 245 mm	-	0.31	4.04	4.44	m	8.48
Skirtings						
14.50 mm x 45 mm; rounded	-	0.10	1.30	0.76	m	2.06
14.50 mm x 70 mm; rounded	-	0.10	1.30	1.05	m	2.35
14.50 mm x 95 mm; rounded	-	0.10	1.30	1.24	m	2.54
14.50 mm x 95 mm; moulded	-	0.10	1.30	1.24	m	2.54
14.50 mm x 120 mm; moulded	-	0.10	1.30	1.44	m	2.74
18 mm x 70 mm; moulded	-	0.10	1.30	1.15	m	2.45
18 mm x 145 mm; moulded	-	0.10	1.30	1.88	m	3.18
Dado rail						
18 mm x 58 mm; moulded	-	0.10	1.30	1.16	m	2.46
Wrought softwood Skirtings, picture rails, dado rails and the like; splayed or moulded						
19 mm x 50 mm; splayed	-	0.10	1.30	1.59	m	2.89
19 mm x 50 mm; moulded	-	0.10	1.30	1.73	m	3.03
19 mm x 75 mm; splayed	-	0.10	1.30	1.87	m	3.17
19 mm x 75 mm; moulded	-	0.10	1.30	2.01	m	3.31
19 mm x 100 mm; splayed	-	0.10	1.30	2.17	m	3.47
19 mm x 100 mm; moulded	-	0.10	1.30	2.31	m	3.61
19 mm x 150 mm; moulded	-	0.12	1.56	2.95	m	4.51
19 mm x 175 mm; moulded	-	0.12	1.56	3.23	m	4.79
22 mm x 100 mm; splayed	-	0.10	1.30	2.39	m	3.69

P BUILDING FABRIC SUNDRIES

	PC £	Labour hours	Labour £	Material £	Unit	Total rate £
Skirtings, picture rails, dado rails and the like; splayed or moulded - cont'd						
25 mm x 50 mm; moulded	-	0.10	1.30	1.92	m	3.22
25 mm x 75 mm; splayed	-	0.10	1.30	2.17	m	3.47
25 mm x 100 mm; splayed	-	0.10	1.30	2.56	m	3.86
25 mm x 150 mm; splayed	-	0.12	1.56	3.36	m	4.92
25 mm x 150 mm; moulded	-	0.12	1.56	3.49	m	5.05
25 mm x 175 mm; moulded	-	0.12	1.56	3.90	m	5.46
25 mm x 225 mm; moulded	-	0.14	1.82	4.69	m	6.51
returned ends	-	0.16	2.08	-	nr	2.08
mitres	-	0.10	1.30	-	nr	1.30
Architraves, cover fillets and the like; half round; splayed or moulded						
13 mm x 25 mm; half round	-	0.12	1.56	1.21	m	2.77
13 mm x 50 mm; moulded	-	0.12	1.56	1.56	m	3.12
16 mm x 32 mm; half round	-	0.12	1.56	1.32	m	2.88
16 mm x 38 mm; moulded	-	0.12	1.56	1.51	m	3.07
16 mm x 50 mm; moulded	-	0.12	1.56	1.64	m	3.20
19 mm x 50 mm; splayed	-	0.12	1.56	1.59	m	3.15
19 mm x 63 mm; splayed	-	0.11	1.43	1.74	m	3.17
19 mm x 75 mm; splayed	-	0.12	1.56	1.87	m	3.43
25 mm x 44 mm; splayed	-	0.12	1.56	1.79	m	3.35
25 mm x 50 mm; moulded	-	0.12	1.56	1.92	m	3.48
25 mm x 63 mm; splayed	-	0.12	1.56	1.97	m	3.53
25 mm x 75 mm; splayed	-	0.12	1.56	2.17	m	3.73
32 mm x 88 mm; moulded	-	0.12	1.56	3.01	m	4.57
38 mm x 38 mm; moulded	-	0.12	1.56	1.91	m	3.47
50 mm x 50 mm; moulded	-	0.12	1.56	2.56	m	4.12
returned ends	-	0.16	2.08	-	nr	2.08
mitres	-	0.10	1.30	-	nr	1.30
Stops; screwed on						
16 mm x 38 mm	-	0.10	1.30	1.20	m	2.50
16 mm x 50 mm	-	0.10	1.30	1.31	m	2.61
19 mm x 38 mm	-	0.10	1.30	1.27	m	2.57
25 mm x 38 mm	-	0.10	1.30	1.41	m	2.71
25 mm x 50 mm	-	0.10	1.30	1.62	m	2.92
Glazing beads and the like						
13 mm x 16 mm	-	0.05	0.65	0.88	m	1.53
13 mm x 19 mm	-	0.05	0.65	0.91	m	1.56
13 mm x 25 mm	-	0.05	0.65	0.97	m	1.62
13 mm x 25 mm; screwed	-	0.05	0.65	1.04	m	1.69
13 mm x 25 mm; fixing with brass cups and screws	-	0.10	1.30	1.27	m	2.57
16 mm x 25 mm; screwed	-	0.05	0.65	1.10	m	1.75
16 mm quadrant	-	0.05	0.65	1.06	m	1.71
19 mm quadrant or scotia	-	0.05	0.65	1.15	m	1.80
19 mm x 36 mm; screwed	-	0.05	0.65	1.29	m	1.94
25 mm x 38 mm; screwed	-	0.05	0.65	1.42	m	2.07
25 mm quadrant or scotia	-	0.05	0.65	1.29	m	1.94
38 mm scotia	-	0.05	0.65	1.83	m	2.48
50 mm scotia	-	0.05	0.65	2.47	m	3.12
Isolated shelves, worktops, seats and the like						
19 mm x 150 mm	-	0.17	2.21	2.79	m	5.00
19 mm x 200 mm	-	0.22	2.86	3.34	m	6.20
25 mm x 150 mm	-	0.17	2.21	3.34	m	5.55
25 mm x 200 mm	-	0.22	2.86	4.10	m	6.96
32 mm x 150 mm	-	0.17	2.21	3.97	m	6.18
32 mm x 200 mm	-	0.22	2.86	4.97	m	7.83

Prices for Measured Works - Major Works
P BUILDING FABRIC SUNDRIES

	PC £	Labour hours	Labour £	Material £	Unit	Total rate £
P20 UNFRAMED ISOLATED TRIMS/ SKIRTINGS/SUNDRY ITEMS - cont'd						
Wrought softwood - cont'd						
Isolated shelves, worktops, seats and the like; cross-tongued joints						
19 mm x 300 mm	-	0.29	3.78	12.99	m	16.77
19 mm x 450 mm	-	0.34	4.43	14.85	m	19.28
19 mm x 600 mm	-	0.41	5.34	21.60	m	26.94
25 mm x 300 mm	-	0.29	3.78	14.40	m	18.18
25 mm x 450 mm	-	0.34	4.43	16.74	m	21.17
25 mm x 600 mm	-	0.41	5.34	24.18	m	29.52
32 mm x 300 mm	-	0.29	3.78	15.79	m	19.57
32 mm x 450 mm	-	0.34	4.43	18.82	m	23.25
32 mm x 600 mm	-	0.41	5.34	27.00	m	32.34
Isolated shelves, worktops, seats and the like; slatted with 50 mm wide slats at 75 mm centres						
19 mm thick	-	1.35	17.58	9.80	m	27.38
25 mm thick	-	1.35	17.58	11.23	m	28.81
32 mm thick	-	1.35	17.58	12.50	m	30.08
Window boards, nosings, bed moulds and the like; rebated and rounded						
19 mm x 75 mm	-	0.19	2.47	2.12	m	4.59
19 mm x 150 mm	-	0.20	2.60	3.06	m	5.66
19 mm x 225 mm; in one width	-	0.27	3.52	3.94	m	7.46
19 mm x 300 mm; cross-tongued joints	-	0.31	4.04	13.07	m	17.11
25 mm x 75 mm	-	0.19	2.47	2.41	m	4.88
25 mm x 150 mm	-	0.20	2.60	3.63	m	6.23
25 mm x 225 mm; in one width	-	0.27	3.52	4.80	m	8.32
25 mm x 300 mm; cross-tongued joints	-	0.31	4.04	14.41	m	18.45
32 mm x 75 mm	-	0.19	2.47	2.75	m	5.22
32 mm x 150 mm	-	0.20	2.60	4.28	m	6.88
32 mm x 225 mm; in one width	-	0.27	3.52	5.76	m	9.28
32 mm x 300 mm; cross-tongued joints	-	0.31	4.04	15.81	m	19.85
38 mm x 75 mm	-	0.19	2.47	3.06	m	5.53
38 mm x 150 mm	-	0.20	2.60	4.81	m	7.41
38 mm x 225 mm; in one width	-	0.27	3.52	6.62	m	10.14
38 mm x 300 mm; cross-tongued joints	-	0.31	4.04	17.07	m	21.11
returned and fitted ends	-	0.16	2.08	-	nr	2.08
Handrails; mopstick						
50 mm diameter	-	0.26	3.39	3.69	m	7.08
Handrails; rounded						
44 mm x 50 mm	-	0.26	3.39	3.69	m	7.08
50 mm x 75 mm	-	0.28	3.65	4.51	m	8.16
63 mm x 87 mm	-	0.31	4.04	6.10	m	10.14
75 mm x 100 mm	-	0.36	4.69	6.84	m	11.53
Handrails; moulded						
44 mm x 50 mm	-	0.23	2.99	4.11	m	7.10
50 mm x 75 mm	-	0.25	3.25	4.94	m	8.19
63 mm x 87 mm	-	0.28	3.65	6.52	m	10.17
75 mm x 100 mm	-	0.32	4.17	7.26	m	11.43
Add 5% to the above material prices for selected softwood for staining						

Prices for Measured Works - Major Works

P BUILDING FABRIC SUNDRIES

	PC £	Labour hours	Labour £	Material £	Unit	Total rate £
Selected West African Mahogany						
Skirtings, picture rails, dado rails and the like; splayed or moulded						
19 mm x 50 mm; splayed	3.40	0.14	1.82	3.92	m	5.74
19 mm x 50 mm; moulded	3.60	0.14	1.82	4.15	m	5.97
19 mm x 75 mm; splayed	4.08	0.14	1.82	4.69	m	6.51
19 mm x 75 mm; moulded	4.27	0.14	1.82	4.90	m	6.72
19 mm x 100 mm; splayed	4.78	0.14	1.82	5.48	m	7.30
19 mm x 100 mm; moulded	4.98	0.14	1.82	5.71	m	7.53
19 mm x 150 mm; moulded	6.38	0.17	2.21	7.29	m	9.50
19 mm x 175 mm; moulded	7.33	0.17	2.21	8.36	m	10.57
22 mm x 100 mm; splayed	5.20	0.14	1.82	5.95	m	7.77
25 mm x 50 mm; moulded	4.03	0.14	1.82	4.63	m	6.45
25 mm x 75 mm; splayed	4.76	0.14	1.82	5.46	m	7.28
25 mm x 100 mm; splayed	5.66	0.14	1.82	6.47	m	8.29
25 mm x 150 mm; splayed	7.79	0.17	2.21	8.88	m	11.09
25 mm x 150 mm; moulded	7.98	0.17	2.21	9.09	m	11.30
25 mm x 175 mm; moulded	8.89	0.17	2.21	10.12	m	12.33
25 mm x 225 mm; moulded	10.05	0.19	2.47	11.43	m	13.90
returned end	-	0.22	2.86	-	nr	2.86
mitres	-	0.16	2.08	-	nr	2.08
Architraves, cover fillets and the like; half round; splayed or moulded						
13 mm x 25 mm; half round	2.50	0.17	2.21	2.91	m	5.12
13 mm x 50 mm; moulded	3.14	0.17	2.21	3.63	m	5.84
16 mm x 32 mm; half round	2.77	0.17	2.21	3.21	m	5.42
16 mm x 38 mm; moulded	3.07	0.17	2.21	3.55	m	5.76
16 mm x 50 mm; moulded	3.36	0.17	2.21	3.88	m	6.09
19 mm x 50 mm; splayed	3.40	0.17	2.21	3.92	m	6.13
19 mm x 63 mm; splayed	3.77	0.17	2.21	4.34	m	6.55
19 mm x 75 mm; splayed	4.08	0.17	2.21	4.69	m	6.90
25 mm x 44 mm; splayed	3.84	0.17	2.21	4.42	m	6.63
25 mm x 50 mm; moulded	4.03	0.17	2.21	4.63	m	6.84
25 mm x 63 mm; splayed	4.31	0.17	2.21	4.95	m	7.16
25 mm x 75 mm; splayed	4.76	0.17	2.21	5.46	m	7.67
32 mm x 88 mm; moulded	6.70	0.17	2.21	7.65	m	9.86
38 mm x 88 mm; moulded	4.11	0.17	2.21	4.72	m	6.93
50 mm x 50 mm; moulded	5.66	0.17	2.21	6.47	m	8.68
returned end	-	0.22	2.86	-	nr	2.86
mitres	-	0.16	2.08	-	nr	2.08
Stops; screwed on						
16 mm x 38 mm	2.70	0.16	2.08	3.05	m	5.13
16 mm x 50 mm	3.00	0.16	2.08	3.39	m	5.47
19 mm x 38 mm	2.90	0.16	2.00	3.27	m	5.36
25 mm x 38 mm	3.23	0.16	2.08	3.65	m	5.73
25 mm x 50 mm	3.67	0.16	2.08	4.14	m	6.22
Glazing beads and the like						
13 mm x 16 mm	2.12	0.05	0.65	2.39	m	3.04
13 mm x 19 mm	2.17	0.05	0.65	2.45	m	3.10
13 mm x 25 mm	2.31	0.05	0.65	2.61	m	3.26
13 mm x 25 mm; screwed	2.12	0.08	1.04	2.54	m	3.58
13 mm x 25 mm; fixing with brass cups and screws	2.31	0.16	2.08	2.84	m	4.92
16 mm x 25 mm; screwed	2.41	0.08	1.04	2.87	m	3.91
16 mm quadrant	2.40	0.07	0.91	2.71	m	3.62
19 mm quadrant or scotia	2.54	0.07	0.91	2.87	m	3.78
19 mm x 36 mm; screwed	2.90	0.07	0.91	3.42	m	4.33
25 mm x 38 mm; screwed	3.23	0.07	0.91	3.80	m	4.71
25 mm quadrant or scotia	2.92	0.07	0.91	3.30	m	4.21
38 mm scotia	4.11	0.07	0.91	4.64	m	5.55
50 mm scotia	5.66	0.07	0.91	6.39	m	7.30

P BUILDING FABRIC SUNDRIES

	PC £	Labour hours	Labour £	Material £	Unit	Total rate £
P20 UNFRAMED ISOLATED TRIMS/ SKIRTINGS/SUNDRY ITEMS - cont'd						
Selected West African Mahogany - cont'd						
Isolated shelves; worktops, seats and the like						
19 mm x 150 mm	6.59	0.22	2.86	7.44	m	10.30
19 mm x 200 mm	7.93	0.31	4.04	8.95	m	12.99
25 mm x 150 mm	7.93	0.22	2.86	8.95	m	11.81
25 mm x 200 mm	9.67	0.31	4.04	10.92	m	14.96
32 mm x 150 mm	9.40	0.22	2.86	10.61	m	13.47
32 mm x 200 mm	11.72	0.31	4.04	13.23	m	17.27
Isolated shelves, worktops, seats and the like; cross-tongued joints						
19 mm x 300 mm	22.11	0.39	5.08	24.96	m	30.04
19 mm x 450 mm	26.36	0.45	5.86	29.75	m	35.61
19 mm x 600 mm	38.22	0.56	7.29	43.14	m	50.43
25 mm x 300 mm	25.20	0.39	5.08	28.44	m	33.52
25 mm x 450 mm	30.78	0.45	5.86	34.74	m	40.60
25 mm x 600 mm	44.10	0.56	7.29	49.78	m	57.07
32 mm x 300 mm	28.27	0.39	5.08	31.91	m	36.99
32 mm x 450 mm	35.29	0.45	5.86	39.83	m	45.69
32 mm x 600 mm	50.38	0.56	7.29	56.87	m	64.16
Isolated shelves, worktops, seats and the like; slatted with 50 mm wide slats at 75 mm centres						
19 mm thick	24.72	1.79	23.31	28.22	m^2	51.53
25 mm thick	27.89	1.79	23.31	31.80	m^2	55.11
32 mm thick	30.83	1.79	23.31	35.12	m^2	58.43
Window boards, nosings, bed moulds and the like; rebated and rounded						
19 mm x 75 mm	4.54	0.24	3.12	5.32	m	8.44
19 mm x 150 mm	6.66	0.28	3.65	7.71	m	11.36
19 mm x 225 mm; in one width	8.74	0.37	4.82	10.06	m	14.88
19 mm x 300 mm; cross-tongued joints	21.99	0.41	5.34	25.01	m	30.35
25 mm x 75 mm	5.22	0.24	3.12	6.08	m	9.20
25 mm x 150 mm	8.05	0.28	3.65	9.28	m	12.93
25 mm x 225 mm; in one width	10.71	0.37	4.82	12.28	m	17.10
25 mm x 300 mm; cross-tongued joints	24.95	0.41	5.34	28.35	m	33.69
32 mm x 75 mm	6.00	0.24	3.12	6.96	m	10.08
32 mm x 150 mm	9.51	0.28	3.65	10.93	m	14.58
32 mm x 225 mm; in one width	13.03	0.37	4.82	14.90	m	19.72
32 mm x 300 mm; cross-tongued joints	28.03	0.41	5.34	31.83	m	37.17
38 mm x 75 mm	6.66	0.24	3.12	7.71	m	10.83
38 mm x 150 mm	10.95	0.28	3.65	12.55	m	16.20
38 mm x 225 mm; in one width	15.02	0.37	4.82	17.14	m	21.96
38 mm x 300 mm; cross-tongued joints	30.81	0.41	5.34	34.97	m	40.31
returned and fitted ends	-	0.23	2.99	-	nr	2.99
Handrails; rounded						
44 mm x 50 mm	7.59	0.33	4.30	8.57	m	12.87
50 mm x 75 mm	9.49	0.37	4.82	10.71	m	15.53
63 mm x 87 mm	12.27	0.41	5.34	13.85	m	19.19
75 mm x 100 mm	14.92	0.45	5.86	16.84	m	22.70
Handrails; moulded						
44 mm x 50 mm	8.13	0.33	4.30	9.18	m	13.48
50 mm x 75 mm	10.05	0.37	4.82	11.34	m	16.16
63 mm x 87 mm	12.83	0.41	5.34	14.48	m	19.82
75 mm x 100 mm	15.47	0.45	5.86	17.46	m	23.32

P BUILDING FABRIC SUNDRIES

	PC £	Labour hours	Labour £	Material £	Unit	Total rate £
American White Ash						
Skirtings, picture rails, dado rails and the like; splayed or moulded						
19 mm x 50 mm; splayed	3.81	0.14	1.82	4.39	m	6.21
19 mm x 50 mm; moulded	4.00	0.14	1.82	4.60	m	6.42
19 mm x 75 mm; splayed	4.76	0.14	1.82	5.46	m	7.28
19 mm x 75 mm; moulded	4.93	0.14	1.82	5.65	m	7.47
19 mm x 100 mm; splayed	5.66	0.14	1.82	6.47	m	8.29
19 mm x 100 mm; moulded	5.85	0.14	1.82	6.69	m	8.51
19 mm x 150 mm; moulded	7.67	0.17	2.21	8.74	m	10.95
19 mm x 175 mm; moulded	8.86	0.17	2.21	10.09	m	12.30
22 mm x 100 mm; splayed	6.24	0.14	1.82	7.13	m	8.95
25 mm x 50 mm; moulded	4.61	0.14	1.82	5.29	m	7.11
25 mm x 75 mm; splayed	5.61	0.14	1.82	6.42	m	8.24
25 mm x 100 mm; splayed	6.82	0.14	1.82	7.78	m	9.60
25 mm x 150 mm; splayed	9.51	0.17	2.21	10.82	m	13.03
25 mm x 150 mm; moulded	9.69	0.17	2.21	11.02	m	13.23
25 mm x 175 mm; moulded	10.88	0.17	2.21	12.37	m	14.58
25 mm x 225 mm; moulded	13.29	0.19	2.47	15.09	m	17.56
returned ends	-	0.22	2.86	-	nr	2.86
mitres	-	0.16	2.08	-	nr	2.08
Architraves, cover fillets and the like; half round; splayed or moulded						
13 mm x 25 mm; half round	2.65	0.17	2.21	3.08	m	5.29
13 mm x 50 mm; moulded	3.43	0.17	2.21	3.96	m	6.17
16 mm x 32 mm; half round	2.97	0.17	2.21	3.44	m	5.65
16 mm x 38 mm; moulded	3.38	0.17	2.21	3.90	m	6.11
16 mm x 50 mm; moulded	3.75	0.17	2.21	4.32	m	6.53
19 mm x 50 mm; splayed	3.81	0.17	2.21	4.39	m	6.60
19 mm x 63 mm; splayed	4.32	0.17	2.21	4.96	m	7.17
19 mm x 75 mm; splayed	4.76	0.17	2.21	5.46	m	7.67
25 mm x 44 mm; splayed	4.42	0.17	2.21	5.07	m	7.28
25 mm x 50 mm; moulded	4.61	0.17	2.21	5.29	m	7.50
25 mm x 63 mm; splayed	5.02	0.17	2.21	5.75	m	7.96
25 mm x 75 mm; splayed	5.61	0.17	2.21	6.42	m	8.63
32 mm x 88 mm; moulded	8.15	0.17	2.21	9.28	m	11.49
38 mm x 88 mm; moulded	4.78	0.17	2.21	5.48	m	7.69
50 mm x 50 mm; moulded	6.82	0.17	2.21	7.78	m	9.99
returned ends	-	0.22	2.86	-	nr	2.86
mitres	-	0.16	2.08	-	nr	2.08
Stops; screwed on						
16 mm x 38 mm	3.01	0.16	2.08	3.40	m	5.48
16 mm x 50 mm	3.38	0.16	2.08	3.82	m	5.90
19 mm x 38 mm	3.21	0.16	2.08	3.62	m	5.70
25 mm x 38 mm	3.64	0.16	2.08	4.11	m	6.19
25 mm x 50 mm	4.24	0.16	2.08	4.79	m	6.87
Glazing beads and the like						
13 mm x 16 mm	2.23	0.05	0.65	2.52	m	3.17
13 mm x 19 mm	2.30	0.05	0.65	2.60	m	3.25
13 mm x 25 mm	2.47	0.05	0.65	2.79	m	3.44
13 mm x 25 mm; screwed	2.47	0.08	1.04	2.94	m	3.98
13 mm x 25 mm; fixing with brass cups and screws	2.47	0.16	2.08	3.02	m	5.10
16 mm x 25 mm; screwed	2.59	0.08	1.04	3.07	m	4.11
16 mm quadrant	2.51	0.07	0.91	2.83	m	3.74
19 mm quadrant or scotia	2.71	0.07	0.91	3.06	m	3.97
19 mm x 36 mm; screwed	3.21	0.07	0.91	3.77	m	4.68
25 mm x 38 mm; screwed	3.64	0.07	0.91	4.26	m	5.17
25 mm quadrant or scotia	3.23	0.07	0.91	3.65	m	4.56
38 mm scotia	4.78	0.07	0.91	5.40	m	6.31
50 mm scotia	6.82	0.07	0.91	7.70	m	8.61

Prices for Measured Works - Major Works
P BUILDING FABRIC SUNDRIES

	PC £	Labour hours	Labour £	Material £	Unit	Total rate £
P20 UNFRAMED ISOLATED TRIMS/ SKIRTINGS/SUNDRY ITEMS - cont'd						
American White Ash - cont'd						
Isolated shelves, worktops, seats and the like						
19 mm x 150 mm	7.94	0.22	2.86	6.96	m	11.82
19 mm x 200 mm	9.67	0.31	4.04	10.92	m	14.96
25 mm x 150 mm	9.67	0.22	2.86	10.92	m	13.78
25 mm x 200 mm	11.95	0.31	4.04	13.49	m	17.53
32 mm x 150 mm	11.59	0.22	2.86	13.08	m	15.94
32 mm x 200 mm	14.65	0.31	4.04	16.54	m	20.58
Isolated shelves, worktops, seats and the like; cross-tongued joints						
19 mm x 300 mm	25.00	0.39	5.08	28.22	m	33.30
19 mm x 450 mm	30.58	0.45	5.86	34.52	m	40.38
19 mm x 600 mm	43.98	0.56	7.29	49.64	m	56.93
25 mm x 300 mm	29.03	0.39	5.08	32.77	m	37.85
25 mm x 450 mm	36.33	0.45	5.86	41.01	m	46.87
25 mm x 600 mm	51.69	0.56	7.29	58.34	m	65.63
32 mm x 300 mm	33.07	0.39	5.08	37.33	m	42.41
32 mm x 450 mm	42.30	0.45	5.86	47.75	m	53.61
32 mm x 600 mm	59.95	0.56	7.29	67.67	m	74.96
Isolated shelves, worktops, seats and the like; slatted with 50 mm wide slats at 75 mm centres						
19 mm thick	27.90	1.79	23.31	31.81	m²	55.12
25 mm thick	31.91	1.79	23.31	36.34	m²	59.65
32 mm thick	36.15	1.79	23.31	41.12	m²	64.43
Window boards, nosings, bed moulds and the like; rebated and rounded						
19 mm x 75 mm	5.19	0.24	3.12	6.05	m	9.17
19 mm x 150 mm	7.95	0.28	3.65	9.16	m	12.81
19 mm x 225 mm; in one width	10.68	0.37	4.82	12.25	m	17.07
19 mm x 300 mm; cross-tongued joints	24.92	0.41	5.34	28.32	m	33.66
25 mm x 75 mm	6.10	0.24	3.12	7.08	m	10.20
25 mm x 150 mm	9.80	0.28	3.65	11.25	m	14.90
25 mm x 225 mm; in one width	13.26	0.37	4.82	15.16	m	19.98
25 mm x 300 mm; cross-tongued joints	28.78	0.41	5.34	32.68	m	38.02
32 mm x 75 mm	7.09	0.24	3.12	8.19	m	11.31
32 mm x 150 mm	11.71	0.28	3.65	13.41	m	17.06
32 mm x 225 mm; in one width	-	0.37	4.82	18.65	m	23.47
32 mm x 300 mm; cross-tongued joints	32.82	0.41	5.34	37.24	m	42.58
38 mm x 75 mm	7.95	0.24	3.12	9.16	m	12.28
38 mm x 150 mm	13.47	0.28	3.65	15.39	m	19.04
38 mm x 225 mm; in one width	18.94	0.37	4.82	21.57	m	26.39
38 mm x 300 mm; cross-tongued joints	36.48	0.41	5.34	41.37	m	46.71
returned and fitted ends	-	0.23	2.99	-	nr	2.99
Handrails; rounded						
44 mm x 50 mm	8.72	0.33	4.30	9.84	m	14.14
50 mm x 75 mm	11.14	0.37	4.82	12.57	m	17.39
63 mm x 87 mm	16.06	0.41	5.34	18.13	m	23.47
75 mm x 100 mm	18.34	0.45	5.86	20.70	m	26.56
Handrails; moulded						
44 mm x 50 mm	9.28	0.33	4.30	10.47	m	14.77
50 mm x 75 mm	11.70	0.37	4.82	13.21	m	18.03
63 mm x 87 mm	16.61	0.41	5.34	18.75	m	24.09
75 mm x 100 mm	18.88	0.45	5.86	21.31	m	27.17

P BUILDING FABRIC SUNDRIES

	PC £	Labour hours	Labour £	Material £	Unit	Total rate £
Pin-boards; medium board Sundeala "A" pin-board or other equal and approved; fixed with adhesive to backing (not included); over 300 mm wide						
6.40 mm thick	-	0.61	7.94	13.14	m²	21.08
Sundries on softwood/hardwood						
Extra over fixing with nails for						
gluing and pinning	-	0.02	0.26	0.06	m	0.32
masonry nails	-	-	-	-	m	0.30
steel screws	-	-	-	-	m	0.28
self-tapping screws	-	-	-	-	m	0.29
steel screws; gluing	-	-	-	-	m	0.48
steel screws; sinking; filling heads	-	-	-	-	m	0.61
steel screws; sinking; pellating over	-	-	-	-	m	1.34
brass cups and screws	-	-	-	-	m	1.64
Extra over for						
countersinking	-	-	-	-	m	0.26
pellating	-	-	-	-	m	1.16
Head or nut in softwood						
let in flush	-	-	-	-	nr	0.61
Head or nut; in hardwood						
let in flush	-	-	-	-	nr	0.91
let in over; pellated	-	-	-	-	nr	2.13
Metalwork; mild steel						
Angle section bearers; for building in						
150 mm x 100 mm x 6 mm	-	0.33	4.22	13.10	m	17.32
150 mm x 150 mm x 8 mm	-	0.36	4.61	20.97	m	25.58
200 mm x 200 mm x 12 mm	-	0.41	5.25	39.33	m	44.58
Metalwork; mild steel; galvanized						
Water bars; groove in timber						
6 mm x 30 mm	-	0.51	6.64	5.61	m	12.25
6 mm x 40 mm	-	0.51	6.64	6.56	m	13.20
6 mm x 50 mm	-	0.51	6.64	8.76	m	15.40
Angle section bearers; for building in						
150 mm x 100 mm x 6 mm	-	0.33	4.22	17.30	m	21.52
150 mm x 150 mm x 8 mm	-	0.36	4.61	27.53	m	32.14
200 mm x 200 mm x 12 mm	-	0.41	5.25	55.05	m	60.30
Dowels; mortice in timber						
8 mm diameter x 100 mm long	-	0.05	0.65	0.10	nr	0.75
10 mm diameter x 50 mm long	-	0.05	0.65	0.26	nr	0.91
Cramps						
25 mm x 3 mm x 230 mm girth; one end bent, holed and screwed to softwood; other end fishtailed for building in	-	0.07	0.91	0.63	nr	1.54
Metalwork; stainless steel						
Angle section bearers; for building in						
150 mm x 100 mm x 6 mm	-	0.33	4.22	48.76	m	52.98
150 mm x 150 mm x 8 mm	-	0.36	4.61	78.64	m	83.25
200 mm x 200 mm x 12 mm	-	0.41	5.25	162.54	m	167.79

Prices for Measured Works - Major Works
P BUILDING FABRIC SUNDRIES

	PC £	Labour hours	Labour £	Material £	Unit	Total rate £
P21 IRONMONGERY						
NOTE: Ironmongery is largely a matter of selection and prices vary considerably, indicative prices for reasonable quantities of "standard" quality ironmongery are given below.						
Ironmongery; NT Laidlaw Ltd or other equal and approved; standard ranges; to softwood						
Bolts						
barrel; 100 mm x 32 mm; PAA	-	0.31	4.04	5.80	nr	9.84
barrel; 150 mm x 32 mm; PAA	-	0.39	5.08	6.64	nr	11.72
flush; 152 mm x 25 mm; SCP	-	0.56	7.29	13.48	nr	20.77
flush; 203 mm x 25 mm; SCP	-	0.56	7.29	14.85	nr	22.14
flush; 305 mm x 25 mm; SCP	-	0.56	7.29	20.11	nr	27.40
indicating; 76 mm x 41 mm; SAA	-	0.62	8.07	9.86	nr	17.93
indicating; coin operated; SAA	-	0.62	8.07	13.48	nr	21.55
panic; single; SVE	-	2.31	30.08	62.20	nr	92.28
panic; double; SVE	-	3.24	42.18	84.72	nr	126.90
necked tower; 152 mm; BJ	-	0.31	4.04	2.74	nr	6.78
necked tower; 203 mm; BJ	-	0.31	4.04	3.69	nr	7.73
mortice security; SCP	-	0.56	7.29	17.89	nr	25.18
garage door bolt; 305 mm	-	0.56	7.29	6.67	nr	13.96
monkey tail with knob; 305 mm x 19 mm; BJ	-	0.62	8.07	8.15	nr	16.22
monkey tail with bow; 305 mm x 19 mm; BJ	-	0.62	8.07	10.27	nr	18.34
Butts						
63 mm; light steel	-	0.23	2.99	0.67	pr	3.66
100 mm; light steel	-	0.23	2.99	1.17	pr	4.16
102 mm; SC	-	0.23	2.99	3.76	pr	6.75
102 mm double flap extra strong; SC	-	0.23	2.99	2.34	pr	5.33
76 mm x 51 mm; SC	-	0.23	2.99	2.72	pr	5.71
51 mm x 22 mm narrow suite S/D	-	0.23	2.99	2.27	pr	5.26
76 mm x 35 mm narrow suite S/D	-	0.23	2.99	3.16	pr	6.15
51 mm x 29 mm broad suite S/D	-	0.23	2.99	2.27	pr	5.26
76 mm x 41 mm broad suite S/D	-	0.23	2.99	3.05	pr	6.04
102 mm x 60 mm broad suite S/D	-	0.23	2.99	6.41	pr	9.40
102 mm x 69 mm lift-off (R/L) hand	-	0.23	2.99	3.41	pr	6.40
high security heavy butts F/pin; 109 mm x 94 mm; ball bearing; SSS	-	0.37	4.82	20.13	pr	24.95
"Hi-load" butts; 125 mm x 102 mm	-	0.28	3.65	46.14	pr	49.79
S/D rising butts (R/L hand); 102 mm x 67 mm; BRS	-	0.23	2.99	38.55	pr	41.54
ball bearing butts; 102 mm x 67 mm; SSS	-	0.23	2.99	9.02	pr	12.01
Catches						
ball catch; 13 mm diameter; BRS	-	0.31	4.04	0.55	nr	4.59
double ball catch; 50 mm; BRS	-	0.37	4.82	0.72	nr	5.54
57 mm x 38 mm cupboard catch; SCP	-	0.28	3.65	6.27	nr	9.92
14 mm x 35 mm magnetic catch; WHT	-	0.16	2.08	0.63	nr	2.71
adjustable nylon roller catch; WNY	-	0.23	2.99	1.67	nr	4.66
"Bales" catch; Nr 4; 41 mm x 16 mm; self-colour brass; mortice	-	0.31	4.04	0.98	nr	5.02
"Bales" catch; Nr 8; 66 mm x 25 mm; self-colour brass; mortice	-	0.32	4.17	2.17	nr	6.34
Door closers and furniture						
standard concealed overhead door closer (L/R hand); SIL	-	1.16	15.10	160.96	nr	176.06
light duty surface fixed door closer (L/R hand); SIL	-	0.93	12.11	68.43	nr	80.54
"Perkomatic" concealed door closer; BRS	-	0.62	8.07	79.16	nr	87.23

Prices for Measured Works - Major Works
P BUILDING FABRIC SUNDRIES

	PC £	Labour hours	Labour £	Material £	Unit	Total rate £
Door closers and furniture - cont'd						
"Softline" adjustable size 2 - 4 overhead door closer; SIL	-	1.39	18.10	75.87	nr	93.97
"Centurion II" size 3 overhead door closer; SIL	-	1.16	15.10	35.28	nr	50.38
"Centurion II" size 4 overhead door closer; SIL	-	1.39	18.10	43.54	nr	61.64
backcheck door closer size 3; SAA	-	1.16	15.10	90.38	nr	105.48
backcheck door closer size 4; SAA	-	1.39	18.10	94.48	nr	112.58
door selector; face fixing; SAA	-	0.56	7.29	67.74	nr	75.03
finger plate; 300 mm x 75 mm x 3 mm	-	0.16	2.08	3.90	nr	5.98
kicking plate; 1000 mm x 150 mm x 30 mm; PAA	-	0.23	2.99	11.34	nr	14.33
floor spring; single and double action; ZP	-	2.31	30.08	166.37	nr	196.45
lever furniture; 280 mm x 40 mm	-	0.23	2.99	22.35	pr	25.34
pull handle; 225 mm; back fixing; PAA	-	0.16	2.08	5.98	nr	8.06
pull handle; f/fix with cover rose; PAA	-	0.31	4.04	32.07	nr	36.11
letter plate; 330 mm x 76 mm; aluminium finish	-	1.23	16.01	8.12	nr	24.13
Latches						
102 mm mortice latch; SCP	-	1.16	15.10	6.83	nr	21.93
cylinder rim night latch; SC7	-	0.69	8.98	19.34	nr	28.32
Locks						
drawer or cupboard lock; 63 mm x 32 mm; SC	-	0.39	5.08	8.51	nr	13.59
mortice dead lock; 63 mm x 108 mm; SSS	-	0.69	8.98	10.96	nr	19.94
12 mm rebate conversion set to mortice dead lock	-	0.46	5.99	10.87	nr	16.86
rim lock; 140 mm x 73 mm; GYE	-	0.39	5.08	8.34	nr	13.42
rim dead lock; 92 mm x 74.50 mm; SCP	-	0.39	5.08	22.04	nr	27.12
upright mortice lock; 103 mm x 82 mm; 3 lever	-	0.77	10.03	11.30	nr	21.33
upright mortice lock; 103 mm x 82 mm; 5 lever	-	0.77	10.03	35.29	nr	45.32
Window furniture						
casement stay; 305 mm long; 2 pin; SAA	-	0.16	2.08	7.39	nr	9.47
casement fastener; standard; 113 mm; SAA	-	0.16	2.08	4.78	nr	6.86
cockspur fastener; ASV	-	0.31	4.04	8.08	nr	12.12
sash fastener; 65 mm; SAA	-	0.23	2.99	2.72	nr	5.71
Sundries						
numerals; SAA	-	0.07	0.91	0.76	nr	1.67
rubber door stop; SAA	-	0.07	0.91	1.14	nr	2.05
medium hot pressed cabin hook; 102 mm; CP	-	0.16	2.08	9.33	nr	11.41
medium hot pressed cabin hook; 203 mm; CP	-	0.16	2.08	12.33	nr	14.41
coat hook; SAA	-	0.07	0.91	1.26	nr	2.17
toilet roll holder; CP	-	0.16	2.08	16.35	nr	18.43
Iromongery; NT Laidlaw Ltd or other equal and approved; standard ranges; to hardwood						
Bolts						
barrel; 100 mm x 32 mm; PAA	-	0.41	5.34	5.80	nr	11.14
barrel; 150 mm x 32 mm; PAA	-	0.52	6.77	6.64	nr	13.41
flush; 152 mm x 25 mm; SCP	-	0.74	9.63	13.48	nr	23.11
flush; 203 mm x 25 mm; SCP	-	0.74	9.63	14.85	nr	24.48
flush; 305 mm x 25 mm; SCP	-	0.74	9.63	20.11	nr	29.74
indicating; 76 mm x 41 mm; SAA	-	0.82	10.68	9.86	nr	20.54
indicating; coin operated; SAA	-	0.82	10.68	13.48	nr	24.16
panic; single; SVE	-	3.08	40.10	62.20	nr	102.30
panic; double; SVE	-	4.32	56.25	84.72	nr	140.97
necked tower; 152 mm; BJ	-	0.41	5.34	2.74	nr	8.08
necked tower; 203 mm; BJ	-	0.41	5.34	3.69	nr	9.03
mortice security; SCP	-	0.74	9.63	17.89	nr	27.52
garage door bolt; 305 mm	-	0.74	9.63	6.67	nr	16.30
monkey tail with knob; 305 mm x 19 mm; BJ	-	0.79	10.29	8.15	nr	18.44
monkey tail with bow; 305 mm x 19 mm; BJ	-	0.79	10.29	10.27	nr	20.56

Prices for Measured Works - Major Works
P BUILDING FABRIC SUNDRIES

	PC £	Labour hours	Labour £	Material £	Unit	Total rate £
P21 IRONMONGERY - cont'd						
Ironmongery; NT Laidlaw Ltd or other equal and approved; standard ranges; to hardwood - cont'd						
Butts						
63 mm; light steel	-	0.32	4.17	0.67	pr	4.84
100 mm; light steel	-	0.32	4.17	1.17	pr	5.34
102 mm; SC	-	0.32	4.17	3.76	pr	7.93
102 mm double flap extra strong; SC	-	0.32	4.17	2.34	pr	6.51
76 mm x 51 mm; SC	-	0.32	4.17	2.72	pr	6.89
51 mm x 22 mm narrow suite S/D	-	0.32	4.17	2.27	pr	6.44
76 mm x 35 mm narrow suite S/D	-	0.32	4.17	3.16	pr	7.33
51 mm x 29 mm broad suite S/D	-	0.32	4.17	2.27	pr	6.44
76 mm x 41 mm broad suite S/D	-	0.32	4.17	3.05	pr	7.22
102 mm x 60 mm broad suite S/D	-	0.32	4.17	6.41	pr	10.58
102 mm x 69 mm lift-off (R/L) hand	-	0.32	4.17	3.41	pr	7.58
high security heavy butts F/pin; 109 mm x 94 mm; ball bearing; SSS	-	0.42	5.47	20.13	pr	25.60
"Hi-load" butts; 125 mm x 102 mm	-	0.37	4.82	46.14	pr	50.96
S/D rising butts (R/L hand); 102 mm x 67 mm; BRS	-	0.32	4.17	38.55	pr	42.72
ball bearing butts; 102 mm x 67 mm; SSS	-	0.32	4.17	9.02	pr	13.19
Catches						
ball catch; 13 mm diameter; BRS	-	0.42	5.47	0.55	nr	6.02
double ball catch; 50 mm; BRS	-	0.43	5.60	0.72	nr	6.32
57 mm x 38 mm cupboard catch; SCP	-	0.37	4.82	6.27	nr	11.09
14 mm x 35 mm magnetic catch; WHT	-	0.21	2.73	0.63	nr	3.36
adjustable nylon roller catch; WNY	-	0.31	4.04	1.67	nr	5.71
"Bales" catch; Nr 4; 41 mm x 16 mm; self-colour brass; mortice	-	0.41	5.34	0.98	nr	6.32
"Bales" catch; Nr 8; 66 mm x 25 mm; self-colour brass; mortice	-	0.43	5.60	2.17	nr	7.77
Door closers and furniture						
standard concealed overhead door closer (L/R hand); SIL	-	1.54	20.05	160.96	nr	181.01
light duty surface fixed door closer (L/R hand); SIL	-	1.20	15.62	68.43	nr	84.05
"Perkomatic" concealed door closer; BRS	-	0.82	10.68	79.16	nr	89.84
"Softline" adjustable size 2 - 4 overhead door closer; SIL	-	1.85	24.09	75.87	nr	99.96
"Centurion II" size 3 overhead door closer; SIL	-	1.62	21.09	35.28	nr	56.37
"Centurion II" size 4 overhead door closer; SIL	-	1.85	24.09	43.54	nr	67.63
backcheck door closer size 3; SAA	-	1.62	21.09	90.38	nr	111.47
backcheck door closer size 4; SAA	-	1.85	24.09	94.48	nr	118.57
door selector; face fixing; SAA	-	0.74	9.63	67.74	nr	77.37
finger plate; 300 mm x 75 mm x 3 mm	-	0.21	2.73	3.90	nr	6.63
kicking plate; 1000 mm x 150 mm x 30 mm; PAA	-	0.41	5.34	11.34	nr	16.68
floor spring; single and double action; ZP	-	3.08	40.10	166.37	nr	206.47
lever furniture; 280 mm x 40 mm	-	0.31	4.04	22.35	pr	26.39
pull handle; 225 mm; back fixing; PAA	-	0.21	2.73	5.98	nr	8.71
pull handle; f/fix with cover rose; PAA	-	0.41	5.34	32.07	nr	37.41
letter plate; 330 mm x 76 mm; aluminium finish	-	1.64	21.35	8.12	nr	29.47
Latches						
102 mm mortice latch; SCP	-	1.62	21.09	6.83	nr	27.92
cylinder rim night latch; SC7	-	0.93	12.11	19.34	nr	31.45
Locks						
drawer or cupboard lock; 63 mm x 32 mm; SC	-	0.52	6.77	8.51	nr	15.28
mortice dead lock; 63 mm x 108 mm; SSS	-	0.93	12.11	10.96	nr	23.07
12 mm rebate conversion set to mortice dead lock	-	0.69	8.98	10.87	nr	19.85
rim lock; 140 mm x 73 mm; GYE	-	0.52	6.77	8.34	nr	15.11
rim dead lock; 92 mm x 74.50 mm; SCP	-	0.52	6.77	22.04	nr	28.81
upright mortice lock; 103 mm x 82 mm; 3 lever	-	1.03	13.41	11.30	nr	24.71
upright mortice lock; 103 mm x 82 mm; 5 lever	-	1.03	13.41	35.29	nr	48.70

Prices for Measured Works - Major Works
P BUILDING FABRIC SUNDRIES

	PC £	Labour hours	Labour £	Material £	Unit	Total rate £
Window furniture						
casement stay; 305 mm long; 2 pin; SAA	-	0.21	2.73	7.39	nr	10.12
casement fastener; standard; 113 mm; SAA	-	0.21	2.73	4.78	nr	7.51
cockspur fastener; ASV	-	0.41	5.34	8.08	nr	13.42
sash fastener; 65 mm; SAA	-	0.31	4.04	2.72	nr	6.76
Sundries						
numerals; SAA	-	0.10	1.30	0.76	nr	2.06
rubber door stop; SAA	-	0.10	1.30	1.14	nr	2.44
medium hot pressed cabin hook; 102 mm; CP	-	0.21	2.73	9.33	nr	12.06
medium hot pressed cabin hook; 203 mm; CP	-	0.21	2.73	12.33	nr	15.06
coat hook; SAA	-	0.10	1.30	1.26	nr	2.56
toilet roll holder; CP	-	0.23	2.99	16.35	nr	19.34
Sliding door gear; Hillaldam Coburn Ltd or other equal and approved; Commercial/Light industrial; for top hung timber/metal doors, weight not exceeding 365 kg						
Sliding door gear						
bottom guide; fixed to concrete in groove	17.27	0.46	5.99	18.56	m	24.55
top track	24.19	0.23	2.99	26.00	m	28.99
detachable locking bar and padlock	41.46	0.31	4.04	44.57	nr	48.61
hangers; timber doors	44.75	0.46	5.99	48.11	nr	54.10
hangers; metal doors	31.90	0.46	5.99	34.29	nr	40.28
head brackets; open, side fixing; bolting to masonry	4.78	0.46	5.99	8.39	nr	14.38
head brackets; open, soffit fixing; screwing to timber	3.61	0.32	4.17	3.93	nr	8.10
door guide to timber door	8.97	0.23	2.99	9.64	nr	12.63
door stop; rubber buffers; to masonry	24.58	0.69	8.98	26.42	nr	35.40
drop bolt; screwing to timber	46.04	0.46	5.99	49.49	nr	55.48
bow handle; to timber	9.95	0.23	2.99	10.70	nr	13.69
Sundries						
rubber door stop; plugged and screwed to concrete	5.17	0.09	1.17	5.56	nr	6.73
P30 TRENCHES/PIPEWAYS/PITS FOR BURIED ENGINEERING SERVICES						
Excavating trenches; by machine; grading bottoms; earthwork support; filling with excavated material and compacting; disposal of surplus soil; spreading on site average 50 m						
Services not exceeding 200 mm nominal size						
average depth of run 0.50 m	-	0.30	2.23	1.39	m	3.62
average depth of run 0.75 m	-	0.45	3.35	2.32	m	5.67
average depth of run 1.00 m	-	0.90	6.70	4.24	m	10.94
average depth of run 1.25 m	-	1.33	9.90	5.77	m	15.67
average depth of run 1.50 m	-	1.71	12.72	7.56	m	20.28
average depth of run 1.75 m	-	2.08	15.48	9.68	m	25.16
average depth of run 2.00 m	-	2.46	18.30	11.09	m	29.39
Excavating trenches; by hand; grading bottoms; earthwork support; filling with excavated material and compacting; disposal; of surplus soil; spreading on site average 50 m						
Services not exceeding 200 mm nominal size						
average depth of run 0.50 m	-	1.06	7.89	-	m	7.89
average depth of run 0.75 m	-	1.60	11.90	-	m	11.90
average depth of run 1.00 m	-	2.34	17.41	1.33	m	18.74
average depth of run 1.25 m	-	3.29	24.48	1.83	m	26.31
average depth of run 1.50 m	-	4.52	33.63	2.23	m	35.86
average depth of run 1.75 m	-	5.96	44.34	2.70	m	47.04
average depth of run 2.00 m	-	6.81	50.67	2.97	m	53.64

P BUILDING FABRIC SUNDRIES

	PC £	Labour hours	Labour £	Material £	Unit	Total rate £
P30 TRENCHES/PIPEWAYS/PITS FOR BURIED ENGINEERING SERVICES - cont'd						
Stop cock pits, valves chmabers and the like; excavating; half brick thick walls in common bricks in cement mortar (1:3); on in situ concrete designated C20 - 20 mm aggregate bed; 100 mm thick						
Pits						
100 mm x 100 mm x 750 mm deep; internal holes for one small pipe; polypropylene hinged box cover; bedding in cement mortar (1:3)	-	4.49	73.01	28.10	nr	101.11
P31 HOLES/CHASES/COVERS/SUPPORTS FOR SERVICES						
Builders' work for electrical installations; cutting away for and making good after electrician; including cutting or leaving all holes, notches, mortices, sinkings and chases, in both the structure and its coverings, for the following electrical points						
Exposed installation						
lighting points	-	0.32	2.99	-	nr	2.99
socket outlet points	-	0.54	5.20	-	nr	5.20
fitting outlet points	-	0.54	5.20	-	nr	5.20
equipment points or control gear points	-	0.75	7.40	-	nr	7.40
Concealed installation						
lighting points	-	0.43	4.07	-	nr	4.07
socket outlet points	-	0.75	7.40	-	nr	7.40
fitting outlet points	-	0.75	7.40	-	nr	7.40
equipment points or control gear points	-	1.06	10.26	-	nr	10.26
Builders' work for other services installations						
Cutting chases in brickwork						
for one pipe; not exceeding 55 mm nominal size; vertical	-	0.43	3.13	-	m	3.13
for one pipe; 55 mm - 110 mm nominal size; vertical	-	0.74	5.39	-	m	5.39
Cutting and pinning to brickwork or blockwork; ends of supports						
for pipes not exceeding 55 mm nominal size	-	0.21	3.41	-	nr	3.41
for cast iron pipes 55 mm - 110 mm nominal size	-	0.35	5.69	-	nr	5.69
Cutting holes for pipes or the like; not exceeding 55 mm nominal size						
half brick thick	-	0.35	2.88	-	nr	2.88
one brick thick	-	0.58	4.77	-	nr	4.77
one and a half brick thick	-	0.95	7.81	-	nr	7.81
100 mm blockwork	-	0.32	2.63	-	nr	2.63
140 mm blockwork	-	0.43	3.53	-	nr	3.53
215 mm blockwork	-	0.53	4.36	-	nr	4.36
Cutting holes for pipes or the like; 55 mm - 110 mm nominal size						
half brick thick	-	0.43	3.53	-	nr	3.53
one brick thick	-	0.74	6.08	-	nr	6.08
one and a half brick thick	-	1.17	9.62	-	nr	9.62
100 mm blockwork	-	0.37	3.04	-	nr	3.04
140 mm blockwork	-	0.53	4.36	-	nr	4.36
215 mm blockwork	-	0.64	5.26	-	nr	5.26

P BUILDING FABRIC SUNDRIES

	PC £	Labour hours	Labour £	Material £	Unit	Total rate £
Cutting holes for pipes or the like; over 110 mm nominal size						
half brick thick	-	0.53	4.36	-	nr	4.36
one brick thick	-	0.91	7.48	-	nr	7.48
one and a half brick thick	-	1.43	11.75	-	nr	11.75
100 mm blockwork	-	0.48	3.95	-	nr	3.95
140 mm blockwork	-	0.64	5.26	-	nr	5.26
215 mm blockwork	-	0.80	6.58	-	nr	6.58
Add for making good fair face or facings one side						
pipe; not exceeding 55 mm nominal size	-	0.08	1.30	-	nr	1.30
pipe; 55 mm - 110 mm nominal size	-	0.11	1.79	-	nr	1.79
pipe; over 110 mm nominal size	-	0.13	2.11	-	nr	2.11
Add for fixing sleeve (supply not included)						
for pipe; small	-	0.16	2.60	-	nr	2.60
for pipe; large	-	0.21	3.41	-	nr	3.41
for pipe; extra large	-	0.32	5.20	-	nr	5.20
Cutting or forming holes for ducts; girth not exceeding 1.00 m						
half brick thick	-	0.64	5.26	-	nr	5.26
one brick thick	-	1.06	8.71	-	nr	8.71
one and a half brick thick	-	1.70	13.97	-	nr	13.97
100 mm blockwork	-	0.53	4.36	-	nr	4.36
140 mm blockwork	-	0.74	6.08	-	nr	6.08
215 mm blockwork	-	0.95	7.81	-	nr	7.81
Cutting or forming holes for ducts; girth 1.00 m - 2.00 m						
half brick thick	-	0.74	6.08	-	nr	6.08
one brick thick	-	1.28	10.52	-	nr	10.52
one and a half brick thick	-	2.02	16.60	-	nr	16.60
100 mm blockwork	-	0.64	5.26	-	nr	5.26
140 mm blockwork	-	0.85	6.99	-	nr	6.99
215 mm blockwork	-	1.06	8.71	-	nr	8.71
Cutting or forming holes for ducts; girth 2.00 m - 3.00 m						
half brick thick	-	1.17	9.62	-	nr	9.62
one brick thick	-	2.02	16.60	-	nr	16.60
one and a half brick thick	-	3.19	26.22	-	nr	26.22
100 mm blockwork	-	1.01	8.30	-	nr	8.30
140 mm blockwork	-	1.38	11.34	-	nr	11.34
215 mm blockwork	-	1.76	14.47	-	nr	14.47
Cutting or forming holes for ducts; girth 3.00 m - 4.00 m						
half brick thick	-	1.60	13.15	-	nr	13.15
one brick thick	-	2.66	21.87	-	nr	21.87
one and a half brick thick	-	4.26	35.02	-	nr	35.02
100 mm blockwork	-	1.17	9.62	-	nr	9.62
140 mm blockwork	-	1.60	13.15	-	nr	13.15
215 mm blockwork	-	2.02	16.60	-	nr	16.60
Mortices in brickwork						
for expansion bolt	-	0.21	1.73	-	nr	1.73
for 20 mm diameter bolt; 75 mm deep	-	0.16	1.32	-	nr	1.32
for 20 mm diameter bolt; 150 mm deep	-	0.27	2.22	-	nr	2.22
Mortices in brickwork; grouting with cement mortar (1:1)						
75 mm x 75 mm x 200 mm deep	-	0.32	2.63	-	nr	2.63
75 mm x 75 mm x 300 mm deep	-	0.43	3.53	-	nr	3.53

Prices for Measured Works - Major Works
P BUILDING FABRIC SUNDRIES

	PC £	Labour hours	Labour £	Material £	Unit	Total rate £
P31 HOLES/CHASES/COVERS/SUPPORTS FOR SERVICES - cont'd						
Builders' work for other services installations - cont'd						
Holes in softwood for pipes, bars, cables and the like						
12 mm thick	-	0.04	0.52	-	nr	0.52
25 mm thick	-	0.06	0.78	-	nr	0.78
50 mm thick	-	0.11	1.43	-	nr	1.43
100 mm thick	-	0.16	2.08	-	nr	2.08
Holes in hardwood for pipes, bars, cables and the like						
12 mm thick	-	0.06	0.78	-	nr	0.78
25 mm thick	-	0.09	1.17	-	nr	1.17
50 mm thick	-	0.16	2.08	-	nr	2.08
100 mm thick	-	0.23	2.99	-	nr	2.99
"SFD Screeduct" or other equal and approved; MDT Ducting Ltd; laid within floor screed; galvanised mild steel						
Floor ducting						
100 mm wide; 40 mm deep; ref SFD40/100/00	5.34	0.21	2.73	6.03	m	8.76
Extra for						
45 degree bend	5.81	0.11	1.43	6.56	nr	7.99
90 degree bend	4.64	0.11	1.43	5.24	nr	6.67
tee section	4.64	0.11	1.43	5.24	nr	6.67
cross over	6.04	0.11	1.43	6.82	nr	8.25
reducer; 100 mm - 50 mm	10.92	0.11	1.43	12.33	nr	13.76
connector / stop end	0.46	0.11	1.43	0.52	nr	1.95
divider	1.39	0.11	1.43	1.57	nr	3.00
ply cover 15 mm/16 mm thick WBP exterior grade	0.54	0.11	1.43	0.61	m	2.04
200 mm wide; 65 mm deep; ref SFD65/200/00	6.27	0.21	2.73	7.08	m	9.81
Extra for						
45 degree bend	8.13	0.11	1.43	9.18	nr	10.61
90 degree bend	6.27	0.11	1.43	7.08	nr	8.51
tee section	6.27	0.11	1.43	7.08	nr	8.51
cross over	8.59	0.11	1.43	9.70	nr	11.13
reducer; 200 mm - 100 mm	10.92	0.11	1.43	12.33	nr	13.76
connector / stop end	0.60	0.12	1.56	0.68	nr	2.24
divider	1.39	0.11	1.43	1.57	nr	3.00
ply cover 15 mm/16 mm thick WBP exterior grade	0.91	0.11	1.43	1.03	m	2.46

Q PAVING/PLANTING/FENCING/SITE FURNITURE

Q10 KERBS/EDGINGS/CHANNELS/PAVING ACCESSORIES

	PC £	Labour hours	Labour £	Material £	Unit	Total rate £
Excavating; by machine						
Excavating trenches; to receive kerb foundations; average size						
300 mm x 100 mm	-	0.02	0.15	0.35	m	0.50
450 mm x 150 mm	-	0.02	0.15	0.70	m	0.85
600 mm x 200 mm	-	0.04	0.30	0.97	m	1.27
Excavating curved trenches; to receive kerb foundations; average size						
300 mm x 100 mm	-	0.01	0.07	0.58	m	0.65
450 mm x 150 mm	-	0.03	0.22	0.81	m	1.03
600 mm x 200 mm	-	0.05	0.37	1.04	m	1.41
Excavating; by hand						
Excavating trenches; to receive kerb foundations; average size						
150 mm x 50 mm	-	0.02	0.15	-	m	0.15
200 mm x 75 mm	-	0.07	0.52	-	m	0.52
250 mm x 100 mm	-	0.12	0.89	-	m	0.89
300 mm x 100 mm	-	0.15	1.12	-	m	1.12
Excavating curved trenches; to receive kerb foundations; average size						
150 mm x 50 mm	-	0.04	0.30	-	m	0.30
200 mm x 75 mm	-	0.08	0.60	-	m	0.60
250 mm x 100 mm	-	0.13	0.97	-	m	0.97
300 mm x 100 mm	-	0.16	1.19	-	m	1.19
Plain in situ ready mixed designated concrete; C7.5 - 40 mm aggregate; poured on or against earth or unblinded hardcore						
Foundations	63.09	1.33	10.93	76.56	m³	87.49
Blinding beds						
not exceeding 150 mm thick	63.09	1.97	16.19	76.56	m³	92.75
Plain in situ ready mixed designated concrete; C10 - 40 mm aggregate; poured on or against earth or unblinded hardcore						
Foundations	63.51	1.33	10.93	77.07	m³	88.00
Blinding beds						
not exceeding 150 mm thick	63.51	1.97	16.19	77.07	m³	93.26
Plain in situ ready mixed designated concrete; C20 - 20 mm aggregate; poured on or against earth or unblinded hardcore						
Foundations	65.69	1.33	10.93	79.71	m³	90.64
Blinding beds						
not exceeding 150 mm thick	65.69	1.97	16.19	79.71	m³	95.90
Precast concrete kerbs, channels, edgings, etc.; BS 340; bedded, jointed and pointed in cement mortar (1:3); including haunching up one side with in situ ready mixed designated concrete C10 - 40 mm aggregate; to concrete base						
Edgings; straight; square edge, fig 12						
50 mm x 150 mm	-	0.27	3.46	2.92	m	6.38
50 mm x 200 mm	-	0.27	3.46	3.55	m	7.01
50 mm x 255 mm	-	0.27	3.46	3.78	m	7.24

	PC £	Labour hours	Labour £	Material £	Unit	Total rate £
Q10 KERBS/EDGINGS/CHANNELS/PAVING ACCESSORIES - cont'd						
Precast concrete kerbs, channels, edgings, etc.; BS 340; bedded, jointed and pointed in cement mortar (1:3); including haunching up one side with in situ ready mixed designated concrete C10 - 40 mm aggregate; to concrete base - cont'd						
Kerbs; straight						
125 mm x 255 mm; fig 7	-	0.35	4.48	5.50	m	9.98
150 mm x 305 mm; fig 6	-	0.35	4.48	7.98	m	12.46
Kerbs; curved						
125 mm x 255 mm; fig 7	-	0.53	6.78	7.11	m	13.89
150 mm x 305 mm; fig 6	-	0.53	6.78	12.66	m	19.44
Channels; 255 mm x 125 mm; fig 8						
straight	-	0.35	4.48	4.92	m	9.40
curved	-	0.53	6.78	6.48	m	13.26
Quadrants; fig 14						
305 mm x 305 mm x 150 mm	-	0.37	4.74	7.72	nr	12.46
305 mm x 305 mm x 255 mm	-	0.37	4.74	8.30	nr	13.04
457 mm x 457 mm x 150 mm	-	0.43	5.50	9.11	nr	14.61
457 mm x 457 mm x 255 mm	-	0.43	5.50	9.69	nr	15.19
Q20 HARDCORE/GRANULAR/CEMENT BOUND BASES/SUB-BASES TO ROADS/PAVINGS						
Filling to make up levels; by machine						
Average thickness not exceeding 0.25 m						
obtained off site; hardcore	-	0.32	2.38	25.85	m^3	28.23
obtained off site; granular fill type one	-	0.32	2.38	35.46	m^3	37.84
obtained off site; granular fill type two	-	0.32	2.38	34.16	m^3	36.54
Average thickness exceeding 0.25 m						
obtained off site; hardcore	-	0.28	2.08	22.23	m^3	24.31
obtained off site; granular fill type one	-	0.28	2.08	35.31	m^3	37.39
obtained off site; granular fill type two	-	0.28	2.08	34.01	m^3	36.09
Filling to make up levels; by hand						
Average thickness not exceeding 0.25 m						
obtained off site; hardcore	-	0.70	5.21	26.31	m^3	31.52
obtained off site; sand	-	0.82	6.10	36.26	m^3	42.36
Average thickness exceeding 0.25 m						
obtained off site; hardcore	-	0.58	4.32	22.50	m^3	26.82
obtained off site; sand	-	0.69	5.13	35.89	m^3	41.02
Surface treatments						
Compacting						
filling; blinding with sand	-	0.05	0.37	1.36	m^2	2.23
Q21 IN SITU CONCRETE ROADS/PAVINGS						
Reinforced in situ ready mixed designated concrete; C10 - 40 mm aggregate						
Roads; to hardcore base						
thickness not exceeding 150 mm	63.51	2.17	17.84	73.39	m^3	91.23
thickness 150 mm - 450 mm	63.51	1.53	12.58	73.39	m^3	85.97

Q PAVING/PLANTING/FENCING/SITE FURNITURE

	PC £	Labour hours	Labour £	Material £	Unit	Total rate £
Reinforced in situ ready mixed designated concrete; C20 - 20 mm aggregate						
Roads; to hardcore base						
thickness not exceeding 150 mm	65.69	2.17	17.84	75.91	m³	93.75
thicness 150 mm - 450 mm	65.69	1.53	12.58	75.91	m³	88.49
Reinforced in situ ready mixed designated concrete; C25 - 20 mm aggregate						
Roads; to hardcore base						
thickness not exceeding 150 mm	66.19	2.17	17.84	76.49	m³	94.33
thickness 150 mm - 450 mm	66.19	1.53	12.58	76.49	m³	89.07
Formwork; sides of foundations; basic finish						
Plain vertical						
height not exceeding 250 mm	-	0.44	5.66	1.42	m	7.08
height 250 mm - 500 mm	-	0.66	8.49	2.47	m	10.96
height 500 mm - 1.00 m	-	0.95	12.22	4.89	m	17.11
add to above for curved radius 6.00 m	-	0.04	0.51	0.19	m	0.70
Reinforcement; fabric; BS 4483; lapped; in roads, footpaths or pavings						
Ref A142 (2.22 kg/m²)						
400 mm minimum laps	0.78	0.16	1.65	0.92	m²	2.57
Ref A193 (3.02 kg/m²)						
400 mm minimum laps	-	0.16	1.65	1.25	m²	2.90
Formed joints; Fosroc Expoandite "Flexcell" impregnated joint filler or other equal and approved						
Width not exceeding 150 mm						
12.50 mm thick	-	0.16	2.06	1.76	m	3.82
25 mm thick	-	0.21	2.70	2.65	m	5.35
Width 150 - 300 mm						
12.50 mm thick	-	0.21	2.70	2.89	m	5.59
25 mm thick	-	0.21	2.70	4.32	m	7.02
Width 300 - 450 mm						
12.50 mm thick	-	0.27	3.47	4.33	m	7.80
25 mm thick	-	0.27	3.47	6.47	m	9.94
Sealants; Fosroc Expandite "Pliastic N2" hot poured rubberized bituminous compound or other equal and approved						
Width 25 mm						
25 mm depth	-	0.22	2.83	1.60	m	4.43
Concrete sundries						
Treating surfaces of unset concrete; grading to cambers; tamping with a 75 mm thick steel shod tamper	-	0.27	2.22	-	m²	2.22

Prices for Measured Works - Minor Works
Q PAVING/PLANTING/FENCING/SITE FURNITURE

	PC £	Labour hours	Labour £	Material £	Unit	Total rate £
Q22 COATED MACADAM/ASPHALT ROADS/PAVINGS						
In situ finishings; Associated Asphalt or other equal and approved						
NOTE: The prices for all in situ finishings to roads and footpaths include for work to falls, crossfalls or slopes not exceeding 15 degrees from horizontal; for laying on prepared bases (prices not included) and for rolling with an appropriate roller. The following rates are based on black bitumen macadam except where stated. Red bitumen macadam rates are approximately 50% dearer.						
Fine graded wearing course; BS 4987; clause 2.7.7, tables 34 - 36; 14 mm nominal size pre-coated igneous rock chippings; tack coat of bitumen emulsion						
19 mm work to roads; one coat						
igneous aggregate	-	-	-	-	m²	15.00
Close graded bitumen macadam; BS 4987; 10 mm nominal size graded aggregate to clause 2.7.4 tables 34 - 36; tack coat of bitumen emulsion						
30 mm work to roads; one coat						
limestone aggregate	-	-	-	-	m²	14.35
igneous aggregate	-	-	-	-	m²	14.48
Bitumen macadam; BS 4987; 45 mm thick base course of 20 mm open graded aggregate to clause 2.6.1 tables 5 - 7; 20 mm thick wearing course of 6 mm nominal size medium graded aggregate to clause 2.7.6 tables 32 - 33						
65 mm work to pavements/footpaths; two coats						
limestone aggregate	-	-	-	-	m²	16.75
igneous aggregate	-	-	-	-	m²	16.94
add to last for 10 mm nominal size chippings; sprinkled into wearing course	-	-	-	-	m²	0.39
Bitumen macadam; BS 4987; 50 mm nominal size graded aggregate to clause 2.6.2 tables 8 - 10						
75 mm work to roads; one coat						
limestone aggregate	-	-	-	-	m²	17.73
igneous aggregate	-	-	-	-	m²	18.05
Dense bitumen macadam; BS 4987; 50 mm thick base course of 20 mm graded aggregate to clause 2.6.5 tables 15 - 16; 200 pen. binder; 30 mm wearing course of 10 mm nominal size graded aggregate to clause 2.7.4 tables 26 - 28						
80 mm work to roads; two coats						
limestone aggregate	-	-	-	-	m²	17.86
igneous aggregate	-	-	-	-	m²	18.05

Prices for Measured Works - Minor Works
Q PAVING/PLANTING/FENCING/SITE FURNITURE

	PC £	Labour hours	Labour £	Material £	Unit	Total rate £
Bitumen macadam; BS 4987; 50 mm thick base course of 20 mm nominal size graded aggregate to clause 2.6.1 tables 5 - 7; 25 mm thick wearing course of 10 mm nominal size graded aggregate to clause 2.7.2 tables 20 - 22						
75 mm work to roads; two coats						
limestone aggregate	-	-	-	-	m²	18.57
igneous aggregate	-	-	-	-	m²	18.76
Dense bitumen macadam; BS 4987; 70 mm thick base course of 20 mm nominal size graded aggregate to clause 2.6.5 tables 15 - 16; 200 pen. binder; 30 mm wearing course of 10 mm nominal size graded aggregate to clause 2.7.4 tables 26 - 28						
100 mm work to roads; two coats						
limestone aggregate	-	-	-	-	m²	20.12
igneous aggregate	-	-	-	-	m²	20.45
Dense bitumen macadam; BS 4987; 70 thick road base of 28 mm nominal size graded aggregate to clause 2.5.2 tables 3 - 4; 50 mm thick base course of 20 mm nominal size graded aggregate to clause 2.6.5 tables 15 - 16; 30 mm wearing course of 10 mm nominal size graded aggregate to clause 2.7.4 tables 26 - 28						
150 mm work to roads; three coats						
limestone aggregate	-	-	-	-	m²	32.58
igneous aggregate	-	-	-	-	m²	33.11
Red bitumen macadam; BS 4987; 70 thick base course of 20 mm nominal size graded aggregate to clause 2.6.5 tables 15 - 16; 30 mm wearing course of 10 mm nominal size graded aggregate to clause 2.7.4 tables 26 - 28						
100 mm work to roads; two coats						
limestone base; igneous wearing course	-	-	-	-	m²	24.02
igneous aggregate	-	-	-	-	m²	24.34
Q23 GRAVEL/HOGGIN/WOODCHIP ROADS/PAVINGS						
Two coat gravel paving; level and to falls; first layer course clinker aggregate and wearing layer fine gravel aggregate						
Pavings; over 300 mm wide						
50 mm thick	-	0.08	1.02	1.80	m²	2.82
63 mm thick	-	0.10	1.28	2.36	m²	3.64

Prices for Measured Works - Minor Works
Q PAVING/PLANTING/FENCING/SITE FURNITURE

	PC £	Labour hours	Labour £	Material £	Unit	Total rate £
Q25 SLAB/BRICK/BLOCK/SETT/COBBLE PAVINGS						
Artificial stone paving; Redland Aggregates "Texitone" or other equal and approved; to falls or crossfalls; bedding 25 mm thick in cement mortar (1:3); staggered joints; jointing in coloured cement mortar (1:3), brushed in; to sand base						
Pavings; over 300 mm wide						
450 mm x 600 mm x 50 mm thick; grey or coloured	3.49	0.48	6.14	17.28	m²	23.42
600 mm x 600 mm x 50 mm thick; grey or coloured	4.05	0.44	5.63	15.40	m²	21.03
750 mm x 600 mm x 50 mm thick; grey or coloured	4.72	0.42	5.38	14.54	m²	19.92
900 mm x 600 mm x 50 mm thick; grey or coloured	5.40	0.38	4.86	13.99	m²	18.85
Brick paviors; 215 mm x 103 mm x 65 mm rough stock bricks; PC £479.30/1000; to falls or crossfalls; bedding 10 mm thick in cement mortar (1:3); jointing in cement mortar (1:3); as work proceeds; to concrete base						
Pavings; over 300 mm wide; straight joints both ways						
bricks laid flat	-	0.85	13.82	22.58	m²	36.40
bricks laid on edge	-	1.19	19.35	33.54	m²	52.89
Pavings; over 300 mm wide; laid to herringbone pattern						
bricks laid flat	-	1.06	17.24	22.58	m²	39.82
bricks laid on edge	-	1.49	24.23	33.54	m²	57.77
Add or deduct for variation of £10.00/1000 in PC of brick paviours						
bricks laid flat	-	-	-	0.45	m²	0.45
bricks laid on edge	-	-	-	0.68	m²	0.68
River washed cobble paving; 50 mm - 75 mm; PC £89.88/t; to falls or crossfalls; bedding 13 mm thick in cement mortar (1:3); jointing to a height of two thirds of cobbles in dry mortar (1:3); tightly butted, washed and brushed; to concrete						
Pavings; over 300 mm wide						
regular	-	4.26	54.53	20.57	m²	75.10
laid to pattern	-	5.32	68.10	20.57	m²	88.67
Concrete paving flags; BS 7263; to falls or crossfalls; bedding 25 mm thick in cement and sand mortar (1:4); butt joints straight both ways; jointing in cement and sand (1:3); brushed in; to sand base						
Pavings; over 300 mm wide						
450 mm x 600 mm x 50 mm thick; grey	1.84	0.48	6.14	9.00	m²	15.14
450 mm x 600 mm x 60 mm thick; coloured	2.12	0.48	6.14	10.17	m²	16.31
600 mm x 600 mm x 50 mm thick; grey	2.09	0.44	5.63	7.86	m²	13.49
600 mm x 600 mm x 50 mm thick; coloured	2.41	0.44	5.63	8.87	m²	14.50
750 mm x 600 mm x 50 mm thick; grey	2.46	0.42	5.38	7.48	m²	12.86
750 mm x 600 mm x 50 mm thick; coloured	2.83	0.42	5.38	8.41	m²	13.79
900 mm x 600 mm x 50 mm thick; grey	2.71	0.38	4.86	6.98	m²	11.84
900 mm x 600 mm x 50 mm thick; coloured	3.17	0.38	4.86	7.94	m²	12.80

Q PAVING/PLANTING/FENCING/SITE FURNITURE

	PC £	Labour hours	Labour £	Material £	Unit	Total rate £
Concrete rectangular paving blocks; to falls or crossfalls; bedding 50 mm thick in dry sharp sand; filling joints with sharp sand brushed in; on earth base						
Pavings; "Keyblock" or other equal and approved; over 300 mm wide; straight joints both ways						
200 mm x 100 mm x 65 mm thick; grey	7.09	0.80	10.24	10.27	m²	20.51
200 mm x 100 mm x 65 mm thick; coloured	7.78	0.80	10.24	11.05	m²	21.29
200 mm x 100 mm x 80 mm thick; grey	7.75	0.85	10.88	11.21	m²	22.09
200 mm x 100 mm x 80 mm thick; coloured	9.00	0.85	10.88	12.62	m²	23.50
Pavings; "Keyblock" or other equal and approved; over 300 mm wide; laid to herringbone pattern						
200 mm x 100 mm x 60 mm thick; grey	-	1.00	12.80	10.27	m²	23.07
200 mm x 100 mm x 60 mm thick; coloured	-	1.00	12.80	11.05	m²	23.85
200 mm x 100 mm x 80 mm thick; grey	-	1.06	13.57	11.21	m²	24.78
200 mm x 100 mm x 80 mm thick; coloured	-	1.06	13.57	12.62	m²	26.19
Extra for two row boundary edging to herringbone paved areas; 200 wide; including a 150 mm high ready mixed designated concrete C10 - 40 mm aggregate haunching to one side; blocks laid breaking joint						
200 mm x 100 mm x 65 mm; coloured	-	0.32	4.10	2.08	m	6.18
200 mm x 100 mm x 80 mm; coloured	-	0.32	4.10	2.18	m	6.28
Pavings; "Mount Sorrel" or other equal and approved; over 300 mm wide; straight joints both ways						
200 mm x 100 mm x 60 mm thick; grey	7.26	0.80	10.24	10.46	m²	20.70
200 mm x 100 mm x 60 mm thick; coloured	8.37	0.80	10.24	11.72	m²	21.96
200 mm x 100 mm x 80 mm thick; grey	8.64	0.85	10.88	12.21	m²	23.09
200 mm x 100 mm x 80 mm thick; coloured	9.94	0.85	10.88	13.68	m²	24.56
Pavings; "Pedesta" or other equal and approved; over 300 mm wide; straight joints both ways						
200 mm x 100 mm x 60 mm thick; grey	7.25	0.80	10.24	10.45	m²	20.69
200 mm x 100 mm x 60 mm thick; coloured	7.34	0.80	10.24	10.55	m²	20.79
200 mm x 100 mm x 80 mm thick; grey	8.41	0.85	10.88	11.95	m²	22.83
200 mm x 100 mm x 80 mm thick; coloured	8.92	0.85	10.88	12.53	m²	23.41
Pavings; "Intersett" or other equal and approved; over 300 mm wide; straight joints both ways						
200 mm x 100 mm x 60 mm thick; grey	8.49	0.80	10.24	11.85	m²	22.09
200 mm x 100 mm x 60 mm thick; coloured	9.41	0.80	10.24	12.89	m²	23.13
200 mm x 100 mm x 80 mm thick; grey	10.14	0.85	10.88	13.90	m²	24.78
200 mm x 100 mm x 80 mm thick; coloured	11.27	0.85	10.88	15.18	m²	26.06
Concrete rectangular paving blocks; to falls or crossfalls; 6 mm wide joints; symmetrical layout; bedding in 15 mm semi-dry cement mortar (1:4); jointing and pointing in cement:sand (1:4); on concrete base						
Pavings; "Trafica" or other equal and approved; over 300 mm wide						
400 mm x 400 mm x 65 mm; Standard; natural	1.34	0.51	6.53	10.50	m²	17.03
400 mm x 400 mm x 65 mm; Standard; buff	1.93	0.51	6.53	14.60	m²	21.13
400 mm x 400 mm x 65 mm; Saxon textured; natural	2.64	0.51	6.53	19.54	m²	26.07
400 mm x 400 mm x 65 mm; Saxon textured; buff	3.12	0.51	6.53	22.87	m²	29.40
400 mm x 400 mm x 65 mm; Perfecta; natural	3.15	0.51	6.53	23.08	m²	29.61
400 mm x 400 mm x 65 mm; Perfecta; buff	3.74	0.51	6.53	27.18	m²	33.71
450 mm x 450 mm x 70 mm; Standard; natural	1.95	0.49	6.27	11.91	m²	18.18
450 mm x 450 mm x 70 mm; Standard; buff	2.59	0.49	6.27	15.43	m²	21.70

Prices for Measured Works - Minor Works
Q PAVING/PLANTING/FENCING/SITE FURNITURE

	PC £	Labour hours	Labour £	Material £	Unit	Total rate £
Q25 SLAB/BRICK/BLOCK/SETT/COBBLE PAVINGS - cont'd						
Concrete rectangular paving blocks; to falls or crossfalls; 6 mm wide joints; symmetrical layout; bedding in 15 mm semi-dry cement mortar (1:4); jointing and pointing in cement:sand (1:4); on concrete base - cont'd Pavings; "Trafica" or other equal and approved; over 300 mm wide - cont'd						
450 mm x 450 mm x 70 mm; Saxon textured; natural	3.27	0.49	6.27	19.17	m^2	25.44
450 mm x 450 mm x 70 mm; Saxon textured; buff	3.87	0.49	6.27	22.47	m^2	28.74
450 mm x 450 mm x 70 mm; Perfecta; natural	3.91	0.49	6.27	22.69	m^2	28.96
450 mm x 450 mm x 70 mm; Perfecta; buff	4.60	0.49	6.27	26.49	m^2	32.76
450 mm x 450 mm x 100 mm; Standard; natural	5.19	0.51	6.53	29.74	m^2	36.27
450 mm x 450 mm x 100 mm; Standard; buff	6.43	0.51	6.53	36.56	m^2	43.09
450 mm x 450 mm x 100 mm; Saxon textured; natural	5.86	0.51	6.53	33.42	m^2	39.95
450 mm x 450 mm x 100 mm; Saxon textured; buff	6.92	0.51	6.53	39.25	m^2	45.78
450 mm x 450 mm x 100 mm; Perfecta; natural	6.28	0.51	6.53	35.73	m^2	42.26
450 mm x 450 mm x 100 mm; Perfecta; buff	7.35	0.51	6.53	41.62	m^2	48.15
York stone slab pavings; to falls or crossfalls; bedding 25 mm thick in cement and sand mortar (1:4); 5 wide joints; jointing in coloured cement mortar (1:3); brushed in; to sand base Pavings; over 300 mm wide						
50 mm thick; random rectangular pattern	67.35	0.80	13.01	75.81	m^2	88.82
600 mm x 600 mm x 50 mm thick	80.82	0.44	7.15	90.65	m^2	97.80
600 mm x 900 mm x 50 mm thick	83.61	0.38	6.18	93.72	m^2	99.90
Granite setts; BS 435; 200 mm x 100 mm x 100 mm; PC £153.28/t; standard "C" dressing; tightly butted to falls or crossfalls; bedding 25 mm thick in cement mortar (1:3); filling joints with dry mortar (1:6); washed and brushed; on concrete base Pavings; over 300 mm wide						
straight joints	-	1.70	21.76	48.76	m^2	70.52
laid to pattern	-	2.13	27.26	48.76	m^2	76.02
Two rows of granite setts as boundary edging; 200 mm wide; including a 150 mm high in situ concrete mix 10.00 N/mm² - 40 mm aggregate (1:3.6); haunching to one side; blocks laid breaking joint	-	0.74	9.47	11.19	m	20.66
Q26 SPECIAL SURFACINGS/PAVINGS FOR SPORT/GENERAL AMENITY						
Sundries						
Line marking						
width not exceeding 300 mm	-	0.05	0.46	0.35	m	0.81

Q PAVING/PLANTING/FENCING/SITE FURNITURE

	PC £	Labour hours	Labour £	Material £	Unit	Total rate £
Q30 SEEDING/TURFING						
Vegetable soil						
Selected from spoil heaps; grading; prepared for turfing or seeding; to general surfaces						
average 75 mm thick	-	0.23	1.71	-	m²	1.71
average 100 mm thick	-	0.25	1.86	-	m²	1.86
average 125 mm thick	-	0.27	2.01	-	m²	2.01
average 150 mm thick	-	0.29	2.16	-	m²	2.16
average 175 mm thick	-	0.30	2.23	-	m²	2.23
average 200 mm thick	-	0.31	2.31	-	m²	2.31
Selected from spoil heaps; grading; prepared for turfing or seeding; to cuttings or embankments						
average 75 mm thick	-	0.27	2.01	-	m²	2.01
average 100 mm thick	-	0.29	2.16	-	m²	2.16
average 125 mm thick	-	0.30	2.23	-	m²	2.23
average 150 mm thick	-	0.31	2.31	-	m²	2.31
average 175 mm thick	-	0.33	2.46	-	m²	2.46
average 200 mm thick	-	0.36	2.68	-	m²	2.68
Imported vegetable soil						
Grading; prepared for turfing or seeding; to general surfaces						
average 75 mm thick	-	0.21	1.56	0.82	m²	2.38
average 100 mm thick	-	0.23	1.71	1.07	m²	2.78
average 125 mm thick	-	0.24	1.79	1.56	m²	3.35
average 150 mm thick	-	0.26	1.93	2.06	m²	3.99
average 175 mm thick	-	0.27	2.01	2.31	m²	4.32
average 200 mm thick	-	0.29	2.16	2.55	m²	4.71
Grading; preparing for turfing or seeding; to cuttings or embankments						
average 75 mm thick	-	0.23	1.71	0.82	m²	2.53
average 100 mm thick	-	0.26	1.93	1.07	m²	3.00
average 125 mm thick	-	0.27	2.01	1.56	m²	3.57
average 150 mm thick	-	0.29	2.16	2.06	m²	4.22
average 175 mm thick	-	0.30	2.23	2.31	m²	4.54
average 200 mm thick	-	0.31	2.31	2.55	m²	4.86
Fertilizer; PC £8.58/25kg						
Fertilizer 0.07 kg/m²; raking in						
general surfaces	-	0.03	0.22	0.03	m²	0.25
Selected grass seed; PC £69.74./25 kg						
Grass seed; sowing at a rate of 0.042 kg/m² two applications; raking in						
general surfaces	-	0.07	0.52	0.25	m²	0.77
cuttings or embankments	-	0.08	0.60	0.25	m²	0.85
Preserved turf from stack on site						
Turfing						
general surfaces	-	0.20	1.49	-	m²	1.49
cuttings or embankments; shallow	-	0.22	1.64	-	m²	1.64
cuttings or embankments; steep; pegged	-	0.31	2.31	-	m²	2.31
Imported turf; selected meadow turf						
Turfing						
general surfaces	1.23	0.20	1.49	1.58	m²	3.07
cuttings or embankments; shallow	1.23	0.22	1.64	1.58	m²	3.22
cuttings or embankments; steep; pegged	1.23	0.31	2.31	1.58	m²	3.89

Prices for Measured Works - Minor Works
Q PAVING/PLANTING/FENCING/SITE FURNITURE

	PC £	Labour hours	Labour £	Material £	Unit	Total rate £
Q31 PLANTING						
Planting only						
Hedge plants						
height not exceeding 750 mm	-	0.26	1.93	-	nr	1.93
height 750 mm - 1.50 m	-	0.61	4.54	-	nr	4.54
Saplings						
height not exceeding 3.00 m	-	1.73	12.87	-	nr	12.87
Q40 FENCING						
NOTE: The prices for all fencing include for setting posts in position, to a depth of 0.60 m for fences not exceeding 1.40 m high and of 0.76 m for fences over 1.40 m high. The prices allow for excavating post holes; filling to within 150 mm of ground level with concrete and all necessary backfilling.						
Strained wire fences; BS 1722 Part 3; 4 mm diameter galvanized mild steel plain wire threaded through posts and strained with eye bolts						
Fencing; height 900 mm; three line; concrete posts at 2750 mm centres	-	-	-	-	m	8.74
Extra for						
end concrete straining post; one strut	-	-	-	-	nr	43.95
angle concrete straining post; two struts	-	-	-	-	nr	63.26
Fencing; height 1.07 m; five line; concrete posts at 2750 mm centres	-	-	-	-	m	12.53
Extra for						
end concrete straining post; one strut	-	-	-	-	nr	65.02
angle concrete straining post; two struts	-	-	-	-	nr	82.29
Fencing; height 1.20 m; six line; concrete posts at 2750 mm centres	-	-	-	-	m	13.08
Extra for						
end concrete straining post; one strut	-	-	-	-	nr	67.61
angle concrete straining post; two struts	-	-	-	-	nr	92.07
Fencing; height 1.40 m; seven line; concrete posts at 2750 mm centres	-	-	-	-	m	13.88
Extra for						
end concrete straining post; one strut	-	-	-	-	nr	72.51
angle concrete straining post; two struts	-	-	-	-	nr	99.27
Chain link fencing; BS 1722 Part 1; 3 mm diameter galvanized mild steel wire; 50 mm mesh; galvanized mild steel tying and line wire; three line wires threaded through posts and strained with eye bolts and winding brackets						
Fencing; height 900 mm; galvanized mild steel angle posts at 3.00 m centres	-	-	-	-	m	12.03
Extra for						
end steel straining post; one strut	-	-	-	-	nr	55.58
angle steel straining post; two struts	-	-	-	-	nr	75.25
Fencing; height 900 mm; concrete posts at 3.00 m centres	-	-	-	-	m	11.74
Extra for						
end concrete straining post; one strut	-	-	-	-	nr	50.93
angle concrete straining post; two struts	-	-	-	-	nr	70.36

Prices for Measured Works - Minor Works
Q PAVING/PLANTING/FENCING/SITE FURNITURE

649

	PC £	Labour hours	Labour £	Material £	Unit	Total rate £
Fencing; height 1.20 m; galvanized mild steel angle posts at 3.00 m centres	-	-	-	-	m	15.63
Extra for						
end steel straining post; one strut	-	-	-	-	nr	67.86
angle steel straining post; two struts	-	-	-	-	nr	89.63
Fencing; height 1.20 m; concrete posts at 3.00 m centres	-	-	-	-	m	14.18
Extra for						
end concrete straining post; one strut	-	-	-	-	nr	62.96
angle concrete straining post; two struts	-	-	-	-	nr	82.34
Fencing; height 1.80 m; galvanized mild steel angle posts at 3.00 m centres	-	-	-	-	m	19.62
Extra for						
end steel straining post; one strut	-	-	-	-	nr	89.58
angle steel straining post; two struts	-	-	-	-	nr	123.64
Fencing; height 1.80 m; concrete posts at 3.00 m centres	-	-	-	-	m	17.88
Extra for						
end concrete straining post; one strut	-	-	-	-	nr	82.34
angle concrete straining post; two struts	-	-	-	-	nr	114.50
Pair of gates and gate posts; gates to match galvanized chain link fencing, with angle framing, braces, etc., complete with hinges, locking bar, lock and bolts; two 100 mm x 100 mm angle section gate posts; each with one strut						
2.44 m x 0.90 m	-	-	-	-	nr	753.99
2.44 m x 1.20 m	-	-	-	-	nr	788.95
2.44 m x 1.80 m	-	-	-	-	nr	893.82
Chain link fencing; BS 1722 Part 1; 3 mm diameter plastic coated mild steel wire; 50 mm mesh; plastic coated mild steel tying and line wire; three line wires threaded through posts and strained with eye bolts and winding brackets						
Fencing; height 900 mm; galvanized mild steel angle posts at 3.00 m centres	-	-	-	-	m	12.53
Extra for						
end steel straining post; one strut	-	-	-	-	nr	55.77
angle steel straining post; two struts	-	-	-	-	nr	75.25
Fencing; height 900 mm; concrete posts at 3.00 m centres	-	-	-	-	m	12.29
Extra for						
end concrete straining post; one strut	-	-	-	-	nr	51.23
angle concrete straining post; two struts	-	-	-	-	nr	70.36
Fencing; height 1.20 m; galvanized mild steel angle posts at 3.00 m centres	-	-	-	-	m	18.37
Extra for						
end steel straining post; one strut	-	-	-	-	nr	67.86
angle steel straining post; two struts	-	-	-	-	nr	89.58
Fencing; height 1.20 m; concrete posts at 3.00 m centres	-	-	-	-	m	14.73
Extra for						
end concrete straining post; one strut	-	-	-	-	nr	62.96
angle concrete straining post; two struts	-	-	-	-	nr	82.34
Fencing; height 1.80 m; galvanized mild steel angle posts at 3.00 m centres	-	-	-	-	m	21.82
Extra for						
end steel straining post; one strut	-	-	-	-	nr	89.58
angle steel straining post; two struts	-	-	-	-	nr	123.73

Prices for Measured Works - Minor Works
Q PAVING/PLANTING/FENCING/SITE FURNITURE

	PC £	Labour hours	Labour £	Material £	Unit	Total rate £
Q40 FENCING - cont'd						
Chain link fencing; BS 1722 Part 1; 3 mm diameter plastic coated mild steel wire; 50 mm mesh; plastic coated mild steel tying and line wire; three line wires threaded through posts and strained with eye bolts and winding brackets - cont'd						
Fencing; height 1.80 m; concrete posts at 3.00 m centres	-	-	-	-	m	20.97
Extra for						
end concrete straining post; one strut	-	-	-	-	nr	82.34
angle concrete straining post; two struts	-	-	-	-	nr	114.50
Pair of gates and gate posts; gates to match plastic chain link fencing; with angle framing, braces, etc. complete with hinges, locking bar, lock and bolts; two 100 mm x 100 mm angle section gate posts; each with one strut						
2.44 m x 0.90 m	-	-	-	-	nr	768.98
2.44 m x 1.20 m	-	-	-	-	nr	798.94
2.44 m x 1.80 m	-	-	-	-	nr	913.78
Chain link fencing for tennis courts; BS 1722 Part 13; 2.50 mm diameter galvanised mild wire; 45 mm mesh; line and tying wires threaded through 45 mm x 45 mm x 5 mm galvanised mild steel angle standards, posts and struts; 60 mm x 60 mm x 6 mm straining posts and gate posts; straining posts and struts strained with eye bolts and winding brackets						
Fencing to tennis court 36.00 m x 18.00 m; including gate 1.07 mm x 1.98 m complete with hinges, locking bar, lock and bolts						
height 2745 mm; standards at 3.00 m centres	-	-	-	-	nr	2361.87
height 3660 mm; standards at 2.50 m centres	-	-	-	-	nr	3195.76
Cleft chestnut pale fencing; BS 1722 Part4; pales spaced 51 mm apart; on two lines of galvanized wire; 64 mm diameter posts; 76 mm x 51 mm struts						
Fencing; height 900 mm; posts at 2.50 m centres	-	-	-	-	m	7.29
Extra for						
straining post; one strut	-	-	-	-	nr	13.29
corner straining post; two struts	-	-	-	-	nr	17.73
Fencing; height 1.05 m; posts at 2.50 m centres	-	-	-	-	m	8.08
Extra for						
straining post; one strut	-	-	-	-	nr	14.68
corner straining post; two struts	-	-	-	-	nr	19.52
Fencing; height 1.20 m; posts at 2.25 m centres	-	-	-	-	m	8.54
Extra for						
straining post; one strut	-	-	-	-	nr	16.03
corner straining post; two struts	-	-	-	-	nr	21.32
Fencing; height 1.35 m; posts at 2.25 m centres	-	-	-	-	m	9.14
Extra for						
straining post; one strut	-	-	-	-	nr	17.97
corner straining post; two struts	-	-	-	-	nr	23.86

Prices for Measured Works - Minor Works
Q PAVING/PLANTING/FENCING/SITE FURNITURE

	PC £	Labour hours	Labour £	Material £	Unit	Total rate £
Close boarded fencing; BS 1722 Part 5; 76 mm x 38 mm softwood rails; 89 mm x 19 mm softwood pales lapped 13 mm; 152 mm x 25 mm softwood gravel boards; all softwood "treated"; posts at 3.00 m centres						
Fencing; two rail; concrete posts						
1.00 m	-	-	-	-	m	30.45
1.20 m	-	-	-	-	m	32.30
Fencing; three rail; concrete posts						
1.40 m	-	-	-	-	m	40.89
1.60 m	-	-	-	-	m	42.44
1.80 m	-	-	-	-	m	44.29
Fencing; two rail; oak posts						
1.00 m	-	-	-	-	m	21.92
1.20 m	-	-	-	-	m	25.17
Fencing; three rail; oak posts						
1.40 m	-	-	-	-	m	28.26
1.60 m	-	-	-	-	m	32.21
1 80 m	-	-	-	-	m	35.85
Precast concrete slab fencing; 305 mm x 38 mm x 1753 mm slabs; fitted into twice grooved concrete posts at 1830 mm centres						
Fencing						
height 1.20 m	-	-	-	-	m	46.74
height 1.50 m	-	-	-	-	m	55.37
height 1.80 m	-	-	-	-	m	69.11
Mild steel unclimbable fencing; in rivetted panels 2440 mm long; 44 mm x 13 mm flat section top and bottom rails; two 44 mm x 19 mm flat section standards; one with foot plate, and 38 mm x 13 mm raking stay with foot plate; 20 mm diameter pointed verticals at 120 mm centres; two 44 mm x 19 mm supports 760 mm long with ragged ends to bottom rail; the whole bolted together; coated with red oxide primer; setting standards and stays in ground at 2440 mm centres and supports at 815 mm centres						
Fencing						
height 1.67 m	-	-	-	-	m	110.10
height 2.13 m	-	-	-	-	m	125.83
Pair of gates and gate posts, to match mild steel unclimbable fencing; with flat section framing, braces, etc., complete with locking bar, lock, handles, drop bolt, gate stop and holding back catches; two 102 mm x 102 mm hollow section gate posts with cap and foot plates						
2.44 m x 1.67 m	-	-	-	-	nr	968.71
2.44 m x 2.13 m	-	-	-	-	nr	1133.49
4.88 m x 1.67 m	-	-	-	-	nr	1498.01
4.88 m x 2.13 m	-	-	-	-	nr	1897.48
PVC coated, galvanised mild steel high security fencing; "Sentinel Sterling fencing" or other equal and approved; Twil Wire Products Ltd; 50 mm x 50 mm mesh; 3 mm/3.50 mm gauge wire; barbed edge - 1; "Sentinal Bi-steel" colour coated posts at 2440 mm centres						
Fencing						
height 1.80 m	-	1.02	7.59	34.11	m	41.70
height 2.10 m	-	1.02	7.59	37.46	m	45.05

R DISPOSAL SYSTEMS

R10 RAINWATER PIPEWORK/GUTTERS

	PC £	Labour hours	Labour £	Material £	Unit	Total rate £
Aluminium pipes and fittings; BS 2997; ears cast on; powder coated finish						
63 50 mm diameter pipes; plugged and screwed	10.82	0.41	4.81	12.97	m	17.78
Extra for						
fittings with one end	-	0.24	2.81	6.72	nr	9.53
fittings with two ends	-	0.46	5.39	6.91	nr	12.30
fittings with three ends	-	0.67	7.85	9.20	nr	17.05
shoe	6.62	0.24	2.81	6.72	nr	9.53
bend	7.06	0.46	5.39	6.91	nr	12.30
single branch	9.21	0.67	7.85	9.20	nr	17.05
offset 228 mm projection	16.28	0.46	5.99	16.27	nr	22.26
offset 304 mm projection	18.16	0.46	5.39	18.34	nr	23.73
access pipe	20.11	-	-	19.16	nr	19.16
connection to clay pipes; cement and sand (1:2) joint	-	0.17	1.99	0.10	nr	2.09
76.50 mm diameter pipes; plugged and screwed	12.60	0.44	5.16	14.99	m	20.15
Extra for						
shoe	9.09	0.28	3.28	9.30	nr	12.58
bend	8.92	0.50	5.86	8.82	nr	14.68
single branch	11.09	0.72	8.44	11.06	nr	19.50
offset 228 mm projection	18.00	0.50	5.86	17.83	nr	23.69
offset 304 mm projection	19.92	0.50	5.86	19.95	nr	25.81
access pipe	21.98	-	-	20.73	nr	20.73
connection to clay pipes; cement and sand (1:2) joint	-	0.19	2.23	0.10	nr	2.33
100 mm diameter pipes; plugged and screwed	21.50	0.50	5.86	25.05	m	30.91
Extra for						
shoe	10.95	0.32	3.75	10.56	nr	14.31
bend	12.43	0.56	6.56	11.93	nr	18.49
single branch	14.84	0.83	9.73	14.05	nr	23.78
offset 228 mm projection	20.82	0.56	6.56	18.90	nr	25.46
offset 304 mm projection	23.13	0.56	6.56	21.45	nr	28.01
access pipe	26.05	-	-	22.72	nr	22.72
connection to clay pipes; cement and sand (1:2) joint	-	0.22	2.58	0.10	nr	2.68
Roof outlets; circular aluminium; with flat or domed grating; joint to pipe						
50 mm diameter	50.84	0.67	11.01	54.65	nr	65.66
75 mm diameter	67.46	0.72	11.84	72.52	nr	84.36
100 mm diameter	94.62	0.78	12.82	101.72	nr	114.54
150 mm diameter	117.70	0.83	13.65	126.53	nr	140.18
Roof outlets; d-shaped; balcony; with flat or domed grating; joint to pipe						
50 mm diameter	60.22	0.67	11.01	64.74	nr	75.75
75 mm diameter	69.22	0.72	11.84	74.41	nr	86.25
100 mm diameter	85.01	0.78	12.82	91.39	nr	104.21
Galvanized wire balloon grating; BS 416 for pipes or outlets						
50 mm diameter	1.41	0.07	1.15	1.52	nr	2.67
63 mm diameter	1.43	0.07	1.15	1.54	nr	2.69
75 mm diameter	1.52	0.07	1.15	1.63	nr	2.78
100 mm diameter	1.68	0.09	1.48	1.81	nr	3.29

R DISPOSAL SYSTEMS

	PC £	Labour hours	Labour £	Material £	Unit	Total rate £
Aluminium gutters and fittings; BS 2997; powder coated finish						
100 mm half round gutters; on brackets; screwed to timber	9.92	0.39	5.08	14.12	m	19.20
Extra for						
stop end	2.69	0.18	2.34	5.73	nr	8.07
running outlet	5.96	0.37	4.82	6.51	nr	11.33
stop end outlet	5.30	0.18	2.34	7.71	nr	10.05
angle	5.51	0.37	4.82	5.04	nr	9.86
113 mm half round gutters; on brackets; screwed to timber	10.39	0.39	5.08	14.68	m	19.76
Extra for						
stop end	2.82	0.18	2.34	5.90	nr	8.24
running outlet	6.50	0.37	4.82	7.06	nr	11.88
stop end outlet	6.08	0.18	2.34	8.53	nr	10.87
angle	6.21	0.37	4.82	5.69	nr	10.51
125 mm half round gutters; on brackets; screwed to timber	11.68	0.44	5.73	17.62	m	23.35
Extra for						
stop end	3.44	0.20	2.60	7.93	nr	10.53
running outlet	7.03	0.39	5.08	7.54	nr	12.62
stop end outlet	6.46	0.20	2.60	10.20	nr	12.80
angle	6.89	0.39	5.08	7.52	nr	12.60
100 mm ogee gutters; on brackets; screwed to timber	12.38	0.41	5.34	18.06	m	23.40
Extra for						
stop end	2.83	0.19	2.47	3.78	nr	6.25
running outlet	6.99	0.39	5.08	6.89	nr	11.97
stop end outlet	5.42	0.19	2.47	8.63	nr	11.10
angle	5.89	0.39	5.08	4.46	nr	9.54
112 mm ogee gutters; on brackets; screwed to timber	13.77	0.46	5.99	19.89	m	25.88
Extra for						
stop end	3.04	0.19	2.47	1.01	nr	6.48
running outlet	7.08	0.39	5.08	6.87	nr	11.95
stop end outlet	6.07	0.19	2.47	9.45	nr	11.92
angle	7.01	0.39	5.08	5.44	nr	10.52
125 mm ogee gutters; on brackets; screwed to timber	15.20	0.46	5.99	21.84	m	27.83
Extra for						
stop end	3.31	0.21	2.73	4.32	nr	7.05
running outlet	7.74	0.41	5.34	7.47	nr	12.81
stop end outlet	6.89	0.21	2.73	10.54	nr	13.27
angle	8.19	0.41	5.34	6.48	nr	11.82
Cast iron pipes and fittings; EN 1462; ears cast on; joints						
65 mm pipes; primed; nailed to masonry	16.27	0.57	6.68	18.72	m	25.40
Extra for						
shoe	14.14	0.35	4.10	14.40	nr	18.50
bend	8.66	0.63	7.38	8.36	nr	15.74
single branch	16.69	0.80	9.38	16.66	nr	26.04
offset 225 mm projection	15.43	0.63	7.38	14.72	nr	22.10
offset 305 mm projection	18.06	0.63	7.38	17.25	nr	24.63
connection to clay pipes; cement and sand (1:2) joint	-	0.17	1.99	0.12	nr	2.11

R10 RAINWATER PIPEWORK/GUTTERS - cont'd

	PC £	Labour hours	Labour £	Material £	Unit	Total rate £
Cast iron pipes and fittings; EN 1462; ears cast on; joints - cont'd						
75 mm pipes; primed; nailed to masonry	16.27	0.61	7.15	18.93	m	26.08
Extra for						
shoe	14.14	0.39	4.57	14.47	nr	19.04
bend	10.52	1.33	16.38	10.48	nr	26.86
single branch	18.40	0.83	9.73	18.68	nr	28.41
offset 225 mm projection	15.43	0.67	7.85	14.79	nr	22.64
offset 305 mm projection	18.96	0.67	7.85	18.31	nr	26.16
connection to clay pipes; cement and sand (1:2) joint	-	0.19	2.23	0.12	nr	2.35
100 mm pipes; primed; nailed to masonry	21.85	0.67	7.85	25.45	m	33.30
Extra for						
shoe	18.43	0.44	5.16	18.83	nr	23.99
bend	14.85	0.72	8.44	14.88	nr	23.32
single branch	21.87	0.89	10.43	22.02	nr	32.45
offset 225 mm projection	30.28	0.72	8.44	30.41	nr	38.85
offset 305 mm projection	30.28	0.72	8.44	29.91	nr	38.35
connection to clay pipes; cement and sand (1:2) joint	-	0.22	2.58	0.10	nr	2.68
100 mm x 75 mm rectangular pipes; primed; nailing to masonry	62.52	0.67	7.85	71.37	m	79.22
Extra for						
shoe	52.97	0.44	5.16	53.22	nr	58.38
bend	50.44	0.72	8.44	50.43	nr	58.87
offset 225 mm projection	71.01	0.44	5.16	68.86	nr	74.02
offset 305 mm projection	75.91	0.44	5.16	72.85	nr	78.01
connection to clay pipes; cement and sand (1:2) joint	-	0.22	2.58	0.10	nr	2.68
Rainwater head; rectangular; for pipes						
65 mm diameter	19.30	0.63	7.38	21.56	nr	28.94
75 mm diameter	19.30	0.67	7.85	21.63	nr	29.48
100 mm diameter	26.65	0.72	8.44	29.86	nr	38.30
Rainwater head; octagonal; for pipes						
65 mm diameter	11.05	0.63	7.38	12.47	nr	19.85
75 mm diameter	12.55	0.67	7.85	14.19	nr	22.04
100 mm diameter	27.80	0.72	8.44	31.13	nr	39.57
Copper wire balloon grating; BS 416 for pipes or outlets						
50 mm diameter	1.72	0.07	0.82	1.85	nr	2.67
63 mm diameter	1.74	0.07	0.82	1.87	nr	2.69
75 mm diameter	1.95	0.07	0.82	2.10	nr	2.92
100 mm diameter	2.22	0.09	1.05	2.39	nr	3.44
Cast iron gutters and fittings; EN 1462						
100 mm half round gutters; primed; on brackets; screwed to timber	8.28	0.44	5.73	11.83	m	17.56
Extra for						
stop end	2.09	0.19	2.47	3.43	nr	5.90
running outlet	6.05	0.39	5.08	5.89	nr	10.97
angle	6.21	0.39	5.08	7.22	nr	12.30
115 mm half round gutters; primed; on brackets; screwed to timber	8.62	0.44	5.73	12.25	m	17.98
Extra for						
stop end	2.70	0.19	2.47	4.08	nr	6.55
running outlet	6.60	0.39	5.08	6.46	nr	11.54
angle	6.39	0.39	5.08	7.35	nr	12.43

R DISPOSAL SYSTEMS

	PC £	Labour hours	Labour £	Material £	Unit	Total rate £
125 mm half round gutters; primed; on brackets; screwed to timber	10.10	0.50	6.51	13.92	m	20.43
Extra for						
stop end	2.70	0.22	2.86	4.09	nr	6.95
running outlet	7.54	0.44	5.73	7.34	nr	13.07
angle	7.54	0.44	5.73	8.31	nr	14.04
150 mm half round gutters; primed; on brackets; screwed to timber	17.25	0.56	7.29	21.67	m	28.96
Extra for						
stop end	3.60	0.24	3.12	6.52	nr	9.64
running outlet	13.04	0.50	6.51	12.68	nr	19.19
angle	13.77	0.50	6.51	14.30	nr	20.81
100 mm ogee gutters; primed; on brackets; screwed to timber	9.23	0.46	5.99	13.16	m	19.15
Extra for						
stop end	1.93	0.20	2.60	4.46	nr	7.06
running outlet	6.61	0.41	5.34	6.42	nr	11.76
angle	6.48	0.41	5.34	7.58	nr	12.92
115 mm ogee gutters; primed; on brackets; screwed to timber	10.15	0.46	5.99	14.23	m	20.22
Extra for						
stop end	2.55	0.20	2.60	5.13	nr	7.73
running outlet	7.02	0.41	5.34	6.77	nr	12.11
angle	7.02	0.41	5.34	7.98	nr	13.32
125 mm ogee gutters; primed; on brackets; screwed to timber	10.65	0.52	6.77	15.12	m	21.89
Extra for						
stop end	2.55	0.23	2.99	5.43	nr	8.42
running outlet	7.66	0.46	5.99	7.42	nr	13.41
angle	7.66	0.46	5.99	8.88	nr	14.87
3 mm thick galvanised heavy pressed steel gutters and fittings; joggle joints; BS 1091						
200 mm x 100 mm (400 mm girth) box gutter; screwed to timber	-	0.72	8.44	19.18	m	27.62
Extra for						
stop end	-	0.39	4.57	10.60	nr	15.17
running outlet	-	0.78	9.14	17.33	nr	26.77
stop end outlet	-	0.39	4.57	24.73	nr	29.30
angle	-	0.78	9.14	19.56	nr	28.70
381 mm boundary wall gutters; 900 mm girth; screwed to timber	-	0.72	8.44	31.59	m	40.03
Extra for						
stop end	-	0.44	5.16	18.14	nr	23.30
running outlet	-	0.78	9.14	24.23	nr	33.37
stop end outlet	-	0.39	4.57	34.35	nr	38.92
angle	-	0.78	9.14	28.53	nr	37.67
457 mm boundary wall gutters; 1200 mm girth; screwed to timber	-	0.83	9.73	42.14	m	51.87
Extra for						
stop end	-	0.44	5.16	23.31	nr	28.47
running outlet	-	0.89	10.43	35.12	nr	45.55
stop end outlet	-	0.44	5.16	37.39	nr	42.55
angle	-	0.89	10.43	38.58	nr	49.01

Prices for Measured Works - Minor Works
R DISPOSAL SYSTEMS

	PC £	Labour hours	Labour £	Material £	Unit	Total rate £
R10 RAINWATER PIPEWORK/GUTTERS - cont'd						
uPVC external rainwater pipes and fittings; BS 4576; slip-in joints						
50 mm pipes; fixing with pipe or socket brackets; plugged and screwed	3.38	0.33	3.87	5.12	m	8.99
Extra for						
shoe	2.00	0.22	2.58	2.73	nr	5.31
bend	2.33	0.33	3.87	3.09	nr	6.96
two bends to form offset 229 mm projection	2.33	0.33	3.87	5.01	nr	8.88
connection to clay pipes; cement and sand (1:2) joint	-	0.15	1.76	0.12	nr	1.88
68 mm pipes; fixing with pipe or socket brackets; plugged and screwed	2.71	0.37	4.34	4.67	m	9.01
Extra for						
shoe	2.00	0.24	2.81	3.02	nr	5.83
bend	3.51	0.37	4.34	4.68	nr	9.02
single branch	6.14	0.49	5.74	7.58	nr	13.32
two bends to form offset 229 mm projection	3.51	0.37	4.34	8.03	nr	12.37
loose drain connector; cement and sand (1:2) joint	-	0.17	1.99	2.81	nr	4.80
110 mm pipes; fixing with pipe or socket brackets; plugged and screwed	5.82	0.40	4.69	9.64	m	14.33
Extra for						
shoe	6.37	0.27	3.16	7.87	nr	11.03
bend	8.99	0.40	4.69	10.76	nr	15.45
single branch	13.29	0.54	6.33	15.50	nr	21.83
two bends to form offset 229 mm projection	8.99	0.40	4.69	19.74	nr	24.43
loose drain connector; cement and sand (1:2) joint	-	0.39	4.57	11.09	nr	15.66
65 mm square pipes; fixing with pipe or socket brackets; plugged and screwed	2.98	0.37	4.34	5.06	m	9.40
Extra for						
shoe	2.18	0.24	2.81	3.22	nr	6.03
bend	2.40	0.37	4.34	3.46	nr	7.80
single branch	6.71	0.49	5.74	8.21	nr	13.95
two bends to form offset 229 mm projection	2.40	0.37	4.34	5.70	nr	10.04
drain connector; square to round; cement and sand (1:2) joint	-	0.39	4.57	3.68	nr	8.25
Rainwater head; rectangular; for pipes						
50 mm diameter	10.86	0.50	5.86	13.09	nr	18.95
68 mm diameter	8.77	0.52	6.09	11.36	nr	17.45
110 mm diameter	18.30	0.61	7.15	21.94	nr	29.09
65 mm square	8.77	0.52	6.09	11.36	nr	17.45
uPVC gutters and fittings; BS 4576						
76 mm half round gutters; on brackets screwed to timber	2.71	0.33	4.30	4.38	m	8.68
Extra for						
stop end	0.92	0.15	1.95	1.34	nr	3.29
running outlet	2.59	0.28	3.65	2.61	nr	6.26
stop end outlet	2.57	0.15	1.95	2.91	nr	4.86
angle	2.59	0.28	3.65	3.21	nr	6.86
112 mm half round gutters; on brackets screwed to timber	2.78	0.37	4.82	5.35	m	10.17
Extra for						
stop end	1.44	0.15	1.95	2.08	nr	4.03
running outlet	2.81	0.32	4.17	2.85	nr	7.02
stop end outlet	2.81	0.15	1.95	3.34	nr	5.29
angle	3.15	0.32	4.17	4.22	nr	8.39
170 mm half round gutters; on brackets; screwed to timber	5.39	0.37	4.82	9.72	m	14.54

R DISPOSAL SYSTEMS

	PC £	Labour hours	Labour £	Material £	Unit	Total rate £
Extra for						
stop end	2.43	0.18	2.34	3.63	nr	5.97
running outlet	5.40	0.34	4.43	5.46	nr	9.89
stop end outlet	5.14	0.18	2.34	6.13	nr	8.47
angle	7.04	0.34	4.43	9.17	nr	13.60
114 mm rectangular gutters; on brackets; screwed to timber	3.04	0.37	4.82	5.97	m	10.79
Extra for						
stop end	1.60	0.15	1.95	2.30	nr	4.25
running outlet	3.08	0.34	4.43	3.13	nr	7.56
stop end outlet	3.08	0.15	1.95	3.66	nr	5.61
angle	3.46	0.32	4.17	4.62	nr	8.79
R11 FOUL DRAINAGE ABOVE GROUND						
Cast iron "Timesaver" pipes and fittings or other equal and approved; BS 416						
50 mm pipes; primed; 2.00 m lengths; fixing with expanding bolts; to masonry	12.80	0.61	7.15	22.56	m	29.71
Extra for						
fittings with two ends	-	0.61	7.15	15.31	nr	22.46
fittings with three ends	-	0.83	9.73	25.91	nr	35.64
bends; short radius	10.12	0.61	7.15	15.31	nr	22.46
access bends; short radius	24.93	0.61	7.15	31.63	nr	38.78
boss; 38 BSP	20.96	0.66	7.74	26.89	nr	34.63
single branch	15.18	0.83	9.73	26.56	nr	36.29
access pipe	24.34	0.61	7.15	29.39	nr	36.54
roof connector; for asphalt	27.46	0.61	7.15	34.63	nr	41.78
isolated "Timesaver" coupling joint	5.74	0.33	3.87	6.32	nr	10.19
connection to clay pipes; cement and sand (1:2) joint	-	0.15	1.76	0.10	nr	1.86
75 mm pipes; primed; 3.00 m lengths; fixing with standard brackets; plugged and screwed to masonry	12.66	0.61	7.15	22.10	m	29.25
75 mm pipes; primed; 2.00 m lengths; fixing with standard brackets; plugged and screwed to masonry	12.80	0.67	7.85	22.26	m	30.11
Extra for						
bends; short radius	10.12	0.67	7.85	15.98	nr	23.83
access bends; short radius	24.93	0.61	7.15	32.30	nr	39.45
boss; 38 BSP	20.96	0.67	7.85	27.93	nr	35.78
single branch	15.18	0.94	11.02	27.76	nr	38.78
double branch	25.59	1.22	14.30	46.23	nr	60.53
offset 115 mm projection	12.47	0.67	7.85	16.91	nr	24.76
offset 150 mm projection	12.47	0.67	7.85	16.48	nr	24.33
access pipe	24.34	0.67	7.85	29.77	nr	37.62
roof connector; for asphalt	31.61	0.67	7.85	39.52	nr	47.37
isolated "Timesaver" coupling joint	6.35	0.39	4.57	7.00	nr	11.57
connection to clay pipes; cement and sand (1:2) joint	-	0.17	1.99	0.10	nr	2.09
100 mm pipes; primed; 3.00 m lengths; fixing with standard brackets; plugged and screwed to masonry	15.29	0.67	7.85	31.22	m	39.07
100 mm pipes; primed; 2.00 m lengths; fixing with standard brackets; plugged and screwed to masonry	15.44	0.74	8.67	31.38	m	40.05

R DISPOSAL SYSTEMS

	PC £	Labour hours	Labour £	Material £	Unit	Total rate £
R11 FOUL DRAINAGE ABOVE GROUND - cont'd						
Cast iron "Timesaver" pipes and fittings or other equal and approved; BS 416 - cont'd						
Extra for						
WC bent connector; 450 mm long tail	15.53	0.67	7.85	20.05	nr	27.90
bends; short radius	14.01	0.74	8.67	21.95	nr	30.62
access bends; short radius	29.63	0.74	8.67	39.16	nr	47.83
boss; 38 BSP	25.03	0.74	8.67	34.09	nr	42.76
single branch	21.65	1.11	13.01	38.01	nr	51.02
double branch	26.78	1.44	16.88	52.78	nr	69.66
offset 225 mm projection	20.13	0.74	8.67	26.42	nr	35.09
offset 300 mm projection	22.69	0.74	8.67	28.72	nr	37.39
access pipe	25.59	0.74	8.67	32.09	nr	40.76
roof connector; for asphalt	24.19	0.74	8.67	32.55	nr	41.22
roof connector; for roofing felt	77.55	0.74	8.67	91.35	nr	100.02
isolated "Timesaver" coupling joint	8.28	0.46	5.39	9.12	nr	14.51
transitional clayware socket; cement and sand (1:2) joint	16.47	0.44	5.16	27.38	nr	32.54
150 mm pipes; primed; 3.00 m lengths; fixing with standard brackets; plugged and screwed to masonry	31.94	0.83	9.73	63.18	m	72.91
150 mm pipes; primed; 2.00 m lengths; fixing with standard brackets; plugged and screwed to masonry	31.69	0.93	10.90	62.91	m	73.81
Extra for						
bends; short radius	25.03	0.93	10.90	40.45	nr	51.35
access bends; short radius	42.08	0.93	10.90	59.24	nr	70.14
boss; 38 BSP	40.83	0.93	10.90	56.97	nr	67.87
single branch	53.69	1.33	15.59	85.97	nr	101.56
double branch	75.43	1.78	20.86	126.56	nr	147.42
access pipe	42.58	0.93	10.90	52.49	nr	63.39
roof connector; for asphalt	61.54	0.93	10.90	77.82	nr	88.72
isolated "Timesaver" coupling joint	-	0.56	6.56	18.24	nr	24.80
transitional clayware socket; cement and sand (1:2) joint	28.85	0.57	6.68	50.13	nr	56.81
Cast iron "Ensign" lightweight pipes and fittings or other equal and approved; EN 877						
50 mm pipes; primed 3.00 m lengths; fixing with standard brackets; plugged and screwed to masonry	-	0.31	3.10	13.65	m	16.75
Extra for						
bends; short radius	-	0.27	2.71	10.89	nr	13.60
single branch	-	0.33	3.30	19.05	nr	22.35
access pipe	-	0.27	2.71	22.36	nr	25.07
70 mm pipes; primed 3.00 m lengths; fixing with standard brackets; plugged and screwed to masonry	-	0.34	3.38	15.26	m	18.64
Extra for						
bends; short radius	-	0.30	2.99	12.16	nr	15.15
single branch	-	0.37	3.69	20.46	nr	24.15
access pipe	-	0.30	2.99	23.84	nr	26.83
100 mm pipes; primed 3.00 m lengths; fixing with standard brackets; plugged and screwed to masonry	-	0.37	3.69	18.01	m	21.70
Extra for						
bends; short radius	-	0.32	3.21	14.92	nr	18.13
single branch	-	0.39	3.89	26.33	nr	30.22
double branch	-	0.46	4.59	38.57	nr	43.16
access pipe	-	0.32	3.21	27.10	nr	30.31
connector	-	0.21	2.08	31.64	nr	33.72
reducer	-	0.32	3.21	18.20	nr	21.41

R DISPOSAL SYSTEMS

	PC £	Labour hours	Labour £	Material £	Unit	Total rate £
Polypropylene (PP) waste pipes and fittings; BS 5254; push fit "O" - ring joints						
32 mm pipes; fixing with pipe clips; plugged and screwed	0.67	0.24	2.81	1.25	m	4.06
Extra for						
fittings with one end	-	0.18	2.11	0.67	nr	2.78
fittings with two ends	-	0.24	2.81	0.67	nr	3.48
fittings with three ends	-	0.33	3.87	0.67	nr	4.54
access plug	0.61	0.18	2.11	0.67	nr	2.78
double socket	0.61	0.17	1.99	0.67	nr	2.66
male iron to PP coupling	1.58	0.32	3.75	1.74	nr	5.49
sweep bend	0.61	0.24	2.81	0.67	nr	3.48
spigot bend	0.61	0.28	3.28	0.67	nr	3.95
40 mm pipes; fixing with pipe clips; plugged and screwed	0.81	0.24	2.81	1.41	m	4.22
Extra for						
fittings with one end	-	0.21	2.46	0.67	nr	3.13
fittings with two ends	-	0.33	3.87	0.67	nr	4.54
fittings with three ends	-	0.44	5.16	0.67	nr	5.83
access plug	0.61	0.21	2.46	0.67	nr	3.13
double socket	0.61	0.22	2.58	0.67	nr	3.25
universal connector	0.61	0.28	3.28	0.67	nr	3.95
sweep bend	0.61	0.33	3.87	0.67	nr	4.54
spigot bend	0.61	0.33	3.87	0.67	nr	4.54
reducer 40 mm - 32 mm	0.61	0.33	3.87	0.67	nr	4.54
50 mm pipes; fixing with pipe clips; plugged and screwed	1.36	0.39	4.57	2.38	m	6.95
Extra for						
fittings with one end	-	0.23	2.70	1.20	nr	3.90
fittings with two ends	-	0.39	4.57	1.20	nr	5.77
fittings with three ends	-	0.52	6.09	1.20	nr	7.29
access plug	1.09	0.23	2.70	1.20	nr	3.90
double socket	1.09	0.26	3.05	1.20	nr	4.25
sweep bend	1.09	0.39	4.57	1.20	nr	5.77
spigot bend	1.09	0.39	4.57	1.20	nr	5.77
reducer 50 mm - 40 mm	1.09	0.39	4.57	1.20	nr	5.77
muPVC waste pipes and fittings; BS 5255; solvent welded joints						
32 mm pipes; fixing with pipe clips; plugged and screwed	1.43	0.28	3.28	2.30	m	5.58
Extra for						
fittings with one end	-	0.19	2.23	1.68	nr	3.91
fittings with two ends	-	0.28	3.28	1.73	nr	5.01
fittings with three ends	-	0.37	4.34	2.41	nr	6.75
access plug	1.30	0.19	2.23	1.68	nr	3.91
straight coupling	0.87	0.19	2.23	1.20	nr	3.43
expansion coupling	1.52	0.28	3.28	1.92	nr	5.20
male iron to muPVC coupling	1.30	0.43	5.04	1.55	nr	6.59
union coupling	3.62	0.28	3.28	4.23	nr	7.51
sweep bend	1.35	0.28	3.28	1.73	nr	5.01
spigot/socket bend	-	0.28	3.28	1.94	nr	5.22
sweep tee	1.91	0.37	4.34	2.41	nr	6.75
40 mm pipes; fixing with pipe clips; plugged and screwed	1.77	0.33	3.87	2.81	m	6.68

R DISPOSAL SYSTEMS

	PC £	Labour hours	Labour £	Material £	Unit	Total rate £
R11 FOUL DRAINAGE ABOVE GROUND - cont'd						
muPVC waste pipes and fittings; BS 5255; solvent welded joints - cont'd						
Extra for						
fittings with one end	-	0.21	2.46	1.77	nr	4.23
fittings with two ends	-	0.33	3.87	1.90	nr	5.77
fittings with three ends	-	0.44	5.16	2.96	nr	8.12
fittings with four ends	6.82	0.59	6.91	7.94	nr	14.85
access plug	1.39	0.21	2.46	1.77	nr	4.23
straight coupling	1.01	0.22	2.58	1.36	nr	3.94
expansion coupling	1.83	0.33	3.87	2.26	nr	6.13
male iron to muPVC coupling	1.54	0.43	5.04	1.82	nr	6.86
union coupling	4.75	0.33	3.87	5.48	nr	9.35
level invert taper	1.48	0.33	3.87	1.87	nr	5.74
sweep bend	1.50	0.33	3.87	1.90	nr	5.77
spigot/socket bend	1.71	0.33	3.87	2.13	nr	6.00
sweep tee	2.41	0.44	5.16	2.96	nr	8.12
sweep cross	6.82	0.59	6.91	7.94	nr	14.85
50 mm pipes; fixing with pipe clips; plugged and screwed	2.67	0.39	4.57	4.33	m	8.90
Extra for						
fittings with one end	-	0.23	2.70	2.74	nr	5.44
fittings with two ends	-	0.39	4.57	3.00	nr	7.57
fittings with three ends	-	0.52	6.09	5.14	nr	11.23
fittings with four ends	-	0.69	8.09	9.06	nr	17.15
access plug	2.27	0.23	2.70	2.74	nr	5.44
straight coupling	1.58	0.26	3.05	1.98	nr	5.03
expansion coupling	2.49	0.39	4.57	2.99	nr	7.56
male iron to muPVC coupling	2.22	0.50	5.86	2.57	nr	8.43
union coupling	7.42	0.39	4.57	8.42	nr	12.99
level invert taper	2.06	0.39	4.57	2.51	nr	7.08
sweep bend	2.50	0.39	4.57	3.00	nr	7.57
spigot/socket bend	4.05	0.39	4.57	4.71	nr	9.28
sweep tee	2.41	0.44	5.16	2.96	nr	8.12
sweep cross	7.84	0.69	8.09	9.06	nr	17.15
uPVC overflow pipes and fittings; solvent welded joints						
19 mm pipes; fixing with pipe clips; plugged and screwed	0.75	0.24	2.81	1.45	m	4.26
Extra for						
splay cut end	-	0.02	0.23	-	nr	0.23
fittings with one end	-	0.19	2.23	1.03	nr	3.26
fittings with two ends	-	0.19	2.23	1.20	nr	3.43
fittings with three ends	-	0.24	2.81	1.35	nr	4.16
straight connector	0.82	0.19	2.23	1.03	nr	3.26
female iron to uPVC coupling	-	0.22	2.58	1.61	nr	4.19
bend	0.98	0.19	2.23	1.20	nr	3.43
bent tank connector	1.52	0.22	2.58	1.74	nr	4.32
uPVC pipes and fittings; BS 4514; with solvent welded joints (unless otherwise described)						
82 mm pipes; fixing with holderbats; plugged and screwed	5.21	0.44	5.16	8.39	m	13.55
Extra for						
socket plug	4.12	0.22	2.58	5.15	nr	7.73
slip coupling; push fit	9.00	0.41	4.81	9.92	nr	14.73
expansion coupling	4.33	0.44	5.16	5.38	nr	10.54
sweep bend	7.26	0.44	5.16	8.61	nr	13.77
boss connector	3.97	0.30	3.52	4.98	nr	8.50
single branch	10.15	0.59	6.91	12.16	nr	19.07

Prices for Measured Works - Minor Works
R DISPOSAL SYSTEMS

	PC £	Labour hours	Labour £	Material £	Unit	Total rate £
Extra for - cont'd						
access door	9.67	0.67	7.85	10.96	nr	18.81
connection to clay pipes; caulking ring and cement and sand (1:2) joint	6.80	0.41	4.81	7.89	nr	12.70
110 mm pipes; fixing with holderbats; plugged and screwed	5.30	0.49	5.74	8.76	m	14.50
Extra for						
socket plug	4.99	0.24	2.81	6.29	nr	9.10
slip coupling; push fit	11.26	0.44	5.16	12.41	nr	17.57
expansion coupling	4.43	0.49	5.74	5.67	nr	11.41
W.C. connector	8.04	0.32	3.75	9.29	nr	13.04
sweep bend	8.50	0.49	5.74	10.16	nr	15.90
W.C. connecting bend	13.19	0.32	3.75	14.96	nr	18.71
access bend	23.58	0.51	5.98	26.77	nr	32.75
boss connector	3.97	0.32	3.75	5.17	nr	8.92
single branch	13.65	0.65	7.62	16.26	nr	23.88
single branch with access	19.25	0.67	7.85	22.43	nr	30.28
double branch	29.05	0.81	9.49	33.65	nr	43.14
W.C. manifold	51.21	0.32	3.75	57.64	nr	61.39
access door	-	0.67	7.85	10.96	nr	18.81
access pipe connector	18.06	0.56	6.56	20.69	nr	27.25
connection to clay pipes; caulking ring and cement and sand (1:2) joint	-	0.46	5.39	2.85	nr	8.24
160 mm pipes; fixing with holderbats; plugged and screwed	13.76	0.56	6.56	22.58	m	29.14
Extra for						
socket plug	9.18	0.28	3.28	11.88	nr	15.16
slip coupling; push fit	28.82	0.50	5.86	31.76	nr	37.62
expansion coupling	13.33	0.56	6.56	16.45	nr	23.01
sweep bend	21.18	0.56	6.56	25.10	nr	31.66
boss connector	5.62	0.37	4.34	7.96	nr	12.30
single branch	49.89	0.79	9.26	57.22	nr	66.48
double branch	116.64	0.93	10.90	131.99	nr	142.89
access door	17.28	0.67	7.85	19.34	nr	27.19
access pipe connector	18.06	0.56	6.56	20.69	nr	27.25
connection to clay pipes; caulking ring and cement and sand (1:2) joint	-	0.56	6.56	4.71	nr	11.27
Weathering apron; for pipe						
82 mm diameter	2.05	0.38	4.45	2.56	nr	7.01
110 mm diameter	2.35	0.43	5.04	3.02	nr	8.06
160 mm diameter	7.06	0.46	5.39	8.63	nr	14.02
Weathering slate; for pipe						
110 mm diameter	33.74	1.00	11.72	37.60	nr	49.32
Vent cowl; for pipe						
82 mm diameter	2.05	0.37	4.34	2.56	nr	6.90
110 mm diameter	2.07	0.37	4.34	2.71	nr	7.05
160 mm diameter	5.41	0.37	4.34	6.81	nr	11.15
Polypropylene ancillaries; screwed joint to waste fitting						
Tubular "S" trap; bath; shallow seal						
40 mm diameter	4.94	0.61	7.15	5.44	nr	12.59
Trap; "P"; two piece; 76 mm seal						
32 mm diameter	3.32	0.43	5.04	3.66	nr	8.70
40 mm diameter	3.84	0.50	5.86	4.23	nr	10.09
Trap; "S"; two piece; 76 mm seal						
32 mm diameter	4.22	0.43	5.04	4.65	nr	9.69
40 mm diameter	4.94	0.50	5.86	5.44	nr	11.30
Bottle trap; "P"; 76 mm seal						
32 mm diameter	3.71	0.43	5.04	4.09	nr	9.13
40 mm diameter	4.43	0.45	5.27	4.88	nr	10.15

Prices for Measured Works - Minor Works
R DISPOSAL SYSTEMS

	PC £	Labour hours	Labour £	Material £	Unit	Total rate £
R11 FOUL DRAINAGE ABOVE GROUND - cont'd						
Polypropylene ancillaries; screwed joint to waste fitting - cont'd						
Bottle trap; "S"; 76 mm seal						
32 mm diameter	4.47	0.43	5.04	4.93	nr	9.97
40 mm diameter	5.42	0.50	5.86	5.97	nr	11.83
R12 DRAINAGE BELOW GROUND						
NOTE: Prices for drain trenches are for excavation in "firm" soil and it has been assumed that earthwork support will only be required for trenches 1.00 m or more in depth.						
Excavating trenches; by machine; grading bottoms; earthwork support; filling with excavated material and compacting; disposal of surplus soil on site; spreading on site average 50 m						
Pipes not exceeding 200 mm nominal size						
average depth of trench 0.50 m	-	0.30	2.23	1.85	m	4.08
average depth of trench 0.75 m	-	0.45	3.35	2.78	m	6.13
average depth of trench 1.00 m	-	0.89	6.62	5.27	m	11.89
average depth of trench 1.25 m	-	1.33	9.90	6.00	m	15.90
average depth of trench 1.50 m	-	1.71	12.72	6.80	m	19.52
average depth of trench 1.75 m	-	2.08	15.48	7.53	m	23.01
average depth of trench 2.00 m	-	2.46	18.30	8.56	m	26.86
average depth of trench 2.25 m	-	3.02	22.47	10.72	m	33.19
average depth of trench 2.50 m	-	3.57	26.56	12.54	m	39.10
average depth of trench 2.75 m	-	3.95	29.39	13.97	m	43.36
average depth of trench 3.00 m	-	4.31	32.07	15.40	m	47.47
average depth of trench 3.25 m	-	4.66	34.67	16.36	m	51.03
average depth of trench 3.50 m	-	4.99	37.13	17.32	m	54.45
Pipes exceeding 200 mm nominal size; 225 mm nominal size						
average depth of trench 0.50 m	-	0.30	2.23	1.85	m	4.08
average depth of trench 0.75 m	-	0.45	3.35	2.78	m	6.13
average depth of trench 1.00 m	-	0.89	6.62	5.27	m	11.89
average depth of trench 1.25 m	-	1.33	9.90	6.00	m	15.90
average depth of trench 1.50 m	-	1.71	12.72	6.80	m	19.52
average depth of trench 1.75 m	-	2.08	15.48	7.53	m	23.01
average depth of trench 2.00 m	-	2.46	18.30	8.56	m	26.86
average depth of trench 2.25 m	-	3.02	22.47	10.72	m	33.19
average depth of trench 2.50 m	-	3.57	26.56	12.54	m	39.10
average depth of trench 2.75 m	-	3.95	29.39	13.97	m	43.36
average depth of trench 3.00 m	-	4.31	32.07	15.40	m	47.47
average depth of trench 3.25 m	-	4.66	34.67	16.36	m	51.03
average depth of trench 3.50 m	-	4.99	37.13	17.32	m	54.45
Pipes exceeding 200 mm nominal size; 300 mm nominal size						
average depth of trench 0.75 m	-	0.51	3.79	3.48	m	7.27
average depth of trench 1.00 m	-	1.04	7.74	5.27	m	13.01
average depth of trench 1.25 m	-	1.42	10.56	6.23	m	16.79
average depth of trench 1.50 m	-	1.86	13.84	7.03	m	20.87
average depth of trench 1.75 m	-	2.16	16.07	7.76	m	23.83
average depth of trench 2.00 m	-	2.46	18.30	9.25	m	27.55
average depth of trench 2.25 m	-	3.02	22.47	11.18	m	33.65
average depth of trench 2.50 m	-	3.57	26.56	12.77	m	39.33
average depth of trench 2.75 m	-	3.95	29.39	14.20	m	43.59
average depth of trench 3.00 m	-	4.31	32.07	15.63	m	47.70
average depth of trench 3.25 m	-	4.66	34.67	17.05	m	51.72

R DISPOSAL SYSTEMS

	PC £	Labour hours	Labour £	Material £	Unit	Total rate £
Extra over excavating trenches; irrespective of depth; breaking out materials						
brick	-	2.08	15.48	8.05	m³	23.53
concrete	-	2.91	21.65	11.13	m³	32.78
reinforced concrete	-	4.17	31.02	16.10	m³	47.12
Extra over excavating trenches; irrespective of depth; breaking out existing hard pavings; 75 mm thick						
tarmacadam	-	0.22	1.64	0.85	m²	2.49
Extra over excavating trenches; irrespective of depth; breaking out existing hard pavings; 150 mm thick						
concrete	-	0.45	3.35	1.89	m²	5.24
tarmacadam and hardcore	-	0.30	2.23	0.98	m²	3.21
Excavating trenches; by hand; grading bottoms; earthwork support; filling with excavated material and compacting; disposal of surplus soil on site; spreading on site average 50 m						
Pipes not exceeding 200 mm nominal size; average depth						
average depth of trench 0.50 m	-	1.02	7.59	-	m	7.59
average depth of trench 0.75 m	-	1.53	11.38	-	m	11.38
average depth of trench 1.00 m	-	2.24	16.67	1.33	m	18.00
average depth of trench 1.25 m	-	3.15	23.44	1.83	m	25.27
average depth of trench 1.50 m	-	4.33	32.22	2.23	m	34.45
average depth of trench 1.75 m	-	5.70	42.41	2.67	m	45.08
average depth of trench 2.00 m	-	6.51	48.43	3.00	m	51.43
average depth of trench 2.25 m	-	8.14	60.56	4.00	m	64.56
average depth of trench 2.50 m	-	9.77	72.69	4.66	m	77.35
average depth of trench 2.75 m	-	10.74	79.91	5.16	m	85.07
average depth of trench 3.00 m	-	11.70	87.05	5.66	m	92.71
average depth of trench 3.25 m	-	12.67	94.26	6.16	m	100.42
average depth of trench 3.50 m	-	13.64	101.48	6.66	m	108.14
Pipes exceeding 200 mm nominal size; 225 mm nominal size						
average depth of trench 0.50 m	-	1.02	7.59	-	m	7.59
average depth of trench 0.75 m	-	1.53	11.38	-	m	11.38
average depth of trench 1.00 m	-	2.24	16.67	1.33	m	18.00
average depth of trench 1.25 m	-	3.15	23.44	1.83	m	25.27
average depth of trench 1.50 m	-	4.33	32.22	2.23	m	34.45
average depth of trench 1.75 m	-	5.70	42.41	2.67	m	45.08
average depth of trench 2.00 m	-	6.51	48.43	3.00	m	51.43
average depth of trench 2.25 m	-	8.14	60.56	4.00	m	64.56
average depth of trench 2.50 m	-	9.77	72.69	4.66	m	77.35
average depth of trench 2.75 m	-	10.74	79.91	5.16	m	85.07
average depth of trench 3.00 m	-	11.70	87.05	5.66	m	92.71
average depth of trench 3.25 m	-	12.67	94.26	6.16	m	100.42
average depth of trench 3.50 m	-	13.64	101.48	6.66	m	108.14
Pipes exceeding 200 mm nominal size; 300 mm nominal size						
average depth of trench 0.75 m	-	1.79	13.32	-	m	13.32
average depth of trench 1.00 m	-	2.60	19.34	1.33	m	20.67
average depth of trench 1.25 m	-	3.66	27.23	1.83	m	29.06
average depth of trench 1.50 m	-	4.88	36.31	2.23	m	38.54
average depth of trench 1.75 m	-	5.70	42.41	2.67	m	45.08
average depth of trench 2.00 m	-	6.51	48.43	3.00	m	51.43
average depth of trench 2.25 m	-	8.14	60.56	4.00	m	64.56
average depth of trench 2.50 m	-	9.77	72.69	4.66	m	77.35
average depth of trench 2.75 m	-	10.74	79.91	5.16	m	85.07
average depth of trench 3.00 m	-	11.74	87.35	5.66	m	93.01
average depth of trench 3.25 m	-	12.67	94.26	6.16	m	100.42
average depth of trench 3.50 m	-	13.64	101.48	6.66	m	108.14

Prices for Measured Works - Minor Works
R DISPOSAL SYSTEMS

	PC £	Labour hours	Labour £	Material £	Unit	Total rate £
R12 DRAINAGE BELOW GROUND - cont'd						
Excavating trenches; by hand; grading bottoms; earthwork support; filling with excavated material and compacting; disposal of surplus soil on site; spreading on site average 50 m - cont'd						
Extra over excavating trenches; irrespective of depth; breaking out existing materials						
brick	-	3.05	22.69	5.77	m^3	28.46
concrete	-	4.58	34.08	9.61	m^3	43.69
reinforced concrete	-	6.11	45.46	13.46	m^3	58.92
Extra over excavating trenches; irrespective of depth; breaking out existing hard pavings; 75 mm thick						
tarmacadam	-	0.41	3.05	0.77	m^2	3.82
Extra over excavating trenches; irrespective of depth; breaking out existing hard pavings; 150 mm thick						
concrete	-	0.71	5.28	1.35	m^2	6.63
tarmacadam and hardcore	-	0.51	3.79	0.96	m^2	4.75
Extra for taking up						
precast concrete paving slabs	-	0.33	2.46	-	m^2	2.46
Sand filling						
Beds; to receive pitch fibre pipes						
600 mm x 50 mm thick	-	0.08	0.60	0.71	m	1.31
700 mm x 50 mm thick	-	0.10	0.74	0.83	m	1.57
800 mm x 50 mm thick	-	0.12	0.89	0.95	m	1.84
Granular (shingle) filling						
Beds; 100 mm thick; to pipes						
100 mm nominal size	-	0.10	0.74	1.72	m	2.46
150 mm nominal size	-	0.10	0.74	2.01	m	2.75
225 mm nominal size	-	0.12	0.89	2.30	m	3.19
300 mm nominal size	-	0.14	1.04	2.58	m	3.62
Beds; 150 mm thick; to pipes						
100 mm nominal size	-	0.14	1.04	2.58	m	3.62
150 mm nominal size	-	0.17	1.26	2.87	m	4.13
225 mm nominal size	-	0.19	1.41	3.16	m	4.57
300 mm nominal size	-	0.20	1.49	3.45	m	4.94
Beds and benchings; beds 100 mm thick; to pipes						
100 mm nominal size	-	0.23	1.71	3.16	m	4.87
150 mm nominal size	-	26.00	193.44	3.16	m	196.60
225 mm nominal size	-	0.31	2.31	4.30	m	6.61
300 mm nominal size	-	0.36	2.68	4.88	m	7.56
Beds and benchings; beds 150 mm thick; to pipes						
100 mm nominal size	-	0.26	1.93	3.45	m	5.38
150 mm nominal size	-	0.29	2.16	3.73	m	5.89
225 mm nominal size	-	0.36	2.68	5.17	m	7.85
300 mm nominal size	-	0.45	3.35	6.32	m	9.67
Beds and coverings; 100 mm thick; to pipes						
100 mm nominal size	-	0.37	2.75	4.30	m	7.05
150 mm nominal size	-	0.45	3.35	5.17	m	8.52
225 mm nominal size	-	0.61	4.54	7.17	m	11.71
300 mm nominal size	-	0.73	5.43	8.61	m	14.04
Beds and coverings; 150 mm thick; to pipes						
100 mm nominal size	-	0.55	4.09	6.32	m	10.41
150 mm nominal size	-	0.61	4.54	7.17	m	11.71
225 mm nominal size	-	0.80	5.95	9.18	m	15.13
300 mm nominal size	-	0.94	6.99	10.91	m	17.90

Prices for Measured Works - Minor Works

R DISPOSAL SYSTEMS

	PC £	Labour hours	Labour £	Material £	Unit	Total rate £
Plain in situ ready mixed designated concrete; C10 - 40 mm aggregate						
Beds; 100 mm thick; to pipes						
100 mm nominal size	-	0.21	1.73	3.67	m	5.40
150 mm nominal size	-	0.21	1.73	3.67	m	5.40
225 mm nominal size	-	0.26	2.14	4.40	m	6.54
300 mm nominal size	-	0.29	2.38	5.14	m	7.52
Beds; 150 mm thick; to pipes						
100 mm nominal size	-	0.29	2.38	5.14	m	7.52
150 mm nominal size	-	0.33	2.71	5.87	m	8.58
225 mm nominal size	-	0.37	3.04	6.61	m	9.65
300 mm nominal size	-	0.42	3.45	7.34	m	10.79
Beds and benchings; beds 100 mm thick; to pipes						
100 mm nominal size	-	0.42	3.45	6.61	m	10.06
150 mm nominal size	-	0.47	3.86	7.34	m	11.20
225 mm nominal size	-	0.56	4.60	8.81	m	13.41
300 mm nominal size	-	0.66	5.43	10.28	m	15.71
Beds and benchings; beds 150 mm thick; to pipes						
100 mm nominal size	-	0.47	3.86	7.34	m	11.20
150 mm nominal size	-	0.52	4.27	8.07	m	12.34
225 mm nominal size	-	0.66	5.43	10.28	m	15.71
300 mm nominal size	-	0.84	6.90	13.21	m	20.11
Beds and coverings; 100 mm thick; to pipes						
100 mm nominal size	-	0.63	5.18	8.81	m	13.99
150 mm nominal size	-	0.73	6.00	10.28	m	16.28
225 mm nominal size	-	1.05	8.63	14.68	m	23.31
300 mm nominal size	-	1.25	10.28	17.61	m	27.89
Beds and coverings; 150 mm thick; to pipes						
100 mm nominal size	-	0.93	7.64	13.21	m	20.85
150 mm nominal size	-	1.05	8.63	14.68	m	23.31
225 mm nominal size	-	1.35	11.10	19.08	m	30.18
300 mm nominal size	-	1.62	13.32	22.75	m	36.07
Plain in situ ready mixed designated concrete; C20 - 40 mm aggregate						
Beds 100 mm thick; to pipes						
100 mm nominal size	-	0.21	1.73	3.80	m	5.53
150 mm nominal size	-	0.21	1.73	3.80	m	5.53
225 mm nominal size	-	0.26	2.14	4.55	m	6.69
300 mm nominal size	-	0.29	2.38	5.32	m	7.70
Beds; 150 mm thick; to pipes						
100 mm nominal size	-	0.29	2.38	5.32	m	7.70
150 mm nominal size	-	0.33	2.71	6.07	m	8.78
225 mm nominal size	-	0.37	3.04	6.84	m	9.88
300 mm nominal size	-	0.42	3.45	7.59	m	11.04
Beds and benchings; beds 100 mm thick; to pipes						
100 mm nominal size	-	0.42	3.45	6.84	m	10.29
150 mm nominal size	-	0.47	3.86	7.59	m	11.45
225 mm nominal size	-	0.56	4.60	9.11	m	13.71
300 mm nominal size	-	0.66	5.43	10.63	m	16.06
Beds and benchings; beds 150 mm thick; to pipes						
100 mm nominal size	-	0.47	3.86	7.59	m	11.45
150 mm nominal size	-	0.52	4.27	8.35	m	12.62
225 mm nominal size	-	0.66	5.43	10.63	m	16.06
300 mm nominal size	-	0.84	6.90	13.66	m	20.56
Beds and coverings; 100 mm thick; to pipes						
100 mm nominal size	-	0.63	5.18	9.11	m	14.29
150 mm nominal size	-	0.73	6.00	10.63	m	16.63
225 mm nominal size	-	1.05	8.63	15.18	m	23.81
300 mm nominal size	-	1.25	10.28	18.22	m	28.50

R DISPOSAL SYSTEMS

	PC £	Labour hours	Labour £	Material £	Unit	Total rate £
R12 DRAINAGE BELOW GROUND - cont'd						
Plain in situ ready mixed designated concrete; C20 - 40 mm aggregate - cont'd						
Beds and coverings; 150 mm thick; to pipes						
100 mm nominal size	-	0.93	7.64	13.66	m	21.30
150 mm nominal size	-	1.05	8.63	15.18	m	23.81
225 mm nominal size	-	1.35	11.10	19.74	m	30.84
300 mm nominal size	-	1.62	13.32	23.53	m	36.85
NOTE: The following items unless otherwise described include for all appropriate joints/couplings in the running length. The prices for gullies and rainwater shoes, etc. include for appropriate joints to pipes and for setting on and surrounding accessory with site mixed in situ concrete 10.00 N/mm^2 - 40 mm aggregate (1:3:6)						
Cast iron "Timesaver" drain pipes and fittings or other equal and approved; BS 437; coated; with mechanical coupling joints						
75 mm pipes; laid straight	19.22	0.48	3.50	25.68	m	29.18
75 mm pipes; in runs not exceeding 3.00 m long	18.67	0.64	4.67	32.78	m	37.45
Extra for						
bend; medium radius	19.06	0.53	3.86	29.45	nr	33.31
single branch	26.50	0.74	5.39	47.18	nr	52.57
isolated "Timesaver" joint	10.62	0.32	2.33	11.70	nr	14.03
100 mm pipes; laid straight	20.31	0.53	3.86	27.71	m	31.57
100 mm pipes; in runs not exceeding 3.00 m long	19.26	0.72	5.25	35.82	m	41.07
Extra for						
bend; medium radius	23.88	0.64	4.67	36.95	nr	41.62
bend; medium radius with access	62.73	0.64	4.67	79.75	nr	84.42
bend; long radius	35.60	0.64	4.67	48.71	nr	53.38
rest bend	27.39	0.64	4.67	39.67	nr	44.34
diminishing pipe	19.97	0.64	4.67	31.49	nr	36.16
single branch	31.69	0.80	5.83	57.45	nr	63.28
single branch; with access	73.08	0.91	6.63	103.05	nr	109.68
double branch	50.92	1.01	7.36	90.87	nr	98.23
double branch; with access	86.72	1.01	7.36	130.32	nr	137.68
isolated "Timesaver" joint	12.77	0.37	2.70	14.07	nr	16.77
transitional pipe; for WC	18.18	0.53	3.86	34.10	nr	37.96
150 mm pipes; laid straight	37.60	0.64	4.67	48.23	m	52.90
150 mm pipes; in runs not exceeding 3.00 m long	35.90	0.87	6.34	57.55	m	63.89
Extra for						
bend; medium radius	54.93	0.74	5.39	71.20	nr	76.59
bend; medium radius with access	116.51	0.74	5.39	139.05	nr	144.44
bend; long radius	73.58	0.74	5.39	89.62	nr	95.01
diminishing pipe	31.13	0.74	5.39	42.85	nr	48.24
single branch	68.41	0.91	6.63	76.71	nr	83.34
isolated "Timesaver" joint	15.46	0.44	3.21	17.04	nr	20.25
Accessories in "Timesaver" cast iron or other equal and approved; with mechanical coupling joints						
Gully fittings; comprising low invert gully trap and round hopper						
75 mm outlet	19.97	0.95	6.93	37.51	nr	44.44
100 mm outlet	31.69	1.01	7.36	52.79	nr	60.15
150 mm outlet	78.83	1.38	10.06	108.45	nr	118.51

Prices for Measured Works - Minor Works

R DISPOSAL SYSTEMS

	PC £	Labour hours	Labour £	Material £	Unit	Total rate £
Add to above for; bellmouth 300 mm high; circular plain grating						
100 mm nominal size; 200 mm grating	33.00	0.48	3.50	54.54	nr	58.04
100 mm nominal size; 100 mm horizontal inlet; 200 mm grating	40.34	0.48	3.50	62.63	nr	66.13
100 mm nominal size; 100 mm horizontal inlet; 200 mm grating	41.37	0.48	3.50	63.77	nr	67.27
Yard gully (Deans); trapped; galvanized sediment pan; 267 mm round heavy grating						
100 mm outlet	214.13	3.08	22.45	271.61	nr	294.06
Yard gully (garage); trapless; galvanized sediment pan; 267 mm round heavy grating						
100 mm outlet	218.43	2.88	21.00	259.31	nr	280.31
Yard gully (garage); trapped; with rodding eye, galvanised perforated sediment pan; stopper; 267 mm round heavy grating						
100 mm outlet	406.21	2.88	21.00	509.67	nr	530.67
Grease trap; internal access; galvanized perforated bucket; lid and frame						
100 mm outlet; 20 gallon capacity	372.16	4.26	31.06	449.61	nr	480.67
Cast iron "Ensign" lightweight pipes and fittings or other equal and approved; EN 877; ductile iron couplings						
100 mm pipes; laid straight	-	0.19	1.86	19.43	m	21.29
Extra for						
bend; long radius	-	0.19	1.86	33.14	nr	35.00
single branch	-	0.23	2.26	33.49	nr	35.75
150 mm pipes; laid straight	-	0.22	2.14	38.85	m	40.99
Extra for						
bend; long radius	-	0.22	2.14	87.53	nr	89.67
single branch	-	0.28	2.73	73.68	nr	76.41
Extra strength vitrified clay pipes and fittings; Hepworth "Supersleve" or other equal and approved; plain ends with push fit polypropylene flexible couplings						
100 mm pipes; laid straight	5.21	0.21	1.53	5.88	m	7.41
Extra for						
bend	4.81	0.21	1.53	10.16	nr	11.69
access bend	31.68	0.21	1.53	40.49	nr	42.02
rest bend	11.02	0.21	1.53	17.17	nr	18.70
access pipe	27.52	0.21	1.53	35.39	nr	36.92
socket adaptor	5.10	0.19	1.39	8.32	nr	9.71
adaptor to "HepSeal" pipe	4.15	0.19	1.39	7.25	nr	8.64
saddle	10.21	0.80	5.83	14.49	nr	20.32
single junction	10.40	0.27	1.97	19.04	nr	21.01
single access junction	36.63	0.27	1.97	48.64	nr	50.61
150 mm pipes; laid straight	9.93	0.27	1.97	11.21	m	13.18
Extra for						
bend	9.91	0.26	1.90	19.52	nr	21.42
access bend	2.63	0.26	1.90	50.71	nr	52.61
rest bend	12.74	0.26	1.90	22.71	nr	24.61
taper pipe	14.68	0.26	1.90	22.48	nr	24.38
access pipe	35.72	0.26	1.90	47.84	nr	49.74
socket adaptor	9.61	0.21	1.53	15.42	nr	16.95
adaptor to "HepSeal" pipe	6.83	0.21	1.53	12.28	nr	13.81
saddle	14.53	0.95	6.93	21.78	nr	28.71
single junction	14.56	0.32	2.33	29.34	nr	31.67
single access junction	52.04	0.32	2.33	71.64	nr	73.97

Prices for Measured Works - Minor Works
R DISPOSAL SYSTEMS

	PC £	Labour hours	Labour £	Material £	Unit	Total rate £
R12 DRAINAGE BELOW GROUND - cont'd						
Extra strength vitrified clay pipes and fittings; Hepworth "HepSeal" or other equal and approved; socketted; with push-fit flexible joints						
100 mm pipes; laid straight	9.53	0.29	2.11	10.76	m	12.87
Extra for						
bend	13.77	0.23	1.68	12.32	nr	14.00
rest bend	16.35	0.23	1.68	15.23	nr	16.91
stopper	5.19	0.15	1.09	5.86	nr	6.95
access pipe	28.99	0.26	1.90	28.42	nr	30.32
single junction	19.11	0.29	2.11	17.27	nr	19.38
double collar	12.25	0.19	1.39	13.83	nr	15.22
150 mm pipes; laid straight	12.36	0.34	2.48	13.95	m	16.43
Extra for						
bend	22.69	0.27	1.97	21.43	nr	23.40
rest bend	27.09	0.23	1.68	26.39	nr	28.07
stopper	7.74	0.17	1.24	8.74	nr	9.98
taper reducer	34.02	0.27	1.97	34.21	nr	36.18
access pipe	46.43	0.27	1.97	46.83	nr	48.80
saddle	11.16	0.86	6.27	12.60	nr	18.87
single junction	12.36	0.34	2.48	27.89	nr	30.37
single access junction	56.71	0.34	2.48	59.71	nr	62.19
double junction	57.20	0.51	3.72	57.59	nr	61.31
double collar	20.20	0.22	1.60	22.80	nr	24.40
225 mm pipes; laid straight	24.47	0.44	3.21	27.62	m	30.83
Extra for						
bend	50.74	0.34	2.48	48.99	nr	51.47
rest bend	61.57	0.34	2.48	61.21	nr	63.69
stopper	15.91	0.22	1.60	17.96	nr	19.56
taper reducer	47.22	0.34	2.48	45.01	nr	47.49
access pipe	117.43	0.34	2.48	121.50	nr	123.98
saddle	21.60	1.15	8.38	24.38	nr	32.76
single junction	90.14	0.44	3.21	90.70	nr	93.91
single access junction	128.52	0.44	3.21	134.02	nr	137.23
double junction	128.58	0.65	4.74	131.32	nr	136.06
double collar	44.32	0.29	2.11	50.03	nr	52.14
300 mm pipes; laid straight	37.52	0.57	4.16	42.35	m	46.51
Extra for						
bend	96.38	0.45	3.28	96.08	nr	99.36
rest bend	137.34	0.45	3.28	142.32	nr	145.60
stopper	36.90	0.29	2.11	41.65	nr	43.76
taper reducer	104.55	0.45	3.28	105.31	nr	108.59
saddle	109.93	1.54	11.23	124.08	nr	135.31
single junction	188.73	0.57	4.16	196.09	nr	200.25
double junction	273.13	0.86	6.27	287.12	nr	293.39
double collar	72.03	0.38	2.77	81.30	nr	84.07
400 mm pipes; laid straight	76.72	0.77	5.61	86.60	m	92.21
Extra for						
bend	288.29	0.62	4.52	299.43	nr	303.95
single unequal junction	270.12	0.77	5.61	270.26	nr	275.87
450 mm pipes; laid straight	99.66	0.95	6.93	112.49	m	119.42
Extra for						
bend	379.62	0.77	5.61	394.75	nr	400.36
single unequal junction	323.13	0.95	6.93	319.74	nr	326.67
British Standard quality vitrified clay pipes and fittings; socketted; cement:sand (1:2) joints						
100 mm pipes; laid straight	6.66	0.43	3.13	7.62	m	10.75

R DISPOSAL SYSTEMS

	PC £	Labour hours	Labour £	Material £	Unit	Total rate £
Extra for						
bend (short/medium/knuckle)	13.32	0.34	2.48	5.37	nr	7.85
bend (long/rest/elbow)	10.95	0.34	2.48	10.21	nr	12.69
single junction	12.24	0.43	3.13	10.94	nr	14.07
double junction	20.37	0.64	4.67	19.39	nr	24.06
double collar	8.04	0.29	2.11	9.18	nr	11.29
150 mm pipes; laid straight	10.25	0.48	3.50	11.67	m	15.17
Extra for						
bend (short/medium/knuckle)	10.25	0.38	2.77	8.20	nr	10.97
bend (long/rest/elbow)	18.51	0.38	2.77	17.53	nr	20.30
taper	24.18	0.38	2.77	23.58	nr	26.35
single junction	20.26	0.48	3.50	18.37	nr	21.87
double junction	48.50	0.72	5.25	49.11	nr	54.36
double collar	13.38	0.32	2.33	15.21	nr	17.54
225 mm pipes; laid straight	20.31	0.58	4.23	23.17	m	27.40
Extra for						
bend (short/medium/knuckle)	32.11	0.47	3.43	29.50	nr	32.93
taper	52.53	0.38	2.77	51.86	nr	54.63
double collar	31.31	0.38	2.77	35.45	nr	38.22
300 mm pipes; laid straight	34.23	0.80	5.83	38.89	m	44.72
Extra for						
bend (short/medium/knuckle)	55.85	0.64	4.67	51.58	nr	56.25
double collar	63.53	0.43	3.13	71.89	nr	75.02
400 mm pipes; laid straight	62.68	106.00	772.74	71.10	m	843.84
450 mm pipes; laid straight	81.02	1.33	9.70	91.81	m	101.51
500 mm pipes; laid straight	101.56	1.55	11.30	115.02	m	126.32
Accessories in vitrified clay; set in concrete; with polypropylene coupling joints to pipes						
Rodding point; with oval aluminium plate						
100 mm nominal size	24.73	0.53	3.86	32.35	nr	36.21
Gully fittings; comprising low back trap and square hopper; 150 mm x 150 mm square gully grid						
100 mm nominal size	26.60	0.91	6.63	37.43	nr	44.06
Access gully; trapped with rodding eye and integral vertical back inlet; stopper; 150 mm x 150 mm square gully grid						
100 mm nominal size	34.46	0.69	5.03	43.33	nr	48.36
Inspection chamber; comprising base; 300 mm or 450 mm raising piece; integral alloy cover and frame; 100 mm inlets						
straight through; 2 nr inlets	101.62	2.13	15.53	120.73	nr	136.26
single junction; 3 nr inlets	109.56	2.34	17.06	132.65	nr	149.71
double junction; 4 nr inlets	118.77	2.55	18.59	146.02	nr	164.61
Accessories in polypropylene; cover set in concrete; with coupling joints to pipes						
Inspection chamber; 5 nr 100 mm inlets; cast iron cover and frame						
475 mm diameter x 585 mm deep	169.86	2.44	17.79	198.34	nr	216.13
475 mm diameter x 930 mm deep	207.83	2.66	19.39	241.19	nr	260.58
Accessories in vitrified clay; set in concrete; with cement:sand (1:2) joints to pipes						
Gully fittings; comprising low back trap and square hopper; square gully grid						
100 mm outlet; 150 mm x 150 mm grid	36.40	1.06	7.73	41.47	nr	49.20
150 mm outlet; 225 mm x 225 mm grid	64.29	1.06	7.73	72.94	nr	80.67

R DISPOSAL SYSTEMS

	PC £	Labour hours	Labour £	Material £	Unit	Total rate £
R12 DRAINAGE BELOW GROUND - cont'd						
Accessories in vitrified clay; set in concrete; with cement:sand (1:2) joints to pipes - cont'd						
Yard gully (mud); trapped with rodding eye; galvanized square bucket; stopper; square hinged grate and frame						
100 mm outlet; 225 mm x 225 mm grid	90.62	3.19	23.26	102.74	nr	**126.00**
150 mm outlet; 300 mm x 300 mm grid	161.38	4.26	31.06	182.97	nr	**214.03**
Yard gully (garage); trapped with rodding eye; galvanized perforated round bucket; stopper; round hinged grate and frame						
100 mm outlet; 273 mm grid	90.30	3.19	23.26	129.42	nr	**152.68**
150 mm outlet; 368 mm grid	165.68	4.26	31.06	187.46	nr	**218.52**
Road gully; trapped with rodding eye and stopper (grate not included)						
300 mm x 600 mm x 100 mm outlet	63.07	3.51	25.59	88.17	nr	**113.76**
300 mm x 600 mm x 150 mm outlet	64.58	3.51	25.59	89.88	nr	**115.47**
400 mm x 750 mm x 150 mm outlet	74.91	4.26	31.06	110.34	nr	**141.40**
450 mm x 900 mm x 150 mm outlet	101.35	5.32	38.78	145.33	nr	**184.11**
Grease trap; with internal access; galvanized perforated bucket; lid and frame						
450 mm x 300 mm x 525 mm deep; 100 mm outlet	448.28	3.73	27.19	524.71	nr	**551.90**
600 mm x 450 mm x 600 mm deep; 100 mm outlet	565.83	4.47	32.59	664.04	nr	**696.63**
Interceptor; trapped with inspection arm; lever locking stopper; chain and staple; cement and sand (1:2) joints to pipes; building in, and cutting and fitting brickwork around						
100 mm outlet; 100 mm inlet	82.17	4.26	31.06	93.21	nr	**124.27**
150 mm outlet; 150 mm inlet	115.32	4.78	34.85	130.62	nr	**165.47**
225 mm outlet; 225 mm inlet	313.91	5.32	38.78	354.83	nr	**393.61**
Accessories; grates and covers						
Aluminium alloy gully grids; set in position						
120 mm x 120 mm	2.74	0.11	0.80	3.09	nr	**3.89**
150 mm x 150 mm	2.74	0.11	0.80	3.09	nr	**3.89**
225 mm x 225 mm	8.14	0.11	0.80	9.19	nr	**9.99**
100 mm diameter	2.74	0.11	0.80	3.09	nr	**3.89**
150 mm diameter	4.18	0.11	0.80	4.72	nr	**5.52**
225 mm diameter	9.10	0.11	0.80	10.27	nr	**11.07**
Aluminium alloy sealing plates and frames; set in cement and sand (1:3)						
150 x 150	10.50	0.27	1.97	11.94	nr	**13.91**
225 x 225	19.22	0.27	1.97	21.79	nr	**23.76**
140 diameter (for 100 mm)	8.56	0.27	1.97	9.75	nr	**11.72**
197 diameter (for 150 mm)	12.30	0.27	1.97	13.97	nr	**15.94**
273 diameter (for 225 mm)	19.70	0.27	1.97	22.33	nr	**24.30**
Coated cast iron heavy duty road gratings and frame; BS 497 Tables 6 and 7; bedding and pointing in cement and sand (1:3); one course half brick thick wall in semi-engineering bricks in cement mortar (1:3)						
445 mm x 400 mm; Grade A1, ref GA1-450 (90 kg)	106.83	2.66	19.39	120.00	nr	**139.39**
400 mm x 310 mm; Grade A2, ref GA2-325 (35 kg)	83.61	2.66	19.39	94.42	nr	**113.81**
500 mm x 310 mm; Grade A2, ref GA2-325 (65 kg)	116.13	2.66	19.39	130.25	nr	**149.64**

Prices for Measured Works - Minor Works

R DISPOSAL SYSTEMS

	PC £	Labour hours	Labour £	Material £	Unit	Total rate £
Accessories in precast concrete; top set in with rodding eye and stopper; cement and sand (1:2) joint to pipe						
Concrete road gully; BS 5911; trapped with rodding eye and stopper; cement and sand (1:2) joint to pipe						
450 mm diameter x 1050 mm deep; 100 mm or 150 mm outlet	31.59	5.05	36.81	53.57	nr	90.38
"Osmadrain" uPVC pipes and fittings or other equal and approved; BS 4660; with ring seal joints						
82 mm pipes; laid straight	7.05	0.17	1.24	7.96	m	9.20
Extra for						
bend; short radius	12.40	0.14	1.02	13.66	nr	14.68
spigot/socket bend	10.42	0.14	1.02	11.48	nr	12.50
adaptor	5.43	0.08	0.58	5.98	nr	6.56
single junction	16.13	0.19	1.39	17.77	nr	19.16
slip coupler	7.66	0.08	0.58	8.44	nr	9.02
100 mm pipes; laid straight	6.24	0.19	1.39	7.87	m	9.26
Extra for						
bend; short radius	16.53	0.17	1.24	17.79	nr	19.03
bend; long radius	26.78	0.17	1.24	27.40	nr	28.64
spigot/socket bend	13.98	0.17	1.24	19.31	nr	20.55
socket plug	6.70	0.05	0.36	7.38	nr	7.74
adjustable double socket bend	18.30	0.17	1.24	23.76	nr	25.00
adaptor to clay	14.29	0.10	0.73	15.43	nr	16.16
single junction	19.73	0.23	1.68	19.63	nr	21.31
sealed access junction	36.14	0.20	1.46	37.71	nr	39.17
slip coupler	7.66	0.10	0.73	8.44	nr	9.17
160 mm pipes; laid straight	14.19	0.23	1.68	17.53	m	19.21
Extra for						
bend; short radius	36.37	0.19	1.39	39.11	nr	40.50
spigot/socket bend	32.98	0.19	1.39	43.30	nr	44.69
socket plug	11.91	0.08	0.58	13.12	nr	13.70
adaptor to clay	28.71	0.13	0.95	30.78	nr	31.73
level invert taper	49.30	0.19	1.39	60.32	nr	61.71
single junction	59.56	0.27	1.97	65.63	nr	67.60
sealed access junction	99.32	0.24	1.75	109.44	nr	111.19
slip coupler	21.47	0.12	0.87	23.66	nr	24.53
uPVC Osma "Ultra-Rib" ribbed pipes and fittings or other equal and approved; WIS approval; with sealed ring push-fit joints						
150 mm pipes; laid straight	-	0.21	1.53	7.00	m	8.53
Extra for						
bend; short radius	19.03	0.19	1.39	20.55	nr	21.94
adaptor to 160 diameter upvc	22.27	0.11	0.80	23.70	nr	24.50
adaptor to clay	45.72	0.11	0.80	49.96	nr	50.76
level invert taper	8.31	0.19	1.39	7.90	nr	9.29
single junction	34.23	0.24	1.75	35.62	nr	37.37
225 mm pipes; laid straight	14.23	0.24	1.75	16.06	m	17.81
Extra for						
bend; short radius	68.57	0.22	1.60	74.59	nr	76.19
adaptor to clay	56.97	0.14	1.02	60.85	nr	61.87
level invert taper	11.92	0.22	1.60	10.24	nr	11.84
single junction	101.78	0.30	2.19	107.33	nr	109.52
300 mm pipes; laid straight	21.34	0.36	2.62	24.08	m	26.70
Extra for						
bend; short radius	108.00	0.32	2.33	117.56	nr	119.89
adaptor to clay	149.85	0.16	1.17	162.23	nr	163.40
level invert taper	35.76	0.32	2.33	35.07	nr	37.40
single junction	217.10	0.41	2.99	231.99	nr	234.98

Prices for Measured Works - Minor Works
R DISPOSAL SYSTEMS

	PC £	Labour hours	Labour £	Material £	Unit	Total rate £
R12 DRAINAGE BELOW GROUND - cont'd						
Interconnecting drainage channel; ref N100; ACO Polymer Products Ltd or other equal and approved; reinforced slotted galvanised steel grating ref 423/4; bedding and haunching in in situ concrete (not included)						
100 mm wide						
laid level or to falls	-	0.51	3.72	79.35	m	83.07
extra for sump unit	-	1.53	11.15	142.57	nr	153.72
extra for end caps	-	0.10	0.73	7.42	nr	8.15
Interconnecting drainage channel; "Birco-lite" ref 8012 or other equal and approved; Marshalls Plc; galvanised steel grating ref 8041; bedding and haunching in in situ concrete (not included)						
100 mm wide						
laid level or to falls	-	0.51	3.72	40.03	m	43.75
extra for 100 mm diameter trapped outlet unit	-	1.53	11.15	69.13	nr	80.28
extra for end caps	-	0.10	0.73	4.99	nr	5.72
Accessories in uPVC; with ring seal joints to pipes (unless otherwise described)						
Cast iron access point						
110 mm diameter	27.84	0.82	5.98	30.68	nr	36.66
Rodding eye						
110 mm diameter	39.19	0.47	3.43	46.86	nr	50.29
Universal gulley fitting; comprising gulley trap, plain hopper						
150 mm x 150 mm grate	22.92	1.02	7.44	30.40	nr	37.84
Bottle gulley; comprising gulley with bosses closed; sealed access covers						
217 mm x 217 mm grate	39.07	0.85	6.20	48.19	nr	54.39
Shallow access pipe; light duty screw down access door assembly						
110 mm diameter	33.08	0.85	6.20	41.60	nr	47.80
Shallow access junction; 3 nr 110 mm inlets; light duty screw down access door assembly						
110 mm diameter	59.41	1.22	8.89	86.36	nr	95.25
Shallow inspection chamber; 250 mm diameter; 600 mm deep; sealed cover and frame						
4 nr 110 mm outlets/inlets	82.65	1.41	10.28	109.41	nr	119.69
Universal inspection chamber; 450 mm diameter; single seal cast iron cover and frame; 4 nr 110 mm outlets/inlets						
500 mm deep	156.82	1.49	10.86	191.14	nr	202.00
730 mm deep	156.82	1.76	12.83	216.21	nr	229.04
960 mm deep	156.82	2.04	14.87	241.28	nr	256.15
Equal manhole base; 750 mm diameter						
6 nr 160 mm outlets/inlets	168.01	1.33	9.70	196.14	nr	205.84
Unequal manhole base; 750 mm diameter						
2 nr 160 mm, 4 nr 110 mm outlets/inlets	156.31	1.33	9.70	183.25	nr	192.95
Kerb to gullies; class B engineering bricks on edge to three sides in cement mortar (1:3) rendering in cement mortar (1:3) to top and two sides and skirting to brickwork 230 mm high; dishing in cement mortar (1:3) to gully; steel trowelled						
230 mm x 230 mm internally	-	1.53	11.15	1.18	nr	12.33

R DISPOSAL SYSTEMS

	PC £	Labour hours	Labour £	Material £	Unit	Total rate £
Excavating; by machine						
Manholes						
maximum depth not exceeding 1.00 m	-	0.22	1.64	4.87	m³	6.51
maximum depth not exceeding 2.00 m	-	0.24	1.79	5.33	m³	7.12
maximum depth not exceeding 4.00 m	-	0.29	2.16	6.26	m³	8.42
Excavating; by hand						
Manholes						
maximum depth not exceeding 1.00 m	-	3.52	26.19	-	m³	26.19
maximum depth not exceeding 2.00 m	-	4.16	30.95	-	m³	30.95
maximum depth not exceeding 4.00 m	-	5.32	39.58	-	m³	39.58
Earthwork support (average "risk" prices)						
Maximum depth not exceeding 1.00 m						
distance between opposing faces not exceeding 2.00 m	-	0.16	1.19	3.42	m²	4.61
Maximum depth not exceeding 2.00 m						
distance between opposing faces not exceeding 2.00 m	-	0.19	1.41	6.50	m²	7.91
Maximum depth not exceeding 4.00 m						
distance between opposing faces not exceeding 2.00 m	-	0.24	1.79	9.59	m²	11.38
Disposal; by machine						
Excavated material						
off site; to tip not exceeding 13 km (using lorries) including Landfill Tax based on inactive waste	-	-	-	16.97	m³	16.97
on site depositing; in spoil heaps; average 50 m distance	-	0.16	1.19	3.15	m³	4.34
Disposal; by hand						
Excavated material						
off site; to tip not exceeding 13 km (using lorries) including Landfill Tax based on inactive waste	-	0.81	6.03	25.96	m³	31.99
on site depositing; in spoil heaps; average 50 m distance	-	1.32	9.82	-	m³	9.82
Filling to excavations; by machine						
Average thickness not exceeding 0.25 m						
arising from the exacavations	-	0.16	1.19	2.32	m³	3.51
Filling to excavations; by hand						
Average thickness not exceeding 0.25 m						
arising from the exacavations	-	1.02	7.59	-	m³	7.59
Plain in situ ready mixed designated concrete; C10 - 40 mm aggregate						
Beds						
thickness not exceeding 150 mm	63.51	3.20	26.30	77.07	m³	103.37
thickness 150 mm - 450 mm	63.51	2.40	19.73	77.07	m³	96.80
thickness exceeding 450 mm	63.51	2.02	16.60	77.07	m³	93.67
Plain in situ ready mixed designated concrete; C20 - 20 mm aggregate						
Beds						
thickness not exceeding 150 mm	65.69	3.20	26.30	79.71	m³	106.01
thickness 150 mm - 450 mm	65.69	2.40	19.73	79.71	m³	99.44
thickness exceeding 450 mm	65.69	2.02	16.60	79.71	m³	96.31

Prices for Measured Works - Minor Works
R DISPOSAL SYSTEMS

	PC £	Labour hours	Labour £	Material £	Unit	Total rate £
R12 DRAINAGE BELOW GROUND - cont'd						
Plain in situ ready mixed designated concrete; **C25 · 20 mm aggregate; (small quantities)**						
Benching in bottoms						
150 mm - 450 mm average thickness	66.19	9.57	97.16	76.49	m³	173.65
Reinforced in situ ready mixed designated concrete; **C20 - 20 mm aggregate; (small quantities)**						
Isolated cover slabs						
thickness not exceeding 150 mm	65.69	7.45	61.24	75.91	m³	137.15
Reinforcement; fabric to BS 4483; lapped; in **beds or suspended slabs**						
Ref A98 (1.54 kg/m²)						
400 mm minimum laps	0.75	0.13	1.34	0.89	m²	2.23
Ref A142 (2.22 kg/m²)						
400 mm minimum laps	0.78	0.13	1.34	0.92	m²	2.26
Ref A193 (3.02 kg/m²)						
400 mm minimum laps	1.06	0.13	1.34	1.25	m²	2.59
Formwork; basic finish						
Soffits of isolated cover slabs						
horizontal	-	3.03	38.97	7.39	m²	46.36
Edges of isolated cover slabs						
height not exceeding 250 mm	-	0.90	11.57	2.08	m	13.65
Precast concrete rectangular access and inspection chambers; "Hepworth" chambers or other equal and approved; comprising cover frame to receive manhole cover (not included) intermediate wall sections and base section with cut outs; bedding; jointing and pointing in cement mortar (1:3) on prepared bed						
Drainage chamber; size 600 mm x 450 mm internally; depth to invert						
600 mm deep	-	4.78	34.85	98.43	nr	133.28
900 mm deep	-	6.38	46.51	123.66	nr	170.17
Drainage chamber; 1200 mm x 750 mm reducing to 600 mm x 600 mm; no base unit; depth of invert						
1050 mm deep	-	7.98	58.17	184.48	nr	242.65
1650 mm deep	-	9.57	69.77	312.06	nr	381.83
2250 mm deep	-	11.70	85.29	438.99	nr	524.28
Common bricks; in cement mortar (1:3)						
Walls to manholes						
one brick thick	0.18	2.55	41.46	33.30	m²	74.76
one and a half brick thick	0.18	3.73	60.65	49.95	m²	110.60
Projections of footings						
two brick thick	0.18	5.22	84.88	66.60	m²	151.48
Class A engineering bricks; in cement mortar (1:3)						
Walls to manholes						
one brick thick	0.33	2.88	46.83	57.48	m²	104.31
one and a half brick thick	0.33	4.15	67.48	59.44	m²	126.92
Projections of footings						
two brick thick	0.33	5.85	95.12	114.95	m²	210.07
Class B engineering bricks; in cement mortar (1:3)						
Walls to manholes						
one brick thick	0.23	2.88	46.83	41.95	m²	88.78
one and a half brick thick	0.23	4.15	67.48	62.92	m²	130.40

Prices for Measured Works - Minor Works   675
R DISPOSAL SYSTEMS

	PC £	Labour hours	Labour £	Material £	Unit	Total rate £
Projections of footings						
two brick thick	0.23	5.85	95.12	83.89	m²	**179.01**
Brickwork sundries						
Extra over for fair face; flush smooth pointing						
manhole walls	-	0.21	3.41	-	m²	**3.41**
Building ends of pipes into brickwork; making good fair face or rendering						
not exceeding 55 mm nominal size	-	0.11	1.79	-	nr	**1.79**
55 mm - 110 mm nominal size	-	0.16	2.60	-	nr	**2.60**
over 110 mm nominal size	-	0.21	3.41	-	nr	**3.41**
Step irons; BS 1247; malleable; galvanized; building into joints						
general purpose pattern	-	0.16	2.60	4.49	nr	**7.09**
Cement:sand (1:3) in situ finishings; steel trowelled						
13 mm work to manhole walls; one coat; to brickwork base over 300 mm wide	-	0.74	12.03	1.31	m²	**13.34**
Cast iron inspection chambers; with bolted flat covers; BS 437; bedded in cement mortar (1:3); with mechanical coupling joints						
100 mm x 100 mm						
one branch	99.43	1.11	8.09	124.29	nr	**132.38**
one branch either side	124.41	1.67	12.17	165.88	nr	**178.05**
150 mm x 100 mm						
one branch	158.25	1.34	9.77	175.02	nr	**184.79**
one branch either side	172.87	1.90	13.85	205.86	nr	**219.71**
150 mm x 150 mm						
one branch	181.91	1.43	10.42	218.79	nr	**229.21**
one branch either side	199.01	2.04	14.87	254.67	nr	**269.54**
Access covers and frames; Drainage Systems or other equal and approved; coated; bedding frame in cement and sand (1:3); cover in grease and sand						
Grade A; light duty; rectangular single seal solid top						
450 mm x 450 mm; ref MC1-45/45	50.12	1.62	11.81	56.69	nr	**68.50**
600 mm x 450 mm; ref MC1-60/45	50.40	1.62	11.81	57.14	nr	**68.95**
600 mm x 600 mm; ref MC1-60/60	107.21	1.62	11.81	119.87	nr	**131.68**
Grade A; light duty; rectangular single seal recessed						
600 mm x 450 mm; ref MC1R-60/45	102.79	1.62	11.81	114.87	nr	**126.68**
600 mm x 600 mm; ref MC1R-60/60	142.00	1.62	11.81	158.21	nr	**170.02**
Grade A; light duty; rectangular double seal solid top						
450 mm x 450 mm; ref MC2-45/45	77.11	1.62	11.81	86.43	nr	**98.24**
600 mm x 450 mm; ref MC2-60/45	101.03	1.62	11.81	112.93	nr	**124.74**
600 mm x 600 mm; ref MC2-60/60	142.46	1.62	11.81	158.71	nr	**170.52**
Grade A; light duty; rectangular double seal recessed						
450 mm x 450 mm; ref MC2R-45/45	128.11	1.62	11.81	142.63	nr	**154.44**
600 mm x 450 mm; ref MC2R-60/45	152.68	1.62	11.81	169.84	nr	**181.65**
600 mm x 600 mm; ref MC2R-60/60	171.63	1.62	11.81	190.85	nr	**202.66**
Grade B; medium duty; circular single seal solid top						
300 mm diameter; ref MB2-50	148.41	2.13	15.53	164.98	nr	**180.51**
550 mm diameter; ref MB2-55	113.34	2.13	15.53	126.33	nr	**141.86**
600 mm diameter; ref MB2-60	99.03	2.13	15.53	110.56	nr	**126.09**

R DISPOSAL SYSTEMS

	PC £	Labour hours	Labour £	Material £	Unit	Total rate £
R12 DRAINAGE BELOW GROUND - cont'd						
Access covers and frames; Drainage Systems or other equal and approved; coated; bedding frame in cement and sand (1:3); cover in grease and sand - cont'd						
Grade B; medium duty; rectangular single seal solid top						
600 mm x 450 mm; ref MB2-60/45	88.91	2.13	15.53	99.57	nr	115.10
600 mm x 600 mm; ref MB2-60/60	113.90	2.13	15.53	127.24	nr	142.77
Grade B; medium duty; rectangular singular seal recessed						
600 mm x 450 mm; ref MB2R-60/45	151.66	2.13	15.53	168.71	nr	184.24
600 mm x 600 mm; ref MB2R-60/60	172.56	2.13	15.53	191.88	nr	207.41
Grade B; "Chevron"; medium duty; double triangular solid top						
600 mm x 600 mm; ref MB1-60/60	108.97	2.13	15.53	121.81	nr	137.34
Grade C; "Vulcan" heavy duty; single triangular solid top						
550 mm x 495 mm; ref MA-T	174.14	2.68	19.54	193.48	nr	213.02
Grade C; "Chevron"; heavy duty double triangular solid top						
550 mm x 550 mm; ref MA-55	156.95	3.19	23.26	174.59	nr	197.85
600 mm x 600 mm; ref MA-60	179.20	3.19	23.26	199.20	nr	222.46
British Standard best quality vitrified clay channels; bedding and jointing in cement:sand (1:2)						
Half section straight						
100 mm diameter x 1.00 m long	4.06	0.85	6.20	4.58	nr	10.78
150 mm diameter x 1.00 m long	6.75	1.06	7.73	7.62	nr	15.35
225 mm diameter x 1.00 m long	15.18	1.38	10.06	17.13	nr	27.19
300 mm diameter x 1.00 m long	31.15	1.70	12.39	35.16	nr	47.55
Half section bend						
100 mm diameter	4.37	0.64	4.67	4.93	nr	9.60
150 mm diameter	7.21	0.80	5.83	8.14	nr	13.97
225 mm diameter	24.04	1.06	7.73	27.14	nr	34.87
300 mm diameter	49.02	1.28	9.33	55.33	nr	64.66
Half section taper straight						
150 mm - 100 mm diameter	18.17	0.74	5.39	20.51	nr	25.90
225 mm - 150 mm diameter	40.55	0.95	6.93	45.77	nr	52.70
300 mm - 225 mm diameter	160.85	1.17	8.53	181.56	nr	190.09
Half section taper bend						
150 mm - 100 mm diameter	27.66	0.95	6.93	31.22	nr	38.15
225 mm - 150 mm diameter	79.24	1.22	8.89	89.44	nr	98.33
300 mm - 225 mm diameter	160.85	1.49	10.86	181.56	nr	192.42
Three quarter section branch bend						
100 mm diameter	9.85	0.53	3.86	11.12	nr	14.98
150 mm diameter	17.13	0.80	5.83	19.34	nr	25.17
225 mm diameter	60.31	1.06	7.73	68.08	nr	75.81
300 mm diameter	126.22	1.42	10.35	142.47	nr	152.82
uPVC channels; with solvent weld or lip seal coupling joints; bedding in cement:sand						
Half section cut away straight; with coupling either end						
110 mm diameter	29.99	0.32	2.33	41.05	nr	43.38
160 mm diameter	51.72	0.43	3.13	76.44	nr	79.57
Half section cut away long radius bend; with coupling either end						
110 mm diameter	30.70	0.32	2.33	41.85	nr	44.18
160 mm diameter	75.92	0.43	3.13	103.75	nr	106.88
Channel adaptor to clay; with one coupling						
110 mm diameter	8.59	0.27	1.97	13.30	nr	15.27
160 mm diameter	18.53	0.35	2.55	29.95	nr	32.50

R DISPOSAL SYSTEMS

	PC £	Labour hours	Labour £	Material £	Unit	Total rate £
Half section bend						
110 mm diameter	11.15	0.35	2.55	12.95	nr	15.50
160 mm diameter	16.63	0.53	3.86	19.62	nr	23.48
Half section channel connector						
110 mm diameter	4.42	0.08	0.58	5.72	nr	6.30
160 mm diameter	10.99	0.11	0.80	14.11	nr	14.91
Half section channel junction						
110 mm diameter	12.51	0.53	3.86	14.49	nr	18.35
160 mm diameter	37.72	0.65	4.74	43.43	nr	48.17
Polypropylene slipper bend						
110 mm diameter	16.80	0.43	3.13	19.33	nr	22.46
Glass fibre septic tank; "Klargester" or other equal and approved; fixing lockable manhole cover and frame; placing in position						
3750 litre capacity; 2000 mm diameter; depth to invert						
1000 mm deep; standard grade	766.42	2.50	18.23	906.29	nr	924.52
1500 mm deep; heavy duty grade	950.37	2.82	20.56	1104.04	nr	1124.60
6000 litre capacity; 2300 mm diameter; depth to invert						
1000 mm deep; standard grade	1226.28	2.68	19.54	1422.61	nr	1442.15
1500 mm deep; heavy duty grade	1558.40	3.01	21.94	1779.64	nr	1801.58
9000 litre capacity; 2660 mm diameter; depth to invert						
1000 mm deep; standard grade	1773.00	2.91	21.21	2010.34	nr	2031.55
1500 mm deep; heavy duty grade	2248.18	3.19	23.26	2521.15	nr	2544.41
Glass fibre petrol interceptors; "Klargester" or other equal and approved; placing in position						
2000 litre capacity; 2370 mm x 1300 mm diameter; depth to invert						
1000 mm deep	868.62	2.73	19.90	933.77	nr	953.67
4000 litre capacity; 4370 mm x 1300 mm diameter; depth to invert						
1000 mm deep	1445.99	2.96	21.58	1554.44	nr	1576.02

R13 LAND DRAINAGE

	PC £	Labour hours	Labour £	Material £	Unit	Total rate £
Excavating; by hand; grading bottoms; earthwork support; filling to within 150 mm of surface with gravel rejects; remainder filled with excavated material and compacting; disposal of surplus soil on site; spreading on site average 50 m						
Pipes not exceeding 200 mm nominal size						
average depth of trench 0.75 m	-	1.71	12.72	8.66	m	21.38
average depth of trench 1.00 m	-	2.31	17.19	13.80	m	30.99
average depth of trench 1.25 m	-	3.19	23.73	17.33	m	41.06
average depth of trench 1.50 m	-	5.50	40.92	21.13	m	62.05
average depth of trench 1.75 m	-	6.52	48.51	24.66	m	73.17
average depth of trench 2.00 m	-	7.54	56.10	28.47	m	84.57
Disposal; by machine						
Excavated material						
off site; to tip not exceeding 13 km (using lorries); including Landfill Tax based on inactive waste	-	-		16.97	m³	16.97
Disposal; by hand						
Excavated material						
off site; to tip not exceeding 13 km (using lorries); including Landfill Tax based on inactive waste	-	0.88	6.55	25.96	m³	32.51

R DISPOSAL SYSTEMS

	PC £	Labour hours	Labour £	Material £	Unit	Total rate £
R13 LAND DRAINAGE - cont'd						
Vitrified clay perforated sub-soil pipes; BS 65; Hepworth "Hepline" or other equal and approved						
Pipes; laid straight						
100 mm diameter	5.98	0.23	1.68	6.75	m	**8.43**
150 mm diameter	10.88	0.29	2.11	12.28	m	**14.39**
225 mm diameter	19.99	0.38	2.77	22.56	m	**25.33**

Prices for Measured Works - Minor Works
S PIPED SUPPLY SYSTEMS

	PC £	Labour hours	Labour £	Material £	Unit	Total rate £
S PIPED SUPPLY SYSTEMS						
S12 HOT AND COLD WATER (SMALL SCALE)						
Copper pipes; EN1057:1996; capillary fittings						
15 mm pipes; fixing with pipe clips and screwed	1.03	0.37	4.34	1.23	m	**5.57**
Extra for						
made bend	-	0.17	1.99	-	nr	**1.99**
stop end	1.01	0.12	1.41	1.11	nr	**2.52**
straight coupling	0.17	0.19	2.23	0.19	nr	**2.42**
union coupling	5.42	0.19	2.23	5.97	nr	**8.20**
reducing coupling	1.84	0.19	2.23	2.03	nr	**4.26**
copper to lead connector	1.73	0.24	2.81	1.91	nr	**4.72**
imperial to metric adaptor	2.21	0.24	2.81	2.44	nr	**5.25**
elbow	0.34	0.19	2.23	0.37	nr	**2.60**
backplate elbow	4.15	0.39	4.57	4.57	nr	**9.14**
return bend	0.27	0.19	2.23	0.30	nr	**2.53**
tee; equal	0.61	0.28	3.28	0.67	nr	**3.95**
tee; reducing	4.44	0.28	3.28	4.89	nr	**8.17**
straight tap connector	1.56	0.56	6.56	1.72	nr	**8.28**
bent tap connector	1.57	0.76	8.91	1.73	nr	**10.64**
tank connector	4.84	0.28	3.28	5.33	nr	**8.61**
overflow bend	10.29	0.24	2.81	11.34	nr	**14.15**
22 mm pipes; fixing with pipe clips and screwed	2.06	0.43	5.04	2.39	m	**7.43**
Extra for						
made bend	-	0.22	2.58	-	nr	**2.58**
stop end	1.90	0.15	1.76	2.09	nr	**3.85**
straight coupling	0.47	0.24	2.81	0.52	nr	**3.33**
union coupling	8.68	0.24	2.81	9.56	nr	**12.37**
reducing coupling	1.84	0.24	2.81	2.03	nr	**4.84**
copper to lead connector	2.59	0.34	3.98	2.85	nr	**6.83**
elbow	0.83	0.24	2.81	0.91	nr	**3.72**
backplate elbow	8.74	0.49	5.74	9.63	nr	**15.37**
return bend	11.98	0.24	2.81	13.20	nr	**16.01**
tee; equal	1.95	0.37	4.34	2.15	nr	**6.49**
tee; reducing	1.54	0.37	4.34	1.70	nr	**6.04**
straight tap connector	1.58	0.19	2.23	1.74	nr	**3.97**
tank connector	2.06	0.37	4.34	2.32	nr	**6.66**
28 mm pipes; fixing with pipe clips and screwed	2.66	0.46	5.39	3.07	m	**8.46**
Extra for						
made bend	-	0.28	3.28	-	nr	**3.28**
stop end	3.34	0.17	1.99	3.68	nr	**5.67**
straight coupling	0.95	0.31	3.63	1.05	nr	**4.68**
reducing coupling	2.53	0.31	3.63	2.79	nr	**6.42**
union coupling	8.68	0.31	3.63	9.56	nr	**13.19**
copper to lead connector	3.51	0.43	5.04	3.87	nr	**8.91**
imperial to metric adaptor	3.82	0.43	5.04	4.21	nr	**9.25**
elbow	1.50	0.31	3.63	1.65	nr	**5.28**
return bend	15.31	0.31	3.63	16.87	nr	**20.50**
tee; equal	4.19	0.45	5.27	4.62	nr	**9.89**
tank connector	2.66	0.45	5.27	3.00	nr	**8.27**
35 mm pipes; fixing with pipe clips and screwed	6.67	0.53	6.21	7.60	m	**13.81**
Extra for						
made bend	-	0.33	3.87	-	nr	**3.87**
stop end	7.37	0.19	2.23	8.12	nr	**10.35**
straight coupling	3.07	0.37	4.34	3.38	nr	**7.72**
reducing coupling	5.97	0.37	4.34	6.58	nr	**10.92**

S12 HOT AND COLD WATER (SMALL SCALE) - cont'd

	PC £	Labour hours	Labour £	Material £	Unit	Total rate £
Copper pipes; EN1057:1996; capillary fittings - cont'd						
Extra for - cont'd						
union coupling	16.60	0.37	4.34	18.29	nr	22.63
flanged connector	45.78	0.49	5.74	50.44	nr	56.18
elbow	6.58	0.37	4.34	7.25	nr	11.59
obtuse elbow	9.74	0.37	4.34	10.73	nr	15.07
tee; equal	10.49	0.51	5.98	11.56	nr	17.54
tank connector	12.45	0.51	5.98	13.72	nr	19.70
42 mm pipes; fixing with pipe clips; plugged and screwed	8.18	0.59	6.91	9.31	m	16.22
Extra for						
made bend	-	0.44	5.16	-	nr	5.16
stop end	12.69	0.21	2.46	13.98	nr	16.44
straight coupling	5.02	0.43	5.04	5.53	nr	10.57
reducing coupling	9.99	0.43	5.04	11.01	nr	16.05
union coupling	24.27	0.43	5.04	26.74	nr	31.78
flanged connector	54.72	0.56	6.56	60.29	nr	66.85
elbow	10.87	0.43	5.04	11.98	nr	17.02
obtuse elbow	17.33	0.43	5.04	19.10	nr	24.14
tee; equal	16.85	0.57	6.68	18.57	nr	25.25
tank connector	16.32	0.57	6.68	17.98	nr	24.66
54 mm pipes; fixing with pipe clips; plugged and screwed	10.50	0.65	7.62	11.93	m	19.55
Extra for						
made bend	-	0.61	7.15	-	nr	7.15
stop end	17.71	0.23	2.70	19.51	nr	22.21
straight coupling	9.26	0.49	5.74	10.20	nr	15.94
reducing coupling	16.77	0.49	5.74	18.48	nr	24.22
union coupling	46.16	0.49	5.74	50.86	nr	56.60
flanged connector	82.72	0.56	6.56	91.15	nr	97.71
elbow	22.44	0.49	5.74	24.73	nr	30.47
obtuse elbow	31.35	0.49	5.74	34.54	nr	40.28
tee; equal	33.97	0.63	7.38	37.43	nr	44.81
tank connector	24.93	0.63	7.38	27.47	nr	34.85
Copper pipes; EN1057:1996; compression fittings						
15 mm pipes; fixing with pipe clips; plugged and screwed	1.03	0.42	4.92	1.23	m	6.15
Extra for						
made bend	-	0.17	1.99	-	nr	1.99
stop end	2.07	0.11	1.29	2.28	nr	3.57
straight coupling	1.67	0.17	1.99	1.84	nr	3.83
reducing set	1.77	0.19	2.23	1.95	nr	4.18
male coupling	1.48	0.22	2.58	1.63	nr	4.21
female coupling	1.78	0.22	2.58	1.96	nr	4.54
90 degree bend	2.01	0.17	1.99	2.21	nr	4.20
90 degree backplate bend	4.95	0.33	3.87	5.15	nr	9.32
tee; equal	2.82	0.24	2.81	3.11	nr	5.92
tee; backplate	8.30	0.24	2.81	9.15	nr	11.96
tank coupling	4.89	0.24	2.81	5.39	nr	8.20
22 mm pipes; fixing with pipe clips; plugged and screwed	2.06	0.47	5.51	2.39	m	7.90
Extra for						
made bend	-	0.22	2.58	-	nr	2.58
stop end	3.00	0.13	1.52	3.31	nr	4.83
straight coupling	2.72	0.22	2.58	3.00	nr	5.58
reducing set	1.91	0.06	0.70	2.10	nr	2.80
male coupling	3.19	0.31	3.63	3.52	nr	7.15

S PIPED SUPPLY SYSTEMS

	PC £	Labour hours	Labour £	Material £	Unit	Total rate £
Extra for - cont'd						
female coupling	2.61	0.31	3.63	2.88	nr	6.51
90 degree bend	3.20	0.22	2.58	3.53	nr	6.11
tee; equal	4.66	0.33	3.87	5.13	nr	9.00
tee; reducing	6.79	0.33	3.87	7.48	nr	11.35
tank coupling	5.23	0.33	3.87	5.76	nr	9.63
28 mm pipes; fixing with pipe clips; plugged and screwed	2.66	0.51	5.98	3.07	m	9.05
Extra for						
made bend	-	0.28	3.28	-	nr	3.28
stop end	5.87	0.16	1.88	6.47	nr	8.35
straight coupling	5.62	0.28	3.28	6.19	nr	9.47
male coupling	3.99	0.39	4.57	4.40	nr	8.97
female coupling	5.16	0.39	4.57	5.69	nr	10.26
90 degree bend	7.27	0.28	3.28	8.01	nr	11.29
tee; equal	11.58	0.41	4.81	12.76	nr	17.57
tee; reducing	11.18	0.41	4.81	12.32	nr	17.13
tank coupling	8.99	0.41	4.81	9.91	nr	14.72
35 mm pipes; fixing with pipe clips; plugged and screwed	6.67	0.57	6.68	7.60	m	14.28
Extra for						
made bend	-	0.33	3.87	-	nr	3.87
stop end	9.06	0.18	2.11	9.98	nr	12.09
straight coupling	11.68	0.33	3.87	12.87	nr	16.74
male coupling	8.87	0.44	5.16	9.77	nr	14.93
female coupling	10.67	0.44	5.16	11.76	nr	16.92
tee; equal	20.52	0.46	5.39	22.61	nr	28.00
tee; reducing	20.05	0.46	5.39	22.09	nr	27.48
tank coupling	10.69	0.46	5.39	11.78	nr	17.17
42 mm pipes; fixing with pipe clips; plugged and screwed	8.18	0.64	7.50	9.31	m	16.81
Extra for						
made bend	-	0.44	5.16	-	nr	5.16
stop end	15.06	0.20	2.34	16.59	nr	18.93
straight coupling	15.36	0.39	4.57	16.92	nr	21.49
male coupling	13.32	0.50	5.86	14.68	nr	20.54
female coupling	14.34	0.50	5.86	15.80	nr	21.66
tee; equal	32.25	0.52	6.09	35.54	nr	41.63
tee; reducing	30.98	0.52	6.09	34.14	nr	40.23
54 mm pipes; fixing with pipe clips; plugged and screwed	10.50	0.69	8.09	11.93	m	20.02
Extra for						
made bend	-	0.61	7.15	-	nr	7.15
straight coupling	22.98	0.44	5.16	25.32	nr	30.48
male coupling	19.67	0.56	6.56	21.67	nr	28.23
female coupling	21.04	0.56	6.56	23.18	nr	29.74
tee; equal	51.81	0.57	6.68	57.09	nr	63.77
tee; reducing	51.81	0.57	6.68	57.09	nr	63.77
Copper, brass and gunmetal ancillaries; screwed joints to fittings						
Stopcock; brass/gunmetal capillary joints to copper						
15 mm nominal size	4.15	0.22	2.58	4.57	nr	7.15
22 mm nominal size	7.75	0.30	3.52	8.54	nr	12.06
28 mm nominal size	22.07	0.38	4.45	24.32	nr	28.77
Stopcock; brass/gunmetal compression joints to copper						
15 mm nominal size	5.70	0.20	2.34	6.28	nr	8.62
22 mm nominal size	10.03	0.27	3.16	11.05	nr	14.21
28 mm nominal size	26.14	0.33	3.87	28.40	nr	32.67

Prices for Measured Works - Minor Works
S PIPED SUPPLY SYSTEMS

	PC £	Labour hours	Labour £	Material £	Unit	Total rate £
S12 HOT AND COLD WATER (SMALL SCALE) - cont'd						
Copper, brass and gunmetal ancillaries; screwed joints to fittings - cont'd						
Stopcock; brass/gunmetal compression joints to polyethylene						
15 mm nominal size	15.18	0.29	3.40	16.73	nr	20.13
22 mm nominal size	15.63	0.37	4.34	17.22	nr	21.56
28 mm nominal size	23.47	0.44	5.16	25.86	nr	31.02
Gunmetal "Fullway" gate valve; capillary joints to copper						
15 mm nominal size	13.01	0.22	2.58	14.34	nr	16.92
22 mm nominal size	15.07	0.30	3.52	16.61	nr	20.13
28 mm nominal size	20.98	0.38	4.45	23.12	nr	27.57
35 mm nominal size	46.82	0.45	5.27	51.59	nr	56.86
42 mm nominal size	58.55	0.52	6.09	64.51	nr	70.60
54 mm nominal size	84.92	0.59	6.91	93.57	nr	100.48
Brass gate valve; compression joints to copper						
15 mm nominal size	9.30	0.33	3.87	10.25	nr	14.12
22 mm nominal size	10.84	0.44	5.16	11.94	nr	17.10
28 mm nominal size	18.95	0.56	6.56	20.88	nr	27.44
Chromium plated; lockshield radiator valve; union outlet						
15 mm nominal size	6.50	0.24	2.81	7.16	nr	9.97
Water tanks/cisterns						
Polyethylene cold water feed and expansion cistern; BS 4213; with covers						
ref SC15; 68 litres	52.14	1.39	16.29	56.04	nr	72.33
ref SC25; 114 litres	61.25	1.61	18.87	65.84	nr	84.71
ref SC40; 182 litres	71.65	1.61	18.87	77.01	nr	95.88
ref SC50; 227 litres	99.74	2.16	25.32	107.22	nr	132.54
GRP cold water storage cistern; with covers						
ref 899.10; 30 litres	122.91	1.22	14.30	132.12	nr	146.42
ref 899.25; 68 litres	154.12	1.39	16.29	165.38	nr	181.97
ref 899.40; 114 litres	194.11	1.61	18.87	208.68	nr	227.55
ref 899.70; 227 litres	240.94	2.16	25.32	259.01	nr	284.33
Storage cylinders/calorifiers						
Copper cylinders; single feed coil indirect; BS 1566 Part 2; grade 3						
ref 2; 96 litres	-	2.22	26.02	160.53	nr	186.55
ref 3; 114 litres	95.01	2.50	29.30	102.14	nr	131.44
ref 7; 117 litres	93.75	2.78	32.58	100.78	nr	133.36
ref 8; 140 litres	106.15	3.33	39.03	114.11	nr	153.14
ref 9; 162 litres	135.58	3.89	45.59	145.75	nr	191.34
Combination copper hot water storage units; coil direct; BS 3198; (hot/cold)						
400 mm x 900 mm; 65/20 litres	110.35	3.11	36.45	118.63	nr	155.08
450 mm x 900 mm; 85/25 litres	113.64	4.33	50.75	122.16	nr	172.91
450 mm x 1075 mm; 115/25 litres	125.01	5.44	63.76	134.39	nr	198.15
450 mm x 1200 mm; 115/45 litres	133.08	6.11	71.61	143.06	nr	214.67
Combination copper hot water storage						
450 mm x 900 mm; 85/25 litres	142.51	4.88	57.19	153.20	nr	210.39
450 mm x 1200 mm; 115/45 litres	162.97	6.66	78.06	175.19	nr	253.25

S PIPED SUPPLY SYSTEMS

	PC £	Labour hours	Labour £	Material £	Unit	Total rate £
Thermal insulation						
19 mm thick rigid mineral glass fibre sectional pipe lagging; plain finish; fixed with aluminium bands to steel or copper pipework; including working over pipe fittings						
around 15/15 pipes	5.84	0.07	0.82	6.59	m	7.41
around 20/22 pipes	6.71	0.11	1.29	7.57	m	8.86
around 25/28 pipes	7.27	0.12	1.41	8.21	m	9.62
around 32/35 pipes	8.06	0.13	1.52	9.10	m	10.62
around 40/42 pipes	8.54	0.15	1.76	9.64	m	11.40
around 50/54 pipes	9.90	0.17	1.99	11.17	m	13.16
19 mm thick rigid mineral glass fibre sectional pipe lagging; canvas or class O lacquered aluminium finish; fixed with aluminium bands to steel or copper pipework; including working over pipe fittings						
around 15/15 pipes	8.50	0.07	0.82	9.59	m	10.41
around 20/22 pipes	9.27	0.11	1.29	10.46	m	11.75
around 25/28 pipes	10.16	0.12	1.41	11.47	m	12.88
around 32/35 pipes	11.06	0.13	1.52	12.48	m	14.00
around 40/42 pipes	11.90	0.15	1.76	13.43	m	15.19
around 50/54 pipes	13.79	0.17	1.99	15.57	m	17.56
60 mm thick glass-fibre filled polyethylene insulating jackets for GRP or polyethylene cold water cisterns; complete with fixing bands; for cisterns size						
450 mm x 300 mm x 300 mm (45 litres)	-	0.44	5.16	-	nr	5.16
650 mm x 500 mm x 400 mm (91 litres)	-	0.67	7.85	-	nr	7.85
675 mm x 525 mm x 500 mm (136 litres)	-	0.78	9.14	-	nr	9.14
675 mm x 575 mm x 525 mm (182 litres)	-	0.89	10.43	-	nr	10.43
1000 mm x 625 mm x 525 mm (273 litres)	-	0.94	11.02	-	nr	11.02
1125 mm x 650 mm x 575 mm (341 litres)	-	0.94	11.02	-	nr	11.02
80 mm thick glass-fibre filled insulating jackets in flame retardant PVC to BS 1763; type 1B; segmental type for hot water cylinders; complete with fixing bands; for cylinders size						
400 mm x 900 mm; ref 2	-	0.37	4.34	-	nr	4.34
450 mm x 900 mm; ref 7	-	0.37	4.34	-	nr	4.34
450 mm x 1050 mm; ref 8	-	0.44	5.16	-	nr	5.16
450 mm x 1200 mm	-	0.56	6.56	-	nr	6.56
S13 PRESSURISED WATER						
Blue MDPE pipes; BS 6572; mains pipework; no joints in the running length; laid in trenches						
Pipes						
20 mm nominal size	0.40	0.12	1.41	0.44	m	1.85
25 mm nominal size	0.49	0.13	1.52	0.54	m	2.06
32 mm nominal size	0.84	0.15	1.76	0.93	m	2.69
50 mm nominal size	1.99	0.17	1.99	2.20	m	4.19
63 mm nominal size	3.18	0.18	2.11	3.52	m	5.63
Ductile iron bitumen coated pipes and fittings; EN598; class K9; Stanton's "Tyton" water main pipes or other equal and approved; flexible joints						
100 mm pipes; laid straight	15.29	0.67	4.88	22.93	m	27.81
Extra for						
bend; 45 degrees	35.14	0.67	4.88	51.00	nr	55.88
branch; 45 degrees; socketted	252.59	1.00	7.29	302.11	nr	309.40
tee	54.74	1.00	7.29	78.79	nr	86.08
flanged spigot	37.30	0.67	4.88	47.77	nr	52.65
flanged socket	36.52	0.67	4.88	46.89	nr	51.77

S PIPED SUPPLY SYSTEMS

	PC £	Labour hours	Labour £	Material £	Unit	Total rate £
S13 PRESSURISED WATER - cont'd						
Ductile iron bitumen coated pipes and fittings; EN598; class K9; Stanton's "Tyton" water main pipes or other equal and approved; flexible joints - cont'd						
150 mm pipes; laid straight	23.03	0.78	5.69	32.17	m	37.86
Extra for						
bend; 45 degrees	60.02	0.78	5.69	80.10	nr	85.79
branch; 45 degrees; socketted	322.39	1.17	8.53	382.42	nr	390.95
tee	113.75	1.17	8.53	146.92	nr	155.45
flanged spigot	44.50	0.78	5.69	56.40	nr	62.09
flanged socket	58.11	0.78	5.69	71.77	nr	77.46
200 mm pipes; laid straight	31.12	1.11	8.09	43.78	m	51.87
Extra for						
bend; 45 degrees	108.31	1.11	8.09	139.55	nr	147.64
branch; 45 degrees; socketted	366.13	1.67	12.17	439.21	nr	451.38
tee	156.23	1.67	12.17	202.28	nr	214.45
flanged spigot	97.03	1.11	8.09	118.17	nr	126.26
flanged socket	91.92	1.11	8.09	112.40	nr	120.49
S32 NATURAL GAS						
Ductile iron bitumen coated pipes and fittings; BS 4772; class K9; Stanton's "Stanlock" gas main pipes or other equal and approved; bolted gland joints						
100 mm pipes; laid straight	30.67	0.74	5.39	48.53	m	53.92
Extra for						
bend; 45 degrees	46.34	0.74	5.39	73.18	nr	78.57
tee	70.71	1.12	8.16	114.61	nr	122.77
flanged spigot	38.38	0.74	5.39	57.24	nr	62.63
flanged socket	37.88	0.74	5.39	56.67	nr	62.06
isolated "Stanlock" joint	12.33	0.37	2.70	13.92	nr	16.62
150 mm pipes; laid straight	46.70	0.95	6.93	72.56	m	79.49
Extra for						
bend; 45 degrees	66.91	0.95	6.93	105.29	nr	112.22
tee	118.38	1.43	10.42	183.23	nr	193.65
flanged spigot	44.50	0.95	6.93	70.07	nr	77.00
flanged socket	55.85	0.95	6.93	82.88	nr	89.81
isolated "Stanlock" joint	17.58	0.48	3.50	19.84	nr	23.34
200 mm pipes; laid straight	62.17	1.38	10.06	96.75	m	106.81
Extra for						
bend; 45 degrees	108.98	1.38	10.06	149.58	nr	159.64
tee	164.47	2.07	15.09	265.36	nr	280.45
flanged spigot	97.03	1.38	10.06	136.09	nr	146.15
flanged socket	86.41	1.38	10.06	124.11	nr	134.17
isolated "Stanlock" joint	23.54	0.69	5.03	26.57	nr	31.60

T MECHANICAL HEATING ETC. SYSTEMS

	PC £	Labour hours	Labour £	Material £	Unit	Total rate £
T MECHANICAL HEATING ETC. SYSTEMS						
T10 GAS/OIL FIRED BOILERS						
Boilers						
Gas fired floor standing domestic boilers; cream or white; enamelled casing; 32 mm diameter BSPT female flow and return tappings; 102 mm diameter flue socket 13 mm diameter BSPT male draw-off outlet						
13.19 kW output (45,000 Btu/Hr)	-	5.55	63.10	656.67	nr	719.77
23.45 kW output (80,000 Btu/Hr)	-	5.55	63.10	844.30	nr	907.40
T31 LOW TEMPERATURE HOT WATER HEATING						
NOTE: The reader is referred to section "S12 Hot and Cold Water (Small Scale)" for rates for copper pipework which will equally apply to this section of work. For further and more detailed information the reader is advised to consult *Spon's Mechanical and Electrical Services Price Book.*						
Radiators; Myson Heat Emitters or other equal and approved						
"Premier HE"; single panel type; steel 690 mm high; wheelhead and lockshield valves						
540 mm long	25.92	2.22	25.24	42.33	nr	67.57
1149 mm long	52.30	2.50	28.43	70.69	nr	99.12
2165 mm long	101.08	2.78	31.61	123.13	nr	154.74
2978 mm long	154.49	3.05	34.68	180.62	nr	215.30

V ELECTRICAL SYSTEMS

V21/V22 GENERAL LIGHTING AND LV POWER

NOTE: The following items indicate approximate prices for wiring of lighting and power points complete, including accessories and socket outlets, but excluding lighting fittings. Consumer control units are shown separately. For a more detailed breakdown of these costs and specialist costs for a complete range of electrical items, reference should be made to *Spon's Mechanical and Electrical Services Price Book*.

	PC £	Labour hours	Labour £	Material £	Unit	Total rate £
Consumer control units						
8-way 60 amp SP&N surface mounted insulated consumer control units fitted with miniature circuit breakers including 2 m long 32 mm screwed welded conduit with three runs of 16 mm² PVC cables ready for final connections	-	-	-	-	nr	161.79
extra for current operated ELCB of 30 mA tripping current	-	-	-	-	nr	69.91
As above but 100 amp metal cased consumer unit and 25 mm² PVC cables	-	-	-	-	nr	184.76
extra for current operated ELCB of 30 mA tripping current	-	-	-	-	nr	149.80
Final circuits						
Lighting points						
wired in PVC insulated and PVC sheathed cable in flats and houses; insulated in cavities and roof space; protected where buried by heavy gauge PVC conduit	-	-	-	-	nr	59.92
as above but in commercial property	-	-	-	-	nr	72.91
wired in PVC insulated cable in screwed welded conduit in flats and houses	-	-	-	-	nr	124.84
as above but in commercial property	-	-	-	-	nr	149.80
as above but in industrial property	-	-	-	-	nr	174.76
wired in MICC cable in flats and houses	-	-	-	-	nr	104.87
as above but in commercial property	-	-	-	-	nr	124.84
as above but in industrial property with PVC sheathed cable	-	-	-	-	nr	144.81
Single 13 amp switched socket outlet points						
wired in PVC insulated and PVC sheathed cable in flats and houses on a ring main circuit; protected where buried by heavy gauge PVC conduit	-	-	-	-	nr	54.92
as above but in commercial property	-	-	-	-	nr	69.91
wired in PVC insulated cable in screwed welded conduit throughout on a ring main circuit in flats and houses	-	-	-	-	nr	94.87
as above but in commercial property	-	-	-	-	nr	109.85
as above but in industrial property	-	-	-	-	nr	124.84
wired in MICC cable on a ring main circuit in flats and houses	-	-	-	-	nr	92.37
as above but in commercial property	-	-	-	-	nr	114.84
as above but in industrial property with PVC sheathed cable	-	-	-	-	nr	144.81

V ELECTRICAL SYSTEMS

	PC £	Labour hours	Labour £	Material £	Unit	Total rate £
Cooker control units						
45 amp circuit including unit wired in PVC insulated and PVC sheathed cable; protected where buried by heavy gauge PVC conduit	-	-	-	-	nr	**129.83**
as above but wired in PVC insulated cable in screwed welded conduit	-	-	-	-	nr	**189.75**
as above but wired in MICC cable	-	-	-	-	nr	**209.72**

W SECURITY SYSTEMS

	PC £	Labour hours	Labour £	Material £	Unit	Total rate £
W SECURITY SYSTEMS						
W20 LIGHTNING PROTECTION						
Lightning protection equipment						
Copper strip roof or down conductors fixed with bracket or saddle clips						
20 mm x 3 mm flat section	-	-	-	-	m	15.73
25 mm x 3 mm flat section	-	-	-	-	m	17.97
Aluminium strip roof or down conductors fixed with bracket or saddle clips						
20 mm x 3 mm flat section	-	-	-	-	m	12.23
25 mm x 3 mm flat section	-	-	-	-	m	12.99
Joints in tapes	-	-	-	-	nr	9.59
Bonding connections to roof and structural metalwork	-	-	-	-	nr	54.92
Testing points	-	-	-	-	nr	26.96
Earth electrodes						
16 mm diameter driven copper electrodes in 1220 mm long sectional lengths (minimum 2440 mm long overall)	-	-	-	-	nr	144.81
first 2440 mm length driven and tested 25 mm x 3 mm copper strip electrode in 457 mm deep prepared trench	-	-	-	-	m	10.99

SPON'S PRICE BOOKS 2002

with free CD-ROM

Free CD-ROM when you order any Spon's 2002 Price Book.
Use the CD-ROM to:
- produce tender documents
- customise data
- keyword search
- export to other major packages
- perform simple calculations.

Spon's Architects' and Builders' Price Book 2002
Davis Langdon & Everest

"Spon's Price Books have always been a 'Bible' in my work - now they have got even better! The CDs are not only quick but easy to use. The CD ROMs will really help me to get the most from my Spon's in my role as a Freelance Surveyor."
Martin Taylor, Isle of Lewis

New Features for 2002 Include:
- A new section on Captial Allowances
- Inclusion of new items within a seperate Measured Works section

September 2001: 1024 pages
Hb & CD-ROM: 0-415-26216-X: £110.00

Spon's Landscapes and External Works Price Book 2002
Davis Langdon & Everest, in association with Landscape Projects

New Features for 2002 Include:
- Fees for professional services
- Revised and updated sections on Cost Information and how to use this book
- Revisions and expansions of the Approximate Estimating section, together with direct links into the Measured Works Section

September 2001: 484 pages
Hb & CD-ROM: 0-415-26220-8: £80.00

Spon's Mechanical and Electrical Services Price Book 2002
Mott Green & Wall

"An essential reference for everybody concerned with the calculation of costs of mechanical and electrical works." *Cost Engineer*

New Features for 2002 Include:
- New sections on modular wiring, emergency lighting, lighting control, sprinkler pre fabricated pipework, UPVC rainwater and gutters, carbon steel pipework and fittings

September 2001: 584 pages
Hb & CD-ROM: 0-415-26222-4: £110.00

Spon's Civil Engineering and Highway Works Price Book 2002
Davis Langdon & Everest

New Features for 2002 Include:
- A revised and extended section on Land Remediation
- The Rail Track section now includes data on Permanent Way work with fully reviewed pricing
- Fully reviewed pricing for the Geotextiles section

September 2001: 688 pages
Hb & CD-ROM: 0-415-26218-6: £120.00

updates available to download from the web
www.pricebooks.co.uk

Return your orders to: Spon Press Customer Service Department, ITPS, Cheriton House, North Way, Andover, Hampshire, SP10 5BE · Tel: +44 (0) 1264 343071 · Fax: + 44 (0) 1264 343005 · Email: book.orders@tandf.co.uk
Postage & Packing: 5% of order value (min. charge £1, max. charge £10) for 3–5 days delivery · Option of next day delivery at an additional £6.50.

SPON PRESS — Taylor & Francis Group

NEW FROM SPON PRESS

Housing Design Quality
THROUGH POLICY, GUIDANCE AND REVIEW

Matthew Carmona,
University College London, UK

Housing Design Quality directly addresses the major planning debate of our time - the delivery and quality of new housing development.

As pressure for new housing development in England increases, a widespread desire to improve the design of the resulting residential environments becomes ever more apparent with increasing condemnation of the standard products of the volume house builders. In recent years central government has come to accept the need to deliver higher quality living environments, and the important role of the planning system in helping to raise design standards. *Housing Design Quality* focuses on this role and in particular on how the various policy instruments available to public authorities can be used in a positive manner to deliver higher quality residential developments.

March 2001: 246x189: 368pp
90 b+w illustrations
Pb: 0-419-25650-4: £35.00

To Order: Tel: +44 (0) 8700 768853, or +44 (0) 1264 343071 Fax: +44 (0) 1264 343005, or Post: Spon Press Customer Services, ITPS Andover, Hants, SP10 5BE, UK Email: book.orders@tandf.co.uk.

Postage & Packing: UK: 5% of order value (min. charge £1, max. charge £10) for 3-5 days delivery. Option of next day delivery at an additional £6.50. Europe: 10% of order value (min. charge £2.95, max. charge £20) for delivery surface post. Option of airmail at an additional £6.50. ROW: 15% of order value (min.charge£6.50, max. charge £30) for airmail delivery.

For a complete listing of all our titles visit: www.sponpress.com

NEW FROM SPON PRESS

Inclusive Design
Designing and Developing Accessible Environments

Rob Imrie and **Peter Hall**, both from the Royal Holloway College, University of London, UK

'This is a well written and informative book on an important topic. The authors have taken an interesting perspective on access and the built environment by focussing on the role of the development industry. Their findings are shocking. Anyone interested in urban issues will find a wealth of insightful material in this book.' Nick Oatley, University of the West of England, UK

The reality of the built environment for disabled people is one of social, physical and attitudinal barriers which prevent their ease of mobility, movement and access. In the United Kingdom, most homes cannot be accessed by wheelchair, while accessible transport is the exception rather than the rule. Pavements are littered with street furniture, while most public and commercial buildings provide few design features to permit disabled people ease of access.

Inclusive Design is a documentation of the attitudes, values, and practices of property professionals, including developers, surveyors and architects, in responding to the building needs of disabled people. It looks at the way in which pressure for accessible building design is influencing the policies and practices of property companies and professionals, with a primary focus on commercial developments in the UK. The book also provides comments on, and references to, other countries, particularly Sweden, New Zealand, and the USA.

June 2001: 246x189: 240pp
5 line figures, 34 b+w photos
Pb: 0-419-25620-2: £29.99

To Order: Tel: +44 (0) 8700 768853, or +44 (0) 1264 343071 Fax: +44 (0) 1264 343005, or
Post: Spon Press Customer Services, ITPS Andover, Hants, SP10 5BE, UK Email: book.orders@tandf.co.uk.

Postage & Packing: UK: 5% of order value (min. charge £1, max. charge £10) for 3-5 days delivery. Option of next day delivery at an additional £6.50. Europe: 10% of order value (min. charge £2.95, max. charge £20) for delivery surface post. Option of airmail at an additional £6.50. ROW: 15% of order value (min.charge£6.50, max. charge £30) for airmail delivery.

For a complete listing of all our titles visit: www.sponpress.com

NEW FROM SPON PRESS

Dictionary of Property and Construction Law

Edited by J. Rostron, School of the Built Environment, Liverpool John Moores University, UK

This is a new dictionary containing over 6,000 entries. It provides clear and concise explanations of the terms used in land, property and construction law and management. The four key areas of coverage are: planning/construction law, land law, equity/trusts and finance and administration. It will be a useful reference for property and building professionals and for students of property and construction law. Jack Rostron is an experienced author and editor whose 1997 Spon title *Sick Building Syndrome* has been well received and widely reviewed.

August 2001: 234x156: 160pp:
Pb: 0-419-26110-9: £19.99
Hb: 0-419-26100-1: £50.00

To Order: Tel: +44 (0) 8700 768853, or +44 (0) 1264 343071 Fax: +44 (0) 1264 343005, or
Post: Spon Press Customer Services, ITPS Andover, Hants, SP10 5BE, UK Email: book.orders@tandf.co.uk.

Postage & Packing: UK: 5% of order value (min. charge £1, max. charge £10) for 3-5 days delivery. Option of next day delivery at an additional £6.50. Europe: 10% of order value (min. charge £2.95, max. charge £20) for delivery surface post. Option of airmail at an additional £6.50. ROW: 15% of order value (min.charge£6.50, max. charge £30) for airmail delivery.

For a complete listing of all our titles visit: www.sponpress.com

PART IV

Approximate Estimating

This part of the book contains the following sections:

Building Costs and Tender Prices, *page 691*
Building Prices per Functional Units, *page 697*
Building Prices per Square Metre, *page 703*
Approximate Estimates (incorporating Comparative Prices), *page 711*
Cost Limits and Allowances, *page 831*
Property Insurance, *page 877*
Capital Allowance, *page 881*

NEW FROM SPON PRESS

The Architectural Expression of Environmental Control Systems

George Baird, Victoria University of Wellington, New Zealand

*The Architectural Expression of Environmental Control System*s examines the way project teams can approach the design and expression of both active and passive thermal environmental control systems in a more creative way. Using seminal case studies from around the world and interviews with the architects and environmental engineers involved, the book illustrates innovative responses to client, site and user requirements, focusing upon elegant design solutions to a perennial problem.

This book will inspire architects, building scientists and building services engineers to take a more creative approach to the design and expression of environmental control systems - whether active or passive, whether they influence overall building form or design detail.

March 2001: 276x219: 304pp
135 b+w photos, 40 colour, 90 line illustrations
Hb: 0-419-24430-1: £49.95

To Order: Tel: +44 (0) 8700 768853, or +44 (0) 1264 343071 Fax: +44 (0) 1264 343005, or
Post: Spon Press Customer Services, ITPS Andover, Hants, SP10 5BE, UK Email: book.orders@tandf.co.uk.

Postage & Packing: UK: 5% of order value (min. charge £1, max. charge £10) for 3-5 days delivery. Option of next day delivery at an additional £6.50. Europe: 10% of order value (min. charge £2.95, max. charge £20) for delivery surface post. Option of airmail at an additional £6.50. ROW: 15% of order value (min.charge£6.50, max. charge £30) for airmail delivery.

For a complete listing of all our titles visit: www.sponpress.com

Building Costs and Tender Prices

The tables which follow show the changes in building costs and tender prices since 1976.

To avoid confusion it is essential that the terms "building costs" and "tender prices" are clearly defined and understood. "Building costs" are the costs incurred by the builder in the course of his business, the principal ones being those for labour and materials. "Tender Price" is the price for which a builder offers to erect a building. This includes "building costs" but also takes into account market considerations such as the availability of labour and materials, and prevailing economic situation. This means that in "boom" periods, when there is a surfeit of building work to be done, "tender prices" may increase at a greater rate than "building costs", whilst in a period when work is scarce, "tender prices" may actually fall when "building costs" are rising.

Building costs

This table reflects the fluctuations since 1976 in wages and materials costs to the builder. In compiling the table, the proportion of labour to material has been assumed to be 40:60. The wages element has been assessed from a contract wages sheet revalued for each variation in labour costs, whilst the changes in the costs of materials have been based upon the indices prepared by the Department of Trade and Industry. No allowance has been made for changes in productivity, plus rates or hours worked which may occur in particular conditions and localities.

1976 = 100

Year	First quarter	Second quarter	Third quarter	Fourth quarter	Annual average
1976	93	97	104	107	100
1977	109	112	116	117	114
1978	118	120	127	129	123
1979	131	135	149	153	142
1980	157	161	180	181	170
1981	182	185	195	199	190
1982	203	206	214	216	210
1983	217	219	227	229	223
1984	230	232	239	241	236
1985	243	245	252	254	248
1986	256	258	266	267	262
1987	270	272	281	282	276
1988	284	286	299	302	293
1989	305	307	322	323	314
1990	326	329	346	347	337
1991	350	350	360	360	355
1992	361	302	367	368	365
1993	370	371	373	374	372
1994	376	379	385	388	382
1995	392	397	407	407	401
1996	407	408	414	414	411
1997	416	417	423	429	421
1998	430	431	448	447	439
1999	446	443	473	478	460
2000	480	482	497	497 (P)	489 (P)
2001	498 (P)	499 (F)	517 (F)	518 (F)	508 (F)

Note: P = Provisional   F = Forecast   R = Revised

Tender Prices

This table reflects the changes in tender prices since 1976. It indicates the level of pricing contained in the lowest competitive tenders for new work in the Greater London area (over £500,000 in value).

1976 = 100

Year	First quarter	Second quarter	Third quarter	Fourth quarter	Annual average
1976	97	98	102	103	100
1977	105	105	109	107	107
1978	113	116	126	139	124
1979	142	146	160	167	154
1980	179	200	192	188	190
1981	199	193	190	195	194
1982	191	188	195	195	192
1983	198	200	198	200	199
1984	205	206	214	215	210
1985	215	219	219	220	218
1986	221	226	234	234	229
1987	242	249	265	279	258
1988	289	299	321	328	309
1989	341	335	340	345	340
1990	320	315	312	290	309
1991	272	262	261	254	262
1992	250	248	241	233	243
1993	227	242	233	239	235
1994	239	247	266	256	252
1995	258	265	266	270	265
1996	265	262	270	270	267
1997	275	287	284	287	283
1998	305	312	318	318	313
1999	325	332	330	342	332
2000	348	353	362	375	360
2001	373 (P)	374 (F)	379 (F)	383 (F)	377 (F)

Note: P = Provisional   F = Forecast

Tender prices in the year to 2000 increased by 8% with the second and the third quarter showing a significance rise of 2.50% and 3.60% over the previous quarter.

After five successive quarterly increases, the first quarter of 2001 saw a halt in the trend with prices falling 0.50% but still 7.00% higher over the first quarter of 2000.

Demand for labour, the largest single factor responsible for a rise in tender prices over the past three years has stabilised. Pressure on the supply of skilled workers has eased. Bricklayer wage rates have been static since the beginning of 2000 and Carpenter rates have now been stabilised at around £100 - £105 a day.

Construction output rose by 1.50% last year. This is 10.00% higher than five years ago. Over this period the volume of new build work grew by 17.00% and this is likely to be the main driving force behind the 35.00% rise in tender prices over the same time span.

New orders increase by 13.00% in the first quarter of 2001 over the fourth quarter of 2000 according to DETR statistics. This is 9.00% higher on the first quarter of 2000.

In the Construction Trends Survey, for the fourth quarter of 2000, a large majority of Construction Confereration members were still relatively optimistic about trading conditions, reporting increases in output, margins, tendering success and new enquiries.

Our forecast for tender prices is a rise of 3.00% to 5.00% this year and 4.00% to 5.00% in the year to 2002.

Readers will be kept abreast of tender price movements in the free *Spon's Updates* and also in the "Tender Price Forecast" and "Cost Update" published quarterly in *Building* magazine.

TENDER PRICES / BUILDING COSTS / RETAIL PRICES

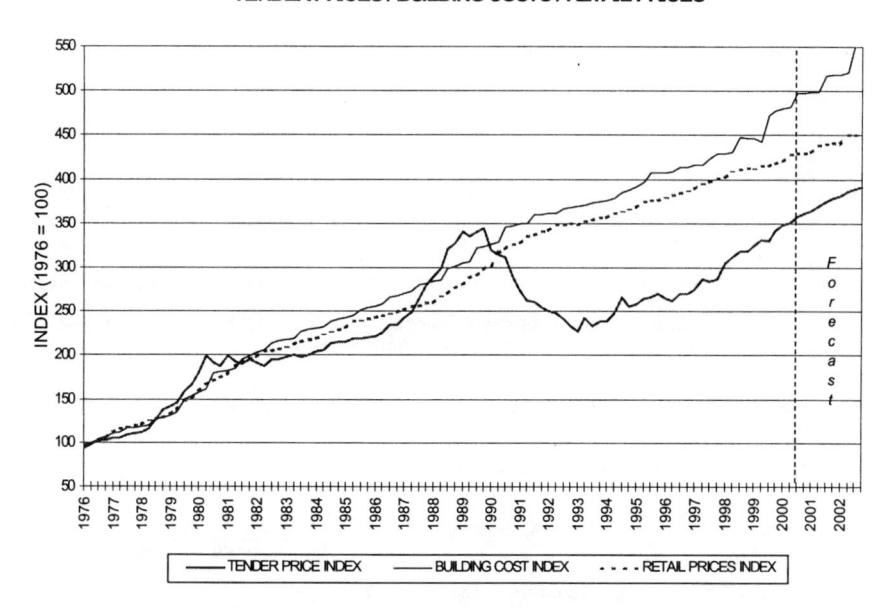

[Chart: Index (1976 = 100) plotted from 1976 to 2002 showing Tender Price Index, Building Cost Index, and Retail Prices Index, with Forecast region at the right.]

Regional Variations

As well as being aware of inflationary trends when preparing an estimate, it is also important to establish the appropriate price level for the project location.

Prices throughout this book reflect prices up to first quarter 2002 in Outer London. Prices in Inner London boroughs can be upto 15% higher while prices in the North and Yorkshire and Humberside can be upto 23% lower. Broad regional adjustment factors to assist with the preparation of initial estimates are shown in the table on the next page.

Over time, price differentials can change depending on regional workloads and local "hot spots". Workloads and prices have risen more strongly in Greater London and the South East over the last year or so than in most other regions. In the year to the first quarter 2001 prices in Greater London rose by 7% but in much of the North only by an average of 5%. *Spon's Updates* and the "Tender Price Forecast" and "Cost Update" features in *Building* magazine will keep readers informed of the latest regional developments and changes as they occur.

The regional variations shown in the table on the next page are based on our forecast of price differentials in each of the economic planning regions in the fourth quarter 2001. The table shows the forecast fourth quarter 2001 tender price index for each region plus the recommended percentage adjustments required to the Major Works section of the Prices for Measured Work. (Prices in the book are at a Tender Price Index level of 388 for Outer London).

Building Costs and Tender Prices Index

Region	Forecast fourth quarter 2001 tender price index	Percentage adjustment to *Major Works* section
Outer London	383	-1
Inner London	413	+7
East Anglia	318	-17
East Midlands	318	-20
Northern	310	-18
Northern Ireland	242	-39
North West	306	-22
Scotland	318	-17
South East	364	-7
South West	338	-18
Wales	295	-22
West Midlands	325	-20
Yorkshire and Humberside	295	-23

Special further adjustment to the above percentages may be necessary when considering city centre or very isolated locations.

The following example illustrates the adjustment of an estimate prepared using *Spon's A&B 2002*, to a price level that reflects the forecast Outer London market conditions for competitive tenders in the fourth quarter 2001:

				£
A	Value of items priced using Spon's A&B 2002 i.e. Tender Price Index 388			1,075,000
B	Adjustment to reduce value of A to forecast price level for fourth quarter 2001 i.e. Forecast Tender Price Index 383 (383 - 388) x 100 = -1.29%			
	388	deduct 1.29%	say	-14,000
				1,061,000
C	Value of items priced using competitive quotations that reflect the market conditions in the fourth quarter 2001			500,000
				1,561,000
D	Allowance for preliminaries +11%		say	172,000
E	Total value of estimate at fourth quarter 2001 price levels		£	1,733,000

Alternatively, for a similar estimate in Scotland:

				£
A	Value of items priced using Spon's A&B 2002 i.e. Tender Price Index 388			1,075,000
B	Adjustment to reduce value of A to forecast price level for fourth quarter 2001 for Scotland (from regional variation table) i.e. Tender Price Index 318 (318 - 388) x 100 = -18.04%			
	388	deduct 18.04%	say	-144,000
				881,000
C	Value of items priced using competitive quotations that reflect the market conditions in the fourth quarter 2001			500,000
				1,381,000
D	Allowance for preliminaries +11%		say	152,000
E	Total value of estimate at fourth quarter 2000 price levels		£	1,533,000

Spons Architects' and
Builders' Price Book
2002Edition

Tender Index 388

Scotland 318

Northern 310

Northern Ireland 242

Yorkshire & Humberside 295

North West 306

East Midlands 318

East Anglia 318

Wales 295

West Midlands 325

Southeast 364 (excl Land)

Southwest 338

Inner London 413
Outer London 383

SPON'S PRICE BOOKS 2002

with free CD-ROM

Free CD-ROM when you order any Spon's 2002 Price Book.
Use the CD-ROM to:
- produce tender documents
- customise data
- keyword search
- export to other major packages
- perform simple calculations.

Spon's Architects' and Builders' Price Book 2002
Davis Langdon & Everest

"Spon's Price Books have always been a 'Bible' in my work - now they have got even better! The CDs are not only quick but easy to use. The CD ROMs will really help me to get the most from my Spon's in my role as a Freelance Surveyor."
Martin Taylor, Isle of Lewis

New Features for 2002 include:
- A new section on Captial Allowances
- Inclusion of new items within a seperate Measured Works section

September 2001: 1024 pages
Hb & CD-ROM: 0-415-26216-X: £110.00

Spon's Landscapes and External Works Price Book 2002
Davis Langdon & Everest, in association with Landscape Projects

New Features for 2002 include:
- Fees for professional services
- Revised and updated sections on Cost Information and how to use this book
- Revisions and expansions of the Approximate Estimating section, together with direct links into the Measured Works Section

September 2001: 484 pages
Hb & CD-ROM: 0-415-26220-8: £80.00

Spon's Mechanical and Electrical Services Price Book 2002
Mott Green & Wall

"An essential reference for everybody concerned with the calculation of costs of mechanical and electrical works." *Cost Engineer*

New Features for 2002 include:
- New sections on modular wiring, emergency lighting, lighting control, sprinkler pre fabricated pipework, UPVC rainwater and gutters, carbon steel pipework and fittings

September 2001: 584 pages
Hb & CD-ROM: 0-415-26222-4: £110.00

Spon's Civil Engineering and Highway Works Price Book 2002
Davis Langdon & Everest

New Features for 2002 include:
- A revised and extended section on Land Remediation
- The Rail Track section now includes data on Permanent Way work with fully reviewed pricing
- Fully reviewed pricing for the Geotextiles section

September 2001: 688 pages
Hb & CD-ROM: 0-415-26218-6: £120.00

updates available to download from the web
www.pricebooks.co.uk

Return your orders to: Spon Press Customer Service Department, ITPS, Cheriton House, North Way, Andover, Hampshire, SP10 5BE · Tel: +44 (0) 1264 343071 ·
Fax: + 44 (0) 1264 343005 · Email: book.orders@tandf.co.uk
Postage & Packing: 5% of order value (min. charge £1, max. charge £10) for 3–5 days delivery · Option of next day delivery at an additional £6.50.

SPON PRESS · Taylor & Francis Group

Building Prices per Functional Units

Prices given under this heading are average prices, on a *fluctuating basis*, for typical buildings based on a tender price level index of 388 (1976 = 100). Prices includes for Preliminaries at 11% and overheads and profit. Unless otherwise stated, prices do not allow for external works, furniture, loose or special equipment and are, of course, exclusive of fees for professional services.

On certain types of buildings there exists a close relationship between its cost and the number of functional units that it accommodates. During the early stages of a project therefore an approximate estimate can be derived by multiplying the proposed unit of accommodation (i.e. hotel bedrooms, car parking spaces etc.) by an appropriate cost.

The following indicative unit areas and costs have been derived from historic data. It is emphasized that the prices must be treated with reserve, as they represent the average of prices from our records and cannot provide more than a rough guide to the cost of a building. There are limitations when using this method of estimating, for example, the functional areas and costs of football stadia are strongly influenced by the extent of front and back of house facilities housed within it, and these areas can vary considerably from scheme to scheme.

The areas may also be used as a "rule of thumb" in order to check on economy of designs. Where we have chosen not to show indicative areas, this is because either ranges are extensive or such figures may be misleading.

Costs have been expressed within a range, although this is not to suggest that figures outside this range will not be encountered, but simply that the calibre of such a type of building can itself vary significantly.

For assistance with the compilation of a closer estimate, or of a Cost Plan, the reader is directed to the *"Building Prices per Square Metre"* and *"Approximate Estimates"* sections. As elsewhere in this edition, prices do not include V.A.T., which is generally applied at the current rate to all non-domestic building (except those with charitable status).

Function:	Indicative functional unit area:	Indicative functional unit cost:
Utilities, civil engineering facilities (Cl/SfB 1)		
Car parking — surface level	20 - 22 m²/car	£900 - £1,325/car
— ground level (under buildings)	22 - 24 m²/car	£1,700 - £3,300/car
— multi-storey	23 - 27 m²/car	£3,900 - £8,250/car
— semi-basement	27 - 30 m²/car	£9,000 - £12,000/car
— basement	28 - 37 m²/car	£12,000 - £20,000/car
Administrative, commercial protective service facilities (Cl/SfB 3)		
Offices - air conditioned — low density cellular	15 m²/person	£8,250 - £16,500/person
— high density open plan	20 m²/person	£11,000 - £22,500/person
Health and welfare facilities (Cl/SfB 4)		
Hospitals — district general	65 - 85 m²/bed	£67,000 - £100,000/bed
— teaching	120 + m²/bed	£88,500 - £122,000/bed
— private	75 - 100 m²/bed	£94,500 - £150,000/bed
Nursing Homes — residential home	-	£20,000 - £38,500/bed
— nursing home	-	£27,500 - £77,500/bed
Recreational facilities (Cl/SfB 5)		
Football Stadia — basic stand	0.50 - 0.85 m²/seat	£530 - £800/seat
— stand plus basic facilities	0.85 - 1.28 m²/seat	£800 - £1,325/seat
— stand plus extensive facilities	1.00 - 1.30 m²/seat	£1,200 - £1,650/seat
— national stadia plus extensive facilities	1.20 - 1.80 m²/seat	£2,250 - £3,350/seat

Function:	Indicative functional unit area:	Indicative functional unit cost:
Recreational facilities (Cl/SfB 5) - cont'd		
Theatres - theatre refurbishment	-	£7,750 - £11,150/seat
- workshop (fewer than 500 seats)	-	£3,300 - £8,850/seat
- more than 500 seats	-	£16,500 - £22,500/seat
Educational, scientific, information facilities (Cl/SfB 7)		
Schools - nursery	3 - 5 m²/child	£3,300 - £7,750/child
- secondary	6 - 10 m²/child	£4,500 - £9,000/child
- boarding	10 - 12 m²/child	£6,700 - £13,250/child
- special	18 - 20 m²/child	£20,000 - £30,750/child
Residential facilities (Cl/SfB 8)		
Housing (Private developer) - terraced; two bedroom	55 - 65 m²/gifa	£33,500 - £38,750/house
- semi-detached; three bedroom	70 - 90 m²/gifa	£50,000 - £55,000/house
- detached; four bedroom	90 - 100 m²/gifa	£67,000 - £100,000/house
- low rise flats; two bedroom	55 - 65 m²/gifa	£36,500 - £53,000/flat
- medium rise flat; two bedroom	55 - 65 m²/gifa	£44,500 - £61,000/flat
Hotels - budget roadside hotel - two to three storey lodge, excluding dining facilities	28 - 35 m²/bedroom	£27,500 - £36,500/bedroom
- budget city-centre hotel (office conversion) - four to six storeys, excluding dining facilities	32 - 38 m²/bedroom	£25,000 - £50,000/bedroom

Function:	Indicative functional unit area:	Indicative functional unit cost:
Residential facilities (CI/SfB 8) - cont'd		
Hotels - cont'd - mid-range provincial hotel - two to three storeys, bedroom extension	33 - 40 m²/bedroom	£37,500 - £53,000/bedroom
- budget city-centre hotel (new build) - four to six storeys, dining bar and facilities	35 - 45 m²/bedroom	£31,250 - £50,000/bedroom
- city centre aparthotel - 4 to seven storeys, apartments with self-catering facilities	50 - 60 m²/bedroom	£50,000 - £77,500/bedroom
- mid-range provincial hotel - two to three storeys, conference and leisure facilities	50 - 60 m²/bedroom	£61,000 - £95,000/bedroom
- business town centre provincial hotel - four to six storeys, conference and wet leisure facilities	70 - 100 m²/bedroom	£105,000 - £150,000/bedroom
- luxury city-centre hotel - multi-storey conference and wet leisure facilities	70 - 130 m²/bedroom	£148,500 - £222,500/bedroom
Hotel furniture, fittings and equipment - budget hotel	-	£3,350 - £6,700/bedroom
- mid-range hotel	-	£11,150 - £16,500/bedroom
- luxury hotel	-	£27,500 - £67,000/bedroom
Students Residences - large turnkey budget schemes (200 + units), simple design, open site; en suite accommodation	18 - 20 m²/bedroom	£11,150 - £20,150/bedroom

Function:	Indicative functional unit area:	Indicative functional unit cost:
Residential facilities (CI/SfB 8) - cont'd		
Students Residences - cont'd - smaller schemes (40 - 100 units) with mid range specifications, some with en suite bathroom and kitchen facilities	19 - 24 m²/bedroom	£20,000 - £33,500/bedroom
- smaller high quality courtyard schemes of collegiate style in restricted city centre sites	24 - 28 m²/bedroom	£33,500 - £55,500/bedroom

NEW FROM SPON PRESS

Dictionary of Property and Construction Law

Edited by J. Rostron, School of the Built Environment, Liverpool John Moores University, UK

This is a new dictionary containing over 6,000 entries. It provides clear and concise explanations of the terms used in land, property and construction law and management. The four key areas of coverage are: planning/construction law, land law, equity/trusts and finance and administration. It will be a useful reference for property and building professionals and for students of property and construction law. Jack Rostron is an experienced author and editor whose 1997 Spon title *Sick Building Syndrome* has been well received and widely reviewed.

August 2001: 234x156: 160pp:
Pb: 0-419-26110-9: £19.99
Hb: 0-419-26100-1: £50.00

To Order: Tel: +44 (0) 8700 768853, or +44 (0) 1264 343071 Fax: +44 (0) 1264 343005, or
Post: Spon Press Customer Services, ITPS Andover, Hants, SP10 5BE, UK Email: book.orders@tandf.co.uk.

Postage & Packing: UK: 5% of order value (min. charge £1, max. charge £10) for 3-5 days delivery. Option of next day delivery at an additional £6.50. Europe: 10% of order value (min. charge £2.95, max. charge £20) for delivery surface post. Option of airmail at an additional £6.50. ROW: 15% of order value (min. charge£6.50, max. charge £30) for airmail delivery.

For a complete listing of all our titles visit: www.sponpress.com

Building Prices per Square Metre

Prices given under this heading are average prices, on a *fluctuating basis*, for typical buildings based on a tender price level index of 388 (1976 = 100). Prices allow for Preliminaries at 11% and overheads and profit. Unless otherwise stated, prices do not allow for external works, furniture, loose or special equipment and are, of course, exclusive of fees for professional services.

Prices are based upon the total floor area of all storeys, measured between external walls and without deduction for internal walls, columns, stairwells, liftwells and the like.

As in previous editions it is emphasized that the prices must be treated with reserve, as they represent the average of prices from our records and cannot provide more than a rough guide to the cost of a building.

In many instances normal commercial pressures together with a limited range of available specifications ensure that a single rate is sufficient to indicate the prevailing average price. However, where such restrictions do not apply a range has been given; this is not to suggest that figures outside this range will not be encountered, but simply that the calibre of such a type of building can itself vary significantly.

For assistance with the compilation of a closer estimate, or of a Cost Plan, the reader is directed to the *"Approximate Estimates"* sections. As elsewhere in this edition, prices do not include V.A.T., which is generally applied at the current rate to all non-domestic building (except those with charitable status).

Utilities, civil engineering facilities (Cl/SfB 1)		Square metre excluding VAT £		
Surface car parking		45	to	65
Surface car parking; landscaped		75	to	100
Multi-storey car parks				
split level		215	to	280
split level with brick facades		240	to	330
flat slab		260	to	335
warped		265	to	305
Underground car parks				
partially underground under buildings; naturally ventilated		345	to	405
completely underground under buildings		445	to	675
completely underground with landscaped roof		730	to	870
Railway stations		1430	to	2425
Bus and coach stations		725	to	1215
Bus garages		695	to	760
Petrol stations				
minor refurbishment	(£23,850 to £116,500 total cost)			
major refurbishment	(£116,500 to £475,000 total cost)			
rebuild	(£475,000 to £715,000 total cost)			
Garage showrooms		535	to	855
Garages, domestic		295	to	475
Airport facilities (excluding aprons)				
airport terminals		1315	to	2870
airport piers/satellites		1605	to	3580
apron/runway - varying infrastructure content		75	to	145
Airport campus facilities				
cargo handling bases		475	to	775
distribution centres		240	to	475

Utilities, civil engineering facilities (CI/SfB 1) - cont'd

Square metre excluding VAT
£

Airport campus facilities - cont'd			
hangars (type C and D aircraft)	1010	to	1190
hangars (type E aircraft)	1190	to	2980
TV, radio and video studios	880	to	1430
Telephone exchanges	720	to	1155
Telephone engineering centres	610	to	760
Branch Post Offices	760	to	1030
Postal Delivery Offices/Sorting Offices	610	to	905
Mortuaries	1385	to	1925
Sub-stations	1030	to	1550

Industrial facilities (CI/SfB 2)

Agricultural storage buildings	365	to	490
Factories			
for letting (incoming services only)	255	to	365
for letting (including lighting, power and heating)	340	to	470
nursery units (including lighting, power and heating)	415	to	625
workshops	475	to	770
maintenance/motor transport workshops	475	to	825
owner occupation-for light industrial use	430	to	625
owner occupation-for heavy industrial use	815	to	935
Factory/office buildings - high technology production			
for letting (shell and core only)	465	to	625
for letting (ground floor shell, first floor offices)	760	to	985
for owner occupation (controlled environment, fully finished)	985	to	1310
Laboratory workshops and offices	890	to	1120
Bi Light industrial/offices buildings			
economical shell, and core with heating only	415	to	730
medium shell and core with heating and ventilation	645	to	960
high quality shell and core with air-conditioning	850	to	1550
developers Category A fit out	340	to	585
tenants Category B fit out	150	to	465
High technology laboratory workshop centres, air conditioned	2065	to	2645
Distribution centres			
low bay; speculative	260	to	370
low bay; owner occupied	390	to	685
low bay; owner occupied and chilled	565	to	1135
high bay; owner occupied	635	to	925
Warehouses			
low bay (6 - 8 m high) for letting (no heating)	255	to	320
low bay for owner occupation (including heating)	305	to	540
high bay (9 - 18 m high) for owner occupation (including heating)	460	to	655
Cold stores, refrigerated stores	540	to	1235

Administrative, commercial protective service facilities (CI/SfB 3)

Embassies	1430	to	2065
County Courts	1265	to	1575
Magistrates Courts	965	to	1215
Civic offices			
non air conditioned	965	to	1215
fully air conditioned	1215	to	1430
Probation/Registrar Offices	705	to	1010
Offices for letting			
low rise, non air conditioned	680	to	970
low rise, air conditioned	850	to	1145

Administrative, commercial protective service facilities (Cl/SfB 3) - cont'd

Square metre excluding VAT
£

Owners for letting - cont'd			
medium rise, non air conditioned	815	to	1070
medium rise, air conditioned	990	to	1335
high rise, non air conditioned	1035	to	1335
high rise, air conditioned	1270	to	1865
Offices for owner occupation			
low rise, non air conditioned	780	to	1020
low rise, air conditioned	970	to	1335
medium rise, non air conditioned	970	to	1215
medium rise, air conditioned	1265	to	2000
high rise, air conditioned	1620	to	2665
Offices, prestige			
medium rise	1590	to	2310
high rise	2310	to	2970
Large trading floors in medium rise offices	2150	to	2565
Two storey ancillary office accommodation to warehouses/factories	695	to	925
Fitting out offices			
basic fitting out including carpets, decorations, partitions and services	220	to	270
good quality fitting out including carpets, decorations, partitions, ceilings, furniture and services	445	to	530
high quality fitting out including raised floors and carpets, decorations, partitions, ceilings, furniture, air conditioning and electrical services	725	to	945
Office refurbishment			
basic refurbishment	370	to	540
good quality, including air conditioning	725	to	980
high quality, including air conditioning	1335	to	1790
Banks			
local	1070	to	1335
city centre/head office	1535	to	1985
Building Society Branch Offices	980	to	1270
refurbishment	540	to	910
Shop shells			
small	460	to	595
large including department stores and supermarkets	395	to	555
Fitting out shell for small shop (including shop fittings)			
simple store	475	to	570
fashion store	900	to	1120
Fitting out shell for department store or supermarket			
excluding shop fittings	540	to	710
including shop fittings	815	to	1850
Retail Warehouses			
shell	305	to	450
fitting out	215	to	250
Supermarkets			
shell	295	to	715
supermarket fit-out	1075	to	1430
hypermarket fit-out	475	to	955
Shopping centres			
malls including fitting out			
comfort cooled	2630	to	4175
air-conditioned	2980	to	4530
food court	2980	to	4295
factory outlet centre mall - enclosed	2385	to	3580
factory outlet centre mall - open	415	to	715
retail area shells, capped off services	600	to	955

Administrative, commercial protective service facilities (Cl/SfB 3) - cont'd

Square metre excluding VAT
£

Shopping centres - cont'd
  landlord's back-up areas, management offices, plant rooms
non air conditioned	690	to	765
excluding shop fittings	540	to	710
including shop fittings	815	to	1850
refurbishment			
mall; limited scope	835	to	1250
mall; comprehensive	1190	to	1790
*Ambulance stations	650	to	965
Ambulance controls centre	905	to	1335
Fire stations	970	to	1275
Police stations	870	to	1325
Prisons	1265	to	1545

Health and welfare facilities (Cl/SfB 4)

*District hospitals	990	to	1335
Refurbishment	475	to	970
Hospice	1035	to	1275
Private hospitals	960	to	1470
Pharmacies	920	to	1235
Hospital laboratories	1275	to	1865
Ward blocks	970	to	1215
Refurbishment	450	to	725
Geriatric units	990	to	1335
Psychiatric units	985	to	1145
Psycho-geriatric units	940	to	1370
Maternity units	985	to	1335
Operating theatres	1060	to	1645
Outpatients/casualty units	1085	to	1430
Hospital teaching centres	815	to	1155
*Health centres	930	to	1170
Welfare centres	990	to	1175
*Day centres	850	to	1175
Group practice surgeries	750	to	945
*Homes for the physically handicapped - houses	965	to	1155
*Homes for the mentally handicapped	780	to	1080
Geriatric day hospital	850	to	1195
Accommodation for the elderly			
residential homes	600	to	955
nursing homes	835	to	1135
*Children's homes	635	to	970
*Homes for the aged	725	to	970
Refurbishment	280	to	680
*Observation and assessment units	720	to	1215

Recreational facilities (Cl/SfB 5)

Public houses	870	to	1175
Kitchen blocks (including fitting out)	1430	to	1645
Dining blocks and canteens in shop and factory	850	to	1215
Restaurants	920	to	1335
Community centres	720	to	995
General purpose halls	720	to	995
Visitors' centres	965	to	1495
Youth clubs	680	to	990
Arts and drama centres	970	to	1120

Recreational facilities (Cl/SfB 5) - cont'd

	Square metre excluding VAT £
Galleries	
refurbishment of historic building to create international standard gallery	2745 to 4530
international-standard art gallery	2275 to 3125
national-standard art gallery	1840 to 2275
independent commercial art gallery	990 to 1275
Arts and drama centres	970 to 1120
Theatres, including seating and stage equipment	
large - over 500 seats	2150 to 2985
studio/workshop - less than 500 seats	1075 to 1790
refurbishment	1190 to 2390
Concert halls, including seating and stage equipment	1790 to 2925
Cinema	
shell	560 to 730
multiplex; shell only	715 to 1010
fitting out including all equipment, air-conditioned	615 to 1120
Exhibition centres	1115 to 1495
Swimming pools	
international standard	1335 to 2405
local authority standard	1080 to 1430
school standard	905 to 1070
leisure pools, including wave making equipment	2150 to 2745
Ice rinks	1000 to 1175
Rifle ranges	780 to 990
Leisure centres	
dry	715 to 1010
extension to hotels (shell and fit-out - including pool)	1490 to 2090
wet and dry	1675 to 2150
Sports halls including changing	560 to 995
School gymnasiums	675 to 780
Squash courts	675 to 945
Indoor bowls halls	415 to 800
Bowls pavilions	675 to 825
Health and fitness clubs	
out-of-town (shell and fit-out - including pool)	955 to 1430
town centre (fit-out - excluding pool)	955 to 1495
Sports pavilions	
changing only	850 to 1155
social and changing	720 to 1240
Grandstands	
simple stands	540 to 635
first class stands with ancillary accommodation	965 to 1215
Football Stadia	
basic stand	415 to 835
medium quality including basic facilities	715 to 1315
high quality including extensive facilities	1315 to 1550
national stadium including extensive facilities	2390 to 3580
Clubhouses	715 to 970
Golf clubhouses	
up to 500 m²	1025 to 1420
500 m² to 2000 m²	620 to 975
over 2000 m²	1620 to 2160

Religious facilities (Cl/SfB 6)

	Square metre excluding VAT £
Temples, mosques, synagogues	1015 to 1430
Churches	870 to 1250
Mission halls, meeting houses	1015 to 1335
Convents	945 to 1070
Crematoria	1175 to 1365

Educational, scientific, information facilities (Cl/SfB 7)

Nursery Schools	870 to 1280
*Primary/junior school	800 to 1170
*Secondary/middle schools	730 to 1170
*Secondary school and further education college buildings	
classrooms	765 to 955
laboratories	785 to 1190
craft design and technology	775 to 1190
music	1170 to 1370
*Extensions to schools	
classrooms	780 to 925
residential	825 to 925
laboratories	1085 to 1175
School refurbishment	215 to 675
Sixth form colleges	815 to 1020
*Special schools	715 to 1020
*Polytechnics	
Students Union buildings	780 to 915
arts buildings	700 to 825
scientific laboratories	940 to 1175
*Training colleges	675 to 1020
Management training centres	965 to 1310
*Universities	
arts buildings	870 to 1080
science buildings	965 to 1245
College/University Libraries	815 to 1155
Laboratories and offices, low level servicing	870 to 1090
Laboratories (specialist, controlled environment)	1335 to 2500
Computer buildings	1145 to 1715
Museums and art galleries	
national-standard museum	2275 to 3345
national-standard independent specialist museum (excluding fit-out)	1550 to 1915
regional, including full air conditioning	1205 to 1990
local, air conditioned	965 to 1365
conversion of existing warehouse to regional standard	940 to 1430
conversion of existing warehouse to local standard	815 to 1215
Learning resource centre	
economical	955 to 1185
high quality	1190 to 1790
Libraries	
branch	815 to 1090
city centre	1085 to 1280
collegiate; including fittings	2090 to 2500

Building Prices per Square Metre

Residential facilities (Cl/SfB 8)

Square metre excluding VAT £

Item	Low		High
*Local authority and housing association schemes			
Bungalows			
semi-detached	615	to	725
terraced	520	to	650
Two storey housing			
detached	570	to	695
semi-detached	505	to	615
terraced	450	to	555
Three storey housing			
semi-detached	535	to	735
terraced	470	to	700
Flats			
low rise excluding lifts	595	to	735
medium rise excluding lifts	610	to	825
Sheltered housing with wardens' accommodation	610	to	880
Terraced blocks of garages	395	to	460
Private developments			
Single detached houses	695	to	970
Houses - two or three storey	455	to	635
High-quality apartments			
three-to-four storey villa	955	to	1135
multi-storey	1550	to	1790
Prestige-quality apartments			
multi-storey	2030	to	2685
Flats			
standard	595	to	725
Warehouse conversions to apartments	695	to	1025
Rehabilitation			
housing	270	to	460
flats	415	to	675
Hotels (including fittings, furniture and equipment)			
luxury city-centre hotel - multi-storey conference and wet leisure facilities	1670	to	2030
business town centre/provincial hotel - 4-to-6 storeys, conference and wet leisure facilities	1190	to	1430
mid range provincial hotel - 2-to-3 storeys, conference and leisure facilities	1190	to	1490
mid-range provincial hotel - 2-to-3 storeys, bedroom extension	890	to	1315
city centre aparthotel - 4-to-7 storeys, apartments with self-catering facilities	955	to	1250
budget city-centre hotel - 4-to-6 storeys, dining and bar facilities	1010	to	1135
budget city-centre hotel (office conversion) - 4-to-6 storeys, excluding dining facilities	775	to	1250
budget roadside hotel - 2-to-3 storey lodge, excluding dining facilities	955	to	1075
Hotel accommodation facilities (excluding fittings, furniture and equipment)			
bedroom areas	650	to	930
front of house and reception	860	to	1135
restaurant areas	955	to	1430
bar areas	835	to	1315
function rooms/conference facilities	715	to	1315
dry leisure	650	to	775
wet leisure	1370	to	1790
*Students' residences			
large budget schemes with en-suite accommodation	775	to	1010
smaller schemes (40 - 100 units) with mid-range specifications, some with en-suite bathroom and kitchen facilities	1010	to	1315
smaller high-quality courtyard schemes, college style	1315	to	1915
Hostels	650	to	965

Common facilities, other facilities (Cl/SfB 9)

	Square metre excluding VAT £
Conference centres	1355 to 1850
Public conveniences	1500 to 2135

* Refer also to *"Cost Limits and Allowances"* in following section.

Approximate Estimates

(incorporating Comparative Prices)

Estimating by means of priced approximate quantities is always more accurate than by using overall prices per square metre. Prices given in this section, which is arranged in elemental order, are derived from "Prices for Measured Works - Major Works" section, but also include for all incidental items and labours which are normally measured separately in Bills of Quantities. As in other sections, they have been established with a tender price index level of 388 (1976 = 100). They include overheads and profit but do not include for preliminaries, details of which are given in Part II and which in the current tendering climate currently amount to 11% of the value of measured work or fees for professional services.

Whilst every effort has been made to ensure the accuracy of these figures, they have been prepared for approximate estimating purposes and on no account should be used for the preparation of tenders.

Unless otherwise described, units denoted as m^2 refer to appropriate area unit (rather than gross floor) areas.

As elsewhere in this edition, prices do not include Value Added Tax, which should be applied at the current rate to all non-domestic building.

Item nr.	SPECIFICATIONS		
1.0	**SUBSTRUCTURE**		
	ground floor plan area (unless otherwise described)		
	comprising:		
	Trench fill foundations		
	Strip foundations		
	Column bases		
	Pile caps		
	Strip or base foundations		
	Raft foundations		
	Piled foundations		
	Underpinning		
	Other foundations/Extras		
	Basements		
	Trench fill foundations		
	Machine excavation, disposal, plain in situ concrete 20.00 N/mm^2 - 20 mm aggregate (1:2:4) trench fill, 300 mm high brickwork in cement mortar (1:3), pitch polymer damp proof course	Unit	450 x 760 mm £
	With common bricks PC £120.00/1000 in		
1.0.1	half brick wall	m	37.30
1.0.2	one brick wall	m	43.40
1.0.3	cavity wall	m	44.90
	With engineering bricks PC £210.00/1000 in		
1.0.4	one brick wall	m	49.50
1.0.5	cavity wall	m	51.00
	With facing bricks PC £130.000 in		
1.0.6	one brick wall	m	47.90
1.0.7	cavity wall	m	49.50
	With facing bricks PC £275.00/1000 in		
1.0.8	one brick wall	m	59.30
1.0.9	cavity wall	m	60.90
	Strip foundations		
	Excavate trench, partial backfill, partial disposal, earthwork support (risk item), compact base of trench, plain insitu concrete 20.00 N/mm^2 - 20 mm aggregate (1:2:4) 250 mm thick, brickwork in cement mortar (1:3), pitch polymer damp proof course		
	A = Wall thickness	A	Half brick
	B = Concrete footing width	B	350 mm wide
	C = Brick bonding	C	Stretcher £
	Hand excavation, depth of wall		
1.0.10	600 mm deep	m	42.60
1.0.11	900 mm deep	m	50.20
1.0.12	1200 mm deep	m	65.40
1.0.13	1500 mm deep	m	77.60
	Machine excavation, depth of wall		
1.0.14	600 mm deep	m	38.80
1.0.15	900 mm deep	m	46.40
1.0.16	1200 mm deep	m	57.80
1.0.17	1500 mm deep	m	65.40
	Extra over for three courses of facing bricks		
1.0.18	PC £110.00/1000	m	-
1.0.19	PC £150.00/1000	m	-
1.0.20	PC £275.00/1000	m	-

Approximate Estimates

450 x 1000 mm £	600 x 760 mm £	600 x 1200 mm £	750 x 1200 mm £	750 x 1500 mm £	Item nr
44.90	44.90	65.40	81.40	101.20	1.0.1
52.50	52.50	74.60	89.00	108.80	1.0.2
54.00	54.00	76.10	89.00	108.80	1.0.3
57.10	57.10	77.60	92.80	108.80	1.0.4
59.30	59.30	81.40	92.80	116.40	1.0.5
55.50	57.10	77.60	92.80	108.80	1.0.6
57.10	60.10	81.40	92.80	116.40	1.0.7
68.50	68.50	89.00	105.00	124.00	1.0.8
70.00	70.00	92.80	105.00	124.00	1.0.9

COMMON BRICKS PC £100.00/1000		ENGINEERING BRICKS PC £175.00/1000		FACING BRICKS PC £110.00/1000		
One brick 600 mm wide English £	Cavity wall 750 mm wide Stretcher £	One brick 600 mm wide English £	Cavity wall 750 mm wide Stretcher £	One brick 600 mm wide English £	Cavity wall 750 mm wide Stretcher £	
66.20	65.20	73.80	77.60	70.00	71.90	1.0.10
81.40	97.40	92.80	108.80	83.70	99.70	1.0.11
104.99	124.00	124.00	140.00	110.40	129.30	1.0.12
124.00	147.60	147.60	170.40	140.00	162.80	1.0.13
57.80	66.20	66.20	77.60	66.20	73.80	1.0.14
77.60	85.20	91.30	101.20	85.20	91.30	1.0.15
92.80	108.80	109.60	124.00	108.80	124.00	1.0.16
108.80	127.80	131.60	147.60	124.00	136.90	1.0.17
0.15	0.23	-	-	-	-	1.0.18
0.76	1.37	-	-	-	-	1.0.19
2.28	4.64	1.37	2.66	-	-	1.0.20

Item nr.	SPECIFICATIONS	Unit	900 mm deep
1.0	**SUBSTRUCTURE - cont'd**		
	ground floor plan area (unless otherwise described)		
	Column bases		
	Excavate pit in firm ground, partial backfill, partial disposal, earthwork support, compact base of pit, plain insitu concrete 20.00 N/mm² - 20 mm aggregate (1:2:4), formwork		
	Hand excavation, base size		
1.0.21	600 mm x 600 mm x 300 mm	nr	47.20
1.0.22	900 mm x 900 mm x 450 mm	nr	86.70
1.0.23	1200 mm x 1200 mm x 450 mm	nr	143.00
1.0.24	1500 mm x 1500 mm x 600 mm	nr	253.30
	Machine excavation, base size		
1.0.25	600 mm x 600 mm x 300 mm	nr	39.60
1.0.26	900 mm x 900 mm x 450 mm	nr	79.10
1.0.27	1200 mm x 1200 mm x 450 mm	nr	119.90
1.0.28	1500 mm x 1500 mm x 600 mm	nr	222.10
	Excavate pit in firm ground by machine, partial backfill, partial disposal, earthwork support, compact base of pit, reinforced in situ concrete 20.00 N/mm² - 20 mm aggregate (1:2:4), formwork		
	Reinforcement at 50 kg/m³ concrete, base size		
1.0.29	1750 mm x 1750 mm x 500 mm	nr	253.00
1.0.30	2000 mm x 2000 mm x 500 mm	nr	316.00
1.0.31	2200 mm x 2200 mm x 600 mm	nr	475.00
1.0.32	2400 mm x 2400 mm x 600 mm	nr	539.00
	Reinforcement at 75 kg/m³ concrete, base size		
1.0.33	1750 mm x 1750 mm x 500 mm	nr	270.00
1.0.34	2000 mm x 2000 mm x 500 mm	nr	332.00
1.0.35	2200 mm x 2200 mm x 600 mm	nr	491.00
1.0.36	2400 mm x 2400 mm x 600 mm	nr	570.00
	Pile caps		
	Excavate pit in firm ground by machine, partial backfill partial disposal, earthwork support, compact base of pit, cut off top of pile and prepare reinforcement, reinforced insitu concrete 25.00 N/mm² - 20 mm aggregate (1:2:4), formwork		
	Reinforcement at 50 kg/m³ concrete, cap size		
1.0.37	900 mm x 900 mm x 1400 mm; one pile	nr	-
1.0.38	2700 mm x 900 mm x 1400 mm; two piles	nr	-
1.0.39	2700 mm x 1475 mm x 1400 mm; three piles*	nr	-
1.0.40	2700 mm x 2700 mm x 1400 mm; four piles	nr	-
1.0.41	3700 mm x 2700 mm x 1400 mm; six piles	nr	-
	Reinforcement at 75 kg/m³ concrete, cap size		
1.0.42	900 mm x 900 mm x 1400 mm; one pile	nr	-
1.0.43	2700 mm x 900 mm x 1400 mm; two piles	nr	-
1.0.44	2700 mm x 1475 mm x 1400 mm; three piles*	nr	-
1.0.45	2700 mm x 2700 mm x 1400 mm; four piles	nr	-
1.0.46	3700 mm x 2700 mm x 1400 mm; six piles	nr	-
	* = triangular on plan, overall dimensions given		
1.0.47	Additional cost of alternative strength concrete	m³	-

Approximate Estimates

1200 mm deep £	1500 mm deep £	DEPTH OF PIT 1800 mm deep £	2100 mm deep £	2500 mm deep £	Item nr
50.20	54.80	63.90	71.50	79.10	1.0.21
102.70	110.30	119.40	143.00	158.20	1.0.22
158.20	173.50	198.60	229.80	253.30	1.0.23
285.30	316.50	332.50	395.60	435.90	1.0.24
39.60	47.20	49.50	54.80	59.30	1.0.25
86.70	86.70	94.30	102.70	110.30	1.0.26
127.10	134.70	143.00	158.20	165.90	1.0.27
237.40	252.60	258.70	285.30	293.70	1.0.28
269.00	285.00	301.00	324.00	348.00	1.0.29
332.00	357.00	373.00	403.00	434.00	1.0.30
491.00	515.00	539.00	570.00	603.00	1.0.31
570.00	594.00	618.00	664.00	714.00	1.0.32
285.00	301.00	316.00	332.00	357.00	1.0.33
357.00	380.00	396.00	427.00	444.00	1.0.34
515.00	539.00	554.00	603.00	633.00	1.0.35
603.00	633.00	649.00	697.00	728.00	1.0.36

1500 mm deep £	1800 mm deep £	2100 mm deep £	2400 mm deep £		Item nr
253.00	256.00	269.00	276.00	-	1.0.37
633.00	628.00	649.00	664.00	-	1.0.38
1149.00	1153.00	1188.00	1226.00	-	1.0.39
1465.00	1506.00	1515.00	1583.00	-	1.0.40
1980.00	2018.00	2095.00	2134.00	-	1.0.41
253.00	259.00	268.00	284.00	-	1.0.42
633.00	644.00	672.00	695.00	-	1.0.43
1164.00	1203.00	1265.00	1304.00	-	1.0.44
1582.00	1613.00	1601.00	1661.00	-	1.0.45
2095.00	2142.00	2219.00	2250.00	-	1.0.46

30.00 N/mm² 20 mm aggregate		40.00 N/mm² 20 mm aggregate			
1.10	-	2.70	-	-	1.0.47

Approximate Estimates

Item nr.	SPECIFICATIONS	Unit	RESIDENTIAL £ range £
1.0	**SUBSTRUCTURE - cont'd**		
	ground floor plan area (unless otherwise described)		
	Strip or base foundations		
	Foundations in good ground; reinforced concrete bed: for one storey development		
1.0.48	shallow foundations	m²	- -
1.0.49	deep foundations	m²	- -
	Foundations in good ground; reinforced concrete bed: for two storey development		
1.0.50	shallow foundations	m²	- -
1.0.51	deep foundations	m²	- -
	Extra for		
1.0.52	each additional storey	m²	- -
	Raft foundations		
	Raft on poor ground for development		
1.0.53	one storey high	m²	- -
1.0.54	two storey high	m²	- -
	Extra for		
1.0.55	each additional storey	m²	- -
	Piled foundations		
	Foundations in poor ground; reinforced concrete slab and ground beams; for two storey residential development		
1.0.56	short bore piled	m²	135.00 - 181.00
1.0.57	fully piled	m²	166.00 - 207.00
	Foundations in poor ground; hollow ground floor; timber and boarding; for two storey residential		
1.0.58	short bore piled	m²	158.00 - 189.00
	Foundations in poor ground; reinforced concrete slab; for one storey commercial development		
1.0.59	short bore piles to columns only	m²	- -
1.0.60	short bore piles	m²	- -
1.0.61	fully piled	m²	- -
	Foundations in poor ground; reinforced concrete slab and ground beams; for two storey commercial development		
1.0.62	short bore piles	m²	- -
1.0.63	fully piled	m²	- -
	Extra for		
1.0.64	each additional storey	m²	- -
	Foundations in bad ground; inner city redevelopment; reinforced concrete slab and ground beams; for two storey commercial development		
1.0.65	fully piled	m²	- -
	Underpinning		
	In stages not exceeding 1500 mm long from one side of existing wall and foundation, excavate preliminary trench by machine and underpinning pit by hand, partial backfill, partial disposal, earthwork support (open boarded), cutting away projecting foundations, prepare underside of existing, compact base of pit, plain in situ concrete 20.00 N/mm² - 20 mm aggregate (1:2:4), formwork, brickwork in cement mortar (1:3), pitch polymer damp proof course, wedge and pin to underside of existing with slates		
	With common bricks PC £100.00/1000, depth of underpinning		
1.0.66	900 mm high, one brick wall	m	-
1.0.67	900 mm high, two brick wall	m	-
1.0.68	1200 mm high, one brick wall	m	-
1.0.69	1200 mm high, two brick wall	m	-
1.0.70	1500 mm high, one brick wall	m	-
1.0.71	1500 mm high, two brick wall	m	-
1.0.72	1800 mm high, one brick wall	m	-
1.0.73	1800 mm high, two brick wall	m	-

Approximate Estimates

INDUSTRIAL £ range £	RETAILING £ range £	LEISURE £ range £	OFFICES £ range £	HOTELS £ range £	Item nr.
47.20 - 63.50	54.80 - 79.10	55.20 - 79.10	-	-	1.0.48
79.10 - 94.70	86.70 - 111.10	79.10 - 94.70	-	-	1.0.49
55.20 - 86.70	63.50 - 94.70	55.90 - 94.70	63.50 - 94.70	63.50 - 94.70	1.0.50
86.70 - 135.00	94.70 - 135.00	94.70 - 135.00	94.70 - 135.00	94.70 - 135.00	1.0.51
19.00 - 22.10	16.00 - 20.50	16.00 - 20.50	16.00 - 20.50	16.00 - 20.50	1.0.52
63.50 - 110.30	71.50 - 110.30	79.10 - 125.50	-	-	1.0.53
102.30 - 150.60	117.90 - 166.20	117.90 - 166.20	117.90 - 162.80	117.90 - 162.80	1.0.54
19.00 - 22.10	16.00 - 20.50	16.00 - 20.50	16.00 - 20.50	16.00 - 20.50	1.0.55
-	-	-	-	-	1.0.56
-	-	-	-	-	1.0.57
-	-	-	-	-	1.0.58
79.10 - 119.40	86.70 - 119.40	79.10 - 119.40	-	-	1.0.59
102.30 - 135.00	102.70 - 142.60	102.30 - 142.60	102.30 - 142.60	102.30 - 142.60	1.0.60
135.00 - 189.40	150.60 - 206.20	150.60 - 206.20	150.60 - 206.20	150.60 - 206.20	1.0.61
-	119.40 - 150.60	119.40 - 149.10	119.40 - 150.60	119.40 - 150.60	1.0.62
-	189.40 - 260.60	189.40 - 260.60	189.40 - 260.60	189.40 - 260.60	1.0.63
19.00 - 22.10	11.00 - 12.60	11.00 - 12.60	11.00 - 12.60	11.00 - 12.60	1.0.64
-	221.80 - 324.50	-	221.80 - 380.40	221.80 - 324.40	1.0.65

EXCAVATION COMMENCING AT

1.00 m below ground level	2.00 m below ground level	3.00 m below ground level	4.00 m below ground level		
233.00	274.00	323.00	351.00	-	1.0.66
277.00	318.00	368.00	396.00	-	1.0.67
291.00	336.00	390.00	417.00	-	1.0.68
350.00	396.00	449.00	476.00	-	1.0.69
333.00	385.00	441.00	469.00	-	1.0.70
405.00	455.00	511.00	539.00	-	1.0.71
377.00	429.00	493.00	522.00	-	1.0.72
461.00	516.00	575.00	604.00	-	1.0.73

Item nr.	SPECIFICATIONS	Unit		
1.0	**SUBSTRUCTURE - cont'd**			
	ground floor plan area (unless otherwise described)			
	Underpinning - cont'd			
	With engineering bricks PC £210/1000, depth of underpinning			
1.0.74	900 mm high; one brick wall	m	-	
1.0.75	900 mm high; two brick wall	m	-	
1.0.76	1200 mm high; one brick wall	m	-	
1.0.77	1200 mm high; two brick wall	m	-	
1.0.78	1500 mm high; one brick wall	m	-	
1.0.79	1500 mm high; two brick wall	m	-	
1.0.80	1800 mm high; one brick wall	m	-	
1.0.81	1800 mm high; two brick wall	m	-	
	Ground slabs			
	Mechanical excavation to reduce levels, disposal, level and compact, hardcore bed blinded with sand, 1200 gauge polythene damp proof membrane concrete 20.00 N/mm^2 - 20 mm aggregate (1:2:4) ground slab, tamped finish		150mm £	
	Plain insitu concrete ground slab, thickness of hardcore bed			
1.0.82	150 mm thick	m^2	30.30	
1.0.83	175 mm thick	m^2	30.40	
1.0.84	200 mm thick	m^2	31.80	
	Add to the foregoing prices for			
	Fabric reinforcement B.S. 4483, lapped			
1.0.85	A142 (2.22kg/m^2) 1 layer	m^2	3.20	
1.0.86	2 layers	m^2	4.70	
1.0.87	A193 (3.02kg/m^2) 1 layer	m^2	3.20	
1.0.88	2 layers	m^2	5.50	
1.0.89	A252 (3.95kg/m^2) 1 layer	m^2	3.20	
1.0.90	2 layers	m^2	6.40	
1.0.91	A393 (6.16kg/m^2) 1 layer	m^2	4.70	
1.0.92	2 layers	m^2	9.40	
	High yield steel bar reinforcement B.S. 4449 straight or bent, at a rate of			
1.0.93	25 kg/m^3	m^2	3.20	
1.0.94	50 kg/m^3	m^2	5.50	
1.0.95	75 kg/m^3	m^2	7.90	
1.0.96	100 kg/m^3	m^2	10.30	
	Alternative concrete mixes in lieu of 20.00 N/mm^2 20 mm aggregate (1:2:4)			
1.0.97	25.00 N/mm^2	m^2	0.30	
1.0.98	30.00 N/mm^2	m^2	0.43	
1.0.99	40.00 N/mm^2	m^2	0.73	
			RESIDENTIAL £ range £	
	Other foundations/alternative slabs/extras			
1.0.100	Cantilevered foundations in good ground; reinforced concrete slab for two storey commercial development	m^2	-	-
1.0.101	Underpinning foundations of existing buildings abutting site	m	-	-
	Extra to substructure rates for			
1.0.102	watertight pool construction	m^2	-	-
1.0.103	ice pad	m^2	-	-
	Reinforced concrete bed including excavation and hardcore under			
1.0.104	150 mm thick	m^2	20.50 -	28.50
1.0.105	200 mm thick	m^2	28.50 -	36.50
1.0.106	300 mm thick	m^2	-	-

Approximate Estimates

EXCAVATION COMMENCING AT

1.00 m below ground level	2.00 m below ground level	3.00 m below ground level	4.00 m below ground level		Item nr.
243.00	286.00	335.00	364.00	-	1.0.74
300.00	341.00	390.00	417.00	-	1.0.75
306.00	353.00	405.00	434.00	-	1.0.76
377.00	423.00	478.00	507.00	-	1.0.77
355.00	402.00	460.00	487.00	-	1.0.78
441.00	505.00	548.00	575.00	-	1.0.79
400.00	455.00	516.00	543.00	-	1.0.80
505.00	558.00	619.00	647.00	-	1.0.81

THICKNESS OF CONCRETE SLAB

200 mm thick £	250 mm thick £	300 mm thick £	375 mm thick £	450 mm thick £	
32.00	36.50	38.00	44.10	49.50	1.0.82
33.50	36.50	39.60	45.60	51.00	1.0.83
33.50	38.00	39.60	47.20	51.00	1.0.84
3.20	3.20	3.20	3.20	3.20	1.0.85
4.70	4.70	5.50	5.50	5.50	1.0.86
3.20	3.20	3.20	3.20	4.00	1.0.87
5.50	6.40	6.40	6.40	6.40	1.0.88
4.00	4.00	4.00	4.70	4.70	1.0.89
6.40	7.20	7.20	7.20	7.90	1.0.90
4.70	4.70	4.70	5.50	5.50	1.0.91
9.40	9.40	9.40	9.40	9.40	1.0.92
4.00	4.70	6.40	7.90	8.70	1.0.93
7.20	9.40	11.10	14.30	16.60	1.0.94
11.10	13.50	16.00	20.50	23.70	1.0.95
14.30	17.50	20.50	26.80	31.80	1.0.96
0.40	0.50	0.59	0.76	0.90	1.0.97
0.56	0.70	0.84	1.05	1.25	1.0.98
0.99	1.22	1.46	1.83	2.19	1.0.99

INDUSTRIAL £ range £	RETAILING £ range £	LEISURE £ range £	OFFICES £ range £	HOTELS £ range £	
- -	268.00 - 350.00	268.00 - 350.00	268.00 - 350.00	268.00 - 350.00	1.0.100
- -	633.00 - 869.00	633.00 - 869.00	633.00 - 869.00	633.00 - 869.00	1.0.101
- -	- -	102.00 - 151.00	- -	- -	1.0.102
- -	- -	120.00 - 181.00	- -	- -	1.0.103
28.50 - 36.50	28.50 - 36.50	28.50 - 36.50	28.50 - 36.50	28.50 - 36.50	1.0.104
38.00 - 49.10	35.00 - 45.60	35.00 - 42.60	35.00 - 42.60	35.00 - 42.60	1.0.105
47.20 - 66.60	42.60 - 60.10	42.60 - 54.00	42.60 - 54.00	42.60 - 54.00	1.0.106

Item nr.	SPECIFICATIONS	Unit	RESIDENTIAL £ range £
1.0	**SUBSTRUCTURE - cont'd**		
	ground floor plan area (unless otherwise described)		
	Other foundations/alternative slabs/extras - cont'd		
	Hollow ground floor with timber and boarding, including excavation, concrete and hardcore under		
1.0.107	300 mm deep	m^2	34.20 - 43.40
	Extra for		
1.0.108	sound reducing quilt in screed	m^2	3.00 - 4.20
1.0.109	50 mm insulation under slab and at edges	m^2	3.00 - 5.30
1.0.110	75 mm insulation under slab and at edges	m^2	- -
1.0.111	suspended precast concrete slabs in lieu of in situ slab	m^2	5.30 - 8.00
	Basement (excluding bulk excavation costs)		
	basement floor/wall area (as appropriate)		
	Reinforced concrete basement floors		
1.0.112	non-waterproofed	m^2	- -
1.0.113	waterproofed	m^2	- -
	Reinforced concrete basement walls		
1.0.114	non-waterproofed	m^2	- -
1.0.115	waterproofed	m^2	- -
1.0.116	sheet piled	m^2	- -
1.0.117	diaphragm walling	m^2	- -
	Extra for		
1.0.118	each additional basement level	%	- -
2.0	**SUPERSTRUCTURE**		
2.1	**FRAME AND UPPER FLOORS**		
	Roof plan area (unless otherwise described)		
	comprising:		
	Reinforced concrete frame		
	Steel frame		
	Other frames/extras		
	Reinforced Concrete Frame		$20N/mm^2$ - 20mm Formwork Basic £
	Reinforced insitu concrete column, bar reinforcement, formwork (assumed four uses)		
	Reinforcement rate 180 kg /m^3, column size		
2.1.1	225 mm x 225 mm	m	37.00
2.1.2	300 mm x 300 mm	m	51.00
2.1.3	300 mm x 450 mm	m	72.00
2.1.4	300 mm x 600 mm	m	88.00
2.1.5	450 mm x 450 mm	m	91.00
2.1.6	450 mm x 600 mm	m	113.00
2.1.7	450 mm x 900 mm	m	157.00
	Reinforcement rate 200 kg /m^3, column size		
2.1.8	225 mm x 225 mm	m	38.00
2.1.9	300 mm x 300 mm	m	53.00
2.1.10	300 mm x 450 mm	m	72.00
2.1.11	300 mm x 600 mm	m	90.00
2.1.12	450 mm x 450 mm	m	93.00
2.1.13	450 mm x 600 mm	m	115.00
2.1.14	450 mm x 900 mm	m	160.00

Approximate Estimates

INDUSTRIAL £ range £	RETAILING £ range £	LEISURE £ range £	OFFICES £ range £	HOTELS £ range £	Item nr.
- -	- -	- -	- -	- -	1.0.107
3.00 - 5.30	3.00 - 5.30	3.00 - 5.30	3.00 - 5.30	3.00 - 5.30	1.0.108
4.90 - 7.20	4.90 - 7.20	4.90 - 7.20	4.90 - 7.20	4.90 - 7.20	1.0.109
6.10 - 9.10	6.10 - 9.10	6.10 - 9.10	6.10 - 9.10	6.10 - 9.10	1.0.110
12.90 - 16.40	10.70 - 12.90	10.70 - 12.90	10.70 - 12.90	10.70 - 12.90	1.0.111
- -	50.00 - 67.00	50.00 - 67.00	50.00 - 67.00	50.00 - 67.00	1.0.112
- -	67.00 - 91.00	67.00 - 91.00	67.00 - 91.00	67.00 - 91.00	1.0.113
- -	149.00 - 190.00	149.00 - 190.00	149.00 - 190.00	149.00 - 190.00	1.0.114
- -	173.00 - 216.00	173.00 - 216.00	173.00 - 216.00	173.00 - 216.00	1.0.115
- -	307.00 - 374.00	307.00 - 374.00	307.00 - 374.00	307.00 - 374.00	1.0.116
- -	339.00 - 399.00	339.00 - 399.00	339.00 - 399.00	339.00 - 399.00	1.0.117
- -	+ 20%	+ 20%	+ 20%	+ 20%	1.0.118

ALTERNATIVE MIXES AND FINISHES

aggregate (1:2:4), finish	30.00 N/mm² - 20 mm aggregate (1:1:2), Formwork finish		40.00 N/mm² - 20 mm aggregate Formwork finish		
Smooth £	Basic £	Smooth £	Basic £	Smooth £	
40.00	37.00	40.00	38.00	42.00	2.1.1
56.00	52.00	56.00	55.00	60.00	2.1.2
76.00	72.00	76.00	75.00	79.00	2.1.3
93.00	88.00	93.00	93.00	98.00	2.1.4
96.00	93.00	98.00	94.00	101.00	2.1.5
119.00	113.00	121.00	119.00	126.00	2.1.6
166.00	158.00	167.00	166.00	173.00	2.1.7
42.00	38.00	42.00	38.00	42.00	2.1.8
59.00	52.00	59.00	55.00	60.00	2.1.9
78.00	72.00	78.00	75.00	79.00	2.1.10
94.00	90.00	94.00	93.00	100.00	2.1.11
98.00	93.00	100.00	96.00	103.00	2.1.12
122.00	116.00	122.00	121.00	129.00	2.1.13
169.00	161.00	170.00	169.00	177.00	2.1.14

Item nr.	SPECIFICATIONS	Unit	20N/mm² -20mm Formwork Basic £
2.1	**FRAME AND UPPER FLOORS - cont'd** roof plan area (unless otherwise described)		
	Reinforced Concrete Frame - cont'd		
	Reinforcement rate 220 kg /m³, column size		
2.1.15	225 mm x 225 mm	m	38.00
2.1.16	300 mm x 300 mm	m	53.00
2.1.17	300 mm x 450 mm	m	73.00
2.1.18	300 mm x 600 mm	m	90.00
2.1.19	450 mm x 450 mm	m	93.00
2.1.20	450 mm x 600 mm	m	116.00
2.1.21	450 mm x 900 mm	m	163.00
	Reinforcement rate 240 kg /m³, column size		
2.1.22	225 mm x 225 mm	m	38.00
2.1.23	300 mm x 300 mm	m	55.00
2.1.24	300 mm x 450 mm	m	75.00
2.1.25	300 mm x 600 mm	m	93.00
2.1.26	450 mm x 450 mm	m	94.00
2.1.27	450 mm x 600 mm	m	119.00
2.1.28	450 mm x 900 mm	m	166.00
	In situ concrete casing to steel column, formwork (assumed four uses), column size		
2.1.29	225 mm x 225 mm	m	35.00
2.1.30	300 mm x 300 mm	m	48.00
2.1.31	300 mm x 450 mm	m	64.00
2.1.32	300 mm x 600 mm	m	76.00
2.1.33	450 mm x 450 mm	m	79.00
2.1.34	450 mm x 600 mm	m	96.00
2.1.35	450 mm x 900 mm	m	133.00
	Reinforced in situ concrete isolated beams, bar reinforcement, formwork (assumed four uses)		
	Reinforcement rate 200 kg /m³, beam size		
2.1.36	225 mm x 450 mm	m	64.00
2.1.37	225 mm x 600 mm	m	81.00
2.1.38	300 mm x 600 mm	m	93.00
2.1.39	300 mm x 900 mm	m	129.00
2.1.40	300 mm x 1200 mm	m	166.00
2.1.41	450 mm x 600 mm	m	119.00
2.1.42	450 mm x 900 mm	m	164.00
2.1.43	450 mm x 1200 mm	m	208.00
2.1.44	600 mm x 600 mm	m	146.00
2.1.45	600 mm x 900 mm	m	197.00
2.1.46	600 mm x 1200 mm	m	251.00
	Reinforcement rate 220 kg /m³, beam size		
2.1.47	225 mm x 450 mm	m	65.00
2.1.48	225 mm x 600 mm	m	81.00
2.1.49	300 mm x 600 mm	m	93.00
2.1.50	300 mm x 900 mm	m	132.00
2.1.51	300 mm x 1200 mm	m	167.00
2.1.52	450 mm x 600 mm	m	121.00
2.1.53	450 mm x 900 mm	m	167.00
2.1.54	450 mm x 1200 mm	m	212.00
2.1.55	600 mm x 600 mm	m	148.00
2.1.56	600 mm x 900 mm	m	200.00
2.1.57	600 mm x 1200 mm	m	256.00
	Reinforcement rate 240 kg /m³, beam size		
2.1.58	225 mm x 450 mm	m	67.00
2.1.59	225 mm x 600 mm	m	84.00
2.1.60	300 mm x 600 mm	m	94.00
2.1.61	300 mm x 900 mm	m	133.00
2.1.62	300 mm x 1200 mm	m	170.00
2.1.63	450 mm x 600 mm	m	122.00
2.1.64	450 mm x 900 mm	m	170.00
2.1.65	450 mm x 1200 mm	m	216.00

Approximate Estimates

ALTERNATIVE MIXES AND FINISHES

aggregate (1:2:4) finish	30.00 N/mm² - 20 mm aggregate (1:1:2), Formwork finish		40.00 N/mm² - 20 mm aggregate Formwork finish		Item nr.
Smooth £	Basic £	Smooth £	Basic £	Smooth £	
42.00	38.00	42.00	40.00	43.00	2.1.15
59.00	55.00	59.00	56.00	60.00	2.1.16
78.00	73.00	78.00	76.00	81.00	2.1.17
96.00	91.00	96.00	94.00	101.00	2.1.18
101.00	94.00	101.00	98.00	104.00	2.1.19
124.00	118.00	126.00	124.00	132.00	2.1.20
172.00	164.00	173.00	172.00	182.00	2.1.21
42.00	38.00	42.00	40.00	43.00	2.1.22
60.00	55.00	60.00	56.00	62.00	2.1.23
79.00	75.00	79.00	78.00	84.00	2.1.24
98.00	93.00	98.00	96.00	101.00	2.1.25
101.00	96.00	103.00	101.00	109.00	2.1.26
126.00	119.00	127.00	126.00	133.00	2.1.27
173.00	169.00	176.00	173.00	186.00	2.1.28
38.00	35.00	38.00	37.00	40.00	2.1.29
51.00	49.00	51.00	51.00	55.00	2.1.30
70.00	65.00	70.00	67.00	72.00	2.1.31
84.00	78.00	85.00	81.00	88.00	2.1.32
87.00	79.00	87.00	85.00	91.00	2.1.33
105.00	98.00	104.00	103.00	112.00	2.1.34
142.00	133.00	143.00	142.00	151.00	2.1.35
68.00	65.00	70.00	67.00	72.00	2.1.36
87.00	81.00	88.00	85.00	91.00	2.1.37
98.00	93.00	100.00	96.00	103.00	2.1.38
137.00	132.00	138.00	135.00	143.00	2.1.39
173.00	167.00	176.00	173.00	185.00	2.1.40
126.00	119.00	127.00	126.00	133.00	2.1.41
173.00	166.00	173.00	173.00	183.00	2.1.42
219.00	211.00	221.00	219.00	233.00	2.1.43
155.00	148.00	155.00	155.00	161.00	2.1.44
210.00	199.00	211.00	211.00	221.00	2.1.45
262.00	255.00	266.00	267.00	280.00	2.1.46
70.00	65.00	70.00	67.00	72.00	2.1.47
88.00	84.00	90.00	87.00	91.00	2.1.48
99.00	94.00	100.00	97.00	103.00	2.1.49
140.00	134.00	142.00	137.00	146.00	2.1.50
179.00	169.00	180.00	175.00	186.00	2.1.51
128.00	122.00	129.00	128.00	134.00	2.1.52
175.00	169.00	178.00	175.00	186.00	2.1.53
222.00	215.00	227.00	225.00	236.00	2.1.54
157.00	151.00	158.00	157.00	164.00	2.1.55
213.00	205.00	216.00	216.00	224.00	2.1.56
268.00	259.00	272.00	275.00	285.00	2.1.57
72.00	67.00	72.00	68.00	72.00	2.1.58
90.00	85.00	90.00	87.00	93.00	2.1.59
100.00	96.00	102.00	99.00	105.00	2.1.60
142.00	135.00	143.00	140.00	148.00	2.1.61
181.00	172.00	183.00	180.00	189.00	2.1.62
132.00	124.00	134.00	129.00	137.00	2.1.63
180.00	172.00	181.00	180.00	189.00	2.1.64
228.00	218.00	231.00	230.00	239.00	2.1.65

Item nr.	SPECIFICATIONS	Unit	20N/mm² -20mm Formwork Basic £
2.1	**FRAME AND UPPER FLOORS - cont'd**		
	roof plan area (unless otherwise described)		
	Reinforced Concrete Frame - cont'd		
	Reinforcement rate 240 kg /m³, beam size - cont'd		
2.1.66	600 mm x 600 mm	m	151.00
2.1.67	600 mm x 900 mm	m	208.00
2.1.68	600 mm x 1200 mm	m	262.00
	In situ concrete casing to steel attached beams, formwork (assumed four uses), beam size		
2.1.69	225 mm x 450 mm	m	59.00
2.1.70	225 mm x 600 mm	m	75.00
2.1.71	300 mm x 600 mm	m	85.00
2.1.72	300 mm x 900 mm	m	116.00
2.1.73	300 mm x 1200 mm	m	146.00
2.1.74	450 mm x 600 mm	m	103.00
2.1.75	450 mm x 900 mm	m	142.00
2.1.76	450 mm x 1200 mm	m	177.00
2.1.77	600 mm x 600 mm	m	126.00
2.1.78	600 mm x 900 mm	m	166.00
2.1.79	600 mm x 1200 mm	m	208.00
	Steel		
	Fabricated steelwork B.S. 4360 grade 40 erected on site with bolted connections		
2.1.80	universal beams	tonne	-
2.1.81	rectangular section beams	tonne	-
2.1.82	composite beams	tonne	-
2.1.83	lattice beams	tonne	-
2.1.84	rectangular section columns	tonne	-
2.1.85	composite columns	tonne	-
2.1.86	roof trusses	tonne	-
2.1.87	smaller sections	tonne	-

Item nr.	SPECIFICATIONS	Unit	RESIDENTIAL £ range £	
	Other frames/extras			
	Space deck on steel frame			
2.1.88	unprotected	m²	-	-
2.1.89	Exposed steel frame for tent/mast structures	m²	-	-
	Columns and beams to 18.00 m high bay warehouse			
2.1.90	unprotected	m²	-	-
	Columns and beams to mansard			
2.1.91	protected	m²	-	-
	Feature columns and beams to glazed atrium roof			
2.1.92	unprotected	m²	-	-

ALTERNATIVE MIXES AND FINISHES

aggregate (1:2:4) finish Smooth £	30.00 N/mm² - 20 mm aggregate (1:1:2), Formwork finish Basic £	Smooth £	40.00 N/mm² - 20 mm aggregate Formwork finish Basic £	Smooth £	Item nr.
160.00	152.00	161.00	160.00	167.00	2.1.66
216.00	210.00	219.00	219.00	230.00	2.1.67
275.00	263.00	278.00	278.00	291.00	2.1.68
64.00	59.00	65.00	62.00	67.00	2.1.69
79.00	75.00	81.00	78.00	85.00	2.1.70
91.00	85.00	91.00	88.00	94.00	2.1.71
125.00	117.00	125.00	122.00	132.00	2.1.72
157.00	148.00	158.00	155.00	164.00	2.1.73
111.00	103.00	113.00	111.00	119.00	2.1.74
151.00	143.00	152.00	151.00	161.00	2.1.75
187.00	180.00	190.00	189.00	201.00	2.1.76
135.00	128.00	137.00	135.00	142.00	2.1.77
177.00	169.00	180.00	180.00	189.00	2.1.78
219.00	211.00	222.00	224.00	237.00	2.1.79

STEEL FRAME £	With galvanising £				Item nr.
1585.00	1776.00	-	-	-	2.1.80
2051.00	2274.00	-	-	-	2.1.81
1951.00	2173.00	-	-	-	2.1.82
1850.00	2072.00	-	-	-	2.1.83
2110.00	2333.00	-	-	-	2.1.84
1908.00	2142.00	-	-	-	2.1.85
1908.00	2142.00	-	-	-	2.1.86
1452.00	1680.00	-	-	-	2.1.87

INDUSTRIAL £ range £	RETAILING £ range £	LEISURE £ range £	OFFICES £ range £	HOTELS £ range £	Item nr.
119.00 - 237.00	128.00 - 237.00	128.00 - 237.00	128.00 - 237.00	- -	2.1.88
204.00 - 304.00	204.00 - 304.00	204.00 - 304.00	-	-	2.1.89
128.00 - 177.00	-	-	-	-	2.1.90
-	85.00 - 118.00	88.00 - 128.00	88.00 - 128.00	88.00 - 128.00	2.1.91
-	-	-	97.00 - 152.00	-	2.1.92

Item nr.	SPECIFICATIONS	Unit	RESIDENTIAL £ range £
2.2	**FRAME & UPPER FLOORS (COMBINED)** **upper floor area** (unless otherwise described)		
	comprising:		
	Softwood floors; no frame		
	Softwood floors; steel frame		
	Reinforced concrete floors; no frame		
	Reinforced concrete floors and frame		
	Reinforced concrete floors; steel frame		
	Precast concrete floors; no frame		
	Precast concrete floors; reinforced concrete frame		
	Precast concrete floors and frame		
	Precast concrete floors; steel frame		
	Other floor and frame constructions/extras		
	Softwood floors; no frame Joisted floor; supported on layers; 22 mm thick chipboard t&g flooring; herringbone strutting; no coverings or finishes		
2.2.1	150 mm x 50 mm thick joists	m²	22.00 - 26.00
2.2.2	175 mm x 50 mm thick joists	m²	25.00 - 30.00
2.2.3	200 mm x 50 mm thick joists	m²	26.00 - 30.00
2.2.4	225 mm x 50 mm thick joists	m²	28.00 - 31.00
2.2.5	250 mm x 50 mm thick joists	m²	30.00 - 33.00
2.2.6	275 mm x 50 mm thick joists	m²	33.00 - 37.00
2.2.7	Joisted floor; average depth; plasterboard; skim; emulsion; vinyl flooring and painted softwood skirtings	m²	52.00 - 67.00
	Softwood construction; steel frame		
2.2.8	Joisted floor; average depth; plasterboard; skim; emulsion; vinyl flooring and painted softwood skirtings	m²	- -
	Reinforced concrete floors; no frame Suspended slab; no coverings or finishes		
2.2.9	3.65 m span; 3.00 kN/m² loading	m²	45.00 - 46.00
2.2.10	4.25 m span; 3.00 kN/m² loading	m²	51.00 - 55.00
2.2.11	2.75 m span; 8.00 kN/m² loading	m²	- -
2.2.12	3.35 m span; 8.00 kN/m² loading	m²	- -
2.2.13	4.25 m span; 8.00 kN/m² loading	m²	- -
	Suspended slab; no coverings or finishes		
2.2.14	150 mm thick	m²	51.00 - 65.00
2.2.15	225 mm thick	m²	76.00 - 91.00
	Reinforced concrete floors and frame Suspended slab; no coverings or finishes		
2.2.16	up to six storeys	m²	- -
2.2.17	seven to twelve storeys	m²	- -
2.2.18	thirteen to eighteen storeys	m²	- -
	Extra for		
2.2.19	section 20 fire regulations	m²	- -
	Wide span suspended slab		
2.2.20	up to six storeys	m²	- -
	Reinforced concrete floors; steel frame Suspended slab; average depth; "Holorib" permanent steel shuttering; protected steel frame; no coverings or finishes		
2.2.21	up to six storeys	m²	- -
	Extra for		
2.2.22	spans 7.50 m to 15.00 m	m²	- -
2.2.23	seven to twelve storeys	m²	- -
	Extra for		
2.2.24	section 20 fire regulations	m²	- -

Approximate Estimates

INDUSTRIAL £ range £	RETAILING £ range £	LEISURE £ range £	OFFICES £ range £	HOTELS £ range £	Item nr.
- -	- -	- -	- -	- -	2.2.1
- -	- -	- -	- -	- -	2.2.2
- -	- -	- -	- -	- -	2.2.3
- -	- -	- -	- -	- -	2.2.4
- -	- -	- -	- -	- -	2.2.5
- -	- -	- -	- -	- -	2.2.6
- -	- -	- -	- -	- -	2.2.7
84.00 - 91.00	91.00 - 125.00	91.00 - 125.00	91.00 - 125.00	91.00 - 125.00	2.2.8
- -	- -	43.00 - 51.00	43.00 - 51.00	43.00 - 51.00	2.2.9
- -	- -	51.00 - 54.00	51.00 - 62.00	51.00 - 62.00	2.2.10
43.00 - 51.00	43.00 - 51.00	43.00 - 51.00	43.00 - 51.00	43.00 - 51.00	2.2.11
49.00 - 58.00	49.00 - 58.00	49.00 - 58.00	49.00 - 58.00	49.00 - 58.00	2.2.12
64.00 - 72.00	64.00 - 72.00	64.00 - 72.00	64.00 - 75.00	64.00 - 75.00	2.2.13
58.00 - 84.00	49.00 - 84.00	49.00 - 76.00	49.00 - 73.00	49.00 - 84.00	2.2.14
91.00 - 103.00	76.00 - 103.00	76.00 - 96.00	76.00 - 96.00	76.00 - 103.00	2.2.15
111.00 - 149.00	91.00 - 134.00	96.00 - 142.00	91.00 - 117.00	91.00 - 125.00	2.2.16
- -	- -	- -	117.00 - 183.00	- -	2.2.17
- -	- -	- -	183.00 - 233.00	- -	2.2.18
- -	- -	- -	8.00 - 12.00	- -	2.2.19
125.00 - 154.00	99.00 - 146.00	111.00 - 149.00	99.00 - 134.00	99.00 - 142.00	2.2.20
149.00 - 183.00	125.00 - 173.00	125.00 - 166.00	125.00 - 154.00	125.00 - 161.00	2.2.21
17.00 - 46.00	22.00 - 51.00	17.00 - 46.00	17.00 - 46.00	17.00 - 46.00	2.2.22
- -	- -	- -	142.00 - 216.00	- -	2.2.23
- -	- -	- -	17.00 - 22.00	- -	2.2.24

Item nr.	SPECIFICATIONS	Unit	RESIDENTIAL £ range £	
2.2	**FRAME & UPPER FLOORS (COMBINED) - cont'd** **upper floor area** (unless otherwise described)			
	Reinforced concrete floors; steel frame - cont'd Suspended slab; average depth; protected steel frame; no coverings or finishes			
2.2.25	up to six storeys	m²	-	-
2.2.26	seven to twelve storeys	m²	-	-
	Extra for			
2.2.27	section 20 fire regulations	m²	-	-
	Precast concrete floors; no frame Suspended slab; 75 mm thick screed; no coverings or finishes			
2.2.28	3.00 m span; 5.00 kN/m² loading	m²	37.00 -	46.00
2.2.29	6.00 m span; 5.00 kN/m² loading	m²	37.00 -	48.00
2.2.30	7.50 m span; 8.50 kN/m² loading	m²	40.00 -	50.00
2.2.31	3.00 m span; 8.50 kN/m² loading	m²	-	-
2.2.32	6.00 m span; 8.50 kN/m² loading	m²	-	-
2.2.33	7.50 m span; 8.50 kN/m² loading	m²	-	-
2.2.34	3.00 m span; 12.50 kN/m² loading	m²	-	-
2.2.35	6.00 m span; 12.50 kN/m² loading	m²	-	-
2.2.36	Suspended slab; average depth; no coverings or finishes	m²	33.00 -	54.00
	Precast concrete floors; reinforced concrete frame			
2.2.37	Suspended slab; average depth; no coverings or finishes	m²	-	-
	Precast concrete floors and frame			
2.2.38	Suspended slab; average depth; no coverings or finishes	m²	-	-
	Precast concrete floors; steel frame Suspended slabs; average depth; unprotected steel frame; no coverings or finishes			
2.2.39	up to three storeys	m²	-	-
	Suspended slabs; average depth; protected steel frame; no coverings or finishes			
2.2.40	up to six storeys	m²	-	-
2.2.41	seven to twelve storeys	m²	-	-
	Other floor and frame construction/extras			
2.2.42	Reinforced concrete cantilevered balcony	nr	1560 -	2092
2.2.43	Reinforced concrete cantilevered walkways	m²	-	-
2.2.44	Reinforced concrete walkways and supporting frame	m²	-	-
	Reinforced concrete core with steel umbrella frame			
2.2.45	twelve to twenty four storeys	m²	-	-
	Extra for			
2.2.46	wrought formwork	m²	3.10 -	3.90
2.2.47	sound reducing quilt in screed	m²	3.50 -	4.60
2.2.48	insulation to avoid cold bridging	m²	3.50 -	5.90

Approximate Estimates 729

INDUSTRIAL £ range £	RETAILING £ range £	LEISURE £ range £	OFFICES £ range £	HOTELS £ range £	Item nr.
142.00 - 190.00	125.00 - 183.00	142.00 - 183.00	125.00 - 166.00	125.00 - 166.00	2.2.25
-	-	-	142.00 - 233.00	-	2.2.26
-	-	-	17.00 - 21.00		2.2.27
-	45.00 - 51.00	45.00 - 51.00	45.00 - 51.00	45.00 - 51.00	2.2.28
-	45.00 - 53.00	45.00 - 53.00	45.00 - 51.00	49.00 - 53.00	2.2.29
	46.00 - 58.00	46.00 - 58.00	46.00 - 51.00	53.00 - 58.00	2.2.30
46.00 - 53.00	-	-	-	-	2.2.31
49.00 - 56.00	-	-	-	-	2.2.32
51.00 - 58.00	-	-	-	-	2.2.33
49.00 - 55.00	-	-	-	-	2.2.34
55.00 - 61.00	-	-	-	-	2.2.35
45.00 - 65.00	-	-	-	-	2.2.36
73.00 - 96.00	65.00 - 103.00	65.00 - 96.00	65.00 - 93.00	65.00 - 103.00	2.2.37
73.00 - 138.00	65.00 - 142.00	65.00 - 138.00	65.00 - 138.00	65.00 - 158.00	2.2.38
81.00 - 146.00	76.00 - 132.00	76.00 - 123.00	65.00 - 123.00	76.00 - 132.00	2.2.39
132.00 - 169.00	111.00 - 158.00	123.00 - 158.00	111.00 - 149.00	111.00 - 149.00	2.2.40
-	-	-	142.00 - 216.00	-	2.2.41
-	-	-	-	-	2.2.42
-	114.00 - 142.00	-	-	-	2.2.43
-	129.00 - 158.00	-	-	-	2.2.44
-	-	-	265.00 - 358.00	-	2.2.45
3.10 - 3.70	3.10 - 6.60	3.10 - 6.60	3.10 - 6.60	3.10 - 8.10	2.2.46
3.30 - 5.90	3.30 - 5.90	3.30 - 5.90	3.30 - 5.90	3.30 - 5.90	2.2.47
5.90 - 8.10	5.90 - 8.10	5.90 - 8.10	5.90 - 8.10	5.90 - 8.10	2.2.48

Approximate Estimates

Item nr.	SPECIFICATIONS	Unit	RESIDENTIAL £ range £
2.3	**ROOF**		
	roof plan area (Unless otherwise described)		
	comprising:		
	Softwood flat roofs		
	Softwood trussed pitched roofs		
	Steel trussed pitch roofs		
	Concrete flat roofs		
	Flat roof decking and finishes		
	Roof claddings		
	Rooflights/patent glazing and glazed roofs		
	Comparative over/underlays		
	Comparative tiling and slating finishes/perimeter treatments		
	Comparative cladding finishes/perimeter treatments		
	Comparative waterproofing finishes/perimeter treatments		
	Softwood flat roofs		
	Structure only comprising roof joists; 100 mm x 50 mm wall plates; herringbone strutting; 50 mm woodwool slabs; no coverings or finishes		
2.3.1	150 mm x 50 mm joists	m^2	28.80 - 33.00
2.3.2	200 mm x 50 mm joists	m^2	33.50 - 37.00
2.3.3	250 mm x 50 mm joists	m^2	36.50 - 40.00
	Structure only comprising roof joists; 100 mm x 50 mm wall plates; herringbone strutting; 25 mm softwood boarding; no coverings or finishes		
2.3.4	150 mm x 50 mm joists	m^2	32.00 - 37.00
2.3.5	200 mm x 50 mm joists	m^2	37.00 - 39.00
2.3.6	250 mm x 50 mm joists	m^2	39.00 - 43.00
	Roof joists; average depth; 25 mm softwood boarding; PVC rainwater goods; plasterboard; skim and emulsion		
2.3.7	three layer felt and chippings	m^2	70.00 - 88.00
2.3.8	two coat asphalt and chippings	m^2	67.00 - 97.00
	Softwood trussed pitched roofs		
	Structure only comprising 75 mm x 50 mm Fink roof trusses at 600 mm centres (measured on plan)		
2.3.9	22.50° pitch	m^2	14.00 - 17.00
	Structure only comprising 100 mm x 38 mm Fink roof trusses at 600 mm centres (measured on plan)		
2.3.10	30° pitch	m^2	15.00 - 19.00
2.3.11	35° pitch	m^2	16.00 - 19.00
2.3.12	40° pitch	m^2	18.00 - 22.00
	Structure only comprising 100 mm x 50 mm Fink roof trusses at 375 mm centres (measured on plan)		
2.3.13	30° pitch	m^2	27.00 - 30.00
2.3.14	35° pitch	m^2	27.00 - 30.00
2.3.15	40° pitch	m^2	29.00 - 33.00
	Extra for		
2.3.16	forming dormers	nr	390.00 - 545.00
	Structure only for "Mansard" roof comprising 100 mm x 50 mm roof trusses at 600 mm centres		
2.3.17	70° pitch	m^2	20.00 - 27.00
	Fink roof trusses; narrow span; 100 mm thick insulation; uPVC rainwater goods; plasterboard; skim and emulsion		
2.3.18	concrete interlocking tile coverings	m^2	65.00 - 87.00
2.3.19	clay pantile coverings	m^2	73.00 - 93.00
2.3.20	composition slate coverings	m^2	76.00 - 96.00
2.3.21	plain clay tile coverings	m^2	93.00 - 107.00
2.3.22	natural slate coverings	m^2	93.00 - 123.00
2.3.23	reconstructed stone coverings	m^2	78.00 - 131.00

Approximate Estimates

INDUSTRIAL £ range £	RETAILING £ range £	LEISURE £ range £	OFFICES £ range £	HOTELS £ range £	Item nr.
- -	29.00 - 33.00	- -	29.00 - 33.00	29.00 - 33.00	2.3.1
- -	33.00 - 37.00	- -	33.00 - 37.00	33.00 - 37.00	2.3.2
- -	37.00 - 40.00	- -	37.00 - 40.00	37.00 - 40.00	2.3.3
- -	32.00 - 37.00	- -	32.00 - 37.00	32.00 - 37.00	2.3.4
- -	37.00 - 39.00	- -	37.00 - 39.00	37.00 - 39.00	2.3.5
- -	37.00 - 57.00	- -	37.00 - 57.00	37.00 - 44.00	2.3.6
- -	70.00 - 88.00	- -	70.00 - 88.00	70.00 - 88.00	2.3.7
- -	68.00 - 97.00	- -	68.00 - 97.00	68.00 - 97.00	2.3.8
- -	14.00 - 19.00	- -	14.00 - 19.00	14.00 - 19.00	2.3.9
- -	17.00 - 20.00	- -	17.00 - 20.00	17.00 - 21.00	2.3.10
- -	18.00 - 20.00	- -	18.00 - 20.00	18.00 - 22.00	2.3.11
- -	19.00 - 23.00	- -	19.00 - 23.00	19.00 - 24.00	2.3.12
- -	27.00 - 30.00	- -	27.00 - 30.00	27.00 - 32.00	2.3.13
- -	29.00 - 32.00	- -	29.00 - 32.00	29.00 - 33.00	2.3.14
- -	29.00 - 35.00	- -	29.00 - 35.00	29.00 - 37.00	2.3.15
- -	391.00 - 545.00	- -	391.00 - 545.00	391.00 - 545.00	2.3.16
- -	20.00 - 29.00	- -	20.00 - 29.00	20.00 - 29.00	2.3.17
- -	72.00 - 97.00	- -	72.00 - 103.00	79.00 - 108.00	2.3.18
- -	78.00 - 103.00	- -	78.00 - 108.00	84.00 - 114.00	2.3.19
- -	82.00 - 107.00	- -	82.00 - 113.00	88.00 - 120.00	2.3.20
- -	97.00 - 123.00	- -	97.00 - 129.00	103.00 - 135.00	2.3.21
- -	103.00 - 129.00	- -	103.00 - 134.00	108.00 - 142.00	2.3.22
- -	84.00 - 134.00	- -	84.00 - 140.00	90.00 - 146.00	2.3.23

Approximate Estimates

Item nr.	SPECIFICATIONS	Unit	RESIDENTIAL £ range £
2.3	**ROOF - cont'd**		
	roof plan area (unless otherwise described)		
	Softwood trussed pitched roofs - cont'd		
	Monopitch roof trusses; 100 mm thick insulation; PVC rainwater goods; plasterboard; skim and emulsion		
2.3.24	concrete interlocking tile coverings	m^2	73.00 - 90.00
2.3.25	clay pantile coverings	m^2	78.00 - 96.00
2.3.26	composition slate coverings	m^2	82.00 - 100.00
2.3.27	plain clay tile coverings	m^2	96.00 - 114.00
2.3.28	natural slate coverings	m^2	103.00 - 123.00
2.3.29	reconstructed stone coverings	m^2	82.00 - 134.00
	Dormer roof trusses; 100 mm thick insulation; PVC rainwater goods; plasterboard; skim and emulsion		
2.3.30	concrete interlocking tile coverings	m^2	97.00 - 123.00
2.3.31	clay pantile coverings	m^2	103.00 - 131.00
2.3.32	composition slate coverings	m^2	107.00 - 148.00
2.3.33	plain clay tile coverings	m^2	122.00 - 148.00
2.3.34	natural slate coverings	m^2	129.00 - 151.00
2.3.35	reconstructed stone coverings	m^2	107.00 - 164.00
	Extra for		
2.3.36	end of terrace semi/detached configuration	m^2	26.00 - 29.00
2.3.37	hipped roof configuration	m^2	26.00 - 33.00
	Steel trussed pitched roofs		
	Fink roof trusses; wide span; 100 mm thick insulation		
2.3.38	concrete interlocking tile coverings	m^2	- -
2.3.39	clay pantile coverings	m^2	- -
2.3.40	composition slate coverings	m^2	- -
2.3.41	clay plain tile coverings	m^2	- -
2.3.42	natural slate coverings	m^2	- -
2.3.43	reconstructed stone coverings	m^2	- -
	Steel roof trusses and beams; thermal and acoustic insulation		
2.3.44	aluminium profiled composite cladding	m^2	- -
2.3.45	copper roofing on boarding	m^2	- -
	Steel roof and glulam beams; thermal and acoustic insulation		
2.3.46	aluminium profiled composite cladding	m^2	- -
	Concrete flat roofs		
	Structure only comprising reinforced concrete suspended slab; no coverings or finishes		
2.3.47	3.65 m span; 3.00 kN/m^2 loading	m^2	46.00 - 48.00
2.3.48	4.25 m span; 3.00 kN/m^2 loading	m^2	55.00 - 58.00
2.3.49	3.65 m span; 8.00 kN/m^2 loading	m^2	- -
2.3.50	4.25 m span; 8.00 kN/m^2 loading	m^2	- -
	Precast concrete suspended slab; average depth; 100 mm thick insulation; uPVC rainwater goods		
2.3.51	two coat asphalt coverings and chippings	m^2	- -
2.3.52	polyester roofing	m^2	- -
	Reinforced concrete or waffle suspended slabs; average depth; 100 mm thick insulation; uPVC rainwater goods		
2.3.53	two coat asphalt coverings and chippings	m^2	- -
2.3.54	two coat asphalt covering and paving slabs	m^2	- -
	Reinforced concrete slabs; on "Holorib" permanent steel shuttering; average depth; 100 mm thick insulation; uPVC rainwater goods		
2.3.55	two coat asphalt coverings and chippings	m^2	- -

Approximate Estimates 733

INDUSTRIAL £ range £	RETAILING £ range £	LEISURE £ range £	OFFICES £ range £	HOTELS £ range £	Item nr.
- -	78.00 - 103.00	- -	78.00 - 107.00	85.00 - 114.00	2.3.24
- -	85.00 - 107.00	- -	85.00 - 114.00	90.00 - 123.00	2.3.25
- -	90.00 - 114.00	- -	90.00 - 123.00	93.00 - 129.00	2.3.26
- -	103.00 - 131.00	- -	103.00 - 134.00	107.00 - 140.00	2.3.27
- -	107.00 - 134.00	- -	107.00 - 140.00	114.00 - 148.00	2.3.28
- -	88.00 - 140.00	- -	88.00 - 148.00	96.00 - 151.00	2.3.29
- -	103.00 - 134.00	- -	103.00 - 140.00	107.00 - 148.00	2.3.30
- -	107.00 - 140.00	- -	107.00 - 148.00	114.00 - 151.00	2.3.31
- -	110.00 - 148.00	- -	110.00 - 151.00	119.00 - 158.00	2.3.32
- -	129.00 - 158.00	- -	129.00 - 164.00	134.00 - 169.00	2.3.33
- -	134.00 - 164.00	- -	134.00 - 169.00	140.00 - 177.00	2.3.34
- -	114.00 - 169.00	- -	114.00 - 177.00	122.00 - 183.00	2.3.35
- -	- -	- -	- -	- -	2.3.36
- -	- -	- -	- -	- -	2.3.37
- -	122.00 - 158.00	119.00 - 151.00	122.00 - 166.00	131.00 - 170.00	2.3.38
- -	131.00 - 166.00	122.00 - 158.00	131.00 - 170.00	137.00 - 180.00	2.3.39
- -	119.00 - 169.00	131.00 - 166.00	134.00 - 173.00	143.00 - 183.00	2.3.40
- -	151.00 - 186.00	146.00 - 177.00	149.00 - 190.00	157.00 - 201.00	2.3.41
- -	157.00 - 190.00	151.00 - 186.00	157.00 - 201.00	166.00 - 207.00	2.3.42
- -	146.00 - 207.00	138.00 - 201.00	146.00 - 215.00	151.00 - 221.00	2.3.43
186.00 - 221.00	169.00 - 207.00	166.00 - 201.00	169.00 - 189.00	- -	2.3.44
- -	186.00 - 228.00	189.00 - 228.00	199.00 - 221.00	- -	2.3.45
- -	186.00 - 246.00	177.00 - 240.00	186.00 - 237.00	- -	2.3.46
- -	- -	45.00 - 53.00	45.00 - 53.00	45.00 - 53.00	2.3.47
- -	- -	51.00 - 55.00	51.00 - 62.00	51.00 - 62.00	2.3.48
49.00 - 58.00	49.00 - 58.00	49.00 - 58.00	49.00 - 58.00	49.00 - 58.00	2.3.49
62.00 - 72.00	62.00 - 72.00	61.00 - 72.00	61.00 - 72.00	61.00 - 72.00	2.3.50
84.00 - 117.00	76.00 - 126.00	76.00 - 120.00	76.00 - 117.00	76.00 - 126.00	2.3.51
88.00 - 103.00	85.00 - 117.00	85.00 - 111.00	85.00 - 102.00	85.00 - 114.00	2.3.52
91.00 - 120.00	85.00 - 132.00	85.00 - 126.00	85.00 - 117.00	85.00 - 132.00	2.3.53
126.00 - 158.00	120.00 - 169.00	120.00 - 161.00	120.00 - 158.00	120.00 - 172.00	2.3.54
85.00 - 102.00	76.00 - 114.00	76.00 - 111.00	76.00 - 102.00	61.00 - 114.00	2.3.55

Item nr.	SPECIFICATIONS	Unit	RESIDENTIAL £ range £
2.3	**ROOF - cont'd**		
	roof plan area (unless otherwise described)		
	Flat roof decking and finishes		
	Woodwool roof decking		
2.3.56	50 mm thick; two coat asphalt coverings to BS 5925 and chippings	m²	- -
	Galvanised steel roof decking; 100 mm thick insulation; three layer felt roofing and chippings		
2.3.57	0.70 mm thick; 2.38 m span	m²	- -
2.3.58	0.70 mm thick; 2.96 m span	m²	- -
2.3.59	0.70 mm thick; 3.74 m span	m²	- -
2.3.60	0.70 mm thick; 5.13 m span	m²	- -
	Aluminium roof decking; 100 mm thick insulation; three layer felt roofing and chippings		
2.3.61	0.90 mm thick; 1.79 m span	m²	- -
2.3.62	0.90 mm thick; 2.34 m span	m²	- -
	Metal decking; 100 mm thick insulation; on wood/steel open lattice beams		
2.3.63	three layer felt roofing and chippings	m²	- -
2.3.64	two layer high performance felt roofing and chippings	m²	- -
2.3.65	two coats asphalt coverings and chippings	m²	- -
	Metal decking to "Mansard"; excluding frame; 50 mm thick insulation; two coat asphalt coverings and chippings on decking		
2.3.66	natural slate covering to "Mansard" faces	m²	- -
	Roof claddings		
	Non-asbestos profiled cladding		
2.3.67	"Profile 3"; natural	m²	- -
2.3.68	"Profile 3"; coloured	m²	- -
2.3.69	"Profile 6"; natural	m²	- -
2.3.70	"Profile 6"; coloured	m²	- -
2.3.71	"Profile 6"; natural; insulated; inner lining panel	m²	- -
	Non-asbestos profiled cladding on steel purlins		
2.3.72	insulated	m²	- -
2.3.73	insulated; with 10% translucent sheets	m²	- -
2.3.74	insulated; plasterboard inner lining of metal tees	m²	- -
	Asbestos cement profiled cladding on steel purlins		
2.3.75	insulated	m²	- -
2.3.76	insulated; with 10% translucent sheets	m²	- -
2.3.77	insulated; plasterboard inner lining on metal tees	m²	- -
2.3.78	insulated; steel sheet liner on metal tees	m²	- -
	PVF2 coated galvanised steel profiled cladding		
2.3.79	0.72 mm thick; "profile 20 B"	m²	- -
2.3.80	0.72 mm thick; "profile TOP 40"	m²	- -
2.3.81	0.72 mm thick "profile 45"	m²	- -
	Extra for		
2.3.82	80 mm thick insulation and 0.40 mm thick coated inner lining sheet	m²	- -
	PVF2 coated galvanised steel profiled cladding on steel purlins		
2.3.83	insulated	m²	- -
2.3.84	insulated; plasterboard inner lining on metal tees	m²	- -
2.3.85	insulated; plasterboard inner lining on metal tees; with 1% fire vents	m²	- -
2.3.86	insulated; plasterboard inner lining on metal tees; with 2.50% fire vents	m²	- -
2.3.87	insulated; coloured inner lining panel	m²	- -
2.3.88	insulated; coloured inner lining panel; with 1% fire vents	m²	- -
2.3.89	insulated; coloured inner lining panel; with 2.50% fire vents	m²	- -
2.3.90	insulated; sandwich panel	m²	- -

Approximate Estimates

INDUSTRIAL £ range £	RETAILING £ range £	LEISURE £ range £	OFFICES £ range £	HOTELS £ range £	Item nr.
35.00 - 49.00	38.00 - 48.00	40.00 - 50.00	35.00 - 50.00	40.00 - 50.00	2.3.56
39.00 - 50.00	39.00 - 50.00	39.00 - 50.00	39.00 - 50.00	39.00 - 50.00	2.3.57
40.00 - 52.00	40.00 - 52.00	40.00 - 52.00	40.00 - 52.00	40.00 - 52.00	2.3.58
42.00 - 52.00	42.00 - 52.00	42.00 - 52.00	42.00 - 52.00	42.00 - 52.00	2.3.59
43.00 - 54.00	43.00 - 54.00	43.00 - 54.00	43.00 - 54.00	43.00 - 54.00	2.3.60
47.00 - 57.00	47.00 - 57.00	47.00 - 57.00	47.00 - 57.00	47.00 - 57.00	2.3.61
49.00 - 60.00	49.00 - 60.00	49.00 - 60.00	49.00 - 60.00	49.00 - 60.00	2.3.62
73.00 - 90.00	78.00 - 91.00	78.00 - 91.00	78.00 - 91.00	78.00 - 94.00	2.3.63
76.00 - 91.00	78.00 - 91.00	78.00 - 97.00	78.00 - 97.00	78.00 - 97.00	2.3.64
73.00 - 96.00	73.00 - 97.00	73.00 - 97.00	73.00 - 97.00	78.00 - 105.00	2.3.65
- -	134.00 - 218.00	122.00 - 201.00	122.00 - 201.00	132.00 - 233.00	2.3.66
12.60 - 15.60	-	-	-	-	2.3.67
14.00 - 17.10	-	-	-	-	2.3.68
14.00 - 17.10	-	-	-	-	2.3.69
15.60 - 17.10	-	-	-	-	2.3.70
24.70 - 32.70	-	-	-	-	2.3.71
24.70 - 31.20	-	-	-	-	2.3.72
27.80 - 32.70	-	-	-	-	2.3.73
40.30 - 47.20	-	-	-	-	2.3.74
24.70 - 29.70	-	-	-	-	2.3.75
27.80 - 32.70	-	-	-	-	2.3.76
38.80 - 47.20	-	-	-	-	2.3.77
38.00 - 47.20	-	-	-	-	2.3.78
17.10 - 22.40	23.20 - 31.20	23.20 - 31.20	-	-	2.3.79
17.10 - 20.90	22.40 - 29.30	22.40 - 29.30	-	-	2.3.80
18.60 - 24.70	24.70 - 32.70	24.70 - 32.70	-	-	2.3.81
11.20 - 12.60	11.20 - 12.60	11.20 - 12.60	-	-	2.3.82
30.10 - 40.30	34.20 - 44.90	34.20 - 44.90	-	-	2.3.83
44.00 - 58.20	48.70 - 63.90	48.70 - 63.90	-	-	2.3.84
52.50 - 73.00	59.00 - 79.10	59.00 - 79.10	-	-	2.3.85
65.40 - 85.20	73.00 - 91.30	73.00 - 91.30	-	-	2.3.86
46.40 - 59.70	52.50 - 65.40	52.50 - 65.40	-	-	2.3.87
52.50 - 65.40	59.00 - 73.00	59.00 - 73.00	-	-	2.3.88
65.40 - 85.20	71.50 - 85.20	71.50 - 85.20	-	-	2.3.89
97.40 - 150.60	124.80 - 170.40	124.80 - 170.40	-	-	2.3.90

Item nr.	SPECIFICATIONS	Unit	RESIDENTIAL £ range £	
2.3	**ROOF - cont'd**			
	roof plan area (unless otherwise described)			
	Roof claddings - cont'd			
	Pre-painted "Rigidal" aluminium profiled cladding			
2.3.91	0.70 mm thick; type WA6	m²	-	-
2.3.92	0.90 mm thick; type A7	m²	-	-
	PVF2 coated aluminium profiled cladding on steel purlins			
2.3.93	insulated; plasterboard inner lining on metal tees	m²	-	-
2.3.94	insulated; coloured inner lining panel	m²	-	-
	Rooflights/patent glazing and glazed roofs			
	Rooflights			
2.3.95	standard PVC	m²	-	-
2.3.96	feature/ventilating	m²	-	-
	Patent glazing; including flashings			
	standard aluminium georgian wired			
2.3.97	single glazed	m²	-	-
2.3.98	double glazed	m²	-	-
	purpose made polyester powder coated aluminium;			
2.3.99	double glazed low emissivity glass	m²	-	-
2.3.100	feature; to covered walkways	m²	-	-
	Glazed roofing on framing; to covered walkways			
2.3.101	feature; single glazed	m²	-	-
2.3.102	feature; double glazed barrel vault	m²	-	-
2.3.103	feature; very expensive	m²	-	-
	Comparative over/underlays			
	Roofing felt; unreinforced			
2.3.104	sloping (measured on face)	m²	1.25 -	1.55
	Roofing felt; reinforced			
2.3.105	sloping (measured on face)	m²	1.55 -	1.78
	Sloping (measured on plan)			
2.3.106	20° pitch	m²	1.78 -	2.25
2.3.107	30° pitch	m²	1.95 -	2.51
2.3.108	35° pitch	m²	2.51 -	2.74
2.3.109	40° pitch	m²	2.51 -	2.81
2.3.110	Building paper	m²	1.08 -	2.25
2.3.111	Vapour barrier	m²	1.63 -	4.87
	Insulation quilt; laid over ceiling joists			
2.3.112	80 mm thick	m²	3.04 -	3.35
2.3.113	100 mm thick	m²	3.65 -	3.88
2.3.114	150 mm thick	m²	5.02 -	5.55
2.3.115	200 mm thick	m²	6.69 -	7.30
	Wood fibre insulation boards; impregnated; density 220 - 350 kg/m³			
2.3.116	12.70 mm thick	m²	-	-
	Polystyrene insulation boards; fixed vertically with adhesive			
2.3.117	12 mm thick	m²	-	-
2.3.118	25 mm thick	m²	-	-
2.3.119	50 mm thick	m²	-	-
2.3.120	Limestone ballast	m²	-	-
	Polyurethane insulation boards; density 32 kg/m³			
2.3.121	30 mm thick	m²	-	-
2.3.122	35 mm thick	m²	-	-
2.3.123	50 mm thick	m²	-	-
	Cork insulation boards; density 112 - 125 kg/m³			
2.3.124	60 mm thick	m²	-	-
	Glass fibre insulation boards; density 120 - 130 kg/m²			
2.3.125	60 mm thick	m²	-	-
	Extruded polystyrene foam boards			
2.3.126	50 mm thick	m²	-	-
2.3.127	50 mm thick; with cement topping	m²	-	-
2.3.128	75 mm thick	m²	-	-

INDUSTRIAL £ range £	RETAILING £ range £	LEISURE £ range £	OFFICES £ range £	HOTELS £ range £	Item nr.
23.00 - 30.00	30.00 - 37.00	30.00 - 37.00	- -	- -	2.3.91
26.00 - 33.00	33.00 - 40.00	33.00 - 40.00	- -	- -	2.3.92
47.00 - 59.00	52.00 - 65.00	52.00 - 65.00	- -	- -	2.3.93
49.00 - 60.00	56.00 - 68.00	56.00 - 68.00	- -	- -	2.3.94
109.00 - 195.00	109.00 - 195.00	109.00 - 195.00	109.00 - 195.00	109.00 - 195.00	2.3.95
- -	195.00 - 358.00	195.00 - 358.00	195.00 - 358.00	195.00 - 358.00	2.3.96
140.00 - 195.00	148.00 - 195.00	148.00 - 195.00	148.00 - 195.00	155.00 - 218.00	2.3.97
163.00 - 210.00	178.00 - 225.00	178.00 - 225.00	178.00 - 225.00	195.00 - 248.00	2.3.98
195.00 - 233.00	210.00 - 248.00	210.00 - 248.00	210.00 - 248.00	218.00 - 272.00	2.3.99
- -	218.00 388.00	218.00 388.00	218.00 - 388.00	218.00 - 428.00	2.3.100
- -	303.00 - 450.00	303.00 - 450.00	303.00 - 450.00	333.00 - 496.00	2.3.101
- -	450.00 - 669.00	450.00 - 669.00	450.00 - 669.00	450.00 - 730.00	2.3.102
- -	669.00 - 837.00	669.00 - 837.00	669.00 - 837.00	669.00 - 890.00	2.3.103
- -	1.25 - 1.63	- -	1.25 - 1.63	1.25 - 1.63	2.3.104
- -	1.63 - 1.78	- -	1.63 - 1.78	1.63 - 1.78	2.3.105
- -	1.78 - 2.25	- -	1.78 - 2.25	1.78 - 2.25	2.3.106
- -	1.95 - 2.51	- -	1.95 - 2.51	1.95 - 2.51	2.3.107
- -	2.51 - 2.74	- -	2.51 - 2.74	2.51 - 2.74	2.3.108
- -	2.51 - 2.74	- -	2.51 - 2.74	2.51 - 2.81	2.3.109
- -	1.17 - 2.25	- -	1.17 - 2.25	1.17 - 2.25	2.3.110
- -	1.63 - 4.87	- -	1.63 - 4.87	1.63 - 4.87	2.3.111
- -	3.04 - 3.35	- -	3.04 - 3.35	3.04 - 3.35	2.3.112
- -	3.65 - 3.88	- -	3.65 - 3.88	1.25 - 3.88	2.3.113
- -	4.87 - 5.63	- -	4.87 - 5.63	1.25 - 5.63	2.3.114
- -	6.69 - 7.30	- -	6.69 - 7.30	1.25 - 7.30	2.3.115
- -	4.49 - 6.24	- -	- -	- -	2.3.116
4.90 - 6.00	4.90 - 6.00	4.90 - 6.00	4.90 - 6.00	4.90 - 6.00	2.3.117
6.00 - 7.00	6.00 - 7.00	6.00 - 7.00	6.00 - 7.00	6.00 - 7.00	2.3.118
7.00 - 8.40	7.00 - 8.40	7.00 - 8.40	7.00 - 8.40	7.00 - 8.40	2.3.119
4.90 - 8.40	4.90 - 8.40	4.90 - 8.40	4.90 - 8.40	4.90 - 8.40	2.3.120
7.00 - 8.40	7.00 - 8.40	7.00 - 8.40	7.00 - 8.40	7.00 - 8.40	2.3.121
8.40 - 8.70	8.40 - 8.70	8.40 - 8.70	8.40 - 8.70	8.40 - 8.70	2.3.122
9.70 - 11.20	9.70 - 11.20	9.70 - 11.20	9.70 - 11.20	9.70 - 11.20	2.3.123
8.70 - 12.20	8.70 - 12.20	8.70 - 12.20	8.70 - 12.20	8.70 - 12.20	2.3.124
12.90 - 15.60	12.90 - 15.60	12.90 - 15.60	12.90 - 15.60	12.90 - 15.60	2.3.125
12.20 - 14.00	12.20 - 14.00	12.20 - 14.00	12.20 - 14.00	12.20 - 14.00	2.3.126
19.80 - 22.40	19.80 - 22.40	19.80 - 22.40	19.80 - 22.40	19.80 - 22.40	2.3.127
16.00 - 17.90	16.00 - 17.90	16.00 - 17.90	16.00 - 17.90	16.00 - 17.90	2.3.128

Approximate Estimates

Item nr.	SPECIFICATIONS	Unit	RESIDENTIAL £ range £
2.3	**ROOF - cont'd**		
	roof plan area (unless otherwise described)		
	Comparative over/underlays - cont'd		
	"Perlite" insulation board; density 170 - 180 kg/m^3		
2.3.129	60 mm thick	m^2	- -
	Foam glass insulation board; density 125 - 135 kg/m^3		
2.3.130	60 mm thick	m^2	- -
	Screeds to receive roof coverings		
2.3.131	50 mm thick cement and sand screed	m^2	8.10 - 9.10
2.3.132	60 mm thick (average) "Isocrete K" screed; density 500 kg/m^3	m^2	8.70 - 9.40
2.3.133	75 mm thick lightweight bituminous screed and vapour barrier	m^2	13.60 - 15.60
2.3.134	100 mm thick lightweight bituminous screed and vapour barrier	m^2	16.70 - 18.60
	50 mm thick woodwool slabs; unreinforced		
2.3.135	sloping (measured on face)	m^2	8.70 - 11.20
	sloping (measured on plan)		
2.3.136	20° pitch	m^2	9.70 - 12.20
2.3.137	30° pitch	m^2	12.20 - 14.40
2.3.138	35° pitch	m^2	13.60 - 16.00
2.3.139	40° pitch	m^2	14.00 - 16.00
2.3.140	50 mm thick woodwool slabs; unreinforced; on and including steel purlins at 600 mm thick centres	m^2	15.10 - 18.30
2.3.141	sloping (measure on face)	m^2	9.70 - 12.20
	sloping (measured on plan)		
2.3.142	20° pitch	m^2	11.20 - 12.90
2.3.143	30° pitch	m^2	13.20 - 15.60
2.3.144	35° pitch	m^2	15.10 - 17.50
2.3.145	40° pitch	m^2	15.60 - 17.90
	25 mm thick "Tanalised" softwood boarding		
2.3.146	sloping (measured on face)	m^2	11.90 - 13.60
	sloping (measured on plan)		
2.3.147	20° pitch	m^2	12.90 - 14.80
2.3.148	30° pitch	m^2	15.60 - 17.90
2.3.149	35° pitch	m^2	17.10 - 19.40
2.3.150	40° pitch	m^2	18.60 - 20.20
	18 mm thick external quality plywood boarding		
2.3.151	sloping (measured on face)	m^2	15.10 - 18.30
	sloping (measured on plan)		
2.3.152	20° pitch	m^2	16.70 - 19.40
2.3.153	30° pitch	m^2	20.20 - 24.00
2.3.154	35° pitch	m^2	23.20 - 26.60
2.3.155	40° pitch	m^2	24.00 - 27.00
	Comparative tiling and slating finishes/perimeter treatments (including underfelt, battening, eaves courses and ridges)		
	Concrete troughed interlocking tiles; 413 mm x 300 mm; 75 mm lap		
2.3.156	sloping (measured on face)	m^2	14.40 - 17.90
	sloping (measured on plan)		
2.3.157	30° pitch	m^2	18.60 - 22.10
2.3.158	35° pitch	m^2	20.90 - 23.60
2.3.159	40° pitch	m^2	22.40 - 24.00
	Concrete interlocking slates; 430 mm x 330 mm; 75 mm lap		
2.3.160	sloping (measured on face)	m^2	15.10 - 18.30
	sloping (measured on plan)		
2.3.161	30° pitch	m^2	16.40 - 22.80
2.3.162	35° pitch	m^2	21.70 - 24.70
2.3.163	40° pitch	m^2	22.40 - 25.10

Approximate Estimates

INDUSTRIAL £ range £	RETAILING £ range £	LEISURE £ range £	OFFICES £ range £	HOTELS £ range £	Item nr.
13.20 - 14.40	13.20 - 14.40	13.20 - 14.40	13.20 - 14.40	13.20 - 14.40	2.3.129
16.70 - 19.40	16.70 - 19.40	16.70 - 19.40	16.70 - 19.40	16.70 - 19.40	2.3.130
-	8.10 - 9.10	-	8.10 - 9.10	8.10 - 9.10	2.3.131
-	8.70 - 9.50	-	8.70 - 9.40	8.70 - 9.50	2.3.132
-	14.00 - 15.60	-	14.00 - 15.60	14.00 - 15.60	2.3.133
-	16.00 - 18.60	-	16.00 - 18.60	16.00 - 18.60	2.3.134
8.70 - 11.20	-	-	8.70 - 11.20	-	2.3.135
9.80 - 12.20	-	-	9.80 - 12.20	-	2.3.136
12.60 - 14.80	-	-	12.60 - 14.80	-	2.3.137
14.00 - 16.00	-	-	14.00 - 16.00	-	2.3.138
14.00 - 16.40	-	-	14.00 - 16.40	-	2.3.139
15.10 - 18.30	-	-	15.10 - 18.30	-	2.3.140
-	9.80 - 12.20	-	9.80 - 12.20	9.80 - 12.20	2.3.141
-	11.20 - 13.30	-	11.20 - 13.30	11.20 - 13.30	2.3.142
-	13.30 - 15.60	-	13.30 - 15.60	13.30 - 15.60	2.3.143
-	15.10 - 17.10	-	15.10 - 17.10	15.10 - 17.10	2.3.144
-	15.60 - 17.90	-	15.60 - 17.90	15.60 - 17.90	2.3.145
-	11.60 - 13.60	-	11.60 - 13.60	11.60 - 13.60	2.3.146
-	12.60 - 14.80	-	12.60 - 14.80	12.60 - 14.80	2.3.147
-	15.60 - 17.90	-	15.60 - 17.90	15.60 - 17.90	2.3.148
-	17.50 - 19.40	-	17.50 - 19.40	17.50 - 19.40	2.3.149
-	18.30 - 20.20	-	18.30 - 20.20	18.30 - 20.20	2.3.150
-	15.10 - 18.30	-	15.10 - 18.30	15.10 - 18.30	2.3.151
-	16.70 - 19.40	-	16.70 - 19.40	16.70 - 19.40	2.3.152
-	20.20 - 23.20	-	20.20 - 23.20	20.20 - 23.20	2.3.153
-	23.20 - 26.20	-	23.20 - 26.20	23.20 - 26.20	2.3.154
-	23.20 - 26.60	-	23.20 - 26.60	23.20 - 26.60	2.3.155
-	14.40 - 17.90	-	14.40 - 17.90	14.40 - 17.90	2.3.156
-	18.60 - 22.40	-	18.60 - 22.40	18.60 - 22.40	2.3.157
-	20.90 - 24.00	-	20.90 - 24.00	20.90 - 24.00	2.3.158
-	22.40 - 24.70	-	22.40 - 24.70	22.40 - 24.70	2.3.159
-	15.10 - 17.90	-	15.10 - 17.90	15.10 - 17.90	2.3.160
-	19.40 - 22.40	-	19.40 - 22.40	19.40 - 22.40	2.3.161
-	21.70 - 24.70	-	21.70 - 24.70	21.70 - 24.70	2.3.162
-	22.40 - 24.70	-	22.40 - 24.70	22.40 - 24.70	2.3.163

Approximate Estimates

Item nr.	SPECIFICATIONS	Unit	RESIDENTIAL £ range £
2.3	**ROOF - cont'd**		
	roof plan area (unless otherwise described)		
	Comparative tiling and slating finishes/perimeter treatments - cont'd		
	Glass fibre reinforced bitumen slates; 900 mm x 300 mm thick; fixed to boarding (not included)		
2.3.164	sloping (measured on face)	m^2	14.00 - 16.70
	sloping (measured on plan)		
2.3.165	30° pitch	m^2	17.90 - 20.20
2.3.166	35° pitch	m^2	19.40 - 23.20
2.3.167	40° pitch	m^2	20.90 - 25.50
	Concrete bold roll interlocking tiles; 418 mm x 332 mm; 75 mm lap		
2.3.168	sloping (measured on face)	m^2	14.40 - 17.90
	sloping (measured on plan)		
2.3.169	30° pitch	m^2	19.00 - 22.40
2.3.170	35° pitch	m^2	20.90 - 24.00
2.3.171	40° pitch	m^2	21.30 - 24.30
	Tudor clay pantiles; 470 mm x 285 mm; 100 mm lap		
2.3.172	sloping (measured on face)	m^2	20.20 - 24.30
	sloping (measured on plan)		
2.3.173	30° pitch	m^2	25.50 - 31.20
2.3.174	35° pitch	m^2	28.50 - 34.60
2.3.175	40° pitch	m^2	30.10 - 36.50
	Natural red pantiles; 337 mm x 241 mm; 76 mm head and 38 mm side laps		
2.3.176	sloping (measured on face)	m^2	24.00 - 28.90
	sloping (measured on plan)		
2.3.177	30° pitch	m^2	31.20 - 36.10
2.3.178	35° pitch	m^2	33.90 - 38.80
2.3.179	40° pitch	m^2	35.40 - 40.30
	Blue composition (non-asbestos) slates; 600 mm x 300 mm; 75 mm lap		
2.3.180	sloping (measured on face)	m^2	24.70 - 30.40
2.3.181	sloping to mansard (measured on face)	m^2	35.00 - 40.30
	sloping (measured on plan)		
2.3.182	30° pitch	m^2	33.50 - 38.00
2.3.183	35° pitch	m^2	36.10 - 41.10
2.3.184	40° pitch	m^2	36.90 - 41.80
2.3.185	vertical to "Mansard"; including 18 mm thick blockboard (measured on face)	m^2	50.20 - 60.10
2.3.186	sloping (measured on face)	m^2	33.50 - 40.30
	sloping (measured on plan)		
2.3.187	30° pitch	m^2	41.50 - 50.20
2.3.188	35° pitch	m^2	46.40 - 55.90
2.3.189	40° pitch	m^2	50.20 - 60.10
	Machine made clay plain tiles; 267 mm x 165 mm; 64 mm lap		
2.3.190	sloping (measured on face)	m^2	37.70 - 44.90
	sloping (measured on plan)		
2.3.191	30° pitch	m^2	46.80 - 55.90
2.3.192	35° pitch	m^2	52.50 - 62.40
2.3.193	40° pitch	m^2	56.70 - 66.90
	Black "Sterreberg" glazed interlocking pantiles; 355 mm x 240 mm; 76 mm head and 38 mm side laps		
2.3.194	sloping (measured on face)	m^2	33.50 - 41.80
	sloping (measured on plan)		
2.3.195	30° pitch	m^2	41.80 - 51.70
2.3.196	35° pitch	m^2	47.20 - 58.60
2.3.197	40° pitch	m^2	51.70 - 63.10

Approximate Estimates 741

INDUSTRIAL £ range £	RETAILING £ range £	LEISURE £ range £	OFFICES £ range £	HOTELS £ range £	Item nr.
-	-	13.60 - 16.70	-	-	13.60 - 16.70	13.60 - 16.70	2.3.164
-	-	17.90 - 19.80	-	-	17.90 - 19.80	13.60 - 19.80	2.3.165
-	-	19.00 - 23.20	-	-	19.00 - 23.20	17.90 - 23.20	2.3.166
-	-	20.90 - 25.10	-	-	20.90 - 25.10	20.90 - 25.10	2.3.167
-	-	14.40 - 17.90	-	-	14.40 - 17.90	14.40 - 17.90	2.3.168
-	-	18.60 - 22.40	-	-	18.60 - 22.40	18.60 - 22.40	2.3.169
-	-	20.90 - 24.00	-	-	20.90 - 24.00	20.90 - 24.00	2.3.170
-	-	21.30 - 24.70	-	-	21.30 - 24.70	21.30 - 24.70	2.3.171
-	-	20.50 - 24.70	-	-	20.50 - 24.70	20.50 - 24.70	2.3.172
-	-	25.50 - 31.20	-	-	25.50 - 31.20	25.50 - 31.20	2.3.173
-	-	28.50 - 34.60	-	-	28.50 - 34.60	28.50 - 34.60	2.3.174
-	-	30.10 - 36.50	-	-	30.10 - 36.50	30.10 - 36.50	2.3.175
-	-	24.00 - 29.30	-	-	24.00 - 29.30	24.00 - 29.30	2.3.176
-	-	31.20 - 36.50	-	-	31.20 - 36.50	31.20 - 36.50	2.3.177
-	-	33.90 - 39.20	-	-	33.90 - 39.20	33.90 - 39.20	2.3.178
-	-	35.00 - 39.90	-	-	35.00 - 39.90	35.00 - 39.90	2.3.179
-	-	24.70 - 30.40	-	-	24.70 - 30.40	24.70 - 30.40	2.3.180
-	-	35.00 - 40.30	-	-	35.00 - 40.30	35.00 - 40.30	2.3.181
-	-	32.70 - 38.00	-	-	32.70 - 38.00	32.70 - 38.00	2.3.182
-	-	35.80 - 41.10	-	-	35.80 - 41.10	35.80 - 41.10	2.3.183
-	-	36.90 - 41.80	-	-	36.90 - 41.80	36.90 - 41.80	2.3.184
-	-	50.20 - 60.10	-	-	50.20 - 60.10	50.20 - 60.10	2.3.185
-	-	33.50 - 40.30	-	-	33.50 - 40.30	33.50 - 40.30	2.3.186
-	-	41.50 - 50.60	-	-	41.50 - 50.60	41.50 - 50.60	2.3.187
-	-	46.40 - 55.90	-	-	46.40 - 55.90	46.40 - 55.90	2.3.188
-	-	50.20 - 60.10	-	-	50.20 - 60.10	50.20 - 60.10	2.3.189
-	-	37.70 - 44.50	-	-	37.70 - 44.50	37.70 - 44.50	2.3.190
-	-	46.80 - 55.90	-	-	46.80 - 55.90	46.80 - 55.90	2.3.191
-	-	52.50 - 62.40	-	-	52.50 - 62.40	52.50 - 62.40	2.3.192
"	-	56.70 - 66.90	-	-	56.70 - 66.90	56.70 - 66.90	2.3.193
-	-	35.00 - 41.80	-	-	35.00 - 41.80	35.00 - 41.80	2.3.194
-	-	41.80 - 51.70	-	-	41.80 - 51.70	41.80 - 51.70	2.3.195
-	-	47.20 - 58.60	-	-	47.20 - 58.60	47.20 - 58.60	2.3.196
-	-	51.70 - 63.10	-	-	51.70 - 63.10	51.70 - 63.10	2.3.197

Item nr.	SPECIFICATIONS	Unit	RESIDENTIAL £ range £
2.3	**ROOF - cont'd**		
	roof plan area (unless otherwise described)		
	Comparative tiling and slating finishes/perimeter treatments - cont'd		
	Red cedar sawn shingles; 450 mm long; 125 mm lap		
2.3.198	sloping (measured on face)	m^2	38.80 - 47.90
	sloping (measured on plan)		
2.3.199	30° pitch	m^2	48.30 - 59.00
2.3.200	35° pitch	m^2	54.40 - 65.40
2.3.201	40° pitch	m^2	60.50 - 70.00
	Welsh natural slates; 510 mm x 255 mm; 76 mm lap		
2.3.202	sloping (measured on face)	m^2	41.80 - 49.50
	sloping (measured on plan)		
2.3.203	30° pitch	m^2	54.40 - 65.40
2.3.204	35° pitch	m^2	61.60 - 68.50
2.3.205	40° pitch	m^2	63.90 - 70.00
	Welsh slates; 610 mm x 305 mm; 76 mm lap		
2.3.206	sloping (measured on face)	m^2	43.40 - 49.50
	sloping (measured on plan)		
2.3.207	30° pitch	m^2	54.40 - 63.90
2.3.208	35° pitch	m^2	63.90 - 70.00
2.3.209	40° pitch	m^2	63.90 - 71.50
	Reconstructed stone slates; random slates; 80 mm lap		
2.3.210	sloping (measured on face)	m^2	26.20 - 49.50
	sloping (measured on plan)		
2.3.211	30° pitch	m^2	32.70 - 63.90
2.3.212	35° pitch	m^2	34.20 - 70.00
2.3.213	40° pitch	m^2	38.80 - 73.00
	Handmade sandfaced plain tiles; 267 mm x 165 mm; 64 mm lap		
2.3.214	sloping (measured on face)	m^2	54.40 - 66.90
	sloping (measured on plan)		
2.3.215	30° pitch	m^2	70.00 - 83.70
2.3.216	35° pitch	m^2	76.10 - 92.80
2.3.217	40° pitch	m^2	83.70 - 98.90
	Westmoreland green slates; random sizes; 76 mm lap		
2.3.218	sloping (measured on face)	m^2	111.10 - 129.30
	sloping (measured on plan)		
2.3.219	30° pitch	m^2	132.40 - 161.30
2.3.220	35° pitch	m^2	161.30 - 175.00
2.3.221	40° pitch	m^2	161.30 - 188.70
		perimeter length	
	Verges to sloping roofs; 250 mm x 25 mm painted softwood bargeboard		
2.3.222	6 mm thick "Masterboard" soffit lining 150 mm wide	m	13.60 - 16.40
2.3.223	19 mm x 150 mm painted soffit softwood	m	16.40 - 18.30
	Eaves to sloping roofs; 200 mm x 25 mm painted softwood fascia; 6 mm thick "Masterboard" soffit lining 225 mm wide		
2.3.224	100 mm uPVC gutter	m	18.60 - 25.10
2.3.225	150 mm uPVC gutter	m	23.60 - 30.40
2.3.226	100 mm cast iron gutter; decorated	m	29.70 - 35.40
2.3.227	150 mm cast iron gutter; decorated	m	36.10 - 41.80
	Eaves to sloping roofs; 200 mm x 25 mm painted softwood fascia; 19 mm x 225 mm painted softwood		
2.3.228	100 mm uPVC gutter	m	22.40 - 27.80
2.3.229	150 mm uPVC gutter	m	27.80 - 33.50
2.3.230	100 mm cast iron gutter; decorated	m	33.50 - 38.00
2.3.231	150 mm cast iron gutter; decorated	m	39.20 - 44.90

Approximate Estimates

INDUSTRIAL £ range £	RETAILING £ range £	LEISURE £ range £	OFFICES £ range £	HOTELS £ range £	Item nr.
- -	38.80 - 47.90	- -	38.80 - 47.90	38.80 - 47.90	2.3.198
- -	47.90 - 59.00	- -	47.90 - 59.00	47.90 - 59.00	2.3.199
- -	54.40 - 65.40	- -	54.40 - 65.40	54.40 - 65.40	2.3.200
- -	59.00 - 71.50	- -	59.00 - 71.50	59.00 - 71.50	2.3.201
- -	41.80 - 49.50	- -	47.90 - 54.40	47.90 - 54.40	2.3.202
- -	54.40 - 65.40	- -	54.40 - 65.40	54.40 - 65.40	2.3.203
- -	62.40 - 68.50	- -	62.40 - 68.50	62.40 - 68.50	2.3.204
- -	63.90 - 70.00	- -	63.90 - 70.00	63.90 - 70.00	2.3.205
- -	43.40 - 49.50	- -	43.40 - 59.00	43.40 - 58.60	2.3.206
- -	54.40 - 63.90	- -	54.40 - 63.90	54.40 - 63.90	2.3.207
- -	62.40 - 70.00	- -	62.40 - 70.00	62.40 - 70.00	2.3.208
- -	63.90 - 71.50	- -	63.90 - 71.50	63.90 - 71.50	2.3.209
- -	26.20 - 49.50	- -	26.20 - 49.50	26.20 - 49.50	2.3.210
- -	32.70 - 63.90	- -	32.70 - 63.90	32.70 - 63.90	2.3.211
- -	34.20 - 70.00	- -	34.20 - 70.00	34.20 - 70.00	2.3.212
- -	38.80 - 73.00	- -	38.80 - 73.00	38.80 - 73.00	2.3.213
- -	54.40 - 66.90	- -	54.40 - 66.90	54.40 - 66.90	2.3.214
- -	68.50 - 85.20	- -	68.50 - 85.20	68.50 - 85.20	2.3.215
- -	77.60 - 94.30	- -	77.60 - 94.30	77.60 - 94.30	2.3.216
- -	82.20 - 97.40	- -	82.20 - 97.40	82.20 - 97.40	2.3.217
- -	111.10 - 129.30	- -	111.10 - 129.30	111.10 - 129.30	2.3.218
- -	132.40 - 161.30	- -	132.40 - 161.30	132.40 - 161.30	2.3.219
- -	161.30 - 175.00	- -	161.30 - 175.00	161.30 - 175.00	2.3.220
- -	167.40 - 188.70	- -	167.40 - 188.70	167.40 - 188.70	2.3.221
- -	13.20 - 16.40	- -	13.20 - 16.40	13.20 - 16.40	2.3.222
- -	16.00 - 18.30	- -	16.00 - 18.30	16.00 - 18.30	2.3.223
- -	19.00 - 25.10	- -	19.00 - 25.10	19.00 - 25.10	2.3.224
- -	23.60 - 30.40	- -	23.60 - 30.40	23.60 - 30.40	2.3.225
- -	29.70 - 35.40	- -	29.70 - 35.40	29.70 - 35.40	2.3.226
- -	36.10 - 41.80	- -	36.10 - 41.80	36.10 - 41.80	2.3.227
- -	22.40 - 28.10	- -	22.40 - 28.10	22.40 - 28.10	2.3.228
- -	27.80 - 33.50	- -	27.80 - 33.50	27.80 - 33.50	2.3.229
- -	33.50 - 37.70	- -	33.50 - 37.70	33.50 - 37.70	2.3.230
- -	39.20 - 45.30	- -	39.20 - 45.30	39.20 - 45.30	2.3.231

Item nr.	SPECIFICATIONS	Unit	RESIDENTIAL £ range £	
2.3	**ROOF - cont'd**			
	roof plan area (unless otherwise described)			
	Comparative tiling and slating finishes/perimeter treatments - cont'd			
	Rainwater pipes; fixed to backgrounds; including offsets and shoe			
2.3.232	68 mm diameter uPVC	m	6.60 -	8.90
2.3.233	110 mm diameter uPVC	m	9.90 -	12.10
2.3.234	75 mm diameter cast iron; decorated	m	23.60 -	27.80
2.3.235	100 mm diameter cast iron; decorated	m	27.80 -	33.50
	Ridges			
2.3.236	concrete half round tiles	m	13.20 -	16.40
2.3.237	machine-made clay half round tiles	m	16.40 -	19.40
2.3.238	hand-made clay half round tiles	m	16.40 -	27.00
	Hips; including cutting roof tiles			
2.3.239	concrete half round tiles	m	17.10 -	22.40
2.3.240	machine-made clay half round tiles	m	24.30 -	27.00
2.3.241	hand-made clay half round tiles	m	24.30 -	28.50
2.3.242	hand-made clay bonnet hip tiles	m	39.20 -	44.50
	Comparative cladding finishes (including underfelt, labours, etc.)			
	0.91 mm thick aluminium roofing; commercial grade			
2.3.243	flat	m^2	-	-
	0.91 mm thick aluminium roofing; commercial grade; fixed to boarding			
2.3.244	sloping (measured on face)	m^2	-	-
	sloping (measured on plan)			
2.3.245	20º pitch	m^2	-	-
2.3.246	30º pitch	m^2	-	-
2.3.247	35º pitch	m^2	-	-
2.3.248	40º pitch	m^2	-	-
	0.81 mm thick zinc roofing			
2.3.249	flat	m^2	-	-
	0.81 mm thick zinc roofing; fixed to boarding (included)			
2.3.250	sloping (measured on face)	m^2	-	-
	sloping (measured on plan)			
2.3.251	20º pitch	m^2	-	-
2.3.252	30º pitch	m^2	-	-
2.3.253	35º pitch	m^2	-	-
2.3.254	40º pitch	m^2	-	-
	Copper roofing			
2.3.255	0.56 mm thick; flat	m^2	-	-
2.3.256	0.61 mm thick; flat	m^2	-	-
	Copper roofing; fixed to boarding (included)			
2.3.257	0.56 mm thick; sloping (measured on face)	m^2	-	-
	0.56 mm thick; sloping (measured on plan)			
2.3.258	20º pitch	m^2	-	-
2.3.259	30º pitch	m^2	-	-
2.3.260	35º pitch	m^2	-	-
2.3.261	40º pitch	m^2	-	-
2.3.262	0.61 mm thick; sloping (measured on face)	m^2	-	-
	0.61 mm thick; sloping (measured on plan)			
2.3.263	20º pitch	m^2	-	-
2.3.264	30º pitch	m^2	-	-
2.3.265	35º pitch	m^2	-	-
2.3.266	40º pitch	m^2	-	-
	Lead roofing			
2.3.267	code 4 sheeting; flat	m^2	-	-
2.3.268	code 5 sheeting; flat	m^2	-	-
2.3.269	code 6 sheeting; flat	m^2	-	-

Approximate Estimates

INDUSTRIAL £ range £	RETAILING £ range £	LEISURE £ range £	OFFICES £ range £	HOTELS £ range £	Item nr.
- -	6.50 - 9.30	- -	6.50 - 9.30	6.50 - 9.30	2.3.232
- -	9.70 - 12.60	- -	9.70 - 12.60	9.70 - 12.60	2.3.233
- -	23.60 - 28.10	- -	23.60 - 28.10	23.60 - 28.10	2.3.234
- -	28.10 - 33.50	- -	28.10 - 33.50	28.10 - 33.50	2.3.235
- -	13.20 - 16.40	- -	13.20 - 16.40	13.20 - 16.40	2.3.236
- -	17.10 - 19.40	- -	17.10 - 19.40	17.10 - 19.40	2.3.237
- -	16.40 - 27.00	- -	16.40 - 27.00	16.40 - 27.00	2.3.238
- -	17.10 - 22.40	- -	17.10 - 22.40	17.10 - 22.40	2.3.239
- -	24.30 - 27.40	- -	24.30 - 27.40	24.30 - 27.40	2.3.240
- -	24.30 - 28.50	- -	24.30 - 28.50	24.30 - 28.50	2.3.241
- -	39.20 - 44.50	- -	39.20 - 44.50	39.20 - 44.50	2.3.242
- -	38.00 - 43.00	38.00 - 43.00	38.00 - 43.00	38.00 - 43.00	2.3.243
- -	40.00 - 47.00	40.00 - 47.00	40.00 - 47.00	40.00 - 47.00	2.3.244
- -	45.00 - 52.00	45.00 - 52.00	45.00 - 52.00	45.00 - 52.00	2.3.245
- -	54.00 - 62.00	54.00 - 62.00	54.00 - 62.00	54.00 - 62.00	2.3.246
- -	60.00 - 70.00	60.00 - 70.00	60.00 - 70.00	60.00 - 70.00	2.3.247
- -	64.00 - 70.00	64.00 - 70.00	64.00 - 70.00	64.00 - 70.00	2.3.248
- -	48.00 - 56.00	48.00 - 59.00	48.00 - 59.00	48.00 - 59.00	2.3.249
- -	52.00 - 60.00	52.00 - 60.00	52.00 - 60.00	52.00 - 60.00	2.3.250
- -	59.00 - 65.00	59.00 - 65.00	59.00 - 65.00	59.00 - 65.00	2.3.251
- -	73.00 - 78.00	73.00 - 78.00	73.00 - 78.00	73.00 - 78.00	2.3.252
- -	79.00 - 87.00	79.00 - 87.00	79.00 - 87.00	79.00 - 87.00	2.3.253
- -	84.00 - 91.00	84.00 - 91.00	84.00 - 91.00	84.00 - 91.00	2.3.254
- -	56.00 - 65.00	56.00 - 65.00	56.00 - 65.00	56.00 - 65.00	2.3.255
- -	60.00 - 67.00	60.00 - 67.00	60.00 - 67.00	60.00 - 67.00	2.3.256
- -	60.00 - 67.00	60.00 - 67.00	60.00 - 67.00	60.00 - 67.00	2.3.257
- -	67.00 - 73.00	67.00 - 73.00	67.00 - 73.00	67.00 - 73.00	2.3.258
- -	79.00 - 87.00	79.00 - 87.00	79.00 - 87.00	79.00 - 87.00	2.3.259
- -	91.00 - 97.00	91.00 - 97.00	91.00 - 97.00	91.00 - 97.00	2.3.260
- -	91.00 - 99.00	91.00 - 99.00	91.00 - 99.00	91.00 - 99.00	2.3.261
- -	64.00 - 72.00	64.00 - 72.00	64.00 - 72.00	64.00 - 72.00	2.3.262
- -	70.00 - 78.00	70.00 - 78.00	70.00 - 78.00	70.00 - 78.00	2.3.263
- -	85.00 - 91.00	85.00 - 91.00	85.00 - 91.00	85.00 - 91.00	2.3.264
- -	94.00 - 102.00	94.00 - 102.00	94.00 - 102.00	94.00 - 102.00	2.3.265
- -	97.00 - 105.00	97.00 - 105.00	97.00 - 105.00	97.00 - 105.00	2.3.266
- -	62.00 - 68.00	62.00 - 68.00	62.00 - 68.00	62.00 - 68.00	2.3.267
- -	73.00 - 79.00	73.00 - 79.00	73.00 - 79.00	73.00 - 79.00	2.3.268
- -	78.00 - 85.00	78.00 - 85.00	78.00 - 85.00	78.00 - 85.00	2.3.269

Item nr.	SPECIFICATIONS	Unit	RESIDENTIAL £ range £	
2.3	**ROOF - cont'd**			
	roof plan area (unless otherwise described)			
	Comparative cladding finishes (including underfelt, labours, etc.) - cont'd			
	Lead roofing; fixed to boarding (included)			
2.3.270	code 4 sheeting; sloping (measured on face)	m^2	-	-
	code 4 sheeting; sloping (measured on plan)			
2.3.271	20° pitch	m^2	-	-
2.3.272	30° pitch	m^2	-	-
2.3.273	35° pitch	m^2	-	-
2.3.274	40° pitch	m^2	-	-
2.3.275	code 6 sheeting; sloping (measured on face)	m^2	-	-
	code 6 sheeting; sloping (measured on plan)			
2.3.276	20° pitch	m^2	-	-
2.3.277	30° pitch	m^2	-	-
2.3.278	35° pitch	m^2	-	-
2.3.279	40° pitch	m^2	-	-
	code 6 sheeting; vertical to mansard; including			
2.3.280	insulation (measured on face)	m^2	-	-
	Comparative waterproof finishes/perimeter treatments			
	Liquid applied coatings			
2.3.281	solar reflective paint	m^2	-	-
2.3.282	spray applied bitumen	m^2	-	-
2.3.283	spray applied co-polymer	m^2	-	-
2.3.284	spray applied polyurethane	m^2	-	-
	20 mm two coat asphalt roofing; laid flat; on felt underlay			
2.3.285	to BS 6925	m^2	-	-
2.3.286	to BS 6577	m^2	-	-
	Extra for			
2.3.287	solar reflective paint	m^2	-	-
2.3.288	limestone chipping finish	m^2	-	-
2.3.289	grp tiles in hot bitumen	m^2	-	-
	20 mm two coat reinforced asphaltic compound; laid flat on felt underlay			
2.3.290	to BS 6577	m^2	-	-
	Built-up bitumen felt roofing; laid flat			
2.3.291	three layer glass fibre roofing	m^2	-	-
2.3.292	three layer asbestos based roofing	m^2	-	-
	Extra for			
2.3.293	granite chipping finish	m^2	-	-
	Built-up self-finished asbestos based bitumen felt roofing; laid sloping			
2.3.294	two layer roofing (measured on face)	m^2	20.90 -	25.10
2.3.295	35° pitch	m^2	31.20 -	35.00
2.3.296	40° pitch	m^2	32.00 -	35.00
2.3.297	three layer roofing (measured on face)	m^2	27.80 -	33.50
	three layer roofing (measured on plan)			
2.3.298	20° pitch	m^2	41.80 -	46.40
2.3.299	30° pitch	m^2	43.40 -	47.90
	Elastomeric single ply roofing; laid flat			
2.3.300	"EPDM" membrane; laid loose	m^2	-	-
2.3.301	butyl rubber membrane; laid loose	m^2	-	-
	Extra for			
2.3.302	ballast	m^2	-	-

Approximate Estimates

INDUSTRIAL £ range £	RETAILING £ range £	LEISURE £ range £	OFFICES £ range £	HOTELS £ range £	Item nr.
- -	65.00 - 72.00	65.00 - 72.00	65.00 - 72.00	65.00 - 72.00	2.3.270
- -	72.00 - 79.00	72.00 - 79.00	72.00 - 79.00	72.00 - 79.00	2.3.271
- -	87.00 - 93.00	87.00 - 93.00	87.00 - 93.00	87.00 - 93.00	2.3.272
- -	97.00 - 105.00	97.00 - 105.00	97.00 - 105.00	97.00 - 105.00	2.3.273
- -	99.00 - 105.00	99.00 - 105.00	99.00 - 105.00	99.00 - 105.00	2.3.274
- -	82.00 - 91.00	82.00 - 91.00	82.00 - 91.00	82.00 - 91.00	2.3.275
- -	88.00 - 99.00	88.00 - 99.00	88.00 - 99.00	88.00 - 99.00	2.3.276
- -	111.00 - 120.00	111.00 - 120.00	111.00 - 120.00	111.00 - 120.00	2.3.277
- -	122.00 - 131.00	122.00 - 131.00	122.00 - 131.00	122.00 - 131.00	2.3.278
- -	128.00 - 137.00	128.00 - 137.00	128.00 - 137.00	128.00 - 137.00	2.3.279
- -	111.00 - 161.00	111.00 - 161.00	111.00 - 161.00	111.00 - 161.00	2.3.280
- -	1.40 - 2.50	1.40 - 2.50	1.40 - 2.50	1.40 - 2.50	2.3.281
- -	5.60 - 9.10	5.60 - 9.10	5.60 - 9.10	5.60 - 9.10	2.3.282
- -	6.60 - 10.10	6.60 - 10.10	6.60 - 10.10	6.60 - 10.10	2.3.283
- -	10.90 - 14.80	12.20 - 14.80	12.20 - 14.80	12.20 - 14.80	2.3.284
- -	11.60 - 15.10	11.60 - 15.10	11.60 - 15.10	11.60 - 15.10	2.3.285
- -	16.40 - 20.20	16.40 - 20.20	16.40 - 20.20	16.40 - 20.20	2.3.286
- -	1.90 - 2.70	1.90 - 2.70	1.90 - 2.70	1.90 - 2.70	2.3.287
- -	2.40 - 6.20	2.40 - 6.20	2.40 - 6.20	2.40 - 6.20	2.3.288
- -	26.20 - 32.70	26.20 - 32.70	26.20 - 32.70	26.20 - 32.70	2.3.289
- -	17.90 - 22.40	17.90 - 22.40	17.90 - 22.40	17.90 - 22.40	2.3.290
- -	15.60 - 19.40	15.60 - 19.40	15.60 - 19.40	15.60 - 19.40	2.3.291
- -	19.40 - 22.40	19.40 - 22.40	19.40 - 22.40	19.40 - 22.40	2.3.292
- -	2.40 - 6.10	2.40 - 6.10	2.40 - 6.10	2.40 - 6.10	2.3.293
- -	- -	- -	- -	- -	2.3.294
- -	- -	- -	- -	- -	2.3.295
- -	- -	- -	- -	- -	2.3.296
- -	- -	- -	- -	- -	2.3.297
- -	- -	- -	- -	- -	2.3.298
- -	- -	- -	- -	- -	2.3.299
18.30 - 20.90	18.30 - 20.90	18.30 - 20.90	18.30 - 20.90	18.30 - 20.90	2.3.300
18.30 - 20.90	18.30 - 20.90	18.30 - 20.90	18.30 - 20.90	18.30 - 20.90	2.3.301
5.60 - 9.10	5.60 - 9.10	5.60 - 9.10	5.60 - 9.10	5.60 - 9.10	2.3.302

Approximate Estimates

Item nr.	SPECIFICATIONS	Unit	RESIDENTIAL £ range £	
2.3	**ROOF - cont'd**			
	roof plan area (unless otherwise described)			
	Comparative waterproof finishes/perimeter treatments - cont'd			
	Thermoplastic single ply roofing; laid flat			
	uPVC membrane			
2.3.303	laid loose	m²	-	-
2.3.304	mechanically fixed	m²	-	-
2.3.305	fully adhered	m²	-	-
2.3.306	"CPE" membrane; laid loose	m²	-	-
2.3.307	"CPSG" membrane; fully adhered	m²	-	-
2.3.308	"PIB" membrane; laid loose	m²	-	-
	Extra for			
2.3.309	ballast	m²	-	-
	High performance built-up felt roofing; laid flat			
	three layer "Ruberglas 120 GP" felt roofing			
2.3.310	granite chipping finish	m²	-	-
	"Andersons" three layer self-finished polyester			
2.3.311	based bitumen felt roofing	m²	-	-
	Three layer polyester based modified bitumen felt roofing			
2.3.312	roofing	m²	-	-
	"Andersons" three layer polyester based bitumen			
2.3.313	felt roofing; granite chipping finish	m²	-	-
	Three layer metal faced glass cloth reinforced			
2.3.314	bitumen roofing	m²	-	-
	Three layer "Ruberfort HP 350" felt roofing; granite			
2.3.315	chipping finish	m²	-	-
	Three layer "Hyload 150E" elastomeric roofing; granite			
2.3.316	chipping finish	m²	-	-
	Three layer "Polybit 350" elastomeric roofing; granite			
2.3.317	chipping finish	m²	-	-
	Torch on roofing; laid flat			
2.3.318	three layer polyester-based modified bitumen roofing	m²	-	-
2.3.319	two layer polymeric isotropic roofing	m²	-	-
	Extra for			
2.3.320	granite chipping finish	m²	-	-
	Edges to flat felt roofs; softwood splayed fillet;			
	280 mm x 25 mm painted softwood fascia; no gutter			
2.3.321	aluminium edge trim	m	26.20 -	28.10
	Edges to flat roofs; code 4 lead drip dressed into gutter; 230 mm x 25 mm painted softwood fascia			
2.3.322	100 mm uPVC gutter	m	24.70 -	32.70
2.3.323	150 mm uPVC gutter	m	31.20 -	37.30
2.3.324	100 mm cast iron gutter; decorated	m	37.30 -	46.40
2.3.325	150 mm cast iron gutter; decorated	m	46.40 -	55.90
	Landcaped roofs			
	Vapour barrier, polyester-based elastomeric bitumen waterproofing and vapour equalisation layer, copper lined bitumen membrane root barrier and waterproofing layer, separation and slip layers, protection layer, 50 mm-thick			
2.3.326	thick drainage board, filter fleece, insulation, top soil and seed	m²	127.00 -	155.00

Approximate Estimates 749

INDUSTRIAL £ range £	RETAILING £ range £	LEISURE £ range £	OFFICES £ range £	HOTELS £ range £	Item nr.
17.50 - 20.90	17.50 - 20.90	17.50 - 20.90	17.50 - 20.90	17.50 - 20.90	2.3.303
22.40 - 26.20	22.40 - 26.20	22.40 - 26.20	22.40 - 26.20	22.40 - 26.20	2.3.304
24.70 - 28.10	24.70 - 28.10	24.70 - 28.10	24.70 - 28.10	24.70 - 28.10	2.3.305
20.20 - 24.70	20.20 - 24.70	20.20 - 24.70	20.20 - 24.70	20.20 - 24.70	2.3.306
20.20 - 24.70	20.20 - 24.70	20.20 - 24.70	20.20 - 24.70	20.20 - 24.70	2.3.307
23.20 - 28.10	23.20 - 28.10	23.20 - 28.10	23.20 - 28.10	23.20 - 28.10	2.3.308
5.60 - 9.10	5.60 - 9.10	5.60 - 9.10	5.60 - 9.10	5.60 - 9.10	2.3.309
24.70 - 27.00	24.70 - 27.00	24.70 - 27.00	24.70 - 27.00	24.70 - 27.00	2.3.310
24.70 - 28.10	24.70 - 28.10	24.70 - 28.10	24.70 - 28.10	24.70 - 28.10	2.3.311
25.50 - 28.10	25.50 - 28.10	25.50 - 28.10	25.50 - 28.10	25.50 - 28.10	2.3.312
28.10 - 31.20	28.10 - 31.20	28.10 - 31.20	28.10 - 31.20	28.10 - 31.20	2.3.313
31.20 - 35.00	31.20 - 35.00	31.20 - 35.00	31.20 - 35.00	31.20 - 35.00	2.3.314
31.20 - 35.00	31.20 - 35.00	31.20 - 35.00	31.20 - 35.00	31.20 - 35.00	2.3.315
32.70 - 36.50	32.70 - 36.50	32.70 - 36.50	32.70 - 36.50	32.70 - 36.50	2.3.316
35.00 - 38.80	35.00 - 38.80	35.00 - 38.80	34.60 - 38.80	34.60 - 38.80	2.3.317
-	22.40 - 26.60	22.40 - 26.60	22.40 - 26.60	22.40 - 26.60	2.3.318
-	22.40 - 26.60	22.40 - 26.60	22.40 - 26.60	22.40 - 26.60	2.3.319
-	2.10 - 5.60	2.10 - 5.60	2.10 - 5.60	2.10 - 5.60	2.3.320
-	26.20 - 28.10	26.20 - 28.10	26.20 - 28.10	26.20 - 28.10	2.3.321
-	24.70 - 32.70	-	24.70 - 32.70	24.70 - 32.70	2.3.322
-	31.20 - 38.00	-	31.20 - 38.00	31.20 - 38.00	2.3.323
-	37.30 - 46.40	-	37.30 - 46.40	37.30 - 46.40	2.3.324
-	46.40 - 59.00	-	46.40 - 59.00	46.40 - 59.00	2.3.325
-	-	127.00 - 155.00	-	-	2.3.326

2.4 STAIRS

storey (unless otherwise described)

Item nr.	SPECIFICATIONS	Unit	RESIDENTIAL £ range £
	comprising:		
	Timber construction		
	Reinforced concrete construction		
	Metal construction		
	Comparative finishes/balustrading		
	Timber construction		
	Softwood staircase; softwood balustrades and hardwood handrail; plasterboard; skim and emulsion to soffit		
2.4.1	2.60 m rise; standard; straight flight	nr	558.00 - 846.00
2.4.2	2.60 m rise; standard; top three treads winding	nr	683.00 - 939.00
2.4.3	2.60 m rise; standard; dogleg	nr	784.00 - 994.00
	Oak staircase; balusters and handrails; plasterboard; skim and emulsion to soffit		
2.4.4	2.60 m rise; purpose-made; dogleg	nr	- -
	plus or minus for		
2.4.5	each 300 mm variation in storey height	nr	- -
	Reinforced concrete construction		
	Escape staircase; granolithic finish; mild steel balustrades and handrails		
2.4.6	3.00 m rise; dogleg	nr	2914 - 3728
	plus or minus for		
2.4.7	each 300 mm variation in storey height	nr	283.00 - 365.00
	Staircase; terrazzo finish; mild steel balustrades and handrails; plastered soffit; balustrades and staircase soffit decorated		
2.4.8	3.00 m rise; dogleg	nr	- -
	plus or minus for		
2.4.9	each 300 mm variation in storey height	nr	- -
	Staircase; terrazzo finish; stainless steel balustrades and handrails; plastered and decorated soffit		
2.4.10	3.00 m rise; dogleg	nr	- -
	plus or minus for		
2.4.11	each 300 mm variation in storey height	nr	- -
	Staircase; high quality finishes; stainless steel and glass balustrades; plastered and decorated soffit		
2.4.12	3.00 m rise; dogleg	nr	- -
	plus or minus for		
2.4.13	each 300 mm variation in storey height	nr	- -
	Metal construction		
	Steel access/fire ladder		
2.4.14	3.00 m high	nr	- -
2.4.15	4.00 m high; epoxide finished	nr	- -
	Light duty metal staircase; galvanised finish; perforated treads; no risers; balustrades and handrails; decorated		
2.4.16	3.00 m rise; spiral; 1548 mm diameter	nr	- -
	plus or minus for		
2.4.17	each 300 mm variation in storey height	nr	- -
2.4.18	3.00 m rise; spiral; 1936 mm diameter	nr	- -
	plus or minus for		
2.4.19	each 300 mm variation in storey height	nr	- -
2.4.20	3.00 m rise; spiral; 2072 mm diameter	nr	- -
	plus or minus for		
2.4.21	each 300 mm variation in storey height	nr	- -

Approximate Estimates 751

INDUSTRIAL £ range £	RETAILING £ range £	LEISURE £ range £	OFFICES £ range £	HOTELS £ range £	Item nr.
- -	- -	- -	- -	- -	2.4.1
- -	- -	- -	- -	- -	2.4.2
- -	- -	- -	- -	- -	2.4.3
- -	- -	- -	- -	5044 - 6870	2.4.4
- -	- -	- -	- -	814.00 - 890.00	2.4.5
4154 - 5158	2914 - 5158	3994 - 4656	3994 - 4656	2914 - 5820	2.4.6
418.00 525.00	280.00 - 525.00	388.00 - 464.00	388.00 - 464.00	280.00 - 586.00	2.4.7
5820 - 7296	5820 - 7798	5470 - 7296	5470 - 7486	4496 - 8460	2.4.8
586.00 - 730.00	586.00 - 784.00	555.00 - 730.00	555.00 - 746.00	449.00 - 844.00	2.4.9
- -	5820 - 9472	6634 - 8726	6634 - 9548	5820 - 10126	2.4.10
- -	586.00 - 951.00	662.00 - 867.00	662.00 - 951.00	586.00 - 997.00	2.4.11
11640 - 11925	12880 - 12835	11640 - 11925	11252 - 11300	13694 - 13740	2.4.12
1164 - 1620	1286 - 1750	1164 - 1620	1126 - 1537	1369 - 1872	2.4.13
441.00 - 616.00	- -	449.00 - 464.00	- -	- -	2.4.14
- -	- -	616.00 - 1057.0	- -	- -	2.4.15
2168 - 2640	- -	- -	- -	- -	2.4.16
218.00 - 266.00	- -	- -	- -	- -	2.4.17
2488 - 3104	- -	- -	- -	- -	2.4.18
248.00 - 312.00	- -	- -	- -	- -	2.4.19
2640 - 3104	- -	- -	- -	- -	2.4.20
263.00 - 310.00	- -	- -	- -	- -	2.4.21

Item nr.	SPECIFICATIONS	Unit	RESIDENTIAL £ range £
2.4	**STAIRS - cont'd**		
	storey (unless otherwise described)		
	Metal construction - cont'd		
	Light duty metal staircase; galvanised finish; perforated treads; no risers; balustrades and handrails; decorated - cont'd		
2.4.22	3.00 m rise; straight; 760 mm wide	nr	- -
	plus or minus for		
2.4.23	each 300 mm variation in storey height	nr	- -
2.4.24	3.00 m rise; straight; 900 mm wide	nr	- -
	plus or minus for		
2.4.25	each 300 mm variation in storey height	nr	- -
2.4.26	3.00 m rise; straight; 1070 mm height	nr	- -
	plus or minus for		
2.4.27	each 300 mm variation in storey height	nr	- -
	Heavy duty cast iron staircase; perforated treads; no risers; balustrades and hand rails; decorated		
2.4.28	3.00 m rise; spiral; 1548 mm diameter	nr	- -
	plus or minus for		
2.4.29	each 300 mm variation in storey height	nr	- -
2.4.30	3.00 m rise; straight	nr	- -
	plus or minus for		
2.4.31	each 300 mm variation in storey height	nr	- -
	Feature metal staircase; galvanised finish perforated treads; no risers; decorated		
2.4.32	3.00 m rise; spiral; balustrades and handrails	nr	- -
	plus or minus for		
2.4.33	each 300 mm variation in storey height	nr	- -
2.4.34	3.00 m rise; dogleg; hardwood balustrades and handrails	nr	- -
2.4.35	3.00 m rise; dogleg; stainless steel balustrades and handrails	nr	- -
	Feature metal staircase; galvanised finish; concrete treads; balustrades and handrails; decorated		
2.4.36	3.00 m rise; dogleg	nr	- -
	Feature metal staircase to water chute; steel springers; stainless steel treads in-filled with tiling; landings every one metre rise		
2.4.37	9.00 m rise; spiral	nr	- -
	Galvanised steel catwalk; nylon coated balustrading		
2.4.38	450 mm wide	m	- -
	Comparative finishes/balustrading		
	Finishes to treads and risers		
2.4.39	uPVC floor tiles including screeds	per storey	- -
2.4.40	granolithic	per storey	- -
2.4.41	heavy duty carpet	per storey	- -
2.4.42	terrazzo	per storey	- -
	Wall handrails		
2.4.43	uPVC covered mild steel rail on brackets	per storey	- -
2.4.44	hardwood handrail on brackets	per storey	- -
2.4.45	stainless steel handrail on brackets	per storey	- -
	Balustrading and handrails		
2.4.46	mild steel balustrades and PVC covered handrails	per storey	- -
2.4.47	mild steel balustrades and hardwood handrails	per storey	- -
2.4.48	stainless steel balustrades and handrails	per storey	- -
2.4.49	stainless steel and glass balustrades	per storey	- -

Approximate Estimates

INDUSTRIAL £ range £	RETAILING £ range £	LEISURE £ range £	OFFICES £ range £	HOTELS £ range £	Item nr.
4892 - 3104	- -	- -	- -	- -	2.4.22
248.00 - 310.00	- -	- -	- -	- -	2.4.23
2640 - 3340	- -	- -	- -	- -	2.4.24
280.00 - 335.00	- -	- -	- -	- -	2.4.25
2792 - 3340	- -	- -	- -	- -	2.4.26
280.00 - 333.00	- -	- -	- -	- -	2.4.27
3416 - 4504	- -	- -	- -	- -	2.4.28
341.00 - 449.00	- -	- -	- -	- -	2.4.29
3728 - 5044	- -	- -	- -	- -	2.4.30
373.00 - 502.00	- -	- -	- -	- -	2.4.31
- -	- -	4892 - 5280	- -	- -	2.4.32
- -	- -	488.00 - 528.00	- -	- -	2.4.33
- -	- -	5044 - 6984	- -	- -	2.4.34
- -	- -	5744 - 9315	- -	- -	2.4.35
- -	- -	8612 - 11640	- -	- -	2.4.36
- -	- -	27936 - 34144	- -	- -	2.4.37
240.00 - 303.00	- -	256.00 - 327.00	- -	- -	2.4.38
- -	586.00 - 822.00	586.00 - 822.00	586.00 - 822.00	586.00 - 822.00	2.4.39
- -	890.00 - 1012.0	890.00 - 1012.0	890.00 - 1012.0	890.00 - 1012.0	2.4.40
- -	1240 - 1552	1240 - 1552	1240 - 1552	1240 - 1552	2.4.41
- -	2488 - 3256	2488 - 3256	2488 - 3256	2488 - 3256	2.4.42
- -	233.00 - 373.00	233.00 - 373.00	233.00 - 373.00	233.00 - 373.00	2.4.43
- -	636.00 - 1057.0	636.00 - 1057.0	636.00 - 1057.0	636.00 - 1057.0	2.4.44
- -	3028 - 4032	3028 - 4032	3028 - 4032	3028 - 4032	2.4.45
- -	730.00 - 936.00	730.00 - 936.00	730.00 - 936.00	730.00 - 936.00	2.4.46
- -	1316 - 1864	1316 - 1864	1316 - 1864	1316 - 1864	2.4.47
- -	5592 - 6771	5592 - 6771	5592 - 4450	5592 - 6771	2.4.48
- -	4900 - 12401	7456 - 12401	7456 - 12401	7456 - 12401	2.4.49

Approximate Estimates

Item nr.	SPECIFICATIONS	Unit	RESIDENTIAL £ range £
2.5	**EXTERNAL WALLS**		
	external wall area (unless otherwise described)		
	comprising:		
	Timber framed walling		
	Brick/block walling		
	Reinforced concrete walling		
	Panelled walling		
	Wall claddings		
	Curtain/glazed walling		
	Comparative external finishes		
	Timber framed walling		
	Structure only comprising softwood studs at 400 mm x 600 mm centres; head and sole plates		
2.5.1	125 mm x 50 mm	m^2	14.40 - 17.90
2.5.2	125 mm x 50 mm; one layer of double sided building paper	m^2	17.10 - 21.30
	Softwood stud wall; vapour barrier and plasterboard inner lining; decorated		
2.5.3	uPVC weatherboard outer lining	m^2	59.70 - 77.60
2.5.4	tile hanging on battens outer lining	m^2	71.50 - 85.20
	Brick/block walling		
	Autoclaved aerated lightweight block walls		
2.5.5	100 mm thick	m^2	17.90 - 19.40
2.5.6	140 mm thick	m^2	20.90 - 25.50
2.5.7	190 mm thick	m^2	28.90 - 33.50
	Dense aggregate block walls		
2.5.8	100 mm thick	m^2	16.40 - 17.90
2.5.9	140 mm thick	m^2	24.00 - 27.00
	Coloured dense aggregate masonry block walls		
2.5.10	100 mm thick; hollow	m^2	30.40 - 33.50
2.5.11	100 mm thick; solid	m^2	33.50 - 36.50
2.5.12	140 mm thick; hollow	m^2	35.00 - 40.30
2.5.13	140 mm thick; solid	m^2	46.40 - 49.50
	Common brick solid walls; bricks PC £120.00/1000		
2.5.14	half brick thick	m^2	27.00 - 28.90
2.5.15	one brick thick	m^2	46.40 - 49.50
2.5.16	one and a half brick thick	m^2	65.40 - 71.50
	Add or deduct for		
	each variation of £10.00/1000 in PC value		
2.5.17	half brick thick	m^2	0.70 - 1.10
2.5.18	one brick thick	m^2	1.40 - 1.70
2.5.19	one and a half brick thick	m^2	2.10 - 2.60
	Extra for		
2.5.20	fair face one side	m^2	1.40 - 1.90
	Engineering brick walls; class B; bricks PC £175.00/1000		
2.5.21	half brick thick	m^2	30.40 - 35.00
2.5.22	one brick thick	m^2	57.40 - 63.90
	Facing brick walls; sand faced facings; bricks PC £130.00/1000		
2.5.23	half brick thick; pointed one side	m^2	30.40 - 35.00
2.5.24	one brick thick; pointed both sides	m^2	55.90 - 63.90
	Facing brick walls; machine-made facings; bricks PC £320.00/1000		
2.5.25	half brick thick; pointed one side	m^2	46.40 - 49.50
2.5.26	half brick thick; built against concrete	m^2	47.90 - 52.50
2.5.27	one brick thick; pointed both sides	m^2	88.30 - 98.90
	Facing bricks solid walls; hand-made facings; bricks PC £525.00/1000		
2.5.28	half brick thick; pointed one side	m^2	60.50 - 68.50
2.5.29	one brick thick; pointed both sides	m^2	119.40 - 135.40

Approximate Estimates

INDUSTRIAL £ range £	RETAILING £ range £	LEISURE £ range £	OFFICES £ range £	HOTELS £ range £	Item nr.
- -	- -	- -	- -	- -	2.5.1
- -	- -	- -	- -	- -	2.5.2
- -	- -	- -	- -	- -	2.5.3
- -	- -	- -	- -	- -	2.5.4
16.40 - 19.40	16.40 - 19.40	16.40 - 19.40	16.40 - 19.40	19.40 - 20.90	2.5.5
20.20 - 24.00	20.20 - 24.00	20.20 - 24.00	20.20 - 24.00	22.40 - 26.20	2.5.6
28.60 - 32.70	28.10 - 32.70	28.10 - 32.70	28.10 - 32.70	30.40 - 34.20	2.5.7
15.60 - 17.90	15.60 - 17.90	15.60 - 17.90	15.60 - 17.90	17.90 - 20.20	2.5.8
22.40 - 26.20	22.40 - 26.20	22.40 - 26.20	22.40 - 26.20	24.70 - 26.20	2.5.9
28.50 - 32.00	28.50 - 32.00	28.50 - 32.00	28.50 - 32.00	32.00 - 35.00	2.5.10
32.00 - 35.00	32.00 - 35.00	32.00 - 35.00	32.00 - 35.00	35.00 - 38.00	2.5.11
35.80 - 38.80	35.80 - 38.80	35.80 - 38.80	35.80 - 38.80	38.80 - 41.80	2.5.12
41.80 - 47.90	41.80 - 47.90	41.80 - 47.90	41.80 - 47.90	46.40 - 52.90	2.5.13
24.00 - 27.00	24.00 - 27.00	24.00 - 27.00	24.00 - 27.00	27.40 - 29.70	2.5.14
43.40 - 47.90	43.40 - 47.90	43.40 - 47.90	43.40 - 47.90	46.40 - 52.90	2.5.15
62.40 - 70.00	62.40 - 70.00	62.40 - 70.00	62.40 - 70.00	66.90 - 73.00	2.5.16
0.70 - 1.00	0.70 - 1.00	0.70 - 1.00	0.70 - 1.00	0.70 - 1.00	2.5.17
1.40 - 1.60	1.40 - 1.60	1.40 - 1.60	1.40 - 1.60	1.40 - 1.60	2.5.18
2.10 - 2.60	2.10 - 2.60	2.10 - 2.60	2.10 - 2.60	2.10 - 2.60	2.5.19
1.60 - 2.10	1.60 - 2.10	1.60 - 2.10	1.60 - 2.10	1.60 - 2.10	2.5.20
29.30 - 33.50	29.30 - 33.50	29.30 - 33.50	29.30 - 33.50	29.30 - 33.50	2.5.21
55.90 - 63.90	55.90 - 63.90	55.90 - 63.90	55.00 - 63.90	60.50 - 66.90	2.5.22
29.30 - 33.50	29.30 - 33.50	29.30 - 33.50	29.30 - 33.50	32.00 - 36.50	2.5.23
55.90 - 60.90	55.90 - 60.90	55.90 - 60.90	55.90 - 60.90	59.00 - 65.40	2.5.24
43.40 - 49.50	43.40 - 49.50	43.40 - 49.50	43.40 - 49.50	46.40 - 52.50	2.5.25
46.40 - 51.00	47.90 - 52.50	47.90 - 52.50	47.90 - 52.50	49.50 - 55.90	2.5.26
85.20 - 98.90	85.20 - 98.90	85.20 - 98.90	85.20 - 98.90	89.80 - 100.40	2.5.27
59.00 - 65.40	60.50 - 66.90	60.50 - 66.90	60.50 - 66.90	63.90 - 73.00	2.5.28
114.10 - 127.80	118.70 - 129.30	118.70 - 129.30	118.70 - 129.30	124.80 - 135.40	2.5.29

Item nr.	SPECIFICATIONS	Unit	RESIDENTIAL £ range £
2.5	**EXTERNAL WALLS - cont'd**		
	external wall area (unless otherwise described)		
	Brick/block walling - cont'd		
	Add or deduct for		
	each variation of £10.00/100 in PC value		
2.5.30	half brick thick	m^2	0.70 - 1.10
2.5.31	one brick thick	m^2	1.40 - 1.60
	Composite solid walls; facing brick on outside; bricks PC £320.00/1000 and common brick on inside; bricks PC £120.00/1000		
2.5.32	one brick thick; pointed one side	m^2	70.00 - 78.00
	Extra for		
2.5.33	one and a half brick thick; pointed one side	m^2	3.80 - 6.10
2.5.34	weather pointing as a separate operation	m^2	91.00 - 85.00
	Composite cavity wall; block outer skin; 50 mm thick insulation; lightweight block inner skin		
2.5.35	outer block rendered	m^2	49.00 - 64.00
	outer block rendered; no insulation; inner skin		
2.5.36	insulating	m^2	1.00 - 1.90
2.5.37	outer block roughcast	m^2	0.50 - 1.90
2.5.38	coloured masonry outer block	m^2	1.00 - 1.90
	Extra for		
2.5.39	heavyweight block inner skin	m^2	1.90 - 3.00
2.5.40	fair face one side	m^2	9.80 - 12.00
2.5.41	75 mm thick cavity insulation	m^2	49.00 - 67.00
2.5.42	100 mm thick cavity insulation	m^2	56.00 - 72.00
2.5.43	plaster and emulsion	m^2	61.00 - 75.00
	Composite cavity wall; facing brick outer skin; 50 mm thick insulation; plasterboard on stud inner skin; emulsion		
2.5.44	sand-faced facings; PC £130.00/1000	m^2	59.00 - 70.00
2.5.45	machine-made facings; PC £320.00/1000	m^2	73.00 - 82.00
2.5.46	hand-made facings; PC £525.00/1000	m^2	87.00 - 97.00
	As above but with plaster on lightweight block inner skin; emulsion		
2.5.47	sand-faced facings; PC £130.00/1000	m^2	52.00 - 67.00
2.5.48	machine-made facings; PC £320.00/1000	m^2	70.00 - 81.00
2.5.49	hand-made facings; PC £525.00/1000	m^2	81.00 - 96.00
	Add or deduct for		
2.5.50	each variation of £10.00/1000 in PC value	m^2	0.70 - 1.00
	Extra for		
2.5.51	heavyweight block inner skin	m^2	1.00 - 1.90
2.5.52	insulating block inner skin	m^2	1.90 - 4.90
2.5.53	30 mm thick cavity wall slab	m^2	2.30 - 5.60
2.5.54	50 mm thick cavity insulation	m^2	2.70 - 3.30
2.5.55	75 mm thick cavity insulation	m^2	3.90 - 4.50
2.5.56	100 mm thick cavity insulation	m^2	4.90 - 5.60
2.5.57	weather-pointing as a separate operation	m^2	3.90 - 6.20
2.5.58	purpose made feature course to windows	m^2	4.90 - 10.00
	Composite cavity wall; facing brick outer skin; 50 mm thick insulation; common brick inner skin; fair face on inside		
2.5.59	sand-faced facings; PC £130.00/1000	m^2	57.00 - 72.00
2.5.60	machine-made facings; PC £320.00/1000	m^2	73.00 - 85.00
2.5.61	hand-made facings; PC £525.00/1000	m^2	88.00 - 100.00
	Composite cavity wall; facing brick outer skin; 50 mm thick insulation; common brick inner skin; plaster and emulsion		
2.5.62	sand-faced facings; PC £130.00/1000	m^2	65.00 - 78.00
2.5.63	machine-made facings; PC £320.00/1000	m^2	82.00 - 93.00
2.5.64	hand-made facings; PC £525.00/1000	m^2	96.00 - 111.00
2.5.65	Composite cavity wall; coloured masonry block; outer and inner skins; fair faced both sides	m^2	91.00 - 123.00

Approximate Estimates

INDUSTRIAL £ range £	RETAILING £ range £	LEISURE £ range £	OFFICES £ range £	HOTELS £ range £	Item nr.
0.70 - 1.10	0.70 - 1.10	0.70 - 1.10	0.70 - 1.10	0.70 - 1.10	2.5.30
1.40 - 1.60	1.40 - 1.60	1.40 - 1.60	1.40 - 1.60	1.40 - 1.60	2.5.31
68.00 - 76.00	72.00 - 78.00	72.00 - 78.00	72.00 - 78.00	75.00 - 81.00	2.5.32
3.70 - 6.10	3.70 - 6.10	3.70 - 6.10	3.70 - 6.10	3.70 - 6.10	2.5.33
85.00 - 96.00	91.00 - 100.00	91.00 - 100.00	91.00 - 100.00	96.00 - 107.00	2.5.34
48.00 - 60.00	49.00 - 64.00	49.00 - 64.00	49.00 - 64.00	52.00 - 67.00	2.5.35
1.00 - 1.90	1.00 - 1.90	1.00 - 1.90	1.00 - 1.90	1.00 - 1.90	2.5.36
0.50 - 1.90	0.50 - 1.90	0.50 - 1.90	0.50 - 1.90	0.50 - 1.90	2.5.37
1.00 - 1.90	1.00 - 1.90	1.00 - 1.90	1.00 - 1.90	1.00 - 1.90	2.5.38
1.90 - 3.00	1.90 - 3.00	1.90 - 3.00	1.90 - 3.00	1.90 - 3.00	2.5.39
9.00 - 12.80	9.00 - 12.80	9.00 - 12.80	9.00 - 12.80	10.50 - 13.60	2.5.40
48.00 - 65.00	49.00 - 67.00	49.00 - 67.00	49.00 - 67.00	52.00 - 68.00	2.5.41
52.00 - 70.00	56.00 - 72.00	56.00 - 72.00	56.00 - 72.00	57.00 - 73.00	2.5.42
59.00 - 73.00	61.00 - 76.00	61.00 - 76.00	61.00 - 76.00	64.00 - 78.00	2.5.43
57.00 - 68.00	59.00 - 70.00	59.00 - 70.00	59.00 - 70.00	61.00 - 72.00	2.5.44
72.00 - 81.00	75.00 - 82.00	75.00 - 82.00	75.00 - 82.00	76.00 - 85.00	2.5.45
82.00 - 93.00	87.00 - 97.00	87.00 - 97.00	87.00 - 97.00	90.00 - 100.00	2.5.46
51.00 - 64.00	52.00 - 67.00	52.00 - 67.00	52.00 - 67.00	57.00 - 70.00	2.5.47
65.00 - 78.00	70.00 - 82.00	70.00 - 82.00	70.00 - 82.00	73.00 - 84.00	2.5.48
78.00 - 96.00	81.00 - 96.00	81.00 - 96.00	81.00 - 96.00	84.00 - 100.00	2.5.49
0.70 - 1.00	0.70 - 1.00	0.70 - 1.00	0.70 - 1.00	0.70 - 1.00	2.5.50
1.00 - 1.90	1.00 - 1.90	1.00 - 1.90	1.00 - 1.90	1.00 - 1.90	2.5.51
1.90 - 4.90	1.90 - 4.90	1.90 - 4.90	1.90 - 4.90	1.90 - 4.90	2.5.52
2.30 - 5.60	2.30 - 5.60	2.30 - 5.60	2.30 - 5.60	2.30 - 5.60	2.5.53
3.00 - 3.70	3.00 - 3.70	3.00 - 3.70	3.00 - 3.70	3.00 - 3.70	2.5.54
3.90 - 4.50	3.90 - 4.50	3.90 - 4.50	3.90 - 4.50	3.90 - 4.50	2.5.55
4.90 - 5.60	4.90 - 5.60	4.90 - 5.60	4.90 - 5.60	4.90 - 5.60	2.5.56
3.90 - 6.20	3.90 - 6.20	3.90 - 6.20	3.90 - 6.20	3.90 - 6.20	2.5.57
4.90 - 10.10	4.90 - 10.10	4.90 - 10.10	4.90 - 10.10	4.90 - 10.10	2.5.58
54.00 - 70.00	56.00 - 72.00	56.00 - 72.00	56.00 - 72.00	59.00 - 73.00	2.5.59
72.00 - 82.00	75.00 - 87.00	75.00 - 87.00	75.00 - 87.00	76.00 - 88.00	2.5.60
84.00 - 96.00	88.00 - 100.00	88.00 - 100.00	88.00 - 100.00	91.00 - 107.00	2.5.61
64.00 - 76.00	65.00 - 78.00	65.00 - 78.00	65.00 - 78.00	68.00 - 84.00	2.5.62
79.00 - 91.00	82.00 - 93.00	82.00 - 93.00	82.00 - 93.00	87.00 - 96.00	2.5.63
91.00 - 105.00	96.00 - 111.00	96.00 - 111.00	96.00 - 111.00	100.00 - 117.00	2.5.64
87.00 - 117.00	91.00 - 123.00	91.00 - 123.00	91.00 - 123.00	94.00 - 126.00	2.5.65

Item nr.	SPECIFICATIONS	Unit	RESIDENTIAL £ range £
2.5	**EXTERNAL WALLS - cont'd**		
	external wall area (unless otherwise described)		
	Reinforced concrete walling		
	In situ reinforced concrete 25.00 N/mm^2; 13 kg/m^2 reinforcement; formwork both sides		
2.5.66	150 mm thick	m^2	76.00 - 93.00
2.5.67	225 mm thick	m^2	93.00 - 103.00
	Panelled walling		
	Precast concrete panels; including insulation; lining and fixings		
2.5.68	standard panels	m^2	- -
2.5.69	standard panels; exposed aggregate finish	m^2	- -
2.5.70	brick clad panels	m^2	- -
2.5.71	reconstructed stone faced panels	m^2	- -
2.5.72	natural stone faced panels	m^2	- -
2.5.73	marble or granite faced panels	m^2	- -
	GRP/laminate panels; including battens; back-up walls; plaster and emulsion		
2.5.74	melamine finished solid laminate panels	m^2	- -
2.5.75	single skin panels	m^2	- -
2.5.76	double skin panels	m^2	- -
2.5.77	insulated sandwich panels	m^2	- -
	Wall claddings		
	Non-asbestos profiled cladding		
2.5.78	"Profile 3"; natural	m^2	- -
2.5.79	"Profile 3"; coloured	m^2	- -
2.5.80	"Profile 6"; natural	m^2	- -
2.5.81	"Profile 6"; coloured	m^2	- -
2.5.82	insulated; inner lining of plasterboard	m^2	- -
2.5.83	"Profile 6"; natural; insulated; inner lining panel	m^2	- -
2.5.84	insulated; with 2.80 m high block inner skin; emulsion	m^2	- -
2.5.85	insulated; with 2.80 m high block inner skin; plasterboard lining on metal tees; emulsion	m^2	- -
	Asbestos cement profiled cladding on steel rails		
2.5.86	insulated; with 2.80 m high block inner skin; emulsion	m^2	- -
2.5.87	Insulated; with 2.80 m high block inner skin plasterboard lining on metal tees; emulsion	m^2	- -
	PVF2 coated galvanised steel profiled cladding		
2.5.88	0.60 mm thick; "profile 20B"; corrugations vertical	m^2	- -
2.5.89	0.60 mm thick; "profile 30"; corrugations vertical	m^2	- -
2.5.90	0.60 mm thick; "profile TOP 40"; corrugations vertical	m^2	- -
2.5.91	0.60 mm thick; "profile 60B"; corrugations vertical	m^2	- -
2.5.92	0.60 mm thick; "profile 30"; corrugations horizontal	m^2	- -
2.5.93	0.60 mm thick; "profile 60B"; corrugations horizontal	m^2	- -
	Extra for		
2.5.94	80 mm thick insulation and 0.40 mm thick coated inner lining	m^2	- -
	PVF2 coated galvanised steel profiled cladding on steel rails		
2.5.95	insulated	m^2	- -
2.5.96	2.80 m high insulating block inner skin; emulsion	m^2	- -
2.5.97	2.80 m high insulating block inner skin; plasterboard lining on metal tees; emulsion	m^2	- -
2.5.98	insulated; coloured inner lining panel	m^2	- -
2.5.99	insulated; full-height insulating block inner skin; plaster and emulsion	m^2	- -
2.5.100	insulated; metal sandwich panel system	m^2	- -

Approximate Estimates

INDUSTRIAL £ range £	RETAILING £ range £	LEISURE £ range £	OFFICES £ range £	HOTELS £ range £	Item nr.
76.00 - 93.00	76.00 - 93.00	76.00 - 93.00	76.00 - 93.00	76.00 - 93.00	2.5.66
93.00 - 103.00	93.00 - 103.00	93.00 - 103.00	93.00 - 103.00	93.00 - 103.00	2.5.67
99.00 - 149.00	102.00 - 149.00	102.00 - 149.00	102.00 - 149.00	102.00 - 161.00	2.5.68
149.00 - 210.00	163.00 - 210.00	163.00 - 210.00	163.00 - 210.00	163.00 - 225.00	2.5.69
195.00 - 248.00	210.00 - 248.00	210.00 - 248.00	210.00 - 248.00	210.00 - 263.00	2.5.70
- -	333.00 - 411.00	333.00 - 411.00	333.00 - 411.00	333.00 - 450.00	2.5.71
- -	464.00 - 586.00	464.00 - 586.00	464.00 - 586.00	464.00 - 624.00	2.5.72
- -	533.00 - 647.00	533.00 - 647.00	533.00 - 647.00	533.00 - 708.00	2.5.73
125.00 - 163.00	125.00 - 163.00	125.00 - 163.00	-	-	2.5.74
140.00 - 178.00	148.00 - 178.00	148.00 - 178.00	117.00 - 148.00	-	2.5.75
195.00 - 233.00	202.00 - 233.00	202.00 - 233.00	178.00 - 225.00	-	2.5.76
- -	233.00 - 297.00	233.00 - 297.00	225.00 - 281.00	225.00 - 281.00	2.5.77
16.40 - 18.60	-	-	-	-	2.5.78
17.90 - 20.20	-	-	-	-	2.5.79
17.90 - 20.20	-	-	-	-	2.5.80
18.60 - 20.90	-	-	-	-	2.5.81
32.00 - 40.30	-	-	-	-	2.5.82
32.00 - 40.30	-	-	-	-	2.5.83
27.80 - 32.00	-	-	-	-	2.5.84
38.00 - 44.90	-	-	-	-	2.5.85
27.80 - 31.20	-	-	-	-	2.5.86
36.50 - 43.40	-	-	-	-	2.5.87
22.40 - 27.80	27.80 - 35.00	27.80 - 35.00	-	-	2.5.88
22.40 - 27.80	27.80 - 35.00	27.80 - 35.00	-	-	2.5.89
20.90 - 26.20	26.20 - 33.50	26.20 - 33.50	-	-	2.5.90
26.20 - 32.70	32.70 - 38.00	32.70 - 38.00	-	-	2.5.91
23.20 - 29.70	29.70 - 36.50	29.70 - 36.50	-	-	2.5.92
26.20 - 33.50	33.50 - 40.30	33.50 - 40.30	-	-	2.5.93
12.60 - 13.20	12.60 - 13.20	12.60 - 13.20	-	-	2.5.94
37.00 - 47.00	-	-	-	-	2.5.95
46.00 - 56.00	-	-	-	-	2.5.96
52.00 - 67.00	-	-	-	-	2.5.97
52.00 - 67.00	59.00 - 78.00	59.00 - 78.00	-	-	2.5.98
65.00 - 88.00	73.00 - 94.00	73.00 - 94.00	73.00 - 94.00	-	2.5.99
132.00 - 202.00	155.00 - 210.00	148.00 - 202.00	170.00 - 202.00	-	2.5.100

Approximate Estimates

Item nr.	SPECIFICATIONS	Unit	RESIDENTIAL £ range £
2.5	**EXTERNAL WALLS - cont'd**		
	external wall area (unless otherwise described)		
	Wall claddings - cont'd		
	PVF2 coated aluminium profiled cladding on steel rails		
2.5.101	insulated	m²	- -
2.5.102	insulated; plasterboard lining on metal tees; emulsion	m²	- -
2.5.103	insulated; coloured inner lining panel	m²	- -
2.5.104	insulated; full-height insulating block inner skin; plaster and emulsion	m²	- -
	Extra for		
2.5.105	heavyweight block inner skin	m²	- -
2.5.106	insulated; aluminium sandwich panel system	m²	- -
	insulated; aluminium sandwich panel system; on framing;		
2.5.107	on block inner skin; fair face one side	m²	- -
	Other cladding systems		
	vitreous enamelled insulated steel sandwich panel system; with non-		
2.5.108	asbestos fibre insulating board on inner face	m²	- -
2.5.109	Formalux sandwich panel system; with coloured lining tray; on steel cladding rails	m²	- -
2.5.110	aluminium over-cladding system rain screen	m²	- -
2.5.111	natural stone cladding on full-height insulating block inner skin; plaster and emulsion	m²	- -
	Curtain/glazed walling		
	12 mm Toughened glass single glazed polyester powder coated aluminium site-constructed 'stick' standard curtain walling system to		
2.5.112	internal trim	m²	- -
	Extra over for		
2.5.113	single door; including hardware	nr	- -
2.5.114	half-hour fire-resistant glazing	m²	- -
2.5.115	6 mm Toughened glass single glazed polyester powder coated aluminium site-construced 'stick' standard curtain walling system to internal trim	m²	- -
2.5.116	6 mm Clear float glass double glazed polyester powder coated aluminium site-constructed 'stick' economical quality standard curtain walling system	m²	- -
	Extra over for		
2.5.117	tinted glass	m²	- -
2.5.118	high performance glass	m²	- -
2.5.119	low E glass	m²	- -
2.5.120	laminated clear float glass	m²	- -
2.5.121	clear solar control glass	m²	- -
2.5.122	translucent thermal insulating glass	m²	- -
2.5.123	screen printing (fritting)	m²	- -
2.5.124	sand-blasting opaque glass	m²	- -
2.5.125	acid etched opaque glass	m²	- -
2.5.126	'look-a-like' steel faced insulating spandrel panels	m²	- -
2.5.127	'look-a-like' aluminium insulating spandrel panels	m²	- -
2.5.128	'look-a-like' laminate opaque insulating spandrel panels	m²	- -
2.5.129	block back-up wall including plaster and emulsion	m²	- -
2.5.130	opening lights; manual (actual area)	m²	- -
2.5.131	opening lights; mechanically-operated (actual area)	m²	- -
2.5.132	single door; including hardware	nr	- -
2.5.133	6 mm Clear float glass double glazed polyester powder coated aluminium site-constructed 'stick' medium quality standard curtain walling system, including opaque insulated spandrel panels	m²	- -
2.5.134	6 mm Clear float glass double glazed polyester powder coated aluminium site-erected 'stick' medium quality standard curtain walling system, including opaque insulated spandrel panels	m²	- -

Approximate Estimates

INDUSTRIAL £ range £	RETAILING £ range £	LEISURE £ range £	OFFICES £ range £	HOTELS £ range £	Item nr.
42.00 - 53.00	- -	- -	- -	- -	2.5.101
56.00 - 65.00	- -	- -	- -	- -	2.5.102
59.00 - 73.00	65.00 - 82.00	65.00 - 82.00	- -	- -	2.5.103
78.00 - 97.00	82.00 - 105.00	82.00 - 105.00	82.00 - 111.00	- -	2.5.104
1.10 - 2.00	1.10 - 2.30	1.10 - 2.30	1.10 - 2.30	- -	2.5.105
111.10 - 147.60	147.60 - 185.60	147.60 - 202.40	155.20 - 202.40	- -	2.5.106
- -	170.00 - 288.00	- -	- -	- -	2.5.107
126.00 - 155.00	- -	- -	- -	- -	2.5.108
148.00 - 178.00	163.00 - 186.00	163.00 - 186.00	163.00 - 186.00	- -	2.5.109
186.00 - 210.00	- -	- -	- -	- -	2.5.110
- -	- -	- -	335.00 - 479.00	350.00 - 495.00	2.5.111
- -	175.00 - 225.00	175.00 - 225.00	175.00 - 225.00	175.00 - 225.00	2.5.112
- -	800.00 - 1000	800.00 - 1000	800.00 - 1000	800.00 - 1000	2.5.113
- -	150.00 - 200.00	150.00 - 200.00	150.00 - 200.00	150.00 - 200.00	2.5.114
- -	225.00 - 275.00	225.00 - 275.00	225.00 - 275.00	225.00 - 275.00	2.5.115
- -	250.00 - 300.00	250.00 - 300.00	250.00 - 300.00	250.00 - 300.00	2.5.116
- -	8.00 - 10.00	8.00 - 10.00	8.00 - 10.00	8.00 - 10.00	2.5.117
- -	20.00 - 30.00	20.00 - 30.00	20.00 - 30.00	20.00 - 30.00	2.5.118
- -	25.00 - 30.00	25.00 - 30.00	25.00 - 30.00	25.00 - 30.00	2.5.119
- -	20.00 - 30.00	20.00 - 30.00	20.00 - 30.00	20.00 - 30.00	2.5.120
- -	50.00 - 60.00	50.00 - 60.00	50.00 - 60.00	50.00 - 60.00	2.5.121
- -	80.00 - 100.00	80.00 - 100.00	80.00 - 100.00	80.00 - 100.00	2.5.122
- -	30.00 - 35.00	30.00 - 35.00	30.00 - 35.00	30.00 - 35.00	2.5.123
- -	15.00 - 20.00	15.00 - 20.00	15.00 - 20.00	15.00 - 20.00	2.5.124
- -	50.00 - 70.00	50.00 - 70.00	50.00 - 70.00	50.00 - 70.00	2.5.125
- -	30.00 - 40.00	30.00 - 40.00	30.00 - 40.00	30.00 - 40.00	2.5.126
- -	20.00 - 30.00	20.00 - 30.00	20.00 - 30.00	20.00 - 30.00	2.5.127
- -	60.00 - 70.00	60.00 - 70.00	60.00 - 70.00	60.00 - 70.00	2.5.128
- -	30.00 - 40.00	30.00 - 40.00	30.00 - 40.00	30.00 - 40.00	2.5.129
- -	150.00 - 250.00	150.00 - 250.00	150.00 - 250.00	150.00 - 260.00	2.5.130
- -	400.00 - 550.00	400.00 - 550.00	400.00 - 550.00	400.00 - 550.00	2.5.131
- -	800.00 - 1000	800.00 - 1000	800.00 - 1000	800.00 - 1000	2.5.132
- -	300.00 - 500.00	300.00 - 500.00	300.00 - 500.00	300.00 - 500.00	2.5.133
- -	350.00 - 500.00	350.00 - 500.00	350.00 - 500.00	350.00 - 500.00	2.5.134

Item nr.	SPECIFICATIONS	Unit	RESIDENTIAL £ range £
2.5	**EXTERNAL WALLS - cont'd**		
	external wall area (unless otherwise described)		
	Curtain/glazed walling - cont'd		
2.5.135	6 mm Laminate glass double glazed polyester powder coated aluminium site-constructed 'stick' high quality bespoke curtain walling system, including opaque insulated spandrel panels	m^2	- -
	Extra over for		
2.5.136	triple glazing	m^2	- -
2.5.137	6 mm Laminate glass double glazed polyester powder coated aluminium factory produced 'unitised/panelled' high quality standard curtain walling system, including opaque insulated spandrel panels	m^2	- -
	Extra over for		
2.5.138	opening lights; manual	m^2	- -
2.5.139	opening lights; mechanically-opened	m^2	- -
2.5.140	10 mm Laminate glass double glazed polyester powder coated aluminium factory produced 'unitised/panelled' medium quality bespoke curtain walling system, including opaque insulated spandrel panels	m^2	- -
	Extra over for		
2.5.141	stainless steel clad curtain walling	m^2	- -
2.5.142	PVF2 finished curtain walling	m^2	- -
2.5.143	brise soleil	m	- -
2.5.144	stone spandrel panels (actual area)	m^2	- -
2.5.145	triple glazing	m^2	- -
2.5.146	triple glazing with wide cavity	m^2	- -
2.5.147	interstitial blind	m^2	- -
2.5.148	bomb-proof glazing	m^2	- -
2.5.149	one-hour fire protection	m^2	- -
2.5.150	facetted glazing	%	- -
2.5.151	curved glazing	%	- -
2.5.152	15 mm Clear and laminate triple glazed polyester powder coated aluminium factory produced modular 'unitised/panelled' prestigious quality bespoke curtain walling system, including opaque insulated spandrel panels	m^2	- -
2.5.153	15 mm Clear and laminate triple glazed bronze finished aluminium factory produced large span 'unitised/panelled' prestigious quality bespoke curtain walling system, including stone faced spandrel panels	m^2	- -
2.5.154	10 mm and 6 mm Clear and laminate structural siliconed double glazed standard 'stick and panel' assembly, with aluminium spacer bar and carrier frame, site-constructed curtain walling system	m^2	- -
2.5.155	10 mm and 6 mm Clear and laminate structural siliconed double glazed standard 'unitised/panelled' assembly on aluminium frame, factory produced curtain walling system	m^2	- -
2.5.156	10 mm and 6 mm Clear and laminate structural siliconed double glazed bespoke 'unitised/panelled' assembly on aluminium frame, factory produced curtain walling system	m^2	- -
	Extra over for		
2.5.157	triple glazing with wide cavity	m^2	- -
2.5.158	12 mm Toughened structural siliconed single glazed standard 'unitised/panelled' assembly factory produced curtain walling system excluding support	m^2	- -
2.5.159	10 mm and 6 mm Clear and laminate structural siliconed double glazed standard 'unitised/panelled' assembly factory produced curtain walling system excluding support	m^2	- -
	Extra over for		
2.5.160	secondary steel supports	m^2	- -
2.5.161	supporting structure for glass vertical fins	m^2	- -
2.5.162	10 mm Clear and laminate structural siliconed inverted suspended double glazed standard 'unitised/panelled' assembly factory produced curtain walling system	m^2	- -
2.5.163	Lift surround of double glazed or laminated glass with aluminium or stainless steel framing	m^2	- -

Approximate Estimates

INDUSTRIAL £ range £	RETAILING £ range £	LEISURE £ range £	OFFICES £ range £	HOTELS £ range £	Item nr.
- -	475.00 - 635.00	475.00 - 635.00	475.00 - 635.00	475.00 - 635.00	2.5.135
- -	85.00 - 130.00	85.00 - 130.00	85.00 - 130.00	85.00 - 130.00	2.5.136
- -	475.00 - 585.00	475.00 - 585.00	475.00 - 585.00	475.00 - 585.00	2.5.137
- -	160.00 - 265.00	160.00 - 265.00	160.00 - 265.00	160.00 - 265.00	2.5.138
- -	425.00 - 585.00	425.00 - 585.00	425.00 - 585.00	425.00 - 585.00	2.5.139
- -	585.00 - 740.00	585.00 - 740.00	585.00 - 740.00	585.00 - 740.00	2.5.140
- -	65.00 - 95.00	65.00 - 95.00	65.00 - 95.00	65.00 - 95.00	2.5.141
- -	85.00 - 105.00	85.00 - 105.00	85.00 - 105.00	85.00 - 105.00	2.5.142
- -	235.00 - 265.00	235.00 - 265.00	235.00 - 265.00	235.00 - 265.00	2.5.143
- -	210.00 - 320.00	210.00 - 320.00	210.00 - 320.00	210.00 - 320.00	2.5.144
- -	85.00 - 105.00	85.00 - 105.00	85.00 - 105.00	85.00 - 105.00	2.5.145
- -	105.00 - 150.00	105.00 - 150.00	105.00 - 150.00	105.00 - 150.00	2.5.146
- -	55.00 - 160.00	55.00 - 160.00	55.00 - 160.00	55.00 - 160.00	2.5.147
- -	95.00 - 130.00	95.00 - 130.00	95.00 - 130.00	95.00 - 130.00	2.5.148
- -	425.00 - 635.00	425.00 - 635.00	425.00 - 635.00	425.00 - 635.00	2.5.149
- -	16.00 - 55.00	16.00 - 55.00	16.00 - 55.00	16.00 - 55.00	2.5.150
- -	105.00 - 115.00	105.00 - 115.00	105.00 - 115.00	105.00 - 115.00	2.5.151
- -	850.00 - 1060	850.00 - 1060	850.00 - 1060	850.00 - 1060	2.5.152
- -	1270 - 1590	1270 - 1590	1270 - 1590	1270 - 1590	2.5.153
- -	370.00 - 475.00	370.00 - 475.00	370.00 - 475.00	370.00 - 475.00	2.5.154
- -	475.00 - 585.00	475.00 - 585.00	475.00 - 585.00	475.00 - 585.00	2.5.155
- -	635.00 - 740.00	635.00 - 740.00	635.00 - 740.00	635.00 - 740.00	2.5.156
- -	105.00 - 150.00	105.00 - 150.00	105.00 - 150.00	105.00 - 150.00	2.5.157
- -	320.00 - 530.00	320.00 - 530.00	320.00 - 530.00	320.00 - 530.00	2.5.158
- -	475.00 - 690.00	475.00 - 690.00	475.00 - 690.00	475.00 - 690.00	2.5.159
- -	105.00 - 210.00	105.00 - 210.00	105.00 - 210.00	105.00 - 210.00	2.5.160
- -	80.00 - 160.00	80.00 - 160.00	80.00 - 160.00	80.00 - 160.00	2.5.161
- -	740.00 - 850.00	740.00 - 850.00	740.00 - 850.00	740.00 - 850.00	2.5.162
- -	475.00 - 795.00	475.00 - 795.00	475.00 - 795.00	475.00 - 795.00	2.5.163

Item nr.	SPECIFICATIONS	Unit	RESIDENTIAL £ range £	
2.5	**EXTERNAL WALLS - cont'd**			
	external wall area (unless otherwise described)			
	Curtain/glazed walling - cont'd			
2.5.164	Patent glazing systems; excluding opening lights and lead flashings etc, 7 mm Georgian wired cast glass, aluminium glazing bars spanning up to 3 m at 600 - 626 mm spacing	m^2	-	-
2.5.165	Patent glazing systems; excluding opening lights and lead flashings etc, 6.4 mm Laminate glass, aluminium glazing bars spanning up to 3 m at 600 - 626 mm spacing	m^2	-	-
	Comparative external finishes			
	Comparative concrete wall finishes			
2.5.166	wrought formwork one side including rubbing down	m^2	-	-
2.5.167	shotblasting to expose aggregate	m^2	-	-
2.5.168	bush hammering to expose aggregate	m^2	-	-
	Comparative insitu finishes			
2.5.169	two coats "Sandtex Matt" cement paint	m^2	-	-
2.5.170	cement and sand plain face rendering	m^2	-	-
2.5.171	three coat "Tyrolean" rendering; including backing	m^2	-	-
2.5.172	"Mineralite" decorative rendering; including backing	m^2	-	-
	Comparative claddings			
2.5.173	25 mm thick tongued and grooved "Tanalised" softwood boarding; including battens	m^2	-	-
2.5.174	25 mm thick tongued and grooved western red cedar boarding including battens	m^2	-	-
2.5.175	machine-made tiles; including battens	m^2	-	-
2.5.176	best hand-made sand-faced tiles; including battens	m^2	-	-
2.5.177	20 mm x 20 mm thick mosaic glass or ceramic; in common colours; fixed on prepared surface	m^2	-	-
2.5.178	75 mm thick Portland stone facing slabs and fixing; including clamps	m^2	-	-
2.5.179	75 mm thick Ancaster stone facing slabs and fixing; including clamps	m^2	-	-
	Comparative curtain wall finishes			
	Extra over aluminium mill finish for			
2.5.180	natural anodising	m^2	-	-
2.5.181	polyester powder coating	m^2	-	-
2.5.182	bronze anodising	m^2	-	-
2.6	**WINDOWS AND EXTERNAL DOORS**			
	window and external door area (unless otherwise described)			
	comprising:			
	Softwood windows and external doors			
	Steel windows and external doors			
	Steel roller shutters			
	Hardwood windows and external doors			
	uPVC windows and external doors			
	Aluminium windows, entrance screens and doors			
	Stainless steel entrance screens and doors			
	Shop fronts, shutters and grilles			
	Softwood windows and external doors			
	Standard windows; painted			
2.6.1	single glazed	m^2	178.00	- 233.00
2.6.2	double glazed	m^2	233.00	- 280.00
	Purpose-made windows; painted			
2.6.3	single glazed	m^2	233.00	- 297.00
2.6.4	double glazed	m^2	297.00	- 373.00

Approximate Estimates

INDUSTRIAL £ range £	RETAILING £ range £	LEISURE £ range £	OFFICES £ range £	HOTELS £ range £	Item nr.
- -	90.00 - 110.00	90.00 - 110.00	90.00 - 110.00	90.00 - 110.00	2.5.164
- -	280.00 - 330.00	280.00 - 330.00	280.00 - 330.00	280.00 - 330.00	2.5.165
- -	2.60 - 5.20	2.60 - 5.20	2.60 - 5.20	2.60 - 5.20	2.5.166
- -	3.30 - 6.60	3.30 - 6.60	3.30 - 6.60	3.30 - 6.60	2.5.167
- -	10.60 - 14.60	10.60 - 14.60	10.60 - 14.60	10.60 - 14.60	2.5.168
- -	6.20 - 8.70	6.20 - 8.70	6.20 - 8.70	6.20 - 8.70	2.5.169
- -	10.90 - 16.40	10.90 - 16.40	10.90 - 16.40	10.90 - 16.40	2.5.170
- -	26.20 - 31.20	26.20 - 31.20	26.20 - 31.20	26.20 - 35.80	2.5.171
- -	52.50 - 60.90	52.50 - 60.90	52.50 - 60.90	55.90 - 68.50	2.5.172
- -	24.70 - 31.20	24.70 - 31.20	24.70 - 31.20	24.70 - 31.20	2.5.173
- -	29.30 - 37.30	29.30 - 37.30	29.30 - 37.30	29.30 - 37.30	2.5.174
- -	31.20 - 36.50	31.20 - 36.50	31.20 - 36.50	31.20 - 36.50	2.5.175
- -	38.00 - 41.80	38.00 - 41.80	38.00 - 41.80	38.00 - 41.80	2.5.176
- -	81.00 - 94.00	81.00 - 94.00	81.00 - 94.00	81.00 - 94.00	2.5.177
- -	- -	- -	312.00 - 388.00	327.00 - 426.00	2.5.178
- -	- -	- -	342.00 - 426.00	- -	2.5.179
- -	14.80 - 19.40	14.80 - 19.40	14.80 - 19.40	14.80 - 19.40	2.5.180
- -	19.40 - 32.70	19.40 - 32.70	19.40 - 32.70	19.40 - 31.20	2.5.181
- -	27.00 - 40.30	27.00 - 40.30	27.00 - 40.30	27.00 - 40.30	2.5.182
170.00 - 218.00	163.00 - 225.00	178.02 - 233.00	178.00 - 233.00	178.00 - 248.00	2.6.1
210.00 - 272.00	225.00 - 272.00	232.80 - 297.00	233.00 - 297.00	233.00 - 297.00	2.6.2
233.00 - 297.00	233.00 - 233.00	248.02 - 304.00	240.00 - 304.00	240.00 - 320.00	2.6.3
289.00 - 342.00	289.00 - 342.00	246.00 - 335.00	289.00 - 335.00	304.00 - 358.00	2.6.4

Approximate Estimates

Item nr.	SPECIFICATIONS	Unit	RESIDENTIAL £ range £
2.6	**WINDOWS AND EXTERNAL DOORS - cont'd** window and external door area (unless otherwise described)		
	Softwood windows and external doors - cont'd Standard external softwood doors and hardwood frames; doors painted; including ironmongery		
2.6.5	two panelled door; plywood panels	nr	304.00 - 365.00
2.6.6	solid flush door	nr	342.00 - 396.00
2.6.7	two panelled door; glazed panels	nr	647.00 - 723.00
	heavy-duty solid flush door		
2.6.8	single leaf	nr	- -
2.6.9	double leaf	nr	- -
	Extra for		
2.6.10	emergency fire exit door	nr	- -
	Steel windows and external doors Standard windows		
2.6.11	single glazed; galvanised; painted	m²	186.00 - 240.00
2.6.12	single glazed; powder-coated	m²	195.00 - 248.00
2.6.13	double glazed; galvanised; painted	m²	240.00 - 288.00
2.6.14	double glazed; powder coated	m²	248.00 - 304.00
	Purpose-made windows		
2.6.15	double glazed; powder coated	m²	- -
	Steel roller shutters Shutters; galvanised		
2.6.16	manual	m²	- -
2.6.17	electric	m²	- -
2.6.18	manual; insulated	m²	- -
2.6.19	electric; insulated	m²	- -
2.6.20	electric; insulated; fire-resistant	m²	- -
	Hardwood windows and external doors Standard windows; stained or uVC coated		
2.6.21	single glazed	m²	248.00 - 350.00
2.6.22	double glazed	m²	320.00 - 403.00
	Purpose-made windows; stained or uPVC coated		
2.6.23	single glazed	m²	288.00 - 388.00
2.6.24	double glazed	m²	350.00 - 449.00
	uPVC windows and external doors Purpose-made windows		
2.6.25	double glazed	m²	502.00 - 624.00
	Extra for		
2.6.26	tinted glass	m²	- -
	Aluminium windows, entrance screens and doors Standard windows; anodised finish		
2.6.27	single glazed; horizontal sliding sash	m²	- -
2.6.28	single glazed; vertical sliding slash	m²	- -
2.6.29	single glazed; casement; in hardwood sub-frame	m²	263.00 - 350.00
2.6.30	double glazed; vertical sliding slash	m²	- -
2.6.31	double glazed; casement; in hardwood sub-frame	m²	320.00 - 426.00
	Purpose-made windows		
2.6.32	single glazed	m²	- -
2.6.33	double glazed	m²	- -
2.6.34	double glazed; feature; with precast concrete surrounds m²	m²	- -
	Purpose-made entrance screens and doors		
2.6.35	double glazed	m²	- -
	Purpose-made revolving door		
2.6.36	2000 mm diameter; double glazed	nr	- -

Approximate Estimates

INDUSTRIAL £ range £	RETAILING £ range £	LEISURE £ range £	OFFICES £ range £	HOTELS £ range £	Item nr.
- -	- -	- -	- -	- -	2.6.5
449.00 - 791.00	358.00 - 403.00	358.00 - 403.00	358.00 - 403.00	365.00 - 418.00	2.6.6
- -	- -	- -	- -	- -	2.6.7
517.00 - 791.00	586.00 - 730.00	586.00 - 715.00	-	-	2.6.8
913.00 - 1301.00	1111.00 - 1240.00	1111.00 - 1240.00	-	-	2.6.9
- -	195.00 - 304.00	195.00 - 304.00	-	-	2.6.10
170.00 - 225.00	170.00 - 233.00	186.00 - 233.00	186.00 - 233.00	186.00 - 256.00	2.6.11
178.00 - 233.00	178.00 - 233.00	195.00 - 248.00	195.00 - 248.00	195.00 - 263.00	2.6.12
225.00 - 280.00	225.00 - 280.00	233.00 - 295.00	233.00 - 295.00	240.00 - 312.00	2.6.13
225.00 - 280.00	225.00 - 295.00	240.00 - 304.00	240.00 - 304.00	248.00 - 320.00	2.6.14
- -	320.00 - 403.00	320.00 - 403.00	320.00 - 403.00	230.00 - 426.00	2.6.15
186.00 - 240.00	170.00 - 225.00	170.00 - 233.00	-	-	2.6.16
225.00 - 320.00	210.00 - 312.00	210.00 - 312.00	-	-	2.6.17
312.00 - 365.00	288.00 - 350.00	288.00 - 350.00	-	-	2.6.18
350.00 - 441.00	335.00 - 426.00	335.00 - 426.00	-	-	2.6.19
844.00 - 1012.00	814.00 - 1012.00	814.00 - 1012.00	-	-	2.6.20
320.00 - 388.00	320.00 - 388.00	320.00 - 388.00	320.00 - 388.00	335.00 - 403.00	2.6.21
403.00 - 487.00	403.00 - 487.00	403.00 - 487.00	403.00 - 487.00	403.00 - 418.00	2.6.22
388.00 - 449.00	388.00 - 449.00	388.00 - 449.00	388.00 - 449.00	388.00 - 487.00	2.6.23
449.00 - 548.00	449.00 - 548.00	449.00 - 548.00	449.00 - 548.00	449.00 - 571.00	2.6.24
464.00 - 517.00	464.00 - 517.00	464.00 - 517.00	464.00 - 517.00	464.00 - 517.00	2.6.25
23.00 - 32.00	23.00 - 32.00	23.00 - 32.00	23.00 - 32.00	23.00 - 32.00	2.6.26
225.00 - 272.00	225.00 - 272.00	225.00 - 272.00	225.00 - 272.00	225.00 - 288.00	2.6.27
350.00 - 418.00	350.00 - 418.00	350.00 - 418.00	350.00 - 418.00	350.00 - 426.00	2.6.28
-	-	-	-	-	2.6.29
388.00 - 449.00	434.00 - 548.00	434.00 - 548.00	434.00 - 548.00	434.00 - 578.00	2.6.30
-	-	-	-	-	2.6.31
225.00 - 320.00	225.00 - 320.00	225.00 - 320.00	225.00 - 320.00	225.00 - 320.00	2.6.32
502.00 - 609.00	502.00 - 609.00	502.00 - 609.00	502.00 - 609.00	502.00 - 624.00	2.6.33
- -	- -	1263.00 - 1902.00	-	-	2.6.34
-	624.00 - 1012.00	624.00 - 1012.00	624.00 - 1012.00	-	2.6.35
-	22367 - 28605	22367 - 28605	22017 - 28605	22017 - 28605	2.6.36

Approximate Estimates

Item nr.	SPECIFICATIONS	Unit	RESIDENTIAL £ range £
2.6	**WINDOWS AND EXTERNAL DOORS - cont'd** **window and external door area** (unless otherwise described)		
	Stainless steel entrance screens and doors Purpose-made screen; double glazed		
2.6.37	with manual doors	m²	- -
2.6.38	with automatic doors	m²	- -
	Purpose-made revolving door		
2.6.39	2000 mm diameter; double glazed	nr	- -
	Shop fronts, shutters and grilles **shop front length**		
2.6.40	Temporary timber shop fronts	m	- -
2.6.41	Grilles or shutters	m	- -
2.6.42	Fire shutters; power-operated	m	- -
	Shop front		
2.6.43	flat facade; glass in aluminium framing; manual centre doors only	m	- -
2.6.44	flat facade; glass in aluminium framing; automatic centre doors only	m	- -
2.6.45	hardwood and glass; including high enclosed window beds	m	- -
2.6.46	high quality; marble or granite plasters and stair risers; window beds and backings; illuminated signs	m	- -
2.7	**INTERNAL WALLS, PARTITIONS AND DOORS** **internal wall area** (unless otherwise described)		
	comprising:		
	Timber or metal stud partitions and doors Brick/block partitions and doors Reinforced concrete walls Solid partitioning and doors Glazed partitioning and doors Special partitioning and doors WC/Changing cubicles Comparative doors/door linings/frames Perimeter treatments		
	Timber or metal stud partitions and doors Structure only comprising softwood studs at 400 mm x 600 mm centres; head and sole plates		
2.7.1	100 mm x 38 mm	m²	11.00 - 14.00
	Softwood stud and plasterboard partitions		
2.7.2	57 mm thick "Paramount" dry partition	m²	20.50 - 22.80
2.7.3	65 mm thick "Paramount" dry partition	m²	22.80 - 25.10
2.7.4	50 mm thick laminated partition	m²	22.10 - 24.30
2.7.5	65 mm thick laminated partition	m²	25.50 - 27.80
2.7.6	65 mm thick laminated partition; emulsion both sides	m²	31.20 - 34.20
2.7.7	100 mm thick partition; taped joints; emulsion both sides	m²	34.20 - 39.60
2.7.8	100 mm thick partition; skim and emulsion both sides	m²	39.60 - 44.90
2.7.9	150 mm thick partition as party wall; skim and emulsion both sides	m²	47.20 - 55.90
	Metal stud and plasterboard partitions		
2.7.10	170 mm thick partition; one hour; taped joints; emulsion both sides	m²	- -
2.7.11	200 mm thick partition; two hour; taped joints; emulsion both sides	m²	- -
2.7.12	325 mm thick two layer partition; cavity insulation	m²	- -
	Extra for		
2.7.13	curved work	m²	- -

Approximate Estimates

INDUSTRIAL £ range £	RETAILING £ range £	LEISURE £ range £	OFFICES £ range £	HOTELS £ range £	Item nr.
- -	1012 - 1522	- -	- -	- -	2.6.37
- -	1255 - 1788	- -	- -	- -	2.6.38
- -	30736 - 38496	28978 - 36294	28978 - 36294	28978 - 36294	2.6.39
- -	41.10 - 58.20	- -	- -	- -	2.6.40
- -	502.00 - 1012.0	- -	- -	- -	2.6.41
- -	890.00 - 1271.0	- -	- -	- -	2.6.42
- -	1012 - 2328	- -	- -	- -	2.6.43
- -	1263 - 2830	- -	- -	- -	2.6.44
- -	3576 - 4336	- -	- -	- -	2.6.45
- -	4032 - 5592	- -	- -	- -	2.6.46
- -	- -	- -	- -	- -	2.7.1
- -	- -	- -	- -	- -	2.7.2
- -	- -	- -	- -	- -	2.7.3
- -	- -	- -	- -	- -	2.7.4
- -	- -	- -	- -	- -	2.7.5
- -	- -	- -	- -	- -	2.7.6
- -	- -	- -	- -	- -	2.7.7
- -	- -	- -	- -	- -	2.7.8
- -	- -	- -	- -	- -	2.7.9
40.00 - 49.00	43.00 - 52.00	43.00 - 52.00	43.00 - 52.00	44.00 - 52.00	2.7.10
56.00 - 62.00	59.00 - 64.00	59.00 - 64.00	59.00 - 64.00	61.00 - 65.00	2.7.11
94.00 - 132.00	100.00 - 137.00	100.00 - 137.00	100.00 - 137.00	100.00 - 137.00	2.7.12
+50%	+50%	+50%	+50%	+50%	2.7.13

Approximate Estimates

Item nr.	SPECIFICATIONS	Unit	RESIDENTIAL £ range £
2.7	**INTERNAL WALLS, PARTITIONS AND DOORS** - cont'd		
	internal wall area (unless otherwise described)		
	Timber or metal stud partitions and doors - cont'd		
	Metal stud and plasterboard partitions; emulsion both sides; softwood doors and frames; painted		
2.7.14	170 mm thick partition	m²	- -
2.7.15	200 mm thick partition; insulated	m²	- -
	Stud or plasterboard partitions; softwood doors and frames; painted		
2.7.16	partition; plaster and emulsion both sides	m²	67.00 - 85.00
	Extra for		
2.7.17	vinyl paper in lieu of emulsion	m²	4.40 - 6.60
2.7.18	hardwood doors and frames in lieu of softwood	m²	19.00 - 22.10
2.7.19	partition; plastered and vinyl both sides	m²	79.00 - 96.00
	Stud or plasterboard partitions; hardwood doors and frames		
2.7.20	partition; plaster and emulsion both sides	m²	85.00 - 107.00
2.7.21	partition; plaster and vinyl both sides	m²	97.00 - 120.00
	Brick/block partitions and doors		
	Autoclaved aerated/lightweight block partitions		
2.7.22	75 mm thick	m²	12.80 - 15.20
2.7.23	100 mm thick	m²	17.10 - 20.50
2.7.24	130 mm thick; insulating	m²	20.90 - 23.20
2.7.25	150 mm thick	m²	23.20 - 24.70
2.7.26	190 mm thick	m²	27.80 - 32.00
	Extra for		
2.7.27	fair face both sides	m²	2.30 - 4.50
2.7.28	curved work		+10% - +20%
2.7.29	average thickness; fair face both sides	m²	22.40 - 27.80
2.7.30	average thickness; fair face and emulsion both sides	m²	26.20 - 33.50
2.7.31	average thickness; plaster and emulsion both sides	m²	40.30 - 47.90
	Concrete block partitions		
2.7.32	to retail units	m²	- -
	Dense aggregate block partitions		
2.7.33	average thickness; fair face both sides	m²	26.20 - 32.00
2.7.34	average thickness; fair face and emulsion both sides	m²	30.40 - 36.50
2.7.35	average thickness; plaster and emulsion both sides	m²	44.90 - 51.00
	Coloured dense aggregate masonry block partition		
2.7.36	fair face both sides	m²	- -
	Common brick partitions; bricks PC £120.00/1000		
2.7.37	half brick thick	m²	24.00 - 26.20
2.7.38	half brick thick; fair face both sides	m²	26.20 - 32.00
2.7.39	half brick thick; fair face and emulsion both sides	m²	30.40 - 36.50
2.7.40	half brick thick; plaster and emulsion both sides	m²	43.40 - 55.90
2.7.41	one brick thick	m²	44.90 - 51.00
2.7.42	one brick thick; fair face and emulsion both sides	m²	47.90 - 55.90
2.7.43	one brick thick; fair face and emulsion both sides	m²	52.50 - 60.50
2.7.44	one brick thick; plaster and emulsion both sides	m²	65.40 - 77.60
2.7.45	Block partitions; softwood doors and frames; painted partition	m²	37.30 - 49.50
2.7.46	partition; fair faced both sides	m²	38.80 - 52.50
2.7.47	partition; fair face and emulsion both sides	m²	44.90 - 57.40
2.7.48	partition; plastered and emulsioned both sides	m²	59.00 - 74.60
	Block partitions; hardwood doors and frames;		
2.7.49	partition	m²	56.00 - 72.00
2.7.50	partition; plaster and emulsion both sides	m²	76.00 - 96.00
	Reinforced concrete walls		
	Walls		
2.7.51	150 mm thick	m²	78.00 - 93.00
2.7.52	150 mm thick; plaster and emulsion both sides	m²	103.50 - 124.80

Approximate Estimates

INDUSTRIAL £ range £	RETAILING £ range £	LEISURE £ range £	OFFICES £ range £	HOTELS £ range £	Item nr.
52.00 - 72.00	56.00 - 75.00	56.00 - 75.00	56.00 - 75.00	56.00 - 76.00	2.7.14
72.00 - 87.00	73.00 - 90.00	73.00 - 90.00	73.00 - 90.00	73.00 - 96.00	2.7.15
65.00 - 82.00	68.00 - 85.00	68.00 - 85.00	68.00 - 85.00	68.00 - 90.00	2.7.16
- -	5.00 - 7.00	5.00 - 7.00	5.00 - 7.00	5.00 - 11.30	2.7.17
19.40 - 22.10	19.40 - 22.10	19.40 - 22.10	19.40 - 22.10	19.40 - 29.70	2.7.18
50.00 - 93.00	78.00 - 96.00	78.00 - 96.00	78.00 - 96.00	85.00 - 113.00	2.7.19
84.00 - 102.00	85.00 - 107.00	85.00 - 107.00	85.00 - 107.00	85.00 - 113.00	2.7.20
113.00 - 111.00	97.00 - 120.00	97.00 - 120.00	97.00 - 120.00	97.00 - 135.00	2.7.21
12.80 - 14.50	12.80 - 15.20	12.80 - 15.20	12.80 - 16.40	14.50 - 16.70	2.7.22
15.60 - 19.40	16.40 - 20.20	16.40 - 20.20	16.40 - 20.20	17.90 - 20.90	2.7.23
19.40 - 22.40	20.90 - 24.00	20.90 - 24.00	20.90 - 24.00	22.40 - 25.50	2.7.24
21.70 - 24.00	22.40 - 24.70	22.40 - 24.70	22.40 - 24.70	24.00 - 24.70	2.7.25
26.20 - 30.40	27.80 - 32.00	27.80 - 32.00	27.80 - 32.00	27.80 - 32.70	2.7.26
2.30 - 5.00	2.30 - 5.00	2.30 - 5.00	2.30 - 5.00	2.30 - 5.00	2.7.27
+10% - +20%	+10% - +20%	+10% - +20%	+10% - +20%	+10% - +20%	2.7.28
22.40 - 26.20	15.80 - 27.80	24.00 - 27.80	24.00 - 27.80	24.70 - 29.70	2.7.29
26.20 - 33.50	26.20 - 33.50	26.20 - 33.50	26.20 - 33.50	28.10 - 35.00	2.7.30
39.60 - 46.40	40.30 - 47.90	40.30 - 47.90	40.30 - 47.90	41.80 - 51.00	2.7.31
- -	35.80 - 40.30	- -	- -	- -	2.7.32
26.20 - 30.40	26.60 - 32.00	26.60 - 32.00	26.60 - 32.00	26.60 - 33.50	2.7.33
29.70 - 36.50	30.40 - 36.50	30.40 - 36.50	30.40 - 36.50	32.00 - 38.00	2.7.34
43.40 - 47.90	44.90 - 51.00	44.90 - 51.00	44.90 - 51.00	44.90 - 52.50	2.7.35
- -	- -	49.50 - 60.50	- -	- -	2.7.36
22.40 - 25.50	24.00 - 26.20	24.00 - 26.20	24.00 - 26.20	24.70 - 27.80	2.7.37
24.70 - 30.40	26.20 - 32.00	26.20 - 32.00	26.20 - 32.00	26.20 - 33.50	2.7.38
29.30 - 36.50	30.40 - 36.50	30.40 - 36.50	30.40 - 36.50	32.00 - 38.00	2.7.39
41.80 - 54.40	43.40 - 55.90	43.40 - 55.90	43.40 - 55.90	44.90 - 60.50	2.7.40
43.40 - 47.90	44.90 - 51.00	44.90 - 51.00	44.90 - 51.00	47.90 - 52.50	2.7.41
46.40 - 52.50	47.90 - 55.90	47.90 - 55.90	47.90 - 55.90	49.50 - 55.90	2.7.42
49.50 - 57.40	52.50 - 60.50	52.50 - 60.50	52.50 - 60.50	55.90 - 63.50	2.7.43
63.50 - 73.00	65.00 - 76.10	65.00 - 76.10	65.00 - 76.10	70.00 - 80.60	2.7.44
35.80 - 47.90	37.30 - 49.50	37.30 - 49.50	37.30 - 49.50	38.80 - 51.00	2.7.45
37.30 - 49.50	38.80 - 52.50	38.80 - 52.50	38.80 - 52.50	41.80 - 52.50	2.7.46
41.80 - 55.90	44.90 - 57.40	44.90 - 57.40	44.90 - 57.40	46.40 - 57.40	2.7.47
57.40 - 71.50	59.00 - 74.60	59.00 - 74.60	59.00 - 74.60	59.00 - 78.00	2.7.48
54.40 - 68.50	55.90 - 71.50	55.90 - 71.50	55.90 - 71.50	60.50 - 73.00	2.7.49
73.00 - 93.00	76.00 - 96.00	76.00 - 96.00	76.00 - 82.00	65.00 - 82.00	2.7.50
78.00 - 91.00	78.00 - 91.00	78.00 - 91.00	78.00 - 91.00	78.00 - 91.00	2.7.51
103.00 - 125.00	103.00 - 125.00	103.00 - 125.00	103.00 - 125.00	103.00 - 125.00	2.7.52

Item nr.	SPECIFICATIONS	Unit	RESIDENTIAL £ range £
2.7	**INTERNAL WALLS, PARTITION AND DOORS - cont'd**		
	internal wall area (unless otherwise described)		
	Solid partitioning and doors		
	Patent partitioning; softwood doors		
2.7.53	frame and sheet	m^2	- -
2.7.54	frame and panel	m^2	- -
2.7.55	panel to panel	m^2	- -
2.7.56	economical	m^2	- -
	Patent partitioning; hardwood doors		
2.7.57	economical	m^2	- -
	Demountable partitioning; hardwood doors		
2.7.58	medium quality; vinyl faced	m^2	- -
2.7.59	high quality; vinyl faced	m^2	- -
	Glazed partitioning and doors		
	Aluminium internal patent glazing		
2.7.60	single glazed	m^2	- -
2.7.61	double glazed	m^2	- -
	Demountable steel partitioning and doors		
2.7.62	medium quality	m^2	- -
2.7.63	high quality	m^2	- -
	Demountable aluminium/steel partitioning and doors		
2.7.64	high quality	m^2	- -
2.7.65	high quality; sliding	m^2	- -
	Stainless steel glazed manual doors and screens		
2.7.66	high quality; to inner lobby of malls	m^2	- -
	Special partitioning and doors		
	Demountable fire partitions		
2.7.67	enamelled steel; half hour	m^2	- -
2.7.68	stainless steel; half hour	m^2	- -
	Soundproof partitions; hardwood doors		
2.7.69	luxury veneered	m^2	- -
	Folding screens		
2.7.70	gym divider; electronically operated	m^2	- -
2.7.71	bar divider	m^2	- -
2.7.72	Squash court glass back wall and door	m^2	- -
	WC/Changing cubicles each (unless otherwise described)		
2.7.73	WC cubicles	nr	- -
	Changing cubicles		
2.7.74	aluminium	nr	- -
2.7.75	aluminium; textured glass and bench seating	nr	- -
	Comparative doors/door linings/frames		
	Standard softwood doors; excluding ironmongery; linings and frames		
	40 mm thick flush; hollow core; painted		
2.7.76	726 mm x 2040 mm	nr	37.00 - 44.00
2.7.77	826 mm x 2040 mm	nr	37.00 - 44.00
	40 mm thick flush; hollow core; plywood faced; painted		
2.7.78	726 mm x 2040 mm	nr	44.00 - 56.00
2.7.79	826 mm x 2040 mm	nr	44.00 - 56.00
	40 mm thick flush; hollow core; sapele veneered hard board faced		
2.7.80	726 mm x 2040 mm	nr	37.00 - 44.00
2.7.81	826 mm x 2040 mm	nr	37.00 - 44.00
	40 mm thick flush; hollow core; teak veneered hard board faced		
2.7.82	726 mm x 2040 mm	nr	87.00 - 107.00
2.7.83	826 mm x 2040 mm	nr	87.00 - 107.00

Approximate Estimates

INDUSTRIAL £ range £	RETAILING £ range £	LEISURE £ range £	OFFICES £ range £	HOTELS £ range £	Item nr.
49.00 - 97.00	56.00 - 100.00	56.00 - 100.00	56.00 - 100.00	56.00 - 123.00	2.7.53
42.00 - 78.00	45.00 - 81.00	45.00 - 81.00	45.00 - 81.00	45.00 - 100.00	2.7.54
62.00 - 132.00	67.00 - 138.00	67.00 - 138.00	67.00 - 138.00	67.00 - 155.00	2.7.55
67.00 - 85.00	72.00 - 91.00	72.00 - 91.00	72.00 - 91.00	72.00 - 113.00	2.7.56
73.00 - 90.00	73.00 - 96.00	73.00 - 96.00	73.00 - 96.00	73.00 - 100.00	2.7.57
96.00 - 125.00	100.00 - 125.00	100.00 - 125.00	100.00 - 125.00	100.00 - 135.00	2.7.58
126.00 - 177.00	126.00 - 177.00	126.00 - 177.00	126.00 - 177.00	132.00 - 192.00	2.7.59
85.00 - 117.00	-	-	-	-	2.7.60
145.00 - 180.00	-	-	-	-	2.7.61
155.00 - 195.00	-	-	155.00 - 210.00	-	2.7.62
195.00 - 240.00	-	-	210.00 - 248.00	-	2.7.63
-	-	233.00 - 418.00	-	333.00 - 418.00	2.7.64
-	-	558.00 - 683.00	-	-	2.7.65
-	333.00 - 893.00	-	-	-	2.7.66
349.00 - 528.00	403.00 - 566.00	403.00 - 566.00	403.00 - 566.00	-	2.7.67
-	706.00 - 893.00	706.00 - 893.00	706.00 - 893.00	-	2.7.68
178.00 - 240.00	195.00 - 272.00	195.00 - 272.00	195.00 - 272.00	195.00 - 295.00	2.7.69
-	-	132.00 - 148.00	-	-	2.7.70
-	-	350.00 - 396.00	-	-	2.7.71
-	-	233.00 - 275.00	-	-	2.7.72
248.00 - 388.00	256.00 - 388.00	256.00 - 388.00	248.00 - 551.00	310.00 - 551.00	2.7.73
-	-	303.00 - 551.00	-	-	2.7.74
-	-	504.00 - 660.00	-	-	2.7.75
37.00 - 44.00	37.00 - 44.00	37.00 - 45.00	37.00 - 44.00	37.00 - 44.00	2.7.76
37.00 - 44.00	37.00 - 44.00	37.00 - 44.00	37.00 - 44.00	37.00 - 44.00	2.7.77
44.00 - 56.00	44.00 - 56.00	44.00 - 56.00	44.00 - 56.00	44.00 - 56.00	2.7.78
44.00 - 56.00	44.00 - 56.00	44.00 - 56.00	44.00 - 56.00	44.00 - 56.00	2.7.79
37.00 - 44.00	37.00 - 44.00	37.00 - 44.00	37.00 - 44.00	24.00 - 44.00	2.7.80
37.00 - 44.00	37.00 - 44.00	37.00 - 44.00	37.00 - 44.00	37.00 - 44.00	2.7.81
87.00 - 105.00	87.00 - 105.00	87.00 - 105.00	87.00 - 105.00	87.00 - 105.00	2.7.82
87.00 - 105.00	87.00 - 105.00	87.00 - 105.00	87.00 - 105.00	87.00 - 105.00	2.7.83

2.7 INTERNAL WALLS, PARTITION AND DOORS - cont'd

each (unless otherwise described)

Comparative doors/door linings/ frames - cont'd
Standard softwood fire doors; excluding ironmongery; lining and frames
44 mm thick flush; half hour fire check; plywood faced; painted

Item nr.	SPECIFICATIONS	Unit	RESIDENTIAL £ range £
2.7.84	726 mm x 2040 mm	nr	56.00 - 62.00
2.7.85	826 mm x 2040 mm	nr	59.00 - 67.00
	44 mm thick flush; half hour fire check; plywood faced; painted		
2.7.86	726 mm x 2040 mm	nr	73.00 - 81.00
2.7.87	826 mm x 2040 mm	nr	75.00 - 84.00
	54 mm thick flush; one hour fire check; sapele veneered hardboard faced		
2.7.88	726 mm x 2040 mm	nr	225.00 - 248.00
2.7.89	826 mm x 2040 mm	nr	233.00 - 248.00
	54 mm thick flush; one hour fire check; sapele veneered hardboard faced		
2.7.90	726 mm x 2040 mm	nr	240.00 - 272.00
2.7.91	826 mm x 2040 mm	nr	248.00 - 280.00
	Purpose-made softwood doors; excluding ironmongery; linings and frames 44 mm thick four panel door; painted		
2.7.92	726 mm x 2040 mm	nr	125.00 - 140.00
2.7.93	826 mm x 2040 mm	nr	132.00 - 148.00
	50 mm thick four door panel; wax polished		
2.7.94	726 mm x 2040 mm	nr	240.00 - 272.00
2.7.95	826 mm x 2040 mm	nr	256.00 - 280.00
	Purpose-made softwood door frames/linings; painted including grounds 32 mm x 100 mm lining		
2.7.96	726 mm x 2040 mm opening	nr	75.00 - 84.00
2.7.97	826 mm x 2040 mm opening	nr	78.00 - 87.00
	32 mm x 140 mm lining		
2.7.98	726 mm x 2040 mm opening	nr	84.00 - 96.00
2.7.99	826 mm x 2040 mm opening	nr	85.00 - 97.00
	32 mm x 250 mm cross-tongued lining		
2.7.100	726 mm x 2040 mm opening	nr	110.00 - 126.00
2.7.101	826 mm x 2040 mm opening	nr	111.00 - 128.00
	32 mm x 375 mm cross-tongued lining		
2.7.102	726 mm x 2040 mm opening	nr	143.00 - 163.00
2.7.103	826 mm x 2040 mm opening	nr	148.00 - 163.00
	75 mm x 100 mm rebated frame		
2.7.104	726 mm x 2040 mm opening	nr	97.00 - 103.00
2.7.105	826 mm x 2040 mm opening	nr	102.00 - 108.00
	Purpose-made mahogany door frames/linings; wax polished including grounds 32 mm x 100 mm lining		
2.7.106	726 mm x 2040 mm opening	nr	100.00 - 108.00
2.7.107	826 mm x 2040 mm opening	nr	102.00 - 111.00
	32 mm x 140 mm lining		
2.7.108	726 mm x 2040 mm opening	nr	111.00 - 126.00
2.7.109	826 mm x 2040 mm opening	nr	114.00 - 128.00
	32 mm x 250 mm cross-tongued lining		
2.7.110	726 mm x 2040 mm opening	nr	128.00 - 137.00
2.7.111	826 mm x 2040 mm opening	nr	131.00 - 142.00
	32 mm x 375 mm cross-tongued lining		
2.7.112	726 mm x 2040 mm opening	nr	195.00 - 210.00
2.7.113	826 mm x 2040 mm opening	nr	195.00 - 218.00
	75 mm x 100 mm rebated frame		
2.7.114	726 mm x 2040 mm opening	nr	140.00 - 148.00
2.7.115	826 mm x 2040 mm opening	nr	140.00 - 148.00

Approximate Estimates 775

INDUSTRIAL £ range £	RETAILING £ range £	LEISURE £ range £	OFFICES £ range £	HOTELS £ range £	Item nr.
56.00 - 62.00	56.00 - 62.00	56.00 - 62.00	56.00 - 62.00	56.00 - 62.00	2.7.84
59.00 - 67.00	59.00 - 67.00	59.00 - 67.00	59.00 - 67.00	59.00 - 67.00	2.7.85
72.00 - 81.00	72.00 - 81.00	72.00 - 81.00	72.00 - 81.00	72.00 - 81.00	2.7.86
75.00 - 84.00	75.00 - 84.00	75.00 - 84.00	75.00 - 84.00	75.00 - 84.00	2.7.87
225.00 - 248.00	225.00 - 248.00	225.00 - 248.00	225.00 - 248.00	225.00 - 248.00	2.7.88
233.00 - 248.00	233.00 - 248.00	233.00 - 248.00	233.00 - 248.00	233.00 - 248.00	2.7.89
240.00 - 272.00	240.00 - 272.00	240.00 - 272.00	240.00 - 272.00	240.00 - 272.00	2.7.90
248.00 - 280.00	248.00 - 280.00	248.00 - 280.00	248.00 - 280.00	248.00 - 280.00	2.7.91
125.00 - 140.00	125.00 - 140.00	125.00 - 140.00	125.00 - 140.00	125.00 - 140.00	2.7.92
132.00 - 148.00	132.00 - 148.00	132.00 - 148.00	132.00 - 148.00	132.00 - 148.00	2.7.93
240.00 - 272.00	240.00 - 272.00	240.00 - 272.00	240.00 - 272.00	240.00 - 272.00	2.7.94
248.00 - 280.00	248.00 - 280.00	248.00 - 280.00	248.00 - 280.00	248.00 - 280.00	2.7.95
75.00 - 84.00	75.00 - 84.00	75.00 - 84.00	75.00 - 84.00	75.00 - 84.00	2.7.96
78.00 - 87.00	78.00 - 87.00	78.00 - 87.00	78.00 - 87.00	78.00 - 87.00	2.7.97
82.00 - 96.00	82.00 - 96.00	82.00 - 96.00	82.00 - 96.00	82.00 - 96.00	2.7.98
85.00 - 99.00	85.00 - 99.00	85.00 - 99.00	85.00 - 99.00	85.00 - 99.00	2.7.99
110.00 - 126.00	110.00 - 126.00	110.00 - 126.00	110.00 - 126.00	110.00 - 126.00	2.7.100
111.00 - 128.00	111.00 - 128.00	111.00 - 128.00	111.00 - 128.00	111.00 - 128.00	2.7.101
151.00 - 157.00	151.00 - 157.00	151.00 - 157.00	151.00 - 157.00	151.00 - 157.00	2.7.102
151.00 - 161.00	151.00 - 161.00	151.00 - 161.00	151.00 - 161.00	151.00 - 161.00	2.7.103
97.00 - 103.00	97.00 - 103.00	97.00 - 103.00	97.00 - 103.00	97.00 - 103.00	2.7.104
103.00 - 110.00	103.00 - 110.00	103.00 - 110.00	103.00 - 110.00	103.00 - 110.00	2.7.105
102.00 - 110.00	102.00 - 110.00	102.00 - 110.00	102.00 - 110.00	102.00 - 110.00	2.7.106
103.00 - 111.00	103.00 - 111.00	103.00 - 111.00	103.00 - 111.00	103.00 - 111.00	2.7.107
111.00 - 126.00	111.00 - 126.00	111.00 - 126.00	111.00 - 126.00	111.00 - 126.00	2.7.108
114.00 - 128.00	114.00 - 128.00	114.00 - 128.00	114.00 - 128.00	114.00 - 128.00	2.7.109
126.00 - 137.00	126.00 - 137.00	126.00 - 137.00	126.00 - 137.00	126.00 - 137.00	2.7.110
132.00 - 145.00	132.00 - 145.00	132.00 - 145.00	132.00 - 145.00	132.00 - 145.00	2.7.111
186.00 - 210.00	186.00 - 210.00	186.00 - 210.00	186.00 - 210.00	186.00 - 210.00	2.7.112
195.00 - 218.00	195.00 - 218.00	195.00 - 218.00	195.00 - 218.00	195.00 - 218.00	2.7.113
140.00 - 148.00	140.00 - 148.00	140.00 - 148.00	140.00 - 148.00	140.00 - 148.00	2.7.114
140.00 - 148.00	140.00 - 148.00	140.00 - 148.00	140.00 - 148.00	140.00 - 148.00	2.7.115

Item nr.	SPECIFICATIONS	Unit	RESIDENTIAL £ range £
2.7	**INTERNAL WALLS, PARTITION AND DOORS - cont'd**		
	each (unless otherwise described)		
	Comparative doors/door linings/frames - cont'd		
	Standard softwood doors and frames; including ironmongery and painting		
2.7.116	flush; hollow core	nr	186.00 - 233.00
2.7.117	flush; hollow core; hardwood faced	nr	186.00 - 248.00
	flush; solid core		
2.7.118	single leaf	nr	218.00 - 280.00
2.7.119	double leaf	nr	318.00 - 428.00
2.7.120	flush; solid core; hardwood faced	nr	225.00 - 295.00
2.7.121	four panel door	nr	318.00 - 388.00
	Purpose-made softwood doors and hardwood frames; including ironmongery; painting and polishing flush		
	solid core; heavy duty		
2.7.122	single leaf	nr	- -
2.7.123	double leaf	nr	- -
	flush solid core; heavy duty; plastic laminate faced		
2.7.124	single leaf	nr	- -
2.7.125	double leaf	nr	- -
	Purpose-made softwood fire doors and hardwood frames; including ironmongery; painting and polishing		
	flush; one hour fire resisting		
2.7.126	single leaf	nr	- -
2.7.127	double leaf	nr	- -
	flush; one hour fire resisting; plastic laminate faced		
2.7.128	single leaf	nr	- -
2.7.128	double leaf	nr	- -
	Purpose-made softwood doors and pressed steel frames;		
2.7.130	flush; half hour fire check; plastic laminate faced	nr	- -
	Purpose-made mahogany doors and frames; including ironmongery and polishing		
2.7.131	four panel door	nr	- -
	Perimeter treatments		
	Precast concrete lintels; in block walls		
2.7.132	75 mm wide	nr	9.90 - 16.30
2.7.133	100 mm wide	nr	16.30 - 20.20
	Precast concrete lintels; in brick walls		
2.7.134	half brick thick	nr	16.30 - 20.20
2.7.135	one brick thick	nr	25.60 - 32.70
	Purpose-made softwood architraves; painted; including grounds		
	25 mm x 50 mm; to both sides of openings		
2.7.136	726 mm x 2040 mm opening	nr	73.00 - 79.00
2.7.137	826 mm x 2040 mm opening	nr	75.00 - 81.00
	Purpose-made mahogany architraves; wax polished; including grounds		
	25 mm x 50 mm; to both sides of openings		
2.7.138	726 mm x 2040 mm opening	nr	120.00 - 132.00
2.7.139	826 mm x 2040 mm opening	nr	120.00 - 132.00

Approximate Estimates 777

INDUSTRIAL £ range £	RETAILING £ range £	LEISURE £ range £	OFFICES £ range £	HOTELS £ range £	Item nr.
186.00 - 233.00	186.00 - 233.00	186.00 - 233.00	186.00 - 233.00	186.00 - 233.00	2.7.116
186.00 - 248.00	186.00 - 248.00	186.00 - 248.00	186.00 - 248.00	186.00 - 248.00	2.7.117
218.00 - 280.00	218.00 - 280.00	218.00 - 280.00	218.00 - 280.00	218.00 - 280.00	2.7.118
318.00 - 428.00	318.00 - 428.00	209.00 - 428.00	318.00 - 428.00	318.00 - 428.00	2.7.119
225.00 - 295.00	225.00 - 295.00	225.00 - 295.00	225.00 - 295.00	225.00 - 295.00	2.7.120
318.00 - 388.00	318.00 - 388.00	318.00 - 388.00	318.00 - 388.00	318.00 - 388.00	2.7.121
536.00 - 628.00	536.00 - 628.00	352.00 - 628.00	536.00 - 628.00	536.00 - 628.00	2.7.122
729.00 - 931.00	729.00 - 931.00	729.00 - 931.00	729.00 - 931.00	729.00 - 931.00	2.7.123
651.00 - 738.00	651.00 - 738.00	651.00 - 738.00	651.00 - 738.00	651.00 - 738.00	2.7.124
901.00 - 1009.0	901.00 - 1009.0	901.00 - 1009.0	901.00 - 1009.0	901.00 - 1009.00	2.7.125
706.00 - 791.00	706.00 - 791.00	706.00 - 791.00	706.00 - 791.00	706.00 - 791.00	2.7.126
901.00 - 1102.0	901.00 - 1102.0	901.00 - 1102.0	901.00 - 1102.0	901.00 - 1102.00	2.7.127
876.00 - 962.00	876.00 - 962.00	876.00 - 962.00	876.00 - 962.00	876.00 - 962.00	2.7.128
1126.0 - 1217.0	1126.0 - 1217.0	1126.0 - 1217.0	1126.0 - 1217.0	1126.0 - 1217.00	2.7.129
814.00 - 978.00	814.00 - 978.00	814.00 - 978.00	814.00 - 978.00	814.00 - 978.00	2.7.130
676.00 - 791.00	676.00 - 791.00	676.00 - 791.00	676.00 - 791.00	676.00 - 791.00	2.7.131
7.50 - 11.70	7.50 - 11.70	7.50 - 11.70	7.50 - 11.70	7.50 - 11.70	2.7.132
9.30 - 13.20	9.30 - 13.20	9.30 - 13.20	9.30 - 13.20	9.30 - 13.20	2.7.133
9.30 - 13.20	9.30 - 13.20	9.30 - 13.20	9.30 - 13.20	9.30 - 13.20	2.7.134
14.30 - 17.50	14.30 - 17.50	14.30 - 17.50	14.30 - 17.50	14.30 - 17.50	2.7.135
73.00 - 79.00	73.00 - 79.00	73.00 - 79.00	73.00 - 79.00	73.00 - 79.00	2.7.136
75.00 - 81.00	75.00 - 81.00	75.00 - 81.00	75.00 - 81.00	75.00 - 69.00	2.7.137
120.00 - 132.00	120.00 - 132.00	120.00 - 132.00	120.00 - 132.00	120.00 - 132.00	2.7.138
122.00 - 132.00	122.00 - 132.00	122.00 - 132.00	122.00 - 132.00	122.00 - 132.00	2.7.139

Approximate Estimates

Item nr.	SPECIFICATIONS	Unit	RESIDENTIAL £ range £
3.1	**WALL FINISHES** **wall finish area** (unless otherwise described) comprising: **Sheet/board finishes** **In situ wall finishes** **Rigid tile/panel finishes**		
	Sheet/board finishes Dry plasterboard lining; taped joints; for direct decoration		
3.1.1	9.50 mm thick Gyproc Wallboard	m^2	7.00 - 10.10
	Extra for		
3.1.2	insulating grade	m^2	0.50 - 0.50
3.1.3	insulating grade; plastic faced	m^2	1.60 - 1.90
3.1.4	12.50 mm thick Gyproc Wallboard (half-hour fire-resisting)	m^2	7.80 - 11.30
	Extra for		
3.1.5	insulating grade	m^2	0.50 - 0.70
3.1.6	insulating grade; plastic faced	m^2	1.50 - 1.60
3.1.7	two layers of 12.50 mm thick Gyproc Wallboard (one-hour fire-resisting)	m^2	14.00 - 17.90
3.1.8	9 mm thick Supalux (half-hour fire-resisting)	m^2	14.00 - 17.90
	Dry plasterboard lining; taped joints; for direct decoration; fixed to wall on dabs		
3.1.9	9.50 mm thick Gyproc Wallboard	m^2	7.80 - 11.30
	Dry plasterboard lining; taped joints; for direct decoration; including metal tees		
3.1.10	9.50 mm thick Gyproc Wallboard	m^2	- -
3.1.11	12.50 mm thick Gyproc Wallboard	m^2	- -
	Dry lining/sheet panelling; including battens; plugged to wall		
3.1.12	6.40 mm thick hardboard	m^2	8.50 - 10.00
3.1.13	9.50 mm thick Gyproc Wallboard	m^2	12.50 - 17.10
3.1.14	6 mm thick birch faced plywood	m^2	14.00 - 16.40
3.1.15	6 mm thick WAM plywood	m^2	17.10 - 20.20
3.1.16	15 mm thick chipboard	m^2	12.50 - 14.00
3.1.17	15 mm thick melamine faced chipboard	m^2	22.40 - 27.40
3.1.18	13.20 mm thick "Formica" faced chipboard	m^2	30.10 - 44.90
	Timber boarding/panelling; on and including battens; plugged to wall		
3.1.19	12 mm thick softwood boarding	m^2	19.40 - 25.10
3.1.20	25 mm thick softwood boarding	m^2	27.80 - 31.20
3.1.21	hardwood panelling; t&g & v-jointed	m^2	45.00 - 101.00
	In situ wall finishes Extra over common brickwork for		
3.1.22	fair face and pointing both sides	m^2	2.80 - 3.90
	Comparative finishes		
3.1.23	one mist and two coats emulsion paint	m^2	2.30 - 3.30
3.1.24	multi-coloured gloss paint	m^2	3.90 - 5.00
3.1.25	two coats of lightweight plaster	m^2	7.80 - 10.10
3.1.26	9.50 mm thick Gyproc Wallboard and skim coat	m^2	9.70 - 12.60
3.1.27	12.50 mm thick Gyproc Wallboard and skim coat	m^2	11.30 - 14.00
3.1.28	two coats of "Thistle" plaster	m^2	9.70 - 13.20
3.1.29	plaster and emulsion	m^2	9.70 - 15.20
	Extra for		
3.1.30	gloss paint in lieu of emulsion	m^2	1.70 - 2.00
3.1.31	two coat render and emulsion	m^2	17.10 - 23.20
3.1.32	plaster and vinyl	m^2	14.00 - 20.20
3.1.33	plaster and fabric	m^2	14.00 - 30.40
3.1.34	squash court plaster "including markings"	m^2	- -
3.1.35	6 mm thick terrazzo wall lining; including backing	m^2	- -
3.1.36	glass reinforced gypsum	m^2	- -

Approximate Estimates

INDUSTRIAL £ range £	RETAILING £ range £	LEISURE £ range £	OFFICES £ range £	HOTELS £ range £	Item nr.
6.50 - 10.10	6.50 - 10.10	6.50 - 10.10	6.50 - 10.10	6.50 - 10.10	3.1.1
0.50 - 0.60	0.50 - 0.60	0.50 - 0.60	0.50 - 0.60	0.50 - 0.60	3.1.2
1.60 - 1.90	1.60 - 1.90	1.60 - 1.90	1.60 - 1.90	1.60 - 1.90	3.1.3
7.80 - 10.50	7.80 - 11.30	7.80 - 11.30	7.80 - 11.30	8.50 - 13.20	3.1.4
0.50 - 0.70	0.50 - 0.70	0.50 - 0.70	0.50 - 0.70	0.50 - 0.70	3.1.5
1.40 - 1.60	1.40 - 1.60	1.40 - 1.60	1.40 - 1.60	1.40 - 1.60	3.1.6
13.20 - 17.90	14.00 - 17.90	14.00 - 17.90	14.00 - 17.90	14.80 - 20.90	3.1.7
13.20 - 17.90	14.00 - 17.90	14.00 - 17.90	14.00 - 17.90	14.80 - 20.90	3.1.8
7.00 - 10.50	7.80 - 11.30	7.80 - 11.30	7.80 - 11.30	8.50 - 12.00	3.1.9
17.10 - 19.80	- -	- -	- -	- -	3.1.10
17.90 - 20.90	- -	- -	- -	- -	3.1.11
8.50 - 10.10	8.50 - 10.10	8.50 - 10.10	8.50 - 10.10	9.30 - 11.30	3.1.12
12.50 - 16.40	12.50 - 17.10	12.50 - 17.10	12.50 - 17.10	14.00 - 17.90	3.1.13
13.20 - 15.60	14.00 - 16.40	14.00 - 16.40	14.00 - 16.40	14.80 - 17.90	3.1.14
16.40 - 20.20	17.10 - 20.90	17.10 - 20.90	17.10 - 20.90	17.90 - 20.90	3.1.15
12.50 - 14.00	12.50 - 14.00	12.50 - 14.00	12.50 - 14.00	14.00 - 15.60	3.1.16
20.90 - 25.90	22.40 - 27.40	22.40 - 27.40	22.40 - 27.40	23.20 - 28.10	3.1.17
28.50 - 43.40	30.40 - 44.90	30.40 - 44.90	30.40 - 44.90	30.40 - 49.50	3.1.18
19.40 - 25.10	19.40 - 25.10	19.40 - 25.10	19.40 - 25.10	20.90 - 26.60	3.1.19
27.00 - 29.70	27.80 - 31.20	27.80 - 31.20	27.80 - 31.20	28.50 - 32.70	3.1.20
45.00 - 100.00	45.00 - 100.00	45.00 - 100.00	30.00 - 100.00	49.00 - 112.00	3.1.21
2.70 - 3.90	2.70 - 3.90	2.70 - 3.90	2.70 - 3.90	2.70 - 3.90	3.1.22
2.20 - 3.30	2.20 - 3.30	2.20 - 3.30	2.20 - 3.30	2.20 - 3.30	3.1.23
3.90 - 5.00	3.90 - 5.00	3.90 - 5.00	3.90 - 5.00	3.90 - 5.00	3.1.24
7.00 - 10.10	7.80 - 10.10	7.80 - 10.10	7.80 - 10.10	8.50 - 11.30	3.1.25
9.30 - 11.60	10.10 - 12.50	10.10 - 12.50	10.10 - 12.50	10.50 - 13.20	3.1.26
10.50 - 13.20	11.30 - 13.20	11.30 - 13.20	11.30 - 13.20	11.60 - 13.20	3.1.27
10.00 - 13.20	10.00 - 13.20	10.00 - 13.20	10.00 - 13.20	10.50 - 13.20	3.1.28
9.30 - 14.00	10.00 - 15.20	10.00 - 15.20	10.00 - 15.20	11.30 - 17.90	3.1.29
1.70 - 1.90	1.70 - 1.90	1.70 - 1.90	1.70 - 1.90	1.70 - 1.90	3.1.30
16.40 - 22.40	17.10 - 23.20	17.10 - 23.20	17.10 - 23.20	17.90 - 24.70	3.1.31
- -	14.00 - 20.20	14.00 - 20.90	14.00 - 20.90	15.20 - 20.90	3.1.32
- -	14.00 - 30.40	14.00 - 30.40	14.00 - 30.40	15.60 - 32.70	3.1.33
- -	- -	19.00 - 25.00	- -	- -	3.1.34
- -	140.00 - 178.00	140.00 - 178.00	140.00 - 178.00	148.00 - 195.00	3.1.35
- -	125.00 - 195.00	125.00 - 195.00	125.00 - 195.00	140.00 - 202.00	3.1.36

Approximate Estimates

Item nr.	SPECIFICATIONS	Unit	RESIDENTIAL £ range £
3.1	**WALL FINISHES - cont'd**		
	wall finish area (unless otherwise described)		
	Rigid tile/panel finishes		
	Ceramic wall tiles; including backing		
3.1.37	economical	m²	18.60 - 34.20
3.1.38	medium quality	m²	34.20 - 60.50
3.1.39	high quality; to toilet blocks; kitchens and first aid rooms	m²	- -
3.1.40	high quality; to changing areas, toilets, showers and fitness areas	m²	- -
	Porcelain mosaic tiling; including backing to swimming pool lining; walls and		
3.1.41	floors	m²	- -
	"Roman Travertine" marble wall linings; polished		
3.1.42	19 mm thick	m²	- -
3.1.43	40 mm thick	m²	- -
3.1.44	Metal mirror cladding panels	m²	- -
3.2	**FLOOR FINISHES**		
	floor finish area (unless otherwise described)		
	comprising:		
	Sheet/board flooring		
	In situ screed and floor finishes		
	Rigid tile/slab finishes		
	Parquet/Wood block finishes		
	Flexible tiling/sheet finishes		
	Carpet tiles/Carpeting		
	Access floors and finishes		
	Perimeter treatments and sundries		
	Sheet/board flooring		
	Chipboard flooring; t&g joints		
3.2.1	18 mm thick	m²	7.00 - 8.50
3.2.2	22 mm thick	m²	8.50 - 10.10
	Wrought softwood flooring		
3.2.3	25 mm thick; butt joints	m²	15.60 - 17.10
3.2.4	25 mm thick; butt joints; cleaned off and polished	m²	17.90 - 20.90
3.2.5	25 mm thick; t&g joints	m²	16.40 - 20.20
3.2.6	25 mm thick; t&g joints; cleaned off and polished	m²	19.40 - 24.00
3.2.7	Wrought softwood t&g strip flooring; 25 mm thick; polished; including fillets	m²	24.00 - 30.40
3.2.8	maple	m²	- -
3.2.9	gurjun	m²	- -
3.2.10	iroko	m²	- -
3.2.11	american oak	m²	- -
	Wrought hardwood t&g strip flooring; 25 mm thick; polished; including fillets		
3.2.12	maple	m²	- -
3.2.13	gurjun	m²	- -
3.2.14	iroko	m²	- -
3.2.15	american oak	m²	- -
	Wrought hardwood t&g strip flooring; 25 mm thick; polished; including rubber pads		
3.2.16	maple	m²	- -
	In situ screed and floor finishes		
	Extra over concrete floor for		
3.2.17	power floating	m²	- -
3.2.18	power floating; surface hardener	m²	- -

Approximate Estimates

INDUSTRIAL £ range £	RETAILING £ range £	LEISURE £ range £	OFFICES £ range £	HOTELS £ range £	Item nr.
17.90 - 31.20	19.40 - 34.20	19.40 - 34.20	19.40 - 34.20	22.40 - 40.30	3.1.37
31.20 - 55.90	34.20 - 62.40	34.20 - 62.40	34.20 - 62.40	38.80 - 68.50	3.1.38
- -	55.90 - 71.50	55.90 - 71.50	55.90 - 71.50	55.90 - 71.50	3.1.39
- -	- -	71.50 - 85.20	- -	- -	3.1.40
- -	- -	44.90 - 60.50	- -	- -	3.1.41
- -	225.00 - 303.00	225.00 - 303.00	225.00 - 303.00	225.00 - 303.00	3.1.42
- -	310.00 - 396.00	310.00 - 396.00	310.00 - 396.00	310.00 - 396.00	3.1.43
- -	233.00 - 396.00	225.00 - 380.00	- -	248.00 - 403.00	3.1.44
- -	- -	- -	- -	- -	3.2.1
- -	- -	- -	- -	- -	3.2.2
- -	- -	- -	- -	- -	3.2.3
- -	- -	- -	- -	- -	3.2.4
- -	- -	- -	- -	- -	3.2.5
- -	- -	- -	- -	- -	3.2.6
- -	- -	- -	- -	- -	3.2.7
- -	- -	41.80 - 47.90	41.80 - 47.90	41.80 - 49.50	3.2.8
- -	- -	- -	37.30 - 41.80	37.30 - 37.10	3.2.9
- -	- -	- -	41.80 - 46.40	41.80 - 41.20	3.2.10
- -	- -	- -	44.90 - 49.50	44.90 - 51.00	3.2.11
- -	- -	46.40 - 52.50	46.40 - 52.50	47.90 - 54.40	3.2.12
- -	- -	- -	41.80 - 46.40	43.40 - 47.90	3.2.13
- -	- -	- -	46.40 - 52.50	47.90 - 52.50	3.2.14
- -	- -	- -	47.90 - 55.90	49.50 - 59.00	3.2.15
- -	- -	55.90 - 66.90	- -	- -	3.2.16
3.90 - 7.80	2.70 - 7.80	- -	- -	- -	3.2.17
7.80 - 11.30	5.90 - 10.10	- -	- -	- -	3.2.18

Approximate Estimates

Item nr.	SPECIFICATIONS	Unit	RESIDENTIAL £ range £
3.2	**FLOOR FINISHES - cont'd**		
	floor finish area (unless otherwise described)		
	Insitu screed and floor finishes - cont'd		
	Latex cement screeds		
3.2.19	3 mm thick; one coat	m^2	- -
3.2.20	5 mm thick; two coats	m^2	- -
3.2.21	Rubber latex non-slip solution and epoxy sealant	m^2	- -
	Cement and sand (1:3) screeds		
3.2.22	25 mm thick	m^2	7.80 - 8.50
3.2.23	50 mm thick	m^2	9.30 - 10.90
3.2.24	75 mm thick	m^2	13.20 - 14.80
	Cement and sand (1:3) paving		
3.2.25	paving	m^2	7.00 - 9.40
3.2.26	32 mm thick; surface hardener	m^2	8.50 - 13.20
3.2.27	screed only (for subsequent finish)	m^2	10.90 - 16.40
3.2.28	screed only (for subsequent finish); allowance for skirtings	m^2	12.40 - 19.40
	Mastic asphalt paving		
3.2.29	20 mm thick; BS 1076; black	m^2	- -
3.2.30	20 mm thick; BS 1471; red	m^2	- -
	Granolithic		
3.2.31	20 mm thick	m^2	- -
3.2.32	25 mm thick	m^2	- -
3.2.33	25 mm thick; including screed	m^2	- -
3.2.34	38 mm thick; including screed	m^2	- -
	"Synthanite"; on and including building paper		
3.2.35	25 mm thick	m^2	- -
3.2.36	50 mm thick	m^2	- -
3.2.37	75 mm thick	m^2	- -
	Acrylic polymer floor finish		
3.2.38	10 mm thick	m^2	- -
	Epoxy floor finish		
3.2.39	1.50 mm - 2.00 mm thick	m^2	- -
3.2.40	5.00 mm - 6.00 mm thick	m^2	- -
	Polyester resin floor finish		
3.2.41	5.00 mm - 9.00 mm thick	m^2	- -
	Terrazzo paving; divided into squares with ebonite strip; polished		
3.2.42	16 mm thick	m^2	- -
3.2.43	16 mm thick; including screed	m^2	- -
	Rigid Tile/slab finishes		
	Quarry tile flooring		
3.2.44	150 mm x 150 mm x 12.50 mm thick; red	m^2	- -
3.2.45	150 mm x 150 mm x 12.50 mm thick; brown	m^2	- -
3.2.46	200 mm x 200 mm x 19 mm thick; brown	m^2	- -
3.2.47	average tiling	m^2	- -
3.2.48	tiling; including screed	m^2	- -
3.2.49	tiling; including screed and allowance for skirtings	m^2	- -
	Brick paving		
3.2.50	paving	m^2	- -
3.2.51	paving; including screed	m^2	- -
	Glazed ceramic tiled flooring		
3.2.52	100 mm x 100 mm x 9 mm thick; red	m^2	- -
3.2.53	150 mm x 150 mm x 12 mm thick; red	m^2	- -
3.2.54	100 mm x 100 mm x 9 mm thick; black	m^2	- -
3.2.55	150 mm x 150 mm x 12 mm thick; black	m^2	- -
3.2.56	150 mm x 150 mm x 12 mm thick; anti-slip	m^2	- -
3.2.57	fully vitrified	m^2	- -
3.2.58	fully vitrified; including screed	m^2	- -
3.2.59	fully vitrified; including screed and allowance for skirtings	m^2	- -
3.2.60	high quality; to service areas; kitchen and toilet blocks; including screed	m^2	- -

Approximate Estimates

INDUSTRIAL £ range £	RETAILING £ range £	LEISURE £ range £	OFFICES £ range £	HOTELS £ range £	Item nr.
3.90 - 4.50	- -	- -	3.90 - 4.50	- -	3.2.19
5.20 - 5.90	- -	- -	5.20 - 5.90	- -	3.2.20
6.60 - 15.60	- -	- -	- -	- -	3.2.21
7.80 - 8.50	7.80 - 8.50	7.80 - 8.50	7.80 - 8.50	7.80 - 8.50	3.2.22
9.40 - 10.90	9.40 - 10.90	9.40 - 10.90	9.40 - 10.90	9.40 - 10.90	3.2.23
13.20 - 14.80	13.20 - 14.80	13.30 - 13.70	13.30 - 13.70	13.30 - 13.70	3.2.24
6.60 - 8.50	7.00 - 9.40	7.00 - 9.40	7.00 - 9.40	7.00 - 9.40	3.2.25
7.80 - 12.40	9.40 - 13.20	9.40 - 13.20	9.40 - 13.20	9.40 - 13.20	3.2.26
10.10 - 15.60	10.90 - 15.60	10.90 - 15.60	10.90 - 15.60	10.90 - 15.60	3.2.27
12.40 - 17.90	13.20 - 19.40	13.20 - 19.40	13.20 - 19.40	13.20 - 20.20	3.2.28
15.60 - 18.60	16.40 - 19.40	16.40 - 19.40	- -	- -	3.2.29
18.60 - 20.90	19.40 - 22.40	19.40 - 22.40	- -	- -	3.2.30
8.50 - 13.20	9.40 - 13.90	9.40 - 13.90	9.40 - 13.90	9.40 - 13.90	3.2.31
11.60 - 15.60	12.40 - 15.60	12.40 - 15.60	12.40 - 15.60	12.40 - 17.10	3.2.32
19.40 - 20.90	19.80 - 22.40	19.80 - 22.40	19.80 - 22.40	19.80 - 21.70	3.2.33
27.40 - 31.20	28.50 - 32.70	28.50 - 32.70	28.50 - 32.70	30.10 - 33.50	3.2.34
17.90 - 20.90	17.90 - 22.40	17.90 - 22.40	17.90 - 22.40	- -	3.2.35
24.70 - 28.50	26.20 - 30.40	26.20 - 30.40	26.20 - 30.40	- -	3.2.36
30.40 - 35.80	32.70 - 36.50	32.70 - 36.50	32.70 - 36.50	- -	3.2.37
17.90 - 22.40	18.60 - 24.00	18.60 - 24.00	18.60 - 24.00	- -	3.2.38
17.90 - 22.40	18.60 - 24.70	18.60 - 24.70	18.60 - 24.70	- -	3.2.39
36.50 - 41.80	37.30 - 42.60	37.30 - 42.60	37.30 - 42.60	- -	3.2.40
41.10 - 46.40	41.80 - 47.90	41.80 - 47.90	41.80 - 47.90	- -	3.2.41
- -	41.10 - 46.40	41.10 - 46.40	41.10 - 46.40	44.90 - 55.90	3.2.42
- -	60.50 - 70.00	60.50 - 70.00	60.50 - 70.00	65.00 - 80.60	3.2.43
22.40 - 24.70	23.20 - 25.50	23.20 - 25.50	23.20 - 25.50	23.20 - 27.00	3.2.44
27.00 - 31.20	28.50 - 32.00	28.50 - 32.00	28.50 - 32.00	30.10 - 33.50	3.2.45
33.50 - 37.30	34.20 - 38.00	34.20 - 38.00	34.20 - 38.00	35.80 - 39.60	3.2.46
22.40 - 37.30	23.20 - 38.00	23.20 - 38.00	23.20 - 38.00	23.20 - 39.60	3.2.47
32.00 - 47.90	33.50 - 49.50	33.50 - 49.50	33.50 - 49.50	34.20 - 49.50	3.2.48
42.60 - 59.00	44.90 - 60.50	44.90 - 60.50	45.30 - 60.50	44.90 - 62.00	3.2.49
- -	22.00 - 47.90	33.50 - 47.90	33.50 - 47.90	33.50 - 47.90	3.2.50
- -	41.80 - 60.50	41.80 - 60.50	41.80 - 60.50	41.80 - 60.50	3.2.51
- -	30.10 - 34.20	30.10 - 34.20	30.10 - 34.60	31.20 - 35.80	3.2.52
- -	26.20 - 30.10	26.20 - 30.10	26.20 - 30.10	27.00 - 32.00	3.2.53
- -	33.50 - 36.50	33.50 - 36.50	33.50 - 36.50	34.20 - 38.00	3.2.54
- -	27.00 - 33.50	27.00 - 33.50	27.00 - 33.50	27.80 - 34.20	3.2.55
- -	33.50 - 35.00	33.50 - 35.00	33.50 - 35.00	33.50 - 36.50	3.2.56
- -	35.00 - 49.50	35.00 - 49.50	35.00 - 49.50	35.80 - 52.90	3.2.57
- -	41.80 - 63.50	41.80 - 63.50	41.80 - 63.50	44.10 - 65.40	3.2.58
- -	47.90 - 76.10	47.90 - 76.10	47.90 - 76.10	47.90 - 80.60	3.2.59
- -	- -	66.60 - 77.60	- -	- -	3.2.60

Item nr.	SPECIFICATIONS	Unit	RESIDENTIAL £ range £	
3.2	**FLOOR FINISHES - cont'd**			
	floor finish area (unless otherwise described)			
	Rigid Tile/slab finishes - cont'd			
	Glazed ceramic tile flooring - cont'd			
3.2.61	high quality; to foyer; fitness and bar areas; including screed	m^2	-	-
3.2.62	high quality; to pool surround, bottoms, steps and changing room; including screed	m^2	-	-
3.2.63	Porcelain mosaic paving; including screed to swimming pool lining; walls and floors	m^2	-	-
	Extra for			
3.2.64	non-slip finish to pool lining; beach and changing areas and showers	m^2	-	-
	Terrazzo tile flooring			
3.2.65	28 mm thick white "Sicilian" marble aggregate tiling	m^2	-	-
3.2.66	tiling; including screed	m^2	-	-
	York stone			
3.2.67	50 mm thick paving	m^2	-	-
3.2.68	paving; including screed	m^2	-	-
	Slate			
3.2.69	200 mm x 400 mm x 10 mm thick ; blue - grey	m^2	-	-
3.2.70	otta riven	m^2	-	-
3.2.71	otta boned (polished)	m^2	-	-
	Portland stone			
3.2.72	50 mm thick paving	m^2	-	-
	Fined sanded "Roman Travertine" marble			
3.2.73	20 mm thick paving	m^2	-	-
3.2.74	paving; including screed	m^2	-	-
3.2.75	paving; including screed and allowance for skirtings	m^2	-	-
	Granite			
3.2.76	20 mm thick paving	m^2	-	-
	Parquet/wood block finishes			
	Parquet flooring; polished			
3.2.77	8 mm thick gurjun "Feltwood"	m^2	22.40 -	25.50
	Wrought hardwood block floorings; 25 mm thick; polished; t&g joints herringbone pattern			
3.2.78	merbau	m^2	-	-
3.2.79	iroko	m^2	-	-
3.2.80	iroko; including screed	m^2	-	-
3.2.81	oak	m^2	-	-
3.2.82	oak; including screed	m^2	-	-
	Composition block flooring			
3.2.83	174 mm x 57 mm blocks	m^2	-	-
	Flexible tiling			
	Thermoplastic tile flooring			
3.2.84	2 mm thick (series 2)	m^2	6.60 -	7.80
3.2.85	2 mm thick (series 4)	m^2	6.60 -	7.80
3.2.86	2 mm thick; including screed	m^2	14.80 -	17.10
	Cork tile flooring			
3.2.87	3.20 mm thick	m^2	11.60 -	14.80
3.2.88	3.20 mm thick; including screed	m^2	20.90 -	27.80
3.2.89	6.30 mm thick	m^2	-	-
3.2.90	6.30 mm thick; including screed	m^2	-	-
	Vinyl floor tiling			
3.2.91	2 mm thick; semi-flexible tiles	m^2	7.00 -	10.10
3.2.92	2 mm thick; fully flexible tiles	m^2	6.60 -	9.40
3.2.93	2.50 mm thick; semi-flexible tiles	m^2	8.50 -	11.30
3.2.94	tiling; including screed	m^2	17.90 -	22.40
3.2.95	tiling; including screed and allowance for skirtings	m^2	19.40 -	26.20
3.2.96	tiling; anti-static	m^2	-	-
3.2.97	tiling; anti-static; including screed	m^2	-	-

Approximate Estimates

INDUSTRIAL £ range £	RETAILING £ range £	LEISURE £ range £	OFFICES £ range £	HOTELS £ range £	Item nr.
- -	- -	70.00 - 81.00	- -	- -	3.2.61
- -	- -	73.00 - 85.00	- -	- -	3.2.62
- -	- -	44.90 - 60.50	- -	- -	3.2.63
- -	- -	1.20 - 2.30	- -	- -	3.2.64
- -	65.40 - 73.00	65.00 - 73.00	65.00 - 73.00	67.00 - 78.00	3.2.65
- -	83.70 - 94.00	84.00 - 94.00	84.00 - 94.00	87.00 - 100.00	3.2.66
- -	85.20 - 113.00	85.00 - 110.00	85.00 - 110.00	90.00 - 122.00	3.2.67
- -	92.80 - 123.00	93.00 - 123.00	93.00 - 123.00	99.00 - 137.00	3.2.68
- -	105.00 - 119.00	105.00 - 119.00	105.00 - 119.00	- -	3.2.69
- -	118.70 - 129.00	119.00 - 129.00	119.00 - 129.00	- -	3.2.70
- -	129.30 - 140.00	129.00 - 140.00	129.00 - 140.00	- -	3.2.71
- -	178.00 - 210.00	178.00 - 210.00	178.00 - 210.00	- -	3.2.72
- -	225.20 - 272.00	225.00 - 272.00	225.00 - 272.00	233.00 - 288.00	3.2.73
- -	232.80 - 288.00	233.00 - 286.00	233.00 - 288.00	240.00 - 295.00	3.2.74
- -	287.60 - 358.00	288.00 - 358.00	288.00 - 358.00	295.00 - 373.00	3.2.75
- -	310.40 - 365.00	310.00 - 365.00	310.00 - 365.00	310.00 - 396.00	3.2.76
- -	- -	- -	- -	- -	3.2.77
- -	46.40 - 54.40	46.40 - 54.40	46.40 - 54.40	47.90 - 54.40	3.2.78
- -	49.50 - 55.90	49.50 - 55.90	49.50 - 55.90	52.50 - 59.00	3.2.79
- -	59.00 - 71.50	59.00 - 71.50	59.00 - 71.50	59.00 - 71.50	3.2.80
- -	47.90 - 62.00	47.90 - 62.00	47.90 - 62.00	49.50 - 63.50	3.2.81
- -	57.40 - 70.00	57.40 - 70.00	57.40 - 70.00	59.00 - 70.00	3.2.82
- -	55.90 - 62.00	55.90 - 62.00	55.90 - 62.00	59.00 - 65.00	3.2.83
- -	- -	- -	- -	- -	3.2.84
- -	- -	- -	- -	- -	3.2.85
- -	- -	- -	- -	- -	3.2.86
- -	- -	- -	- -	- -	3.2.87
- -	- -	- -	- -	- -	3.2.88
- -	20.90 - 24.00	20.90 - 24.00	20.90 - 24.00	22.40 - 24.70	3.2.89
- -	27.80 - 36.50	27.80 - 36.50	27.80 - 36.50	28.50 - 38.00	3.2.90
- -	7.80 - 10.10	7.80 - 10.10	7.80 - 10.10	7.80 - 10.10	3.2.91
- -	7.00 - 9.40	7.00 - 9.40	7.00 - 9.40	7.00 - 9.40	3.2.92
- -	7.80 - 11.30	7.80 - 11.30	7.80 - 11.30	7.80 - 11.30	3.2.93
- -	17.90 - 22.40	17.90 - 22.40	17.90 - 22.40	17.90 - 22.40	3.2.94
- -	19.40 - 26.20	19.40 - 26.20	19.40 - 26.20	19.40 - 26.20	3.2.95
- -	32.70 - 37.30	32.70 - 37.30	32.70 - 37.30	32.70 - 37.30	3.2.96
- -	40.30 - 49.80	40.30 - 49.80	40.30 - 49.80	40.30 - 49.80	3.2.97

Item nr.	SPECIFICATIONS	Unit	RESIDENTIAL £ range £	
3.2	**FLOOR FINISHES - cont'd**			
	floor finish area (unless otherwise described)			
	Flexible tiling/sheet finishes - cont'd			
	Vinyl sheet flooring; heavy duty			
3.2.98	2 mm thick	m²	-	-
3.2.99	2.50 mm thick	m²	-	-
3.2.100	3 mm thick; needle felt backed	m²	-	-
3.2.101	3 mm thick; foam backed	m²	-	-
	Sheeting; including screed and allowance for skirtings			
3.2.102	Altro "Safety" flooring	m²	-	-
3.2.103	2 mm thick; Altro "Marine T20" flooring	m²	-	-
3.2.104	2.50 mm thick; Altro "Classic D25" flooring	m²	-	-
3.2.105	3.50 thick; Altro "Stronghold" flooring	m²	-	-
3.2.106	flooring	m²	-	-
3.2.107	flooring; including screed	m²	-	-
	Linoleum tile flooring			
3.2.108	3.20 mm thick; coloured	m²	-	-
3.2.109	3.20 mm thick; coloured; including screed	m²	-	-
	Linoleum sheet flooring			
3.2.110	3.20 mm thick; coloured	m²	-	-
3.2.111	3.20 mm thick; marbled; including screed	m²	-	-
	Rubber tile flooring; smooth; ribbed or studded tiles			
3.2.112	2.50 mm thick	m²	-	-
3.2.113	5 mm thick	m²	-	-
3.2.114	5 mm thick; including screed	m²	-	-
	Carpet tiles/Carpeting			
3.2.115	Underlay	m²	3.90 -	5.50
	Carpet tiles			
3.2.116	nylon needlepunch (stick down)	m²	10.10 -	12.40
3.2.117	80% animal hair; 20% wool cord	m²	-	-
3.2.118	100% wool	m²	-	-
3.2.119	80% wool; 20% nylon antistatic economical; including screed and allowance for	m²	-	-
3.2.120	skirtings	m²	-	-
3.2.121	good quality	m²	-	-
3.2.122	good quality; including screed	m²	-	-
3.2.123	good quality; including screed and allowance for skirtings	m²	-	-
	Carpet; including underlay			
3.2.124	nylon needlepunch	m²	-	-
3.2.125	100% acrylic; light duty	m²	13.20 -	16.40
3.2.126	80% animal hair; 20% wool cord	m²	-	-
3.2.127	open-weave matting poolside carpet (no underlay)	m²	-	-
3.2.128	80% wool; 20% acrylic; light duty	m²	-	-
3.2.129	100% acrylic; heavy duty	m²	-	-
3.2.130	cord	m²	-	-
3.2.131	100% wool	m²	-	-
3.2.132	good quality; including screed	m²	-	-
	good quality (grade 5); including screed and allowance			
3.2.133	for skirtings	m²	-	-
3.2.134	80% wool; 20% acrylic; heavy duty	m²	-	-
3.2.135	"Wilton"/pile carpet	m²	-	-
3.2.136	high quality; including screed	m²	-	-
3.2.137	high quality; including screed and allowance for skirtings	m²	-	-
	Access floors and finishes			
	Shallow void block and battened floors			
3.2.138	chipboard on softwood battens; partial access	m²	19.40 -	24.70
3.2.139	chipboard on softwood cradles (for uneven floors)	m²	23.20 -	27.00
	chipboard on softwood battens and cross battens;			
3.2.140	full access	m²	-	-

Approximate Estimates

INDUSTRIAL £ range £	RETAILING £ range £	LEISURE £ range £	OFFICES £ range £	HOTELS £ range £	Item nr.
- -	12.40 - 13.90	12.40 - 13.90	12.40 - 13.90	12.40 - 13.90	3.2.98
- -	12.40 - 15.60	12.40 - 15.60	12.40 - 15.60	12.40 - 15.60	3.2.99
- -	8.50 - 11.30	8.50 - 11.30	8.50 - 11.30	8.50 - 11.30	3.2.100
- -	12.40 - 15.60	12.40 - 15.60	12.40 - 15.60	12.40 - 15.60	3.2.101
- -	23.20 - 28.10	23.20 - 28.10	23.20 - 28.10	23.20 - 28.10	3.2.102
- -	17.90 - 22.40	17.90 - 22.40	17.90 - 22.40	- -	3.2.103
- -	22.40 - 26.20	22.40 - 26.20	22.40 - 26.20	- -	3.2.104
- -	28.10 - 32.70	28.10 - 32.70	28.10 - 32.70	- -	3.2.105
- -	17.90 - 33.50	17.90 - 33.50	17.90 - 33.50	- -	3.2.106
- -	26.20 - 44.90	26.20 - 44.90	26.20 - 44.90	- -	3.2.107
- -	13.90 - 16.40	13.90 - 16.40	13.90 - 16.40	- -	3.2.108
- -	22.40 - 30.40	22.40 - 30.40	22.40 - 30.40	- -	3.2.109
- -	14.80 - 16.40	14.80 - 19.40	14.80 - 19.40	- -	3.2.110
- -	23.20 - 30.40	23.20 - 30.40	23.20 - 30.40	- -	3.2.111
- -	22.40 - 26.20	22.40 - 26.20	22.40 - 26.20	- -	3.2.112
- -	26.20 - 30.40	26.20 - 30.40	26.20 - 30.40	- -	3.2.113
- -	35.30 - 44.90	35.80 - 44.90	35.80 - 44.90	- -	3.2.114
- -	3.90 - 5.40	3.90 - 5.40	3.90 - 5.40	- -	3.2.115
- -	10.10 - 12.40	10.10 - 12.40	10.10 - 12.40	- -	3.2.116
- -	17.90 - 20.20	17.90 - 20.20	17.90 - 20.20	- -	3.2.117
- -	24.70 - 30.40	24.70 - 30.40	24.70 - 30.40	- -	3.2.118
- -	26.20 - 36.50	26.20 - 36.50	26.20 - 36.50	- -	3.2.119
- -	27.00 - 30.40	27.00 - 30.40	27.00 - 30.40	- -	3.2.120
- -	24.70 - 36.50	24.70 - 36.50	24.70 - 36.50	24.70 - 40.30	3.2.121
- -	33.50 - 44.90	33.50 - 44.90	33.50 - 44.90	33.50 - 49.50	3.2.122
- -	37.30 - 49.50	37.30 - 49.50	37.30 - 49.50	37.30 - 55.90	3.2.123
- -	13.20 - 16.40	13.20 - 16.40	13.20 - 16.40	- -	3.2.124
- -	17.10 - 20.20	17.10 - 20.20	17.10 - 20.20	- -	3.2.125
- -	20.90 - 26.20	20.90 - 26.20	20.90 - 26.20	- -	3.2.126
- -	- -	17.90 - 22.40	- -	- -	3.2.127
- -	26.20 - 34.20	26.20 - 34.20	26.20 - 34.20	- -	3.2.128
- -	27.00 - 32.70	27.00 - 32.70	27.00 - 32.70	- -	3.2.129
- -	- -	32.70 - 36.50	- -	- -	3.2.130
- -	32.70 - 42.60	32.70 - 42.60	32.70 - 42.60	32.70 - 45.60	3.2.131
- -	40.30 - 55.90	40.30 - 55.90	40.30 - 55.90	41.80 - 60.50	3.2.132
- -	41.80 - 59.00	41.80 - 59.00	41.80 - 59.00	45.00 - 02.40	3.2.133
- -	42.60 - 50.60	42.60 - 50.60	42.60 - 50.60	45.60 - 53.60	3.2.134
- -	46.40 - 51.40	46.40 - 51.40	46.40 - 51.40	47.90 - 53.60	3.2.135
- -	55.90 - 65.40	55.90 - 65.40	55.90 - 65.40	59.00 - 66.90	3.2.136
- -	59.00 - 66.90	59.00 - 66.90	59.00 - 66.90	60.90 - 73.00	3.2.137
- -	- -	- -	19.40 - 24.70	- -	3.2.138
- -	- -	- -	23.20 - 27.00	- -	3.2.139
- -	- -	- -	28.90 - 32.00	- -	3.2.140

3.2 FLOOR FINISHES - cont'd

floor finish area (unless otherwise described)

Item nr.	SPECIFICATIONS	Unit	RESIDENTIAL £ range £	
	Access floors and finishes - cont'd			
	Shallow void block and battened floors - cont'd			
3.2.141	fibre and particle board; on lightweight concrete pedestal blocks	m²	-	-
	Shallow void block and battened floors; including carpet-tile finish			
3.2.142	fibre and particle board; on lightweight concrete pedestal blocks	m²	-	-
	Access floors; excluding finish			
	600 mm x 600 mm chipboard panels faced both sides with galvanised steel sheet; on adjustable steel/aluminium pedestals; cavity 100 mm - 300 mm high			
3.2.143	light grade duty	m²	-	-
3.2.144	medium grade duty	m²	-	-
3.2.145	heavy grade duty	m²	-	-
3.2.146	extra heavy grade duty	m²	-	-
	600 mm x 600 mm chipboard panels faced both sides with galvanised steel sheet; on adjustable steel/aluminium pedestals; cavity 300 mm - 600 mm high			
3.2.147	medium grade duty	m²	-	-
3.2.148	heavy grade duty	m²	-	-
3.2.149	extra heavy grade duty	m²	-	-
	Access floor with medium quality carpeting			
3.2.150	"Durabella" suspended floors	m²	-	-
3.2.151	"Buroplan" partial access raised floor	m²	-	-
3.2.152	"Pedestal" partial access raised floor	m²	-	-
3.2.153	modular floor; 100% access raised floor	m²	-	-
	Access floor with high quality carpeting			
3.2.154	computer loading; 100% access raised floor	m²	-	-
	Common floor coverings bonded to access floor panels			
3.2.155	heavy-duty fully flexible vinyl; to BS 3261; type A	m²	-	-
3.2.156	fibre-bonded carpet	m²	-	-
3.2.157	high-pressure laminate; to BS 2794; class D	m²	-	-
3.2.158	anti-static grade fibre-bonded carpet	m²	-	-
3.2.159	anti-static grade sheet PVC; to BS 3261	m²	-	-
3.2.160	low loop tufted carpet	m²	-	-
	Perimeter treatments and sundries			
	Comparative skirtings			
3.2.161	25 mm x 75 mm softwood skirting; painted; including grounds	m	8.10 -	9.70
3.2.162	25 mm x 100 mm mahogany skirting; polished; including grounds	m	11.50 -	12.90
3.2.163	12.50 mm x 150 mm quarry tile skirting; including backing	m	11.60 -	14.40
3.2.164	13 mm x 75 mm granolithic skirting; including backing	m	16.40 -	19.40
3.2.165	6 mm x 75 mm terrazzo; including backing	m	28.50 -	33.00
3.2.166	Entrance matting in aluminium-framed matwell	m²	-	-

Approximate Estimates

INDUSTRIAL £ range £	RETAILING £ range £	LEISURE £ range £	OFFICES £ range £	HOTELS £ range £	Item nr.
- -	- -	- -	27.00 - 33.50	- -	3.2.141
- -	- -	- -	40.30 - 46.40	- -	3.2.142
- -	- -	- -	36.50 - 43.40	- -	3.2.143
- -	- -	- -	41.80 - 48.70	- -	3.2.144
- -	- -	- -	52.50 - 66.20	- -	3.2.145
- -	- -	- -	59.00 - 66.20	- -	3.2.146
- -	- -	- -	48.00 - 54.00	- -	3.2.147
- -	- -	- -	54.00 - 67.00	- -	3.2.148
- -	- -	- -	60.00 - 67.00	- -	3.2.149
- -	- -	- -	52.00 - 57.00	- -	3.2.150
- -	- -	- -	59.00 - 67.00	- -	3.2.151
- -	- -	- -	67.00 - 82.00	- -	3.2.152
- -	- -	- -	87.00 - 110.00	- -	3.2.153
- -	- -	- -	97.00 - 145.00	- -	3.2.154
- -	- -	- -	6.60 - 18.60	- -	3.2.155
- -	- -	- -	7.00 - 13.60	- -	3.2.156
- -	- -	- -	7.00 - 20.90	- -	3.2.157
- -	- -	- -	8.50 - 14.40	- -	3.2.158
- -	- -	- -	12.00 - 18.60	- -	3.2.159
- -	- -	- -	14.40 - 20.90	- -	3.2.160
7.40 - 8.80	8.10 - 9.70	8.10 - 9.70	8.10 - 9.70	9.70 - 10.30	3.2.161
11.10 - 12.60	11.50 - 12.90	11.50 - 12.90	11.50 - 12.90	12.20 - 13.60	3.2.162
10.90 - 13.60	11.60 - 14.40	11.60 - 14.40	11.60 - 14.40	12.00 - 14.40	3.2.163
15.60 - 19.40	16.40 - 19.40	16.40 - 19.40	16.40 - 19.40	17.10 - 20.90	3.2.164
- -	28.90 - 33.50	28.90 - 33.50	28.90 - 33.50	30.40 - 35.00	3.2.165
- -	272.00 - 388.00	263.00 - 358.00	263.00 - 365.00	263.00 - 365.00	3.2.166

Approximate Estimates

Item nr.	SPECIFICATIONS	Unit	RESIDENTIAL £ range £
3.3	**CEILING FINISHES**		
	ceiling finish area (unless otherwise described)		
	comprising:		
	In situ board finishes		
	Suspended and integrated ceilings		
	In situ/board finishes		
	Decoration only to soffits		
3.3.1	to exposed steelwork	m²	- -
3.3.2	to concrete soffits	m²	2.30 - 3.30
3.3.3	one mist and two coats emulsion paint; to plaster/plasterboard	m²	2.30 - 3.30
	Plaster to soffits		
3.3.4	lightweight plaster	m²	7.80 - 10.50
3.3.5	plaster and emulsion	m²	10.50 - 15.60
	Extra for		
3.3.6	gloss paint in lieu of emulsion	m²	1.70 - 2.00
	Plasterboard to soffits		
3.3.7	9.50 mm Gyproc lath and skim coat	m²	11.30 - 13.20
3.3.8	9.50 mm Gyproc insulating lath and skim coat	m²	11.60 - 13.20
3.3.9	plasterboard, skim and emulsion	m²	13.20 - 16.40
	Extra for		
3.3.10	gloss paint in lieu of emulsion	m²	1.70 - 2.00
3.3.11	plasterboard and "Artex"	m²	9.40 - 11.30
3.3.12	plasterboard; "Artex" and emulsion	m²	11.30 - 14.00
3.3.13	plaster and emulsion; including metal lathing	m²	17.90 - 24.70
	Other board finishes; with fire-resisting properties; excluding decoration		
3.3.14	12.50 mm thick Gyproc "Fireline"; half hour	m²	- -
3.3.15	6 mm thick "Supalux"; half hour	m²	- -
3.3.16	two layers of 12.50 mm thick Gyproc "Wallboard"; half hour	m²	- -
3.3.17	two layers of 12.50 mm thick Gyproc "Fireline"; one hour	m²	- -
3.3.18	9 mm thick "Supalux"; one hour; on fillets	m²	- -
	Specialist plasters; to soffits		
3.3.19	sprayed acoustic plaster; self-finished	m²	- -
3.3.20	rendering; "Tyrolean" finish	m²	- -
	Other ceiling finishes		
3.3.21	50 mm thick wood wool slabs as permanent lining	m²	- -
3.3.22	12 mm thick pine t&g boarding	m²	14.00 - 17.10
3.3.23	16 mm thick softwood t&g boardings	m²	17.10 - 20.20
	Suspended and integrated ceilings		
	Suspended ceiling		
3.3.24	economical; exposed grid	m²	- -
3.3.25	jointless; plasterboard	m²	- -
3.3.26	semi-concealed grid	m²	- -
3.3.27	medium quality; "Minatone"; concealed grid	m²	- -
3.3.28	high quality; "Travertone"; concealed grid	m²	- -
	Other suspended ceilings		
3.3.29	metal linear strip; "Dampa"/"Luxalon"	m²	- -
3.3.30	metal tray	m²	- -
3.3.31	egg-crate	m²	- -
3.3.32	open grid; "Formalux"/"Dimension"	m²	- -
	Integrated ceilings		
3.3.33	coffered; with steel surfaces	m²	- -
	Acoustic suspended ceilings		
3.3.34	on anti vibration mountings	m²	- -

Approximate Estimates

INDUSTRIAL £ range £	RETAILING £ range £	LEISURE £ range £	OFFICES £ range £	HOTELS £ range £	Item nr.
2.70 - 3.90	2.30 - 3.30	-	-	-	3.3.1
2.30 - 3.30	2.30 - 3.30	2.30 - 3.30	2.30 - 3.30	2.30 - 3.30	3.3.2
2.30 - 3.30	2.30 - 3.30	2.30 - 3.30	2.30 - 3.30	2.30 - 3.30	3.3.3
7.80 - 10.10	7.80 - 10.10	7.80 - 10.10	7.80 - 10.10	7.80 - 10.10	3.3.4
10.10 - 15.10	10.50 - 16.00	10.50 - 16.00	10.50 - 16.00	12.00 - 17.90	3.3.5
1.70 - 2.00	1.70 - 2.00	1.70 - 2.00	1.70 - 2.00	1.70 - 2.00	3.3.6
10.50 - 13.20	11.30 - 13.20	11.30 - 13.20	11.30 - 13.20	12.00 - 13.20	3.3.7
11.30 - 13.20	12.00 - 13.20	12.00 - 13.20	12.00 - 13.20	12.00 - 14.00	3.3.8
12.00 - 16.00	13.20 - 16.40	13.20 - 16.40	13.20 - 16.40	13.20 - 17.10	3.3.9
1.70 - 2.00	1.70 - 2.00	1.70 - 2.00	1.70 - 2.00	1.70 - 2.00	3.3.10
7.80 - 10.50	8.50 - 11.30	8.50 - 11.30	8.50 - 11.30	9.40 - 12.00	3.3.11
10.50 - 13.20	11.30 - 14.00	11.30 - 14.00	11.30 - 14.00	12.00 - 14.80	3.3.12
17.10 - 24.70	17.90 - 24.70	17.90 - 24.70	17.90 - 24.70	17.90 - 24.70	3.3.13
8.50 - 11.30	9.40 - 11.30	9.40 - 11.30	9.40 - 11.30	9.70 - 12.50	3.3.14
11.30 - 12.50	12.00 - 13.20	12.00 - 13.20	12.00 - 13.20	14.80 - 15.60	3.3.15
12.50 - 14.40	13.20 - 15.20	13.20 - 15.20	13.20 - 15.20	13.20 - 15.60	3.3.16
14.40 - 17.10	15.20 - 17.90	15.20 - 17.90	15.20 - 17.90	15.60 - 17.90	3.3.17
16.40 - 20.20	16.40 - 20.20	16.40 - 20.20	16.40 - 20.20	17.90 - 20.90	3.3.18
22.40 - 31.20	-	-	-	-	3.3.19
-	-	23.20 - 33.50	-	-	3.3.20
-	-	-	10.50 - 13.20	-	3.3.21
-	14.00 - 17.10	14.00 - 17.10	14.00 - 17.10	14.40 - 17.90	3.3.22
-	17.90 - 20.20	17.90 - 20.20	17.90 - 20.20	17.90 - 20.20	3.3.23
17.90 - 24.00	19.40 - 24.70	19.40 - 24.70	19.40 - 24.70	19.40 - 28.10	3.3.24
22.40 - 28.10	23.20 - 30.40	23.20 - 30.40	23.20 - 30.40	23.20 - 33.50	3.3.25
24.00 - 30.40	26.20 - 33.50	26.20 - 33.50	26.20 - 33.50	26.20 - 35.80	3.3.26
26.20 - 35.80	27.00 - 38.00	27.00 - 38.00	27.00 - 38.00	27.00 - 40.30	3.3.27
30.10 - 38.80	32.70 - 40.30	32.70 - 40.30	32.70 - 40.30	32.70 - 44.10	3.3.28
33.00 - 42.00	36.00 - 45.00	36.00 - 45.00	36.00 - 45.00	-	3.3.29
34.00 - 45.00	37.00 - 45.00	37.00 - 45.00	37.00 - 45.00	-	3.3.30
37.00 - 79.00	40.00 - 81.00	40.00 - 81.00	40.00 - 81.00	-	3.3.31
68.00 - 84.00	70.00 - 87.00	70.00 - 87.00	70.00 - 87.00	-	3.3.32
78.00 - 129.00	78.00 - 129.00	78.00 - 129.00	78.00 - 129.00	-	3.3.33
-	-	40.00 50.00	-	-	3.3.34

Approximate Estimates

Item nr.	SPECIFICATIONS	Unit	RESIDENTIAL £ range £
3.4	**DECORATIONS** **surface area** (unless otherwise described) comprising:		
	Comparative wall and ceiling finishes		
	Comparative steel/metalwork finishes		
	Comparative woodwork finishes		
	Comparative wall and ceiling finishes		
	Emulsion		
3.4.1	two coats	m²	1.60 - 2.00
3.4.2	one mist and two coats	m²	2.00 - 2.70
	"Artex" plastic compound		
3.4.3	one coat; textured	m²	2.60 - 3.50
3.4.4	Wall paper	m²	3.50 - 5.70
3.4.5	Hessian wall coverings	m²	- -
	Gloss		
3.4.6	primer and two coats	m²	3.00 - 4.10
3.4.7	primer and three coats	m²	4.10 - 5.10
	Comparative steel/metalwork finishes		
	Primer		
3.4.8	only	m²	- -
3.4.9	grit blast and one coat zinc chromate primer	m²	- -
3.4.10	touch up primer and one coat of two pack epoxy zinc phosphate primer	m²	- -
	Gloss		
3.4.11	three coats	m²	4.10 - 5.00
	Sprayed mineral fibre		
3.4.12	one hour	m²	- -
3.4.13	two hour	m²	- -
	Sprayed vermiculite cement		
3.4.14	one hour	m²	- -
3.4.15	two hour	m²	- -
	Intumescent coating with decorative top seal		
3.4.16	half hour	m²	- -
3.4.17	one hour	m²	- -
	Comparative woodwork finishes		
	Primer		
3.4.18	only	m²	1.20 - 1.20
	Gloss		
3.4.19	two coats; touch up primer	m²	2.30 - 2.60
3.4.20	three costs; touch up primer	m²	3.00 - 3.80
3.4.21	primer and two coat	m²	3.50 - 4.10
3.4.22	primer and three coat	m²	4.60 - 5.00
	Polyurethene lacquer		
3.4.23	two coats	m²	2.30 - 2.80
3.4.24	three coats	m²	3.50 - 4.10
	Flame-retardant paint		
3.4.25	three coats	m²	5.20 - 6.60
	Polish		
3.4.26	wax polish; seal	m²	5.70 - 7.80
3.4.27	wax polish; stain and body-in	m²	9.50 - 11.00
3.4.28	french polish; stain and body-in	m²	14.10 - 17.10

1.60 - 2.00	1.60 - 2.00	1.60 - 2.00	1.60 - 2.00	1.60 - 2.00	3.4.1
2.00 - 2.70	2.00 - 2.70	2.00 - 2.70	2.00 - 2.70	2.00 - 2.70	3.4.2
2.60 - 3.50	2.60 - 3.50	2.60 - 3.50	2.60 - 3.50	2.60 - 3.50	3.4.3
3.50 - 6.10	3.50 - 6.10	3.50 - 6.10	3.50 - 6.10	3.60 - 9.10	3.4.4
-	7.80 - 11.30	7.80 - 11.30	7.80 - 11.30	8.80 - 13.80	3.4.5
3.00 - 4.10	3.00 - 4.10	3.00 - 4.10	3.00 - 4.10	3.00 - 4.10	3.4.6
4.10 - 5.00	4.10 - 5.00	4.10 - 5.00	4.10 - 5.00	4.10 - 5.00	3.4.7
0.60 - 1.20	-	-	-	-	3.4.8
1.30 - 2.00	-	-	-	-	3.4.9
1.80 - 2.30	-	-	-	-	3.4.10
-	-	-	-	-	3.4.11
7.80 - 12.00	-	-	-	-	3.4.12
13.00 - 15.60	-	-	-	-	3.4.13
8.80 - 13.10	-	-	-	-	3.4.14
10.60 - 15.60	-	-	-	-	3.4.15
14.10 - 15.60	14.10 - 15.60	14.10 - 15.60	-	-	3.4.16
22.40 - 27.40	22.40 - 27.40	22.40 - 27.40	-	-	3.4.17
1.10 - 1.20	1.10 - 1.20	1.10 - 1.20	1.10 - 1.20	1.10 - 1.20	3.4.18
2.30 - 2.60	2.30 - 2.60	2.30 - 2.60	2.30 - 2.60	2.30 - 2.60	3.4.19
3.00 - 3.90	3.00 - 3.90	3.00 - 3.90	3.00 - 3.90	3.00 - 3.90	3.4.20
3.50 - 4.10	3.50 - 4.10	3.50 - 4.10	3.50 - 4.10	3.50 - 4.10	3.4.21
4.60 - 5.00	4.60 - 5.00	4.60 - 5.00	4.60 - 5.00	4.60 - 5.00	3.4.22
2.30 - 2.70	2.30 - 2.60	2.30 - 2.70	2.30 - 2.70	2.30 - 2.70	3.4.23
3.50 - 4.10	3.50 - 4.10	3.50 - 4.10	3.50 - 4.10	3.50 - 4.10	3.4.24
5.20 - 6.30	5.20 - 6.30	5.20 - 6.30	5.20 - 6.30	5.20 - 6.30	3.4.25
5.20 - 6.30	5.20 - 7.80	5.20 - 7.80	5.30 - 7.80	5.20 - 7.80	3.4.26
9.50 - 11.00	9.50 - 11.00	9.50 - 11.00	9.50 - 11.00	9.50 - 11.00	3.4.27
14.10 - 17.10	14.10 - 17.10	14.10 - 17.10	14.10 - 17.10	14.10 - 17.10	3.4.28

Item nr.	SPECIFICATIONS	Unit	RESIDENTIAL £ range £
4.1	**FITTINGS AND FURNISHINGS** gross internal area or individual units		
	comprising:		
	Residential fittings **Comparative fittings/sundries** **Industrial/Office furniture, fittings and equipment** **Retail furniture, fittings and equipment** **Leisure furniture, fittings and equipment** **Hotel furniture, fittings and equipment**		
	Residential fittings Kitchen fittings for residential units		
4.1.1	one person flat/bed-sit	nr	504.00 - 1013.0
4.1.2	two person flat/house	nr	679.00 - 1712
4.1.3	three person flat/house	nr	811.00 - 2353
4.1.4	four person house	nr	863.00 - 3026
4.1.5	five person house	nr	1129.00 - 5592
	Comparative residential fittings/sundries Individual kitchen fittings		
4.1.6	600 mm x 600 mm x 300 mm wall unit	nr	62.40 - 74.60
4.1.7	1200 mm x 900 mm x 300 mm wall unit	nr	98.90 - 114.10
4.1.8	500 mm x 900 mm x 600 mm floor unit	nr	98.90 - 114.10
4.1.9	600 mm x 500 mm x 195 mm store cupboard	nr	147.60 - 162.80
4.1.10	1200 mm x 900 mm x 600 mm sink unit (excluding top)	nr	147.60 - 162.80
	Comparative wrought softwood shelving		
4.1.11	25 mm x 225 mm including black japanned brackets	m	9.60 - 11.10
4.1.12	25 mm thick slatted shelving; including bearers	m^2	39.60 - 44.10
4.1.13	25 mm thick cross-tongued shelving; including bearers	m^2	50.20 - 55.90
	Industrial/Office furniture, fittings and equipment Reception desk, shelves and cupboards for general areas		
4.1.14	economical	m^2	- -
4.1.15	medium quality	m^2	- -
4.1.16	high quality	m^2	- -
	Extra for		
4.1.17	high quality finishes to reception areas	m^2	- -
4.1.18	full kitchen equipment (one cover/20 m^2)	m^2	- -
	Furniture and fittings to general office areas		
4.1.19	economical	m^2	- -
4.1.20	medium quality	m^2	- -
4.1.21	high quality	m^2	- -
	Retail fitting out, furniture, fittings and equipment Mall furniture, etc.		
4.1.22	minimal provision	m^2	- -
4.1.23	good provision	m^2	- -
4.1.24	internal planting	m^2	- -
4.1.25	glazed metal balustrades to voids	m^2	- -
4.1.26	feature pond and fountain	m^2	- -
4.1.27	Fitting out a retail warehouse	m^2	- -
	Fitting out shell for small shop (including shop fittings)		
4.1.28	simple store	m^2	- -
4.1.29	fashion store	m^2	- -
	Fitting out shell for department store or supermarket		
4.1.30	excluding shop fittings	m^2	- -
4.1.31	including shop fittings	m^2	- -

Approximate Estimates

INDUSTRIAL £ range £	RETAILING £ range £	LEISURE £ range £	OFFICES £ range £	HOTELS £ range £	Item nr.
- -	- -	- -	- -	- -	4.1.1
- -	- -	- -	- -	- -	4.1.2
- -	- -	- -	- -	- -	4.1.3
- -	- -	- -	- -	- -	4.1.4
- -	- -	- -	- -	- -	4.1.5
- -	- -	- -	- -	- -	4.1.6
- -	- -	- -	- -	- -	4.1.7
- -	- -	- -	- -	- -	4.1.8
- -	- -	- -	- -	- -	4.1.9
- -	- -	- -	- -	- -	4.1.10
9.70 - 11.10	9.70 - 11.10	9.70 - 11.10	9.70 - 11.10	9.70 - 11.10	4.1.11
37.30 - 44.10	37.30 - 44.10	37.30 - 44.10	37.30 - 44.10	37.30 - 44.10	4.1.12
50.20 - 55.90	50.20 - 55.90	50.20 - 55.90	50.20 - 55.90	50.20 - 55.90	4.1.13
3.10 - 7.00	- -	- -	5.20 - 10.80	- -	4.1.14
5.60 - 11.10	- -	- -	8.10 - 14.80	- -	4.1.15
8.90 - 17.10	- -	- -	12.60 - 19.90	- -	4.1.16
- -	- -	- -	5.60 - 8.10	- -	4.1.17
8.10 - 9.70	- -	- -	8.90 - 11.10	- -	4.1.18
- -	- -	- -	6.60 - 8.90	- -	4.1.19
- -	- -	- -	8.90 - 14.00	- -	4.1.20
- -	- -	- -	15.60 - 22.90	- -	4.1.21
- -	8.10 - 15.60	- -	- -	- -	4.1.22
- -	34.20 - 40.30	- -	- -	- -	4.1.23
- -	17.90 - 22.90	- -	- -	- -	4.1.24
- -	601.00 - 760.80	- -	- -	- -	4.1.25
- -	33368 - 51216	- -	- -	- -	4.1.26
- -	209.20 - 240.40	- -	- -	- -	4.1.27
- -	524.90 - 601.00	- -	- -	- -	4.1.28
- -	791.20 - 1126.0	- -	- -	- -	4.1.29
- -	555.40 - 753.20	- -	- -	- -	4.1.30
- -	791.20 - 1126.0	- -	- -	- -	4.1.31

Approximate Estimates

Item nr.	SPECIFICATIONS	Unit	RESIDENTIAL £ range £
4.1	**FITTINGS AND FURNISHINGS - cont'd**		
	gross internal area or internal units (unless otherwise described)		
	Retail fitting out, furniture, fittings and equipment - cont'd		
	Special fittings/equipment		
4.1.32	refrigerated installation for cold stores; display fittings in food stores	m^2	- -
4.1.33	food court furniture; fittings and special finishes (excluding catering display units)	m^2	- -
4.1.34	bakery ovens	nr	- -
4.1.35	refuse compactors	nr	- -
	Leisure furniture, fittings and equipment		
	General fittings		
4.1.36	internal planting	m^2	- -
4.1.37	signs, notice-boards, shelving, fixed seating, curtains and blinds	m^2	- -
4.1.38	electric hand-dryers, incinerators, mirrors	m^2	- -
	Specific fittings		
4.1.39	lockers, coin return locks	nr	- -
4.1.40	kitchen units; excluding equipment	nr	- -
4.1.41	folding sun bed	nr	- -
4.1.42	security grille	nr	- -
4.1.43	entrance balustrading and control turnstile	nr	- -
4.1.44	sports nets, screens etc in a medium sized sports hall reception counter, fittings and reception counter	nr	- -
4.1.45	screen	nr	- -
4.1.46	bar and fittings	nr	- -
4.1.47	telescopic seatings	nr	- -
	Swimming pool fittings		
4.1.48	metal balustrades to pool areas	m	- -
4.1.49	skimmer grilles to pool edge	nr	- -
	stainless steel pool access ladder		
4.1.50	1700 mm high	nr	- -
4.1.51	2500 mm high	nr	- -
4.1.52	stainless steel tube ladders and fixing sockets to pool	nr	- -
4.1.53	level deck starting blocks	nr	- -
4.1.54	turning boards and brackets	nr	- -
4.1.55	backstroke warning set and infill tubes	nr	- -
4.1.56	false start equipment set including infill tubes	nr	- -
4.1.57	set of four 25 m lane rope sets; storage trolley and flush deck level adaptors	nr	- -
4.1.58	electrically operated cover to pool	nr	- -
	Leisure pool fittings		
4.1.59	stainless steel lighting post	nr	- -
4.1.60	water cannons	nr	- -
4.1.61	fountain or water sculpture	nr	- -
4.1.62	loudspeaker tower	nr	- -
4.1.63	grp water chute; 65.00 m - 80.00 m long; steel supports; spiral stairs and balustrading	nr	- -
	Sports halls		
4.1.64	administration areas	m^2	- -
4.1.65	aerobic dance studios	m^2	- -
4.1.66	badminton courts	m^2	- -
4.1.67	bowls halls	m^2	- -
4.1.68	cafe/restaurant areas	seat	- -
4.1.69	changing rooms/WCs	m^2	- -
4.1.70	circulation areas	m^2	- -
4.1.71	creche areas	m^2	- -
4.1.72	fitness areas	m^2	- -
4.1.73	indoor cricket	m^2	- -
4.1.74	rifle ranges	m^2	- -
4.1.75	reception areas	m^2	- -
4.1.76	school gymnasium (no changing)	m^2	- -
4.1.77	sports halls (with changing)	m^2	- -
4.1.78	squash courts	m^2	- -

Approximate Estimates 797

INDUSTRIAL £ range £	RETAILING £ range £	LEISURE £ range £	OFFICES £ range £	HOTELS £ range £	Item nr.
- -	22.10 - 67.70	- -	- -	- -	4.1.32
- -	715.00 - 776.00	- -	- -	- -	4.1.33
- -	9008 - 13580	- -	- -	- -	4.1.34
- -	9624 - 17688	- -	- -	- -	4.1.35
- -	- -	12.60 - 17.10	- -	- -	4.1.36
- -	- -	9.70 - 11.10	- -	- -	4.1.37
- -	- -	2.50 - 3.70	- -	- -	4.1.38
- -	- -	95.00 - 155.00	- -	- -	4.1.39
- -	- -	1864 - 2488	- -	- -	4.1.40
- -	- -	4504 - 4960	- -	- -	4.1.41
- -	- -	6208 - 8856	- -	- -	4.1.42
- -	- -	11336 - 12416	- -	- -	4.1.43
- -	- -	15520 - 18624	- -	- -	4.1.44
- -	- -	18624 - 31040	- -	- -	4.1.45
- -	- -	31040 - 37278	- -	- -	4.1.46
- -	- -	43456 - 49664	- -	- -	4.1.47
- -	- -	280.00 - 502.00	- -	- -	4.1.48
- -	- -	248.00 - 318.00	- -	- -	4.1.49
- -	- -	936.00 - 1029.0	- -	- -	4.1.50
- -	- -	1019.0 - 1170	- -	- -	4.1.51
- -	- -	415.00 - 470.00	- -	- -	4.1.52
- -	- -	581.00 - 638.00	- -	- -	4.1.53
- -	- -	386.00 - 443.00	- -	- -	4.1.54
- -	- -	443.00 - 498.00	- -	- -	4.1.55
- -	- -	221.00 - 277.00	- -	- -	4.1.56
- -	- -	2771 - 3104	- -	- -	4.1.57
- -	- -	6651 - 11086	- -	- -	4.1.58
- -	- -	125.00 - 155.00	- -	- -	4.1.59
- -	- -	852.00 - 1126.0	- -	- -	4.1.60
- -	- -	3424 - 17118	- -	- -	4.1.61
- -	- -	4352 - 5128	- -	- -	4.1.62
- -	- -	1e+05 - 117515	- -	- -	4.1.63
- -	- -	128.00 - 256.00	- -	- -	4.1.64
- -	- -	65.00 - 88.00	- -	- -	4.1.65
- -	- -	26.00 - 38.00	- -	- -	4.1.66
- -	- -	14.00 - 26.00	- -	- -	4.1.67
- -	- -	1386 - 1607	- -	- -	4.1.68
- -	- -	64.00 - 126.00	- -	- -	4.1.69
- -	- -	6.00 - 14.00	- -	- -	4.1.70
- -	- -	52.00 - 78.00	- -	- -	4.1.71
- -	- -	510.00 - 765.00	- -	- -	4.1.72
- -	- -	14.00 - 26.00	- -	- -	4.1.73
- -	- -	14.00 - 26.00	- -	- -	4.1.74
- -	- -	128.00 - 254.00	- -	- -	4.1.75
- -	- -	52.00 - 64.00	- -	- -	4.1.76
- -	- -	26.00 - 52.00	- -	- -	4.1.77
- -	- -	6.00 - 14.00	- -	- -	4.1.78

Item nr.	SPECIFICATIONS	Unit	RESIDENTIAL £ range £
4.1	**FITTINGS AND FURNISHINGS - cont'd**		
	gross internal area or internal units		
	(unless otherwise described)		
	Leisure furniture, fittings and equipment - cont'd		
	Sports halls - cont'd		
4.1.79	table tennis	m²	- -
4.1.80	tennis courts	m²	- -
4.1.81	viewing areas	m²	- -
	Football stadia		
4.1.82	facilities to basic stand	m²	- -
	Theatres		
	Stage engineering equipment to large-scale modern touring theatre		
4.1.83	flying	nr	- -
4.1.84	safety curtains and fire doors	nr	- -
4.1.85	lighting access	nr	- -
4.1.86	stage/pit lifts	nr	- -
4.1.87	hoists	nr	- -
4.1.88	curtains and masking	nr	- -
	Stage engineering equipment to medium-scale modern repertory touring theatre		
4.1.89	flying	nr	- -
4.1.90	safety curtains and fire doors	nr	- -
4.1.91	lighting access	nr	- -
4.1.92	stage/pit lifts	nr	- -
4.1.93	hoists	nr	- -
4.1.94	curtains and masking	nr	
	Museums and art galleries	nr	- -
	Display and retail areas		
4.1.95	display cases and lighting, finishes	m²	- -
4.1.96	display fittings and power	m²	- -
4.1.97	interactive display	m²	- -
4.1.98	ticketing and retail	m²	- -
4.1.99	display case (to 1.50 m x 1.50 m)	nr	- -
4.1.100	fibre optic lighting to display case	nr	- -
4.1.101	interactive "hands on" feature	nr	- -
4.1.102	interactive video installations		
4.1.103	video wall	nr	- -
4.1.104	video monitor hardware	nr	- -
4.1.105	AV installation to cinema	nr	- -
	Aviation		
	Airport passenger lounges		
4.1.106	international airport	m²	- -
4.1.107	domestic airport	m²	- -
4.1.108	allowance for furniture	m²	- -
4.1.109	baggage handling facilities	m²	- -
	Education		
	Schools and colleges		
4.1.110	general teaching	m²	- -
4.1.111	laboratories	m²	- -
4.1.112	information technology and business studies	m²	- -
4.1.113	design and technology	m²	- -
4.1.114	art and design	m²	- -
4.1.115	library resource centre	m²	- -
4.1.116	circulation	m²	- -
4.1.117	laboratory preparation	m²	- -
4.1.118	staff accommodation	m²	- -

Approximate Estimates

INDUSTRIAL £ range £	RETAILING £ range £	LEISURE £ range £	OFFICES £ range £	HOTELS £ range £	Item nr.
- -	- -	7.00 - 13.00	- -	- -	4.1.79
- -	- -	7.00 - 13.00	- -	- -	4.1.80
- -	- -	13.00 - 127.00	- -	- -	4.1.81
- -	- -	255.00 - 399.00	- -	- -	4.1.82
- -	- -	432334 - 504391	- -	- -	4.1.83
- -	- -	387994 - 460048	- -	- -	4.1.84
- -	- -	260509 - 304851	- -	- -	4.1.85
- -	- -	288223 - 360279	- -	- -	4.1.86
- -	- -	216169 - 243880	- -	- -	4.1.87
- -	- -	332565 - 224751	- -	- -	4.1.88
- -	- -	188452 - 219420	- -	- -	4.1.89
- -	- -	88683 - 127483	- -	- -	4.1.90
- -	- -	216169 - 243880	- -	- -	4.1.91
- -	- -	99769 - 144111	- -	- -	4.1.92
- -	- -	44342 - 72055	- -	- -	4.1.93
- -	- -	27715 - 72055	- -	- -	4.1.94
- -	- -	2660 - 5321	- -	- -	4.1.95
- -	- -	1330 - 2395	- -	- -	4.1.96
- -	- -	1995 - 3325	- -	- -	4.1.97
- -	- -	1330 - 1995	- -	- -	4.1.98
- -	- -	3990 - 1995	- -	- -	4.1.99
- -	- -	931 - 1331	- -	- -	4.1.100
- -	- -	9312 - 13299	- -	- -	4.1.101
- -	- -	332566 - 465593	- -	- -	4.1.102
- -	- -	4654 - 6657	- -	- -	4.1.103
- -	- -	133026 - 465593	- -	- -	4.1.104
- -	- -	1374 - 1374	- -	- -	4.1.106
- -	- -	776 - 1374	- -	- -	4.1.107
- -	- -	216 - 1374	- -	- -	4.1.108
- -	- -	2034 - 1374	- -	- -	4.1.109
- -	- -	127 - 199	- -	- -	4.1.110
- -	- -	166 - 277	- -	- -	4.1.111
- -	- -	361 - 510	- -	- -	4.1.112
- -	- -	350 - 393	- -	- -	4.1.113
- -	- -	233 - 277	- -	- -	4.1.114
- -	- -	393 - 441	- -	- -	4.1.115
- -	- -	40 - 67	- -	- -	4.1.116
- -	- -	361 - 418	- -	- -	4.1.117
- -	- -	420 - 441	- -	- -	4.1.118

Approximate Estimates

Item nr.	SPECIFICATIONS	Unit	RESIDENTIAL £ range £
5.1	**SANITARY AND DISPOSAL INSTALLATIONS** **gross internal area** (unless otherwise described)		
	comprising:		
	Comparative sanitary fittings/sundries **Sanitary and disposal installations**		
	Comparative sanitary fittings/sundries Note: Material prices vary considerably, the following composite rates are based on "average" prices for mid priced fittings: Individual sanitary appliances (including fittings) lavatory basins; vitreous china; chromium plated taps; waste; chain and plug; cantilever brackets		
5.1.1	white	nr	171.00 - 194.00
5.1.2	coloured	nr	194.00 - 233.00
	low level WC's; vitreous china pan and cistern; black plastic seat; low pressure ball valve; plastic flush pipe; fixing brackets - on ground floor		
5.1.3	white	nr	148.00 - 171.00
5.1.4	coloured	nr	186.00 - 202.00
	- one of a range; on upper floors		
5.1.5	white	nr	280.00 - 318.00
5.1.6	coloured	nr	318.00 - 350.00
	Extra for bowl type wall urinal; white glazed vitreous china flushing cistern; chromium plated flush pipes and spreaders; fixing brackets		
5.1.7	white	nr	- -
	shower tray; glazed fireclay; chromium plated waste; chain and plug; riser pipe; rose and mixing valve		
5.1.8	white	nr	- -
5.1.9	coloured	nr	388.00 - 450.00
	sink; glazed fireclay; chromium plated waste; chain and plug; fixing		
5.1.10	white	nr	435.00 - 481.00
	sink; stainless steel; chromium plated waste; chain and self coloured		
5.1.11	single drainer	nr	- -
5.1.12	double drainer	nr	186.00 - 233.00
	bath; reinforced acrylic; chromium plated taps; overflow; waste; chain and plug; "P" trap and overflow connections		
5.1.13	white	nr	218.00 - 342.00
5.1.14	coloured	nr	280.00 - 342.00
	bath; enamelled steel; chromium plated taps; overflow; waste; chain and plug; "P" trap and overflow connections		
5.1.15	white	nr	310.00 - 373.00
5.1.16	coloured	nr	342.00 - 403.00
	Soil waste stacks; 3.15 m storey height; branch and connection to drain		
5.1.17	110 mm diameter PVC	nr	248.00 - 280.00
	Extra for		
5.1.18	additional floors	nr	124.00 - 140.00
5.1.19	100 mm diameter cast iron; decorated	nr	- -
	Extra for		
5.1.20	additional floors	nr	- -

Approximate Estimates

INDUSTRIAL £ range £	RETAILING £ range £	LEISURE £ range £	OFFICES £ range £	HOTELS £ range £	Item nr.
148.00 - 186.00	171.00 - 233.00	171.00 - 233.00	171.00 - 233.00	171.00 - 280.00	5.1.1
- -	- -	194.00 - 274.00	194.00 - 274.00	194.00 - 274.00	5.1.2
140.00 - 171.00	148.00 - 171.00	148.00 - 171.00	148.00 - 171.00	148.00 - 202.00	5.1.3
- -	- -	186.00 - 233.00	186.00 - 233.00	186.00 - 256.00	5.1.4
272.00 - 320.00	280.00 - 318.00	280.00 - 318.00	280.00 - 318.00	280.00 - 342.00	5.1.5
- -	- -	318.00 - 358.00	318.00 - 373.00	318.00 - 396.00	5.1.6
124.00 - 148.00	140.00 - 171.00	140.00 - 171.00	140.00 - 171.00	140.00 - 171.00	5.1.7
365.00 - 411.00	396.00 - 449.00	396.00 - 449.00	396.00 - 449.00	396.00 - 449.00	5.1.8
- -	- -	396.00 - 479.00	396.00 - 502.00	396.00 - 738.00	5.1.9
186.00 - 233.00	186.00 - 318.00	- -	186.00 - 318.00	- -	5.1.10
- -	- -	- -	- -	- -	5.1.11
- -	- -	- -	- -	- -	5.1.12
- -	- -	- -	- -	280.00 - 403.00	5.1.13
- -	- -	- -	- -	280.00 - 403.00	5.1.14
- -	- -	- -	- -	310.00 - 434.00	5.1.15
- -	- -	- -	- -	341.00 - 449.00	5.1.16
240.00 - 280.00	248.00 - 280.00	248.00 - 280.00	248.00 - 280.00	248.00 - 280.00	5.1.17
124.00 - 148.00	124.00 - 148.00	124.00 - 148.00	124.00 - 148.00	125.00 - 148.00	5.1.18
502.00 - 533.00	502.00 - 533.00	502.00 - 533.00	502.00 - 533.00	502.00 - 533.00	5.1.19
248.00 - 280.00	248.00 - 280.00	248.00 - 280.00	248.00 - 280.00	248.00 - 280.00	5.1.20

Approximate Estimates

Item nr.	SPECIFICATIONS	Unit	RESIDENTIAL £ range £
5.1	**SANITARY AND DISPOSAL INSTALLATIONS** **gross internal area** (unless otherwise described)		
	Sanitary and disposal installations		
	Residential units		
5.1.21	range including WC; wash handbasin; bath	nr	1240 - 2092
5.1.22	range including WC; wash handbasin; bidet; bath	nr	1552 - 2488
5.1.23	and kitchen sink	nr	1864 - 2762
	range including two WC's; two wash handbasins; bidet		
5.1.24	bath and kitchen sink	nr	2326 - 3223
	Extra for		
5.1.25	rainwater pipe per storey	nr	55.90 - 70.00
5.1.26	soil pipe per storey	nr	124.00 - 140.00
5.1.27	shower over bath	nr	310.00 - 441.00
	Industrial buildings		
	warehouse		
5.1.28	minimum provision	m²	- -
5.1.29	high provision	m²	- -
	production unit		
5.1.30	minimum provision	m²	- -
5.1.31	minimum provision; area less than 1000 m²	m²	- -
5.1.32	high provision	m²	- -
	Retailing outlets		
5.1.33	to superstore	m²	- -
	to shopping centre malls; public conveniences; branch		
5.1.34	connections to shop shells	m²	- -
5.1.35	fitting out public conveniences in shopping mall block	m²	- -
5.1.36	Leisure buildings	m²	- -
	Office and industrial office buildings		
5.1.37	speculative; low rise; area less than 1000 m²	m²	- -
5.1.38	speculative; low rise	m²	- -
5.1.39	speculative; medium rise; area less than 1000 m²	m²	- -
5.1.40	speculative; medium rise	m²	- -
5.1.41	speculative; high rise	m²	- -
5.1.42	owner-occupied; low rise; area less than 1000 m²	m²	- -
5.1.43	owner-occupied; low rise	m²	- -
5.1.44	owner-occupied; medium rise; area less than 1000 m²	m²	- -
5.1.45	owner-occupied; medium rise	m²	- -
5.1.46	owner-occupied; high rise	m²	- -
	Hotels		
5.1.47	WC; bath; shower; basin to each bedroom; sanitary accommodation to public areas	m²	- -
5.20	**WATER INSTALLATIONS** **gross internal area**		
	Hot and cold water installations		
5.2.1	Complete installations	m²	13.20 - 23.30
5.2.2	To mall public conveniences; branch connections to shop shells	m²	- -

Approximate Estimates

INDUSTRIAL £ range £	RETAILING £ range £	LEISURE £ range £	OFFICES £ range £	HOTELS £ range £	Item nr.
- -	- -	- -	- -	- -	5.1.21
- -	- -	- -	- -	- -	5.1.22
- -	- -	- -	- -	- -	5.1.23
- -	- -	- -	- -	- -	5.1.24
- -	- -	- -	- -	- -	5.1.25
- -	- -	- -	- -	- -	5.1.26
- -	- -	- -	- -	- -	5.1.27
8.10 - 11.30	- -	- -	- -	- -	5.1.28
11.30 - 17.10	- -	- -	- -	- -	5.1.29
11.30 - 17.10	- -	- -	- -	- -	5.1.30
13.20 - 22.80	- -	- -	- -	- -	5.1.31
12.50 - 20.20	- -	- -	- -	- -	5.1.32
- -	2.90 - 7.00	- -	- -	- -	5.1.33
- -	5.60 - 9.10	- -	- -	- -	5.1.34
- -	4230 - 6794	- -	- -	- -	5.1.35
- -	- -	9.70 - 12.40	- -	- -	5.1.36
4.20 - 10.60	- -	- -	4.20 - 10.60	- -	5.1.37
7.80 - 14.00	- -	- -	7.40 - 13.90	- -	5.1.38
- -	- -	- -	8.10 - 14.80	- -	5.1.39
- -	- -	- -	11.00 - 17.10	- -	5.1.40
- -	- -	- -	11.00 - 17.10	- -	5.1.41
- -	- -	- -	7.00 - 13.90	- -	5.1.42
7.00 - 13.90	- -	- -	11.00 - 17.10	- -	5.1.43
11.00 - 17.10	- -	- -	11.00 - 17.10	- -	5.1.44
- -	- -	- -	13.20 - 19.40	- -	5.1.45
- -	- -	- -	15.60 - 21.70	- -	5.1.46
- -	- -	- -	- -	19.40 - 47.90	5.1.47
5.60 - 14.80	- -	17.90 - 25.50	10.10 - 14.80	28.10 - 34.20	5.2.1
- -	4.00 - 7.00	- -	- -	- -	5.2.2

Item nr.	SPECIFICATIONS	Unit	RESIDENTIAL £ range £	
5.3	**HEATING, AIR-CONDITIONING AND VENTILATING INSTALLATIONS** gross internal area serviced (unless otherwise described)			
	comprising:			
	Solid fuel radiator heating			
	Gas or oil-fired radiator heating			
	Gas or oil-fired convector heating			
	Electric and under floor heating			
	Hot air systems			
	Ventilation systems			
	Heating and ventilation systems			
	Comfort cooling systems			
	Full air-conditioning systems			
	Solid fuel radiator heating			
5.3.1	Chimney stack; hearth and surround to independent residential unit	nr	1712 -	2016
	Chimney; hot water service and central heating for			
5.3.2	two radiators	nr	2488 -	2944
5.3.3	three radiators	nr	2944 -	3271
5.3.4	four radiators	nr	3271 -	3606
5.3.5	five radiators	nr	3606 -	3956
5.3.6	six radiators	nr	3956 -	4306
5.3.7	seven radiators	nr	4306 -	5044
	Gas or oil-fired radiator heating			
	Gas-fired hot water service and central heating for			
5.3.8	three radiators	nr	1864 -	2640
5.3.9	four radiators	nr	2640 -	2876
5.3.10	five radiators	nr	2876 -	3066
5.3.11	six radiators	nr	3066 -	3271
5.3.12	seven radiators	nr	3416 -	3880
	Oil-fired hot water service, tank and central heating for			
5.3.13	seven radiators	nr	3048 -	4012
5.3.14	ten radiators	nr	3398 -	3329
5.3.15	LPHW radiator system	m²	25.00 -	39.00
5.3.16	speculative; area less than 1000 m²	m²	-	-
5.3.17	speculative	m²	-	-
5.3.18	owner-occupied; area less than 1000 m²	m²	-	-
5.3.19	owner-occupied	m²	-	-
5.3.20	LPHW radiant panel system			
5.3.21	speculative; less than 1000 m²	m²	-	-
5.3.22	speculative	m²	-	-
5.3.23	LPHW fin tube heating	m²	-	-
	Gas or oil-fired convector heating			
5.3.24	LPHW convector system	m²		
5.3.25	speculative; area less than 1000 m²	m²	-	-
5.3.26	speculative	m²	-	-
5.3.27	owner-occupied; area less than 1000 m²	m²	-	-
5.3.28	owner-occupied	m²	-	-
	LPHW sill-line convector system			
5.3.29	owner-occupied	m²	-	-
	Electric and under floor heating			
5.3.30	Panel heaters	m²	-	-
5.3.31	Skirting heaters	m²	-	-
5.3.32	Storage heaters	m²	-	-
5.3.33	Underfloor heating in changing areas	m²	-	-

Approximate Estimates

INDUSTRIAL £ range £	RETAILING £ range £	LEISURE £ range £	OFFICES £ range £	HOTELS £ range £	Item nr.
- -	- -	- -	- -	- -	5.3.1
- -	- -	- -	- -	- -	5.3.2
- -	- -	- -	- -	- -	5.3.3
- -	- -	- -	- -	- -	5.3.4
- -	- -	- -	- -	- -	5.3.5
- -	- -	- -	- -	- -	5.3.6
- -	- -	- -	- -	- -	5.3.7
- -	- -	- -	- -	- -	5.3.8
- -	- -	- -	- -	- -	5.3.9
- -	- -	- -	- -	- -	5.3.10
- -	- -	- -	- -	- -	5.3.11
- -	- -	- -	- -	- -	5.3.12
- -	- -	- -	- -	- -	5.3.13
- -	- -	- -	- -	- -	5.3.14
37.30 - 52.90	37.30 - 52.90	46.40 - 60.50	- -	55.90 - 77.60	5.3.15
- -	- -	- -	45.60 - 57.40	- -	5.3.16
- -	- -	- -	49.50 - 68.50	- -	5.3.17
- -	- -	- -	49.50 - 66.90	- -	5.3.18
- -	- -	- -	52.50 - 74.60	- -	5.3.19
- -	44.90 - 68.50	55.90 - 68.50	- -	- -	5.3.20
49.50 - 59.00	- -	- -	- -	- -	5.3.21
51.00 - 60.50	- -	- -	- -	- -	5.2.22
65.40 - 96.20	- -	- -	- -	- -	5.2.23
38.80 - 51.00	37.70 - 71.50	49.50 - 63.90	- -	65.40 - 82.20	5.3.24
- -	- -	- -	51.00 - 60.50	- -	5.3.25
- -	- -	- -	52.50 - 71.50	- -	5.3.26
- -	- -	- -	55.90 - 68.50	- -	5.3.27
- -	- -	- -	62.40 - 77.60	- -	5.3.28
- -	- -	- -	80.60 - 112.60	- -	5.3.29
- -	- -	- -	11.00 - 14.80	- -	5.3.30
- -	- -	- -	17.10 - 22.80	- -	5.3.31
- -	- -	- -	19.80 - 26.60	- -	5.3.32
- -	- -	42.60 - 54.40	- -	- -	5.3.33

Item nr.	SPECIFICATIONS	Unit	RESIDENTIAL £ range £
5.3	**HEATING, AIR-CONDITIONING AND VENTILATING INSTALLATIONS - cont'd**		
	gross internal area serviced (unless otherwise described)		
	Hot air systems		
	Hot water service and ducted hot air heating to		
5.3.34	three rooms	nr	1712 - 2206
5.3.35	five rooms	nr	2282 - 2602
5.3.36	"Elvaco" warm air heating	m²	- -
5.3.37	Gas-fired hot air space heating	m²	- -
5.3.38	Warm air curtains; 1.60 m long electrically-operated	nr	- -
5.3.39	Hot water-operated; including supply pipework	nr	- -
	Ventilation system		
	Local ventilation to		
5.3.40	WC's	nr	186.40 - 248.00
5.3.41	toilet areas	m²	- -
5.3.42	bathroom and toilet areas	m²	- -
5.3.43	Air extract system	m²	- -
5.3.44	Air supply and extract system	m²	- -
5.3.45	Service yard vehicle extract system	m²	- -
	Heating and ventilation systems		
5.3.46	Space heating and ventilation - economical	m²	- -
5.3.47	Heating and ventilation	m²	- -
5.3.48	Warm air heating and ventilation	m²	- -
5.3.49	Hot air heating and ventilation to shopping malls; including automatic remote vents in rooflights	m²	- -
	Extra for		
5.3.50	comfort cooling	m²	- -
5.3.51	full air-conditioning	m²	- -
	Comfort cooling systems		
5.3.52	Fan coil/induction systems	m²	- -
5.3.53	speculative; area less than 1000 m²	m²	- -
5.3.54	speculative	m²	- -
5.3.55	owner-occupied; area less than 1000 m²	m²	- -
5.3.56	owner-occupied	m²	- -
5.3.57	"VAV" system	m²	- -
5.3.58	speculative; area less than 1000 m²	m²	- -
5.3.59	speculative	m²	- -
5.3.60	owner-occupied; area less than 1000 m²	m²	- -
5.3.61	owner-occupied	m²	- -
	Stand-alone air-conditioning unit systems		
5.3.62	air supply and extract	m²	- -
5.3.63	air supply and extract; including heat re-claim	m²	- -
	Extra for		
5.3.64	automatic control installation	m²	- -
5.3.65	Full air-conditioning with dust and humidity control	m²	- -
5.3.66	Fan coil/induction systems	m²	- -
5.3.67	speculative; area less than 1000 m²	m²	- -
5.3.68	speculative	m²	- -
5.3.69	owner-occupied; area less than 1000 m²	m²	- -
5.3.70	owner-occupied	m²	- -
5.3.71	"VAV" system	m²	- -
5.3.72	speculative; area less than 1000 m²	m²	- -
5.3.73	speculative	m²	- -
5.3.74	owner-occupied; area less than 1000 m²	m²	- -
5.3.75	owner-occupied	m²	- -

Approximate Estimates

INDUSTRIAL £ range £	RETAILING £ range £	LEISURE £ range £	OFFICES £ range £	HOTELS £ range £	Item nr.
- -	- -	- -	- -	- -	5.3.34
- -	- -	- -	- -	- -	5.3.35
- -	- -	- -	60.50 - 77.60	- -	5.3.36
22.80 - 44.90	- -	- -	- -	- -	5.3.37
- -	1864 - 2526	- -	- -	- -	5.3.38
- -	3766 - 4260	- -	- -	- -	5.3.39
- -	- -	- -	- -	- -	5.3.40
- -	- -	- -	3.40 - 8.40	- -	5.3.41
- -	- -	- -	- -	19.00 - 17.50	5.3.42
- -	31.20 - 41.80	31.20 - 41.80	31.20 - 41.80	34.20 - 44.90	5.3.43
- -	- -	- -	44.90 - 68.50	- -	5.3.44
- -	31.20 - 41.80	- -	- -	- -	5.3.45
19.00 - 43.40	20.50 - 37.70	- -	- -	- -	5.3.46
49.50 - 62.40	- -	- -	- -	- -	5.3.47
- -	93.00 - 119.00	93.00 - 119.00	93.00 - 119.00	100.00 - 140.00	5.3.48
- -	81.00 - 105.00	- -	- -	- -	5.3.49
- -	70.00 - 100.00	- -	- -	- -	5.3.50
- -	81.00 - 148.00	- -	- -	- -	5.3.51
- -	- -	- -	- -	140.00 - 194.00	5.3.52
- -	- -	- -	140.00 - 126.00	140.00 - 194.00	5.3.53
- -	- -	- -	148.00 - 178.00	- -	5.3.54
- -	- -	- -	148.00 - 186.00	- -	5.3.55
- -	- -	- -	155.00 - 194.00	- -	5.3.56
- -	140.00 - 194.00	- -	- -	171.00 - 248.00	5.3.57
- -	- -	- -	155.00 - 233.00	- -	5.3.58
- -	- -	- -	163.00 - 233.00	- -	5.3.59
- -	- -	- -	171.00 - 240.00	- -	5.3.60
- -	- -	- -	171.00 - 248.00	- -	5.3.61
- -	- -	116.00 - 155.00	- -	- -	5.3.62
- -	- -	132.00 - 171.00	- -	- -	5.3.63
- -	- -	11.00 - 34.20	- -	- -	5.3.64
140.00 - 233.00	- -	- -	- -	- -	5.3.65
- -	155.00 - 233.00	- -	- -	163.00 - 248.00	5.3.66
- -	- -	- -	148.00 - 186.00	- -	5.3.67
- -	- -	- -	148.00 - 210.00	- -	5.3.68
- -	- -	- -	155.00 - 233.00	- -	5.3.69
- -	- -	- -	163.00 - 248.00	- -	5.3.70
- -	- -	- -	- -	186.00 - 280.00	5.3.71
- -	- -	- -	178.00 - 233.00	- -	5.3.72
- -	- -	- -	186.00 - 248.00	- -	5.3.73
- -	- -	- -	186.00 - 271.00	- -	5.3.74
- -	- -	- -	186.00 - 280.00	- -	5.3.75

Item nr.	SPECIFICATIONS	Unit	RESIDENTIAL £ range £
5.4	**ELECTRICAL INSTALLATIONS** gross internal area serviced (unless otherwise described)		
	comprising:		
	Mains and sub-mains switchgear and distribution **Lighting installation** **Lighting and power installation** **Comparative fittings/rates per point**		
	Mains and sub-mains switchgear and distribution		
5.4.1	Mains intake only	m²	- -
5.4.2	Mains switchgear only	m²	2.20 - 3.70
	Mains and sub-mains distribution		
5.4.3	to floors only	m²	- -
5.4.4	to floors; including small power and supplies to equipment	m²	- -
5.4.5	to floors; including lighting and power to landlords areas and supplies to equipment	m²	- -
5.4.6	to floors; including power, communication and supplies to equipment	m²	- -
5.4.7	to shop units; including fire alarms and telephone distribution	m²	- -
	Lighting installation Lighting to		
5.4.8	warehouse area	m²	- -
5.4.9	production area	m²	- -
5.4.10	General lighting; including luminaries	m²	- -
5.4.11	Emergency lighting	m²	- -
5.4.12	standby generators only	m²	- -
5.4.13	Underwater lighting	m²	- -
	Lighting and power installations Lighting and power to residential units		
5.4.14	one person flat/bed-sit	nr	852 - 1324
5.4.15	two person flat/house	nr	1012 - 1864
5.4.16	three person flat/house	nr	1164 - 2252
5.4.17	four person house	nr	1400 - 3416
5.4.18	five/six person house	nr	1712 - 3416
	Extra for		
5.4.19	intercom	nr	340.80 - 388.00
	Lighting and power to industrial buildings		
5.4.20	warehouse area	m²	- -
5.4.21	production area	m²	- -
5.4.22	production area; high provision	m²	- -
5.4.23	office area	m²	- -
5.4.24	office area; high provision	m²	- -
	Lighting and power to retail outlets		
5.4.25	shopping mall and landlords' areas	m²	- -
	Lighting and power to offices		
5.4.26	speculative office areas; average standard	m²	- -
5.4.27	speculative office areas; high standard	m²	- -
5.4.28	owner-occupied office areas; average standard	m²	- -
5.4.29	owner-occupied office areas; high standard	m²	- -
	Comparative fittings/rates per point		
5.4.30	Consumer control unit	nr	140.00 - 155.20
	Fittings; excluding lamps or light fittings		
5.4.31	lighting point; PVC cables	nr	44.10 - 47.90
5.4.32	lighting point; PVC cables in screwed conduits	nr	94.30 - 103.50
5.4.33	lighting point; MICC cables	nr	80.60 - 86.70

Approximate Estimates

INDUSTRIAL £ range £	RETAILING £ range £	LEISURE £ range £	OFFICES £ range £	HOTELS £ range £	Item nr.
1.40 - 2.80	- -	- -	- -	- -	5.4.1
3.70 - 7.40	3.40 - 7.80	- -	- -	- -	5.4.2
4.30 - 8.10	- -	17.10 - 31.20	12.50 - 21.70	- -	5.4.3
- -	- -	12.40 - 14.80	- -	- -	5.4.4
8.10 - 19.40	- -	- -	31.20 - 51.00	- -	5.4.5
- -	- -	- -	- -	51.00 - 73.00	5.4.6
- -	5.00 - 12.40	- -	- -	- -	5.4.7
17.90 - 34.20	- -	- -	- -	- -	5.4.8
21.70 - 38.00	- -	- -	- -	- -	5.4.9
- -	- -	21.70 - 37.30	- -	19.40 - 31.20	5.4.10
- -	- -	8.10 - 12.40	- -	3.40 - 6.20	5.4.11
2.20 - 8.90	2.20 - 8.90	2.20 - 8.90	2.20 - 8.90	2.20 - 8.90	5.4.12
- -	- -	5.60 - 11.00	- -	- -	5.4.13
- -	- -	- -	- -	- -	5.4.14
- -	- -	- -	- -	- -	5.4.15
- -	- -	- -	- -	- -	5.4.16
- -	- -	- -	- -	- -	5.4.17
- -	- -	- -	- -	- -	5.4.18
- -	- -	- -	- -	- -	5.4.19
34.20 - 55.90	- -	- -	- -	- -	5.4.20
38.80 - 62.40	- -	- -	- -	- -	5.4.21
55.90 - 73.00	- -	- -	- -	- -	5.4.22
83.70 - 100.40	- -	- -	- -	- -	5.4.23
111.80 - 133.90	- -	- -	- -	- -	5.4.24
- -	55.90 - 95.10	- -	- -	- -	5.4.25
- -	- -	- -	74.60 - 105.00	- -	5.4.26
- -	- -	- -	94.30 - 111.80	- -	5.4.27
- -	- -	- -	100.40 - 133.90	- -	5.4.28
- -	- -	- -	117.90 - 147.60	- -	5.4.29
- -	- -	- -	- -	- -	5.4.30
- -	49.50 - 55.90	49.50 - 55.90	49.50 - 55.90	49.50 - 55.90	5.4.31
133.90 - 147.60	110.30 - 125.50	110.30 - 125.50	110.30 - 125.50	110.30 - 125.50	5.4.32
110.30 - 125.50	96.60 - 105.70	96.60 - 105.70	96.60 - 105.70	96.60 - 105.70	5.4.33

Approximate Estimates

Item nr.	SPECIFICATIONS	Unit	RESIDENTIAL £ range £
5.4	**ELECTRICAL INSTALLATIONS - cont'd**		
	gross internal area serviced		
	Comparative fittings/rates per point - cont'd		
	Switch socket outlet; PVC cables		
5.4.34	single	nr	46.00 - 52.00
5.4.35	double	nr	56.00 - 64.00
	Switch socket outlet; PVC cables in screwed conduit		
5.4.36	single	nr	73.00 - 78.00
5.4.37	double	nr	76.00 - 85.00
	Switch socket outlet; MICC cables		
5.4.38	single	nr	73.00 - 78.00
5.4.39	double	nr	78.00 - 87.00
5.4.40	Immersion heater point (excluding heater)	nr	68.00 - 78.00
5.4.41	Cooker point; including control unit	nr	96.60 - 155.00
5.5	**GAS INSTALLATION**		
5.5.1	Connection charge	nr	502.00 - 639.00
	Supply to heaters within shopping mall and capped off		
5.5.2	Supply to shop shells	m²	- -
5.6	**LIFT AND CONVEYOR INSTALLATIONS**		
	lift or escalator (unless otherwise described)		
	comprising:		
	Passenger lifts		
	Escalators		
	Goods lift		
	Dock levellers		
	Electro-hydraulic passenger lifts		
	Eight to twelve person lifts		
5.6.1	8 person; 0.30 m/sec ; 2 - 3 levels	nr	- -
5.6.2	8 person; 0.63 m/sec ; 3 - 5 levels	nr	- -
5.6.3	8 person; 1.50 m/sec ; 3 levels	nr	- -
5.6.4	8 person; 1.60 m/sec ; 6 - 9 levels	nr	- -
5.6.5	10 person; 0.63 m/sec ; 2 - 4 levels	nr	- -
5.6.6	10 person; 1.00 m/sec ; 2 levels	nr	- -
5.6.7	10 person; 1.00 m/sec ; 3 - 6 levels	nr	- -
5.6.8	10 person; 1.00 m/sec ; 3 - 6 levels	nr	- -
5.6.9	10 person; 1.60 m/sec ; 4 levels	nr	- -
5.6.10	10 person; 1.60 m/sec ; 10 levels	nr	- -
5.6.11	10 person; 1.60 m/sec ; 13 levels	nr	- -
5.6.12	10 person; 2.50 m/sec ; 18 levels	nr	- -
5.6.13	12 person; 1.00 m/sec ; 2 - 3 levels	nr	- -
5.6.14	12 person; 1.00 m/sec ; 3 - 6 levels	nr	- -
	Thirteen person lifts		
5.6.15	13 person; 0.30 m/sec ; 2 - 3 levels	nr	- -
5.6.16	13 person; 0.63 m/sec ; 4 levels	nr	- -
5.6.17	13 person; 1.00 m/sec ; 2 levels	nr	- -
5.6.18	13 person; 1.00 m/sec ; 2 - 3 levels	nr	- -
5.6.19	13 person; 1.00 m/sec ; 3 - 6 levels	nr	- -
5.6.20	13 person; 1.60 m/sec ; 7 - 11 levels	nr	- -
5.6.21	13 person; 2.50 m/sec ; 7 - 11 levels	nr	- -
5.6.22	13 person; 2.50 m/sec ; 12 - 15 levels	nr	- -
	Sixteen person lifts		
5.6.23	16 person; 1.00 m/sec ; 2 levels	nr	- -
5.6.24	16 person; 1.00 m/sec ; 3 levels	nr	- -
5.6.25	16 person; 1.00 m/sec ; 3 - 6 levels	nr	- -
5.6.26	16 person; 1.60 m/sec ; 3 - 6 levels	nr	- -
5.6.27	16 person; 1.60 m/sec ; 7 - 11 levels	nr	- -
5.6.28	16 person; 2.50 m/sec ; 7 - 11 levels	nr	- -
5.6.29	16 person; 2.50 m/sec ; 12 - 15 levels	nr	- -
5.6.30	16 person; 3.50 m/sec ; 12 - 15 levels	nr	- -

Approximate Estimates

INDUSTRIAL £ range £	RETAILING £ range £	LEISURE £ range £	OFFICES £ range £	HOTELS £ range £	Item nr.
- -	57.00 - 62.00	57.00 - 62.00	57.00 - 62.00	57.00 - 62.00	5.4.34
107.00 - 121.00	68.00 - 75.00	68.00 - 75.00	68.00 - 75.00	68.00 - 75.00	5.4.35
88.00 - 98.00	81.00 - 84.00	81.00 - 84.00	81.00 - 84.00	81.00 - 84.00	5.4.36
100.00 - 109.00	87.00 - 97.00	87.00 - 97.00	87.00 - 97.00	87.00 - 97.00	5.4.37
94.00 - 99.00	81.00 - 85.00	81.00 - 85.00	81.00 - 85.00	81.00 - 85.00	5.4.38
100.00 - 126.00	87.00 - 97.00	87.00 - 97.00	87.00 - 97.00	87.00 - 97.00	5.4.39
- -	- -	- -	- -	- -	5.4.40
- -	- -	- -	- -	- -	5.4.41
- -	- -	- -	- -	- -	5.5.1
- -	6.24 - 9.51	- -	- -	- -	5.5.2
- -	- -	30701 - 42680	28712 - 42680	- -	5.6.1
- -	- -	- -	53179 - 73720	- -	5.6.2
64417 - 88479	- -	- -	- -	- -	5.6.3
- -	- -	- -	95060 - 114118	- -	5.6.4
- -	- -	- -	63221 - 82621	- -	5.6.5
- -	- -	- -	53179 - 58961	- -	5.6.6
- -	- -	- -	72198 - 91598	- -	5.6.7
- -	- -	- -	76839 - 95859	- -	5.6.8
- -	- -	- -	76839 - 98446	- -	5.6.9
- -	- -	- -	106282 - 121041	- -	5.6.10
- -	- -	- -	121878 - 140057	- -	5.6.11
- -	- -	- -	213780 - 233342	- -	5.6.12
71410 - 87513	- -	- -	- -	- -	5.6.13
- -	- -	- -	- -	84762 - 100530	5.6.14
- -	- -	- -	33779 - 44278	- -	5.6.15
- -	- -	- -	72190 - 90077	- -	5.6.16
- -	84427 - 94812	- -	- -	- -	5.6.17
71410 - 93306	- -	- -	- -	- -	5.6.18
- -	- -	- -	72198 - 95859	84762 - 105496	5.6.19
- -	- -	- -	94337 - 117922	1e+05 - 117426	5.6.20
- -	- -	- -	151700 - 155103	- -	5.6.21
- -	- -	- -	155103 - 170368	- -	5.6.22
- -	80214 - 102035	- -	- -	- -	5.6.23
- -	94812 - 109408	- -	- -	- -	5.6.24
- -	- -	- -	82621 - 106358	- -	5.6.25
- -	- -	- -	88479 - 112140	- -	5.6.26
- -	- -	- -	121878 - 140137	- -	5.6.27
- -	- -	- -	159080 - 164539	- -	5.6.28
- -	- -	- -	163180 - 190740	- -	5.6.29
- -	- -	- -	171055 - 197324	- -	5.6.30

Approximate Estimates

Item nr.	SPECIFICATIONS	Unit	RESIDENTIAL £ range £
5.6	**LIFT AND CONVEYOR INSTALLATIONS** -cont'd		
	lift or escalator (unless otherwise described)		
	Electro-hydraulic passenger lifts - cont'd		
	Twenty one person lifts		
5.6.31	21 person/bed; 0.40 m/sec ; 3 levels	nr	- -
5.6.32	21 person; 0.60 m/sec ; 2 levels	nr	- -
5.6.33	21 person; 1.00 m/sec ; 4 levels	nr	- -
5.6.34	21 person; 1.60 m/sec ; 4 levels	nr	- -
5.6.35	21 person; 1.60 m/sec ; 7 - 11 levels	nr	- -
5.6.36	21 person; 1.60 m/sec ; 10 levels	nr	- -
5.6.37	21 person; 1.60 m/sec ; 13 levels	nr	- -
5.6.38	21 person; 2.50 m/sec ; 7 - 11 levels	nr	- -
5.6.39	21 person; 2.50 m/sec ; 12 - 15 levels	nr	- -
5.6.40	21 person; 2.50 m/sec ; 18 levels	nr	- -
5.6.41	21 person; 3.50 m/sec ; 12 - 15 levels	nr	- -
	Extra for		
5.6.42	enhanced finish to car	nr	- -
5.6.43	glass backed observation car	nr	- -
	Ten person wall climber lifts		
5.6.44	10 person; 0.50 m/sec ; 2 levels	nr	- -
	Escalators		
	30⁰ Escalator; 0.50 m/sec; enamelled steel glass balustrades		
5.6.45	3.50 m rise; 600 mm step width	nr	- -
5.6.46	3.50 m rise; 800 mm step width	nr	- -
5.6.47	3.50 m rise; 1000 mm step width	nr	- -
	Extra for		
5.6.48	enhanced finish	nr	- -
5.6.49	4.40 m rise; 800 mm step width	nr	- -
5.6.50	4.40 m rise; 1000 mm step width	nr	- -
5.6.51	5.20 m rise; 800 mm step width	nr	- -
5.6.52	5.20 m rise; 1000 mm step width	nr	- -
5.6.53	6.00 m rise; 800 mm step width	nr	- -
5.6.54	6.00 m rise; 1000 mm step width	nr	- -
	Extras (per escalator)		
5.6.55	under step lighting	nr	- -
5.6.56	under handrail lighting	nr	- -
5.6.57	stainless steel balustrades	nr	- -
5.6.58	mirror glass cladding to sides and soffits	nr	- -
5.6.59	heavy duty chairs	nr	- -
	Goods lifts		
5.6.60	Hoist	nr	- -
	Kitchen service hoist		
5.6.61	50 kg; 2 levels	nr	- -
	Electric heavy duty goods lifts		
5.6.62	500 kg; 2 levels	nr	- -
5.6.63	500 kg; 2 - 3 levels	nr	- -
5.6.64	500 kg; 5 levels	nr	- -
5.6.65	1000 kg; 2 levels	nr	- -
5.6.66	1000 kg; 2 - 3 levels	nr	- -
5.6.67	1000 kg; 5 levels	nr	- -
5.6.68	1500 kg; 3 levels	nr	- -
5.6.69	1500 kg; 4 levels	nr	- -
5.6.70	1500 kg; 7 levels	nr	- -
5.6.71	2000 kg; 2 levels	nr	- -
5.6.72	2000 kg; 3 levels	nr	- -
5.6.73	3000 kg; 2 levels	nr	- -
5.6.74	3000 kg; 3 levels	nr	- -
	Oil hydraulic heavy duty goods lifts		
5.6.75	500 kg; 3 levels	nr	- -
5.6.76	1000 kg; 3 levels	nr	- -
5.6.77	2000 kg; 3 levels	nr	- -
5.6.78	3500 kg; 4 levels	nr	- -

Approximate Estimates

INDUSTRIAL £ range £	RETAILING £ range £	LEISURE £ range £	OFFICES £ range £	HOTELS £ range £	Item nr.
- -	- -	- -	60482 - 67900	- -	5.6.31
- -	- -	- -	86920 - 106320	- -	5.6.32
- -	- -	- -	103200 - 121916	- -	5.6.33
- -	- -	- -	91484 - 106320	- -	5.6.34
- -	- -	- -	121916 - 150483	- -	5.6.35
- -	- -	- -	137322 - 136561	- -	5.6.36
- -	- -	- -	150483 - 166612	- -	5.6.37
- -	- -	- -	173839 - 206173	- -	5.6.38
- -	- -	- -	195902 - 228616	- -	5.6.39
- -	- -	- -	235843 - 257906	- -	5.6.40
- -	- -	- -	221008 - 243070	- -	5.6.41
- -	- -	- -	- -	11640 - 12401	5.6.42
- -	- -	- -	- -	13162 - 23584	5.6.43
- -	191718 - 240408	- -	191718 - 243071	191718 - 243071	5.6.44
- -	- -	- -	63297 - 77980	- -	5.6.45
- -	- -	- -	67900 - 82621	- -	5.6.46
- -	91598 - 120965	73796 - 88403	73796 - 88403	76078 - 107118	5.6.47
- -	27160 - 38420	19020 - 26323	19020 - 26323	27160 - 41158	5.6.48
- -	76078 - 95859	- -	- -	- -	5.6.49
- -	86920 - 103086	- -	- -	- -	5.6.50
- -	81024 - 99663	- -	- -	- -	5.6.51
- -	91675 - 108792	- -	- -	- -	5.6.52
- -	85969 - 103086	- -	- -	- -	5.6.53
- -	95859 - 114118	- -	- -	- -	5.6.54
- -	426 - 583	- -	- -	- -	5.6.55
- -	1712 - 3347	- -	- -	- -	5.6.56
- -	1864 - 3690	- -	- -	- -	5.6.57
- -	13200 - 16281	- -	- -	- -	5.6.58
- -	16281 - 22900	- -	- -	- -	5.6.59
6619 - 23584	- -	- -	- -	- -	5.6.60
- -	- -	7760 - 8521	- -	- -	5.6.61
- -	57439 - 65199	- -	55918 - 65199	55918 - 65199	5.6.62
25106 - 35300	- -	- -	- -	- -	5.6.63
- -	- -	- -	67862 - 85360	- -	5.6.64
- -	54320 - 73796	- -	60482 - 73796	62080 - 70601	5.6.65
32562 - 44278	- -	- -	- -	- -	5.6.66
- -	- -	- -	- -	73796 - 84599	5.6.67
67862 - 81100	- -	- -	76839 - 91598	- -	5.6.68
- -	- -	- -	82621 - 98902	- -	5.6.69
- -	- -	- -	98902 - 113661	- -	5.6.70
- -	- -	- -	76839 - 91598	- -	5.6.71
- -	- -	- -	91598 - 106358	- -	5.6.72
- -	- -	- -	82621 - 98902	- -	5.6.73
- -	- -	- -	103162 - 121878	- -	5.6.74
76839 - 91598	- -	- -	- -	- -	5.6.75
81100 - 95859	- -	- -	- -	- -	5.6.76
94337 - 10909	- -	- -	- -	- -	5.6.77
84143 - 98902	- -	- -	- -	- -	5.6.78

Item nr.	SPECIFICATIONS	Unit	RESIDENTIAL £ range £
5.6	**LIFT AND CONVEYOR INSTALLATIONS - cont'd** **lift or escalator** (unless otherwise described)		
	Dock levellers		
5.6.79	Dock levellers	nr	- -
5.6.80	Dock leveller and canopy	nr	- -
5.7	**PROTECTIVE, COMMUNICATION AND SPECIAL INSTALLATIONS** **gross internal area served** (unless otherwise described)		
	comprising:		
	Fire fighting/protective installations **Security/communication installations** **Special installations**		
	Fire fighting/protective installations Fire alarms/appliances		
5.7.1	loose fire fighting equipment	m²	- -
5.7.2	smoke detectors; alarms and controls	m²	- -
5.7.3	hosereels; dry risers and extinguishers	m²	- -
	Sprinkler installations		
5.7.4	landlords areas; supply to shop shells; including fire alarms; appliances etc.	m²	- -
5.7.5	single level sprinkler systems, alarms and smoke detectors; low hazard	m²	- -
5.7.6	Extra for ordinary hazard	m²	- -
5.7.7	single level sprinkler systems; alarms and smoke detectors; ordinary hazard	m²	- -
5.7.8	double level sprinkler systems; alarms and smoke detectors; high hazard	m²	- -
	Smoke vents		
5.7.9	automatic smoke vents over glazed shopping mall	m²	- -
5.7.10	smoke control ventilation to atria	m²	- -
5.7.11	Lightning protection		
	Security/communication installations		
5.7.12	Clock installation	m²	- -
5.7.13	Security alarm system	m²	- -
5.7.14	Telephone system	m²	- -
5.7.15	Public address, television aerial and clocks	m²	- -
5.7.16	Closed-circuit television	m²	- -
5.7.17	Public address system	m²	- -
5.7.18	Closed-circuit television and public address system	m²	- -
	Special installations Window cleaning equipment		
5.7.19	twin track	m	- -
5.7.20	manual trolley/cradle	nr	- -
5.7.21	automatic trolley/cradle	nr	- -
5.7.22	Refrigeration installation for ice rinks	ice area m²	- -
5.7.23	Pool water treatment installation	pool area m²	- -
5.7.24	Laundry chute	nr	- -
5.7.25	Sauna	nr	- -
5.7.26	Jacuzzi installation	nr	- -
5.7.27	Wave machine; four chamber wave generation equipment	nr	- -
5.7.28	Swimming pool; size 13.00 m x 6.00 m Extra over cost including structure; finishings, ventilation; heating and filtration	m²	- -

Approximate Estimates

INDUSTRIAL £ range £	RETAILING £ range £	LEISURE £ range £	OFFICES £ range £	HOTELS £ range £	Item nr.
8102 - 19400	8262 - 19400	- -	- -	8102 - 19400	5.6.79
11640 - 26780	11640 - 26780	- -	- -	- -	5.6.80
- -	- -	0.18 - 0.30	- -	- -	5.7.1
2.80 - 5.55	2.80 - 5.55	5.55 - 7.00	2.80 - 5.55	6.24 - 11.18	5.7.2
4.34 - 10.42	4.95 - 10.42	- -	4.95 - 10.42	4.79 - 9.66	5.7.3
- -	8.14 - 14.76	- -	- -	- -	5.7.4
10.27 - 15.60	11.11 - 15.60	- -	- -	- -	5.7.5
4.34 - 5.55	4.34 - 5.55	- -	- -	- -	5.7.6
14.76 - 20.54	15.60 - 20.54	- -	13.16 - 19.40	11.64 - 18.64	5.7.7
22.06 - 28.15	24.73 - 29.29	- -	22.06 - 28.15	- -	5.7.8
- -	15.60 - 25.11	- -	- -	- -	5.7.9
- -	- -	- -	47.93 - 55.92	47.93 - 60.48	5.7.10
0.55 - 0.70	1.08 - 2.18	0.55 - 0.70	0.78 - 1.48	0.78 - 1.48	5.7.11
- -	- -	0.14 - 1.08	- -	- -	5.7.12
- -	- -	1.55 - 2.25	1.55 - 2.25	- -	5.7.13
- -	- -	0.85 - 1.86	0.85 - 1.86	- -	5.7.14
2.18 - 3.41	- -	2.48 - 3.73	- -	2.18 - 4.73	5.7.15
- -	3.42 - 4.11	3.42 - 4.11	- -	- -	5.7.16
- -	8.83 - 10.42	8.83 - 10.42	- -	- -	5.7.17
- -	20.54 - 40.32	12.40 - 14.76	- -	- -	5.7.18
- -	126.00 - 148.00	- -	110.00 - 132.00	121.00 - 140.00	5.7.19
- -	8110 - 9700	- -	8110 - 9700	7760 - 9312	5.7.20
- -	17878 - 22900	- -	19020 - 22900	18639 - 22900	5.7.21
- -	- -	435.00 - 505.00	- -	- -	5.7.22
- -	- -	435.00 - 472.00	- -	- -	5.7.23
- -	- -	11069 - 13580	- -	- -	5.7.24
- -	- -	- -	- -	10897 - 13369	5.7.25
- -	- -	12425 - 16853	- -	6529 - 13369	5.7.26
- -	- -	41193 - 58048	- -	- -	5.7.27
- -	- -	- -	- -	843.00 - 959.00	5.7.28

Approximate Estimates

Item nr.	SPECIFICATIONS	Unit	RESIDENTIAL £ range £	
5.8	**BUILDERS' WORK IN CONNECTION WITH SERVICES** gross internal area			
	General builders work to			
5.8.1	main supplies, lighting and power to landlords areas	m²	-	-
5.8.2	central heating and electrical installation	m²	-	-
5.8.3	space heating and electrical installation	m²	-	-
5.8.4	central heating, electrical and lift installations	m²	-	-
5.8.5	space heating, electrical and ventilation installations	m²	-	-
5.8.6	air-conditioning	m²	-	-
5.8.7	air-conditioning and electrical installation	m²	-	-
5.8.8	air conditioning, electrical and lift installations	m²	-	-
	General builders work, including allowance for plant rooms; to			
5.8.9	central heating and electrical installations	m²	-	-
5.8.10	central heating, electrical and lift installations	m²	-	-
5.8.11	air-conditioning	m²	-	-
5.8.12	air-conditioning and electrical installation	m²	-	-
5.8.13	air-conditioning, electrical and lift installations	m²	-	-

Keep your figures up to date, free of charge
Download updates from the web: www.pricebooks.co.uk

This section, and most of the other information in this Price Book, is brought up to date every three months with the Price Book Updates, until the next annual edition. The updates are available free to all Price Book purchasers.

To ensure you receive your free copies, either complete the reply card from the centre of the book and return it to us or register via the website www.pricebooks.co.uk

Approximate Estimates

INDUSTRIAL £ range £	RETAILING £ range £	LEISURE £ range £	OFFICES £ range £	HOTELS £ range £	Item nr.
1.60 - 4.50	- -	- -	- -	- -	5.8.1
3.20 - 10.70	5.20 - 11.90	5.20 - 11.90	7.80 - 11.90	7.80 - 11.90	5.8.2
4.20 - 11.90	7.00 - 12.40	7.00 - 12.40	9.00 - 12.40	9.00 - 12.40	5.8.3
4.60 - 12.40	7.80 - 13.20	7.80 - 13.20	10.50 - 14.00	10.50 - 14.00	5.8.4
7.00 - 16.40	10.50 - 17.90	10.50 - 19.40	14.80 - 19.40	14.80 - 19.40	5.8.5
11.80 - 18.30	14.00 - 19.40	14.00 - 19.40	16.70 - 21.70	16.70 - 21.70	5.8.6
13.20 - 19.40	16.40 - 21.70	16.40 - 21.70	18.60 - 24.00	18.60 - 24.00	5.8.7
14.80 - 21.70	17.90 - 24.00	17.90 - 24.00	20.90 - 25.60	20.90 - 25.60	5.8.8
- -	20.90 - 26.40	20.90 - 26.40	26.40 - 32.70	26.40 - 32.70	5.8.9
- -	26.40 - 32.70	26.40 - 32.70	32.70 - 37.30	32.70 - 37.30	5.8.10
- -	37.30 - 43.40	37.30 - 43.40	47.90 - 52.50	47.90 - 52.50	5.8.11
- -	47.90 - 52.50	47.90 - 52.50	59.00 - 65.40	59.00 - 65.40	5.8.12
- -	59.00 - 65.40	59.00 - 65.40	68.50 - 76.10	68.50 - 76.10	5.8.13

Visit www.pricebooks.co.uk

Approximate Estimates

Item nr.	SPECIFICATIONS		Unit	ALL AREAS £ range £	
6.1	**SITE WORK** surface area (unless otherwise described)				
	comprising:				
	Preparatory excavation and sub-bases	Roadbridges			
	Seeded and planted areas	Underpasses			
	Sports Pitches	Roundabouts			
	Parklands	Guard rails and parking bollards, etc.			
	Paved areas				
	Car parking alternatives	Street furniture			
	Roads and barriers	Playground equipment			
	Road Crossing	Fencing and screen walls, ancillary buildings etc.			
	Footbridges				
	Preparatory excavation and sub-bases				
	Excavating				
6.1.1	top soil; average 150 mm deep		m^2	0.85 -	1.55
	top soil; average 225 mm deep; preserving in spoil heaps				
6.1.2	by machine		m^2	1.55 -	2.10
6.1.3	by hand		m^2	3.80 -	5.93
	to reduce levels; not exceeding 0.25 m deep				
6.1.4	by machine		m^3	2.40 -	2.66
6.1.5	by hand		m^3	10.50 -	11.64
6.1.6	to form new foundation levels and contours		m^3	2.63 -	12.10
	Filling; imported top soil				
6.1.7	150 mm thick; spread and levelled for planting		m^2	4.87 -	5.63
	Comparative sub-bases/beds				
6.1.8	50 mm thick; sand		m^2	1.70 -	1.86
6.1.9	75 mm thick; ashes		m^2	1.48 -	1.70
6.1.10	75 mm thick; sand		m^2	2.40 -	2.63
6.1.11	50 mm thick; blinding concrete (1:8)		m^2	2.80 -	3.12
6.1.12	100 mm thick; granular fill		m^2	2.80 -	3.12
6.1.13	150 mm thick; granular fill		m^2	4.41 -	4.64
6.1.14	75 mm thick; blinding concrete (1:3:6)		m^2	4.18 -	4.64
	Seeded and planted areas				
	Plant supply, planting, maintenance and 12 months guarantee				
6.1.15	seeded areas		m^2	2.80 -	5.63
6.1.16	turfed areas		m^2	3.65 -	7.30
	Planted areas (per m^2 of surface area)				
6.1.17	herbaceous plants		m^2	3.12 -	4.18
6.1.18	climbing plants		m^2	4.18 -	7.30
6.1.19	general planting		m^2	9.36 -	19.02
6.1.20	woodland		m^2	14.15 -	28.30
6.1.21	shrubbed planting		m^2	18.64 -	52.11
6.1.22	dense planting		m^2	23.28 -	46.41
6.1.23	shrubbed area including allowance for small trees		m^2	28.00 -	65.43
	Trees				
6.1.24	advanced nursery stock trees (12 - 20 cm girth)		tree	115.00 -	140.00
	semi-mature trees; 5 - 8 m high				
6.1.25	coniferous		tree	373.00 -	939.00
6.1.26	deciduous		tree	566.00 -	1552.00

Approximate Estimates

Item nr.	SPECIFICATIONS	Unit	ALL AREAS £ range £
6.1	**SITE WORK - cont'd**		
	surface area (unless otherwise described)		
	Sports pitches		
	Costs include for cultivating ground, bringing to appropriate levels for the specified game, applying fertiliser, weedkiller, seeding and rolling and white line marking with nets, posts, etc. as required		
6.1.27	football pitch (114 m x 72 m)	nr	13504
6.1.28	cricket outfield (160 m x 142 m)	nr	45951
6.1.29	cricket square (20 m x 20m) including imported marl or clay loam, bringing to accurate levels, seeding with cricket square type grass	nr	4184
6.1.30	bowling green (38 m x 38 m) rink including French drain and gravel path on four sides	nr	16623
6.1.31	grass tennis courts 1 court (35 m x 17 m) including bringing to accurate levels, chain link perimeter fencing and gate, tennis posts and net	nr	16927
6.1.32	two grass tennis courts (35 m x 32 m) ditto	pair	28720
6.1.33	artificial surface tennis courts (35 m x 17 m) including chain link fencing, gate, posts and net	nr	13352
6.1.34	two courts (45 m x 32 m) ditto	pair	24992
6.1.35	artificial football pitch, including sub-base, bitumen macadam open textured base and heavy duty "Astroturf" type carpet	nr	263840
6.1.36	golf-putting green	hole	1354
6.1.37	pitch and putt course	hole	4184 - 6353
6.1.38	full length golf course, full specifications including watering system	hole	14802 - 28044
6.1.39	championship golf course	hole	up to 104075
	Parklands		
	NOTE: Work on parklands will involve different techniques of earth shifting and cultivation. The following rates include for normal surface excavation, they include for the provision of any land drainage.		
6.1.40	Parklands, including cultivating ground, applying fertiliser, etc and seeding with parks type grass	ha	12895
6.1.41	General sports field	ha	15368
	Lakes including excavation average 10 m deep, laying 1.50 mm thick butyl rubber sheet and spreading top soil evenly on top 300 mm deep		
6.1.42	under 1 hectare in area	ha	282403
6.1.43	between 1 and 5 hectare in area	ha	263840
6.1.44	Extra for planting aquatic plants in lake top soil	m^2	42.00
	Land drainage		
	NOTE: If land drainage is required on a project, the propensity of the land to flood will decide the spacing of the land drains. Costs include for excavation and backfilling of trenches and laying agricultural clay drain pipes with 75 mm diameter lateral runs average 600 mm deep, and 100 mm diameter main runs average 750 mm deep.		
	land drainage to parkland with laterals at		
6.1.45	30 m centres and main runs at 100 m centres	ha	2663
	land drainage to sportsfields with laterals at		
6.1.46	10 m centres and main runs at 33 m centres	ha	7760

Approximate Estimates

Item nr.	SPECIFICATIONS	Unit	ALL AREAS £ range £	
6.1	**SITE WORK - cont'd**			
	surface area (unless otherwise described)			
	Paved areas			
	Gravel paving rolled to falls and cambers			
6.1.47	50 mm thick	m²	1.90 -	2.60
6.1.48	paving on sub-base; including excavation	m²	7.30 -	10.50
	Cold bitumen emulsion paving; in three layers			
6.1.49	25 mm thick	m²	3.70 -	4.60
	Tarmacadam paving; two layers; limestone or igneous chipping finish			
6.1.50	65 mm thick	m²	5.60 -	7.00
6.1.51	paving on sub-base; including excavation	m²	13.50 -	19.40
	Precast concrete paving slabs			
6.1.52	50 mm thick	m²	8.10 -	17.10
6.1.53	50 mm thick "Texitone" slabs	m²	11.60 -	16.00
6.1.54	slabs on sub-base; including excavation	m²	18.60 -	26.60
	Precast concrete block paviours			
6.1.55	65 mm thick "Keyblock" grey paving	m²	15.60 -	19.40
6.1.56	65 mm thick "Mount Sorrel" grey paving	m²	15.60 -	18.60
6.1.57	65 mm thick "Intersett" paving	m²	16.40 -	19.40
6.1.58	60 mm thick "Pedesta" paving	m²	11.80 -	19.40
6.1.59	paviours on sub-base; including excavation	m²	22.40 -	32.00
	Brick paviours			
	229 mm x 114 mm x 38 mm paving bricks			
6.1.60	laid flat	m²	25.50 -	32.00
6.1.61	laid to herringbone pattern	m²	41.80 -	46.40
6.1.62	paviours on sub-base; including excavation	m²	47.90 -	55.90
	Granite setts			
6.1.63	200 mm x 100 mm x 100 mm setts	m²	65.40 -	71.50
6.1.64	setts on sub-base; including excavation	m²	79.10 -	85.20
	York stone slab paving			
6.1.65	paving on sub-base; including excavation	m²	85.20 -	101.20
	Cobblestone paving			
6.1.66	50 mm - 75 mm diameter	m²	47.90 -	59.00
6.1.67	cobblestones on sub-base; including excavation	m²	55.90 -	73.00
	Car Parking alternatives			
	Surface level parking; including lighting and drainage			
6.1.68	tarmacadam on sub-base	car	887 -	1220
6.1.69	concrete interlocking blocks	car	998 -	1331
6.1.70	"Grasscrete" precast concrete units filled with top soil and grass seed	car	543 -	743
6.1.71	at ground level with deck or building over	car	4551 -	5266
	Garages etc			
6.1.72	single car park	nr	596 -	936
6.1.73	single; traditional construction; in a block	nr	1862 -	2660
6.1.74	single; traditional construction; pitched roof	nr	1838 -	5432
6.1.75	double; traditional construction; pitched roof	nr	6086 -	7797
	Multi-storey parking; including lighting and drainage on roof of two storey shopping centre; including			
6.1.76	ramping and strengthening structure	car	3880 -	5896
6.1.77	split level/parking ramp	car	5362 -	6041
6.1.78	multi-storey flat slab	car	6491 -	7797
6.1.79	multi-storey warped slab	car	6847 -	8314
	Extra for			
6.1.80	feature cladding and pitched roof	car	831 -	1414
	Underground parking; including lighting and drainage			
6.1.81	partially underground; natural ventilation; no sprinklers	car	8868 -	12195
6.1.82	completely underground; mechanical ventilation and sprinklers	car	12195 -	16281
6.1.83	completely underground; mechanical ventilation; sprinklers and landscaped roof	car	14329 -	19955

Approximate Estimates

Item nr.	SPECIFICATIONS	Unit	ALL AREAS £ range £
6.1	**SITE WORK - cont'd**		
	surface area (unless otherwise described)		
	Roads and barriers		
	Tarmacadam or reinforced concrete roads, including all earthworks, drainage, pavements, lighting, signs, fencing and safety barriers (where necessary); average cut 1.50 m		
6.1.84	two lane road 7.30 m wide-rural location	m	1377 - 1575
6.1.85	two lane road 7.30 m wide-urban location	m	1552 - 1864
6.1.86	two lane road 10.00 m wide-rural location	m	1491 - 1712
6.1.87	two lane road 10.00 m wide-urban location	m	1712 - 1864
	Road crossings		
	NOTE: Costs include road markings, beacons, lights, signs, advance danger signs etc.		
6.1.88	Zebra crossing	nr	3576 - 4184
6.1.89	Pelican crossing	nr	14607 - 15520
	Footbridges		
	Footbridge of either precast concrete or steel construction 4.00 m wide, 6.00 m high including deck, access stairs and ramp, parapets etc.		
6.1.90	15.00 m - 20.00 m span to two lane road	nr	145272 - 161438
6.1.91	30.00 m span to four lane dual carriageway	nr	161438 - 193696
	Roadbridges		
	Roadbridges including all excavation, reinforcement, formwork, concrete, bearings, expansion joints, deck water proofing and finishings, parapets etc.		
	RC bridge with precast beams		
6.1.92	10.00 m span	deck area m^2	852.00
6.1.93	15.00 m span	deck area m^2	806.00
	RC bridge with prefabricated steel beams		
6.1.94	20.00 m span	deck area m^2	844.00
6.1.95	30.00 m span	deck area m^2	791.00
	Underpass		
	Provision of underpasses to new roads, constructed as part of a road building programme		
	Precast concrete pedestrian underpass		
6.1.96	3.00 m wide x 2.50 m high	m	3104 - 3652
	Precast concrete vehicle underpass		
6.1.97	7.00 m wide x 5.00 m high	m	8673 - 10864
6.1.98	14.00 m wide x 5.00 m high	m	22101
	Bridge type structure vehicle underpass		
6.1.99	7.00 m wide x 5.00 m high	m	15368 - 18944
6.1.100	14.00 m wide x 5.00 m high	m	37583
	Roundabouts		
6.1.101	Roundabout on existing dual carriageway; including perimeter road, drainage and lighting, signs and disruption while under construction	nr	282479 - 415921
	Guard rails and parking bollards etc.		
6.1.102	Open metal post and rail fencing 1.00 m high	m	104.00 - 121.00
6.1.103	Galvanised steel post and rail fencing 2.00 m high	m	116.00 - 152.00
6.1.104	Steel guard rails and vehicle barriers	m	37.00 - 56.00
	Parking bollards		
6.1.105	precast concrete	nr	84.00 - 99.00
6.1.106	steel	nr	135.00 - 178.00
6.1.107	cast iron	nr	158.00 - 218.00
6.1.108	Vehicle control barrier; manual pole	nr	730.00 - 784.00
6.1.109	Galvanised steel cycle stand	nr	31.00 - 40.00
6.1.110	Galvanised steel flag staff	nr	822.00 - 1042.00

Approximate Estimates

Item nr.	SPECIFICATIONS	Unit	ALL AREAS £ range £
6.1	**SITE WORK - cont'd**		
	surface area (unless otherwise described)		
	Street Furniture		
	Reflectorised traffic signs 0.25 m² area		
6.1.111	on steel post	nr	75.00 - 135.00
	Internally illuminated traffic signs		
6.1.112	dependent on area	nr	161.00 - 218.00
	Externally illuminated traffic signs		
6.1.113	dependent on area	nr	396.00 - 1042.00
	Lighting to pedestrian areas and estates		
6.1.114	roads on 4.00 m - 6.00 m columns with up to 70 W lamps	nr	179.00 - 263.00
	Lighting to main roads		
6.1.115	10.00 m -12.00 m columns with 250 W lamps	nr	411.00 - 502.00
6.1.116	12.00 m -15.00 m columns with 400 W high pressure sodium lighting	nr	525.00 - 631.00
6.1.117	Benches - hardwood and precast concrete	nr	154.00 - 210.00
	Litter bins		
6.1.118	precast concrete	nr	154.00 - 178.00
6.1.119	hardwood slatted	nr	62.00 - 84.00
6.1.120	cast iron	nr	264.00
6.1.121	large aluminium	nr	464.00
6.1.122	Bus stops	nr	280.00
6.1.123	Bus stops including basic shelter	nr	647.00
6.1.124	Pillar box	nr	233.00
6.1.125	Telephone box	nr	2556.00
	Playground equipment		
	Modern swings with flat rubber safety seats:		
6.1.126	four seats; two bays	nr	1039.00
6.1.127	Stainless steel slide, 3.40 m long	nr	1149.00
6.1.128	Climbing frame - igloo type 3.20 m x 3.75 m on plan x 2.00 m high	nr	1194.00
6.1.129	Seesaw comprising timber plank on sealed ball bearings 3960 mm x 230 mm x 70 mm thick	nr	829.00
6.1.130	Wicksteed "Tumbleguard" type safety surfacing around play equipment	m²	70.00
6.1.131	Bark particles type safety surfacing 150 mm thick on hardcore bed	m²	9.40
	Fencing and screen walls, ancillary buildings etc		
	Chain link fencing; plastic coated		
6.1.132	1.20 m high	m	12.90 - 15.60
6.1.133	1.80 m high	m	18.60 - 20.90
	Timber fencing		
6.1.134	1.20 m high chestnut pale facing	m	14.50 - 17.10
6.1.135	1.80 m high close-boarded fencing	m	37.30 - 46.40
	Screen walls; one brick thick; including foundations etc.		
6.1.136	1.80 m high facing brick screen wall	m	186.00 - 233.00
6.1.137	1.80 m high coloured masonry block boundary wall	m	209.00 - 263.00
6.1.138	Squash courts, independent building, including shower	nr	56488 - 66112
	Demolish existing buildings		
6.1.139	of brick construction	m³	3.90 - 7.00

Approximate Estimates

Item nr.	SPECIFICATIONS	Unit	ALL AREAS £ range £
6.20	**EXTERNAL SERVICES** **gross internal area** (unless otherwise described)		
	Service runs All laid in trenches including excavation Water main		
6.2.1	75 mm uPVC main in 225 mm diameter ductile iron pipe as duct	m	38.80
	Electric main		
6.2.2	600/1000 volt cables. Two core 25 mm diameter cable including 100 mm diameter clayware duct	m	25.50
	Gas main		
6.2.3	150 mm diameter ductile or cast iron gas pipe	m	38.80
	Telephone		
6.2.4	British Telecom installation in 100 mm diameter uPVC duct	m	15.60
6.2.5	External lighting (per m² of lighted area)	m²	1.90 - 3.00
	Connection areas The privatisation of telephone, water, gas, and electricity has complicated the assessment of service connection charges. Typically, service connection charges will include the actual cost of the direct connection plus an assessment of distribution costs from the main. The latter cost is difficult to estimate as it depends on the type of scheme and the distance from the mains. In addition, service charges are complicated by discounts that may be offered. For instance, the electricity boards will charge less for housing connections if the house is all electric. However, typical charges for an estate of 200 houses might be as follows		
6.2.6	Water	house	426.00 - 867.00
	Electric		
6.2.7	all electric	house	218.00
6.2.8	gas/electric	house	426.00
	Extra cost of		
6.2.9	sub-station	nr	10879 - 16129
6.2.10	Gas	house	426.00 - 548.00
	Extra cost of		
6.2.11	governing station	nr	10879
6.2.12	Telephone	house	140.00
6.2.13	Sewerage	house	327.00 - 426.00

Item nr.	SPECIFICATIONS	Unit	ALL AREAS £ range £
6.3	**DRAINAGE**		
	gross internal area (unless otherwise described)		
	comprising:		
	Overall £/m² allowances		
	Comparative pipework		
	Comparative manholes		
	Overall £/m² allowances		
6.3.1	Site drainage (per m² of paved area)	m²	5.60 - 14.40
6.3.2	Building drainage (per m² of gross floor area)	m²	5.60 - 12.60
6.3.3	Drainage work beyond the boundary of the site and final connection	nr	1245 - 8369
	Drains; hand excavation, grade bottom, earthwork support, laying and jointing pipe, backfill and compact, disposal of surplus soil; land drain, clay field drain, pipes, nominal size		0.75m £
6.3.4	75 mm diameter	m	20.90
6.3.5	100 mm diameter	m	22.40
6.3.6	150 mm diameter	m	23.60
	Land drain, vitrified clay perforated sub-soil pipes, nominal size		
6.3.7	100 mm diameter	m	23.60
6.3.8	150 mm diameter	m	28.00
6.3.9	225 mm diameter	m	33.50
	Machine excavation, grade bottom, earthwork support, laying and jointing pipes and accessories, backfill and compact, disposal of surplus soil (excluding beds, benchings and coverings)		
	Vitrified clay pipes and fittings, socketted, cement and sand joints, nominal size		1.00m £
6.3.10	100 mm diameter	m	17.10
6.3.11	150 mm diameter	m	19.40
6.3.12	225 mm diameter	m	29.50
6.3.13	300 mm diameter	m	44.90
6.3.14	450 mm diameter	m	82.20
	Vitrified clay pipes and fittings, "Hepseal" socketted with push fit flexible joints, nominal size		
6.3.15	100 mm diameter	m	19.40
6.3.16	150 mm diameter	m	23.60
6.3.17	225 mm diameter	m	36.50
6.3.18	300 mm diameter	m	55.90
6.3.19	450 mm diameter	m	113.40
	Class M tested concrete centrifugally spun pipes and fittings, flexible joints, nominal size		
6.3.20	300 mm diameter	m	35.00
6.3.21	450 mm diameter	m	47.90
6.3.22	600 mm diameter	m	62.40
6.3.23	900 mm diameter	m	-
6.3.24	1200 mm diameter	m	-
	Cast iron "Timesaver" drain pipes and fittings, mechanical coupling joints, nominal size		
6.3.25	75 mm diameter	m	40.30
6.3.26	100 mm diameter	m	43.40
6.3.27	150 mm diameter	m	65.40
	uPVC pipes and fittings, lip seal coupling joints, nominal size		
6.3.28	100 mm diameter	m	14.00
6.3.29	160 mm diameter	m	17.90

Approximate Estimates

Item nr

AVERAGE DEPTH

1.00 m £	1.25 m £	1.50 m £	1.75 m £	2.00 m £	
27.80	36.50	52.50	62.40	71.50	6.3.4
29.50	37.30	54.40	63.90	73.00	6.3.5
31.20	38.80	55.90	65.40	76.10	6.3.6
31.20	38.80	55.90	65.40	76.10	6.3.7
34.20	41.80	60.50	68.50	77.60	6.3.8
38.80	49.50	65.40	76.10	83.70	6.3.9

AVERAGE DEPTH OF DRAIN

1.50 m £	2.00 m £	2.50 m £	3.00 m £	3.50 m £	
20.90	25.10	33.50	38.80	52.50	6.3.10
23.60	29.70	36.50	41.80	55.90	6.3.11
33.50	38.80	46.40	52.90	55.90	6.3.12
50.60	54.40	60.50	66.90	71.50	6.3.13
86.70	91.30	102.30	107.30	113.40	6.3.14
23.60	29.70	36.50	41.80	55.90	6.3.15
28.10	32.00	38.80	44.90	60.50	6.3.16
40.30	44.90	52.50	59.00	82.20	6.3.17
59.70	65.40	73.00	77.60	82.20	6.3.18
117.90	122.50	132.00	138.50	144.50	6.3.19
38.80	44.90	51.00	57.40	62.40	6.3.20
55.90	59.00	68.50	74.60	79.10	6.3.21
70.00	74.60	83.70	91.30	97.40	6.3.22
119.40	125.50	138.50	146.10	153.70	6.3.23
-	192.50	205.40	216.10	223.70	6.3.24
44.90	50.20	57.40	62.40	76.10	6.3.25
47.20	51.00	60.50	65.40	79.10	6.3.26
71.50	74.60	82.20	88.30	101.90	6.3.27
18.30	23.60	31.20	35.80	49.50	6.3.28
23.60	27.80	35.80	40.30	55.90	6.3.29

Approximate Estimates

Item nr.	SPECIFICATIONS	Unit	1.00 m £
6.3	**DRAINAGE - cont'd**		
	gross internal area (unless otherwise described)		
	uPVC "Ultra-Rib" ribbed pipes and fittings, sealed ring push fit joints, nominal size		
6.3.30	150 mm diameter	m	15.60
6.3.31	225 mm diameter	m	24.00
6.3.32	300 mm diameter	m	33.50

Item nr.	SPECIFICATIONS	Unit	100 mm thick £
	Pipe beds, benching and coverings granular filling, pipe size		
6.3.33	100 mm diameter	m	1.63
6.3.34	150 mm diameter	m	1.80
6.3.35	225 mm diameter	m	2.08
6.3.36	300 mm diameter	m	2.39
6.3.37	450 mm diameter	m	2.97
6.3.38	600 mm diameter	m	3.24
	In situ concrete 10.00 N/mm^2 - 40 mm aggregate (1:3:6), pipe size		
6.3.39	100 mm diameter	m	3.29
6.3.40	150 mm diameter	m	3.29
6.3.41	225 mm diameter	m	3.86
6.3.42	300 mm diameter	m	4.60
6.3.43	450 mm diameter	m	5.92
6.3.44	600 mm diameter	m	6.53
6.3.45	900 mm diameter	m	7.85
6.3.46	1200 mm diameter	m	10.48
	In situ concrete 20.00 N/mm^2 - 20 mm aggregate (1:2:4), pipe size		
6.3.47	100 mm diameter	m	3.35
6.3.48	150 mm diameter	m	3.35
6.3.49	225 mm diameter	m	4.02
6.3.50	300 mm diameter	m	4.70
6.3.51	450 mm diameter	m	6.06
6.3.52	600 mm diameter	m	6.71
6.3.53	900 mm diameter	m	8.05
6.3.54	1200 mm diameter	m	10.74
	Brick Manholes		
	Excavate pit in firm ground, partial backfill, partial disposal, earthwork support, compact base of pit, plain in situ concrete 20.00 N/mm^2 - 20 mm aggregate (1:2:4) base, formwork, one brick wall of engineering bricks PC £175.00/1000 in cement mortar (1:3) finished fair face, vitrified clay channels, plain insitu concrete 25.00 N/mm^2 - 20 mm aggregate (1:1:5:3) benchings, reinforced insitu concrete 20.00 N/mm^2 - 20 mm aggregate (1:2:4) cover and reducing slabs, fabric reinforcement, formwork step irons, cast iron cover and frame, depth from cover to invert		600mm x450mm £
6.3.55	0.75 m	nr	266.00
6.3.56	1.00 m	nr	326.00
6.3.57	1.25 m	nr	391.00
6.3.58	1.50 m	nr	450.00
6.3.59	1.75 m	nr	513.00
	with reducing slab and brick shaft internal size 600 mm x 450 mm; depth from cover to invert		
6.3.60	2.00 m	nr	-
6.3.61	2.50 m	nr	-
6.3.62	3.00 m	nr	-
6.3.63	3.50 m	nr	-
6.3.64	4.00 m	nr	-

Approximate Estimates

	AVERAGE DEPTH OF DRAIN				Item nr
1.50 m £	2.00 m £	2.50 m £	3.00 m £	3.50 m £	
19.40	23.60	32.00	38.00	51.70	6.3.30
28.10	33.50	40.30	46.40	49.50	6.3.31
38.80	41.80	49.50	55.90	60.50	6.3.32

BED	BED AND BENCHING		BED AND COVERING		
150 mm thick £	100 mm thick £	150 mm thick £	100 mm thick £	150 mm thick £	
2.39	3.24	3.56	4.70	6.95	6.3.33
2.66	3.36	3.90	5.75	7.84	6.3.34
2.97	4.35	5.17	7.84	10.10	6.3.35
3.24	4.98	6.44	9.40	12.01	6.3.36
4.29	7.55	9.01	13.41	16.89	6.3.37
5.05	9.81	11.26	17.24	20.36	6.3.38
4.60	6.13	6.86	8.60	12.87	6.3.39
5.25	6.86	7.58	10.01	14.30	6.3.40
5.92	8.23	9.59	14.30	18.59	6.3.41
6.53	9.59	12.32	17.15	22.17	6.3.42
8.52	14.42	17.15	24.30	30.77	6.3.43
9.84	18.52	21.91	31.47	37.19	6.3.44
12.45	30.13	34.98	47.91	60.79	6.3.45
15.05	44.55	49.33	65.79	85.82	6.3.46
4.70	6.28	7.01	8.78	13.16	6.3.47
5.39	7.01	7.76	10.24	14.64	6.3.48
6.06	8.43	9.83	14.64	19.08	6.3.49
6.71	9.83	12.63	17.56	22.67	6.3.50
8.73	14.76	17.57	24.88	31.47	6.3.51
10.07	18.96	22.44	32.49	38.05	6.3.52
12.75	30.87	35.82	49.01	62.19	6.3.53
15.44	45.63	50.50	67.31	87.79	6.3.54

INTERNAL SIZE OF MANHOLE

750mm x 450mm £	900mm x 600mm £	900mm x 900mm £	900mm x 1500mm £	1200mm x 1800mm £	
291.00	358.00	414.00	554.00	695.00	6.3.55
356.00	434.00	502.00	665.00	829.00	6.3.56
425.00	514.00	595.00	781.00	968.00	6.3.57
491.00	592.00	685.00	890.00	1100.00	6.3.58
557.00	669.00	773.00	1003.00	1236.00	6.3.59
-	-	855.00	1089.00	1324.00	6.3.60
-	-	1024.00	1287.00	1554.00	6.3.61
-	-	1155.00	1423.00	1698.00	6.3.62
-	-	1278.00	1557.00	1838.00	6.3.63
-	-	1404.00	1687.00	1977.00	6.3.64

Item nr.	SPECIFICATIONS	Unit
6.3	**DRAINAGE - cont'd**	
	gross internal area (unless otherwise described)	
	Concrete manholes	
	Excavate pit in firm ground, disposal, earthwork support, compact base of pit, plain insitu concrete 20.00 N/mm^2 - 20 mm aggregate (1:2:4) base, formwork, reinforced precast concrete chamber and shaft rings, taper pieces and cover slabs bedded jointed and pointed in cement:mortar (1:3), weak mix concrete filling to working space, vitrified clay channels, plain insitu concrete 25.00 N/mm^2 - 20 mm aggregate (1:1.5:3) benchings, step irons, cast iron cover and frame; depth from cover to invert	
6.3.65	0.75 m	nr
6.3.66	1.00 m	nr
6.3.67	1.25 m	nr
6.3.68	1.50 m	nr
6.3.69	1.75 m	nr
	with taper piece and shaft 675 mm diameter; depth from cover to invert	
6.3.70	2.00 m	nr
6.3.71	2.50 m	nr
6.3.72	3.00 m	nr
6.3.73	3.50 m	nr
6.3.74	4.00 m	nr

INTERNAL DIAMETER OF MANHOLE

1350 mm £	1500 mm £	1800 mm £	Item nr
450.00	533.00	698.00	6.3.65
507.00	593.00	773.00	6.3.66
563.00	657.00	854.00	6.3.67
621.00	720.00	930.00	6.3.68
676.00	781.00	1004.00	6.3.69
828.00	978.00	1175.00	6.3.70
945.00	1070.00	1337.00	6.3.71
1048.00	1184.00	1485.00	6.3.72
1146.00	1293.00	1627.00	6.3.73
1242.00	1404.00	1767.00	6.3.74

NEW FROM SPON PRESS

Inclusive Design
Designing and Developing Accessible Environments

Rob Imrie and **Peter Hall**, both from the Royal Holloway College, University of London, UK

'This is a well written and informative book on an important topic. The authors have taken an interesting perspective on access and the built environment by focussing on the role of the development industry. Their findings are shocking. Anyone interested in urban issues will find a wealth of insightful material in this book.' *Nick Oatley, University of the West of England, UK*

The reality of the built environment for disabled people is one of social, physical and attitudinal barriers which prevent their ease of mobility, movement and access. In the United Kingdom, most homes cannot be accessed by wheelchair, while accessible transport is the exception rather than the rule. Pavements are littered with street furniture, while most public and commercial buildings provide few design features to permit disabled people ease of access.

Inclusive Design is a documentation of the attitudes, values, and practices of property professionals, including developers, surveyors and architects, in responding to the building needs of disabled people. It looks at the way in which pressure for accessible building design is influencing the policies and practices of property companies and professionals, with a primary focus on commercial developments in the UK. The book also provides comments on, and references to, other countries, particularly Sweden, New Zealand, and the USA.

June 2001: 246x189: 240pp
5 line figures, 34 b+w photos
Pb: 0-419-25620-2: £29.99

To Order: Tel: +44 (0) 8700 768853, or +44 (0) 1264 343071 Fax: +44 (0) 1264 343005, or
Post: Spon Press Customer Services, ITPS Andover, Hants, SP10 5BE, UK Email: book.orders@tandf.co.uk.

Postage & Packing: UK: 5% of order value (min. charge £1, max. charge £10) for 3-5 days delivery. Option of next day delivery at an additional £6.50. Europe: 10% of order value (min. charge £2.95, max. charge £20) for delivery surface post. Option of airmail at an additional £6.50. ROW: 15% of order value (min.charge£6.50, max. charge £30) for airmail delivery.

For a complete listing of all our titles visit: www.sponpress.com

Cost Limits and Allowances

Information given under this heading is based upon the cost targets currently in force for buildings financed out of public funds, i.e., hospitals, schools, universities and public authority housing. The information enables the cost limit for a scheme to be calculated and is not intended to be a substitute for estimates prepared from drawings and specifications.

The cost limits are generally set as target costs based upon the user accommodation. However ad-hoc additions can be agreed with the relevant Authority in exceptional circumstances.

The documents setting out cost targets are almost invariably complex and cover a range of differing circumstances. They should be studied carefully before being applied to any scheme; this study should preferably be undertaken in consultation with a Chartered Quantity Surveyor.

The cost limits for Public Authority housing generally known as the "Housing Cost Yardstick" have been replaced by a system of new procedures. The cost criteria prepared by the Department of the Environment that have superseded the Housing Cost Yardstick are intended as indicators and not cost limits.

HOSPITAL BUILDINGS

Cost Allowances for all health buildings are now set at a Median Index of Public Sector Building Tender Prices (MIPS) Variation of Price (VOP) index level of 310 with a Firm Price (FP) differential allowance of 5.10% increase on the VOP level, that is, an FP index level of 325. The effective date for these increased allowances was 14 December 2000.

Both the Cost Allowances (DCAGs) and the Equipment Cost Allowance Guides (ECAGs) are now included in version 1.0 of the Health Capital Investment (HCI) document, issued in March 1997. This supersedes previous publications including version 13 of the Concise 4D/DCAG database.

"Quarterly Briefing" is now the official document for the notification of all new/revised DCAGs and ECAGs. Users should check Briefings issued from and including volume 7 no. 3 to ensure that all revisions to the HCI have been identified.

The NHS Estates has recently completed research into alternative indices for its Equipment Price Index, this was due to concerns that the index did not reflect market conditions as it had not been effectively maintained over the past few years. The result has been the development of a new equipment price index which reflects more accurately the market indicators for equipment.

The new index is a combination of various sources of information. This includes taking into consideration equipment-specific indices issued by the NHS Executive, namely the Health Service Cost Index, the retail price index and a general construction materials index. The combination of all this produces a new index series which reflects price movement and inflationary effects specifically associated with equipment.

The new equipment price index is set at 1997 = 100.

The ECAGs in the HCI document result from a complete revision of existing cost guidance and therefore supersede all previous equipment cost guidance.

During the calculation of capital costs for business cases, consideration must be given to the date when the equipment is likely to be purchased in relation to the construction tender index base date. This is the base date that the capital costs are adjusted to once a business case has been approved. Most equipment is likely to be purchased near the end of a building contract and will therefore have a later base date than the construction date.

Further advice can be obtained from NHS Estates (see Useful Addresses for Further Information on page 973).

UNIVERSITY BUILDINGS

The Universities Funding Council and the Polytechnics & Colleges Funding Council merged in April 1993 to form the Higher Education Funding Council for England. Procedures for appraising funding requirements for University buildings have been under review for a number of years and due to public spending constraints the Government have encouraged the use of Private Finance Initiatives for proposed estates projects. For guidance on techniques to be used by higher education institutions when appraising proposals that involve the development, ownership, leasing and occupation of land and buildings contact:

Higher Education Funding Council for England
Northavon House
Coldharbour Lane
Bristol
BS16 1QD
Tel: 0207 931 7317

Information regarding Scotland and Wales can be obtained from:

Scottish Higher Education Funding Council
Donaldson House
97 Haymarket Terrace
Edinburgh
EH12 5HD
Tel: 0131 313 6500

Higher Education Funding Council for Wales
Lambourne House
Cardiff Business Park
Llanishen
Cardiff
CF4 5GL
Tel: 02920 761 861

EDUCATIONAL BUILDINGS (Procedures and Cost Guidance)

In January 1986 the then Department of Education and Science introduced new procedures for the approval of educational building projects.

In its Administrative Memorandum 1/86 and subsequent letter AB13/12/028, dated 12 November 1991, the Department currently requires detailed individual approval only for projects costing £2 million and over. Minor works i.e., costing less than £200,000, require no approval except where the authority requires such or where statutory notices are involved. Projects of £200,000, up to £2 million require approval which will be given automatically upon the submission of particulars of the scheme and performance data relating to it.

Capital works and repair projects supported by grant from the Department for Education will continue to require individual approval.

EDUCATIONAL BUILDINGS

The Department for Education and Employment publishes information on costs and performance data for school building projects.

Building Guide Costs

The Building Guide Costs are intended to provide reasonable and achievable target costs for use in the cost planning of new schools and extensions. The values relate to median values between middle and lower quartile costs in LEA projects.

The Building Guide Costs relate to Q1 2000 national average price levels. The March 1999 edition of the Building Guide Costs gave an indicative index of 115 @ 1Q1999 which was subsequently found to be too optimistic. Tender prices during late 1998 and 1999 do not appear to have increased as forecast. The current indicative index of 116 for 1Q2000 is based on the DETR Publication dated December 1999.

The values *exclude* the cost of:

- 'Abnormal' substructure costs (i.e. the proportion of the total substructure costs over £57/m² at Q1 2000 prices).
- External works.
- Furniture, fittings and equipment, whether built-in or loose.
- Fees and VAT.

If used as the basis for project cost planning, the values should be adjusted to take account of tender date, location and contract size factors. Other factors affecting costs will include design, specification and site conditions. The guide values for primary and secondary extensions are intended to relate to projects providing an average mix of accommodation types. Actual costs may therefore be higher in schemes providing disproportionately more specialist accommodation.

The Building Guide Costs may be adjusted to reflect tender price movements by reference to DETR's PUBSEC index. It is suggested that adjustments for location and contract size may be made by reference to factors published by the RICS's Building Cost Information Service.

The table below provides indicative unit costs for four common types of project. The Department intends to use these values for reference when scrutinising cost standards in grant-aided school projects.

Building Guide Costs per m² at Q1 2000 prices
(Relating to DETR PUBSEC indicative index 116, Q1 2000)
(NOTE: DETR PUBSEC indicative index 1995 = 100)

New primary school	£780/m²
Primary extension	£745/m²
New secondary school	£698/m²
Secondary extension	£685/m²

HOUSING ASSOCIATION SCHEMES

Grant rates 2001/2002 and other current literature is available from The Housing Corporation, 149 Tottenham Court Road, London, W1P 0BN. Tel: 0207 292 4400. For the benefit of readers, the TCI system instigated by Housing Corporation Circular HC 33/91 has been reproduced here in full and amended to accord with the cost information in F2 - 22/99 and Guidance Notes effective from 1 April 2001.

Advises associations of:

(i) the Total Cost Indicators, and grant rates effective from 1 April 2001.

(ii) the grant rates applicable to 2001/2002 approvals for mixed and publicly funded schemes for rent and housing for sale schemes approved under the Housing Act 1988 and funded by the Housing Corporation or by local authorities.

The Association Investment Profile system, and the associated Guidance **(last updated in September 1989)**, has now been withdrawn and replaced by a new, simpler system entitled Performance Assessment and Investment Summary (PAIS). This brings together the various assessments of associations made by the Corporation in the following areas, to inform its decisions on making allocation to associations:

- Development Programme Delivery
- Development Function
- Scheme Audit
- Control and Conduct
- Landlord
- Financial Control
- Financial Health

The further reduction in grant rates for **1995/96** led the Corporation to develop improved tools for satisfying itself that associations developing for rent or sale using mixed funding will be in a position to raise the necessary private finance. The Corporation has a duty to safeguard the public investment in associations and the interests of the existing occupiers of their dwellings. As part of the Financial Health element of the PAIS, the Corporation will therefore appraise associations' financial position against typical private lender tests of gearing and interest cover.

The Corporation's objective in carrying out these tests is to satisfy itself before making mixed funded allocations that associations will in due course be able to obtain the necessary private finance. If an association has difficulty at present in satisfying the Corporation tests it will discuss with it its ability, over an agreed timescale, to satisfy typical private lender tests. It will not be debarred from receiving allocations; if it can show to the Corporation's satisfaction that it will be able to secure the necessary private finance when it is required, its bids for ADP resources will be considered.

Conversely, satisfying the tests will not carry an automatic entitlement to an allocation.

From 1 April 1995, special needs schemes receiving approval for Special Needs Management Allowance (SNMA) will no longer necessarily receive a grant rate of 100%. The grant rate will be determined on a scheme by scheme basis at 100% or such lower level as is embodied in the association's bid for funding.

Enquiries about the contents of this Circular to Regional offices of the Corporation.

Annex to Circular HC 27/93

PERFORMANCE ASSESSMENT AND INVESTMENT SUMMARY (PAIS)

Introduction

(i) The Performance Assessment and Investment Summary (PAIS) is designed to assist accountable investment decisions by the Corporation by ensuring that all relevant aspects of an association's performance are taken into account. The PAIS therefore fulfils the same function as, and supersedes, the previous system of Association Investment Profiles.

(ii) A PAIS will be produced annually for each association seeking development funding from the corporation. (Associations seeking only TIS or Miscellaneous Works allocations will not normally have a PAIS). Each association will receive a single PAIS from the Corporation, produced by the lead Regional Director for the association.

Contents of the PAIS

(iii) The PAIS brings together the Corporation's most recent assessments of associations under the following general headings.

- Development Programme Delivery
- Development Function
- Scheme Audit
- Control and Conduct (embracing control and propriety, accountability, agency services, and equal opportunities).
- Landlord (embracing access to housing; housing management and maintenance)
- Financial Control
- Financial Health

(iv) The sources for the above are:

- Performance audit reports
- Scheme audit reports
- Programme Delivery assessments (see paragraph (viii) below)
- Annual accounts review
- Quarterly financial returns.

Associations will previously have been informed of any areas of concern in all of these areas; the process of producing the PAIS does not involve the Corporation in making any fresh judgements of the association's performance.

(v) **It is essential that associations provide both audited annual accounts and quarterly financial returns to the published timetables.**

(vi) As well as recording the assessments previously made, the PAIS also includes a brief narrative summarising the performance of the association and setting out the implications, as seen by the Corporation, for future ADP-funded development by the association. This narrative will take into account the association response to any Corporation concerns expressed in the earlier assessments. If any new assessment which substantively alters the position is made by the Corporation between the issue of a PAIS and allocation decisions being made, the Corporation will issue an updated PAIS.

(vii) The format and content of the PAIS as initially used in the current year will be further developed by the Corporation in the light of experience and feedback, and to take account of changes in the Corporation's regulatory arrangements. Comments on the PAIS issued this year will therefore be welcome, and should be sent to the lead Regional Director for the association.

Programme Delivery Assessment

(viii) The PDA summarises the association's performance on the ADP in the following areas:

> Cash Spend Performance
> Allocation performance for each programme heading
> Cost Control/Development Finance (embracing cost movement between allocation and scheme approval; cost overruns post-approval; prompt payment of debts to the Corporation; prompt submission of HAG recovery forms to the Corporation).

(ix) Programme Delivery Assessments (PDA) will in future be produced during the first quarter of each financial year, covering the association's performance during the previous year.

(x) A separate PDA will be sent to the association by each Corporation Region in which it had a Cash Planning Target (CPT) in the relevant year.

TOTAL COST INDICATORS 2001/2002 GUIDANCE NOTES

1. Introduction

1.1 These guidance notes are designed to explain the use of Total Cost Indicators (TCI), grant rates and rent caps and administrative allowances in the Social Housing Grant (SHG) funding framework. They are intended to help registered social landlords (RSLs) complete scheme submissions. The notes are effective from 1 April 2001.

1.2 Enquiries about the contents of this guidance should be directed to the regional offices of the Housing Corporation.

2. Main changes from the 2000/2001 TCI

2.1 The main change from the 2000/2001 Total Cost Indicators is that a number of local authority areas have changed TCI cost groups, reflecting latest evidence of relative costs.

2.2 The works cost element within the TCI base table has been updated to include a broader range of archetypes within the model. The archetypes and costings derive from work commissioned by the DETR and carried out by the Building Cost Information Service (BCIS). As a result, the distribution of costs within the range of the base table has changed. Generally, works costs have increased greater than average above the mid-band but with increasing cost reductions below the mid-band;

2.3 The existing supplementary multipliers for sheltered, frail older persons, supported and shared housing have been adjusted to reflect the redistribution of costs within the base table;

2.4 The existing supplementary multipliers for Rural Areas and National Parks have been enhanced in recognition of the increased difficulties of developing schemes in such areas;

2.5 A new listing has been incorporated to assist RSLs in the identification of 'Areas of Outstanding Beauty' (AOBs) and National Parks;

2.6 A new supplementary multiplier has been introduced in order to encourage greater sustainability;

2.7 A new supplementary multiplier has been introduced to encourage RSLs to design homes with lofts that may be readily adapted to provide future additional habitable space;

2.8 A new supplementary multiplier has been introduced as an encouragement for RSLs to obtain 'Chartered Client' status from the 'Condfederation of Construction Clients' and in recognition of the additional expense involved in attaining such status;

2.9 The 'standard house type' supplementary multiplier deflator has been withdrawn in order to encourage RSLs to pursue standardisation within the design process;

TOTAL COST INDICATORS 2001/2002 GUIDANCE NOTES - cont'd

2. **Main changes from the 2000/2001 TCI** - cont'd

   2.10 The basic on-cost percentage has been increased by 0.50% within relevant scheme types in recognition of the increased costs associated with:

   * DETR signboard requirements outlined in Circular HC F2-7/00;
   * Housing Quality Indictors (HQI) assessment of new schemes;
   * Submission of HQI data to a national HQI database; and

   Note: This increase is absorbed in most cost groups by a decrease arising from a change in the Acquisition: Works relationship. The higher the proportion of acquisition; the lower the on-cost percentage (and vice versa), since the percentage of fees on acquisition related items is lower than on works related items.

3. **Explanation of TCI**

   3.1 A key objective of the funding system is to achieve value for money in return for grant, and to ensure the correct level of grant is paid. TCI form the basis of this system, and are divided into unit type and cost group area categories.

   3.2 TCI apply equally to units funded with Social Housing Grant (SHG) by the Housing Corporation, or those sponsored by a local authority.

   3.3 TCI represent the basis for a cost evaluation of SHG funded units. TCI are also used to calculate the maximum level of grant or other public subsidy payable. Further details on this inter-relationship between TCI and grant levels are given in the separate grant rate guidance notes on page 865.

   3.4 Key and supplementary multipliers are applied to the base TCI figures to allow for scheme variations as outlined in the multiplier tables on pages 846 to 853. Thus, there is a relationship between the base norm cost of a unit and its unit type.

4. **Types of accommodation**

   4.1 Different types of accommodation other than self-contained housing for general needs are classified as follows:

   4.2 **Accommodation for older people**

   (i) Category 1 - self-contained accommodation for the more active older person, which may include an element of support and/or additional communal facilities;

   (ii) Category 2 - self-contained accommodation for the less active older person, which includes an element of support and the full range of communal facilities;

   The term "sheltered" is used generally to describe Category 1 and Category 2 schemes;

   (iii) Frail older people - supported extra care accommodation, which may be either shared or self-contained, for the frail older person. Includes the full range of communal facilities, plus additional special features, including wheelchair user environments and supportive management.

   4.3 **Shared accommodation**

   Accommodation predominantly for single persons, which includes a degree of sharing between tenants of some facilities (e.g. kitchens, bathrooms, living room) and may include an element of support and/or additional communal facilities.

   4.4 **Supported housing**

   Accommodation, which may be either shared or self-contained, designed to meet the needs of particular user groups for intensive housing management (see the Housing Corporation's Guide to Supported Housing). Such accommodation may also include additional communal facilities.

4. Types of accommodation - cont'd

4.5 Accommodation for wheelchair users

Accommodation, which may be either shared or self-contained, designed for independent living by people with physical disabilities and wheelchair users. Where such accommodation is incorporated within schemes containing communal facilities, these facilities should be wheelchair accessible.

4.6 Communal facilities

Ancillary communal accommodation, the range of which comprises:

(i) Common room - consisting of common room/s of adequate size to accommodate tenants and occasional visitors, chair store and kitchenette for tea-making;

(ii) Associated communal facilities - consisting of warden's office, laundry room and guest room.

5. Treatment of combined supported housing and general needs schemes

5.1 Arrangements exist which allow the combination of supported housing and general needs units within a single scheme. Further guidance with regards to supported housing schemes is set out on page 847.

6. Temporary housing

6.1 Temporary Social Housing (TSH) Grant is a term used to describe SHG paid to RSLs to cover the cost of bringing properties into temporary use.

6.2 Properties are eligible for TSH Grant if they are available for use by the RSL for a period of time covered by a lease or licence for not less than two years and not more than 29 years.

6.3 A capital grant contribution will be available towards initial acquisition costs or periodic lease charges up to a grant maximum. The maxima are based upon the Housing Corporation's own assessment of what constitutes a reasonable contribution to the capitalised lease value. This is calculated using capitalised lease premium factors.

6.4 The *TSH multipliers and capitalised lease premium factor tables for 2001/2002* are included in this guidance on pages 853 and 854.

7. The composition of TCI

7.1 TCI comprise the following elements:

Acquisition

(i) Purchase price of land/property.

Works

(i) Main works contract costs (including where applicable adjustments for additional claims and fluctuations, but excluding any costs defined as on-costs) (see page 839);

(ii) major site development works (where applicable). These include piling, soil stabilisation, road/sewer construction, major demolition;

(iii) major pre-works (rehabilitation) where applicable;

(iv) statutory agreements, associated bonds and party wall agreements (including all fees and charges directly attributable to such works) where applicable;

(v) additional costs associated with complying with archaeological works and party wall agreement awards (including all fees, charges and claims attributable to such works) where applicable;

7. The composition of TCI - cont'd

Works - cont'd

   (vi) home loss and associated costs. This applies to new build only;

   (vii) VAT on the above, where applicable.

On-Costs

   (i) Legal fees, disbursements and expenses;

   (ii) stamp duty;

   (iii) net gains/losses via interest charges on development period loans;

   (iv) building society or other valuation and administration fees;

   (v) fees for building control and planning permission;

   (vi) fees and charges associated with compliance with European Community directives, and the Housing Corporation's requirements relating to energy rating of dwellings and Housing Quality Indicators;

   (vii) in-house or external consultants' fees, disbursements and expenses (where the development contract is a design and build contract) (see below);

   (viii) insurance premiums including building warranty and defects/liability insurance (except contract insurance included in works costs);

   (ix) contract performance bond premiums;

   (x) borrowing administration charges (including associated legal and valuation fees);

   (xi) an appropriate proportion of the RSL's development and administration costs (formerly Acquisition and Development allowances), excluding Co-operative Promotional Allowance (CPA) and Special Projects Promotion Allowances (SPPA) and including an appropriate proportion of any abortive scheme costs;

   (xii) furniture, loose fittings and furnishings;

   (xiii) home loss and disturbance payments for rehabilitation;

   (xiv) preliminary minor site development works (new build), pre-works (rehabilitation), and minor works (off-the-shelf) and minor works and repairs in connection with existing satisfactory purchases;

   (xv) marketing costs - for sale schemes only;

   (xvi) post completion interest - for sale schemes only;

   (xvii) legal, administrative and related fees and costs associated with negotiating and arranging leases - for TSH only;

   (xviii) VAT on the above, where applicable.

   Note: Where the development contract is design and build, the on-costs are deemed to include the builder's design fee element of the contract sum. Therefore the amount included by the builder for design fees should be deducted from the works cost element submitted by the RSL to the Housing Corporation.

   Similarly, other non-works costs that may be included by the builder such as fees for building and planning permission, building warranty and defects/liability insurance, contract performance bond and energy rating of dwellings should also be deducted from the works cost element submitted by the RSL to the Housing Corporation.

   The Housing Corporation will subsequently check compliance through its compliance audit framework.

8. Explanation of on-costs

8.1 TCI are inclusive of on-costs contained in the relevant on-cost table. The on-costs vary according to TCI cost group and the general purpose of the scheme. TCI levels are set with the assumption that the RSL's development and administrative costs will be contained within the percentages in the relevant on-cost table.

8.2 In order to allow a proper comparison between the total eligible costs and the relevant TCI it is necessary to add the percentage on-cost (from the relevant table) to the estimated eligible final costs of acquisition and works.

8.3 Major repairs schemes and adaptation schemes are not measured against TCI. However, on-costs (from the relevant table) should be added to such schemes. Supplementary on-costs may not be used in any circumstances in connection with major repairs or adaptation schemes.

9. Selection of on-costs

9.1 One key on-cost will apply per scheme. To this should be added any appropriate supplementary on-cost.

9.2 Supplementary on-costs may be used when the accommodation is designed to meet the relevant standards set out in the Housing Corporation publication Scheme Development Standards (latest version, August 2000).

9.3 The appropriate key or supplementary on-cost is determined by the predominant dwelling type in a scheme. Predominance is established, where necessary, by the largest number of persons in total.

9.4 Where two key on-costs or two supplementary on-costs are equally applicable (e.g. supported housing and shared), the higher should be used.

Keep your figures up to date, free of charge
Download updates from the web: www.pricebooks.co.uk

This section, and most of the other information in this Price Book, is brought up to date every three months with the Price Book Updates, until the next annual edition. The updates are available free to all Price Book purchasers.

To ensure you receive your free copies, either complete the reply card from the centre of the book and return it to us or register via the website www.pricebooks.co.uk

KEY ON-COSTS 2001/2002 BY COST GROUP
Note: only one of the following to be used.

Key on-costs			A %	B %	Cost group C %	D %	E %
a)	**New Build**						
	(i)	acquisition and works	13	13	14	15	16
	(ii)	off-the-shelf	8	8	8	8	8
	(iii)	works only	17	17	17	17	17
b)	**Rehabilitation**						
	(i)	acquisition and works					
		- vacant	12	12	12	13	15
		- tenanted	15	14	15	18	18
	(ii)	existing satisfactory	10	10	10	10	10
	(iii)	purchase and repair	8	8	8	8	8
	(iv)	works only					
		- vacant	22	22	22	22	22
		- tenanted	26	26	26	26	26
	(v)	re-improvements	22	22	22	22	22
	(vi)	TSH					
		- unimproved vacant	21	21	21	21	21
		- improved vacant	9	9	9	9	9
c)	**Major repairs and miscellaneous works**		25	25	25	25	25
d)	**Adaptation works**		13	13	13	13	13

SUPPLEMENTARY ON-COSTS 2001/2002

	Supplementary on-costs (all cost groups)	Purchase & Repair and Acquisition & Works %	Works only and reimprovements %	Off-the-Shelf existing satisfactory %
a)	Sheltered with common room or communal facilities	+2	+3	+1
b)	Frail older persons	+4	+6	+3
c)	Supported housing: i) supported housing ii) with common room or communal facilities	+4 +5	+6 +7	+3 +4
d)	Shared	+5	+7	+5
e)	Construction Clients Charter	+1	+1	0
f)	Housing for Sale	+8	+11	+5
g)	TSH: i) shared unimproved +7 ii) shared improved +5			

10. Tranches

10.1 For units developed under the Housing Act 1996, a set percentage of approved grant can be paid once a unit reaches certain key development stages. These grant payments are known as tranches.

10.2 The key stages are:

    (i)   exchange of purchase contracts (ACQ);

    (ii)   start on site of main contract works (SOS); this is deemed to be the date when the contractor took possession of the site/property in accordance with the signed main building contract;

    (iii)   practical completion of the scheme (PC).

10.3 Where a *public subsidy* is given by way of discounted land, (acquisition public subsidy), the whole subsidy is deducted from the first tranche; any excess balance should be deducted from the second tranche. This is to ensure that grant is not paid in advance of need. In other circumstances any other public subsidy will be deducted from each tranche on a pro-rata basis.

10.4 In the case of *land inclusive packages* the first and second tranches are paid together at start on site stage.

10.5 Tranche details for Special Project Promotion Allowances (SPPA) and Co-op Promotion Allowances (CPA), together with further guidance relating to tranches for management contracting arrangements, can be found in the Capital Funding System Procedure Guide.

10.6 The grant on outstanding mortgages for reimprovement schemes will be paid in accordance with the tranche percentage for the scheme.

10.7 All tranche payments may be paid directly to RSLs rather than via solicitors.

GRANT PAYMENTS FOR TRANCHES 2001/2002

Tranche percentages

Housing for Rent	Scheme	Cost Groups	Acquisition & Works			Works Only & Reimprovements		Off-Shelf/ Existing Satisfactory	Purchase & Repair	
			(i)	(ii)	(iii)	(ii)	(iii)	(i)	(i)+(ii)	(iii)
Mixed Funded	New Build	ALL	40%	40%	20%	65%	35%	100%		
	Rehab.	ALL	50%	30%	20%	60%	40%	100%	80%	20%
100% SHG	New Build	A, B & C	30%	35%	35%	50%	50%	100%		
		D & E	20%	35%	45%	45%	55%	100%		
	Rehab.	A, B & C	50%	20%	30%	40%	60%	100%	80%	20%
		D & E	45%	20%	35%	35%	65%	100%	80%	20%
TSH	Un-improved					65%	35%			
	Improved							100%		
Housing for Sale Sale										
Mixed Funded	New Build & Rehab.	ALL	50%	45%	5%	95%	5%	100%	95%	5%

11. **The use of the TCI base table**

   11.1 The unit size in square metres shown in the TCI *base table 20012002 self-contained accommodation* (on page 845) relates to the total floor area of the unit. The probable occupancy figure in the tables is only a guideline figure. The number of occupants is derived from the total number of bedspaces provided.

   11.2 The TCI for a unit where the total floor area exceeds 120 m^2 will be the cost of a unit of 115 - 120 m^2 plus for each additional 5 m^2 or part thereof, the difference between the cost of a 110 - 115 m^2 unit, and the cost of a 115 - 120 m^2 unit.

   11.3 In calculating the appropriate TCI floor area band the relevant floor area should be rounded to nearest whole number.

   For self-contained accommodation

   11.4 Self-contained units provide each household, defined as a tenancy, with all their basic facilities behind their own lockable front door.

   11.5 For self-contained units the base TCI is determined by its total floor area and the cost group in which it is located. The dwelling floor area is determined by the area of each unit for the private use of a single household. Communal areas or any facilities shared by two or more households should be excluded.

   11.6 The total floor area of self-contained accommodation is measured to the finished internal faces of the main containing walls on each floor of the accommodation and includes the space, on plan, taken up by private staircases, partitions, internal walls (but not "party" or similar walls), chimney breasts, flues and heating appliances. It includes the area of tenants' internal and/or external essential storage space. It excludes:

For self-contained accommodation - cont'd

(i) any space where the height to the ceiling is less than 1.50 m (e.g. areas in rooms with sloping ceilings, external dustbin enclosures);

(ii) any porch, covered way; etc., open to the air;

(iii) all balconies (private, escape and access) and decks;

(iv) non-habitable basements, attics thermal buffer zones, conservatories or sheds;

(v) external storage space in excess of 2.50 m^2;

(vi) all space for purposes other than housing (e.g. garages, commercial premises etc.).

11.7 The TCI for Frail older persons dwellings should always be calculated as *self-contained* units even if they have some characteristics which are more typical of shared accommodation.

For shared accommodation

11.8 Shared accommodation is defined as one household (i.e. one tenancy or licence) which shares facilities (i.e. bathroom, kitchen) with other households. Each household sharing such accommodation may comprise more than one person. The base TCI for shared accommodation should be calculated separately to any self-contained accommodation in the scheme. The base TCI for shared accommodation is calculated on a per bedspace basis and may include any staff with a residential tenancy for shared accommodation. Staff sleep over accommodation which is not subject to a tenancy is not regarded as a bedspace for TCI purposes. The relevant base TCI cost for the cost group from the *TCI base table 2001/2002: shared accommodation* on page 845 is used.

For TSH shared accommodation

11.9 The relevant floor area, for TCI and capitalised lease premium calculation purposes only, should be taken as the overall building gross floor area measured to the finished internal faces of the main containing walls all as otherwise described for self-contained units above, divided by the number of people sharing. This calculation will result in the correct band size per person sharing which will then be used to select the base figure from the *TCI base table: self-contained accommodation* on page 845. This figure should then be multiplied by the number of people sharing prior to applying the relevant TSH key and supplementary multipliers or capitalised lease premium factor.

TCI BASE TABLE 2001/2002: SELF CONTAINED ACCOMMODATION

Total unit costs

All self-contained accommodation
(including all frail older persons and TSH)
£ per unit
Cost Group

Unit Floor Area (m²)	Probable occupancy (persons)	A	B	C	D	E
Up to 25	1	53,800	48,000	40,800	34,300	31,300
Exceeding/not exceeding						
25/30	1	60,600	54,000	45,500	37,800	34,300
30/35	1 and 2	67,400	60,000	50,100	41,300	37,300
35/40	1 and 2	74,300	66,000	54,800	44,800	40,300
40/45	2	81,100	72,000	59,500	48,400	43,300
45/50	2	87,900	78,000	64,200	51,900	46,300
50/55	2 and 3	94,800	84,000	68,800	55,400	49,300
55/60	2 and 3	101,600	90,000	73,500	58,900	52,300
60/65	3 and 4	108,400	95,900	78,200	62,400	55,200
65/70	3 and 4	115,300	101,900	82,900	65,900	58,200
70/75	3, 4 and 5	122,100	107,900	87,600	69,400	61,200
75/80	3, 4 and 5	129,000	113,900	92,200	72,900	64,200
80/85	4, 5 and 6	135,800	119,900	96,900	76,400	67,200
85/90	4, 5 and 6	142,600	125,900	101,600	79,900	70,200
90/95	5 and 6	149,500	131,900	106,300	83,400	73,200
95/100	5 and 6	156,300	137,900	111,000	87,000	76,200
100/105	6 and 7	163,100	143,900	115,600	90,500	79,100
105/110	6 and 7	170,000	149,900	120,300	94,000	82,100
110/115	6, 7 and 8	176,800	155,800	120,000	97,500	85,100
115/120	6, 7 and 8	183,600	161,800	129,700	101,000	88,100

TCI BASE TABLE 2001/2002: SHARED ACCOMMODATION

All shared accommodation (except TSH)
£ per person sharing
Cost Group

Total costs per person	A	B	C	D	E
Each person bedspace	74,000	65,400	53,100	42,100	37,100

12. Selection of key multipliers

12.1 Only one key multiplier can be used per unit.

12.2 *New build acquisition and works* is the basic key multiplier, hence its neutral value.

12.3 The *off-the-shelf* multiplier is used where new dwellings to a standard suitable for social housing letting are purchased, following inspection, from contractors/developers or their agents. The cost of any minor works required should be set against the on-cost allowance.

12.4 The *existing satisfactory* multiplier is used where existing dwellings of a standard and in a condition suitable for social housing letting are purchased, following inspection, from the second-hand property market. The cost of any minor works required should be set against the on-cost allowance.

12.5 The *purchase and repair* multiplier is used where existing dwellings are purchased, following inspection, from the property market which necessitate a degree of repair to bring them to a standard and condition suitable for social housing but not full rehabilitation. Purchase and repair classification will apply where the estimated repair/improvement costs of each dwelling exceed £1,500 but are less than £10,000 (exclusive of VAT).

12.6 The *works only* multiplier is used for accommodation which involves the development of land or property already in the RSL's ownership and for which no acquisition costs (other than basic legal charges) apply.

12. Selection of key multipliers - cont'd

12.7 The *reimprovement* multiplier is used for dwellings which have already received some form of public/grant subsidy and are now being rehabilitated. Reimprovement schemes are generally expected to be submitted no less than 15 years after a rehabilitation or 30 years after construction. It should be noted that the outstanding mortgage of the original works can be considered an eligible cost and will attract the grant rate applicable to the new scheme as a whole. The reimprovement key multiplier is linked to that of the Rehabilitation works only (vacant) multiplier.

12.8 The *TSH improved and unimproved vacant* multiplier is used for accommodation with a lease of between two and 29 years.

KEY MULTIPLIERS 2001/2002 BY COST GROUP
Note: only one of the following to be used.

	Scheme Type		TCI calc. form line No.	A	B	Cost group C	D	E
a)	New build							
	(i)	acquisition and works	0010	1.00	1.00	1.00	1.00	1.00
	(ii)	off-the-shelf	0020	1.00	0.96	0.95	0.90	0.88
	(iii)	works only	0090	0.59	0.61	0.69	0.82	0.89
b)	Rehabilitation							
	(i)	acquisition and works						
		- vacant	0100	1.24	1.16	1.12	1.12	1.07
		- tenanted	0105	1.16	1.08	1.05	1.04	1.00
	(ii)	existing satisfactory	0120	1.02	0.97	0.96	0.92	0.90
	(iii)	purchase and repair	0110	1.00	0.96	0.95	0.90	0.88
	(iv)	works only						
		- vacant	0125	0.55	0.51	0.53	0.60	0.65
		- tenanted	0130	0.57	0.52	0.55	0.61	0.67
	(v)	re-improvements	0135	0.55	0.51	0.53	0.60	0.65
c)	TSH							
	(i)	vacant improved	0030	1.24	1.16	1.12	1.12	1.07
	(ii)	vacant unimproved	0035	1.24	1.16	1.12	1.12	1.07

Cost Limits and Allowances   847

13. Selection of supplementary multipliers

13.1 The table on page 850 *is used for acquisition and works, off-the-shelf, existing satisfactory and purchase and repair schemes. For other scheme types use the table on page 851. None of these supplementary multipliers are applicable to TSH schemes.*

13.2 Supplementary multipliers can be applied to new build and rehabilitation units when the accommodation is designed to meet the relevant standards set out in the Housing Corporation publication Scheme Development Standards.

13.3 More than one supplementary multiplier can be used per unit. However certain combinations of multipliers are invalid. Multipliers for sheltered, frail older people, supported housing and extended families cannot be combined. The matrix on page 852 gives a comprehensive list of valid combinations of multipliers.

13.4 The main reason for combinations of multipliers being invalid is that the combination of those multipliers would lead to a duplication of the financial provision for the facilities accounted for in the multipliers. e.g. category 2 includes allowance for new lifts or single storey. This means that in these circumstances the other relevant multiplier does not apply i.e. a new lift or single storey multiplier cannot be used with a category 2 multiplier.

13.5 The *supported housing* multiplier and on-costs will only apply to schemes approved within the funding framework introduced in 1995. A scheme which is developed within this framework must be eligible to receive SHMG whether or not SHMG is actually being claimed for the scheme. Applications for approval of capital only supported housing schemes which utilise the supported housing multiplier must be accompanied by form TS1 (revenue budget). This multiplier should only be used where it is the RSL's plan to use the accommodation to provide supported housing in the long term.

13.6 Supported housing schemes with shared facilities cannot be combined with either of the three common room multipliers in *part (h)* of the supplementary multiplier table.

13.7 The supported housing multiplier cannot be used for staff units.

13.8 The appropriate *shared* supplementary multiplier is determined by the total number of bedspaces provided within all of the households sharing facilities in the scheme. Where a cluster (i.e. more than one) of shared accommodation is provided the appropriate shared supplementary multiplier is determined by the total number of bedspaces contained within the households comprising each independent and self sufficient shared accommodation arrangement. The *TCI base table: shared accommodation* on page 845 is deemed to include all communal and ancillary facilities.

13.9 The shared multiplier does not apply to either sheltered or frail older people schemes.

13.10 The *extended families* multiplier is used when a self-contained dwelling is to cater for eight or more persons and additional or duplicate facilities i.e. kitchen and/or sanitary fittings/equipment are provided. The additional space requirement is accounted for in the size band selection. Where significant additional facilities are not provided the RSL should contact the relevant regional office of the Corporation to receive confirmation that the extended families multiplier may be applied.

13.11 The *served by new lifts* multiplier is used when new vertical passenger lift provision is incorporated for access to dwelling entrances and communal accommodation. The multiplier does not apply to dwellings with entrances at ground floor level unless, exceptionally, the scheme includes essential communal accommodation provided at a level other than at ground floor level.

13.12 The *wheelchair with individual carport* multiplier should be used only with individual self-contained dwellings designed in accordance with the relevant standards set out in the Housing Corporation publication Scheme Development Standards. Where non-individual self-contained wheelchair user dwellings are provided e.g. on some category 2 schemes, or where a waiver for non-compliance with the carport provisions has been obtained from the regional office, the wheelchair without individual carport multiplier should be used.

13.13 The *wheelchair* multiplier does not allow for any fixed additional equipment for people with disabilities; these needs should be met via the adaptations funding framework.

13. Selection of supplementary multipliers - cont'd

13.14 The *parent with children refuges* multiplier applies to shared accommodation specifically designed to meet the needs of parents with children, including vulnerable women with babies and women and children at risk of domestic violence. The TCI is calculated on a per bedspace basis with children counted as full bedspaces. The provision of additional bunk-bedspaces should be disregarded for TCI calculation purposes.

13.15 The *housing for sale* multiplier is used for all sale schemes. It cannot be used with the supported housing or shared multipliers.

13.16 The *rehabilitation to pre-1919 properties* multiplier is used where the scope of the refurbishment or conversion work is carried out on a property or properties originally constructed prior to 1919. It cannot be used in connection with existing satisfactory or purchase and repair multipliers.

13.17 The *no VAT rehabilitation* multiplier is used when the scope of the refurbishment or conversion work is such that the relevant local Customs and Excise office determines that VAT is not chargeable on the works.

13.18 The *rural housing* multiplier is used for schemes identified by rural investment codes R, G and F. Rural areas mainly comprise those with 1,000 or less inhabitants and a minimum of 60% of the rural programme is targeted to relieve housing need in these settlements. Exceptionally, the limit may be extended to 3,000 inhabitants on a case by case basis by the relevant regional office of the Housing Corporation. (See the Housing Corporation's Guide to the Allocations Process, for details of population settlements, the Housing Corporation's Rural Settlements Gazetteer, last published in 1998).

13.19 The *national park* multiplier may be applied to any scheme within a formally designated national park or formally designated area of outstanding beauty. It may be used in conjunction with the rural housing multiplier. Please see list on page 855.

13.20 The *sustainability* multiplier has been introduced to encourage RSLs towards greater sustainability. In order to qualify for the 1.01 multiplier, two separate aspects must be addressed:

(i) greening - in pursuit of the government's stared policy of increasing energy efficiency and reducing levels of CO_2 in the atmosphere, an 'Eco-Home' rating of 'Good' (second level) must be certified;

(ii) security - Secured by Design certification must be obtained for the scheme.

13.21 The *adaptability* multiplier has been introduced has been introduced to encourage RSLs to design homes which lofts that may be readily adapted to provide additional habitable space. In order to qualify for the 1.02 multiplier the following features must be present:

(i) a clear loft area of a size and height that will satisfy planning requirementr=s to accommodate a single bedroom;

(ii) loft floor joists size for floor loading;

(iii) loft floor trimmed for new staircase;

(iv) landing layout and size that will accommodate a new access to the converted loft;

(v) a bedroom sized gable window, dormer window or opening rooflight.

13.22 The *construction clients' charter* mulitplier has been introduced in recognition of additional expense involved in setting up and maintaining 'Chartered' client status as supported by Government and encouraged for all agencies and associated organisations in receipt of public funds. In order to qualify for the supplementary on-cost of 1% and the resultant supplementary mulitplier of 1.01, the RSL must be able to satisfy and comply with the 'Chartered Client' conditions set by the Confederation of Construction Clients (www.construction-clients.org).

13.23 The *TSH* supplementary multiplier is used for all schemes with a lease length of between two and 29 years requiring works, (see table on page 853) and it can only be used with the TSH shared supplementary multiplier. It does not apply to improved TSH schemes.

13. Selection of supplementary multipliers - cont'd

13.24 *Note:* Where a RSL is in any doubt concerning the appropriateness of the application of any supplementary multiplier, it is advised to seek advice from the relevant regional office of the Housing Corporation. RSLs should also ensure that relevant information relating to the appropriateness of the application of supplementary multipliers is maintained on file for compliance audit purposes.

SUPPLEMENTARY MULTIPLIERS FOR ACQUISITION AND WORKS, OFF-THE-SHELF, EXISTING SATISFACTORY AND PURCHASE AND REPAIR SCHEMES ONLY 2001/2002

	SCHEME TYPE	TCI calc. form line no.	COST GROUP A	B	C	D	E
a)	Sheltered:						
	(i) category 1	0140	1.04	1.04	1.05	1.06	1.06
	(ii) category 2 (includes new lifts/single storey)	0170	1.30	1.31	1.35	1.40	1.42
b)	Frail older person (use alone only)	0180	1.47	1.48	1.54	1.61	1.66
c)	Supported housing	0186	1.12	1.13	1.14	1.16	1.17
d)	Extended general families	0195	1.09	1.10	1.11	1.14	1.15
e)	Shared (not used with Frail older person) Bedspaces per scheme/cluster						
	i) 2 to 3	0200	1.00	1.00	1.00	1.00	1.00
	ii) 4 to 6	0210	0.88	0.88	0.88	0.88	0.88
	iii) 7 to 10	0220	0.79	0.79	0.79	0.79	0.79
	iv) 11 and over	0235	0.74	0.74	0.74	0.74	0.74
f)	Served by new lifts (not used with Frail older person and Sheltered category 2)	0245	1.07	1.07	1.08	1.09	1.10
g)	Single Storey (New build only not used with Frail older person and Sheltered category 2 or Wheelchair with carport)	0095	1.21	1.21	1.18	1.14	1.11
h)	Common room etc. (Sheltered category 1 and Self-contained supported housing only)						
	i) common room only	0250	1.08	1.08	1.08	1.08	1.09
	ii) associated communal facilities only	0260	1.05	1.05	1.06	1.07	1.07
	iii) common room and communal facilities	0270	1.13	1.13	1.14	1.16	1.16
i)	Rehabilitation to pre-1919 properties (not used with existing satisfactory and purchase and repair)	0310	1.07	1.07	1.07	1.08	1.09
j)	No VAT rehabilitation	0320	0.92	0.93	0.93	0.92	0.91
k)	Wheelchair (except Single storey and where included as above)						
	i) with individual carport (not used with Frail older person or Single storey)	0380	1.33	1.33	1.30	1.24	1.21
	ii) without individual carport (not used with Frail older person) (Note: provision (ii) requires a waiver from the regional office)	0385	1.07	1.07	1.08	1.10	1.11
l)	Parent with children refuges (Shared general needs and Shared supported housing only)	0420	0.48	0.48	0.48	0.48	0.48
m)	Housing for sale	5100	1.05	1.05	1.05	1.04	1.04
n)	Rural housing	0430	-	1.12	1.14	1.16	1.18
o)	National parks	0440	1.10	1.10	1.12	1.14	1.15
o)	National parks	0440	1.10	1.10	1.12	1.14	1.15
p)	Sustainability	0450	1.01	1.01	1.01	1.02	1.02
q)	Construction Clerks Charter except off-the-shelf and existing satisfactory	0460	1.01	1.01	1.01	1.01	1.01
r)	Adaptability: Lofts	0470	1.02	1.02	1.03	1.03	1.04

SUPPLEMENTARY MULTIPLIERS FOR WORKS ONLY AND REIMPROVEMENT SCHEMES 2001/2002

	SCHEME TYPE	TCI calc. form line no.	COST GROUP A	B	C	D	E
a)	Sheltered						
	i) category 1	0140	1.07	1.07	1.07	1.07	1.07
	ii) category 2 (includes new lifts/ single storey)	0170	1.51	1.51	1.50	1.49	1.48
b)	Frail older person (use alone only)	0180	1.80	1.79	1.77	1.75	1.74
c)	Supported housing	0186	1.21	1.21	1.20	1.19	1.19
d)	Extended general families	0195	1.16	1.16	1.16	1.17	1.17
e)	Shared (not used with Frail older person) Bedspaces per scheme/cluster						
	i) 2 to 3	0200	1.00	1.00	1.00	1.00	1.00
	ii) 4 to 6	0210	0.80	0.80	0.83	0.85	0.86
	iii) 7 to 10	0220	0.64	0.65	0.70	0.74	0.76
	iv) 11 and over	0235	0.56	0.57	0.63	0.68	0.71
f)	Served by new lifts (not used with Frail older person and Sheltered category 2)	0245	1.12	1.12	1.12	1.11	1.11
g)	Single storey (New build only not used with Frail older person and Sheltered category 2 or Wheelchair with carport)	0095	1.07	1.07	1.07	1.07	1.07
h)	Common room etc. (Sheltered category 1 and Self-contained supported housing only)						
	i) common room only	0250	1.07	1.07	1.07	1.07	1.07
	ii) associated communal facilities only	0260	1.08	1.08	1.08	1.08	1.08
	iii) common room and communal facilities	0270	1.15	1.15	1.15	1.15	1.15
i)	Rehabilitation to pre-1919 properties (not used with existing satisfactory and purchase and repair)	0310	1.13	1.13	1.13	1.13	1.13
j)	No VAT rehabilitation	0320	0.86	0.86	0.86	0.86	0.86
k)	Wheelchair (except Single storey and where included as above)						
	i) with individual carport (not used with Frail older person or Single Storey)	0380	1.15	1.15	1.16	1.16	1.16
	ii) without individual carport (not used with Frail older person)	0385	1.12	1.12	1.12	1.12	1.12
	(Note: provision (ii) requires a waiver from the regional office)						
l)	Parent with children refuges (Shared general needs and Shared supported housing only)	0420	0.11	0.14	0.25	0.36	0.41
m)	Housing for sale	5100	1.09	1.09	1.07	1.05	1.05
n)	Rural housing	0430	-	1.19	1.19	1.20	1.20
o)	National parks	0440	1.16	1.16	1.17	1.17	1.17
p)	Sustainability	0450	1.02	1.02	1.02	1.02	1.02
o)	Construction Clerks Charter except off-the-shelf and existing satisfactory	0460	1.02	1.01	1.01	1.01	1.01
r)	Adaptability: Lofts	0470	1.04	1.04	1.04	1.04	1.04

VALID COMBINATION OF SUPPLEMENTARY MULTIPLIERS 2001/2002

[Matrix table showing valid combinations of supplementary multipliers. Rows and columns both list the following codes: CAT1 0140, CAT2 0170, FE 0180, SN 0186, ExFm 0195, SH23 0200, SH46 0210, SH710 0220, SH11+ 0235, LIFT 0245, SS 0095, CR 0250, FAC 0260, CR&FC 0270, 1919 0310, VAT 0320, WC+C 0380, WC 0385, PWCR 0420, HFS 5100, RH 0430, NP 0440, SUS 0450, CCC 0460, ADA 0470. Black dots indicate valid combinations; shaded cells indicate invalid combinations; the diagonal is blacked out.]

● indicates a valid combination

Key:

Code	No.	Description	Code	No.	Description	Code	No.	Description
CAT1	0140	Category 1	SS	0095	Single storey	VAT	0320	No VAT rehabilitation
CAT2	0170	Category 2	SUS	0450	Sustainability	WC+C	0380	Wheel chair and individual carport
FE	0180	Frail older person	CCC	0460	Construction Clients' Charter			
SN	0186	Supported housing	ADA	0470	Adaptability	WC	0385	Wheelchair without individual carport
ExFm	0195	Extended families	CR	0250	Common room			
SH23	0200	Shared 2-3 bedspace per unit	FAC	0260	Common room associated communal facilities only	PWCR	0420	Parent with children refuges
SH46	0210	Shared 4-6 bedspace per unit				HFS	5100	Housing for sale
SH710	0220	Shared 7-10 bedspace per unit	CR&FC	0270	Common room and communal facilities	RH	0430	Rural housing
SH11+	0235	Shared 11+ bedspace per unit				NP	0440	National parks
LIFT	0245	Served by new lifts	1919	0310	Rehab pre-1919 properties			

Notes:

It is valid to combine each supplementary multiplier singularly with each key multiplier with the following exceptions:
    the Housing for Sale multiplier cannot be combined with the Rehabilitation reimprovement key multiplier;
    the Single storey multiplier can only be combined with the New build acquisition and works, New build off-the-shelf and New build works only key multipliers;
    the Standard house type multiplier cannot be combined with the New build off-the-shelf, Rehabilitation existing satisfactory or Purchase and repair key multipliers;
    the Rehabilitation pre-1919 multiplier cannot be combined with the Existing satisfactory or Purchase and repair key multipliers.
TSH is not included in this matrix. Refer to guidance on page 848.

SUPPLEMENTARY MULTIPLIERS (INCLUSIVE OF ON-COSTS) FOR TEMPORARY HOUSING (TSH) 2001/2002

		TCI calc. form line no.	Cost group A	B	C	D	E
TSH		4000					
term:	29 years		0.44	0.43	0.46	0.52	0.59
	28 years		0.43	0.42	0.46	0.52	0.58
	27 years		0.42	0.42	0.45	0.51	0.58
	26 years		0.42	0.41	0.45	0.50	0.57
	25 years		0.41	0.41	0.44	0.50	0.56
	24 years		0.40	0.39	0.43	0.48	0.55
	23 years		0.39	0.38	0.41	0.47	0.53
	22 years		0.38	0.37	0.40	0.46	0.52
	21 years		0.37	0.36	0.39	0.44	0.50
	20 years		0.36	0.35	0.38	0.43	0.49
	19 years		0.35	0.34	0.37	0.42	0.47
	18 years		0.34	0.33	0.36	0.40	0.46
	17 years		0.33	0.32	0.35	0.39	0.45
	16 years		0.32	0.31	0.34	0.38	0.43
	15 years		0.31	0.30	0.33	0.37	0.42
	14 years		0.27	0.26	0.28	0.32	0.37
	13 years		0.23	0.23	0.25	0.28	0.32
	12 years		0.20	0.20	0.22	0.24	0.28
	11 years		0.18	0.17	0.19	0.21	0.24
	10 years		0.15	0.15	0.16	0.19	0.21
	9 years		0.15	0.15	0.16	0.18	0.21
	8 years		0.15	0.15	0.16	0.18	0.20
	7 years		0.15	0.14	0.16	0.18	0.20
	6 years		0.15	0.14	0.15	0.17	0.20
	5 years		0.12	0.12	0.13	0.15	0.17
	4 years		0.11	0.10	0.11	0.13	0.14
	3 years		0.08	0.08	0.09	0.10	0.11
	2 years		0.07	0.06	0.07	0.08	0.09
TSH (shared)		4001					
term:	2 - 29 years		1	1	1	1	1

CAPITALISED LEASE PREMIUM FACTORS FOR TEMPORARY HOUSING (TSH) 2001/2002

		TCI calc. form line no.	Cost group A	B	C	D	E
TSH		n/a					
term:	29 years		0.23	0.24	0.22	0.20	0.16
	28 years		0.23	0.23	0.22	0.19	0.16
	27 years		0.22	0.22	0.21	0.19	0.16
	26 years		0.22	0.22	0.21	0.18	0.15
	25 years		0.21	0.21	0.20	0.18	0.15
	24 years		0.21	0.21	0.20	0.17	0.14
	23 years		0.20	0.20	0.19	0.17	0.14
	22 years		0.19	0.20	0.19	0.16	0.14
	21 years		0.19	0.19	0.18	0.16	0.13
	20 years		0.18	0.19	0.18	0.16	0.13
	19 years		0.18	0.18	0.17	0.15	0.13
	18 years		0.17	0.18	0.17	0.15	0.12
	17 years		0.17	0.17	0.18	0.14	0.12
	16 years		0.16	0.17	0.16	0.14	0.11
	15 years		0.16	0.16	0.15	0.13	0.11
	14 years		0.15	0.15	0.15	0.13	0.11
	13 years		0.15	0.15	0.14	0.12	0.10
	12 years		0.14	0.14	0.14	0.12	0.10
	11 years		0.14	0.14	0.13	0.12	0.10
	10 years		0.13	0.13	0.13	0.11	0.09
	9 years		0.13	0.13	0.12	0.11	0.09
	8 years		0.12	0.12	0.12	0.10	0.09
	7 years		0.12	0.12	0.11	0.10	0.08
	6 years		0.11	0.11	0.11	0.09	0.08
	5 years		0.09	0.09	0.08	0.07	0.06
	4 years		0.07	0.07	0.07	0.06	0.05
	3 years		0.05	0.05	0.05	0.05	0.04
	2 years		0.04	0.04	0.04	0.04	0.03

14. Calculation of maximum grant contribution to lease costs

14.1 The above factors are used to calculate maximum grant contributions for lease premiums outside the normal TCI framework. Whilst the factors are applied to the appropriate figures in the TCI base table and the relevant TSH key multiplier to determine the maximum contribution, any grant paid is additional to that paid in respect of works costs, which is subject to separate value for money assessment using the TCI multipliers set out on page 853. The grant payment will be calculated according to the actual cost of the lease premium.

14.2 The above factors should be used only in conjunction with the *TCI base table for 2001/2002: self-contained accommodation* (see table on page 845) and the relevant TSH key multiplier. The resultant value reflects the maximum grant contribution towards any capitalised lease premium payable to acquire the lease. No other factors or supplementary multipliers apply. The on-costs associated with setting up the lease are included within the TCI element of eligible costs. The different principles underlying the calculation of TCI for self-contained accommodation and for shared accommodation as outlined on page 843 to 844 apply equally to the calculation of capitalised lease premiums.

LIST OF NATIONAL PARKS IN ENGLAND

The Broads	Dartmoor
Exmoor	Lake District
The New Forest	North Yorkshire Moor
Northumberland	Peak District
Yorkshire Dales	

DIRECTORY OF AREAS OF OUTSTANDING BEAUTY

Arnside & Silverdale	Blackdown Hills
Cannock Chase	Chichester Harbour
Chilterns	Cornwall
Cotswolds	Cranbourne Chase and West Wiltshire Downs
Dedham Vale	Dorset
East Devon	East Hampshire
Forest of Bowland	High Weald
Howardian Hills	Isle of Wight
Isles of Scilly	Kent Downs
Lincolnshire Wolds	Malvern Hills
Mendip Hills	Nidderdale
Norfolk Coast	North Devon
North Pennines	Northumberland Coast
North Wessex Downs	Quantock Hills
Shropshire Hills	Solway Coast
South Devon	South Hampshire Coast
Suffolk Coast and Heaths	Surrey Hills
Sussex Downs	Tamar Valley
Wye Valley	

TCI COST GROUP TABLES

TCI shall apply to schemes in cost groups as follows:

GROUP A
Comprises:

The following inner London boroughs:
Brent	Camden
City of London	Hackney
Hammersmith and Fulham	Islington
Greenwich	Kensington and Chelsea
Lambeth	Southwark
Tower Hamlets	Wandsworth
Westminster	

The following outer London boroughs:
Barnet	Ealing
Enfield	Haringey
Harrow	Hillingdon
Hounslow	Kingston upon Thames
Merton	Richmond upon Thames

The following unitary authority:

Windsor and Maidenhead

The following local authorities in the counties of:

Buckinghamshire
Chiltern
South Buckinghamshire

Cornwall and Isles of Scilly
Isles of Scilly

Essex
Epping Forest

Hertfordshire
Dacorum
Hertsmere
Three Rivers
Watford

GROUP B
Comprises:

The following inner London boroughs:
Havering
Lewisham

The following outer London boroughs:
Barking and Dagenham
Bromley
Newham
Sutton
Bexley
Croydon
Redbridge
Waltham Forest

The following unitary authorities:
Bracknell Forest
Reading
West Berkshire
Brighton and Hove
Slough
Wokingham

The following local authorities in the counties of:

Bedfordshire:
South Bedfordshire

Buckinghamshire
Aylesbury Vale
Wycombe

Cambridgeshire
Cambridge

Essex
Basildon
Chelmsford
Uttlesford
Brentwood
Harlow

Hampshire
Basingstoke and Deane
Hart
Winchester
East Hampshire
Rushmoor

Hertfordshire
Broxbourne
St. Albans
East Hertfordshire
Welwyn Hatfield

Kent
Sevenoaks
Tunbridge Wells
Tonbridge and Malling

Oxfordshire
Oxford
South Oxfordshire
Vale of White Horse
West Oxfordshire

Surrey
Elmbridge
Guildford
Reigate and Banstead
Spelthorne
Tandridge
Woking
Epsom and Ewell
Mole Valley
Runnymede
Surrey Heath
Waverley

West Sussex
Arun
Chichester
Mid Sussex
Crawley
Horsham

GROUP C
Comprises:

The following unitary authorities:

Bath and North East Somerset
Bristol
Milton Keynes
Poole
South Gloucestershire
Southend-on-Sea
The Meadway Towns

Bournemouth
Luton
North Somerset
Portsmouth
Southampton
Swindon
Thurrock

The following local authorities in the counties of:

Bedfordshire
Mid Bedfordshire   Bedford

Cambridgeshire
South Cambridgeshire

Cheshire
Macclesfield

Dorset
Christchurch   East Dorset
North Dorset   Purbeck
West Dorset   Weymouth and Portland

East Sussex
Eastbourne   Hastings
Lewes   Rother
Wealdon

Essex
Basildon   Castle Point
Colchester   Maldon
Rochford

Gloucestershire
Cheltenham   Cotswold

Greater Manchester
Stockport

Hampshire
Eastleigh   Fareham
Gosport   Havant
New Forest   Test Valley

Hertfordshire
North Hertfordshire   Stevenage

Kent
Ashford   Canterbury
Dartford   Dover
Gravesham   Maidstone
Shepway   Swale
Thanet

Northamptonshire
South Northamptonshire

GROUP C - cont'd
Comprises:

North Yorkshire
Harrogate

Oxfordshire
Cherwell

Somerset
Taunton Deane

Warwickshire
Stratford-on-Avon   Warwick

West Midlands
Birmingham   Solihull

West Sussex
Adur   Worthing

Wiltshire
Kennet   North Wiltshire
Salisbury   West Wiltshire

Worcestershire
Malvern Hills District   Worcester City
Wychavon

GROUP D
Comprises:

The following unitary authorities:

Blackburn
Darlington
Halton
Isle of Wight
Leicester
Peterborough
Rutland
The Wrekin
York

Blackpool
East Riding
Herefordshire
Kingston-upon-Hull
Nottingham City
Plymouth
Torbay
Warrington

The following local authorities in the counties of:

Cambridgeshire
East Cambridgeshire

Huntingdonshire

Cheshire
Chester
Vale Royal

Congleton

Cornwall and Isles of Scilly
Caradon
Kerrier
Penwith

Carrick
North Cornwall
Restormel

Cumbria
Barrow-in-Furness
South Lakeland

Eden

Derbyshire
Derbyshire Dales
High Peak

Erewash
North East Derbyshire

Devon
East Devon
Mid Devon
South Hams
Teignbridge

Exeter
North Devon
Torridge
West Devon

Durham
Chester-le-Street
Durham

Derwentside

Essex
Tendring

Gloucestershire
Forest of Dean
Stroud

Gloucester
Tewkesbury

Greater Manchester
Bolton
Manchester
Rochdale
Tameside
Wigan

Bury
Oldham
Salford
Trafford

GROUP D - cont'd
Comprises:

Lancashire
Chorley
Lancaster
Ribble Valler
West Lancashire

Flyde
Preston
South Ribble
Wyre

Leicestershire
Blaby
Harborough
Melton
Oadby and Wigston

Charnwood
Hinckley and Bosworth
North West Leicestershire

Merseyside
Knowsley
Sefton

Liverpool
St. Helens

Norfolk
Broadland
Norwich

King's Lynn and West Norfolk

North Yorkshire
Craven
Richmondshire
Scarborough

Hambleton
Ryedale
Selby

Northamptonshire
Corby
East Northamptonshire
Northampton

Daventry
Kettering
Wellingborough

Northumberland
Castle Morpeth

Tynedale

Nottinghamshire
Basset Law
Newark and Sherwood

Gedling
Rushcliffe

Shropshire
Bridgnorth
Shrewsbury and Atcham

Oswestry
South Shropshire

Somerset
Mendip
South Somerset

Sedgemoor
West Somerset

South Yorkshire
Doncaster

Sheffield

Staffordshire
Cannock Chase
South Staffordshire
Staffordshire Moorlands

Lichfield
Stafford
Tamworth

Suffolk
Babergh
Ipswich
St Edmundsbury

Forest Heath
Mid Suffolk
Suffolk Coastal

GROUP D - cont'd
Comprises:

Tyne and Wear
Gateshead
North Tyneside
Sunderland

Newcastle upon Tyne
South Tyneside

Warwickshire
North Warwickshire
Rugby

Nuneaton and Bedworth

West Midlands
Coventry
Sandwell
Wolverhampton

Dudley
Walsall

West Yorkshire
Bradford
Leeds

Kirklees

Worcestershire
Bromsgrove
Wyre Forest

Redditch

Cost Limits and Allowances 863

GROUP E
Comprises:

The following unitary authorities:
City of Derby   Hartlepool
Middlesbrough   North East Lincolnshire
North Lincolnshire   Recar and Cleveland
Stockton-on-Tees   Stoke-on-Trent

The following local authorities in the counties of:

Cambridgeshire
Fenland

Cheshire
Crewe and Nantwich   Ellesmere Port and Neston

Cumbria
Allderdale   Carlisle
Copeland

Derbyshire
Amber Valley   Bolsover
Chesterfield   South Derbyshire

Durham
Easington   Sedgefield
Teesdale   Wear Valley

Lancashire
Burnley   Hyndburn
Pendle   Rossendale

Lincolnshire
Boston   East Lindsey
Lincoln   North Kesteven
South Holland   South Kesteven
West Lindsey

Merseyside
Wirral

Norfolk
Breckland   Great Yarmouth
North Norfolk   South Norfolk

Northumberland
Alnwick   Berwick-upon-Tweed
Blyth Valley   Wansbeck

Nottinghamshire
Ashfield   Broxtowe
Mansfield

Shropshire
North Shropshire

South Yorkshire
Barnsley   Rotherham

Staffordshire
East Staffordshire   Newcastle-under-Lyme

GROUP E - cont'd
Comprises:

Suffolk
Waveney

West Yorkshire
Calderdale   Wakefield

Keep your figures up to date, free of charge
Download updates from the web: www.pricebooks.co.uk

This section, and most of the other information in this Price Book, is brought up to date every three months with the Price Book Updates, until the next annual edition. The updates are available free to all Price Book purchasers.

To ensure you receive your free copies, either complete the reply card from the centre of the book and return it to us or register via the website www.pricebooks.co.uk

GRANT RATES 2001/2002 GUIDANCE NOTES

1. Introduction

1.1 These guidance notes are designed to explain the use of grant rates in the Social Housing Grant (SHG) funding framework, and help registered social landlords (RSLs) complete scheme submissions. The notes are effective from 1 April 2001; they are issued together with separate Total Cost Indicator (TCI) guidance notes.

1.2 Enquiries about the contents of this guidance should be directed to regional offices of the Housing Corporation.

2. Explanation of grant rates

2.1 A key objective of the funding system is to achieve value for money in return for grant that allows an affordable rent, and to ensure the correct level of grant is paid. Grant rates are an essential part of this system and are published by cost group area categories and unit types.

2.2 Grant rates apply equally to units funded with SHG by the Housing Corporation and to those sponsored by a local authority.

2.3 Grant rates represent the maximum proportion of scheme costs which will be funded by any form of public subsidy including SHG. (The definition of public subsidy in this context is contained in appendix 6 of the Capital Funding System Procedure Guide). Where a scheme gets public subsidy from sources other than SHG, the maximum amount of SHG payable is reduced pound for pound.

2.4 The grant percentages published in this booklet represent the maxima which may be paid. Grant paid to RSLs will be based upon the lower of the amount allocated in response to bids or the amount produced by the calculations set out in this part of the guidance notes.

3. Treatment of combined supported housing and general needs schemes

3.1 The Housing Corporation has arrangements which allow the submission of schemes containing a combination of supported housing and general needs units within a single scheme.

3.2 Consequently such schemes may have two grant rates, one for the supported homes and one for the other homes.

3.3 The calculation of the grant rate for capital funded supported housing schemes follows the procedures set out in the Capital Funding System Procedure Guide.

4. Cost group areas

4.1 The local authority area in which a scheme is located defines the cost group for the scheme. A list of the local authority areas in each cost group is included on pages 856 to 864. Different grant rates apply in each cost group area.

5. Unit type

5.1 Because of the different capital and revenue costs of different dwelling types, separate grant rates are published for a range of units. The TCI multipliers which are applied to a unit (see pages 836 to 864) define the dwelling type for grant rate purposes.

6. The grant rate tables

6.1 Grant rates for schemes for rent - Appendix 1
Set out the grant rates which apply to schemes for rent by TCI cost group area.

6.3 Grant rates for schemes for sale - Appendices 2 and 3
Set out the grant rates which apply to schemes for sale; together with the Low Cost Home Ownership (LCHO) factors which are applied for housing for sale schemes to the appropriate grant rates in the table *Grant rates for schemes for rent* (Appendix 1). The value limits and discount amounts applicable to home ownership schemes, including Homebuy, are included here.

7. Using the grant rate tables - Schemes for rent

7.1 To use the tables (Appendix 1) to identify the appropriate grant rate, the cost group and gross TCI multiplier must be identified.

7.2 The Total Cost Indicator Guidance Notes list the TCI cost groups into which local authority areas fall. Those notes also list the current TCI multipliers and an explanation of their use.

7.3 For housing for rent units, the gross multiplier and maximum grant rate is calculated for each unit type on Form SFN1, which is also used to derive the aggregate maximum grant rate for the scheme on a weighted basis according to the mix of units.

7.4 To identify the grant rate for units for rent:

   i) *find the TCI gross multiplier* (rounded to two decimal places) in one of the ranges established by the "Lowest" and "Highest" columns at the left of the table

   ii) *determine the appropriate grant column* by referring to the column headings and the key at the foot of the table

   iii) *read off the grant rate* from the appropriate column for the particular unit type.

   iv) where appropriate *apply any necessary Standard Percentage Adjustment* calculation to the grant percentage taken from the table, using the formula set out on form SFN1.

7.5 "Standard" grant rates for rented projects are shown in bold. These have been calculated as follows:

   i) the standard rate is the rate for the unit type at the top of the column in the absence of any other design or need characteristics which, by attracting additional key or supplementary multipliers, would produce an adjusted TCI and a different gross multiplier. For example, a category 1 acquisition and works unit in TCI cost group area A with no additional characteristics has a standard grant rate at a gross multiplier of 1.04, reflecting the fact that category 1 dwellings attract a supplementary multiplier of 1.04 which is applied to the basic new build TCI.

   ii) the standard rate for shared housing is calculated by applying to the basic TCI the adjustment for acquisition and works schemes with between 4 and 6 bedspaces. Schemes with higher or lower numbers of bedspaces attract different supplementary multipliers leading to grant rates above or below the standard.

7.6 Grant rates are shown for a wide range of gross multipliers, but it is possible that a scheme may attract a multiplier outside these ranges. In such cases, the grant rate appropriate to the nearest published range should be used.

8. Using the grant rate tables - Schemes for sale

8.1 For all shared ownership schemes, the first stage is to use the table for schemes for rent (appendix 1) in exactly the same way as described above for schemes for rent. Then turn to the *LCHO factors* (appendix 2) and apply the appropriate factor to the result from tables in appendix 1 to calculate the grant appropriate to the scheme.

8.3 *Homebuy value limits and on-cost allowance, Voluntary Purchase Grant and Right to Acquire discount amounts and Local Authority Do-It-Yourself Shared Ownership value limits, grant rates and on-cost allowances* are on pages 873 to 874.

9. Standard percentage adjustments (SPAs)

9.1 Standard Percentage Adjustments (SPAs) can be set by the Housing Corporation to reflect particular high or low cost locations within a single TCI cost group area. These adjustments are intended to make the TCI more sensitive to local circumstances and will be fixed in a range between 80 - 120% of TCI. SPAs have the effect of:

   i) *amending the norm cost level used in the assessment of schemes*. For example, in a location for which an SPA is set at 105%, only schemes with costs greater than 105% of TCI will be treated as over TCI cases for scheme appraisal purposes. If an SPA is set in a location at 85% of TCI, all schemes with costs of more than 85% of TCI will be regarded as over TCI.

   ii) *modifying the total grant rate for schemes approved in these areas*, so that the cost of operating the acknowledged high or low cost areas is only partly translated into loan requirements and thereby rents.

9. Standard percentage adjustments (SPAs) - cont'd

9.2 The SPA grant adjustment calculation (shown on Form SFN1) only partially compensates for the effects of operating in areas with high or low norm costs. In evaluating their development options RSLs will need to consider the potential for achieving lower rents on schemes produced at norm cost in low cost areas within TCI areas and conversely, the disadvantages in terms of rent levels of working in particularly high cost areas.

10. Grant calculation and cost over-runs

All schemes

10.1 Projects above TCI for rent or sale will be subject to value for money assessment. Schemes will not be approved at more than 130% of TCI.

Schemes for rent

10.2 For schemes for rent the maximum grant percentage is fixed at grant confirmation stage. Maximum grant entitlement calculated at this stage is estimated eligible cost x grant rate. The costs are reviewed at practical completion, and grant eligibility recalculated as eligible outturn cost x the original grant rate. For RSLs following the scheme contract funding route grant will not be paid on an outturn cost greater than 110% of the estimate at grant confirmation or 130% of TCI (excluding any SPA), whichever is the lower. RSLs following the programme contract funding route will calculate grant eligibility on cost over-runs as above, but these will be funded from their grant pot or from their own resources, rather than by allocation of additional Housing Corporation resources.

Schemes for sale

10.3 For housing for sale schemes maximum grant eligibility is calculated at grant confirmation as:

Estimated eligible cost x grant rate (modified by any SPA) x LCHO factor = grant in £

10.4 The grant is fixed at grant confirmation; it is not recalculated at practical completion.

GRANT RATE TABLES   **"STANDARD" GRANT RATES SHOWN IN BOLD**

UNIT TYPE KEY:

GEN, SN* NB S/C	New build housing for general needs or non-framework, supported housing, self contained.
GEN, CAT½#, SN* RH S/C	Rehabilitation housing for general needs or CAT½# older people or for non-framework supported housing, self contained.
CAT 1 NB S/C	New build Category 1 housing for older people, self contained.
CAT 2# NB S/C	New build Category 2# housing for older people, self contained.
GEN, SN* NB SHD	New build housing for general needs or non-framework supported housing, shared.
GEN, SN* RH SHD	Rehabilitation housing for general needs or non-framework supported housing, shared.
EXT FAM NB S/C	New build housing for extended families, self contained.
EXT FAM RH S/C	Rehabilitation housing for extended families, self contained.

*   To be used for appropriate supported housing units if not under the supported housing framework.

\#   All self-contained units meeting wheelchair design requirements or units for frail older people, but which are not under the supported housing framework, use CAT2 (NB or RH) grant rates.

GRANT RATES FOR SCHEMES FOR RENT: TCI COST GROUP A
APPENDIX 1

GROSS MULTIPLIER		GEN SN* NB S/C G010	GEN, CAT ½# SN* RH S/C G020	CAT 1 NB S/C G030	CAT 2# NB S/C G040	GEN, SN* NB SHD G050	GEN, SN* RH SHD G060	EXT FAM NB S/C G070	EXT FAM RH S/C G080
LOW	HIGH								
1.96	2.00	86.2	91.8	90.6	94.8	85.6	90.6	90.7	94.8
1.91	1.95	81.2	86.6	85.4	89.5	80.6	85.4	85.6	89.6
1.86	1.90	80.6	86.2	85.0	89.2	80.0	85.0	85.1	89.2
1.81	1.85	80.1	85.8	84.5	88.8	79.4	84.5	84.7	88.9
1.76	1.80	79.4	85.3	84.0	88.5	78.8	84.0	84.2	88.5
1.71	1.75	78.8	84.9	83.5	88.1	78.1	83.5	83.7	88.1
1.66	1.70	78.1	84.4	82.9	87.7	77.4	83.0	83.1	87.7
1.61	1.65	77.3	83.8	82.3	87.2	76.7	82.4	82.5	87.3
1.56	1.60	76.5	83.2	81.7	86.7	75.9	81.8	81.9	86.8
1.51	1.55	75.7	82.6	81.0	86.2	75.0	81.1	81.3	86.3
1.46	1.50	74.8	82.0	80.3	85.7	74.1	80.4	80.6	85.8
1.41	1.45	73.9	81.3	79.5	85.1	73.2	79.6	79.8	85.3
1.36	1.40	72.8	80.5	78.7	84.5	72.1	78.8	79.0	84.7
1.31	1.35	71.7	79.7	77.8	83.9	71.0	78.0	78.1	84.0
1.26	1.30	70.5	78.9	76.9	83.1	69.8	77.0	77.2	83.3
1.21	1.25	69.2	77.9	75.8	82.4	68.5	76.0	76.2	82.6
1.18	1.20	68.1	77.1	74.9	81.7	67.4	75.2	75.3	81.9
1.15	1.17	67.2	76.5	74.2	81.2	66.5	74.5	74.6	81.4
1.12	1.14	66.3	75.8	73.5	80.6	65.5	73.7	73.9	80.9
1.09	1.11	65.3	75.1	72.7	80.1	64.6	73.0	73.1	80.3
1.06	1.08	64.3	74.4	71.9	79.4	63.5	72.2	72.3	79.7
1.03	1.05	63.2	73.6	71.0	78.8	62.4	71.3	71.4	79.1
1.00	1.02	62.0	72.7	70.1	78.1	61.3	70.4	70.5	78.4
0.97	0.99	60.8	71.9	69.1	77.4	60.0	69.4	69.6	77.7
0.94	0.96	59.4	70.9	68.0	76.6	58.7	68.4	68.5	76.9
0.91	0.93	58.0	69.9	66.9	75.7	57.3	67.3	67.4	76.1
0.88	0.90	56.5	68.8	65.7	74.9	55.8	66.1	66.3	75.3
0.85	0.87	54.9	67.7	64.4	73.9	54.2	64.9	65.0	74.3
0.82	0.84	53.2	66.5	63.0	72.9	52.5	63.6	63.7	73.4
0.79	0.81	51.3	65.1	61.6	71.8	50.7	62.1	62.2	72.3
0.76	0.78	49.3	63.7	60.0	70.6	47.8	60.6	60.7	71.2
0.73	0.75	47.2	62.2	58.2	69.3	46.5	58.9	59.0	69.9
0.70	0.72	44.8	60.5	56.4	68.0	44.2	57.1	57.2	68.6
0.67	0.69	42.3	58.7	54.3	66.5	41.7	55.1	55.2	67.2
0.64	0.66	39.5	56.7	52.1	64.8	39.0	53.0	53.0	65.6
0.61	0.63	36.4	54.6	49.6	63.0	36.0	50.6	50.6	63.8
0.58	0.60	33.0	52.2	46.9	61.0	32.7	48.0	48.0	62.0
0.55	0.57	29.2	49.6	43.9	58.8	29.0	45.1	45.1	59.9
0.52	0.54	25.0	46.6	40.6	56.4	24.9	41.9	41.8	57.5
0.49	0.51	20.3	43.4	36.8	53.6	20.4	38.3	38.2	54.9
0.46	0.48	15.0	39.7	32.5	50.5	15.3	34.3	34.1	52.0
0.43	0.45	8.9	35.5	27.7	47.0	9.5	29.7	29.4	48.7
0.40	0.42	1.9	30.7	22.1	43.0	2.9	24.4	24.1	44.9
0.37	0.39	1.0	25.2	15.7	38.3	1.0	18.4	17.9	40.5
0.34	0.36	1.0	18.8	8.0	32.9	1.0	11.2	10.6	35.5
0.00	0.33	1.0	1.0	1.0	1.0	1.0	1.0	1.0	1.0

* To be used for appropriate special needs units if not under the special needs framework.
\# All self-contained units meeting wheelchair design requirements or units for frail older people, but which are not under the special needs framework, use CAT2 (NB or RH) grant rates.

GRANT RATES FOR SCHEMES FOR RENT, TCI COST GROUP B
APPENDIX 1

GROSS MULTIPLIER		GEN SN* NB S/C G010	GEN, CAT ½# SN* RH S/C G020	CAT1 NB S/C G030	CAT2# NB S/C G040	GEN, SN* NB SHD G050	GEN, SN* RH SHD G060	EXT FAM NB S/C G070	EXT FAM RH S/C G080
LOW	HIGH								
1.96	2.00	83.5	90.2	88.4	93.1	83.0	89.1	88.6	93.4
1.91	1.95	78.5	85.0	83.3	87.9	78.0	84.0	83.5	88.2
1.86	1.90	77.9	84.6	82.7	87.5	77.4	83.5	83.0	87.8
1.81	1.85	77.2	84.1	82.2	87.1	76.7	83.0	82.4	87.5
1.76	1.80	76.5	83.6	81.6	86.7	76.0	82.5	81.9	87.1
1.71	1.75	75.7	83.1	81.0	86.2	75.3	81.9	81.3	86.6
1.66	1.70	74.9	82.6	80.4	85.7	74.5	81.3	80.7	86.2
1.61	1.65	74.1	82.0	79.7	85.2	73.6	80.7	80.0	85.7
1.56	1.60	73.2	81.3	79.0	84.7	72.7	80.0	79.3	85.2
1.51	1.55	72.2	80.7	78.2	84.1	71.7	79.3	78.5	84.7
1.46	1.50	71.2	80.0	77.4	83.5	70.7	78.6	77.7	84.1
1.41	1.45	70.1	79.2	76.5	82.8	69.6	77.8	76.9	83.5
1.36	1.40	68.9	78.4	75.6	82.1	68.4	76.9	75.9	**82.8**
1.31	1.35	67.7	77.5	74.6	**81.4**	67.2	76.0	74.9	82.1
1.26	1.30	66.3	76.6	73.5	80.6	65.8	74.9	73.9	81.3
1.21	1.25	64.8	**75.6**	72.3	79.7	64.3	73.9	72.7	80.5
1.18	1.20	63.6	74.7	71.3	78.9	63.1	72.9	71.7	79.8
1.15	1.17	62.6	74.0	70.5	78.3	62.1	72.2	70.9	79.3
1.12	1.14	61.5	73.2	69.9	77.7	61.0	71.4	70.1	78.7
1.09	1.11	60.4	72.5	68.7	77.0	59.9	**70.6**	**69.2**	78.0
1.06	1.08	59.2	71.7	67.8	76.3	58.7	69.7	68.3	77.4
1.03	1.05	57.9	70.8	**66.8**	75.5	57.5	68.8	67.3	76.7
1.00	1.02	**56.6**	69.9	65.7	74.7	56.1	67.8	66.2	75.9
0.97	0.99	55.2	68.9	64.6	73.9	54.7	66.7	65.1	75.2
0.94	0.96	53.7	67.9	63.4	73.0	53.3	65.6	64.0	74.3
0.91	0.93	52.1	66.8	62.1	72.1	51.7	64.5	62.7	73.4
0.88	0.90	50.4	65.6	60.7	71.0	**50.0**	63.2	61.4	72.5
0.85	0.87	48.5	64.3	59.3	69.9	48.2	61.9	59.9	71.5
0.82	0.84	46.5	63.0	57.7	68.8	46.2	60.4	58.4	70.4
0.79	0.81	44.4	61.6	56.0	67.5	44.2	58.9	56.7	69.3
0.76	0.78	42.1	60.0	54.2	66.2	41.9	57.2	55.0	68.0
0.73	0.75	39.7	58.3	52.2	64.7	39.5	55.4	53.0	66.7
0.70	0.72	37.0	56.5	50.0	63.1	36.9	53.5	51.0	65.2
0.67	0.69	34.1	54.5	47.7	61.4	34.0	51.3	48.7	63.6
0.64	0.66	30.9	52.4	45.1	59.5	30.9	49.0	46.2	61.9
0.61	0.63	27.3	50.0	42.3	57.4	27.5	46.5	43.5	60.0
0.58	0.60	23.5	47.4	39.2	55.1	23.8	43.7	40.5	57.9
0.55	0.57	19.2	44.5	35.8	52.6	19.6	40.6	37.1	55.6
0.52	0.54	14.4	41.3	32.0	49.8	15.0	37.1	33.4	53.1
0.49	0.51	9.0	37.8	27.7	46.6	9.9	33.2	29.3	50.2
0.46	0.48	2.9	33.7	22.8	43.1	4.1	28.9	24.6	47.0
0.43	0.45	1.0	29.2	17.2	39.0	1.0	24.0	19.2	43.4
0.40	0.42	1.0	24.0	10.9	34.4	1.0	18.3	13.1	39.2
0.37	0.39	1.0	17.9	3.5	29.0	1.0	11.7	6.0	34.4
0.34	0.36	1.0	10.9	1.0	27.7	1.0	4.1	1.0	28.8
0.00	0.33	1.0	1.0	1.0	1.0	1.0	1.0	1.0	1.0

* To be used for appropriate special needs units if not under the special needs framework.
\# All self-contained units meeting wheelchair design requirements or units for frail older people, but which are not under the special needs framework, use CAT2 (NB or RH) grant rates.

GRANT RATES FOR SCHEMES FOR RENT: TCI COST GROUP C

APPENDIX 1

GROSS MULTIPLIER		GEN SN* NB S/C G010	GEN, CAT ½# SN* RH S/C G020	CAT1 NB S/C G030	CAT2# NB S/C G040	GEN, SN NB SHD G050	GEN, SN* RH SHD G060	EXT FAM NB S/C G070	EXT FAM RH S/C G080
LOW	HIGH								
1.96	2.00	80.8	88.0	86.8	92.4	80.7	86.6	86.5	91.6
1.91	1.95	75.9	82.9	81.8	87.2	75.8	81.5	81.4	86.4
1.86	1.90	75.2	82.4	81.2	86.8	75.1	81.0	80.8	86.0
1.81	1.85	74.4	81.8	80.6	86.3	74.3	80.4	80.2	85.5
1.76	1.80	73.6	81.2	79.9	85.8	73.5	79.8	79.5	85.0
1.71	1.75	72.7	80.6	79.2	85.3	72.6	79.1	78.8	84.5
1.66	1.70	71.8	80.0	78.5	84.8	71.7	78.4	78.1	84.0
1.61	1.65	70.8	79.3	77.7	84.2	70.7	77.7	77.3	83.5
1.56	1.60	69.7	78.6	76.9	83.6	69.7	76.9	76.5	82.9
1.51	1.55	68.6	77.8	76.0	82.9	68.6	76.1	75.6	82.2
1.46	1.50	67.4	77.0	75.1	82.3	67.4	75.2	74.7	81.6
1.41	1.45	66.2	76.1	74.1	81.5	66.2	74.3	73.6	80.9
1.36	1.40	64.8	75.2	73.0	80.7	64.9	73.3	72.6	80.1
1.31	1.35	63.4	74.2	71.9	**79.9**	63.4	72.2	71.4	**79.3**
1.26	1.30	61.8	73.1	70.6	78.9	61.9	71.0	70.1	78.4
1.21	1.25	60.1	71.9	69.3	77.9	60.2	69.7	68.8	77.4
1.18	1.20	58.6	**70.9**	68.1	77.1	58.7	68.6	67.6	76.6
1.15	1.17	57.4	70.1	67.2	76.4	57.6	67.8	66.7	75.9
1.12	1.14	56.2	69.2	66.2	75.7	56.4	66.9	65.7	75.2
1.09	1.11	54.0	68.3	65.2	74.9	55.1	65.9	**64.7**	74.5
1.06	1.08	53.5	67.4	64.1	74.1	53.8	64.9	63.6	73.7
1.03	1.05	52.1	66.4	**63.0**	73.3	52.3	**63.8**	62.4	72.9
1.00	1.02	**50.5**	65.3	61.8	72.4	50.8	62.7	61.2	72.0
0.97	0.99	48.9	64.2	60.5	71.4	49.2	61.5	59.9	71.1
0.94	0.96	47.1	63.0	59.1	70.4	47.5	60.2	58.5	70.2
0.91	0.93	45.3	61.7	57.7	69.3	45.7	58.8	57.1	69.1
0.88	0.90	43.3	60.4	56.1	68.2	**43.8**	57.3	55.5	68.0
0.85	0.87	41.2	58.9	54.4	66.9	41.8	55.8	53.8	66.8
0.82	0.84	38.9	57.4	52.6	65.6	39.6	54.1	52.0	65.6
0.79	0.81	36.4	55.7	50.7	64.2	37.2	52.3	50.1	64.2
0.76	0.78	33.8	53.9	48.6	62.7	34.6	50.4	48.0	62.8
0.73	0.75	30.9	52.0	46.4	61.0	31.9	48.3	45.7	61.2
0.70	0.72	27.8	49.9	43.9	59.2	28.9	46.0	43.3	59.5
0.67	0.69	24.4	47.6	41.2	57.3	25.6	43.5	40.6	57.6
0.64	0.66	20.7	45.1	38.3	55.1	22.1	40.8	37.7	55.6
0.61	0.63	16.6	42.3	35.1	52.8	18.2	37.8	34.5	53.4
0.58	0.60	12.2	39.3	31.6	50.2	14.0	34.6	31.0	51.0
0.55	0.57	7.2	36.0	27.7	47.4	9.2	31.0	27.1	48.3
0.52	0.54	1.6	32.3	23.3	44.2	4.0	26.9	22.7	45.3
0.49	0.51	1.0	28.2	18.4	40.6	1.0	22.4	17.9	42.0
0.46	0.48	1.0	23.5	12.9	36.6	1.0	17.4	12.4	38.2
0.43	0.45	1.0	18.2	6.6	32.1	1.0	11.6	6.1	34.0
0.40	0.42	1.0	12.2	1.0	26.9	1.0	5.0	1.0	29.1
0.37	0.39	1.0	5.2	1.0	20.8	1.0	1.0	1.0	23.5
0.34	0.36	1.0	1.0	1.0	13.7	1.0	1.0	1.0	17.0
0.00	0.33	1.0	1.0	1.0	1.0	1.0	1.0	1.0	1.0

* To be used for appropriate special needs units if not under the special needs framework.
\# All self-contained units meeting wheelchair design requirements or units for frail older people, but which are not under the special needs framework, use CAT2 (NB or RH) grant rates.

GRANT RATES FOR SCHEMES FOR RENT: TCI COST GROUP D
APPENDIX 1

GROSS MULTIPLIER		GEN SN* NB S/C G010	GEN, CAT ½# SN* RH S/C G020	CAT1 NB S/C G030	CAT 2# NB S/C G040	GEN, SN* NB SHD G050	GEN, SN* RH SHD G060	EXT FAM NB S/C G070	EXT FAM RH S/C G080
LOW	HIGH								
1.96	2.00	78.4	86.1	85.8	92.3	79.4	86.0	84.3	89.8
1.91	1.95	73.5	81.0	80.7	87.0	74.4	80.9	79.2	84.7
1.86	1.90	72.7	80.5	80.1	86.5	73.6	80.3	78.6	84.2
1.81	1.85	71.8	79.8	79.4	86.0	72.8	79.7	77.1	83.6
1.76	1.80	70.9	79.2	78.7	85.5	71.9	79.0	77.1	83.1
1.71	1.75	69.9	78.5	77.9	84.9	70.9	78.3	76.3	82.5
1.66	1.70	68.8	77.7	77.1	84.3	69.9	77.6	75.4	81.9
1.61	1.65	67.7	77.0	76.2	83.7	68.9	76.8	74.5	81.2
1.56	1.60	66.5	76.1	75.3	83.0	67.7	75.9	73.5	80.5
1.51	1.55	65.3	75.2	74.3	82.3	66.5	75.1	72.5	79.8
1.46	1.50	63.9	74.3	73.3	81.5	65.2	74.1	71.4	79.0
1.41	1.45	62.5	73.3	72.2	80.7	63.8	73.1	70.2	78.2
1.36	1.40	60.9	72.2	71.0	79.8	62.3	72.0	69.0	77.3
1.31	1.35	59.2	71.0	69.7	78.9	60.8	70.8	67.6	76.3
1.26	1.30	57.4	69.8	68.3	77.9	59.0	69.5	66.1	75.2
1.21	1.25	55.5	68.4	66.8	76.8	57.2	68.1	64.5	74.1
1.18	1.20	53.8	67.2	65.5	75.8	55.6	67.0	63.2	73.1
1.15	1.17	52.5	66.3	64.5	75.1	54.3	66.0	62.1	72.4
1.12	1.14	51.1	65.3	63.4	74.3	53.0	65.0	61.0	71.5
1.09	1.11	49.6	64.3	62.2	73.4	51.6	64.0	59.8	70.7
1.06	1.08	48.0	63.2	61.0	72.5	50.1	62.9	58.5	69.8
1.03	1.05	49.3	62.0	59.7	71.6	48.5	61.7	57.1	68.8
1.00	1.02	44.6	60.8	58.4	70.6	46.9	60.5	55.7	67.8
0.97	0.99	42.7	59.5	56.9	69.6	45.1	59.2	54.2	66.7
0.94	0.96	40.7	58.2	55.4	68.4	43.2	57.8	52.6	65.6
0.91	0.93	38.6	56.7	53.8	67.2	41.2	56.3	50.9	64.4
0.88	0.90	36.3	55.1	52.0	66.0	39.1	54.7	49.0	63.1
0.85	0.87	33.9	53.5	50.2	64.6	36.8	53.0	47.1	61.7
0.82	0.84	31.3	51.7	48.2	63.1	34.4	51.2	45.0	60.2
0.79	0.81	28.5	49.8	46.0	61.6	31.8	49.3	42.7	58.6
0.76	0.78	25.5	47.7	43.7	59.9	29.0	47.1	40.3	56.9
0.73	0.75	22.2	45.4	41.2	58.0	25.9	44.9	37.7	55.0
0.70	0.72	18.7	43.0	38.4	56.1	22.6	42.4	34.8	53.0
0.67	0.69	14.8	40.4	35.4	53.9	19.0	39.7	31.7	50.8
0.64	0.66	10.6	37.5	32.2	51.5	15.1	36.8	28.3	48.4
0.61	0.63	6.0	34.4	28.6	49.0	10.8	33.6	24.6	45.8
0.58	0.60	1.0	30.9	24.6	46.1	6.1	30.1	20.5	43.0
0.55	0.57	1.0	27.1	20.3	43.0	1.0	26.1	15.9	39.8
0.52	0.54	1.0	22.8	15.4	39.4	1.0	21.8	10.8	36.3
0.49	0.51	1.0	18.0	9.9	35.5	1.0	16.9	5.2	32.4
0.46	0.48	1.0	12.7	3.7	31.1	1.0	11.4	1.0	27.9
0.43	0.45	1.0	6.6	1.0	26.0	1.0	5.2	1.0	22.9
0.40	0.42	1.0	1.0	1.0	20.2	1.0	1.0	1.0	17.2
0.37	0.39	1.0	1.0	1.0	13.6	1.0	1.0	1.0	10.6
0.34	0.36	1.0	1.0	1.0	5.7	1.0	1.0	1.0	2.9
0.00	0.33	1.0	1.0	1.0	1.0	1.0	1.0	1.0	1.0

* To be used for appropriate special needs units if not under the special needs framework.
\# All self-contained units meeting wheelchair design requirements or units for frail older people, but which are not under the special needs framework, use CAT2 (NB or RH) grant rates.

GRANT RATES FOR SCHEMES FOR RENT: TCI COST GROUP E
APPENDIX 1

GROSS MULTIPLIER		GEN SN* NB S/C G010	GEN, CAT ½# SN* RH S/C G020	CAT1 NB S/C G030	CAT2# NB S/C G040	GEN, SN* NB SHD G050	GEN, SN* RH SHD G060	EXT FAM NB S/C G070	EXT FAM RH S/C G080
LOW	HIGH								
1.96	2.00	75.2	85.4	83.8	91.2	76.2	85.3	81.5	89.2
1.91	1.95	70.3	80.3	78.7	85.9	71.4	80.2	76.5	84.0
1.86	1.90	69.4	79.7	78.0	85.3	70.4	79.5	75.7	83.4
1.81	1.85	68.4	79.0	77.2	84.8	69.5	78.9	74.8	82.9
1.76	1.80	67.3	78.3	76.4	84.2	68.4	78.1	74.0	82.3
1.71	1.75	66.2	77.5	75.5	83.6	67.4	77.4	73.0	81.6
1.66	1.70	65.0	76.7	74.6	82.9	66.2	76.6	72.0	80.9
1.61	1.65	63.7	75.9	73.7	82.2	65.0	75.7	71.0	80.2
1.56	1.60	62.4	75.0	72.6	81.4	63.7	74.8	69.9	79.4
1.51	1.55	61.0	74.0	71.5	80.6	62.3	73.8	68.7	78.6
1.46	1.50	59.4	73.0	70.3	79.8	60.9	72.8	67.4	77.7
1.41	1.45	57.8	71.9	69.1	**78.8**	59.3	71.6	66.1	76.8
1.36	1.40	56.0	70.7	67.7	77.8	57.6	70.4	64.6	75.8
1.31	1.35	54.1	69.4	66.3	76.8	55.8	69.2	63.1	**74.7**
1.26	1.30	52.1	68.1	64.7	75.6	53.9	67.8	61.4	73.6
1.21	1.25	49.9	66.6	63.0	74.4	51.8	66.3	59.6	72.3
1.18	1.20	48.0	65.3	61.6	73.3	50.0	65.0	58.0	71.3
1.15	1.17	46.5	**64.3**	60.4	72.5	48.5	64.0	**56.8**	70.4
1.12	1.14	44.9	63.2	59.2	71.6	47.0	62.9	55.5	69.5
1.09	1.11	43.2	62.1	57.9	70.7	45.4	61.7	54.1	68.5
1.06	1.08	41.4	60.9	**56.6**	69.7	43.7	60.5	52.6	67.5
1.03	1.05	39.5	59.7	55.1	68.6	41.9	**59.3**	51.1	66.5
1.00	1.02	**37.5**	58.3	**53.6**	67.5	40.1	**57.9**	49.4	65.4
0.97	0.99	35.4	56.9	52.0	66.3	38.1	56.5	47.7	64.2
0.94	0.96	33.1	55.4	50.2	65.1	35.9	55.0	45.9	62.9
0.91	0.93	30.7	53.8	48.4	63.8	33.7	53.4	43.9	61.6
0.88	0.90	28.2	52.1	46.4	62.3	**31.2**	51.6	41.8	60.1
0.85	0.87	25.4	50.3	44.3	60.8	28.7	49.8	39.5	58.6
0.82	0.84	22.5	48.4	42.1	59.2	25.9	47.8	37.1	57.0
0.79	0.81	19.3	46.3	39.6	57.4	22.9	45.7	34.5	55.2
0.76	0.78	15.9	44.0	37.0	55.5	19.7	43.4	31.7	53.3
0.73	0.75	12.2	41.6	34.2	53.5	16.3	40.9	28.7	51.3
0.70	0.72	8.1	39.0	31.1	51.2	12.5	38.2	25.4	49.0
0.67	0.69	3.8	36.1	27.8	48.8	8.5	35.3	21.9	46.6
0.64	0.66	1.0	33.0	24.1	46.2	4.0	32.1	18.0	44.0
0.61	0.63	1.0	29.5	20.1	43.3	1.0	28.6	13.7	41.1
0.58	0.60	1.0	25.7	15.6	40.1	1.0	24.8	9.0	37.9
0.55	0.57	1.0	21.6	10.7	36.6	1.0	20.5	3.7	34.4
0.52	0.54	1.0	16.9	5.2	32.6	1.0	15.8	1.0	30.6
0.49	0.51	1.0	11.7	1.0	28.2	1.0	10.5	1.0	26.2
0.46	0.48	1.0	5.9	1.0	23.3	1.0	4.5	1.0	21.3
0.43	0.45	1.0	1.0	1.0	17.6	1.0	1.0	1.0	15.8
0.40	0.42	1.0	1.0	1.0	11.1	1.0	1.0	1.0	9.5
0.37	0.39	1.0	1.0	1.0	3.7	1.0	1.0	1.0	2.2
0.34	0.36	1.0	1.0	1.0	1.0	1.0	1.0	1.0	1.0
0.00	0.33	1.0	1.0	1.0	1.0	1.0	1.0	1.0	1.0

*   To be used for appropriate special needs units if not under the special needs framework.
\#   All self-contained units meeting wheelchair design requirements or units for frail older people, but which are not under the special needs framework, use CAT2 (NB or RH) grant rates.

HOUSING FOR SALE GRANT RATES
APPENDIX 2

LOW COST HOME OWNERSHIP (LCHO) FACTORS

Sale programme	Priority Investment Areas	Other Areas
Shared ownership	58 %	50 %
Shared ownership for the elderly	68 %	60 %
Improvement for outright sale Shared ownership for the elderly	33 %	

HOUSING FOR SALE GRANT RATES
APPENDIX 3

1. HOMEBUY VALUE LIMITS AND ON-COST ALLOWANCE

TCI COST GROUP

	A	B	C	D	E
Homebuy value limits (£)					
- Homes with up to and including two bedrooms	130,900	110,300	89,200	67,700	58,800
- Homes with more than two bedrooms	164,000	137,900	110,50	82,900	71,500
On-cost allowance (%)					
- Applied to actual purchase price	3	3	3	3	3

2. VOLUNTARY PURCHASE GRANT (VPG) AND RIGHT TO ACQUIRE (RTA) DISCOUNT AMOUNTS

TCI COST GROUP

	A	B	C	D	E
- VPG and RTA discount amounts (£)	16,000	13,500	11,000	10,000	9,000

Note: The RTA discount amounts, and the local authorities they apply to, are set by Statutory Instrument (1997 No 626, The Housing (Right to Acquire Discount) Order), and may be varied by subsequent orders. The current intention is that RTA discount levels should continue to reflect those for VPG. Any discrepancies should be reported to the Housing Corportation

HOUSING FOR SALE GRANT RATES
APPENDIX 3

3. LOCAL AUTHORITY DO-IT-YOIURSELF SHARED OWNERSHIP (LA DIYSO) VALUE LIMITS, GRANT RATES AND ON-COST ALLOWANCE

	TCI COST GROUP				
	A	B	C	D	E
LA DIYSO value limits (£)					
- Homes with up to and including two bedrooms	130,900	110,300	89,200	67,700	58,800
- Homes with more than two bedrooms	164,000	137,900	110,50	82,900	71,500
Grant rates (%)					
- Applied to actual purchase price	72.7	68.9	63.0	56.7	52.1
On-cost allowance (%)					
- Applied to actual purchase price	5	5	5	5	5

Note: Value limits are as for Homebuy.
Total grant will be: (value of unsold equity x grant rate) +(total purchase price x on-cost allowance).
LADIYSO programme subject to review by the Department of the Environmnet, Transport and Regions (DETR).

The Housing Corporation registers, promotes, funds and supervises non-profit making housing associations. Its mission is to support social housing in England by working with housing associations and others to provide good homes for those in housing need.

In England, there are some 2,300 registered housing associations, providing half a million homes for rent for people in housing need. They also provide homes for sale through low cost home ownership. The Corporation is also responsible for helping tenants interested in Tenants' Choice, and for the approval and revocation of new landlords in relation to this scheme.

THE HOUSING CORPORATION'S OFFICES

London
Waverley House
7-12 Noel Street
London
W1V 4BA
Tel: (0207) 292 4400

South East
Leon House
High Street
Croydon
Surrey
CR9 1UH
Tel: (0208) 253 1400

South West
Beaufort House,
51 New North Road
Exeter
EX4 4EP
Tel: (01392) 428 200

East
Attenborough House
109/119 Charles Street
Leicester
LE1 1QF
Tel:(0116) 242 4800

West Midlands
31 Waterloo Road
Wolverhampton
WV1 4DJ
Tel: (01902) 795 000

North Eastern
St Paul's House
23 Park Square South
Leeds
LS1 2ND
Tel: (0113) 233 7100

North West and Merseyside
North West office
Elisabeth House
16 St. Peter's Square
Manchester
M2 3DF
Tel: (0161) 242 2000

Merseyside office
Colonial Chambers
3-11 Temple Street
Liverpool
L2 5RH
Tel: (0151) 242 1200

NEW FROM SPON PRESS

Housing Design Quality
THROUGH POLICY, GUIDANCE AND REVIEW

Matthew Carmona,
University College London, UK

Housing Design Quality directly addresses the major planning debate of our time - the delivery and quality of new housing development.

As pressure for new housing development in England increases, a widespread desire to improve the design of the resulting residential environments becomes ever more apparent with increasing condemnation of the standard products of the volume house builders. In recent years central government has come to accept the need to deliver higher quality living environments, and the important role of the planning system in helping to raise design standards. *Housing Design Quality* focuses on this role and in particular on how the various policy instruments available to public authorities can be used in a positive manner to deliver higher quality residential developments.

March 2001: 246x189: 368pp
90 b+w illustrations
Pb: 0-419-25650-4: £35.00

To Order: Tel: +44 (0) 8700 768853, or +44 (0) 1264 343071 Fax: +44 (0) 1264 343005, or
Post: Spon Press Customer Services, ITPS Andover, Hants, SP10 5BE, UK Email: book.orders@tandf.co.uk.

Postage & Packing: UK: 5% of order value (min. charge £1, max. charge £10) for 3-5 days delivery. Option of next day delivery at an additional £6.50. Europe: 10% of order value (min. charge £2.95, max. charge £20) for delivery surface post. Option of airmail at an additional £6.50. ROW: 15% of order value (min.charge£6.50, max. charge £30) for airmail delivery.

For a complete listing of all our titles visit: www.sponpress.com

Property Insurance

The problem of adequately covering by insurance the loss and damage caused to buildings by fire and other perils has been highlighted in recent years by the increasing rate of inflation.

There are a number of schemes available to the building owner wishing to insure his property against the usual risk. Traditionally the insured value must be sufficient to cover the actual cost of reinstating the building. This means that in addition to assessing the current value an estimate has also to be made of the increases likely to occur during the period of the policy and of rebuilding which, for a moderate size building, could amount to a total of three years. Obviously such an estimate is difficult to make with any degree of accuracy, if it is too low the insured may be penalized under the terms of the policy and if too high will result in the payment of unnecessary premiums.

There are variations on the traditional method of insuring which aim to reduce the effects of over estimating and details of these are available from the appropriate offices. For the convenience of readers who may wish to make use of the information contained in this publication in calculating insurance cover required the following may be of interest.

1 PRESENT COST

The current rebuilding costs may be ascertained in a number of ways:

(a) where the actual building cost is known this may be updated by reference to tender prices (page 692);
(b) by reference to average published prices per square metre of floor area (page 703). In this case it is important to understand clearly the method of measurement used to calculate the total floor area on which the rates have been based;
(c) by professional valuation;
(d) by comparison with the known cost of another similar building.

Whichever of these methods is adopted regard must be paid to any special conditions that may apply, i.e., a confined site, complexity of design, or any demolition and site clearance that may be required.

2 ALLOWANCE FOR INFLATION

The "Present Cost" when established will usually, under the conditions of the policy, be the rebuilding cost on the first day of the policy period. To this must be added a sum to cover future increases. For this purpose, using the historical indices on *pages ??? - ???*, as a base and taking account of the likely change in building costs and tender climate the following annual average indices are predicted for the future.

	Cost Index	Tender Index
1991	355	262
1992	365	243
1993	372	235
1994	382	252
1995	401	265
1996	411	267
1997	421	283
1998	439	313
1999	460	332
2000	489 (P)	359
2001	508 (F)	378 (F)
2002	536 (F)	395 (F)

	Cost Index	Tender Index
2003	562 (F)	411 (F)
2004	579 (F)	423 (F)

3 FEES

To the total of 1 and 2 above must be added an allowance for fees.

4 VALUE ADDED TAX (V.A.T.)

To the total of 1 to 3 above must be added Value Added Tax. Historically, relief may have been given to total reconstruction following fire damage etc. Since the 1989 Finance Act, such work, except for self-contained dwellings and other residential buildings and certain non-business charity buildings, has attracted V.A.T. and the limit of insurance cover should be raised to allow for this.

5 EXAMPLE

An assessment for insurance cover is required in the fourth quarter of 2000 for a property which cost £200,000 when completed in 1976.

Present Cost
Known cost at mid 1976 £ 200,000.00

Predicted tender index fourth quarter 2001 = 383
Tender index fourth quarter 1976 = 100
Increase in tender index = 283%
applied to known cost = £ 566,000.00
Present cost (excluding any allowance for demolition) £ 766,000.00

Allowance for inflation
Present cost at day one of policy £ 766,000.00
Allow for changes in tender levels during 12 month currency of policy

Predicted tender index fourth quarter 2002 = 401
Predicted tender index fourth quarter 2001 = 383
Increase in tender index = 4.70%
applied to present cost = say £ 36,000.00
Anticipated cost at expiry of policy £ 802,000.00

Assuming that total damage is suffered on the last day of the currency of the policy and that planning and documentation would require a period of twelve months before re-building could commence, then a further similar allowance must be made.

Predicted tender index fourth quarter 2003 = 416
Predicted tender index fourth quarter 2002 = 401
Increase in tender index = 3.74%
applied to cost at expiry of policy = say £ 30,000.00
Anticipated cost at tender date £ 832,000.00

Assuming that reconstruction would take one year, allowance must be made for the increases in costs which would directly or indirectly be met under a building contract.

Predicted cost index fourth quarter 2004 = 587
Predicted cost index fourth quarter 2003 = 570
Increase in cost index = 2.98%
This is the total increase at the end of the one year period.
The amount applicable to the contract would be about half, say £ 12,400.00
Estimated cost of reinstatement £ 844,400.00

SUMMARY OF EXAMPLE

Estimated cost of reinstatement	£	844,400.00
Add professional fees at, say 16%	£	135,100.00
	s/t £	979,500.00
Add for V.A.T., currently at 17½% say	£	171,400.00
Total insurance cover required	£	**1,150,900.00**

NEW FROM SPON PRESS

The Architectural Expression of Environmental Control Systems

George Baird, Victoria University of Wellington, New Zealand

The Architectural Expression of Environmental Control Systems examines the way project teams can approach the design and expression of both active and passive thermal environmental control systems in a more creative way. Using seminal case studies from around the world and interviews with the architects and environmental engineers involved, the book illustrates innovative responses to client, site and user requirements, focusing upon elegant design solutions to a perennial problem.

This book will inspire architects, building scientists and building services engineers to take a more creative approach to the design and expression of environmental control systems - whether active or passive, whether they influence overall building form or design detail.

March 2001: 276x219: 304pp
135 b+w photos, 40 colour, 90 line illustrations
Hb: 0-419-24430-1: £49.95

To Order: Tel: +44 (0) 8700 768853, or +44 (0) 1264 343071 Fax: +44 (0) 1264 343005, or
Post: Spon Press Customer Services, ITPS Andover, Hants, SP10 5BE, UK Email: book.orders@tandf.co.uk.

Postage & Packing: UK: 5% of order value (min. charge £1, max. charge £10) for 3-5 days delivery. Option of next day delivery at an additional £6.50. Europe: 10% of order value (min. charge £2.95, max. charge £20) for delivery surface post. Option of airmail at an additional £6.50. ROW: 15% of order value (min.charge£6.50, max. charge £30) for airmail delivery.

For a complete listing of all our titles visit: www.sponpress.com

Capital Allowances

What are Capital Allowances ?

Capital Allowances provide tax relief by prescribing a statutory rate of depreciation for tax purposes in place of that used for accounting purposes. They are utilised by government to provide an incentive to invest in capital equipment, including commercial property, by allowing the majority of taxpayers a deduction from taxable profits for certain types of capital expenditure, thereby deferring tax liabilities.

The two commonest types of capital allowances applicable to real estate are those given for capital expenditure on both new and existing industrial buildings, and plant and machinery in all commercial buildings.

Other types of allowances relevant to property are hotel allowances and enterprise zone allowances, which are in fact variants of industrial building allowances. Enhanced rates of allowance are available on certain types of energy saving plant and machinery, whilst reduced rates apply to items with an expected life of more than 25 years.

Plant and Machinery Allowances

Plant and machinery allowances are the most common form of capital allowances in that they are available on virtually every type of commercial building. The legal definition of plant is somewhat imprecise but would typically include items such as heating and ventilation installations, lifts, fire alarms, emergency lighting, sanitary fittings and carpets. A statutory definition of what cannot qualify is included in the Act but there is no definition of what can qualify.

Plant and machinery allowances are available at 25% per annum on a reducing balance basis. For every £100 of qualifying expenditure, £25 is claimable in year 1, £18.75 in year 2 and son on until either all the allowances are claimed or the building sold.

The real value of these allowances is the product of the allowances and the claimant's tax rate.

Industrial Building Allowances

Industrial Building Allowances are available on the entire cost of construction of a building that is used for a qualifying purpose. It is the use of the building that is important. What constitutes a qualifying use is defined in the Capital Allowances Act 2001, and includes such uses as manufacturing, certain types of storage and transport undertakings.

The person who incurs the original construction expenditure can claim the allowances over 25 years (annual allowance = 4% of building cost) from the date of first use. However, a subsequent purchaser can, subject to certain caveats, claim the same value of allowances over the number of years remaining to the 25th anniversary of the date of first use. So, if an investor purchases a 20-year old building then the allowances can be claimed over 5 years (annual allowance = 20% of original building cost). The closer a building gets to its 25th anniversary the more valuable the annual allowance becomes.

Who Can Claim ?

Primarily, the claimant must be liable to income or corporation tax on the profits arising from his or her business activity. The statute does, however, require further criteria to be fulfilled. To qualify for plant and machinery that is owned by him and is used for the purposes of a qualifying activity.

In property terms, claimants can be property investors, owner-occupiers or tenants who have incurred capital expenditure on the acquisition, construction, refurbishment or fitting out of their premises for the purposes of their business.

When a property is acquired it is important that the correct interest has been purchased. The allowances attach to the title upon which the original expenditure was incurred. So, if the property was developed on the freehold then the freehold is the "relevant interest" for the purposes of claiming allowances. The purchase of a lesser interest would mean that the allowances might not be available. A buyer should always be suspicious when offered a property by way of the creation of a new 999-year lease. This probably means that the seller wants to "strip-out" or retain the allowances.

Property traders who buy or build a property with the intention of selling it at a profit are not entitled to claim capital allowances, because their costs are a trading expense rather than capital expenditure.

Is There Time Limit On Claiming These Allowances ?

There is no restriction preventing claims on capital expenditure incurred several years ago. It is, in theory, possible to make a claim on any property you still own regardless of when the expenditure was originally incurred. There is, however, a restriction on when you can start claiming the allowances. Under self-assessment a retrospective claim can effectively only be made in the current year and two preceding accounting periods as long as the asset was owned in those years. The rules vary slightly for individuals and offshore investors paying income tax.

For more information contact:

NBW Crosher & James
Help Line: 0800 526 262
London: 020 7845 0600
Birmingham: 0121 632 3600
Edinburgh: 0131 220 4225
www.nbwcrosherjames.com

PART V

Tables and Memoranda

This part of the book contains the following sections:

Conversion Tables, *page* 885
Formulae, *page* 886
Design Loadings for Buildings, *page* 888
Planning Parameters, *page* 895
Sound Insulation, *page* 913
Thermal Insulation, *page* 914
Weights of Various Materials, *page* 918
Memoranda for each Trade, *page* 919
Useful Addresses for Further Information, *page* 956
Natural Sines, *page* 971
Natural Cosines, *page* 973
Natural Tangents, *page* 975

DAVIS LANGDON & EVEREST
Authors of Spon's Price Books

www.davislangdon.com

Davis Langdon & Everest is an independent practice of Chartered Quantity Surveyors, with some 1000 staff in 20 UK offices and, through Davis Langdon & Seah International, some 2,500 staff in 85 offices worldwide.

DLE manages client requirements, controls risk, manages cost and maximises value for money, throughout the course of construction projects, always aiming to be - and to deliver - the best.

TYPICAL PROJECT STAGES, DLE INTEGRATED SERVICES AND THEIR EFFECT:

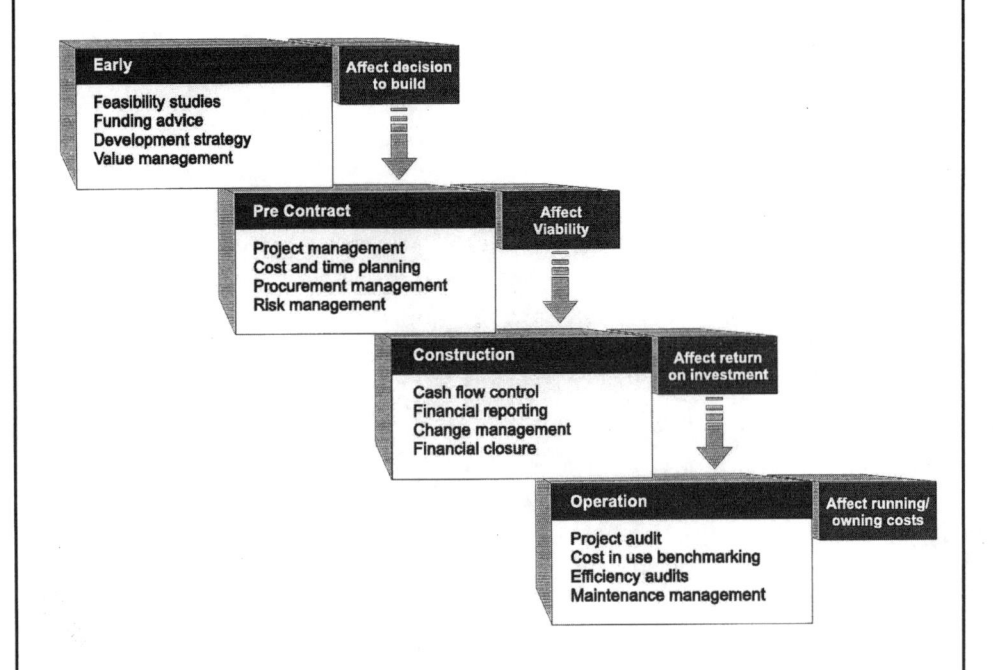

Early
Feasibility studies
Funding advice
Development strategy
Value management
→ Affect decision to build

Pre Contract
Project management
Cost and time planning
Procurement management
Risk management
→ Affect Viability

Construction
Cash flow control
Financial reporting
Change management
Financial closure
→ Affect return on investment

Operation
Project audit
Cost in use benchmarking
Efficiency audits
Maintenance management
→ Affect running/owning costs

EUROPE ⇨ ASIA - AUSTRALIA - AFRICA - AMERICA
GLOBAL REACH - LOCAL DELIVERY

Tables and Memoranda

CONVERSION TABLES

	Unit	Conversion factors			
Length					
Millimetre	mm	1 in	= 25.4 mm	1 mm	= 0.0394 in
Centimetre	cm	1 in	= 2.54 cm	1 cm	= 0.3937 in
Metre	m	1 ft	= 0.3048 m	1 m	= 3.2808 ft
		1 yd	= 0.9144 m		= 1.0936 yd
Kilometre	km	1 mile	= 1.6093 km	1km	= 0.6214 mile

Note: 1 cm = 10 mm   1 ft = 12 in
       1 m = 1 000 mm   1 yd = 3 ft
       1 km = 1 000 m   1 mile = 1 760 yd

Area

Square Millimetre	mm^2	$1\ in^2$	$= 645.2\ mm^2$	$1\ mm^2$	$= 0.0016\ in^2$
Square Centimetre	cm^2	$1\ in^2$	$= 6.4516\ cm^2$	$1\ cm^2$	$= 1.1550\ in^2$
Square Metre	m^2	$1\ ft^2$	$= 0.0929\ m^2$	$1\ m^2$	$= 10.764\ ft^2$
		$1\ yd^2$	$= 0.8361\ m^2$	$1\ m^2$	$= 1.1960\ yd^2$
Square Kilometre	km^2	$1\ mile^2$	$= 2.590\ km^2$	$1\ km^2$	$= 0.3861\ mile^2$

Note: $1\ cm^2 = 100\ mm^2$ $1\ ft^2 = 144\ in^2$
       $1\ m^2 = 10\ 000\ cm^2$ $1\ yd^2 = 9\ ft^2$
       $1\ km^2 = 100$ hectares   1 acre $= 4\ 840\ yd^2$
          $1\ mile^2 = 640$ acres

Volume

Cubic Centimetre	cm^3	$1\ cm^3$	$= 0.0610\ in^3$	$1\ in^3$	$= 16.387\ cm^3$
Cubic Decimetre	dm^3	$1\ dm^3$	$= 0.0353\ ft^3$	$1\ ft^3$	$= 28.329\ dm^3$
Cubic Metre	m^3	$1\ m^3$	$= 35.3147\ ft^3$	$1\ ft^3$	$= 0.0283\ m^3$
		$1\ m^3$	$= 1.3080\ yd^3$	$1\ yd^3$	$= 0.7646\ m^3$
Litre	l	1 l	= 1.76 pint	1 pint	= 0.5683 l
			= 2.113 US pt		= 0.4733 US l

Note: $1\ dm^3 = 1\ 000\ cm^3$ $1\ ft^3 = 1\ 728\ in^3$ 1 pint = 20 fl oz
       $1\ m^3 = 1\ 000\ dm^3$ $1\ yd^3 = 27\ ft^3$ 1 gal = 8 pints
       1 l = 1 dm^3

Neither the Centimetre nor Decimetre are SI units, and as such their use, particularly that of the Decimetre, is not widespread outside educational circles.

Mass

Milligram	mg	1 mg	= 0.0154 grain	1 grain	= 64.935 mg
Gram	g	1 g	= 0.0353 oz	1 oz	= 28.35 g
Kilogram	kg	1 kg	= 2.2046 lb	1 lb	= 0.4536 kg
Tonne	t	1 t	= 0.9842 ton	1 ton	= 1.016 t

Note: 1 g = 1000 mg   1 oz = 437.5 grains   1 cwt = 112 lb
       1 kg = 1000 g   1 lb = 16 oz   1 ton = 20 cwt
       1 t = 1000 kg   1 stone = 14 lb

CONVERSION TABLES

Unit		Conversion factors		

Force

Newton	N	1 lbf	= 4.448 N	1 kgf	= 9.807 N
Kilonewton	kN	1 lbf	= 0.004448 kN	1 ton f	= 9.964 kN
Meganewton	MN	100 tonf	= 0.9964 MN		

Pressure and stress

Kilonewton per square metre	kN/m^2	$1\ lbf/in^2$	$= 6.895\ kN/m^2$
		1 bar	$= 100\ kN/m^2$
Meganewton per square metre	MN/m^2	$1\ tonf/ft^2$	$= 107.3\ kN/m^2 = 0.1073\ MN/m^2$
		$1\ kgf/cm^2$	$= 98.07\ kN/m^2$
		$1\ lbf/ft^2$	$= 0.04788\ kN/m^2$

Coefficient of consolidation (Cv) or swelling

Square metre per year	m^2/year	$1\ cm^2/s$	$= 3\ 154\ m^2$/year
		$1\ ft^2$/year	$= 0.0929\ m^2$/year

Coefficient of permeability

Metre per second	m/s	1 cm/s	= 0.01 m/s
Metre per year	m/year	1 ft/year	= 0.3048 m/year
			$= 0.9651 \times (10)^8$ m/s

Temperature

Degree Celsius °C $°C = \frac{5}{9} \times (°F - 32)$ $°F = \frac{9 \times °C}{5} + 32$

FORMULAE

Two dimensional figures

Figure	Area
Square	$(side)^2$
Rectangle	Length x breadth
Triangle	½ (base x height)
	or $\sqrt{(s(s-a)(s-b)(s-c))}$ where a, b and c are the lengths of the three sides, and $s = \frac{a+b+c}{2}$
	or $a^2 = b^2 + c^2 - (2b\ c\cos A)$ where A is the angle opposite side a
Hexagon	$2.6 \times (side)^2$
Octagon	$4.83 \times (side)^2$
Trapezoid	height x ½ (base + top)
Circle	$3.142 \times radius^2$ or $0.7854 \times diameter^2$ (circumference = $2 \times 3.142 \times radius$ or $3.142 \times diameter$)
Sector of a circle	½ x length of arc x radius
Segment of a circle	area of sector - area of triangle

FORMULAE

Two dimensional figures

Figure	Area
Ellipse	$3.142 \times AB$ (where $A = \frac{1}{2} \times$ height and $B = \frac{1}{2} \times$ length)
Bellmouth	$\frac{3}{14} \times \text{radius}^2$

Three dimensional figures

Figure	Volume	Surface Area
Prism	Area of base $\times$ height	circumference of base $\times$ height
Cube	$(\text{side})^3$	$6 \times (\text{side})^2$
Cylinder	$3.142 \times \text{radius}^2 \times \text{height}$	$2 \times 3.142 \times \text{radius} \times (\text{height} - \text{radius})$
Sphere	$\frac{4}{3} \times 3.142 \times \text{radius}^3$	$4 \times 3.142 \times \text{radius}^2$
Segment of a sphere	$\frac{(3.142 \times h) \times (3 \times r^2 + h^2)}{6}$	$2 \times 3.142 \times r \times h$
Pyramid	$\frac{1}{3}$ of area of base $\times$ height	$\frac{1}{2} \times$ circumference of base $\times$ slant height
Cone	$\frac{1}{3} \times 3.142 \times \text{radius}^2 \times h$	$3.142 \times \text{radius} \times \text{slant height}$
Frustrum of a pyramid	$\frac{1}{3} \times \text{height} [A + B + \sqrt{(AB)}]$ where A is the area of the large end and B is the area of the small end	$\frac{1}{2} \times$ mean circumference $\times$ slant height
Frustrum of a cone	$(\frac{1}{3} \times 3.142 \times \text{height} (R^2 + r^2 + R \times r))$ where R is the radius of the large end and r is the radius of the small end	$3.142 \times \text{slant height} \times (R + r)$

Other formulae

Formula	Description
Pythagoras' theorum	$A^2 = B^2 + C^2$ where A is the hypotenuse of a right-angled triangle and B and C are the two adjacent sides
Simpson's Rule	Volume $= \frac{x}{3} [(y_1 + y_n) + 2(y_3 + y_5) + 4(y_2 + y_4)]$

The volume to be measured must be represented by an odd number of cross-sections ($y_1, -y_n$) taken at fixed intervals (x), the sum of the areas at even numbered intermediate cross-sections (y_2, y_4, etc.) is multiplied by 4 and the sum of the areas at odd numbered intermediate cross-sections (y_3, y_5, etc.) is multiplied by 2, and the end cross-sections (y_1 and y_n) taken once only. The resulting *weighted average* of these areas is multiplied by a of the distance between the cross-sections (x) to give the total volume.

FORMULAE

Other formulae

Formula	Description
Trapezoidal Rule	(0.16 x [Total length of trench] x [area of first section x 4 times area of middle section + area of last section])

Note: Both Simpson's Rule and Trapezoidal Rule are useful in accurately calculating the volume of an irregular trench, or similar longitudinal earthworks movement, e.g. road construction.

DESIGN LOADINGS FOR BUILDINGS

Note: Refer to BS 6399: Part 1: 1996 Code of Practice for Dead and Imposed Loads for minimum loading examples.

Definitions

Dead load: The load due to the weight of all walls, permanent partitions, floors, roofs and finishes, including services and all other permanent construction.

Imposed load: The load assumed to be produced by the intended occupancy or use, including the weight of moveable partitions, distributed, concentrated, impact, inertia and snow loads, but excluding wind loads.

Distributed load: The uniformly distributed static loads per square metre of plan area which provide for the effects of normal use. Where no values are given for concentrated load it may be assumed that the tabulated distributed load is adequate for design purposes.

Note: The general recommendations are not applicable to certain atypical usages particularly where mechanical stacking, plant or machinery are to be installed and in these cases the designer should determine the loads from a knowledge of the equipment and processes likely to be employed.

The additional imposed load to provide for partitions, where their positions are not shown on the plans, on beams and floors, where these are capable of effective lateral distributional of the load, is a uniformly distributed load per square metre of not less than one-third of the weight per metre run by the partitions but not less than 1 kN/m^2.

Floor area usage	Distributed load kN/m^2	Concentrated load kN/300 mm^2
Industrial occupancy class (workshops, factories)		
Foundries	20.0	-
Cold storage	5.0 for each metre of storage height with a minimum of 15.0	9.0
Paper storage, for printing plants	4.0 for each metre of storage height	9.0
Storage, other than types listed separately	2.4 for each metre of storage height	7.0
Type storage and other areas in printing plants	12.5	9.0
Boiler rooms, motor rooms, fan rooms and the like, including the weight of machinery	7.5	4.5

DESIGN LOADINGS FOR BUILDINGS

Floor area usage	Distributed load kN/m²	Concentrated load kN/300 mm²
Industrial occupancy class (workshops, factories) - cont'd		
Factories, workshops and similar buildings	5.0	4.5
Corridors, hallways, foot bridges, etc. subject to loads greater than for crowds, such as wheeled vehicles, trolleys and the like	5.0	4.5
Corridors, hallways, stairs, landings, footbridges, etc.	4.0	4.5
Machinery halls, circulation spaces therein	4.0	4.5
Laboratories (including equipment), kitchens, laundries	3.0	4.5
Workrooms, light without storage	2.5	1.8
Toilet rooms	2.0	-
Cat walks	-	1.0 at 1 m centres
Institutional and educational occupancy class (prisons, hospitals, schools, colleges)		
Dense mobile stacking (books) on mobile trolleys	4.8 for each metre of stack height but with a minimum of 9.6	7.0
Stack rooms (books)	2.4 for each metre of stack height but with a minimum of 6.5	7.0
Stationery stores	4.0 for each metre of storage height	9.0
Boiler rooms, motor rooms, fan rooms and the like, including the weight of machinery	7.5	4.5
Corridors, hallways, etc. subject to loads greater than from crowds, such as wheeled vehicles, trolleys and the like	5.0	4.5
Drill rooms and drill halls	5.0	9.0
Assembly areas without fixed seating, stages gymnasia	5.0	3.6
Bars	5.0	-
Projection rooms	5.0	-

DESIGN LOADINGS FOR BUILDINGS

Floor area usage	Distributed load kN/m²	Concentrated load kN/300 mm²
Institutional and educational occupancy class (prisons, hospitals, schools, colleges) - cont'd		
Corridors, hallways, aisles, stairs, landings, foot-bridges, etc.	4.0	4.5
Reading rooms with book storage, e.g. libraries	4.0	4.5
Assembly areas with fixed seating	4.0	-
Laboratories (including equipment), kitchens, laundries	3.0	4.5
Corridors, hallways, aisles, landings, stairs, etc. not subject to crowd loading	3.0	2.7
Classrooms, chapels	3.0	2.7
Reading rooms without book storage	2.5	4.5
Areas for equipment	2.0	1.8
X-ray rooms, operating rooms, utility rooms	2.0	4.5
Dining rooms, lounges, billiard rooms	2.0	2.7
Dressing rooms, hospital bedrooms and wards	2.0	1.8
Toilet rooms	2.0	-
Bedrooms, dormitories	1.5	1.8
Balconies	same as rooms to which they give access but with a minimum of 4.0	1.5 per metre run concentrated at the outer edge
Fly galleries	4.5 kN per metre run distributed uniformly over the width	-
Cat walks	-	1.0 at 1 m centres
Offices occupancy class (offices, banks)		
Stationery stores	4.0 for each metre of storage height	9.0
Boiler rooms, motor rooms, fan rooms and the like, including the weight of machinery	7.5	4.5

DESIGN LOADINGS FOR BUILDINGS

Floor area usage	Distributed load kN/m²	Concentrated load kN/300 mm²
Offices occupancy class (offices, banks) - cont'd		
Corridors, hallways, etc. subject to loads greater than from crowds, such as wheeled vehicles, trolleys and the like	5.0	4.5
File rooms, filing and storage space	5.0	4.5
Corridors, hallways, stairs, landings, footbridges, etc.	4.0	4.5
Offices with fixed computers or similar equipment	3.5	4.5
Laboratories (including equipment), kitchens, laundries	3.0	-
Banking halls	3.0	4.5
Offices for general use	2.5	2.7
Toilet rooms	2.0	-
Balconies	Same as rooms to which they give access but with a minimum of 4.0	1.5 per metre run concentrated at the outer edge
Cat walks	-	1.0 at 1 m centre
Public assembly occupancy class (halls, auditoria, restaurants, museums, libraries, non-residential clubs, theatres, broadcasting studios, grandstands)		
Dense mobile stacking (books) on mobile trucks	4.8 for each metre of stack height but with a minimum of 9.6	7.0
Stack rooms (books)	2.4 for each metre of stack height but with a minimum of 6.5	7.0
Boiler rooms, motor rooms fan rooms and the like, including the weight of machinery	7.5	4.5
Stages	7.5	4.5
Corridors, hallways, etc. subject to loads greater than from crowds, such as wheeled vehicles, trolleys and the like. Corridors, stairs, and passage ways in grandstands	5.0	4.5
Drill rooms and drill halls	5.0	9.0
Assembly areas without fixed seating dance halls, gymnasia, grandstands	5.0	3.6
Projection rooms, bars	5.0	-

DESIGN LOADINGS FOR BUILDINGS

Floor area usage	Distributed load kN/m²	Concentrated load kN/300 mm²
Public assembly occupancy class (halls, auditoria, restaurants, museums, libraries, non-residential clubs, theatres, broadcasting studios, grandstands) - cont'd		
Museum floors and art galleries for exhibition purposes	4.0	4.5
Corridors, hallways, stairs, landings, footbridges, etc.	4.0	4.5
Reading rooms with book storage, e.g. libraries	4.0	4.5
Assembly areas with fixed seating	4.0	-
Kitchens, laundries	3.0	4.5
Chapels, churches	3.0	2.7
Reading rooms without book storage	2.5	4.5
Grids	2.5	-
Areas for equipment	2.0	1.8
Dining rooms, lounges, billiard rooms	2.0	2.7
Dressing rooms	2.0	1.8
Toilet rooms	2.0	-
Balconies	Same as rooms to which they give access but with a minimun of 4.0	1.5 per metre run concentrated at the outer edge
Fly galleries	4.5 kN per metre run distributed uniformly over the width	
Cat walks	-	1.0 at 1 m centres
Residential occupancy class		
Self contained dwelling units and communal areas in blocks of flats not more than three storeys in height and with not more than four self-contained dwelling units per floor accessible from one staircase		
All usages	1.5	1.4
Boarding houses, lodging houses, guest houses, hostels, residential clubs and communal areas in blocks of flats other than type 1		
Boiler rooms, motor rooms, fan rooms and the like including the weight of machinery	7.5	4.5

DESIGN LOADINGS FOR BUILDINGS

Floor area usage	Distributed load kN/m²	Concentrated load kN/300 mm²
Residential occupancy class - cont'd		
Boarding houses, lodging houses, guest houses, hostels, residential clubs and communal areas in blocks of flats other than type 1 - cont'd		
Communal kitchens, laundries	3.0	4.5
Corridors, hallways, stairs, landings, footbridges etc.	3.0	4.5
Dining rooms, lounges, billiard rooms	2.0	2.7
Toilet rooms	2.0	-
Bedrooms, dormitories	1.5	1.8
Balconies	Same as rooms to which they give access but with a minimum of 3.0	1.5 per metre run concentrated at the outer edge
Cat walks	-	1.0 at 1 m centres
Hotels and Motels		
Boiler rooms, motor rooms, fan rooms and the like, including the weight of machinery	7.5	4.5
Assembly areas without fixed seating, dance halls	5.0	3.6
Bars	5.0	-
Assembly areas with fixed seating	4.0	-
Corridors, hallways, stairs, landings, footbridges, etc.	4.0	4.5
Kitchens, laundries	3.0	4.5
Dining rooms, lounges, billiard rooms	2.0	2.7
Bedrooms	2.0	1.8
Toilet rooms	2.0	-
Balconies	Same as rooms to which they give access but with a minimum of 4.0	1.5 per metre run concentrated at the outer edge
Cat Walks	-	1.0 at 1 m centres
Retail occupancy class (shops, departmental stores, supermarkets)		
Cold storage	5.0 for each metre of storage height with a minimum of 15.0	9.0
Stationery stores	4.0 for each metre of storage height	9.0

DESIGN LOADINGS FOR BUILDINGS

Floor area usage	Distributed load kN/m²	Concentrated load kN/300 mm²
Retail occupancy class (shops, departmental stores, supermarkets) - cont'd		
Storage, other than types separately	2.4 for each metre of storage height	7.0
Boiler rooms, motor rooms, fan rooms and the like, including the weight of machinery	7.5	4.5
Corridors, hallways, etc. subject to loads greater than from crowds, such as wheeled vehicles, trolleys and the like	5.0	4.5
Corridors, hallways, stairs, landings, footbridges, etc.	4.0	4.5
Shop floors for the display and sale of merchandise	4.0	3.6
Kitchens, laundries	3.0	4.5
Toilet rooms	2.0	-
Balconies	Same as rooms to which they give access but with a minimum of 4.0	1.5 per metre run concentrated at the outer edge
Cat walks	-	1.0 at 1 m centres
Storage occupancy class (warehouses)		
Cold storage	5.0 for each metre of storage height with a minimum of 15.0	9.0
Dense mobile stacking (books) on mobile trucks	4.8 for each metre of storage height with a minimum of 15.0	7.0
Paper storage, for printing plants	4.0 for each metre of storage height	9.0
Stationery stores	4.0 for each metre of storage height	9.0
Storage, other than types listed separately, warehouses	2.4 for each metre of storage height	7.0
Motor rooms, fan rooms and the like, including the weight of machinery	7.5	4.5
Corridors, hallways, footbridges, etc. subject to loads greater than for crowds, such as wheeled vehicles, trolleys and the like	5.0	4.5
Cat walks	-	1.0 at 1 m centres

DESIGN LOADINGS FOR BUILDINGS

Floor area usage	Distributed load kN/m²	Concentrated load kN/300 mm²
Vehicular occupancy class (garages, car parks, vehicle access ramps)		
Motor rooms, fan rooms and the like, including the weight of machinery	7.5	4.5
Driveways and vehicle ramps, other than in garages for the parking only of passenger vehicles and light vans not exceeding 2500 kg gross mass	5.0	9.0
Repair workshops for all types of vehicles, parking for vehicles exceeding 2500 kg gross mass including driveways and ramps	5.0	9.0
Footpaths, terraces and plazas leading from ground level with no obstruction to vehicular traffic, pavement lights	5.0	9.0
Corridors, hallways, stairs, landings, footbridges, etc. subject to crowd loading	4.0	4.5
Footpaths, terraces and plazas leading from ground level but restricted to pedestrian traffic only	4.0	4.5
Car parking only, for passenger vehicles and light vans not exceeding 2500 kg gross mass including garages, driveways and ramps	2.5	9.0
Cat walks	-	1.0 at 1 m centres

PLANNING PARAMETERS

Definitions

*   For precise definitions consult the Code of Measuring Practice published by the Royal Institution of Chartered Surveyors and the Incorporated Society of valuers and Auctioneers.

General definitions

Plot ratio *
Ratio of GEA to site area where the site area is expressed as one.

Gross external area (GEA) *
Gross area on each floor including the external walls of all spaces except open balconies and fire escapes, upper levels of atria and areas less than 1.5 m (5ft) such as under roof slopes, open covered ways or minor canopies, open vehicle parking areas, terraces and party walls beyond the centre line. Measured over structural elements and services space such as partitions and plant rooms. Roof level plant rooms may be excluded from the planning area

Site area *
Total area of the site within the site title boundaries measured on the horizontal plane.

PLANNING PARAMETERS

General Definitions - cont'd

Gross site area *
The site area, plus any area of adjoining roads enclosed by extending the side boundaries of the site up to the centre of the road, or to 6 m (20 ft) out from the frontage, whatever is the less.

Gross internal floor area (GIFA) *
Gross area measured on the same basis as GEA, but excluding external wall thickness.

Net internal floor area (NIFA) *
Net usable area measured to the internal finish of the external walls excluding all auxiliary and ancillary spaces such as WC's and lobbies, ducts, lift, tank and plant space etc, staircases, lift wells and major access circulation, fire escape corridors and lobbies, major switchroom space and areas used by external authorities, internal structural walls and columns, car parking and areas with less than 1.5 m headroom, such as under roof slopes, corridors used in common with other occupiers or of a permanent essential nature such as fire corridors, smoke lobbies, space occupied by permanent air-conditioning, heating or cooling apparatus and surface mounted ducting causing space to be unusable.

Cubic content *
The GEA multiplied by the vertical height from the lowest basement floor or average ground to the average height of the roof.

Internal cube
The GIFA of each floor multiplied by its storey height.

Ceiling height *
The height between the floor surface and the underside of the ceiling.

Building frontage *
The measurement along the front of the building from the outside of the external walls or the centre line of party walls.

External wall area
The wall area of all the enclosed spaces fulfilling the functional requirements of the buildings measured on the outer face of the external walls and overall windows and doors etc.

Wall to floor ratio
The factor produced by dividing the external wall area by the GIFA.

Window to external wall ratio
The factor produced by dividing the external windows and door area by the external wall area.

Circulation (C)
Circulation and ancillary area measured on plan on each floor for staircases, lift lobbies, lift wells, lavatories, cleaners' cupboards usually represented as the allowances for circulation and ancillary space as a percentage of NIFA.

Plant area
Plant rooms and vertical duct space.

Retail definitions

Sales area *
NIFA usable for retailing excluding store rooms unless formed by non-structural partitions.

Storage area *
NIFA not forming part of the sales area and usable only for storage.

Shop frontage *
Overall external frontage to shop premises including entrance and return shop frontage, but excluding recesses, doorways and the like of other accommodation.

Overall frontage *
Overall measurement in a straight line across the front of the building and any return frontage, from the outside of external walls and / or the entire line or party walls.

PLANNING PARAMETERS

Retail definitions - cont'd

Shop width *
Internal measurement between inside faces of external walls at shop front or other points of reference.

Shop depth *
Overall measurement from back of pavement or forecourt to back of sales area measured over any non-structural partitions.

Built depth *
Overall external ground level measurement from front to rear walls of building.

Zone A
Front zone of 6 m in standard retail units 6 m x 24 m.

Housing definitions

Number of persons housed
The total number for whom actual bed spaces are provided in the dwellings as designed.

Average number of persons per dwelling
The total number of persons housed divided by the total number of dwellings.

Density
The total number of persons housed divided by the site in hectares or acres.
The total number of units divided by the site area in hectares or acres.

Functional units

As a "rule of thumb" guide to establish a cost per functional unit, or as a check on economy of design in terms of floor area, the following indicative functional unit areas have been derived from historical data. For indicative unit costs see "Building Prices per Functional Units" (Part IV - Approximate Estimating) on page 677.

Car parking	- surface	20 - 22 m^2/car
	- multi storey	23 - 27 m^2/car
	- basement	28 - 37 m^2/car
Concert Halls		8 m^2/seat
Halls of residence	- college/polytechnic	25 - 35 m^2/bedroom
	- university	30 - 50 m^2/bedroom
Hospitals	- district general	65 - 85 m^2/bed
	- teaching	120 + m^2/bed
	- private	75 - 100 m^2/bed
Hotels	- budget	28 - 35 m^2/bedroom
	- luxury city centre	70 - 130 m^2/bedroom
Housing		**Gross internal floor area**
Private developer:	1 Bedroom Flat	45 - 50 m^2
	2 Bedroom Flat	55 - 65 m^2
	2 Bedroom House	55 - 65 m^2
	3 Bedroom House	70 - 90 m^2
	4 Bedroom House	90 - 100 m^2
Offices	- high density open plan	20 m^2/person
	- low density cellular	15 m^2/person

PLANNING PARAMETERS

Functional units - cont'd

Schools
- nursery — 3 - 5 m^2/child
- secondary — 6 - 10 m^2/child
- boarding — 10 -12 m^2/child

Theatres
- small, local — 3 m^2/seat to
- large, prestige — 7 m^2/seat

Typical planning parameters

The following are indicative planning design and functional criteria derived from historical data for a number of major building types.

Gross internal floor areas (GIFA)

Offices

Feasibility assessment of GIFA for:

Curtain wall office	GEA x 0.97
Solid wall office	GEA x 0.95

These measures apply except for thick stone façades - take measurements on site.

Typical dimensions measured on plan between the internal finishes of the external walls for:

Speculative offices	13.75 m
Open plan offices	15.25 m
Open plan / cellular offices	18.3 m

Retail

Typical gross internal floor areas:

Food courts, comprising	232 to 372 m^2
Kiosks	37 m^2
Services - per seat	1.1 to 1.5 m^2
Seating area in mall - per seat	1.2 to 1.7 m^2
Retail Kiosks	56 to 75 m^2
Small specialist shops	465 to 930 m^2
Electrical goods	930 to 1 395 m^2
DIY	930 to 4 645 m^2
Furniture / carpets	1858 to 5 575 m^2
Toys	3715 to 4 645 m^2
Superstores	3715 to 5 575 m^2
Department stores within shopping centres	5575 to 27 870 m^2
Specialist shopping centres	5574 to 9 290 m^2

Leisure

Standard sizes:

Large sports halls	Medium sports halls	Small sports halls
36.5 x 32 x 9.1 m	29 x 26 x 7.6 - 9.1 m	29.5 x 16.5 x 6.7 - 7.6 m
32 x 26 x 7.6 - 9.1 m	32 x 23 x 7.6 - 9.1 m	26 x 16.5 x 6.7 - 7.6 m
	32 x 17 x 6.7 - 7.6 m	22.5 x 16.5 x 6.7 - 7.6 m

Community halls
17.2 x 15.6 x 6.7 m
17 x 8.5 x 6.7 m

PLANNING PARAMETERS
Leisure - cont'd

Court sizes:

badminton	13.4 x 6.1 m	volleyball	18 x 9 m
basketball	26+2 x 14+1 m	tug of war	35 (min) x 5 m (min)
handball	30 - 40 x 17 - 20 m	bowls	4.5 x 32 m (min) per rink
hockey	36 - 44 x 18 - 22 m	cricket nets	3.05 (min) x 33.5 m per net
women's lacrosse	27 - 36 x 15 - 21m	snooker	3.7 x 1.9 m table size
men's lacrosse	46 - 48 x 18 - 24m	ice hockey	56.61 x 26 - 30.5 m
netball	30.50 x 15.25 m	racquets	18.288 x 9.144 m
tennis	23.77 x 10.97 m	squash	9.754 x 6.4 x 5.64 m

Typical swimming pool dimensions:

Olympic standard	50 m x 21 m (8 lanes) water depth 1.8 m (constant)
ASA, national and county championship standard	25 or 33.3 m long with width multiple of 2.1 m wide lanes minimum water depth 900 mm 1 m springboard needs minimum 3 m water depth
Learner pool	width 7.0 - 7.5 m depth 600 - 900 mm
Toddlers pool	450 mm depth
Leisure pool	informal shape: will sometimes encompass 25 m in one direction to accommodate roping-off for swimming lanes; water area from 400 - 750 m²
Splash pool	minimum depth 1.05 m
Changing cubicles	minimum dimensions: 914 x 1057 mm

Note: For 25 m pool developments the ratio of water area to gross floor area may average 1:3. For free form leisure pool developments, a typical ratio is 1:5.5.

Multiplex space planning data:

Ideal number of screens	10 (minimum six)
Average area per screen	325 m²
Typical dimensions:	71 x 45 m (10 screens) 66 x 43 m (8 screens) plus 20 m² food area

Housing

Typical densities	Persons per hectare	Units per hectare
Urban	200	90
Suburban	150	55
Rural	110	35

Typical gross internal floor areas for housing associations / local authorities schemes:

	(m²)
Bungalows	
one-bed	48
two-bed	55 - 65
Houses	
one-bed	44
two-bed	62 - 80
three-bed	75 - 95
four-bed	111 - 145
Flats	
bedsitters	23
one-bed	35 - 63
two-bed	55 - 80
three-bed	75 - 100

PLANNING PARAMETERS

Housing - cont'd

Gross internal floor areas for private developments are much more variable and may be smaller or larger than the indicative areas shown above, depending on the target market. Standards for private housing are set out in the NHBC's Registered House Builders Handbook. There are no floor space minima, but heating, kitchen layout, kitchens and linen storage, WC provisions, and the number of electrical socket outlets are included.

Average housing room sizes - net internal floor areas:

	Living room (m^2)	Kitchen (m^2)	Bathroom (m^2)	Main bedroom (m^2)	Average bedroom size (m^2)
Bungalows					
one-bed	15.0	6.0	3.5	11.0	-
two-bed	17.0	9.0	3.5	12.5	10.0
Houses					
one-bed	14.5	6.5	3.5	11.0	-
two-bed	17.5	9.5	4.5	10.0	9.0
three-bed	17.5	13.5	7.0	13.0	10.5
four-bed	22.5	12.5	8.0	17.5	12.5
Flats					
bedsitters	18.0	-	3.0	-	-
one-bed	13.5	7.5	4.5	10.0	-
two-bed	17.0	10.0	5.5	13.5	11.5
three-bed	23.0	3.5	5.5	14.0	14.0

Storage accommodation for housing

NHBC requirements are that in every dwelling, enclosed domestic storage accommodation shall be provided as follows:

Area of dwelling (m^2)	Minimum volume of storage (m^3)
less than 60	1.3
60 - 80	1.7
over 80	2.3

Hotels

Typical gross internal floor areas per bedroom:

	m^2
Five star, city centre hotel	60+
Four star, city centre / provincial centre hotel	45 to 55
Three star, city / provincial hotel	40 to 45
Three / two star, provincial hotel	33 to 40
Three / two star bedroom extension	26 to 30

Indicative space standards (unit):

Suites including bedroom, living room bathroom and hall (nr)	55 to 65
Double bedrooms including bathroom and lobby (nr)	
large	30 to 35
average	25 to 30
small	20 to 25
disabled	3 to 5 m^2 extra
Restaurant (seat)	
first class	1.85
speciality/grill	1.80

PLANNING PARAMETERS

Hotels - cont'd

Indicative space standards (unit) - cont'd:	m²
Coffee shop (seat)	1.80
Bar (customer standing)	0.40 to 0.45
Food preparation/main kitchen/storage	40% to 50% of restaurant and bar areas
Banquet (seat)	1.40
Catering to banquets	10 % to 25 % of banquet area
Function/meeting rooms (person)	1.50
Staff areas (person)	0.40 to 0.60
Staff restaurant and kitchen (seat)	0.70 to 0.90
Service rooms (floor)	30 to 50
General storage and housekeeping	1.5 to 2% of bedroom and circulation areas
Front hall, entrance areas, lounge	2 to 3% average (up to 5%) of total hotel area
Administrative areas	Allowances based on number of accounts staff. Additional area if self accounting 15 to 25 per cent for bedroom floors depending on number of storeys, layout and operating principles, 20 to 25 per cent for public areas
Plant rooms and ducts	4 to 5 % of total hotel area for non-air-conditioned areas, 7 to 8% for air conditioned areas

Typical internal bedroom dimensions:

Bedroom including bathroom	
five star	8.0 m x 4.0 m
four star	7.5 m x 3.75 m
three/two star	7.0 m x 3.5 m
Typical corridor width	1.4 m to 1.6 m

Circulation (C)

Figures represent net area which is gross area less space to be set aside for staircases, lift lobbies, lift wells, lavatories, cleaners' cupboards, service risers, plant space, etc.

Typical NIFA to GIFA areas:	Percentage of GIFA
Offices	
2 to 4 storey	82 - 87
5 to 9 storey	76 - 82
10 to 14 storey	72 - 76
15 to 19 storey	68 - 72
20 + storeys	65 - 68
Adjustments	
for fancoil air-conditioned offices	deduct 2 - 3
for VAV air-conditioned offices	deduct 6 - 7

PLANNING PARAMETERS

Circulation - cont'd

Percentage of GIFA

Flats
- Staircase access — 85
- Enclosed balcony — 83
- Internal corridor and lobby — 80

Typical sales to gross internal areas

Retail
- Superstores — 45 - 55
- Department stores — 50 - 60
- Retail warehouses — 75 - 85

Wall and Window to floor ratios

Typical ratios based on historic data:

Legend:
- (1) W/F - External wall to gross floor area (GIFA) ratio
- (2) W/W - External window to external wall ratio
- (3) IW/F - Internal wall to gross floor area (GIFA) ratio

Building types	(1) W/F	(2) W/W	(3) IW/F
Industrial			
warehouse	0.45	0.04	-
factory	0.60	0.14	-
nursery	0.70	0.14	-
Offices			
open	0.80	0.35	0.30
cellular	0.80	0.35	1.10

Plant area

Percentage of GIFA

Industrial — 3 - 5

Offices — 4 - 11

Percentage of treated floor area

Leisure

all air, low velocity	4.0 - 6.0
induction	2.0 - 3.0
fan coil	1.5 - 2.5
VAV	3.0 - 4.5
versatemp	1.5 - 2.0
boiler plant (excluding hws cylinders)	0.8 - 1.8
oil tank room	1.0 - 2.0
refrigeration plant (excluding cooling towers)	1.0 - 2.0
supply and extract ventilation	3.0 - 5.0
electrical (excluding input substation or standby generation)	0.5 - 1.5
lift rooms	0.2 - 0.5
toilet ventilation	0.3 - 1.0

PLANNING PARAMETERS

Other key dimensions

Structural grid and cladding rail spacing for industrial buildings

Typical economic dimensions	m
spans	18
column spacing	6 - 7.5
purlin spacing	1.8

Wall to core for offices

	m
Typical dimensions measured on plan between the internal finish of external wall to finish of core	7.3

Floor to floor heights

Typical dimensions, measured on section

	m
Industrial	
top of ground slab to top of first floor slab	3.9 - 4.5
top of first floor slab to underside of beams / eaves	3.4 - 3.7

Minimum dimensions; floor finish to floor finish

	m
Offices	
speculative centrally heated	3.3
speculative air-conditioned	3.8
trading floors air-conditioned	4.7
Hotels	
bedrooms	2.7 - 3
public areas	3.5 - 3.6

Floor to underside of structure heights

	m
Industrial	
Minimum internal clear height	
minimum cost stacking warehouse/light industrial	5 - 5.5
minimum height for storage racking	7.5
turret trucks used for stacking	9
automatic warehouse with stacker cranes	15 - 30

Clearance for structural members, sprinklers and lighting in addition to the above

	m
Retail	
Clear height from floor to underside of beams / eaves:	
shop sales area	3.3 - 3.8
shop non-sales area	3.2 - 3.6
retail warehouse	4.75 - 5.5
Leisure	
Specified by each sport's governing body	
badminton/tennis to county standard	7.6
badminton/tennis/ trampolining to international standard	9.1
pool hall from pool surround	8.4 - 8.9

PLANNING PARAMETERS

Other key dimensions - cont'd m

Floor to underside of structure heights

Industrial floor to eaves height
Typical dimensions measured on section:
    low bay warehouse   6
    high bay warehouse   9 - 18

Floor to ceiling height

Typical dimensions measured on section:
Industrial
    top of ground slab to underside of first floor slab   3.7 - 4.3
    top of first floor finish to ceiling finish   2.75 - 3

Minimum dimensions measured on section from floor finish to ceiling finish:
Offices
    Speculative offices   2.6
    Trading floors   3
Leisure
    Multiple cinemas   6
    Fitness/dance studios   5 - 6
    Snooker room   3
    Projectile room   3
    Changing rooms   3.5
Houses
    Ground floor   2.1 - 2.55
    First floor   2.35 - 2.55
Flats   2.25 - 2.65
Bungalows   2.4
Hotels
    Bedrooms   2.5
    Lounges   2.7
    Meeting rooms   2.8
    Restaurant / coffee shop / bar   3
    Function rooms   4 - 4.8

Raised floor areas mm
Minimum clear void for:
    Speculative offices   100 - 200
    Trading floors   300

Note: one floor box per 9 m^2

Suspended ceilings mm
Minimum clear voids (beneath beams)
    Mechanically ventilated offices   300
    Fan coil air-conditioned offices   450
    VAV air-conditioned offices   550
    Trading floors   760

PLANNING PARAMETERS

Typical floor loadings

For more precise floor loadings according to usage refer to section on *DESIGN LOADINGS FOR BUILDINGS* earlier in this section.

Typical loadings (based on minimum uniformly distributed loads plus 25% for partition loads) are:

	KN/m^2
Industrial	24 - 37
Offices	5 - 7
Retail warehouse / storage	24 - 29
Shop sales areas	6
Shop storage	12
Public assembly areas	6
Residential dwelling units	2 - 2.5
Residential corridor areas	4
Hotel bedrooms	3
Hotel corridor areas	4
Plant rooms	9
Car parks and access ramps	3 - 4

Fire protection and means of escape

BS 5588: Fire Precautions in the Design and Construction of Building: includes details of:
- angle between escape routes
- disposition of fire resisting construction
- permitted travel distances

The Building Regulations fire safety approved document B 1992 provides advice on interpretation of the Building Regulations and is still the relevant controlling legislation for fire regulations, although the Loss Prevention Council have recently produced an advisory note, the *Code of Practice for the Construction of Buildings* which argues for a higher performance than the mandatory regulations.

Some minimum periods of fire resistance in minutes for elements of a structure are reproduced hereafter, based on Appendix A Table A2 of the Building Regulations fire safety approved document B, but refer to the relevant documentation to ensure that the information is current.

PLANNING PARAMETERS

Building group	Minimum fire resistance in minutes						
	Basement storey		Ground and Upper storey				
	<10m deep	>10m deep	<5m high	>20m high	<30m high	>30m high	
Industrial							
not sprinklered	120	90	60	90	120	not allowed	
sprinklered	90	60	30*	60	60	120#	
Offices							
not sprinklered	90	60	30*	60	90	not allowed	
sprinklered	60	60	30*	30*	60	120#	
Shop, commercial and leisure							
not sprinklered	90	60	60	60	90	not allowed	
sprinklered	60	60	30*	60	60	120#	
Residential dwelling houses	-	30*	30*	60	-	-	

  *   Increase to a minimum of 60 minutes for compartment walls separating buildings
  #   Reduce to 90 minutes for elements not forming part of the structural frame

Section 20

Applies to buildings in the Greater London area - refer to *London Building Acts (Amendment) Act 1939: Section 20, Code of Practice*. Major cost considerations include 2 hour fire resistance to reinforced concrete columns, possible requirement for sprinkler installation in offices and / or basement car parks, automatic controls and smoke detection in certain ventilation trucking systems, 4 hour fire resistance to fire fighting lift/stair/lobby enclosures and requirements for ventilated lobbies with a minimum floor area of 5.5 m^2 to fire fighting staircases.

Sprinkler installations

Sprinkler installations should be considered where any of the following are likely to occur:

    rapid fire spread likely, for example warehouses with combustible goods/packaging
    large uncompartmented areas
    high financial or consequential loss arising from fire damage

Refer to BS 5306: Part 2: 1990 for specification of sprinkler systems and associated Technical Bulletins from the Fire Officers Committee.

Sanitary provisions

For the provisions of sanitary appliances refer to BS 6465: Part 1: 1994, which suggests the following minimum requirements (refer to the relevant documentation to ensure information is correct).

Factories (table 5)	Males	Females
WC's	1 per 25 persons or part thereof	1 per 25 persons or part thereof
Urinals	As required	Not applicable
Baths or showers	As required	As required
	Male and Female	
Wash basins	1 per 20 persons; for clean processes 1 per 10 persons; for dirty processes 1 per 5 persons; for injurious processes	

PLANNING PARAMETERS

Sanitary provisions - cont'd

Housing (table 1)	2 - 4 person	5 person	6 person and over
One level, eg. bungalows and flats			
WC's	1	1	2
Bath	1	1	1
Wash basin *	1	1	1
Sink and drainer	1	1	1
On two or more levels, eg. houses and maisonettes			
WC's	1	2	2
Bath	1	1	1
Wash basin *	1	1	1
Sink and drainer	1	1	1

* in addition, allow one extra wash basin in every separate WC compartment which does not adjoin a bathroom.

Tables 2 and 3 deal with sanitary provisions for elderly people

Office building and shops (table 4)	Number per male and per female staff
WC's (no urinals) and wash hand basins	1 for 1 to 15 persons 2 for 16 to 30 persons 3 for 31 to 50 persons 4 for 51 to 75 persons 5 for 76 to 100 persons add 1 for every additional 25 persons or part thereof
Cleaners' sink	At least 1 per floor

For WC's (urinals provided), urinals, incinerators, etc. refer to BS 6465: Part 1: 1984. One unisex type WC and one smaller compartment for each sex on each floor where male and female toilets are provided - refer to BS 5810: 1979 and Building Regulations 1985 Schedule 2 (shortly to be replaced by part M).

PLANNING PARAMETERS

Sanitary provisions - cont'd

Swimming pools (table 11)

	For spectators		For bathers	
	Males	Females	Males	Females
WC's	1 for 1 - 200 persons 2 for 201 - 500 persons 3 for 501 - 1000 persons Over 1000 persons, 3 plus 1 for every additional 500 persons or part thereof	1 for 1 - 100 persons 2 for 101 - 250 persons 3 for 251 - 500 persons Over 500 persons, 3 plus 1 for every additional 400 persons or part thereof	1 per 20 changing places	1 per 10 changing places
Urinals	1 per 50 persons	n/a	1 per 20 changing places	n/a
Wash basins	1 per 60 persons	1 per 60 persons	1 per 15 changing places	1 per 15 changing places
Showers	n/a	n/a	1 per 8 changing places	1 per 8 changing places

Refer also to BS 6465: Part 1: 1994 for sanitary provisions for schools, leisure, hotels and restaurants, etc.

Minimum cooling and ventilation requirements

General offices   40 W/m^2
Trading floors   60 W/m^2
Fresh air supply
    offices/dance halls   8 - 12 litres/person/second
    bars   12 - 18 litres/person/second

Recommended design values for internal environmental temperatures and empirical values for air infiltration and natural ventilation allowances

	Temperature °C (winter	Air infiltration rate (changes per hour)	Ventilation allowance (W/m^3 degrees C)
Warehouses			
working and packing spaces	16	0.5	0.17
storage space	13	0.25	0.08
Industrial			
production	16	0.5	0.17
offices	20	1.0	0.33
Offices	20	1.0	0.33
Shops			
small	18	1.0	0.33
large	18	0.5	0.17
department store	18	0.25	0.08
fitting rooms	21	1.5	0.50
store rooms	15	0.5	0.17

PLANNING PARAMETERS

Recommended design values for internal environmental temperatures and empirical values for air infiltration and natural ventilation allowances - cont'd

	Temperature °C (winter)	Air infiltration rate (changes per hour)	Ventilation allowance (W/m³ degrees C)
Housing			
living rooms	21	1.0	0.33
bedrooms	18	0.5	0.17
bed sitting rooms	21	1.0	0.33
bathrooms	22	2.0	0.67
lavatory, cloakrooms	18	1.5	0.50
entrance halls, staircases, corridors	16	1.5	0.50
Hotels			
bedrooms (standard)	22	1.0	0.33
bedrooms (luxury)	24	1.0	0.33
public rooms	21	1.0	0.33
corridors	18	1.5	0.50
foyers	18	1.5	0.50

Typical design temperatures and mechanical ventilation allowances for leisure buildings

	Air temperature °C	Mechanical airchange rates (changes per hour)
Leisure buildings		
ice rink	below 25 (heating temperature in winter: -8)	6
sports hall	16 - 21	3
squash courts	16 - 21	3
bowls halls	16 - 21	3
activity rooms	16 - 21	3
function room/bar	21 ± 2	2 - 4
fitness / dance studio	16 - 21	3 - 6
snooker room	16 - 21	3 - 6
projectile room	16 - 21	3 - 6
changing rooms	22	10
swimming pools	28	4 - 6
bar and cafe areas	23	2 - 4
administration areas	21	2 - 4

	Pool water temperature °C	
Swimming pools		
main pool	27	Ventilation rates must be related to the control of condensation. The criteria is the water area and the recommended basis is 20 litres/per m² of water surface, plus a margin (say 20 per cent) to allow for the effect of wet surrounds.
splash pool	27	
learners pool	28 - 30	
diving pool	27	
leisure pool	29	
jacuzzi pool	35	

PLANNING PARAMETERS

Typical lighting levels

Lighting levels for a number of common building types are given below. For more precise minimum requirements refer to the IES Code.

	Lux
Industrial building - production/assembly areas	100 - 1000 (varies)
Offices	500
Conventional shops with counters or wall displays and self-service shops	500
Supermarkets	500
Covered shopping precincts and arcades	
main circulation paces	100 - 200
lift, stairs, escalators	150
staff rooms	150
external covered walkways	30
Sports buildings	
multi use sports halls	500
squash courts	500
dance / fitness studio	300
snooker room	500 on table
projectile room	300 generally / 1000 on target
Homes	
living rooms	
general	50
casual reading	150
bedrooms	
general	50
bedhead	150
studios	
desk and prolonged reading	300
kitchens	
working areas	300
bathrooms	100
halls and landings	150
stairs	100
Hotels	
internal corridors	200
guest room sleep area; stair wells	300
guest room activity area; housekeeping areas	500
meeting / banquet facilities	800

Electrical socket outlets (NHBC)

	Desirable provision	Minimum provision
Homes		
working area of kitchen	4	4
dining area	2	1
living area	5	3
first or only double bedroom	3	2
other double bedrooms	2	2
single bedrooms	2	2
hall and landing	1	1
store/workshop/garage	1	-
single study bedrooms	2	2
single bed sitting rooms in family dwellings	3	3
single bed sitting rooms in self contained bed sitting room dwellings	5	5

PLANNING PARAMETERS

Lifts
Performance standard to be not less than BS 5655: Lifts and service Lifts.

Industrial
Typical goods lift - 1000 kg

Offices
Dependant on number of storeys and planning layout, usually based on:

	Number of lifts
< 4 storeys	1
> = 4 storeys and < 10 000m² GIA	2
> = 4 storeys and > 10 000m² GIA	3

Hotels
Dependant on number of bedrooms, number of storeys and planning layout.
Typical examples

120 bed hotel on 3 floors	two 6 - 8 person lifts and service lift
200 bed hotel on 10 floors	four 13 person lifts and fireman's lift and service lift

Car park

Typical car space requirements   **One car space per**

Industrial 45 - 55m² GIA	
Offices	
medium tech	28 - 37 m² GIFA
high tech	19 - 25 m² GIFA
Retail	
superstores	8 - 10 m² GIFA
shopping centres/out of town retailing	18 - 23 m² GIFA
furniture/DIY stores	20 - 30 m² GIFA
Leisure	
swimming pools	
patrons	10 m² pool area
staff	2 nr staff
leisure centres	
patrons	10 m² activity area
Residential	1 - 2 dwellings (depending on garage space, standard of dwelling, etc)

Goods and reception and service vehicles

Typical goods reception bay suitable for two 15 m articulated lorries with 1.5 m clearance either side. Loading bays must be level and have a clear height of 4.73 m. Approach routes should have a clear minimum height of 5.03 m. Minimum articulated lorry turning circle 13 m.

Typical design load for service yard 20 KN/m².

PLANNING PARAMETERS

Recommended sizes of various sports facilities

Archery (Clout)	7.3 m firing area Range 109.728 (Women), 146.304 (Men) 182.88 (Normal range)
Baseball	Overall 60 m x 70 m
Basketball	14 m x 26 m
Camogie	91 - 110 m x 54 - 68 m
Discus and Hammer	Safety cage 2.74 m square Landing area 45 arc (65° safety) 70 m radius
Football, American	Pitch 109.80 m x 48.80 m overall 118.94 m x 57.94 m
Football, Association	NPFA rules Senior pitches 96 - 100 m x 60 - 64 m Junior pitches 90 m x 46 - 55 m International 100 - 110 m x 64 - 75 m
Football, Australian Rules	Overall 135 - 185 m x 110 - 155 m
Football, Canadian	Overall 145.74 m x 59.47 m
Football, Gaelic	128 - 146.40 m x 76.80 - 91.50 m
Football, Rugby League	111 - 122 m x 68 m
Football, Rugby Union	144 m max x 69 m
Handball	91 - 110 m x 55 - 65 m
Hockey	91.50 m x 54.90 m
Hurling	137 m x 82 m
Javelin	Runway 36.50 m x 4.27 m Landing area 80 - 95 m long, 48 m wide
Jump, High	Running area 38.80 m x 19 m Landing area 5 m x 4 m
Jump, Long	Runway 45 m x 1.22 m Landing area 9 m x 2.75 m
Jump, Triple	Runway 45 m x 1.22 m Landing area 7.30 m x 2.75 m
Korfball	90 m x 40 m
Lacrosse	(Mens) 100 m x 55 m (Womens) 110 m x 73 m
Netball	15.25 m x 30.48 m
Pole Vault	Runway 45 m x 1.22 m Landing area 5 m x 5 m
Polo	275 m x 183 m
Rounders	Overall 19 m x 17 m

Recommended sizes of various sports facilities - cont'd

400m Running Track	115.61 m bend length x 2 84.39 m straight length x 2 Overall 176.91 m long x 92.52 m wide
Shot Putt	Base 2.135 m diameter Landing area 65° arc, 25 m radius from base
Shinty	128 - 183 m x 64 - 91.50 m
Tennis	Court 23.77 m x 10.97 m Overall minimum 36.27 m x 18.29 m
Tug-of-war	46 m x 5 m

SOUND INSULATION

Sound reduction requirements as Building Regulations (E1/2/3)

The Building Regulations on airborne and impact sound (E1/2/3) state simply that both airborne and impact sound must be reasonably reduced in floors and walls. No minimum reduction is given but the following tables give example sound reductions for various types of constructions.

Sound reductions of typical walls	Average sound reduction (dB)
13 mm Fibreboard	20
16 mm Plasterboard	25
6 mm Float glass	30
16 mm Plasterboard, plastered both sides	35
75 mm Plastered concrete blockwork (100 mm)	44
110 mm half brick wall, half brick thick, plastered both sides	43
240 mm Brick wall one brick thick, plastered both sides	48
Timber stud partitioning with plastered metal lathing both sides	35
Cupboards used as partitions	30
Cavity block wall, plastered both sides	42
75 mm Breeze block cavity wall, plastered both sides	50
100 mm Breeze block cavity wall, plastered both sides including 50 mm air-gap and plasterboard suspended ceiling	55
As above with 150 mm Breeze blocks	65
19 mm T & G boarding on timber joists including plasterboard ceiling and plaster skim coat	32
As above including metal lash and plaster ceiling	37
As above with solid sound proofing material between joists approx 98 kg per sq metre	55
As above with floating floor of T & G boarding on batten and soundproofing quilt	75

SOUND INSULATION

Impact noise is particularly difficult to reduce satisfactorily. The following are the most efficient methods of reducing such sound.

1) Carpet on underlay of rubber or felt;
2) Pugging between joists (e.g. Slag Wool); and
3) A good suspended ceiling system.

Sound requirements

Housing

NHBC requirements are that any partition between a compartment containing a WC and a living-room or bedroom shell have an average sound insulation index of not less than 35 dB over the frequency range of 100 - 3150 Hz when tested in accordance with BS2750.

Hotels

Bedroom to bedroom or bedroom to corridor 48dB

THERMAL INSULATION

Thermal properties of various building elements

Thickness (mm)	Material	(m^2k/W) R	(W/m^2K) U - Value
n/a	Internal and external surface resistance	0.18	-
	Air-gap cavity	0.18	-
103	Brick skin	0.12	-
	Dense concrete block		
100	ARC conbloc	0.09	11.11
140	ARC conbloc	0.13	7.69
190	ARC conbloc	0.18	5.56
	Lightweight aggregate block		
100	Celcon standard	0.59	1.69
125	Celcon standard	0.74	1.35
150	Celcon standard	0.88	1.14
200	Celcon standard	1.18	0.85
	Lightweight aggregate thermal block		
125	Celcon solar	1.14	0.88
150	Celcon solar	1.36	0.74
200	Celcon solar	1.82	0.55
	Insulating board		
25	Dritherm	0.69	1.45
50	Dritherm	1.39	0.72
75	Dritherm	2.08	0.48
13	Lightweight plaster "Carlite"	0.07	14.29
13	Dense plaster "Thistle"	0.02	50.00
	Plasterboard		
9.5	British gypsum	0.06	16.67
12.7	British gypsum	0.08	12.50
40	Screed	0.10	10.00
150	Reinforced concrete	0.12	8.33
100	Dow roofmate insulation	3.57	0.28

THERMAL INSULATION

Resistance to the passage of heat

Provisions meeting the requirement set out in the Building Regulations (L2/3):

		Minimum U - Value
a)	**Dwellings**	
	Roof	0.35
	Exposed wall	0.60
	Exposed floor	0.60
b)	**Residential, Offices, Shops and Assembly Buildings**	
	Roof	0.06
	Exposed wall	0.60
	Exposed floor	0.60
c)	**Industrial, Storage and Other Buildings**	
	Roof	0.70
	Exposed wall	0.70
	Exposed floor	0.70

TYPICAL CONSTRUCTIONS MEETING THERMAL REQUIREMENTS

External wall, masonry construction:

Concrete blockwork	U - Value
200 mm lightweight concrete block, 25 mm air-gap, 10 mm plasterboard	0.68
200 mm lightweight concrete block, 20 mm EPS slab, 10 mm plasterboard	0.54
200 mm lightweight concrete block, 25 mm air-gap, 25 mm EPS slab, 10 mm plasterboard	0.46

Brick/Cavity/Brick

105 mm brickwork, 50 mm UF foam, 105 mm brickwork, 3 mm lightweight plaster	0.55

Brick/Cavity/Block

105 mm brickwork, 50 mm cavity, 125 mm Thermalite block, 3 mm lightweight plaster	0.59
105 mm brickwork, 50 mm cavity, 130 mm Thermalite block, 3 mm lightweight plaster	0.57
105 mm brickwork, 50 mm cavity, 130 mm Thermalite block, 3 mm dense plaster	0.59
105 mm brickwork, 50 mm cavity, 100 mm Thermalite block, foilbacked plasterboard	0.55
105 mm brickwork, 50 mm cavity, 115 mm Thermalite block, 9.5 mm plasterboard	0.58
105 mm brickwork, 50 mm cavity, 115 mm Thermalite block, foilbacked plasterboard	0.52
105 mm brickwork, 50 mm cavity, 125 mm Theramlite block, 9.5 mm plasterboard	0.55
105 mm brickwork, 50 mm cavity, 100 mm Thermalite block, 25 mm insulating plasterboard	0.53
105 mm brickwork, 50 mm cavity, 125 mm Thermalite block, 25 mm insulating plasterboard	0.47
105 mm brickwork, 25 mm cavity, 25 mm insulation, 100 mm Thermalite block, lightweight plaster	0.47

THERMAL INSULATION

TYPICAL CONSTRUCTIONS MEETING THERMAL REQUIREMENTS - cont'd

Brick/Cavity/Block - cont'd	**U-Value**
105 mm brickwork, 25 mm cavity, 25 mm insulation, 115 mm Thermalite block, lightweight plaster	0.44
Render, 100 mm "Shield" block, 50 mm cavity, 100 mm Thermalite block, lightweight plaster	0.50
Render, 100 mm "Shield" block, 50 mm cavity, 115 mm Thermalite block, lightweight plaster	0.47
Render, 100 mm "Shield" block, 50 mm cavity, 125 mm Thermalite block, lightweight plaster	0.45

Tile hanging

10 mm tile on battens and felt, 150 mm Thermalite block, lightweight plaster	0.57
25 mm insulating plasterboard	0.46
10 mm tile on battens and felt, 190 mm Thermalite block, lightweight plaster	0.47
25 mm insulating plasterboard	0.40
10 mm tile on battens and felt, 200 mm Thermalite block, lightweight plaster	0.45
25 mm insulated plasterboard	0.38
10 mm tile on battens, breather paper, 25 mm air-gap, 50 mm glass fibre quilts, 10 mm plasterboard	0.56
10 mm tile on battens, breather paper, 25 mm air-gap, 75 mm glass fibre quilts, 10 mm plasterboard	0.41
10 mm tile on battens, breather paper, 25 mm air-gap, 100 mm glass fibre quilts, 10 mm plasterboard	0.33

Pitched roofs

Slate or concrete tiles, felt, airspace, Rockwool flexible slabs laid between rafters, plasterboard

Slab	40 mm thick	0.62
	50 mm thick	0.52
	60 mm thick	0.45
	75 mm thick	0.38
	100 mm thick	0.29

Concrete tiles, sarking felt, rollbatts between joists, plasterboard

Insulation	100 mm thick	0.31
	120 mm thick	0.26
	140 mm thick	0.23
	160 mm thick	0.21

THERMAL INSULATION

TYPICAL CONSTRUCTIONS MEETING THERMAL REQUIREMENTS - cont'd

Pitched roofs - cont'd U-Value

Steel frame Rockwool insulation sandwiched between steel exterior profiled sheeting and interior sheet lining

Insulation	60 mm thick	0.53
	80 mm thick	0.41
	100 mm thick	0.34

Steel frame, steel profiled sheeting, Rockwool insulation over purlins and plasterboard lining

Insulation	60 mm thick	0.51
	80 mm thick	0.38
	100 mm thick	0.32
	120 mm thick	0.27
	140 mm thick	0.24
	160 mm thick	0.21

Flat roofs

Asphalt, Rockwool roof slabs, 25 mm timber boarding, timber joists and 9.5 mm plasterboard

Insulation	30 mm thick	0.68
	40 mm thick	0.57
	50 mm thick	0.49
	60 mm thick	0.44
	70 mm thick	0.39
	80 mm thick	0.35
	90 mm thick	0.32
	100 mm thick	0.29

Asphalt, Rockwool roof slabs on 150 mm dense concrete deck and screed with 16 mm plaster finish

Insulation	40 mm thick	0.68
	50 mm thick	0.57
	60 mm thick	0.49
	70 mm thick	0.43
	80 mm thick	0.39
	90 mm thick	0.35
	100 mm thick	0.32

Asphalt, Rockwool roof slabs on 150 mm dense concrete deck and screed with suspended plasterboard ceiling

Insulation	40 mm thick	0.60
	50 mm thick	0.52
	60 mm thick	0.45
	70 mm thick	0.40
	80 mm thick	0.36
	90 mm thick	0.33
	100 mm thick	0.30

THERMAL INSULATION

TYPICAL CONSTRUCTIONS MEETING THERMAL REQUIREMENTS

Flat roofs - cont'd   U-Value

Steel frame, asphalt on insulation slabs on troughed steel decking

Insulation	50 mm thick	0.59
	60 mm thick	0.51
	70 mm thick	0.45
	80 mm thick	0.39
	90 mm thick	0.35
	100 mm thick	0.33

Steel frame, asphalt on insulation slabs on troughed steel decking including suspended plasterboard ceiling

Insulation	40 mm thick	0.67
	50 mm thick	0.57
	60 mm thick	0.49
	70 mm thick	0.43
	80 mm thick	0.38
	90 mm thick	0.34
	100 mm thick	0.32

WEIGHTS OF VARIOUS MATERIALS

Material		kg/m³	Material		kg/m³
Aggregates					
Ashes		610	Lime:	Chalk (lump)	704
Cement	(Portland)	1600		Ground	961
Chalk		2406		Quick	880
Chippings	(stone)	1762	Sand:	Dry	1707
Clinker	(furnace)	800		Wet	1831
	(concrete)	1441	Water		1000
Ballast or stone		2241	Shale/Whinstone		2637
Pumice		640	Broken stone		1709
Gravel		1790	Pitch		1152
Metals					
Aluminium		2559	Lead		11260
Brass		8129	Tin		7448
Bronze		8113	Zinc		7464
Gunmetal		8475			
Iron:	Cast	7207			
	Wrought	7687			
Stone and brickwork					
Blockwork:			Brickwork:		
	Aerated	650		Common Fletton	1822
	Dense concrete	1800		Glazed brick	2080
	Lightweight concrete	1200		Staffordshire Blue	2162
	Pumice concrete	1080		Red Engineering	2240
				Concrete	1841

Material	kg/m³	Material	kg/m³
Stone and brickwork - cont'd			
Stone:			
Artificial	2242	Granite	2642
Bath	2242	Marble	2742
Blue Pennant	2682	Portland	2170
Cragleith	2322	Slate	2882
Darley Dale	2370	York	2402
Forest of Dean	2386	Terra-cotta	2116
Wood			
Blockboard	500 - 700	Jarrah	816
Cork Bark	80	Maple	752
Hardboard:		Mahogany:	
Standard	940 - 1000	Honduras	576
Tempered	940 - 1060	Spanish	1057
Wood chipboard:		Oak:	
Type I	650 - 750	English	848
Type II	680 - 800	American	720
Type III	650 - 800	Austrian & Turkish	704
Type II/III	680 - 800	Pine:	
Laminboard	500 - 700	Pitchpine	800
Timber:		Red Deal	576
Ash	800	Yellow Deal	528
Baltic spruce	480	Spruce	496
Beech	816	Sycamore	530
Birch	720	Teak:	
Box	961	African	961
Cedar	480	Indian	656
Chestnut	640	Moulmein	736
Ebony	1217	Walnut:	
Elm	624	English	496
Greenheart	961	Black	720

MEMORANDA FOR EACH TRADE

EXCAVATION AND EARTHWORK

Transport capacities

Type of vehicle	Capacity of vehicle m³ (solid)
Standard wheelbarrow	0.08
2 ton truck (2.03 t)	1.15
3 ton truck (3.05 t)	1.72
4 ton truck (4.06 t)	2.22
5 ton truck (5.08 t)	2.68
6 ton truck (6.10 t)	3.44
2 cubic yard dumper (1.53 m³)	1.15
3 cubic yard dumper (2.29 m³)	1.72
6 cubic yard dumper (4.59 m³)	3.44
10 cubic yard dumper (7.65 m³)	5.73

MEMORANDA FOR EACH TRADE

EXCAVATION AND EARTHWORK - cont'd

Planking and strutting

Maximum depth of excavation in various soils without the use of earthwork support

Ground conditions	Metres (m)
Compact soil	3.65
Drained loam	1.85
Dry sand	0.30
Gravelly earth	0.60
Ordinary earth	0.90
Stiff clay	3.00

It is important to note that the above table should only be used as a guide. Each case must be taken on its merits and, as the limited distances given above are approached, careful watch must be kept for the slightest signs of caving in.

Baulkage of soils after excavation

Soil type	Approximate bulk of 1m³ after excavation
Vegetable soil and loam	25 - 30%
Soft clay	30 - 40%
Stiff clay	10 - 15%
Gravel	20 - 25%
Sand	40 - 50%
Chalk	40 - 50%
Rock, weathered	30 - 40%
Rock, unweathered	50 - 60%

CONCRETE WORK

Approximate average weights of materials

Materials	Percentage of voids (%)	Weight per m³ (kg)
Sand	39	1660
Gravel 10 - 20 mm	45	1440
Gravel 35 - 75 mm	42	1555
Crushed stone	50	1330
Crushed granite		
(over 15 mm)	50	1345
(n.e. 15 mm)	47	1440
"All-in" ballast	32	1800

MEMORANDA FOR EACH TRADE

CONCRETE WORK - cont'd

Common mixes for various types of work per m³

Recommended mix	Class of work suitable for:	Cement (kg)	Sand (kg)	Coarse Aggregate (kg)	No. of 50 kg bags of cement per m³ of combined aggregate
1:3:6	Roughest type of mass concrete such as footings, road haunchings 300 mm thick	208	905	1509	4.00
1:2.5:5	Mass concrete of better class than 1:3:6 such as bases for machinery, walls below ground, etc.	249	881	1474	5.00
1:2:4	Most ordinary uses of concrete such as mass walls above ground, road slabs etc. and general reinforced concrete work	304	889	1431	6.00
1:1.5:3	Watertight floors, pavements and walls, tanks, pits, steps, paths, surface of two course roads, reinforced concrete where extra strength is required	371	801	1336	7.50
1:1:2	Work of thin section such as fence posts and small precast work	511	720	1206	10.50

Bar reinforcement

Cross-sectional area and mass

Nominal sizes (m)	Cross-sectional area (mm²)	Mass per metre run (kg)
6*	28.3	0.222
8	50.3	0.395
10	78.5	0.616
12	113.1	0.888
16	201.1	1.579
20	314.2	2.466
25	490.9	3.854
32	804.2	6.313
40	1256.6	9.864
50*	1963.5	15.413

* Where a bar larger than 40 mm is to be used the recommended size is 50 mm. Where a bar smaller than 8 mm is to be used the recommended size is 6 mm.

MEMORANDA FOR EACH TRADE

CONCRETE WORK - cont'd

Fabric reinforcement

Preferred range of designated fabric types and stock sheet sizes

Fabric reference	Longitudinal wires			Cross wires			
	Nominal wire size (mm)	Pitch (mm)	Area (mm^2/m)	Nominal wire size (mm)	Pitch (mm)	Area (mm^2/m)	Mass (kg/m^2)
Square mesh							
A393	10	200	393	10	200	393	6.16
A252	8	200	252	8	200	252	3.95
A193	7	200	193	7	200	193	3.02
A142	6	200	142	6	200	142	2.22
A98	5	200	98	5	200	98	1.54
Structural mesh							
B1131	12	100	1131	8	200	252	10.90
B785	10	100	785	8	200	252	8.14
B503	8	100	503	8	200	252	5.93
B385	7	100	385	7	200	193	4.53
B283	6	100	283	7	200	193	3.73
B196	5	100	196	7	200	193	3.05
Long mesh							
C785	10	100	785	6	400	70.8	6.72
C636	9	100	636	6	400	70.8	5.55
C503	8	100	503	5	400	49.0	4.34
C385	7	100	385	5	400	49.0	3.41
C283	6	100	283	5	400	49.0	2.61
Wrapping mesh							
D98	5	200	98	5	200	98	1.54
D49	2.5	100	49	2.5	100	49	0.77

Stock sheet size 4.8 m x 2.4 m, Area 11.52 m^2

Average weight kg/m^3 of steelwork reinforcement in concrete for various building elements

Substructure	kg/m^3 concrete		
Pile caps	110 - 150	Plate slab	150 - 220
Tie beams	130 - 170	Cant slab	145 - 210
Ground beams	230 - 330	Ribbed floors	130 - 200
Bases	125 - 180	Topping to block floor	30 - 40
Footings	100 - 150	Columns	210 - 310
Retaining walls	150 - 210	Beams	250 - 350
Raft	60 - 70	Stairs	130 - 170
Slabs - one way	120 - 200	Walls - normal	40 - 100
Slabs - two way	110 - 220	Walls - wind	70 - 125

Note: For exposed elements add the following % :

Walls 50%, Beams 100%, Columns 15%

MEMORANDA FOR EACH TRADE

BRICKWORK AND BLOCKWORK

Number of bricks required for various types of work per m² of walling

Description	Brick size	
	215 x 102.5 x 50 mm	215 x 102.5 x 65 mm
Half brick thick		
Stretcher bond	74	59
English bond	108	86
English garden wall bond	90	72
Flemish bond	96	79
Flemish garden wall bond	83	66
One brick thick and cavity wall of two half brick skins		
Stretcher bond	148	119

Quantities of bricks and mortar required per m² of walling

Standard bricks	Unit	No of bricks required	Mortar required (cubic metres)		
			No frogs	Single frogs	Double frogs
Brick size 215 x 102.5 x 50 mm					
half brick wall (103 mm)	m²	72	0.022	0.027	0.032
2 x half brick cavity wall (270 mm)	m²	144	0.044	0.054	0.064
one brick wall (215 mm)	m²	144	0.052	0.064	0.076
one and a half brick wall (322 mm)	m²	216	0.073	0.091	0.108
Mass brickwork	m³	576	0.347	0.413	0.480
Brick size 215 x 102.5 x 65 mm					
half brick wall (103 mm)	m²	58	0.019	0.022	0.026
2 x half brick cavity wall (270 mm)	m²	116	0.038	0.045	0.055
one brick wall (215 mm)	m²	116	0.046	0.055	0.064
one and a half brick wall (322 mm)	m²	174	0.063	0.074	0.088
Mass brickwork	m³	464	0.307	0.360	0.413
Metric modular bricks			Perforated		
Brick size 200 x 100 x 75 mm					
90 mm thick	m²	67	0.016	0.019	
190 mm thick	m²	133	0.042	0.048	
290 mm thick	m²	200	0.068	0.078	
Brick size 200 x 100 x 100 mm					
90 mm thick	m²	50	0.013	0.016	
190 mm thick	m²	100	0.036	0.041	
290 mm thick	m²	150	0.059	0.067	
Brick size 300 x 100 x 75 mm					
90 mm thick	m²	33	-	0.015	
Brick size 300 x 100 x 100 mm					
90 mm thick	m²	44	0.015	0.018	

Note: Assuming 10 mm thick joints.

MEMORANDA FOR EACH TRADE

BRICKWORK AND BLOCKWORK - cont'd

Mortar required per m² blockwork (9.88 blocks/m²)

Wall thickness	75	90	100	125	140	190	215
Mortar m³/m²	0.005	0.006	0.007	0.008	0.009	0.013	0.014

Standard available block sizes

Block	Length x height		
	Co-ordinating size	Work size	Thicknesses (work size)
A	400 x 100	390 x 90	(75, 90, 100,
	400 x 200	440 x 190	(140 & 190 mm
	450 x 225	440 x 215	(75, 90, 100 (140, 190, & 215 mm
B	400 x 100	390 x 90	(75, 90, 100
	400 x 200	390 x 190	(140 & 190 mm
	450 x 200	440 x 190	(
	450 x 225	440 x 215	(75, 90, 100
	450 x 300	440 x 290	(140, 190, & 215 mm
	600 x 200	590 x 190	(
	600 x 225	590 x 215	(
C	400 x 200	390 x 190	(
	450 x 200	440 x 190	(
	450 x 225	440 x 215	(60 & 75 mm
	450 x 300	440 x 290	(
	600 x 200	590 x 190	(
	600 x 225	590 x 215	(

ROOFING

Total roof loadings for various types of tiles/slates

		Slate/Tile	Roof load (slope) kg/m² Roofing underlay and battens	Total dead load kg/m²
Asbestos cement slate (600 x 300)		21.50	3.14	24.64
Clay tile	interlocking	67.00	5.50	72.50
	plain	43.50	2.87	46.37
Concrete tile	interlocking	47.20	2.69	49.89
	plain	78.20	5.50	83.70
Natural slate (18" x 10")		35.40	3.40	38.80
			Roof load (plan) kg/m²	
Asbestos cement slate (600 x 300)		28.45	76.50	104.95
Clay tile	interlocking	53.54	76.50	130.04
	plain	83.71	76.50	60.21
Concrete tile	interlocking	57.60	76.50	134.10
	plain	96.64	76.50	173.14
Natural slate (18" x 10")		44.80	76.50	121.30

MEMORANDA FOR EACH TRADE

ROOFING - cont'd

Tiling data

Product		Lap (mm)	Gauge of battens	No. slates per m^2	Battens (m/m^2)	Weight as laid (kg/m^2)
CEMENT SLATES						
Eternit slates	600 x 300 mm	100	250	13.4	4.00	19.50
(Duracem)		90	255	13.1	3.92	19.20
		80	260	12.9	3.85	19.00
		70	265	12.7	3.77	18.60
	600 x 350 mm	100	250	11.5	4.00	19.50
		90	255	11.2	3.92	19.20
	500 x 250 mm	100	200	20.0	5.00	20.00
		90	205	19.5	4.88	19.50
		80	210	19.1	4.76	19.00
		70	215	18.6	4.65	18.60
	400 x 200 mm	90	155	32.3	6.45	20.80
		80	160	31.3	6.25	20.20
		70	165	30.3	6.06	19.60
CONCRETE TILES/SLATES						
Redland Roofing						
Stonewold slate	430 x 380 mm	75	355	8.2	2.82	51.20
Double Roman tile	418 x 330 mm	75	355	8.2	2.91	45.50
Grovebury pantile	418 x 332 mm	75	343	9.7	2.91	47.90
Norfolk pantile	381 x 227 mm	75	306	16.3	3.26	44.01
		100	281	17.8	3.56	48.06
Renown inter-locking tile	418 x 330 mm	75	343	9.7	2.91	46.40
"49" tile	381 x 227 mm	75	306	16.3	3.26	44.80
		100	281	17.8	3.56	48.95
Plain, vertical tiling	265 x 165 mm	35	115	52.7	8.70	62.20
Marley Roofing						
Bold roll tile	420 x 330 mm	75	344	9.7	2.90	47.00
		100	-	10.5	3.20	51.00
Modern roof tile	420 x 330 mm	75	338	10.2	3.00	54.00
		100	-	11.0	3.20	58.00
Ludlow major	420 x 330 mm	75	338	10.2	3.00	45.00
		100	-	11.0	3.20	49.00
Ludlow plus	387 x 229 mm	75	305	16.1	3.30	47.00
		100	-	17.5	3.60	51.00
Mendip tile	420 x 330 mm	75	338	10.2	3.00	47.00
		100	-	11.0	3.20	51.00
Wessex	413 x 330 mm	75	338	10.2	3.00	54.00
		100	-	11.0	3.20	58.00
Plain tile	267 x 165 mm	65	100	60.0	10.00	76.00
		75	95	64.0	10.50	81.00
		85	90	68.0	11.30	86.00
Plain vertical tiles (feature)	267 x 165 mm	35	110	53.0	8.70	67.00
		34	115	56.0	9.10	71.00
CLAY TILES						
Redland Roofing						
Old English pantile	342 x 241 mm	75	267	18.5	3.76	47.00
Bold roll Roman tiles	342 x 266 mm	75	267	17.8	3.76	50.00

MEMORANDA FOR EACH TRADE

ROOFING - cont'd

Slate nails, quantity per kilogram

	Type			
Length	Plain wire	Galvanised wire	Copper nail	Zinc nail
28.5 mm	325	305	325	415
34.4 mm	286	256	254	292
50.8 mm	242	224	194	200

Metal sheet coverings

Thicknesses and weights of sheet metal coverings

Lead to BS 1178

BS Code No	3	4	5	6	7	8
Colour Code	Green	Blue	Red	Black	White	Orange
Thickness (mm)	1.25	1.80	2.24	2.50	3.15	3.55
kg/m^2	14.18	20.41	25.40	28.36	35.72	40.26

Copper to BS 2870

Thickness (mm)	0.60	0.70
Bay width		
Roll (mm)	500	650
Seam (mm)	525	600
Standard width to form bay	600	750
Normal length of sheet	1.80	1.80
Density kg/m^2		

Zinc to BS 849

Zinc Gauge (Nr)	9	10	11	12	13	14	15	16
Thickness (mm)	0.43	0.48	0.56	0.64	0.71	0.79	0.91	1.04
Density kg/m^2	3.1	3.2	3.8	4.3	4.8	5.3	6.2	7.0

Aluminium to BS 4868

Thickness (mm)	0.5	0.6	0.7	0.8	0.9	1.0	1.2
Density kg/m^2	12.8	15.4	17.9	20.5	23.0	25.6	30.7

MEMORANDA FOR EACH TRADE

ROOFING - cont'd

Type of felt	Nominal mass per unit area (kg/10m)	Nominal mass per unit area of fibre base (g/m²)	Nominal length of roll (m)
Class 1			
1B fine granule surfaced bitumen	14	220	10 or 20
	18	330	10 or 20
	25	470	10
1E mineral surfaced bitumen	38	470	10
1F reinforced bitumen	15	160 (fibre) 110 (hessian)	15
1F reinforced bitumen, aluminium faced	13	160 (fibre) 110 (hessian)	15
Class 2			
2B fine granule surfaced bitumen asbestos	18	500	10 or 20
2E mineral surfaced bitumen asbestos	38	600	10
Class 3			
3B fine granule surfaced bitumen glass fibre	18	60	20
3E mineral surfaced bitumen glass fibre	28	60	10
3E venting base layer bitumen glass fibre	32	60*	10
3H venting base layer bitumen glass fibre	17	60*	20

* Excluding effect of perforations

MEMORANDA FOR EACH TRADE

WOODWORK

Conversion tables (for timber only)

Inches	Millimetres	Feet	Metres
1	25	1	0.300
2	50	2	0.600
3	75	3	0.900
4	100	4	1.200
5	125	5	1.500
6	150	6	1.800
7	175	7	2.100
8	200	8	2.400
9	225	9	2.700
10	250	10	3.000
11	275	11	3.300
12	300	12	3.600
13	325	13	3.900
14	350	14	4.200
15	375	15	4.500
16	400	16	4.800
17	425	17	5.100
18	450	18	5.400
19	475	19	5.700
20	500	20	6.000
21	525	21	6.300
22	550	22	6.600
23	575	23	6.900
24	600	24	7.200

Planed softwood

The finished end section size of planed timber is usually 3/16" less than the original size from which it is produced. This however varies slightly depending upon availability of material and origin of the species used.

Standards (timber) to cubic metres and cubic metres to standards (timber)

Cubic metres	Cubic metres standards	Standards
4.672	1	0.214
9.344	2	0.428
14.017	3	0.642
18.689	4	0.856
23.361	5	1.070
28.033	6	1.284
32.706	7	1.498
37.378	8	1.712
42.050	9	1.926
46.722	10	2.140
93.445	20	4.281
140.167	30	6.421
186.890	40	8.561
233.612	50	10.702
280.335	60	12.842
327.057	70	14.982
373.779	80	17.122
420.502	90	19.263
467.224	100	21.403

MEMORANDA FOR EACH TRADE

WOODWORK - cont'd

Standards (timber) to cubic metres and cubic metres to standards (timber)

1 cu metre = 35.3148 cu ft = 0.21403 std

1 cu ft = 0.028317 cu metres

1 std = 4.67227 cu metres

Basic sizes of sawn softwood available (cross sectional areas)

Thickness (mm)	Width (mm)							
	75	100	125	150	175	200	225	250	300
16	X	X	X	X					
19	X	X	X	X					
22	X	X	X	X					
25	X	X	X	X	X	X	X	X	
32	X	X	X	X	X	X	X	X	
36	X	X	X	X					
38	X	X	X	X	X	X	X		
44	X	X	X	X	X	X	X	X	
47*	X	X	X	X	X	X	X	X	X
50	X	X	X	X	X	X	X	X	X
63	X	X	X	X	X	X	X		
75		X	X	X	X	X	X	X	X
100		X		X		X		X	X
150				X		X			X
200						X			
250								X	
300									X

* This range of widths for 47 mm thickness will usually be found to be available in construction quality only.

Note: The smaller sizes below 100 mm thick and 250 mm width are normally but not exclusively of European origin. Sizes beyond this are usually of North and South American origin.

MEMORANDA FOR EACH TRADE

WOODWORK - cont'd

Basic lengths of sawn softwood available (metres)

1.80	2.10	3.00	4.20	5.10	6.00	7.20
	2.40	3.30	4.50	5.40	6.30	
	2.70	3.60	4.80	5.70	6.60	
		3.90			6.90	

Note: Lengths of 6.00 m and over will generally only be available from North American species and may have to be recut from larger sizes.

Reductions from basic size to finished size by planing of two opposed faces

Purpose		Reductions from basic sizes for timber			
		15 - 35 mm	36 - 100 mm	101 - 150 mm	over 150 mm
a)	constructional timber	3 mm	3 mm	5 mm	6 mm
b)	Matching interlocking boards	4 mm	4 mm	6 mm	6 mm
c)	Wood trim not specified in BS 584	5 mm	7 mm	7 mm	9 mm
d)	Joinery and cabinet work	7 mm	9 mm	11 mm	13 mm

Note: The reduction of width or depth is overall the extreme size and is exclusive of any reduction of the face by the machining of a tongue or lap joints.

Maximum spans for various roof trusses

Maximum permissible spans for rafters for Fink trussed rafters

Basic size (mm)	Actual size (mm)	Pitch (degrees)								
		15 (m)	17.5 (m)	20 (m)	22.5 (m)	25 (m)	27.5 (m)	30 (m)	32.5 (m)	35 (m)
38 x 75	35 x 72	6.03	6.16	6.29	6.41	6.51	6.60	6.70	6.80	6.90
38 x 100	35 x 97	7.48	7.67	7.83	7.97	8.10	8.22	8.34	8.47	8.61
38 x 125	35 x 120	8.80	9.00	9.20	9.37	9.54	9.68	9.82	9.98	10.16
44 x 75	41 x 72	6.45	6.59	6.71	6.83	6.93	7.03	7.14	7.24	7.35
44 x 100	41 x 97	8.05	8.23	8.40	8.55	8.68	8.81	8.93	9.09	9.22
44 x 125	41 x 120	9.38	9.60	9.81	9.99	10.15	10.31	10.45	10.64	10.81
50 x 75	47 x 72	6.87	7.01	7.13	7.25	7.35	7.45	7.53	7.67	7.78
50 x 100	47 x 97	8.62	8.80	8.97	9.12	9.25	9.38	9.50	9.66	9.80
50 x 125	47 x 120	10.01	10.24	10.44	10.62	10.77	10.94	11.00	11.00	11.00

MEMORANDA FOR EACH TRADE

WOODWORK- cont'd

Sizes of internal and external doorsets

Description	Internal Size (mm)	Permissible deviation	External Size (mm)	Permissible deviation
Co-ordinating dimension: height of door leaf height sets	2100		2100	
Co-ordinating dimension: height of ceiling height set	2300 2350 2400 2700 3000		2300 2350 2400 2700 3000	
Co-ordinating dimension: width of all door sets S = Single leaf set D = Double leaf set	600 S 700 S 800 S&D 900 S&D 1000 S&D 1200 D 1500 D 1800 D 2100 D		900 S 1000 S 1200 D 1500 D 1800 D 2100 D	
Work size: height of door leaf height set	2090	± 2.0	2095	± 2.0
Work size: height of ceiling height set	2285) 2335) 2385) ± 2.0 2685) 2985)		2295) 2345) 2395) ± 2.0 2695) 2995)	
Work size: width of all door sets S = Single leaf set D = Double leaf set	590 S) 690 S) 790 S&D) 890 S&D) 990 S&D) ± 2.0 1190 D) 1490 D) 1790 D) 2090 D)		895 S) 995 S) 1195 D) 1495 D) ± 2.0 1795 D) 2095 D)	
Width of door leaf in single leaf sets F = Flush leaf P = Panel leaf	526 F) 626 F) 726 F&P) ± 1.5 826 F&P) 926 F&P)		806 F&P) 906 F&P)	± 1.5

MEMORANDA FOR EACH TRADE

WOODWORK - cont'd

Description	Internal Size (mm)	Internal Permissible deviation	External Size (mm)	External Permissible deviation
Width of door leaf in double leaf sets F = Flush leaf P = Panel leaf	362 F) 412 F) 426 F) 562 F&P) ± 1.5 712 F&P) 826 F&P) 1012 F&P)		552 F&P) 702 F&P) ± 1.5 852 F&P) 1002 F&P)	
Door leaf height for all door sets	2040	± 1.5	1994	± 1.5

STRUCTURAL STEELWORK

Tables showing the mass and surface area per metre run for various steel members

Size (mm)	Mass (kg/m)	Surface area per m run (m²)
Universal beams		
914 x 419	388	3.404
	343	3.382
914 x 305	289	2.988
	253	2.967
	224	2.948
	201	2.932
838 x 292	226	2.791
	194	2.767
	176	2.754
762 x 267	197	2.530
	173	2.512
	147	2.493
686 x 254	170	2.333
	152	2.320
	140	2.310
	125	2.298
610 x 305	238	2.421
	179	2.381
	149	2.361
610 x 229	140	2.088
	125	2.075
	113	2.064
	101	2.053
533 x 210	122	1.872
	109	1.860
	101	1.853
	92	1.844
	82	1.833

MEMORANDA FOR EACH TRADE

STRUCTURAL STEELWORK - cont'd

Tables showing the mass and surface area per metre run for various steel members - cont'd

Size (mm)	Mass (kg/m)	Surface area per m run (m²)
Universal beams - cont'd		
457 x 191	98	1.650
	89	1.641
	82	1.633
	74	1.625
	67	1.617
457 x 152	82	1.493
	74	1.484
	67	1.474
	60	1.487
	52	1.476
406 x 178	74	1.493
	67	1.484
	60	1.476
	54	1.468
406 x 140	46	1.332
	39	1.320
356 x 171	67	1.371
	57	1.358
	51	1.351
	45	1.343
356 x 127	39	1.169
	33	1.160
305 x 165	54	1.245
	46	1.235
	40	1.227
305 x 127	48	1.079
	42	1.069
	37	1.062
305 x 102	33	1.006
	28	0.997
	25	0.988
254 x 146	43	1.069
	37	1.060
	31	1.050
254 x 102	28	0.900
	25	0.893
	22	0.887
203 x 133	30	0.912
	25	0.904

MEMORANDA FOR EACH TRADE

STRUCTURAL STEELWORK - cont'd

Tables showing the mass and surface area per metre run for various steel members - cont'd

Size (mm)	Mass (kg/m)	Surface area per m run (m^2)
Universal columns		
356 x 406	634	2.525
	551	2.475
	467	2.425
	393	2.379
	340	2.346
	287	2.312
	235	2.279
356 x 368	202	2.187
	177	2.170
	153	2.154
	129	2.137
305 x 305	283	1.938
	240	1.905
	198	1.872
	158	1.839
	137	1.822
	118	1.806
	97	1.789
254 x 254	167	1.576
	132	1.543
	107	1.519
	89	1.502
	73	1.485
203 x 203	86	1.236
	71	1.218
	60	1.204
	52	1.194
	46	1.187
152 x 152	37	0.912
	30	0.900
	23	0.889
Joists		
254 x 203	81.85	1.193
254 x 114	37.20	0.882
203 x 152	52.09	0.911
152 x 127	37.20	0.722
127 x 114	29.76	0.620
127 x 114	26.79	0.635
114 x 114	26.79	0.600
102 x 102	23.07	0.528

MEMORANDA FOR EACH TRADE

STRUCTURAL STEELWORK - cont'd

Tables showing the mass and surface area per metre run for various steel members - cont'd

Size (mm)	Mass (kg/m)	Surface area per m run (m²)	
Joists - cont'd			
89 x 89	19.35	0.460	
76 x 76	12.65	0.403	

Circular hollow sections - outside dia (mm)	Mass	Surface area per m run (m²)	Thickness (mm)
21.30	1.43	0.067	3.20
26.90	1.87	0.085	3.20
33.70	1.99	0.106	2.60
	2.41	0.106	3.20
	2.93	0.106	4.00
42.40	2.55	0.133	2.60
	3.09	0.133	3.20
	3.79	0.133	4.00
48.30	3.56	0.152	3.20
	4.37	0.152	4.00
	5.34	0.152	5.00
60.30	4.51	0.189	3.20
	5.55	0.189	4.00
	6.82	0.189	5.00
76.10	5.75	0.239	3.20
	7.11	0.239	4.00
	8.77	0.239	5.00
88.90	6.76	0.279	3.20
	8.38	0.279	4.00
	10.30	0.279	5.00
114.30	9.83	0.359	3.60
	13.50	0.359	5.00
	16.80	0.359	6.30
139.70	16.60	0.439	5.00
	20.70	0.439	6.30
	26.00	0.439	8.00
	32.00	0.439	10.00
168.30	20.10	0.529	5.00
	25.20	0.529	6.30
	31.60	0.529	8.00
	39.00	0.529	10.00

MEMORANDA FOR EACH TRADE

STRUCTURAL STEELWORK - cont'd

Tables showing the mass and surface area per metre run for various steel members - cont'd

Size (mm)	Mass (kg/m)	Surface area per m run (m²)	Thickness (mm)
Circular hollow sections - outside diameter (mm) - cont'd			
193.70	23.30	0.609	5.00
	29.10	0.609	6.30
	36.60	0.609	8.00
	45.30	0.609	10.00
	55.90	0.609	12.50
	70.10	0.609	16.00
219.10	33.10	0.688	6.30
	41.60	0.688	8.00
	51.60	0.688	10.00
	63.70	0.688	12.50
	80.10	0.688	16.00
	98.20	0.688	20.00
273.00	41.40	0.858	6.30
	52.30	0.858	8.00
	64.90	0.858	10.00
	80.30	0.858	12.50
	101.00	0.858	16.00
	125.00	0.858	20.00
	153.00	0.858	25.00
323.90	62.30	1.020	8.00
	77.40	1.020	10.00
	96.00	1.020	12.50
	121.00	1.020	16.00
	150.00	1.020	20.00
	184.00	1.020	25.00
406.40	97.80	1.280	10.00
	121.00	1.280	12.50
	154.00	1.280	16.00
	191.00	1.280	20.00
	235.00	1.280	25.00
	295.00	1.280	32.00
457.00	110.00	1.440	10.00
	137.00	1.440	12.50
	174.00	1.440	16.00
	216.00	1.440	20.00
	266.00	1.440	25.00
	335.00	1.440	32.00
	411.00	1.440	40.00

MEMORANDA FOR EACH TRADE

STRUCTURAL STEELWORK - cont'd

Tables showing the mass and surface area per metre run for various steel members - cont'd

Size (mm)	Mass (kg/m)	Surface area per m run (m^2)	Thickness (mm)
Square hollow sections			
20 x 20	1.12	0.076	2.00
	1.35	0.074	2.50
30 x 30	2.14	0.114	2.50
	2.51	0.113	3.00
40 x 40	2.92	0.155	2.50
	3.45	0.154	3.00
	4.46	0.151	4.00
50 x 50	4.66	0.193	3.20
	5.72	0.191	4.00
	6.97	0.189	5.00
60 x 60	5.67	0.233	3.20
	6.97	0.231	4.00
	8.54	0.229	5.00
70 x 70	7.46	0.272	3.60
	10.10	0.269	5.00
80 x 80	8.59	0.312	3.60
	11.70	0.309	5.00
	14.40	0.306	6.30
90 x 90	9.72	0.352	3.60
	13.30	0.349	5.00
	16.40	0.346	6.30
100 x 100	12.00	0.391	4.00
	14.80	0.389	5.00
	18.40	0.386	6.30
	22.90	0.383	8.00
	27.90	0.379	10.00
120 x 120	18.00	0.469	5.00
	22.30	0.466	6.30
	27.90	0.463	8.00
	34.20	0.459	10.00
150 x 150	22.70	0.589	5.00
	28.30	0.586	6.30
	35.40	0.583	8.00
	43.60	0.579	10.00
	53.40	0.573	12.50
	66.40	0.566	16.00

MEMORANDA FOR EACH TRADE

STRUCTURAL STEELWORK - cont'd

Tables showing the mass and surface area per metre run for various steel members - cont'd

Size (mm)	Mass (kg/m)	Surface area per m run (m²)	Thickness (mm)
Square hollow sections - cont'd			
180 x 180	34.20	0.706	6.30
	43.00	0.703	8.00
	53.00	0.699	10.00
	65.20	0.693	12.50
	81.40	0.686	16.00
200 x 200	38.20	0.786	6.30
	48.00	0.783	8.00
	59.30	0.779	10.00
	73.00	0.773	12.50
	91.50	0.766	16.00
250 x 250	48.10	0.986	6.30
	60.50	0.983	8.00
	75.00	0.979	10.00
	92.60	0.973	12.50
	117.00	0.966	16.00
300 x 300	90.70	1.180	10.00
	112.00	1.170	12.50
	142.00	1.170	16.00
350 x 350	106.00	1.380	10.00
	132.00	1.370	12.50
	167.00	1.370	16.00
400 x 400	122.00	1.580	10.00
	152.00	1.570	12.50
Rectangular hollow sections			
50 x 30	2.92	0.155	2.50
	3.66	0.153	3.20
60 x 40	4.66	0.193	3.20
	5.72	0.191	4.00
80 x 40	5.67	0.232	3.20
	6.97	0.231	4.00
90 x 50	7.46	0.272	3.60
	10.10	0.269	5.00
100 x 50	6.75	0.294	3.00
	7.18	0.293	3.20
	8.86	0.291	4.00
100 x 60	8.59	0.312	3.60
	11.70	0.309	5.00
	14.40	0.306	6.30

MEMORANDA FOR EACH TRADE

STRUCTURAL STEELWORK - cont'd

Tables showing the mass and surface area per metre run for various steel members - cont'd

Size (mm)	Mass (kg/m)	Surface area per m run (m²)	Thickness (mm)
Rectangular hollow sections - cont'd			
120 x 60	9.72	0.352	3.60
	13.30	0.349	5.00
	16.40	0.346	6.30
120 x 80	14.80	0.389	5.00
	18.40	0.386	6.30
	22.90	0.383	8.00
	27.90	0.379	10.00
150 x 100	18.70	0.489	5.00
	23.30	0.486	6.30
	29.10	0.483	8.00
	35.70	0.479	10.00
160 x 80	18.00	0.469	5.00
	22.30	0.466	6.30
	27.90	0.463	8.00
	34.20	0.459	10.00
200 x 100	22.70	0.589	5.00
	28.30	0.586	6.30
	35.40	0.583	8.00
	43.60	0.579	10.00
250 x 150	38.20	0.786	6.30
	48.00	0.783	8.00
	59.30	0.779	10.00
	73.00	0.773	12.50
	91.50	0.766	16.00
300 x 200	48.10	0.986	6.30
	60.50	0.983	8.00
	75.00	0.979	10.00
	92.60	0.973	12.50
	117.00	0.966	16.00
400 x 200	90.70	1.180	10.00
	112.00	1.170	12.50
	142.00	1.170	16.00
450 x 250	106.00	1.380	10.00
	132.00	1.370	12.50
	167.00	1.370	16.00

MEMORANDA FOR EACH TRADE

STRUCTURAL STEELWORK - cont'd

Tables showing the mass and surface area per metre run for various steel members - cont'd

Size (mm)	Mass (kg/m)	Surface area per m run (m²)
Channels		
432 x 102	65.54	1.217
381 x 102	55.10	1.118
305 x 102	46.18	0.966
305 x 89	41.69	0.920
254 x 89	35.74	0.820
254 x 76	28.29	0.774
229 x 89	32.76	0.770
229 x 76	26.06	0.725
203 x 89	29.78	0.720
203 x 76	23.82	0.675
178 x 89	26.81	0.671
178 x 76	20.84	0.625
152 x 89	23.84	0.621
152 x 76	17.88	0.575
127 x 64	14.90	0.476

Angles - sum of leg lengths	Thickness (mm)	Mass (kg/m)	Surface area per m run (m²)
50	3	1.11	0.10
	4	1.45	0.10
	5	1.77	0.10
80	4	2.42	0.16
	5	2.97	0.16
	6	3.52	0.16
90	4	2.74	0.18
	5	3.38	0.18
	6	4.00	0.18
100	5	3.77	0.20
	6	4.47	0.20
	8	5.82	0.20
115	5	4.35	0.23
	6	5.16	0.23
	8	6.75	0.23

MEMORANDA FOR EACH TRADE

STRUCTURAL STEELWORK - cont'd

Tables showing the mass and surface area per metre run for various steel members - cont'd

Angles - sum of leg lengths - cont'd	Thickness (mm)	Mass (kg/m)	Surface area per m run (m^2)
120	5	4.57	0.24
	6	5.42	0.24
	8	7.09	0.24
	10	8.69	0.24
125	6	5.65	0.25
	8	7.39	0.25
200	8	12.20	0.40
	10	15.00	0.40
	12	17.80	0.40
	15	21.90	0.40
225	10	17.00	0.45
	12	20.20	0.45
	15	24.80	0.45
240	8	14.70	0.48
	10	18.20	0.48
	12	21.60	0.48
	15	26.60	0.48
300	10	23.00	0.60
	12	27.30	0.60
	15	33.80	0.60
	18	40.10	0.60
350	12	32.00	0.70
	15	39.60	0.70
	18	47.10	0.70
400	16	48.50	0.80
	18	54.20	0.80
	20	59.90	0.80
	24	71.10	0.80

MEMORANDA FOR EACH TRADE

PLUMBING AND MECHANICAL INSTALLATIONS

Dimensions and weights of tubes

Outside diameter (mm)	Internal dia (mm)	Weight per m (kg)	Internal dia (mm)	Weight per m (kg)	Internal dia (mm)	Weight per m (kg)
Copper to EN 1057:1996						
	Table X		Table Y		Table Z	
6	4.80	0.0911	4.40	0.1170	5.00	0.0774
8	6.80	0.1246	6.40	0.1617	7.00	0.1054
10	8.80	0.1580	8.40	0.2064	9.00	0.1334
12	10.80	0.1914	10.40	0.2511	11.00	0.1612
15	13.60	0.2796	13.00	0.3923	14.00	0.2031
18	16.40	0.3852	16.00	0.4760	16.80	0.2918
22	20.22	0.5308	19.62	0.6974	20.82	0.3589
28	26.22	0.6814	25.62	0.8985	26.82	0.4594
35	32.63	1.1334	32.03	1.4085	33.63	0.6701
42	39.63	1.3675	39.03	1.6996	40.43	0.9216
54	51.63	1.7691	50.03	2.9052	52.23	1.3343
76.1	73.22	3.1287	72.22	4.1437	73.82	2.5131
108	105.12	4.4666	103.12	7.3745	105.72	3.5834
133	130.38	5.5151	-	-	130.38	5.5151
159	155.38	8.7795	-	-	156.38	6.6056

PLUMBING AND MECHANICAL INSTALLATIONS - cont'd

Dimensions and weights of tubes - cont'd

Nominal size (mm)	Outside diameter max (mm)	Outside diameter min (mm)	Wall thickness (mm)	Weight (kg/m)	Weight screwed and socketted (kg/m)
Steel pipes to BS 1387					
Light gauge					
6	10.1	9.7	1.80	0.361	0.364
8	13.6	13.2	1.80	0.517	0.521
10	17.1	16.7	1.80	0.674	0.680
15	21.4	21.0	2.00	0.952	0.961
20	26.9	26.4	2.35	1.410	1.420
25	33.8	33.2	2.65	2.010	2.030
32	42.5	41.9	2.65	2.580	2.610
40	48.4	47.8	2.90	3.250	3.290
50	60.2	59.6	2.90	4.110	4.180
65	76.0	75.2	3.25	5.800	5.920
80	88.7	87.9	3.25	6.810	6.980
100	113.9	113.0	3.65	9.890	10.200
Medium gauge					
6	10.4	9.8	2.00	0.407	0.410
8	13.9	13.3	2.35	0.650	0.654
10	17.4	16.8	2.35	0.852	0.858
15	21.7	21.1	2.65	1.220	1.230
20	27.2	26.6	2.65	1.580	1.590
25	34.2	33.4	3.25	2.440	2.460
32	42.9	42.1	3.25	3.140	3.170
40	48.8	48.0	3.25	3.610	3.650
50	60.8	59.8	3.65	5.100	5.170
65	76.6	75.4	3.65	6.510	6.630
80	89.5	88.1	4.05	8.470	8.640

PLUMBING AND MECHANICAL INSTALLATIONS - cont'd

Nominal size (mm)	Outside diameter max (mm)	min (mm)	Wall thickness (mm)	Weight (kg/m)	Weight screwed and socketted (kg/m)
Medium gauge - cont'd					
100	114.9	113.3	4.50	12.100	12.400
125	140.6	138.7	4.85	16.200	16.700
150	166.1	164.1	4.85	19.200	19.800
Heavy gauge					
6	10.4	9.8	2.65	0.493	0.496
8	13.9	13.3	2.90	0.769	0.773
10	17.4	16.8	2.90	1.020	1.030
15	21.7	21.1	3.25	1.450	1.460
20	27.2	26.6	3.25	1.900	1.910
25	34.2	33.4	4.05	2.970	2.990
32	42.9	42.1	4.05	3.840	3.870
40	48.8	48.0	4.05	4.430	4.470
50	60.8	59.8	4.50	6.170	6.240
65	76.6	75.4	4.50	7.900	8.020
80	89.5	88.1	4.85	10.100	10.300
100	114.9	113.3	5.40	14.400	14.700
125	140.6	138.7	5.40	17.800	18.300
150	166.1	164.1	5.40	21.200	21.800
Stainless steel pipes to BS 4127					
8	8.045	7.940	0.60	0.1120	
10	10.045	9.940	0.60	0.1419	
12	12.045	11.940	0.60	0.1718	
15	15.045	14.940	0.60	0.2174	
18	18.045	17.940	0.70	0.3046	
22	22.055	21.950	0.70	0.3748	
28	28.055	27.950	0.80	0.5469	
35	35.070	34.965	1.00	0.8342	

PLUMBING AND MECHANICAL INSTALLATIONS - cont'd

Maximum distances between pipe supports

Pipe material	BS nominal pipe size		Pipes fitted vertically support distances in metres	Pipes fitted horizontally onto low gradients support distances in metres
	inch	mm		
Copper	0.50	15.0	1.90	1.3
	0.75	22.0	2.50	1.9
	1.00	28.0	2.50	1.9
	1.25	35.0	2.80	2.5
	1.50	42.0	2.80	2.5
	2.00	54.0	3.90	2.5
	2.50	67.0	3.90	2.8
	3.00	76.1	3.90	2.8
	4.00	108.0	3.90	2.8
	5.00	133.0	3.90	2.8
	6.00	159.0	3.90	2.8
muPVC	1.25	32.0	1.20	0.5
	1.50	40.0	1.20	0.5
	2.00	50.0	1.20	0.6
Polypropylene	1.25	32.0	1.20	0.5
	1.50	40.0	1.20	0.5
uPVC	-	82.4	1.20	0.5
	-	110.0	1.80	0.9
	-	160.0	1.80	1.2

Litres of water storage required per person in various types of building

Type of building	Storage per person (litres)
Houses and flats	90
Hostels	90
Hotels	135
Nurse's home and medical quarters	115
Offices with canteens	45
Offices without canteens	35
Restaurants, per meal served	7
Boarding school	90
Day schools	30

PLUMBING AND MECHANICAL INSTALLATIONS - cont'd

Cold water plumbing - thickness of insulation required against frost

Bore of tube	Pipework within buildings declared thermal conductivity (W/m degrees C)		
	Up to 0.040	0.041 to 0.055	0.056 to 0.070
(mm)	Minimum thickness of insulation (mm)		
15	32	50	75
20	32	50	75
25	32	50	75
32	32	50	75
40	32	50	75
50	25	32	50
65	25	32	50
80	25	32	50
100	19	25	38

Cisterns

Capacities and dimensions of galvanised mild steel cisterns from BS 417

Capacity (litres)	BS type	Dimensions (mm)		
		length	width	depth
18	SCM 45	457	305	305
36	SCM 70	610	305	371
54	SCM 90	610	406	371
68	SCM 110	610	432	432
86	SCM 135	610	457	482
114	SCM 180	686	508	508
159	SCM 230	736	559	559
191	SCM 270	762	584	610
227	SCM 320	914	610	584
264	SCM 360	914	660	610
327	SCM 450/1	1220	610	610
336	SCM 450/2	965	686	686
423	SCM 570	965	762	787
491	SCM 680	1090	864	736
709	SCM 910	1170	889	889

Capacities of cold water polypropylene storage cisterns from BS 4213

Capacity (litres)	BS type	Maximum height (mm)
18	PC 4	310
36	PC 8	380
68	PC 15	430
91	PC 20	510
114	PC 25	530
182	PC 40	610
227	PC 50	660
273	PC 60	660
318	PC 70	660
455	PC 100	760

HEATING AND HOT WATER INSTALLATIONS

Storage capacity and recommended power of hot water storage boilers

Type of building	Storage at 65°C (litres per person)	Boiler power to 65°C (kW per person)
Flats and dwellings		
(a) Low rent properties	25	0.5
(b) Medium rent properties	30	0.7
(c) High rent properties	45	1.2
Nurses homes	45	0.9
Hostels	30	0.7
Hotels		
(a) Top quality - upmarket	45	1.2
(b) Average quality - low market	35	0.9
Colleges and schools		
(a) Live-in accommodation	25	0.7
(b) Public comprehensive	5	0.1
Factories	5	0.1
Hospitals		
(a) General	30	1.5
(b) Infectious	45	1.5
(c) Infirmaries	25	0.6
(d) Infirmaries (inc. laundry facilities)	30	0.9
(e) Maternity	30	2.1
(f) Mental	25	0.7
Offices	5	0.1
Sports pavilions	35	0.3

Thickness of thermal insulation for heating installations

Size of tube (mm)	Up to 0.025	Declared thermal conductivity		
		0.026 to 0.040	0.041 to 0.055	0.056 to 0.070
		Minimum thickness of insulation		
LTHW Systems				
15	25	25	38	38
20	25	32	38	38
25	25	38	38	38
32	32	38	38	50
40	32	38	38	50
50	38	38	50	50
65	38	50	50	50
80	38	50	50	50

HEATING AND HOT WATER INSTALLATIONS - cont'd

Minimum thickness of insulation

Size of tube (mm)	Up to 0.025	0.026 to 0.040	0.041 to 0.055	0.056 to 0.070
LTHW Systems - cont'd				
100	38	50	50	63
125	38	50	50	63
150	50	50	63	63
200	50	50	63	63
250	50	63	63	63
300	50	63	63	63
Flat surfaces	50	63	63	63
MTHW Systems and condensate				

Declared thermal conductivity

15	25	38	38	38
20	32	38	38	50
25	38	38	38	50
32	38	50	50	50
40	38	50	50	50
50	38	50	50	50
65	38	50	50	50
80	50	50	50	63
100	50	63	63	63
125	50	63	63	63
150	50	63	63	63
200	50	63	63	63
250	50	63	63	75
300	63	63	63	75
Flat surfaces	63	63	63	75
HTHW Systems and steam				
15	38	50	50	50
20	38	50	50	50
25	38	50	50	50
32	50	50	50	63
40	50	50	50	63
50	50	50	75	75
65	50	63	75	75
80	50	63	75	75
100	63	63	75	100
125	63	63	100	100
150	63	63	100	100
200	63	63	100	100
250	63	75	100	100
300	63	75	100	100
Flat surfaces	63	75	100	100

HEATING AND HOT WATER INSTALLATIONS - cont'd

Capacities and dimensions of copper indirect cylinders (coil type) from BS 1566

Capacity (litres)	BS Type	External diameter (mm)	External height over dome (mm)
96	0	300	1600
72	1	350	900
96	2	400	900
114	3	400	1050
84	4	450	675
95	5	450	750
106	6	450	825
117	7	450	900
140	8	450	1050
162	9	450	1200
206	9 E	450	1500
190	10	500	1200
245	11	500	1500
280	12	600	1200
360	13	600	1500
440	14	600	1800

Capacity (litres)	BS Type	Internal diameter (mm)	Height (mm)
109	BSG 1M	457	762
136	BSG 2M	457	914
159	BSG 3M	457	1067
227	BSG 4M	508	1270
273	BSG 5M	508	1473
364	BSG 6M	610	1372
455	BSG 7M	610	1753
123	BSG 8M	457	838

Energy costs (July 2001)

GAS SUPPLIES

The last year has seen the wholesale gas market remain somewhat volatile. Suppliers source their gas from this market unless they have a related company producing gas when they can purchase using the transfer pricing mechanism, which is again market based and equally voltaile. This volatility has invoked a continuing increase in the wholesale price of gas and such increases continue to be reflected in the price that the end-user has to pay.

The reasons suggested for the presentsituation include heavy buying by suppliers in European markets and the link of such markets with oil prices. Another reason is that a new system has been introduced to allocate and price the capacity suppliers require to pout their gas into the network from the producers. This capacity has been restricted due to maintenance of the system which has forced entry capacity prices to rise. The suppliers are indicating that rates have reached their peak, but there would seem no real prospect of them falling off.

Domestic Markets

This sector refers to individual supply points, which do not consume more 73,250 (2,500 therms) per annum. Suppliers must supply at their published rates, although there are exceptions to this for bulk purchasing schemes, which can reduce prices by approximately 3%. By contrasting with an independent supplier, savings can still be achieved over British Gas Tariffs. Care must be exercised when selecting a supplier, look beyond the savings as some supply contracts contain onerous risk clauses. A typical "all in" supply rate would be 1.3p/kw hr for a domestic property. This shows an increase on last year.

Energy costs (July 2001) - cont'd

GAS SUPPLIES - cont'd

Commercial

This sector refers to all other gas supplies. During the last 12 months the price of gas in this sector has continued to increase. For a typical supply consuming 1,000,000 kw/annum, rates approaching 1.2 p/kw are not uncommon.

ELECTRICITY SUPPLIES

In contrast with the significant downward trend last year the market has somewhat levelled out.

Over 100 KVA Supplies

For supplies in this sector of the market there are many options to choose from regarding the charging structure. A typical contract for a supply site with an annual expenditure of £50,000 can expect an "all-in" rate in the region of 4.5 p/kw hr dependent on the load factor. Supplies in this sector require half hourly meters with the associated telephone line in order to collect the half hourly consumption data.

Under 100 KVA Supplies (Non Domestic)

This sector of the market completed its deregulation in 1999. All consumers can purchase their electricity from any authorised supplier, generally a Regional Electricty Company (REC) or Generator, although there are other independent companies in the market place. As suppliers in this market have established themselves, their pricing structures have matured, many no longer just offer discounts off the REC's tariff, but offer a pricing structure to meet the consumer's needs. However, the process of changing supplier has not in some cases been the smooth process that was intendedwith some supplies experiencing extreme difficulties in managing the transfer process. In extreme cases the industry regulator OFGEM has suspended some suppliers from taking on further business until they, OFGEM, are satisfied the companies in question have the ability to manage the process.

Supply rates achieved vary from region to region, but a typical average rate for a day night supply with an annual expenditure of £500 remains at around 6 p/kw hr.

Domestic Tariff Supplies

Again this sector of the market completed its deregulation process in 1999. Generally the principal is the same as the "Under 100 KVA" market except that the typical discounts are lower, a typical discount being 15% off the host REC tariff.

GENERALLY

For users who are able to group purchase their fuels (e.g. schools, health trusts, local authorities, housing associations and any other organisation with multiple supplie) further savings can be achieved. Advice on how to go about this or energy purchasing in general can be obtained from the editor's Davis Langdon & Everest's, Cambridge office, Tel: 01223 351 258, Fax: 01223 321 002 who have considerable experience in both purchasing energy and the establishment of bulk purchasing schemes.

CLIMATE CHANGE LEVY

This levy, which is a tax on industrial and commercial use of energy is designed to encourage businesses to use less energy and so reduce carbon dioxide emissions. It came into effect on 1st April 2001 and applies to electricity, natural gas, coal, coke and liquid petroleum gas (LPG) but is not levied on standing charges.

The rates for 2001 - 2002 are as follows:

Electricity	0.43 p/kWh
Natural Gas, Coal and Coke	0.15 p/kWh
Liquid Petroleum Gas (LPG)	0.07 p/kWh

which could add 8% - 15% to the energy bills of most businesses. VAT is charged on the levy. Energy supplies are responsible for collecting this levy from customers.

Energy costs (July 2001) - cont'd

CLIMATE CHANGE LEVY - cont'd

National Insurance contributions

The Government has reduced the level of employer's National Insurance contributions by the same amount it expects the levy to raise - so, supposedly, there will be no increase in taxation, but the impact is likely to vary company to company, or even sector by sector. The reduction in employers' National Insurance contributions is 0.30%.

Exemptions

Where a taxable commodity (electricty, gas, coal etc.) Is used for a non energy purpose, e.g. coal is used as a raw material to make carbon filters, the levy is not due. Additionally, where an organisation uses a taxable commodity to produce another taxable commodity, this is also exempt from the levy, e.g. burning gas in a power station to produce electricty. Further, in certain circumstances combined Heat and Power Plants (CHP) are exempt from the levy, and if VAT is paid at the reduced level, i.e. 5% (domestic rate) on any supplies these are not levied.

For further information you may care to access the Customs and Excise website http://www.hmce.gov.uk. A Climate Change Levy Helpdesk also exists on Tel: 0161 827 0332, Fax: 0161 827 0356. Again the Cambridge Office of Davis Langdon & Everest are happy to advise.

VENTILATION AND AIR-CONDITIONING

Typical fresh air supply factors in typical situations

Building type	Litres of fresh air per second per person	Litres of fresh air per second per m² floor area
General offices	5 - 8	1.30
Board rooms	18 - 25	6.00
Private offices	5 - 12	1.20 - 2.00
Dept. stores	5 - 8	3.00
Factories	20 - 30	0.80
Garages	-	8.00
Bars	12 - 18	-
Dance halls	8 - 12	-
Hotel rooms	8 - 12	1.70
Schools	14	-
Assembly halls	14	-
Drawing offices	16	-

Note: As a global figure for fresh air allow per 1000 m² 1.20 m³/second.

Typical air-changes per hour in typical situations

Building type	Air changes per hour
Residences	1 - 2
Churches	1 - 2
Storage buildings	1 - 2
Libraries	3 - 4
Book stacks	1 - 2
Banks	5 - 6
Offices	4 - 6
Assembly halls	5 - 10
Laboratories	4 - 6
Internal bathrooms	5 - 6
Laboratories - internal	6 - 8
Restaurants/cafes	10 - 15
Canteens	8 - 12
Small kitchens	20 - 40
Large kitchens	10 - 20
Boiler houses	15 - 30

GLAZING

Float and polished plate glass

Nominal thickness (mm)	Tolerance on thickness (mm)	Approximate weight (kg/m^2)	Normal maximum size (mm)
3	+ 0.2	7.50	2140 x 1220
4	+ 0.2	10.00	2760 x 1220
5	+ 0.2	12.50	3180 x 2100
6	+ 0.2	15.00	4600 x 3180
10	+ 0.3	25.00)	
12	+ 0.3	30.00)	6000 x 3300
15	+ 0.5	37.50	3050 x 3000
19	+ 1.0	47.50)	
25	+ 1.0	63.50)	3000 x 2900

Clear sheet glass

Nominal thickness (mm)	Tolerance on thickness (mm)	Approximate weight (kg/m^2)	Normal maximum size (mm)
2 *	+ 0.2	5.00	1920 x 1220
3	+ 0.3	7.50	2130 x 1320
4	+ 0.3	10.00	2760 x 1220
5 *	+ 0.3	12.50)	
6 *	+ 0.3	15.00)	2130 x 2400

Cast glass

Nominal thickness (mm)	Tolerance on thickness (mm)	Approximate weight (kg/m^2)	Normal maximum size (mm)
3	+ 0.4 / - 0.2	6.00)	
4	+ 0.5	7.50)	2140 x 1280
5	+ 0.5	9.50	2140 x 1320
6	+ 0.5	11.50)	
10	+ 0.8	21.50)	3700 x 1280

Wired glass

(Cast wired glass)

Nominal thickness (mm)	Tolerance on thickness (mm)	Approximate weight (kg/m^2)	Normal maximum size (mm)
6	+ 0.3 / - 0.7	-))	3700 x 1840
7	+ 0.7	-)	

(Polished wire glass)

Nominal thickness (mm)	Tolerance on thickness (mm)	Approximate weight (kg/m^2)	Normal maximum size (mm)
6	+ 1.0	-	330 x 1830

    * The 5 mm and 6 mm thickness are known as *thick drawn sheet*. Although 2 mm sheet glass is available it is not recommended for general glazing purposes.

DRAINAGE

Width required for trenches for various diameters of pipes

Pipe diameter (mm)	Trench n.e. 1.50 m deep	Trench over 1.50 m deep
n.e. 100 mm	450 mm	600 mm
100 - 150 mm	500 mm	650 mm
150 - 225 mm	600 mm	750 mm
225 - 300 mm	650 mm	800 mm
300 - 400 mm	750 mm	900 mm
400 - 450 mm	900 mm	1050 mm
450 - 600 mm	1100 mm	1300 mm

Weights and dimensions of typically sized uPVC pipes

Nominal size	Mean outside diameter (mm) min	Mean outside diameter (mm) max	Wall thickness	Weight kg per metre
Standard pipes				
82.40	82.40	82.70	3.20	1.20
110.00	110.00	110.40	3.20	1.60
160.00	160.00	160.60	4.10	3.00
200.00	200.00	200.60	4.90	4.60
250.00	250.00	250.70	6.10	7.20

Perforated pipes

Heavy grade as above

Thin wall

82.40	82.40	82.70	1.70	-
110.00	110.00	110.40	2.20	-
160.00	160.00	160.60	3.20	-

Vitrified clay pipes

Product	Nominal diameter (mm)	Effective pipe length (mm)	Limits of bore load per metre length min	Limits of bore load per metre length max	Crushing strength (kN/m)	Weight kg/pipe (/m)
Supersleve	100	1600	96	105	35.00	15.63 (9.77)
Hepsleve	150	1600	146	158	22.00 (normal)	36.50 (22.81)
Hepseal	150	1500	146	158	22.00	37.04 (24.69)
	225	1750	221	235	28.00	95.24 (54.42)
	300	2500	295	313	34.00	196.08 (78.43)
	400	2500	394	414	44.00	357.14 (142.86)

DRAINAGE - cont'd

Vitrified clay pipes - cont'd

Product	Nominal diameter (mm)	Effective pipe length (mm)	Limits of bore load per metre length min	max	Crushing strength (kN/m)	Weight kg/pipe (/m)
Hepseal - cont'd	450	2500	444	464	44.00	500.00 (200.00)
	500	2500	494	514	48.00	555.56 (222.22)
	600	3000	591	615	70.00	847.46 (282.47)
	700	3000	689	719	81.00	1111.11 (370.37)
	800	3000	788	822	86.00	1351.35 (450.35)
	1000	3000	985	1027	120.00	2000.00 (666.67)
Hepline	100	1250	95	107	22.00	15.15 (12.12)
	150	1500	145	160	22.00	32.79 (21.86)
	225	1850	219	239	28.00	74.07 (40.04)
	300	1850	292	317	34.00	105.26 (56.90)
Hepduct (Conduit)	90	1500	-	-	28.00	12.05 (8.03)
	100	1600	-	-	28.00	14.29 (8.93)
	125	1250	-	-	22.00	21.28 (17.02)
	150	1250	-	-	22.00	28.57 (22.86)
	225	1850	-	-	28.00	64.52 (34.88)
	300	1850	-	-	34.00	111.11 (60.06)

USEFUL ADDRESSES FOR FURTHER INFORMATION

ACOUSTICAL INVESTIGATION & RESEARCH ORGANISATION LTD (AIRO)
Duxon's Turn
Maylands Avenue
Hemel Hempstead
Hertfordshire
HP2 4SB
Tel: 01442 247 146
Fax: 01442 256 749

ALUMINIUM FEDERATION LTD (ALFED)
Broadway House
Calthorpe Road
Five Ways
Birmingham
West Midlands
B15 1TN
Tel: 0121 456 1103
Fax: 0121 456 2274

ALUMINIUM FINISHING ASSOCIATION
Broadway House
Calthorpe Road
Five Ways
Birmingham
West Midlands
B15 1TN
Tel: 0121 456 1103
Fax: 0121 456 2274

ALUMINIUM ROLLED PRODUCTS MANUFACTURERS ASSOCIATION
Broadway House
Calthorpe Road
Five Ways
Birmingham
West Midlands
B15 1TN
Tel: 0121 456 1103
Fax: 0121 456 2274

AMERICAN HARDWOOD EXPORT COUNCIL (AHEC)
10 Throgmorton Avenue
London
EC2N 2DL
Tel: 0207 588 8811
Fax: 0207 588 8855

ANCIENT MONUMENTS SOCIETY (AMS)
Saint Ann's Vestry Hall
2 Church Entry
London
EC4V 5HB
Tel: 0207 236 3934
Fax: 0207 329 3677

APA - THE ENGINEERED WOOD ASSOCIATION
MWB Business Exchange
Hinton Road
Bournemouth
BH1 2EF
Tel: 01202 201 007
Fax: 01202 201 008

ARCHITECTS AND SURVEYING INSTITUTE (ASI)
Saint Mary House
15 Saint Mary Street
Chippenham
Wiltshire
SN15 3WD
Tel: 01249 444 505
Fax: 01249 443 602

ARCHITECTURAL ADVISORY SERVICE CENTRE (POWDER/ANODIC METAL FINISHES)
Spendale House
The Runway
South Ruislip
Middlesex
HA4 6SJ
Tel: 0208 842 3111
Fax: 0208 845 9798

ARCHITECTURAL ASSOCIATION (AA)
34 - 36 Bedford Square
London
WC1B 3ES
Tel: 0207 887 4000
Fax: 0207 414 0782

ARCHITECTURAL CLADDING ASSOCIATION (ACA)
60 Charles Street
Leicester
Leicestershire
LE1 1FB
Tel: 0116 253 6161
Fax: 0116 251 4568

ASBESTOS INFORMATION CENTRE (AIC)
PO Box 69
Widnes
Cheshire
WA8 9GW
Tel: 0151 420 5866
Fax: 0151 420 5853

ASBESTOS REMOVAL CONTRACTORS' ASSOCIATION (ARCA)
Friars House
6 Parkway
Chelmsford
Essex
CM2 0NF
Tel: 01245 259 744
Fax: 01245 490 722

USEFUL ADDRESSES FOR FURTHER INFORMATION

ASSOCIATION OF INTERIOR SPECIALISTS
Olton Bridge
245 Warwick Road
Solihull
West Midlands
B92 7AH
Tel: 0121 707 0077
Fax: 0121 706 1949

BOX CULVERT ASSOCIATION (BCA)
60 Charles Street
Leicester
Leicestershire
LE1 1FB
Tel: 0116 253 6161
Fax: 0116 251 4568

BRITISH ADHESIVES & SEALANTS ASSOCIATION (BASA)
33 Fellowes Way
Stevenage
Hertfordshire
SG2 8BW
Tel: 01438 358 514
Fax: 01438 742 565

BRITISH AGGREGATE CONSTRUCTION MATERIALS INDUSTRIES LTD (BACMI)
156 Buckingham Palace Road
London
SW1W 9TR
Tel: 0207 730 8194
Fax: 0207 730 4355

BRITISH APPROVALS FOR FIRE EQUIPMENT (BAFE)
Neville House
55 Eden Street
Kingston upon Thames
Surrey
KT1 1BW
Tel: 0208 541 1950
Fax: 0208 547 1564

BRITISH APPROVALS SERVICE FOR CABLES (BASEC)

23 Presley Way
Crownhill
Milton Keynes
Buckinghamshire
MK8 0ES
Tel: 01908 267 300
Fax: 01908 267 255

BRITISH ARCHITECTURAL LIBRARY (BAL)
Royal Institute of British Architects
66 Portland Place
London
W1N 4AD
Tel: 0207 580 5533
Fax: 0207 631 1802

BRITISH ASSOCIATION OF LANDSCAPE INDUSTRIES (BALI)
Landscape House
National Agricultural Centre
Stoneleigh Park
Warwickshire
CV8 2LG
Tel: 0247 669 0333
Fax: 0247 669 0077

BRITISH BATHROOM COUNCIL
Federation House
Station Road
Stoke-on-Trent
Staffordshire
ST4 2RT
Tel: 01782 747 123
Fax: 01782 747 161

BRITISH BOARD OF AGREMENT (BBA)
PO Box 195
Bucknalls Lane
Garston
Watford
Hertfordshire
WD2 7NG
Tel: 01923 665 300
Fax: 01923 665 301

BRITISH CABLES ASSOCIATION (BCA)
37a Walton Road
East Molesey
Surrey
KT8 0DH
Tel: 0208 941 4079
Fax: 0208 783 0104

BRITISH CARPET MANUFACTURERS ASSOCIATION LTD (BCMA)
PO Box 1155
MCF Complex
60 New Road
Kidderminster
Worcestershire
DY10 2ZH
Tel: 01562 747 351
Fax: 01562 747 359

BRITISH CEMENT ASSOCIATION (BCA), CENTRE FOR CONCRETE INFORMATION
Century House
Telford Avenue
Crowthorne
Berkshire
RG45 6YS
Tel: 01344 762 676
Fax: 01344 761 214

USEFUL ADDRESSES FOR FURTHER INFORMATION

BRITISH CERAMIC CONFEDERATION (BCC)
Federation House
Station Road
Stoke-on-Trent
Staffordshire
ST4 2SA
Tel: 01782 744 631
Fax: 01782 744 102

BRITISH CERAMIC RESEARCH LTD (BCR)
Queens Road
Penkhull
Stoke-on-Trent
Staffordshire
ST4 7LQ
Tel: 01782 845 431
Fax: 01782 412 331

BRITISH CERAMIC TILE COUNCIL (BCTC)
Forum Court
83 Copers Cope Road
Beckenham
Kent
BR3 1NR
Tel: 0208 663 0946
Fax: 0208 663 0949

BRITISH COMBUSTION EQUIPMENT MANUFACTURERS ASSOCIATION (BCEMA)
58 London Road
Leicester
LE2 0QD
Tel: 0116 275 7111
Fax: 0116 275 7222

BRITISH CONCRETE MASONRY ASSOCIATION (BCMA)
Grove Crescent House
18 Grove Place
Bedford
MK40 3JJ
Tel: 01234 353 745
Fax: 01234 353 745

BRITISH CONSTRUCTIONAL STEELWORK ASSOCIATION LTD (BCSA)
4 Whitehall Court
Westminster
London
SW1A 2ES
Tel: 0207 839 8566
Fax: 0207 976 1634

BRITISH CONTRACT FURNISHING ASSOCIATION (BCFA)
Suite 214
The Business Design Centre
52 Upper Street
Islington Green
London
N1 0QH
Tel: 0207 226 6641
Fax: 0207 288 6190

BRITISH DECORATORS ASSOCIATION (BDA)
32 Coton Road
Nuneaton
Warwickshire
CV11 5TW
Tel: 0247 635 3776
Fax: 0247 635 4513

BRITISH ELECTRICAL SYSTEMS ASSOCIATION (BESA)
Granville Chambers
2 Radford Street
Stone
Staffordshire
ST15 8DA
Tel: 01785 812 426
Fax: 01785 818 157

BRITISH ELECTROTECHNICAL APPROVALS BOARD (BEAB)
1 Station View
Guildford
Surrey
GU1 4JY
Tel: 01483 455 466
Fax: 01483 455 477

BRITISH FIRE PROTECTION SYSTEMS ASSOCIATION LTD (BFPSA)
Neville House
55 Eden Street
Kingston-upon-Thames
Surrey
KT1 1BW
Tel: 0208 549 5855
Fax: 0208 547 1564

BRITISH FLAT ROOFING COUNCIL (BFRC)
186 Beardall Street
Hucknall
Nottingham
Nottinghamshire
NG15 7JU
Tel: 0115 956 6666
Fax: 0115 963 3444

USEFUL ADDRESSES FOR FURTHER INFORMATION

BRITISH FURNITURE MANUFACTURERS FEDERATION LTD (BFM Ltd)
30 Harcourt Street
London
W1H 2AA
Tel: 0207 724 0851
Fax: 0207 706 1924

BRITISH GEOLOGICAL SURVEY (BGS)
Keyworth
Nottingham
Nottinghamshire
NG12 5GG
Tel: 0115 936 3100
Fax: 0115 936 3200

BRITISH INDEPENDENT PLASTIC EXTRUDERS ASSOCIATION (BIPE)
89 Cornwall Street
Birmingham
West Midlands
B3 3BY
Tel: 0121 236 1866
Fax: 0121 200 1389

BRITISH INSTITUTE OF ARCHITECTURAL TECHNOLOGISTS (BIAT)
397 City Road
London
EC1V 1NH
Tel: 0207 278 2206
Fax: 0207 837 3194

BRITISH LAMINATED FABRICATORS ASSOCIATION
6 Bath Place
Rivington Street
London
EC2A 3JE
Tel: 0207 457 5025
Fax: 0207 457 5038

BRITISH LIBRARY BIBLIOGRAPHIC SERVICE AND DOCUMENT SUPPLY
Boston Spa
Wetherby
West Yorkshire
LS23 7BQ
Tel: 01937 546 585
Fax: 01937 546 586

BRITISH LIBRARY ENVIRONMENTAL INFORMATION SERVICE
96 Euston Road
London
NW1 2DB
Tel: 0207 012 7955
Fax: 0207 412 7954

BRITISH LIBRARY, SCIENCE REFERENCE AND INFORMATION SERVICE
25 Southampton Buildings
Chancery Lane
London
WC2A 1AW
Tel: 0207 323 7494
Fax: 0207 323 7495

BRITISH NON-FERROUS METALS FEDERATION
10 Greenfield Crescent
Edgbaston
Birmingham
West Midlands
B15 3AU
Tel: 0121 456 3322
Fax: 0121 456 1394

BRITISH PLASTICS FEDERATION (BPF)
Plastics & Rubber Advisory Service
6 Bath Place
Rivington Street
London
EC2A 3JE
Tel: 0207 457 5000
Fax: 0207 457 5045

BRITISH PRECAST CONCRETE FEDERATION LTD
60 Charles Street
Leicester
Leicestershire
LE1 1FB
Tel: 0116 253 6161
Fax: 0116 251 4568

BRITISH PROPERTY FEDERATION (BPF)
1 Warwick Row
7th Floor
London
SW1E 5ER
Tel: 0207 828 0111
Fax: 0207 834 3442

BRITISH RESILIENT FLOORING MANUFACTURERS ASSOCIATION
4 Queen Square
Brighton
East Sussex
BN1 3FD
Tel: 01273 727 906
Fax: 01273 206 217

BRITISH RUBBER MANUFACTURERS' ASSOCIATION LTD (BRMA)
6 Bath Place
Rivington Street
London
EC2A 3JE
Tel: 0207 457 5040
Fax: 0207 972 9008

USEFUL ADDRESSES FOR FURTHER INFORMATION

BRITISH STANDARDS INSTITUTION (BSI)
389 Chiswick High Road
London
W4 4AL
Tel: 0208 996 9000
Fax: 0208 996 7400

CORUS RESEARCH DEVELOPMENT AND TECHNOLOGY
Swinden Technology Centre
Moorgate
Rotherham
South Yorkshire
S60 3AR
Tel: 01709 820 166
Fax: 01709 825 337

BRITISH WATER
1 Queen Anne's Gate
London
SW1H 9BT
Tel: 0207 957 4554
Fax: 0207 957 4565

BRITISH WELDED STEEL TUBE ASSOCIATION
6 Brookhouse Mill
Painswick Stroud
Gloucestershire
GL6 6SE
Tel: 01452 814 514
Fax: 01452 814 514

BRITISH WOOD PRESERVING & DAMP PROOFING ASSOCIATION (BWPDA)
1 Gleneagles House
Vernon Gateoad
South Street
Derby
DE1 1UP
Tel: 01332 225 101
Fax: 01332 225 101

BRITISH WOODWORKING FEDERATION
56 - 64 Leonard Steet
London
EC2A 4JX
Tel: 0207 608 5050
Fax: 0207 608 5051

BUILDING CENTRE
The Building Centre
26 Store Street
London
WC1E 7BT
Tel: 0207 692 4000
Fax: 0207 580 9641

BUILDING COST INFORMATION SERVICE LTD (BCIS)
Royal Institution of Chartered Surveyors
12 Great George Street
Parliament Square
London
SW1P 3AD
Tel: 0207 222 7000
Fax: 0207 695 1501

BUILDING EMPLOYERS CONFEDERATION (BEC)
56 - 64 Leonard Street
London
EC2A 4JX
Tel: 0207 608 5000
Fax: 0207 608 5001

BUILDING MAINTENANCE INFORMATION (BMI)
Royal Institution of Chartered Surveyors
12 Great George Street
Parliament Square
London
SW1P 3AD
Tel: 0207 695 1516
Fax: 0207 695 1501

BUILDING RESEARCH ESTABLISHMENT (BRE)
Bucknalls Lane
Garston
Watford
Hertfordshire
WD2 7JR
Tel: 01923 664 000
Fax: 01923 664 010

BUILDING RESEARCH ESTABLISHMENT: SCOTLAND (BRE)
Kelvin Road
East Kilbride
Glasgow
G75 0RZ
Tel: 01355 233 001
Fax: 01355 241 895

BUILDING SERVICES RESEARCH AND INFORMATION ASSOCIATION
Old Bracknell Lane West
Bracknell
Berkshire
RG12 7AH
Tel: 01344 426 511
Fax: 01344 487 575

BUILT ENVIRONMENT RESEARCH GROUP
c/o Disabled Living Foundation
380 - 384 Harrow Road
London
W9 2HU
Tel: 0207 289 6111
Fax: 0207 273 4340

USEFUL ADDRESSES FOR FURTHER INFORMATION

CATERING EQUIPMENT MANUFACTURERS ASSOCIATION (CEMA)
Carlyle House
235/237 Vauxhall Bridge Road
London
SW1V 1EJ
Tel: 0207 233 7724
Fax: 0207 828 0667

CAVITY FOAM BUREAU
PO Box 79
Oldbury
Warley
West Midlands
B69 4PW
Tel: 0121 544 4949
Fax: 0121 544 3569

CHARTERED INSTIRUTE OF ARBITRATORS (CIArb)
24 Angel Gate
City Road
London
EC1V 2RS
Tel: 0207 837 4483
Fax: 0207 837 4185

CHARTERED INSTITUTE OF BUILDING (CIOB)
Englemere
Kings Ride
Ascot
Berkshire
SL5 8BJ
Tel: 01344 630 700
Fax: 01344 630 888

CHARTERED INSTITUTION OF BUILDING SERVICES ENGINEERS (CIBSE)
Delta House
222 Balham High Road
London
SW12 9BS
Tel: 0208 675 5211
Fax: 0208 675 5449

CLAY PIPE DEVELOPMENT ASSOCIATION (CPDA)
Copsham House
53 Broad Street
Chesham
Buckinghamshire
HP5 3EA
Tel: 01494 791 456
Fax: 01494 792 378

CLAY ROOF TILE COUNCIL
Federation House
Station Road
Stoke-on-Trent
Staffordshire
ST4 2SA
Tel: 01782 744 631
Fax: 01782 744 102

COLD ROLLED SECTIONS ASSOCIATION (CRSA)
National Metal Forming Centre
Birmingham Road
West Bromwich
West Midlands
B70 6PY
Tel: 0121 601 6350
Fax: 0121 601 6378

COMMONWEALTH ASSOCIATION OF ARCHITECTS (CAA)
66 Portland Place
London
W1N 4AD
Tel: 0207 490 3024
Fax: 0207 253 2592

CONCRETE ADVISORY SERVICE
37 Cowbridge Road
Pontyclun
Cardiff
South Glamorgan
CF72 9EB
Tel: 01433 237 210
Fax: 01433 237 271

CONCRETE BRIDGE DEVELOPMENT GROUP
Century House
Telford Avenue
Old Wokingham Road
Crowthorne
Berkshire
RG45 6YS
Tel: 01344 725 727
Fax: 01344 727 204

CONCRETE LINTEL ASSOCIATION
60 Charles Street
Leicester
Leicestershire
LE1 1FB
Tel: 0116 253 6161
Fax: 0116 251 4568

CONCRETE PIPE ASSOCIATION (CPA)
60 Charles Street
Leicester
Leicestershire
LE1 1FB
Tel: 0116 253 6161
Fax: 0116 251 4568

CONCRETE REPAIR ASSOCIATION (CRA)
Association House
235 Ash Road
Aldershot
Hampshire
GU12 4DD
Tel: 01252 321 302
Fax: 01252 333 901

USEFUL ADDRESSES FOR FURTHER INFORMATION

CONCRETE SOCIETY
Century House
Telford Avenue
Crowthorne
Berkshire
RG45 6YS
Tel: 01344 466 007
Fax: 01344 466 008

CONFEDERATION OF BRITISH INDUSTRY (CBI)
Centre Point
103 New Oxford Street
London
WC1A 1DU
Tel: 0207 395 8020
Fax: 0207 240 0988

CONSTRUCT - CONCRETE STRUCTURES GROUP LTD
Century House
Telford Avenue
Crowthorne
Berkshire
RG11 6YS
Tel: 01344 725 744
Fax: 01344 761 214

CONSTRUCTION EMPLOYERS FEDERATION LTD (CEF)
143 Malone Road
Belfast
Northern Ireland
BT9 6SU
Tel: 02890 661 711
Fax: 02890 666 323

CONSTRUCTION INDUSTRY RESEARCH & INFORMATION ASSOCIATION (CIRIA)
6 Storey's Gate
Westminster
London
SW1P 3AU
Tel: 0207 222 8891
Fax: 0207 222 1708

CONSTRUCTION PLANT-HIRE ASSOCIATION (CPA)
52 Rochester Row
London
SW1P 1JU
Tel: 0207 630 6868
Fax: 0207 630 6765

CONTRACT FLOORING ASSOCIATION (CFA)
4c Saint Mary's Place
The Lace Market
Nottingham
Nottinghamshire
NG1 1PH
Tel: 0115 941 1126
Fax: 0115 941 2238

CONTRACTORS MECHANICAL PLANT ENGINEERS (CMPE)
3 Hillview
Hornbeam
Waterlooville
Hampshire
PO8 9EYH
Tel: 023 9236 5829
Fax: 023 9236 5829

COPPER DEVELOPMENT ASSOCIATION
Verulam Industrial Estate
224 Londons Road
Saint Albans
Hertfordshire
AL1 1AQ
Tel: 01727 731 200
Fax: 01727 731 216

COUNCIL FOR ALUMINIUM IN BUILDING (CAB)
191 Cirencester Road
Charlton Kings
Cheltenham
Gloucestershire
GL53 8DF
Tel: 01242 578 278
Fax: 01242 578 283

DRY STONE WALLING ASSOCIATION OF GREAT BRITAIN (DSWA)
PO Box 8615
Sutton Coldfield
West Midlands
B75 7HR
Tel: 0121 378 0493
Fax: 0121 378 0493

DUCTILE IRON PIPE ASSOCIATION (DIPA)
The McLaren Building
35 Dale End
Birmingham
West Midlands
B4 7LN
Tel: 0121 200 2100
Fax: 0121 200 1306

ELECTRICAL CONTRACTORS ASSOCIATION (ECA)
ESCA House
34 Palace Court
Bayswater
London
W2 4HY
Tel: 0207 313 4800
Fax: 0207 221 7344

USEFUL ADDRESSES FOR FURTHER INFORMATION

ELECTRICAL CONTRACTORS ASSOCIATION OF SCOTLAND (SELECT)
Bush House
Bush Estate
Midlothian
Scotland
EH26 0SB
Tel: 0131 445 5577
Fax: 0131 445 5548

ELECTRICAL INSTALLATION EQUIPMENT MANUFACTURERS ASSOCIATION LTD (EIEMA)
Westminster Tower
3 Albert Embankment
London
SE1 7SL
Tel: 0207 793 3013
Fax: 0207 735 4158

ELECTRICITY ASSOCIATION
30 Millbank
London
SW1P 4RD
Tel: 0207 963 5817
Fax: 0207 963 5957

ENERGY SYSTEMS TRADE ASSOCIATION LTD (ESTA)
Palace Chambers
41 London Road
Stroud
Gloucestershire
GL5 2AJ
Tel: 01453 767 373
Fax: 01453 767 376

EUROPEAN LIQUID ROOFING ASSOCIATION (ELRA)
Fields House
Gower Road
Haywards Heath
West Sussex
RH16 4PL
Tel: 01444 417 458
Fax: 01444 415 616

FABRIC CARE RESEARCH ASSOCIATION
Forest House Laboratories
Knaresborough Road
Harrogate
North Yorkshire
HG2 7LZ
Tel: 01423 885 977
Fax: 01423 880 045

FACULTY OF BUILDING
Central Office
35 Hayworth Road
Sandiacre
Nottingham
Nottinghamshire
NG10 5LL
Tel: 0115 949 0641
Fax: 0115 949 1664

FEDERATION OF MANUFACTURERS OF CONSTRUCTION EQUIPMENT & CRANES
Ambassador House
Brigstock Road
Thornton Heath
Surrey
CR7 7JG
Tel: 0208 665 5727
Fax: 0208 665 6447

FEDERATION OF MASTER BUILDERS
Gordon Fisher House
14 - 15 Great James Street
London
WC1N 3DP
Tel: 0207 242 7583
Fax: 0207 404 0296

FEDERATION OF PILING SPECIALISTS
Forum Court
83 Coppers Cope Road
Beckenham
Kent
BR3 1NR
Tel: 0208 663 0947
Fax: 0208 663 0949

FEDERATION OF PLASTERING & DRYWALL CONTRACTORS
Construction House
56 - 64 Leonard Street
London
EC2A 4JX
Tel: 0207 608 5092
Fax: 0207 608 5081

FEDERATION OF THE ELECTRONICS INDUSTRY (FEI)
Russell Square House
10 - 12 Russell Square
London
WC1B 5EE
Tel: 0207 331 2000
Fax: 0207 331 2040

FENCING CONTRACTORS ASSOCIATION
Warren Road
Trellech
Monmouthshire
NP5 4PQ
Tel: 07000 560 722
Fax: 01600 860 614

USEFUL ADDRESSES FOR FURTHER INFORMATION

FINNISH PLYWOOD INTERNATIONAL
Stags End House
Gaddesden Row
Hemel Hempstead
Hertfordshire
HP2 6HN
Tel: 01582 794 661
Fax: 01582 792 755

FLAT GLASS MANUFACTURERS ASSOCIATION
Prescot Road
Saint Helens
Merseyside
WA10 3TT
Tel: 01744 288 82
Fax: 01744 692 660

FLAT ROOFING ALLIANCE
Fields House
Gower Road
Haywards Heath
West Sussex
RH16 4PL
Tel: 01444 440 027
Fax: 01444 415 616

FURNITURE INDUSTRY RESEARCH ASSOCIATION (FIRA INTERNATIONAL LTD)
Maxwell Road
Stevenage
Hertfordshire
SG1 2EW
Tel: 01438 777 700
Fax: 01438 777 800

GLASS & GLAZING FEDERATION (GGF)
44 - 48 Borough High Street
London
SE1 1XB
Tel: 0207 403 7777
Fax: 0207 357 7458

HEATING & VENTILATING CONTRACTORS' ASSOCIATION
ESCA House
34 Palace Court
Bayswater
London
W2 4JG
Tel: 0207 229 2488
Fax: 0207 727 9268

HOUSING CORPORATION HEADQUARTERS
149 Tottenham Court Road
London
W1P 0BN
Tel: 0207 393 2000
Fax: 0207 393 2111

INSTITUTE OF ACOUSTICS
77A Saint Peter' Street
Saint Albans
Hertfordshire
AL1 3BN
Tel: 01727 848 195
Fax: 01727 850 553

INSTITUTE OF ASPHALT TECHNOLOGY
Office 5
Trident House
Clare Road
Stanwell
Middlesex
TW19 7QU
Tel: 01784 423 444
Fax: 01784 423 888

INSTITUTE OF MAINTENANCE AND BUILDING MANAGEMENT
Keets House
30 East Steet
Farnham
Surrey
GU9 7SW
Tel: 01252 710 994
Fax: 01252 737 741

INSTITUTE OF MATERIALS
1 Carlton House Terrace
London
SW1Y 5DB
Tel: 0207 451 7300
Fax: 0207 839 1702

INSTITUTE OF PLUMBING
64 Station Lane
Hornchurch
Essex
RM12 6NB
Tel: 01708 472 791
Fax: 01708 448 987

INSTITUTE OF SHEET METAL ENGINEERING
Unit 16
Greenlands Business Centre
Studley Road
Redditch
Worcestershire
B98 7HD
Tel: 01527 515 351
Fax: 01527 514 745

INSTITUTE OF WASTES MANAGEMENT
9 Saxon Court
Saint Peter's Gardens
Northampton
Northamptonshire
NN1 1SX
Tel: 01604 620 426
Fax: 01604 621 339

USEFUL ADDRESSES FOR FURTHER INFORMATION

INSTITUTE OF WOOD SCIENCE
Stocking Lane
Hughenden Valley
High Wycombe
Buckinghamshire
HP14 4NU
Tel: 01494 565 374
Fax: 01494 565 395

INSTITUTION OF BRITISH ENGINEERS
The Royal Liver Building
6 Hampton Place
Brighton
East Sussex
BN1 3DD
Tel: 01273 734 274

INSTITUTION OF CIVIL ENGINEERS (ICE)
1 - 7 Great George Street
London
SW1P 3AA
Tel: 0207 222 7722
Fax: 0207 222 7500

INSTITUTION OF INCORPORATED ENGINEERS
Savoy Hill House
Savoy Hill
London
WC2R 0BS
Tel: 0207 836 3357
Fax: 0207 497 9006

INSTITUTION OF STRUCTURAL ENGINEERS (ISE)
11 Upper Belgrave Street
London
SW1X 8BH
Tel: 0207 235 4535
Fax: 0207 235 4294

INTERPAVE (THE PRECAST CONCRETE PAVING & KERB ASSOCIATION)
60 Charles Street
Leicester
Leicestershire
LE1 1FB
Tel: 0116 253 6161
Fax: 0116 251 4568

JOINT CONTRACTS TRIBUNAL LTD
66 Portland Place
London
W1N 4AD
Tel: 0207 580 5533
Fax: 0207 255 1541

KITCHEN SPECIALISTS ASSOCIATION
PO Box 311
Worcester
Worcestershire
WR1 1DR
Tel: 01905 619 922
Fax: 01905 726 469

LIGHTING ASSOCIATION LTD
Stafford Park 7
Telford
Shropshire
TF3 3BQ
Tel: 01952 290 905
Fax: 01952 290 906

MASTIC ASPHALT COUNCIL LTD
Claridge House
5 Elwick Road
Kent
TN23 1PD
Tel: 01233 634 411
Fax: 01233 634 466

METAL CLADDING & ROOFING MANUFACTURERS ASSOCIATION
18 Mere Farm Road
Prenton
Birkenhead
Merseyside
CH43 9TT
Tel: 0151 652 3846
Fax: 0151 653 4080

NATIONAL HOUSE-BUILDING COUNCIL (NHBC)
Technical Department
Buildmark House
Chiltern Avenue
Amersham
Buckinghamshire
HP6 5AP
Tel: 01494 434 477
Fax: 01494 728 521

NATURAL SLATE QUARRIES ASSOCIATION
26 Store Street
London
WC1E 7BT
Tel: 0207 323 3770
Fax: 0207 323 0307

NHS ESTATES
Departments of Health
1 Trevelyan Square
Boar Lane
Leeds
West Yorkshire
LS1 6AE
Tel: 0113 254 7000
Fax: 0113 254 7299

ORDNANCE SURVEY
Romsey Road
Maybush
Southampton
SO16 4GU
Tel: 08456 050 505
Fax: 02380 792 452

USEFUL ADDRESSES FOR FURTHER INFORMATION

PAINT & POWDER FINISHING ASSOCIATION
Federation House
10 Vyse Street
Birmingham
B18 6LT
Tel: 0121 237 1123
Fax: 0121 237 1124

PIPELINE INDUSTRIES GUILD
14 - 15 Belgrave Square
London
SW1X 8PS
Tel: 0207 235 7938
Fax: 0207 235 0074

PLASTIC PIPE MANUFACTURERS SOCIETY
89 Cornwall Street
Birmingham
West Midlands
B3 3BY
Tel: 0121 236 1866
Fax: 0121 200 1389

PRECAST FLOORING FEDERATION
60 Charles Street
Leicester
Leicestershire
LE1 1FB
Tel: 0116 253 6161
Fax: 0116 251 4568

PRESTRESSED CONCRETE ASSOCIATION
60 Charles Street
Leicester
Leicestershire
LE1 1FB
Tel: 0116 253 6161
Fax: 0116 251 4568

PROPERTY CONSULTANTS SOCIETY LTD
107a Tarrant Street
Arundel
West Sussex
BN18 9DP
Tel: 01903 883 787
Fax: 01903 889 590

QUARRY PRODUCTS ASSOCIATION
156 Buckingham Palace Road
London
SW1W 9TR
Tel: 0207 730 8194
Fax: 0207 730 4355

READY-MIXED CONCRETE BUREAU
Century House
Telford Avenue
Crowthorne
Berkshire
RG45 6YS
Tel: 01344 725 732
Fax: 01344 774 976

REINFORCED CONCRETE COUNCIL
Century House
Telford Avenue
Crowthorne
Berkshire
RG45 6YS
Tel: 01344 725 733
Fax: 01344 761 214

ROYAL INCORPORATION OF ARCHITECTS IN SCOTLAND (RIAS)
15 Rutland Square
Edinburgh
Scotland
EH1 2BE
Tel: 0131 229 7545
Fax: 0131 228 2188

ROYAL INSTITUTE OF BRITISH ARCHITECTS (RIBA)
66 Portland Place
London
W1N 4AD
Tel: 0207 580 5533
Fax: 0207 255 1541

ROYAL INSTITUTION OF CHARTERED SURVEYORS (RICS)
12 Great George Street
London
SW1P 3AD
Tel: 0207 222 7000
Fax: 0207 334 3800

ROYAL TOWN PLANNING INSTITUTE (RTPI)
26 Portland Place
London
W1N 4BE
Tel: 0207 636 9107
Fax: 0207 323 1582

RURAL DESIGN AND BUILDING ASSOCIATION
6 Mill House
Cock Road
Cotton
Stowmarket
Suffolk
IP14 4QH
Tel: 01449 781 307
Fax: 01449 780 327

RURAL DEVELOPMENT COMMISSION
Headquarters
141 Castle Street
Salisbury
Wiltshire
SP1 3TP
Tel: 01722 336 255
Fax: 01722 332 769

USEFUL ADDRESSES FOR FURTHER INFORMATION

SCOTTISH BUILDING CONTRACT COMMITTEE
c/o MacRoberts Solicitors
27 Melville Street
Edinburgh
Scotland
EH3 7JF
Tel: 0131 226 2552
Fax: 0131 226 2501

SCOTTISH BUILDING EMPLOYERS FEDERATION
Carron Grange
Carron Grange Avenue
Stenhousemuir
Scotland
FK5 3BQ
Tel: 01324 555 550
Fax: 01324 555 551

SCOTTISH HOMES
Thistle House
91 Haymarket Terrace
Edinburgh
Scotland
EH12 5HE
Tel: 0131 313 0044
Fax: 0131 313 2680

SCOTTISH NATURAL HERITAGE
Communications Directorate
12 Hope Terrace
Edinburgh
EH9 2AS
Tel: 0131 447 4784
Fax: 0131 446 2279

SINGLE PLY ROOFING ASSOCIATION
Association House
186 Beardall Street
Hucknall
Nottingham
Nottinghamshire
NG15 7JU
Tel: 0115 956 6666
Fax: 0115 963 3444

SMOKE CONTROL ASSOCIATION
Henley Road
Medmenham
Marlow
Buckinghamshire
SL7 2ER
Tel: 01491 578 674
Fax: 01491 575 024

SOCIETY FOR THE PROTECTION OF ANCIENT BUILDINGS (SPAB)
37 Spital Square
London
E1 6DY
Tel: 0207 377 1644
Fax: 0207 247 5296

SOCIETY OF GLASS TECHNOLOGY
Don Valley House
Saville Street East
Sheffield
South Yorkshire
S4 7UQ
Tel: 0114 263 4455
Fax: 0114 263 4411

SOIL SURVEY AND LAND RESEARCH CENTRE
Cranfield University
Silsoe Campus
Bedford
Bedfordshire
MK45 4DT
Tel: 01525 863 263
Fax: 01525 863 253

SOLAR ENERGY SOCIETY
192 Franklin Road
Birmingham
West Midlands
B30 2HE
Tel: 0121 459 4826
Fax: 0121 459 8206

SPECIALIST ENGINEERING CONTRACTORS' GROUP
ESCA House
34 Palace Court
Bayswater
London
W2 4JG
Tel: 0207 243 4919
Fax: 0207 727 9268

SPORT ENGLAND
16 Upper Woburn Place
London
WC1H 0QP
Tel: 0207 273 1500
Fax: 0207 383 5740

SPORT SCOTLAND
Caledonia House
South Gyle
Edinburgh
Scotland
EH12 9DQ
Tel: 0131 317 7200
Fax: 0131 317 7202

SPORTS COUNCIL FOR WALES
Welsh Institute of Sport
Sophia Gardens
Cardiff
CF11 9SW
Tel: 02920 300 500
Fax: 02920 300 600

USEFUL ADDRESSES FOR FURTHER INFORMATION

SPORTS TURF RESEARCH INSTITUTE (STRI)
Saint Ives Estate
Bingley
West Yorkshire
BD16 1AU
Tel: 01274 565 131
Fax: 01274 561 891

SPRAYED CONCRETE ASSOCIATION
Association House
235 Ash Road
Aldershot
Hampshire
GU12 4DD
Tel: 01252 321 302
Fax: 01252 333 901

STAINLESS STEEL ADVISORY SERVICE
Room 2041
The Innovation Centre
217 Portobello
Sheffield
South Yorkshire
S1 4DP
Tel: 0114 224 2240
Fax: 0114 273 0444

STEEL CONSTRUCTION INSTITUTE
Silwood Park
Ascot
Berkshire
SL5 7QN
Tel: 01344 623 345
Fax: 01344 622 944

STEEL WINDOW ASSOCIATION
The Building Centre
26 Store Street
London
WC1E 7BT
Tel: 0207 637 3571
Fax: 0207 637 3572

STONE FEDERATION GREAT BRITAIN
Construction House
56 - 64 Leonard Street
London
EC2A 4JX
Tel: 0207 608 5094
Fax: 0207 608 5081

SUSPENDED ACCESS EQUIPMENT MANUFACTURERS ASSOCIATION
Construction House
56 - 64 Leonard Street
London
EC2A 4JX
Tel: 0207 608 5098
Fax: 0207 608 5081

SWIMMING POOL & ALLIED TRADES ASSOCIATION (SPATA)
Spata House
Junction Road
Andover
Hampshire
SP10 3QT
Tel: 01264 356 210
Fax: 01264 332 628

TAR INDUSTRIES SERVICES
BCRA Scientific & Technical Service
Mill Lane
Wingerworth
Chesterfield
Derbyshire
S42 6NG
Tel: 01246 209 654
Fax: 01246 272 247

THERMAL INSULATION CONTRACTORS ASSOCIATION
Charter House
450 High Road
Ilford
Essex
IG1 1UF
Tel: 0208 514 2120
Fax: 0208 478 1256

TIMBER AND BRICK INFORMATION COUNCIL
Gable House
40 High Street
Rickmansworth
Hertfordshire
WD3 1ES
Tel: 01923 778 136
Fax: 01923 720 724

TIMBER RESEARCH & DEVELOPMENT ASSOCIATION (TRADA)
Stocking Lane
Hughenden Valley
High Wycombe
Buckinghamshire
HP14 4ND
Tel: 01494 563 091
Fax: 01494 565 487

TIMBER TRADE FEDERATION
4th Floor
Clareville House
26-27 Oxenden Street
London
SW1Y 4EL
Tel: 0207 839 1891
Fax: 0207 930 0094

USEFUL ADDRESSES FOR FURTHER INFORMATION

TOWN & COUNTRY PLANNING ASSOCIATION (TCPA)
17 Carlton House Terrace
London
SW1Y 5AS
Tel: 0207 930 8903
Fax: 0207 930 3280

TREE COUNCIL
51 Catherine Place
London
SW1E 6DY
Tel: 0207 828 9928
Fax: 0207 828 9060

TRUSSED RAFTER ASSOCIATION
31 Station Road
Sutton
Retford
Nottinghamshire
DN22 8PZ
Tel: 01777 869 281
Fax: 01777 869 281

TWI (FORMERLY THE WELDING INSTITUTE)
Granada Park
Great Abington
Cambridge
Cambridgeshire
CB1 6AL
Tel: 01223 891 162
Fax: 01223 892 588

UK STEEL ASSOCIATION: REINFORCEMENT MANUFACTURING PRODUCT GROUP
Millbank Tower
21 - 24 Millbank
London
SW1P 4QP
Tel: 0207 343 3150
Fax: 0207 343 3190

UNDERFLOOR HEATING MANUFACTURERS' ASSOCIATION
Belhaven House
67 Walton Road
East Moseley
Surrey
KT8 0DB
Tel: 0208 941 7177
Fax: 0208 941 815

VERMICULITE INFORMATION SERVICE
c/o Mandoval Ltd
170 Priestley Road
Surrey Research Park
Guildford
Surrey
GU2 5RQ
Tel: 01483 883 550
Fax: 01483 883 555

WALLCOVERING MANUFACTURERS ASSOCIATION
James House
Bridge Street
Leatherhead
Surrey
KT22 7EP
Tel: 01372 360 660
Fax: 01372 376 069

WASTE MANAGEMENT INFORMATION BUREAU
AEA Technology Environment
F6 Culham
Abingdon
Oxfordshire
OX14 3DB
Tel: 01235 463 162
Fax: 01235 463 004

WATER RESEARCH CENTRE
PO Box 16
Henley Road
Medmenham
Marlow
Buckinghamshire
SL7 2HD
Tel: 01491 571 531
Fax: 01491 579 094

WATER SERVICES ASSOCIATION
1 Queen Anne's Gate
London
SW1H 9BT
Tel: 0207 957 4567
Fax: 0207 957 4666

WATERHEATER MANUFACTURERS ASSOCIATION
Milverton Lodge
Hanson Road
Victoria Park
Manchester
M14 5BZ
Tel: 0161 224 4009
Fax: 0161 257 2166

WELDING MANUFACTURERS' ASSOCIATION
Westminster Tower
3 Albert Embankment
London
SE1 7SL
Tel: 0207 793 3041
Fax: 0207 582 8020

WIMLAS LTD
Saint Peter's House
6 - 8 High Street
Iver
Buckinghamshire
SL0 9NG
Tel: 01753 737 744
Fax: 01753 792 321

USEFUL ADDRESSES FOR FURTHER INFORMATION

WOOD PANEL INDUSTRIES FEDERATION
28 Market Place
Grantham
Lincolnshire
NG31 6LR
Tel: 01476 563 707
Fax: 01476 579 314

WOOD WOOL SLAB MANUFACTURERS ASSOCIATION
26 Store Street
London
WC1E 7BT
Tel: 0207 323 3770
Fax: 0207 323 0307

ZINC DEVELOPMENT ASSOCIATION
42 Weymouth Street
London
W1N 3LQ
Tel: 0207 499 6636
Fax: 0207 493 1555

Visit www.pricebooks.co.uk

NATURAL SINES

Degrees	0' 0°.0	6' 0.1	12' 0.2	18' 0.3	24' 0.4	30' 0.5	36' 0.6	42' 0.7	48' 0.8	54' 0.9	Mean Differences 1 2 3 4
0	0.0000	0.0017	0.0035	0.0052	0.0070	0.0087	0.0105	0.0122	0.0140	0.0157	3 6 9 12
1	0.0175	0.0192	0.0209	0.0227	0.0244	0.0262	0.0279	0.0297	0.0314	0.0332	3 6 9 12
2	0.0349	0.0366	0.0384	0.0401	0.0419	0.0436	0.0454	0.0471	0.0488	0.0506	3 6 9 12
3	0.0523	0.0541	0.0558	0.0576	0.0593	0.0610	0.0628	0.0645	0.0663	0.0680	3 6 9 12
4	0.0698	0.0715	0.0732	0.0750	0.0767	0.0785	0.0802	0.0819	0.0837	0.0854	3 6 9 12
5	0.0872	0.0889	0.0906	0.0924	0.0941	0.0958	0.0976	0.0993	0.1011	0.1028	3 6 9 12
6	0.1045	0.1063	0.1080	0.1097	0.1115	0.1132	0.1149	0.1167	0.1184	0.1201	3 6 9 12
7	0.1219	0.1236	0.1253	0.1271	0.1288	0.1305	0.1323	0.1340	0.1357	0.1374	3 6 9 12
8	0.1392	0.1409	0.1426	0.1444	0.1461	0.1478	0.1495	0.1513	0.1530	0.1547	3 6 9 12
9	0.1564	0.1582	0.1599	0.1616	0.1633	0.1650	0.1668	0.1685	0.1702	0.1719	3 6 9 12
10	0.1736	0.1754	0.1771	0.1788	0.1805	0.1822	0.1840	0.1857	0.1874	0.1891	3 6 9 12
11	0.1908	0.1925	0.1942	0.1959	0.1977	0.1994	0.2011	0.2028	0.2045	0.2062	3 6 9 11
12	0.2079	0.2096	0.2113	0.2130	0.2147	0.2164	0.2181	0.2198	0.2215	0.2233	3 6 9 11
13	0.2250	0.2267	0.2284	0.2300	0.2317	0.2334	0.2351	0.2368	0.2385	0.2402	3 6 8 11
14	0.2419	0.2436	0.2453	0.2470	0.2487	0.2504	0.2521	0.2538	0.2554	0.2571	3 6 8 11
15	0.2588	0.2605	0.2622	0.2639	0.2656	0.2672	0.2689	0.2706	0.2723	0.2740	3 6 8 11
16	0.2756	0.2773	0.2790	0.2807	0.2823	0.2840	0.2857	0.2874	0.2890	0.2907	3 6 8 11
17	0.2924	0.2940	0.2957	0.2974	0.2990	0.3007	0.3024	0.3040	0.3057	0.3074	3 6 8 11
18	0.3090	0.3107	0.3123	0.3140	0.3156	0.3173	0.3190	0.3206	0.3223	0.3239	3 6 8 11
19	0.3256	0.3272	0.3289	0.3305	0.3322	0.3338	0.3355	0.3371	0.3387	0.3404	3 5 8 11
20	0.3420	0.3437	0.3453	0.3469	0.3486	0.3502	0.3518	0.3535	0.3551	0.3567	3 5 8 11
21	0.3584	0.3600	0.3616	0.3633	0.3649	0.3665	0.3681	0.3697	0.3714	0.3730	3 5 8 11
22	0.3746	0.3762	0.3778	0.3795	0.3811	0.3827	0.3843	0.3859	0.3875	0.3891	3 5 8 11
23	0.3907	0.3923	0.3939	0.3955	0.3971	0.3987	0.4003	0.4019	0.4035	0.4051	3 5 8 11
24	0.4067	0.4083	0.4099	0.4115	0.4131	0.4147	0.4163	0.4179	0.4195	0.4210	3 5 8 11
25	0.4226	0.4242	0.4258	0.4274	0.4289	0.4305	0.4321	0.4337	0.4352	0.4368	3 5 8 11
26	0.4384	0.4399	0.4415	0.4431	0.4446	0.4462	0.4478	0.4493	0.4509	0.4524	3 5 8 10
27	0.4540	0.4555	0.4571	0.4586	0.4602	0.4617	0.4633	0.4648	0.4664	0.4679	3 5 8 10
28	0.4695	0.4710	0.4726	0.4741	0.4756	0.4772	0.4787	0.4802	0.4818	0.4833	3 5 8 10
29	0.4848	0.4863	0.4879	0.4894	0.4909	0.4924	0.4939	0.4955	0.4970	0.4985	3 5 8 10
30	0.5000	0.5015	0.5030	0.5045	0.5060	0.5075	0.5090	0.5105	0.5120	0.5135	3 5 8 10
31	0.5150	0.5165	0.5180	0.5195	0.5210	0.5225	0.5240	0.5255	0.5270	0.5284	2 5 7 10
32	0.5299	0.5314	0.5329	0.5344	0.5358	0.5373	0.5388	0.5402	0.5417	0.5432	2 5 7 10
33	0.5446	0.5461	0.5476	0.5490	0.5505	0.5519	0.5534	0.5548	0.5563	0.5577	2 5 7 10
34	0.5592	0.5606	0.5621	0.5635	0.5650	0.5664	0.5678	0.5693	0.5707	0.5721	2 5 7 10
35	0.5736	0.5750	0.5764	0.5779	0.5793	0.5807	0.5821	0.5835	0.5850	0.5864	2 5 7 10
36	0.5878	0.5892	0.5906	0.5920	0.5934	0.5948	0.5962	0.5976	0.5990	0.6004	2 5 7 9
37	0.6018	0.6032	0.6046	0.6060	0.6074	0.6088	0.6101	0.6115	0.6129	0.6143	2 5 7 9
38	0.6157	0.6170	0.6184	0.6198	0.6211	0.6225	0.6239	0.6252	0.6266	0.6280	2 5 7 9
39	0.6293	0.6307	0.6320	0.6334	0.6347	0.6361	0.6374	0.6388	0.6401	0.6414	2 4 7 9
40	0.6428	0.6441	0.6455	0.6468	0.6481	0.6494	0.6508	0.6521	0.6534	0.6547	2 4 7 9
41	0.6561	0.6574	0.6587	0.6600	0.6613	0.6626	0.6639	0.6652	0.6665	0.6678	2 4 7 9
42	0.6691	0.6704	0.6717	0.6730	0.6743	0.6756	0.6769	0.6782	0.6794	0.6807	2 4 6 9
43	0.6820	0.6833	0.6845	0.6858	0.6871	0.6884	0.6896	0.6909	0.6921	0.6934	2 4 6 8
44	0.6947	0.6959	0.6972	0.6984	0.6997	0.7009	0.7022	0.7034	0.7046	0.7059	2 4 6 8

NATURAL SINES

Degrees	0' 0°.0	6' 0.1	12' 0.2	18' 0.3	24' 0.4	30' 0.5	36' 0.6	42' 0.7	48' 0.8	54' 0.9	Mean Differences 1 2 3 4
45	0.7071	0.7083	0.7096	0.7108	0.7120	0.7133	0.7145	0.7157	0.7169	0.7181	2 4 6 8
46	0.7193	0.7206	0.7218	0.7230	0.7242	0.7254	0.7266	0.7278	0.7290	0.7302	2 4 6 8
47	0.7314	0.7325	0.7337	0.7349	0.7361	0.7373	0.7385	0.7396	0.7408	0.7420	2 4 6 8
48	0.7431	0.7443	0.7455	0.7466	0.7478	0.7490	0.7501	0.7513	0.7524	0.7536	2 4 6 8
49	0.7547	0.7559	0.7570	0.7581	0.7593	0.7604	0.7615	0.7627	0.7638	0.7649	2 4 6 8
50	0.7660	0.7672	0.7683	0.7694	0.7705	0.7716	0.7727	0.7738	0.7749	0.7760	2 4 6 7
51	0.7771	0.7782	0.7793	0.7804	0.7815	0.7826	0.7837	0.7848	0.7859	0.7869	2 4 5 7
52	0.7880	0.7891	0.7902	0.7912	0.7923	0.7934	0.7944	0.7955	0.7965	0.7976	2 4 5 7
53	0.7986	0.7997	0.8007	0.8018	0.8028	0.8039	0.8049	0.8059	0.8070	0.8080	2 3 5 7
54	0.8090	0.8100	0.8111	0.8121	0.8131	0.8141	0.8151	0.8161	0.8171	0.8181	2 3 5 7
55	0.8192	0.8202	0.8211	0.8221	0.8231	0.8241	0.8251	0.8261	0.8271	0.8281	2 3 5 7
56	0.8290	0.8300	0.8310	0.8320	0.8329	0.8339	0.8348	0.8358	0.8368	0.8377	2 3 5 6
57	0.8387	0.8396	0.8406	0.8415	0.8425	0.8434	0.8443	0.8453	0.8462	0.8471	2 3 5 6
58	0.8480	0.8490	0.8499	0.8508	0.8517	0.8526	0.8536	0.8545	0.8554	0.8563	2 3 5 6
59	0.8572	0.8581	0.8590	0.8599	0.8607	0.8616	0.8625	0.8634	0.8643	0.8652	1 3 4 6
60	0.8660	0.8669	0.8678	0.8686	0.8695	0.8704	0.8712	0.8721	0.8729	0.8738	1 3 4 6
61	0.8746	0.8755	0.8763	0.8771	0.8780	0.8788	0.8796	0.8805	0.8813	0.8821	1 3 4 6
62	0.8829	0.8838	0.8846	0.8854	0.8862	0.8870	0.8878	0.8886	0.8894	0.8902	1 3 4 5
63	0.8910	0.8918	0.8926	0.8934	0.8942	0.8949	0.8957	0.8965	0.8973	0.8980	1 3 4 5
64	0.8988	0.8996	0.9003	0.9011	0.9018	0.9026	0.9033	0.9041	0.9048	0.9056	1 3 4 5
65	0.9063	0.9070	0.9078	0.9085	0.9092	0.9100	0.9107	0.9114	0.9121	0.9128	1 2 4 5
66	0.9135	0.9143	0.9150	0.9157	0.9164	0.9171	0.9178	0.9184	0.9191	0.9198	1 2 3 5
67	0.9205	0.9212	0.9219	0.9225	0.9232	0.9239	0.9245	0.9252	0.9259	0.9265	1 2 3 4
68	0.9272	0.9278	0.9285	0.9291	0.9298	0.9304	0.9311	0.9317	0.9323	0.9330	1 2 3 4
69	0.9336	0.9342	0.9348	0.9354	0.9361	0.9367	0.9373	0.9379	0.9385	0.9391	1 2 3 4
70	0.9397	0.9403	0.9409	0.9415	0.9421	0.9426	0.9432	0.9438	0.9444	0.9449	1 2 3 4
71	0.9455	0.9461	0.9466	0.9472	0.9478	0.9483	0.9489	0.9494	0.9500	0.9505	1 2 3 4
72	0.9511	0.9516	0.9521	0.9527	0.9532	0.9537	0.9542	0.9548	0.9553	0.9558	1 2 3 3
73	0.9563	0.9568	0.9573	0.9578	0.9583	0.9588	0.9593	0.9598	0.9603	0.9608	1 2 2 3
74	0.9613	0.9617	0.9622	0.9627	0.9632	0.9636	0.9641	0.9646	0.9650	0.9655	1 2 2 3
75	0.9659	0.9664	0.9668	0.9673	0.9677	0.9681	0.9686	0.9690	0.9694	0.9699	1 1 2 3
76	0.9703	0.9707	0.9711	0.9715	0.9720	0.9724	0.9728	0.9732	0.9736	0.9740	1 1 2 3
77	0.9744	0.9748	0.9751	0.9755	0.9759	0.9763	0.9767	0.9770	0.9774	0.9778	1 1 2 3
78	0.9781	0.9785	0.9789	0.9792	0.9796	0.9799	0.9803	0.9806	0.9810	0.9813	1 1 2 2
79	0.9816	0.9820	0.9823	0.9826	0.9829	0.9833	0.9836	0.9839	0.9842	0.9845	1 1 2 2
80	0.9848	0.9851	0.9854	0.9857	0.9860	0.9863	0.9866	0.9869	0.9871	0.9874	0 1 1 2
81	0.9877	0.9880	0.9882	0.9885	0.9888	0.9890	0.9893	0.9895	0.9898	0.9900	0 1 1 2
82	0.9903	0.9905	0.9907	0.9910	0.9912	0.9914	0.9917	0.9919	0.9921	0.9923	0 1 1 2
83	0.9925	0.9928	0.9930	0.9932	0.9934	0.9936	0.9938	0.9940	0.9942	0.9943	0 1 1 1
84	0.9945	0.9947	0.9949	0.9951	0.9952	0.9954	0.9956	0.9957	0.9959	0.9960	0 1 1 1
85	0.9962	0.9963	0.9965	0.9966	0.9968	0.9969	0.9971	0.9972	0.9973	0.9974	0 0 1 1
86	0.9976	0.9977	0.9978	0.9979	0.9980	0.9981	0.9982	0.9983	0.9984	0.9985	0 0 1 1
87	0.9986	0.9987	0.9988	0.9989	0.9990	0.9990	0.9991	0.9992	0.9993	0.9993	0 0 0 1
88	0.9994	0.9995	0.9995	0.9996	0.9996	0.9997	0.9997	0.9997	0.9998	0.9998	0 0 0 0
89	0.9998	0.9999	0.9999	0.9999	0.9999	1.0000	1.0000	1.0000	1.0000	1.0000	0 0 0 0
90	1.0000										

NATURAL COSINES

Degrees	0' 0°.0	6' 0.1	12' 0.2	18' 0.3	24' 0.4	30' 0.5	36' 0.6	42' 0.7	48' 0.8	54' 0.9	Mean Differences 1 2 3 4
0	1.0000	1.0000	1.0000	1.0000	1.0000	1.0000	0.9999	0.9999	0.9999	0.9999	0 0 0 0
1	0.9998	0.9998	0.9998	0.9997	0.9997	0.9997	0.9996	0.9996	0.9995	0.9995	0 0 0 0
2	0.9994	0.9993	0.9993	0.9992	0.9991	0.9990	0.9990	0.9989	0.9988	0.9987	0 0 0 1
3	0.9986	0.9985	0.9984	0.9983	0.9982	0.9981	0.9980	0.9979	0.9978	0.9977	0 0 1 1
4	0.9976	0.9974	0.9973	0.9972	0.9971	0.9969	0.9968	0.9966	0.9965	0.9963	0 0 1 1
5	0.9962	0.9960	0.9959	0.9957	0.9956	0.9954	0.9952	0.9951	0.9949	0.9947	0 1 1 1
6	0.9945	0.9943	0.9942	0.9940	0.9938	0.9936	0.9934	0.9932	0.9930	0.9928	0 1 1 1
7	0.9925	0.9923	0.9921	0.9919	0.9917	0.9914	0.9912	0.9910	0.9907	0.9905	0 1 1 2
8	0.9903	0.9900	0.9898	0.9895	0.9893	0.9890	0.9888	0.9885	0.9882	0.9880	0 1 1 2
9	0.9877	0.9874	0.9871	0.9869	0.9866	0.9863	0.9860	0.9857	0.9854	0.9851	0 1 1 2
10	0.9848	0.9845	0.9842	0.9839	0.9836	0.9833	0.9829	0.9826	0.9823	0.9820	1 1 2 2
11	0.9816	0.9813	0.9810	0.9806	0.9803	0.9799	0.9796	0.9792	0.9789	0.9785	1 1 2 2
12	0.9781	0.9778	0.9774	0.9770	0.9767	0.9763	0.9759	0.9755	0.9751	0.9748	1 1 2 3
13	0.9744	0.9740	0.9736	0.9732	0.9728	0.9724	0.9720	0.9715	0.9711	0.9707	1 1 2 3
14	0.9703	0.9699	0.9694	0.9690	0.9686	0.9681	0.9677	0.9673	0.9668	0.9664	1 1 2 3
15	0.9659	0.9655	0.9650	0.9646	0.9641	0.9636	0.9632	0.9627	0.9622	0.9617	1 2 2 3
16	0.9613	0.9608	0.9603	0.9598	0.9593	0.9588	0.9583	0.9578	0.9573	0.9568	1 2 2 3
17	0.9563	0.9558	0.9553	0.9548	0.9542	0.9537	0.9532	0.9527	0.9521	0.9516	1 2 3 3
18	0.9511	0.9505	0.9500	0.9494	0.9489	0.9483	0.9478	0.9472	0.9466	0.9461	1 2 3 4
19	0.9455	0.9449	0.9444	0.9438	0.9432	0.9426	0.9421	0.9415	0.9409	0.9403	1 2 3 4
20	0.9397	0.9391	0.9385	0.9379	0.9373	0.9367	0.9361	0.9354	0.9348	0.9342	1 2 3 4
21	0.9336	0.9330	0.9323	0.9317	0.9311	0.9304	0.9298	0.9291	0.9285	0.9278	1 2 3 4
22	0.9272	0.9265	0.9259	0.9252	0.9245	0.9239	0.9232	0.9225	0.9219	0.9212	1 2 3 4
23	0.9205	0.9198	0.9191	0.9184	0.9178	0.9171	0.9164	0.9157	0.9150	0.9143	1 2 3 5
24	0.9135	0.9128	0.9121	0.9114	0.9107	0.9100	0.9092	0.9085	0.9078	0.9070	1 2 4 5
25	0.9063	0.9056	0.9048	0.9041	0.9033	0.9026	0.9018	0.9011	0.9003	0.8996	1 3 4 5
26	0.8988	0.8980	0.8973	0.8965	0.8957	0.8949	0.8942	0.8934	0.8926	0.8918	1 3 4 5
27	0.8910	0.8902	0.8894	0.8886	0.8878	0.8870	0.8862	0.8854	0.8846	0.8838	1 3 4 5
28	0.8829	0.8821	0.8813	0.8805	0.8796	0.8788	0.8780	0.8771	0.8763	0.8755	1 3 4 6
29	0.8746	0.8738	0.8729	0.8721	0.8712	0.8704	0.8695	0.8686	0.8678	0.8669	1 3 4 6
30	0.8660	0.8652	0.8643	0.8634	0.8625	0.8616	0.8607	0.8599	0.8590	0.8581	1 3 4 6
31	0.8572	0.8563	0.8554	0.8545	0.8536	0.8526	0.8517	0.8508	0.8499	0.8490	2 3 5 6
32	0.8480	0.8471	0.8462	0.8453	0.8443	0.8434	0.8425	0.8415	0.8406	0.8396	2 3 5 6
33	0.8387	0.8377	0.8368	0.8358	0.8348	0.8339	0.8329	0.8320	0.8310	0.8300	2 3 5 6
34	0.8290	0.8281	0.8271	0.8261	0.8251	0.8241	0.8231	0.8221	0.8211	0.8202	2 3 5 7
35	0.8192	0.8181	0.8171	0.8161	0.8151	0.8141	0.8131	0.8121	0.8111	0.8100	2 3 5 7
36	0.8090	0.8080	0.8070	0.8059	0.8049	0.8039	0.8028	0.8018	0.8007	0.7997	2 3 5 7
37	0.7986	0.7976	0.7965	0.7955	0.7944	0.7934	0.7923	0.7912	0.7902	0.7891	2 4 5 7
38	0.7880	0.7869	0.7859	0.7848	0.7837	0.7826	0.7815	0.7804	0.7793	0.7782	2 4 5 7
39	0.7771	0.7760	0.7749	0.7738	0.7727	0.7716	0.7705	0.7694	0.7683	0.7672	2 4 6 7
40	0.7660	0.7649	0.7638	0.7627	0.7615	0.7604	0.7593	0.7581	0.7570	0.7559	2 4 6 8
41	0.7547	0.7536	0.7524	0.7513	0.7501	0.7490	0.7478	0.7466	0.7455	0.7443	2 4 6 8
42	0.7431	0.7420	0.7408	0.7396	0.7385	0.7373	0.7361	0.7349	0.7337	0.7325	2 4 6 8
43	0.7314	0.7302	0.7290	0.7278	0.7266	0.7254	0.7242	0.7230	0.7218	0.7206	2 4 6 8
44	0.7193	0.7181	0.7169	0.7157	0.7145	0.7133	0.7120	0.7108	0.7096	0.7083	2 4 6 8

NATURAL COSINES

Degrees	0' 0°.0	6' 0.1	12' 0.2	18' 0.3	24' 0.4	30' 0.5	36' 0.6	42' 0.7	48' 0.8	54' 0.9	Mean Differences 1 2 3 4
45	0.7071	0.7059	0.7046	0.7034	0.7022	0.7009	0.6997	0.6984	0.6972	0.6959	2 4 6 8
46	0.6947	0.6934	0.6921	0.6909	0.6896	0.6884	0.6871	0.6858	0.6845	0.6833	2 4 6 8
47	0.6820	0.6807	0.6794	0.6782	0.6769	0.6756	0.6743	0.6730	0.6717	0.6704	2 4 6 9
48	0.6691	0.6678	0.6665	0.6652	0.6639	0.6626	0.6613	0.6600	0.6587	0.6574	2 4 7 9
49	0.6561	0.6547	0.6534	0.6521	0.6508	0.6494	0.6481	0.6468	0.6455	0.6441	2 4 7 9
50	0.6428	0.6414	0.6401	0.6388	0.6374	0.6361	0.6347	0.6334	0.6320	0.6307	2 4 7 9
51	0.6293	0.6280	0.6266	0.6252	0.6239	0.6225	0.6211	0.6198	0.6184	0.6170	2 5 7 9
52	0.6157	0.6143	0.6129	0.6115	0.6101	0.6088	0.6074	0.6060	0.6046	0.6032	2 5 7 9
53	0.6018	0.6004	0.5990	0.5976	0.5962	0.5948	0.5934	0.5920	0.5906	0.5892	2 5 7 9
54	0.5878	0.5864	0.5850	0.5835	0.5821	0.5807	0.5793	0.5779	0.5764	0.5750	2 5 7 9
55	0.5736	0.5721	0.5707	0.5693	0.5678	0.5664	0.5650	0.5635	0.5621	0.5606	2 5 7 10
56	0.5592	0.5577	0.5563	0.5548	0.5534	0.5519	0.5505	0.5490	0.5476	0.5461	2 5 7 10
57	0.5446	0.5432	0.5417	0.5402	0.5388	0.5373	0.5358	0.5344	0.5329	0.5314	2 5 7 10
58	0.5299	0.5284	0.5270	0.5255	0.5240	0.5225	0.5210	0.5195	0.5180	0.5165	2 5 7 10
59	0.5150	0.5135	0.5120	0.5105	0.5090	0.5075	0.5060	0.5045	0.5030	0.5015	3 5 8 10
60	0.5000	0.4985	0.4970	0.4955	0.4939	0.4924	0.4909	0.4894	0.4879	0.4863	3 5 8 10
61	0.4848	0.4833	0.4818	0.4802	0.4787	0.4772	0.4756	0.4741	0.4726	0.4710	3 5 8 10
62	0.4695	0.4679	0.4664	0.4648	0.4633	0.4617	0.4602	0.4586	0.4571	0.4555	3 5 8 10
63	0.4540	0.4524	0.4509	0.4493	0.4478	0.4462	0.4446	0.4431	0.4415	0.4399	3 5 8 10
64	0.4384	0.4368	0.4352	0.4337	0.4321	0.4305	0.4289	0.4274	0.4258	0.4242	3 5 8 11
65	0.4226	0.4210	0.4195	0.4179	0.4163	0.4147	0.4131	0.4115	0.4099	0.4083	3 5 8 11
66	0.4067	0.4051	0.4035	0.4019	0.4003	0.3987	0.3971	0.3955	0.3939	0.3923	3 5 8 11
67	0.3907	0.3891	0.3875	0.3859	0.3843	0.3827	0.3811	0.3795	0.3778	0.3762	3 5 8 11
68	0.3746	0.3730	0.3714	0.3697	0.3681	0.3665	0.3649	0.3633	0.3616	0.3600	3 5 8 11
69	0.3584	0.3567	0.3551	0.3535	0.3518	0.3502	0.3486	0.3469	0.3453	0.3437	3 5 8 11
70	0.3420	0.3404	0.3387	0.3371	0.3355	0.3338	0.3322	0.3305	0.3289	0.3272	3 5 8 11
71	0.3256	0.3239	0.3223	0.3206	0.3190	0.3173	0.3156	0.3140	0.3123	0.3107	3 6 8 11
72	0.3090	0.3074	0.3057	0.3040	0.3024	0.3007	0.2990	0.2974	0.2957	0.2940	3 6 8 11
73	0.2924	0.2907	0.2890	0.2874	0.2857	0.2840	0.2823	0.2807	0.2790	0.2773	3 6 8 11
74	0.2756	0.2740	0.2723	0.2706	0.2689	0.2672	0.2656	0.2639	0.2622	0.2605	3 6 8 11
75	0.2588	0.2571	0.2554	0.2538	0.2521	0.2504	0.2487	0.2470	0.2453	0.2436	3 6 8 11
76	0.2419	0.2402	0.2385	0.2368	0.2351	0.2334	0.2317	0.2300	0.2284	0.2267	3 6 8 11
77	0.2250	0.2233	0.2215	0.2198	0.2181	0.2164	0.2147	0.2130	0.2113	0.2096	3 6 9 11
78	0.2079	0.2062	0.2045	0.2028	0.2011	0.1994	0.1977	0.1959	0.1942	0.1925	3 6 9 11
79	0.1908	0.1891	0.1874	0.1857	0.1840	0.1822	0.1805	0.1788	0.1771	0.1754	3 6 9 11
80	0.1736	0.1719	0.1702	0.1685	0.1668	0.1650	0.1633	0.1616	0.1599	0.1582	3 6 9 12
81	0.1564	0.1547	0.1530	0.1513	0.1495	0.1478	0.1461	0.1444	0.1426	0.1409	3 6 9 12
82	0.1392	0.1374	0.1357	0.1340	0.1323	0.1305	0.1288	0.1271	0.1253	0.1236	3 6 9 12
83	0.1219	0.1201	0.1184	0.1167	0.1149	0.1132	0.1115	0.1097	0.1080	0.1063	3 6 9 12
84	0.1045	0.1028	0.1011	0.0993	0.0976	0.0958	0.0941	0.0924	0.0906	0.0889	3 6 9 12
85	0.0872	0.0854	0.0837	0.0819	0.0802	0.0785	0.0767	0.0750	0.0732	0.0715	3 6 9 12
86	0.0698	0.0680	0.0663	0.0645	0.0628	0.0610	0.0593	0.0576	0.0558	0.0541	3 6 9 12
87	0.0523	0.0506	0.0488	0.0471	0.0454	0.0436	0.0419	0.0401	0.0384	0.0366	3 6 9 12
88	0.0349	0.0332	0.0314	0.0297	0.0279	0.0262	0.0244	0.0227	0.0209	0.0192	3 6 9 12
89	0.0175	0.0157	0.0140	0.0122	0.0105	0.0087	0.0070	0.0052	0.0035	0.0017	3 6 9 12
90	0.0000										

NATURAL TANGENTS

Degrees	0' 0°.0	6' 0.1	12' 0.2	18' 0.3	24' 0.4	30' 0.5	36' 0.6	42' 0.7	48' 0.8	54' 0.9	Mean Differences 1 2 3 4
0	0.0000	0.0017	0.0035	0.0052	0.0070	0.0087	0.0105	0.0122	0.0140	0.0157	3 6 9 12
1	0.0175	0.0192	0.0209	0.0227	0.0244	0.0262	0.0279	0.0297	0.0314	0.0332	3 6 9 12
2	0.0349	0.0367	0.0384	0.0402	0.0419	0.0437	0.0454	0.0472	0.0489	0.0507	3 6 9 12
3	0.0524	0.0542	0.0559	0.0577	0.0594	0.0612	0.0629	0.0647	0.0664	0.0682	3 6 9 12
4	0.0699	0.0717	0.0734	0.0752	0.0769	0.0787	0.0805	0.0822	0.0840	0.0857	3 6 9 12
5	0.0875	0.0892	0.0910	0.0928	0.0945	0.0963	0.0981	0.0998	0.1016	0.1033	3 6 9 12
6	0.1051	0.1069	0.1086	0.1104	0.1122	0.1139	0.1157	0.1175	0.1192	0.1210	3 6 9 12
7	0.1228	0.1246	0.1263	0.1281	0.1299	0.1317	0.1334	0.1352	0.1370	0.1388	3 6 9 12
8	0.1405	0.1423	0.1441	0.1459	0.1477	0.1495	0.1512	0.1530	0.1548	0.1566	3 6 9 12
9	0.1584	0.1602	0.1620	0.1638	0.1655	0.1673	0.1691	0.1709	0.1727	0.1745	3 6 9 12
10	0.1763	0.1781	0.1799	0.1817	0.1835	0.1853	0.1871	0.1890	0.1908	0.1926	3 6 9 12
11	0.1944	0.1962	0.1980	0.1998	0.2016	0.2035	0.2053	0.2071	0.2089	0.2107	3 6 9 12
12	0.2126	0.2144	0.2162	0.2180	0.2199	0.2217	0.2235	0.2254	0.2272	0.2290	3 6 9 12
13	0.2309	0.2327	0.2345	0.2364	0.2382	0.2401	0.2419	0.2438	0.2456	0.2475	3 6 9 12
14	0.2493	0.2512	0.2530	0.2549	0.2568	0.2586	0.2605	0.2623	0.2642	0.2661	3 6 9 12
15	0.2679	0.2698	0.2717	0.2736	0.2754	0.2773	0.2792	0.2811	0.2830	0.2849	3 6 9 13
16	0.2867	0.2886	0.2905	0.2924	0.2943	0.2962	0.2981	0.3000	0.3019	0.3038	3 6 9 13
17	0.3057	0.3076	0.3096	0.3115	0.3134	0.3153	0.3172	0.3191	0.3211	0.3230	3 6 10 13
18	0.3249	0.3269	0.3288	0.3307	0.3327	0.3346	0.3365	0.3385	0.3404	0.3424	3 6 10 13
19	0.3443	0.3463	0.3482	0.3502	0.3522	0.3541	0.3561	0.3581	0.3600	0.3620	3 7 10 13
20	0.3640	0.3659	0.3679	0.3699	0.3719	0.3739	0.3759	0.3779	0.3799	0.3819	3 7 10 13
21	0.3839	0.3859	0.3879	0.3899	0.3919	0.3939	0.3959	0.3979	0.4000	0.4020	3 7 10 13
22	0.4040	0.4061	0.4081	0.4101	0.4122	0.4142	0.4163	0.4183	0.4204	0.4224	3 7 10 14
23	0.4245	0.4265	0.4286	0.4307	0.4327	0.4348	0.4369	0.4390	0.4411	0.4431	3 7 10 14
24	0.4452	0.4473	0.4494	0.4515	0.4536	0.4557	0.4578	0.4599	0.4621	0.4642	4 7 11 14
25	0.4663	0.4684	0.4706	0.4727	0.4748	0.4770	0.4791	0.4813	0.4834	0.4856	4 7 11 14
26	0.4877	0.4899	0.4921	0.4942	0.4964	0.4986	0.5008	0.5029	0.5051	0.5073	4 7 11 15
27	0.5095	0.5117	0.5139	0.5161	0.5184	0.5206	0.5228	0.5250	0.5272	0.5295	4 7 11 15
28	0.5317	0.5340	0.5362	0.5384	0.5407	0.5430	0.5452	0.5475	0.5498	0.5520	4 8 11 15
29	0.5543	0.5566	0.5589	0.5612	0.5635	0.5658	0.5681	0.5704	0.5727	0.5750	4 8 12 15
30	0.5774	0.5797	0.5820	0.5844	0.5867	0.5890	0.5914	0.5938	0.5961	0.5985	4 8 12 16
31	0.6009	0.6032	0.6056	0.6080	0.6104	0.6128	0.6152	0.6176	0.6200	0.6224	4 8 12 16
32	0.6249	0.6273	0.6297	0.6322	0.6346	0.6371	0.6395	0.6420	0.6445	0.6469	4 8 12 16
33	0.6494	0.6519	0.6544	0.6569	0.6594	0.6619	0.6644	0.6669	0.6694	0.6720	4 8 13 17
34	0.6745	0.6771	0.6796	0.6822	0.6847	0.6873	0.6899	0.6924	0.6950	0.6976	4 9 13 17
35	0.7002	0.7028	0.7054	0.7080	0.7107	0.7133	0.7159	0.7186	0.7212	0.7239	4 9 13 18
36	0.7265	0.7292	0.7319	0.7346	0.7373	0.7400	0.7427	0.7454	0.7481	0.7508	5 9 14 18
37	0.7536	0.7563	0.7590	0.7618	0.7646	0.7673	0.7701	0.7729	0.7757	0.7785	5 9 14 18
38	0.7813	0.7841	0.7869	0.7898	0.7926	0.7954	0.7983	0.8012	0.8040	0.8069	5 9 14 19
39	0.8098	0.8127	0.8156	0.8185	0.8214	0.8243	0.8273	0.8302	0.8332	0.8361	5 10 15 20
40	0.8391	0.8421	0.8451	0.8481	0.8511	0.8541	0.8571	0.8601	0.8632	0.8662	5 10 15 20
41	0.8693	0.8724	0.8754	0.8785	0.8816	0.8847	0.8878	0.8910	0.8941	0.8972	5 10 16 21
42	0.9004	0.9036	0.9067	0.9099	0.9131	0.9163	0.9195	0.9228	0.9260	0.9293	5 11 16 21
43	0.9325	0.9358	0.9391	0.9424	0.9457	0.9490	0.9523	0.9556	0.9590	0.9623	6 11 17 22
44	0.9657	0.9691	0.9725	0.9759	0.9793	0.9827	0.9861	0.9896	0.9930	0.9965	6 11 17 23

NATURAL TANGENTS

Degrees	0' 0°.0	6' 0.1	12' 0.2	18' 0.3	24' 0.4	30' 0.5	36' 0.6	42' 0.7	48' 0.8	54' 0.9	Mean Differences 1 2 3 4
45	1.0000	1.0035	1.0070	1.0105	1.0141	1.0176	1.0212	1.0247	1.0283	1.0319	6 12 18 24
46	1.0355	1.0392	1.0428	1.0464	1.0501	1.0538	1.0575	1.0612	1.0649	1.0686	6 12 18 25
47	1.0724	1.0761	1.0799	1.0837	1.0875	1.0913	1.0951	1.0990	1.1028	1.1067	6 13 19 25
48	1.1106	1.1145	1.1184	1.1224	1.1263	1.1303	1.1343	1.1383	1.1423	1.1463	7 13 20 27
49	1.1504	1.1544	1.1585	1.1626	1.1667	1.1708	1.1750	1.1792	1.1833	1.1875	7 14 21 28
50	1.1918	1.1960	1.2002	1.2045	1.2088	1.2131	1.2174	1.2218	1.2261	1.2305	7 14 22 29
51	1.2349	1.2393	1.2437	1.2482	1.2527	1.2572	1.2617	1.2662	1.2708	1.2753	8 15 23 30
52	1.2799	1.2846	1.2892	1.2938	1.2985	1.3032	1.3079	1.3127	1.3175	1.3222	8 16 24 31
53	1.3270	1.3319	1.3367	1.3416	1.3465	1.3514	1.3564	1.3613	1.3663	1.3713	8 16 25 33
54	1.3764	1.3814	1.3865	1.3916	1.3968	1.4019	1.4071	1.4124	1.4176	1.4229	9 17 26 34
55	1.4281	1.4335	1.4388	1.4442	1.4496	1.4550	1.4605	1.4659	1.4715	1.4770	9 18 27 36
56	1.4826	1.4882	1.4938	1.4994	1.5051	1.5108	1.5166	1.5224	1.5282	1.5340	10 19 29 38
57	1.5399	1.5458	1.5517	1.5577	1.5637	1.5697	1.5757	1.5818	1.5880	1.5941	10 20 30 40
58	1.6003	1.6066	1.6128	1.6191	1.6255	1.6319	1.6383	1.6447	1.6512	1.6577	11 21 32 43
59	1.6643	1.6709	1.6775	1.6842	1.6909	1.6977	1.7045	1.7113	1.7182	1.7251	11 23 34 45
60	1.7321	1.7391	1.7461	1.7532	1.7603	1.7675	1.7747	1.7820	1.7893	1.7966	12 24 36 48
61	1.8040	1.8115	1.8190	1.8265	1.8341	1.8418	1.8495	1.8572	1.8650	1.8728	13 26 38 51
62	1.8807	1.8887	1.8967	1.9047	1.9128	1.9210	1.9292	1.9375	1.9458	1.9542	14 27 41 55
63	1.9626	1.9711	1.9797	1.9883	1.9970	2.0057	2.0145	2.0233	2.0323	2.0413	15 29 44 58
64	2.0503	2.0594	2.0686	2.0778	2.0872	2.0965	2.1060	2.1155	2.1251	2.1348	16 31 47 63
65	2.1445	2.1543	2.1642	2.1742	2.1842	2.1943	2.2045	2.2148	2.2251	2.2355	17 34 51 68
66	2.2460	2.2566	2.2673	2.2781	2.2889	2.2998	2.3109	2.3220	2.3332	2.3445	18 37 55 73
67	2.3559	2.3673	2.3789	2.3906	2.4023	2.4142	2.4262	2.4383	2.4504	2.4627	20 40 60 79
68	2.4751	2.4876	2.5002	2.5129	2.5257	2.5386	2.5517	2.5649	2.5782	2.5916	22 43 65 87
69	2.6051	2.6187	2.6325	2.6464	2.6605	2.6746	2.6889	2.7034	2.7179	2.7326	24 47 71 95
70	2.7475	2.7625	2.7776	2.7929	2.8083	2.8239	2.8397	2.8556	2.8716	2.8878	26 52 78 104
71	2.9042	2.9208	2.9375	2.9544	2.9714	2.9887	3.0061	3.0237	3.0415	3.0595	29 58 87 116
72	3.0777	3.0961	3.1146	3.1334	3.1524	3.1716	3.1910	3.2106	3.2305	3.2506	32 64 96 129
73	3.2709	3.2914	3.3122	3.3332	3.3544	3.3759	3.3977	3.4197	3.4420	3.4646	36 72 108 144
74	3.4874	3.5105	3.5339	3.5576	3.5816	3.6059	3.6305	3.6554	3.6806	3.7062	41 81 122 163
75	3.7321	3.7583	3.7848	3.8118	3.8391	3.8667	3.8947	3.9232	3.9520	3.9812	46 93 139 186
76	4.0108	4.0408	4.0713	4.1022	4.1335	4.1653	4.1976	4.2303	4.2635	4.2972	53 107 160 213
77	4.3315	4.3662	4.4015	4.4373	4.4737	4.5107	4.5483	4.5864	4.6252	4.6646	
78	4.7046	4.7453	4.7867	4.8288	4.8716	4.9152	4.9594	5.0045	5.0504	5.0970	Mean
79	5.1446	5.1929	5.2422	5.2924	5.3435	5.3955	5.4486	5.5026	5.5578	5.6140	
80	5.6713	5.7297	5.7894	5.8502	5.9124	5.9758	6.0405	6.1066	6.1742	6.2432	differences
81	6.3138	6.3859	6.4596	6.5350	6.6122	6.6912	6.7720	6.8548	6.9395	7.0264	
82	7.1154	7.2066	7.3002	7.3962	7.4947	7.5958	7.6996	7.8062	7.9158	8.0285	
83	8.1443	8.2636	8.3863	8.5126	8.6427	8.7769	8.9152	9.0579	9.2052	9.3572	cease to be
84	9.5144	9.6768	9.8448	10.0187	10.1988	10.3854	10.5789	10.7797	10.9882	11.2048	
85	11.4301	11.6645	11.9087	12.1632	12.4288	12.7062	12.9962	13.2996	13.6174	13.9507	sufficiently
86	14.3007	14.6685	15.0557	15.4638	15.8945	16.3499	16.8319	17.3432	17.8863	18.4645	
87	19.0811	19.7403	20.4465	21.2049	22.0217	22.9038	23.8593	24.8978	26.0307	27.2715	
88	28.6363	30.1446	31.8205	33.6935	35.8006	38.1885	40.9174	44.0661	47.7395	52.0807	accurate
89	57.2900	63.6567	71.6151	81.8470	95.4895	114.5887	143.237	190.984	286.478	572.957	
90	0.0000										

New Items

This edition of Spon's Architects' and Builders' Price Book introduces a number of new items and sections which have been designed to compliment the existing updated information that makes Spon's the most detailed, professionally relevant source of construction price data.

A new section is included on Capital Allowances on page 881. Many clients lose thousands of pounds each year as a direct result of lost or under-claimed Capital Allowances. This new section provides a broad outline of those items which qualify for allowances, which will assist clients and practitioners involved with commercial construction contracts, PFI projects and refurbishment projects where it is not uncommon to find that more than 75% of cost can qualify.

The following pages highlight those new items contained within the Measured Works section of the book for Major Works. Particular attention this year has been focused upon introducing 'accessible' items. The Disability Discrimination Act 1995 imposed a duty on employers to make reasonable adjustments to any physical feature of his premises which placed a disabled person at a substantial disadvatange to those who were not disabled. It also imposed a duty on those designing public buildings to make them accessible as failure to do so may constitute a breach of the duty to exercise reasonable care and skill. The Act prescribed that compliance must be made by 2004. Accordingly, we have introduced a number of items relating to automatic entrance doors, grab rails and accessible sanitary items required under the Act.

Other items include the revised RICS Schedule of Basic Plant Charges, lightweight concrete, intumescent paint fire protection to structural steelwork and roofing slates.

Prices for Measured Works - Major Works
NEW ITEMS

	PC £	Labour hours	Labour £	Material £	Unit	Total rate £
E IN SITU CONCRETE/LARGE PRECAST CONCRETE						
E05 IN SITU CONCRETE CONSTRUCTION GENERALLY						
Mixed concrete prices						
Lightweight concrete						
grade 15; lytag medium and natural sand; 40 mm aggregate	-	-	-	80.10	m³	80.10
grade 20; lytag medium and natural sand; 20 mm aggregate	-	-	-	81.05	m³	81.05
grade 25; lytag medium and natural sand; 20 mm aggregate	-	-	-	81.45	m³	81.45
grade 30; lytag medium and natural sand; 20 mm aggregate	-	-	-	83.65	m³	83.65
G STRUCTURAL/CARCASSING METAL/TIMBER						
G10 STRUCTURAL STEEL FRAMING						
Surface treatment						
At works						
intumescent paint fire protection, 30 minutes; spray applied	-	-	-	-	m²	8.00
intumescent paint fire protection, 60 minutes; spray applied	-	-	-	-	m²	12.00
Extra over intumescent paint for; decorative sealer top	-	-	-	-	m²	2.00
On site						
intumescent paint fire protection, 30 minutes; spray applied	-	-	-	-	m²	6.00
intumescent paint fire protection, 60 minutes; spray applied	-	-	-	-	m²	8.00
Extra over intumescent paint for; decorative sealer top	-	-	-	-	m²	2.00
H CLADDING/COVERING						
H60 PLAIN ROOF TILING						
Concrete interlocking slates; Redland "Richmond" smooth finish tiles or other equal and approved; PC £753.70/1000; 430 x 380; to 75 mm lap; on 25 mm x 38 mm battens and type 1F reinforced underlay						
Roof coverings	-	0.32	5.66	9.02	m²	14.68
Extra over coverings for						
fixing every tile	-	0.02	0.35	0.45	m²	0.80
eaves; eaves filler	-	0.02	0.35	0.50	m	0.85
verges; extra single undercloak course of plain tiles	-	0.23	4.06	2.69	m	6.75
ambi-dry verge system	-	0.19	3.36	7.36	m	10.72
ambi-dry verge eave/ridge end piece	-	0.02	0.35	2.64	m	2.99
valley trough tiles; cutting both sides	-	0.56	9.90	20.99	m	30.89
"Delta" ridge tiles	-	0.46	8.13	7.83	m	15.96
"Delta" hip tiles; cutting both sides	-	0.60	10.60	10.21	m	20.81
"Delta" mono-ridge tiles	-	0.46	8.13	15.41	m	23.54
gas ridge terminal	-	0.46	8.13	61.77	nr	69.90
ridge vent with 110 mm diameter flexible adaptor	-	0.46	8.13	54.31	nr	62.44
holes for pipes and the like	-	0.19	3.36	-	nr	3.36

Prices for Measured Works - Major Works

NEW ITEMS

	PC £	Labour hours	Labour £	Material £	Unit	Total rate £
L WINDOWS/DOORS/STAIRS						
L20 DOORS/SHUTTERS/HATCHES						
Doorsets; factory assembled; Dorma Entrance Systems or equal and approved; space saver automatic double doors; powder coated aluminium framing and base plates; alumnium entrance fanlights; laminated safety glazing; weatherstops; fixing by bolting						
Doorsets						
1485 mm x 2985 mm (co-ordinated size)	-	-	-	-	nr	**7000.00**
N FURNITURE/EQUIPMENT						
N13 SANITARY APPLIANCES/FITTINGS						
Sinks; Armitage Shanks or equal and approved						
Sinks; white glazed fireclay; BS 1206; pointing all round with Dow Corning Hansil silicone sealant ref 785						
Belfast sink; 455 mm x 380 mm x 205 mm ref 350016S; Nimbus ½" inclined bib taps ref 6610400; ½" wall mounts ref 81460PR; 1½" slotted waste, chain and plug, screw stay ref 70668M8; aluminium alloy build-in fixing brackets ref 7931WD0	-	2.78	33.98	117.33	nr	**151.31**
Belfast sink; 610 mm x 455 mm x 255 mm ref 350086S; Nimbus ½" inclined bib taps ref 6610400; ½" wall mounts ref 81460PR; 1½" slotted waste, chain and plug, screw stay ref 70668M8; aluminium alloy build-in fixing brackets ref 7931VE0	-	2.78	33.98	147.84	nr	**181.82**
Belfast sink; 760 mm x 455 mm x 255 mm ref 3500A6S; Nimbus ½" inclined bib taps ref 6610400; ½" wall mounts ref 81460PR; 1½" slotted waste, chain and plug, screw stay ref 70668M8; aluminium alloy build-in fixing brackets ref 7931VE0	-	2.78	33.98	215.69	nr	**249.67**
Lavatory basins; Armitage Shanks or equal and approved						
Basins; white vitreous china; BS 5506 Part 3; pointing all round with Dow Corning Hansil silcone sealant ref 785						
Portman basin 400 mm x 365 mm ref 117913J; Nuastyle 2 ½" pillar taps with anti-vandal indices ref 6973400; 1½" bead chain waste and plug, 80 mm slotted tail, bolt stay ref 90547N1; 1½" plastics bottle trap with 75 mm seal ref 70237Q4; concealed fixing bracket ref 790002Z; Isovalve servicing valve ref 9060400; screwing	-	2.13	26.03	91.61	nr	**117.64**
Portman basin 500 mm x 420 mm ref 117923S; Nuastyle 2 ½" pillar taps with anti-vandal indices ref 6973400; 1½" bead chain waste and plug, 80 mm slotted tail, bolt stay ref 90547N1; 1½" plastics bottle trap with 75 mm seal ref 70237Q4; concealed fixing bracket ref 790002Z; Isovalve servicing valve ref 9060400; screwing	-	2.13	26.03	111.29	nr	**137.32**

Prices for Measured Works - Major Works
NEW ITEMS

	PC £	Labour hours	Labour £	Material £	Unit	Total rate £
N13 SANITARY APPLIANCES/FITTINGS - cont'd						
Lavatory basins; Armitage Shanks or equal and approved - cont'd						
Basins; white vitreous china; BS 5506 Part 3; pointing all round with Dow Corning Hansil silcone sealant ref 785 - cont'd						
Tiffany basin 560 mm x 455 mm ref 121614A; Tiffany pedestal ref 132408S; Millenia ½" monobloc mixer tap with non-return valves and 1½" pop-up waste ref 6104XXX; Universal Porcelain handwheels ref 63614XXX; Isovalve servicing valve ref 9060400; screwing	-	2.13	26.03	124.24	nr	150.27
Montana basin 510 mm x 410 mm ref 120814E; Montana pedestal ref 132408S; Millenia ½" monobloc mixer tap with non-return valves and 1½" pop-up waste ref 6104XXX; Millenia metal handwheels ref 63664XX; Isovalve servicing valve ref 9060400; screwing	-	2.31	28.23	129.27	nr	157.50
Basins; white vitreous china; BS 5506 Part 3; pointing all round with Dow Corning Hansil silcone sealant ref 785 - cont'd						
Montana basin 580 mm x 475 mm ref 120824E; Montana pedestal ref 132408S; Millenia ½" monobloc mixer tap with non-return valves and 1½" pop-up waste ref 6104XXX; Millenia metal handwheels ref 63664XX; Isovalve servicing valve ref 9060400; screwing	-	2.31	28.23	132.43	nr	160.66
Portman basin 600 mm x 480 mm ref 117933S; Nuastyle 2 ½" pillar taps with anti-vandal indices ref 6973400; 1½" bead chain waste and plug, 80 mm slotted tail, bolt stay ref 90547N1; 1½" plastics bottle trap with 75 mm seal ref 70237Q4; concealed fixing bracket ref 790002Z; Isovalve servicing valve ref 9060400; screwing	-	2.13	26.03	145.53	nr	171.56
Cottage basin 560 mm x 430 mm ref 121013S; Cottage pedestal ref 132008S; Cottage ½" pillar taps ref 9290400; 1½" bead chain waste and plug, 80 mm slotted tail, bolt stay ref 90547N1' Isovalve servicing valve ref 9060400; screwing	-	2.31	28.23	206.11	nr	234.34
Cottage basin 625 mm x 500 mm ref 121023S; Cottage pedestal ref 132008S; Cottage ½" pillar taps ref 9290400; 1½" bead chain waste and plug, 80 mm slotted tail, bolt stay ref 90547N1' Isovalve servicing valve ref 9060400; screwing	-	2.31	28.23	243.74	nr	271.97
Cliveden basin 620 mm x 525 mm ref 121223S; Cliveden pedestal ref 132308S; Cliveden ½" mixer tap with 1½" pop-up waste ref 9456400; Isovalve servicing vale ref 9060400; screwing	-	2.31	28.23	532.18	nr	560.41
Drinking fountains; Armitage Shanks or equal and approved						
White vitreous china fountains; pointing all round with Dow Corning Hansil silicone selant ref 785						
Aqualon drinking fountain ref 200100; ½" self closing non-conclussive valve with flow control, plastics strainer waste, concealed hanger ref 0275A00; 1½" plastics bottle trap with 75 mm seal; screwing	-	2.31	28.23	173.04	nr	201.27

NEW ITEMS

	PC £	Labour hours	Labour £	Material £	Unit	Total rate £
Stainless steel fountains; pointing all round with Dow Corning Hansil silicone selant ref 785						
Purita drinking fountain, self closing non-conclussive valve with push button operation, flow control and 1½" strainer waste ref 53400Z5; screwing	-	2.31	28.23	154.99	nr	183.22
Purita drinking fountain, self closing non-conclussive valve with push button operation, flow control and 1½" strainer waste ref 53400Z5; pedestal shroud ref 5341000; screwing	-	2.78	33.98	343.59	nr	377.57
Baths; Armitage Shanks or equal and approved						
Bath; reinforced acrylic rectangular pattern; chromium plated overflow chain and plug; 40 mm diameter chromium plated waste; cast brass "P" trap with plain outlet and overflow connection; pair 20 mm diameter chromium plated easy clean pillar taps to BS 1010						
1700 mm long; white	171.87	3.50	42.78	184.76	nr	227.54
1700 mm long; coloured	190.44	3.50	42.78	204.72	nr	247.50
Bath; enamelled steel; medium gauge rectangular pattern; 40 mm diameter chromium plated overflow chain and plug; 40 mm diameter chromium plated waste; cast brass "P" trap with plain outlet and overflow connection; pair 20 mm diameter chromium plated easy clean pillar taps to BS 1010						
1700 mm long; white	-	3.50	42.78	284.62	nr	327.40
1700 mm long; coloured	-	3.50	42.78	299.60	nr	342.38
Water closets; Armitage Shanks or equal and approved						
White vitreous china pans and cisterns; pointing all round base with Dow Corning Hansil silicone sealant ref 785						
Seville close coupled washdown pan ref 147001A; Seville plastics seat and cover ref 68780B1; Panekta WC pan P trap connector ref 9013000; Seville 6 litres capacity cistern and cover, bottom supply ball valve, bottom overflow and close coupling fitment ref 17156FR; Seville modern lever ref 7959STR	-	3.05	37.28	156.71	nr	193.99
Extra over for; Panekta WC S trap connector ref 9014000	-	-	-	1.40	nr	1.40
Wentworth close coupled washdown pan ref 150601A; Orion III plastics seat and cover; Panekta WC pan P trap connector ref 9013000; Group 7½ litres capacity cistern and cover, bottom supply ballvalve, bottom overflow, close coupling fitment and lever ref 17650FB	-	3.05	37.28	184.66	nr	221.94

Prices for Measured Works - Major Works
NEW ITEMS

	PC £	Labour hours	Labour £	Material £	Unit	Total rate £
N13 SANITARY APPLIANCES/FITTINGS - cont'd						
Water closets; Armitage Shanks or equal and approved - cont'd						
White vitreous china pans and cisterns; pointing all round base with Dow Corning Hansil silicone sealant ref 785 - cont'd						
Tiffany back to wall washdown pan ref 154601A; Saturn plastics seat and cover ref 68980B1; Panekta WC pan P trap connector ref 9013000; Conceala 7½ litres capacity cistern and cover, side supply ball valve, side overflow, flushbend and extended lever ref 42350JE	-	3.05	37.28	200.28	nr	237.56
Extra over for; Panekta WC pan S trap connector ref 9014000	-	-	-	1.46	nr	1.46
Tiffany close coupled washdown pan ref 154301A; Saturn plastics seat and cover ref 68980B1; Panekta WC pan P trap connector ref 9013000; Tiffany 7l litres capacity cistern and cover, bottom supply ball valve, bottom overflow, close coupling fitment and side lever ref 17751FN	-	3.05	37.28	213.50	nr	250.78
Extra over for Panekta WC pan S trap connector ref 9014000	-	-	-	1.46	nr	1.46
White vitreous china pans and cisterns; pointing all round base with Dow Corning Hansil silicone sealant ref 785 - cont'd						
Cameo close coupled washdown pan ref 154301A; Cameo plastics seat and cover ref 6879NB2; Panekta WC pan P trap connector ref 9013000; Cameo 7½ litres capactity cistern and cover, bottom supply ball valve, bottom overflow and close coupling fitment ref 17831KR; luxury metal lever ref 7968000	-	3.05	37.28	275.34	nr	312.62
Extra over for; Panekta WC pan S trap connector ref 9014000	-	-	-	1.46	nr	1.46
Cottage close coupled washdown pan ref 152301A; mahogany seat and cover ref 68970B2; Panekta WC pan P trap connector ref 9013000; 7½ lites capacity cistern and cover, bottom supply ball valve, bottom overflow and close coupling fitment ref 17700FR; mahogany lever assembly ref S03SN23	-	3.05	37.28	431.13	nr	468.41
Extra over for; Panekta WC pan S trap connector ref 9014000	-	-	-	1.46	nr	1.46
Cliveden close coupled washdown pan ref 153201A; mahogany seat and cover ref 68970B2; Panekta WC pan P trap connector ref 9013000; 7½ litres capacity cistern and cover, bottom supply ball valve, bottom overflow and close coupling fitment ref 17720FR; brass level assembly ref S03SN01	-	3.05	37.28	479.09	nr	516.37
Extra over for; Panekta WC pan S trap connector ref 9014000	-	-	-	1.46	nr	1.46
Concept back to wall washdown pan ref 153301A; Concept plastics seat and cover ref 6896AB2; Panekta WC pan P trap connector ref 9013000; Conceala 7½ litres capacity cistern and cover, side supply ball valve, side overflow, flushbend and extended lever ref 42350JE	-	3.05	37.28	566.60	nr	603.88

Prices for Measured Works - Major Works

NEW ITEMS

	PC £	Labour hours	Labour £	Material £	Unit	Total rate £
Wall urinals; Armitage Shanks or equal and approved White vitreous china bowls and cisterns; pointing all round with Dow Corning Hansil silicone sealant ref 785						
single Sanura 400 mm bowls ref 261119E; top inlet spreader ref 74344A1; concealed steel hangers ref 7220000; 1½" plastics domed strainer waste ref 90568N0; 1½" plastics bottle trap with 75 mm seal ref 70238Q4; Conceala 4½ litres capacity cistern and cover ref 4225100; polished stainless steel exposed flushpipes ref 74450A1PO; screwing	-	3.70	45.22	208.21	nr	253.43
single Sanura 500 mm bowls ref 261129E; top inlet spreader ref 74344A1; concealed steel hangers ref 7220000; 1½" plastics domed strainer waste ref 90568N0; 1½" plastics bottle trap with 75 mm seal ref 70238Q4; Conceala 4½ litres capacity cistern and cover ref 4225100; polished stainless steel exposed flushpipes ref 74450A1PO; screwing	-	3.70	45.22	269.77	nr	314.99
range of 2 nr Sanura 400 mm bowls ref 261119E; top inlet spreaders ref 74344A1; concealed steel hangers ref 7220000; 1½" plastics domed strainer wastes ref 90568N0; 1½" plastics bottle traps with 75 mm seal ref 70238Q4; Conceala 9 litres capacity cistern and cover ref 4225200; polished stainless steel exposed flushpipes ref 74450B1PO; screwing	-	6.94	84.83	338.03	nr	422.86
range of 2 nr Sanura 500 mm bowls ref 261129E; top inlet spreaders ref 74344A1; concealed steel hangers ref 7220000; 1½" plastics domed strainer wastes ref 90568N0; 1½" plastics bottle traps with 75 mm seal ref 70238Q4; Conceala 9 litres capacity cistern and cover ref 4225200; polished stainless steel exposed flushpipes ref 74450B1PO; screwing	-	6.94	84.83	461.15	nr	545.98
range of 3 nr Sanura 400 mm bowls ref 261119E; top inlet spreaders ref 74344A1; concealed steel hangersref 7220000; 1½" plastics domed strainer wastes ref 90568N0; 1½" plastics bottle traps with 75 mm seal ref 70238Q4; Conceala 9 litres capacity cistern and cover ref 4225200; polished stainless steel flushpipes ref 74450C1PO; screwing	-	10.18	124.43	456.35	nr	580.78
range of 3 nr Sanura 500 mm bowls ref 261129E; top inlet spreaders ref 74344A1; concealed steel hangers ref 7220000; 1½" plastics domed strainer wastes ref 90568N0; 1½" plastics bottle traps with 75 mm seal ref 70238Q4; Conceala 9 litres capacity cistern and cover ref 4225200; polished stainless steel flushpipes ref 74450C1PO; screwing	-	10.18	124.43	640.82	nr	765.25

NEW ITEMS

	PC £	Labour hours	Labour £	Material £	Unit	Total rate £
N13 SANITARY APPLIANCES/FITTINGS - cont'd						
Wall urinals; Armitage Shanks or equal and approved - cont'd						
White vitreous china bowls and cisterns; pointing all round with Dow Corning Hansil silicone sealant ref 785 - cont'd						
range of 4 nr Sanura 400 mm bowls ref 261119E; top inlet spreaders ref 74344A1; concealed steel hangers ref 7220000; 1½" plastics domed strainer wastes ref 90568N0; 1½" plastics bottle traps with 75 mm seal ref 70238Q4; Conceala 13.60 litres capacity cistern and cover ref 4225300; polished stainless steel flushpipes ref 74450D1PO; screwing	-	13.41	163.91	592.52	nr	756.43
range of 4 nr Sanura 500 mm bowls ref 261129E; top inlet spreaders ref 74344A1; concealed steel hangers ref 7220000; 1½" plastics domed strainer wastes ref 90568N0; 1½" plastics bottle traps with 75 mm seal ref 70238Q4; Conceala 13.60 litres capacity cistern and cover ref 4225300; polished stainless steel flushpipes ref 74450D1PO; screwing	-	13.41	163.91	838.75	nr	1002.66
White vitreous china division panels; pointing all round with Dow Corning Hansil silicone sealant ref 785						
625 mm long ref 2605000; screwing	-	0.69	8.43	53.47	nr	61.90
Bidets; Armitage Shanks or equal and approved						
Bidet; vitreous china; chromium plated waste; mixer tap with hand wheels						
600 mm x 400 mm x 400 mm; white	-	3.50	42.78	184.76	nr	227.54
Shower trays and fittings; Armitage Shanks or equal and approved						
Shower tray; glazed fireclay with outlet and grated waste; chain and plug; bedding and pointing in waterproof cement mortar						
760 mm x 760 mm x 180 mm; white	-	3.00	36.67	114.85	nr	151.52
760 mm x 760 mm x 180 mm; coloured	-	3.00	36.67	174.76	nr	211.43
Shower fitting; riser pipe with mixing valve and shower rose; chromium plated; plugging and screwing mixing valve and pipe bracket						
15 mm diameter riser pipe; 127 mm diameter shower rose	-	5.00	61.11	224.71	nr	285.82
Miscellaneous fittings; Pressalit or equal and approved						
Raised seats						
Ergosit; 50 mm high; ref R19000	-	0.50	4.92	76.50	nr	81.42
Dania; 50 mm high; ref R23000	-	0.50	4.92	84.50	nr	89.42
Dania; 100 mm high; ref R24000	-	0.50	4.92	129.50	nr	134.42
Dania; sloping 100 mm - 50 mm high; ref R25000	-	0.50	4.92	123.00	nr	127.92
Raised seats and covers						
Ergosit; 50 mm high; ref R20000	-	0.50	4.92	104.50	nr	109.42
Dania; 50 mm high; ref R33000	-	0.50	4.92	91.50	nr	96.42
Dania; 100 mm high; ref R34000	-	0.50	4.92	151.00	nr	155.92
Dania; sloping 100 mm - 50 mm high; ref R35000	-	0.50	4.92	145.50	nr	150.42

NEW ITEMS

	PC £	Labour hours	Labour £	Material £	Unit	Total rate £
Miscellaneous fittings; Pressalit Ltd or equal and approved						
Grab rails						
300 mm long ref RT100000; screwing	-	0.50	4.92	37.43	nr	**42.35**
450 mm long ref RT101000; screwing	-	0.50	4.92	43.77	nr	**48.69**
600 mm long ref RT102000; screwing	-	0.50	4.92	50.26	nr	**55.18**
800 mm long ref RT103000; screwing	-	0.50	4.92	56.25	nr	**61.17**
1000 mm long ref RT104000; screwing	-	0.50	4.92	64.74	nr	**69.66**
Angled grab rails						
900 mm long, angled 135 ref RT110000; screwing	-	0.50	4.92	81.22	nr	**86.14**
1300 mm long, angled 90 ref RT119000; screwing	-	0.75	7.38	127.32	nr	**134.70**
Hinged grab rails						
600 mm long ref R3016000 ; screwing	-	0.35	3.44	131.71	nr	**135.15**
600 mm long with spring counter balance ref RF016000 ; screwing	-	0.35	3.44	182.64	nr	**186.08**
800 mm long ref R3010000 ; screwing	-	0.35	3.44	157.68	nr	**161.12**
800 mm long with spring counter balance ref RF010000 ; screwing	-	0.35	3.44	194.12	nr	**197.56**

SPON'S PRICE BOOKS 2002

with free CD-ROM

Free CD-ROM when you order any Spon's 2002 Price Book.
Use the CD-ROM to:
- produce tender documents
- customise data
- keyword search
- export to other major packages
- perform simple calculations.

updates available to download from the web
www.pricebooks.co.uk

Spon's Architects' and Builders' Price Book 2002
Davis Langdon & Everest

"Spon's Price Books have always been a 'Bible' in my work - now they have got even better! The CDs are not only quick but easy to use. The CD ROMs will really help me to get the most from my Spon's in my role as a Freelance Surveyor."
Martin Taylor, Isle of Lewis

New Features for 2002 include:
- A new section on Captial Allowances
- Inclusion of new items within a seperate Measured Works section

September 2001: 1024 pages
Hb & CD-ROM: 0-415-26216-X: £110.00

Spon's Landscapes and External Works Price Book 2002
Davis Langdon & Everest, in association with Landscape Projects

New Features for 2002 include:
- Fees for professional services
- Revised and updated sections on Cost Information and how to use this book
- Revisions and expansions of the Approximate Estimating section, together with direct links into the Measured Works Section

September 2001: 484 pages
Hb & CD-ROM: 0-415-26220-8: £80.00

Spon's Mechanical and Electrical Services Price Book 2002
Mott Green & Wall

"An essential reference for everybody concerned with the calculation of costs of mechanical and electrical works." *Cost Engineer*

New Features for 2002 include:
- New sections on modular wiring, emergency lighting, lighting control, sprinkler pre fabricated pipework, UPVC rainwater and gutters, carbon steel pipework and fittings

September 2001: 584 pages
Hb & CD-ROM: 0-415-26222-4: £110.00

Spon's Civil Engineering and Highway Works Price Book 2002
Davis Langdon & Everest

New Features for 2002 include:
- A revised and extended section on Land Remediation
- The Rail Track section now includes data on Permanent Way work with fully reviewed pricing
- Fully reviewed pricing for the Geotextiles section

September 2001: 688 pages
Hb & CD-ROM: 0-415-26218-6: £120.00

Return your orders to: Spon Press Customer Service Department, ITPS, Cheriton House, North Way, Andover, Hampshire, SP10 5BE · Tel: +44 (0) 1264 343071 · Fax: + 44 (0) 1264 343005 · Email: book.orders@tandf.co.uk
Postage & Packing: 5% of order value (min. charge £1, max. charge £10) for 3–5 days delivery · Option of next day delivery at an additional £6.50.

SPON PRESS
Taylor & Francis Group

Index

References in brackets after page numbers refer to SMM7 and the Common Arrangement of Work sections.

A PRELIMINARIES/GENERAL/CONDITIONS, 105, 410
Access
    and inspection chambers, 394 (R12), 674 (R12)
    covers and frames, 396 (R12), 675 (R12)
    flooring, 266 (K41), 786
"Accoflex" vinyl tile flooring, 312 (M50), 590 (M50)
Acrylic polymer flooring 782
Addresses for further information, 956
Afrormosia,
    278 (L20), 559 (L20)
Agba, basic price, 206 (G20)
Air conditioning systems, 804
Air bricks, 190 (F30), 489 (F30)
Albi products, 321 (M60), 600 (M60)
Alterations, 411 (C20)
Alternative material prices
    blocks, 177 (F10)
    facing bricks, 166 (F10)
    insulation, 340 (P10)
    sheet linings and casings, 250 (K10)
    slate or tile roofing, 229 (H60)
    tile slab and block finishings, 307 (M40)
    timber, 206 (G20)
Altro
    "Altroflow 3000" epoxy grout, 298 (M12)
    "Mondopave" tile flooring, 313 (M50), 592 (M50)
    safety flooring, 311 (M50), 590 (M50), 786
Aluminium
    covering, 239 (H72), 522 (H72), 744
    eaves trim, 244 (J21), 248 (J41), 527 (J21), 531 (J41)
    flashings, 239 (H72), 522 (H72)
    grates and covers, 390 (R12), 670 (R12)
    gutters, 370 (R10), 653 (R10)
    "Kal-Zip" cladding, 222 (H31)
    paint, 323, (M60), 604, (M60)
    patio doors, 281 (L20), 562 (L20)
    pipes, rainwater, 369 (R10), 652 (R10)
    polyester powder coated flashings, 240 (H72), 522 (H72)
    roof decking, 249 (J43), 734
    windows, 766
American White Ash
    270 (L10), 279, 284 (L20), 347 (P20), 551 (L10), 561, 565 (L20), 629 (P20)
    basic price, 206 (G20)
Anchor slots, 157 (E42), 461 (E42)
Anchors, 192 (F30), 228 (H51), 491 (F30)
Andersons roofing, 246 (J41), 529 (J41)

Angles, 195 (G10)
    galvanised steel, 349 (P20), 631 (P20)
    mild steel, 195 (G10), 349 (P20), 631 (P20)
    stainless steel, 349 (P20), 631 (P20)
"Anticon", sarking membrane, 235 (H60), 517 (H60)
Antivlam board, 257 (K11), 540 (K11)
APPROXIMATE ESTIMATES, 711
"Aquaseal" timber treatments, 206 (G20)
Arch frames, steel, 305 (M30), 585 (M30)
Arches, brick, 163 (F10), 168 (F10), 171 (F10), 174 (F10), 465 (F10), 468 (F10), 472 (F10) 475 (F10)
ARCHITECTS' FEES, 4
Architraves
    hardwood, 345 (P20), 347 (P20), 627 (P20), 629 (P20)
    softwood, 343 (P20), 625 (P20)
"Arlon" tile flooring, 312 (M50), 591 (M50)
"Armstrong" vinyl sheet flooring, 312 (M50), 590 (M50)
"Artex", 317 (M60), 596 (M60), 792
Artificial stone paving, 361 (Q25), 644 (Q25)
Asbestos-free
    accessories, "Eternit", 235 (H61), 518 (H61)
    cladding, "Eternit", 221 (H30), 510 (H30)
    corrugated sheets, "Eternit", 221 (H30), 510 (H30)
    slates, "Eternit", 243 (H61), 526 (H61)
Asphalt
    acid-resisting, 297 (M11), 578 (M11)
    damp-proofing, 243 (J20), 526 (J20)
    flooring, 297 (M11), 578 (M11)
    roofing, 244 (J21), 527 (J21), 746
    tanking. 243 (J20), 526 (J20)
"Astra-Glaze" blocks, 183 (F10), 482 (F10)
Atrium, 724
Attendance upon electrician, 354 (P31), 636 (P31)

Balloon gratings, 369 (R10), 652 (R10)
Balustrades
    metal, 290 (L30), 571 (L30), 750
    stainless steel, 290 (L30)
    timber, 286 (L30), 567 (L30)
Bar reinforcement, 129 (D40), 132 (D50), 149 (E30), 443 (D50), 453 (E30)
Basements
    approximate estimates, 720

Basic material prices
  concrete, 135 (E05)
  glazing, 291 (L40)
  mortar, 161 (F10)
  paints, 315 (M60)
  stonework, 226 (H51)
  structural steelwork, 194 (G10)
  timber, 206 (G20)
Basins, lavatory, 334 (N13), 615 (N13)
Bath panel
  glazed hardboard, 258 (K11), 541 (K11)
Baths, 336 (N13), 617 (N13)
Beams, laminated, 211 (G20), 500 (G20), 732
Bedding and pointing frames, 285 (L20), 566 (L20)
Beds, 295 (M10), 576 (M10), 782
Beds, benching and coverings
  concrete, 383 (R12), 665 (R12)
  granular fill, 383 (R12), 664 (R12)
Beech
  basic price, 206 (G20)
  strip flooring, 264 (K20), 547 (K20)
"Bentonite" slurry, 129 (D40)
Bidet, 339 (N13), 620 (N13)
Bitumen
  felt roofing, 245 (J41), 748
  macadam
    pavings, 820
    roads, 360, (Q22), 642 (Q22), 821
  strip slates, 242 (H76), 525 (H76), 740
"Bitumetal" roof decking, 249 (J43)
"Bituthene"
  fillet, 245 (J40), 528 (J40)
  sheeting, 245 (J40), 528 (J40)
Blasting concrete, 154 (E41), 458 (E41)
Blinds
  roller, 333 (N10/11), 614 (N10/11)
  vertical louvre, 333 (N10/11), 614 (N10/11)
Blockboard
  alternative sheet material prices, 250 (K10)
  american white ash, 342 (P20), 624 (P20)
  birch faced, 256 (K11), 539 (K11)
  linings, 256 (K11), 539 (K11)
  sapele veneered, 342 (P20), 624 (P20)
Block flooring, wood, 310 (M42), 589 (M42), 784
Blocks
  alternative material prices, 177 (F10)
  paviors, 362 (Q25), 645 (Q25), 820
  walling, 177 (F10), 477 (F10), 754
Blockwork
  approximate estimates, 754, 770
  "Astra-Glaze", 183 (F10), 482 (F10)
  Celcon, 177 (F10)
  dense aggregate, 181 (F10), 481 (F10)
  Durox "Supablocs", 177 (F10)
  Fenlite, 177 (F10)
  Forticrete "Bathstone", 182 (F10), 482 (F10)
  Forticrete, 177 (F10)
  Hanson "Conbloc", 181 (F10), 481 (F10)
  Hemelite, 177 (F10)
  Tarmac "Topblock", 177 (F10)
  Thermalite, 177 (F10), 477 (F10)

Blockwork - cont'd
  Toplite, 177 (F10)
Boilers, 405 (T10), 685 (T10)
Bollards, parking, 821
"Bradstone"
  architectural dressings, 185 (F22), 484 (F22)
  roof slates, 237 (H63), 520 (H63)
  walling, 184 (F22), 483 (F22)
Breaking out, existing materials, 121, 123 (D20), 433, 435 (D20)
Brick
  cladding systems, 228 (H51)
  manholes, 395 (R12), 674 (R12)
  paviors, 362 (Q25), 644 (Q25), 820
  reinforcement, 188 (F30), 487 (F30)
  sill, 169, 172, 175 (F10), 469, 473 476 (F10)
  slips, 176 (F10), 477 (F10)
"Brickforce" reinforcement, 188 (F30), 487 (F30)
Bricks
  alternative material prices, 166 (F10)
  Breacon Hill Brick Company, 166 (F10)
  common, 133 (D50), 161 (F10), 443 (D50), 463 (F10), 754, 770, 778
  engineering, 133 (D50), 164, 165 (F10), 443 (D50), 466 (F10)
  facings, 166 (F10), 467 (F10), 754
  hand made, 173 (F10), 474 (F10), 754
  Ibstock, 166 (F10)
  London brick, 166 (F10)
  machine made, 170 (F10), 471 (F10), 754
  sand faced, 167 (F10), 467 (F10), 754
  sandlime, 170 (F10), 470 (F10)
Brickwork and blockwork
  removal of, 413 (C20)
  repairs to, 415 (C41)
Bridges
  foot, 821
  road, 821
Builder's work, 816
  electrical, 354 (P31), 636 (P31)
  other services, 354 (P31), 636 (P31)
Building
  costs index, 691
  fabric sundries, 340 (P10), 622 (P10)
  hourly rates/operatives' earnings, 100
  paper, 144 (E05), 313 (M50), 340 (P10), 448 (E05), 586 (M50), 622 (P10), 736
  PRICES PER SQUARE METRE, 703
  PRICES PER FUNCTIONAL UNITS, 697
BUILDING REGULATIONS FEES, 65
Burn off, existing, 594 (M60), 603 (M60)
Bush hammering, 764

C DEMOLITION/ALTERATION/RENOVATION, 411
Calorifiers, 402 (S10/11), 682 (S12)
Carcassing, sawn softwood, 206 (G20), 495 (G20)
"Carlite" plaster, 299, 301 (M20), 580, 582 (M20)
Car Parking, 820
Carpentry/First fixing, timber, 206 (G20), 495 (G20)

Carpet, 314 (M51), 592 (M51), 786
  gripper, 314 (M51), 593 (M51)
  tiles, 314 (M50, 592 (M50), 786
  underlay, 314 (M51), 593 (M51), 786
Cart away from site, 125 (D20), 132 (D50), 393 (R12), 398 (R13), 438 (D20), 442 (D50), 673 (R12), 677 (R13)
Casings, 260 (K11), 543 (K11)
Cast iron
  access covers and frames, 396 (R12), 675 (R12)
  accessories/gullies, 386 (R12), 666 (R12)
  drain pipes, 386 (R12), 666 (R12), 814
  "Ensign" lightweight pipes, 375 (R11), 387 (R12), 658 (R11), 667 (R12)
  gutters, 371 (R10), 654 (R10)
  inspection chambers, 396 (R12), 675 (R12)
  lightweight pipes, 375 (R11), 387 (R12), 658 (R11), 667 (R12)
  rainwater pipes, 370 (R10), 653 (R10), 744
  "Timesaver" waste pipes, 374 (R11), 386 (R12), 657 (R11), 666 (R12)
Cast stonework, 184 (F22), 483 (F22)
Cat ladders, 752
Catnic steel lintels, 192 (F30), 491 (F30)
Cat walk, 752
Cavity
  closers, 190 (F30), 490 (F30)
  closing, 162, 163, 178, 179 (F10), 463, 465, 478, 490 (F10)
  flashings, "Bithu-thene", 191 (F30), 490 (F30)
  forming, 186 (F30), 485 (F30)
  trays, polypropylene, 191 (F30), 490 (F30)
  wall insulation, 340 (P10), 622 (P10), 756
Cedar
  boarding, 764
  shingles, 237 (H64), 520 (H64), 742
Ceiling finishes, 780
Ceiling
  suspended, 265 (K40), 790
  integrated, 790
"Celcon" blocks, 177 (F10)
"Celgard CF" wood preservatives, 208, 211 (G20), 497, 499 (G20)
"Celuka" PVC cladding, 225 (H32), 511 (H32)
Cement paint, 323 (M60), 603 (M60)
Cement:sand
  beds/backings, 295 (M10), 298 (M20), 576 (M10), 579 (M20), 782
  paving, 295 (M10), 576 (M10), 782
  rendering, 298 (M20), 579 (M20), 782
  screeds, 295 (M10), 570 (M10), 738, 780
Central heating, 804
Ceramic wall and floor tiles, 308 (M40), 587 (M40), 780, 782
Chain link fencing, 366 (Q40), 648 (Q40), 822
Changing cubicles, 772
Channels
  clay, 397 (R12), 676 (R12)
  precast concrete 358 (Q10), 640 (Q10)
  uPVC, 397 (R12), 676 (R12)
Chemical anchors, 217 (G20), 506 (G20)
Chequer plate flooring, 290 (L30), 571 (L30)
Chimney pots, 192 (F30), 491 (F30)

Chipboard
  boarding, 257 (K11), 540 (K11)
  fittings, 328 (N10/11), 609 (N10/11)
  flame retardant, 257 (K11), 540 (K11)
  melamine faced, 257 (K11), 540 (K11), 778
  flooring, 257 (K11), 540 (K11), 780
  linings, 256 (K11), 539 (K11) 778
Cisterns, 402 (S10/11), 682 (S12)
CITB levy, 100
Cladding/Covering, 220 (H), 509 (H)
Cladding
  aluminium, 223 (H31), 736, 760
  asbestos free, 221 (H30), 510 (H30), 734
  brick, insulated, 228 (H51)
  granite, 308 (M40)
  "Kal-Zip", 223 (H31)
  "Plannja", 223 (H31)
  steel, 222 (H31), 734, 758
  support systems, 158 (E42)
  translucent, 225 (H41), 511 (H41)
  vermiculite gypsum "Vicuclad", 255 (K10), 538 (K10)
"Classic" vinyl sheet flooring, 311, (M50), 590 (M50)
Clay
  accessories, 389 (R12), 669 (R12)
  channels, 397 (R12), 676 (R12)
  land drains, 398 (R13), 677 (R13)
  gullies, 389 (R12), 669 (R12)
  pantiles, 730, 732, 740
    Langley, 229 (H60)
    Sandtoft, 229 (H60), 512 (H60)
    William Blyth, 227 (H60), 512 (H60)
  pipes, 824
  plain tiles, 730, 732, 740
    Hinton, Perry and Davenhill, 230 (H60), 513 (H60)
    Keymer, 230 (H60), 513 (H60)
Clean out gutters, 422 (C90)
Clearing the site, 120 (D20), 432 (D20)
Close boarded fencing, 368 (Q40), 651 (Q40), 812
Cobblestone paving, 362 (Q25), 644 (Q25), 820
Coffered slabs, 146 (E20), 450 (E20)
Column guards, 158 (E42), 462 (E42)
Comfort cooling system, 806
Common bricks, 133 (D50), 161 (F10), 443 (D50), 463 (F10), 754, 770, 778
Comparative
  finishes
    balustrading, 752
    ceiling, 792
    cladding, 744
    external, 764
    fittings, 794
    staircases, 750
    steel/metalwork, 792
    wall, 778, 792
    weatherproof, 746
    woodwork, 792
Composite floor, 160 (E60)
Compressor, plant rates, 104
Computation of labour rates, 102

Concrete
   air-entrained, 140 (E05)
   basic prices, 135 (E05)
   beds and backings, 295 (M10), 576 (M10)
   beds, benchings and coverings, 384 (R12), 665 (R12)
   blasting surfaces, 154 (E41), 458 (E41)
   blocks
      approximate estimates, 754, 770
      Celcon, 177 (F10)
      dense aggregate, 181 (F10), 481 (F10)
      Durox "Supabloc", 177 (F10)
      Fenlite, 177 (F10)
      Forticrete, 177 (F10)
      Hanson "Conbloc", 181 (F10), 481 (F10)
      standard dense, 177 (F10)
      Tarmac "Topblock", 177 (F10)
      Thermalite, 177 (F10), 477 (F10)
      Toplite, 177 (F10)
   diaphragm walling, 129 (D40)
   filling, 141 (E05), 445 (E05)
   floors, 720, 726
   foundations, 141 (E05), 357 (Q10), 445 (E05), 639 (Q10), 712
   frames, 720
   hacking, 154 (E41), 458 (E41)
   lintels, 193 (F31), 492 (F31), 776
   manhole rings, 395 (R12)
   mixer, plant rates, 104
   pavement lights, 220 (H14)
   paving
      blocks, 362 (Q25), 645 (Q25), 820
      flags, 362 (Q25), 644 (Q25), 820
   piling, 127 (D30)
   pipes, 390 (R12)
   poured against excavation faces, 141 (E05), 445 (E05)
   precast, 158 (E60), 193 (F31), 460 (E60), 492 (F31), 728
   ready mixed, 135 (E)
   repairs to, 415 (C41)
   roads/paving/bases, 358 (Q21), 640 (Q21), 820
   rooflights, 220 (H14)
   roofs, approximate estimates, 730
   roof tiles
      approximate estimates, 738
      Marley, 229, 231, 232 (H60), 513, 514, 515 (H60)
      Redland, 229, 232, 233, 234 (H60), 515, 516 (H60)
   shot-blasting, 154 (E41), 458 (E41)
   shuttering, 145 (E20), 449 (E20)
   site mixed, 139 (E)
   sundries, 359 (Q21), 641 (Q21)
   tamping, 154 (E41), 458 (E41),
   walls, 758
   water repellent, 140 (E)
Connection charges, 823
Construction Industry Training Board Levy, 100
CONSULTING ENGINEERS' FEES, 45
Conversion tables, 885
Cooker control units, 407 (V21), 687 (V21), 808
Copings concrete, precast, 193 (F31), 492 (F31)

Copper
   coverings, 240 (H73), 523 (H73), 744
   flashings, 240 (H73), 523 (H73)
   pipes, 399 (S10/11), 679 (S12)
   roofing, 240 (H73), 523 (H73), 744
Cordek
   "Claymaster" formwork, 144 (E20), 448 (E20)
   trough former, 146 (E20), 450 (E20)
Cork
   insulation boards, 248 (J41), 531 (J41), 736
   tiling, 313 (M50), 592 (M50), 784
Cost limits and allowances
   educational buildings, 833
   hospital buildings, 831
   housing association schemes, 834
   university buildings, 832
Costs index, 691
Coverings
   aluminium, 239 (H72), 522 (H72), 744
   copper, 240 (H73), 523 (H73), 744
   felt roof, 245 (J41), 528 (J41), 746
   lead, 238 (H71), 520 (H71), 744
   stainless steel, 241 (H75), 524 (H75)
   zinc, 240 (H74), 523 (H74), 744
"Coxdome" rooflights, 275 (L10), 556 (L10)
Cramps, 192 (F30), 228 (H51), 349 (P20), 491 (F30), 631 (P20)
Creosote, 327 (M60), 607 (M60)
"Crown Wool" insulation, 340 (P10)
Cubicles
   changing, 772
   WC, 264 (K32), 772
"Cullamix" Tyrolean rendering, 302 (M20), 583 (M20)
Curtains, warm air, 806
Cylinders
   combination, 402 (S10/11), 682 (S12)
   hot water, 402 (S10/11), 682 (S12)
   insulation, 402 (S10/11), 683 (S12)
   storage, 402 (S10/11), 682 (S12)

D GROUNDWORK, 120, 432
Dado rails
   mdf, 343 (P20), 624 (P20)
Damp proof courses
   bitumen-based, lead-cored, 134 (D50), 187 (F30), 444 (D50), 487 (F30)
   fibre based, 187 (F30), 486 (F30)
   hessian based, 133 (D50), 187 (F30), 444 (D50), 486 (F30)
   lead, 188 (F30), 487 (F30)
   lead cored, 134 (D50), 187 (F30), 444 (D50), 487 (F30)
   "Peter Cox" chemical transfusion system, 487 (F30)
   pitch polymer, 133 (D50), 187 (F30), 444 (D50), 486 (F30)
   polyethylene, 187 (F30), 486 (F30)
   silicone injection, 487 (F30)
   slate, 134 (D50), 188 (F30), 444 (D50), 487 (F30)
Damp proof membranes
   asphalt, 243 (J20), 526 (J20)
   liquid applied, 244 (J30), 527 (J30),
DAYWORK AND PRIME COST
   building indusry, 71

Decorative
  plywood lining, 259 (K11), 542 (K11)
  wall coverings, 314 (M52), 593 (M52)
Decra lightweight steel roof tiles, 223 (H31), 511 (H31)
Defective timber, cut out, 418 (C51)
Demolition/Alteration/Renovation, 411 (C)
Design loadings for buildings, 888
Diamond drilling, 155 (E41), 459 (E41)
Diaphragm walling, 129 (D40), 720
Discounts, 104
Disposal
  hand, 126 (D20), 132 (D50), 393 (R12), 439 (D20), 442 (D50), 673 (R12)
  mechanical, 125 (D20), 393 (R12), 438 (D20), 673 (R12)
  systems, 369 (R), 652 (R)
Division strips, 309 (M41), 589 (M41)
Dock levellers, 814
Dogleg staircase, 286 (L30), 567 (L30), 750
Door
  frames
    hardwood, 284, 285 (L20), 564, 565, 566 (L20)
    linings, comparative, 772
    softwood, 282 (L20), 563 (L20)
  mats, 333 (N10/11), 614 (N10/11)
  sills, oak, 285 (L20), 566 (L20)
  stops, 345, 347 (P21), 633, 635 (P21)
Doors
  approximate estimates, 768
  comparative prices, 772
  fire resisting, 277 (L20), 558 (L20), 774
  flush, 276 (L20), 557 (L20), 772
  garage, 280 (L20), 561 (L20)
  half-hour firecheck, 277 (L20), 558 (L20)
  hardwood, 279 (L20), 560, 561 (L20), 774
  "Leaderflush" type B30, 277 (L20), 559 (L20)
  matchboarded, 276 (L20), 559 (L20)
  one-hour firecheck, 277 (L20), 559 (L20)
  panelled
    hardwood, 279 (L20), 560, 561 (L20), 776
    softwood, 278 (L20), 553 (L20), 776
  patio, aluminium, 281 (L20), 562 (L20)
  softwood, 276 (L20), 557 (L20), 772
  steel, 280 (L20), 561 (L20)
  "Weatherbeater", 279 (L20), 556 (L20)
Dormer roofs, 732
Dowels, 349 (P20), 631 (P20)
Drainage, 824
  beds, etc., 383 (R12), 664 (R12)
  channels, 397 (R12), 676 (R12)
Drains
  cast iron, "Timesaver" pipes, 374(R11), 386 (R12), 657 (R11), 666 (R12),
  clay, 387 (R12), 667 (R12), 824
  land, 398 (R13), 677 (R13)
  unplasticized PVC, 391 (R12), 671 (R12), 824
Drinking fountains, 335 (N13), 616 (N13)
Dry
  linings, 250 (K10), 533 (K10), 778
  partitions, 250 (K10), 533 (K10), 768
  ridge tiles, 231 (H60), 512 (H60)
  risers, 814
  verge system, 231 (H60), 512 (H60)

"Dri-Wall top coat", 255 (K10), 538 (K10)
Dubbing out, 423 (C56)
Duct covers, 356 (P31), 638 (P31)
Ducting, floor, 356 (P31), 638 (P31)
Dulux "Weathershield" paint, 323 (M60), 603 (M60)
"Duplex" insulating board, 252 (K10), 535 (K10)
Durabella "Westbourne" flooring, 257 (K11), 540 (K11)
Durox, "Supablocs" , 177 (F10)
Dust-proof screen, 431 (C90)

E IN SITU CONCRETE/LARGE PRECAST CONCRETE, 135, 445
Earnings, guaranteed minimum weekly, 101
Earthwork support, 123 (D20), 131 (D50), 393 (R12), 436 (D20), 441 (D50), 673 (R12)
  inside existing building, 125 (D20), 438 (D20)
Eaves
  verges and fascia boarding, 212, 215 (G20), 500, 504 (G20)
  trim, 244 (J21), 248 (J41), 527 (J21), 531 (J41)
Edgings, precast concrete, 357 (Q10), 639 (Q10)
EDUCATIONAL BUILDINGS, 833
Electrical installations, 808
Electrical systems, 406 (V), 686 (V)
"Elvaco" warm air heating, 806
Emulsion paint, 316 (M60), 595 (M60), 778
Engineering bricks and brickwork, 164, (D50), 165 (F10), 443 (D50), 466 (F10)
ENGINEERS' FEES, 45
"Ensign" pipes,
  drain, 387 (R12), 667 (R12)
  soil, 375 (R11), 658 (R11)
Entrance matting, 788
Epoxy
  damp proof membrane, 244 (J30), 527 (J30)
  flooring, 782
  resin grout, 143 (E10), 447 (E10)
  resin sealer, 320 (M60), 600 (M60)
Escalators, 812
"Eurobrick" brick cladding system, 228 (H51)
Evode "Flashband", 242 (H76), 525 (H76)
Excavation
  constants for alternative soils, 120 (D20), 432 (D20)
  drain trenches, 378 (R12), 662 (R12)
  generally, 120 (D20), 432 (D20)
  hand, 122 (D20), 130 (D50), 353 (P30), 357 (Q10), 380 (R12), 434 (D20), 441 (D50), 635 (P30), 639 (Q10), 663, 673 (R12)
  manholes, 393 (R12), 673 (R12)
  mechanical, 120 (D20), 130 (D50), 353 (P30), 357 (Q10), 378 (R12), 432 (D20), 440 (D50), 635 (P30), 639 (Q10), 662, 673 (R12)
  services trenches, 353 (P30), 635 (P30)
  underpinning, 130 (D50), 441 (D50)
  valve pits, 354 (P30), 636 (P30)
Excavators, plant rates, 104
Existing surfaces, preparation of, 593 (M60), 602 (M60)
Expamet
  beads, 303 (M20), 584 (M20)
  foundation bolt boxes, 157 (E42), 461 (E42)
  wall starters, 191 (F30), 491 (F30)
Expanded metal lathing, 305 (M30), 585 (M30)

Expanding bolts, 216 (G20), 505 (G20)
Expandite "Flexcell" joint filler, 152 (E40), 359 (Q21), 456 (E40), 641 (Q21)
Expansion joints, fire-resisting, 189 (F30), 488 (F30)
External lighting, 822
External services, 823
External walls, 754
External water and gas installations, 403 (S13), 404 (S32), 683 (S13), 684 (S32)
"Extr-aqua-vent", 244 (J21), 248 (J41), 527 (J21), 531 (J41)
Extruded polystyrene foam boards, 248 (J41), 531 (J41), 736

F MASONRY, 161, 463
Fabricated steelwork, 194 (G10), 493 (G10)
Fabric reinforcement, 150 (E30), 359 (Q21), 394 (R12), 454 (E30), 641 (Q21), 674 (R12)
Facing bricks, 166 (F10), 467 (F10), 754
    alternative material prices, 166 (F10)
    Breacon Hill Brick Company, 166 (F10)
    hand made, 173 (F10), 474 (F10), 754
    Ibstock, 166 (F10)
    London brick, 166 (F10)
    machine made, 170 (F10), 471 (F10), 754
    sand faced, 167 (F10), 467 (F10), 754
"Famlex" tanking, 245 (J40), 528 (J40)
FEES
    architects', 4
    building (local authority charges) regulations, 65
    consulting engineers, 45
    planning, 58
    quantity surveyors', 9
Felt roofing, 245 (J41), 528 (J41), 746
Fencing, 822
    chain link, 366 (Q40), 648 (Q40), 822
    security, 368 (Q40), 651 (Q40)
    timber, 367 (Q40), 650 (Q40), 822
"Fenlite" blocks, 177 (F10)
Fertilizer, 365 (Q30), 647 (Q30)
"Fibreglass" insulation, 340 (P10), 622 (P10)
Fibrous plaster, 306 (M31), 586 (M31)
Fillets/pointing/wedging and pinning, etc., 188 (F30), 487 (F30)
Fill existing openings, 429 (C90)
Filling
    hand, 126 (D20), 132 (D50), 358 (Q20), 394 (R12), 439 (D20), 442 (D50), 640 (Q20), 673 (R12)
    mechanical, 126 (D20), 358 (Q20), 394 (R12), 439 (D20), 640 (Q20), 673 (R12)
Filter membrane, 127 (D20), 440 (D20)
Finishes,
    approximate estimates, 778
    ceiling, 792
    external wall, 764
    floor, 780
    stairs/balustrading, 752
    steel/metalwork, 792
    wall, 778
    wall/ceiling, 792
    woodwork, 792

Fire
    barrier, 341 (P10), 623 (P10)
    check channel, 341 (P10), 622 (P10)
    doors, 277 (L20), 558 (L20)
    fighting equipment, 814
    insurance, 108, 877
    protection
        compound, 341 (P10), 623 (P10)
        Mandolite "CP2" coatings, 304 (M22), 585 (M22)
        "Unitherm" intumescent paint, 320 (M60), 600 (M60)
        "Vermiculite" claddings, 255 (K10), 538 (K10)
    resisting
        coatings, 792
        expansion joints, 189 (F30), 488 (F30)
        glass, 293 (L40), 574 (L40)
        retardant paint/coatings, 321 (M60), 600 (M60), 792
        seals, 341 (P10), 618 (P10)
        shutters, 768
        stops, 341 (P10), 618 (P10)
Fireline board lining, 250 (K10), 533 (K10)
Firrings, timber, 213, 214 (G20), 495, 497 (G20)
First fixings, woodwork, 206 (G20), 495 (G20)
Fitted carpeting, 314 (M51), 592 (M51), 786
Fittings, 328 (N10/11), 609 (N10/11)
Fittings and furnishings, 794
Fittings, kitchen, 331 (N10/11), 612 (N10/11), 794
Fixing
    cramps, 158 (E42), 462 (E42)
    ironmongery
        to hardwood, 351 (P21), 633 (P21)
        to softwood, 350 (P21), 632 (P21)
    kitchen fittings, 331 (N10/11), 612 (N10/11)
Flame
    proofing timber, 208, 211 (G20), 497, 499 (G20)
    retardant surface coating, 321 (M60), 600 (M60)
Flashings
    aluminium, 239 (H72), 522 (H72)
    copper, 240 (H73), 523 (H73)
    "Flashband Original", 242 (H76), 525 (H76)
    lead, 238 (H71), 521 (H71)
    stainless steel, 241 (H75), 524 (H75)
    zinc, 241 (H74), 523 (H74)
Flat roofs
    approximate estimates, 730, 732
"Flettons", see common bricks
Flexible sheet coverings, 786
Floor
    finishes, 780
    hardener, 320 (M60), 600 (M60)
Flooring
    access, 266 (K41), 786
    acrylic polymer, 782
    block, 310 (M42), 589 (M42), 784
    brick, 782
    carpet, 314 (M51), 592 (M51), 786
    carpet tiles, 314 (M50), 592 (M50), 786
    cement:sand, 295 (M10), 576 (M10), 782
    ceramic tile, 308 (M40), 587 (M40), 782
    chequer plate, 290 (L30), 571 (L30)

Flooring - cont'd
  chipboard, 257 (K11), 539 (K11), 780
  cork tiles, 313 (M50), 592 (M50), 784
  epoxy, 782
  granite, 308 (M40), 784
  granolithic, 295 (M10), 576 (M10), 782
  hardwood, 264 (K20), 547 (K20), 780
  linoleum sheet, 311 (M50), 590 (M50), 786
  linoleum tiles, 313 (M50), 591 (M50), 786
  marble, 784
  mosiac, 784
  open grid steel, 290 (L30), 571 (L30)
  plywood, 257 (K11), 540 (K11)
  polyester resin, 782
  precast concrete, 158 (E60), 462 (E60)
  quarry tiles, 307 (M40), 587 (M40), 782
  rubber tiles, 312 (M50), 590 (M50), 786
  safety flooring, 311 (M50), 590 (M50)
  slate, 309 (M40), 588 (M40), 784
  slip resistant 311 (M50), 590 (M50)
  softwood boarded, 263 (K20), 545 (K20), 780
  stone, 784
  terrazzo, 309 (M41), 589 (M41), 782, 784
  thermoplastic tiles, 313 (M50), 591 (M50), 784
  underlay, 313 (M50), 314 (M51), 592(M50), 593 (M51), 786
  vinyl sheet, 311 (M50), 590 (M50), 786
  vinyl tiles, 312 (M50), 590 (M50), 784
  wood blocks, 310 (M42), 589 (M42), 784
Floors
  approximate estimates, 726
  composite, 160 (E60), 462 (E60)
  raised, 266 (K41), 786
Fluctuations, 109
Flue
  blocks, 190 (F30), 489 (F30)
  linings, 189 (F30), 489 (F30)
Flush doors, 276 (L20), 557 (L20), 772
Folding screens, 772
Formed joint, 152 (E40), 359 (Q21), 456 (E40), 641 (Q21)
Formulae, 886
Formwork, 132 (D50), 144 (E20), 359 (Q21), 394 (R12), 443 (D50), 448 (E20), 641 (Q21), 674 (R12)
"Forticrete" blocks, 177 (F10), 481 (F10)
Foundation boxes, 157 (E42), 461 (E42)
Foundations, 712
Frame and upper floors, 720
Frame to roof, 720
Frames
  approximate estimates, 720, 726
  concrete, 720
  door
    hardwood, 284, 285 (L20), 560, 561 (L20)
    softwood, 282 (L20), 559 (L20)
    steel, 768
  window
    hardwood, 271, 272 (L10), 552, 553 (L10)
    softwood, 271 (L10), 552 (L10)
Framework, timber, 213, 214 (G20), 502, 503 (G20)
French polish, 322 (M60), 602 (M60), 792
"Frodingham" sheet piling, 128 (D32)
Fungus/Beetle eradication, 420 (C52)
Furniture/equipment, 328 (N), 612 (N)

G STRUCTURAL/CARASSING METAL/TIMBER, 194, 493
Galvanized steel
  angles, 349 (P20), 631 (P20)
  gutters, 372 (R10), 655 (R10)
  joint reinforcement, 188 (F30), 487 (F30)
  lintels, 192 (F30), 491 (F30)
  profile sheet cladding, 222 (H31), 734, 758
  roof decking, 249 (J43), 734
  windows, 273 (L10), 554 (L10)
Galvanizing, 205 (G10), 494 (G10)
Garage doors, steel, 280 (L20), 561 (L20)
Gas
  flue linings, 189 (F30), 489 (F30)
  installations, 810
  pipework, 404 (S32), 684 (S32)
Gates, 366, 367, 368 (Q40), 650, 651 (Q40)
Gate valves, 402 (S10/11), 682 (S12)
"Glasal" sheeting, 221 (H20), 509 (H20)
"Glasroc" GRC boards, 262 (K14), 545 (K14)
Glass
  basic material prices, 291 (L40)
  block walling, 183 (F11)
  fibre insulation, 340 (P10), 622 (P10)
  fire resisting, 293 (L40), 574 (L40)
  generally, 291 (L40), 572 (L40)
  louvres, 294 (L40), 574 (L40)
  mirrors, 332 (N10/11), 613 (N10/11)
  pavement/rooflights, 220 (H14)
  security, 291, 293 (L40), 574 (L40)
Glazed
  block walling, 183 (F10), 482 (F10)
  ceramic wall and floor tiles, 308 (M40), 587 (M40), 782
  hardboard, 258 (K11), 541 (K11)
Glazing, 291 (L40), 572 (L40)
  basic material prices, 291 (L40)
  beads
    hardwood, 346 (P20), 348 (P20), 627 (P20), 629 (P20)
    softwood, 344 (P20), 625 (P20)
  patent, 220 (H10), 509 (H10)
  systems, 764
Gloss paint, 317 (M60), 596 (M60), 778
Glulam timber beams, 211 (G20), 500 (G20), 722
Goods lifts, 812
Granite
  flooring, 308 (M40), 784
  setts, 364 (Q25), 646 (Q25), 820
  wall linings, 308 (M40)
Granolithic
  paving, 295 (M10), 576 (M10), 782
  skirtings, 296 (M10), 577 (M10), 788
Granular beds, benches and coverings, 383 (R12), 664 (R12)
Granular fill
  type 1, 126 (D20), 358 (Q20), 439 (D20), 640 (Q20)
  type 2, 126 (D20), 358 (Q20), 439 (D20), 640 (Q20)
Grass seed, 365 (Q30), 647 (Q30), 818
Grating
  cast iron, road, 390 (R12), 670 (R12)
  wire balloon, 369 (R10), 652 (R10)

Gravel pavings, 361 (Q23), 643 (Q23), 820
Grilles roller, 282 (L20), 562 (L20), 768
Ground floor, 720
Groundwater level, extending below, 121, 122 (D20), 433 434 (D20)
Groundwork, 120 (D), 432 (D)
Grouting
   cement mortar, 143 (E05), 447 (E05)
   cementitious conbex, 143 (E05), 447 (E05)
   epoxy resin, 143 (E05), 447 (E05)
GRP panels, 758
Guard rails, 821
Gullies
   cast iron, 386 (R12), 666 (R12)
   uPVC, 392 (R12), 672 (R12)
Gurjun, strip flooring, 264 (K20), 557 (K20)
Gutter boarding, 212, 214, 215 (G20), 501, 502, 504 (G20)
Gutters
   aluminium, 370 (R10), 653 (R10)
   cast iron, 371 (R10), 654 (R10)
   clean out, 422 (C90)
   pressed steel, 372 (R10), 655 (R10)
   uPVC, 373 (R10), 656 (R10), 748
"Gyproc"
   dry linings, 255 (K10), 537 (K10)
   laminated partitions, 248 (K10), 527 (K10), 758
   M/F dry linings, 255 (K10), 538 (K11)
Gypsum
   plank, 252 (K10), 535 (K10)
   plasterboard, 251 (K10), 534 (K10)
   thermal board, 253 (K10), 535 (K10)

H CLADDING/COVERING, 220, 509
Hacking concrete, 154 (E41), 458 (E41)
Hacking brickwork, 189 (F30), 488 (F30)
Half-hour fire doors, 277 (L20), 558 (L20)
"Halfen" support systems, 158 (E42)
Hand made facing bricks, 173 (F10), 474 (F10), 755
Handrails, 752
Handrail bracket, metal, 290 (L30), 571 (L30)
Hardboard,
   bath panels, 258 (K11), 541 (K11)
   linings, 258 (K11), 539 (K11), 778
   underlay, 313 (M50), 592 (M50)
Hardcore, 126, (D20), 358 (Q20), 439 (D20), 640 (Q20)
Hardener, surface, 296 (M10), 577 (M10)
"Hardrow" roof slates, 236 (H63), 519 (H63)
Hardwood
   architraves, 345, 347 (P20), 627, 629 (P20)
   basic prices, 206 (G20)
   doors, 279, (L20), 561, 562 (L20), 774
   door frames, 284, 285 (L20), 564, 565, 566 (L20)
   skirtings, 345, 347 (P20), 627, 629 (P20), 778
   staircase, 286 (L30), 567 (L30)
   strip flooring, 264 (K20), 547 (K20), 780
   windows, 269, 270 (L10), 551, 552, 553 (L10), 756
   window frames, 271, 272 (L10), 552, 553 (L10)
Heating, air conditioning and ventilating installations, 804
Heating
   "Elvaco" warm air, 806
   underfloor, 804
"Hepline", 398 (R13), 678 (R13)

"HepSeal" vitrified clay pipes, 387 (R12), 668 (R12)
Herringbone strutting, 208, 210 (G20), 497, 499 (G20), 726
Hessian-based damp proof course, 133 (D50), 187 (F30), 444 (D50), 486 (F30)
Hessian wall coverings, 792
Hip irons, 234 (H60), 517 (H60)
Hips, comparative prices, 744
Holes in
   brickwork, 354 (P31), 636 (P31)
   concrete,
      cutting, 156 (E41), 460 (E41)
      forming, 149 (E20), 453 (E20)
   hardwoood, 356 (P31), 638 (P31)
   metal, 290 (L30), 572 (L30)
   softwood, 355 (P31), 638 (P31)
Holidays with pay, 100
"Holorib" steel shuttering, 146 (E20), 160 (E60), 450 (E20), 726, 732
HOSPITAL BUILDINGS, 831
HOUSING ASSOCIATION SCHEMES, 834
"Hyload" 150E roofing, 247 (J41), 531 (J41), 748
"Hyrib" permanent shuttering, 145 (E20), 449 (E20)

Immersion heater, 804
Imported soil, 126, 127 (D20), 439, 440 (D20)
Incoming mains services, 823
Index
   Building cost, 691
   Tender prices, 692
Industrial board lining, 250 (K10)
Inserts, concrete, 157 (E42), 462 (E42)
In situ concrete/Large precast concrete, 135 (E), 445 (E), 720, 758
Inspection chambers
   cast iron, 396 (R12), 675 (R12)
   concrete, 394 (R12), 674 (R12)
   polypropylene, 389 (R12), 659 (R12)
   unplasticised PVC, 393 (R12), 672 (R12)
Insulated roof screeds, 208 (M12), 579 (M12)
Insulating jackets 403 (S10/11), 683 (S12)
Insulation, 340 (P10), 622 (P10)
   alternative material prices, 340 (P10)
   boards to roofs, 248 (J41), 531 (J41), 736
   cold bridging, 341 (P10), 623 (P10)
   "Crown Wool" glass fibre, 340 (P10)
   "Isowool" glass fibre, 340 (P10), 622 (P10)
   "Jablite", 341 (P10), 622 (P10)
   pipe, 402 (S10/11), 683 (S12)
   quilt 340 (P10), 622 (P10), 736
   sound-deadening quilt, 340 (P10), 728
   "Styrofoam" Floormate, 341 (P10), 622 (P10)
   tank lagging, 402 (S10/11), 683 (S12)
Insurance, 108
   national, 100
   property, 877
Integrated ceilings, 790
Intercepting traps, 390 (R12), 670 (R12)

Interlocking
　roof tiles, 231, (H60), 513, (H60), 732
　sheet piling, 128 (D32), 716
　woodwool slabs, 219 (G32), 508 (G32)
Internal walls, partitions and doors, 768
"Intersett" paving blocks, 363 (Q25), 645 (Q25)
INTRODUCTION
　Major works, 99
　Minor works, 409
Intumescent
　paste, 293 (L40)
　plugs, 341 (P10), 623 (P10)
　seals, 341 (P10), 623 (P10)
Iroko
　basic price, 206 (G20)
　block flooring, 310 (M42), 589 (M42), 784
　fittings, 330 (N10/11), 611 (N10/11)
　strip flooring, 264 (K20), 547 (K20), 780
Ironmongery, 350 (P21), 632 (P21)

J WATERPROOFING, 243, 526
Jacuzzi installation, 814
Joint filler,
　"Expandite", 152 (E40), 359 (Q21), 456 (E40), 641 (Q21)
　"Kork-pak", 152 (E40), 456 (E40)
Joint reinforcement, 188 (F30), 487 (F30)
Joints, expansion, 152 (E40), 189 (F30), 456 (E40), 488 (F30)
Joists/beams
　hangers, 216 (G20), 504 (G20)
　steel, 194 (G10)
　struts, 208, 210 (G20), 497, 499 (G20)
"Junckers" wood strip, 264 (K20)

K LININGS/SHEATHING/DRY PARTITIONING, 250, 533
"Kal-Zip" cladding, 223 (H31)
"Kawneer" patent glazing, 220 (H10), 509 (H10)
Kerbs
　foundations, 357 (Q10), 639 (Q10)
　precast concrete, 357 (Q10), 639 (Q10)
"Keyblock" rectangular paving blocks, 362 (Q25), 645 (Q25), 820
Kitchen fittings, 331 (N10/11), 612 (N10/11), 794
"Kork -Pak" joint filler, 152 (E40), 456 (E40)

L WINDOWS/DOORS/STAIRS, 267, 548
Labour rates, computation of, 102
Lacquers, 322 (M60), 602 (M60)
Lagging pipes, 402 (S10/11), 683 (S12)
"Lamatherm"
　fire barriers, 341 (P10), 623 (P10)
Laminated
　chipboard, 329 (N10/11), 610 (N10/11), 778
　roof beams, 211 (G20), 500 (G20), 732
Lamp post, 822
Land drainage, 398 (R13), 677 (R13), 824
Land remediation, xi
Landfill Tax, ix
Landscaped roofs, 748
Latex screeds, 297 (M12), 579 (M12), 782
Lathing,
　metal, 305 (M30), 585 (M30)
Lavatory basins, 334 (N13), 615 (N13), 800

Lead
　coverings, 238 (H71), 520 (H71), 744
　damp proof course, 188 (F30), 487 (F30)
　flashings, 238 (H71), 521 (H71)
"Leaderflush" type B30 doors, 277 (L20), 559 (L20)
Lift and conveyor installations, 810
Lighting and power for the works, 112
Lighting
　emergency, 808
　external, 822
　installation, 808
　points, 406 (V21-22), 686 (V21-22), 808
　temporary, 112
Lightning protection, 408 (W20), 688 (W20), 814
Lightweight
　cast iron pipes, 375 (R11), 387 (R12), 657 (R11), 667 (R12)
　plaster, 299 (M20), 580 (M20), 778
　roof screeds, 298 (M12), 579 (M12), 738
"Limelite" plaster, 301 (M20), 581 (M20)
Limestone
　chippings, 244 (J21), 527 (J21)
　Guiting random walling, 183 (F20), 483 (F20)
　Portland facework, 226 (H51)
　reconstructed walling, 184 (F22), 483 (F22)
Lining paper, 314 (M52), 593 (M52)
Linings
　blockboard, 256 (K11), 539 (K11)
　chipboard, 256 (K11), 539 (K11), 778
　fireline board, 250, 251 (K10), 533 (K10)
　granite, 308 (M40), 784
　hardboard, 258 (K11), 541 (K11), 778
　industrial board, 250 (K10)
　insulation board, 258 (K11), 541 (K11)
　Laminboard, 258 (K11), 541 (K11)
　marble, 309 (M40)
　"Masterboard", 258 (K11), 541 (K11)
　Melamine faced, 257 (K11), 543 (K11), 778
　"Monolux 40", 259 (K11), 542 (K11)
　plasterboard, 251 (K10), 534 (K10)
　plywood, 259 (K11), 542 (K11), 778
　"Supalux", 258 (K11), 542 (K11), 778
　veneered plywood, 261 (K11), 544 (K11)
Linings/Sheathing/Dry partitioning, 250 (K), 533 (K)
Linoleum
　floor tiles, 313 (M50), 591 (M50), 786
　sheet flooring, 311 (M50), 590 (M50), 786
Linseed oil, 600 (M60)
Lintels
　precast, 776
　prestressed concrete, 193 (F31), 492 (F31)
　steel, 192 (F30), 491 (F30)
Liquid applied
　coatings, 746
　damp proof, 244 (J30), 527 (J30)
　tanking, 244 (J30), 527 (J30)
Lockers
　clothes, 331 (N10/11), 612 (N10/11)
　wet area, 331 (N10/11), 613 (N10/11)
Loft ladders, aluminium, folding, 290 (L30), 571 (L30)
Lost time, 100

Louvres
    aluminium, 275 (L10), 557 (L10)
    glass, 294 (L40), 574 (L40)
Luxcrete rooflights, pavement lights, 220 (H14)

M SURFACE FINISHES, 295, 576
Macadam, bitumen, 360 (Q22), 642 (Q22)
Mahogany
    basic price, 206 (G20)
    fittings, 330 (N10/11), 611 (N10/11)
    West African, 270 (L10), 279 (L20), 284 (L20), 286 (L30), 560 (L10), 564 (L20), 567 (L20), 774, 776
Mains pipework, 403 (S13), 683 (S13)
"Mandolite CP2", fire protection, 304 (M22), 585 (M22)
Manhole covers, 396 (R12), 675 (R12)
Manholes
    brick, 395 (R12), 674 (R12)
    concrete, 394 (R12), 674 (R12)
Mansard roofs, 730
Maple, basic price, 206 (G20)
Maple, Canadian, strip flooring, 264 (K20), 547 (K20), 780
Marble
    flooring, 309 (M40), 784
    wall linings, 309 (M40), 780
"Marley HD"
    vinyl sheet flooring, 311 (M50), 590 (M50)
    vinyl tile flooring, 312 (M50), 590 (M50)
"Marleyflex" vinyl tile flooring, 312 (M50), 591 (M50)
"Marmoleum" linoleum sheet flooring, 311 (M50), 590 (M50)
Masonry, 161 (F), 463 (F)
    basic material prices, 226 (H51)
    sundries, 186 (F30), 485 (F30)
Masonry slots, 157 (E42), 461 (E42)
"Masterboard", 212 (G20), 258 (K11), 500 (G20), 541 (K11)
Mastic asphalt
    damp proofing/tanking, 243 (J20), 526 (J20)
    flooring, 297 (M11), 578 (M11)
Mat or quilt insulation, 340 (P10), 622 (P10), 736
Mats, door, 333 (N10/11), 614 (N10/11)
Matting, entrance, 788
Matchboarded door, 276 (L20), 557 (L20)
Materials, weights of, 918
Matwells, 333 (N10/11), 614 (N10/11)
Mechanical plant cost, 104
Mechanical heating/Cooling/Refrigeration systems, 405 (T), 685 (T)
Medium density fireboard
    dado rails, 343 (P20), 624 (P20)
    skirtings, 343 (P20), 624 (P20)
    window boards, 342 (P20), 624 (P20)
Melamine faced chipboard, 328 (N10/11), 610 (N10/11)
Membranes, damp proof, 243 (J20), 526 (J20)
Metal
    balustrades, 290 (L30), 571 (L30)
    beads, 303 (M20), 584 (M20)
    handrail bracket, 290 (L30), 571 (L30)
    holes in, 290 (L30), 572 (L30)
    inserts, 157 (E42), 462 (E42)
    lathing, 305 (M30), 585 (M30)

Metal - cont'd
    roof decking, 249 (J43), 734
    windows, 273 (L10), 554 (L10), 766
Metalwork
    removal of, 413 (C20)
    repairs to, 418 (C50)
"Metsec" beams, 206 (G12), 495 (G12)
Mild steel
    angles, 349 (P20), 631 (P20)
    bar reinforcement 129 (D40), 132 (D50), 149 (E30), 443 (D50), 453 (E30)
    fabric, 150 (E30), 359 (Q21), 394 (R12), 453 (E30), 641 (Q21), 674 (R12)
Milled lead, 238 (H71), 520 (H71), 744
Mirrors, 332 (N10/11), 613 (N10/11)
    panels, 294 (L40), 574 (L40)
    cladding panels, 780
Mixed concrete prices, 135 (E05)
"Moistop" vapour barrier, 340 (P10), 622 (P10)
"Mondopave" rubber tiling, 313 (M50), 592 (M50)
"Monks Park" stone, 226 (H51)
Mortar
    basic material prices, 161 (F10)
    coloured, 161 (F10)
    plasticiser, 161 (F10)
Mosaic
    flooring, 784
    tiling, 780
"Mount Sorrel" paving blocks, 363 (Q25), 645 (Q25), 820
muPVC waste pipes, 376 (R11), 659 (R11)

N FURNITURE/EQUIPMENT, 328, 609
National insurance contributions, 100
Natural slates, 235 (H62), 518 (H62), 742
Natural stone walling, 226 (H51)
"Nitoflor Lithurin" surface hardener, 320 (M60), 600 (M60)
Non-asbestos
    cladding, 221 (H30), 510 (H30), 734
    roof slates, 235 (H61), 518 (H61)
"Nordic" roof windows, 275 (L10), 556 (L10)
Nosings, 313 (M50), 592 (M50)
"Nullifire" fire protection system, 203 (G10), 488 (G10)

Oak
    basic price, 206 (G20)
    block flooring, 310 (M42), 589 (M42), 784
    door sills, 285 (L20), 566 (L20)
    strip flooring, 264 (K20), 547 (K20), 780
Obeche, basic price, 206 (G20)
Office, temporary, 111
One-hour fire doors, 277 (L20), 559 (L20)
"Osmadrain" upvc pipes and fittings, 391 (R12), 671 (R12)
Overhead charges and profit,
    interpretation of, 99

P BUILDING FABRIC SUNDRIES, 340, 622
Padstones, precast concrete, 193 (F31), 492 (F31)

Paint
  aluminium 323 (M60), 604 (M60)
  anti-graffiti, 323 (M60), 604 (M60)
  basic prices, 315 (M60)
  bituminous, 318 (M60), 599 (M60)
  burning off, 588 (M60), 603 (M60)
  cement "Sandtex Matt", 323(M60), 603 (M60)
  cementitious, sprayed, 305 (M22), 585 (M22)
  eggshell, 318 (M60), 597 (M60)
  emulsion, 316 (M60), 595 (M60)
  flame retardent, 321 (M60), 600 (M60)
  gloss, 317 (M60), 596 (M60)
  "Hammerite", 327 (M60), 607 (M60)
  knotting, 317 (M60), 597 (M60)
  polyurethane, 321 (M60), 600 (M60)
  "Weathershield", 323 (M60), 603 (M60)
Panelled doors
  hardwood, 279 (L20), 560 (L20), 776
  softwood, 278 (L20), 559 (L20), 776
Panelled screens
  softwood stud, 772
  stainless steel, 772
  steel, 772
  terazzo faced, 264 (K33)
Pantiles, clay, 740
  Langley, 229 (H60)
  Sandtoft, 229 (H60), 512 (H60)
Paperhanging, 314 (M52), 593 (M52), 792
Parking bollards, 821
Parklands, 819
Parquet flooring, 310 (M42), 589 (M42), 784
Partitions
  approximate estimates, 768
  demountable, 772
  fire, 772
  glazed, 772
  "Gyproc"
    laminated, 250 (K10), 533 (K10), 768
    metal stud, 251 (K10), 533 (K10), 768
Patent glazing, 220 (H10), 509 (H10), 736, 764
Patio doors, aluminium, 281 (L20), 562 (L20)
Pavement lights, concrete, 220 (H14)
Paving
  artificial stone, 361 (Q25), 644 (Q25), 820
  bitumen emulsion, 820
  bitumen macadam, 360 (Q22), 642 (Q22), 820
  block, 362 (Q25), 645 (Q25), 810
  cement:sand, 293 (M10), 576 (M10), 782
  cobble, 362 (Q25), 644 (Q25), 820
  concrete flags, 362 (Q25), 644 (Q25), 820
  granite setts, 364 (Q25), 646 (Q25), 820
  granolithic, 295 (M10), 576 (M10), 782
  gravel, 361 (Q23), 643 (Q23), 820
  "Intersett" paving blocks, 363 (Q25), 645 (Q25), 820
  "Keyblock" rectangular concrete blocks, 362 (Q25), 645 (Q25), 820
  marble, 786
  mastic asphalt, 297 (M11), 578 (M11), 782
  "Mount Sorrel" blocks, 363 (Q25), 645 (Q25), 820

Paving - cont'd
  "Pedesta" blocks, 363 (Q25), 645 (Q25), 820
  quarry tiles, 307 (M40), 587 (M40), 782
  terrazzo, 309 (M41), 589 (M41), 782
  terrazzo tiles, 309 (M41), 589 (M41), 782
  "Trafica" blocks, 363 (Q25), 645 (Q25)
  York stone slab, 363 (Q25), 646 (Q25)
Paving/Planting/Fencing/Site furniture, 357 (Q), 639 (Q)
Paviors
  block, 362 (Q25), 645 (Q25)
  brick, 362 (Q25), 644 (Q25), 782, 820
"PC" value,
  interpretation of, 104
Pebble dash, 302 (M20), 583 (M20)
"Pedesta" paving blocks, 363 (Q25), 645 (Q25), 820
Pellating, 349 (P20), 631 (P20)
Pensions, 107
Performance bonds, 109
Perimeter treatments, roofing, 738
Perlite insulation boards, 248 (J41), 531 (J41), 738
Permanent shuttering
  concrete, 145 (E20), 449 (E20)
  "Holorib", 146 (E20), 160 (E60), 450 (E20) 726, 732
  "Hyrib", 145 (E20), 449 (E20)
  "Ribdeck", 146 (E20), 450 (E20)
"Peter Cox"
  chemical transfusion dpc, 487 (F30)
Piled foundations, 716
Piling
  concrete, 127 (D30), 716
  steel, 128 (D32), 716
Pin-board, "Sundeala A", 349 (P20), 631 (P20)
Pipe
  casings, plywood, 259 (K11), 542 (K11)
  insulation, 402 (S10/11), 683 (S12)
  trenches, 353 (P30), 635 (P30)
Piped supply systems, 399 (S), 679 (S)
Pipework
  aluminium, 369 (R10), 652 (R10)
  cast iron, 370 (R10), 653 (R10), 744
  gas mains, 404 (S32), 685 (S32)
  "Timesaver", 374 (R11), 386 (R12), 657 (R11), 666 (R12)
  clay, 387 (R12), 667 (R12), 824
  "HepSeal", 387 (R12), 668 (R12)
  "SuperSleve", 387 (R12), 667 (R12)
  concrete, 390 (R12)
  copper, 399 (S10/11), 679 (S12)
  mains, 403 (S13), 683 (S13)
  MDPE, 403 (S13), 683 (S13)
  muPVC waste, 376 (R11), 659 (R11)
  overflow, 377 (R11), 660 (R11)
  polypropylene, 375 (R11), 659 (R11)
  PVC, see "uPVC"
  rainwater, 369 (R10), 652 (R10), 744
  soil, 377 (R11), 660 (R11), 802
  uPVC (unplasticised PVC)
    drain, 391 (R12), 671 (R12), 824
    drain, ribbed pipe, 392 (R12), 671 (R12)
    overflow, 377 (R11), 660 (R11)
    rainwater, 372 (R10), 656 (R10), 744
    soil/waste, 377 (R11), 660 (R11), 802

uPVC (unplasticised PVC) - cont'd
   vitrified clay, 387 (R12), 667 (R12), 824
Pitch polymer damp proof course, 133, (D50), 187 (F30), 444 (D50), 486 (F30)
Pitched roofs, 730
Plain roof tiles, 229, 230, 234 (H60), 513, 517 (H60), 740
PLANNING FEES, 58
Planning parameters, 895
"Plannja" metal cladding/decking, 223 (H31), 249 (J43)
Planting, 365 (Q31), 648 (Q31), 818
Plaster, 298 (M20), 579 (M20), 778
   beads, 303 (M20), 584 (M20)
   "Carlite", 299, 301 (M20), 580, 582 (M20), 778
   fibrous, 306 (M31), 586 (M31)
   "Hardwall", 300 (M20), 581 (M20)
   lightweight, 299 (M20), 580 (M20), 778
   "Renovating", 301(M20), 423 (C90), 581 (M20)
   "Snowplast", 300 (M20), 581 (M20)
   squash court, 302 (M20), 778
   "Thistle", 300 (M20), 581 (M20), 778
   "Thistle" board finish, 302 (M20), 583 (M20)
   "Thistle Multi Finish", 300 (M20), 581 (M20)
   "Thistle X-ray", 301 (M20), 582 (M20)
Plasterboard, 250 (K10), 534 (K10), 778, 790
   accessories, 303 (M20), 584 (M20)
   cove, 306 (M31), 586 (M31)
   dry linings, 255 (K10), 537 (K10), 778
   "Duplex" insulating board, 252 (K10), 303 (M20), 535 (K10), 583 (M20)
   plank, 252 (K10), 535 (K10)
   plastic faced, 253 (K10), 536 (K10)
   thermal board, 253 (K10), 535 (K10)
   vermiculite cladding, 255 (K10), 538 (K10)
   wallboard, 251 (K10), 534 (K10), 778
Playground equipment, 822
Plumbers
   earnings, 103
   hourly rates, 103
Plumbing operatives, 103
Plywood, 212 (G20), 257 (K11), 500 (G20), 540 (K11)
   marine quality, 212 (G20), 501 (G20)
   linings, 259 (K11), 542 (K11), 778
   pipe casings, 259 (K11), 542 (K11)
   roofing, 212 (G20), 258 (K11), 500 (G20), 540 (K11)
   underlay, 313 (M50), 592 (M50)
Pointing
   expansion joints, 152 (E40), 456 (E40)
   wood frames, etc., 285 (L20), 566 (L20)
Polishing
   floors, 311 (M42), 589 (M42)
   French, 322 (M60), 602 (M60), 792
   wax, 321 (M60), 601 (M60), 792
"Polybit 350" roofing, 247 (J41), 530 (J41), 748
Polycarbonate sheet, 291 (L40)
Polyester resin flooring, 782
Polyethylene
   cistern, 402 (S10/11), 682 (S12)
   damp proof course, 187 (F30), 486 (F30)
"Polyflex" vinyl tile flooring, 312 (M50), 591 (M50)
Polypropylene
   accessories, 389 (R12), 669 (R12)
   pipes, waste, 375 (R11), 659 (R11)
   traps, 378 (R11), 661 (R11)

Polystyrene board, 341 (P10), 622 (P10), 736
Polythene damp proof membrane, 144 (E05), 448 (E05)
Polyurethane
   boards, 248 (J41), 531 (J41), 726
   sealer/lacquer, 321 (M60), 589 (M42), 600 (M60)
"Portabond"/"Portaflek", 317 (M60), 596 (M60)
Portland
   cement, 140 (E05)
   stone cladding, 226 (H51)
Portland stone, 226 (H51)
"Powerlon 250 BM", 235 (H60), 517 (H60)
Powered float finish, 154 (E41), 458 (E41), 780
Precast concrete
   access or inspection chambers, 394 (R12), 674 (R12)
   channels, 359 (Q10), 639 (Q10)
   copings, 193 (F31), 492 (F31)
   edgings, 357 (Q10), 639 (Q10)
   fencing, 368 (Q40), 651 (Q40)
   flooring, 158 (E60), 462 (E60)
   floors, 728
   kerbs, 358 (Q10), 639 (Q10)
   lintels, 776
   manhole units, 395 (R12)
   padstones, 193 (F31), 492 (F31)
   panels, 758
   paving blocks, 362 (Q25), 645 (Q25), 820
   paving flags, 362 (Q25), 644 (Q25), 820
   pipes, 390 (R12)
   sills/lintels/copings, 193 (F31), 492 (F31)
Preliminaries/General conditions, 105 (A), 410 (A)
Preparation for redecoration
   external, 602 (M60)
   internal, 593 (M60)
Preformed plywood casings, 260 (K11), 543 (K11)
Preservatives, wood, 315, 327 (M60), 607 (M60)
Pressed steel gutters, 372 (R10), 655 (R10)
Prestressed concrete, 158 (E60), 193 (F31), 462 (E60), 492 (F31)
Prime cost, building industry, 71
Priming steelwork, 205 (G10), 494 (G10)
PROPERTY INSURANCE, 877
Protection board, 245 (J40), 528 (J40)
Protective, communication and special installations, 814
PVC
   channels, 397 (R12), 676 (R12)
   gutters, 373 (R10), 656 (R10), 748
   pipes, 372 (R10), 377 (R11), 391 (R12), 656 (R10), 659 (R11), 671 (R12), 824
   rooflights, 275 (L10), 556 (L10)
   weatherboard, 754
   windows, 274 (L12), 555 (L12), 766
"Pyran" fire-resisting glass, 293 (L40), 574 (L40)
"Pyrostop" fire-resisting glass, 294 (L40), 574 (L40)

Q PAVING/PLANTING/FENCING/SITE FURNITURE, 357, 639
QUANTITY SURVEYORS' FEES, 9
Quarry tiles, 307 (M40), 587 (M40), 782
Quoin jambs, 424 (C90)

R DISPOSAL SYSTEMS, 369, 652
Radiators, 405 (T31), 685 (T31), 804
Rafters, trussed, 211 (G20), 499 (G20), 730
Rainwater pipes, 744
Raised floors, 266 (K41), 786
RATES OF WAGES, 91
Rates on temporary buildings, 111
"Rawlbolts", 216 (G20), 505 (G20)
Reconstructed
    stone walling, 184 (F22), 483 (F22)
    stone slates, 236 (H63), 519 (H63), 742
"Redrib" lathing, 306 (M30), 585 (M30)
Refractory
    flue linings, 189 (F30), 489 (F30)
Refrigeration installation, 814
Reinforcement
    bars, 129 (D40), 132 (D50), 149 (E30), 443 (D50), 453 (E30)
    "Brickforce", 188 (F30), 487 (F30)
    fabric, 150 (E30), 359 (Q21), 394 (R12), 454 (E30), 641 (Q21), 674 (R12)
Reinstating concrete, 415 (C41)
Removal of existing
    brickwork and blockwork, 412 (C20)
    claddings, 425 (C90)
    metalwork, 413 (C20)
    outbuildings, 411 (C20)
    pipework etc., 422 (C90)
    stonework, 413 (C20)
    surface finishes, 423 (C90)
    timber, 413 (C20)
    trees, 118 (D20), 432 (D20)
Removing from site, see "Disposal"
Rendering
    cement:sand, 298 (M20), 579 (M20), 778
    "Cemrend" self coloured, 302 (M20)
    "Sika", waterproof, 243 (J10), 526 (J10)
Renovations, 415 (C41), 417 (C50), 417 (C51)
Repairs to
    brickwork and blockwork, 415 (C41)
    claddings, 425 (C90)
    concrete work, 415 (C41)
    metalwork, 418 (C50)
    pipework, 422 (C90)
    surface finishes, 423 (C90)
    timber, 418 (C51)
    waterproof finishes, 420 (C90)
"Resoplan" sheeting, 221 (H20), 509 (H20)
"Rhino Contract" sheet flooring, 312 (M50), 590 (M50)
"Riblath", lathing, 305 (M30), 586 (M30)
"Ribdeck" permanent shuttering, 146 (E20), 450 (E20)
Ridges
    comparative prices, 744
"Rigifix" column guard, 158 (E42), 462 (E42)
Road crossings, 821
Road gratings, 390 (R12), 670 (R12)
Road markings, 364 (Q26), 646 (Q26)
Roads
    and barriers, 821
    and footpaths, concrete, 358 (Q21), 640 (Q21), 821
    bitumen macadam, 360 (Q22), 642 (Q22), 821

"Rockwool"
    fire barriers, 341 (P10), 623 (P10)
    fire stops, 341 (P10), 622 (P10)
Roller shutters, 766
Roof
    approximate estimates, 730
    boarding, 262 (K20), 545 (K20), 736
    claddings/coverings
        aluminium, 223 (H31), 736
        "Andersons" HT, 246 (J41), 529 (J41)
        asbestos-free
            corrugated sheets, "Eternit", 221 (H30), 510 (H30)
            fibre cement slates, 235 (H61), 518 (H61)
        clay
            pantiles, 229 (H60), 512 (H60), 740
            plain tiles, 230 (H60), 513 (H60), 740
        concrete
            interlocking tiles, 231 (H60), 513 (H60), 738
            plain tiles, 234 (H60), 517 (H60),
        copper, 240 (H73), 523 (H73), 744
        CPE membrane, 748
        CPSG memrane, 748
        Elastomeric single ply, 746
            EPDM membrane, 746
            Butyl rubber membrane, 746
        felt, 245 (J41), 528 (J41), 746
            "Hyload 150 E", 247 (J41), 531 (J41), 748
        "Kal-Zip", 223 (H31)
        lead, 744
        "Marley" tiles, 229, 231, 232 (H60), 513, 514, 515 (H60)
        mastic asphalt, 244 (J21), 527 (J21), 746
        natural slates, 235 (H62), 518 (H62), 742
            Westmorland green, 236 (H62), 519 (H62), 742
        PIB membrane, 748
        polyester-based, 246 (J41), 529 (J41), 748
        reconstructed stone slates, 236 (H63), 519 (H63), 732
        red cedar shingles, 237 (H64), 520 (H64), 742
        "Redland" tiles, 232, 233, 234 (H60), 515, 516 (H60)
        "Ruberfort HP" 350, 247 (J41), 530 (J41), 745
        "Rubergias 120 GP", 246 (J41), 530 (J41), 745
        stainless steel, 241 (H75), 524 (H75)
        steel troughed sheeting, 222 (H31),
        thermoplastic single ply, 748
            PVC membrane, 748
        torch-on, 748
        zinc, 240 (H74), 523 (H74), 744
    decking,
        "Plannja", 249 (J43)
        woodwool, 219 (G32), 508 (G32)
    dormer, 732
    eaves, 742
    finishes, 734
    glazing, 736
    hips, 744
    insulation boards, 248 (J41), 531 (J41), 736

Index

Roof - cont'd
   lights, 736
      concrete, 220 (H14)
      PVC, 275 (L10), 556 (L10)
      Velux, 275 (L10), 556 (L10)
   outlets, aluminium, 369 (R10), 652 (R10)
   ridges, 744
   screeds, 297 (M12), 579 (M12), 738
   screed ventilators, 244 (J21), 527 (J21)
   trusses, timber, 211 (G20), 499 (G20), 730
   underlay, 235 (H60), 517 (H60), 736
   verges, 742
   windows, 275 (L10), 556 (L10)
Roundabouts, 821
Rubber floor tiles, 786
Rubble walling, 183 (F20), 483 (F20)
"Ruberfort HP350" roofing, 247 (J41), 530 (J41), 748
"Ruberglas 120 GP" roofing, 246 (J41), 529 (J41), 748

S PIPED SUPPLY SYSTEMS, 399, 679
"Sadolin", 315, 321, 322 (M60), 601, 608 (M60)
Safety, health and welfare of work people, 111
"Safety" vinyl sheet flooring, 311 (M50), 590 (M50), 786
Sand
   blinding/beds, 127 (D20), 440 (D20), 820
   faced facing bricks, 167 (F10), 467 (F10), 754
   filling, 126 (D20), 358 (Q20), 383 (R12), 439 (D20), 640 (Q20), 664 (R12)
Sandlime bricks, 170 (F10), 470 (F10)
"Sandtex Matt", cement paint, 323 (M60), 603 (M60), 764
Sanitary and disposal installations, 800
Sanitary fittings, 334 (N13), 615 (N13), 800
Sapele, basic price, 200 (G20)
Sawn softwood
   tanalised, 209 (G20), 497 (G20)
   untreated, 207 (G20), 495 (G20)
Scaffolding, 117
Scaffolding sheeting, 117
Scarfed joint, 420 (C51)
Screeds
   cement:sand, 295 (M10), 576 (M10), 782
   "Isocrete", 298 (M12), 579 (M12), 738
   latex, 297 (M12), 579 (M12), 782
   roof, 298 (M12), 579 (M12), 738
"Screeduct", 356 (P31), 638 (P31)
Screen walls, 822
Screens
   dust-proof, 431 (C90)
   folding, 772
   stainless steel, 768, 772
   temporary, 431 (C90)
Sealers, polyurethane, 321 (M60), 600 (M60)
"Sealmaster" fire and smoke seals, 341 (P10), 623 (P10)
Security fencing, 368 (Q40), 651 (Q40)
Security systems, 408 (W), 688 (W)
Security/communications installations, 814
Seeding, 358 (Q30), 637 (Q30), 818
"Sentinel Sterling", fencing, 368 (Q40), 651 (Q40)
Septic tanks, 397 (R12), 677 (R12)
Service runs, 823
"Servi-pak" protection board, 245 (J40), 528 (J40)
Setting out the site, 106
Setts, granite, 364 (Q25), 646 (Q25), 820
Shelving systems, 332 (N10/N11), 613 (N10/N11)

Sheet
   flooring
      "Altro Safety", 311 (M50), 590 (M50), 786
      "Armstrong", 312 (M50), 590 (M50)
      "Classic", 311 (M50), 590 (M50)
      "Forbo-Nairn", 311 (M50), 590 (M50)
      "Gerflex", 311 (M50), 590 (M50)
      linoleum, 311 (M50), 590 (M50), 786
      "Marley", 311 (M50), 590 (M50)
      "Marley HD", 311 (M50), 590 (M50)
      "Rhino Contract", 312 (M50), 590 (M50)
      vinyl, 311 (M50), 590 (M50), 786
   linings and casings, 256 (K11), 539 (K11)
   metal roofing, flashings and gutters, 238 (H71), 239 (H72), 240 (H73), 241 (H74), 241 (H75), 520 (H71), 522 (H72), 523 (H73), 523 (H74), 524 (H75), 734
   piling, 128 (D32), 716
Shingles, cedar, 237 (H64), 520 (H64), 742
Shop fronts, 768
Shoring and strutting, 414 (C30)
Shot blasting, 154 (E41), 205 (G10), 458 (E41), 494 (G10)
Shower fitting, 339(N13), 620 (N13),
Shuttering, permanent
   "Holorib", 146 (E20), 450 (E20),
   "Hyrib", 145 (E20), 449 (E20)
   "Ribdeck", 146 (E20), 450 (E20)
Shutters,
   counter, 281 (L20), 562 (L20)
   fire, 768
   roller, 281 (L20), 562 (L20), 766
   vertically opening, 281 (L20), 556 (L20)
Sick pay, 100
Signs/Notices, 339 (N15), 621 (N15)
"Sika" waterproof rendering, 243 (J10), 526 (J10)
"Sikkens Cetol", 315, 316, 322 (M60), 595, 601, 608 (M60)
Sill
   brick, 169, 172, 175 (F10), 469, 473, 476 (F10)
   glazed wall tile, 308 (M40), 588 (M40)
   hardwood, 271, 272 (L10), 284, 285 (L20), 552, 553 (L10), 565, 566 (L20)
   quarry tile, 307 (M40), 587 (M40)
   softwood, 271 (L10), 552 (L10)
   steel, 273 (L10), 554 (L10)
Sills and tile creasing, 189 (F30), 489 (F30)
Sinks, 334 (N13), 615 (N13), 800
Site
   preparation, 120 (D20), 432 (D20)
   vegetation, clearing, 120 (D20), 432 (D20)
Skirtings
   asphalt, 244 (J21), 527 (J21)
   bitumen felt, etc., 245, 246, 247 (J41), 528, 530 (J41)
   granolithic, 296 (M10), 577 (M10), 788
   hardwood, 345, 347 (P20), 627, 629 (P20), 788
   MDF, 343 (P20), 624 (P20)
   quarry tile, 307 (M40), 587 (M40), 782
   softwood, 343 (P20), 624 (P20), 788
   terrazzo, 310 (M41), 788
Slab
   coffered, 146 (E05), 438, 450 (E05)
   suspended, 726

Slates
    asbestos-free "Eternit", 235 (H61), 518 (H61)
    concrete interlocking, 232 (H60), 515 (H60), 738
    damp proof course, 134 (D50), 188 (F30), 444 (D50), 487 (F30)
    flooring, 309 (M40), 588 (M40), 784
    natural, 235 (H62), 518 (H62), 742
    reconstructed, 236 (H63), 519 (H63), 742
    Welsh, 236 (H62), 518 (H62), 742
    Westmorland green, 236 (H62), 519 (H62), 742
Sliding door gear, 353 (P21), 635 (P21)
Slots, masonry, 157 (E42), 461 (E42)
Smoke vents, 814
"Snowplast", 300 (M20), 581 (M20)
Softwood
    architraves, 343 (P20), 625 (P20)
    basic prices, carcassing, 206 (G20)
    boarding, 262 (K20), 545 (K20), 742, 778, 790
    doors, 276 (L20), 557 (L20), 772
    fittings, 329 (N10/11), 610 (N10/11)
    flooring, 263 (K20), 545 (K20), 780
    floors, 726
    g.s. grade, 206, 208, 211 (G20), 497, 499 (G20)
    joinery quality, basic prices, 206 (G20)
    repairs, 418 (C51)
    roof boarding, 262 (K20), 545 (K20), 730, 738
    s.s. grade, 206, 208, 211 (G20), 497, 499 (G20)
    skirtings, 343 (P20), 624 (P20), 788
    staircase, 286 (L30), 566 (L30), 750
    trussed rafters, 211 (G20), 499 (G20), 730
    windows, 267, 268, 269 (L10), 548, 549, 550 (L10)
    window frames, 271 (L10), 552 (L10)
Soil
    imported, 126, 127 (D20), 439, 440 (D20)
    vegetable, 364 (Q30), 647 (Q30)
Solid strutting, 208, 210 (G20), 497, 499 (G20)
"Solignum", wood preservatives, 315, 327 (M60), 607 (M60)
Sound-deadening quilt insulation, 340 (P10)
Space decks, 724
Space heating, 806
Special equipment, 796,
Sports pitches, 819
Spot items, 421 (C90)
"Spraylath" lathing, 305 (M30), 586 (M30)
Sprinkler installations, 814
Squash court, 796
    plaster, 778
Staff costs, 110
Stains
    preservatives, 315, 327 (M60), 607 (M60)
    varnishes, 321 (M60), 600 (M60)
    wood, 322 (M60), 601, 608 (M60)
Stainless steel
    angle beads/stop beads, 303 (M20), 584 (M20)
    angles, 349 (P20), 631 (P20)
    balustrades, 290 (L30)
    column guards, 158 (E42), 462 (E42)
    cramps, etc., 228 (H51)

Stainless steel - cont'd
    entrance screens, 768
    joint reinforcement, 188 (F30), 487 (F30)
    reinforcement, 150 (E30), 454 (E30)
    roofing, 241 (H75), 524 (H75)
    sheet coverings/flashings, 241 (H75), 521 (H75)
    ties, 191 (F30), 491 (F30)
    wall starters, 191 (F30), 491 (F30)
Staircases
    cast iron, 752
    concrete, 142, 143 (E05), 446, 447 (E05), 750
    feature, 752
    formwork, 148 (E20), 453 (E20)
    hardwood, 286 (L30), 567 (L30), 750
    metal, 750
    softwood, 286 (L30), 566 (L30), 750
    spiral, 289 (L30)
    steel, 750
Stair nosings, 313 (M50), 592 (M50)
Stairs, 750
"Stanlock" joints, 404 (S32), 684 (S32)
Steel
    arch-frames, 305 (M30), 585 (M30)
    balustrades, handrail, etc., 290 (L30), 571 (L30), 750
    cladding, 222 (H31), 734, 758
    deck/shuttering,
        "Holorib", 146 (E20), 450 (E20), 732
        "Ribdeck", 146 (E20), 450 (E20)
    door/frame, 281 (L21), 562 (L21)
    light joisted floors, 726
    lintels, 192 (F30), 491 (F30)
    metalwork finishes, 792
    piling, 128 (D32), 716
    reinforcement
        bars, 129 (D40), 133 (D50), 149 (E30), 443 (D50), 453 (E30)
        fabric, 150 (E30), 359 (Q21), 394 (R12), 454 (E30), 641 (Q21), 674 (R12)
    roof tiles, lightweight, 223 (H31), 511 (H31)
    space decks, 724
    staircases, 750
    troughed sheeting, 222 (H31),
    windows, 273 (L10), 554 (L10), 766
Steelwork
    basic prices, 194 (G10)
    fabricated and erected, 204 (G10), 493 (G10)
    unfabricated and erected, 205 (G12), 494 (G12)
Step irons, 396 (R12), 675 (R12)
"Sto External Render System", 299 (M20)
"StoTherm" mineral external wall insulation system, 304 (M21)
Stone
    basic material prices, 226 (H51)
    cladding, 226 (H51)
    flooring, 784
    natural, 226 (H51),
    reconstructed roofing slates, 236 (H63), 519 (H63), 742
    reconstructed walling, 184 (F22), 483 (F22)
    rubble walling, 183 (F20), 483 (F20)
    York, 363 (Q25), 646 (Q25)

Stonework
    cast, 184 (F22), 483 (F22)
    removal of, 413 (C20)
Stop beads, 303 (M20), 584 (M20)
Stopcocks, 401 (S10/11), 681 (S12)
Stop valve pit, 354 (P30), 636 (P30)
Straps, 216 (G20), 504 (G20)
Street furniture, 822
Strip flooring, hardwood, 264 (K20), 547 (K20), 780
Stripping wallpaper, 423 (C90)
Structural/Carcassing metal/timber, 194 (G), 493 (G)
"Styrofoam" insulation, 341 (P10), 622 (P10)
Substructure, 712
"Sundeala A" pinboard, 349 (P20), 631 (P20)
Sundries, Building Fabric, 340 (P), 622 (P)
"Supalux" lining, 258 (K11), 541 (K11)
"SuperSleve" vitrified clay pipes, 387 (R12), 667 (R12)
Superstructure, 720
Supervision, 102, 106
Surface
    finishes, 315 (M), 576 (M)
    removal of, 423 (C90)
    repairs to, 423 (C90)
    treatments, 127 (D20), 132 (D50), 358 (Q20), 440 (D20), 442 (D50), 640 (Q20)
Suspended ceilings, 265 (K40), 790
Switch socket/outlet points, 406 (V21-22), 686 (V21-22),
"Synthaprufe", 188 (F30), 244 (J30), 487 (F30), 527 (J30)

T MECHANICAL HEATING/COOLING/REFRIGERATION SYSTEMS, 405, 685
TABLES AND MEMORANDA, 883
    for each trade, 919
"Tacboard" fire resisting boards, 261 (K11), 544 (K11)
"Tacfire" fire resisting boards, 261 (K11), 544 (K11)
Tanking
    liquid applied, 244 (J30), 527 (J30)
    mastic asphalt, 243 (J20), 526 (J20)
Tanks, 402 (S10/11), 682 (S12)
Teak
    basic price, 206 (G20)
    fittings, 330 (N10/11), 612 (N10/11)
Telephones, 113
Temporary
    accommodation, 111, 112
    fencing, hoarding and screens, 118
    lighting, 112
    roads, 116
Tender prices index, 692
Tennis court
    fencing, 367 (Q40), 650 (Q40), 819
Terne coated stainless steel, 241 (H75), 524 (H75)
"Terram" filter membrane, 127 (D20), 440 (D20)
Terrazzo
    faced partitions, 264 (K33)
    paving, 309 (M41), 589 (M41), 782
"Texitone" paving slabs, 361 (Q25), 644 (Q25), 820
Thermal boards, 253 (K10), 535 (K10)

Thermal insulation
    cavity wall, 340 (P10), 622 (P10), 756
    "Crown Wool", 340 (P10)
    cylinders, 403 (S10/11), 682 (S12)
    pipes, 402 (S10/11), 683 (S12)
    "Styrofoam", 341 (P10), 622 (P10)
    tanks, 403 (S10/11), 682 (S12)
"Thermabate" cavity closer, 190 (F30), 490 (F30)
"Thermalite" blocks, 177 (F10), 477 (F10)
Thermoplastic floor tiles, 313 (M50), 591 (M50), 784
"Thistle" plaster, 300 (M20), 581 (M20), 778
"Thistle X-ray" plaster, 301 (M20), 582 (M20)
Tile
    flooring
        "Accoflex", 312 (M50), 590 (M50)
        alternative material prices, 307 (M40)
        "Altro", 311 (M50), 590 (M50), 786
        anti-static, 312 (M50), 591 (M50)
        "Arlon", 312 (M50), 591 (M50)
        "Armstrong", 312 (M50), 590 (M50)
        carpet, 314 (M50), 592 (M50),
        cork, 313 (M50), 592 (M50), 784
        "Forbo-Nairn", 311 (M50), 590 (M50)
        linoleum, 313 (M50), 591 (M50), 786
        "Marley", 312 (M50), 591 (M50)
        "Marleyflex", 312 (M50), 591 (M50)
        "Marley HD", 312 (M50), 591 (M50)
        "Mondopave", 313 (M50), 592 (M50)
        "Polyflex", 312 (M50), 591 (M50)
        quarry, 307 (M40), 587 (M40), 782
        studded rubber, 313 (M50), 592 (M50), 786
        terrazzo, 309 (M41), 589 (M41),
        thermoplastic, 313 (M50), 591 (M50), 784
        vinyl, 312 (M50), 590 (M50), 784
        "Vylon", 312 (M50), 591 (M50)
        "Wicanders", 313 (M50), 592 (M50)
    roofing
        alternative material prices, 229 (H60)
        clay
            pantiles, 229 (H60), 512 (H60), 740
            plain tiles, 230 (H60), 513 (H60), 740
        concrete
            interlocking tiles, 231 (H60), 513 (H60)
            plain tiles, 234 (H60), 517 (H60), 730
    wall, 308 (M40), 587 (M40), 780
"Tilene 200P" underlay, 235 (H60), 517 (H60)
Timber
    "Aquaseal" treatments, 206 (G20)
    basic material prices, 206 (G20)
    board flooring, 263 (K20), 546 (K20),
    connectors, 216 (G20), 505 (G20)
    cut out defective, 420 (C51)
    doors, 276 (L20), 557 (L20)
    fencing, 822
    flame proofing, 208, 211 (G20), 497, 499 (G20)
    floors, 726
    framework, 213, 214 (G20), 502, 503 (G20)

Timber - cont'd
  holes in, 355 (P31), 638 (P31)
  linings, 262 (K20), 544 (K20)
  removal of, 413 (C20)
  repairs to, 418 (C51)
  roofs, 730
  shingling, 237 (H64), 520 (H64), 742
  staircases, 286 (L30), 566 (L30)
  stress grading, 206, 208, 211 (G20), 497, 499 (G20)
  studding, 768
  treatment, 206 (G20)
  windows, 267 (L10), 548 (L10), 764
  wrot faces, 209, 211 (G20), 497, 499(G20)
"Timesaver" pipes,
  drain, 386 (R12), 666 (R12)
  soil, 374 (R11), 657 (R11)
"Topblock" blocks, 177 (F10)
"Toplite" blocks, 177 (F10)
"Torvale" woodwool slabs, 219 (G32), 261 (K11), 508 (G32), 544 (K11)
Trap
  clay (intercepting), 390 (R12), 670 (R12)
  polypropylene, 378 (R11), 661 (R11)
Treated sawn softwood, 209, 214 (G20), 497, 502 (G20),
Treatment prices, 206 (G20)
Trees,
  removing, 120 (D20), 432 (D20)
  planting, 818
"Tretolastex 202T", 245 (J30), 528 (J30)
Trocal PVC roofing, 249 (J42), 532 (J42)
Troughed slabs, 142, 143 (E05), 446, 447 (E05)
Trough formers
  "Cordek", 146 (E20), 450 (E20)
Truss clips, 217 (G20), 506 (G20)
Trussed rafters, 211 (G20), 499 (G20), 730
"Tuftiguard" door mats, 333 (N10/11), 614 (N10/11)
Turf, 365 (Q30), 647 (Q30), 818
Turf, lifting, 120 (D20), 432 (D20)
Tyrolean decorative rendering, 302 (M20), 583 (M20), 790
"Tyton" joints, 403 (S13), 683 (S13)

"Uginox AE", 241 (H75), 524 (H75)
"Uginox AME", 241 (H75), 524 (H75)
"Ultra-Rib" UPVC drain pipes, 392 (R12), 671 (R12)
Unclimbable fencing, 368 (Q40), 651 (Q40)
Underfloor heating, 804
Underlay, to
  carpet, 314 (M51), 593 (M51), 786
  floor finishings, 313 (M50), 592 (M50), 786
  roofs, 235 (H60), 517 (H60), 736
Underpass, 821
Underpinning, 130 (D50), 440 (D50)
Unfabricated steelwork, 205 (G12), 494 (G12)
Universal
  beams, 194 (G10)
  columns, 194 (G10)
UNIVERSITY BUILDINGS, 832
Unplasticised PVC (UPVC) pipes
  accessories, 392 (R12), 672 (R12)
  drain, 391 (R12), 671 (R12), 824
    ribbed, 392 (R12), 671 (R12)
  gutters, 373 (R10), 656 (R10), 748

Unplasticised PVC (UPVC) pipes - cont'd
  overflow pipes, 377 (R11), 660 (R11)
  rainwater, 372 (R10), 656 (R10), 744
  soil/waste, 377 (R11), 659 (R11), 800
UPVC windows, 274 (L10), 555 (L10), 766
Upper floors, 726
Urinals, 337 (N13), 618 (N13), 800
Utile, basic price, 206 (G20)

V ELECTRICAL SYSTEMS, 406, 686
Value Added Tax, 99, 104, 107, 878
Valves, 402 (S10/11), 682 (S12)
"Vandex" slurry, 245 (J30), 528 (J30)
Vapour barrier, 248 (J41), 531 (J41)
Varnish, 321 (M60), 600 (M60)
Vegetable soil
  imported, 364 (Q30), 647 (Q30)
  selected, 365 (Q30), 647 (Q30)
Vehicle control barrier, 821
"Velux" roof windows, 275 (L10), 556 (L10)
Ventilation systems, 806
Vents, 814
  fascia, 234 (H60), 517 (H60)
  soffit, 234 (H60), 517 (H60)
"Vermiculite" gypsum cladding, Viculad, 255 (K10), 538 (K10)
"Vigerflex" wood block flooring, 310 (M42), 589 (M42)
Vinyl
  floor tiles, 312 (M50), 590 (M50), 784
  sheet flooring, 311 (M50), 590 (M50), 786
"Visqueen" sheeting, 144 (E05), 448 (E05)
Vitrified clay, see "Clay"
"Vylon" vinyl tile flooring, 312 (M50), 591 (M50)

W SECURITY SYSTEMS, 408, 688
WAGE RATES
  building industry, 91, 101
  plumbing industry, 95, 103
  road haulage workers, 94
Walkways, 728
Wall
  finishes
    approximate estimates, 778
    granite, 309 (M40)
    marble, 309 (M40)
  kickers, 148 (E20), 452 (E20)
  linings,
    sheet, 250 (K), 533 (K)
    paper, 314 (M52), 593 (M52)
  ties
    butterfly type, 158 (E42), 191 (F30), 485 (F30)
    twisted, 192 (F30), 228 (H51), 491 (F30)
  tiles, 308 (M40), 587 (M40), 780
  units, 331 (N10/11), 612 (N10/11), 794
"Wallforce" reinforcement, 188 (F30), 487 (F30)
Walling, diaphragm, 129 (D40), 720
Walls, 754
"Wareite Xcel" laminated chipboard, 329 (N10/11), 610 (N10/11)
Warm air curtains, 806
Wash down existing, 593, 602 (M60)
Waterbars, 349 (P20), 631 (P20)

Water
    for the works, 113
    installations, 802
    mains, 403 (S13), 683 (S13)
    tanks, 402 (S10/11), 682 (S12)
    treatment installations, 814
Waterproofing, 243 (J), 526 (J)
Waterstops, 153 (E40), 457 (E40)
Wax polish, 321 (M60), 601 (M60), 792
WC
    suites, 336 (N13), 800
"Weathershield" emulsion, 323 (M60), 603 (M60)
Wedging and pinning, 134 (D50), 444 (D50)
Weights of various materials, 918
Welsh slates, 236 (H62), 518 (H62), 742
"Westbrick" cavity closer, 190 (F30), 490 (F30)
"Westmoreland" green slates, 236 (H62), 519 (H62), 742
Western Red Cedar boarding, 764
White sandlime bricks, 170 (F10), 470 (F10)
Window boards, MDF, 342 (P20), 624 (P20)
Window frames
    hardwood, 271, 272 (L10), 552, 553 (L10)
    softwood, 271 (L10), 552 (L10)
Windows
    aluminium, 766
    and external doors, 764
    cleaning equipment, 814
    hardwood, 269, 270 (L10), 550, 551 (L10), 766
    roof, 275 (L10), 556 (L10)
    softwood, 267, 268, 269 (L10), 548, 549, 550 (L10)
    steel, 273 (L11), 554 (L11), 766
    uPVC, 274 (L12), 555 (L12), 766
Windows/Doors/Stairs, 267 (L), 548 (L)
Wood
    block flooring, 310 (M42), 589 (M42), 784
    fibre boards, 248 (J41), 531 (J41), 736
    preservatives, "Solignum", 315, 327 (M60), 607 (M60)
    strip flooring, 264 (K20), 547 (K20), 780
    "Vigerflex" block flooring, 310 (M42), 589 (M42)
Woodwool roof decking
    reinforced slabs, 219 (G32), 508 (G32)
    unreinforced slabs, 261 (K11), 738
Woodwork, see "Timber" and "Softwood"
Working space, allowance, 121 (D20), 123 (D20), 433 (D20), 435 (D20)
Wrot faces,
    timber, 209, 211 (G20), 497, 499 (G20)

York stone, 363 (Q25), 646 (Q25)

Zinc
    coverings, 240 (H74), 523 (H74), 744
    flashings, 241 (H74), 523 (H74)

CD-Rom Single-User Licence Agreement

We welcome you as a user of this Spon Press CD-ROM and hope that you find it a useful and valuable tool. Please read this document carefully. **This is a legal agreement** between you (hereinafter referred to as the "Licensee") and Taylor and Francis Books Ltd., under the imprint of Spon Press (the "Publisher"), which defines the terms under which you may use the Product. **By breaking the seal and opening the package containing the CD-ROM you agree to these terms and conditions outlined herein. If you do not agree to these terms you must return the Product to your supplier intact, with the seal on the CD case unbroken.**

1. Definition of the Product

The product which is the subject of this Agreement, *Spon's Architects' and Builders' Price Book on CD-ROM* (the "Product") consists of:

1.1　Underlying data comprised in the product (the "Data")
1.2　A compilation of the Data (the "Database")
1.3　Software (the "Software") for accessing and using the Database
1.4　A CD-ROM disk (the "CD-ROM")

2. Commencement and Licence

2.1　This Agreement commences upon the breaking open of the package containing the CD-ROM by the Licensee (the "Commencement Date").
2.2　This is a licence agreement (the "Agreement") for the use of the Product by the Licensee, and not an agreement for sale.
2.3　The Publisher licenses the Licensee on a non-exclusive and non-transferable basis to use the Product on condition that the Licensee complies with this Agreement. The Licensee acknowledges that it is only permitted to use the Product in accordance with this Agreement.

3. Installation and Use

3.1　The Licensee may provide access to the Product for individual study in the following manner: The Licensee may install the Product on a secure local area network on a single site for use by one user. For more than one user or for a wide area network or consortium, use is only permissible with the express permission of the Publisher in writing and requires the payment of the appropriate fee as specified by the Publisher, and signature by the Licensee of a separate multi-user licence agreement.
3.2　The Licensee shall be responsible for installing the Product and for the effectiveness of such installation.
3.3　Text from the Product may be incorporated in a coursepack. Such use is only permissible with the express permission of the Publisher in writing and requires the payment of the appropriate fee as specified by the Publisher and signature of a separate licence agreement.

4. Permitted Activities

4.1　The Licensee shall be entitled:
4.1.1　to use the Product for its own internal purposes;
4.1.2　to download onto electronic, magnetic, optical or similar storage medium reasonable portions of the Database provided that the purpose of the Licensee is to undertake internal research or study and provided that such storage is temporary;
4.1.3　to make a copy of the Database and/or the Software for back-up/archival/disaster recovery purposes.
4.2　The Licensee acknowledges that its rights to use the Product are strictly as set out in this Agreement, and all other uses (whether expressly mentioned in Clause 5 below or not) are prohibited.

5. Prohibited Activities

The following are prohibited without the express permission of the Publisher:

5.1　The commercial exploitation of any part of the Product.
5.2　The rental, loan (free or for money or money's worth) or hire purchase of the product, save with the express consent of the Publisher.
5.3　Any activity which raises the reasonable prospect of impeding the Publisher's ability or opportunities to market the Product.
5.4　Any networking, physical or electronic distribution or dissemination of the product save as expressly permitted by this Agreement.
5.5　Any reverse engineering, decompilation, disassembly or other alteration of the Product save in accordance with applicable national laws.
5.6　The right to create any derivative product or service from the Product save as expressly provided for in this Agreement.
5.7　Any alteration, amendment, modification or deletion from the Product, whether for the purposes of error correction or otherwise.

6. General Responsibilities of the Licensee

6.1 The Licensee will take all reasonable steps to ensure that the Product is used in accordance with the terms and conditions of this Agreement.

6.2 The Licensee acknowledges that damages may not be a sufficient remedy for the Publisher in the event of breach of this Agreement by the Licensee, and that an injunction may be appropriate.

6.3 The Licensee undertakes to keep the Product safe and to use its best endeavours to ensure that the product does not fall into the hands of third parties, whether as a result of theft or otherwise.

6.4 Where information of a confidential nature relating to the product or the business affairs of the Publisher comes into the possession of the Licensee pursuant to this Agreement (or otherwise), the Licensee agrees to use such information solely for the purposes of this Agreement, and under no circumstances to disclose any element of the information to any third party save strictly as permitted under this Agreement. For the avoidance of doubt, the Licensee's obligations under this sub-clause 6.4 shall survive termination of this Agreement.

7. Warrant and Liability

7.1 The Publisher warrants that it has the authority to enter into this Agreement, and that it has secured all rights and permissions necessary to enable the Licensee to use the Product in accordance with this Agreement.

7.2 The Publisher warrants that the CD-ROM as supplied on the Commencement Date shall be free of defects in materials and workmanship, and undertakes to replace any defective CD-ROM within 28 days of notice of such defect being received provided such notice is received within 30 days of such supply. As an alternative to replacement, the Publisher agrees fully to refund the Licensee in such circumstances, if the Licensee so requests, provided that the Licensee returns the Product to the Publisher. The provisions of this sub-clause 7.2 do not apply where the defect results from an accident or from misuse of the product by the Licensee.

7.3 Sub-clause 7.2 sets out the sole and exclusive remedy of the Licensee in relation to defects in the CD-ROM.

7.4 The Publisher and the Licensee acknowledge that the Publisher supplies the Product on an "as is" basis. The Publisher gives no warranties:

7.4.1 that the Product satisfies the individual requirements of the Licensee; or

7.4.2 that the Product is otherwise fit for the Licensee's purpose; or

7.4.3 that the Data are accurate or complete or free of errors or omissions; or

7.4.4 that the Product is compatible with the Licensee's hardware equipment and software operating environment.

7.5 The Publisher hereby disclaims all warranties and conditions, express or implied, which are not stated above.

7.6 Nothing in this Clause 7 limits the Publisher's liability to the Licensee in the event of death or personal injury resulting from the Publisher's negligence.

7.7 The Publisher hereby excludes liability for loss of revenue, reputation, business, profits, or for indirect or consequential losses, irrespective of whether the Publisher was advised by the Licensee of the potential of such losses.

7.8 The Licensee acknowledges the merit of independently verifying Data prior to taking any decisions of material significance (commercial or otherwise) based on such data. It is agreed that the Publisher shall not be liable for any losses which result from the Licensee placing reliance on the Data or on the Database, under any circumstances.

7.9 Subject to sub-clause 7.6 above, the Publisher's liability under this Agreement shall be limited to the purchase price.

8. Intellectual Property Rights

8.1 Nothing in this Agreement affects the ownership of copyright or other intellectual property rights in the Data, the Database or the Software.

8.2 The Licensee agrees to display the Publisher's copyright notice in the manner described in the Product.

8.3 The Licensee hereby agrees to abide by copyright and similar notice requirements required by the Publisher, details of which are as follows:

"© 2002 Spon Press. All rights reserved. All materials in *Spon's Architects' and Builders' Price Book on CD-ROM* are copyright protected. © 2001 Adobe Systems Incorporated. All rights reserved. No such materials may be used, displayed, modified, adapted, distributed, transmitted, transferred, published or otherwise reproduced in any form or by any means now or hereafter developed other than strictly in accordance with the terms of the licence agreement enclosed with the CD-ROM. However, text and images may be printed and copied for research and private study within the preset program limitations. Please note the copyright notice above, and that any text or images printed or copied must credit the source."

8.4 This Product contains material proprietary to and copyrighted by the Publisher and others. Except for the licence granted herein, all rights, title and interest in the Product, in all languages, formats and media throughout the world, including all copyrights therein, are and remain the property of the Publisher or other copyright owners identified in the Product.

9. Non-assignment

This Agreement and the licence contained within it may not be assigned to any other person or entity without the written consent of the Publisher.

10. Termination and Consequences of Termination

10.1 The Publisher shall have the right to terminate this Agreement if:

10.1.1		the Licensee is in material breach of this Agreement and fails to remedy such breach (where capable of remedy) within 14 days of a written notice from the Publisher requiring it to do so; or
10.1.2		the Licensee becomes insolvent, becomes subject to receivership, liquidation or similar external administration; or
10.1.3		the Licensee ceases to operate in business.
10.2		The Licensee shall have the right to terminate this Agreement for any reason upon two months' written notice. The Licensee shall not be entitled to any refund for payments made under this Agreement prior to termination under this sub-clause 10.2.
10.3		Termination by either of the parties is without prejudice to any other rights or remedies under the general law to which they may be entitled, or which survive such termination (including rights of the Publisher under sub-clause 6.4 above).
10.4		Upon termination of this Agreement, or expiry of its terms, the Licensee must:
10.4.1		destroy all back up copies of the product; and
10.4.2		return the Product to the Publisher.

11. General

11.1 *Compliance with export provisions*
The Publisher hereby agrees to comply fully with all relevant export laws and regulations of the United Kingdom to ensure that the Product is not exported, directly or indirectly, in violation of English law.

11.2 *Force majeure*
The parties accept no responsibility for breaches of this Agreement occurring as a result of circumstances beyond their control.

11.3 *No waiver*
Any failure or delay by either party to exercise or enforce any right conferred by this Agreement shall not be deemed to be a waiver of such right.

11.4 *Entire agreement*
This Agreement represents the entire agreement between the Publisher and the Licensee concerning the Product. The terms of this Agreement supersede all prior purchase orders, written terms and conditions, written or verbal representations, advertising or statements relating in any way to the Product.

11.5 *Severability*
If any provision of this Agreement is found to be invalid or unenforceable by a court of law of competent jurisdiction, such a finding shall not affect the other provisions of this Agreement and all provisions of this Agreement unaffected by such a finding shall remain in full force and effect.

11.6 *Variations*
This Agreement may only be varied in writing by means of variation signed in writing by both parties.

11.7 *Notices*
All notices to be delivered to: Spon Press, an imprint of Taylor & Francis Books Ltd., 11 New Fetter Lane, London EC4P 4EE, UK.

11.8 *Governing law*
This Agreement is governed by English law and the parties hereby agree that any dispute arising under this Agreement shall be subject to the jurisdiction of the English courts.

If you have any queries about the terms of this licence, please contact:

Spon's Price Books
Spon Press
an imprint of Taylor & Francis Books Ltd.
11 New Fetter Lane
London EC4P 4EE
United Kingdom
Tel: +44 (0) 20 7583 9855
Fax: +44 (0) 20 7842 2298
www.sponpress.com

CD-Rom Installation Instructions

System requirements

Minimum

- 66 MhZ processor
- 12 MB of RAM
- 10 MB available hard disk space
- Quad speed CD-ROM drive
- Microsoft Windows 95/98/2000/NT
- VGA or SVGA monitor (256 colours)
- Mouse

Recommended

- 133 MhZ (or better) processor
- 16 MB of RAM
- 10 MB available hard disk space or more
- 12x speed CD-ROM drive
- Microsoft Windows 95/98/2000/NT
- SVGA monitor (256 colours) or better
- Mouse

Microsoft ® is a registered trademark and Windows™ is a trademark of the Microsoft Corporation.

Installation

How to install *Spon's Architects' and Builders' Price Book 2002 CD-ROM*

Windows 95/98/2000/NT

Spon's Architects' and Builders' Price Book 2002 CD-ROM should run automatically when inserted into the CD-ROM drive. If it fails to run, follow the instructions below.

- Click the **Start** button and choose **Run**.
- Click the **Browse** button.
- Select your CD-ROM drive.
- Select the Setup file [setup.exe] then click **Open**.
- Click the OK button.
- Follow the instructions on screen.
- The installation process will create a folder containing an icon for *Spon's Architects' and Builders' Price Book CD-ROM* and also an icon on your desktop.

How to run the *Spon's Architects' and Builders' Price Book CD-ROM*

- Double click the icon (from the folder or desktop) installed by the Setup program.
- Follow the instructions on screen.

© COPYRIGHT ALL RIGHTS RESERVED

All materials in *Spon's Architects' and Builders' Price Book 2002 CD-ROM* are copyright protected. No such materials may be used, displayed, modified, adapted, distributed, transmitted, transferred, published or otherwise reproduced in any form or by any means now or hereafter developed other than strictly in accordance with the terms of the above licence agreement.

The software used in *Spon's Architects' and Builders' Price Book 2002 CD-ROM* is furnished under a licence agreement. The software may be used only in accordance with the terms of the licence agreement.